Single-Stranded RNA Phages

Single-Stranded RNA Phages
From Molecular Biology to Nanotechnology

Paul Pumpens, Dr. habil. biol.
Professor of Molecular Biology
University of Latvia, Riga

CRC Press
Taylor & Francis Group
Boca Raton London New York

CRC Press is an imprint of the
Taylor & Francis Group, an **informa** business

CRC Press
Taylor & Francis Group
6000 Broken Sound Parkway NW, Suite 300
Boca Raton, FL 33487-2742

First issued in paperback 2021

© 2020 by Taylor & Francis Group, LLC
CRC Press is an imprint of Taylor & Francis Group, an Informa business

No claim to original U.S. Government works

ISBN-13: 978-0-367-02858-9 (hbk)
ISBN-13: 978-1-03-217503-4 (pbk)
DOI: 10.1201/9780429001208

To my family
To my teacher, Elmārs Grēns
To my friends

Contents

Preface xv
Author xix
Introduction xxi

Chapter 1 History and First Look 1

Classical Reviews 1
Male Specificity 1
RNA as the Genome 1
Growth Cycle 11
Coliphages 12
Worldwide 13
Non-Coli Phages 13
Diversity of Pili 14
Newcomers 15
Dangerous Similarity 15
Antibodies 15
Portraits of Icosahedra 16
Composition 18
Genome 19
Translation 20
Replication 22
Regulation 24
Reconstitution 25
Phage Tales 26

Chapter 2 Taxonomy and Grouping 29

Entries in Prestigious Books of Reference and Encyclopedias 29
Official ICTV Taxonomy 29
Relatedness to Other Taxa 31
NCBI Classification 32
Serogrouping 35
Affinities to Millipore Filters 37
Template Specificity of Replicases 38
Hybrid Particles 39
Physicochemical Parameters 39
Genogrouping 40
Metaviromics 42
Urgent Necessity for Improvements in the Taxonomy 44

Chapter 3 Pili and Hosts 45

Pili: Current Definition 45
F Pili: Ascending to Clarity 45
How Pili Work 47
F Type Pili: R Factors 49
Genetics of F Pili 50
Chromosomal Mutations 52
F Pili in the *Enterobacteriaceae* Other Than *E. coli* 53
Conjugation and RNA Phages 53
Elimination of Sex Factors 54
Non-F-Encoded Pili 55

Incompatibility Groups: Definition 57
Incompatibility Groups and RNA Phages 58
Fine Mapping of F Pili 61
Fine Mapping of Polar Pili 61
Caulobacter and *Acinetobacter* Pili 64
Hosts 66
    Enterobacteriaceae 66
    Pseudomonadaceae 66
    Caulobacteraceae 66
    Moraxellaceae 67
Spheroplasts 67

Chapter 4   Physiology and Growth 69

Attachment or Adsorption 69
Ejection or Eclipse 71
Penetration 73
Replication Cycle 75
Crystalline Inclusions 78
Release 79
Plaque Assay 81
Phage Effect on Host 81
Host Mutants 84
Inhibition of Phage Growth 86
Rifamycin 89
Mixed Infections and Superinfections 93
Preparative Growth and Purification 95
Cell-Free Synthesis 96
Lysogeny 97

Chapter 5   Distribution 99

Geography 99
Sources 102
Human Beings 105
Enumeration 105

Chapter 6   Ecology 111

Measure of Pollution 111
Water Quality 112
Indicators of Fecal Pollution 113
Surveillance of Ground and Surface Waters 113
Water Surveillance 114
    United States 114
    Canada 115
    Africa 115
    New Zealand and Australia 115
    Israel 115
    Europe 116
    Central and South America 116
    Japan 116
    China 117
    India 117
    Southeast Asia Countries 117
    Antarctica 117
Different Survival in Water 117

Surrogates of Human Viruses 118
Inactivation 122
Ultraviolet Irradiation 122
Photocatalytic Disinfection 127
UV-LEDs 129
Photosensitizing Agents 130
Photosensitization of Fullerol Nanoparticles 130
Graphene and Carbon Nanomaterials 131
Other Nanoparticles 131
Ionizing Radiation 132
Plasma 132
Selective Photonic Disinfection 133
Metal Compounds 133
Nitrous Acid 134
Alkylating Agents 134
Bisulfite 135
Hydroxylamine 135
Chlorine, Iodine, and Bromine 135
Ammonia 137
Alkaline Stabilization 137
Peracetic Acid 137
Ozone 138
Other Inactivating Agents 138
Heat 140
Sunlight 141
High Pressure 143
Ultrasound 143
CO_2 143
Desiccation, Humidity, and Aerosols 143
Biofilms 144
Dyed Phage Particles 145
Different Resistance to Inactivation 145
Adsorption and Coagulation 147
Fracture Flow and Transport 150
Filtration by Soil and Sand Columns 152
Membrane Filtration 154
Oysters and Mussels 157
Fruits and Vegetables 159
Food Total 161
Patient Care 163
Pseudomonaphage PP7 165

Chapter 7 Genome 167

Genetic Makeup 167
Full-Length Sequences 167
Peculiarities of the Genome 171
Similarities and Dissimilarities 173

Chapter 8 Genetics and Mutants 175

Conditionally Lethal Mutations 175
Polar Mutations 176
Gene Order 176
Frequencies of Mutations 176
Complementation 177
Function of Mutant Genes 177

Azure Mutants 181
Op-3 Mutant 183
Streptomycin Suppression 183
Mutants of the Phage f2 183
Mutants of the Phage MS2 184
Mutants of the Phage fr 186
Mutants of the Phage R17 186
Mutants of the Phage GA 186
Mutants of the Phages Qβ, SP, and FI.. 187
Mutants of the Phage PP7 188

Chapter 9   Particles 189

Molecular Mass.. 189
Size 190
Composition 191
Symmetry 192
Morphology 192
Coat-RNA Contact 193
Defective Particles 193
Reconstitution... 195
Monitoring.. 198
Aggregation 201
Standards and Internal Controls... 202

Chapter 10 RNA 205

Biophysics... 205
Gel Electrophoresis 208
General Properties... 208
Hybridization...210
Preparations for Sequencing...210
MS2 Sequencing: Walter Fiers' Team 211
MS2 Sequencing: Other Groups 216
By-Products and Direct Consequences of the MS2 Sequencing....................................217
R17 Sequencing...219
f2 Sequencing... 221
Qβ Sequencing 222
Sequencing of Other Phage RNAs 224
Electron Microscopy 226
Secondary Structure: Jan van Duin's Team 227
Secondary Structure: Further Inquiry... 231
MS2 RNA and Programmed Cell Death 232
MS2 RNA as a Substrate and Interaction Probe 233
MS2 RNA as a Control 234

Chapter 11 Maturation Protein 237

Mutant Allusions.. 237
Gel Electrophoresis 237
Isolation 238
Chemical Studies.. 238
Fate of the Maturation Protein 238
Sequencing 241
Homology 242
Spatial Structure... 242
Lysis.. 243

Chapter 12 Coat Protein...247

    Composition and Sequencing..247
    N-Terminal Modification...249
    Isolation...249
    Aggregation...249
    Physical Properties 250
    Homology..250
    Spatial Structure 251
    Lysis..251
    Readthrough Protein 252

Chapter 13 Replicase 257

    Activity...257
    Replicase Subunit 258
    Homology..258
    Qβ Replicase 258
    Subunits of the Qβ Replicase 262
    EF Tu-Ts 265
    Ribosomal Protein S1 267
    Fine Structure of the Qβ Replicase 268
    Host Factors..272
    Host Factor I, or HF, or Hfq 278
    The *hfq* Gene 279
    Fine Structure of the Hfq 280
    Some Special Features of the Qβ Replicase Enzyme 283
    Non-*Allolevivirus* Replicases 286
    f2 Replicase 286
    MS2 Replicase..288
    GA Replicase 288
    R17 Replicase 289

Chapter 14 Lysis Protein...291

    Discovery..291
    Homology..291
    Natural Expression of the Lysis Gene 293
    Lysis Mechanism..295
    Protein Antibiotics...298
    Other Applications 299

Chapter 15 Immunology...301

    Serological Criteria 301
    Serogrouping...301
    Immunoglobulins 302
    Phylogeny of Antibodies 304
    Target of Antibodies...306
    Mapping..309
    Immunodetection 310

Chapter 16 Translation...313

    The Essentials...313
    Translation *In Vivo* 315
    Translation *In Vitro* 316
    Heterologous Translation *In Vitro* 318

Polarity 319
Nascent Chains 322
Initiation 323
Initiation Sites 326
Shine-Dalgarno 327
Elongation 330
Termination 331
Intergenic Sequences 333
RNA Fragments as Messengers 333
Miscellaneous by the *In Vitro* Translation 334
Secondary and Tertiary Structure 334
Ribosomal Frameshifting 338
Reinitiation 338
Large-Scale Translation 339
Genetic Code 339
Non-Coliphages 343
Repressor Complex I 344
Repressor Complex I: f2 346
Repressor Complex I: R17 346
Repressor Complex I: MS2 349
Repressor Complex I: fr 356
Repressor Complex I: GA 358
Repressor Complex I: Qβ 359
Repressor Complex I: PP7 360
Repressor Complex I: PRR1 361
Repressor Complex II 362

Chapter 17 Replication 365

The Essentials 365
Early Studies 366
Replicative Intermediate 367
Regulation of the Replication Cycle 371
Induction of Interferon 375
Larifan 378

Chapter 18 Evolution 381

Darwinian Experiments 381
6S RNA 383
MDV-1: A Midi-Variant 384
Micro- and Nanovariants 386
De Novo Synthesis? 387
Molecular Parasites 390
Molecular Colonies 390
Autocatalytic Synthesis of Heterologous RNAs 391
RQ Variants 393
Mapping of Qβ Replicase 395
Diagnostics 396
Mutagenesis System 399
Coupled Replication-Translation 399
RNA World 400
Phylogeny 404
RNA-Dependent Polymerases 406
Recombination 410
Quasispecies and Evolution 413

Chapter 19 Expression 419

    Era of Gene Engineering...419

    Generation of Mutants.. 421

    Chimeric Phages.. 422

    Maturation Gene... 423

    Lysis Gene 424

    Replicase Gene 425

    Replicase Fusions 427

    Coat Gene 431

Chapter 20 Chimeric VLPs... 435

    Protein Engineering and VLP Nanotechnology 435

    Genetic Fusions of Coats.. 438

        fr 438

        MS2 439

        Qβ 440

        PP7 442

        AP205.. 444

        GA 445

    Peptide Display... 446

    Chemical Coupling.. 447

        Martin F. Bachmann's Team 447

        Bryce Chackerian and David S. Peabody.. 447

    Functionalization.. 448

        MS2 448

        Qβ 452

    Plug-and-Display: AP205... 458

    Stability 460

    Packaging 461

Chapter 21 3D Particles 467

    Viruses and/or VLPs?... 467

    RNA Aptamers... 470

    Packing of Genomic RNA 471

    Maturation Protein 476

    Self-Assembly.. 480

    Lattices 484

    AP205... 486

    Chimeric VLPs... 487

Chapter 22 VLP Vaccines... 489

    General 489

    Basic Immunology 489

    Genetic Fusions 493

    Chemical Coupling... 501

    Chimeric Phages... 514

Chapter 23 Non-Vaccine VLPs 515

    Armored RNA.. 515

    Nanocontainers: Drug Delivery 518

    Nanomaterials and Imaging 529

    Nanoreactors... 529

Chapter 24 Non-VLPs..535

   Gene Therapy...535
   Tethering..536
   Imaging..540
   New Challenge: PP7..546
   CRISPR-Cas9..549

Chapter 25 Epilogue...553

   Statistics..553
   Goal Achievements..553
   Virome...554
   Novel VLP Models..556
   Three-Dimensional Structure...556
   Immediate Future...557
   Bacterial Immunity..558
   Phage Therapy...559
   PHAGE: Therapy, Applications, and Research..559

Author Index...561

Subject Index...607

References: Can Be Found on the Book's Website: www.crcpress.com/9780367028589

Preface

When a person tries in good faith, even if he fails, he cannot be blamed.

William Shakespeare
Antony and Cleopatra, *Act 2, Scene 5*

This book is a first humble attempt to compose a complete guide to all important species of the *Leviviridae* family, the first discovered RNA bacteriophages, which were found in 1961. It is designed as a comprehensive account of the history of single-stranded RNA phages during the 1961–2019 period. The book presents a maximally full outline of the discovery, history, taxonomy, genome structure, physical, structural, and functional properties of the phage RNA and encoded proteins, of their virions and virus-like particles (VLPs), as well as the general scientific importance and numerous practical applications of the *Leviviridae* phages. The current state of knowledge demonstrates the unique role of the RNA phages in the scientific establishment of the central molecular biology principles such as genetic code, translation, replication, regulation of gene expression, and suppression of mutations. Special attention is devoted to the broad modern applications of the RNA phages and their products in nanotechnology, vaccinology, gene tethering and imaging, and evolutionary and environmental studies.

First, the rapidly developing field of RNA phage-derived VLPs in the context of immunology, virology, microbiology, biochemistry, and chemistry is described. The book offers a complete overview of the generation of novel vaccines, gene therapy vectors, drug delivery, and diagnostic tools on the basis of RNA phage VLPs. Next, the great role of RNA phage-derived products is outlined in the revolutionary progress of bioimaging, namely, in the protocols intended to visualize biological processes in real time. The book then demonstrates the marked contribution of RNA phages to environmental sciences by enabling improved water and food quality control and by systematic studies on the inactivation of infectious agents. Last but not least, future perspectives for novel ideas and applications of the growing number and role of the *Leviviridae* family members are presented.

Books generally summarize only the most important facts in the field due to the enormous amount of data. The disadvantage of this approach, unfortunately, is the fact that the data are presented according to the author's taste, but tastes tend to differ. The advantage of the RNA phage topic is that the field, although broad, still retains the ability to quickly summarize all collected evidence, to provide readers with further selection and operation with maximal information, including totally forgotten—not always justifiably—papers. This is the last time for summarizing, since the number of publications is constantly growing.

After the discovery of single-stranded RNA phages about 60 years ago, the first book devoted exclusively to them, *Regulatory Mechanisms of RNA-Containing Phage Replication*, was published in Riga, Latvia. This book was written by my teacher and mentor, Professor Elmārs Grēns, in 1974. The book gained durable popularity among intellectual scientists and students in the then Soviet Union, but remained practically unrecognized by the broader scientific community because of its Russian language. In 1975, the comprehensive book *RNA Phages* (Cold Spring Harbor Laboratory Press, Cold Spring Harbor, NY) was edited by one of the RNA phage discoverers, Norton D. Zinder, the father of RNA phages, with leading RNA phage experts as contributors. This famous green-covered book is still available on the internet.

After these two books, single-stranded RNA phages did not attract attention from publishers for more than 40 years. Recognizing the tremendous amount of inspiring knowledge obtained in recent studies of RNA phages, I felt it was my duty and responsibility to the scientific community to publish the entire RNA phage story and to draw attention to the enormous progress occurring today in this broad and still unpredictable field. Having begun my scientific career in Elmārs Grēns' lab in 1970 with my academic theses on the regulation of the RNA phage replication, I then finished my scientific career in 2014 with the RNA phage-derived nanotechnological tools at the Biomedical Research and Study Centre, which was established and headed by Elmārs Grēns, I felt ready to undertake this book project.

Norton D. Zinder
(1928–2012)
Father of RNA phages
(Adapted from the
New York Times.)

Elmārs Grēns
b. 1935
*Father of molecular biology
in Latvia*
(Courtesy of F64 Photo
Agency/Girts Ozoliņš.)

This book contains the full story of RNA phages—not only the successful undertakings, but also how these undertakings have been successful and the contributions that have been made to global science, from molecular biology to nanotechnology. The book includes not only what happened, but also how it happened and who did it. Thus, as much as possible, I have presented original data, including reports that might not be regarded as important. In any case, these data are a part of the RNA phage history and their potency may not have been maximally explored.

This book would not have been possible without the constant support of many people during my lucky scientific career. First and foremost, I have to thank my lifelong teacher and mentor, Professor Elmārs Grēns, who introduced me to the topic, conveyed a constant spirit of adventure in regard to research, and has taught me more than I could ever give him credit for here.

I am greatly indebted to my dear closest colleagues Dr. Andris Dišlers, Dr. habil. biol. Tatyana Kozlovska, Dr. habil. biol. Velta Ose, Dr. Peter Pushko (now Frederick), Dr. Regīna Renhofa, Dr. Indulis Cielēns, and Dr. habil. biol. Alexander Tsimanis (now Rehovot) who have been extremely supportive and who worked enthusiastically in the RNA phage field in Riga and readily shared their unpublished data that are presented in this book. Many thanks to Professor Vadim I. Agol (Moscow) for his serious attention to my RNA phage PhD thesis and to Professor Eduard I. Budowsky (Moscow) for my first training abroad. All of us in Riga are thankful to the great RNA phage classicists Professor Charles Weissmann (Zürich) and Professor Walter Fiers (Ghent) for their friendly attitude and for providing us with RNA phages and their mutants.

I am highly indebted to Uppsala University professors Bror Strandberg and Lars Liljas, who introduced our team to 3D RNA phage studies, to Dr. Maija Bundule who started the 3D studies in Riga, and to Professor Kaspars Tārs, a well-known 3D expert, who started in Riga and after 11 years, left Uppsala to return to Riga.

I am also grateful to all of those with whom I have had the pleasure to work in the RNA phage field in Riga: Ms. Ināra Akopjana, Dr. Elita Avota (now Würzburg), Dr. Andris Avots (now Würzburg), Mr. Artis Brede, Dr. Dzidra Dreiliņa, Dr. Anda Dreimane, Dr. Guna Feldmane, Dr. Jānis Freivalds, Dr. Esther Grinstein (now Berlin), Mr. Indulis Gusārs, Dr. Inta Jansone, Dr. Juris Jansons, Dr. Andris Kazāks, Dr. Jānis Kloviņš, Dr. Viktors Kumpiņš, Dr. Normunds Līcis, Ms. Valerie Markushevich, Dr. Tatjana Nikitina, Dr. Ivars Petrovskis, Dr. Laima Šēriņa, Dr. Dace Skrastiņa, Dr. Dagnija Sniķere, Dr. Irina Sominskaya, Dr. Juris Šteinbergs, Dr. Anna Strelnikova, Dr. Arnis Strods, Mr. Valdis Tauriņš, Dr. Inta Vasiljeva, Dr. Tatjana Voronkova-Kazāka, Dr. Andris Zeltiņš. I am highly indebted to Dr. Jānis Rūmnieks for the excellent idea concerning the structure of the prolonged RNA phage icosahedra and for the nice illustration of this phenomenon.

I would especially like to thank Professor Martin F. Bachmann (Zürich / Bern / Oxford) who converted RNA phages into real nanotechnological tools, started many amazing innovative projects, and remains active and successful in the RNA phage field. In the nanotechnological connection,

I am highly indebted to Dr. Alain C. Tissot (now Penzberg), Dr. Gary T. Jennings (Zürich), and Dr. Stephan Oehen (now Basel). Warm gratitude to the Nobelist Rolf M. Zinkernagel (Zürich) for his friendly attitude and valuable guidance at the beginning of the deep immunological studies of RNA phage VLPs.

I am highly indebted to professors David S. Peabody and Bryce Chackerian at the University of New Mexico in Albuquerque, who represent both theory and praxis of the RNA phages. Many thanks to Professor David Klatzmann (Paris) who was involved in two joint biotechnological projects on the RNA phages.

I am grateful to Professor Dr. Rüdiger Schmitt (Regensburg) for the principal and unique contribution at the very beginning of our studies on RNA phage AP205 that unfortunately remain unpublished.

I am highly indebted to the famous pioneers of RNA phages Professor Sankar Mitra (Houston), Dr. Henri Pierre-Marcel Fromageot (West Caldwell), Professor Margarita Salas (Madrid). Also to Dr. Maarten de Smit (Leiden) and Professor Stephen H. Zinder (Ithaca) and Ms. Lenny Gemski (Asheville) for providing me with the necessary copyright permissions.

I am very thankful to Professor R. Anthony Crowther (Cambridge) for his collaboration and constant guidance in 3D structures and to Professor David J. Rowlands (Leeds) for long-term collaboration on VLP technologies. Many thanks to the VLP classicists Professor David R. Milich (San Diego), Professor Michael Nassal (Freiburg), Professor Adam Zlotnick (Bloomington), Professor Michael Roggendorf (Essen), Professor George P. Lomonossoff (Norwich), Professor Trevor Douglas (Bloomington), Professor Polly Roy (London), Professor Bogdan Dragnea (Bloomington), Professor Bettina Böttcher (Würzburg), and Professor Reinhold Schirmbeck (Ulm).

A warm thank you to my old colleagues and friends Dr. Vadim Bichko (now in San Diego), Professor Kęstutis Sasnauskas (Vilnius), Professor Matti Sällberg (Stockholm), Professor Henrick Garoff (Stockholm), and Professor Kenneth Lundstrom (Lutry). I am very thankful to Essen's Professors Sergei Viazov and Stefan Roß, to Professor Mikhail Mikhailov (Moscow), and to Professor Michael Kann (Göteborg). My warm thanks to Dr. Rainer G. Ulrich (Greifswald-Insel Riems), Dr. Helga Meisel (Berlin), Professor Detlev H. Krüger (Berlin), to Tartu Professors Richard Villems and Mart Ustav, to Professor Mart Saarma (Helsinki), and to Professor Joseph Holoshitz (Ann Arbor).

I am highly indebted to my friends Professor Indriķis Muižnieks, Professor Yury Dekhtyar, Professor Alexander Rapoport, Professor Katrīna Ērenpreisa, Dr. Edith Carron (now Leonardtown, Maryland), Dr. Valentīna Sondore, Professor Jāzeps Keišs, Dr. Rūta Brūvere, Professor Uga Dumpis, and Dr. Uldis Bērziņš for their constant understanding, friendship, and support.

I wish to thank Professor Jānis Krastiņš, a famous expert in art history, who taught me how to write books. My cordial gratitude to Mr. Alexander Kiselev, my oldest friend.

My thanks to Dr. Pavel Zayakin for his professional help and advice on IT applications.

I am highly indebted to my late colleagues and friends Professor Viesturs Baumanis, Professor Valdis Bērziņš, Dr. Galina Borisova, Dr. Alıse Jalinska, Professor Valts Loža, Mr. Juris Ozols, Dr. Oleg Plotnikov, Dr. Guntis Rozentāls, Dr. Anatoly Sharipo, Dr. Zinaida Shomstein, Professor Eva Stankeviča, Dr. Laima Tihomirova, Professor Lev Kisselev (Moscow), Professor Jan van Duin (Leiden), and Professors Hans-Alfred and Sinaida Rosenthal (Berlin), who will never see this book. I am extremely thankful to Professor Peter H. Hofschneider (München), one of the RNA phage discoverers, whom I recognized during his (and my) hepatitis B stage of our scientific career.

Last but not least, I would like to express my deepest appreciation to Professor Wolfram H. Gerlich (Giessen) for his constant friendly support and advice, and many thanks to Dr. Yury E. Khudyakov (Atlanta) for his extensive friendly help. Yury is the one who encouraged me to take the opportunity to start work on this RNA phage book.

Finally, I wish to thank Dr. Charles "Chuck" Crumly, Ms. Ana Lucia Eberhart, Ms. Linda Leggio, Ms. Kelsey Barham, Christian Munoz (cover designer), and the great CRC Press/Taylor & Francis team for the creation of this volume.

A complete and searchable Reference list is available for Free Download at: www.crcpress.com/9780367028589.

Author

Paul Pumpens, Dr. habil. biol., graduated from the Chemical Department of the University of Latvia in 1970 and earned his PhD in molecular biology from the Latvian Academy of Sciences, Riga; and DSc from the Institute of Molecular Biology of the USSR Academy of Sciences, Moscow, USSR. Dr. Pumpens started his career as a research fellow at the Institute of Organic Synthesis, Riga, where he conducted research from 1973 to 1989. He served as head of the Laboratory of Protein Engineering at the Institute of Organic Synthesis (1989–1990), as head of the Department of Protein Engineering at the Institute of Molecular Biology of the Latvian Academy of Sciences (since 1993, the Biomedical Research and Study Centre) in Riga (1990–2002), and as scientific director of the Biomedical Research and Study Centre (2002–2014). He served as a professor of the Biological Department of the University of Latvia from 1999 until 2013.

Dr. Pumpens pioneered genetic engineering research in Latvia. He was one of the first in the world to perform successful cloning of the hepatitis B virus genome and the expression of hepatitis B virus genes in bacterial cells. His major scientific interests are in designing novel recombinant vaccines and diagnostic reagents and development of tools for gene therapy on the basis of virus-like particles.

Dr. Pumpens is the author of more than 300 scientific papers and issues or pending patents. He has edited, together with Dr. Yury E. Khudyakov, the CRC Press/Taylor & Francis book *Viral Nanotechnology* (2015).

Introduction

Faust: Nun gut, wer bist du denn?
Mephistopheles: Ein Teil von jener Kraft,
Die stets das Böse will und stets das Gute schafft.

Johann Wolfgang von Goethe
Faust: Eine Tragödie
Kapitel 6

Faust: Well, what are you then?
Mephistopheles: Part of the Power that would
Always wish Evil, and always works the Good.

Johann Wolfgang von Goethe
Faust, Part I Scene III: The Study
Translation by A.S. Kline

The world of bacteriophages, or simply "phages," the creatures that devour bacteria and archaea, is a substantial part of the huge virus world. The phages are here, there, and everywhere that bacteria may exist. Undoubtedly, the phages are the most numerous organisms on Earth, considering their number by more than 10^{31} units, and are therefore the most impressive reservoir of the genetic diversity in the Earth's biosphere. The total contribution of the phages to very different fields of human knowledge in life sciences is keeping up well with their number. The single-stranded positive-sense RNA phages of the *Leviviridae* family, the subject of this book, are unique by their features within the total phage world and impersonate rather widely distributed single-stranded positive-sense RNA eukaryotic viruses. First, only 2 out of 21 known phage families possess RNA genomes: *Leviviridae*, with the linear single-stranded RNA genome, and *Cystoviridae*, with the segmented double-stranded RNA genome.

Second, the positive-sense RNA viruses that are unique among phages are widely distributed in eukaryotes, including humans and animals, and are responsible for a large number of dangerous diseases. The *Leviviridae* is the only single-stranded RNA virus family that infects bacteria within the huge *Riboviria* realm that includes one large *Negarnaviricota* phylum and three large orders (*Nidovirales*, *Picornavirales*, and *Tymovirales*) as well as 40 families and 8 genera that have been not assigned to a higher order. Norton D. Zinder, the "father of the RNA phages," regarded the RNA phages to be the most populous organism in the world, considering their small size and large yield.

Third, the RNA phages are pili-specific, i.e., they recognize pili, namely, filamentous proteinaceous structures that extend from the cell surface in Gram-negative bacteria and in the vast majority of cases perform the conjugative function.

The RNA phages are nonenveloped, spherical viruses with predominant T = 3 icosahedral symmetry and diameters ranging mostly from approximately 28−30 nm. The particles of the most-studied RNA phages are composed of 178 chemically identical coat protein molecules, or 89 dimers, and 1 copy of the maturation, or A, protein, which replaces a single-coat protein dimer. The phage's monopartite, positive-sense, single-stranded, or plus-ssRNA genome is approximately 4 kb in size and serves as messenger RNA for the synthesis of the capsid-forming coat protein as well as three other viral proteins: the maturation, or A protein, replicase protein, and lysis protein. The first and classical *Escherichia coli*-infecting RNA phages f2, MS2, R17, fr, M12, and Qβ that played an extraordinary role in the buildup of molecular biology and still play a leading role in modern bionanotechnology were identified in the early 1960s, starting in 1961.

The RNA phages came to scientific consideration only 3 years after the central dogma of molecular biology was first stated by Francis Crick in 1958. It so happened that the RNA phages contributed strongly to the definitive establishment of the central dogma, and the method of RNA phage studies was virtually identical to the general route to the global creation of molecular biology. One of the main targets of the book is the gradual unveiling of the enthusiastic spirit of the revolutionary 1960s, with novel ideas and novel discoveries brought about by young and ambitious molecular biologists.

Chapter 1 presents not only a general introduction to the subject, but also a short story that is aimed to replace the need to read the entire book if the reader's time is limited. It is devoted first to the discovery process and first inventions of the RNA phages as "instrumental in the making of molecular biology," per Jérôme Pierrel. This chapter explains briefly how and why the RNA phages provided the scientific community with purified RNA and markedly contributed to the genome sequencing and decryption of the genetic code, to the understanding of RNA translation and replication mechanisms, and to the elucidation of the virus-host interactions and self-regulation in biological systems. This chapter is intended to present a full list of the currently described RNA phages and also to pay tribute to the great personalities of this unforgettable time: Norton D. Zinder, Robert L. Sinsheimer, Peter Hans Hofschneider, Hartmut Hoffmann-Berling, Charles Weissmann, Sol Spiegelman, Paul Kaesberg, Peter Knolle, Severo Ochoa, William Paranchych, Itaru Watanabe, and other great pioneers of the RNA phage field.

The next 23 chapters fall into three large logical blocks: Chapters 2–6 deal with the RNA phages in microbiological, virological, and ecological terms. Chapters 7–18 are devoted to the classical molecular biology and biophysics of the RNA phages. And finally, Chapters 19–24 introduce gene engineering, high-resolution structure, and broad nanotechnological applications of the RNA phages. Chapter 25 serves as an epilogue and a bridge to the future. The chapters are written in a consecutive manner where novel facts sequentially expand the information described before. Nevertheless, each chapter is meant to be self-sufficient without the need for preliminary information.

Thus, as noted above, the RNA phages are members of the family *Leviviridae* (from Latin *levis*, light, i.e., not heavy, also trivial, swift, gentle, frivolous, mild, rapid, superficial, undependable, questionable, unsubstantial, unsure, changeable, soft, unstable, unreliable). This family is officially assigned to the huge *Riboviria* realm, but not to any higher order, class, subphylum, or phylum. The *Leviviridae* family comprises two genera: *Levivirus* and *Allolevivirus*. The place of the *Leviviridae* family in the present official taxonomies (ICTV, NCBI), the detailed structure of the *Leviviridae* family, its possible interconnections with other viral families, as well as numerous attempts to classify the single-stranded RNA phages by different structural and functional parameters, together with drawbacks of all existing classification and typing systems, are the subject of Chapter 2.

The presence of pili (the male specificity of bacteria), appeared as one of the basic physiological features of the single-stranded RNA phages and stimulated active studies of bacterial pili that led to broadening of the lists of existing pili, their classification, and clear understanding of their structure and function. Thus, Chapter 3 concentrates on microbiology in general and describes the RNA phages from the point of view of the bacterial hosts and their pili. This chapter unveils the interaction of the RNA phages with different pili, presenting their genetics and variability. Moreover, it explains how pili work, what incompatibility groups are, and why only four bacterial families, *Enterobacteriaceae*, *Pseudomonadaceae*, *Caulobacteraceae*, and *Moraxellaceae*, are known as hosts for the RNA phages. In this connection, the intriguing interplay of the RNA phage infection with bacterial conjugation is highlighted.

Chapter 4 continues the microbiological line and presents up-to-date information on the general mechanisms of the early infection steps: attachment, or *adsorption*, ejection of RNA, or *eclipse*, penetration of the phage RNA into the bacterial cell, and phage release by the infected cells. The next aim of this chapter is to acquaint the reader with the basic concept of the replication cycle and the approaches to study it. Then, the phage effects on the host, and vice versa the effects of the host mutations on the infection outcome, as well as inhibition of phage growth by different agents and outcomes of mixed infections, are presented. Finally, historical and current data on the preparative RNA phage growth and purification are summarized, considering the great demand for the large-scale cultivation and purification protocols that are appearing nowadays because of the cost pressure and growing Good Manufacturing Practice requirements in the bionanotechnological applications.

Chapter 5 shows the RNA phage geography with localization of the more popular and less popular representatives of the family on the geographical map, as well as the source of the RNA phages, general methodologies of their enumeration, and the present state of metaviromics, or modern approaches by the search for novel RNA phage family members in the next-generation sequencing era.

The huge Chapter 6 surveys the RNA phage ecology as a part of the total phage ecology, first emphasizing the striking role of the RNA phages by the surveillance of water quality and general pollution of the surroundings. The rise of the RNA phages to the status of the surrogates of human and animal pathogens is one of the central subjects of this chapter. An overview is given of the huge literature on the very different phage inactivation methods as a part of the global disinfection problem, since the RNA phages not only sped up the rapid development of molecular biology but also attracted a purely chemical interest to the inactivation of genomic nucleic acids—in this case RNA—and pioneered the application of chemistry to the problems of biological purity and safety. Special emphasis is placed on water treatment and source tracking in wastewater, on food production, processing, and storage, as well as on patient care by clinical applications.

By the genome size and number of genes, the RNA phages belong to the smallest organisms, which are nevertheless able to regulate their macromolecular syntheses close to perfection. The genome characteristics and peculiarities of genome organization, including location of the genes and differences in genomic organization among different phage species, are the subject of Chapter 7. It is especially interesting that the genomes of the genera *Levivirus* and *Allolevivirus* demonstrate a strict difference that is generated by new *Allolevivirus* protein, namely, a minor A1 protein. The latter is nothing more than a C-terminally extended coat protein, which appears as a result of ribosomal read-through at a leaky opal termination codon of the *Allolevivirus* coat gene and is found essential for the formation of viable virions *in vivo*. An up-to-date analysis of full-length RNA phage genomes and their phylogenetic relatedness is presented.

The RNA phages not only conclusively solved the central problems in molecular biology and chemistry, but also contributed markedly to the understanding of the mutagenic actions and application of chemical mutagenesis in practice. Genetic studies of the RNA phages have been performed using a large number of temperature-sensitive and suppressor-specific mutations. Chapter 8 deals with genetic studies and practical generation of the numerous RNA phage mutants.

Chapter 9 is concerned with the most general characteristics of the phage particles in a historical retrospective including reconstitution experiments, but without going deeply into three-dimensional structures.

Chapter 10 is devoted entirely to the structure of the phage RNA. Early studies on the physical properties of the phage RNAs are presented. The main attention is devoted, however, to the development of sequencing techniques and elucidation of the primary structure of the MS2 RNA as the first fully sequenced genome in the world, which was achieved by the famous Walter Fiers' team.

The four RNA phage proteins—maturation protein, or A protein (Chapter 11), coat protein (Chapter 12), replicase protein, or replicase subunit—as well as the structure of full replicase enzyme (Chapter 13) and lysis protein, or lysin (Chapter 14) are presented in all available structural and functional detail. The structure and function of phage replicases, first of the Qβ RNA replicase, with a full description of the protein factors required for the replication of phage Qβ RNA, are presented in accordance with up-to-date findings. Chapter 14 presents not only the intriguing method of lysis protein discovery, but also unveils the diversity of the lysis protein structure, location within the genome, and possible interconnections with other functionally related proteins.

The intrinsic antigenicity and immunogenicity data, including historically important evidence on the induction of phage-neutralizing antibodies and mapping of the exposed RNA phage B cell epitopes, are the subject of Chapter 15.

The most important theoretical studies on the RNA phages are connected with the translation mechanisms and deciphering of the genetic code. The appropriate data are summarized in Chapter 16. The in-depth elucidation of phage RNAs as classical translation templates *in vivo* and *in vitro* and the global role of the phage RNAs in clarification of the general mechanisms of ribosome recognition, chain initiation and termination, and the structure of the RNA initiation sites are described. Special attention is devoted to the critical role of both the coat protein and the replicase subunit as specific repressors of the phage RNA translation.

Replication of the RNA phages, which is described in Chapter 17, has served as a prototype for all further studies on virus replication. This chapter presents data on the replication features as steps in the infectious life cycle of the RNA phages, asymmetric synthesis of plus- and minus-RNA, and fine structure of the replicative intermediates. The regulation of the RNA replication by the interconnection with the translation of the RNA plus strands, together with an original model of this regulation, is described in particular detail, since the RNA phages were the first examples of an *operon* mechanism of gene regulation by *self* proteins. This unique mechanism is described as a full-cycle biological feedback system, where gene expression is regulated by two phage proteins: coat protein and replicase. The coat proteins of most RNA phages have been shown to repress translation of the replicase gene by binding to an RNA hairpin as an operator at the start site of the replicase gene. X-ray crystallography led to a breakthrough in understanding the protein-RNA interactions that occur during translational repression and genome encapsidation. This breakthrough was particularly apparent after the first crystal structure of a complex of recombinant MS2 capsids with the 19-nucleotide RNA operator was resolved and the residues responsible for the protein-RNA interactions were localized. Finally, possible medical usage of double-stranded phage RNA as an interferon inducer is presented here.

The RNA phages have also presented substantial background for studies on phylogeny and genome evolution, as described in Chapter 18. In time, the evolutionary experiments *in vitro* and the famous Sol Spiegelman's Monster raised more questions than answers. The problem remains urgent today. This chapter sheds light on the current status quo not only in the self-replication by Spiegelman, but also for the multicomponent structure of Qβ RNA replicase and the fine mechanisms of Qβ RNA synthesis.

Chapter 19 is dedicated to the triumphant entry of the RNA phages into the gene engineering era, when RNA phage genomes and genes were expressed successfully and the appropriate phage proteins—coat and read-through A1

proteins, maturation A and A2 proteins, phage-specific replicase subunit, and lysis proteins—were studied structurally and functionally in the context of global molecular biology. The successful beginning of the RNA phage coat-based virus-like particle (VLP) story is outlined here, since the expression of the coat protein genes in *E. coli* and yeast species *Saccharomyces cerevisiae* and *Pichia pastoris* led to the high-level production of the correctly self-assembled icosahedral capsids that were morphologically and immunologically indistinguishable from virions and paved the way to numerous nanotechnological applications.

Chapter 20 is dedicated to the main principles of the practical realization of the idea of RNA phage VLPs being used as universal carriers of foreign sequences in the form of the so-called chimeric VLPs. The numerous studies have shown on the background that both genetic fusion and chemical coupling have been successfully used for the production of the impressive number of VLP chimeras. The level of VLP tolerance to foreign elements displayed on VLP surfaces and the stability of VLPs are analyzed here in general terms. Moreover, the data that demonstrate the ability of coat proteins to tolerate different organic (such as peptides, oligonucleotides, and carbohydrates) and inorganic (such as metal ions) compounds, either chemically coupled or non-covalently added to the outer and/or inner surfaces of the chimeric VLPs, are presented.

The place of RNA phages, as well as their regular and chimeric VLPs in the global history of viral architecture, is invaluable, since they have been an essential contribution to the substantial development of the spatial characterization and 3D resolution of biological subjects by step-wise introduction of electron microscopy, x-ray crystallography, electron cryomicroscopy, and nuclear magnetic resonance. Chapter 21 deals with the corresponding 3D data. This long way of 3D studies is of special interest, since the capsids of the RNA phages R17 and f2 were among the first observed virions with resolved icosahedral symmetry, after the classical work by Donald Caspar and Aaron Klug on the structural analysis of plant viruses. The novel sensational data on the maturation protein location on the virions as well as the data on the structural basis of the RNA packaging process are discussed here. The recent success with high resolution structures of chimeric VLPs is also highlighted.

Vaccines, the subject of Chapter 22, represent the most advanced field of RNA phage VLP applications due to the excellent and well-established scaffold properties and structural tolerance to the decoration by foreign immunogenic sequences. Such decoration can be performed both genetically and chemically, and the VLP scaffold may provide foreign epitopes with the strong T cell response. Moreover, the RNA phage VLPs can serve as nanocontainers that can encapsulate specific adjuvants, such as immunostimulatory oligodeoxynucleotides, or CpGs, as the TLR9 ligands. RNA phage VLPs can package single-stranded or double-stranded RNA fragments as TLR7 and TLR3 ligands, respectively. Moreover, recombinant RNA phage VLPs contain spontaneously encapsulated bacterial RNA, which may also act as an adjuvant. It is emphasized here that the chemical coupling approach was validated by an impressive line of experimental therapeutic vaccines developed by Martin F. Bachmann's team. Bachmann's idea of therapeutic vaccines was based on the assumption that the VLP carriers can present not only surface-displayed self-antigens but also augment their ability to overcome the natural tolerance of the immune system toward self-proteins, and to induce high levels of specific autoantibodies. This approach was initially planned as a method of replacing host-specific monoclonal antibodies in the treatment of acute and chronic diseases. The experimental vaccines and vaccine prototypes, including therapeutic ones, are listed in this chapter.

Chapter 23 is devoted to VLP applications other than vaccines. First, the popular *armored* RNA technology is described as an excellent example of the broad and successful application of the RNA phage VLPs in modern diagnostic kits. *Armored* nucleic acids as noninfectious, easily available reagents are employed widely for quality control in the diagnosis of many pathogenic viruses, including influenza virus, hepatitis C virus, severe acute respiratory syndrome (SARS) coronavirus, or Ebola virus. Second, a detailed list is given of the prospective VLP-based experimental approaches that could be classified as VLP *nanocontainer* studies, including the *packaging, targeting,* and *delivery* steps. Third, a detailed list of the RNA phage VLP-based bioimaging applications is presented. The use of imaging agents in combination with RNA phage VLPs has contributed recently to the high-resolution and noninvasive visualization of these particles, as well as to the potential treatment of diseases. Finally, the solid contribution of RNA phage VLPs to the general idea of *nanoreactors* is highlighted.

Chapter 24 is devoted to RNA phage non-VLP applications, and begins with the exciting story how the RNA phages paved the way for the antisense-based gene therapy via the generation of the so-called *mRNA-interfering complementary RNA (micRNA) immune system.* Major attention is focused, however, on the tethering and imaging applications, which are based on the ability of the RNA phage coat protein to recognize the corresponding operator stem-loop. This feature led to the development of the pioneering tethering methodology that allowed identification, isolation, and purification of the desired RNA-protein complexes of various origins, when mRNAs were tagged with the operator sequence and were then highly specifically recognized by the RNA phage coat fused to different functional probes. The tethering technique allowed affinity purification of such desired RNA-protein complexes. Moreover, when the RNA phage coat was fused to fluorescent probes, a revolutionary imaging technology arose. This technique enabled imaging of the processing, export, localization, translation, and degradation of operator-tagged mRNAs in live cells and live animals. The tethering/imaging methodology exploited mostly the coat protein-operator composition from the phages MS2 and PP7, including simultaneous MS2 and PP7 two-color labeling. Finally, the tethering technique is applied to the further development of the highly productive CRISPR-Cas9 technology.

A summary and future perspectives of RNA phage applications are presented in Chapter 25. First, the yearly statistics are presented on the papers devoted to the RNA phages, showing the enormous growth of the number of publications. Next, development of the evolutionary studies is highlighted, due to the rapid progress of metagenomic sequencing. In this light, evolutionary studies are recommended as the most direct method for the discovery of novel RNA phage VLP carriers and for the further participation of the RNA phages in the invaluable progress of nanotechnology. Finally, the rapid development of revolutionary bacterial immunity and phage therapy investigations is scrutinized from the point of view of the RNA phage field.

A special feature of this book is the comprehensive list of papers dealing with RNA phages and their applications. To simplify working with these papers, they are grouped by year of publication and arranged in alphabetical order within each year group. This idea was adapted from the first RNA phage book, written by Elmārs Grēns.

In conclusion, the major goal of this book is not only to show the continuous development of RNA phage-connected studies over 60 years, but also to inspire the reader with the enthusiasm of the numerous brilliant scientists who have contributed to this immense field.

1 History and First Look

The RNA phages were discovered accidentally.

Norton N. Zinder

Everything will turn out right, the world is built on that.

Mikhail Bulgakov
The Master and Margarita

CLASSICAL REVIEWS

The early history and first classical applications of RNA phages have been reviewed by pioneers and heroes of the RNA phage discovery and elucidation (Zinder 1963, 1965, 1966–1967, 1980, 1988; Schindler 1964b; Gussin et al. 1966; Bradley 1967; Brinton and Beer 1967; Hoffmann-Berling et al. 1967; Kaesberg 1967; Watanabe and August 1967b; Weissmann 1967, 1974, 1976; Hofschneider 1969, 1972; Valentine et al. 1969b; Hohn and Hohn 1970; Stavis and August 1970; Spiegelman 1971b; Brown and Hull 1973; Weissmann et al. 1973a,b; Zubay 1973; Eoyang and August 1974; Fiers 1974, 1977, 1979; Tikchonenko 1975; Ochoa 1976; van Duin 1988, 2005; Lodish 2012). The first book fully devoted to RNA phage replication was published in Riga by Elmārs Grēns (Grens 1974). The classical *RNA Phages* book of 1975 collected 14 excellent review articles written by the famous RNA phages pioneers. Norton D. Zinder, the father of RNA phages, wrote in the classical "Portraits of Viruses: RNA Phage" review article (Zinder 1980): "The RNA phages were discovered accidentally. A group of phages had been isolated on the basis of their ability to grow on male *Escherichia coli* (Loeb 1960)." This and further citations of the classical Zinder review of 1980 are reprinted here with kind permission of S. Karger AG, Basel. Table 1.1 sets the characterized RNA phages in order by year of their appearance and elucidation.

In the context of this table, it should be noted that the *Gremmeniella abietina* RNA virus MS2 (GaRV-MS2) was described (Tuomivirta and Hantula 2005). *G. abietina* is a species of fungal diseases infecting coniferous forests. The GaRV-MS2 genome is composed of three double-stranded RNA molecules, and the virus belongs to the *Partitiviridae* family (Tuomivirta and Hantula 2005; Zhang T et al. 2013) and has nothing to do with the *Leviviridae* members.

MALE SPECIFICITY

Citing the first exhaustive review of the father of RNA phages (Zinder 1965):

These phages will attach only to male bacteria (Loeb 1960; Loeb and Zinder 1961). This includes Hfr, F+, and F′ strains. Although there is no evidence which shows that only the RNA penetrates the host bacteria. this seems likely considering that the infectivity becomes sensitive to the enzyme ribonuclease

(Zinder 1963). Calcium is required for the phage penetration, but any physiological ionic environment will suffice for phage attachment. Transfer of the *E[scherichia] coli* F agent to other bacteria including *Salmonella*, *Shigella*, and *Proteus* often confers upon them sensitivity to infection (Brinton et al. 1964). (Citations of Zinder's classical review: RNA phages. *Annu Rev Microbiol.* 1965;19:455–472 are printed here and further with permission of Annual Reviews.)

Furthermore, Zinder (1965) concluded that "the phage attaches to some special pili associated with male *E. coli* [Brinton et al. 1964; Crawford and Gesteland 1964]. These pili may be composed of the male antigen which was detected serologically by the Ørskovs [1960]." Only a limited number of such structures have been seen per cell, and they were distinguished from other filamentous appendages by being covered with many RNA phage particles attached to the side of the pili (Crawford and Gesteland 1964), as well as by filamentous single-stranded DNA phages which attached to the end of the pilus. This point has been illustrated by electron micrograph, presented in Figure 1.1.

In contrast to whole RNA phages, the purified phage RNA was able to infect spheroplasts of both male and female *E. coli*, as well as of *Salmonella*, *Proteus*, and *Shigella* (Davis et al. 1961; Fouace and Huppert 1962; Knolle and Kaudewitz 1962; Paranchych 1963; Engelhardt and Zinder 1964). After detection of the normal multiplication of RNA phages in *Proteus* and *Shigella* spheroplasts, converting of *P. mirabilis* (Horiuchi and Adelberg 1965) and *S. flexneri* (Kitano 1966a, 1966b) to F′+ and lac+ by transferring F-*lac* exogenote from *E. coli* K12 allowed adsorption and multiplication of RNA phages MS2 or M12 in intact *P. mirabilis* and *S. flexneri* cells, although efficiency of the adsorption remained low and no plaque-forming ability on agar plates was observed.

RNA AS THE GENOME

RNA phages were the first detected bacteriophages with an RNA molecule as a genome. At that time only one RNA virus was known, namely, tobacco mosaic virus, a member of *Tobamovirus* genus from *Virgaviridae* family. Citing Zinder (1980) on the first identification of RNA as a genomic substance of the RNA phage f2:

One finding more than any other led to an early chemical analysis of these phages: the titer of the raw lysates was about 10^{12}/ml. Purification was rather simple but also surprising. 20% saturated ammonium sulfate solutions precipitated the infectivity quantitatively without loss of titer. However, centrifuging at the usual 60,000 g left the phage in the supernatant. At each step a diphenylamine test for DNA and an orcinol test for RNA were done. 100,000 g for 2 h pelleted the infectivity. A technician was most concerned that the pellets

TABLE 1.1

The Consecutive Introduction of RNA Phage Species into the Scientific Mode of Life

Year	Phage	Source	Genus	Host	Pili, Incompatibility Group	Serotype	Comments	References
1959	MS2	Sewage, Berkeley, CA	*Levivirus*	*E. coli*	FI-FIV	I	Date of isolation cited by Friedman et al. (2009a)	A.J. Clark, unpublished; Davis et al. (1961)
1960	f2 till f7	Sewage, Manhattan, NY	*Levivirus*	*E. coli*	FI-FIV	I	First mention of the group in 1960; f2 was selected for study	Loeb (1960); Loeb and Zinder (1961)
1960	f4	Sewage, Manhattan, NY	*Levivirus*	*E. coli*	FI-FIV	I	Cross-reaction with f2 antisera: an additional methionine residue per molecule	Loeb and Zinder (1961); Modak and Notani (1969)
1961	μ2	Sewage, Milan	*Levivirus*	*E. coli*	FI-FIV	I	Close relationship with R17 and fr; significant differences with Qβ	Maccacaro (1961); Dettori et al. (1961)
1961	Qβ	Human feces, Kyoto	*Allolevivirus*	*E. coli*	FI-FIV	III	The first appearance	Watanabe (1964); Watanabe et al. (1967a)
1962	R17	Sewage, Philadelphia, PA	*Levivirus*	*E. coli*	FI-FIV	I	The first appearance	Paranchych and Graham (1962)
1962	FH5	Unknown, France	*Levivirus*	*E. coli*	FI-FIV	I	Immunologically related to f2, MS2, and M12	Fouace and Huppert (1962)
1962	ft5 (new name fr)	Dung hill, Heidelberg	*Levivirus*	*E. coli*	FI-FIV	I	Related to f2	Hartmut Hoffmann-Berling, unpublished; Knolle and Kaudewitz (1962)
1963	M12	Sewage, Munich	*Levivirus*	*E. coli*	FI-FIV	I	The first appearance	Hofschneider (1963a,b)
1963	fr	Dung hill, Heidelberg	*Levivirus*	*E. coli*	FI-FIV	I	The first appearance under the fr name (former ft5)	Marvin and Hoffmann-Berling (1963a,b)
1963	β	Sewage, Tokyo	*Levivirus*	*E. coli*	FI-FIV	I	Closely related to MS2	Nonoyama et al. (1963)
1963	B1, I	Sewage, Tokyo	*Levivirus*	*E. coli*	FI-FIV	I	Both serologically related; the relationship between MS2 and β is close but that between β (or MS2) and B1 (or I) is far	Nonoyama et al. (1963); Yuki and Ikeda (1966)
1963	7s	Sewage, New Orleans, LA	Unclassified *Levivirus*	*P. aeruginosa*	Polar pili FP, TFP, type IV pili, T4P	–	Serologically related to PP7, but not identical	Feary et al. (1963, 1964)
1963	VK	Sewage, Tokyo	*Allolevivirus*	*E. coli*	FI-FIV	III	A strain of Qβ, according to recent classification	Watanabe et al. (1967a)

(Continued)

TABLE 1.1 (Continued)
The Consecutive Introduction of RNA Phage Species into the Scientific Mode of Life

Year	Phage	Source	Genus	Host	Pili, Incompatibility Group	Serotype	Comments	References
1964	f_can1	Sewage, Canberra	*Levivirus*	*E. coli*	FI-FIV	I	Neutralized with R17 and μ2 antisera	Davern (1964b)
1964	ZIK/1 ZJ/1 ZG	Marquette, Michigan; Basingstoke, Hampshire, UK; Pangbourne, Berkshire, UK	*Levivirus*	*E. coli*	FI-FIV	?	ZIK/1 and ZJ/1 are similar to each other but are serologically dissimilar to the other coliphages	Bradley (1964a,b)
1964	GA	Sewage, Ōokayama, Tokyo area	*Levivirus*	*E. coli*	FI-FIV	II	The GA subgroup within the serogroup II	Watanabe et al. (1967a,b)
1964	MY	Sewage, Matsuyama, Western Japan	*Levivirus*	*E. coli*	FI-FIV	I		Watanabe et al. (1967a,b)
1965	α15	Water, Braid Burn, Edinburgh	?	*E. coli*	FI-FIV	?	Group I ?	Bishop and Bradley (1965)
1965	ZS/3 ZL/3	Sewage, Linlithgow, Stoke-on-Trent, UK	?	*E. coli*	FI-FIV	?	ZG, ZS/3, ZL/3 and α15 are serologically related and dissimilar to ZIK/1 and ZJ/1, but may fall into two subgroups (i.e., ZG + ZS/3 and α15 + ZL/3)	Bishop and Bradley (1965)
1965	10	Sewage, Tokyo?	*Levivirus*	*E. coli*	FI-FIV	?		Taketo et al. (1965)
1965	PP7	Sewage, Pangbourne, Berkshire, UK	Unclassified *Leviviridae*	*P. aeruginosa*	polar pili FP, TFP, type IV pili, T4P	—	No male specificity; serologically related to 7s, but not identical	Bradley (1965b, 1966b)
1965	φCB8r φCB9	Raw sewage, Berkeley?	Unclassified *Leviviridae*	*C. bacteroides*	?	—	Group IV: serologically distinct from groups V and VI	Schmidt and Stanier (1965)
1965	φCB2 φCB4 φCB5 φCB12r φCB15	Chlorinated and raw sewage, Berkeley?	Unclassified *Leviviridae*	*C. vibrioides* and other *Caulobacter* strains	?	—	Group V: all are serologically related; φCB4 could be a subgroup	Schmidt and Stanier (1965)
1965	φCB23r	Raw sewage, Berkeley?	Unclassified *Leviviridae*	*C. fusiformis*	?	—	Group VI: serologically distinct from groups IV and V	Schmidt and Stanier (1965)
1965	ZR	Sewage, Mitaka, Tokyo area	*Levivirus*	*E. coli*	FI-FIV	I	Identical to MS2 and R17 by amino acid sequence of the coat protein; a single amino acid substitution in the coat protein as compared to f2	Sakurai et al. (1967); Watanabe et al. (1967a)
1966	SD	Sewage, Sendai, Northern Japan	*Levivirus*	*E. coli*	FI-FIV	II	The GA subgroup within the serogroup II	Watanabe et al. (1967a,b)
1966	R23	Atlantic Beach, New York	*Levivirus*	*E. coli*	FI-FIV	I	Related to f2 by serology and RNA base composition; strong inhibition of the host RNA and protein synthesis	Watanabe and August (1966, 1967a,b); Watanabe M (1967)

(Continued)

TABLE 1.1 (Continued)

The Consecutive Introduction of RNA Phage Species into the Scientific Mode of Life

Year	Phage	Source	Genus	Host	Pili, Incompatibility Group	Serotype	Comments	References
1966	ST	Sewage, Sendai, Northern Japan	*Allolevivirus*	*E. coli*	FI-FIV	III	Serologically identical to MS2	Watanabe et al. (1967a)
1967	MSO-12, 14	Sewage, Norman, OK	*Levivirus*	*E. coli*	FI-FIV	I	Immunologically unrelated, inhibited slowly by anti-R23 serum; R34 is similar to R23 by inhibition of RNA synthesis	Lancaster (1968)
1968	R34, R40	The same as for the R23 phage?	Unassigned	*E. coli*	FI-FIV	?		Watanabe M et al. (1968); Watanabe and Watanabe (1970a)
1968	SP	Feces of a Siamang Gibbon *Symphalangus syndactylus*	*Allolevivirus*	*E. coli*	FI-FIV	IV	Serologically unrelated to any of MS2 or ZR (group I), GA (group 11), and Qβ (group III)	Sakurai et al. (1968)
1969	FI: FIA, FIB, FIC	Feces of infants, Hachioji, Tokyo	*Allolevivirus*	*E. coli*	FI-FIV	IV	Similar except the host range; mutants originating from the same origin?	Miyake et al. (1969)
1970	G17	Qβ samples	*Levivirus*	*E. coli*	FI-FIV	?	A member of the R17 group, but serologically distinguishable from R17	Strauss and Kaesberg (1970)
1971	Y Z	Sewage, Tokyo	Unassigned	*E. coli*	FI-FIV	?	Serologically distinct from both phage β (related to MS2) and phage I distinct from β; phage Z was inactivated by antisera of I, Y, and β, and vice versa; the presence of at least three types of RNA phages was suggested	Tsuchida et al. (1971a)
1971	06N-58P	Sea water, 31°04′N-130°35′E	Unassigned	*Pseudomonas* sp.	?	–	The host bacterium of 06N-58P is a marine *Pseudomonas* sp.; the phage 06N-58P is small and apparently simple in structure. It shares the property of susceptibility to the action of ribonuclease in the presence of host bacterium; diameter 60 nm	Hidaka (1971, 1972, 1975); Hidaka and Ichida (1976)
1971	Hd	Sewage, Dunedin, New Zealand	*Levivirus*	*E. coli*	FI-FIV	I	Serologically related to f2 and not to Qβ	Bilimoria and Kalmakoff (1971)
1971	TW18	Taiwan?	*Allolevivirus*		FI-FIV	III	Group III, subgroup (d), a strain of Qβ?	Furuse et al. (1978)
1971	TW19 TW28	Sewage, Changhua, Taiwan	*Allolevivirus*	*E. coli*	FI-FIV	IV		Miyake et al. (1971a)
1971	BZ1	Sewage/feces, Recife, Brazil	*Allolevivirus*	*E. coli*	FI-FIV	III	A strain of Qβ?	Miyake et al. (1973)
1972	TH1	Sewage, Thailand	*Levivirus*	*E. coli*	FI-FIV	II	An analogue of GA?	Aoi et al. (1972)
1973	PRR1	Sewage, Kalamazoo, MI	Unclassified *Levivirus*	*P. aeruginosa*	P-1 (R1822 RP1, RP4)	–	The first male-specific *Pseudomonas* phage	Olsen and Shipley (1973); Olsen and Thomas (1973)

(Continued)

TABLE 1.1 (Continued)
The Consecutive Introduction of RNA Phage Species into the Scientific Mode of Life

Year	Phage	Source	Genus	Host	Pili, Incompatibility Group	Serotype	Comments	References
1973	KU1	Sewage, Kuwait	*Levivirus*	*E. coli*	FI-FIV	II	Member of the GA subgroup	Furuse et al. (1975b)
1973?	JP501	Sewage, Japan	*Levivirus*	*E. coli*	FI-FIV	I		Furuse et al. (1973)
1973	JP34, JP500	Japan	*Levivirus*	*E. coli*	FI-FIV	II	Serological intermediates between groups I and II; a subgroup within the group II that is different from the GA subgroup	Furuse et al. (1973)
1973	BZ13	Sewage, Brazil	*Levivirus*	*E. coli*	FI-FIV	II	A member of the GA subgroup	Miyake et al. (1973)
1974	KC	?	*Levivirus*	*E. coli*	FI-FIV	I	Antigenically related to MS2	Dhillon et al. (1974)
1974	HK102	Sewage, Hong Kong	*Allolevivirus*	*E. coli*	FI-FIV	III	Antigenically related to Qβ	Dhillon et al. (1974)
1975	BO1	Sewage, Bolivia	*Levivirus*	*E. coli*	FI-FIV	I	An analogue of fr?	Furuse et al. (1975b)
1975	MX1	Domestic drainage, Mexico	*Allolevivirus*	*E. coli*	FI-FIV	III	A serological intermediate between groups III and IV; inactivated by antisera against the group IV phage SP and the group III phage Qβ, while anti-SP serum inactivates more extensively (about 100-fold) than does anti-Qβ serum	Furuse et al. (1975b); Hirashima et al. (1983)
1975	pilE/R1	?	Unassigned	*E. coli*	T	—		To et al. (1975)
1976	φCp2 φCp18 φCp21 φCp28 φCp32 φCp42	Sewage and river water, Tokyo and Ibaragi	Unassigned	*C. crescentus*		—	A group of φCp2, φCp21, φCp28, φCp32, and φCp42 phages and φCp18 are serologically different from each other and from φCB5 as well; phages φCp2 and φCp18 may be classified in the fifth group (φCB5) of Schmidt and Stanier 1965	Miyakawa et al. (1976); Fujiki et al. (1978)
1976	PP25	Sewage, Philippines	*Levivirus*	*E. coli*	FI-FIV	I	Group I subgroup (a)	Furuse et al. (1978)
1976	PP3	Sewage, Philippines	*Levivirus*	*E. coli*	FI-FIV	II	Group II subgroup (c)	Furuse et al. (1978)
1976	PP4, 6, 7, 8, 11, 12, 15, 18, 19, 23, 24, 26, 27, 28, 30, 31	Sewage, Philippines	*Allolevivirus*	*E. coli*	FI-FIV	III	Group III subgroup (a)	Furuse et al. (1978)
1976	PP2, 9, 13, 14, 16, 17, 21, 22, 29	Sewage, Philippines	*Allolevivirus*	*E. coli*	FI-FIV	III	Group III subgroup (b)	Furuse et al. (1978)
1976	PP1, 5, 10, 20	Sewage, Philippines	*Allolevivirus*	*E. coli*	FI-FIV	III	Group III subgroup (g)	Furuse et al. (1978)
1976	TL1	Sewage, Bangkok, Thailand	*Allolevivirus*	*E. coli*	FI-FIV	III	Group III subgroup (a)	Furuse et al. (1978)

(Continued)

TABLE 1.1 (Continued)
The Consecutive Introduction of RNA Phage Species into the Scientific Mode of Life

Year	Phage	Source	Genus	Host	Pili, Incompatibility Group	Serotype	Comments	References
1976	TL2	Sewage, Bangkok, Thailand	*Levivirus*	*E. coli*	FI-FIV	II	An analogue of GA group II, subgroup (d)	Furuse et al. (1978)
1976	SP5	Tanglin Sewage Pumping Station, Singapore	*Levivirus*	*E. coli*	FI-FIV	II	Group II subgroup (a)	Furuse et al. (1978)
1976	SP1, 2, 3, 4, 6, 7, 8, 9, 10, 11, 12	Tanglin Sewage Pumping Station, Singapore	*Allolevivirus*	*E. coli*	FI-FIV	III	Group III subgroup (a)	Furuse et al. (1978)
1976	IN7	Jakarta, Indonesia	*Levivirus*	*E. coli*	FI-FIV	II	Group II subgroup (d)	Furuse et al. (1978)
1976	IN1, 2, 3, 4, 5, 6	Jakarta, Indonesia	*Allolevivirus*	*E. coli*	FI-FIV	III	Group III, subgroup (a)	Furuse et al. (1978)
1976	ID1	Sewage, India	*Allolevivirus*	*E. coli*	FI-FIV	III	Group III, subgroup (d)	Furuse et al. (1978)
1976	ID2	Sewage, India	*Allolevivirus*	*E. coli*	FI-FIV	IV	Group IV, subgroup (f); neutralized well by the antisera of TW19 (IV) and TW28 (IV), but also by antisera of the group I, II, and III phages, although to a lesser extent; classified under group IV by RNA size; the first isolate to share common characteristics of groups I, II, III, and IV	Furuse et al. (1978, 1979b)
1977	φCR14 φCR28	Ponds, slow-moving streams, and tropical fish tanks, Baltimore, MD; island off the coast, VA	Unassigned	*Caulobacter*	?	–	Similar to φCB5 by salt sensitivity	Johnson et al. (1977)
1979	KR17 KR32 KR69 KR87	Sewage samples from domestic drainage, Seoul and Busan	*Levivirus*	*E. coli*	FI-FIV	I	Six standard antisera were used by isolation: MS2 (group I, GA and JP34 (group II), Qβ and VK (group III), SP (group IV)	Osawa et al. (1981a)
1979	KR: 47 strains	Sewage samples from domestic drainage, Seoul and Busan	*Levivirus*	*E. coli*	FI-FIV	II		Osawa et al. (1981a)
1979	KR: 50 strains	Sewage samples from domestic drainage, Seoul and Busan	*Allolevivirus*	*E. coli*	FI-FIV	IIIa		Osawa et al. (1981a)
1979	KR2 KR52 KR58 KR60 KR97	Sewage samples from domestic drainage, Seoul and Busan	*Allolevivirus*	*E. coli*	FI-FIV	IIIb		Osawa et al. (1981a)
1980	GH1	Sewage samples from domestic drainage, Accra	*Levivirus*	*E. coli*	FI-FIV	I		Osawa et al. (1983b)

(Continued)

TABLE 1.1 (Continued)
The Consecutive Introduction of RNA Phage Species into the Scientific Mode of Life

Year	Phage	Source	Genus	Host	Pili, Incompatibility Group	Serotype	Comments	References
1980	MD3	Sewage samples from domestic drainage, Antananarivo	*Levivirus*	*E. coli*	FI–FIV	I		Osawa et al. (1983b)
1980	GH2 GH3 GH4	Sewage samples from domestic drainage, Accra	*Levivirus*	*E. coli*	FI–FIV	II		Osawa et al. (1983b)
1980	MD2	Sewage samples from domestic drainage, Antananarivo	*Levivirus*	*E. coli*	FI–FIV	II		Osawa et al. (1983b)
1980	SG4 SG6	Sewage samples from domestic drainage, Dakar	*Levivirus*	*E. coli*	FI–FIV	II		Osawa et al. (1983b)
1980	GH5 GH6 GH7	Sewage samples from domestic drainage, Accra, Suhum, Ghana	*Allolevivirus*	*E. coli*	FI–FIV	IIIa		Osawa et al. (1983b)
1980	MD7	Sewage samples from domestic drainage, Moramanga, Madagascar	*Allolevivirus*	*E. coli*	FI–FIV	IIIa		Osawa et al. (1983b)
1980	SG1, 2, 3, 5, 7, 8, 9, II, 12, 13, 14, 16	Sewage samples from domestic drainage, Dakar	*Allolevivirus*	*E. coli*	FI–FIV	IIIa		Osawa et al. (1983b)
1980	SG15	Sewage samples from domestic drainage, Dakar	*Allolevivirus*	*E. coli*	FI–FIV	IIIb		Osawa et al. (1983b)
1980	SG10	Sewage samples from domestic drainage, Dakar	*Allolevivirus*	*E. coli*	FI–FIV	IVf		Osawa et al. (1983b)
1980	MD1, 4, 5, 6	Sewage samples from domestic drainage, Antananarivo	*Allolevivirus*	*E. coli*	FI–FIV	IVg		Osawa et al. (1983b)
1980	UA-6	Sewage, Edmonton	Unassigned	*E. coli*	FV, SI (EDP208)	–	Hosts carrying EDP208 plasmid	Armstrong et al. (1980)
1981	F₀*lac*	Sewage, Pretoria	Unassigned	*E. coli* *S. typhimurium*	FV, SI (EDP208)	–	Hosts carrying EDP208 plasmid; serologically related to the phage UA-6	Bradley et al. (1981a)
1981	t, or τ	Sewage, Pretoria	Unassigned	*E. coli* *S. typhimurium*	T	–	Hosts carrying IncT plasmid	Bradley et al. (1981b)
1981	C-1, later C-1 INW-2012	Sewage, Pretoria	Unclassified *Levivirus*	*E. coli* *S. typhimurium* *P. mirabilis* *Ser. marcescens*	C	–	Hosts carrying IncC plasmid	Sirgel et al. (1981); Sirgel and Coetzee (1983); Kannoly et al. (2012)

(Continued)

TABLE 1.1 (Continued)
The Consecutive Introduction of RNA Phage Species into the Scientific Mode of Life

Year	Phage	Source	Genus	Host	Pili, Incompatibility Group	Serotype	Comments	References
1981	Ri (9 clones)	Sewage? Tbilisi	*Levivirus*	*E. coli*	FI-FIV	I	Neutralized by sera against f2 and fr; attributed to one of the three subgroups within the group I	Darsavelidze and Chanishvili (2005)
1981	LR60	A concentrate of river water, Coventry, UK	Unassigned	Not described	P-1 (R1822 RP1, RP4)	–	Hosts carrying P-group plasmid; resembles the phage PRR1	Primrose et al. (1982)
1982	Iα	Sewage, Windhoek, Namibia	Unassigned	*E. coli* S. typhimurium	I_1 I_ζ	–		Coetzee et al. (1982)
1983	M	Sewage, Pretoria	Unclassified *Levivirus*	*Proteus* and other genera harboring incM plasmids	M	–	Sensitive to chloroform	Coetzee et al. (1983)
1985	pilHα	Sewage, Pretoria	Unassigned	*E. coli, S. typhimurium, Klebsiella, P. morganii* and *Ser. marcescens* strains carrying any one of the H-complex plasmids used	HI HII	–	Temperature-sensitive, sensitive to chloroform	Coetzee et al. (1985a)
1985	D	Sewage, Pretoria-Johannesburg area	Unassigned	*E. coli K 12* strains and strains of *Salmonella typhimurium, Proteus morganii,* and *Klebsiella oxytoca* harboring one of these plasmids	D	–	Sensitive to chloroform; the sera against phage D did not neutralize MS2, t, F_0lac, M, or pilHα.	Coetzee et al. (1985b)
1986	F_0lac h, SR	Sewage, Pretoria-Johannesburg area	Unassigned	*E. coli S. typhimurium*	S	–	Sensitive to chloroform; F_0lac and its mutant F_0lac h are related serologically but are distinct from phages SR, MS2, and D, each of which is serologically unique	Coetzee et al. (1985b, 1986)
1986	NL95	Calf feces, The Netherlands	*Allolevivirus*	*E. coli*	FI-FIV	IV	Low immunological cross-reactivity with five different antisera against subgroup IV phages	Havelaar et al. (1986)
1986?	M11	Bank filtrate, The Netherlands	*Allolevivirus*	*E. coli*	FI-FIV	III	Originally serotyped as the subgroup IV phage (K. Furuse, personal communication)	A. Havelaar, unpublished data; Beekwilder et al. (1995)

(Continued)

TABLE 1.1 (Continued)
The Consecutive Introduction of RNA Phage Species into the Scientific Mode of Life

Year	Phage	Source	Genus	Host	Pili, Incompatibility Group	Serotype	Comments	References
1987	Hgal1 till Hgal27	Sewage, Galway docks, Ireland	Unclassified Levivirus	E. coli and other bacterial strains carrying plasmids of the IncHI or IncHII groups	H	–	Temperature-sensitive; the Hgal phage is closely related to pilHα by a set of properties including resistance to chloroform	Nuttall et al. (1987); Maher et al. (1991)
1988	φ3122, φ7087.1, φ7087.2	Activated sludge, Tampere, Finland	Levivirus?	E. coli	FI-FIV	?	The protein composition and sedimentation behavior of these phages suggest that they are single-stranded RNA-type phages like the phage MS2	Hantula et al. (1991)
1995	AP205	Quebec, Canada	Unclassified Leviviridae	Acinetobacter genospecies 16	?	–		Coffi (1995)
2000	HL4-9	Hog lagoon, Duplin County, NC	Allolevivirus	E. coli	FI-FIV	III	A strain of Qβ?	Friedman et al. (2009a)
2000	J20	Chicken litter, South Carolina	Levivirus	E. coli	FI-FIV	I		Friedman et al. (2009a)
2002	HB-P22	Bird, Talbert Marsh sandflats, Huntington Beach, CA	Allolevivirus	E. coli	FI-FIV	IV	A strain of FI	Friedman et al. (2009a)
2002	HB-P24	Bird, Talbert Marsh sandflats, Huntington Beach, CA	Allolevivirus	E. coli	FI-FIV	IV	A strain of FI	Friedman et al. (2009a)
2002	T72	Bird, Talbert Marsh sandflats, Huntington Beach, CA	Levivirus	E. coli	FI-FIV	II	A strain of BZ13	Friedman et al. (2009a)
2004	DL1	River water, Tijuana River, CA	Levivirus	E. coli	FI-FIV	I		Friedman et al. (2009a)
2004	DL2	Bay water, Delaware Bay, DE	Levivirus	E. coli	FI-FIV	I		Friedman et al. (2009a)
2004	DL13	Oyster, Whiskey Creek (Masonboro Is), NC	Levivirus	E. coli	FI-FIV	I		Friedman et al. (2009a)
2004	DL16	Bay water, Great Bay (Nannie Is), NH	Levivirus	E. coli	FI-FIV	I		Friedman et al. (2009a)
2004	DL52	Bay water, Rachel Carson Reserve, NC	Levivirus	E. coli	FI-FIV	I-JS-like	A possible strain or isolate of MS2	Friedman et al. (2012)
2004	DL54	Bay water, Narragansett Bay, RI	Levivirus	E. coli	FI-FIV	I-JS-like		Friedman et al. (2012)

(Continued)

TABLE 1.1 (Continued)
The Consecutive Introduction of RNA Phage Species into the Scientific Mode of Life

Year	Phage	Source	Genus	Host	Pili, Incompatibility Group	Serotype	Comments	References
2004	WWTP1_50 2G1I3	Wastewater, Massachusetts and South Carolina	*Levivirus*	*E. coli*	FI-FIV	I-JS	Novel group JS proposed, in addition to MS2- and GA-like phages of the groups I and II, respectively; the two JS strains, both isolated from wastewater samples, have 40% sequence diversity with strains from the MS2 or GA groups	Vinjé et al. (2004)
2005	BR1	Water, Guerin Creek, Charleston, SC	*Allolevivirus*	*E. coli*	FI-FIV	IV		Friedman et al. (2009a)
2005	BR8	Water, Bull Creek, Charleston, SC	*Allolevivirus*	*E. coli*	FI-FIV	IV		Friedman et al. (2009a)
2005	BR12	Water, New Market Creek Charleston, SC	*Allolevivirus*	*E. coli*	FI-FIV	III	A strain of Qβ?	Friedman et al. (2009a)
2005	DL10	Mussel, Tijuana River, CA	*Levivirus*	*E. coli*	FI-FIV	II	A strain of BZ13	Friedman et al. (2009a)
2005	DL20	Clam, Narragansett Bay, RI	*Levivirus*	*E. coli*	FI-FIV	II	A strain of BZ13	Friedman et al. (2009a)
2007	XY-1	Sewage pond in hospital, Chongqing, China	Unassigned	*E. coli*	FI-FIV	?	Broad host range among *E. coli* strains (six *E. coli* reference strains killed); related to f2 by morphology; single-stranded RNA genome of ~ 5.0 kb; diameter 40–50 nm	Song et al. (2007)
2012	PaMx54 PaMx60 PaMx61	Environmental and sewage water, Central Mexico	Unassigned	*P. aeruginosa*	?	–	Three novel records in addition to the three previously described *P. aeruginosa* phages	Sepúlveda-Robles et al. (2012)
2016	Strains 1 and 5	Lehtoniemi municipal wastewater treatment plant, Kuopio, Finland	Unassigned	*E. coli* ?	?	–	Chlorine- and UV-resistant	Zyara et al. (2016a,b, 2017)
2017	LeviOr01	Sewage water, Armand Trousseau Hospital, Paris	Unclassified *Leviviridae*	*P. aeruginosa* strain PcyII-10	IV	–	The first RNA phage with the clearly acknowledged carrier state in its host	Pourcel et al. (2017)
Unknown	Raa₄	Unknown	Unassigned	*E. coli* ?	?	–	Used as a standard in the wastewater treatment study	Rajala-Mustonen and Heinonen-Tanski (1995)
Unknown	ST4	Unknown	*Levivirus*	*E. coli*	FI-FIV	I		Friedman et al. (2009a)

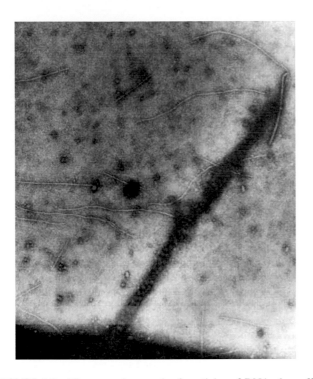

FIGURE 1.1 Electron micrograph of particles of RNA phage f2 attached to an F pilus. The thin filaments in the field are the filamentous phage f1 (800 nm) which attach to the end of a pilus. Negatively stained with phosphotungstic acid, ×125,000. (Reprinted with permission of S. Karger AG, Basel from Zinder ND. *Intervirology*. 1980;13:257–270.)

had little DNA but lots of RNA 'contamination'. It was most tantalizing. However, we had lost about 10-fold in infectivity, some to mechanical loss and some to inactivation of phage. Dependent upon the amount of nucleic acid per particle, the amount of DNA still present could have explained the residual infectivity. We were sure we had a small phage, as there were small pellets and large titers. The year was 1960 and the only small phage known was φX174, described only 1 year before (Sinsheimer 1959a,b). φX174 made large plaques, had a small burst size and not only was it not male-specific but it did not grow at all on *E. coli* K12. Thus, our phage was not φX174. We subjected the phage to CsCl density gradient centrifugation, setting the average density at the 1.5 of the typical phage. A phage band of about 3 mm width was just below the top of the gradient (phage's σ = 1.4). With purified phage at hand, chemical analysis soon revealed that the phage, f2, had to contain an oxypentose sugar and, by precedent, probably RNA. Further analysis showed the sugar to be ribose. We had isolated the first RNA phage (Loeb and Zinder 1961). (Citations of Zinder's classical review: Portraits of viruses: RNA phage. Intervirology. 1980;13:257–270 are printed here and further with permission of S Karger AG, Basel.)

Further, full independence of the RNA phage replication from DNA was demonstrated by the first discoverers (Cooper and Zinder 1962; Lodish and Zinder 1966b).

Although free RNA phages were not inactivated by ribonuclease, a culture did not become infected if small quantities of ribonuclease were added to the medium prior to infection, as reported by Tim Loeb in 1961 (cited by Valentine and Wedel 1965). The complete block of the RNA phage plaque yield

by the ribonuclease pretreatment was published first in the case of the RNA phage fr (Knolle and Kaudewitz 1963). The reduction of titers of the *Pseudomonas* phage 7s in the presence of ribonuclease was also noted (Feary et al. 1964). The same was true also for the *Caulobacter* phages of the φCB series (Schmidt and Stanier 1965). This effect of the ribonuclease treatment has since been used for the simple and rapid estimation of the nucleic acid type of newly isolated phages.

GROWTH CYCLE

The first growth cycle of an RNA phage from the first paper on RNA phages is presented in Figure 1.2. Zinder characterized the growth cycle of the RNA phage f2 as

… not different from that of any other virus. In broth media at 37°C there is a 15 min eclipse period in which no infective particles can be recovered from the infected bacteria. Phage then accumulates intracellularly. The phage is released by lysis of bacteria. There is, however, no evidence of a lysozyme-like enzyme as has been found following infection with other bacteriophages. The first phage is released from bacteria at about 22 min and phages continue to be released for another 30 min (Zinder 1965).

At high cell density, the cultures could be lysis-inhibited and some 20–40,000 particles could be synthesized by each bacterium (Loeb and Zinder 1961). Interestingly, the first calculations that under conditions of lysis inhibition some 5% of the bacterial mass might become virus led to conclusion that

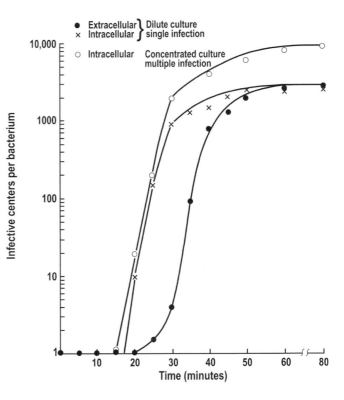

FIGURE 1.2 The first representation of the intra- and extracellular growth of f2. (Redrawn with permission of Stephen H. Zinder from Loeb T, Zinder ND. *Proc Natl Acad Sci U S A*. 1961;47:282–289.)

"the RNA phage is currently the most populous organism in the world" (Zinder 1965).

COLIPHAGES

A first series of bacteriophages, or coliphages, which were specific for male strains of *E. coli* K12, was isolated from New York sewage by Tim Loeb (1960), who was Zinder's first student at Rockefeller University. Originally, the work was designed to search for a bacteriophage capable of distinguishing F⁻ from F⁺ and Hfr mating types of *E. coli* K12. Because of their specificity for strains carrying the fertility agent F, the selected phages were labeled f1 through f7, the numbers designating isolations from different batches of sewage (Loeb 1960). It was concluded that the inability of the selected phages to grow on female bacteria was due to their failure to attach to them.

Of the original set, only f1 did not contain RNA, but was "a unique and strange creature," a filamentous phage with a single-stranded DNA that started another famous line of the bacteriophage history (Zinder et al. 1963). But this is a special long story of the extensive development and brilliant applications, including famous phage display, of the rod-shaped circular single-stranded DNA phages forming the *Inoviridae* family.

One of the RNA phages in the list, namely f2, became not only the first described RNA bacteriophage, but also the first species in the line of highly popular and well-studied family of single-stranded RNA phages (Loeb and Zinder 1961). This revolutionary paper (Loeb and Zinder 1961) presented clear biochemical evidence that the bacteriophage f2 contained RNA and not DNA as its nucleic acid. The evidence was based "primarily on the analysis of purified material but also on the distribution of the two types of nucleic acid synthesized after infection" (Loeb and Zinder 1961). The authors emphasized that "although f2 is an extremely small phage, there is a compensating large yield per bacterium (about 10,000 pfu)" and presented an efficient scheme for the isolation and purification of the phage f2 in large quantities. From the remaining five RNA phages of the first set, only the phage f4 was characterized in further detail and found to be closely related to the RNA phage f2 (Modak and Notani 1969).

The first paper was followed by a long list of novel RNA phage species from around the world. The famous RNA phage MS2, the evident leader by current number of publications and variety of applications, was found in 1961, or even in 1959 by some references, in Berkeley by Alvin John Clark (cited in Davis et al. 1961; Zinder 1965) who discovered, together with Ann Dee Margulies, recombination-less (Rec⁻), mutants of *E. coli* and recA gene (Clark and Margulies 1965).

The phage MS2 was always referenced in papers as a private gift or private communication and never published by its discoverer. This phage was characterized and published first in the short communication of the famous Robert L. Sinsheimer's lab (Davis et al. 1961). This paper noticed for the first time that the isolated nucleic acid, obtained

from MS2 by phenol treatment, is infective to bacterial protoplasts of F⁺ strains, as well as of F⁻ strains, and the infectivity was destroyed by ribonuclease but not by deoxyribonuclease. The RNA phages f2 and MS2 were therefore the first species that formed a genus that was named later *Levivirus*.

Concerning the next related coliphages, the RNA phage µ2 was isolated in Italy from Milan sewage by Giulio Alfredo Maccacaro (Dettori et al. 1961; Maccacaro 1961), who was very popular in Italy, since he had combined his academic work with many public undertakings, e.g., the popular *Medicina Democratica* movement.

The RNA phage Qβ, the first member of the *Allolevivirus* genus and the next popular and the most important now RNA phage after MS2, was isolated from local feces in 1961 at the Institute for Virus Research, Kyoto University, in Western Japan by the pioneering initiative of Itaru Watanabe, and published only 3 years later (Watanabe 1964; Watanabe et al. 1967a).

The RNA phage R17 was found in Philadelphia by Paranchych and Graham (1962). William Paranchych, from the Wistar Institute in Philadelphia (further at the University of Alberta, Canada), continued his work on RNA phages actively and contributed greatly to physiology of the RNA phages.

Many prominent scientists have initiated elucidation of RNA phages in Germany. The popular RNA phage fr (that appeared in the first paper under the ft5 name) was found in Heidelberg by the well-known microbiologist Hartmut Hoffmann-Berling and published together with Donald A. Marvin and other colleagues (Hoffmann-Berling et al. 1963a, 1963b; Marvin et al. 1963a, 1963b). Remarkably, the RNA phage fr was first published by Peter Knolle's group from Berlin-Dahlem (Knolle and Kaudewitz 1962) and characterized further in parallel to Hoffmann-Berling's studies (Kaudewitz and Knolle 1963; Knolle and Kaudewitz 1963). It is noteworthy that another widely-used single-stranded DNA phage fd, an analogue to the phage f1, was found as a pair to the RNA phage fr (Hoffmann-Berling et al. 1963a, 1963b; Marvin and Hoffmann-Berling 1963a, 1963b).

The RNA phage M12 was found in Munich by Peter Hans Hofschneider (1963a, 1963b), a brilliant scientist who contributed enormously to the genetic engineering of hepatitis B by isolating the first cloned HBV genome (Burrell et al. 1979). And again, as in the case of the f1 and f2 pair and the fr and fd pair before, this discovery was accompanied by the finding of another famous single-stranded DNA phage M13 (Hofschneider and Preuss 1963).

In France, the RNA phage FH5 was found by Fouace and Huppert (1962), and in Australia the RNA phage f$_{can1}$ was isolated from Canberra sewage samples by Davern (1964b).

Six RNA phages including ZIK/1 (which remained the most studied phage from this group) were isolated from sewage samples in Edinburgh by David E. Bradley (Bradley 1964a,b; Bishop and Bradley 1965), who also published a large set of brilliant RNA phage electron micrographs. It is noteworthy that raw sewage was obtained from different places whose names were retained in the phage names: Marquette, Michigan, USA (ZIK); Basingstoke, Hampshire,

England (ZG); and Pangbourne Village on the River Thames in the English county of Berkshire (ZJ).

The RNA phage β, together with 12 other species of RNA phages acting upon *E. coli* K12, was isolated from sewage samples of the Arakawa Sewage Disposal Plant in Tokyo by the use of antiserum against the RNA phage MS2 (Nonoyama et al. 1963). From the remaining phages of this isolation, B1 and I were studied further (Yuki and Ikeda 1967) and found physiologically differing from MS2 and β (size of plaques, cultivation at 45°C, lytic response to the host). Moreover, they were distinct from MS2 and β by RNA annealing properties and base composition and also by serological specificity: phage B1 seemed to be more closely related to MS2 and β than phage I did. However, when anti-phage I serum was used for the isolation of RNA phages from sewage in place of anti-MS2 serum, new phages Y and Z were isolated (Tsuchida et al. 1971a). They were serologically distinct from both phage β closely related to MS2 and phage I distinct from β. Phage Z was inactivated by antisera of I, Y, and β, and *vice versa*. Conversely, the coat protein of the novel phage ZR from sewage in the Tokyo area demonstrated no difference in amino acid sequence from R17 and MS2 coats but differed by a single amino acid substitution from that of the f2 phage (Sakurai et al. 1967; Watanabe et al. 1967).

The phages MSO-12 and MSO-14 that have been isolated from the municipal sewage disposal plant at Norman, Oklahoma, were serologically identical to the MS2 phage (Lancaster 1968). The phages that have been only slightly inactivated by MS2 antiserum were also found in this study, but they were never characterized further to identify them as RNA phages.

The next group of findings demonstrated defined distinctions from the most traditional f2- and MS2-like phages presented above. First, the RNA phage R23 was isolated from sewage of Atlantic Beach, New York (Watanabe and August 1966, 1967a,b; Watanabe M 1967) in the laboratory of J. Thomas August, who contributed greatly to the RNA phage story, especially for the RNA phage replication. Contrary to all known RNA phages, the R23 phage markedly inhibited the RNA and protein syntheses of the host cell but demonstrated relatedness to the f2 phage in serology and RNA base composition (Watanabe and Watanabe 1970a). In the same block of studies, the R34 and R40 phages have been described (Watanabe et al. 1968). According to further comparative physiological studies by Hiroko and Mamoru Watanabe, the R34 and R40 phages differed serologically from each other and from the R23 phage, but nevertheless were inactivated to some extent by anti-R23 serum (Watanabe et al. 1970a).

Secondly, the RNA phage GA, the first and the most studied species of the serological group II, appeared in a paper in 1967, together with the phage SD (Watanabe et al. 1967a,b). In fact, the phage GA (together with other possible members of the group II) was isolated from Ōokayama sewage in Tokyo area in 1964, due to great activity of Itaru Watanabe and his colleagues at the Keio University. The next large group of representatives of the serogroup II was isolated in the 1970s by Itaru Watanabe's laboratory at Keio University.

Third, relatives of the Qβ phage, namely, phages VK and ST, novel species of the serogroup III, were found by Itaru Watanabe and his colleagues (Watanabe et al. 1967a).

Fourthly, Itaru Watanabe's laboratory has identified the novel RNA phage serogroup IV that was represented by phages SP (Sakurai et al. 1968), FI (Miyake et al. 1969), TW19 and TW28 (Miyake et al. 1971a), MX1 (Miyake et al. 1975), and ID2 (Furuse et al. 1978, 1979b). Later, biochemical characterization revealed a closer relationship of the MX1 phage to the serogroup III than to the serogroup IV (Hirashima et al. 1983).

In the 1980s, the list of well-characterized RNA phages, e.g., NL95, M11, was enriched in the Netherlands by Arie Hendrik Havelaar.

WORLDWIDE

Together with isolation of numerous RNA phages from sewage in central Japan (Watanabe et al. 1967a; Furuse et al. 1973, 1979a, 1981); Okinawa Island (Aoi et al. 1974); and the islands in the adjacent seas of Japan (Furuse et al. 1975a) performed altogether from 1972 to 1977. Watanabe's team completed an unprecedented study on phage isolation and grouping in Taiwan in 1970 (Miyake et al. 1971a); Arabia, India, and Thailand (Aoi et al. 1972); and Brazil in 1971 (Miyake et al. 1973); Peru, Bolivia, Mexico, Kuwait, France, Australia, and the United States in 1975 (Furuse et al. 1975b); the Philippines, Singapore, Indonesia, India, and Thailand in 1976 (Furuse et al. 1978); Korea (Osawa et al. 1981a); and Senegal, Ghana, and Madagascar (Furuse et al. 1983b). The results of this first worldwide survey were exhaustively reviewed by Furuse (1987).

NON-COLI PHAGES

The phage 7s was the first RNA phage described for a genus other than *Escherichia*. This phage infecting *Pseudomonas* was found in New Orleans by Feary et al. (1963, 1964). The RNA phage 7s was isolated from a lysogenic strain of *P. aeruginosa* designated Ps-7. It was incapable of plaque formation on strain Ps-7 but was adsorbed to cell suspensions of the Ps-7. Morphologically, it was similar to the phage f2.

The *P. aeruginosa* RNA phage PP7 that acquired now great popularity in nanotechnology and bioimaging applications, was isolated by Bradley in Edinburgh, but the sewage influent came from Pangbourne on the River Thames (Bradley 1965b, 1966b).

The next *P. aeruginosa* RNA phage PRR1 was isolated at the University of Michigan, Ann Arbor, and came from the Kalamazoo, Michigan, sewage (Olsen and Shipley 1973). Biochemical and physiological characteristics, such as membrane filter-salt elution patterns, ribonuclease sensitivity, inactivation in low ionic strength solutions, and host range, served to distinguish PRR1 from the coliphage f2 and two other *Pseudomonas* RNA phages, 7s and PP7 (Olsen and Thomas 1973).

Three novel RNA pseudomonaphages, PaMx54, PaMx60, and PaMx61, in addition to the three known, were isolated among 68 *P. aeruginosa* phages from environmental and sewage water samples collected from four states in central Mexico (Sepúlveda-Robles et al. 2012).

Recently, a new *P. aeruginosa* levivirus, vB_PaeL_PcyII-10_LeviOr01, or simply LeviOr01, was isolated from hospital wastewater (Pourcel et al. 2017). This phage demonstrated unique capability to establish a carrier state in a fraction of the infected cells, conferring superinfection immunity.

A set of RNA phages infecting stalked bacteria of the genera *Caulobacter* and designated φCB was isolated from sewage, soil, and pond water in Berkeley (Schmidt and Stanier 1965). No lytic action by the *Caulobacter* RNA phages was found against *E. coli* Hfr strains, as well as against a number of species of *Flavobacterium*, *Erwinia*, *Xanthomonas*, and *Pseudomonas*. Then, the RNA phage φCB5, properties of which were described further by Bendis and Shapiro (1970), attained, thanks to its unique properties, special popularity in the nanotechnology.

Itaru Watanabe's group isolated the next six *C. crescentus* phages, two of which, φCp2 and φCp18, demonstrated serological difference, and both differed from φCB5 by cross-neutralization efficiency by specific antibodies (Miyakawa et al. 1976). Concerning structural characterization of this group of *Caulobacter* phages, the 5'- and 3'-terminal bases of φCp2 RNA were determined as pppG and A_{OH}, respectively, and were the same as in the MS2 RNA (Fujiki et al. 1978). Two next caulophages, φCr14 and φCr28, were isolated by Johnson et al. (1977) and demonstrated a typical pattern of high salt sensitivity identical to that previously described for the phage φCB5.

The first *Acinetobacter* RNA phage AP205, a real star of the present viral nanotechnology, was isolated from Quebec sewage by Hortense Coffi (1995). The *Acinetobacter* genospecies 16 that was originally isolated from urine was used as a host bacterium.

Lucille Shapiro and Ina Bendis (1975) presented the first global review of the RNA phages infecting bacteria other than *E. coli*.

DIVERSITY OF PILI

As we know now, sex pili are a diverse group of nonflagellar filamentous appendages that function as conjugational tubes in bacterial mating and are found on many strains of bacteria. It is noteworthy to emphasize here that prior to the f2 experiments there was no hint of the existence of sex pili. It was the absence of F pili on female bacteria that accounted for the male specificity of f2 and not some other intrinsic property (Zinder 1965).

The *Pseudomonas* phage PRR1 was the first RNA phage that was found in 1973 to be dependent on transferable plasmid other than F plasmid (and therefore on conjugational tubes other than F pili) and namely R1822 plasmid (Olsen and Shipley 1973; Olsen and Thomas 1973). According to the authors' statement, "the potential for promiscuity shown for the R1822 plasmid, then, may make PRR1 phage one of the most ubiquitous viruses in the microbial world" (Olsen and Thomas 1973). A short information on a new RNA phage pilE/R1 specific for the new epiviral pilus, the E pilus (determined by plasmids of group T), appeared in 1975 (To et al. 1975).

In 1980, William Paranchych's team isolated RNA phage UA-6, which was dependent on EDP208 pili and later chosen as a representative of several such phages isolated from local sewage in Edmonton (Armstrong et al. 1980). The EDP208 pili were encoded by a derepressed derivative of a naturally occurring *lac* plasmid, F_0lac (incompatibility group FV), originally isolated from *Salmonella typhi*. Although EDP208 pili were serologically unrelated to F pili and did not promote infection by F-specific RNA phages, it was concluded that F and EDP208 pili are closely related structures (Armstrong et al. 1980). By the authors' measurements, the UA-6 group phages were somewhat smaller in diameter than the F-specific RNA phages: purified preparations consisted of spherical particles approximately 20 nm in diameter versus 26 nm for the F-specific RNA phages.

The next studies deciphering the role of pili encoded by plasmids of different incompatibility groups were performed by Bradley and his colleagues. Thus, the RNA phage F_0lac specific to EDP208 pili was isolated from sewage derived from a Pretoria hospital (Bradley et al. 1981a). The phage was described as having "hexagonal outline with a diameter of 28 nm, containing RNA, resistant to chloroform, and probably adsorbed preferentially to the sides of EDP208 pili very near the tip" (Bradley et al. 1981a). Multiplication of the phage was demonstrated on *E. coli* or *S. typhimurium* strains carrying the plasmid F_0lac. In spite of widely separated geographical locations, the phage F_0lac was found to be serologically related to the phage UA-6, Paranchych's EDP208 pili-specific isolate found in Canada.

Further, an impressive set of RNA phages specific for the different pili was isolated from sewage samples of South Africa: the phage t that infected *E. coli* and *S. typhimurium* cells carrying plasmids belonging to the incompatibility group T (Bradley et al. 1981b); the phage C-1 infecting bacterial strains carrying various IncC plasmids (Sirgel et al. 1981; Sirgel and Coetzee 1983); the phage Iα specific for the I_1 or I_γ pili (Coetzee et al. 1982); the phage M specific for bacterial strains, of various genera, harboring plasmids of the M incompatibility group (Coetzee et al. 1983); the temperature-sensitive phage pilHα specific for bacterial strains, of various genera, harboring plasmids of the HI and HII incompatibility groups (Coetzee et al. 1985a); the phage D specific for pili encoded by the incompatibility group D plasmids (Coetzee et al. 1985b); the phage F_0lac h, a host range mutant of the phage F_0lac; and the phage SR which adsorbed to pili encoded by plasmids of the S-complex (Coetzee et al. 1986). Unfortunately, these unique phages were never characterized structurally.

Finally, a comprehensive method for the direct isolation of IncH plasmid-dependent phages and a set of 27 temperature-sensitive phages, Hgal 1 till Hgal 27, were presented (Nuttall et al. 1987).

NEWCOMERS

In the early 2000s, the increased interest in the modern genotyping of the RNA phages was initiated, first of all, because of the growing use of male-specific coliphages as indicators of microbial inputs to source waters and tools for microbial source tracking (Vinjé et al. 2004; Friedman et al. 2009a, 2009b). A revolutionary contribution to the discovery, characterization, and classification of old and novel coliphages was made by Jan Vinjé, Stephanie D. Friedman, and their colleagues from a group of American institutions. First, a novel technology, namely, reverse line blot hybridization (RLB), using subgroup-specific oligonucleotides, was developed and applied for the successful genotyping of more than 100 of the enriched RNA phage samples from animal feces and wastewater (Vinjé et al. 2004). The technique was strongly improved by introduction of highly precise, forward and reverse genogroup-specific, RT-PCR primers based on a total of 30 RNA phages of several strains from all four genogroups (Friedman et al. 2009a), including 19 newly sequenced RNA coliphages (Friedman et al. 2009b). Full-length sequencing and detailed analysis of two phages, DL52 and DL54 (a pair of genomes that belonged to the specific subgroup JS within the genotype I by analogy to a pair of the WWTP1_50 and 2GI13 genomes from the previous study (Vinjé et al. 2004), led to unexpected evidence for possible recombination in environmental RNA phage strains (Friedman et al. 2012). Previous classical genetic studies have concluded that RNA phages would not undergo recombination (Horiuchi 1975).

It is worthy of notice also that a strange RNA coliphage was isolated from a sewage pond in a Chongqing hospital in the southwest China (Song et al. 2007). The XY-1 phage was declared by the authors as an analogue of the phage f2. However, it demonstrated an unusually broad host range of infected *E. coli* strains, genome size of 5.0 kb, and diameter of 40–50 nm. Unfortunately, the unusual phage XY-1 did not favor further structural studies.

Finally, the metagenomic sequencing era opened new opportunities for the recovery of novel RNA phage information. Thus, two genomes, namely EC and MB, that have been assembled from 1 liter of San Francisco wastewater, formed two novel genera within the *Leviviridae* family (Greninger and DeRisi 2015).

DANGEROUS SIMILARITY

In 1972, seven phages MAC-1, 1′, 2, 4, 4′, 5, and 7 attacking and lysing saprophytic strains of *Bdellovibrio bacteriovorus* were isolated from raw sewage of Lexington, Kentucky (Althauser et al. 1972). The phages were tailless, had a regular hexagonal outline, were approximately 25 nm in diameter, and for these reasons were attributed by authors to the "Bradley's group E" that was previously assigned to the RNA phages (Bradley 1967). However, further characterization of one of the discovered phages, MAC-1, led to the conclusion that the phage genome is made up by a circular

single-stranded DNA (Roberts et al. 1987). Therefore, this phage appeared as a member of the *Microviridae*, a family of the smallest spherical somatic DNA phages typified by the phage φX174 (Sinsheimer 1959).

ANTIBODIES

It is noteworthy that helpful serological criteria were used in the first paper on RNA phages (Loeb and Zinder 1961). In his first review article on the RNA phages, Zinder wrote that the "neutralization by antiserum is critical as an indicator of the relationship of the different male-specific RNA phages" (Zinder 1965). Anti-phage sera were prepared by traditional immunization methods with rabbits as hosts. Phage-specific antibodies allowed scientists to indicate different RNA phages and to distinguish them from each other, as well as from unrelated phages—specifically, in the case of the first RNA phage paper, to easily distinguish the f2 group phages from the DNA phage f1 (Loeb and Zinder 1961).

After the f2 phage story, the serological approach was used efficiently for the characterization and standardization of other early discovered phages such as β (Nonoyama et al. 1963), f_{can1} (Davern 1964b). fr (Hoffmann-Berling et al. 1963a, 1963b; Knolle and Kaudewitz 1963; Hoffmann-Berling and Mazé 1964), M12 (Hofschneider 1963a), MS2 (Davis et al. 1964), 7s (Feary et al. 1964), ZIK/l, ZG, ZJ/1, α15, ZS/3, and ZL/3 (Bishop and Bradley 1965), φCB5 and other φCB phage group members (Schmidt and Stanier 1965), Qβ (Spiegelman et al. 1965), R17 (Enger and Kaesberg 1965), PP7 (Bradley 1966b), PRR1 (Stanisich 1974), D (Coetzee et al. 1985b), and F_0lac, F_0lac h, and SR (Coetzee 1986). For example, because of the serological characterization, the two first isolated *Pseudomonas* phages, 7s and PP7, were found to be nonidentical, although a definite serological relationship among them did exist (Bradley 1966b).

The first systematic serological study on RNA phages was performed in 1965 by David William Scott. He collected an almost complete set of the different isolates known at that time: MS2, f2 and f4, M12, R17, fr, ft5 (new name fr), FH5, and β, and has shown by comparative neutralization tests that none of the phages differed by more than a factor of 10 in sensitivity of a particular anti-phage serum (Scott 1965). It was therefore most reasonable to assume that these phages should be considered mutational derivatives of a common ancestor. With this in mind, it was "assumed that any fact developed for one of these phages is essentially true for all of them" (Zinder 1965).

Serological properties and relationships between different RNA phages remained a top priority also for further studies on RNA phages. Moreover, serological grouping of RNA phages appeared as a first systematic taxonomical approach for RNA phage classification. The serogroup affiliation of the RNA phages is indicated in the Table 1.1 and, when possible, provided with specific comments on the serological peculiarities of a phage in question. The full history and principles of RNA phage serogrouping, as well as the content of the four serogroups I through IV, are described in Chapter 2.

PORTRAITS OF ICOSAHEDRA

The first electron microscopy portrait of an RNA phage from the first paper published on RNA phages (Loeb and Zinder 1961) is shown in Figure 1.3. On the basis on this first micrograph, the phage particles were characterized in the first review article as "small polyhedra, some 200 Å in diameter" where "no capsomeric structures have been seen" (Zinder 1965).

The first portrait was followed by a stream of high-quality micrographs of other RNA phages. Thus, for the phage M12, a diameter of ~270 Å was found (Hofschneider 1963a). For the phage fr, the "particle diameter measured on electron micrographs is 210 Å + 10%. The nature of the virus fine structure is not clear" (Marvin and Hoffmann-Berling 1963b). The phage R17 was presented as "a spherical virus 200 Å in diameter, the surface of the particle being composed of hollow capsomeres" and "morphologically similar to the RNA phage f2 isolated by Loeb and Zinder" (Crawford and Gesteland 1964). A set of excellent micrographs was published in 1964 by Bradley to illustrate his papers on the phages ZIK/l, ZJ/1, and ZG/l, which all "appear to be morphologically similar… are 225 Å in diameter and show a hexagonal outline. They also pack into a hexagonal array indicating that they are in the form of regular polyhedra" (Bradley 1964a,b). An example of Bradley's excellent EM micrographs is presented in Figure 1.4.

Further, the phage 7s was portrayed as a 25 nm particle (Feary et al. 1964) and the phage φCB5 as a "simple structure, 21–23 nm in diameter" (Schmidt and Stanier 1965). In 1965, Bradley published a portrait of the phage PP7 (Bradley 1965b). He used his excellent photos of the phage ZIK/1 to conclude that the RNA phages are icosahedra and follow the prediction of Crick and Watson (1956, 1957) that this form would be preferred by small "spherical" virions, very common amongst animal and plant viruses. Therefore, the RNA phages were characterized as "the only phages with a morphology close to

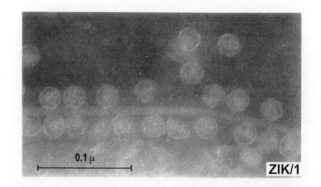

FIGURE 1.4 An example of David E. Bradley's EM micrographs. The RNA phage ZIK/1 adsorbed on an associated filamentous type, phosphotungstate negative stain, ×360,000. (Reprinted from *J Ultrastruct Res.*, 10, Bradley DE, Some preliminary observations on filamentous and RNA bacteriophages, 385–389, Copyright 1964, with permission from Elsevier.)

that of non-bacterial viruses" (Bradley 1965b). A straightforward attempt of Bradley to collineate the phage photo with an icosahedron is shown in Figure 1.5.

It is of special interest that the first systematic electron microscopy elucidation of the most popular RNA phage, namely MS2, was published in Moscow by one of the leaders of Soviet virology, Tomas Iosifovich Tikhonenko and his coauthors (Tikhonenko et al. 1965). Earlier, Strauss and Sinsheimer (1963) mentioned only that "in the electron microscope MS2 appears to be a polyhedral object about 26 mμ in diameter (A. Hodge, in preparation)." In 1966, the phage MS2 was photographed together with the second popular RNA phage Qβ by Sol Spiegelman's group and both were characterized as "polyhedrons with a diameter of about 25 mμ" (Overby et al. 1966a). Nice electron micrographs of the phage MS2 attached to F pili were published by Caro and Schnös (1966) and Lawn (1966, 1967).

The portrait of the *Pseudomonas* phage PRR1 revealed close similarity with the f2 group phages and an average diameter of 25 nm (Olsen and Thomas 1973).

Electron microscopy evaluation of the *Caulobacter* phages φCp2 and φCp18 (Miyakawa 1976) led to the diameters of 29.2 and 28.5 nm, respectively, larger therefore than that of 23 nm for the phage φCB5 (Bendis and Shapiro, 1970).

Electron microscopy evaluation of the nanotechnologically highly-prominent phage AP205 by Jan van Duin's group revealed the size of 27–30 nm, and the phage was seen attached to the sides of *Acinetobacter* pili of about 6 nm in diameter (Klovins et al. 2002).

Thanks to the extraordinary experience of Bradley in electron microscopy technique, the story of different pili-dependent RNA phages was always illustrated by excellent electron micrographs. Thus, the phage M demonstrated a diameter of 27 nm (Coetzee et al. 1983), the phage $F_0 lac$ – 28 nm (Bradley et al. 1981a), the phage t – 25 nm (Bradley et al. 1981b), the phage C-1–27±2 nm (Sirgel et al. 1982); the phage Iα – 24 nm (Coetzee et al. 1982), the phage pilHα – 25 nm (Coetzee et al. 1985a), the phage D – 27 nm (Coetzee et al. 1985b), and both $F_0 lac$ h and SR – 28 nm (Coetzee et al. 1986).

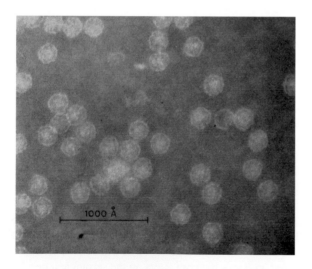

FIGURE 1.3 The first electron micrograph of an RNA phage, namely f2. The phage was negatively stained by embedding in neutral phosphotungstate. This micrograph was kindly taken by Dr. W. Stoeckenius. (Reprinted with permission of Stephen H. Zinder from Loeb T, Zinder ND. *Proc Natl Acad Sci U S A.* 1961;47:282–289.)

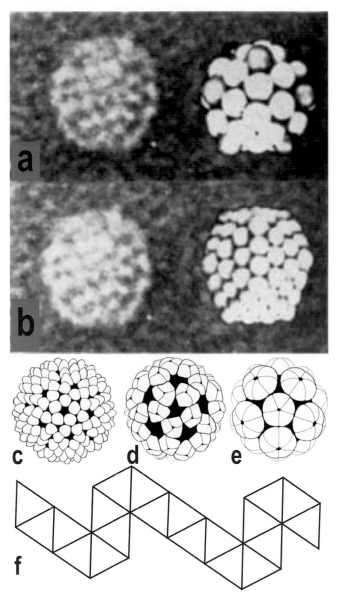

FIGURE 1.5 A compilation of early structural attempts to model the subunit structure of the RNA phages: (upper row) interpretation of electron micrographs: (a) ZIK/l in phosphotungstate showing capsomeres and model of icosahedron with 42 subunits; (b) ZIK/l in phosphotungstate and model having 92 subunits (both models are taken from Bradley (1965b); (middle row) models for the icosahedral arrangement of 180 identical subunits (triangulation number T = 3), (c) no clustering, (d) clustering in trimers, (e) clustering in hexamers and pentamers (Hohn and Hohn 1970); (lower row) construction of a model from a strip of paper (f) that can easily be folded into an icosahedron (Hohn and Hohn 1970). (Adapted with permission of Elsevier and Wiley-Blackwell from the cited papers.)

The cytological investigation of f2-infected *E. coli* cells by electron microscopy revealed extensive intracellular crystallization of the phage at a time when the intracellular phage content could be about 20,000 particles per cell (Schwartz and Zinder 1963). The presence of a paracrystallinic area in R17-infected cells was also found by high resolution autoradiography (Granboulan and Franklin 1966b).

The term *icosahedral*, in connection with RNA phages, was used for the first time by Bradley (Bradley 1964b, 1965b) with reference to the Crick and Watson (1956, 1957) hypothesis on virus structure, and then by Gary N. Gussin, with a personal reference to John T. Finch and Aaron Klug, in the classical review on RNA phages published together with James D. Watson and other highly prominent scientists (Gussin et al. 1966).

Serious structural estimations were performed at that time by Richard M. Franklin and his colleagues. Thus, the morphology of the phage R17 was studied in detail by electron microscopy of negatively stained virions: "the hexagonal shape, the presence of a maximum of 10 units at the periphery, and especially the observation of central fivefold points of symmetry with neighboring five and six coordinated units indicated icosahedral symmetry with 32 morphological units" (Vazquez et al. 1966). It was proposed for the first time that the morphological units of the phage particle were probably situated on a triangulation number T = 3 icosahedral surface lattice (Vazquez et al. 1966) in accordance with the quasi-equivalence theory of Caspar and Klug (1962), who pointed out that the assembly of viruses was related to crystallization and was governed by the law of statistical mechanics. According to the idea, a maximum number of the most stable intersubunit bonds was achieved in the final structure if all subunits were in equivalent position. Therefore, not more than 60 subunits could be arranged in strictly equivalent position in isometric viruses.

The fact that the RNA phage might appear to be clustered in small units, probably trimers, that could be arranged on the surface of a regular icosahedron of the triangulation T = 3, with the total number of subunits being 180, was discussed for the first time in the exhaustive and nicely illustrated review on the RNA phage assembly and disassembly published by Thomas and Barbara Hohn (Hohn and Hohn 1970). The early models, including the strip of paper that can easily be folded into an icosahedron, are presented in Figure 1.5.

Further development of the quasiequivalence principle for the self-assembly of the RNA phages was performed by A. Keith Dunker and William Paranchych by arranging subunits in bonding domains (Dunker 1974; Dunker and Paranchych 1975).

A novel highly prospective approach to three-dimensional image reconstruction from electron micrographs was elaborated by R. Anthony Crowther and Aaron Klug (Crowther and Klug 1975) and used for the RNA phages R17 and f2 (Crowther et al. 1975). It showed that the protein subunits were located at the 2-fold positions and extended towards the quasi-3-fold positions of the T = 3 icosahedral surface lattice, leaving holes on the 5-fold and 3-fold axes. The particles were reconstructed as roughly circular of diameter about 240 Å, where protein was concentrated into a relatively thin shell some 20−30 Å thick, and superposition patterns characteristic of views down 2-fold, 3-fold, and 5-fold axes of symmetry were recognized (Crowther et al. 1975). The course of the image reconstruction is illustrated in Figure 1.6.

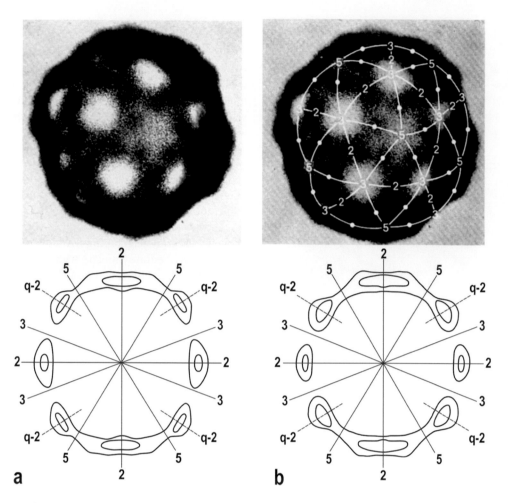

FIGURE 1.6 A compilation of the data on the classical structural resolution of the RNA phages f2 and R17. Upper row: A density plot of a reconstruction of f2 phage, in which high density represents exclusion of stain. On the right, part of the T = 3 icosahedral surface lattice has been superimposed on the density. Strict 2-fold, 3-fold, and 5-fold axes of symmetry are indicated and the white dots mark the positions of the quasi-2-fold axes. Lower row: Equatorial sections normal to 2-fold axes of reconstructions of (a) phage f2, (b) phage R17. The axes of 2-fold, 3-fold, and 5-fold symmetry lying in the plane are indicated. The best mirror lines at the quasi-2-fold positions are drawn as dotted lines and labelled q-2. Note that the two reconstructions are extremely similar, although there are small differences which we attribute to (b) being a preliminary reconstruction of lower resolution. (Reprinted from *J Mol Biol.*, 98, Crowther RA et al. Three-dimensional image reconstructions of bacteriophages R17 and f2, 631–635, Copyright 1975, with permission from Elsevier.)

COMPOSITION

According to early measurements, the RNA phage particles demonstrated a sedimentation constant $S_{20,w}$ of 76–80, banding in CsCl at 1.43 gm/mL, and a total particle weight of $3–3.5 \times 10^6$ (Davis and Sinsheimer 1963; Enger et al. 1963; Marvin and Hoffmann-Berling 1963a; Strauss and Sinsheimer 1963; Zinder 1963; Gesteland and Boedtker 1964; Overby et al. 1966a). The molecular mass of the f2 phage particle was later estimated at 4.0×10^6, "not much larger than a 70 S ribosome" (Zinder 1980).

In first classical review (Zinder 1965), RNA phage consisted of two major components: (i) a single RNA molecule of about 10^6 molecular mass units enclosed by (ii) a single protein species, namely, coat protein that was calculated of a molecular mass of 15,000 Da for the monomer consisting of some 133 amino acid residues. The single particle was estimated therefore to contain about 150 such coat protein units

(Zinder 1965), which was in fact not very far from the final precise data: 129 amino acid residues of the f2 coat protein, 13,700 Da for the f2 coat monomer and 180 subunits within the particle (for review, Zinder 1980). One striking compositional feature of the coat protein of the f2-like phage group was its lack of the amino acid histidine.

In 1968, Joan Argetsinger Steitz found that attachment of the phage to the bacterial pili is accomplished by a protein called A (for the order of the phage genes within their genome or for the attachment) protein, or maturation protein, or sometimes maturase, which appeared in a single copy within the phage particle, probably, on its surface and in a tight complex with RNA (Steitz 1968a,b,c). The presence of the A protein (its molecular mass was calculated as about 40,000) was absolutely required for infectivity (Roberts and Steitz 1967). Spatial models to explain the presence of the A protein on the phage surface have been proposed by the William Paranchych's team (O'Callaghan et al. 1973a; Dunker and Paranchych 1975).

Finally, a C-terminally extended coat protein known as an A1 protein (IIb in early studies) was found in the phage Qβ particles: Alan M. Weiner and Klaus Weber identified this protein, which appeared as a result of ribosomal read-through of a leaky opal termination codon of the CP gene, as a minor component of the particles (Weiner and Weber 1971, 1973). The read-through nature of this protein was postulated at the same time also by Moore et al. (1971). The A1 protein was incorporated in 3–10 copies per virion, or in 12 copies in accordance with a more recent study (Skamel et al. 2014), and was essential for the formation of viable Qβ particles *in vivo* (Hofstetter et al. 1974; Engelberg-Kulka et al. 1977, 1979a), although its precise function remains unclear. The A1 proteins are a trademark of the *Allolevirus* genus. After Qβ, the analogous read-through proteins were found by Takeshi Aoi and Paul Kaesberg for the phages TW19 and TW28 (Aoi and Kaesberg 1976).

GENOME

All phage genomic RNA was found in one single-stranded positive-sense, or plus strand, molecule. The RNA phage R17 minus strands failed to bind ribosomes and did not stimulate any protein synthesis in the cell-free extracts of *E. coli* (Schwartz et al. 1969).

RNA of the phages MS2 (Strauss and Sinsheimer 1963), fr (Marvin and Hoffman-Berling 1963a), R17 (Mitra et al. 1963; Gesteland and Boedtker 1964; Sinha et al. 1965a,b), and M12 (Sinha et al. 1965a) demonstrated a molecular mass of 1.1×10^6. This mass value was consistent with estimates based on the chemically determined fraction (about 30%–32%) of whole phage that is RNA (Strauss and Sinsheimer 1963; Enger et al. 1963; Hoffman-Berling et al. 1963). It is noteworthy that somewhat lower molecular mass, 0.9×10^6, was determined for Qβ RNA, in parallel experiment with MS2 RNA which demonstrated 1.0×10^6 in this case (Overby et al. 1966b).

According to Zinder's apt remark, "one million daltons of RNA can code for polypeptides with a total of 1.200 amino acids. This is scarcely larger than the number of amino acids in the enzyme β-galactosidase, whose function is the cleavage of a simple glycosidic linkage. Few functions can be encoded in so small a genome" (Zinder 1980).

Since the RNA phages contained so little nucleic acid, they must have only a few genes. Ohtaka and Spiegelman (1963) have found three MS2-specific electrophoretically separable proteins and anticipated existence of three genes in the phage genome. In fact, there are currently four genes identified that are packed into the small RNA genome.

Needless to say, it seemed possible to obtain mutations in each of these genes, which resulted in the introduction of (i) temperature-sensitive (*ts*) mutations and (ii) host suppressor (*amber*)-dependent mutations that could both be qualified as conditionally lethal mutations. The appropriate mutants were obtained for the following phages: f2 (Engelhardt and Zinder 1964; Lodish et al. 1964, 1965; Valentine et al. 1964; Valentine and Zinder 1964a,b; Zinder and Cooper 1964; Notani et al. 1965; Horiuchi et al. 1966; Lodish and Zinder 1966a; Engelhardt et al. 1967; Model et al. 1979), fr

(Kaudewitz and Knolle 1963; Heisenberg and Blessing 1965; Heisenberg 1966, 1967; Knolle 1966a,b,c), f_{can1} (Davern 1964a), MS2 (Pfeifer et al. 1964; Van Montagu 1966, 1968; Gillis et al. 1967; Van Montagu et al. 1967; Nathans et al. 1969; Sugiyama et al. 1969; Vandekerckhove et al. 1969, 1971; Van Assche et al. 1972; Vandamme et al. 1972; Van Assche and Van Montagu 1974; Van Assche et al. 1974), R17 (Argetsinger and Gussin 1966; Gussin 1966; Gussin et al. 1966; Roberts and Gussin 1967; Tooze and Weber 1967; Igarashi et al. 1970a,b), GA (Aoi 1973; Aoi et al. 1973), β (Tsuchida et al. 1966), and Qβ (Horiuchi and Matsuhashi 1970; Palmenberg and Kaesberg 1973; Domingo et al. 1975; Gupta et al. 1975). Remarkably, the phage has never been known to mutate to infect F^--bacteria (Zinder 1965, 1980).

At first, the *ts* and *amber* mutants revealed three complementation groups for the following phages: f2 (Horiuchi et al. 1966), R17 (Gussin 1966; Gussin et al. 1966) and Qβ (Horiuchi and Matsuhashi 1970). Each of the groups was correlated with a particular phage function: one gene specified a coat protein, one gene a maturation protein, and another an RNA replicase.

A fourth complementation group was defined by a mutant that grew only in UGA-suppressing strains. It failed to lyse the nonpermissive hosts while filling them with thousands of normal phage particles, since it did not produce lysis protein (Atkins et al. 1979; Beremand and Blumenthal 1979; Model et al. 1979). According to Zinder, "many years passed before this mutation was discovered to be in a gene whose reading frame was different from other genes and overlapped two of the known genes" (Zinder 1980). Thus, the lysis gene was discovered in the *Levivirus* genus, but the *Allolevivirus* genus members did not encode the lysis gene and used maturation protein to perform lysis function.

While functions of the coat and replicase genes were clearly defined just by the gene names, the maturation, or A protein gene, remained for some time as an abstraction that was in some way required for the infectivity of the phage (Heisenberg and Blessing 1965; Lodish et al. 1965).

Since the RNA phages did not undergo molecular recombination, the order of the genes as well as positions of mutations were obtained by structural rather than genetic mapping for the following RNA phages: R17 (Gesteland and Spahr 1969; Jeppesen et al. 1970b), f2 (Lodish and Robertson 1969b; Robertson and Lodish 1970), M12 (Konings et al. 1970), and Qβ (Hindley et al. 1970; Staples et al. 1971). Earlier attempts of gene ordering led to an erroneous assumption that the coat protein gene could be located close to the 5'-end of the genome (Ohtaka and Spiegelman 1963; Engelhardt et al. 1968; Lodish 1968d; Robertson and Zinder 1969; Webster et al. 1969).

Direct sequencing of the phage genome was initiated by Frederick Sanger, who twice won the Nobel Prize in Chemistry. Sanger's eighth Hopkins Memorial Lecture gave an account of his group's studies on the sequences in the RNA of the R17 phage (Sanger 1971). In the course of these studies, the RNA fragments that remained relatively resistant to ribonuclease were isolated, sequenced, and contributed strongly to the correct ordering of the phage R17 genes (Jeppesen et al. 1970b). The same gene order was independently found

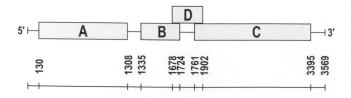

FIGURE 1.7 Genetic map of group I RNA phages. Genes (rectangles) are drawn more or less in proportion to their size. A = maturation gene; B = coat gene; C = replicase gene; D = lysis gene. The numbers on the line below sum the nucleotides from the 5′ end of the RNA molecule to the significant sites on the genetic map above. (Adapted with permission of S. Karger AG, Basel from Zinder ND. *Intervirology.* 1980;13:257–270; originally drawn after Fiers W et al. *Nature.* 1976;260:500–507; Atkins JF et al. *Cell.* 1979;18:247–256; Beremand MN, Blumenthal T. *Cell.* 1979;18:257–266.)

by studies on the *in vitro* translation of nascent RNA strands derived from replicative intermediates as a function of their chain length (Konings et al. 1970). Further, the famous Walter Fiers' team performed full gene-by-gene sequencing of MS2 RNA and successfully published the first decoded genome in the world (Fiers et al. 1976).

As just mentioned above, the fourth gene, namely, the lysis gene, appeared on the RNA phage genetic map only in 1979. It was localized for the phages f2 (Beremand and Blumenthal 1979), MS2 (Atkins et al. 1979a,b; Kastelein et al. 1982, 1983a,b), fr (Adhin and van Duin 1989), KU1 (Groeneveld et al. 1996), AP205 (Klovins et al. 2002), and φCB5 (Kazaks et al. 2011). The lysis proteins for a large group of RNA phages were compared by their primary structure and searched for putative homology (Nishihara et al. 2006). Generally, the outstanding contribution of Jan van Duin and his laboratory to our knowledge in the RNA phage lysis proteins must be emphasized.

Figure 1.7 presents the first genetic and structural map of this first known whole genome in the world.

TRANSLATION

Genomic RNAs of the RNA phages have been the first natural messengers and, moreover, the only real messenger RNAs available for a long time, which were used to decipher genetic code and to elucidate the translational mechanisms. The pioneering role of the RNA phage messengers was highlighted by the sole definition of the mRNA term (Singer and Leder 1966). The R17, MS2, and Qβ genes became in 1970s the first messenger RNAs of known sequence.

The polygenic, or polycistronic, phage messenger demonstrated many interesting features. The messenger possessed noncoding regions at both ends of the molecule. Each gene was separated from the next by the first intergenic spaces of about 40 nucleotides. Two of the three genes that were known in the mid-1970s started with AUG and one with GUG while they terminated with UAG or UAA. All of the synonymous codons were used and with frequencies that might be expected on a more or less random basis. The structure of the three initiation regions brought up the idea of the

regulated replication and mechanisms of its realization. The first one, that of the maturation gene, was surrounded by secondary structure, whereas that of the coat gene was open and accessible and that of the replicase gene was base paired to the proximal sequence of the coat gene. This last finding explained why the coat gene must be translated prior to the effective translation of the replicase gene (Zinder 1980). It was clarified therefore that ribosomes did not glide from gene to gene; they entered each of the ribosome-binding sites independently, but the sites differed both in accessibility and in their affinity for ribosomes, and regulated the time and amount of the various proteins that were synthesized (Webster and Zinder 1969a,b).

Fifteen years before the full-length genome, in early 1961, Marshall Warren Nirenberg and J. Heinrich Matthaei showed that that *E. coli* extracts containing the ribosomes and supernatant factors might synthesize protein when template RNA was added (Nirenberg and Matthaei 1961). From statistical analysis of various mixed synthetic polyribonucleotides which have been used as messengers, certain genetic code words have been assigned to nucleotide triplets. Appearance of the easily accessible natural messengers, namely, phage RNAs, at a rather propitious time stimulated enormously the progress in the principles of the protein biosynthesis. Daniel Nathans, who received the 1978 Nobel Prize for the restriction enzyme, has shown, together with his colleagues, that phage f2 RNA directed the *in vitro* synthesis of a product giving rise to tryptic peptides which co-chromatographed with those from authentic f2 coat protein (Nathans et al. 1962a,b). It is remarkable that the f2 RNA translation studies were done in the laboratories of Norton D. Zinder and James D. Watson in parallel, and caused challenging competition. As the studies revealed, 70% of the protein synthesized on f2 RNA was its coat protein (Nathans et al. 1962a). Replicase or its fragments amounted to 30% of the protein synthesized (Lodish 1968d). No maturation protein could be found. These findings were in good agreement with the idea of the optimal regulation of the RNA phage replication. It is pertinent to note the vital role of Severo Ochoa, the 1959 Nobelist, in the translation studies in tight connection with the RNA phage replication, together with Charles Weissmann and other colleagues (Weissmann et al. 1963a,b; Ochoa et al. 1964; Weissmann et al. 1964a,b; Weissmann et al. 1966; Feix et al. 1967a; Weissmann and Ochoa 1967; Viñuela et al. 1968; Garwes et al. 1969; Ochoa 1969), and more specifically in the application of phage messengers (Stanley et al. 1966; Ochoa 1967; Viñuela et al. 1967a,b,c; Iwasaki et al. 1968; Ochoa et al. 1968; Miller et al. 1969; Sabol et al. 1970; Lee-Huang and Ochoa 1971, 1972, 1973; Sabol and Ochoa 1971; Lee-Huang et al. 1973, 1974; Sabol et al. 1973; Ochoa 1976). In September 1975, at an international symposium held in Barcelona and Madrid, when Ochoa's students and colleagues celebrated his 70th birthday and his outstanding contributions to biochemistry spanning nearly half a century, Weissmann gave an exclusive talk on the discovery of RNA phages, which was published further in the collection of essays *Reflections on Biochemistry: In Honour of Severo*

FIGURE 1.8 The RNA synthetase research team at New York University (1964). Left to right: Piet Borst, (Ted Abbot), Severo Ochoa, Charles Weissmann, Martin Billeter, and Roy Burdon. On the table: the wooden mallet used in the first step of synthetase purification and the scheme of phage RNA replication. (Reprinted from A phage in New York, Weissmann C, (Eds: Kornberg A, Horecker BL, Cornudella L, Oro J.), *Reflections on Biochemistry in Honour of Severo Ochoa.*, pp. 283–292, Pergamon Press, Oxford-New York-Toronto-Sydney-Paris-Frankfurt, Copyright 1976, with permission from Elsevier.)

Ochoa (Weissmann 1976). Figure 1.8, which is reprinted from these *Reflections*, is intended to carry back to the great individuals of the early RNA phage days.

The elucidation of the host suppressor—dependent mutants of RNA phages led to the clarification of the *amber* (Capecchi and Gussin 1965; Engelhardt et al. 1965; Notani et al. 1965; Weber et al. 1966; Zinder et al. 1966) and *opal* (Chan et al. 1971) suppression mechanisms. This helped to explain such basic translation mechanism elements as the chain termination at the UAG codon (Webster et al. 1967), chain termination—involved release of a peptide from tRNA (Engelhardt et al. 1965), suppression of mutants by mutant tRNAs which inserted such amino acids as serine, tyrosine, and glutamine at the *amber* codon (Capecchi and Gussin 1965; Engelhardt et al. 1965), and dissociation of ribosomes from the phage RNA and the tRNA from the peptide at the site of a chain-terminating event (Webster and Zinder 1969a,b).

As for the initiation of protein chain synthesis, the phage RNA-directed process *in vitro* allowed natural initiation in a precise and proper order in low Mg^{2+} concentration (Nathans et al. 1962a,b), while the initiation on polyribonucleotides has occurred at high Mg^{2+} concentration which causes spurious initiation (Nirenberg and Matthaei 1961). This is described in Zinder's classical review,

> … a clue to faithful initiation appeared when the prematurely terminated coat protein hexapeptide was found to have a blocked amino terminus. Removal of the blocking group by

mild acid hydrolysis indicated that it was a formyl group. We remembered that, a year before, Marcker and Sanger (1964) had found a formylated methionine tRNA. Before we had finished the experiment designed to test the role of N-formyl methionine, James Watson came flying into the laboratory and told us that Capecchi (*Mario Ramberg Capecchi is the future Nobelist 2007 – PP*) had found N-formyl methionine on the amino terminus of *in vitro* synthesized phage protein. Within a month the experiments were finished and two papers sent off to the Proceedings of the National Academy of Sciences of the USA (Adams and Capecchi 1966; Webster et al. 1966). Remarkably, of the myriad of known and unknown factors required for protein synthesis in these cell extracts, only the deformylase had died, leaving the formyl group on the peptides ready to be discovered (Zinder 1980).

A quick perusal of the composition of the *E. coli* proteins led to the conclusion that all *E. coli* proteins were started by N-formyl methionine (Capecchi 1966b), when "the addition of a formyl group seemed to be a clever ploy by nature to form the first peptide bond and provide protein synthesis with unique direction" (Zinder 1980). In the same year, 1966, the same conclusions were reached by the *in vitro* translation of the phage μ2 RNA (Clark and Marcker 1966).

The next crucial contribution of the RNA phages to the translation mechanisms consisted of the identification of the ribosome binding sites. Steitz isolated ribosome-bound phage RNA that was ribonuclease-resistant by a combination of two techniques: (i) binding of phage RNA to ribosomes (Gupta et al. 1969, 1970, 1971) and (ii) Sanger's sequencing small amounts of radioactively labeled RNA (Barrell and Sanger 1969). Of the three ribosome-binding sites found (Steitz 1969a,b), the ribosome-binding site of the coat protein gene appeared in much larger amounts than those for the replicase and maturation genes. However, no common feature that would identify ribosome-binding sites was apparent (Steitz 1972). It was further established that in addition to AUG (or GUG), a ribosome-binding site was provided with 3–7 purines, so-called Shine-Dalgarno sequence, some 5–10 bases to the left of the AUG (Shine and Dalgarno 1974). This oligopurine tract was capable of base pairing with the nucleotides at the 3′-end of the 30S ribosomal RNA (Steitz and Jakes 1975).

The *in vitro* translation system revealed the sizes of the three major phage MS2 proteins which were defined by their mobility when electrophoresed in a polyacrylamide gel and compared with the *in vivo* synthesized products (Nathans et al. 1966; Eggen et al. 1967; Viñuela et al. 1967a,b,c). A bit later, Qβ proteins were compared with the MS2 proteins by the double-labeling technique in the polyacrylamide gel (Garwes et al. 1969), and the Qβ proteins synthesized *in vitro* were compared with the Qβ proteins synthesized *in vivo* by Paul Kaesberg's group (Jockusch et al. 1970). Figure 1.9 presents one of the first examples of the polyacrylamide gels used for the identification of RNA phage proteins.

Since the three proteins produced *in vitro* were the same as the proteins appearing *in vivo*, there was no doubt that the RNA in the virion must be the messenger strand, or plus-sense

RNA. Remarkably, the three proteins were made *in vitro* in different amounts, somewhat in proportion to what was found in infected cells (Lodish 1975).

Summarizing, the application of the phage RNA messengers (i) verified the fidelity of the *in vitro* translation system, (ii) contributed crucially to the mechanisms of chain initiation and chain termination, including determination of the mechanism of nonsense suppression and identification of punctuation signals in protein synthesis, which were not readily studied by the use of synthetic polyribonucleotides, (iii) initiated studies of ribosome recognition by correct polypeptide initiation sites on mRNA, and (iv) established therefore a solid basis for the modern understanding of protein biosynthesis, as reviewed substantially by Lucas-Lenard and Lipmann (1971). Moreover, the sequences of the phage RNAs have confirmed the strong accuracy of the genetic code.

REPLICATION

Just after the discovery of RNA phages, Roy H. Doi and Sol Spiegelman showed by newly applied nucleic acid hybridization that there is no homology between the viral RNA and the bacterial DNA from either uninfected or infected cells (in their experiments, by MS2 infection) and therefore excluded participation of a possible DNA intermediate during the RNA phage replication (Doi and Spiegelman 1962). The early involvement of Spiegelman, who generated the technique of nucleic acid hybridization which helped to lay the

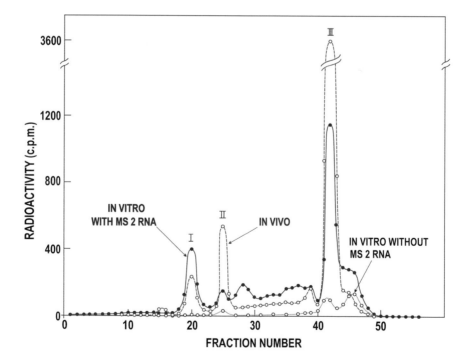

FIGURE 1.9 One of the first examples of an electrophoretic pattern of proteins synthesized *in vitro* and *in vivo*; in this case, by RNA phage MS2. Proteins from *in vitro* (C14-leucine label) and *in vivo* (H3-leucine label) incubations were mixed and subjected to polyacrylamide gel electrophoresis. The radioactivity of the fractions was measured simultaneously in two channels of the Packard liquid scintillation spectrometer. Solid line and circles (•-•, *blue*), *in vitro* incubation with MS2 RNA; dotted line and open circles (o-o, *black*), *in vitro* incubation without MS2 RNA; dashed line and open circles (o-o, *red*), *in vivo* incubation (MS2-infected spheroplasts). Positions of the replicase, A protein, and coat protein are indicated by roman numbers I, II, and III, respectively. (Redrawn [in color] with permission of Margarita Salas from Viñuela E et al. *Proc Natl Acad Sci U S A*. 1967;57:729–734.)

groundwork for further advances in recombinant DNA technology, was very fruitful for the RNA phages at that moment. Spiegelman's role in the story of the nucleic acid hybridization is reviewed exhaustively by Giacomoni (1993).

Furthermore, from the early RNA phage studies, the work on the conservation of the parental phage strands during full MS2 replication cycle must be mentioned (Doi and Spiegelman 1963). More generally, Spiegelman's name is associated with self-reproducing phage Qβ RNA-derived structures, called Spiegelman's Monster, and with the famous extracellular Darwinian experiments. Spiegelman worked on trying to establish that retroviruses cause human cancers, but died in 1983 just before the cause of AIDS was identified as a human retrovirus.

The participation of DNA synthesis in the phage replication (in this case, f2) was excluded also by use of mitomycin C, an inhibitor of DNA synthesis (Cooper and Zinder 1962). It is worth mentioning in this connection that the reverse transcription was found 10 years later for retroviruses (Temin and Baltimore 1972).

An obligate necessity of the parental phage RNA's translation was confirmed by the fact that chloramphenicol added at the moment of infection inhibited the synthesis of viral RNA, while addition of the drug within 10–15 min after infection did allow normal synthesis of viral RNA even though protein synthesis is inhibited, in the case of phages f2 (Cooper and Zinder 1963) and R17 (Paranchych and Ellis 1964).

By the early phage R17 elucidation, it was well established that the RNA phages were synthesized completely *de novo* and no host protein or RNAs were involved as parts in the RNA phage particles, while synthesis of the host RNA was partially inhibited (Ellis and Paranchych 1963). Replication of R17 was studied also by the Richard M. Franklin's team (Erikson et al. 1964; Fenwick et al. 1964; Erikson et al. 1965).

The pioneering studies on the MS2 replication that led to the conclusion that there was no parent to progeny transfer of RNA were performed by Robert L. Sinsheimer's team (Davis and Sinsheimer 1963; Strauss and Sinsheimer 1963). Intensive replication studies were performed at that time on the phage M12 by Peter H. Hofschneider's team (Ammann et al. 1964a,b; Delius and Hofschneider 1964) and on the phage fr by Hartmut Hoffmann-Berling's team (Kaerner and Hoffmann-Berling 1964a,b) and by Peter Knolle (1964a).

Addition of chloramphenicol at the time of infection prevented conversion of parental MS2 phage RNA molecules into a "double-stranded" form, a first sign of the ongoing replication (Kelly and Sinsheimer 1964). It was therefore "concluded that a new phage-induced enzyme, which is involved in the synthesis of viral RNA, was needed; i.e., phage RNA made phage RNA" (Zinder 1965).

The three groups concurrently discovered the new enzymatic activity, i.e., the appearance of an RNA-dependent RNA replicase after infection of *E. coli* cells with the phages f2 (August et al. 1963, 1965; Shapiro and August 1965a, 1966) and MS2 (Haruna et al. 1963; Spiegelman and Doi 1963; Weissmann and Borst 1963; Weissmann et al. 1963a,b;

Borst and Weissmann 1965; Haywood and Sinsheimer 1965; Weissmann 1965).

Concerning RNA synthesis *in vivo*, its asymmetric character was beyond question: 5–10 plus-strands were made per minus strand (Weissmann et al. 1968a). The mechanism of phage RNA synthesis was studied *in vitro* due to unique properties; first, the extraordinary stability of the RNA replicase of Qβ, the phage from the serological group III. The isolation and characterization of the famous Qβ replicase was reported first in 1965 by Ichiro Haruna and Sol Spiegelman (Haruna and Spiegelman 1965a,b,c; Spiegelman et al. 1965).

The enzyme contained four components: the phage-specific replicase, S1, a protein from the small ribosome subunit, and the factors required for elongation in protein synthesis, EF-Tu and EF-Ts (Kamen 1970; Kondo et al. 1970; Blumenthal et al. 1972; Groner et al. 1972c; Inouye et al. 1974; Wahba et al. 1974). It was intriguing that three of four proteins in the Qβ replicase were cell proteins involved in protein biosynthesis (Zinder 1980). Their precise role in RNA remained not fully clear at that time, although it could be connected with RNA recognition. Thus, S1 was necessary for the proper messenger binding (Noller et al. 1971). EF-Tu and EF-Ts carried the charged tRNAs from the supernatant to the ribosomes (Lucas-Lenard and Lipmann 1971). In this sense, it seemed significant that the structure of the 3'-end of the phage RNA resembled the cloverleaf of tRNAs. In addition to S1 and EF-Tu and EF-Ts, two other factors were identified as necessary for full replicase activity (August et al. 1968; Franze de Fernandez and August 1968; Franze de Fernandez et al. 1968; Shapiro et al. 1968). One of these two factors, Hfq, became famous during recent studies.

The Qβ replicase, when primed with Qβ plus strands, first synthesized minus strands and later predominantly plus strands (Weissmann and Feix 1966), but when primed with Qβ minus strands produced plus strands from the very outset of the reaction (Weissmann et al. 1967).

A titanic work was performed on the purification and characterization of the f2 replicase (Fedoroff and Zinder 1971a,b; 1972a,b; 1973). The f2 replicase purified by Nina V. Fedoroff appeared as the first efficacious replicase from the phage that did not belong to the *Allolevivirus* genus.

The next important finding for phage replication was the fact that the phage-specific intermediate in RNA synthesis is not double-stranded. In all stages of synthesis, the nascent RNA strands appeared as single-stranded tails and were associated only at the growing point (Spiegelman et al. 1968; Weissmann et al. 1968b). The presence of the double-stranded intermediate would be dangerous for phage replication, since *E. coli* contained a ribonuclease specific for double-stranded RNA (Robertson et al. 1968a), although some double-stranded RNA fragments were found in the case of the replicase overexpression (Lodish and Zinder 1966a). Generally, it was concluded that the strands hybridize to each other during deproteinization and extraction procedures (Erikson and Franklin 1966; Robertson and Zinder 1969).

Figure 1.10 presents the original outline of the RNA replication cycle from Charles Weissmann's famous review by data as of 1968.

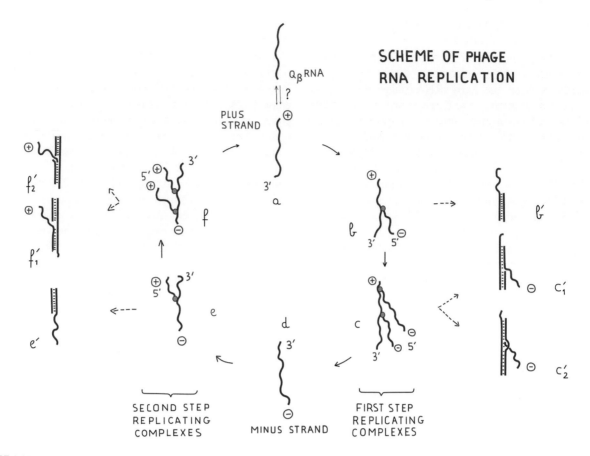

FIGURE 1.10   Scheme of phage RNA replication. In the first stage of replication a single-stranded minus strand is synthesized on a plus strand template; the intermediate, the first-step-replicating complex (b or c), is an 'open' structure in which template and product do not form a double helix. In the second step of replication the single-stranded minus strand (d) is used as a template for plus strand synthesis; the intermediate, the second-step-replicating complex (e or f), is analogous to the first, but the full-length template is a minus strand. Replicating complexes may collapse, spontaneously or under the influence of external agents, to yield double-helical structures (b′, c′, e′, f′). (Reprinted from A phage in New York, 465, Weissmann C, (Eds: A. Kornberg, BL Horecker, L. Cornudella, J. Oro.), In: *Reflections on Biochemistry in honour of Severo Ochoa.*, pp. 283-292, Pergamon Press, Oxford-New York-Toronto-Sydney-Paris-Frankfurt, Copyright 1976, with permission from Elsevier.)

REGULATION

Since RNA phages are the simplest organisms, or because of this prominent simplicity, they must possess a strong and efficient self-regulation mechanism that would work on all steps of the replication cycle, i.e., the time and amount of synthesis of the different four gene products. For example, one of the earliest observations of the phage f2 life cycle disclosed the fact that the synthesis of viral polymerase ceased long before much virus RNA, coat, or particles have been produced (Lodish et al. 1964). While DNA genomes perform much of this regulation by controlling the rate of gene translation into messenger RNA, the RNA genomes needed other special mechanisms to get minus and plus RNA strands in time and in proper ratio and to ensure the needed amounts of the four different proteins. As Harvey F. Lodish, one of the main players in the discovery of the translational regulation in RNA phages, wrote in the introduction to his *Nature* paper: "One way in which an organism can regulate the synthesis of specific proteins is by translation of the different genes on a polygenic mRNA at different rates and different times" (Lodish 1968d).

The first level of the special RNA phage regulation was established in the specific exposition of the three ribosome binding sites to the ribosomes, when only that of the coat protein gene is accessible for ribosomes but those of the replicase and maturation genes are hidden appear accessible only when the RNA is broken by radioactive decay of the ^{32}P label (Steitz 1969a,b).

The limited accessibility of the ribosome binding site of the replicase gene was ensured by the mechanism of so-called polarity control. Pleiotropic mutations were called polar, when genes on the operator-distal side of the mutation were affected. In the case of RNA phages, the polarity control was operated by coat protein gene over distal replicase gene: the ribosome binding site of the replicase gene opened only when translation of the proximal region of the coat protein gene was started on the parental RNA after its injection bacterial cell.

The polar mutants with *amber* mutation in the coat protein gene at glutamine residue position 6 have been isolated and characterized in the RNA phages f2 (Notani et al. 1965; Lodish and Zinder 1966; Weber et al. 1966; Lodish 1968a), R17 (Gussin 1966; Tooze and Weber 1967), and MS2 (Fiers et al. 1969). The polar effect was unveiled by comparison both *in vivo* and *in vitro* of polar *amber* mutations with non-polar at glutamine residue positions 50, 54, and 70 in the coat protein gene. The *amber* mutation at position 6 was polar *in vivo*, i.e., fully prevented further translation of the parental RNA, while the polar mutant grew well on supA strains, and the amino acid serine was found at the respective position (Zinder et al. 1966). *In vitro*, the position 6 mutant directed the synthesis of an amino terminal hexapeptide in *E. coli* Su⁻ extract (Engelhardt et al. I965) and did not allow the incorporation of any other amino acids. It was also polar *in vitro* both in f2 (Zinder et al. 1966) and R17 (Capecchi 1967b). The position 50 and longer mutants directed in Su⁻ cell extract the synthesis of the appropriate amino terminal fragments of the coat and of the replicase protein: they were not polar. The polarity phenomenon was described also for the phage Oβ mutants (Ball and Kaesberg 1973).

It was clearly shown that the non-polar mutations in the coat gene resulted in the loss of the coat protein control over replicase synthesis and enabled excess production of the replicase protein (Lodish and Zinder 1966a), the phenomenon used further for the production and purification of the replicase.

Further investigations confirmed the idea that the expression of viral genes is controlled by phage-specific proteins which are bound to the RNA template and form translational repressor complexes. A complex of a few coat protein subunits per R17 RNA was first observed by Capecchi and Gussin (1965) as newly *in vitro* synthesized protein subunits cosedimenting with RNA. The repressor complex I was formed by the coat protein and prevented initiation of the replicase gene, as was found in the case of the phages f2 (Lodish and Zinder 1966a; Robertson et al. 1968; Ward et al. 1967, 1968), M12 (Ward et al. 1970), MS2 (Eggen and Nathans 1967, 1969; Sugiyama and Nakada 1967a,b, 1968, 1970; Sugiyama et al. 1967), R17 (Spahr et al. 1969), fr (Hohn 1969a,c; Kaerner 1970), and Qβ (Robertson et al. 1968). Once again, the *in vitro* protein-synthesizing system clarified a problem, since addition of purified coat protein *in vitro* inhibited replicase synthesis. The added coat protein occupied the initiation sequence of the replicase gene, as was shown clearly in the case of the phage R17 (Bemardi and Spahr 1972a,b). Therefore, the coat protein blocked ribosome attachment and consequently replicase synthesis by a very simple but the efficient control mechanism.

The repressor complex II formed by the replicase presumably abolished initiation of both coat protein and replicase synthesis (Kolakofsky and Weissmann 1971a,b). The model of replicase-controlled regulation for coat protein and replicase synthesis on Qβ RNA, by means of repressor complex II, was put forward by Daniel Kolakofsky and Charles Weissmann. According to this model, the function of repressor complex II consisted of removing ribosomes from the plus-strand templates and switching the latter from translation to replication. The replicase in the repressor complex II did bind to the coat protein ribosome binding site, as was shown *in vitro* for Qβ replicase and Qβ RNA (Weber et al. 1972; Meyer et al. 1975; Vollenweider et al. 1976).

Further findings demonstrated that the repressor complex II was indeed functional *in vivo* and, together with the repressor complex I, was necessary and sufficient to regulate production of viral RNA, including the proper ratio of plus and minus strands (Bauman et al. 1978; Pumpen et al. 1978a,b). Such control might be possible if replicase and RNA syntheses would be somehow coupled. According to the idea of the Elmārs Grēns' laboratory, the initiation of RNA minus strands required unfolding of the 3′-terminal part of the RNA plus template, which was probably effected by the ribosome during translation of the replicase cistron. The initiation of the plus-strand synthesis at the same time appeared to be independent of the replicase gene translation. If RNA minus strands were functionally unstable, the rate of the plus-strand synthesis decreased after a period equal to the lifetime of an RNA minus strand in the absence of replicase synthesis. Thus, the regulation of replicase synthesis would be sufficient in such a case to exert effective control over viral RNA replication. The idea was supported by data on the phages f2, fr, MS2, R17, Qβ, and their suppressor-dependent mutants (Bauman et al. 1978; Pumpen et al. 1978a,b). Figure 1.11 presents a summarizing picture of the RNA phage replication with all regulatory links.

A last example of regulation, which is not illustrated in Figure 1.11, would result in the small amount of lysis protein made. It could be controlled by a very modest affinity of its ribosome-binding site for ribosomes.

RECONSTITUTION

In the 1960s, the idea of the *in vitro* reconstitution of RNA phages from their structural subunits was rather theoretical and met the challenge to produce infectious phages in a test tube and repeat (or model) therefore the natural process of life. Nowadays, this is a popular technique of virus-like particle (VLP) nanotechnology which serves for purely practical motives. For this reason, the early approach to the problem is included in the first acquisition with the RNA phages.

Thomas and Barbara Hohn provided the first reconstitution attempts, with a solid theoretical and experimental background, which was exhaustively presented in their comprehensive and well-illustrated review (Hohn and Hohn 1970). For simplicity, they introduced a summarizing phage name fII, the same name as the first isolated species but written with a Roman numeral for distinction, for the f2 and f2-like phages fr, MS2, and R17. These authors' term fII is used in this section only.

Thus, the first reconstitutions of the fII capsid were performed using coat protein isolated by the acetic acid method and phage RNA isolated by phenol extraction (Hohn 1967; Sugiyama et al. 1967). The major products of these reconstitution experiments were particles of size and general electron microscopic appearance similar to an RNA phage. They contained a normal amount of RNA, but they do not adsorb to bacteria and were noninfectious. The low infectivity was argued by the absence of the A protein in the reconstructed particles. The formation of the fII-defective particles was also observed in an *in vitro* protein synthesizing system from infected or uninfected *E. coli* directed by fII RNA (Knolle 1969).

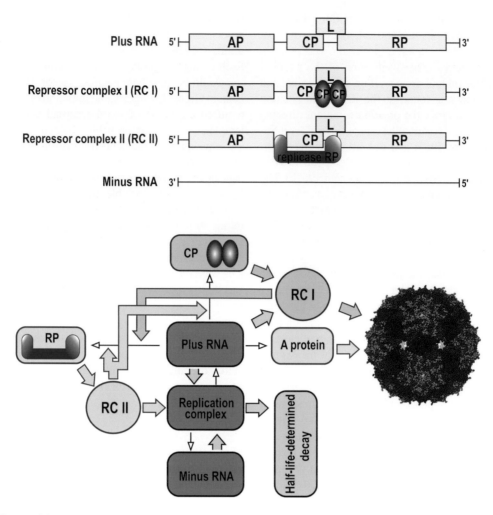

FIGURE 1.11 Summarizing scheme of the RNA phage replication with inclusion of regulation mechanisms. The details are explained in Chapters 16 and 17.

Defective particles similar in most of their properties to the ones in the fII system were obtained for the phage Qβ by reconstitution of the Qβ protein obtained by the guanidinium-HCl method and the Qβ RNA obtained by phenol extraction (Hung and Overby 1969).

It is remarkable, especially considering current activities, that a number of different RNAs, including the fII RNA fragments, plant virus RNA, Qβ RNA, and even polyuridylic acid and polyvinyl sulfate, were proved to be active in inducing the formation of particles in size and appearance similar to the fII phages, while RNAs smaller than 20 nucleotides were inactive (Hohn 1969c). Moreover, the double-sized TMV and TYMV RNA could be shared by two growing capsids so that a high proportion of "twinned" and "tailed" particles was synthesized (Hohn and Hohn 1970).

Roberts and Argetsinger-Steitz (1967) were the first to perform reconstitution from three components: RNA, coat protein, and A protein, and therefore improved the infectivity of the resulting product to some extent. Further, addition of the A protein to the system resulted in the improved infectivity of the reconstituted particles, although the yield of infectious particles still remained low (Herrmann R., thesis, Univ. of Tubingen, 1970, cited by Hohn and Hohn 1970).

The assembly of phage-like particles, by modern talking VLPs, occurred also in the absence of RNA by special conditions in the fII group phages (Herrmann et al. 1968; Hohn 1969b; Matthews 1970; Rohrmann and Krueger 1970c; Schubert and Frank 1970a,b, 1971; Zelazo and Haschemeyer 1970; Zipper et al. 1971b, 1973). The reconstitution of the RNA phages was reviewed by Heinz Ludwig Fraenkel-Conrat (1970). Later, Peter Knolle and Thomas Hohn (1975) observed the RNA phage morphogenesis and self-assembly problems from all possible sides. The next extensive reviews at that time concentrated on the comparison of the morphogenesis of the RNA phages and other viruses (Hohn 1976; Hung 1976).

PHAGE TALES

To complete the first look at RNA phages, let us refer to Sidney Brenner, the 2002 Nobelist, who made significant contributions to work on the genetic code and other areas of molecular biology while working in the Medical Research Council (MRC) Laboratory of Molecular Biology in Cambridge:

A correspondent has asked me about the origins of a legend which has become known as *The phage in the letter*. Common to all variants of this story is the following: one scientist, called X, sends a request to another, Z, for a particular bacteriophage strain which Z has discovered. Z replies saying he is not sending it out. X thereupon plates out the letter and retrieves the phage from it. Some versions include a third scientist, Y, to whom the outraged X takes the letter; it is Y who advises X on how to recover the bacteriophage. Now Z, in this story, is Norton Zinder, who discovered an RNA phage, f2, in the sewers of New York and who did not send out the phage to the large number of scientists who requested it. He also found a single-stranded DNA filamentous phage, f1, in the same sewage. Both f1 and f2 would only grow on bacteria with a sex factor. It has been suggested that I was either X or Y in the legend; that is, I either plated out the letter myself or got somebody else to do it. In fact, this is an invented story and … I do not know that it has ever been attempted (Brenner 1997).

Disclosing the origins of this popular legend in 1997, Brenner wrote:

When Norton Zinder published his paper on the RNA phage, I thought of writing to ask him for a sample of it (or f1, for that matter) because I wanted to use it to test bacteria for the presence of sex factor. However, I knew that if I gave this reason he would not believe me, and would suspect that I had made it up, simply to get the phage and work on RNA replication. Many people complained to me, and I am sure to others, about his not turning over his discovery to the world instantly, and there was much talk about how everything should be made available once it is published. I did suggest to more than one of these people that they should try to recover the phage by plating out the letter. Perhaps I was guilty of hinting that I had successfully done this myself.

I obtained many of these phages myself simply by going to the Cambridge sewage plant and bringing back a bottle of their best vintage and plating it on bacteria with a sex factor. Prodigious as New York sewage might be, we could still match it. Other people also went to their particular sewers and found their own phages, some of which, such as Qβ, an RNA phage, and M13, a DNA phage, became more famous than the original finds. I didn't want to add to the confusion, but to this day still use my isolates to test bacteria for their sex.

François Jacob told me that whenever he wanted a phage he would take himself off to the nearest *pharmacie*, where all kinds of phage were sold as remedies for intestinal complaints. I traced the history of these and found that many had been isolated from the Paris sewers by two characters called Sertic and Boulgakov. The X in phage φX174 is not the letter *eks* but the Roman numeral ten, indicating that it was the 174th isolate from a particularly good sewer in the *Xe arrondissement* (Brenner 1997). (Reprinted from *Curr Biol.*, 7, Brenner S., Bacteriophage tales, R736, Copyright 1997, with permission from Elsevier.)

To round off the first look at RNA phages, it remains only to stress that one of the characters mentioned by Sidney Brenner above, namely, Boulgakov, is, with a great likelihood, Nikolaï Boulgakov, a French microbiologist from the Institut Pasteur. Nikolaï Boulgakov is a brother of the famous Russian writer Mikhail Bulgakov: the second epigraph to the present chapter is taken from his immortal novel *The Master and Margarita*. This novel is called one of the masterpieces of the twentieth century, as also are the RNA phages.

2 Taxonomy and Grouping

The beginning of wisdom is a definition of terms.

Socrates

Crude classifications and false generalizations are the curse of organized life.

George Bernard Shaw

ENTRIES IN PRESTIGIOUS BOOKS OF REFERENCE AND ENCYCLOPEDIAS

From the early years of RNA phage investigation, they have been presented in prominent virological paper collections and reference books by the pioneers of the field. Thus, a set of thorough chapters (Fraenkel-Conrat 1968a; Fraenkel-Conrat and Weissmann 1968; Hofschneider and Hausen 1968; Kaper 1968) has been written for the historically highly important collection *The Small RNA Viruses of Plants, Animals and Bacteria* as a part of classical *Molecular Basis of Virology book* edited by Heinz Fraenkel-Conrat (1968b) and published by Reinhold Book Corporation. Further, the exhaustive general reviews written by the field leaders appeared systematically in the *Comprehensive Virology* collections guided by Heinz Fraenkel-Conrat together with other well-known virologists (Eoyang and August 1974; Fiers 1979; Murphy and Gordon 1981).

Walter Fiers prepared a general entry on RNA phages for the *Handbook of Genetics* (Fiers 1974), which was reprinted by Springer in 2013. Jan van Duin contributed to the two editions of the classical *The Bacteriophages* book edited by Stephen T. Abedon and Richard Lane Calendar (van Duin 1988; van Duin and Tsareva 2005) and to the modern and highly useful *eLS (Encyclopedia of Life Sciences)* edition which started in 2002 and is now published online since 2011 in a continually updated form (van Duin 2002; Olsthoorn and van Duin 2007, 2011). This unique project, spanning the entire spectrum of the life sciences, is also available as an enormous 32-volume set. Figure 2.1 shows the cover of the ambitious *Encyclopedia of Life Sciences* project. Of note, David J. Rowlands and Peter G. Stockley contributed to this project on virus assembly (Rowlands and Stockley 2001).

Stockley also contributed with a general entry on the RNA phages to the comprehensive *Desk Encyclopedia Animal and Bacterial Virology* edited by Brian W.J. Mahy and Marc H.V. van Regenmortel in 2010. Leonard Mindich wrote a general entry on the RNA phages in the *Encyclopedia of Genetics* edited by Eric C.R. Reeve (Mindich 2001). In the *Encyclopedia of Microbiology* of 2009, the RNA phages are mentioned rather briefly in more general entries on phages and phage ecology (Abedon et al. 2009; Hyman and Abedon 2009). The informative entries on the RNA phages to the 8th

and 9th reports of the Virus taxonomy commission were written by van Duin and van den Worm (2005) and van Duin and Olsthoorn (2012), respectively.

OFFICIAL ICTV TAXONOMY

According to the current (10th) release 2018b, ratified on February 2019, *Report on Virus Taxonomy* of the International Committee on Taxonomy of Viruses (ICTV), which is published as a freely available online resource (https://talk.ictvonline.org/ictv-reports/ictv_online_report/) and introduced by a summarizing paper (King et al. 2018), the current totals at each taxonomic level now stand at 14 orders, 150 families, 79 subfamilies, 1019 genera, and 5560 species. By the previous ICTV standards, an "Order," the use of which was optional, was the highest taxonomic level into which virus species could be classified. A "Family" was the next level in the taxonomic hierarchy, followed by a "Genus" and then "Species," which was the lowest taxonomic level in the hierarchy approved by the ICTV. While subspecies levels of classification might exist for some species, the ICTV did not discuss or approve the classification of viruses below the species level. One "Type Species" was chosen for each Genus to serve as an example of a well characterized species for that Genus.

The 2018b release introduced a set of novel taxonomic levels, such as "Realm," "Subrealm," "Kingdom," "Subkingdom," "Phylum," "Subphylum," "Class," and "Subclass," which are followed now by the formerly higher "Order" level. The "Realm" is defined now as the highest taxonomic rank into which virus species can be classified. The only currently defined realm *Riboviria* covers RNA viruses: altogether 1 phylum, 3 orders, 40 families, and 8 genera. The *Leviviridae* family is included therefore into the huge *Riboviria* realm.

Although such a crucial parameter as genome composition is not regarded as an official taxonomy term, the currently available ICTV chapters on the families (https://talk.ictvonline.org/ictv-reports/ictv_online_report/) are placed in groups by accordance with the nature, i.e., molecular and genetic composition, of the virus genome packaged into the virion. As described by the latest update (EC 50, Washington, DC, July 2018) of the ICTV Master Species List (MSL34) which is published online on March 8, 2019, the possible values of the genome composition packaged into the virion are: dsDNA, ssDNA, ssDNA(−), ssDNA(+), ssDNA(±), dsDNA-reverse transcription (RT), ssRNA-RT, dsRNA, ssRNA, ssRNA(−), ssRNA(+), and ssRNA(±).

In this respect, the ICTV classification continues and develops traditions of the classification suggested by David Baltimore, the 1975 Nobelist, and based on the genome structure and mechanism of mRNA production by the viruses (Baltimore 1974). In accordance with the Baltimore

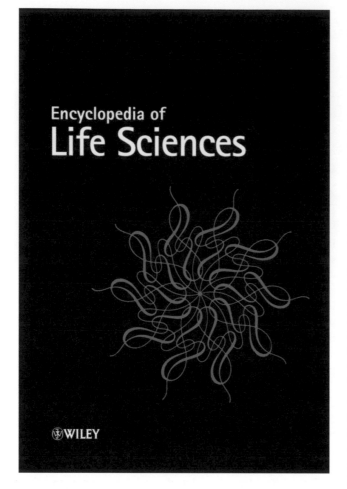

FIGURE 2.1 The cover of the *eLS* (*Encyclopedia of Life Sciences*) edition, which is now published online and covers the entire spectrum of the life sciences including comprehensive entries on the RNA phages. (Printed with permission of Wiley-Blackwell.)

classification, all viruses were placed into seven groups in the following order: I: double-stranded DNA (dsDNA) viruses, II: single-stranded (ssDNA) viruses with plus-strand or sense DNA; III: double-stranded RNA (dsRNA) viruses; IV: positive single-stranded RNA, or (+)ssRNA, viruses with plus-strand or sense RNA; V: negative single-stranded RNA, or (−)ssRNA, viruses with minus-strand or antisense RNA; VI: single-stranded RNA viruses using reverse transcription (ssRNA-RT) with plus-strand or sense RNA and DNA intermediate in life cycle; VII: double-stranded DNA viruses using reverse transcription (dsDNA-RT) with intermediate pregenomic RNA in life cycle.

Therefore, according to the Baltimore classification, the *Leviviridae* family covering the RNA phages falls into the (+)ssRNA, or single-stranded positive-sense RNA genome, group IV. Before 2018, the *Leviviridae* family was not assigned to any higher order. Now, it is placed in the *Riboviria* realm, together with other 39 RNA virus families which are not assigned to any order.

The latest ICTV online information (https://talk.ictvonline. org/taxonomy/) declares that the *Leviviridae* family falls into genera *Allolevivirus* and *Levivirus*, each of which is represented by two species: Escherichia virus FI and Escherichia virus Qβ for the *Allolevivirus* genus and Escherichia virus BZ13 and Escherichia virus MS2 for the *Levivirus* genus, where Qβ and MS2 are marked as "type species."

In the preceding 9th report of the Virus Taxonomy, the *Leviviridae* family was described in the short version of the general table (King et al. 2011, p. 13) as an unassigned family, ssRNA(+) by the nature of genome, icosahedral by morphology, 26 nm by virion size, one linear segment by genome configuration, 3.5–4.3 kb by genome size, and bacteria as a host. The *Leviviridae* family contained two genera: *Allolevivirus* and *Levivirus*, which were represented by type species *Enterobacteria phages* MS2 and Qβ, respectively (King et al. 2011, p. 34).

Full description of the *Leviviridae* family in the 9th report (King et al. 2011, pp. 1035–1043) was contributed by Jan van Duin and René C.L. Olsthoorn. René Olsthoorn was a chair of the *Leviviridae* Study Group (King et al. 2011, p. 1267). In the full description, the lists of both genera species (King et al. 2011, pp. 1039 and 1041) did not coincide fully with the short information provided by the ICTV site and were very close to the historical unofficial lists of RNA phages (see Table 1.1) and NCBI information discussed below. Two lists of species of the related viruses (King et al. 2011, pp. 1039 and 1041) which could be members of the *Leviviridae* family and of the *Levivirus* genus but have not been approved as species reflected fully the information given the NCBI online. Of note, the 9th report explained derivation of names "levi" from Latin levis, "light," and "allo" from Greek ἄλλο, "other."

The 10th report is not finished at the time of this writing, and the latest description of the *Leviviridae* family is still not available on the ICTV site. However, the introduction and some other parts of the 10th report are available online (https://talk.ictvonline.org/ictv-reports/ictv_online_report/).

It is remarkable that the concept of the virus species as the lowest taxon (group) in a branching hierarchy of viral taxa was formalized for the first time by the 7th ICTV Report (Van Regenmortel et al. 2000). As defined therein, "a virus species is a polythetic class of viruses that constitute a replicating lineage and occupy a particular ecological niche." A "polythetic class" is one whose members have several properties in common, although they do not necessarily all share a single common defining property. In other words, the members of a virus species are defined collectively by a consensus group of properties. Virus species thus differ from the higher viral taxa, which are "universal" classes and as such are defined by properties that are necessary for membership. The seventh ICTV report (Van Regenmortel et al. 2000) stated the general classification principles that are presented at the ICTV homepage (https://talk.ictvonline.org/taxonomy/w/ictv-taxonomy): "(1) viruses are real physical entities produced by biological evolution and genetics, whereas virus species and higher taxa are abstract concepts produced by rational thought and logic. The virus/species relationship thus represents the front line of the interface between biology and logic; (2) viruses (including virus isolates, strains, variants, types, sub-types, serotypes, etc.) should wherever possible be assigned as members of the appropriate virus species, although many viruses remain unassigned because they are inadequately characterized; (3) all virus species must be represented by at least one virus isolate; (4) almost all virus species are members of recognized genera. A few species remain unassigned in their families although they have been clearly identified as new species; (5) Some genera are members of recognized subfamilies; (6) all subfamilies and most genera are members of recognized families. Some genera are not yet assigned to a family; in the future they may either join an existing family or constitute a new family with other unassigned genera; (7) some families are members of the following recognized orders; (8) only the aforementioned taxa are recognized by the ICTV; (9) other groupings (from clade to superfamily), may communicate useful descriptive information in some circumstances but they have no formally recognized taxonomic meaning. Similarly, the term "quasispecies," although it captures an important concept, has no recognized taxonomic meaning". Therefore, the current hierarchy of recognized viral taxa was established in 1996 and published in 2000.

Two genera in the *Leviviridae* family were established by the 5th ICTV report (Francki et al. 1991). According to the Minutes of the 8th plenary meeting of the ICTV, Berlin, 29 August 1990, it was decided there "(1) to establish two genera within the *Leviviridae* family; (2) to retain the existing generic name *Levivirus* to provide a name for a new genus that recalls an earlier designation (supergroup A); (3) to designate the MS2 phage group as the type species of this genus; (4) to name the second genus *Allolevivirus*; (5) to designate the Qβ group as the type species of this genus."

According to the 4th ICTV report (Matthews 1982), the genus including the MS2 phage group was named *Levivirus*.

By the 3rd ICTV report (Matthews 1979), a phage R40 was abolished from the *Leviviridae* as not fully recognized member but phages B6, B7, ZIK/1, φCB5, φCB8r, φCB12r, φCB23r, PPR1, PP7, and 7s were listed as new species.

The 2nd ICTV report (Fenner 1975) listed families of viruses "primarily of interest to bacterial virologists," among them the *Leviviridae*, or f2 phage group with a reference to Wildy (1971, p. 66). The group was shortly defined as having "linear single-stranded DNA [*an unfortunate misprint* – PP], mol. wt. 1.2×10^6, small icosahedral virion." Remarkably from the current point of view, it was concluded that comparative studies of bacterial viruses as such have hardly begun by the ICTV, "since they are of minor economic importance" (Fenner 1975). As a result, The Bacterial Virus Subcommittee has suggested family names for six groups defined in the 1st report, including *Leviviridae* family (Wildy 1971). However, no definitive family or generic names were proposed for approval at the Madrid meeting in 1975 and it was decided that "the families and names suggested for them should be exposed to working virologists to determine their usefulness" (Fenner 1975).

Therefore, we see that the 1st ICTV report stood at the very beginning of the official *Leviviridae* history (Wildy 1971), when it appeared for the first time among 43 families and groups after discussion in 1968 at the International Congress of Virology in Helsinki.

Currently, the ICTV site allows easy follow-up of the entry history for the four assigned members: MS2, BZ13, Qβ, and FI, and a number of unassigned members of the *Leviviridae* family.

RELATEDNESS TO OTHER TAXA

Although similarity with other taxa is stated as "not reported" in the description of the *Leviviridae* family, the 9th ICTV report identifies the *Leviviridae* family as distantly related to narnaviruses and mitoviruses of the *Narnaviridae* family by the structure of their RNA dependent RNA polymerases, or simply replicases (King et al. 2011, pp. 1058–1059). Furthermore, the 3′-end secondary structures of members of the genus *Narnavirus* resemble those of coliphages in the family *Leviviridae*. In a neighbor-joining phylogenetic tree of families of fungus viruses and related viruses in other taxa, based on aa sequences of the putative replicase proteins, the families *Narnaviridae* and *Leviviridae* form a cluster with 69.2% bootstrap support. Figure 2.2 presents the corresponding phylogenetic tree.

Figure 2.3 presents an illustration from the introduction to the 10th ICTV report, where virus particles of the taxa infecting bacteria and archaea are drawn in the size scale. It is noteworthy that only 2 out of 21 families of phages that infect bacteria and archaea have RNA genomes: *Leviviridae* (linear ssRNA) and *Cystoviridae* (segmented dsRNA).

On the contrary, positive-sense RNA viruses are widely distributed in eukaryotes including humans and animals and account for a large fraction of dangerous pathogens such as the hepatitis C virus, West Nile virus, dengue virus, SARS and MERS coronaviruses, as well as less clinically serious pathogens such as rhinoviruses that cause the common cold, and many others. The positive-sense RNA viruses form the most abundant group of viruses. It is highly remarkable that *Leviviridae* is the only positive-strand RNA virus family employing bacteria as a host against the huge number of the *Riboviria* realm members that do not infect bacteria. Recently, it was hypothesized however that the bisegmented double-stranded RNA viruses of the *Picobirnaviridae* family are in fact prokaryotic (Krishnamurthy and Wang 2018).

Figure 2.4 is intended to show the most recent phylogenetic analysis and the evolutionary connections among viral taxa that were elaborated on the basis of the three-dimensional structures of the viral RNA dependent RNA polymerases, or replicases, resolved by x-ray crystallography and available as of September 2017 (Venkataraman et al. 2018). Thus, the 3D hierarchy of replicases revealed the representatives of dsRNA virus family *Cystoviridae* as a nearest neighbor of the phage Qβ from the *Leviviridae* family. The next relatives of the latter are dsRNA virus families *Permutotetraviridae* and *Birnaviridae* and finally, the representatives of the (+) strand virus family *Flaviviridae*.

Some relatedness of the *Leviviridae* family to other positive-sense RNA viruses led to an idea of the *Flavivirata,*

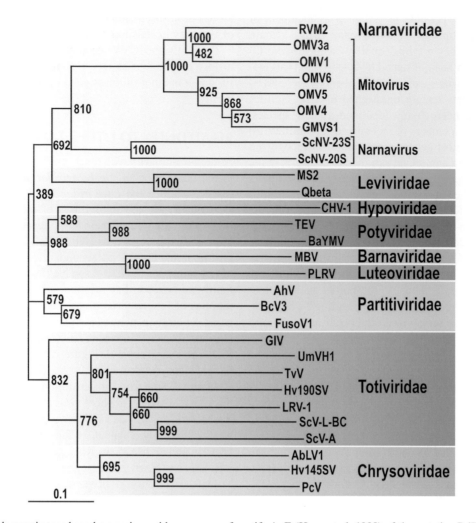

FIGURE 2.2 Phylogenetic tree based on amino acid sequences of motifs A–E (Hong et al. 1998) of the putative RdRp, or RNA dependent RNA polymerase, proteins of members of the family *Narnaviridae*, other families of RNA viruses of fungi, and related viruses in other host taxa, and the family *Leviviridae* of RNA bacteriophages. Sequence alignments and the neighbor-joining tree were made using the Clustal X program. Bootstrap numbers (1000 replicates) are shown on the nodes. Abbreviations and sequence acquisition numbers are: AbVL1, Agaricus bisporus virus L1 X94361; AhV, Atkinsonella hypoxylon virus L39126; BaYMV, Barley yellow mosaic virus D01091; BcV3, Beet cryptic virus 3 S63913; CHV1, Cryphonectria hypovirus 1 M57938; CMV1, Cryphonectria mitovirus 1 L31849; FusoV1, Fusarium solani virus 1 D55668; GlV, Giardia lamblia virus L13218; GMVS1, Gremmeniella mitovirus S1 AF534641; Hv145SV, Helminthosporium victoriae 145S virus AF297176; Hv190SV, Helminthosporium victoriae 190S virus U41345; LRV1, Leishmania RNA virus 1-1 M92355; MBV, Mushroom bacilliform virus U07551; MS2, Enterobacteria phage MS2 GB-PH:MS2CG; OMV3a, Ophiostoma mitovirus 3a AJ004930; OMV4, Ophiostoma mitovirus 4 AJ132754; OMV5, Ophiostoma mitovirus 5 AJ132755; OMV6, Ophiostoma mitovirus 6 AJ132756; PcV, Penicillium chrysogenum virus AF296439; PLRV, Potato leafroll virus X14600; Qbeta, Enterobacteria phage Qβ AY099114; RVM2, Rhizoctonia virus M2 U51331; ScV-L-A, Saccharomyces cerevisiae virus L-A J04692; ScV-L-BC, Saccharomyces cerevisiae virus L-BC U01060; ScNV-20S, Saccharomyces 20S RNA narnavirus M63893; ScNV-23S, Saccharomyces 23S RNA narnavirus M86595; TEV, Tobacco etch virus M15239; TvV, Trichomonas vaginalis virus U08999; UmVH1, Ustilago maydis virus H1 U01059. (Reprinted from *Ninth Report of the International Committee on Taxonomy of Viruses*, King AMQ et al. Virus taxonomy. Classification and nomenclature of viruses, 1338 p., p. 1059, Copyright 2011, with permission from Elsevier.)

or flavi-like group, where *Leviviridae* would be combined together with carmoviruses, dianthoviruses, flaviviruses, pestiviruses, statoviruses, tombusviruses, hepatitis C virus, and a subset of luteoviruses (barley yellow dwarf virus) (Koonin 1991b).

NCBI CLASSIFICATION

The National Center for Biotechnology Information (NCBI) of the U.S. National Library of Medicine (NLM) is located

in Bethesda. As defined at the official site, the NCBI (1) conducts research on fundamental biomedical problems at the molecular level using mathematical and computational methods; (2) maintains collaborations with several NIH institutes, academia, industry, and other governmental agencies; (3) fosters scientific communication by sponsoring meetings, workshops, and lecture series; (4) supports training on basic and applied research in computational biology for postdoctoral fellows through the NIH Intramural Research Program; (5) engages members of the international scientific community

Virus Taxa Infecting Bacteria and Archaea

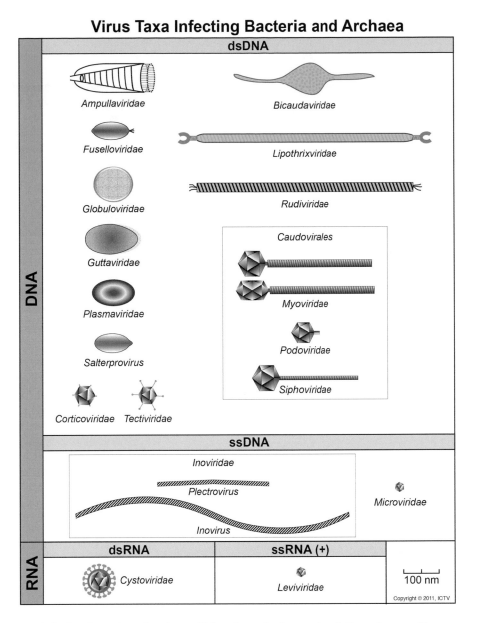

FIGURE 2.3 Virus taxa infecting bacteria and archaea. (Taken from the International Committee on Taxonomy of Viruses [ICTV] report: https://talk.ictvonline.org/ictv-reports/ictv_online_report/introduction/w/introduction-to-the-ictv-online-report/422/hosts-bacteria-and-archaea. Printed with permission of ICTV.)

in informatics research and training through the Scientific Visitors Program; (6) develops, distributes, supports, and coordinates access to a variety of databases and software for the scientific and medical communities, and (7) develops and promotes standards for databases, data deposition and exchange, and biological nomenclature (https://www.ncbi.nlm.nih.gov/home/about/mission/). In fact, the NCBI ensures simple and easy access of scientists to the most advanced databases for taxonomy (by the Taxonomy Browser), genome information (by Viral Genome Browser), sequences and their analysis (by GenBank, BLAST, etc.), and 3D structure (by MMDB, [Molecular Modeling Database] of 3D protein structures). Although the disclaimer on the NCBI Taxonomy browser says that "the NCBI taxonomy database is not an authoritative source for nomenclature or classification," it is

in reality highly authoritative and most advanced by provided size and quality of information and reflects and develops in principle the ICTV classification when unassigned information within the latter is taken into account.

Figure 2.5 demonstrates the current status (on June 6, 2019) of the *Leviviridae* phage in the NCBI Taxonomy browser.

According to the NCBI Taxonomy browser information (https://www.ncbi.nlm.nih.gov/Taxonomy/Browser/wwwtax.cgi?mode=Undef&id=11989&lvl=3&keep=1&srchmode=1&unlock), the *Levivirus* genus consists of the BZ13 and MS2 species, as it is approved by the ICTV classification. The BZ13 species includes BO1, GA, BZ13, JP34, JP500, KU1, SD, TH1, and TL2 as "no rank" members. Inclusion of the phage BO1 here could be regarded as a simple mistake, since BO1 belongs to the group I phages, in contrast to other

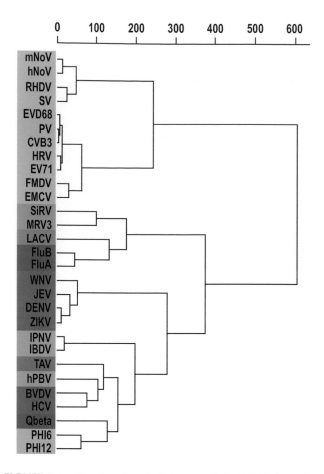

FIGURE 2.4 Structure-based phylogeny of viral RNA dependent replicases. Groups of ss (+) RNA viruses are shown in shades of blue: pale blue indicating *Caliciviridae*: murine norovirus (mNoV), human norovirus (hNoV), rabbit hemorrhagic disease virus (RHDV), Sapporo virus (SV); medium blue representing *Picornaviridae*: enterovirus D68 (EVD68), poliovirus type I (PV), coxsackievirus B3 (CVB3), human rhinovirus 16 (HRV), enterovirus A71 (EV71), foot-and-mouth disease virus (FMDV), encephalomyocarditis virus 1 (EMCV), and dark blue indicating *Flaviviridae*: West Nile virus (WNV), Japanese encephalitis virus (JEV), dengue virus (DENV), Zika virus (ZIKV), bovine viral diarrhea virus (BVDV), hepatitis C virus (HCV). dsRNA viruses are shown in shades of green: *Reoviridae* in dark green: simian rotavirus SA11 (SiRV), mammalian orthoreovirus 3 (MRV3), *Birnaviridae* in medium green: infectious pancreatic necrosis virus (IPNV), infectious bursal disease virus (IBDV), *Picobirnaviridae* in light green: human picobirnavirus (hPBV), and *Cystoviridae* in lime green: *Pseudomonas* phage φ6 (PHI6), *Pseudomonas* phage φ12 (PHI12). *Permutotetraviridae* Thosea asigna virus (TAV) and *Leviviridae* coliphage Qβ from the group IV are presented in light and medium purple, respectively. ss (−) RNA viruses are colored in shades of red, *Orthomyxoviridae* in dark red: influenza A virus (FluA), influenza B virus (FluB) and *Bunyaviridae* in light red: La Crosse virus (LACV). (Redrawn from Venkataraman S et al. *Viruses.* 2018;10:E76.)

BZ13 species members. In the same manner, the MS2 species includes the phages f2, fr, JP501, M12, R17, ZR, and MS2. Remarkably, the BZ13 and MS2 species are including the "no rank" BZ13 and MS2 entries. Furthermore, the *Levivirus* genus also includes four unclassified members: C-1 INW-2012, HgalI, M, and the pseudomonaphage PRR1.

Lineage (full): Viruses; Riboviria

- **Leviviridae** *Click on organism name to get more information.*
  - **Allolevivirus**
    - **Enterobacteria phage FI sensu lato**
    - **Enterobacteria phage Qbeta sensu lato**
    - **Escherichia virus FI**
      - **Enterobacteria phage ID2**
      - **Enterobacteria phage NL95**
      - **Enterobacteria phage SP**
      - **Enterobacteria phage TW19**
      - **Enterobacteria phage TW28**
    - **Escherichia virus Qbeta**
      - **Enterobacteria phage M11**
      - **Enterobacteria phage MX1**
      - **Enterobacteria phage ST**
      - **Enterobacteria phage VK**
  - **Levivirus**
    - **Escherichia virus BZ13**
      - **Enterobacteria phage BO1**
      - **Enterobacteria phage GA**
      - **Enterobacteria phage isolate BZ13**
      - **Enterobacteria phage JP34**
      - **Enterobacteria phage JP500**
      - **Enterobacteria phage KU1**
      - **Enterobacteria phage SD**
      - **Enterobacteria phage TH1**
      - **Enterobacteria phage TL2**
    - **Escherichia virus MS2**
      - **Enterobacteria phage f2**
      - **Enterobacteria phage fr**
      - **Enterobacteria phage JP501**
      - **Enterobacteria phage M12**
      - **Enterobacteria phage MS2**
      - **Enterobacteria phage R17**
      - **Enterobacteria phage ZR**
    - **unclassified Levivirus**
      - **Enterobacteria phage C-1 INW-2012**
      - **Enterobacteria phage Hgal1**
      - **Enterobacteria phage M**
      - **Pseudomonas phage PRR1**
  - **unclassified Leviviridae**
    - **Acinetobacter phage AP205**
    - **Bdellovibrio bacteriovirus**
    - **Caulobacter phage phiCb5**
    - **LeviOr01 phage**
    - **Pseudomonas phage PP7**
    - **Leviviridae sp.**

FIGURE 2.5 The *Leviviridae* page in the NCBI Taxonomy browser. (Printed with permission of NLM.)

Based on the NCBI Taxonomy classification of the *Allolevivirus* genus, the latter contains two species FI and Qβ, in accordance with the ICTV classification. However, in contrast to the content of the *Levivirus* genus, the "no rank" FI and Qβ entries do not appear under the appropriate FI and Qβ species but are defined as two independent species "Enterobacteria phage FI sensu lato" and "Enterobacteria phage Qβ sensu lato." Therefore, the FI species includes phages ID2, NL95, SP, TW19, TW28, but not FI. The Qβ species includes phages M11, MX1, ST, VK, but not Qβ.

The indication "sensu lato" ("in the broad sense") would be attributed to the FI and Qβ species per se and the introduction of the "sensu lato" species is rather misleading. Moreover, the structural information that appears under both "sensu lato" species is not clearly specified and repeats the same information given under the FI and Qβ species. The source and history of the appearance of both "sensu lato" species remains unclear.

The list of unclassified *Leviviridae* members includes the acinetophage AP205, the pseudomonaphages PP7 and LeviOr01, and the caulophage φCB5. The appearance of the entry "Bdellovibrio bacteriovirus" could be explained as an aftereffect of the misinterpretation connected with *Bdellovibrio bacteriovorus* phages which is described in the Dangerous similarity section in Chapter 1. The *Leviviridae* sp. directory contains sequences acquired metagenomically by Krishnamurthy et al. (2016) and therefore opens a way to novel species discovered by novel approaches and not by the traditional search for the infectious phage particles.

However, the major role of the NCBI Taxonomy page consists of the collection and classification of structural entries. Table 2.1 presents a number of entries devoted to the structural data concerning RNA phages. Nevertheless, it is necessary to take into account that the entries are not fully systematized: some entries are doubled by repetition of the same information under different GenBank accession numbers, and some older information is missing, e.g., the first full-length Qβ genome (Mekler 1981) and the first amino acid sequence of MS2 coat protein deduced by protein sequencing (Lin et al. 1967).

The genomic information on the *Leviviridae* family is presented now also by the viruSITE (http://www.virusite.org), an integrated database for viral genomics, which contains information on virus taxonomy, host range, genome features, and sequential relatedness as well as the properties and functions of viral genes and proteins (Stano et al. 2016).

SEROGROUPING

From the very beginning, it is necessary to acknowledge that the current official taxonomy of RNA phages that has been described above has grown up from their immunological characteristics and remains structured up to now in a general accordance with the early serological grouping.

Serological relationship, namely "the partial or complete neutralization of the infectivity of one virus by the antiserum to a second virus," was suggested as one of the most important taxonomic criteria, together with morphological relationship, for the classification of viruses by Mark H. Adams (1953). Moreover, he stated in this early paper that the "absence of detectable neutralization, however, does not preclude a close biological relationship, since circumstantial evidence indicates that serological specificity may alter through mutations."

A first attempt to discover the immunological differences among RNA phages known at that time: f2, f4, fr, ft5 (new name fr), FH5, M12, MS2, R17, and β led to the clear conclusion that they all are serologically related and "may all

be serological mutants of the same phage" (Scott 1965). The following year, the phage β and two novel phages isolated in Japan, B1 and I, were found serologically related but not identical to the phage MS2 (Yuki and Ikeda 1966).

The broad step toward the serological classification was made by the finding that the phage Qβ from Japan (Watanabe 1964) is not serologically cross-reactive with the MS2, and therefore with all other immunologically related phages studied before (Overby et al. 1966a). It is noteworthy that Robert G. Krueger was not so categorical in his detailed immunological study: he concluded that the phages MS2 and Qβ could be inactivated by anti-Qβ and anti-MS2 serum, respectively, and therefore could appear to be distantly related (Krueger 1969a).

Furthermore, thanks to serious efforts in the search for novel RNA phages by Itaru Watanabe and his colleagues (30 strains were isolated in 3 years after 1961), three serological groups of RNA phages were established (Sakurai et al. 1967; Watanabe et al. 1967a). A weak cross-reaction was observed by authors between group I and group II, but group III showed virtually no cross-reaction to groups I and II. As a result, the popular phages MS2, f2, and R17 were combined into group I together with phages MY, ZR, and some other species isolated in Japan but never mentioned in more detailed studies: MB, AB, SZ, MC, CC, and KM. The phages GA and SD formed group II together with the less-known phages EI, SS, KJ, SW, UW, SB, FM, and YG. The phages Qβ and VK belonged to group III together with the less-known phages NM, NH, AG, GM, ON, SO, HI, CF, Qα, and KF (Watanabe et al. 1967a).

Figure 2.6, which is redrawn from this innovative paper (Watanabe et al. 1967a), demonstrates inactivation of the RNA phages representing three serological groups by the appropriate three antisera.

Isolation of the phage SP that differed from the three existing group members led to the addition of serological group IV to the serological classification (Sakurai et al. 1968). Moreover, isolation of the phage FI announced a putative group V (Miyake et al. 1969). However, the subsequent grouping of RNA phages based on the template specificity of their RNA replicases placed the phage FI into group IV, which was divided into the two subgroups SP and FI (Miyake et al. 1971b). Not long after, two novel representatives of group IV, namely TW19 and TW28, were isolated in Taiwan (Miyake et al. 1971a) and characterized by the synthesis of "read-through" proteins typical for the group III and IV phages (Aoi and Kaesberg 1976).

More detailed serological analysis led to identification of subgroups within the basic serological groups I–IV. Thus, three subgroups were identified in group III with the phages Qβ, NH, SG belonging to subgroup I, the phages VK and SO to subgroup II, and the phages NM and ST to subgroup III (Miyake et al. 1968). Further subdivision of all four groups was elaborated by the fundamental phage survey in South and East Asia (Furuse et al. 1978). Figure 2.7 presents the general outlook of groups I–IV.

Furthermore, some RNA phages were defined as serological intermediates: the phages JP34 and JP500 (Furuse et al. 1973) between groups I and II and the phage MX1 (Furuse et al. 1975) between groups III and IV. In the case of the phage

TABLE 2.1

The RNA Phage Entries at the NCBI Taxonomy Browser (Status: June 6, 2019)

| Genus | Phage | Number of Entries | | | Protein Sequencing Data |
		Nucleotide (Direct Links)	Protein (Direct Links)	Full-Length Genomes	
Allolevivirus	FI	25	37	3	–
	ID2	–	–	–	
	NL95	1	4	1	
	SP	10	41	5	
	TW19	1	2	–	
	TW28	1	2	–	
	Qβ	108	269	12	CAPSD_BPQBE Maita and Konigsberg (1971)
	M11	4	7	1	
	MX1	1	8	1	
	ST	1	2	–	
	VK	2	3	–	
Levivirus	BZ13	20	33	3	
	BO1	2	4	–	
	GA	3	61	1	
	JP34	2	9	–	
	JP500	1	3	–	
	KU1	3	8	1	
	SD	1	3	–	
	TH1	2	6	–	
	TL2	1	3	–	
	MS2	6 (subtree links: 532)	33 (subtree links: 1465)	5	
	f2	1	3	–	CAPSD_BPF2 Weber and Konigsberg (1967)
	fr	5	26	1	CAPSD_BPFR Wittmann-Liebold and Wittmann (1967)
	JP501	2	6	–	
	M12	1	4	–	
	R17	13	10	–	CAPSD_BPR17 Weber (1967)
	ZR	1	4	–	CAPSD_BPZR Nishihara et al. (1970)
Unclassified *Levivirus*	C-1 INW-2012	2	8	1	
	Hgal1	2	8	1	
	M	2	8	1	
	PRR1	4	15	1	COAT_BPPRR Dhaese et al. (1979)
Unclassified *Leviviridae*	AP205	3	14	1	
	φCB5	2	8	1	
	LeviOr01	1	3	1	
	PP7	6	26	1	COAT_BPPP7 Dhaese et al. (1980)
	Leviviridae sp.	20	47	1	

JP34, the discrepancy between serological and biophysical classification was resolved by full-length sequencing of the JP34 RNA that demonstrated high nucleotide similarity (more than 95%) with group II and low (less than 45%) with group I (Adhin et al. 1989) genome sequences. Therefore, the phage JP34 was classified as a member of group II but the altered serotype of JP34 was most likely due to the change of three critical amino acids of the coat protein to residues present in group I phage MS2 at the homologous positions (Adhin et al. 1989). The phage MX1 was attended to group III by biochemical criteria (Hirashima et al. 1983).

Remarkably, the phage ID2 was neutralized by the antisera of group IV phages, but also inactivated significantly by those of phage groups I, II, and III, although to a lesser extent (Furuse et al. 1978). Thus, the phage ID2 was seen to share serological characteristics common to groups I, II, III, and IV. Unfortunately, this intriguing phage was never subjected to the genome sequencing.

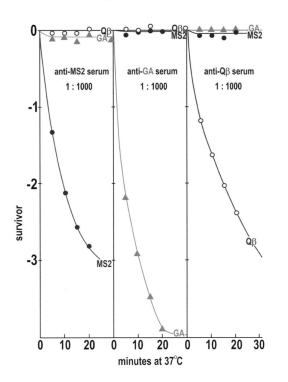

FIGURE 2.6 Inactivation of RNA phages by various antisera. (Redrawn in color from Watanabe M et al. Methods for selecting RNA bacteriophage. In: *Methods in Virology*, vol. 3. Maramorosch K, Koprowski H (Eds). Academic Press, New York, 1967, pp. 337–350 with permission from the *Proceedings of the Japan Academy, Ser B.*)

The maximally full list of the group members, where such serological intermediates are also indicated, is presented in Table 2.2. Exhaustive review of the RNA phage serogrouping is given by the first author of many of the above cited papers, Kohsuke Furuse (1987).

After this comprehensive serological investigation, in consistency with the results of grouping by structural and physicochemical criteria, it was advised that groups I and II could be assembled into one large group A and groups III and IV into another large group B (Ando et al. 1976; Furuse et al. 1978).

Thus, as a logical result of the use and completion of the serological classification, the current *Leviviridae* taxonomy

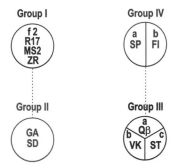

FIGURE 2.7 A schema of grouping of *E. coli* RNA phages. This schema is based on properties of phage particles and the template specificities of RNA replicases. Broken lines indicate a similarity between the groups. (Redrawn with permission from Miyake T et al. *Proc Natl Acad Sci USA* 1971;68:2022–2024.)

was arranged: the group I and II members formed the *Levivirus* genus, while the group III and IV members were attributed to the *Allolevivirus* genus. Since then, the phages MS2, GA, Qβ, and SP are traditionally recognized as reference strains for the groups (not only sero-, but also genogroups) I, II, III, and IV, respectively. However, an important point is that the four groups are not arranged fully on the serological ground: inclusion of the phage FI into group IV occurred after the template specificity of the RNA phage replicases has been determined. Moreover, the serogroup reference strains do not coincide with the taxonomic RNA phage references described above.

Unfortunately, the well-elaborated serogrouping into groups I–IV can describe the RNA coliphages only. Non-coli RNA phages remain untouched by this classification.

The first discovered *Pseudomonas* RNA phages 7s (Feary et al. 1963, 1964) and PP7 (Bradley 1965b, 1966) were serologically related to each other but not identical. No serological relationship of these phages to coliphages was studied. The next *Pseudomonas* phage PRR1 (Olsen and Shipley 1973; Olsen and Thomas 1973) was not compared serologically not only to coliphages, but also to the pseudomonaphages 7s and PP7.

The *Caulobacter* RNA phages fell into three different groups: IV (φCB8r, φCB 9), V (φCB2, φCB4, φCB5, φCB12r, φCB15), and VI (φCB23r) (Schmidt and Stanier 1965) that have never been confronted with the coliphage groups I–IV. The same is true for two other *Caulobacter* RNA phages, φCp2 and φCp18, that were different from each other serologically and different from the most-studied caulophage, φCB5 (Miyakawa et al. 1976). Nevertheless, Miyakawa et al. (1976) concluded that the phages φCp2 and φCp18 might represent a subgroup within Schmidt and Stanier's group V, which included the most popular *Caulobacter* phage φCB5.

Poor serological data are available on the numerous RNA phages that are specific for hosts carrying pili-encoding plasmids of different incompatibility groups. Thus, no serological characteristics have been published for the phage M (Coetzee et al. 1983) and many others. However, a solid and highly important serological investigation (Maher et al. 1991) was performed with the RNA phages pilHα (Coetzee et al. 1985a) and Hgal (Nuttall et al. 1987) specific for incompatibility group HI and HII plasmids. These RNA phages did not demonstrate any cross-reactivity with the group I phages MS2, BO1, and JP501; with the group II phages GA, BZ13, TH1, KU1, and JP34; with the group III phages Qβ, VK, ST, TW18, and MX1; and with the group IV phages SP, FI, TW19, TW18, and ID2. This study was undertaken by Kohsuke Furuse (Maher et al. 1991).

AFFINITIES TO MILLIPORE FILTERS

In parallel with the first serological grouping scheme (Watanabe et al. 1967a), the Itaru Watanabe laboratory elaborated a simple and easy grouping method based on the Millipore filtration and elution (Miyake et al. 1967). This grouping approach was based on the earlier observation that the RNA phage f2 could be attached to nitrocellulose Millipore filters of an HA WP type (pore size 0.45 μ) at

TABLE 2.2

Serological Grouping of RNA Phages

I	I/II	II	III	III/IV	IV
AB B1 BO1 CC f2 f4	JP34→	BZ13 FM GA EI	AG CF GM HI ID1 IN1 IN2 IN3 IN4 IN5 IN6 KF NH	←MX1	FI ID2 SP
FH5 fr ft5 I JP501 KM	JP500→	IN7 KJ KU1 PP3	NM ON PP1 PP2 PP4 PP5 PP6 PP7 PP8 PP9 PP10		TW19
M12 MB MC MS2 MY		SB SD SP5 SS SW	PP11 PP12 PP13 PP14 PP15 PP16 PP17 PP18 PP19		TW28
PP25 R17 R23 R34		TH1 TL2 UW YG	PP20 PP21 PP22 PP23 PP24 PP26 PP27 PP28 PP29		
R40 SZ ZR β μ2			PP30 PP31 Qα Qβ SG SO SP1 SP2 SP3 SP4 SP6		
			SP7 SP8 SP9 SP10 SP11 SP12 ST TL1 TW18 VK		

Source: The data of the following papers are compiled: Scott DW. *Virology.* 1965;26:85–88; Yuki A and Ikeda Y. *J Gen Appl Microbiol.* 1966;12:79–89; Watanabe I. J. *Keio Med. Soc.* 1967;44:622–631; Miyake T et al. *Jpn J Microbiol.* 1968;12:167–170; Krueger RG. *J Virol.* 1969;4:567–573; Watanabe H and Watanabe M. *Can J Microbiol.* 1970;16:859–864; Piffaretti JC and Pitton JS. *J Virol.* 1976;20:314–318; Furuse K. et al. *Appl. Environ. Microbiol.* 1978;35:995–1002.

certain salt concentration (Lodish and Zinder 1965a). Only a short time before, a similar finding was reported in the case of ribosomes (Nirenberg and Leder 1964). It was beyond question that this phenomenon was related to the chemical nature of the phage coats (or ribosomes) and must reflect differences in the phage coat proteins by quantitative analysis of the phage attachment to the filters.

The F (filtration) method used in the first experiments which required a series of Millipore filters, each eluted with a different concentration of NaCl salt, proved impractical. A modification, the filtration-elution (F-E) method, in which the phages bound to one filter were eluted successively with different salt concentrations, yielded results similar to the F method and was used routinely (Miyake et al. 1967). As a result, the F-E patterns of the various RNA phages fell into three distinct groups, I, II, and III, in good agreement with the serological classification of the same phages (Watanabe et al. 1967a). Therefore, the exact fit of both grouping approaches indicated clearly that the differences in the F-E patterns also reflected differences in coat structures, as did anti-phage antibodies. Furthermore, the F-E method was able to divide group III in two subgroups: IIIa (phages Qβ, CF, HI, NH, NM, and SG) and IIIb (phages VK, ST, and SO). These differences have not been observed by the serological grouping. It is noteworthy that some mutants of Qβ, a member of the IIIa subgroup, showed the F-E patterns of the IIIb type (Miyake et al. 1967).

By the lack of the serological cross-investigation of the coli- and non-coli phages, it was reasonable that the Millipore filtering was used for direct comparison of the pseudomonaphages PRR1, 7s, and PP7 with the coliphage f2 (Olsen and Thomas 1973). Interestingly, the phages 7s and PP7 were found similar to f2, whereas the phage PRR1 was markedly different from two other pseudomonaphages or coliphage f2, as well as from the other phages tested by Millipore filtering (Miyake et al. 1967).

TEMPLATE SPECIFICITY OF REPLICASES

The template specificity of replicases was the first example when a functional criterion was applied for the grouping of

the RNA coliphages instead of the traditional structural ones before (Haruna et al. 1967; Miyake et al. 1971). Based on the new criterion, namely the template specificity of the phage RNA replicases, Miyake et al. (1971) showed that (a) Qβ, VK, and ST belonged to one group, group III, and SP and FI belonged to another group, group IV, and (b) some similarity existed between groups III and IV. As noted previously in the Serogrouping section, the template specificity criterion helped to establish the final outlook of the four major groups I–IV (see Figure 2.7). The subgroups *a*, *b*, and *c* in group III (Qβ, VK, ST, respectively) appeared in this outlook due to the serological properties, filtration and elution patterns, and sensitivity to the enzyme Pronase, although the template specificities of Qβ (IIIa), VK (IIIb), and ST (IIIc) replicases were similar. The subgroups *a* and *b* (SP and FI) in group IV were, however, the result of the introduction of the template specificity criterion.

In fact, the introduction of the template specificity criterion was possible due to the isolation and characterization of RNA replicases induced by SP and FI phages (Haruna et al. 1971) as a necessary addition to the previous success with the isolation of the Qβ replicase and characterization of its specific template requirements (Haruna and Spiegelman 1965a,b,c).

According to the template specificity measurements (Miyake et al. 1971), the Qβ replicase utilized its own RNA as a template at a high efficiency, but it could not utilize the RNAs of the phages MS2 and GA. The RNAs from SP and FI were certainly utilized by Qβ replicase, but to a significantly lesser extent. The replicases of SP and FI could not utilize MS2 and GA RNA, but could utilize SP and FI RNA at a similarly high efficiency. The RNAs from the group III phages (Qβ and ST) were certainly utilized by the SP and FI enzymes, but the efficiency was much lower than that with their own RNAs. Importantly, no difference in the template specificity of SP and FI replicases was observed, although the serological properties of SP and FI particles were remarkably different and led to the announcement of the phage FI as a member of a putative serology group V (Miyake et al. 1969).

A definite drawback of the replicase specificity grouping criterion consisted of the fact that the replicases of the group I and II phages were difficult to isolate and purify, for purely technical reasons (Miyake et al. 1971). Nevertheless, definite progress was achieved 10 years later by the isolation of the GA, a member of the group II, replicase (Yonesaki and Aoyama 1981; Yonesaki and Haruna 1981). This success allowed measurement among all four groups of RNA coliphages based on the template specificity, this time, of the GA replicase (Yonesaki et al. 1982). As expected, the latter made it possible to distinguish the group I and II from the group III and IV representatives. RNAs from the group I phages f2 and MS2 showed high template activities with the GA replicase as well as those from the group II phages GA and SD. RNAs from the group III phages Qβ and ST revealed no template activities even at higher RNA concentrations, while remarkably, those from the group IV phages SP and FI showed significant template activities, though lower than those of the group I and II phage RNAs (Yonesaki et al. 1982).

As one more result of this investigation, the phage MX1, which was serologically grouped as a group IV member (Furuse et al. 1975b), was attributed finally to group III, since the MX1 RNA showed almost the same template activity as Qβ RNA with Qβ replicase but low template activity with the SP replicase (Yonesaki et al. 1982). An additional argument for placement into group III was the fact that the phage MX1 RNA demonstrated an S value typical for group III by sucrose density gradient centrifugation (Furuse et al. 1979b).

Therefore, the template specificity criterion, together with physiochemical data (see later), was used to improve the original serological grouping of RNA phages and establish the final structure of the classical groups I–IV and supergroups A and B, aka *Levivirus* and *Allolevivirus* genera.

HYBRID PARTICLES

The ability of RNA phages to form hybrid particles by their replication *in vivo* is a highly challenging criterion for assessment of grouping quality. Thus, simultaneous infection with Qβ and MS2 did not result in phenotypic mixing, or mixed coat particles, and genomic masking, or formation of hybrid particles (Ling et al. 1970). However, hybrid particles appeared in *E. coli* cells which were simultaneously infected with the group IVa phage SP-222 (an electrophoretic mutant of the RNA phage SP) and the group IVb phage FIC (a host-range variant of the RNA phage FI) (Miyake and Shiba 1971). The altered character of these hybrid particles was not genetic, but rather phenotypic, and resulted in phenotypically mixed particles, or mixed coat particles, and depended therefore on the ratio of two coat proteins present in particles. The particles reverted to their parental type corresponding to their genotype after multiplication in an appropriate host.

Furthermore, after a complete plasmid-based complementation system for RNA phage Qβ was established, it was found that three proteins: A2, or maturation, protein; A1, or read-through, protein, and replicase of phages Qβ (group III) and SP (group IV) can be interchanged.

However, the very low chance of the appearance of hybrid particles among phages that were not closely related markedly diminished the importance of the hybrid particle approach.

PHYSICOCHEMICAL PARAMETERS

Large datasets on physicochemical properties of the RNA phage virions, RNA, and proteins were used to support the serology-based classification of RNA phages into four groups, I–IV.

At the very beginning, in 1966, by direct comparison of MS2 and Qβ, clear differences were described for the general virion characteristics: $S_{20,w}$ value of 79 and 84, molecular mass of 3.6×10^6 and 4.2×10^6, density of 1.422 and 1.439, and pH 3.9 and 5.3 as an isoelectric point, respectively (Overby et al. 1966a). Furthermore, RNA molecules of MS2 and Qβ were characterized as showing an $S_{20,w}$ of 25.8 and 28.9, a molecular mass by light scattering of 10^6 and 0.9×10^6, and an adenine-uracil ratio 0.95 and 0.75, respectively; the two RNA preparations were readily separated by chromatography on columns of methylated albumin but both gave identical buoyant densities in cesium sulfate of 1.64 g/mL (Overby et al. 1966b). The molecular mass of coat protein subunits was measured as 15,500 for Qβ and 14,000 for MS2; both proteins differed, however, in that the Qβ coat lacked tryptophan and histidine, whereas the MS2 coat lacked only histidine (Overby et al. 1966b).

Next, the initial grouping of RNA phages into the three groups, I–III, by serological and Millipore filtration-elution, demonstrated a good agreement with the results of chemical and physical studies: virion size and buoyant density, sedimentation constants, RNA size and base ratios, coat protein tryptophan content, and UV and pH sensitivity of representatives of the three groups (Watanabe et al. 1967b).

A detailed comparison of the virions of groups I–III for the buoyant densities by CsCl equilibrium density gradient centrifugation led to the clear conclusion that the buoyant densities of the RNA phages are not different within the same group but are different and specific for each group (Nishihara and Watanabe 1969). In good accordance with the three-group classification, the RNA phages were physically separated at three distinct densities (g/cm³): 1.46 for group I (MS2, ZR, MY, R17, f2, GR), 1.44 for group II (GA, SD, SW, KJ, EI) and 1.47 for group III (Qβ, VK, ST, NM, NH) (Nishihara and Watanabe 1969). For groups I–III, the $S_{20,w}$ was determined for phage particles as 79S, 76S, and 83S (Watanabe et al. 1967b) and RNA base ratios as 0.94–0.98, 0.84–0.86, and 0.78–0.79 (Osawa et al. 1968; Nishihara 1969), respectively. The different effect of the ultraviolet irradiation on various RNA phages was described for the first time by Furuse et al. (1967).

A thorough elucidation of the four RNA phage groups by the electrophoretic mobility of their representatives on cellulose acetate membranes led to a general conclusion that the electrophoretic phage properties are closely related to the serological ones (Harigai et al. 1981).

The typical group I–IV representatives—f2, GA, Qβ, and SP—showed different group-specific resistance to high hydrostatic pressure that was evaluated with respect to pressure magnitude, treatment temperature, and suspending medium (Guan et al. 2006, 2007).

In order to elucidate the intergroup relationships among the four RNA phage groups by physicochemical characteristics of their RNAs, the sizes of the latter were studied by measuring the sedimentation velocity of RNAs in a sucrose density gradient, and the electrophoretic mobility of the RNAs and that of proteins in polyacrylamide gel (Furuse et al. 1979b). Thus, the RNAs of the group I–IV phages were found to have sedimentation coefficients of 24, 23, 27, and 28 S by sucrose density gradient centrifugation analysis and to have average molecular mass of 1.21, 1.20, 1.39, and 1.42×10^6 daltons by gel electrophoretic analysis, respectively. In the virions of the group I and II phages, there were two kinds of protein: A, or maturation, protein and coat protein. In those of the group III and IV phages, an additional protein, read-through (A1 or IIb) protein (average molecular mass: 3.85×10^4 for group III, and 3.90×10^4 for group IV phages) was detected. The average molecular mass of coat protein from groups I, II, III, and IV were 1.40, 1.29, 1.69, and 1.73×10^4, respectively.

That of maturation protein were 4.48, 4.45, 4.50, and 4.8×10^4, respectively (Furuse et al. 1979b). It is noteworthy that the coat proteins from groups I–III (Nishihara et al. 1969) and group IV (Hirashima et al. 1982) were compared carefully by the group specificity of their amino acid composition.

Studies on the size and properties of the RNA phage components indicated for the first time that a distinct difference (about 20%) in molecular size of RNA exists between groups I and II and groups III and IV, which reflected the presence of read-through protein in groups III and IV. The presented structural data not only showed good agreement with the previous serological grouping but also offered a highly useful idea for the genogrouping of RNA phages (Furuse et al. 1979b).

GENOGROUPING

Strong input toward upcoming phylogenetic analysis and genogrouping was first given by the full-length sequencing of the MS2 genome, which was reported for the first time in 1975 in Madrid and published in 1976 (Min Jou et al. 1972a,b; Vandenberghe et al. 1975; Fiers et al. 1975a,b, 1976).

The full-length MS2 sequence allowed its comparison (Min Jou and Fiers 1976b) with the previously sequenced fragments of the R17 and f2 genomes from the same group I. The overall degree of variation amounts was estimated to be 3.9% (MS2-R17), 3.4% (MS2-f2), and 3.7% (R17-f2) and was accounted for by single-base substitutions (Min Jou and Fiers 1976b).

Full-length genomes were achieved for the coliphages of different groups: Qβ (Mekler 1981) from group III, GA (Inokuchi et al. 1986), JP34 (Adhin et al. 1989), and KU1 (Groeneveld et al. 1996) from group II, SP (Inokuchi et al. 1988) from group IV.

The first full-length genome of non-coli phages, namely, of the pseudomonaphage PP7 (Olsthoorn et al. 1995a), demonstrated no significant nucleotide sequence identity between PP7 and the coliphages except for a few regions. It was concluded that PP7 is related to the coliphages but branched off before the coliphages diverged into separate groups (Olsthoorn et al. 1995a).

Strong differences from the coliphages were found in the full-length genomes of the acinetophage AP205 (Klovins et al. 2002), the broad host range P-pili–specific phage PRR1 (Ruokoranta et al. 2006), the caulophage φCB5 (Kazaks et al. 2011), the M-pili–specific phage M (Rumnieks and Tars 2012), and the R-plasmid–dependent phages C-1 INW-2012 and HgaII (Kannoly et al. 2012). These sequences evoked a strong interest in the problem of RNA phage evolution and improvement of the classical four-group-based classification of the two *Leviviridae* genera.

A solid theoretical background for the practical genotyping was laid out by the first systematic study on the RNA phage phylogeny and genome evolution (Bollback and Huelsenbeck 2001), although it was performed before massive sequencing of the non-coli RNA phage genomes was started: only the pseudomonaphage PP7 has been included for the comparative analysis in this study.

Meanwhile, a genotyping scheme by plaque hybridization with oligonucleotide probes was elaborated for the RNA phage classification in accordance with the existing taxonomic structure and in order to replace the laborious and sometimes inconclusive serotyping procedure (Hsu et al. 1995; Beekwilder et al. 1996a; Griffin et al. 2000). The main goal of this genotyping consisted of simple distinguishing between human and animal waste by genotyping, instead of serotyping, of the F pili-specific RNA coliphage isolates. Thus oligoprobes I, II, III, IV, A, and B were selected to detect group I, II, III, IV, I plus II, and III plus IV phages, respectively.

Furthermore, the RT-PCR approach was introduced for the genotyping of F pili-specific RNA phages in wastewater and polluted marine environment (Limsawat and Ohgaki 1997; Rose et al. 1997). Both plaque hybridization and RT-PCR were compared for their efficiency in water samples (Schaper and Jofre 2000). The plaque hybridization was found to be more sensitive for that time and was applied successfully to stool samples and a wide variety of fecally polluted waters in order to investigate the extent to which the genotypes of F pili-specific RNA phages reflect fecal pollution of human and animal origin in water environments (Schaper et al. 2002b).

A highly useful reverse line blot (RLB) hybridization assay which allowed for the simultaneous detection and genotyping of both F pili-specific RNA as well as F pili-specific DNA coliphages was elaborated (Vinjé et al. 2004). This approach led to identification of a potentially new *Levivirus* group, JS, which included strains having more than 40% nucleotide sequence diversity with the known *Levivirus* groups I and II represented by the phages MS2 and GA, respectively (Vinjé et al. 2004).

Further refining of the hybridization and RT-PCR genotyping techniques improved specificity of classification, e.g., it allowed discrimination of the M11-like phages from the Qβ-like phages within group III (Stewart et al. 2006).

Since traditional genotyping methods suffered from the time-consuming need to isolate viruses, the RT-PCR (Dryden et al. 2006) and reverse quantitative (RT-qPCR) techniques

(Ogorzaly and Gantzer, 2006; Kirs and Smith 2007; Wolf et al. 2008) were generated without the need for viral isolation and membrane hybridization.

Meanwhile, a CLAT (FRNA culture, latex agglutination, and typing) method, i.e., an antibody-coated polymeric bead agglutination assay, was proposed as a simple and rapid 180-min procedure to detect the FRNA coliphage groups with antibody-coated particles (Love and Sobsey 2007). The CLAT assay was performed on a cardboard card by mixing a drop of coliphage enrichment culture with a drop of antibody-coated polymeric beads as the detection reagent and visual agglutination or clumping of positive samples occurred in <60 seconds. The CLAT method successfully classified FRNA coliphages into four serogroups, in similar proportions to those obtained with a nucleic acid hybridization assay (Love and Sobsey 2007).

A major step toward highly precise genotyping was taken by Friedman et al. (2009a) who designed a set of the forward and reverse genogroup-specific, RT-PCR primers based on a total of 30 F pili-specific RNA phages of several strains from all four genogroups (Friedman et al. 2009b): 19 novel full-length sequences (10 *Levivirus* strains and 9 *Allolevivirus* strains) in addition to the 11 full-length phage sequences available at that time in GenBank.

Figure 2.8 shows an unrooted phylogenetic analysis of full-length nucleotide sequences that appeared as a result of the above-mentioned study (Friedman et al. 2009a,b). This alignment makes it possible to localize most of the popular RNA coliphages that were mentioned in Chapter 1. It is noteworthy that this study confirmed previous localization of some markedly differing strains within the groups. Thus, although the group I strain fr and the group III strains MX1 and M11 shared only 70%–78% sequence identity with strains in their respective groups, phylogenetic analyses of the complete genome and the individual genes suggested that the phage fr should be grouped in the *Levivirus* group I and that the phages MX1 and M11 belong to the *Allolevivirus* group III (Friedman et al. 2009b).

During the same period of time, a multiplexed real-time TaqMan RT-PCR assay with a sample process control was elaborated for the detection of the FRNA coliphage genogroups I and IV (Jones et al. 2009b). In this study, the

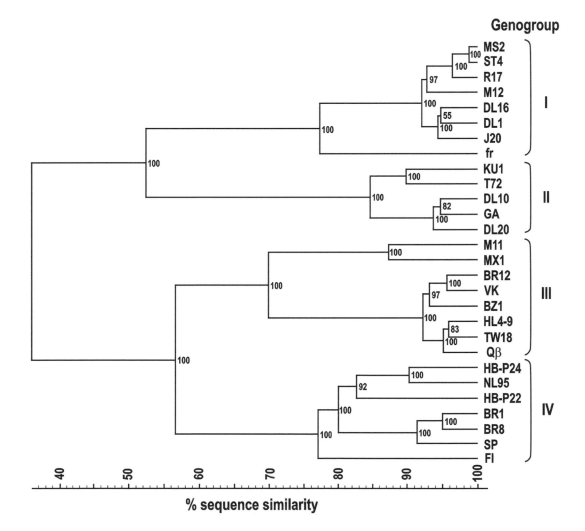

FIGURE 2.8 Unrooted phylogenetic analysis of full-length nucleotide sequences. Nucleotide percent similarity between each male-specific FRNA strain (family *Leviviridae*) is represented on the horizontal axis. The numbers on each node indicate bootstrap values expressed as percentages. (Redrawn with permission of American Society for Microbiology from Friedman SD et al. *J Virol.* 2009b;83:11233–11243.)

individual real-time TaqMan RT-PCR assays for the FRNA phage replicase gene were multiplexed with a real-time TaqMan RT-PCR assay for feline calicivirus as a sample process control for the simultaneous detection and enumeration of the genogroup I and IV FRNA coliphages.

On the basis of this impressive set of novel and previously sequenced full-length coliphage genomes, a leading-edge real-time RT-qPCR assay was introduced (Friedman et al. 2011). This assay was based on the most optimal target regions that were identified to develop RT-qPCR primers and probes specific for each coliphage group. Due to this developed technique, the authors identified two novel coliphages representing the JS group: DL52 and DL54, which demonstrated for the first time evidence for possible RNA recombination (Friedman et al. 2012).

The detection of the FRNA phage genogroups I–IV was included in a viral multiplex RT-qPCR toolbox for tracking sources of fecal contamination, which detected a set of noroviruses and adenoviruses excreted by infected humans, pigs, cattle, sheep, deer, and goats (Wolf et al. 2010). It is noteworthy in this connection that treatment of water samples with an alkaline protein-rich solution and optimization of RNA extraction for RT-qPCR minimized the interference of aluminum, zirconium, and chitosan coagulants with the phage MS2 detection (Christensen et al. 2018), while ferrihydrite treatment mitigated inhibition of the RT-qPCR detection of the phage Qβ from large-volume environmental water samples (Canh et al. 2019).

METAVIROMICS

In early studies involving the traditional search for viruses, the marine and coastal RNA virus communities demonstrated very few if any RNA phages (Hidaka 1975). The RNA phages were not detected by the early metagenomic analysis of coastal RNA virus communities in British Columbia, Canada (Culley et al. 2006). Therefore, the argument that most marine phages have DNA genomes (Weinbauer 2004) was supported. The phage MS2 was detected by the metagenomic analysis of RNA viruses in freshwater in Lake Needwood, Maryland (Djikeng et al. 2009). The first extensive RNA sequencing of gene expression profiles from the oral phage community in men did not reveal any traces of the RNA phages (Santiago-Rodriguez et al. 2015). A study on the early life dynamics of the human gut virome and bacterial microbiome in infants did not yield any RNA phages (Lim et al. 2015). In the later metaviromic datasets, the small fraction of phage diversity was represented, and double-stranded DNA phage genomes outnumbered the RNA phage genomes (Bruder et al. 2016). The freshwater metaviromics, including phages, remained under scrutiny at that time, as followed in an authoritative review (Bruder et al. 2016). This review emphasized the point that, even when sequenced, the RNA phages could be especially difficult to identify, either due to their relative scarcity in a particular sample (Zhang T et al. 2006), or due to biases in the analysis itself (Krishnamurthy et al. 2016). The next extensive review of the present situation in the human intestinal virome

fixed the rare appearance of the RNA phages in human (Carding et al. 2017).

Despite the rather disappointing first attempts, the metagenomic sequencing opened a new method of RNA phage search and contributed first to the RNA phage list by marine RNA phages EC and MB, which sequences have been assembled from the metagenomic sequencing data of marine organisms (Greninger and DeRisi 2015). In contrast with the previous genotyping data concentrated on the RNA coliphages, the EC and MB phages showed only moderate translated amino acid identity to other enterobacteria phages and appeared to constitute novel members of the *Leviviridae* family as well as two novel possible genera within the *Leviviridae* family.

Therefore, the first metagenomically attributed RNA phage genomes were assembled for the phages EC and MB (Greninger and DeRisi 2015), but the first massive metagenomic search for the RNA phage genomes was performed by Krishnamurthy et al. (2016). The main goal of the authors consisted of the idea of overcoming the limited number of known RNA phage species and, as a result, to overcome the predominant concentration on the RNA coliphages. This survey of metagenomic databases revealed 158 partial single-stranded RNA phage genome sequences belonging to 122 distinct phylotypes, 66 of which possessed a putative open reading frame predicted to be the coat gene. These novel RNA phage sequences were present in samples collected from a range of ecological niches worldwide, including invertebrates and extreme microbial sediment, demonstrating that the RNA phages are more widely distributed than previously recognized. The genomic analyses of these novel phages yielded multiple novel genome organizations. Thus, one RNA phage was detected in the transcriptome of a pure culture of *Streptomyces avermitilis*, suggesting for the first time that the known tropism of RNA phages may include Gram-positive bacteria (Krishnamurthy et al. 2016). This extraordinary study identified new dimensions of RNA phage biology, including phages with novel genome organizations, numerous open reading frames that contained novel genes with no detectable homology to the known phage genes, RNA phage presence in novel ecological niches, and the first data in support of an RNA phage infection of a Gram-positive bacterium. These results critically illuminated an unexamined dimension of molecular and ecological phage diversity and fundamentally established a necessary framework that enabled a more accurate dissection of RNA phage modulation of microbial populations.

Figure 2.9 presents a phylogenetic analyses of novel RNA phages discovered in the metagenomic sequencing datasets in light of current ICTV taxonomic classification for RNA phages and eukaryotic viruses (Krishnamurthy et al. 2016). It is noteworthy that only 20 genomes from this study are present in the NCBI taxonomy browser under the *Leviviridae* sp. entry, under *unclassified Leviviridae* members.

During the same period of time, a great survey of RNA virus sequences from invertebrates resulted in 67 additional levi-like RNA phage genome sequences, among newly discovered 1445 RNA viruses (Shi et al. 2016). The identified

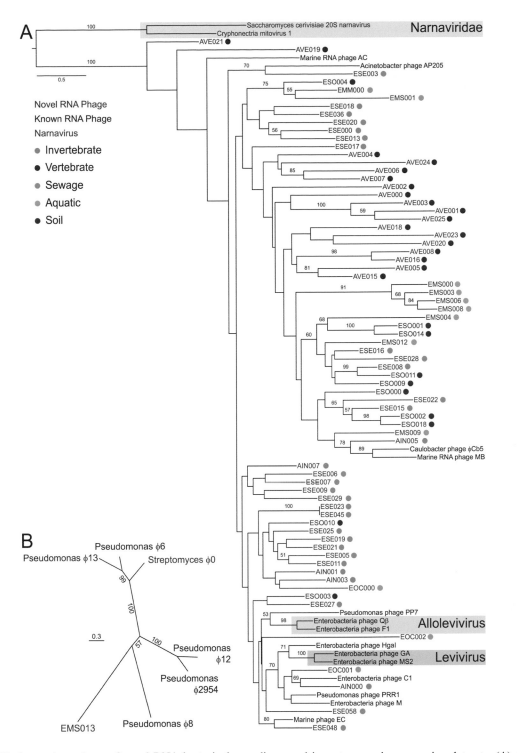

FIGURE 2.9 Phylogenetic analyses of novel RNA bacteriophages discovered in metagenomic sequencing datasets. (A) ssRNA RdRp domain-based tree. Colored dots represent the ecological niche from which each partial genome was identified. Bootstrap values under 50 are not displayed. The current ICTV taxonomic classification for RNA bacteriophages and eukaryotic viruses is shaded. (B) Phylogenetic analysis based on RdRp of EMS013 and *Streptomyces* bacteriophage φ0 with known cystoviruses. (Reprinted from Krishnamurthy SR et al. *PLOS Biol.* 2016;14:e1002409.)

viruses filled major gaps in the RNA virus phylogeny and revealed an evolutionary history that was characterized by both host switching and co-divergence. The authors concluded with good reason that the RNA virosphere is more phylogenetically and genomically diverse than that depicted in current classification schemes and provides a more solid foundation

for studies in virus ecology and evolution. Remarkably, none of these levi-like genomes appears in the NCBI taxonomy browser under *Leviviridae* family, but they are included under *unclassified Riboviria*.

The growing interest in RNA phages in the current decade of the RNA virus metagenomics was pointed out recently by

Greninger (2018). Although the RNA phages were a surprisingly late arrival to the metagenomic discovery party, they rapidly contributed to the open reading frames that had no similarity to known proteins and led to discovery of a set of levi-like viruses (Krishnamurthy et al. 2016; Shi et al. 2016; Greninger 2018). Meanwhile, the phage MS2 appeared as a reliable internal quality control by the metagenomic next-generation sequencing of RNA and DNA viruses from clinical respiratory samples from patients with acute respiratory infections (Bal et al. 2018).

It is noteworthy that the levi-like genomes were found by the metagenomic elucidation of the fecal virome of captive and wild Tasmanian devils (Chong et al. 2018).

Adriaenssens et al. (2018) revealed a diverse assemblage of RNA viruses, including the levi-like phages, by viromic analysis of wastewater input to the Conwy River catchment area in Wales. Remarkably, the *Leviviridae* members were among the most represented families or groupings. The authors observed a striking co-occurrence of the leviviruses with norovirus signatures. The most commonly observed viruses in this sample were pseudomonaphage PRR1 and coliphages FI and M11 of the *Allolevivirus* genus. The significant correlation between the presence of leviviruses and noroviruses in water samples seemed intriguing and made it possible to propose that the higher abundance of alloleviviruses than of MS2-like viruses could indicate that the former might be more relevant as model systems for noroviruses (for detailed outline, see Chapter 6). The latest technologies of the phage metagenomics were presented by Owen et al. (2019).

Finally, the Kaspars Tārs team performed an unprecedent similarity analysis of the coat proteins of all known levi-like RNA genomes, deciphered mostly from the metagenomic studies (Liekniņa et al. 2019). Full analysis of the data are presented in Chapter 12.

Appraising the great contribution of the metagenomic studies that have greatly expanded our familiarity with the levi-like RNA phages and their diversity, it must be acknowledged that it is very difficult, if possible, to resurrect infectious phages from the partial genome sequences. Moreover, metaviromics does not provide us with highly important information on host bacteria, along with other aspects of the phage biology.

URGENT NECESSITY FOR IMPROVEMENTS IN THE TAXONOMY

In view of the latest studies, and especially the growing contribution of metagenomics, it is necessary to conclude that both ICTV and NCBI classifications comprising two RNA phage genera that are accompanied by lists of unassigned and unclassified species at two levels (in the NCBI page they are unclassified *Leviviridae* and unclassified *Levivirus*) need strong reconsideration. The ICTV and NCBI classifications are adequate and work well in the case of coliphages only, since they have been based historically on the large data bodies dealing with the coliphages. Both

classifications were elaborated upon in detail on the basis of the traditional serogrouping, which resulted, after some improvements and corrections from non-serological arguments, in the four groups I–IV, which fell logically into two large blocks: A (groups I and II) and B (groups III and IV). These two blocks served as a historical background for the current *Leviviridae* subdivision into two *Levivirus* and *Allolevivirus* genera. The recent attempts to assign some non-coli phages to *Levivirus*, *Allolevivirus*, or leave them as "unassigned" or "unclassified" members may serve only as a temporary solution.

Nowadays, the classical RNA phage taxonomy and grouping seem rather artificial and lack clear rationale aside from historical considerations. A new classification system based solely on sequence similarities should be employed, e.g., the most-conserved replicase sequences could be used as a rationale for such sequence-based classification. For the new classification, the *Levivirus* and *Allolevivirus* genera could be left predominantly for coliphages but new genera could be introduced based on the sequence data, including results of metagenomic sequencing. Many computing approaches have been proposed recently to contribute directly to the phage classification needs. Thus, a Host Taxon Predictor software has been written in order to predict taxon of the host of a newly discovered virus (Gałan et al. 2019). Chibani et al. (2019a) elaborated the ClassiPhage, which uses a set of phage-specific hidden Markov models (HMMs) generated from clusters of related proteins. However, direct success with the classification of the potential *Leviviridae* species was achieved when HMMs were used to scan phage datasets and the resulting Phage Family-proteome to Phage-derived-HMMs scoring matrix was used to develop and train an artificial neural network to find patterns for phage classification into one of the phage families (Chibani et al. 2019b). The performance of the neural network demonstrated near-perfect prediction for the *Leviviridae* family.

The next unsolved problem that is directly related to taxonomy is the identity of phages. The existing phage names have been given by authors and are very far from any systematization. As Kropinski et al. (2009) wrote, "most phage names do not result in immediate recognition, even among phage researchers," since phage naming has been based on different strategies including meaningless assemblies of letters or simple numeric codes, morphological peculiarities, and references to persons, and the origin of the names is often difficult to trace. To bring prokaryotic virus nomenclature into the twenty-first century, Kropinski et al. (2009) designed a proposal by which a name has two components: the first part contains a preface, identifying what follows as a bacterial virus: the isolation host and virus familial relationship, while the second part provides the specific laboratory designation (freely chosen but with proposed guidelines). The pseudomonaphage vB_PaeL_PcyII-10_LeviOr01 was the first that received its name in accordance with these recommendations (Pourcel et al. 2017). Nevertheless, the short name LeviOr01 seems more useful from a practical point of view.

3 Pili and Hosts

Men who wish to know about the world must learn about it in its particular details.

Heraclitus

All animals are equal, but some animals are more equal than others.

George Orwell
Animal Farm

PILI: CURRENT DEFINITION

A pilus (Latin for *hair*; plural: pili). According to the current Medical Subject Headings (MeSH) definition at the appropriate NLM page (https://meshb.nlm.nih.gov/record/ui?name=Sex%20Pilus):

> *pili, sex*, are filamentous or elongated proteinaceous structures which extend from the cell surface in gram-negative bacteria that contain certain types of conjugative plasmid. These pili are the organs associated with genetic transfer and have essential roles in conjugation. Normally, only one or a few pili occur on a given donor cell. (Singleton P, Sainsbury D: Dictionary of Microbiology and Molecular Biology, 2nd ed. p. 675. 1993. Copyright Wiley-VCH Verlag GmbH & Co. KGaA. Reproduced with permission.)

The third revised edition of the above-cited dictionary, published in 2006, specified this a bit definition to the

> *pili (singular: pilus) (conjugative pili; sex pili)* are elongated or filamentous, proteinaceous, plasmid-encoded structures which extend from the surface of those (gram-negative) bacteria which contain an (expressed) conjugative plasmid. Some types of pili have been shown to play an essential role in conjugation, and it is generally believed (though not proven) that all pili have essential roles in conjugation. Commonly, only a few pili (sometimes only one) occur on a given donor cell. At least some types of pili appear to be retractable, but the mechanism of retraction is unknown. (Singleton P, Sainsbury D: Dictionary of Microbiology and Molecular Biology, 3rd ed. p. 587. 2006. Copyright Wiley-VCH Verlag GmbH & Co. KGaA. Reproduced with permission.)

Therefore, according to the information given in the third edition of the above dictionary, such thin, flexible pili mediate conjugation which can occur equally well within a body of liquid (e.g., broth) or on a moist (but not submerged) solid surface (e.g., an agar plate or membrane filter).

For example, the most well-studied F pili encoded by the F plasmid are one to several micrometers in length, ca. 8 nm in diameter; each F pilus has an axial channel, ca. 2 nm diameter, with a hydrophilic lumen. It has been assumed that the base of a pilus is associated with an adhesion site or similar structure. F pili are composed mainly or solely of a single type of subunit, pilin; the 70-amino acid pilin molecule is a single polypeptide containing one residue of D-glucose and two phosphate residues.

Two major theories regarding the operation mode of the pili were proposed: the first *conduction* theory, which suggested that the pili are acting as tubes within which a nucleic acid molecule can travel during conjugation and phage infection (Brinton 1965), and the second F pili *retraction* theory, saying that, on contact with a recipient cell or phage, the pilus would retract, being depolymerized by a mechanism at its base (Curtiss 1969; Marvin and Hohn 1969). The strongest objection to the first theory was that these thin filaments must be capable of transporting different kinds of nucleic acids in different directions.

F PILI: ASCENDING TO CLARITY

Norton D. Zinder wrote in his first classical review that "prior to the f2 experiments there was no hint of the existence of sex pili. It is the absence of pili on female bacteria that accounts for the male specificity of f2 and not some other intrinsic property; phage RNA can infect females as well as males" (Zinder 1965).

Just before the discovery of the male-specific RNA phages by Loeb and Zinder (1961), it was known that "the pili usually cover most of the bacterial surface and are continually growing out in a manner quite analogous to hair or fur" (Brinton 1959). Charles C. Brinton stated in this exhaustive review paper that "a possible correlation of piliation with fertility was investigated, but both F+ and F− bacteria of one strain (*E. coli* K-12, W677) were piliated while neither F+ nor F− bacteria of another strain (*E. coli* K-12, 58–161) were piliated. A strain of *E. coli* showing high frequency of recombination (Hfr of Hayes) was non-piliated." Therefore, regardless of serious suspicions, no correlation of piliation with maleness was found at that time.

At the same time, Ida and Frits Ørskov (1960) found that F+, Hfr, and F− strains could be distinguished by serological methods. They identified an antigen termed f+ with the male status of the appropriate *E. coli* strains. Sneath and Lederberg (1961) detected a "mating substance" of a probably periodate-reactive polysaccharide nature on the surface of male bacteria. Tsutomu Watanabe et al. (1962) were the first to assume that the antigen f+, the "mating substance," and the receptor substance for the F-specific phages could be identical.

The full clarity was established after E. Margaret Crawford and Raymond F. Gesteland (1964) found by electron microscopy that the RNA phage R17 adsorbed to pili of an Hfr and an F+ strain of *E. coli* but not to pili of an F− strain. This led the authors to suggest that the attachment to pili might represent the initial step in the RNA phage replication. The authors also observed that in conjugating mixtures of Hfr and F⁻ bacteria, the RNA phage attached to the pili of only one of the two mating partners. Therefore, these observations implied the existence of two or more (?) different kinds of pili, only one of which could be connected with the RNA phage adsorption and male specificity. The absorption of the phages ZIK/1 and ZJ/1 to the "fimbriae" structures of the male *E. coli* cells was reported at that time also by Bradley (1964b).

A bit later, these observations were clearly confirmed with another RNA phage, M12, which also adsorbed to some of the pili present on the male bacteria but not to pili on the female strains (Brinton et al. 1964). That the synthesis of pili capable of adsorbing M12 was determined by the *E. coli* F factor was shown by the correlation of F piliation with the presence of the F factor in derivatives of *E. coli* K12, the disappearance of F pili when the F factor was removed, and the appearance of F pili when the F factor was added (Brinton et al. 1964). F pili were much less numerous than other pili and could not be distinguished from them in electron micrographs in the absence of male phage adsorption. It led to a highly productive idea, which was used further in a large number of microbiological studies, that the phage attachment could be used to visualize F pili, when negatively stained in electron micrographs, from other types of pili occurring on the same cell (Brinton et al. 1964). Moreover, due to the discovery of RNA phages, the reason earlier observations have been unable to correlate piliation with fertility was understood (Brinton 1959).

The data on RNA phage adsorption (Brinton et al. 1964; Crawford and Gesteland 1964) strongly supported the identity of F pili to the f⁺ surface antigen found by Ørskovs (1960). In fact, the f⁺ antigen occurred only on Hfr and F⁺ strains and not on F⁻ strains. The fact that the average number of F pili per cell could be about one explained why the f⁺ agglutination reaction was weak. Strong experimental argumentation for the identity of the f⁺ antigen and the cellular receptor for the phage fr was presented by Peter Knolle and Ida Ørskov (1967) not long after the RNA phage receptor function of an *F fimbriae* was proposed (Knolle 1964b) and published (Knolle 1967b). Moreover, specific aggregation of male *E. coli* (or of isolated F pili) by anti-f⁺ antiserum was inhibited when the phage M12 was mixed with the male culture before the addition of anti-f⁺ antiserum. Therefore, it was concluded that the F pilus is the structure of the f⁺ antigen and is responsible for f⁺ agglutination (Ishibashi 1967).

Furthermore, mutant *E. coli* strains were isolated with a changed F factor which were still fertile but resistant to male-specific phages (Nishimura and Hirota 1962; Brinton et al. 1964). However, no fertile mutants have been isolated which have been demonstrated to lack F pili or the f⁺ antigen. As for the role of F pili in the conjugation process, preliminary experiments have provided direct evidence that F pili are necessary for chromosome transfer (Brinton et al. 1964). A complete review on the status of the piliation problem was published shortly thereafter by Brinton (1965). It is noteworthy that the term "F pili" appeared for the first time in the papers of the Brinton's team (Brinton et al. 1964; Brinton

1965). Brinton et al. (1964) were the ones who acknowledged for the first time that the phage receptive pili were of a special type, wider and longer than *common pili* (*fimbriae* by Duguid and Wilkinson 1961; *type I pili* by Brinton 1965) and named them F pili.

It is also important that the RNA phage particles attached to the sides and completely covered F pili, in contrast to F-specific DNA phages, namely the phage f1, that attached to the tip of F pili, and that each break of the pilus, creating a new tip, also created a new site of attachment (Caro and Schnös 1966). Therefore, the F-specific DNA phages attaching to the tip and the RNA phages, which attached laterally along the pilus, did not compete for attachment sites.

Figure 3.1, reprinted from Brinton et al. (1964) presents a typical electron microscopic visualization of F pili after attachment of the RNA phage M12 and verifies full coincidence of the phage attachment with the presence of the F plasmid.

Moreover, Raymond C. Valentine and Mette Strand (1965) found that F pili, which were sheared from the surface of *E. coli* cells by treatment in a Waring blender, retained capability to bind the RNA phage f2 particles in the presence of Ca^{2+} ions. Such complexes of F pili and phage were easily assayed because of their retention by membrane filter pads, in contrast to non-attached phage. The rate of phage adsorption to F pili was relatively insensitive to temperature and took place readily even when incubation is at 0°C. Disruption of F pili−phage complexes by blending resulted in release of viable phages from the complex, indicating that their RNA was not injected into the core of the F pilus. Figure 3.2 shows electron micrographs of F pili-RNA phage f2 complexes formed *in vitro* from this paper (Valentine and Strand 1965).

In parallel, the mechanical shearing technique was used to discover the fate of the R17 phage components after the attachment of the ³⁵S- or ³²P-labeled phage to F pili: the protein ³⁵S-label was separable from the host cell by shear, while the viability and the nucleic acid ³²P-label remained insensitive to shear and associated with the host (Edgell and Ginoza 1965).

The simple shearing technique using a radioactive ³²P-labeled phage paved the way to precise quantitative measurements of the phage attachment to F pili, or "male substance" by the previous nomenclature of the authors (Ippen and Valentine 1965). For example, this method made it possible to show definitively that the phage adsorption did not appear at any time during the growth of the female F⁻ culture (Ippen and Valentine 1965). Moreover, it was assumed that the female → male conversion of *E. coli* cells was a relatively rapid process taking place as little as 30 min after infection of the female cell with the fertility factor (Valentine 1966). The reverse conversion, male → female, was regarded in this study as a relatively rare event which occurs occasionally in the male population apparently due to the spontaneous loss of the F factor from a male cell. Such isolation of F pili by blending led to the tentative assessment of their structural characteristics (Wendt et al. 1966). It was concluded that F pili varied considerably in length, the longer segments being readily broken by shearing into smaller active fragments still

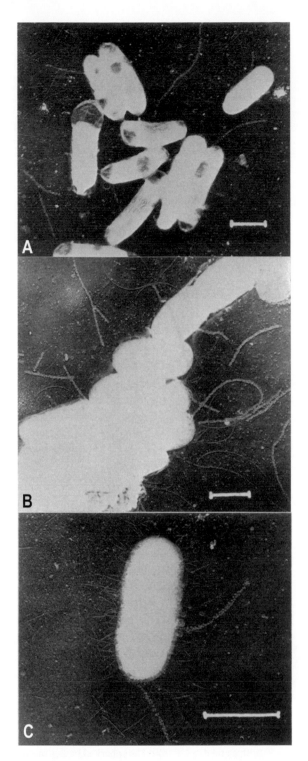

FIGURE 3.1 F pili and RNA phages. (A) *E. coli* K12, W1895, Hfr (Cavalli) + M12 phage. F pili are distinguishable from Type I pili by the presence of adsorbed M12 phage. (B) *E. coli* K12 58–161 AO-1, Mg13, F+ + M12 phage. This strain, originally F+ and having both F pili and Type I pili, had lost its F pili by acridine orange elimination of the fertility factor. Upon reintroduction of the fertility factor by infection, the F pili reappeared. (C) *Proteus mirabilis* F-lac+ + M12 phage. This strain, originally F− and lacking F pili but having Type I pili, acquired F pili and fertility upon introduction of the *E. coli* fertility factor. White lines are 1 μ. (Reprinted with permission of Lenny Gemski from Brinton CC Jr et al. *Proc Natl Acad Sci U S A.* 1964;52:776–783.)

capable of adsorbing phage. F pili demonstrated low buoyant density in CsCl and high sensitivity to lipid solvents and heat.

It is important to note that the ^{32}P-labeled phage assay measured only simple phage adsorption to F pili, subsequent "injection stages" being prevented by the incubation at 0°C, while formation of the F pili–phage complex occurred readily at this temperature (Ippen and Valentine 1965). The filtration data also confirmed the previous electron microscopic observations (Brinton et al. 1964) that F pili were observed in preparations of not only *E. coli*, but also of other grown Gram-negative bacteria including *Salmonella*, *Shigella*, *Serratia*, and *Proteus*.

The direct role of F pili in the RNA phage infection was confirmed once again by experiments which involved (i) full removal of F pili by blending and (ii) rapid regeneration of the latter after a short lag period of 1–2 min (Valentine et al. 1965). As a result, it was proposed that F pili, being sex hairs of male cells, may serve as "tails" for male-specific phages (Valentine and Wedel 1965).

Thus, the notion of a viral "nucleic acid pump" to account for observations on the penetration of male-specific phage nucleic acids into male cells was introduced in order to stress the active role of the male cell in this process and the function of the F pilus in nucleic acid transport (Ippen and Valentine 1966).

In further studies on the possible role of F pilus as a "sensory fiber, conjugation tube, or mating arm," the RNA phage f2, as well as the F-specific DNA phage f1 were defined as specific mating inhibitors that functioned as a "side inhibitor" and a "tip inhibitor," respectively (Ippen and Valentine 1967).

HOW PILI WORK

Brinton et al. (1964) initially suggested that pili may conduct DNA between mating bacteria, and later it was proposed that pili could conduct male phage nucleic acid (Brinton 1965; Brinton and Beer 1967). Next, it was suggested that the pilus may be a retractable structure that attaches to a female and draws her to the surface of the male cell as it retracts (Curtiss 1969). A similar retraction mechanism was proposed to explain how filamentous phage can attach to the tip of the pilus and then penetrate the cell intact (Marvin and Hohn 1969).

Some evidence that pili may conduct bacterial DNA was provided by the demonstration that genetic transfer can take place between mating pairs that never make wall-to-wall contact (Ou and Anderson 1970).

F pili production was found to be very sensitive to temperature: (i) when cells were grown at 18°C they did not adsorb the phage fr but regained this ability when the temperature was raised to 37°C; (ii) the yield of the phage fr plaques decreased at temperatures below 34°C (Knolle and Ørskov 1967).

Definite progress in the pili studies was achieved by the elaboration of a method for assaying F pili based on serum-blocking power (Novotny and Lavin 1971). It has been shown before that serum from rabbits immunized with either whole *E. coli* male cells or preparations of sex pili contained an

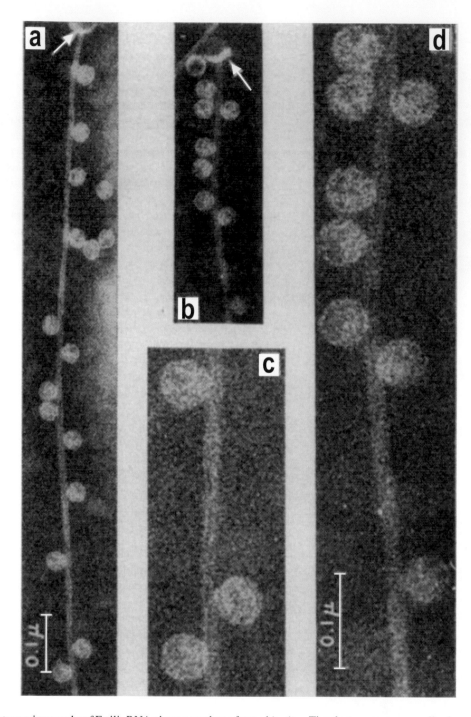

FIGURE 3.2 Electron micrographs of F pili–RNA phage complexes formed *in vitro*. The phages appear as small spheres (diameter ~ 250 Å) attached to the surface of the long filaments (F pilus). The preparation was freed of unattached phage by several cycles of sedimentation and washings. Complexes were negatively (phosphotungstic acid) stained. a and b, ×70,000; c and d, ×110,000. The spherical bodies marked by arrows in the micrograph are apparently not phage particles but may represent a membranal base or root of the F pili. (From Valentine RC and Strand M. Complexes of F-pili and RNA bacteriophage. *Science*. 1965;148:511–513. Reprinted with permission of AAAS.)

antibody that bound along F pili and prevented conjugation and male phage infection (Ishibashi 1967; Lawn et al. 1967). The serum-blocking assay was based on the ability of a known amount of antibody to inhibit male phage infection before and after adsorption to F pili (Novotny and Lavin 1971). This assay was used successfully to quantify the effect of temperature on the F pili production by *E. coli* B/r F$^+$ strain: maximum number of pili per cell occurred between 37° and 42°C;

below 37°C the number decreased, reaching zero at about 25°C; when cells were grown at 37°C, blended, and resuspended in fresh media at 25°C, they made F pili, probably, by assembling from a pool of subunits that were synthesized during growth at 37°C (Novotny and Lavin 1971).

Detailed studies of the effects of chloramphenicol, nalidixic acid, mitomycin C, NaCN, and ultraviolet irradiation at 253.7 nm on F pili production showed that *E. coli* cells contained pools of pili protein, F pilin, and that assembly did not require synthesis of protein or DNA (Novotny et al. 1972). The phage R17 was used permanently to label F pili for electron microscopy. It was observed that NaCN did not prevent the phage attachment to free F pili and concluded, in accordance with the earlier finding that the phages adsorbed to pili at 0°C (Brinton 1965; Ippen and Valentine 1965; Valentine and Strand 1965), that an effect of the energy poisons on the phage attachment was not likely (Novotny et al. 1972). However, the disappearance of F pili on *E. coli* cells in the presence of cyanide was qualified as a retraction process, since the pili which disappeared from the cell did not appear as free pili in the culture medium, suggesting that the pili had retracted into the cell (Novotny and Fives-Taylor 1974). Remarkably, the adsorption of either F pili antibody or R17 bacteriophage to the sides of pili prevented retraction. New pili were produced at a normal rate approximately 3 min after NaCN was removed. In contrast to R17, the DNA phage M13 retained ability to adsorb, suggesting that the tips of retracted pili remained exposed (Novotny and Fives-Taylor 1974).

By maximal thymine → 5-bromuracil substitution in the cell DNA, the modal length of the pilus doubled and the number of pili per cell reached approximately 50% that of thymine-grown cells, as well as the 50% ability of 5-bromuracil-grown cells to form mating pairs and to be infected by the phage R17 and also by DNA phage M13 (Fives-Taylor and Novotny 1974).

Under certain specific conditions, namely in *E. coli* cultures grown in glycerol-containing medium, addition of arsenate by R17 eclipse led to a dramatic loss in the cell-associated F pili: upon removal of the inhibitor, the cells were found to regain F pili rapidly and to become sensitive once again to the RNA phage infection (O'Callaghan et al. 1973c). A similar effect was detected earlier by the addition of phenethyl alcohol to *E. coli* cultures during the attachment step of the phage f2 (Wendt and Mobach 1969).

Natural retraction of F pili was first observed as the result of infection with a filamentous phage Ff, when the loss of pili paralleled the appearance of infected cells (Marvin and Hohn 1969).

The authors proposed that the attachment of the phage Ff to the pilus tip triggered retraction of the pilus, which brought the phage Ff to the cell surface where penetration occurred. Subsequently, it was shown that F pili with the filamentous M13 phage on their tips became shorter with time, as predicted by the "retraction hypothesis," and, as pili disappeared, M13 phage appeared attached directly to the cell surface (Jacobson 1972).

F pili were retracted rapidly at high temperatures (46°C–50°C) but the adsorption of either R17 phage or F pili antibody to the sides of pili prevented retraction (Novotny and Fives-Taylor 1978).

The infection of *E. coli* with low multiplicities of M13 phage (10 plaque-forming units per cell) caused no detectable loss of the cell-associated F pili as measured by electron microscopy or the ability to attach the phage R17. However, larger multiplicities of M13 infection led to a rapid loss of F pili, since large amounts of M13 caused a breakage or release of F pili from the cell (O'Callaghan et al. 1973b).

F TYPE PILI: R FACTORS

The resistance factors (R) responsible for multiple drug resistance, as well as the colicin factors (Col), were shown to have many properties similar to those of the F factor. Elinor Meynell and Naomi Datta (1965) established functional homology of the resistance transfer factors (R plasmids) with the sex factors (F and F' plasmids). However, the R plasmids differed from the classical F sex factors by lower conjugation activity due to a specific repression mechanism. Considering the restrictions produced by the repression mechanism, the capability of the *E. coli* R+ cells to support infection by the phages MS2 and μ2 was demonstrated successfully (Meynell and Datta 1965). One large class of R factors, which have been named *fi+* (for fertility inhibition), was closely related to the F factor and inhibited fertility of *E. coli* K12 strains carrying the F factor. The other class of factors which did not repress conjugation was named *fi-* (Meynell and Datta 1966a,b). The loss of F fertility caused by *fi+* factors was accompanied by the loss of sensitivity to male-specific phages, while *fi-* factors did not affect this sensitivity.

According to these findings, the sex factor F was itself exceptional among conjugation factors in being freely expressed in all the bacteria of a culture, while related factors which are largely unexpressed were widespread. As a result, only a small proportion of cells was sensitive to F-specific phages in the R plasmid−carrying strains.

Nevertheless, the R+ *E. coli* strains carrying R1, R124, and R237 factors could produce the F type pili which were similar to those found on F+ bacteria and were recognized by the RNA phages (Datta et al. 1966). Moreover, the efficiency of the F type piliation correlated fully with the sensitivity to the F-specific RNA phages. Furthermore, derepressed mutants of the R1 factor were isolated: they demonstrated high and stable sensitivity to the phage MS2 (Meynell and Datta 1967). A derepressed variant of the R100 factor, namely R100-1, produced pili which were morphologically like F pili, but which could be distinguished from them by a lesser affinity for the F-specific RNA phages. Some attachment of the phages MS2, μ2, and f2 to the R100-1 pili could be demonstrated, but it was markedly less than with an F pilus (Nishimura et al. 1967). The R100-1 pili also differed from the F pili immunologically (Frost et al. 1985).

Systematic morphological studies of the F type pili were performed by A.M. Lawn (1966). Thus, the phage MS2 was used to differentiate the specific F type pili, which adsorbed the phage, from common pili, which did not. It was concluded that the F type pili were wider and generally longer than common pili, but the F type pili formed by *E. coli* carrying the different R factors could not be morphologically distinguished from each other.

Morphological studies led to a highly important improvement in the electron microscopic evaluation technique:

immunological labeling of the subject with the specific anti-bodies was introduced just for characterization of the pili–phage complexes (Lawn 1967).

Furthermore, it was assembled that the F-like class comprised F, ColV, and ColB factors and about half of known R factors (Lawn et al. 1967). Thus, cultures of *E. coli* or *S. typhimurium* were examined by electron microscopy and their plasmids fell into two groups: the first, related to F, included F itself (both autonomous in an F+ strain, or as F'13 or F*lac*, and integrated in the strains HfrH and HfrC), colV-K94, F₀*lac* and the *fi+* R factors, R1 (both wild type and dere-pressed), R100-1 (a derepressed mutant of R100), and HFT cultures of R124 and R237. The second group, which was responsible for the type I pili, was related to colI and included derepressed mutants of the fi⁻ factors, R64 and R144, as well as HFT cultures of colIb and colEIa-16 (Lawn et al. 1967).

The sex pili of members of the first, F-like, group, whether determined by F, colV, or F₀*lac*, or by R1, R100-1, R124, or R237, could not be distinguished from one another by appearance and adsorbed the phage MS2 and the DNA phage M13. Those of the second, I-like group, comprising colIb-P9, colEIa-16, R64, and R166, were also indistinguishable from one another except for the occasional presence of unusually thin filaments associated with colIb, and adsorbed specifically the filamentous DNA phage I. There were, however, certain definite differences between the sex pili of the two groups (Lawn et al. 1967).

The first impressive systematization of drug-resistance factors and other transmissible bacterial plasmids with the active role of the RNA phages in their evaluation was published by Elinor Meynell et al. (1968). It was concluded that gross differences distinguished the two major groups of F-like and I-like pili and their factors, while minor differences are evident within the F-like group, as with the drug-resistance factor, R 100-1, and the plasmid, F₀*lac*.

Next, the MS2 phage contributed as a derepression marker of the male properties to the study on the superinfection immunity and repressor susceptibility of the F-like R factors (Frydman and Meynell 1969).

The phage μ2 was used as a marker in order to characterize and classify into three classes the *fi+* property of an impressing set of transfer factors (Pitton and Anderson 1970).

A systematic study on the MS2- and M13-based distinction between the pili encoded by the F-like (fi+) and I-like (fi⁻) R factors, when the R factors are derepressed and most cells produce both F-like and I-like sex pili, was published by Salzman (1971).

The phage MS2 was used as a measure of the degree of sex piliation in bacterial populations by the serious investigation of the R factor repressor functions during anaerobic growth of *E. coli* (Burman 1975).

The phages MS2 and Qβ, as well as the DNA phage f1, were used by a massive screening of a large number of sero-logically typed *E. coli* strains in order to detect F-like factors (Shchipkov et al. 1977).

The phage R17 was involved as a marker in the large and authoritative study that was devoted to the fine mechanisms of conjugation and compared proteins involved in pilus synthesis

and mating pair stabilization from the related plasmids F and R100-1 (Anthony et al. 1999).

GENETICS OF F PILI

In the early studies, it was defined therefore that the natural resistance to the phage fr might arise from (i) loss of the F factor, (ii) mutation in the F factor, and (iii) mutation in the bacterial chromosome (Schnegg and Kaudewitz 1968).

Looking for the F factor, or the "nucleic acid pump," mechanism, a remarkable class of mutants of male bacteria with altered sex hairs was isolated after *N*-methyl-*N'*-nitro-*N*-nitrosoguanidine, or shorter nitrosoguanidine, treatment of an *E. coli* HfrC strain (Silverman et al. 1967a). These mutants produced normal levels of F pili and readily adsorbed the phage f2 but restricted the infection of the latter at a later extracellular stage, possibly at a transport or injection, or pen-etration, step. At the same time, the mutants were infected normally by the male-specific DNA phage fl and by the sero-logically distinct RNA phage Qβ.

Detailed elucidation of these mutants showed that the intact phage f2 RNA could penetrate the mutant cells at a low frequency, especially when the phage RNA was forced into these mutants by raising the multiplicity of infection, but most of the penetrated f2 appeared fragmented and noninfectious (Silverman et al. 1967b).

It is noteworthy that a special study was undertaken on the nitrosoguanidine mutagenesis assay that demonstrated non-random distribution of mutations, namely the f1-, f2-, and Qβ-resistant mutations appeared more often than expected in Hfr strains but not in F+ strains (Lloveres and Cerdá-Olmedo 1973).

Next, all potential F pili mutants were classified into four classes by blocking one of the following functions: (a) adsorp-tion of the phage onto the pilus, (b) injection of the phage nucleic acid into the pilus, (c) transport of the nucleic acid from the point of injection on the F pilus to the cell surface, and (d) penetration of the nucleic acid into the cell proper (Silverman et al. 1968). The first mutants that were described above fulfilled the requirements to the class (a). After screen-ing of several thousand additional nitrosoguanidine induced mutants of an *E. coli* Hfr strain for the f2 resistance, two new mutant groups belonging to classes (b) and (c) were found (Silverman et al. 1968). Selection of these mutants was reviewed by Valentine et al. (1969a).

The isolation of one mutant whose transfer ability has been affected, but which retained full sensitivity to the male spe-cific phage infection, has also been briefly reported (Cuzin and Jacob 1967).

Novel mutants of the derepressed F-like R factor, namely R1*drd*-16, were selected after ethylmethane sulfonate muta-genesis and demonstrated resistance to the phage MS2, as a result of poor adsorption and normal sensitivity to the phage Qβ, as well as to the F-specific DNA phage fd (Meynell and Aufreiter 1969). The novel mutants allowed conjugation and gene transfer, although poor adsorption of the phage MS2 also resulted in the structural instability of the pilus.

Walker and Pittard (1969) described an *E. coli* mutant that was insensitive to the phages MS2 and R17 but sensitive to the phage Qβ at 42°C and suggested that the structure of the F pilus was altered at high temperature in some way similar to that by mutants described before (Silverman et al. 1967a).

A prominent study on the isolation and characterization of transfer-defective (*tra*) mutants of the F′ and Hfr strains and on the involvement of the *tra* genes in genetic transfer was performed by active exploitation of the RNA phages M12, MS2, f2, and Qβ (Ohtsubo 1970; Ohtsubo et al. 1970).

A large series of amber-suppressible *tra* mutants of F*lac* was isolated by nitrosoguanidine or ethyl methane sulfonate treatment and tested on phage sensitivity, including f2 and Qβ: again, one of the mutant groups demonstrated resistance to the phage f2 and sensitivity to the phage Qβ (Achtman et al. 1971). Further, genetic analysis of these mutants revealed existence of at least eleven genes, *traA* through *traK*, necessary for conjugational DNA transfer. Mutants in *traI* and *traD* and some in *traG* still made F pili, although *traD* mutants were resistant to the phage f2 (but sensitive to the phage Qβ); therefore, their products might be involved in conjugational DNA metabolism.

Remarkably, the Tra⁻fl^sf2^RQβ^S phenotype of the *traD* mutants was observed for sex factors without any mutations, but in obstacles when expression of the *traD* gene was inhibited by the FinC system (van de Pol et al. 1979; Willetts 1980). Moreover, further studies on temperature-sensitive *traD* mutants, which were temperature sensitive for the group I RNA phages MS2, f2, and R17 but not to the group III phage Qβ (Schoulaker and Engelberg-Kulka 1978), led to the conclusion that the mutant cells permitted the penetration and translation of the phage MS2 RNA but did not permit the MS2 RNA replication (Schoulaker-Schwarz and Engelberg-Kulka 1981, 1983; Engelberg-Kulka and Schoulaker-Schwarz 1983). The effect was explained by the marked *traD* mutation-induced changes in the membrane, where *traD* gene product could be either directly or indirectly involved in the replication of the group I phages, possibly due to the lability of their replicases, in contrast to the stable Qβ replicase.

Other mutants in *traG* and all mutants in the remaining eight genes did not make F pili. One of these, *traJ* was a control gene, and the others specified a biosynthetic pathway responsible for synthesis and modification of the F pilin subunit protein and its assembly into the F pilus (Achtman et al. 1972; Willetts and Achtman 1972). It was detected at that time that F pili consists of one protein, namely F pilin, which contains two moles of phosphate and one mole of glucose per molecule (Brinton 1971). Finally, the order of *tra* genes was established to be: … *traJ traA traE traK traB traC traF traH traG traD traL* (Ippen-Ihler et al. 1972).

In parallel, six novel F pili mutants were isolated after mutagenesis of an *E.coli* F′8 (F*gal*) strain by UV irradiation and nitrosoguanidine treatment: the mutants retained conjugal fertility and synthesized F pili, but three of them had lost sensitivity to the phage MS2 (but continued to adsorb it and supposed to be *traD* mutants), and the other three were insensitive to the phages Qβ and MS2 but did not adsorb MS2 and did not correlate with any of identified *tra* genes

(Tomoeda et al. 1972). Similar mutants were isolated further by Paranchych (1975), as well as by Orosz and Wootton (1977) in the case of the F-like plasmid R192-7.

The function of the F transfer gene *traL* was the first to be studied in detail (Willetts 1973). The *traL* gene was intimately connected with the F-specific phage infection, as well as the *traA* product. The phages MS2, M12, and Qβ were used in this study. The *traL* gene was needed for conjugational DNA transfer and for pilus formation but not for surface exclusion; it was not plasmid-specific, being indistinguishable from the *traL* products of the F-like plasmids ColV2, ColYB*trp*, R100-1 and Rl-19. The above-mentioned order of *tra* genes was corrected to … *traA traL traE* … and the possibility that the *traL* product could specify the structural pili subunit, or F pilin, was mentioned (Willetts 1973).

Furthermore, 12 genes that are required for pilus formation were identified within the single *tra* operon (Helmuth and Achtman 1975) by taking into account 4 new *tra* genes: *traN*, *traU*, *traV*, and *tra W* (Miki et al. 1978) and with the final order being *traA, L, E, K, B, V, W, C, U, F, H, G* (Willetts and Skurray 1980). While new *traU*, *traV*, and *traW* mutants were resistant to the male-specific phages f1, f2 (or MS2), and Qβ, but *traN* mutants were sensitive against them, it was concluded that, while *traV*, *traW*, and *traU* were required for pilus formation, *traN* was not (Miki et al. 1978).

In order to prevent confusion with the *tra* gene nomenclature, an explanation from an authoritative review (Willetts and Skurray 1980) is included here, according to which the 19 known transfer genes were conveniently divided into 4 groups. The first group included the above-mentioned genes *traA, L, E, K, D, V; W. C, U, F, H, G*, which were directly required for pilus formation and hence for recipient cell recognition and mating pair formation, as well as for infection by the F-specific phages. The second group contained *traN* and *traG*, required for stabilization of mating pairs, and the third included *traM, Y. G, D, L Z*, which were concerned with conjugal DNA metabolism. Mutants in the genes belonging to these two groups still synthesized the pilus and consequently retained sensitivity to the F-specific phages, except that *traD* mutants, though sensitive to filamentous DNA phages and to the phage Qβ, were resistant to the group I phages f2, R17, and MS2 because of inability of the phage RNA to penetrate the cell envelope (Achtman et al. 1971; Paranchych 1975). The sole gene in the fourth group, *traJ*, controlled expression of most if not all of the other transfer genes (Finnegan and Willetts 1973).

Finally, the *traA* gene was identified as the structural gene for F pilin, the protein subunit comprising F pili (Minkley et al. 1976). For this reason, the number of radioactively labeled ^125^I-tyrosine peptides obtained from purified pili, which were produced by serine and tyrosine suppressed *tra* amber mutants, were analyzed. The other genes were presumed to determine a biochemical pathway for the conversion of F pilin to the extracellular pilus.

Most of the above-mentioned RNA phage-resistant F mutants were transfer-deficient and did not form an extracellular pilus (Ohtsubo et al. 1970; Achtman et al. 1971, 1972). However, a few F mutants partially or completely resistant to one or more

F-specific phages, but still largely transfer-proficient, have been obtained (Tomoeda et al. 1972; Paranchych 1975).

A number of the R17-resistant mutants of *E. coli* with an altered F pilus structure and function were reported by Burke et al. (1979): some of the mutants produced unusually long pili, displayed wide variations in the number of pili per cell, or were deficient in pilus retraction and synthesis.

These studies investigated the phenotype but not the genotype of the mutants. In an attempt to further delineate the dual role of the pilus in conjugation and F-specific phage infection, Willetts et al. (1980) isolated a series of F*lac* mutants that were deficient mainly in the latter function and determined both their pilus-related and genetic properties. This paper was therefore a first attempt to combine functional and genetic information on the *tra* mutants.

The most common type of mutant pili have almost complete sensitivity to f1 but total or almost total resistance to both R17 and Qβ. Similar mutants have been isolated by R. Weppelman, D. Popkin, K. Ippen-Ihler, and C.C. Brinton, and two were shown to carry mutations in the *traA* gene (unpublished, cited by Willetts et al. 1980). All these mutants adsorbed the phage R17 well, and the defect in R17 infection was presumably due to a failure of the ejection or penetration stages (Willets 1980). Some mutants of this group made two- to fivefold more pili than normal, and both cyanide treatment and temperature-shift experiments suggested that this resulted from defects in the abilities of these pili to retract, or from an increase in their rate of outgrowth. Other mutants showed other combinations of altered pilus properties: one mutant was resistant to f1, R17, and Qβ, but demonstrated almost normal R17 adsorption, although the cells were apparently piliated to only about 10% of normal. Another mutant was resistant to Qβ and f1 and partially resistant to R17: piliation and R17 adsorption were similar to normal. One more mutant retained about one-quarter of the normal transfer and piliation levels but was resistant to all three classes of the F-specific phages and adsorbed R17 very poorly. Remarkably, all these altered phenotypes seemed to result from mutations in the *traA* gene. The authors suggested that specific alterations in the amino-acid sequence of the pilin precursor protein, encoded by *traA*, gave rise to particular combinations of altered pilus properties.

It is noteworthy that another protein, namely the *traJ* gene product, was found, at almost the same time, to be immunologically related to F pilin and regarded as a precursor to F pilin (Achtman et al. 1979).

The problem of the gene encoding F pilin was solved once and for all when direct sequencing made it possible to locate the *traA* reading frame and to compare it with the amino acid composition and sequence of N- and C-terminal peptides isolated from the purified F pilin (Frost et al. 1984).

An intriguing temperature-sensitive *traC* mutant was found (Schandel et al. 1987). Being resistant to the phage f2, this mutant retained sensitivity to the filamentous bacteriophage f1 in the absence of expression of extended F pili, since the f1 attachment sites were still present at the cell surface (Schandel et al. 1987).

Furthermore, whole genome analysis of an *E. coli* strain partially resistant to the phage Qβ suggested that a single mutation in the *traQ* gene was responsible for the partially resistant phenotype (Kashiwagi et al. 2015). The TraQ product interacted with propilin, encoded by the *traA* gene: overexpression of the wild-type TraQ in the partially resistant *E. coli* strain resulted in recovery of both TraA protein content, including propilin and pilin, and Qβ amplification to levels comparable to those observed in the susceptible strain (Kashiwagi et al. 2015). This study represented an example of how *E. coli* can become partially resistant to the RNA phage infection via changes in a protein involved in maturation of a receptor rather than in the receptor itself.

CHROMOSOMAL MUTATIONS

Besides *tra* genes connected directly with the sex plasmids described above, chromosomal *E. coli* genes might play a decisive role in male functions, including sensitivity to the RNA phages. Thus, two chromosomal *E. coli* K12 Hfr genes required for the expression of F plasmid functions in the presence of normal plasmid DNA were found by selection of Qβ-resistant *E. coli* mutants (McEwen and Silverman 1980) in accordance with the previously elaborated methodology for the isolation of such mutants (Mandal and Silverman 1977). Mutants in the two newly designated genes *cpxA* and *cpxB* (*cpx = conjugative plasmid expression*) were resistant to the phage Qβ and defective as conjugal donors. These characteristics were attributed to the inability of mutant Hfr cells to elaborate F pili (McEwen et al. 1980). Furthermore, chromosomal mutants of *E. coli* deficient in the expression of F plasmid functions were selected by mutagenizing F⁻ cells, introducing an F' plasmid into the mutagenized cells by conjugation, and identifying transconjugants resistant to the phage Qβ: such mutants were defective in an extracellular stage of Qβ infection, suggesting that they failed to elaborate F pili (Silverman et al. 1980).

Similar chromosomal mutants in the two genes designated *sfrA* and *sfrB* were described by Beutin and Achtman (1979). These mutants exhibited pleiotropic effects on the expression of F factor *tra* genes and provoked resistance to the phages M12 and f2 in the cells carrying normal F factor.

A subsequent pair of chromosomal mutants in the genes *fexA* and *fexB* (*fex = F expression*) was presented (Lerner and Zinder 1979, 1982). Together, the *fexA* and *fexB* exerted a pleiotropic effect on the expression of the F *tra* genes, prevented appearance of F pili, and were therefore found resistant to the phage f2 (Lerner and Zinder 1982). An analog of the *E. coli* gene *sfrB*, namely gene *rfaH*, which makes lipopolysaccharides but is also required for the expression of the F factor functions including piliation and attachment of the phage f2, was localized in *S. typhimurium* (Sanderson and Stocker 1981).

Furthermore, cyclic AMP and its receptor protein appeared as being required for the expression of *tra* operon in *E. coli* Hfr cells: a number of *cya* and *crp* mutants demonstrated inability to adsorb the phages MS2 and Qβ, but cAMP supplementation suppressed this defect in *cya* but not in *crp* mutants (Kumar and Srivastava 1983). This inability was caused by the simple fact that the *cya* and *crp* mutants did not produce F pili in normal amounts.

Mutation in a novel bacterial locus *fii* did not allow the filamentous bacteriophage f1 to infect bacteria, although the cell capabilities of conjugation or infection by the phage f2 were not affected (Sun and Webster 1986). Furthermore, the mutation in the *tolQ* (previously *fii*) gene, a member of the *tolQRAB* cluster encoding proteins that transport large molecules across the bacterial envelope, was shown to affect only tip-adsorbing filamentous phages and not shaft-adsorbing ones, e.g., RP4-specific filamentous phage Pf3 (Bradley and Whelan 1989). The sensitivity of the *tolQ* mutants to the shaft-adsorbing RNA phages was confirmed by infection with the phages $Q\beta$, F_0lac, $I\alpha$, t, and pilHα (Bradley and Whelan 1989).

F PILI IN THE *ENTEROBACTERIACEAE* OTHER THAN *E. COLI*

Shortly after Brinton et al. (1964) demonstrated with an electron microscope that the phage M12 adsorbed to the pili of *P. mirabilis* cells which harbor the F*lac* plasmid, but not to those of the F$^-$ cells of *P. mirabilis*, the RNA phage MS2 was shown to infect and grow in *P. mirabilis* strains harboring the F genotes derived from *E. coli* (Horiuchi and Adelberg 1965). Although a burst size of 2000–3000 phage particles per cell was similar to that in *E. coli* K12 and no host-controlled modification of MS2 was detected, the *P. mirabilis* strains failed to show plaque formation by the phage MS2 on agar plates. This failure was explained by low efficiency of phage adsorption. Remarkably, both *E. coli* and *P. mirabilis* strains harboring a different sex plasmid, P*lac*, failed to permit MS2 growth (Horiuchi and Adelberg 1965).

Transferring F*lac* exogenote from *E. coli* K12 F$^{/+}$ strain to *Shigella flexneri* 2a and 3a, of which lactose inferment-ability is one of the important markers, converted it to *lac*$^+$ and also to F$^{/+}$ (Kitano 1966a). Such *S. flexneri* F$^{/+}$ was able to adsorb the MS2 or M12 phages, although the efficiency of adsorption was very low (Kitano 1966b). However, once the phage RNA was injected into the *Shigella* bacteria, phage replication occurred therein as well as in *E. coli* K12: the burst size was about 2000 and the latent period was about 20 min, in both *E. coli* and *S. flexneri*. The phages MS2, M12, and $Q\beta$ did not form any plaque on intact F$^{/+}$ *Shigella*, but the phage RNA could infect spheroplasts of *S. flexneri* with a similar plating efficiency as in its infection of *E. coli*. No affirmative evidence for the multiplication and continuous release of RNA phages without killing the host bacterial cells, *Shigella* F$^{/+}$, has been obtained (Kitano 1966b). Remarkably, transferring F*lac* exogenote back from *S. flexneri* F$^{/+}$ to *E. coli* F$^-$ strain resulted in the progeny that demonstrated full sensitivity to the phage MS2 infection (Kitano 1966a). The low efficiency of *S. flexneri* F$^{/+}$ infection was explained by the following assumptions: (i) *S. flexneri* F$^{/+}$ may be able to form only labile or incomplete F pili, (ii) the fine structure of F pili may be modified under the genetic control mechanism of *Shigella*, so that the receptor site is insufficient for accepting the RNA phage, or (iii) only a minor population of *S. flexneri* F$^{/+}$ may be able to form complete F pili (Kitano 1966b).

The variants sensitive to the phage MS2 appeared among MS2 non-sensitive *S. flexneri* cultures in the course of liquid medium passages, where sensitive variants possessed pili and electron microscopy showed adsorption of the MS2 phages on the F-like pili of these variants (Kerekes et al. 1983a,b).

The phage R17 was used as a marker of the male properties in the attempts to perform chromosome transfer between *E. coli* Hfr strains and *P. mirabilis* (Gemski et al. 1967).

When either the F*lac* or the F$'$*Cm* plasmid was transferred from *E. coli* into *Pasteurella pseudotuberculosis*, the *P. pseudotuberculosis* (F$'$) strains isolated formed plaques with both RNA and DNA male-specific phages (Molnar and Lawton 1971). In contrast, strains of *P. pestis* harboring *E. coli* (F$'$) plasmids did not form plaques with the male-specific phages, although such strains permitted limited multiplication of the phage MS2. The adsorption and burst size of MS2 were approximately the same in both species of *Pasteurella*, but the percent of adsorbed MS2 that produced infective centers was much lower in *P. pestis* than it was in *P. pseudotuberculosis*. Nevertheless, the authors isolated a single clone of *P. pestis* F$'$ cells that could form not only MS2, but also $Q\beta$, f2, and R17 plaques (Molnar and Lawton 1971).

An unusual sex factor variant, F*lac*S, was found in a strain of *S. typhimurium* which carried the temperature-sensitive F$_{ts}$*lac* plasmid (Macrina and Balbinder 1973). Since *S. typhimurium* repressed the synthesis of pili, the F*lac*S plasmid was examined in *E. coli*. The F*lac*S plasmid carried a mutation that conferred not only extreme genetic stability and resistance to curing by acridine orange, but also resistance to the phages MS2, $Q\beta$, and f2. The sex pili synthesized by *E. coli* strains carrying F*lac*S were altered in some manner that resulted in a very inefficient adsorption of the male-specific phages (Macrina and Balbinder 1973).

CONJUGATION AND RNA PHAGES

Although some early reports (Ippen and Valentine 1967; Knolle 1967b) indicated the existence of the interference among phage infection and conjugation, a later study (Novotny et al. 1968) led to the final conclusion that the F-specific RNA, as well as DNA, phages did inhibit the bacterial mating process of *E. coli*. A step in the RNA phage infection, prior to RNA penetration, prevented the formation of mating pairs and, in addition, prevented a fraction of existing mating pairs from completing the mating process. In contrast, the DNA phages prevented the formation of mating pairs but had no effect on mating pairs once they were formed. These findings supported all earlier arguments that donor cells have a single surface structure involved in both conjugation and male-phage adsorption and that this element is the F pilus (Novotny et al. 1968).

The RNA phages R17, $\mu2$, f_{can1}, MS2, and $Q\beta$ were involved in a classical study of the initial events during bacterial conjugation, in order to elaborate optimal conditions for achieving high mating efficiencies (Curtiss et al. 1969). Once again, the authors concluded that the F pili were essential for specific pair formation but also found that the presence of the F pili was not sufficient for display of donor ability, nor was the absence of F pili enough for cells to exhibit recipient ability (Curtiss et al. 1969). Figure 3.3, which is reprinted from this excellent paper, demonstrates the involvement of F pili in specific pair formation, where F pili are "stained" with the phage MS2.

To relate quantitatively the loss of visible cell-surface structures (including F pili, type I pili and flagella) from Hfr and RTF donor cells in electron microscopy with blending speed, the improved blending procedure was elaborated with the use of the phage M12 as a marker (Novotny et al. 1969a). These precise measurements formed a background for the accompanying paper where the mechanisms of conjugation and male phage infection were studied in detail by comparing quantitative changes in the ability of donor cells to form mating pairs and adsorb male phage with the morphological surface changes which occurred during and after blending, with the use of the phage R17 in this specific case (Novotny et al. 1969b).

The effect of the phage MS2, as well as of the DNA phage f1, on the formation of mating pairs in *E. coli* conjugation was demonstrated directly in the Coulter counter (Ou 1973). In contrast to f1, when the formation of mating pairs was inhibited immediately and completely, the phage MS2, at a relatively high multiplicity of infection (MOI), also inhibited the formation of mating pairs significantly although not completely: the inhibitory effect of MS2 phage was dependent on the time of addition and the MOI used. At relatively low MOI (<20), the MS2 phage showed some inhibitory effect when added to a male culture prior to mixing with females, whereas

no effect was observed when phages were added after mating pair formation had already commenced. At a high MOI (>400), the phage MS2 disrupted the mating pairs already formed. Some preformed mating pairs were resistant to the high MOI of the phage MS2, however, and the "sensitive" (to high MOI) mating pairs seemed to mature into "resistant" mating pairs as a function of time. It was concluded that the tip of an F pilus was the specific attachment site for mating, whereas weak contact of the sides of F pili was also a necessary step in the mating pair formation (Ou 1973). These data reflected an earlier finding that Zn^{2+} at 10^{-3} M greatly reduced the formation of mating pairs (Ou and Anderson 1972a), as well as inhibited the adsorption of the DNA phage to the tips of F pili but not the adsorption of the phages MS2 and R17 to the sides of F pili (Ou and Anderson 1972b).

Overall, *E. coli* Hfr cells that were engaged in conjugation with F⁻ cells were killed by the phage MS2 with the same kinetics as were Hfr cells that were not conjugating (Schreil and Christensen 1974). Preincubation with large numbers of MS2 did interfere with the ability of Hfr cells to form recombinants but this interference did not have certain properties that might be expected if it were due to competition for nucleic acid transport (Schreil and Christensen 1974).

The ability of the phage MS2 to decrease both mating aggregates and transconjugants in a mating mixture was also observed during conjugation directed by an F-like R factor in *E. coli* (Eckerson and Reynard 1977).

RNA phages were used traditionally as markers in numerous studies of a machinery involved in the bacterial conjugation, e.g., the phages f2 (Sanderson 1996), MS2 (Buyanova et al. 1991; Grishina and Pekhov 1991; Grishina et al. 1993; Nosova et al. 2005), R17 (Bayer et al. 1995, 2000, 2001; Pölzleitner et al. 1997; Beranek et al. 2004; Gruber et al. 2016), and PRR1 (Lessl et al. 1993). A modern review of the interplay of the conjugative plasmids with the phage infection was published by Getino and de la Cruz (2018).

ELIMINATION OF SEX FACTORS

Usually, cells were cured from sex factors by a classical method using acridine orange (Hirota and Iijima 1957). A simple and efficient method for the elimination of sex F and drug resistance R factors from *E. coli* cells was elaborated by treatment with sodium dodecyl sulfate (SDS) (Tomoeda et al. 1968). By such treatment of F⁺ cells with SDS, unusual F⁺ cells which retained mating ability but showed resistance to the phage M12 were isolated, together with mutants of another type which have lost mating ability but retained sensitivity to M12 (Tomoeda et al. 1968). Remarkably, the treatment with SDS of *E. coli* K-12 Hfr cells, where F factor was integrated, resulted in the appearance of variants of several types which have lost part or all the F factor (Inuzuka et al. 1969).

Furthermore, the elimination of F factors in *E. coli* by urea was reported: growth of *E. coli* harboring F or F′8 (F*gal*) factors in broth containing urea led to the loss of these genetic elements and yielded F⁻ cells (Tomoeda et al. 1970). Remarkably, unusual F⁺ or F′8 cells which retained the ability

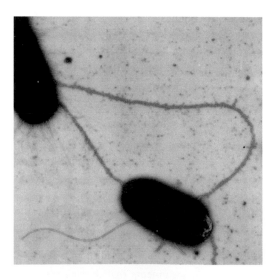

FIGURE 3.3 Electron micrograph of presumed specific pair between an Hfr cell (bottom) and an F⁻ cell. The phage MS2 was used to "stain" F pili. The specimen was prepared soon after mixing the Hfr and F⁻ cells. Two F pili have been used to make contact with a single F⁻ cell. ×28,000. (Reprinted with permission of ASM from Curtiss R 3rd. *Microbiol.* 1969;23:69–127.)

of genetic transfer but showed resistance to the phage M12 were also isolated.

NON-F-ENCODED PILI

F pili encoded by F and F-like plasmids or provided by a chromosome of the *E. coli* Hfr strains retained the central position in the RNA phage studies for a long time. For this reason, the RNA phages were frequently described, and are still described now, as F pili-dependent, or simply F-dependent, RNA phages (FRNA phages). In a broader sense, the RNA phages remained connected with the *Enterobacteriaceae* family of Gram-negative bacteria including such genera as *Escherichia*, *Klebsiella*, *Proteus*, *Salmonella*, *Shigella*, and *Yersinia*.

However, just at the beginning of the RNA phage era, the RNA phages infecting bacteria representing other than *Enterobacteriaceae* families of the Gram-negative bacteria kingdom were discovered. The RNA phage 7s of *P. aeruginosa* from the *Pseudomonadaceae* family, unable to infect *E. coli*, was found by Feary et al. (1963). The phage PP7, which was to become very popular in the future, after more than 50 years, followed soon (Bradley 1965b). It was shown by excellent negative contrasting electron microscopy that the phage PP7 infected the cell via polar pili of about 45 Å in diameter (Bradley 1966b). After 30 years, nucleotide sequence of the PP7 genomic RNA was determined (Olsthoorn et al. 1995).

A large group of the RNA phages infecting stalked bacteria of the *Caulobacter* genus from the *Caulobacteraceae* family (for details, see Table 1.1) was discovered by Schmidt and Stanier (1965). Although the paper did not address the problem of the phage adsorption mechanism to the bacteria, Ippen and Valentine (1965) noticed from the electron micrographs published there that the F pili-like structures "might be synthesized even by various *Caulobacter* strains." The following year, the attachment of the *Caulobacter* RNA phages to pili of their hosts was described in detail (Schmidt 1966). It was found that the removal of the pili from the host by shear treatment before phage adsorption resulted in effective inhibition of phage adsorption. Efficient adsorption occurred on swarmer rather than on stalked caulobacter populations, while the swarmer caulobacters frequently possessed pili that were located at the flagellated pole of the organism, but stalked caulobacters were almost always devoid of these structures (Schmidt 1966). Three serologically distinct groups of caulophages demonstrated non-overlapping specificity to three *Caulobacter* strains: *C. crescentus*, *C. bacteroides*, and *C. fusiformis*, always accompanied by the pili—phage complex formation. The pili emanated from the flagellated pole of the swarmer, which eventually gave rise to stalk development, and were not found at the opposite non-flagellated pole, nor did they occur at peritrichous sites of origin. The pili were several microns in length and approximately 40 Å in diameter (Schmidt 1966). By means of their selective adsorption to only one form of the host cell, namely to the swarmer cells, the RNA phage φCB5 was used as a tool to monitor the morphogenic changes occurring in synchronized cultures of

C. crescentus (Shapiro and Agabian-Keshishian 1970). Such specific susceptibility of the *Caulobacter* host cell to RNA phage infection during only a limited time in its life cycle contributed strongly to the studies on the prokaryotic differentiation (Shapiro et al. 1971). The specific role of the *Caulobacter* pili was confirmed by isolation of a *C. crescentus* mutant that did not produce pili and was resistant to the phage φCB5 (Kurn et al. 1974). The same effect, loss of pili and resistance to the phage φCB5, was induced by growth in the presence of cyclic GMP derivatives, which repressed surface structure differentiation in *C. crescentus* (Kurn and Shapiro 1976). At the same time, a large set of pleiotropic *C. crescentus* mutants both deprived of pili and resistant to the phages φCp2 and φCp18 was isolated (Fukuda et al. 1976) and characterized in detail (Fukuda et al. 1977). Then, a subsequent series of nonmotile *C. crescentus* mutants showed decrease of sensitivity to the phage φCB5 (Johnson and Ely 1979). Furthermore, Sommer and Newton (1988) demonstrated that pilus formation is indeed periodic in the *C. crescentus* cell cycle and that this structure is assembled in the swarmer cell immediately after cell division. Nevertheless, the function of *C. crescentus* pili remained unknown: they have not been implicated in bacterial conjugation, and it was possible that they might serve a novel function in this differentiating organism (Sommer and Newton 1988). By using the phage φCB5, an interesting mutant of *C. crescentus* was isolated, namely a mutant of the *tacA* gene encoding a putative activator of the alternative sigma factor σ^{54}, which played an important role in the expression of late flagellar genes of *C. crescentus* (Marques et al. 1997).

Bradley (1965b) was the first to raise the urgent question of whether or not RNA phages might indicate sexuality in their *Pseudomonas* and *Caulobacter* hosts, as they did in the case of *E. coli* where they were exclusively male specific. Both the genera *Pseudomonas* and *Caulobacter* were characterized by polar flagellation, but no connection to the possible sexual role of these chromosomally-encoded filaments did exist. Nevertheless, since an infectious fertility factor (FP) has been demonstrated for *P. aeruginosa* (Holloway and Jennings 1958), it was speculated that the pili observed by the *Pseudomonas* phage infection could probably be the same sex pili encoded by the FP factor (Bradley 1966b). Negative staining of uninfected cells did not reveal the pili (Bradley 1966b), and the RNA phages thus had the effect of delineating them as is the case with *E. coli* F pili (Brinton 1965). However, the pili were later found readily by negative contrast staining of *P. aeruginosa* which had not been deliberately infected with phage (Fuerst and Hayward 1969). The *P. aeruginosa* pili were rather thinner than those of *E. coli*: about 45−54−60 Å (Bradley 1966b; Fuerst and Hayward 1969; Bradley 1972a) as opposed to 85–95 Å (Brinton 1965; Lawn 1966) and resembled the filamentous phages of *E. coli*. The designation FP pili, (*fertility factor* pili) was suggested for the *Pseudomonas* pili by Bradley (1966b).

Nevertheless, further elucidation led to the conclusion that the RNA phages of *P. aeruginosa* have no sexual specificity (Holloway 1969). Unlike the *E. coli* RNA phages, they were sex-indifferent, plating equally well on male and female strains and showing adsorption characteristics different from those of comparable phages in *E. coli*: the absence of sex-specific pili might also explain the failure to isolate sex-specific *P. aeruginosa* RNA phages (Holloway 1969 and Holmes, unpublished data, cited in this reference). In fact, conjugation between strains of *P. aeruginosa* depended upon the sex factor FP being present in at least one of the parents. However, there was no evidence that the presence of the FP factor induced the formation of a structure analogous to the sex pilus of the *Enterobacteriaceae*. Electron microscopic observations (Holmes, personal communication; Veitch, personal communication, cited by Holloway 1969) of a range of donor and recipient strains of *P. aeruginosa* have revealed no difference in the surface anatomy of these two types. Moreover, no true Hfr form of *P. aeruginosa* was isolated (Holloway 1969).

At the same time, clear evidence was obtained in favor of the proposal that the chromosomally encoded polar FP pili are the adsorption organelle for the *P. aeruginosa* phage PP7: a pili-less mutant derived from the phage's piliated host, was found to be phage resistant and unable to adsorb phage (Weppelman and Brinton 1971). The ability of the mutant to maintain phage replication was confirmed by the fact that the spheroplasts prepared from either the mutant or the original strain could be infected by naked PP7 RNA with equal efficiency, demonstrating that the phage resistance of the mutant was due to a defect in the adsorption-penetration mechanism and not due to an inability to replicate the phage (Weppelman and Brinton 1971).

Fine properties of the *Pseudomonas* pili were studied further in a series of papers by David E. Bradley. Thus, it was confirmed that these pili are not involved in the transfer of the FP sex factor and supposed that the *P. aeruginosa* pili could be coded for by a plasmid other than FP (Bradley 1972a). Remarkably, about 25 times more pili per cell were present after PP7 adsorption than before it. This result supported rather the pilus retraction theory (Marvin and Hohn 1969) instead of acting as simple tubes for the transfer of genetic material (Brinton 1965). Furthermore, the phage-stimulated pilus synthesis was demonstrated (Bradley 1972d). Strong experimental support for the pilus retraction model was found by electron microscopic measuring the lengths of polar *P. aeruginosa* pill of before and after RNA phage adsorption, where expected averaging 50% length reduction was observed, if one phage per pilus would adsorb and stop further retraction (Bradley 1972b,c, 1974b; Bradley and Pitt 1974).

Unlike *Enterobacteriaceae*, where RNA phages and filamentous DNA phages shared the same pili, differing only in the attachment site, the filamentous *P. aeruginosa* phage Pf used other pili that did not adsorb the phage PP7 and appeared on a serologically different strain from that infected by the RNA phage (Bradley 1972a). The designation *FP pili* was further fully replaced by the term *type IV pili*, or simplified T4P, or just TFP, which followed the definition suggested by Ottow (1975) for the "flexible, rodlike, polarly inserted fimbriae which were observed on members of the genus

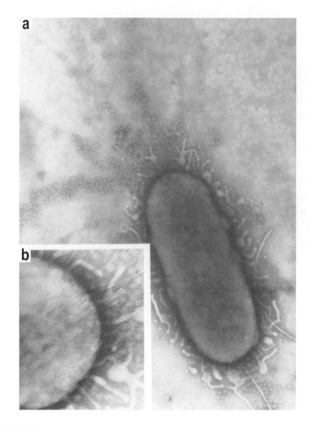

FIGURE 3.4 *Pseudomonas aeruginosa* strain 1 (PAO 1) with PP7 phage virions adsorbed to polar pili, ×60,000 (a). (b) The upper pole enlarged to show pili entering the cell, ×130,000. (Reprinted with permission of Cambridge Journals from Bradley DE. *Genet Res.* 1972;19:39–51.)

Pseudomonas and *Vibrio*." Figure 3.4 shows how PP7 phage virions are adsorbed to the polar *P. aeruginosa* pili.

The phage PRR1, isolated by enrichment from sewage, demonstrated specificity to the hosts carrying the R1822 factor and therefore might be regarded as the first male-specific *Pseudomonas* phage (Olsen and Shipley 1973). The R1822, or RP4 (Datta et al. 1971), or RP1 (Grinsted et al. 1972) by other designations, a plasmid specifying multiple drug resistances and regarded originally as a *Pseudomonas* factor, demonstrated high potential of promiscuity and was transferred to a variety of species including *Enterobacteriaceae*, soil saprophytes as *Acinetobacter*, *Neisseria perflava*, and photosynthetic bacteria. With the acquisition of drug resistances, these strains became sensitive to the phage PRR1 (Olsen and Shipley 1973). However, not all strains harboring the R1822 plasmid supported productive infection by PRR1 (Olsen and Thomas 1973). The phage PRR1 differed from coliphages and from two other *Pseudomonas* RNA phages, 7s and PP7, by host range and biochemical properties: membrane filter-salt elution patterns, RNase sensitivity, and inactivation in low ionic strength solutions (Olsen and Thomas 1973). As other RNA phages, PRR1 required pili for the adsorption (C.H. To and C. Brinton, unpublished data cited by Olsen and Thomas 1973; Bradley 1974b). Thus, short thick pili were found by electron microscopy on bacteria carrying the RP1 and R1822 plasmids, and the phage PRR1 was seen to adsorb to the bases of the pili (Bradley 1974a). The phage PRR1 was specific for a number of *Pseudomonas* RP strains but did not propagate on FP strains (Stanisich 1974). Moreover, the phage PRR1 was the only phage specific for the *Pseudomonas* RP sex pili from a group of tested phages including filamentous one, namely Pf3 (Bradley et al. 1974a, Stanisich 1974). The P-1 pilus encoded for by the RP1 plasmid was declared finally as a receptor for the phage PRR1 (Bradley and Cohen 1977). The P-1 (or P) is a designation of one of the shared enterobacterial and pseudomonad incompatibility groups (the plasmids R1822, RP1, and RP4 are members of this group) and further narrative on the RNA phage pili will be connected with the incompatibility nomenclature.

INCOMPATIBILITY GROUPS: DEFINITION

Following Singleton and Sainsbury (*Dictionary of Microbiology and Molecular Biology*, 3rd ed, Wiley, pp. 391–392), *Incompatibility (in plasmids)* is

the inability of plasmids to coexist stably, within the same cell when they have similar or identical systems for replication. Two incompatible plasmids which occupy the same cell will (in the absence of a selective pressure for both plasmids) tend to segregate to different cells during cell division. The stable intracellular co-existence of one plasmid with another requires that each plasmid be able to control, independently of the other, its own replication/partition such that it can establish and maintain a stable copy number; the inability of a given plasmid to maintain a stable copy number in the presence of another plasmid is the characteristic feature of incompatibility.

It must be stressed that the idea of incompatibility was put forward first by Naomi Datta and R.W. Hedges (Datta and Hedges 1971; Hedges and Datta 1971) and described in a classical review by Richard P. Novick (1987).

The incompatibility or compatibility measure was used for the classification of plasmids, since it provided a useful classifying criterion in early steps, before bulk sequencing era. Thus, the plasmids were classified into so-called incompatibility groups (Inc groups) in such a way that all the plasmids in a given Inc group expressed mutual incompatibility (Singleton and Sainsbury, p. 391).

The Inc groups were designated by the prefix "Inc" followed by a capital Latin letter and, sometimes, by a Roman numeral or a Greek letter, e.g., IncFII, IncIα. The letter, e.g., F, I, indicated the type of conjugation system specified by the transfer operon; thus, e.g., an IncF plasmid specified a conjugative system like that of the F plasmid or of an F-like plasmid, and encoded therefore either F pili or F-like pili. The Roman numeral or Greek letter indicated a particular type of the plasmid replication system; thus e.g., IncFI and IncFII plasmids have different replication systems (i.e., they are compatible) though they specify similar conjugation systems. Some examples (including plasmids, which are mentioned in this chapter) of enterobacterial Inc

groups are IncFI including the F plasmid, the colicin plasmid ColV-K94, and the R plasmid R386; IncFII including the R1 plasmid, R6, and R100; IncIα (sometimes written IncI1 or IncI₁) including ColIb-P9, and R64; IncN including N3, R46, and R269N-1; IncX including the R6K plasmid (Singleton and Sainsbury, p. 392).

Pseudomonas Inc groups were designated IncP-1 to IncP-13, the numbers referring to each of the 13 different types of replication system.

Shared enterobacterial and pseudomonad Inc groups were represented (the enterobacterial Inc group being given first) by: IncC ($\equiv$ IncP-3); IncP ($\equiv$ IncP-1) including R68, R751, RK2, RP1, and RP4; IncQ ($\equiv$ IncP-4) including a popular RSF1010 plasmid (Singleton and Sainsbury, p. 392).

In 2009, altogether 27 Inc groups were recognized in *Enterobacteriaceae* by the Plasmid Section of the National Collection of Type Cultures in London, including six IncF (FII–VII) and three IncI (I1, Iγ, I2) variants. It is remarkable that the first Inc groups were defined in tight connection with the susceptibility to the phage infection: IncI, plasmids producing type I pili susceptible to the DNA phage Ifl; IncN, N3-related plasmids susceptible to the DNA phage IKe; IncF, plasmids producing type F pili susceptible to the DNA phage Ff; and IncP, RP4-related plasmids, susceptible to the phage PRR1 (for review see Carattoli 2009).

These conjugative plasmids from Gram-negative bacteria directed the synthesis of the extracellular pili which had an essential role in the recognition of recipient cells and the establishment of cell-to-cell contact, as well in the adsorption of the RNA phages. The pilin subunits, namely the structural units of the pili, often shared no similarity. For simplicity, pili were classified into two broad morphological groups: long flexible (1 μm) and short rigid (0.1 μm). Long pili were like those expressed by cells carrying the F plasmid. Short pili were expressed by plasmids of the IncN, IncP, and IncW incompatibility groups. Some plasmids (Inc groups I₁, I₂, I₅, B, K, and Z) encoded both long and short pili (*Encyclopedia of Life Sciences*, 2008, p. 2208).

Where plasmids with distinct compatibility specificities determined serologically related pili, it was the practice to describe them as constituting a complex encompassing several incompatibility groups, e.g., the F complex (Hedges and Datta 1972), the H complex (Bradley et al. 1982), and the I complex (Bradley 1984). Following to the Inc abbreviations, which could be sometimes misleading, it is necessary to keep in mind that, for example, the F complex includes the Inc groups I, II, III, and IV, but not an IncFV group, which was represented by a popular plasmid EDP208, a pilus-derepressed derivative of the plasmid F₀*lac*; both of these were included later in an S complex (Coetzee et al. 1986). This was an attempt to prevent the possible misunderstanding by the exhaustive comments in Table 1.1.

Generally, the incompatibility classification scheme was complicated by the presence of two or more different replicons in some plasmids, and has become obsolete nowadays owing to the availability of rapid methods for determining and comparing DNA sequences. Novel and much more universal sequence-based classification schemes are elaborated now in the genomic era.

Thus, the iterons, or multiple small repeats adjacent to the origin of replication in forward or reverse orientations to one another, could be the structural basis for the Inc phenotype whereby closely related plasmids, i.e., ones with the same iteron sequences, are unable to be stably inherited in the same cell line (Thomas et al. 2017).

A computer algorithm (Carattoli et al. 2014) can classify *Enterobacteriaceae* plasmids to a putative Inc group by comparison to known replicons. However, it should be stressed that while new classification schemes occur, incompatibility assays are still required to demonstrate the phenotype of incompatibility (Thomas et al. 2017).

It is noteworthy that the extent of sequence identity among Inc group members varies markedly between different Inc groups: the similarity within the IncP, IncN, or IncW groups extends to most of the plasmid backbone, while in the IncF group the similarity among members is limited to the conjugation region (reviewed by Fernandez-Lopez et al. 2017).

To set up an operational plasmid taxonomy, it would be necessary to construct phylogenies using some conserved genetic marker. The conjugative relaxase, the protein required to initiate plasmid mobilization through conjugation, seems to be the most feasible classification unit, which allows division of plasmids in eight mobility (MOB) classes with clades, or families, within, with true phylogenetic meaning (see Fernandez-Lopez et al. 2017 for review).

More generally, the plasmids could be regarded as an element of mobile genetic elements (MGEs) that would include plasmids, phages, and their integrated counterparts: integrative and conjugative elements ICEs and prophages (de Toro et al. 2014).

Nevertheless, many urgent questions remain unanswered; for example, why IncF and IncI plasmids and not others occur so frequently in *E. coli* (de Toro et al. 2014).

INCOMPATIBILITY GROUPS AND RNA PHAGES

The most studied group of the RNA phages, the coliphages, attached to the pili encoded for by the IncF group, one group of the large number of the Inc groups present in nature and mentioned above.

The first example of the non-IncF-specific phage was the above-described phage PRR1 which adsorbed exclusively to the IncP plasmid-encoded pili (Olsen and Thomas 1973). Moreover, Olsen and Thomas (1973) tested for the first time the susceptibility of members of many other plasmid Inc groups including F, I, N, W, C, A, J, and T for the PRR1 infection and found no evidence of the PRR1 plaque formation after spotting of surface lawns, and therefore supported high specificity of the RNA phage attachment. After more than 30 years, genomic sequence of the phage PRR1 was determined (Ruokoranta et al. 2006).

After the PRR1 phage, an impressive set of the RNA phages specific for other than the IncF group was isolated, due to enormous activity of David E. Bradley in Canada and J.N. Coetzee and colleagues in South Africa. Bradley also

introduced the first highly informative morphological and serological classification of pili into three groups: thin flexible, thick flexible, and rigid filaments or rods (Bradley 1980b). Morphologically identical thin flexible pili were determined by plasmids of the I complex, IncB, and IncK. Thick flexible pili were determined by plasmids of Inc groups C, D, the F complex, H1, H2, J, T, V, X, com9, the single plasmid F_0lac, and the unclassified plasmid R687. Serological tests showed that C pili were related to J pili, H1 pili to H2 pili, com9 pili to F_0lac pili, and R687 pili to D pili, the remainder being unrelated. Rigid pili were determined by plasmids of Inc groups M, N, P, W, and by the unclassified plasmids R775, RA3, and pAr-32. No cross-reactions were found between pili of the three different morphological groups (Bradley 1980b).

From the novel F pili unrelated phages, the phage C-1 (termed now C-1 INW-2012 at the NCBI), which adsorbed to the shafts, or sides, of pili coded for by IncC plasmids, formed plaques on *S. typhimurium* strains carrying IncC plasmids but failed to multiply on strains lacking plasmids of this group. It also plated on *P. mirabilis* and *Serratia marcescens* strains carrying various IncC plasmids but failed to form plaques on *E. coli* strains harboring most of these plasmids, although in all cases, phage multiplication on the IncC-carrying strains was demonstrated. However, no phage increase occurred in any of the strains which lacked an IncC plasmid or contained plasmids of other incompatibility groups (Sirgel et al. 1981). Electron microscopy of the C-1 phage attachment to the C pili supported the above-presented pili retraction model, since single adsorbed phages were found at the pilus bases where they were thought to have stopped retraction (Bradley 1989). Biochemical comparison with the coliphages f2 and R17, double-stranded RNA phage φ6 and some DNA phages allowed to number the C-1 phage among *Leviviridae* family members (Sirgel and Coetzee 1983). Later, genomic sequence of the phage C-1 was determined (Kannoly et al. 2012).

The non-F pili-specific phage t recognized serologically unique pili encoded by the IncT plasmid (Bradley et al. 1981b). The phage t formed plaques only on *E. coli* and *S. typhimurium* strains that carried plasmids belonging to the IncT group. However, a *P. mirabilis* strain failed to support phage growth, although it transferred the plasmid and concomitant phage sensitivity to a *E. coli* strain.

Another RNA phage pilE/R1 was shown to adsorb to pili determined by plasmids of the IncT group (To et al. 1975). The phage adsorbed along the entire length of the pili (cited by Bradley et al. 1981b).

The phage F_0lac was dependent on the plasmid EDP208, a constitutive pilus-producing derivative of plasmid F_0lac, which belonged to the IncFV group, and plated on *E. coli* or *S. typhimurium* strains harboring the plasmid EDP208 but not on isogenic strains without the plasmid (Bradley et al. 1981a). It is necessary to emphasize that the cells carrying EDP208 were resistant to the F-specific RNA phages, e.g., to the phages f2, MS2, and Qβ (Bradley and Meynell 1978). Bradley (1978) found that F_0lac pili were serologically unrelated to those of F-like plasmids and the plasmid F_0lac has been removed from the F complex and simply classified as unique (Datta 1979).

The F_0lac phage adsorbed preferentially to the sides of EDP208 pili very near the tip but no increase in titer occurred on *E. coli* strains carrying plasmids of the F complex. Moreover, results of the F-specific RNA phage multiplication experiments on strains carrying the derepressed pilus-producing plasmids R71 or TP224-Tc, which determined pili serologically related to those of EDP208, were inconclusive (Bradley et al. 1981a).

Attachment of another EDP208 pili-dependent phage UA-6, serologically similar to the F_0lac phage, was much less efficient than that of the phage R17 to its host (Armstrong et al. 1980). Nevertheless, the attachment was highly specific: the phage UA-6 did not bind at all to F⁻ cells or cells carrying an F plasmid but its attachment to the EDP208 pili was inhibited by anti-EDP208 pilus antiserum (Armstrong et al. 1980).

The phage Iα formed negative plaques on *E. coli* and *S. typhimurium* strains carrying I_1 plasmids or the Iγ plasmid R621a, whereas the phage adsorbed to thin flexible I_1 pili of the morphological class 1 (Bradley 1980b) along the length of their shafts (Coetzee et al. 1982). The pili coded for by Iγ plasmid were serologically related to the I_1 type. Originally, the I complex including the incompatibility groups 1_1 ($\equiv$ Iα), I_2, Iγ, Iζ, B, and K was found identical with the set of plasmids determining pili of the morphological class 1 (Bradley 1980b). The major part of the I complex was represented by the I_1 complex, plasmids of which (typical are R64, R144, and ColIb-P9) showed no DNA homology with the I_2 group (for references see Coetzee et al. 1982). High specificity of the phage Iα attachment was illustrated by the fact that the pili encoded by the B and K plasmids were serologically related to the I1 pili (Bradley 1980b) but incapable of absorbing the phage Iα (Coetzee et al. 1982). Thus, these pili, though similar, were not identical. Generally, shaft-adsorbing phages tended to be more specific than those adsorbing to pilus tips: in this respect the host range of the phage Iα resembled that of the phage PRR1, which only adsorbed to the shafts of the P-1 pili (Olsen and Thomas 1973), the phage C-1, which adsorbed only to shafts of the C pili (Sirgel et al. 1981), and the phages t and F_0lac (Bradley et al. 1981a,b), which adsorbed to the terminal shafts of the pili produced by the T plasmids and the unique plasmid F_0lac, respectively (Coetzee et al. 1982). The I-specific phages were rare, since the I pili were probably only rarely available in nature, and the range of phages adapted to adsorption to such receptors was likely to be limited (Coetzee et al. 1982). The high host specificity of the phage Iα was used, together with that of the phages t and R17, as well as of some DNA phages, for the incompatibility group identification for repressed plasmids using host cell lysis by these specific phages (Bradley and Fleming 1983).

The phage M was specific for bacterial strains of various genera: *Escherichia*, *Salmonella*, *Klebsiella*, *Proteus*, and *Serratia*, harboring plasmids of the IncM group and carrying therefore the M pili, which were rigid and belonged to Bradley's morphological group 3 (Coetzee et al. 1983). The phage M adsorbed along the length of shafts of the M pili. Although the M plasmids have been identified in a wide range of bacterial species, they seemed to be most closely associated

with the *Klebsiella* strains and formed a high proportion of the R plasmids of this tribe (for references see Coetzee et al. 1983). According to this study, the phage M was adapted to exploit the host range of the M plasmids like the phage PRR1 which lysed many hosts of the wide host-range IncP-1 plasmids (Olsen and Thomas 1973). It is noteworthy that the M pili were the only rigid pili (Bradley 1980b) which possessed the RNA phage adsorption sites, except for the PRR1-attaching P-1 pili, which are however thinner than the M pili (Bradley 1980b). The phage M was used as a control in a special study on the specificity of the phage C-1 to the IncC plasmid-encoded pili (Bradley 1989).

The phage pilHα was also specific for bacterial strains of various genera: *Klebsiella*, *E. coli*, *S. typhimurium*, *P. morganii*, and *Ser. marcescens* strains carrying any one of the H-complex plasmids from IncHI and IncHII groups and was the first wild-type phage reported to be temperature sensitive for plaque formation (Coetzee et al. 1985a). Plaque formation was temperature sensitive in that plaques formed at 26°C but not at 37°C. The pilHα phage adsorbed along the length of the shafts of the IncHI and HII plasmid-coded pili (Coetzee et al. 1985a), which were of the thick flexible variety, belonged to the morphological group 1, and were serologically related (Bradley 1980b). Furthermore, the phage pilHα was used for the mapping of transfer and H pilus coding regions of the IncHII plasmid pHH1508a (Yan and Taylor 1989).

A reliable method for the direct isolation of IncHI and IncHII plasmid-dependent phages by use of an isogenic plasmid-lacking host to absorb the majority of nonspecific phages allowed recovery of a total of 27 IncH-dependent RNA phages, namely the phages Hgal1 till Hgal27, all temperature-sensitive and failing to produce plaques above 30°C (Nuttall et al. 1987). The H-plasmid-dependent Hgal phages differed (Nuttall et al. 1987) by some physiological parameters (plaque turbidity, resistance to chloroform) from the first temperature-sensitive H-complex-dependent phage pilHα (Coetzee et al. 1985a).

Two independently isolated IncH-dependent temperature-sensitive phages, pilHα and Hgal, were compared for their physiological, biochemical, and immunological properties and no differences between the phages were observed for any of the features analyzed (Maher et al. 1991). Furthermore, the ribonuclease-inactivated phage Hgal was used as a labeling unit for examination of the kinetics of the H-pilus outgrowth and assembly, as well as their morphology and stability by transmission electron microscopy and field emission scanning electron microscopy (Maher et al. 1993). Meanwhile, it was detected that, after pili were removed by vortexing, the outgrowth of full-length 2-μ-long pili required 20 min. Figure 3.5 presents an impressive picture from Maher et al. (1993), which demonstrates a view of the phage−pili complex by field emission scanning electron microscopy.

After many decades of obscurity, the phage Hgal1 was selected, together with the phage C-1, for full-length sequencing, mainly because of the wide host range of both phages (Kannoly et al. 2012). The sequencing and phylogenetic analyses showed that both C-1 and Hgal1 were most closely

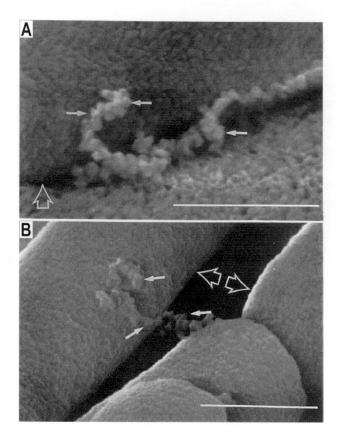

FIGURE 3.5 FESEM (Field emission scanning electron microscopy) of donor strain DT1944 showing uniform adsorption of bacteriophage Hgal to H-pilus accelerating potentials of 3.5 kV (A) and 6.0 kV (B). Open arrows indicate host strain DT1944; closed arrows indicate Hgal particles adsorbed to pilus (coiled morphology). Bars, 0.5 μm. (Reprinted with permission of ASM from Maher D et al. *J Bacteriol.* 1993;175:2175–2183.)

related to the above-described IncP-plasmid-dependent phage PRR1, which used *P. aeruginosa* and *E. coli* harboring the plasmid R1822 (Kannoly et al. 2012).

The isolation of the phage D allowed successful delineation of the IncD group (Coetzee et al. 1985b). The temperature sensitive phage D adsorbed specifically to the very distal ends of the shafts of IncD plasmid-encoded pili produced by *E. coli*, *S. typhimurium*, *P. morganii*, and *K. oxytoca* harboring one of the IncD plasmids but did not plate and propagate on IncD plasmid-carrying strains of *Providencia* or *Ser. marcescens*. The IncD-encoded pili were thick flexible (Bradley 1980b), similar to those coded for by IncF subgroups I-IV, F₀*lac*, and IncC plasmids carried by the same hosts but differed from them serologically. It is remarkable that the phage MS2 was able to adsorb to the D pili but not propagate in the IncD plasmid-carrying cells (Bradley and Meynell 1978). The authors explained this phenomenon by differences in the pili recognition by phages: whereas formalin-fixed preparations showed the phage MS2 adsorbed uniformly along the entire length of the shafts of the D pili, the phage D only adsorbed to the very distal sides of the shafts. Meanwhile, no serological relationship was demonstrated among the phages D and MS2 (Coetzee et al. 1985b).

Two RNA phages—F₀*lac* h, a mutant of the F₀*lac* phage specific to the EDP208 pili and described above, and SR, a newly isolated phage antigenically distinct from the F₀*lac* and F₀*lac* h—adsorbed unevenly to the shafts of the thick flexible pili (by classification of Bradley 1980b) encoded by plasmids of the newly suggested complex, designated S (Coetzee et al. 1986). The phages F₀*lac* h and SR had similar host ranges and plated with similar efficiencies on the *E. coli*, *Salmonella*, *Klebsiella*, and *Serratia* strains harboring plasmids of the S complex representatives from SI to SIV subgroups (Coetzee et al. 1986). It is interesting to note that the phages F₀*lac* h and SR adsorbed along the length of the EDP208 pilus shafts (Coetzee et al. 1986), whereas phage F₀*lac* attached mainly to the sides of the tips of these shafts (Bradley et al. 1981a). This could be an indication of a difference existing between the distal and more proximal areas of the shafts of the pili and raises the possibility that other phages might exist which adsorb to the more proximal areas of shafts of pili for which only distal-adsorbing RNA-containing phages have been discovered to date (Coetzee et al. 1986). Figure 3.6, taken from Kannoly et al. (2012),

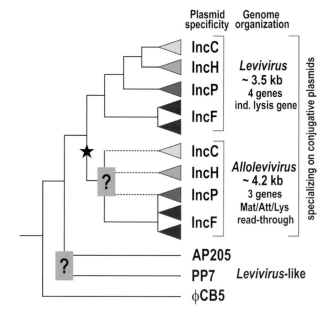

FIGURE 3.6 A hypothetical phylogenetic tree of ssRNA phages. All ssRNA phages that depend on conjugative plasmids for attachment and cell entry belong to one of the two main clades, *Levivirus* and *Allolevivirus*, with the later lineage thought to have evolved from a *Levivirus*-like ancestral phage (indicated by the star). Plasmid dependency evolved independently within each clade, although not necessarily with the same branching order (a question mark indicates an uncertain branching order, and dashed lines indicate hypothetical non-F plasmid-dependent *Allolevivirus*). However, within each clade, phage strains that depend on the same Inc plasmid (represented by the shaded triangles) are thought to be monophyletic. Phages AP205, PP7, and φCB5 share some but not all genomic traits with *Levivirus*. The most notable difference is that the placement of the lysis gene is different for AP205 (before the maturation/attachment gene) and φCB5 (after the replicase gene). (Redrawn with permission of ASM from Kannoly S et al. *J Bacteriol.* 2012;194:5073–5079.)

demonstrates a hypothetical phylogenetic tree of the RNA phages in a connection with the Inc groups of the responsive pili-encoding factors.

FINE MAPPING OF F PILI

Concerning F pili structure, the data on the *traA* gene sequencing (Frost et al. 1984) showed that there was a leader peptide of 51 amino acids and that F pilin contained 70 amino acids, giving molecular weights of 13,200 for F propilin and 7200 for mature F pilin. At the same time, the pilin, or *traA*, gene of the conjugative plasmid ColB2 was cloned and sequenced: the pilin protein of ColB2 was identical to F, except at the amino terminus, where Ala-Gln of ColB2 pilin corresponded to Ala-Gly-Ser-Ser of F pilin (Finlay et al. 1984). The cloning and sequencing technique allowed comparison of the primary structures of the next eight F-like pilins (Frost et al. 1985). For the first time, all known pilins were divided here into six groups on the basis of their primary structures, and the latter were correlated with the specific peculiarities in the pili behavior, such as their antigenicity and phage sensitivity. Summarizing, it was concluded that the F pilus consisted of a single subunit protein, pilin, which had a molecular weight of 10,800–11,800 and carried two phosphate groups and one glucose residue per molecule (Brinton 1971; Date et al. 1977; Helmuth and Achtman 1978). This "global sequencing" stage of the pili studies was excellently reviewed by William Paranchych and Laura S. Frost (1988), and a bit later by Frost (1993), Frost et al. (1994), and Silverman (1997).

Therefore, gene sequencing opened a broad path for mutant pili analysis by localization of missense and nonsense point mutations in the *traA* gene (Frost and Paranchych 1988). It made it possible to reveal two domains in the F pilin subunit exposed on the surface of the F pilus which mediate phage attachment. These two domains included residues 14–17 (approximately) and the last few residues at the carboxy-terminus of the pilin protein (Frost and Paranchych 1988).

Remarkably, the phage MS2 labeled with fluorescent dye, showing the best results with rhodamine B, allowed light-microscopic visualization of F pili, including semi-quantitative determination of the F pili amount (Biebricher and Düker 1984). Furthermore, the phage R17 conjugated with a fluorescent dye, Alexa 488, was used to make F pili visible by immunofluorescence microscopy and to study pilin mutations for recognition of the labeled phage (Daehnel et al. 2005).

Extensive mutational analysis of *traA*, the F pilin gene, revealed specific regions that could be associated with pilus assembly and DNA transfer, as well as with sensitivity to the R17 phage infection (Manchak et al. 2002). Thus, mutations involving lysines and phenylalanines within residues 45–60 suggested that these residues might participate in transmitting a signal down the length of the pilus that initiates DNA transfer or R17 eclipse. Figure 3.7 presents a summary of the positions of mutations that affect pili-specific processes.

The first x-ray diffraction studies of bacterial pili were performed with the I pili of *E. coli* in a way similar to that which was first employed for tobacco mosaic virus elucidation

(Mitsui et al. 1973). The first high resolution study on F pili combined both x-ray diffraction and electron microscopy approaches (Folkhard et al. 1979) with a further reassessment (Marvin and Folkhardt 1986). All available knowledge on the F pili structure and function, with a strong emphasis on the structure, was reviewed at that time by Ippen-Ihler and Minkley (1986).

A serious contribution to our knowledge on the packing geometry of F pili was offered by Wang et al. (2009) by electron cryomicroscopy and single-particle methods. Moreover, the authors considered the data that the F pilus, in addition to establishing contact between mating cells and retracting (as discussed above), might serve as a channel for passing DNA (or maybe RNA also?) (Babic et al. 2008). Wang et al. (2009) showed that the F pilus has an entirely different symmetry from any of the known bacterial pili as well as any of the filamentous bacteriophages, which have been suggested to be structural homologs. Two subunit packing schemes were identified: one has stacked rings of four subunits axially spaced by approximately 12.8 Å, while the other has a one-start helical symmetry with an axial rise of approximately 3.5 Å per subunit and a pitch of approximately 12.2 Å. Both structures have a central lumen of approximately 30 Å diameter, which is more than large enough to allow for the passage of single-stranded DNA. Remarkably, both schemes appear to coexist within the same filaments, in contrast to filamentous phages that have been described as belonging to one of two possible symmetry classes (Wang et al. 2009).

Clarke et al. (2008) applied live-cell imaging to characterize the dynamics of F pili with the green fluorescent phage R17 and *E. coli* strain HfrH expressing a red fluorescent cytoplasmic marker.

It was established that the F pili normally underwent cycles of extension and retraction in the absence of any obvious triggering event, such as contact with a recipient cell. When made, such contacts were able to survive the shear forces felt by bacteria in liquid media. Additionally and unexpectedly, the extension and retraction of the F pili were accompanied by rotation about the long axis of the filament.

Silverman and Clarke (2010) published an excellent review by integrating long-established facts about the F pili biology during 40-year studies with modern tools of fluorescence and electron microscopy. The authors outlined an interesting idea that one function for F pili is to search a large volume around donor cells in liquid culture for the presence of other cells.

Recently, the F-encoded pili structures were visualized in the native bacterial cell envelope by *in situ* cryoelectron tomography (Hu et al. 2019).

FINE MAPPING OF POLAR PILI

The functional evaluation of F pili has always developed by structural comparison of the F pili with the polar pili of *Pseudomonas*. Figure 3.8 presents an example of such comparison.

During these studies in the 1970s–1980s, the attention was more and more concentrated on the polar *Pseudomonas* pili,

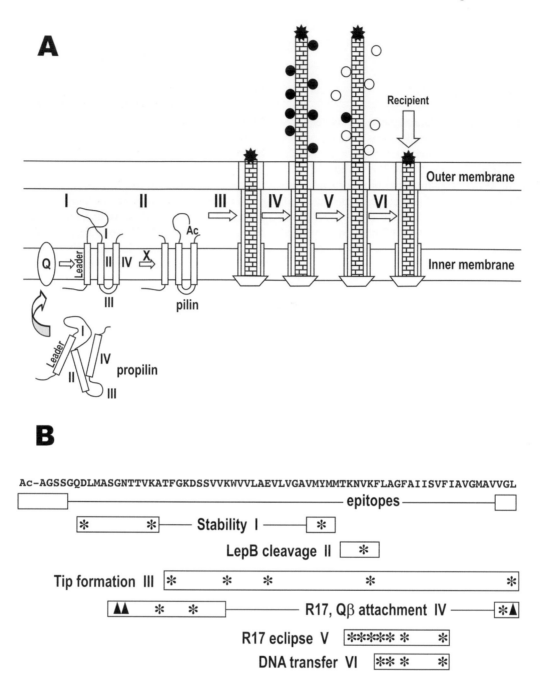

FIGURE 3.7 Pili and RNA phages. (A) The steps in pilus assembly affected by various mutations in pilin. The various mutations are divided into classes I–VI; class I affects the stability of pilin perhaps by affecting the insertion of propilin into the inner membrane via a TraQ-dependent step (D7A, K17A, Y42D); class II affects processing of propilin to pilin by LepB peptidase (ANVA); class III affects the assembly of the pilus tip at the cell surface possibly resulting in the release of unassembled pilin subunits into the medium (F20V, K28I, E34A, E34G, ANKA); class IV affects attachment of R17 phage [K17E (Frost and Paranchych 1988), A18E, D23G, K28A]; class V affects the eclipse of R17 phage (K46R, K46V, V48A, K49A, K49T, F50A, F54V, F60V, AKVA, RNVR); class VI affects DNA transfer at the level of either signaling that transfer should begin or DNA transport itself (K49A, K49T, F50A, F54V, F60V). The solid circles are intact R17 particles, whereas the open circles are eclipsed particles. The star represents the unknown structure at the pilus tip. Q and X refer to the propilin chaperone, TraQ, and the pilin acetylase, TraX respectively. (B) A summary of the positions of mutations that affect each of the six classes of mutations. * refers to positions of mutations isolated in this study; a closed triangle refers to the positions of important amino acid changes in F-like pilins. (Manchak J, Anthony KG, Frost LS: Mutational analysis of F-pilin reveals domains for pilus assembly, phage infection and DNA transfer. Mol Microbiol. 2002. 43. 195–205. Copyright Wiley-VCH Verlag GmbH & Co. KGaA. Reproduced with permission.)

first named FP pili, then sometimes in 1970s PSA pili, but later preferably the type IV, or TFP, pili. This structural success was based on the isolation of a multi-piliated and non-retractile mutant of *P. aeruginosa* K, namely the PAK/2Pfs

mutant (Bradley 1974a), that was many times more piliated than the wild-type strain and facilitated the preparation of large amounts of pure pili for biochemical studies (Frost and Paranchych 1977). It was found that wild-type and PAK/2PfS

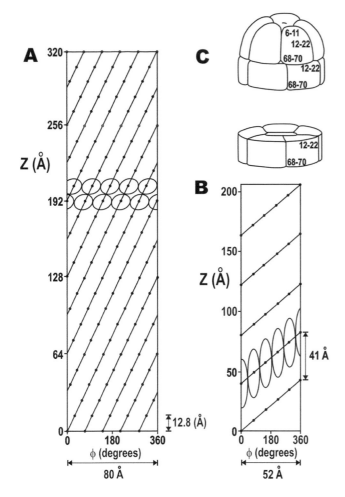

FIGURE 3.8 A presentation of the physical parameters of pili. (A) The F pilus, outer diameter 8 nm, inner diameter 2 nm. The F pilus can be represented as a five-start helix with a rise per subunit of 1.28 nm and a helix symmetry of 25 subunits in two turns of the helix and crystallographic repeat of 32 nm (134). (B) The polar pili of *Pseudomonas* represent a filamentous structure with a single-start helix, five subunits per turn, with a rise per subunit of 4.1 nm. The diameter of these pili is 5.2 nm (81, 200). The sketch of the subunits illustrates the packing of the subunits in the pilus; however, there is no information on the shape of the subunits in either system. (C) A sketch of a single layer of five subunits in the F pilus as well as a representation of a slightly different packing of the subunits at its tip. The numbers 6–11 represent the amino acids in the pilin subunit (70 residues) that form a domain thought to be involved in filamentous phage attachment; numbers 12–22 represent the site of Group I RNA phage attachment, and 68–70 represent the site of Group m phage attachment. Residues 6–11 may be exposed on the upper surface of broken pili, and the acetylated N-terminus may be exposed at the tip. (With kind permission from Springer Science+Business Media: *Bacterial Conjugation*, Conjugative pili and pilus-specific phages, 1993, pp. 189–221, Frost LS.)

pili were identical and that the mutation responsible for producing the multi-piliated state in PAK/2PfS was probably located outside the structural gene for pili (Frost and Paranchych 1977). However, it should be kept in mind that the *P. aeruginosa* strain PAK which gave this enormous start to the structural investigations did not support infection by the RNA phages (Bradley 1974a).

Soon, the amino-terminal amino acid sequence of the *Pseudomonas* pili protein revealed high homology to that of pilins obtained from *Neisseria gonorrhoeae* and *Moraxella nonliquefaciens* (Paranchych et al. 1978). Moreover, all three sequences demonstrated an unusual amino acid, N-monomethylphenylalanine, at the amino terminus (Frost et al. 1978) that paved a way for the elucidation of a special class of the similarly modified proteins involved in large macromolecular structures such as ribosomes, myofibrils, nucleosomes, pilins, or flagella (Stock et al. 1987). The N-terminal sequences of all of the methylated pilins were identical. However, the methylation reaction did not appear to be essential, since the pilin from *E. coli*, which contained a distinct N-terminal sequence, was not methylated (Hermodson et al. 1978).

A strong interest in the polar, or type IV, or N-monomethylphenylalanine pili was multiplied by discovery of an effect named *twitching motility*, as a function of *P. aeruginosa* PAO polar pili. In contrast to the *P. aeruginosa* PAK strain described above, the pili of the PAO strain were sensitive to the phage PP7 and the latter was used in this revolutionary paper (Bradley 1980a). The twitching motility was formulated as a mode of a primitive motion or flagella-independent surface translocation exhibited by *P. aeruginosa* and other bacteria on solid media and needed the presence of retractile, usually polar pili, while strains with no pili or non-retractile pili were unable to move, and anti-pilus serum also prevented twitching motility (Bradley 1980a). The subject of twitching motility was thoroughly reviewed at that time by Jørgen Henrichsen (1983), a pioneer in the field, and later by Kaiser (2000), Mattick (2002), and Daum and Gold (2018).

Furthermore, the type IV pilus was declared as the major virulence-associated adhesin of *P. aeruginosa* that promotes attachment of bacteria to epithelial cells and contributes to the virulence of this pathogen (for review see Hahn 1997). This led to an idea for the construction of the pilin-based anti-*Pseudomonas* vaccine (Hahn et al. 1997).

At the end of the 1970s, it was discovered that the *P. aeruginosa* PAK and PAO strain pili (although behaving differently from RNA phage infection) had very similar amino acid compositions and contained a number of homologous regions including the same sequence of 20 amino acids at the N-terminus, which was extremely hydrophobic and was strikingly homologous, as stated above, to the N-terminus of common pili isolated earlier from *N. gonorrhoeae* and *M. nonliquefaciens* (Paranchych et al. 1979). X-ray diffraction studies showed that the polar pili of *P. aeruginosa* strains PAK and PAO are hollow cylinders with 52 Å outer diameter and 12 Å inner diameter, and they form a helical array with 4.06–4.08 units per turn of a basic helix that had a pitch of 40.8 Å for the strain PAK pili and 41.3 Å for the strain PAO pili at 75% relative humidity (Folkhard et al. 1981). The first amino acid sequence of the PAK strain pilin, consisting of 145 amino acid residues and corresponding to molecular mass 15,082, was determined by Sastry et al. (1983).

Thanks to the known N-terminal sequence of the pilin, the further elucidation of the pilins was stimulated by the cloning and sequencing of the appropriate chromosomal

Pseudomonas gene of the PAK strain (Pasloske et al. 1985) and expression of the cloned gene in *E. coli* without, however, the assembly into pili (Finlay et al. 1986). The sequencing of pilins from PAO (Sastry et al. 1985) and other (Johnson et al. 1986) *Pseudomonas* strains, as well as expression of the PAK pilin mutants with their insertion into cytoplasmatic membrane of *E. coli* (Pasloske et al. 1988), followed.

Finally, a gene, *fimA,* encoding the pilin was located in the *P. aeruginosa* PAK chromosome (Hobbs et al. 1988). Then, three genes: *pilB, pilC,* and *pilD,* which are required for biogenesis of the polar *P. aeruginosa* pili, were identified being adjacent to the *pilA,* or *fimA,* the pilin structural gene (Nunn et al. 1990). All available data made it possible to conclude that the type IV, or the type IV methylphenylalanine pilins, were synthesized as prepilins, containing similar N-terminal six- or seven-amino-acid cationic sequences that were absent from the mature pilin subunit. These leader sequences were notably shorter than the typical signal sequences of *E. coli* pilins. The export and assembly of type IV pilins was accompanied by cleavage of the short leader peptide and methylation of the newly formed N-terminal phenylalanine (Nunn et al. 1990; Strom and Lory 1991).

A bit surprisingly, *E. coli* contained a set of genes homologous to those involved in protein secretion, DNA uptake, and the assembly of the type IV pili (Hobbs and Mattick 1993; Whitchurch and Mattick 1994). However, these genes were not necessary for the assembly of the conjugative F pilus, since they did not influence infection by the phage MS2.

The first high resolution (1.63 Å) crystal structure of the *P. aeruginosa* strain PAK pilin was resolved by Hazes et al. (2000) for the N-terminally truncated molecule. Craig et al. (2003) presented crystal structure of the full-length *P. aeruginosa* strain PAK pilin. An exhaustive review on the type IV pili structure was published simultaneously (Craig et al. 2004).

Furthermore, a new system of nomenclature was proposed, in which *P. aeruginosa* type IV pili, or TFP, were divided into five distinct phylogenetic groups, classifying the above-described PAO, PAK, and some other pilins to group II (Kus et al. 2004). Group II was formed by pilins which were the smallest among the five groups and the only ones lacking associated accessory proteins, making them an exception in the *P. aeruginosa* pilin repertoire (Kus et al. 2004).

Cryoelectron microscopy reconstruction the type IV pili, or T4P, of the *P. aeruginosa* PAK was performed at ~8 Å resolution (Wang F et al. 2017). Figure 3.9, taken from this paper, illustrates these structural findings.

Nowadays, modern studies on the complicated machinery of the type IV secretion systems (T4SSs) revived interest in the RNA phage infection which could be mediated by the T4SS (Christie et al. 2014). Electron cryotomography, which provides 3D views of the cell ultrastructure and macromolecular machines in their native context and unprecedented detail, will play a central role in further structural studies on the subject (Oikonomou and Jensen 2017).

Finally, a pilin region that determined the *P. aeruginosa* host range for the phage PP7 was mapped by Kim et al. (2018). First, it was detected that only the strains with the type IV pili, or TFP, pilin, which belonged to a specific subset of group II (named by authors the group IIa), were susceptible to PP7. Second, the coexpression of the PAO (group IIa) pilin rendered all the strains susceptible to PP7 (with the exception of the strains with group I pilin). Thirdly, site-directed and random mutation analyses of the PAO pilin led to identification of a pilin mutant in the β1−β2 loop (Gly96Ser) that is fully functional but resistant to PP7 infection. This study provided a structural background for the difference in the PAK (phage-resistant) and PAO (phage-sensitive) strains (both belonging to the group II) in their attitude to the PP7 infection (Kim et al. 2018). It was concluded that the responsible determinants are associated with the β1−β2 loop region that is flexibly protruding among the highly conserved structure of group II pilins. Therefore, this was the first study that defined a molecular pilin determinant that is responsible for the host spectrum.

An exhaustive review on the conjugative pili and their phages was published by Arutyunov and Frost (2013). Remarkably, the phage R17 was involved in the recent study that unveiled the novel role of F pili by the import of contact-dependent growth inhibition toxins (Beck et al. 2014).

CAULOBACTER AND *ACINETOBACTER* PILI

The first characterization of the *C. crescentus* pili, which occurred during a short period of the cell cycle and were only present at the flagellar pole of the swarmer cell, revealed their amino acid composition and molecular mass of 8500 (Lagenaur and Agabian 1977). The *C. crescentus* pilin was purified extensively and characterized biochemically (Smit et al. 1981). In fact, two pilins were compared from *C. crescentus* strains CB15 and CB13, both demonstrating similar sensitivity to the caulophage φCB5 and serological relatedness: there were significant differences in the amino acid composition, molecular mass, and other physical properties of the pilins. Moreover, the amino-terminal sequence of about 40% of the CB15 pilin molecule was determined and no homology to any of the other pilin sequences, as well as no indication of an unusual NH$_2$-terminal amino acid in *Caulobacter* pilin, were found (Smit et al. 1981).

A cluster of seven genes, including the major pilin subunit gene *pilA,* was mapped and regulation of their expression was elucidated (Skerker and Shapiro 2000). The pili genes encoding the appropriate components of the *Caulobacter* type IV pili secretion apparatus were described in detail by Harding et al. (2013) and Christen et al. (2016). The latter study employed the pili-specific tailed DNA phage φCBK (Agabian-Keshishian and Shapiro 1970) as a selection driver, but not *Caulobacter* RNA phages. However, it was established before that a piliated *Caulobacter* strain was sensitive to both φCBK and the RNA phage φCB5 (Lagenaur and Agabian 1977). Thus there is a good reason to believe that the susceptible pili are the same for both phages.

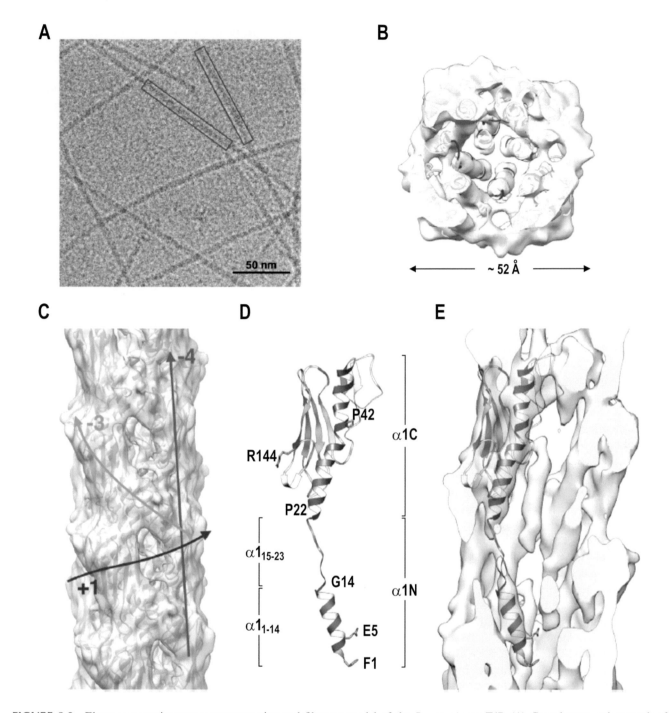

FIGURE 3.9 Electron cryomicroscopy reconstruction and filament model of the *P. aeruginosa* T4P. (A) Cryoelectron micrograph of *P. aeruginosa* T4P. Red rectangles indicate boxed pilus filaments. (B) Slice of the PaK pilus reconstruction and filament model viewed along the filament axis. (C) Side view of the PaK pilus reconstruction and filament model. Colored arrows indicate the paths of the (+)1-, (−)3-, and (−)4-start helices. (D) Ribbon representation of the PilA subunit model generated by fitting the protein into the cryo-EM density in three segments: the globular domain and α1:1–14 were fit separately as rigid bodies from the PilA crystal structure PDB: 1OQW, and α1:15–23 was modeled in an extended conformation. α1 is purple and the remainder of the globular domain is yellow. This subunit model was used to generate the PaK pilus filament model. (E) Slice through the cryo-EM density map showing a single PilA subunit. (Reprinted with permission of Cell Press from Wang F et al. *Structure*. 2017;25:1423–1435.)

Recently, Ellison et al. (2019) performed transposon sequencing of mutant *C. crescentus* libraries infected with the phage φCB5 to identify genes required for the phage infection. They found that the φCB5 infection, in contrast to the φCBK ones, was 75% prevented when pilus retraction was obstructed.

The *Acinetobacter* pili, which are responsible for the acinetophage AP205 uptake, are organized in accordance with the type IV pili principles (Palmen and Hellingwerf 1997; Wilharm et al. 2013; Piepenbrink et al. 2016; Leong et al. 2017). In fact, a gene *comP*, which was essential for the natural transformation of an *Acinetobacter* sp. strain, encoded a 15-kDa polypeptide that displayed significant similarities to the type IV pilins (Porstendörfer et al. 1997). However, no one has yet used the phage AP205 in the studies of the *Acinetobacter* pili.

HOSTS

ENTEROBACTERIACEAE

All *Enterobacteriaceae* family members that possessed machinery to build up the F- or F-like pili, namely by a plasmid from the IncFI-IncFIV group or by the Hfr status resulting from the integration of the plasmid into bacterial chromosome, were able to support the RNA phage infection. Thus, transfer of the *E. coli* F agent to other bacteria including *Salmonella*, *Shigella*, and *Proteus* conferred sensitivity to infection (Brinton et al. 1964). Transmission of F′*lac* plasmid from *E. coli* to bacteria of a large number of strains belonging to the genus *Erwinia* led to the sensitivity of the latter to the phages f2 and Qβ, as well as to the DNA phage f1 (Prokulevich and Fomichev 1978). Nevertheless, *E. coli* remained practically the only classical host for the RNA coliphages.

In the early 1960s, at the very beginning of the RNA phage era, it was recognized that the DNA phages, e.g., the phage λ, were modified in their host range by growth on particular bacterial hosts (Arber and Dussoix 1962; Luria 1962). Thus, one of the first goals of the RNA phage pioneers was the verification of this obstacle. According to Zinder (1965), "the two phage-host systems which modify the phage λ were tested with f2. The instances studied were phage grown on *E. coli* K as compared with phage grown on *E. coli* C or *E. coli* K (P1). In neither case was f2 restricted or modified. Clearly, the specific nucleotide sequences that respond to modifying systems are either not reactive or do not exist in f2."

The next general problem was a specific clarification of the host range term, since two typical situations might occur: (i) a phage is capable of forming negative plaques on bacteria by the soft-agar layer method and multiplies during cultivation of infected cells in a liquid culture, and (ii) a phage is unable to form negative plaques, but multiplies by infection of cells in a liquid culture. Both situations were regarded as necessary and sufficient to nominate a strain in question as a host for RNA phages.

The first systematic study of the RNA phage host range specificities was performed by Dennison and Hedges (1972), who found some phage- and host-defined peculiarities by plating of the representatives of the four serogroups I–IV on *E. coli* harboring F plasmid as a reference point and various derepressed R plasmids. However, the major conclusion of this study was the fact that the pseudomonaphage PP7 failed to form plaques on any of the strains of *E. coli* (Dennison and Hedges 1972).

It is worthy of mention that the phage MS2 was used by the construction of an improved genetic map of *E. coli*, as an instrument to selectively kill donor bacteria (Taylor and Thoman 1964).

PSEUDOMONADACEAE

The pseudomonaphages 7s and PP7 that were dependent on the *Pseudomonas* chromosomally encoded polar pili demonstrated a very narrow host range, which was sensitive to small differences in the pili structure, e.g., among phage-sensitive PAO and phage-resistant PAK strains (Paranchych et al. 1979; Folkhard et al. 1981; Kus et al. 2004).

The above-postulated problem of the host range definition appeared early in the case of the phage 7s that did not form negative plaques on the *P. aeruginosa* Ps-7 strain, although it was isolated from this strain, but was titrated well on the *P. aeruginosa* strain 1 (Feary et al. 1964).

Opposite to the 7s and PP7, the phage PRR1, which was dependent on pili encoded by the IncP-1 group plasmid, remained coliphages by its infection habits. The high promiscuity level of the appropriate IncP plasmids, namely R1822, RP4, RP1, and their derivatives, has demonstrated the ability to be transferred to a variety of species including *Enterobacteriaceae*, soil saprophytes as *Acinetobacter*, *Neisseria perflava*, and photosynthetic bacteria (Olsen and Shipley 1973).

Highly remarkably from the evolutionary point of view, an idea that the phage PP7 might increase *P. aeruginosa* diversity was experimentally approved (Brockhurst et al. 2005). Thus, the phage PP7 drove cycles of morphological diversification in host populations of *P. aeruginosa* due to the *de novo* evolution of small-rough colony variants that coexisted with large diffuse colony morph bacteria, while, in the absence of the phage, bacteria only displayed the large diffuse colony morphology of the wild-type (Brockhurst et al. 2005).

A present view on the diversity of the *Pseudomonas* phages including both PP7 and PRR1, with special attention to emerging studies in coevolutionary and in therapeutic settings, is presented by Ceyssens and Lavigne (2010).

CAULOBACTERACEAE

The generic specificity of the caulophage φCB series was characterized as an absolute, and the RNA caulophages emerged as a completely isolated group with respect to host/phage interactions (Schmidt and Stanier 1965). The caulophages were tested on *E. coli*, *Xanthomonas*, *Erwinia*, *Pseudomonas*, and, conversely, phages active against members of the *Pseudomonas*, *Xanthomonas*, and *Erwinia* groups were tested on a large number of the *Caulobacter* strains with completely negative results (Schmidt and Stanier 1965). However, different RNA caulophages showed specificity to the different *Caulobacter* strains: *C. crescentus*, *C. bacteroides*, or *C. fusiformis*. Moreover, no overlap of the host specificity among serologically distinct caulophages was observed (Schmidt 1966).

It is noteworthy that the RNA caulophages remained highly specific also within the *Caulobacteraceae* family: none of the numerous caulophages was active against any strain of *Asticcacaulis*, a second genus of the *Caulobacteraceae* family (Schmidt and Stanier 1965).

A special feature of the caulophages was the fact that they did not multiply in segregated populations of stalked caulobacters. The very infrequent occurrence of pili on stalked caulobacters and the poor adsorption of the caulophages to the stalked caulobacters provided a possible basis for the explanation of this observation. The stalked caulobacters may represent a form relatively resistant to the caulophages, which were, however, able to infect the piliated swarmer before it had undergone stalk development (Schmidt 1966). This feature was used to elaborate a differentiation assay that was based on the fact that the caulophage receptor sites were synthesized at specific times in the cell's life cycle and the phage φCB5, labeled with radioactive adenine, was retained on Millipore filters only if combined with cells (Shapiro et al. 1971). Since φCB5 specifically combined with the pili present on swarmer cells (Schmidt 1966), the swarmer cell lost the ability to retain labeled phage as it changed into a stalked cell and pili were lost (Shapiro and Agabian-Keshishian 1970). Thus, the RNA caulophage adsorption assay reflected morphogenic changes occurring in *Caulobacter* and was used to monitor simply and accurately the transition in synchronized cultures of swarmer to stalked cells (Shapiro et al. 1971).

The two phages that were tested from the caulophage φCp series (Miyakawa et al. 1976) demonstrated the same high host range specificity, similar to those of the φCB5 group by Schmidt and Stanier (1965).

It should be stressed that *Caulobacter*, an aquatic bacterium, has very wide occurrence and may present more discoveries in the search for novel RNA phages. One of the major supporting arguments for this thesis is the fact that it was relatively easy to isolate caulophages from sewage and river water collected at different locations in Japan (Miyakawa et al. 1976).

MORAXELLACEAE

No special studies have been undertaken on the host range of the *Acinetobacter* phage AP205.

SPHEROPLASTS

It was demonstrated at the very beginning of the RNA phage era that the spheroplasts, sometimes termed protoplasts, resulting from lysozyme-EDTA treatment of bacterial cells (Guthrie and Sinsheimer 1960) and prepared from both male and female *E. coli*, as well as from *Salmonella*, *Proteus*, *Shigella*, and *Aerobacter* cells, were susceptible to infection with the purified phage RNA (Davis et al. 1961; Fouace and Huppert 1962; Knolle and Kaudewitz 1962; Haywood and Sinsheimer 1963; Paranchych 1963; Engelhardt and Zinder 1964; Taketo et al. 1965; Hamon and Peron 1966). The spheroplast infectivity assays for the phages MS2 (Davis et al. 1964; Strauss 1964; Strauss and Sinsheimer 1967; Taketo and Kuno 1969), M12 (Benzinger et al. 1967; Hofschneider and Delius 1968), and Qβ (Pace and Spiegelman 1966b) have been elaborated in detail.

The use of DEAE-dextran and freezing permitted spheroplasts to be stored for periods up to 80 days with constant activity and high efficiencies (Merril and Geier 1970).

Further technical improvements concentrated on the comparison of different methods of the spheroplast preparation (Grabovskaia and Golubkov 1975; Rymar' and Kordium 1976).

The same lysozyme-EDTA treatment rendered *P. aeruginosa* cells susceptible to infection by the phenol-extracted PP7 RNA, where the infected spheroplast yield was directly proportional to the RNA concentration over a tenfold range (Weppelman and Brinton 1970). Then, sequential steps of the spheroplast infection from the ribonuclease-sensitive to the ribonuclease-resistant status of the infecting RNA were identified in this study. Moreover, the spheroplasts prepared from the pili-less *P. aeruginosa* mutant, which was derived from the piliated host and found to be phage-resistant and unable to adsorb phage, demonstrated susceptibility to infection by the naked PP7 RNA with equal efficiency as the spheroplasts prepared from the sensitive strain (Weppelman and Brinton 1971).

After use of the lysozyme-EDTA treatment for spheroplast preparation, a simple treatment of cells with 0.05 M $CaCl_2$ was employed to transfect Qβ RNA into the cells (Taketo 1972). Nevertheless, the author regarded his procedure as inferior to that of the lysozyme-EDTA spheroplasts and suggested regarding both techniques as complementary to each other in transfection assay. Furthermore, cells of *E. coli* were efficiently transfected with Qβ RNA by electroporation and the latter method was found superior to the conventional spheroplast techniques (Taketo 1989). It remains intriguing that the spheroplast-like L-forms of *E. coli* were obtained by penicillin treatment (Androsov and Levashev 1976).

4 Physiology and Growth

Art and science have their meeting-point in method.

Edward George Bulwer-Lytton

Biology is the only science in which multiplication means the same thing as division.

Ancient matter

The height of the special interest in the early steps of phage infection occurred in the second half of the 1960s–first half of the 1970s. It was the time when the three general lines of the RNA phages belonging to the coli-, pseudomona-, and caulophages developed almost simultaneously. For this reason, the available information on their physiology and growth is combined here under general physiological, but not taxonomical, headings, since no evident physiological differences have been observed until now among these three well-studied phage groups.

The three early steps: (i) attachment, or phage adsorption, (ii) ejection, or eclipse, and (iii) penetration of the phage genome into the cell, followed by (iv) nucleic acid and protein synthesis, or replication in the strict sense, and finally followed by the late steps:(v) virion assembly, and (vi) phage release were systematized and classified mostly by William Paranchych's team on the phage R17 (Paranchych and Graham 1962; Paranchych 1966; Danziger and Paranchych 1970a,b; Paranchych et al. 1970, 1971; Paranchych and Krahn 1971; Krahn et al. 1972; Paranchych 1975; Reynolds and Paranchych 1976; Wong and Paranchych 1976a,b,c), by Raymond C. Valentine and colleagues on the phage f2 (Ippen and Valentine 1965; Valentine and Strand 1965; Valentine and Wedel 1965; Valentine et al. 1965; Ippen and Valentine 1966, 1967; Valentine 1966; Wendt et al. 1966; Silverman et al. 1967a,b; Silverman and Valentine 1969; Valentine et al. 1969a,b), and by Peter Knolle (1964b; 1967b,c,e,g) on the phage fr.

William Paranchych, one of the great RNA phage pioneers, wrote in his classical chapter in Norton D. Zinder's *RNA Phages*:

> It is perhaps ironical, in view of our present level of understanding of the RNA phage replicative processes, that the molecular mechanisms involved in the adsorption and penetration stages of RNA phage infection are relatively poorly understood. This irony becomes even more acute when it is realized that the total sequence analysis of all the structural components of the phage is almost at hand, and that the synthesis and assembly of infectious particles in cell-free systems is on the edge of becoming a reality. (Paranchych 1975.)

It becomes more ironic when we need to accept that the situation in the physiological sense has not changed markedly so far, after more than 40 years.

According to Paranchych's opinion, the "early efforts to study the RNA phage attachment and penetration processes were seriously hindered by an inexplicable lack of experimental reproducibility, which seemed to be an inherent aspect of the RNA phage-host system, and the problem that only 5%–10% of phage populations were infectious" (Paranchych 1975). As described in the previous chapter, the specific role of F pili in the adsorption and penetration processes found by Crawford and Gesteland (1964) and then quickly confirmed by Brinton et al. (1964), by Valentine and Strand (1965), and by Knolle (1967b), was at least surprising for that time, since it differed RNA phages from other phages known at that time that adsorbed directly to the bacterial cell wall. It is noteworthy, however, that in their classical paper, Crawford and Gesteland (1964) still assumed that, by a high multiplicity of infection and by a highly reversible adsorption process, the productive infection might occur only after direct attachment of a few phages directly to the cell wall. Extreme fragility of F pili to such routine microbiological procedures as the vigorous aeration of bacterial cultures or the routine washing of bacterial cells led to irreproducibility that did appear in the early studies (Paranchych 1975).

Although William Paranchych, Charles Brinton, and many other prominent scientists contributed enormously to the functional and structural elucidation of the F pili, type IV pili, and other bacterial filaments and, moreover, the structural studies on the subject are running still to date (see Chapter 3), the fine mechanisms of the phage-pilus interaction and of the precise role played by each of the structural components of the phage—coat protein, A-protein, and RNA—remain a bit mysterious in molecular terms.

The early studies concentrated mostly on the physiological aspects of the phage-pilus interaction reviewed below.

ATTACHMENT OR ADSORPTION

Valentine and Wedel (1965) have divided the early infection processes into four steps: (i) simple adsorption of the phage to the F pilus, (ii) injection of phage RNA into the core of the F pilus, (iii) transport of the RNA down the core of the pilus, and (iv) final penetration of the RNA into the cell. The proposed steps (iii) and (iv) were usually combined into one, namely the penetration step. The adsorption step occurred even at 0°C and an integral relationship between cell and F pilus was necessary to proceed further, since free F pili fragments did not initiate subsequent stages after the attachment (Valentine and Wedel 1965). The attachment appeared as a diffusion-limited reaction requiring no energy of activation (Knolle 1967g). The next infection steps after the attachment could be blocked also by Mg^{2+} deficiency, since the latter or other divalent cations were absolutely necessary for the next ribonuclease-sensitive steps (Paranchych 1966).

There was no difference between the two media: broth or phosphate-buffered saline was observed with respect to speed or efficiency of the phage MS2 adsorption at 37°C. Although approximately 90% of the phage particles were adsorbed at the end of 15 min in this study, the adsorption was reversed by series of washes (Rappaport 1965).

Just before F pili were identified as an RNA phage target, it was known that the replication of RNA phages is dependent upon divalent cations, although the function of the divalent cations in the infective process has remained a mystery. Loeb and Zinder (1961) reported that the phage f2 infection is abortive in the absence of added Ca^{2+}, but that calcium is not needed for phage adsorption. Davis and Sinsheimer (1963), on the other hand, indicated that divalent cations are required for the phage MS2 attachment. Furthermore, Paranchych (1966) found that divalent cations are necessary for penetration, but not for the attachment and ejection steps. In contrast to the filamentous DNA phage f1, the adsorption rate of which was markedly reduced by divalent Zn^{2+} ions in concentrations of 10^{-3} M, the latter did not affect the adsorption of the phages R17 and MS2 (Ou and Anderson 1972b). Finally, concerning the role of the ions in phage attachment, the latter was concluded to be independent of ion specificity, requiring only an adequate ionic strength (about 0.1 M or greater) for maximum adsorption, and the addition of a divalent salt, at optimal levels of any monovalent salt, caused no further stimulation of phage-pili complex formation (Danziger and Paranchych 1970a).

The adsorption of the ^{32}P-labeled phage R17 to *E. coli* Hfr cells was shown to be a rapid, temperature-independent, and reversible process in the paper describing discovery of the phage R17 (Paranchych and Graham 1962). Furthermore, the R17 attachment process at 4°C was described as reversible, but at 37°C the rapid interaction of particles with host cells was followed by a very rapid dissociation of the phage-cell complex (Paranchych et al. 1970). Upon separation from the host cell, the phage particles did not dissociate into coat protein subunits, but remained intact as empty, or partially empty, capsids. Totally, about 50% of the phage population was defective in their ability to adsorb to F pili, the final result being that they reacted only weakly with the host cell. Although this brief interaction caused sensitization of most of the phage RNA to ribonuclease, release of the RNA from the particles did not occur. Therefore, the apparent low infectivity (about 10%) of phage preparations was explained almost entirely on the basis of the attachment capability of the various classes of phage particles (Paranchych et al. 1970; Paranchych 1975).

Quantitative measurements of the phage-pili complexes were started after the introduction of the simple membrane filter retention method (Valentine and Strand 1965). The idea began with the observation that cell-free filtrates of F$^+$ strains of *E. coli* "inactivated" RNA phage titer as much as 50%–75% and the active "inactivation" agents have been identified in the electron microscope as F pili. Since then, the phage-pili complexes were easily assayed because of their retention by membrane filters. Contrary to the *in vivo* observations

presented above, the phage-pili complex formation obligatory needed Ca^{2+} ions. The phage-pili attachment was fully reversible, since blending of the "inactivated" phage-pili complex reversed the phage "inactivation," while, on further standing, the "inactivation" again took place (Valentine and Strand 1965). The attachment as such did not liberate phage RNA, since treatment of the free phage-pili complexes with ribonuclease neither destroyed the viability of the phage nor disrupted the complex, while similar treatment of phage-cell complexes resulted in complete hydrolysis of the infecting RNA (Valentine and Strand 1965). Therefore, the adsorption of phage as the first step in phage infection was clearly differentiated from the next step, namely RNA injection, or ejection, or eclipse. Use of a radioactive ^{32}P-labeled phage allowed quantitative measurements of the phage-pili complexes (Ippen and Valentine 1965): the attachment studies, using highly purified radioactive phage, showed that total saturation of F pili usually resulted in the binding of about 1000 physical particles per bacterium (Paranchych 1975).

It is noteworthy that at that time a fraction that increased infectivity, i.e., attachment, of the phage f2 was isolated from uninfected *E. coli* cells (Chroboczek 1978). Although this fraction consisted of proteins, lipids, carbohydrates, and unidentified material, the lipids were suspected to assure the infectivity-restoring activity.

Finally, in order to clarify the phage-pilus interaction from the physical point of view, the kinetics of the binding reaction of the purified MS2 phage to a highly purified F pilus of *E. coli* were studied by the membrane filter assay (Date 1979). The binding reaction was a simple bimolecular reaction and was reversible, as mentioned above. The rate of dissociation k_d of the MS2-phage–F pilus complex was very slow ($k_d = 4.6 \times 10^5$ s^{-1}) and followed first-order kinetics with a half-life of 4.2 h at 30°C in the standard buffer. The dissociation rate was rather insensitive to temperature but became more rapid at high ionic strength or at basic pH. In a 0.25-M ionic strength buffer, the half-life of the complex was about 1.0 min. The rate of association was very fast and followed second-order kinetics with the rate constant for association k_a being 8×10^7 M^{-1} s^{-1} at 30°C. The equilibrium constant K was determined as 6×10^{-13} M. The adsorption of the phage MS2 to the F pilus was a favorable free energy change; the driving force for the binding reaction came from a large entropy increase. The rate of association was almost insensitive to ionic strength but slightly sensitive to pH or temperature. Monovalent cations could also promote the binding reaction as well as divalent cations, but the complex formed with monovalent cation was unstable. The study of the kinetics of dissociation suggested that there were two types of interaction between MS2 phage and F pilus: one was a strong interaction formed with divalent cations and the other was a weak one formed with monovalent cations. The physical nature of the bonds involved in the former and the latter were characterized as mainly electrostatic and non-electrostatic, respectively (Date 1979). Therefore, the possible role of mono- and divalent cations in the phage-pili complex formation studied before and described above was fully clarified.

Not long ago, the adsorption rate constants α for the phages C-1 and Hgal1 were determined to be 8.27 ($\pm$0.997) $\times$ 10^{-11} and 3.44 ($\pm$0.126) $\times$ 10^{-11} cells^{-1} mL^{-1} min^{-1}, respectively (Kannoly et al. 2012), whereas the adsorption rate constant of the phage Qβ attachment to *E. coli*, which was explained in terms of a collision reaction, was estimated to be 4 $\times$ 10^{-10} mL cells^{-1} min^{-1} by Tsukada et al. (2009).

The efficiency of the attachment was dependent on the structure of the pili: Paranchych (1975) compared the attachment capacity of the phage R17 using bacteria harboring representatives of each of the four serological types of F-like plasmid FI–FIV and concluded that the host cells' phage attachment capability was at least partly responsible for determining the phage sensitivity of the host strain. The pili were arranged by decreasing capability to adsorb phage: F > R1-19 $\geq$ R538-1 drd > R100-1. As for the possible specificity of the phage species by the attachment, the efficiency of the attachment was compared directly for five coliphages: R23, f2, Qβ, R34, and R40 by Hiroko Watanabe and Mamoru Watanabe (1970a).

Concerning pseudomonaphages, the phage PP7 attached to the *P. aeruginosa* FP pili which were found by Bradley (1966b) due to his work with this RNA phage and were later defined as polar pili, or type IV pili, or T4P, or TFP, as was described in more detail in the Chapter 3. Further strong evidence in favor of the proposal that FP pili were the genuine adsorption organelle for the RNA phages was presented by Weppelman and Brinton (1971): the spheroplasts that were prepared from pili-less *P. aeruginosa* cells were sensitive to infection by the naked PP7 RNA. Moreover, experiments with synchronously dividing cells demonstrated that the host was susceptible to infection only during a brief phase of the cellular growth cycle, beginning shortly before division (Weppelman and Brinton 1971). In contrast to the attachment of coliphages to cell-free F pili, the phage PP7 did not appear to attach to the FP pili that were removed from the host bacteria, possibly due to some physical denaturation of the free FP pill (Lin and Schmidt 1972).

It is noteworthy that the velocity constant of the adsorption of the pseudomonaphage 7s to the *P. aeruginosa* strain Ps-7 cells has been determined to have a value of 4.1 $\times$ 10^{-8} mL min^{-1} just before the phage attachment site on the cells was mapped (Feary et al. 1964).

The adsorption of the phage PRR1, which attached pili encoded by the promiscuous IncP-1 plasmids, was greatest by using *Pseudomonas* species as hosts, whereby 70%–90% of the phage were adsorbed in 10 min at 37°C with less than 20% adsorbed under similar conditions by using *Enterobacteriaceae* (Olsen and Thomas 1973).

The adsorption of the caulophages appeared to be similar to that of the RNA coli- or pseudomonaphages (Schmidt and Stanier 1965). Further detailed evaluation showed that the removal of the pili from the *C. crescentus* host by shear treatment before phage adsorption resulted in effective inhibition of phage adsorption (Schmidt 1966). While swarmer caulobacters frequently possessed pili which were located at the flagellated pole of the organism, stalked caulobacters were almost always devoid of these structures. The adsorption experiments with segregated populations of caulobacters indicated that stalked caulobacters adsorbed RNA phages very poorly, while in swarmer populations and unsegregated cultures the RNA phages were efficiently adsorbed (Schmidt 1966). Remarkably, adsorption of the phage φCB5 prevented the loss of pili at the proper time during development of *C. crescentus* (Sommer and Newton 1988).

EJECTION OR ECLIPSE

The adsorption, which was necessary to transfer the phage from the appendage to the cell surface but remained reversible, was followed by the next and the first irreversible infection step, namely ejection, or injection, or eclipse. The capability of successful ejection had a strong influence on the phage attachment to the intact cell-bound pili, since all above-described studies on the attachment were complicated by the real observation that all RNA phage preparations were heterogeneous with respect to the attachment function (Paranchych et al. 1970; Krahn and Paranchych 1971; Paranchych 1975). The main reason for this heterogeneity was the presence of a class of noninfectious particles, about 10% of the population, which lacked A protein and were unable to interact with F pili at all. However, the remainder of the phage population contained both infectious and noninfectious particles, showing a further heterogeneity in their affinity for F pili, with about 10% of the phage population being infectious, as stated above (Paranchych 1975). Both infectious and noninfectious particles of the remainder adsorbed to F pili, but the attachment of the noninfectious particles, which constituted approximately 80% of the phage population, was weaker (Krahn and Paranchych 1971). The differential attachment capacities of these two parts of the population remained not understood, with a guess only that the configuration of the A protein in these noninfectious particles was different from that in the infectious particles and less favorable in terms of promoting the phage attachment process (Paranchych 1975). Figure 4.1, taken from this review chapter of the famous *RNA Phages* book illustrates the attachment process of infectious and total phage R17 to *E. coli* HB11 at 4°C.

Thus, the productive ejection was connected with the presence of the functionally wrapped A protein. Jeffrey W. Roberts and Joan E. Argetsinger Steitz (1967) were the first to directly show by phage reconstitution that the A protein, which existed in a tight complex with the coliphage RNA, was absolutely required for phage infectivity. This requirement of the tight A protein–RNA complex must be realized on both the attachment and ejection steps (Steitz 1968a,b,c). A bit earlier, the absolute necessity of the A protein for the production of infective phage was demonstrated by studies of the defective particles synthesized during the growth of the A gene amber mutants of the phages f2 (Lodish et al. 1965), fr (Heisenberg 1966), and R17 (Argetsinger and Gussin 1966) in non-permissive cells. Such A protein–deficient mutants were unable to attach the host bacteria (Heisenberg and Blessing 1965; Lodish et al. 1965; Valentine and Strand 1965). In fact,

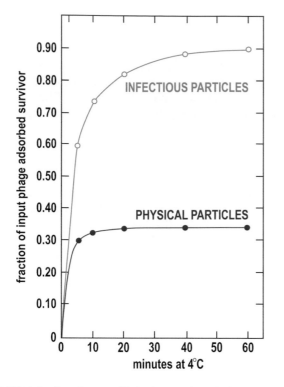

FIGURE 4.1 Attachment of infectious and total phage to *E. coli* HB11 at 4°C. Density of bacteria was 5×10^8/mL. Phage concentration was 5×10^9 particles/mL, of which 10% were infectious particles. The physical particles were detected by use of ^{32}P-labeled R17 phage. The infectious phage was determined by assaying for the remaining plaque-forming units in the supernatant solution of centrifuged samples of culture. (Redrawn with permission of Cold Spring Harbor Laboratory Press from Paranchych W. *RNA Phages.* Cold Spring Harbor Laboratory, NY, USA, 1975, pp. 85–111.)

Gary N. Gussin (1966) and Joan Argetsinger Steitz (1968c) were pioneers who suggested that the A protein might be an attachment organelle.

Furthermore, the A protein–RNA complex, which was reconstituted *in vitro* from purified RNA and A protein of the phage MS2 (Shiba and Miyake 1975; Iglewski 1976) and GA (Shiba and Miyake 1975), or isolated as a complex by acetic acid treatment from the phage R17 (Paranchych 1975; Reynolds and Paranchych 1976) and from the phage M12 (Leipold and Hofschneider 1975) was found infectious for intact *E. coli* cells. A similar complex was isolated also in the case of the phage Qβ (Shiba and Suzuki 1981).

Next, the phage attachment to cell-associated F pili (but not cell-free F pili) at 37°C was followed almost immediately by a loss of phage infectivity and the concomitant sensitization of the phage RNA to ribonuclease (Valentine and Wedel 1965; Knolle 1967e,g; Silverman and Valentine 1969; Danziger and Paranchych 1970b; Paranchych 1975). First, in connection with the ribonuclease sensitivity, the step following phage attachment was described as the ribonuclease-sensitive step (Valentine and Wedel 1965; Paranchych 1966), the injection step (Valentine and Wedel 1965; Silverman and Valentine 1969), the invasion step (Knolle 1967e,g), and the eclipse reaction (Paranchych 1966; Danziger and Paranchych 1970b). The term *ejection* was chosen by Paranchych (1975)

since it "describes the nature of the event more precisely than the other terms mentioned and avoids confusion with later events involving the injection of phage RNA from the F pilus into the host bacterium."

The three experimental approaches that have been devised to measure the RNA ejection reaction were reviewed exhaustively by Paranchych (1975). The simplest method involved the plaque assay procedure to detect the loss of phage infectivity: in order to prevent the detection of infected cells as infectious centers, the method used either introduction of ribonuclease into the culture to prevent cell infection (Paranchych 1966) or destroying the infected cells by lysing them with chloroform (Silverman and Valentine 1969), sonic vibration (Danziger and Paranchych 1970b), or freezing and thawing procedures (Knolle 1967e,g). The second method was based on the measurement of the amount of phage RNA that became converted to a ribonuclease-sensitive state (Valentine and Wedel 1965; Silverman and Valentine 1969; Paranchych et al. 1970; Paranchych et al. 1971). The third method involved the use of sucrose density gradient centrifugation procedures to examine the sedimentation properties of radioactively labeled phage that had been allowed to interact with host bacteria (Silverman and Valentine 1969; Paranchych et al. 1970; Paranchych et al. 1971).

In contrast to the attachment, the ejection was the energy-dependent step, since it occurred only at warm temperatures. By studying the effect of temperature on the rate of the ejection reaction, the values of 36 kcal/mole (Knolle 1967e) and 10 kcal/mole (Danziger and Paranchych 1970b) were obtained for the energy of activation of the RNA ejection process. However, these data, as well as numerous studies on the effect of metabolic inhibitors on the ejection process, must be interpreted with considerable caution. Thus, the loss of the ejection efficiency could be provoked by the loss of pili, since many metabolic poisons, as well as some physiological conditions, might lead to the rapid disappearance of F pili from the cell surface (Paranchych 1975). Some examples of the pili disappearance were presented in the Chapter 3. The same was true for the studies on the effect of temperature on the ejection step, since Novotny and Lavin (1971) have shown that sudden reductions in the temperature of a bacterial culture caused significant losses of cell-associated F pili. For this reason, it was important to study F pilus functions under experimental conditions which did not cause losses of the cell-associated F pili. Thus, the addition of the RNA phage to F-piliated bacteria was found to produce sudden losses in cellular nucleoside triphosphate ATP, GTP, UTP, and CTP levels (Paranchych et al. 1970). Paranchych et al. (1971) found that similar results were obtained under conditions where all stages of infection subsequent to the ejection step were prevented by the omission of divalent cations from the culture medium. Under these conditions, the ejection reaction led to the sensitization of phage particles to ribonuclease, but little or no penetration of phage RNA into the host cell occurred (Paranchych 1966; Silverman and Valentine 1969). Yamazaki (1969) has shown that the ejection reaction caused an abrupt inhibition of amino acid transport into the cell, suggesting that the RNA ejection process and the amino acid transport process might compete for a common membrane-associated energy source (Paranchych 1975).

The sensitivity to extracellular ribonucleases rendered by the ejection process provided the first indication that the uncoating of the phage RNA might be an extracellular event (Zinder 1963). Later, it was detected directly that the phage coat protein remained outside the host cell while the phage RNA was transferred into the cell (Edgell and Ginoza 1965). Moreover, the separation of the phage RNA and coat protein occurred almost immediately after phage attachment to the F pilus, and the coat protein was then released from the F pilus as an intact empty capsid (Silverman and Valentine 1969; Paranchych et al. 1970; Paranchych 1975). It is noteworthy that the characterization of different ribonucleases that would be able to prevent infection by the RNA phages M12, f2, Qβ, and PP7 remained actual until now, and the corresponding novel data were reported very recently (Sharipova et al. 2017).

Summarizing, the ribonuclease sensitivity and the protein A participation appeared as the main attributes of the ejection step. Concerning the fate of the A protein, a definite native linking of the latter to RNA was proposed in the second half of 1960s (Richelson and Nathans 1967; Oriel 1969; Kaerner 1970). Finally, it was found that the A protein and RNA penetrated the host cell together (Kozak and Nathans 1971; Krahn et al. 1972). In this connection, O'Callaghan et al. (1973a) proposed an interesting model for the RNA phage 3D structure which was consistent with the specific role of the A protein. This model, placing the latter on the surface of phage particles, was described in detail by Paranchych (1975). This structural problem was solved only recently when electron cryomicroscopy studies led to the final conclusion that the A protein actually replaces a single coat protein dimer within the symmetric phage particle and the virion actually contains not 180 but 178 coat monomers, or not 90 but 89 coat dimers (Dent et al. 2013; Koning et al. 2016).

As a result of the ejection process, the phage A protein was cleaved into two peptides of molecular mass 15 and 24 kDa (Krahn et al. 1972). It was speculated that this A protein cleavage reaction might play a key role as the triggering mechanism for the ejection process, or as a pili-catalyzed signal to initiate the unfolding of the tightly packed phage RNA (Paranchych 1975). Although some mechanisms involving the interaction of charged groups in the RNA molecule with mono- and divalent cations located within and without the virion were suggested (Valentine et al. 1969b), the nature of the driving force which may cause the unraveling of the long RNA molecule (approximately 1000 nm) remained unknown.

PENETRATION

Delivery of the phage RNA during the penetration step was always one of the most mysterious processes of phage infection. In 1975, Paranchych wrote: "The mechanisms involved in the movement of phage RNA along the F pilus and through the cell membrane are among the least understood in the overall infectious process" (Paranchych 1975). After more than 40

years, the level of understanding of these processes has not changed markedly. It remains unclear how the movement of RNA is accomplished by the pili and how far the pili and/or other cell components are involved in promoting the penetration of phage RNA through the cell membrane. As explained by the ejection, the A protein plays a critical role by the ejection/penetration steps and enters the cell together with phage RNA (Krahn et al. 1972). Nevertheless, successful infection can be achieved by transfection of the cell spheroplasts by pure phage RNA, without any signs of the A protein, although the specific infectivity of the R17 phage RNA-A protein complex was 35-fold greater than that of R17 phage RNA alone, but both preparations were ribonuclease-sensitive (Iglewski 1976).

The critical role of the A protein was supported further by the studies on the sensitivity of the penetration step to ribonuclease (Wong and Paranchych 1976a). It was found that the penetration of the phage RNA is affected by the ribonuclease concentrations as low as 0.1 μg/mL, while the penetration of the A protein was unaffected by ribonuclease concentrations as high as 20 μg/mL. Moreover, a significant fraction of the phage RNA remained resistant to the ribonuclease and gave rise to the penetration of intact RNA molecules that produced the expected number of infectious centers (Wong and Paranchych 1976a). The results were explained by an assumption that the A protein precedes the RNA during its entry into the cell and that a defined segment of RNA is protected from ribonuclease by association with the A-protein, or that certain sites on the F pilus are "protected" and allow the penetration of intact phage RNA even in the presence of ribonuclease. Thus, this interpretation supposed that the ejection and penetration stages could apparently proceed whether phage attachment occurred at the "protected" or "unprotected" sites on the F pilus: if the attachment occurred at an "unprotected" site, which was presumably somewhat removed from the pilus base, the RNA moiety was sensitive to ribonuclease and was completely degraded, while the "protected" sites were actual phage attachment sites at or near the base of the pilus which allowed the A protein–RNA complex to eject from the virion and penetrate into the cell without coming in contact with the extracellular medium (Wong and Paranchych 1976a). Preservation of the secondary structure of the phage RNA while being transported into the cell supported the idea that the RNA and A protein remained attached to the exterior surface of the F pilus during penetration (Wong and Paranchych 1976c). Further studies on the polarity of the R17 RNA penetration showed that the 3′-end of RNA, in conjunction with the A protein, acted as the pilot end during the penetration of phage RNA into the host bacterium (Wong and Paranchych 1976b). The site of interaction of the phage MS2 A protein on MS2 RNA was determined by analyzing the sequence of the RNA fragment which was released from the ribonuclease-resistant complex prepared by ribonuclease digestion of the RNA-A protein complex (Shiba and Suzuki 1981). At least two sites on MS2 RNA were protected by the A protein: one had a sequence which corresponded to the 5′-A-protein region of MS2 RNA at nucleotides 388–398 and the other corresponded to the 3′-untranslated region at nucleotides 3510–3520 (Shiba and Suzuki 1981).

Interestingly, the caulophage PRR1 was characterized as unusually sensitive to ribonuclease in the early stages of infection and behaved rather as coliphages with defective A proteins or other structural anomalies (Olsen Thomas 1973).

Meanwhile, it should be remembered that only 10% of the total phage population succeeded in releasing intact RNA molecules to host bacteria during the course of infection (Paranchych et al. 1970).

It is well established that the penetration, but not the attachment and the ejection steps, cannot occur without divalent cations Mg^{2+}, Ca^{2+}, Sr^{2+}, and Ba^{2+}, which were found equally effective in promoting R17 infection by concentration of 0.7 mM or greater, whereas no phage growth occurred in the presence of the cations Mn^{2+}, Zn^{2+}, Ni^{2+}, or Co^{2+} (Paranchych 1966). The impairment of phage RNA penetration which occurred as a result of a lack of divalent cations during the initial stages of infection cannot be reversed by the later addition of divalent cations to the medium. On the other hand, if divalent cations were removed from the medium subsequent to the initiation of infection, phage replication proceeded at a normal rate (Paranchych 1966). However, there were no experimentally supported ideas why these divalent cations were required, or whether they were required at the level of RNA transport along the F pilus or its penetration through the cell membrane (Paranchych 1975).

One possible clue to the resolution of the penetration mechanisms may be found in the above-mentioned studies (see Chapter 3) with transfer-deficient mutants carrying an amber mutation in the *traD* gene (Achtman et al. 1971). These *traD* mutants produced normal amounts of F pili, which were chemically identical to wild-type F pili (Brinton, personal communication, cited by Paranchych 1975), but were highly sensitive to the phage Qβ (and the F-specific filamentous DNA phages) and completely resistant to the phages of the f2 group (Achtman et al. 1971). Thus, while promoting the adsorption and the ejection reactions with all phages studied, the *traD* mutants demonstrated specificity during the penetration step (Achtman et al. 1971; Paranchych 1975). Thus, the *traD* gene product (as well as *traI* and *traM*) was suspected to participate in the formation of a membrane pore which might permit transfer of nucleic acids (Goldberg 1980). Nevertheless, Schoulaker-Schwarz and Engelberg-Kulka (1978, 1981, 1983) have found that a temperature-sensitive (*ts*) *traD* mutation permitted the penetration and translation of the phage MS2 RNA but did not permit MS2 RNA replication at 42°C, in contrast to the phage Qβ, which replicated well in these obstacles. Again, the phenomenon was explained by the different role of cell membranes in the replication of both phages.

It is established now that the TraD protein belongs to the type IV coupling proteins (T4CPs) that mediate multiple protein–protein interactions with cytoplasmic and inner membrane components of the secretion system and act as substrate receptors at the cytoplasmic entrance of the secretion channel (Lang and Zechner 2012). Generally, the T4CP

cumulatively senses an intracellular signal (substrate docking) and an extracellular signal (pilus bound by phage or a recipient cell) to coordinate a late stage assembly or gating reaction that enables bidirectional transmission of nucleoprotein substrates through the T4SS, or bacterial type IV secretion system (Lang et al. 2011). Figure 4.2, which is taken from Berry and Christie (2011), presents a modern view on the uptake of the phage R17 by F pili.

Historically, the penetration process was explained by functioning of pili: (i) like a hollow tube for the nucleic acid transport (Brinton 1965); (ii) as two parallel protein filaments (Brinton 1971), (iii) by retracting into the cell after receiving an appropriate stimulus and sequential depolymerization of pili subunits, which resulted in RNA phage being pulled to the base of the pilus, where it injected its RNA (Curtiss 1969; Marvin and Hohn 1969). The latter mechanism was supported by Bradley (1972c), who found by electron microscopy studies that the infection of *P. aeruginosa* with the phage PP7 resulted in an average reduction in pilus length of about 50%, and that phage particles were almost always at the bases of the pili. Experiments similar to the above were carried out by Paranchych et al. (1971) with the phage R17 and F-piliated *E. coli*. These studies also showed that the interaction of pili with the RNA phage resulted in an overall shortening of the pilus lengths. However, the shortening of F pili was accompanied by a corresponding increase of cell-free F pilus fragments in the medium, suggesting that the phage interaction with the F pili had resulted in a fragmentation of the F pili rather than retraction. The foregoing observations were also interpretable, however, in terms of retraction of old F pili followed by the outgrowth and release of newly synthesized F pili (Paranchych 1975). Reduction of F pili was observed also by infection with filamentous DNA phages f1 (Jacobson 1972) and M13 (O'Callaghan et al. 1973b). Nevertheless, Ou and Anderson (1970) demonstrated strong evidence against the retraction model by studies where cells were involved in bacterial conjugation and the exchange of genetic information occurred between pairs of cells which were never visibly in close contact, suggesting that the donor DNA is transferred to the recipient via the F pilus in the absence of the F pilus retraction (for qualified discussion see Paranchych 1975).

Concerning kinetics of the phage RNA entry, the penetration was shown to proceed very rapidly during the first 5 min of the infectious period, but that it then slowed down to a very low rate for the duration of the infectious cycle (Brinton and Beer 1967; Krahn et al. 1972). It is highly remarkable in this aspect that F pili underwent an irreversible change during the initial 5-min period of interaction with RNA phage, namely, in the period when both a decrease in the level of cellular nucleoside triphosphates and a release of F pilus fragments into the surrounding medium were occurring (Paranchych et al. 1971). Although the molecular basis of these F pilus changes was not yet understood, the changes were probably responsible for the abrupt decrease of the phage RNA entry into the host cell at approximately 5 min postinfection (Paranchych 1975).

Based on kinetic studies, it was concluded that only one or a few phage particles could be injected into a cell by a single

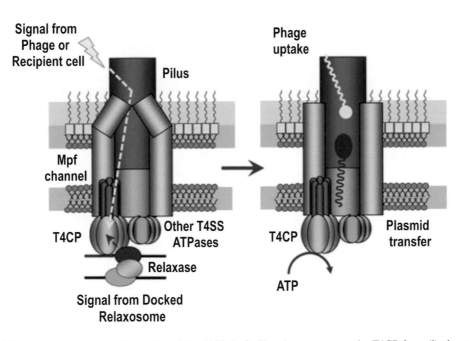

FIGURE 4.2 Model for contact-dependent stimulation of the T4SS. Left: Signals converge on the T4CP from (i) phage binding or contact with a recipient cell (yellow dashed arrow) and (ii) a docked relaxosome (red dashed arrow). Right: The T4CP in turn stimulates channel gating through ATP hydrolysis and energy-induced conformational changes in the T4SS. The gated channel serves as a conduit for nucleoprotein substrates: (i) phage uptake, e.g., R17 protein A bound to RNA (yellow); or (ii) plasmid transfer, e.g., R1 TraI relaxase bound to ssDNA (red). The DNA strand of plasmid R1 translocated through the T4SS is depicted in red in the left panel. (Berry TM, Christie PJ: Caught in the act: The dialogue between bacteriophage R17 and the type IV secretion machine of plasmid R1. *Mol Microbiol.* 2011. 82. 1039–1043. Copyright Wiley-VCH Verlag GmbH & Co. KGaA. Reproduced with permission.)

F pilus (Brinton and Beer 1967). Conversely, Krahn et al. (1972) have found that a maximum of about 30–35 phage equivalents of RNA entered a single cell during the 5-min penetration period, although phage inputs of several thousand physical particles per cell were required to achieve such high levels of penetration. At lower multiplicities of infection (100 physical particles per cell or less), there was roughly a 1:1 relationship between the number of phage equivalents of injected RNA and the input concentration of the potentially infectious particles, suggesting that there was little or no restriction of phage RNA entry up to a level of 10 phage genomes per cell (Paranchych 1975).

Concluding the penetration step, it seems noteworthy to cite Maxime Schwartz who wrote at the end of the most active phage search decade:

> Very little information is available regarding the first irreversible step in phage adsorption. The best-documented cases are those of the pilus-dependent phages, where this step seems to be a cleavage of the adsorption protein on the phage. Conceivably such cleavages, or other covalent modifications, could occur during the adsorption of other types of phages, but they have not been demonstrated. (Schwartz 1980).

REPLICATION CYCLE

The molecular biology of the replication cycle is described in Chapters 16 and 17. Here, only the time frame and general outcome of the total replication process at standard temperatures are documented. Although values of the two main parameters: (i) duration of the replication cycle and (ii) phage burst size varied for different phages in different labs, the variability was not crucial but explainable generally by a variety of the methodology used, including growth conditions, etc.

The first paper on the RNA phages reported the single-cycle growth of the phage f2 of 30 min length with a yield of the infectious phage per bacterium amounting to 2000–4000 in the dilute culture and to greater than 9000 in the dense culture, up to as high as 20,000 plaque-forming units (pfu) in some experiments, when the bacteria were probably lysis-inhibited in dense culture (Loeb and Zinder 1961). Davis et al. (1961) reported that the phage MS2 was lysis-inhibited and provided a burst size of 10,000-20,000. A bit later, replication cycle, or an eclipse period, by the author's terminology, of the phage MS2 was found to be 15–17 min, after which mature phage made their intracellular appearance, but the first bursts occurred 38 min after adsorption and a final yield of 2000 pfu per cell was reached (Rappaport 1965).

The one-step growth curve of the phage R17 demonstrated that the replication cycle continued 25 or 35 min with a phage yield of 350 or 4000 pfu in poor and rich media, respectively, while the burst size was between 10,000 and 20,000 by growth in dense cultures (Graham and Paranchych 1962). For the phage fr, the so-called latent period of 20 min and the burst size of 10,000 pfu per bacterium (much less, namely, 800 ± 430 pfu per cell, by the one-step growth) were found (Hoffmann-Berling et al. 1963a). Similar burst size was reported for the phage M12 (Hofschneider 1963a). The first single-step growth measures by phage replication in *E. coli* spheroplasts were undertaken by Knolle and Kaudewitz (1963) with the phage fr.

An exclusive material was achieved by the direct elucidation of the ultrastructural changes in *E. coli* cells during the phage R17 replication cycle under conditions of one-step growth by parallel use of (i) negative staining electron microscopy (Franklin and Granboulan 1966) and (ii) high resolution autoradiography (Granboulan and Franklin 1966b). The electron microscopy showed that no morphological alterations were seen during the latent period. During the period of rapid viral synthesis, a fibrillar lesion surrounded by ribonucleoprotein particles was observed in a polar region. Late in infection, paracrystalline arrays of virions were found in over 90% of the cells (Franklin and Granboulan 1966). The fine method of high-resolution autoradiography was elaborated, first of all, for the radioactive ^{3}H-uridine labeling of the cellular RNA (Franklin and Granboulan 1965; Granboulan 1967). In the case of the R17 infected *E. coli* cells, tritiated precursors of RNA, DNA, and protein were employed in separate experiments. As a result, all three types of syntheses were localized in the infected cells. It is remarkable that the RNA synthesis in infected cells was predominantly cytoplasmic, but later in the latent period, and during the stage of active viral growth, the label was localized in a polar region, whereas normal RNA synthesis in uninfected cells occurred in the nucleoid. In the late stages of viral growth, the RNA synthesis occurred only around the crystals. The protein synthesis also became localized in a polar region, but the DNA synthesis remained confined to the nucleoid (Granboulan and Franklin 1966b).

Very similar values of the replication length and phage burst size appeared when different coliphages were compared by concurrent evaluation of these parameters, e.g., for the phages f2, Qβ, R23, R34, and R40, when the replication cycle, or an eclipse period, by the authors' terminology, was in the frame of 20–25 min and the burst size reached from 1300 to 10,000 pfu per cell (Watanabe and Watanabe 1970a). The putative differences in the bulk production of the phages MS2 and Qβ were explained by specific physiological reasons and overpowered successfully by optimizing conditions of cultivation (Jalinska et al. 1971; Bauman et al. 1976). The one-step RNA replication cycles of the phages MS2, f2, fr, R17, and Qβ were practically identical at 37°C (Bauman et al. 1978; Pumpen et al. 1978a).

Concerning the group IV phages, the replication cycle length, or an eclipse period, by the authors' terminology, was 25 min for the phage SP and 24 min for the phage FI, while the burst size reached 2000 and 2500, respectively (Miyake et al. 1969). The latent period, or time until the extracellular phage appeared, was determined as 31 and 33 min for the phages SP and FI, respectively (Miyake et al. 1969).

The parameters of the replication cycle were generally similar also for the pseudomonaphages 7s (Feary et al. 1963, 1964; Benson 1974) and PP7 (Bradley 1966b), including the phage PP7 infection of synchronous *P. aeruginosa* culture (Weppelman and Brinton 1971). In the case of the pseudomonaphages, an exclusive observation of the phage PP7 replication cycle was undertaken by electron microscopy for the first time: the sampling occurred along the replication cycle, in parallel with a control uninfected *P. aeruginosa* culture, and the morphological changes during different infection stages were documented on the high-quality electron micrographs (Bradley 1966b). The first definite changes were seen after 15 min of infection, when dense areas appeared at the poles and at the center of the cell as the effect of reducing the area occupied by the nucleoplasm.

Figure 4.3 presents the appearance of an infected cell at 40 min after infection, where the dense patches of undifferentiated viral RNA and protein can be seen at the periphery of the bulge, and at the ends of the two undistorted lengths of cell, often called "rabbit's ears," and at 80 min after infection, when the crystalline phage particle inclusions appeared (Bradley 1966b, 1967).

A bit later, electron microscope observation of the replication cycle was performed with the phage MS2 in *E. coli* cells and a series of impressive micrographs was published (Meywisch et al. 1974). It is noteworthy that the special fixation method was used for electron microscopic visualization of intracellular MS2 particles in thin sections of *E. coli* cells (Teuchy and Meyvisch 1970).

The acridine-orange staining with subsequent examination of the probes under the fluorescence microscope was introduced by David E. Bradley for the identification of the phage RNA and as a confirmatory test on the RNA phage replication (Bradley 1965a, 1966a).

The one-step growth curves of the phage PRR1 showed longer replication cycles than for the cases reviewed above; the phage release started after 50 min and reached maximum after 80 min (Olsen and Thomas 1973). However, this curve described extracellular, but not intracellular phage. The burst size of 50–100 on R$^+$ pseudomonads or 10–20 on *Enterobacteriaceae* was somewhat less for PRR1 than that observed for coliphages. Altogether, it allowed the authors to conclude "that PRR1 differs significantly from f2 in its ability to direct phage progeny synthesis" (Olsen and Thomas 1973).

Concerning the caulophages, the single-step growth curves of the φCB RNA phage series on a *Caulobacter* host demonstrated prolonged character, with the first extracellular phage after 2 hr of infection and reaching a plateau after 4 hr with the burst size from 14 to 33 infectious particles per cell (Schmidt and Stanier 1965). It is necessary to remember, however, that the phage replication occurred at 30°C temperature, in contrast to all above-reviewed data that were obtained at 37°C. Growth at 33°C and optimization of the growth and infection conditions made it possible to markedly improve the yield of the phage φCB5 (Bendis and Shapiro 1970).

The growth temperature was absolutely decisive for the duration of the replication cycle. Thus, the one-step replication studies of the phage Qβ replication indicated that the replicative cycle was prolonged when the temperature decreased, and the replication was ceased below 25°C (Woody and Cliver 1995). The systematic prolongation of the replication cycle by decreasing the growth temperature from 37°C to 25°C was observed for the phages MS2, f2, fr, R17, and Qβ (Andris Dišlers, unpublished observations

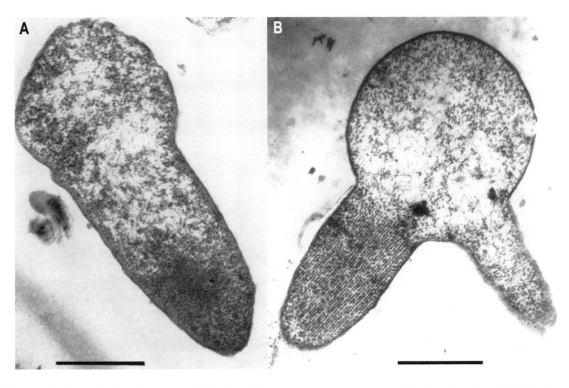

FIGURE 4.3 Lysis of bacteria by RNA phage. (A) Section of *Pseudomonas aeruginosa* cell 40 min after phage PP7 infection, ×50,000, scale 0.5 μ. (B) Section of *P. aeruginosa* cell 80 min after RNA phage PP7 infection, ×37,500, scale 1 μ. (Reprinted with permission of Microbiology Society from Bradley DE. *J Gen Microbiol*. 1966b;45:83–96.)

of 1974–1978). Moreover, the further careful studies on the effect of not only temperature, but also such factors as the host cell number, nutrition, the competition from insusceptible cells and from non-FRNA coliphages, namely, the factors that could influence the phage replication by natural conditions, led to the conclusion that the Qβ replication cannot occur in such nutrient-poor environments as wastewater and groundwater (Woody and Cliver 1997).

The single-step growth curves and phage RNA replication cycle demonstrated an *E. coli* strain-specific character. Thus, the replicative cycles were protracted in *E. coli* Q13, a ribonuclease I–deficient strain, and *E. coli* AB105, deficient in both ribonucleases I and III, in comparison to a wild-type *E. coli* AB259 strain, as well as a number of other *E. coli* strains that possessed a complete ribonuclease set (Bauman et al. 1978). The maximum RNA synthesis in the two mutant strains was reached later, and the rate of RNA accumulation declined more slowly. Nevertheless, the described strain-specific differences were not affected by the growth rate of cells: the two- to fourfold variation in the culture growth rate (by the change of the growth medium at constant 37°C temperature) failed to alter the shape of the RNA synthesis rate curves and the pattern of intracellular phage development (Bauman et al. 1978). The prolongation of the replication cycle was achieved only by changing the temperature conditions. Another way to change the length of the replication cycle was the introduction of so-called regulatory mutations (Pumpen et al. 1978b, 1982; Dishler et al. 1980a).

The replication cycle of RNA phages as a unique complex of synthetic and regulatory events was considered in the context of the famous hypercycle conception (Eigen et al. 1991). In fact, the same idea was used as a starting point for the replication cycle studies performed in the 1970s by Elmārs Grēns' team in Riga (Pumpen and Gren 1975; Bauman et al. 1978; Pumpen et al. 1978a,b, 1982; Dishler et al. 1980a). A set of equations describing all phage-initiated processes and considering putative regulatory links among them was set up by the Grēns team together with the physicist Boris Isaakovich Kaplan in 1975 and solved on the first Hewlett-Packard computer in Riga by the qualified assistance of Velga Sīle from the Institute of Organic Synthesis. The resulting model described well the experimental behavior of RNA phages and their specific "regulation" mutants, which were isolated at that time by the young talented fellow Andris Dišlers (Dishler et al. 1980a), in different growth and suppressing conditions. Unfortunately, these early modeling data were never published.

The same old good idea of the modeling of the phage replication cycle was revived by the first modern quantitative analysis of the phage Qβ infection cycle, which was performed by Tsukada et al. (2009). As mentioned above by the attachment characteristics, the adsorption rate constant of the Qβ interaction with *E. coli* cells was 4×10^{-10} mL cells^{-1} min^{-1}. During the replication cycle, approximately 130 molecules of the phage replicase and 2×10^5 molecules of coat protein were synthesized in 15 min. Replication of Qβ RNA proceeded in two steps: an exponential phase until 20 min and a nonexponential phase after 30 min. Prior to the burst of infected cells, phage RNAs and coat proteins accumulated in the cells at an average of up to 2300 molecules and 5×10^5 molecules, respectively. An average of 90 infectious phage particles per infected cell was released during a single infection cycle up to 105 min (Tsukada et al. 2009).

Nevertheless, the first kinetic model for the intracellular growth of the phage Qβ in *E. coli*, with a special goal to decipher the energetic costs of all template-dependent polymerization reactions, in ATP equivalents, was elaborated by Kim and Yin (2004). This model led to the conclusion that translation dominated phage growth, requiring 85% of the total energy expenditure, and only 10% of the total energy was applied to activities other than the direct synthesis of progeny phage components, reflecting primarily the cost of making the negative-strand RNA template that was needed for replication of phage genomic RNA (Kim and Yin 2004). Summarizing, the model suggested that the phage Qβ has evolved to optimally utilize the finite resources of its host cells, which was in good agreement with the preceding ideas on the highly optimized regulation mode of the phage replication cycle by limitation of not only protein, but also RNA synthesis (Pumpen et al. 1978a). Remarkably, the idea of the numerical simulations of the impact of a phage infection on its *E. coli* host, yielding important insights into the time course of the metabolic demands of a viral infection in total, was developed recently for the infection of a much more complicated phage, namely, for the phage T4 (Mahmoudabadi et al. 2017).

Intriguingly, the length of the phage MS2 replication cycle was influenced by low-frequency 60-Hz electromagnetic fields (Staczek et al. 1998). Thus, a significant delay in phage yield was found by the strength of the field of 5 G, although the final phage concentration was not altered when compared with control cultures. Moreover, the strength of the field of 25 G resulted in both impeded phage replication and increased phage yield. The authors claimed that the phage MS2 was the simplest biological system in which an electromagnetic field-induced effect has been demonstrated (Staczek et al. 1998). However, this effect could be rather secondary by affecting, first of all, bacterial machinery. For example, it was not excluded that bacterial lysis was impeded: the authors did not measure the intracellular phage during the replication cycle.

An interesting more current study showed that the phage MS2 yield can be increased by the presence of chemically benign apatite nanoparticles (Nickel et al. 2010).

Recently, quantitative comparison of the phage Qβ infection cycle in rich and minimal media was performed by Inomata et al. (2012). The adsorption rate constants in both media were almost the same. A difference of 15 min in the latent period and an approximately twofold increase in the rate of phage release were observed, although approximately 10^5 molecules of coat proteins, equivalent to approximately 600–1000 phage particles, accumulated in an infected cell prior to burst. Addition of Mg^{2+} to minimal medium markedly affected the Qβ infection cycle, and Mg^{2+} was found to be required for the stages of the infectious cycle after adsorption.

A novel high-throughput and rapid method for temporal kinetic analysis of the lytic MS2 activity was introduced to replace the classical plaque assay (Davidi et al. 2014). This method, where bacterial growth was monitored using a multiwell plate reader, made it possible to obtain such important

additional information as phage replication rate, progeny size per cycle, and viral propagation during bacterial growth. The data gained were in good agreement with general values reported earlier.

Finally, Yin J and Redovich (2018) published a global review on the kinetic growth modeling of many viruses, with a special chapter for the Qβ RNA replication.

CRYSTALLINE INCLUSIONS

The crystalline, or paracrystalline, aggregates were first detected in the *E. coli* cells infected with the phages f2 (Schwartz and Zinder 1963) and μ2 (De Petris and Nava 1963). In the latter paper, the number of virions per cell was estimated at 10,000 from counts of virions in paracrystals in sectioned bacteria. Figure 4.4 presents a combination of pictures from the Schwartz and Zinder (1963) paper together with an excellent photo published a bit later in a classical review written by Bradley (1967). Such crystalline arrays of virions filled over half of a bacterial section (Schwartz and Zinder 1963).

Excellent micrographs of the paracrystalline arrays were obtained in *E. coli* infected with the phage R17 (Franklin and Granboulan 1966). It is remarkable that the paracrystalline arrays were seen more clearly after "etching" with pepsin, which led to better definition of the crystal regions.

Furthermore, the crystalline inclusions were documented in *P. aeruginosa* cells infected with the phage PP7: the crystals

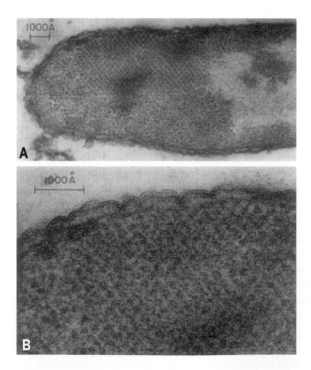

FIGURE 4.4 Electron micrograph of longitudinally sectioned *Escherichia coli*, 50 minutes after infection with f2. (A) Magnification: ×80,000. (B) Enlargement of an area of (A). Total magnification: ×218,500. (Reprinted from *Virology*, 21, Schwartz FM, Zinder ND, Crystalline aggregates in bacterial cells infected with the RNA bacteriophage f2, 276–278, Copyright 1963, with permission from Elsevier.)

continued to increase in size until the spheroplast-like cells ruptured and lysis occurred (Bradley 1966b). These impressive crystals are seen in Figure 4.3. Bradley calculated that there were some 30 rows of them, each containing about 40 phage particles making a total of 1200 virions in this section.

Paracrystalline sheets of the icosahedral virions were observed also *in vitro*, within phage preparations: for the pseudomonaphage 7s (Feary et al. 1964), as well as for the phage f2 group representatives (Hohn and Hohn 1970).

Figure 4.5 presents nice paracrystalline aggregates from the modern twenty-first century paper about the phage R17 infection in view of the *traD* role in the phage RNA transport (Lang et al. 2011).

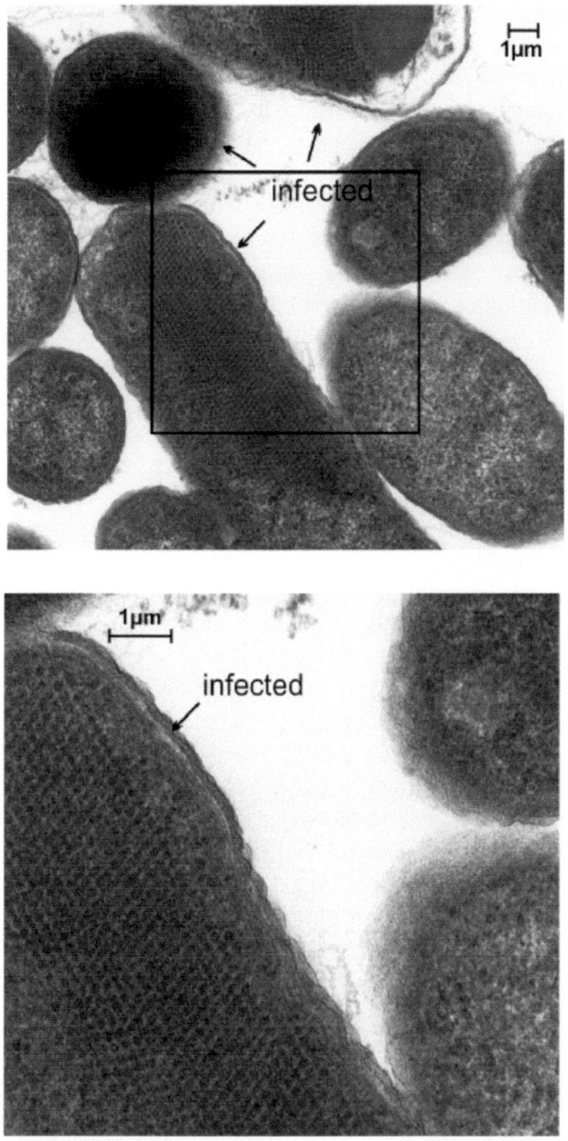

FIGURE 4.5 Electron microscopy of the R17-infected *E. coli* cells demonstrating distinctive honeycomb pattern of phage particles. Upper part presents an enlarged region of the honeycomb pattern. (Lang S et al: An activation domain of plasmid R1 TraI protein delineates stages of gene transfer initiation. *Mol Microbiol.* 2011. 82. 1071–1085. Copyright Wiley-VCH Verlag GmbH & Co. KGaA. Reproduced with permission.)

The *P. aeruginosa* cells examination demonstrated bacteria densely packed with the LeviOr01 phage, but the appearance was different from the crystal areas described above for the coliphages, as well as for the *P. aeruginosa* cells infected with the pseudomonaphage PP7 (Pourcel et al. 2017).

RELEASE

The true molecular mechanisms of the RNA phage-induced lysis of the host cells are presented and discussed in the Chapter 14. Here, the most general outlines of the RNA phage release physiology are summarized.

The first classical review stated the fact that the RNA phages were released by lysis of bacteria, whereas the first phage was released from bacteria at about 22 min and phages continued to be released for another 30 min, but no evidence of a lysozyme-like enzyme was found (Zinder 1965). Moreover, when the bacteria were kept at high cell density, the cultures were lysis-inhibited and some 20,000–40,000 particles were synthesized by each bacterium (Loeb and Zinder 1961). It was calculated that the equivalent of some 5% of the bacterial mass became virus, under conditions of lysis inhibition, and 10–20 mg per liter of readily purified phage could be obtained by the optimum conditions of growth (Zinder 1965). Of note, Zinder assumed further: "It might be mentioned parenthetically that considering its small size and large yield, the RNA phage is currently the most populous organism in the world" (Zinder 1965).

The above remarks on the lysis inhibition illustrated that the early observation that the RNA phage-provoked lysis was not absolute and differed markedly from the deep DNA phage-induced lysis. In contrast to T coliphages, where no surviving bacteria remained after infection, 10%–20% surviving bacteria were observed in liquid cultures infected with high multiplicities of the phage R17: for example, about 20% of the cells appeared as survivors 10 min after infection with multiplicity of 10 phages per cell (Graham Paranchych 1962). Rappaport (1965) mentioned that about one-half of the MS2-infected *E. coli* cells did not lyse and retained their colony-forming ability. Moreover, it was suspected that the entire RNA phages M12 (Hofschneider and Preuss 1963) or fr (Hoffmann-Berling and Mazé 1964) could be released from bacteria without lysis. Later, the release without lysis was attributed to the MS2-infected *E. coli* at 30°C, but not at 37°C, in contrast to the growth at 37°C, where lysis occurred (Engelberg and Soudry 1971b). Nevertheless, further experiments led to the clear conclusion that the phage release from the f2-infected cells was discontinuous, i.e., provoked by cell lysis, at both 30°C and 37°C (Lerner and Zinder 1977).

From the very beginning, chloroform and lysozyme were added at the end of the phage cultivation (Loeb and Zinder 1961) or sonic vibration was used (Graham and Paranchych 1962) to complete lysis and liberate the phage particles. In the first experiments, cyanide was added to stop growth before the addition of the chloroform and lysozyme (Loeb and Zinder 1961).

The lysis process was then clearly documented by electron microscopy for the phage PP7-infected *P. aeruginosa* cells

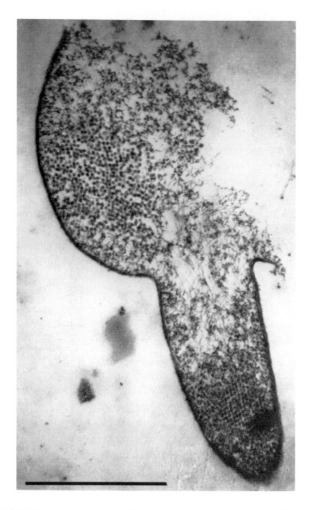

FIGURE 4.6 Section of *Pseudomonas aeruginosa* cell 50 min after RNA phage PP7 infection. Magnification ×50,000, scale 1 μ. (Reprinted with permission of Microbiology Society from Bradley DE. *J Gen Microbiol.* 1966b;45:83–96.)

(Bradley 1965b, 1966b, 1967). Figure 4.6 presents an impressive outline of these studies.

It is noteworthy that the lysis of *P. aeruginosa* was preceded by the formation of spheroplasts, while it was not the case with the infected *E. coli* cells (Bradley 1966b, 1967). As in the case of the coliphages, these studies on the pseudomonaphages also identified the rapid growth of resistant bacteria after the lysis of a broth culture and found that a fairly large proportion of the original bacterial population remained resistant to infection; indeed, it was extremely easy to obtain clones resistant to the phages 7s and PP7 (Bradley 1966b). The novel phage LeviOr01, a PP7-like levivirus, did not initiate cell lysis just after 7 h, although phage was present in the culture medium, but lysis was observed after 20 h, together with the high phage production (Pourcel et al. 2017).

An outstanding electron microscopy report on the phage MS2-induced lysis of *E. coli* cells was published by Meyvisch et al. (1974). The authors distinguished two morphologically different modes of bacterial lysis. First, the bacteria were observed, which had a gap both in the cell wall and membrane, usually located at the side of the cell, through which

small amounts of the cytoplasm and virus particles flew out. When the cell contents had completely leaked out, the cell ghosts, consisting of cell wall and membrane, remained. Second, there were cells where the cell envelope seemed to be broken down more gradually: the interruption appeared first in the outer cell wall through which a gradually increasing portion of the cell contents, still surrounded by the cell membrane, protruded, but finally resulted in a swollen spheroplast, limited only by the membrane (Meyvisch et al. 1974), like in the above-mentioned study with the infected *P. aeruginosa* cells.

Direct comparison of the five phages f2, R40, R34, R23, and Qβ revealed clear differences in their lysing onset and efficiency: lysis of the infected cultures began 30, 50, 50, 60, and 90 min, respectively, and differed in the same order from the deepest to the more moderate level of lysis (Watanabe and Watanabe 1970a).

Propst Ricciuti (1972) found that the *E. coli* cells that were infected during the stationary phase of growth were able to support the phage MS2-directed syntheses and production of virions, but not the phage release. Therefore, the fact of the substantial protein synthesis was not sufficient for the phage release, in contrast to the assumption proposed earlier by Engelberg and Soudry (1971b). In a similar way, the absence of cell lysis and the failure of progeny virus release was observed by MS2 infection of glucose, sulfur- or nitrogen-starved *E. coli* (Propst-Ricciuti 1976).

Addition of glucose or sulfur to the correspondingly starved cells resulted in the normal release of virus within 40–60 min, but return of nitrogen to nitrogen-starved cells did not result in the release of virus, even after 90 min. This phenomenon was connected with the cell division, since the latter demonstrated analogous response to returning of the limiting compound. Therefore, these data supported a hypothesis that lysis by RNA phage was related to cell division and might result at the time of cell division from failure of the cells to divide properly (Haywood 1974). However, it is necessary to add that the MS2 infection of sulfur-starved or glucose-starved *E. coli* resulted in a 100-fold decrease in the yield of plaque-forming units, and the major cause of this phenomenon was the formation of noninfective virions (Propst Ricciuti and Haywood 1974).

The hypothesis about the strong correlation of the phage release and cell division was based on the data that the lysis could be regarded as an aberrant cell division (Haywood 1974). The hypothesis intended correlation of the lysis with the host's doubling time and postulated that the lysis should not occur in cells that were not dividing. Experiments in which *E. coli* cells growing at different rates in different media were infected with the phage MS2 supported the idea: the onset of lysis correlated with doubling time (Haywood 1974). The same idea of the united nature of both phage lysis and cell division was incorporated into the model of the regulation of the phage replication cycle proposed by Grēns' team (Bauman et al. 1978; Pumpen et al. 1978a).

After division inhibitors, the phage release was blocked by the introduction of *fi+*, or *fertility inhibition*, resistance factors and transfer factors into *E. coli* K12 F$^+$ (Pitton and Anderson 1970). Such introduction showed a range of inhibitory activity of lysis by the male-specific phage μ2 and allowed subdivision of the *fi*$^+$ factors into fi^{+1}, fi^{+2}, fi^{+3}, and fi^{+4} classes, according to the degree of inhibition of the phage μ2 lysis (Pitton and Anderson 1970).

Recently, Malekpour et al. (2018) modeled the phage burst size variation by considering the lysis time decisions as a game, where each player in the game is a phage that has initially infected and lysed its host bacterium.

PLAQUE ASSAY

In 2014, as mentioned in the Replication cycle section of this chapter, a novel method was suggested for RNA phage enumeration (Davidi et al. 2014) to replace the classical plaque assay. Of note, the latter served as a satisfying method for more that 50 years in the RNA phage research.

The *double agar layer*, or *soft agar overlay*, method, a classical tool for the determination of the phage titer, namely, plaque-forming units (pfu), was elaborated originally by two great bacteriologists and geneticists: by André Gratia (1936), and independently, by the 1969 Nobelist Alfred Day Hershey et al. (1943). As a laboratory routine, the method was formalized and introduced by Mark H. Adams (1950, 1959). The RNA phage pioneers usually gave the reference of Adams (1950). By the original description, the method is performed in the following way:

> About 2 mL of melted 0.6% agar is cooled to 45°C and inoculated with a drop of a concentrated suspension of the host bacterium. A measured volume of phage suspension is then added, and the entire mixture poured over the surface of a hardened layer of nutrient agar. After the upper agar layer has solidified the plate is incubated. The bacteria grow as a multitude of tiny colonies within the soft agar layer, fed by the layer underneath, forming an opaque background against which plaques are easily seen. Advantages of the agar layer method are: (1) phage samples up to 1 mL. in volume can be plated per petri dish; (2) the time required for plating each sample is less than 30 seconds; (3) the plaque size is larger than that given by the spreading method; and (4) the efficiency of plating is often higher. The agar layer method has permitted accurate kinetic study of reactions that were too rapid to be followed by other methods of phage assay and has been indispensable in genetic studies relying on recognition of different types of plaque. (Adams 1959)

In fact, the method was suitable not only for the enumeration, but also for the characterization of the RNA phages and especially of their mutants by the individual plaque morphology: size, turbidity, presence of a halo, etc. Although such plaque characteristics did not demonstrate absolute character and could be influenced by many factors in different laboratories, they were highly useful and were always referenced in the corresponding papers.

For example, considerable attention has been given to the plaque morphology in the studies on the caulophages (Schmidt and Stanier 1965; Miyakawa 1976). Then, the marked difference in plaque size permitted a simple distinction between the phages MS2 and Qβ (Overby et al. 1966a). A combination of the plaque assay with the radioactive labeling of the infective centers contributed markedly to the understanding of the replication cycle by polar coat protein mutants (Pumpen et al. 1978b). A mutant of the phage MS2 with very specific plaque form, the so-called "thin rim" mutant, was isolated by Tamara Sherban (1969) and further characterized by Dishler et al. (1980b).

Peter Knolle concentrated marked attention on the plaque assay technique (Knolle 1967c). Furthermore, the effect of 21 surfactants was tested on plaque formation by the coliphages M12, f2, and Qβ and by the pseudomonaphage PP7, and anionic detergents proved able to influence plaque formation substantially (Menzel 1985). An effect of pH on the putative aggregation of the phage MS2 and its possible effect on the plaque assay was studied: indeed, the phage MS2 exhibited significant aggregation processes for pH at or below pI$_{MS2}$, which was 3.9, leading to aggregates with sizes of few micrometers and therefore reduced the pfu counts (Langlet et al. 2007). Moreover, in an attempt to improve the visibility and consistency of the current plaque assay for the quantification of the phage MS2, a spread plate technique was introduced instead of the pour plate technique used commonly in the classical methods (Cormier and Janes 2014). Using the spread plate technique resulted in an increase of plaque size by approximately 50% and contributed to better visibility, and it was combined with other improvements: addition of glucose, CaCl$_2$, and thiamine supplements and reduction of agar thickness and hardness (Cormier and Janes 2014).

Generally, the traditional plaque assay, which still retained its current gold standard position, was soon replaced by such modern methods as quantitative real-time polymerase chain reaction (QPCR) and NanoSight (NS) Limited technologies for the rapid (about 5 min) enumeration of phage particles (Anderson 2011). Moreover, Fouts (2018) adapted the single-plaque assay for the whole genome sequencing of a desired phage with the special protocol for the RNA phages.

PHAGE EFFECT ON HOST

The RNA phages were the first discovered bacteriophages that did not kill infected bacteria at once. Citing again the classical Zinder (1965) review:

> After infection with f2, all bacterial metabolic processes continue normally until late in infection when the cell lyses. There is no inhibition of cellular DNA synthesis or RNA synthesis (Loeb and Zinder 1961). Sucrose gradient fractionation of RNA labeled after infection reveals essentially the same pattern as does RNA from uninfected cells. In addition, enzymes such as β-galactosidase and alkaline phosphatase can be induced normally (Zinder 1963) after infection until just before lysis.

Four ribonucleoside monophosphokinases, namely AMP, UMP, GMP, and CMP kinases, as well as dTMP kinase demonstrated the same chromatographic pattern and activity after

40 min infection with the phage MS2 as that from the uninfected cells (Hiraga and Sugino 1966). The same was true for the ribonucleoside mono- (Argyrakis 1968) and diphosphokinases (Argyrakis-Vomvoyannis 1968) in the phage f2-infected cells. The ATP-dependent deoxyribonuclease of *E. coli* was not affected by the Qβ infection, in contrast to the infection with the double-stranded DNA phages that inactivated or markedly reduced the enzyme activity (Sakaki 1974). The phage MS2 infection did not induce any changes in sulfur-labeling of RNA, in contrast to the DNA phages T2 and T4 (Hsu et al. 1967).

The DNA synthesis, although maintained following infection, was not necessary for phage synthesis, since fluorouracil, fluorouracildeoxyribotide, and mitomycin inhibited DNA synthesis and growth of the DNA phages but had no gross effect on the phage f2 yield (Cooper and Zinder 1962). Starvation for thymine of a thymine-requiring strain did inhibit the growth of f2 unless uracil was added (Zinder 1965). The RNA phages were not affected by four systems of host control variation in *E. coli*: C-K, K-B, B-K, and K-K(P1) (Eskridge et al. 1967).

At the early stage of phage studies, Doi and Spiegelman (1962) showed that the host cell did not contain a sequence complementary to the phage MS2 RNA, by a novel revolutionary hybridization test using specific formation of hybrids between DNA and RNA (Hall and Spiegelman 1961). These results clearly implied that DNA is not an intermediate in phage growth. Starvation of histidine auxotrophs for histidine late in infection had little effect on the phage yield, since histidine was not present in the coat protein of the phage f2 (Cooper and Zinder 1963). Cooper and Zinder (1962) showed that all of the phosphorus found in the phage progeny was taken up by the cells after infection, so that the cellular enzymes which make the RNA precursors must be functioning after infection. A bit later, the phage fr replication was also found independent of the bacterial DNA synthesis (Knolle and Kaudewitz 1964).

Therefore, it was generally accepted that the RNA phages did not kill bacteria at once but were going to subvert almost half of their host's metabolism to their own needs (Weissmann 1974). Nevertheless, the role of bacterial RNA and protein synthesis in the growth of phage remained unclear for some period of time. Thus, actinomycin D was first reported to have no effect on the growth of the phage MS2 in infected protoplasts of *E. coli* (Haywood and Sinsheimer 1963), which would imply that the cellular RNA synthesis was not involved in phage replication, but further investigations revealed a marked decrease of the phage f2 yield following the actinomycin D treatment (Zinder 1965). Finally, it was postulated that actinomycin D caused irreversible inhibition of the synthesis of a host structure required for the MS2 replication prior to the synthesis of progeny MS2 RNA, since addition of the drug later than 16 min postinfection did not inhibit the phage MS2 yield (Haywood and Harris 1966).

Similarly, according to Ellis and Paranchych (1963), ribosomal RNA synthesis in *E. coli* was rapidly inhibited during the first 15 min after infection with the phage R17. During this interval, rRNA synthesis was depressed to a rate of about 20% of that prevailing at the time of infection, and then continued at this slower rate for at least 20 minutes. This inhibition seemed similar to that observed in mammalian cells infected with certain of the smaller RNA-containing viruses, such as poliovirus and Mengo virus (for discussion see Ellis and Paranchych 1963). It was excluded, however, that the continuing late synthesis of rRNA would be provided by the "survivors" and would probably decrease as these survivors slowly disappeared (Paranchych and Graham 1962). Again, Ellis and Paranchych (1963) concluded that there was no incorporation of host protein and/or RNA into progeny phage during the course of infection. Marked reduction of DNA, RNA, and protein syntheses during the latent period of the R17 infection was confirmed using autoradiographic techniques (Granboulan and Franklin 1966). Decrease of the host's rRNA synthesis 15 min after infection with the phage ZIK/1 was reported by Bishop (1965, 1966b). Knolle (1967f) determined the phage fr provoked decrease of the host rRNA synthesis by 20%–30% and detected a definite impairment of ribosome formation after infection, in obstacles when only a definite fraction of the host cells was found productively infected (Knolle 1964).

More detailed investigation showed that the R17 infection caused decrease of the host RNA synthesis by 70%–80%, including the same 70%–80% decrease of the rRNA synthesis, and that this inhibition occurred during the first 15–20 min after infection, although the function and properties of the bacterial DNA-dependent RNA polymerase system were not influenced (Hudson and Paranchych 1967). The tRNA synthesis was also not influenced by the R17 infection (Hudson and Paranchych 1968). Hung and Overby (1968) found, however, a codon-specific change in the function of tRNAs from *E. coli* after infection with the phage Qβ.

The marked inhibition of both 16S and 23S rRNA, as well as DNA (but not of tRNA) syntheses with a maximum at 30 min after infection was observed in the R23-infected *E. coli* cells, although infected cells continued to divide for about one generation (45–60 min) after infection (Watanabe and Watanabe 1971). Strong inhibition of the host's rRNA synthesis after R17 infection was confirmed by experiments with a temperature-sensitive phage mutant in the replicase gene (Igarashi et al. 1970a).

By infecting *E. coli* cells with the R17 amber mutants which lacked the ability to synthesize one or more of the viral proteins, the synthesis of the replicase subunit only correlated with the inhibition degree of the host's RNA synthesis, which occurred solely at the transcriptional level (Spangler and Iglewski 1972). The ability of the infecting phage to synthesize either the coat or A proteins was not correlated with the inhibition of the rRNA synthesis.

Concerning protein synthesis in *E. coli* cells, both total and β-galactosidase synthesis were gradually reduced during the progress of the MS2 infection, whereas synthesis of the viral coat protein increased gradually until it accounted for 30%–40% of the total protein synthesized in the cell (Sugiyama and Stone 1968a,b). The distribution of polysomes obtained from the R17-infected cells shifted toward smaller polysomes as early as 10 minutes after infection (Hotham-Iglewski and Franklin 1967). The inhibition of the inducible β-galactosidase synthesis, although not complete, was apparent shortly after infection of *E. coli* with the phage R23 and was maximal after the first 20 min of infection (Watanabe and Watanabe 1968; Watanabe M et al. 1968). The primary effect of this phenomenon was explained by limitation of synthesis of enzyme-specific mRNA, whereas the inhibitory process was separated into two phases: early inhibition, which did not require the expression of the viral genome, and late inhibition, which required the expression of the viral RNA replicase gene (Watanabe and Watanabe 1968, 1970c).

Yamazaki (1969) explained the early inhibition of protein and RNA synthesis in the R17-infected cells by the greatly reduced capacity of the infected cells to transport exogenous amino acids. The intracellular functioning of the viral genome was not required for amino acid transport inhibition, since a UV-inactivated phage behaved similarly to the intact phage. Remarkably, this inhibition resulted in the phage-induced synchronous division of bacterial cells (Potter and Yamazaki 1969). A bit later, the early phase of inhibition of the host protein synthesis was explained by reaction of the cells to tryptone broth, which was often used to suspend the phage R17: the early inhibition was not observed when *E. coli* cultures were inoculated with the purified phage R17 suspended in phosphate buffer (Scott and Iglewski 1974). Therefore, the early inhibition was not an intrinsic feature of the phage R17 infection. In the same context of some uncertainty, it is worth mentioning that Jiresová and Janecek (1977), who studied the β-galactosidase synthesis after infection not only with the phage MS2, but also with the DNA phages T1, T2, T3, and T4, did not find any inhibitory effect of the MS2 infection, in contrast to the T-phages that caused immediate inhibition of the enzyme synthesis.

Another set of studies showed that the synthesis of the host amino acid biosynthetic enzymes was permitted during infection with the phage MS2 and that the phage had evolved so that it benefited from such host biosynthesis (Goldman 1982; Koontz et al. 1983; Rojiani and Goldman 1986). A functional dependence of the phage MS2 growth upon the postinfection host gene function was perhaps the strongest argument that the RNA phage did not shut off host mRNA and protein synthesis (Rojiani and Goldman 1986).

The membrane as the intracellular location site of the MS2 phage syntheses was postulated for the first time by Haywood and Sinsheimer (1965) and studied in more detail by Haywood et al. (1969) and Haywood (1973). Continuing this approach, the same place on the membrane was identified as the site of the rRNA synthesis (Haywood 1971). Logically, the competition of both the phage and rRNA syntheses arose (Haywood and McClellen 1973). As a result of such competition, it appeared that the fate of rRNA synthesis was different in two different *E. coli* host strains by the MS2 infection: (i) in the D-10 strain, the synthesis and turnover of the 16 s rRNA precursor was similar in both infected and uninfected cells, but only one-third of this precursor matured in the infected cell, and the remainder appeared to be degraded; the 23 s

rRNA precursor was synthesized at a slightly slower rate than in the uninfected cell and did not turn over; (ii) in the MRE 600 strain, in contrast to the D-10 strain, 90% reduction of the rRNA precursor synthesis occurred (Haywood and McClellen 1973).

The implication of the bacterial membrane permeability into the phage-induced processes was confirmed by a rapid, transient increase in K^+ efflux from the R23-infected cells, which was terminated 10–15 min after infection (Watanabe and Watanabe 1970b). Moreover, the RNA synthetic capacity of the membrane complex was inhibited by the R23 infection (Hunt et al. 1971). Such inhibition was caused by the release of ribosomes from the membrane complex, as demonstrated *in vitro* on the membrane-DNA-RNA polymerase complex isolated from the R23-infected *E. coli* cells (Hunt and Watanabe 1973). The effect was reversal, since the addition of ribosomes to such a membrane preparation restored the RNA synthetic capacity to uninfected levels. The reversibility of the inhibition of the rRNA synthesis by that of the phage MS2 or Qβ RNAs was demonstrated later *in vivo* (Pumpen and Gren 1977).

By direct comparative analysis, the coliphages differed by their capability to inhibit the host synthesis. Thus, the phage R23 decreased the synthesis of bacterial RNA and protein to a greater extent than the phages f2, Qβ, R34, or R40 (Watanabe and Watanabe 1970a). The phenomenon was explained by the higher yield and longest life cycle of the phage R23 when compared to that of other phages, as noted above. The phage Qβ accounted for the lesser effect on host metabolism due probably to its lower efficiency by the cell infection (Watanabe and Watanabe 1970a).

A complex investigation with an aim to simultaneously fix the changes in the *E. coli* host DNA, RNA, and protein syntheses after infection with the phage MS2 was performed by the Grēns' team (Berzin and Gren 1971, 1972a,b; Rosenthal and Gren 1972; Berzin et al. 1974). As a result, it was concluded that up to 25–30 min after infection only host rRNA and protein syntheses essentially underwent suppression with hardly any detectable effect on DNA, mRNA, and tRNA syntheses. The synthesis of host proteins was gradually displaced by that of viral proteins as early as 10–15 min after infection.

An unexpected interference of the phage Qβ infection with the *E. coli* host cell functions, which led to alteration of the host character, was described by Miyake et al. (1966): the number of colonies formed by *lac⁻* host on minimal lactose agar medium increased 15 times when the host was plated after Qβ infection. This effect was explained by the concomitant liberation of β-galactosidase accumulated in constitutive cells together with the liberation of phage and the increase of the extracellular β-galactosidase. Furthermore, the Qβ infection caused a codon-specific change in the function of tRNA: preparations of the latter from infected cells showed a markedly decreased ability to support translation of polycytidylic acid into polypeptide, as compared to tRNA from uninfected cells, whereas polyuridylic acid−directed phenylalanine incorporation was not influenced (Hung and Overby 1968). This phenomenon was explained by 2−3 times decrease in ribosomal binding of the prolyl-tRNA. Another new function of the Qβ infection consisted in the inhibition of bacterial cell wall mucopeptide synthesis: incorporation of ^{3}H-diaminopimelic acid into the mucopeptide layer of the cell wall was markedly decreased starting at about the midpoint of the phage replication cycle (Ozaki and Valentine 1973).

Brief mention should be made of the nice and popular hypothesis based on the discovery of a ψ_r factor. ψ_r factor was found by Travers et al. (1970a,b) and postulated as a common activity that participated in the synthesis of both rRNA and phage RNA, in the latter case as a hypothetical component of the phage Qβ replicase. Chapter 13 will describe the further development of this hypothesis to the clear observation , but in 1970 it seemed clear enough to explain the competitive character of the rRNA and phage RNA synthesis. Travers et al. (1970b) tried to connect the ψ_r factor with the CAP, or catabolite gene activator protein, regulated by cyclic AMP, or cyclic 3′: 5′-adenosine monophosphate (Zubay et al. 1970). However, the first experimental response to the ψ_r hypothesis indicated that the decreased availability of ψ_r was probably not the cause of the reduced host RNA synthesis which followed infection with the phage R23 (Hunt and Watanabe 1971).

In fact, the idea of the ψ_r factor had its roots in the studies of a relationship between the RNA phage infection and the status, stringent or relaxed, of the so-called *RC* gene(s), or a genetic *rel* locus, that was responsible for the regulation of the rRNA synthesis by the presence of a full complement of required amino acids (Stent and Brenner 1961; Kurland and Maaløe 1962). The stringency of the amino acid requirement for RNA synthesis was greatly relaxed in *E. coli* mutants carrying a "relaxed" allele of the *RC* gene (Stent and Brenner 1961).

James D. Friesen (1965) showed for the first time that the synthesis of the phage f2 RNA was controlled by amino acids in a manner similar to the control of the host RNA, and it was postulated that the phage RNA synthesis is subjected to RC control, as its host RNA. This conclusion was confirmed by further investigations with the phages f2 (Gallant and Cashel 1967; Friesen 1969), fr (Knolle 1967a), R17 (Yamazaki 1969; Khan and Yamazaki 1970), and R23 (Ernberg and Sköld 1976). Surprisingly, no *rel* locus control over the synthesis of Qβ RNA was detected (Siegel and Kjeldgaard 1971). Finally, Watson and Yamazaki (1972) concluded that the phage RNA synthesis was not influenced by the *rel* gene of the host in the same manner as was host RNA itself.

One of the last topics exploiting the stringent or relaxed status of the host's *rel* locus in connection with the RNA phage infection was devoted to polyamine synthesis (Fukuma and Cohen 1973). The presented data indicated a close correlation between the synthesis of RNA and spermidine, suggesting a significant role for this polyamine in the multiplication of the phage R17. Furthermore, the phage R17 and its RNA were found to contain significant amounts of spermidine, but not of putrescine. When isolated at 0.01 M KCl, up to 1000 molecules of spermidine were associated with the R17 virion. The phage R17 RNA isolated with phenol plus sodium lauryl sulfate contained approximately 70–90 molecules of spermidine (Fukuma and Cohen 1975a). Finally, Fukuma and Cohen (1975b) acknowledged the fact that the R17 infection permitted the synthesis of RNA and spermidine in a stringent *E. coli* strain in the absence of the exogenous essential amino acid, namely arginine in this case. Therefore, the fact of the phage infection as such was suspected to alter the availability of the *rel* gene product. This conclusion was close to that mentioned above: although phage RNA synthesis may be controlled by the *rel* locus, this control was different from the host's control over rRNA synthesis (Watson and Yamazaki 1972). In conclusion, the connection of the *rel* locus status with the inhibition of the phage RNA synthesis generally retained its mysterious character.

Mathematical modeling was involved to understand how metabolism could be impacted by the host *E. coli* interaction with the pathogen phage MS2 during a viral infection (Jain et al. 2006a,b; Jain and Srivastava 2009). Employing a metabolic modeling strategy known as "flux balance analysis" coupled with experimental studies, it was predicted how the MS2 infection would alter bacterial metabolism: the cell growth and biosynthesis of the cell wall would be halted, a substantial increase in metabolic activity of the pentose phosphate pathway as a means to enhance viral biosynthesis would occur, while a breakdown in the citric acid cycle and no changes in the glycolytic pathway were predicted (Jain and Srivastava 2009).

The rapid development of microarray technology has provided unique means to monitor host cell responses to viral infection at the level of global changes in mRNA levels: this methodology was applied to investigate the gene expression changes caused by the phage PRR1 infection of *P. aeruginosa* cells (Ravantti et al. 2008). The PRR1 infection resulted in changes in expression levels of less than 4% of *P. aeruginosa* genes. Interestingly, the number of genes affected by the phage infection was significantly lower than the number of genes affected by changes in the growth conditions during the experiment. Compared with a similar study that focused on the double-stranded DNA phage PRD1, it was evident that there were no universal responses to viral infection. Generally, the phage PRR1 infection did not grossly affect cell metabolism until the time of lysis. The genes most affected by PRR1 infection were grouped into three major functional classes: transport, energy production, and protein synthesis (Ravantti et al. 2008).

Unfortunately, microarray investigations have focused mostly on viral infections in animals and plants and not as much on bacteria. Involvement of this modern methodology could solve many phage–bacteria interaction moments, especially in the case of the RNA phages with their "mild" effect on the host, that remained controversially evaluated above.

HOST MUTANTS

A large number of different *E. coli* mutants, besides pili mutants that were described in Chapter 3, were checked for their susceptibility to the RNA coliphage infection. The story of the majority of *E. coli* mutants studied in their connection with the RNA phage infection is presented below in chronological order.

Thus, it was reported that growth of the phage MS2 was not observed to occur in *E. coli* strains lacking ribonuclease I, or RNaseI⁻ mutants (Ohyama et al. 1969, 1970). It should be noted here that one of the strains studied by Ohyama et al. (1969), Q13, was used successfully in the phage replication studies by the Grēns' team with the phages MS2, f2, fr, R17, and Qβ (Pumpen and Gren 1975; Bauman et al. 1978; Pumpen et al. 1978a). Moreover, the strain AB105 lacking both ribonucleases I and III was able to support phage replication (Bauman et al. 1978). This double RNaseI⁻ RNaseIII⁻ mutant, originally named *E. coli* 301–105, was isolated by Kindler et al. (1973) after treating *E. coli* cells of a ribonuclease I-free strain, A19, with nitrosoguanidine and selection of mutants for inability to degrade double-stranded RNA. Although the mutant demonstrated less than 1% of the ribonuclease III activity related to the parental strain, the plating efficiency for the phage M12, as well as M13, was equal for the mutant and the parental strain (Kindler et al. 1973). The *E. coli* 105 strain was characterized in detail by Apirion and Watson (1974) including preparation of its mutant resistant to the phage R17. Meanwhile, Apirion and Watson (1974) proved that the mutation to RNase III⁻ survived in strain AB105 as a compensation for some other defect that did not permit such strains to grow on minimal medium at elevated temperatures.

The susceptibility of the ampicillin-resistant mutants of *E. coli* K-12 with lipopolysaccharide alterations to the phage MS2 was dependent on the resistance level, moderate (class I) or high (class III): mutants from class I were rapidly killed by MS2, whereas the class III mutant remained almost unaffected (Monner et al. 1971).

Special mention should be made of a temperature-sensitive mutation preventing cell division and Qβ RNA replication but susceptible to the phages f2 and R23 at non-permissive temperature (Silverman and Mandal 1972). After this mutant was lost, a variety of other Qβ-resistant mutants were isolated (Mandal and Silverman 1977). One of the latter was capable of supporting replication of the phage Qβ at 33°C, but not at 40°C, while the phages f2 and R23 formed plaques on mutant cells at both temperatures. The phage Qβ replication was blocked in the mutant within the first 20–30 min of infection, but the defect did not prevent translation of the Qβ replicase gene or assembly of catalytically active Qβ replicase molecules. In fact, mutant cells infected at 40°C hyperinduced replicase active both *in vivo* and *in vitro*. However, zone sedimentation of the *in vivo* RNA product showed it to consist of partially double-stranded material sedimenting at 9 S, with little or no viral 32 S RNA. It thus appeared that the temperature-sensitive component was required for viral RNA replication (Mandal and Silverman 1977).

Considering the special role of the translation elongation factors EF-Tu and EF-Ts in RNA phage replication, the appropriate *E. coli* mutants were isolated and checked. First, *E. coli* with a temperature-sensitive elongation factor EF-Ts was selected (Kuwano et al. 1973). The infection of the mutant with a mutant Qβam12, which produced large amounts of replicase at the permissive temperature led to production of the thermolabile Qβ replicase (Kuwano et al. 1973, 1974). The temperature-sensitive EF-Tu mutant of *E. coli* was unable to replicate the phage MS2 at 42° but permitted phage production at 37°C (Lupker et al. 1974a). The new method to select cells unable to propagate the RNA phage MS2 at 42°C, as well as exhaustive experimental details of the elucidation of the EF-Tu mutant were presented further by Lupker (1974b).

Mutations affecting utilization of lactose and resistance to the phages f2, and Qβ, as well as to the DNA phage f1, tended to occur simultaneously more often than expected by chance in Hfr strains whose origin of transfer is close to the genes for lactose utilization, but not in F⁺ strains (Lloveres and Cerdá-Olmedo 1973).

The phage f2 grew poorly in a conditional putrescine auxotroph of *E. coli* during polyamine starvation, while addition of putrescine simultaneously with f2 enhanced phage growth, shortened the latent period, and increased the burst size (Young and Srinivasan 1974). The phenomenon was explained by the ability of polyamines to stimulate the translation of a preformed messenger. In contrast to these data, Hafner et al. (1979) reported that the f2 replication proceeded normally in a polyamine-deficient *E. coli* mutant that did not contain putrescine or spermidine. Although the phage f2 was poorly adsorbed, the burst size was roughly the same in the amine-starved and in the amine-supplemented cultures. It was concluded that polyamines were not required for the phage f2 and Qβ production, but the phage adsorption defects could be attributable to a decrease in the stability of the F pili (Hafner 1981).

Mutation of *E. coli* K12 to resistance to fluorophenylalanine resulted in the changes in the plaque morphology of the phage MS2 on this strain and led to an increased efficiency of propagation of the phage in liquid cultures (Jenkins et al. 1974). Evidence was obtained that the mutation resulted in inhibition of early lysis in infected cells.

Genetic analysis of streptomycin-resistant mutants of *E. coli* K Hfr strain showed that restriction of the phage MS2, derestriction of the female-specific DNA phage T7, and resistance to streptomycin were the pleiotropic effects of a single mutation at the *strA* locus responsible for the ribosomal protein S12 (Chakrabarti and Gorini 1975). In fact, the S12 protein was defined as a product of the *strA* gene earlier (Ozaki et al. 1969) and was later found to be responsible for the restriction of the Qβ read-through protein synthesis (Yates et al. 1977). Another streptomycin-resistant spontaneous *E. coli* mutant was temperature sensitive for the phage Qβ, but not for the group I phages f2, MS2, and R17 (Engelberg-Kulka et al. 1977). The Qβ infection of the mutant at the non-permissive 42°C temperature resulted in the release of a near-normal burst of noninfectious particles. It was assumed that the mutant was defective at elevated temperatures in the suppression of nonsense codons, thereby producing Qβ-like particles which were noninfectious because of the lack of the read-through protein A1 (Engelberg-Kulka et al. 1977, 1979b). The Qβ particles that were produced at 41°C by the temperature-sensitive streptomycin-resistant *E. coli* mutant were deficient in both minor capsid proteins of Qβ: A2, or maturation, or IIa protein, and A1, or read-through, or IIb protein, and did not adsorb to F-piliated bacteria (Engelberg-Kulka et al. 1979a). Both minor proteins A1 and A2 were not produced in Qβ-infected mutant cells at 41°C. Moreover, a shorter RNA, which sedimented mainly at 23 S, instead of the full-length 30 S RNA of Qβ, was found in the defective particles (Engelberg-Kulka et al. 1979a). The nature of the temperature-sensitive mutation was assumed to be related to an alteration in the ribosomal protein S12 (Zeevi et al. 1979).

Deficiency of *E. coli* cells in the enzyme tRNA nucleotidyltransferase in so-called *cca* mutants did not affect infection with the phages f2 and Qβ, while the decrease amounted to as much as 90% in the case of T-even phages, and 50%–65% for T-odd phages (Morse and Deutscher 1976).

A *lpo* mutant of *E. coli* lacking a specific outer membrane lipoprotein remained susceptible to the phages f2 and MS2, as well as to the DNA phage f1 and all other coliphages examined (Hirota et al. 1977).

The temperature-sensitive *E. coli* mutant that inhibited the growth of the phage β but not the growth of the host cell itself at 43°C was isolated and found to have an altered membrane organization which interfered with normal viral replication (Yasuo et al. 1978).

Plasmids harboring the N-terminal part of the DNA phage f1 gene II conferred to bacterial cells partial resistance to infection with the male-specific phages f1 and f2 (Dotto et al. 1981). This effect was due to the production of large amounts of a ~20 Kd polypeptide corresponding to the N-terminal part of the gene II protein.

The development of the phages MS2 and Qβ, but not that of M13, was more efficient in the *pcnB* mutant of *E. coli* relative to an otherwise isogenic *pcnB*⁺ host (Jasiecki and Wegrzyn 2005). The product of the *pcnB* gene was poly(A) polymerase I (PAP I), which was the main enzyme responsible for RNA polyadenylation in *E. coli* and was localized in membrane, where polyadenylated RNA molecules were rapidly degraded by a multiprotein complex called RNA degradosome. Thus, it was shown for the first time that membrane-associated RNA turnover enzymes could be involved in the protection of the cell against the RNA phages: polyadenylation of the phage RNA at the penetration stage might result in its more efficient degradation, thus impairing the infection process (Jasiecki and Wegrzyn 2005).

Intriguingly, expression of the human immunodeficiency virus-1 integrase in *E. coli*, at levels that had no effect on bacterial cell growth, blocked plaque formation by the phages R17, Qβ, and PRR1, as well as M13, while plaque formation by phages having double-stranded DNA genomes was unaffected (Levitz et al. 1994). An early stage in infection was affected by the integrase and it was suggested that the integrase interacted *in vivo* with phage nucleic acid, while the putative effect on the pili was excluded. This conclusion was supported by studies in which the integrase was shown to have a DNA-binding activity in its C-terminal portion. This portion of the integrase was both necessary and sufficient for the interference of plaque formation by the phages R17 and M13 (Qβ and PRR1 were not tested). Expression of the N-terminal portion of the integrase at the same level as the intact integrase had little effect on the phage growth, indicating that the expression of foreign protein in general was not responsible for the inhibitory effect (Levitz et al. 1994).

ALKBH proteins, the homologs of *E. coli* AlkB dioxygenase, which constitute a direct, single-protein repair system, protecting cellular DNA and RNA against the cytotoxic and mutagenic activity of alkylating agents, were found to increase the survival of the phage MS2 after treatment with such alkylating agents as methyl methanesulfonate or chloroacetaldehyde, indicating efficient repair of 1meA/3meC lesions and etheno adducts in the phage RNA (Mielecki et al. 2012).

INHIBITION OF PHAGE GROWTH

The experiments with classical inhibitors revealed first the phage independence of the host DNA synthesis: reproduction of the RNA phages was insensitive to such inhibitors of DNA synthesis as fluorodeoxyuridine in thymine-less bacteria or mitomycin C in the case of the phage f2 (Cooper and Zinder 1962), bromodeoxyuridine or mitomycin C in the case of the phage M12 (Hofschneider 1963b), and mitomycin C in the case of the phage fr (Knolle and Kaudewitz 1964). Growth of the phages β and 10 was insensitive to nalidixic acid, a naphthyridine derivative, which inhibited the DNA synthesis (Taketo and Watanabe 1967). It should be noted, however, that RNA synthesis of the phages β and MS2 was found sensitive to phenetyl alcohol, known at that time as an inhibitor of DNA synthesis (Nonoyama and Ikeda 1964).

Second, the RNA phage infection was found insensitive to actinomycin D which inhibited DNA-dependent RNA synthesis by binding to DNA (Haywood and Sinsheimer 1963). The actinomycin D-treated MS2-infected *E. coli* spheroplasts allowed extraction of the phage-specific syntheses from the bacterial background (Haywood and Sinsheimer 1965). In total, multiplication of the phage MS2 proceeded in *E. coli* spheroplasts at actinomycin D concentrations causing 75%–90% inhibition of the host protein synthesis (Haywood and Sinsheimer 1963), but increased inhibition of the host protein synthesis was associated with a decreased yield of the phage MS2 (Haywood and Sinsheimer 1965). EDTA-treated cells were then adopted for the studies instead of spheroplasts (Haywood and Harris 1966). The ability of EDTA-treated cells to form colonies and to produce MS2 decreased by equal amounts and in proportion to the decrease of host incorporation of ³H-uracil and ³H-lysine, when the cells were treated with increasing doses of actinomycin D. If the cells were incubated with different concentrations of actinomycin D for 15 min, resuspended in growth media without it, and then infected, the colony-forming ability prior to infection and the ability to produce MS2 were reduced to the same degree as when the cells were left in the drug. When actinomycin D was added to the infected cells more than 16 min after infection, it did not inhibit MS2 yield. It was postulated that actinomycin D caused irreversible inhibition of the synthesis of a host structure required for MS2 replication prior to the synthesis of the progeny MS2 RNA (Haywood and Harris 1966). The reason for the suppression of phage production by actinomycin

D, both in spheroplasts and EDTA-treated cells, remained unclear (Lunt and Sinsheimer 1966). Although the efficiency of infection and burst size of the phage MS2 were reduced markedly at high concentrations of actinomycin D, the synthesis of protein and RNA at an optimum concentration was largely dependent on the phage infection and therefore allowed the studies of the phage-specific protein syntheses (Oeschger and Nathans 1966). The technique of the actinomycin D treatment was improved by Fastame and Algranati (1994), in order to follow the phage Qβ replication in the EDTA-permeabilized *E. coli* cells without a definite impact on the phage yield.

Furthermore, novobiocin, the first clinically employed antibiotic that has been shown to inhibit nucleic acid synthesis without direct binding to DNA, blocked replication of the phage MS2 (Smith and Davis 1965). Chromomycin A$_3$, another inhibitor of the host's DNA-dependent RNA-polymerase, reduced the yield of phage MS2 by 90% (Kaziro and Kamiyama 1965; Hasegawa 1966), although purified Qβ replicase was insensitive to the drug (Kaziro and Kamiyama 1967).

Miracil D, a drug suppressing bacterial ribonucleic acid and protein synthesis, was used to observe MS2 phage-specific replication products (Cramer and Sinsheimer 1971, 1972). However, the drug inhibited phage penetration and it was necessary to perform penetration without the drug and further achievements in the presence of chloramphenicol. Then, chloramphenicol was to be removed by washing. Nevertheless, addition of miracil D after 20 min postinfection only did not inhibit the infection process. The phenomenon was explained by the inhibition of the synthesis of an unstable enzyme, namely replicase, in the presence of the drug (Cramer and Sinsheimer 1972).

Rifamycin was the most intriguing inhibitor of the host's DNA-dependent RNA polymerase: it replaced actinomycin D by the separation of the phage-specific syntheses from bacterial background and played a special role in the phage studies. The next section in this chapter, as well as a substantial part of Chapter 17, are devoted to the rifamycin story.

Concerning protein synthesis in the infected cells, the phage f2 RNA synthesis was inhibited if chloramphenicol was added within 2 min after infection (Cooper and Zinder 1963). If chloramphenicol was added at later times, some phage RNA was synthesized. Chloramphenicol added 15 min after infection did not inhibit phage RNA synthesis at all. By using a strain which required histidine and methionine as host bacterium (the f2 coat protein does not contain histidine), it was shown that the synthesis of f2 became independent of the presence of histidine halfway through the latent period but remained dependent on the continued presence of methionine. First, these results suggested that an essential protein other than the coat, namely replicase, was synthesized soon after phage f2 infection. Second, the time frame of the phage dependence on the protein synthesis was identified. The same effect of chloramphenicol on the phage R17 propagation was found by Edgell and Ginoza (1965). Schmidt (1966) reported that pretreatment of the host caulobacters with chloramphenicol

prevented the caulophage adsorption. Generally, the unique role of chloramphenicol in the story of the phage regulation mechanisms will be unveiled in Chapter 17.

Concerning the synthesis of the functional phage RNA, halogenated pyrimidines able to replace analogous bases in nucleic acids were used. Thus, when 5-fluorouracil at a concentration of 33 μg/mL was present from time 0, 5, or 10 min after MS2 infection, no increase in infectious virus was found, whereas when 5-fluorouracil was added at 15, 20, or 25 min after infection, active phage particles were formed, though with a reduced yield (Shimura and Nathans 1964). Furthermore, a method for preparing the phage MS2 containing 5-fluorouracil, or FU phage, with markedly increased buoyant density, was presented in detail (Shimura et al. 1965). Later, a fraction of the phage MS2 with buoyant density lower than normal was identified in the FU phage: this fraction was noninfectious, did not adsorb to pili, and possessed fragmented RNA due to lack of the A protein (Shimura et al. 1967). RNA fragments from these FU phage particles were used further as messengers for specific bacteriophage proteins (Shimura et al. 1968).

Complete inhibition of the R17 RNA and phage synthesis occurred when 5-fluorouracil was added within 2 min after infection (Graham and Kirk 1965). If the addition of 5-fluorouracil was made later than 5 min after infection, infectious RNA synthesis was blocked but infectious phage was still formed; the infectious RNA made before the addition of FU continued to be incorporated into mature phage. These properties of the drug were used to determine the kinetics of phage RNA synthesis and the size of the phage precursor RNA pool. At a concentration of 2.2×10^{-5} M 5-fluorouracil, the yield of phage was reduced to 15% of that in an uninhibited control, 28% of the phage RNA uracil was replaced with 5-fluorouracil, but the specific infectivity of the phage was unaltered (Graham and Kirk 1965).

In the case of the phage f$_{can1}$ infection, 5-fluorouracil was shown to be mutagenic, and production of noninfective phage particles in the presence of the drug was demonstrated (Davern 1964a).

Streptomycin inhibited infection of the phage MS2 using *E. coli* host cells that were completely resistant to the concentrations of antibiotic used (Brock 1962). The drug was able to inhibit when it was present at the time of the adsorption and injection, but not when added shortly after the injection has occurred (Brock 1962). In the streptomycin-sensitive host, the phage protein synthesis was at least as sensitive to inhibition by the antibiotic as host protein synthesis, but the production of infectious RNA continued for a considerable period after both R17 phage and host protein synthesis have stopped and after uninfected bacteria have lost the capacity to divide (Brownstein 1964). Furthermore, streptomycin reduced the average burst size of the phages f2 and μ2 by 60%–80% in streptomycin-resistant cells, as well as in streptomycin-sensitive cells under conditions when neither growth nor protein synthesis was affected (Schindler 1964a). The inhibitory effect of streptomycin was blocked by sodium polymethacrylate, spermine, deoxyribonucleic acid, and adenylic acid. However, spermine itself at concentrations from 0.5 to 200 μg/mL inhibited the development of the phage f2 when added together with the phage (Schindler 1965a). It is necessary to add here that spermine, like histone and protamine, and, just to 10^5 greater extent on a molar basis, lysozyme was able to precipitate the phage f2 (Matthews and Cole 1972c). Streptomycin, as well as neomycin, but not kanamycin, formed reversible complexes with the phage particles, but the inhibition of phage reproduction could not be explained only by the reaction of the antibiotics with the free virions (Schindler 1965b).

Colicine E2, a bacteriocidal agent, inhibited the phage R17 replication not only by prior treatment of the host cells with E2, but even if E2 was added after the infection (Fujimura 1966). However, it took a longer time for inhibition by E2, at least inhibition of viral RNA synthesis, when the addition of the colicine was delayed.

A set of carcinogenic polycyclic aromatic hydrocarbons, such as the carcinogen 7,12-dimethylbenz[a]anthracene (7,12-DMBA), inhibited multiplication of the phage MS2 in *E. coli* spheroplasts but not in intact bacteria (Hsu WT et al. 1965). Moreover, when other closely related hydrocarbons were tested in the system, it became obvious that a correlation existed between the observed carcinogenic activity of a compound and its ability to inhibit plaque formation (Hsu WT et al. 1965, 1967).

The infective activity of the phage f2 and its infectious RNA was inhibited by histone and protamine, when they were added together with the phage, while neither protamine nor histone exerted any inhibitory effect on growing *E. coli* culture (Schindler 1966).

Spermine, known as one of a number of compounds stabilizing spheroplasts to lysis in distilled water, demonstrated its stabilizing effect on the f2-infected cells by complete prevention or interruption of the phage release, although phage reproduction was markedly inhibited (Groman 1966; Groman and Suzuki 1966). The degree of inhibition was constant over the range of concentrations tested but was roughly related to the time of addition: significant inhibition was observed even when spermine was added as late as 30 min after infection, which was just before lysis began in the control (Groman and Suzuki 1966).

Phleomycin, an antibiotic which blocked DNA-dependent RNA polymerase of *E. coli*, also inhibited the phage R23-directed RNA synthesis, whereas synthesis of coat protein, phage assembly, and the formation of infective particles were unaffected (Watanabe and August 1968b). The phenomenon was explained by the binding of the drug to the phage RNA.

Both 5-azadeoxycytidine and 5-azacytidine inhibited replication of the phage f2 in wild-type cells of *E. coli* but had practically no inhibitory effects in deaminase-less mutants (Doskočil and Šorm 1970).

Fluorophenylalanine did not inhibit multiplication of the phage β (Ikeda 1967). It was later found, however, that fluorophenylalanine reduced by 100-fold the number of the phage MS2s produced and increased the latent period of infection (Wray et al. 1970). It was most effective when added concurrent with infection. Addition of a tenfold greater concentration of phenylalanine reversed the inhibition caused by fluorophenylalanine. The exposure of the MS2-infected cells to another amino acid analog, namely threo amino chlorobutyric acid (TACB), an analog of valine, and incorporation of the analog into the MS2-coded proteins prevented phage maturation (Prouty 1975). Abnormal proteins, synthesized from the phage genome, were degraded, presumably by a host catabolic system, more rapidly than the normal gene products.

Levorphanol and levallorphan, structural analogs of morphine, have demonstrated ability to inhibit RNA phage replication. Thus, reproduction of the phages f2, MS2, and Qβ was markedly (>99%) inhibited by levorphanol (Simon et al. 1970). Even at concentrations which were virtually without any effect on the host, the phage yield was decreased by 85%–90%. Levorphanol was most effective when added before or at the time of infection, but no inhibition was observed when it was added 30 min or more after infection. Inhibition of phage reproduction was reversed when, even after exposure to levorphanol for an hour, the infected cells were washed free of drug. The phage-inhibitory action of levorphanol seemed to be unique, since it exerted stronger inhibition on the phage than on the host, in contrast to the actinomycin D or rifamycin action. However, parallel investigation on the levallorphan that demonstrated similar inhibition of the MS2 infection showed that most of the inhibition was due to a failure to infect the host (Raab and Röschenthaler 1970). The inhibition of phage replication by levallorphan was found to be less affected than the phage adsorption.

An antibiotic (-)rugulosin isolated from *Myrothecium verucaria* demonstrated a potent anti-phage effect on the phages MS2, GA and Qβ: the inhibition was not mainly due to a drop in the burst size but rather to a decrease of the phage-producing cells during the early stages of phage infection and replication (Nakamura et al. 1971a). Further studies, predominantly with the phage MS2, indicated that the mechanism of the (-)rugulosin action consisted of at least two parts, i.e., the inhibition of the penetration of phage RNA into the host bacteria and some early intracellular reaction(s) for phage multiplication (Nakamura et al. 1971b).

Requinomycin, an anthracycline antibiotic, inhibited propagation of the phage f2 when the drug was present at the time of infection: as a result, pili were suspected to be the binding sites of requinomycin (Hori et al. 1972). In contrast, another antibiotic, desdanine, known to inhibit the synthesis of nucleic acid in *E. coli*, suppressed the multiplication of the phage Qβ, as well as that of the ssDNA phage f1, but not the adsorption or penetration stages (Tanida et al. 1976).

Large sets of compounds were tested routinely on the effect on the RNA phage propagation. Thus, 26 plant growth regulators including herbicides were investigated for their effect on the bacterial growth and multiplication of phages including the phages M12 and Qβ (Menzel et al. 1975). Some of herbicides inhibited the phage M12 propagation at different stages (Touré and Stenz 1977). Some substituted thioureas were found to produce strong inhibitory effect on the phage M12 replication (Rehnig and Schuster 1978). The number of the M12 and Qβ plaques was diminished or increased depending on the kind of the 1,3,5-triazines tested (Stenz and Menzel 1978). Then, a set of morpholinium compounds was tested on the replication of the phages M12, f2, and Qβ, as well as on that of plant viruses: some drugs significantly enhanced plaque formation by the phages (Kluge et al. 1978). Some triazole compounds reduced plaque formation by the phages M12, f2, and Qβ and retarded the liberation of phages (Menzel and Kluge 1979, 1980). An 1,3,4-thiadiazole derivative inhibited the MS2 propagation (Varvaresou et al. 2000).

Metaupon, a detergent, inhibited the adsorption of the phages f2, M12, and Qβ, but the number of the phage f2 and M12 plaques could be increased, depending on the substance concentration (Menzel and Stenz 1979). Furthermore, 21 anionic detergents were tested on plaque formation by the coliphages f2, M12, and Qβ and by the pseudomonaphage PP7: the effect was dependent on the phage species and on the detergent sort and concentration; some detergents increased the number of the phage plaques (Menzel 1985).

The effect of 14 pyrrol-carboxylic acid derivatives was tested on the RNA phage growth (Yamabe and Shimizu 1977). Enlargements of plaque size were observed with seven compounds but increase in plaque numbers was not registered. These enlargements of plaque size were specific to the phages MS2, GA, and Qβ, but not to the DNA phages tested. The enlargements of plaque size by the compounds was explained by an increase in the rate of the phage particle release (Yamabe and Shimizu 1977).

It is noteworthy that a long list of potential antitumor substances was tested for the inhibition of the phage fr growth by Schering AG (Mujtic 1969). The use of the RNA phages as one of the microbial models for the screening of potential antitumor drugs was reviewed by White (1982).

Finally, some exotic potentially antiviral substances were tested on the anti-RNA phage activity. Thus, an extract obtained from the roots of *Boerhaavia diffusa* plants, which inhibited the infection of several plant RNA viruses, demonstrated no effect with the PP7/*P. aeruginosa* system but exerted some stimulatory effect on the coliphage M12 and Qβ plating (Awasthi and Menzel 1986). Crude extracts of a set of Greek ferns were checked on the anti-phage activity including the anti-MS2 activity in frame of a prescreen for antiviral agents: some extracts inhibited the MS2 plating, with the *Asplenium ceterach* extract as the most active of them (Skaltsa et al. 1991).

Growth of the phage μ2 was not inhibited in *E. coli* cells that were collected from aerosol and survived spraying from distilled water into a nitrogen atmosphere as a function of aerosol age (Cox 1968).

RIFAMYCIN

A group of rifamycin antibiotics was started with the discovery of rifamycin B as a product of *Streptomyces mediterranei* (Sensi et 1959) and followed by many semi-synthetic rifamycins (Sensi et al. 1960, 1962, 1964). Among those provided with *in vivo* activity against Gram-positive and tubercular

infections, rifamycin SV (Bergamini and Fowst 1965) and rifamycin B diethylamide (Maggi et al. 1965) have both found clinical use, but only by parenteral administration. Rifampicin was synthesized (Maggi et al. 1966) as a new representative of the rifamycin group and was highly effective by oral administration in experimental infections induced by Gram-positive and Gram-negative bacteria and in experimental tuberculosis. The chemical structure of rifamycins was determined by Oppolzer et al. (1964).

Hartmann et al. (1967) reported that rifamycins are specific inhibitors of bacterial DNA-directed RNA synthesis. Whereas a bacterial RNA polymerase was very sensitive to rifamycin, mammalian RNA polymerases were not affected or affected only at high concentrations (Wehrli 1968b). In contrast to actinomycin D, which interacted with the DNA, rifamycin acted directly on the enzyme itself (Wehrli 1968b). Using a ^{14}C-labeled rifamycin derivative, it was shown that RNA polymerase forms a very stable complex with the antibiotic with simultaneous loss of activity (Wehrli 1968a). Moreover, RNA polymerase from rifamycin-resistant mutants of *E. coli* did not form a complex with the antibiotic and was not inhibited (Ezekiel and Hutchins 1968; Wehrli 1968a).

In parallel, an antimicrobial substance B44P was isolated from another strain of actinomycetes, *Streptomyces* spp. No. B44-P1 almost identified with *S. spectabilis* (Yamazaki 1968a). This substance belonged to streptovaricins, a set of antituberculosis agents (Siminoff et al. 1957), and was characterized as a potential agent against *Mycobacterium tuberculosis* and *Staphylococcus aureus* (Yamazaki 1968b). The substance B44P, or streptovaricin, which was a mixture of at least five compounds, inhibited the DNA-dependent RNA polymerase reaction in *E. coli* (Mizuno et al. 1968a). It was demonstrated that streptovaricin repressed neither the formation of the DNA-enzyme complex nor the polymerizing process, but it inhibited the initiation of RNA synthesis (Mizuno et al. 1968b). Umezawa et al. (1968) accepted that streptovaricin A had a structural resemblance to rifamycins, and demonstrated that rifamycins B and SV had the same inhibitory action on the DNA-dependent RNA polymerase of *E. coli* as determined before with the streptovaricin (Mizuno et al. 1968a,b).

Figure 4.7 presents the structure of the most popular rifamycin group members. Rifampicin, by its WHO nonproprietary name, or rifampin (US Adopted Names Council), or Rifadin® (Dow Chemical Company), or Rimactane® (Ciba Pharmaceutical Company), properties of which were exhaustively reviewed by Lester (1972), was the most used substance of the rifamycin group in the RNA phage studies. Nevertheless, the basic term rifamycin will be used here for all members of the group.

The highly specific action of rifamycins on the DNA-dependent RNA-polymerase has aroused particular interest from the pioneers of the RNA phage studies, in order to separate phage-specific syntheses from the background of bacterial syntheses and to obtain exact quantitative data about the kinetics of the phage macromolecule turnover *in vivo*. The first attempt showed that the growth of the phages f2

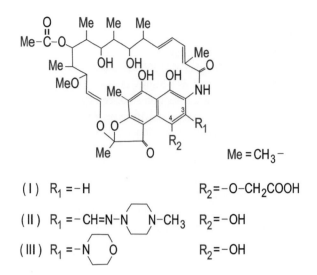

FIGURE 4.7 The naturally occurring rifamycin B (I) and several chemically modified derivatives, such as rifampicin (II) and 3-morpholino-rifamycin SV (III), are known as particularly active agents against Gram-positive microorganisms. (Reprinted from *Biochim Biophys Acta*, 145, Hartmann G et al. The specific inhibition of the DNA-directed RNA synthesis by rifamycin, 843–844, Copyright 1967, with permission from Elsevier.)

and MS2 in the wild-type *E. coli* was sensitive to rifampicin, when added before infection, but the phages grew normally in the presence of rifampicin on the mutant that possessed an altered polymerase (Marino et al. 1968; Tocchini-Valentini et al. 1968). These results were compatible, however, with an early idea that at least a part of the host DNA-dependent RNA polymerase has a role in the replication process of the phage RNA. Subak-Sharpe et al. (1969) noted this hypothesis in their study that demonstrated how rifamycin inhibited the growth of some mammalian viruses, namely poxviruses and adenoviruses. This problem was solved once and for all by Bandle and Weissmann (1970) who showed that the synthesis of both phage-specific plus- and minus-RNA by Qβ replicase *in vitro* were insensitive to high concentrations of rifamycin and no one component of the DNA-dependent RNA polymerase was involved in the phage propagation.

Fromageot and Zinder (1968) were the first to appreciate the great advantages of rifamycin allowing the phage f2 growth to proceed in a near-normal fashion in whole *E. coli* cells, despite the fact that cellular RNA and protein synthesis were almost completely inhibited. It was the first case in which the problem of cell permeability was circumvented, in contrast to the drug actinomycin D when it has been necessary first to modify the host's permeability barrier. The pretreatments distorted the phage macromolecular syntheses in an undefined way. It was found that application of rifamycin prior to phage infection partially inhibited phage synthesis at the step of penetration. When rifamycin was added a few minutes after infection, biosynthesis of cellular components ceased but phage-specific synthesis proceeded, giving rise to a normal yield of progeny particles (Fromageot and Zinder 1968). Simon et al. (1970), in their work on the levorphanol effect, which was referenced in the preceding section

, established, however, that the effect of rifamycin was not as selective in their hands with the phage MS2 as it was in the experiments with the phage f2 of Fromageot and Zinder (1968). When rifamycin was added 15 min after infection with MS2 they obtained the greatest selectivity: host RNA synthesis was decreased 95% while phage RNA synthesis was reduced 70% (Simon et al. 1970).

Garwes et al. (1969) got the excellent pattern of the Qβ-specific protein synthesis in the presence of rifamycin, in parallel with actinomycin D, but they used however spheroplasts of *E. coli*.

Engelberg and Soudry (1971a) reported inhibition of the phage MS2 release from its host by rifamycin, in addition to interfering with intracellular growth. Rifamycin prevented the release of the progeny MS2 phage particles at both 37°C and 30°C temperatures (as postulated by authors before, the release was accompanied by cell lysis at 37°C, whereas at 30°C phage was released from intact cells), while rifamycin did not inhibit phage release from mutant *E. coli* cells possessing a rifamycin-resistant DNA-dependent RNA polymerase (Engelberg and Soudry 1971b). The effect of rifamycin on the phage release was confirmed later by the excellent electron micrographs of Meyvisch et al. (1974) where most of the MS2-infected cells were packed with phage particles, and in more than 50% of the cells the phage particles were condensed into large crystals in the presence of rifamycin from 15 min of infection.

Moreover, in addition to the release, Engelberg (1972) established that the RNA minus strands were synthesized on the parental template at a reduced rate, whereas the synthesis of the progeny plus strands was completely inhibited in the presence of rifamycin. The effect of the drug was not immediate; rather, a period of approximately 15 min was required for the drug to exert a maximal inhibitory effect on both stages of the phage RNA replication. The phenomenon was interpreted by a competition of the phage replicase with the host's DNA-dependent RNA polymerase for a common protein (Engelberg 1972).

Later, Engelberg et al. (1975) found a discriminative effect of rifamycin on RNA replication of various RNA phages. According to this study, rifamycin interfered exclusively with the RNA replication *in vivo* of the group I phages MS2, f2, and R17, whereas Qβ RNA replication was not affected by the drug. In addition, rifamycin was found to have a discriminative effect on the group I phage RNA replication: the antibiotic differentially interfered with the synthesis of minus RNA strands of the phage f2, whereas it had almost no effect on the synthesis of the progeny plus strands. For the phage MS2, the drug differentially arrested the synthesis of the progeny plus strands and almost failed to affect the synthesis of the minus RNA strands. For the phage R17, both steps of its RNA replication were affected by rifamycin, although each step was only partially (approximately 50%) inhibited (Engelberg et al. 1975).

Van de Voorde and Fiers (1971) added rifamycin 3 minutes before the labeled precursor when they performed 5-minute radioactive pulse at 15 or 25 min after the MS2 infection by the search for the place where phage proteins were synthesized. Passent and Kaesberg (1971) studied the effect of rifamycin on the development of the phage Qβ *in vivo*. Rifamycin, when added at the time of infection, inhibited synthesis of the phage Qβ. It was found that both viral RNA and viral proteins were made in nearly the same amount as in the absence of rifamycin, but the rate of assembly into phage particles was low.

The same old idea of the different reaction of the group I phage R17 against Qβ by the rifamycin treatment was supported by Igarashi and Bissonnette (1971b). Moreover, the clear differences were postulated among the group I phages f2, MS2, and R17 in their reaction against rifamycin (Igarashi and Bissonnette 1971b). An impressive work was undertaken to show that the R17 replicase could be dissociated into two major components, one of which would be connected with the DNA-dependent RNA polymerase complex (Igarashi and Bissonnette 1971b; Igarashi 1973). By the work, a set of rifamycin-resistant *E. coli* mutants was isolated, which permitted the growth of the phage R17 (Igarashi 1972). Finally, Igarashi and Elliott (1975) came to the conclusion that the rifamycin-sensitive stage for the R17 growth was the first 20 min of the infectious cycle during which time the synthesis of phage components, but not the assembly, took place, and that the RNA synthesis was definitely sensitive to rifamycin by the R17 infection, in contrast to the other RNA phage systems. Shortly thereafter, an effect of rifamycin on the R17 particle synthesis was studied briefly by Knolle (1972a).

The same idea of a host factor which would be necessary for the R17 phage RNA synthesis and assembly and could be depleted by the presence of rifamycin was supported by Rothwell and Yamazaki (1972). It should be noted that a combination of rifamycin and chloramphenicol treatment, which played an extraordinary role in the replication regulation studies (see Chapter 17), was used in this study for the first time.

A similar conclusion was also achieved by Meier and Hofschneider (1972), who studied replication of the phage M12 in the presence of rifamycin. It was demonstrated that the M12 replication proceeded in a normal way only some 25 min after addition of the drug. Later, both phage-specific RNA and protein synthesis simultaneously came to a halt, leading to the assumption that rifamycin caused a depletion of one or more host-controlled factors necessary for both phage RNA and protein synthesis (Meier and Hofschneider 1972).

Meanwhile, the effect of rifamycin on the MS2 replication was connected with the inhibition of the episomal gene expression in the drug-treated *E. coli* cells and the loss of capacity to the phage adsorption by an alteration or lack of phage receptors (Riva et al. 1972). When the effects of rifamycin and streptolydigin, another inhibitor of bacterial RNA synthesis, on the production of F pili by *E. coli* were studied by electron microscopy, a reduction of the number of new pili produced by depiliated cells was observed (Fives-Taylor and Novotny 1976). However, neither the length or number of F pili that were present at the time of inhibition were affected, and the retraction of these preexisting pili did not occur. It was suggested that the rifamycin-sensitive step might be linked to the establishment of a site for the F pili production (Fives-Taylor and Novotny 1976). The phage R17 was used in this study to label pili for electron microscopy.

In the rifamycin context, an interesting temperature-sensitive RNA synthesis mutant of *E. coli* was described by Patterson et al. (1971). At the non-permissive temperature, the capacity for RNA and protein synthesis decreased logarithmically in the mutant, and the latter was unable to support the growth of the phage f2, even at the permissive temperature.

It should be noted here that Fastame and Algranati (1994) developed an improved method for the actinomycin D treatment that was superior against the rifamycin treatment for the segregation of the phage-directed processes from the bacterial background in that study.

Concluding the negative effects of rifamycin on phage replication, the direct influence of the drug on phage replication was excluded since this effect was absent in a mutant host with the drug-resistant DNA-dependent RNA polymerase (Marino et al. 1968, Passent and Kaesberg 1971; Igarashi 1972). Generally, it was assumed that RNA phage replication required participation of a short-lived host protein which became deficient within 25–30 min postinfection in the presence of rifamycin. Such a hypothetical protein was suggested to be involved in either phage RNA synthesis (Engelberg 1972; Engelberg et al. 1975), or both RNA and protein syntheses (Meier and Hofschneider 1972; Rothwell and Yamazaki 1972), or in assembly (Passent and Kaesberg 1971), or in the phage release (Engelberg and Soudry 1971a). Remarkably, direct treatment of the phage f2 with rifamycin *in vitro* led to a dramatic loss of infectivity by the drug binding to phage RNA at a few specific sites (Naimski and Chroboczek 1977). By this interaction, the phage capsid acted as a protective barrier, since inhibition of phage RNA infectivity occurred at 10–100 times lower molar excess of rifamycin than inhibition of infectivity of the intact phage particles.

Extensive studies on the rifamycin effect on RNA phage replication have been performed by the Grēns' team, who believed that rifamycin met the necessary requirements for phage replication studies to a great extent and did not markedly change the global RNA phage replication pattern (Baumanis et al. 1972; Pumpen et al. 1974; Pumpen and Gren 1975, 1977; Bauman et al. 1978; Pumpen et al. 1978a,b). It was postulated that, regardless of the time of addition, rifamycin failed to alter, at least qualitatively, the general pattern of both MS2 and Qβ replication (Pumpen et al. 1974) and was acceptable for replication regulation studies (Pumpen and Grens 1975; Pumpen et al. 1978a,b). In fact, rifamycin had very little effect on phage replication. Moreover, the drug was even capable of stimulating it in some cases. Rifamycin arrested only those processes which could be either directly or indirectly associated with cell division, namely phage release, or lysis, as found by Engelberg and Soudry (1971a,b). The block of phage release caused by the drug appeared to be one of the main reasons why the interference of rifamycin with phage replication tended to be overestimated in the works referenced above. In fact, no essential differences were found in the replication pattern of the phages from group I, namely MS2, f2, fr, and R17, or the phage Qβ from group III, in response to rifamycin addition (Bauman et al. 1978).

Exposure of infected cells to rifamycin before or early in infection resulted in a sharp reduction in the number of infective centers, whereas neither adsorption nor injection appeared to be affected by the drug (Pumpen et al. 1974). The penetration of ^{32}P-labeled RNA phage was almost independent of the time of rifamycin addition (Bauman et al. 1978). This discrepancy between the amount of penetrated RNA and the observed number of infective centers was attributed to the limitations of the infective center assay based on the assumption that the infectious particles could be released from the cells exposed to rifamycin. However, since lysis was arrested by early presence of the drug, the actual number of infective centers tended to be underestimated in these experiments. Therefore, pre-replicative stages were declared as not affected by the drug.

To delineate the effect of rifamycin on the phage replication cycle, the RNA synthesis rate and single-step growth curves were obtained in the presence of the drug (Bauman et al. 1978). Figure 4.8 presents the curves obtained for three *E. coli* strains varying in RNase content, AB259, Q13 (RNase I⁻, PNPase⁻), and AB105 (RNase I⁻, RNase III⁻).

It is apparent that the rate curves of RNA synthesis and the pattern of intracellular phage development were not influenced by the time of rifamycin addition, not even when the drug was added before infection. It was assumed that the observed differences in the RNA synthesis patterns were due to the different set of ribonucleases, when lower content of the latter in Q13 and AB105 strains resulted in a longer lifetime of the phage RNA and prolonged RNA replication cycle (Pumpen et al. 1978a). The drug appeared to exert only a slight enhancing effect on the single-step growth curve. The phage yield was never less than 40% as compared with the untreated cultures, even under the most unfavorable conditions, i.e., when rifamycin was added 5 min before infection. The phage yield was always in good agreement with the total amount of synthesized RNA.

The shape of the RNA synthesis rate curves and that of the single-step growth curves was strain-specific and characterized by certain peculiarities with respect to the strains used. Thus, the maximal RNA synthesis in *E. coli* Q13 and AB105 was reached later and the rate of RNA accumulation declined more slowly than in the AB259 strain. The described strain-specific differences were not affected by the growth rate of cells. The growth media used for different strains, as seen in Figure 4.8, were chosen to obtain roughly equal values of the generation time (Bauman et al. 1978). Figure 4.9 shows that two- to fourfold variations in the culture growth rate, by the change of the growth medium at constant temperature, failed to alter the shape of the RNA synthesis rate curves and the pattern of intracellular phage development. Both characteristics in the defined strain could only be affected by changing temperature conditions (Bauman et al. 1978).

By the presence of rifamycin from 3 min postinfection, the emergence of extracellular phage was determined by the rate

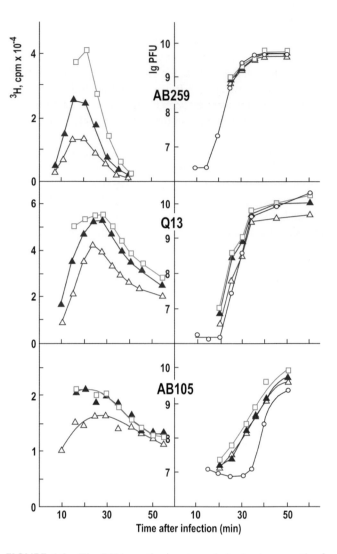

FIGURE 4.8 The RNA synthesis rate and single-step growth of the phage MS2 in different *E. coli* strains with respect to time of the rifamycin addition. The media used were glycerol salt for AB259, TPG for Q13, and tryptone broth for AB105. Total phage and RNA synthesis rate in the presence of rifamycin added: 5 min before (△), 3 min after (▲), and 10 min after (□) infection. (○) Total phage without rifamycin. (Redrawn with permission of ASM from Bauman V et al. *J Virol.* 1978;28:717–724.)

of cell division as opposed to the appearance of intracellular particles. Since rifamycin inhibited cell division, the release of the phage became blocked by the drug. The extent of blocking was determined by both the time of rifamycin addition and the culture growth rate. In slow-growing cultures when generation time exceeded 30 min, addition of rifamycin at 10 min postinfection (or earlier) resulted in complete inhibition of phage release. In fast-growing cultures, addition of rifamycin at 10 min was followed by phage release constituting 10% of intracellular phage in AB259 and 5% in Q13 (as determined at 50 min postinfection). In good agreement with Propst Ricciuti (1976), the phage release was well correlated with the growth rate of cells. The phage release, besides being correlated with the cell growth rate, appeared to be independent of the phage intracellular replication as such. If the rate

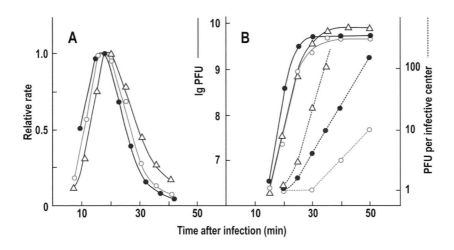

FIGURE 4.9 Phage MS2 RNA synthesis rate and single-step growth in rifamycin-treated *E. coli* AB259 grown on different media. Rifamycin was added at 3 min postinfection. (A) Rate of RNA synthesis; (B) total phage (solid line); extracellular phage in the control without rifamycin (dashed line). Symbols: (o) glycerol salt; (•) TPG + tryptone; (Δ) tryptone broth. (Redrawn with permission of ASM from Bauman V et al. *J Virol.* 1978;28:717–724.)

of bacterial multiplication was low, the phage particles might persist within the cell for a long time after the replication was completed. Since cell division required host protein synthesis, it became blocked by rifamycin after an appropriate lag period. This was probably the reason rifamycin treatment of infected cells for more than 20 min before lysis suppressed the release of phage particles, as was found by Engelberg and Soudry (1971a,b).

The ratio of the RNA plus and minus strands synthesized, as well as the ratio of protein versus RNA synthesis did not depend on the time of rifamycin addition (Bauman et al. 1978). It was concluded that the parameters of the RNA phage replication cycle were not susceptible to rifamycin inhibiting specifically host RNA and protein syntheses. The drug added even before infection failed to cause any qualitative changes in the replication pattern, although a certain reduction in the replication efficiency could be observed (Bauman et al. 1978). This reduction was explained by the cumulative effect of such factors as a certain decrease in the number of infective centers, some disturbances in the macromolecular synthesis caused by exhaustion of the total biosynthetic capacity of the cells due to very early exposure to the drug, etc. Moreover, in the majority of cases the addition of rifamycin accelerated phage replication, thus reducing the latent period of infection. This phenomenon was explained by the elimination of competition between phage replication and host-specific synthetic processes, primarily the accumulation of the host's rRNA (Pumpen and Gren 1977). Certainly, rifamycin did not interfere with phage assembly, and it did not seem to increase the proportion of defective particles, since correlation among the phage yield and the total RNA synthesis was always evident (Bauman et al. 1978).

In conclusion, these results suggested that rifamycin did not interfere with phage replication for two main reasons: (i) high specificity of the drug and (ii) high degree of autonomy of the phage replication from macromolecular syntheses of the host cell. For this reason, rifamycin was used extensively

in the studies of the self-regulation of the phage replication cycle, as it will be described in Chapter 17.

MIXED INFECTIONS AND SUPERINFECTIONS

The problem of the effective mixed infection arose primarily by genetic studies when complementation of different mutants was encountered. Zinder wrote in his first classical review (Zinder 1965) that there have been two reports of interference (Kaudwitz and Knolle, Proc Intern Congr Genet, The Hague 1963; Davern 1964a) in attempts at mixed infection of host bacteria with different mutants, when cells yielded only one type of phage, some yielding one and some the other. However, successful complementation between mutants has been reported (Pfeifer et al. 1964, Valentine et al. 1964), "indicating that mixed infection can sometimes, if not always, be obtained" (Zinder 1965). Details of the complementation studies will be discussed in Chapter 8.

Superinfection of the previously T2-infected cells by the phage f2 was performed by Neubauer and Zavada (1965). In this way, the authors tried to establish the level of independence of the RNA phage growth of the cell syntheses, since it was known that the T-even phage infection immediately blocked the bacterial genome (Nomura et al. 1962). Moreover, this experiment would establish whether and to what extent the capacity for f2 was affected by the preinfection of the host with the UV inactivated phage T2, which was used to avoid its reproduction in treated cells. However, the idea did not work: it appeared that only those cells that survived infection by the UV-irradiated phage T2 were able to reproduce phage f2 (Neubauer and Zavada 1965).

The next superinfection by the phage f2 was made on the induced cultures of *E. coli* K-12 (λ112) F+, where λ112 was a lysis-inhibiting mutant of the phage λ (Groman and Suzuki 1965). When the f2 superinfection occurred within 90 min after induction, lysis was observed in normally lysis-inhibited cultures, but later superinfections produced very little lysis.

Following early superinfection, both λ112 and f2 phages were produced in induced cells. When superinfection occurred during the period in which growth was inhibited, the phage f2 production was totally inhibited. However, the step at which f2 was inhibited remained unspecified (Groman and Suzuki 1965).

The double infection of *E. coli* by the two morphologically and genetically distinct male-specific phages f1 and f2 demonstrated that both phages were competing for a common pathway at the penetration stage, when nucleic acid transport occurred over pili "as a part of an active nucleic acid pump of the male bacterium," but not for the adsorption (Ippen and Valentine 1966).

Mutual phage exclusion was studied by mixed infection of *E. coli* with the phage FH5 and the subsequently superinfecting DNA phages T6 or φX174 (Huppert and Blum-Emerique 1967; Huppert et al. 1967). In this case, the second phage was not excluded and the development of the first phage, namely FH5, was blocked or inhibited, even late in its growth cycle. However, single cells that were capable of producing both phages were isolated when superinfection by the DNA phage was delayed for 15–25 min: the yield of the RNA phage was reduced under these circumstances (Huppert et al. 1967).

The infection of *E. coli* with a strain of the phage T4 interrupted the previous infection with the phages MS2 or R23 (Yarosh and Levinthal 1967) or M12 (Hattman and Hofschneider 1967) when the cultures were superinfected 20 min later with the phage T4. Addition of actinomycin D or chloramphenicol before the T4 superinfection prevented the interruption (Yarosh and Levinthal 1967).

Summarizing the simultaneous M12 and T4 infection, it appeared that the parental M12 RNA remained intact, but no replicative intermediates were formed, whereas, by delayed T4 superinfection, the M12 RNA was synthesized but no M12 progeny phage was produced (Hattman and Hofschneider 1967). Detailed elaboration of the inhibitory effect showed that the production of the replicase activity of the phage M12 was completely blocked by the T4 superinfection and therefore the T4 interfered with translation of the M12 RNA, possibly on the polysomal level (Hattman and Hofschneider 1968a,b). A sharp reduction in the M12 coat production was observed within several minutes after superinfection with the phage T4 (Hattman 1970).

More detailed information on the T4 superinfection effect on the phage f2 propagation was achieved by using rifamycin and peptide-mapping procedures (Goldman and Lodish 1971). It appeared that the exclusion of the phage f2 required the T4 gene function soon after T4 infection, while synthesis of the f2 coat occurred at a reduced level until 4 min after T4 superinfection and then ceased abruptly, preexisting f2 replicative intermediate RNA and f2 single-stranded RNA were degraded to small fragments, and most f2-specific RNA was released from polyribosomes (Goldman and Lodish 1971). As a consequence, it was suggested that the phage T4 induced the synthesis of an endoribonuclease which specifically degraded f2 RNA and bacterial mRNA.

The *in vitro* translation studies indicated that the T4 infection induced an alteration in host ribosomes which restricted the translation of host and other T4-unrelated template RNAs including MS2 RNA but permitted normal translation of T4 RNA (Hsu and Weiss 1969). The factor that was present in the ribosome fraction from the infected cells operated at the level of polypeptide chain initiation and interfered unequally with the initiation of protein synthesis at the three known genes of the phage R17 (Dube and Rudland 1970; Pollack et al. 1970; Lee-Huang and Ochoa 1971).

The situation with the T4 superinfection appeared more complicated when T4 ghosts were involved into the superinfection studies (Goldman and Lodish 1973). Both the phage T4 and DNA-free T4 ghosts inhibited replication of the phage f2, but by fundamentally different mechanisms. Most but not all of the effects by T4 upon f2 growth could be blocked by the addition of rifamycin prior to T4 superinfection; by contrast, the inhibition of the f2 synthesis by T4 ghosts could not be blocked by rifamycin. Therefore, the inhibition by intact T4 required gene function, while inhibition by ghosts did not. However, no new f2 RNA was synthesized after superinfection with either (Goldman and Lodish 1973). Further studies led to the suggestion that messenger RNA competition might be a mechanism by which the T4 superinfection of cells infected with the phage f2 blocked translation of f2 RNA and possibly host mRNA (Goldman and Lodish 1975).

The exclusion phenomenon among related RNA phages was investigated with four phages, β, I, Y, and Z, which could be distinguished from one another by their serological characters and plaque types (Tsuchida et al. 1971b). When *E. coli* cells were infected simultaneously with two serologically related phages, no exclusion was observed, but the exclusion was apparent when the cells were infected with two serologically unrelated phages. A phage having been judged to be highly "virulent" by the simultaneous infection method was excluded by a weakly "virulent" phage when tested by the superinfection method (Tsuchida et al. 1971b).

Simultaneous infection with both Qβ and MS2 phages led to infection of the cells by both phages (Ling et al. 1970). Upon simultaneous addition, the two phages interfered with each other and resulted in reduced infection by each phage; however, preinfection of the cell by one phage did not affect subsequent infection by the other. The phage progenies produced during double infection were authentic Qβ and MS2; phenotypic mixing (mixed-coat particles) and genomic masking (hybrid particles) were not detected. The *in vitro* phage reassembly showed that high species specificity in forming reassembled particles could be achieved when the proteins and RNAs of the two phages freely interacted with one another (Ling et al. 1970). By the coinfection of *E. coli* with the phages fr and GA, a pair more related with each other than the Qβ and MS2 pair mentioned above, only authentic fr and GA virions were formed (Rumnieks et al. 2009).

The superinfection of *E. coli* with amber mutants of the phage R17, which produced high levels of the R17 replicase, inhibited production of the phage Qβ as well as of the DNA phage fd, whereas the inhibition required R17 replicase production and was related to the amount of replicase produced (Scott and Iglewski 1975).

Recently, a coinfection study by the phage MS2 and T7 or φX174 was undertaken, in order to observe the impact of internal competition between phages on their replicative fitness which might lead to establishing the possibility of developing a model for human pathogenic coinfections (Brewster et al. 2012). The chosen alternate phages differed from Qβ in their receptor requirement, to eliminate extracellular competition for the adsorption places, and genome type, to test the impact of different genomes. Generally, it was concluded that the coinfection resulted in an earlier incidence of lysis compared to single infection with one phage alone. Competitive inhibition among phages was explained by mutual requirements for the cellular machinery and metabolites, such as ribosomes, nucleotides, and amino acids (Brewster et al. 2012).

PREPARATIVE GROWTH AND PURIFICATION

The semipreparative isolation of the phage f2 in large quantities was described in the first paper on the RNA phages (Loeb and Zinder 1961). Approximately 2×10^8 bacteria per mL in a 15-liter culture were infected and allowed to go to lysis. Chloroform and lysozyme were added to complete the lysis. The phage yield was 7×10^{11} pfu per mL. The lysate was made 2 M with ammonium sulfate and the precipitate collected by centrifugation. This precipitate was suspended in 200 mL of water, brought to pH 8 with NaOH, and then stirred in a blender. Deoxyribonuclease and ribonuclease were added, and the solution was centrifuged, giving a supernatant containing a total of 8×10^{15} phage. Further purification included ammonium sulfate precipitations, ultracentrifugation at 60,000 g which led to a marked loss of viable titer by about 90%, CsCl density gradient centrifugation that did not show further loss of viable titer, re-centrifugation at 60,000 g, re-suspension and removing debris by low-speed centrifugation, with no further loss of viability. The final solution had a viable titer of 5×10^{13} per mL of a total of 5.5×10^{14} pfu, with a recovery of about 5% of viable particles (Loeb and Zinder 1961).

Strauss and Sinsheimer (1963) proposed a semipreparative MS2 propagation method in 7-l cultures with a detailed description of the purification scheme adopted as a modification from the phage φX174 purification procedure. In order to scale up the MS2 growth and purification, Rushizky et al. (1965a) implemented a large-scale 300-l fermentation by the use of enriched media and inoculation of MS2 into higher *E. coli* densities (4×10^9 cells/mL), which was followed by a simplified purification procedure and final outcome of 65 mg of phage per liter of medium. The further improvement of this method by better aeration and a different medium led to the final yield of 265 mg of the phage MS2 per liter (Rogerson and Rushizky 1975).

The spermine-induced intracellular retention of the phage f2, which was mentioned above, was proposed as a simple method of the phage concentration for its scaled-up production (Groman 1966; Groman and Suzuki 1966).

Virustat, a device in analogy with the bacteria-producing chemostat, was invented by Homer Jacobson and Leslie J. Jacobson (1966). The continuous production of the phage MS2 was accomplished in the Virustat over run periods of 1–6 days. The phage lysates were produced continuously at flow rates of 20–170 mL/hr and at titers ranging nearly to those given by optimal batch cultures. Nevertheless, Nagai et al. (1971) demonstrated that the production of the phage MS2 was tenfold higher in a 2-liter fermenter than in shake flasks. An efficient method for the propagation and purification of the phage MS2 was elaborated at that time by the Grēns' team (Rosenthal et al. 1971).

Later, a Cellstat with a bubble wall-growth scraper was used for the continuous culture of the phage Qβ (Husimi and Keweloh 1987). This Cellstat was developed earlier to culture the DNA phage fd for the study of molecular evolution and was used then for the evolution studies of the phage Qβ (Husimi 1989).

An automated system was used by the phage MS2 growth measurements and followed by the mathematical modeling and theoretical analysis of the growth kinetics (Corman et al. 1986). Then, the continuously cultured host bacteria infected with the phage Qβ promoted the development of computer simulations and further applications of the optimized system to strategies of molecular evolution (Schwienhorst et al. 1996). It is worth mention here the earlier mathematical models for the RNA phage production which were developed by Chen et al. (1971) and Grossi et al. (1977). Finally, the above-described model for the energy-efficient growth of the phage Qβ was elaborated (Kim and Yin 2004).

The striking methodology based on the culture vessels capable of giving maximum oxygen solution rates of the dense bacterial cultures was used successfully for the high-yield production of the phage μ2 (Sargeant and Yeo 1966). Thus, the 3-liter, 20-liter, and 150-liter bacterial cultures were grown in stirred, deep culture vessels to average bacterial cell densities of 71×10^8, 63×10^8, and 43×10^8 viable organisms per milliliter, respectively, and then infected with phage. The average yield of progeny phage in each case was ca. 3000 minimum pfu per cell and the average mass of phage obtained in the 3-liter experiments was not less than 124 mg/L, or about 20-fold higher than it was obtainable by conventional methods in aerated, shaken culture flasks. In fact, the actual phage yields were much higher than the minimum values calculated above from plaque counts. The carbon dioxide evolution rate of cultures was measured and used as a guide to the time at which phage should be added, while cultures grown under conditions of low aeration gave poor yields of phage (Sargeant and Yeo 1966). This advanced methodology was approved later by the cultivation of the phages R17 (Sargeant 1970) and Qβ (Sargeant et al. 1971).

The shake flask and mechanically agitated fermentor cultures were compared by the production of the phage MS2: oxygen supply was the most critical point for the strong advantages of the fermentor cultivation, and dissolved oxygen was found to be a useful indicator of residual glucose (Nagai et al. 1971).

The new steric chromatography technique using beds or columns of powdered glass, the individual particles of which contain pores of closely controlled size, which appeared in 1965, was used opportunely for the isolation of the phage MS2 (Haller 1967). This technique was applied successfully further for the large-scale preparation of the phages M12 and Qβ, as well as of the DNA phages M13, φX174, and T4, from highly concentrated crude lysates without detectable loss of infectivity (Gschwender et al. 1969). To get such highly concentrated phage lysates, Gschwender and Hofschneider (1969) achieved lysis inhibition of the M12- and Qβ-infected *E. coli*, as well as of the φX174-infected bacteria, by magnesium ions, which resulted in two- to threefold higher phage yields. Lysis of the host bacteria was inhibited if the growth medium was adjusted to 0.2 M MgSO₄ 30 min before infection or earlier.

As an alternate method of the phage MS2 purification by chromatography, calcium phosphate columns were proposed (Cernohorský et al. 1968). Then, the molecular sieve chromatography on an agarose column was used for the purification of the phage MS2 (Yoshinaga and Shimomura 1971). The phage MS2 was concentrated also by charge-modified filter chromatography: phage lysates were first clarified by filtration through serum-coated membrane filters, then the clarified lysate was adjusted to pH 6 and passed through a Zeta-plus filter (30 S size); greater than 99% phage adsorption occurred under these conditions and adsorbed phage was successfully eluted (Goyal et al. 1980a). The phages f2 and MS2 were involved, together with a set of the DNA phages, in thorough studies on the potential application of DEAE sepharose and octyl-sepharose column chromatography for the purification of viruses (Shields and Farrah 2002).

A rapid phage concentration from crude lysates in the presence of polyethylene glycol, namely PEG 6000, was applied for the large-scale purification of a long set of the DNA phages, but also of the phage R17 (Yamamoto et al. 1970). This paper was decisive for the further extension of the phase partition technique for the concentration and purification of viruses and nucleic acids.

Large-scale purification of the phage Qβ by differential centrifugation was described by McGregor et al. (1974). Using the phage R17, a tricky and useful improvement was introduced for the sedimentation of viruses at very low concentrations: a "trap" of 5% sucrose placed in the bottom of the tube by high-speed centrifugation prevented the redistribution of the small pelleted material after the rotor came to rest (McNaughton and Matthews 1971).

A special procedure was necessary for the large-scale isolation in milligram quantities of the caulophage φCB5 (Leffler et al. 1971). Isolation of the φCB5 by conventional procedures was impossible because of its extreme sensitivity to high ionic strength that excluded salt precipitation or the CsCl gradient fractions. To achieve the goal, the infected cells containing mature virions were concentrated 5.5 hours after infection before the cells begin to disintegrate. Then, the concentrated infected cells were converted to spheroplasts by lysozyme treatment in the presence of an osmotically protective concentration of polyethylene glycol (Leffler et al. 1971).

The flow field-flow fractionation (FFF), was applied for the separation and purification of the phage Qβ as a model,

although the phages f2 and MS2 were also involved in the study (Giddings et al. 1977). The forced-flow electrophoresis (FFE), was developed later with the use of the phage MS2, among other phages, to characterize the separation process (Mullon et al. 1987).

The phages MS2 and Qβ were used, among other icosahedral viruses, by the novel universal calibration of gel permeation chromatography and determination of its separation principles (Potschka 1987).

The phage MS2 was used as a standard by the elaboration of zonal density gradient electrophoresis for the separation of proteins and phages, as well for the correct estimation of their electrophoretic mobility (Mullon et al. 1986).

Recently, an optimal preparation methodology was published for the phage MS2 in connection with growing role of the latter as an additive or seeding indicator by phage ecology, or virus inactivation, studies (Dunkin et al. 2017c).

Due to the current ecological and nanotechnological challenges, the problem retains the urgent character of the RNA phage propagation and purification. Thus, Shi and Tarabara (2018) evaluated comparatively two phage propagation methods, in broth and on double agar overlay, and three popular purification procedures—PEG precipitation, centrifugal diafiltration, and CsCl density gradient centrifugation—using the phage MS2 as a model, in parallel with the DNA phage P22. The CsCl density gradient centrifugation gave the highest quality phage suspensions, while the phage propagation in broth media resulted in the higher purity phage stocks regardless of the purification method applied. The production and storage methodology of the phage MS2 as a simulant for yellow fever virus and Hantaan virus was presented by Shin et al. (2018). Subsequently, a general method was published on how to produce phages in bioreactors, which could be adapted easily to many host-phage systems and various operating conditions (Agboluaje and Sauvageau 2018). A substantial overview was later devoted to different operation modes of the phage production such as batch, semi-continuous, and especially continuous, with the pros and cons of each (Jurač et al. 2019).

CELL-FREE SYNTHESIS

The conception of the continuous RNA phage RNA-directed cell-free translation system capable of producing polypeptides in high yield was developed by Alexander Sergeevich Spirin's team. This system used a continuous flow of the feeding buffer, including amino acids, ATP, and GTP, through the reaction mixture and continuous removal of a polypeptide product (Spirin et al. 1988). Both prokaryotic *E. coli* and eukaryotic wheat embryo *Triticum* sp. versions of the system were generated. In both cases, the system has proven active for long amounts of time, synthesizing polypeptides at a high constant rate for dozens of hours. Thus, with the MS2 RNA, 100 copies of coat protein per RNA were synthesized for 20 hours (Spirin et al. 1988).

Formation of the phage MS2 infectious units in the cell-free translation system was demonstrated first by Katanaev et al. (1996). Thus, the simple cell-free translation system from *E. coli*, programmed with the phage MS2 RNA, was able to infect male *E. coli* cells. The plaques appearing on the *E. coli* host strain were morphologically indistinguishable from those derived from the normal phage MS2 infection. The cell-free based infection was maximal under conditions favoring the highest synthesis of the A protein. Of note, it was the first example of the infectious unit formation in a cell-free translation system. By the cell-free translation technique, the phage Qβ RNA was used further as a vector, where dihydrofolate reductase mRNA was inserted instead of the coat protein gene (Katanaev et al. 1995). Then, Qβ RNA variants appeared as templates by the coupled replication-translation, as presented in Chapter 18. The large-scale translation process by the cell-free synthesis is outlined in Chapter 16.

Furthermore, *E. coli*-based cell-free synthesis was used for the efficient production of the MS2 virus-like particles (VLPs) and other VLPs, such as hepatitis B core VLPs, and demonstrated capability to overcome many of the limitations of the current VLP production processes (Bundy et al. 2008). Thus, the MS2 VLPs were produced at a yield 1×10^{14} particles per mL, about 14 times the best published production yield.

Recently, a new generation *E. coli*-based cell-free transcription-translation (TXTL) system was capable of synthesizing infectious phages, such as MS2, φX174, and T7 (Garamella et al. 2016; Rustad et al. 2017). The TXTL system, including the phage MS2 RNA-driven one, was improved substantially by the engagement of mass spectrometry to determine the presence of synthesized proteins (Garenne et al. 2019).

Strong contribution of cell-free techniques to modern nanotechnology was achieved recently by Schwarz-Schilling et al. (2018). Gregorio et al. (2019) published an extensive overview of cell-free protein synthesis over 60 years.

LYSOGENY

In contrast to the lytic cycle of the phage reproduction, the lysogenic cycle assumes formation of a prophage that is either integration of the phage genome into the host bacterium's genome or formation of a circular replicon in the bacterial cytoplasm, while the host cell continues to live and reproduce normally. Although absence of the elements complementary to the RNA phage genome in the host's chromosome (Doi and Spiegelman 1962) and full independence of the phage replication from the DNA synthesis (Cooper and Zinder 1962; Knolle and Kaudewitz 1964) were established at the first steps of the RNA phage studies, the idea of the putative lysogeny persisted.

The reason for such persistence consisted first of the RNA phage physiology and their pronounced tendency to coexist with host bacteria. This was evident from the following general features: lack of a strong suppression of the bacterial processes, incompleteness of lysis, and high number of surviving bacteria. These typical RNA phage properties allowed early speculations about the putative "stable lysogenization" or an "unstable carrier state" of the RNA phage-infected bacteria, which could appear predominantly, for example, by the RNA phage fr infection at late log-phase of the host with cell concentration of about 3×10^8 per mL, which led to

a non-lytic outcome of the infection (Hoffmann-Berling et al. 1963a). Addition of an anti-phage serum to the host resulted in the full depletion of the cells from the phage. Nevertheless, the presence of the fr-lysogenic cells, which were growing more slowly than the non-lysogenic ones, was acknowledged (Hoffmann-Berling et al. 1963a). Knolle (1964a) accepted three potential responses to the phage fr infection:

> … a cell may become (a) temporarily immune to phage fr, while some of its descendants return to fr sensitivity; (b) infected, resulting in lysis; (c) infected, resulting in a carrier-property, whereby the cell either liberates phage particles without lysis and produces cellular progeny, or divides after infection, permitting a phage-carrying descendant to lyse and nonphage-carrying descendants to continue to grow," although it was accepted that "so far no cooperation between the fr RNA-genome and host DNA has been demonstrated. (Reprinted from Knolle P. Cellular responses of bacterial hosts to contact with particles of the RNA phage fr. *Virology.* 1964;23:271–273, Copyright 1964, with permission from Elsevier.)

Concerning pseudomonaphages, a potential recognition of the lysogeny was more clearly defined than in the case of the coliphages. As Feary et al. (1964) wrote, "the nature of phage 7s is somewhat of an enigma." It was generally accepted that the relationship between the phage 7s and *P. aeruginosa* strain Ps-7 "could be obtained by application of the sequence complementarity test devised by Hall and Spiegelman (1961) and recently used (Doi and Spiegelman 1962) to show that no homology exists between the RNA of phage MS2 and the DNA of its host strain *E. coli* Hfr strain K-16" (Feary et al. 1964). However, such information was never published. Bradley (1965b) accepted that the "the phage 7s is temperate and causes lysogeny in a strain of *P. aeruginosa* but can be grown on suitable indicator strains of the same species." When a direct test of the phage PP7 for the lysogeny in two *P. aeruginosa* strains was performed, a strong indication for the lysogeny was found in one of them; however, this indication of apparent lysogeny was "cured" after growth in the presence of phage antiserum (Bradley 1972a). Thus, it was concluded that this strain "was not truly lysogenic and phage PP7 existed in some form of carrier state rather than as prophage, which was suggested for phage 7s, an RNA phage described by Feary et al. (1964)" (Bradley 1972a).

Direct arguments for the lysogeny in the case of the phage f2 were published by Vera Zgaga (1977). The reaction of Charles Weissmann (1978) followed immediately and clarified the situation. It was noted that

> the state in which bacterial cultures produce phage without lysing and are resistant to superinfection has been called "carrier state" or "viral persistence" without implying any mechanism… The term "lysogeny" implies that the viral genome is present in the host in a noninfectious form, in particular that it is integrated into the host genome; until evidence for such criteria is adduced in the case of RNA phages it would be preferable to retain the original designation of "carrier state" (Weissmann 1978).

After half a year, Weissmann's lab published a straightforward study that described a hybrid plasmid carrying a complete DNA copy of the phage Qβ genome and capable to elicit, inserted in either orientation, the formation of phage Qβ when introduced into *E. coli* (Taniguchi et al. 1978a,b).

In this context of Weissmann's criteria, the "pseudolysogenization" was studied on the phage Qβ and fairly stable "lysogenic-like" bacteria were established (Watanabe I 1967; Watanabe I et al. 1968b; Sakurai 1969). These cells contained few mature phages intracellularly (less than 10^{-3} pfu per cell), continued to grow with a potential to produce Qβ phages upon downward temperature shift from 44° to 37°C, and showed an immunity-like response against homologous phage infection. Anti-Qβ serum did not influence the growth. Judging from temperature shift experiments, the phage growth seemed to be blocked at some stage in these cells, and this status was designated as a "metaphage," or "pseudolysogenic" state (Watanabe et al. 1979). However, the pseudolysogenic cells obtained in this manner segregated phage-free cells with a relatively high frequency. The fairly stable lysogenic-like bacteria were established between an A2, or maturation, protein amber mutant of the phage Qβ (Qβam205) and its non-permissive *E. coli* host BE110. These bacteria contained few mature phages intracellularly (less than 10^{-3} pfu per cell), continued to grow and produce Qβam205 spontaneously, and showed an immunity-like response against homologous phage infection and no reaction against antiserum presence, as stated above. These cells, which were capable of replicating the phage mutant and to distribute it to daughter cells synchronously with cell division, were regarded as pseudolysogenic bacteria (Watanabe et al. 1979).

In the context of lysogeny and/or pseudolysogeny of the RNA phages, the name of Tamara Pererva, nee Sherban, a clear enthusiast of the putative RNA phage lysogeny, must be mentioned. She participated in the early studies on the action of mutagens on the phage MS2 and its RNA (Krivisky and Sherban 1969, 1970; Sherban 1969; Sherban and Krivisky 1969; Budowsky et al. 1971, 1972, 1973, 1974a,b; Bespalova et al. 1976). Morphologic mutants of the phage MS2 that were isolated during these studies induced an interest to the lysogeny problem, although one of the most prominent morphological mutants, a rim mutant (Sherban 1969) was classified later as a temperature-sensitive coat protein mutant carrying a missense mutation in the coat (Dishler et al. 1980b).

To explain the above-mentioned immunity, or phage-induced resistance of about 1,3% of the *E. coli* host cells to the infection with the same phage, in this case, MS2, it was assumed that a direct interaction of phage products with the episomal DNA coding proteins for F pili might occur (Pererva 1977). Detailed evaluation of these mutants revealed their further segregation by F pili properties (Pererva and Malyuta 1984). As a result of complicated selection experiments, a λ phage able to conduct MS2 genome was isolated (Pererva et al. 1993a). Moreover, cloning of the chromosomal DNA of one of the MS2-induced host mutants led to plasmids that were able to induce synthesis of an MS2-specific RNA in the transduced *E. coli* cells (Pererva et al. 1993b). The size and map of the cloned MS2-specific sequences were determined, and a hypothesis that the plasmid of interest might contain a plus-chain of MS2 RNA-like DNA or even MS2 RNA within a three-stranded structure was proposed (Pererva et al. 1995). Segregation of the MS2-specific information from transducing P1 and λ phages was studied, but mostly by microbiological methodology (Pererva et al. 1996). Furthermore, DNA-containing phages with some MS2-like immunological properties were obtained (Pererva 1998a) and the development of the RNA phage-resistant host forms was described (Pererva 1998b, 1999). The MS2-specific sequences were mapped by hybridization within the physical map of the MS2-induced DNA-containing P23-2 phage genome (Miryuta et al. 2001). Then, inhibition of the spontaneous segregation of new forms of the MS2-induced *E. coli* mutants was achieved by extracts of some medicinal plants, such as *Polyscias filicifolia*, *Rhodiola rosea*, and *Ungernia victoria* (Dvornyk et al. 2004). The MS2-induced *E. coli* mutant was used for the construction of a bacterial test system for the primary screening of anticancer drugs (Pererva et al. 2007). The whole set of Pererva's team studies on the RNA phage lysogeny was self-reviewed thoroughly and is available in English (Pererva et al. 2008).

Finally, the pseudomonaphage LeviOr01 was discovered, which was capable of inducing a stable carrier state in a *P. aeruginosa* host and could be continuously released (Pourcel et al. 2017). Moreover, the establishment of a carrier state was a frequent outcome of infection, since the majority of tolerant bacteria happened to produce phage. Such cells also resisted other phages that used the type IV pili as a receptor. The carrier population was composed of a mixture of the cells producing phage and the susceptible cells that were non-carriers. The carrier cells accumulated phage until they burst, releasing large quantities of virions. Whole-genome sequencing of the LeviOr01-resistant variants revealed the presence of mutations that most probably were selected by the phage. The mutations were situated in the genes involved in the type IV pilus biogenesis, but also in the genes affecting lipopolysaccharide synthesis. Such stable carrier state has not been described previously for leviviruses (Pourcel et al. 2017). It is noteworthy that the phage LeviOr01 was the first RNA phage that received its official name according to the nomenclature proposed by Kropinski et al. (2009): the phage was called vB_PaeL_PcyII-10_LeviOr01, or LeviOr01 for short.

5 Distribution

We are more closely connected to the invisible than to the visible.

Novalis

There are no passengers on Spaceship Earth. Everybody's crew.

Marshall McLuhan

GEOGRAPHY

The two first RNA phages, f2 and MS2, were very similar and appeared by further elucidation as rather variants of the same virus than different species, although they have been isolated at the Atlantic and Pacific coasts of the United States, respectively. Looking at Table 1.1, it is evident that all early isolated coliphages in the United States and Europe have been closely related, belonging to the later established serological group I (this serological group was nominated in 1965, 5 years after the first phages were found). The closely related specimens were isolated in Australia and later in New Zealand. The slight immunological differences that have been detected among them were explained after some years, when the structure of coat protein was determined, by a few amino acid substitutions in their coat proteins.

The first serologically different phages, Qβ, VK, and ST from the future serological group III, and the phages GA and SD from group II, as well as the phages SP and FI, the first representatives of group IV, were discovered in Japan. Nevertheless, the closely related analogs of the group I phages, namely the phages β, B1, I, MY, and ZR, were also identified in Japan.

Remarkably, the serologically related pseudomonaphages 7s and PP7 were discovered in the United States and the UK, respectively. The similar caulophages of the φCB5 group and a pair φCR14 and φCR28 were isolated at the Pacific and Atlantic coasts of the United States, respectively.

Figure 5.1 presents a picture of the geographical location of the most popular RNA phages, which have been listed in Table 1.1, on the world map. It is clear from the latter that the representatives of the serological groups II, III, and IV were found, in the course of time, also in North and South America, as well as in Europe. However, Figure 5.1 does not pretend to be a real demonstration of the RNA phage distribution, since, in the most cases, the phages were searched for special properties (e.g., for pili specificity), but not for pure statistics of their distribution, as in the case of a special set of pili-dependent phages in South Africa, and did not reflect the statistical situation in the region in question.

The first large and unprecedented survey on the geographical distribution of the RNA phages was undertaken by Japanese scientists. Isolation and grouping of RNA phages was performed in Taiwan (Miyake et al. 1971a); Arabia, India, and Thailand (Aoi et al. 1972); Brazil (Miyake et al. 1973); Peru,

Bolivia, Mexico, Kuwait, France, Australia, and the United States (Furuse et al. 1975b). In parallel, the exhaustive survey was also performed in Japan proper (Furuse et al. 1973); Okinawa Island (Aoi et al. 1974); and the islands in the adjacent seas of Japan: Rebun, Rishiri, Hachijojima, Miyakejima, and Niijima Islands (Furuse et al. 1975a). Alterations in the distributional pattern of RNA phages in summer and in winter were considered (Furuse and Watanabe 1973). In general, the distribution of the RNA coliphages in South and East Asia was reviewed systematically by Furuse et al. (1978). Figure 5.2, which is taken from this paper, demonstrates a geographical map of this distribution with the prevalence of the group representatives in different regions.

In total, the survey in the Philippines, Singapore, Indonesia, India, and Thailand was performed by collecting sewage samples from domestic drainage in November 1976 (Furuse et al. 1978). Of the 221 samples collected from domestic drainage, 50 contained RNA phages. By serological analysis, 46 of the 52 strains were found to belong to group III. Thus, the latter was declared as the most prevalent in Southeast Asia, namely in Amamiohshima, mainland Okinawa, Ishigakijima, Iriomotejima Islands, Taiwan, the Philippines, Singapore, and Indonesia. The sewage samples collected from domestic drainage in Japan indicated that the most prevalent RNA phages in the mainland Japan (north of Kyushu) were the group II phages, whereas the group III phages were predominant in the southern part of Japan (south of Amamiohshima Island). Therefore, a borderline between Kyushu and Amamiohshima Island was proposed for the geographical distribution of the RNA coliphages in the domestic drainage of South and East Asia (Furuse et al. 1978). This borderline of the group II and III prevalence regions is shown in Figure 5.2 by a colored dotted line.

The continuous survey over a 5-year period from 1973 to 1977 was undertaken on the RNA phages from domestic drainage in Japan proper and islands in the seas adjacent to Japan (Furuse et al. 1979a). The stability and continuity of the RNA phage distribution in their natural habitats was confirmed by the amount and group types of the RNA phages in the respective domestic drainage samples. The frequencies of the RNA phage isolation remained high and constant. The group types of the RNA phages isolated were also stable, for example, in the three cities, Choshi, Niigata, and Toyama in Japan proper, which gave the average ratio for the groups II:III = 3:1 for the three cities. A survey in Choshi city also revealed a similar result: II:III = 6:1 in 1981 compared with the II:III = 4:1 in 1977 (Furuse 1987). In the islands, the frequency of isolation of the RNA phages and the group types of the isolated RNA phages were fairly high as in the case of the above three cities in Japan proper, and were also stable (Furuse et al. 1979a). The local predominance confirmed substantially the picture that is presented in Figure 5.2. Thus,

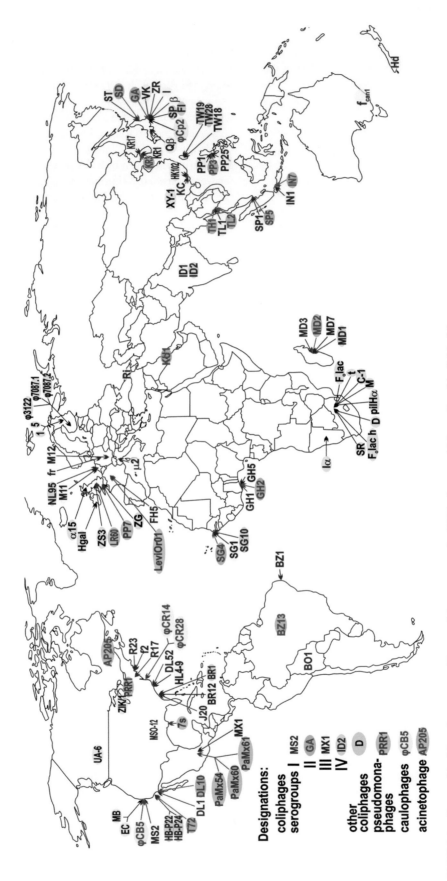

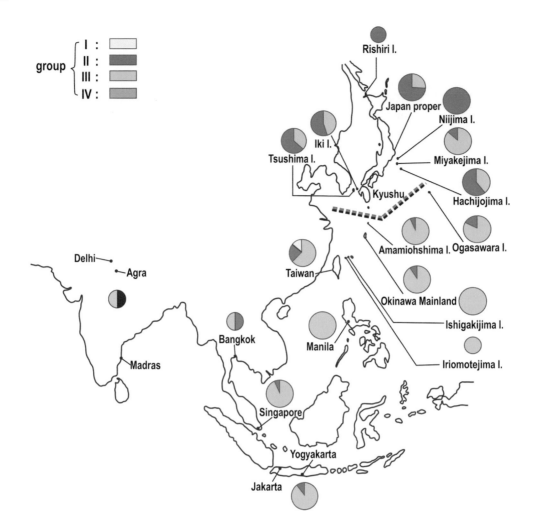

FIGURE 5.2 Distribution of the RNA phages in South and East Asia. Smaller circles indicate a smaller number of RNA phage strains isolated. The dotted line shows a borderline between Kyushu and Amamiohshima Islands. (Redrawn with permission of ASM from Furuse K et al. *Appl Environ Microbiol.* 1978;35:995–1002.)

insofar as domestic drainage was concerned, the appearance of the RNA coliphages was highly stable in terms of their quantity and group types in Japan proper and islands over relatively long periods of time: at least for 5 years in Niigata and Toyama, and 10 years in Choshi (Furuse 1987), and the apparent difference in the geographical distribution of RNA phages in Japan existed between Kyushu and Amamiohshima Islands (Furuse et al. 1979a), in accordance with the previously established borderline (Furuse et al. 1978).

Next, the systematic survey on the geographical RNA phage distribution was performed in Korea by collecting sewage samples from domestic drainage in densely populated urban areas in July through August 1979 (Osawa et al. 1981a). Of the samples, 74 of 132, or 56%, contained 106 strains of RNA phages. The ratio 4:47:55 was established for the I, II, and III groups, respectively, by serological analysis. Based on previous data for Japan with the ratio 3:1 = II:III and Southeast Asia with mostly group III, the distribution pattern of the RNA phages in Korea was of an intermediate type between those of Japan and Southeast Asia (Osawa et al. 1981a). Remarkably, the group II:III = 1.5:1 ratio for Iki and Tsushima Islands, which are located between Japan and

Korea, showed a somewhat different distribution pattern from that of mainland Japan with the groups II:III = 3:1 (Furuse et al. 1979a). A gradual increase in the group III over the group II phages existed, according to separation from Japan proper toward the south as follows: Rebun and Rishiri (mostly group II) → Japan proper (north of Kyushu) (II:III = 3:1) → Iki and Tsushima (II:III = 3:2) → Korea (II:III = 1:1) → Southeast Asia (south of Amamiohshima) (mostly group III) (Furuse 1987). Insofar as domestic drainage was concerned, the frequencies of isolation of the group I and the group IV phages were negligible in Korea, as in Japan (Furuse et al. 1973, 1975a) and Southeast Asia (Furuse et al. 1978).

The evident difference in the group ratio was explained by the local climate influence, since the group II phages, which were the most prevalent RNA phages in Japan proper, had the lowest optimal temperature for growth among the four known groups of the RNA phages in *in vitro* experiments. They could propagate even at 20°C, which was a limiting temperature for the group I, III, and IV phages. Furthermore, the group I, III, and IV phages could propagate almost normally at 40°C, whereas the group II phages could produce any progeny at that temperature (Osawa et al. 1981a). In the temperature sensitivity context, it is necessary to emphasize that the large survey data appeared to reflect the influence of general climate surroundings on the domestic drainage, because the sewage samples were obtained exclusively from the domestic drainage whose physical conditions appeared to be directly controlled by the local climate. Moreover, the domestic drainage seemed to suffer little influence from contamination by animal sources such as human or domestic animal feces, and seemed to represent the most favorable natural habitat for the RNA coliphages (Furuse 1987).

In parallel with the large Japanese survey in the Southeast Asia, the appearance of the RNA coliphages was studied in the sewage from rural and urban areas of Hong Kong by Dhillon et al. (1970). As a result, the coliphage content of sewage collected from 11 different Hong Kong localities was determined. In general, urban sewage tended to be richer than rural sewage both in the pfu count as well as in the plaque morphological variation, and approximately 32% of 77 phage isolates were found to be male specific. It is remarkable that 700 *E. coli* types of colonies were tested, and none was found to be lysed by the RNA phage MS2. On the other hand, when liquid cultures of the same strains were infected with MS2, six cultures showed an increase in phage titer, indicating the presence of some phage-sensitive cells (Dhillon et al. 1970). The RNA phage isolates fell into two distinguishable antigenic classes: members of one class being related to the phage MS2 and those of the other being related to the phage Qβ. It is remarkable that in this study the MS2-related phages, i.e., belonging to group I, were found to be more widely distributed than the Qβ-related phages, i.e., belonging to group III, whereas most habitats sampled were found to yield only one or the other kind of phage (Dhillon and Dhillon 1974). Thus, from the viewpoint of the geographical distribution of the RNA phages, the sewage of Hong Kong appeared to have certain features in contrast with those of other Asian countries.

To verify the generality of such a climate dependent distribution pattern of the RNA phages observed in South and East Asia, the distribution pattern of the RNA phages was examined in two mainland African countries (Senegal and Ghana) and in Madagascar, where the environmental conditions were unique and remarkably different from those of Asia (Furuse et al. 1983b). In Senegal, among 65 sewage samples, 14 or 22%, contained 16 strains of the RNA phages, 13 of which belonged to group III. It was consistent with the distribution pattern of the RNA coliphages in tropical and subtropical regions of Asia as Taiwan (28%) (Miyake et al. 1971a), Japan (20%–70%) (Furuse et al. 1973, 1975a, 1979a), the Philippines (48%), Singapore (35%), Indonesia (21%) (Furuse et al. 1978), and Korea (56%) (Osawa et al. 1981a). The sewage samples from Ghana and Madagascar, however, had only limited numbers of the RNA phages, which was consistent with those for Brazil (5%) (Miyake et al. 1973), Mexico (2%) (Furuse et al. 1975b), India (3%), and Thailand (5%) (Furuse et al. 1978). In Ghana, among 106 samples, only seven, or 7%, contained the RNA phages in the ratio I:II:III = 1:3:3. In Madagascar, among 124 samples, 7, or 6%, contained RNA phages in the ratio I:II:III:IV = 1:1:1:4.

In spite of the low isolation frequency, Madagascar appeared to have a unique distribution pattern that differed from that of any other countries that have been examined before, but the generality of the distribution pattern of the RNA phages in the tropical region, namely abundance of the group III phages, was verified at least in Senegal (Furuse et al. 1983b). In parallel, another study on the distribution of the RNA coliphages in Senegal, Ghana, and Madagascar was published in a local journal (Rasolofonirina and Coulanges 1982).

SOURCES

Japanese scientists were the first to pay profound attention not only to the location, but also to the physical source of the RNA phages. In the course of the systematic surveys on the RNA phage distribution, the sewage samples from domestic drainage and feces of certain domestic animals and of humans were used as sources for the isolation of the phages. Analysis of these three types of material revealed a unique distribution pattern of the RNA phages in each material examined. As stated above, the sewage from domestic drainage in Japan proper contained group II and III phages in a 3:1 ratio (Furuse et al. 1973, 1979a). The sewage from slaughterhouses, feces of domestic animals, and feces of animals in zoological gardens contained the group I phages exclusively, and the feces of humans contained the group II and III phages almost equally, although the frequencies of isolation of these phages in human and animal feces were fairly low: 2% and 6%, respectively (Osawa et al. 1981b). The distribution patterns of the RNA phages in raw sewage collected from treatment plants in Sapporo, Tokyo, and Toyama contained appreciable amounts of the group I phages in addition to the group II and III phages (Furuse et al. 1981). As a whole, raw sewage from treatment plants in Japan contained the RNA phages of the three groups in the ratio I:II:III = 1:2:5. Based on the distribution patterns of the RNA phages in sewage from domestic drainage in Japan proper, namely the group II:III = 3:1 as stated above, in animal feces and sewage from slaughterhouses (mostly group I), and in human feces (group II:III = 1:1), Furuse et al. (1981) were the first to proclaim clearly that the group I phages tend to be introduced from animal sources and group II and III phages tend to be introduced from human sources.

In his exhaustive review on the natural distribution of the RNA phages, Kohsuke Furuse (1987) indicated seven potential phage sources: (i) sewage from domestic drainage, (ii) raw sewage from treatment plants, (iii) animal feces including those of man, cows, pigs, and several animals in zoological gardens, (iv) river water, (v) pond or lake water, (vi) irrigation water, and (vii) seawater, where the first three were the most suitable sources for the isolation of the RNA coliphages. Thus, the coliphages were detected in almost all sewage samples collected from the domestic drainage examined. The amounts of total coliphages in this material were fairly high, ranging from 10 to 10^7 pfu/mL (many between 10^2 and 10^5 pfu/mL) and the relative amounts of the RNA phages in the sewage samples were high, occupying 10%–90% of the total coliphages isolated. As a result, the 1020 strains isolated in the large Japanese survey were classified into one of the four known RNA phage groups (Furuse 1987).

Concerning the distribution pattern of the RNA coliphages in animals and in man, and the extent of their contribution to the propagation and transmission cycles in their natural habitats, the large Japanese survey investigated the numbers and types of RNA coliphages in animal sources such as (i) fecal samples from domestic animals (cows, pigs, horses, and fowl), some other animals in zoological gardens, and humans, (ii) the gastrointestinal contents of cows and pigs, and (iii) sewage samples from treatment plants in slaughterhouses. The concentration of the RNA coliphages from the first and second sources was fairly low ($10-10^3$ pfu/mL of original phage sample), whereas that from the third source was fairly high (10^3-10^5 pfu/g). As for the group types of the RNA phages from the first and second sources, human feces were found to contain the RNA phages of groups II and III in almost equal proportions, the gastrointestinal contents of pigs included those of groups I and II equally, and the feces and gastrointestinal contents of mammals other than man and pigs had those of group I exclusively. From the third source, the group I phages were found most, with a minor fraction of the group II phages (Osawa et al. 1981b; Furuse 1987). Furuse (1987) concluded that the microbial flora demonstrated the prominent feature of the distribution pattern of the RNA coliphages by the existence of preferential relationships between the RNA phage groups and their host animals. Thus, the group III phages were isolated exclusively from human feces, and the group I phages were isolated from the feces and gastrointestinal contents of mammals other than man. The group II phages were thought to belong primarily to humans, although they were also found in the gastrointestinal contents of pigs and in raw sewage collected from treatment plants of slaughterhouses. It remained unclear whether the group II phages observed in pigs were intrinsic to them or had been introduced from human sources by chance. In any case, pigs appeared to be able to support the propagation of the RNA phages of group II as well as of group I under the usual breeding conditions, suggesting some similarities of gastrointestinal conditions between humans and pigs. The group IV phages could be isolated from the feces of animals and humans, sewage from domestic drainage, and raw sewage from treatment plants, showing the broadest habitat among the four RNA phage groups, although the isolation frequencies of these phages were very low. The group I phages could be isolated only from animals such as foxes, elephants, horses, cows, and pigs, and from sewage from slaughterhouses, but not from humans or sewage from domestic drainage. The group I phages were therefore thought to belong primarily to animals (Furuse 1987). The fact that the feces of some domestic animals (pigs and cows) contained fairly large amounts of the RNA phages suggested that the RNA phages could be capable of propagating in the intestines of these animals. Experiments were designed to clarify this point using germfree mice as the test animals (Ando et al. 1979). The propagation of the phage MS2 was detected in the F$^+$-established gnotobiotic

mice, while MS2 was unable to propagate or colonize in the intestines of the germfree or F$^-$-established gnotobiotic mice. At 6–24 hr after phage inoculation, the phage MS2 excreted in the feces of the mice reached a maximum level of $10^{10}-10^{11}$ pfu/g of feces. At 2–4 days after phage inoculation, the titer of the phage MS2 decreased appreciably, but the pfu levels were found to stabilize at between 10^5 and 10^9 pfu/g of feces thereafter. When the F$^+$-established mice were inoculated with a lower and a higher titer of the phages GA, Qβ, or SP, their growth patterns were similar to those of the phage MS2 except with inoculations of a low titer of the phage Qβ. Thus, the serogroup types and inoculum size had no effect on the final, stable pfu levels (10^4-10^{10} pfu/g of feces) except in the case of the phage Qβ (Ando et al. 1979). By double infections, the best colonization response was exhibited by the phage MS2 (over Qβ and SP), which was consistent with the aforementioned data that the feces and gastrointestinal contents of mammals other than man contained the group I phages exclusively (Osawa et al. 1981b; Furuse 1987). By double infection with the phages GA and Qβ, the predominance of GA was demonstrated (Osawa et al. 1981b).

Administration of antibiotics to the MS2-established mice (excreting 10^8 pfu/g of feces) demonstrated clearly that the male bacteria growth or colonization in the intestines of germfree mice was obligatory for the propagation of the RNA phages in the mice (Osawa et al. 1981b).

The data on the serogroup predominance in animal and human specimens, which were collected during the large survey in Japan, were confirmed by survey in the Netherlands by isolation of many RNA phages from fecal samples of various domestic animals (Havelaar et al. 1986). In this study, the RNA phages were detected in appreciable numbers only in feces from pigs, broiler chickens, sheep, and calves but not from dogs, cows, horses, and humans. In tight accordance with the previous data (Furuse et al. 1981), all animal isolates belonged to either group I or IV. In contrast with animal isolates, 19 isolates from hospital wastewater belonged to the groups II or III (Havelaar et al. 1986). Another and most prominent feature of this study was the abundance of the group IV phages isolated from animal feces, in accordance with the observations in Asia (Furuse 1987).

The distribution patterns of the RNA phages in raw sewage, which was thought to manifest the overall distribution pattern but collected from the treatment plants in various localities in Japan and several other countries, were necessary to determine the transmission cycle of the RNA phages in their natural habitats (Furuse et al. 1981). The densities of total coliphages were variable, ranging from 10 to 10^8 pfu/mL. A tendency appeared toward somewhat higher phage densities in summer (10^4-10^6 pfu/mL) than in winter ($10-10^3$ pfu/mL) from the continuous surveys (two to four times per year) over 3- to 5-year periods on raw sewage from several treatment plants in Japan. Almost all of the samples contained the RNA phages at relatively high densities, that is, 5%–90% of the total coliphages present (with many at 30%–60%). Thus, the RNA phages constituted the predominant phage species in raw sewage from treatment plants, as in the case of sewage from domestic drainage (Furuse 1987). By serological analysis, 1832 RNA phage strains isolated from Japan were classified into one of the four known groups (I/II/III/IV:221/487/1122/2), while most of the samples contained only group II and III phages. As a whole, raw sewage from treatment plants in Japan contained RNA phages belonging to the three groups in a ratio of I:II:III = 1:2:5. Two prominent features gathered from this material were the predominance of the group III over the group II phages and the existence of the group I phages in certain treatment plants (Furuse 1987). It is noteworthy in this connection that the treatment plant in Tokyo received sewage from slaughterhouses together with that from human sources.

The continual occurrence of the RNA phages in raw sewage from treatment plants was also confirmed from continuous surveys over 3- to 5-year periods, although the density of the RNA phages fluctuated more widely than that in domestic drainage (Furuse et al. 1979a; Furuse 1987). Most of the sewage samples contained the RNA phages of group II and/or III at fairly high densities comparable with those from the treatment plants in Japan. However, sewage samples from Recife (Brazil) and Würzburg (Germany) contained the group I phages only, although it was impossible to ascertain whether or not the samples included sewage from slaughterhouses.

Based on the distribution patterns of the RNA phages in sewage from domestic drainage in Japan proper (II:III = 3:1), in animal feces and sewage from slaughterhouses (mostly group 1), and in human feces (III:II = 1:1), it was reasonably assumed that the group I phages observed in the raw sewage from treatment plants were most likely introduced from animal sources, and the group II and III phages were from human sources (Furuse 1987).

Concluding the 5-year period of the continuous survey on the distribution pattern of the RNA coliphages in fixed domestic drainages in Japan, the qualitative and quantitative stabilities of these phages in sewage (Furuse et al. 1979a) and in raw sewage from treatment plants (Furuse et al. 1981) were demonstrated. The findings suggested strongly the idea that the sewage of domestic drainage constituted one of the most suitable natural habitats for the RNA coliphages (Furuse 1987).

In the context of the stability of the RNA coliphage distribution, the survey performed in 2007–2008 in the Tonegawa River basin, Japan, was of outstanding interest (Haramoto et al. 2009). Thus, after about 30 years of the surveys described above, distribution of the FRNA coliphage genogroup, which repeated the same RNA phage serogroups, was tested in a total of 18 river water and sediment samples. This time, modern plaque isolation procedures combined with the TaqMan-based genogroup-specific RT-PCR were applied. The FRNA phages of the human genogroups II and III were detected in 32, or 38%, plaques, whereas those of the animal genogroups I and IV were detected in 17, or 20%, plaques (Haramoto et al. 2009). Similar results were obtained by the survey performed in 2010–2011 in the Kofu basin, Japan, where wastewater and river water were tested. This time, the FRNA coliphages were detected in the 30 tested water samples that were collected

from a wastewater treatment plant and a river in the Kofu basin on fine weather days. Of phage plaques isolated, 187 (82%) of 227 were classified into one of the four FRNA coliphage genogroups. In full accordance with the previous survey, the human genogroup II and III FRNA coliphages were more abundant in raw sewage than those of the animal genogroups I and IV. However, an unexpected finding was that the genogroup distribution of the FRNA coliphages in the secondary-treated sewage samples was quite different from that in the raw sewage samples: the group I phages were much more abundant in the secondary-treated sewage, probably because of a less efficient inactivation of them in comparison with other genogroups during wastewater treatment.

The group II phages were abundant in the river water samples that were considered to be contaminated with human feces, independent of rainfall effects. For the first time, the rainfall effect was studied by collecting water samples not only on fine weather days, but also by a sequential sampling of river water during a rainfall event (Haramoto et al. 2012c). The appearance of the FRNA phage genogroup representatives in the raw water samples at drinking water treatment plants over Japan demonstrated different contribution of human and animal fecal contamination depending on the drinking water treatment plant location (Haramoto et al. 2012b). In order to support these studies, a novel prospective method, the electronegative membrane-vortex (EMV) method, was developed for simultaneous concentration of viruses and protozoa from a single water sample (Haramoto et al. 2012a).

The lowest and the highest reduction of the group I and group III FRNA phages, respectively, was confirmed also at a full-scale wastewater treatment plant in Japan by the qPCR and plate count assays (Hata et al. 2012).

The general tendencies of the FRNA phage distribution were confirmed by a methodologically well-supported and modern survey of surface water samples that were collected at three points along the Katsura River and a point on the Furu River, a tributary of the Uji River in Kyoto Prefecture, Japan, in 2014–1015 (Hata et al. 2016). In this study, the number of infectious phages, but not only total number of genomes, was considered. For the characterization of the infectious genotypes, the integrated culture RT-PCR coupled with the most probable number approach was applied to surface water samples. Furthermore, a special method was elaborated to recover low concentrations of the FRNA phage genotypes. Again, the genotypes I and II tended to be predominant at locations impacted by treated and untreated municipal wastewater, respectively. Therefore, the ratio of the group I and II phages could be a good indicator of viral contaminations originating from either untreated or treated municipal wastewater. The group III phages could be a better indicator of the contamination of untreated wastewater, since they were more susceptible to wastewater treatments than the group II phages. The group IV phages appeared as a rare genotype in water environments where human waste was the main contamination source (Hata el al. 2012, 2016). Remarkably, the numbers and proportions of the infectious FRNA phages tended to be higher during the winter season when water temperature decreased.

Because of these "new period" studies in Japan, the group I FRNA phages, as the most stable during treatment, were proposed as an appropriate indicator of virus reduction during wastewater treatment (Haramoto et al. 2015).

The phages from activated sludge as a specific ecosystem were studied for the first time in Tampere, Finland, by Hantula et al. (1991). As a result, three RNA phages, φ3122, φ7087.1, and φ7087, were isolated from 49 studied virus-host systems. They were found similar to the phage MS2 by the protein composition and sedimentation behavior, but no serological or other characterization was performed.

By analogy to the above-described coliphages, the known RNA pseudomona- and caulophages were isolated from sewage samples (see Table 1.1), with two exceptions for the caulophages φCR14 and φCR28 where sources were described as "ponds, slowmoving streams and tropical fish tanks" (Johnson et al. 1977) and for the PRR1-like phage LR60, which was isolated from concentrated river water (Primrose et al. 1982). However, no special studies on the geographical and/or natural source distribution of the pseudomona- and caulophages have been performed yet.

It is of great interest and importance that there is currently only one reported marine RNA phage, as confirmed by highly authoritative reviewers (Lang et al. 2009; Culley 2018). This RNA phage, 06N-58P, isolated from coastal Japanese waters and infecting a marine strain of *Pseudomonas* (Hidaka 1971, 1972, 1975), demonstrated, however, a diameter of 60 nm, more than expected for the leviviruses (Hidaka and Ichida 1976). Unfortunately, nothing new has appeared about the properties of this phage 50 years after discovery.

Even more intriguingly, metagenomic sequencing revealed the presence of the *Leviviridae* sequences on Livingston Island (Antarctic Peninsula), especially on the lysis plaque-like macroscopic blighted patches within the predominant microbial mats (Velázquez et al. 2016). Those blighting circles were associated with decay in physiological traits in the spatial microstructure and were evidenced at a time of unprecedented rates of local warming in the Antarctic Peninsula area.

Without any doubt, future developments by the search for the RNA phage sources relate to the coming of the next-generation sequencing era. By performing weekly metagenomic sequencing of organisms in San Francisco wastewater, full-length genomes of two novel RNA phages, EC and MB, were assembled (Greninger and DeRisi 2015). The EC and MB phages comprised the typical *Leviviridae* genome organization. Absolutely novel RNA phage sequences were found by Krishnamurthy et al. (2016). These novel sequences were present in samples collected from a range of ecological niches worldwide, including invertebrates and extreme microbial sediment, demonstrating that the RNA phages were more widely distributed than was recognized previously. One RNA phage was detected in the transcriptome of a pure culture of *Streptomyces avermitilis*, suggesting for the first time that the known tropism of the RNA phages may include Gram-positive bacteria. Then, two RNA phages found in stool samples from a longitudinal cohort of macaques, suggesting

that they were generally acutely present rather than persistent (Krishnamurthy et al. 2016).

HUMAN BEINGS

As presented above, geographical and source studies on the RNA phage distribution triggered the onset and rapid development of a novel discipline known as "phage ecology." The next logical question of phage ecology was directed to the ecological role of the phages in human health, first of all, in the human intestine. For this reason, fecal samples were collected from healthy individuals and from patients with certain intestinal diseases (Furuse et al. 1983a). However, in contrast with the widespread distribution of the RNA coliphages in sewage samples from domestic drainage or raw sewage from treatment plants, the human fecal samples contained only limited numbers of the RNA phages: only 12 RNA phage strains appeared among 747 fecal samples (Furuse 1987). When the individuals whose feces had been shown to contain RNA phage at the time of the first survey were followed, the RNA phages were not detected in the samples obtained from the same subjects on the second and third occasions (Furuse 1987). Accordingly, no RNA coliphages were found in feces of cows, pigs, and humans in Hong Kong (Dhillon et al. 1976). Therefore, the RNA coliphages seemed to constitute only a minor fraction of the total coliphages in the feces of humans and animals (Osawa et al. 1981b; Furuse 1987).

Although the RNA phages were making a minimal contribution to the diversity of the intestinal virome (Zhang T et al. 2006), the attempt to characterize virus-like particles that were associated with fecal and cecal microbiota is worthy of mention (Hoyles et al. 2014).

ENUMERATION

Logically, the permanent discovery of the male-specific phages (or F-specific RNA phages, or FRNA phages for short) in the wastewater was connected with the growing control of water quality. It was well-established before that human "enteric" viruses can survive water treatment processes, especially disinfection, better than pathogenic or indicator bacteria (Berg et al. 1978). Thus, it was postulated that the water regarded as hygienically safe, because conventional bacterial indicators of fecal pollution such as coliform organisms and fecal streptococci could not be detected, might still contain infectious viruses (Havelaar and Hogeboom 1984). Since it was technically difficult to detect those viruses that appeared to be most significant epidemiologically such as hepatitis A virus, Norwalk-like viruses, rotavirus, and other still unclassified and unknown gastroenteritis viruses, the FRNA phages were proposed as a surrogate instrument for the solution of the water quality problem. For this reason, in order to operate quantitatively with the RNA phage presence in the different sources, a special method for the enumeration of male-specific bacteriophages in sewage was developed (Havelaar and Hogeboom 1984). For the first time, a special F *S. typhimurium* strain, namely, WG49, was designed for the enumeration of the FRNA phages in fecally polluted waters. Provided that the number of the F-specific phages in sewage was at least one order of magnitude greater than that of somatic *Salmonella* phages, the interference by the latter could be ignored from a practical point of view. The typical representatives of the four serogroups—MS2, GA, Qβ, and SP—were used to confirm the capabilities of the constructed enumeration strain. Almost all the phages detected by the enumeration had a host range restricted to male *Salmonella* or *Escherichia coli* strains, were resistant to chloroform, and their infectivity was inhibited by ribonuclease, while electron microscopy revealed phage particles that were morphologically identical to the RNA phages (Havelaar and Hogeboom 1984).

The development of this simple and highly selective method, based on the specially constructed host strain of *S. typhimurium* WG49 which ensured more reliable and higher FRNA phage counts from natural sewage samples than those obtained on *E. coli* hosts, was described further with all necessary arguments and technical details (Havelaar et al. 1985, 1986; Havelaar 1987a,b).

As a result, the FRNA phages together with the simple and specific enumeration method were chosen for the virologic analysis of water as a necessary and sufficient surrogate model (Havelaar 1987b). Remarkably, the FRNA phages were seldom found in non-fecally contaminated wastewater, but surprisingly low numbers were found in feces. The FRNA phages in wastewater effluent were found to be highly resistant to chloramines and relatively resistant to UV inactivation. Thus, the FRNA phages were involved effectively in the evaluation of disinfection processes for water treatment plants as indicator organisms for human pathogenic viruses (Havelaar 1987a,b). The high specificity of the special *S. typhimurium* WG49 strain for the FRNA phages was confirmed by studies from many laboratories over a long period of time (Rhodes and Kator 1991; Bonadonna et al. 1993; Handzel et al. 1993; Payment and Franco 1993; Stetler and Williams 1996; Pallin et al. 1997; Maiello et al. 1999; Lucena et al. 2004; Pusch et al. 2005; Ottoson et al. 2006; Yousefi and Zazouli 2008; Hata et al. 2014). However, some interference with the assay by somatic *Salmonella* phages was also reported by the detection of the FRNA phages in groundwater (Williams and Stetler 1994).

Nevertheless, the choice of a good host strain for phage enumeration remained one of the most important questions that influenced the usage of the FRNA phages as indicators of enteric viruses during the 1990s. The above-mentioned paper of Hantula et al. (1991), where three new RNA phages appeared, was devoted generally to the search for the appropriate host. The search for the appropriate *E. coli* hosts was performed in order to detect phages in effluents of Finnish and Nicaraguan wastewater treatment plants. Meanwhile, 13 new RNA phages, but without any given names, were isolated out of 38 phages that were found in this parallel survey in Finland and Nicaragua, countries where the climate was very different (Rajala-Mustonen and Heinonen-Tanski 1994). Furthermore, the hosts, as well as procedures for the FRNA phage detection, were standardized by the International Organization for Standardization (ISO; Geneva, Switzerland). An official ISO standard for the detection and enumeration of F-specific RNA bacteriophages, ISO 10705-1 was issued (ISO 1995). As the appropriate link https://www.iso.org/standard/18794.html informs, this standard was last reviewed and confirmed in 2017. Therefore, this version remains current.

A serious evaluation of the ISO 10705-1 was carried out during the European project "Bacteriophages in bathing waters" from January 1996–June 1999, mainly on optimization of the different steps for culturing the host strain *S. typhimurium* WG49 (Mooijman et al. 2002). It was concluded that all steps described in the ISO 10705-1 were necessary and, if followed carefully, using a culture of the WG49 strain of good quality, reliable results could be obtained for the enumeration of the FRNA phages. Further, conservation of the reference materials and water samples containing the FRNA phages was optimized (Mendez et al. 2002).

A conversion of the *E. coli* strain WG5 recommended by the ISO and by the U.S. Environmental Protection Agency (USEPA) to detect somatic coliphages into an F+ strain capable of simultaneous detection of both somatic and F-specific coliphages was presented by Guzmán et al. (2008). Furthermore, *E. coli* strain CB390 was proposed for the simultaneous detection of the FRNA and somatic phages (Agulló-Barceló et al. 2016; Bailey et al. 2017). The whole-genome sequence of the classical *S. typhimurium* strain WG49 and *E. coli* strain WG5 were published recently (Bothma et al. 2018).

An alternate *E. coli* host strain was proposed for the enumeration of the FRNA phages by Debartolomeis and Cabelli (1991). This strain had three antibiotic resistance markers: ampicillin on the Famp plasmid, which coded for F pili production, and streptomycin and nalidixic acid on the chromosome, and more than 95% of the phages from environmental samples which plaqued on the host strain were F-specific.

Simple membrane filtration/elution methods to concentrate and enumerate the FRNA phages in raw and finished drinking water were elaborated in the 1990s (Sobsey et al. 1990; Sinton et al. 1996; Li et al. 1998). The purpose of these methods was to use the FRNA phages as a viral indicator, an approach which became very popular in the near future. Moreover, the phage MS2 played a central role in the use of a method for the detection of animal and bacterial viruses adsorbed to biomedical waste crush (Anderson et al. 1996).

The integrated virus detection system (IVDS) was developed with the phage MS2 as a standard (Wick and McCubbin 1999), while purification of the phage MS2 from complex growth media by ultrafiltration and resulting analysis by the IVDS approach were presented (Wick and Patrick 1999b).

The recovery and quantitation of the FRNA phages from marine waters and sediments was improved by Reynolds et al. (1993). A comparative study of techniques used to recover the FRNA phages from residual urban sludge was performed (Mignotte et al. 1999). Later, a VIRADEN procedure, which combined concentration and detection and offered the opportunity of having realistic measurements of viruses in volumes of up to 10 liters of tested water, was applied for the FRNA phage detection (Mocé-Llivina et al. 2005).

Further development of the FRNA phage enumeration technique, which was achieved by the introduction of such modern methodological approaches as plaque and line blot hybridization with oligonucleotide probes, CLAT, or antibody-coated polymeric bead agglutination assay, RT-PCR, RT-qPCR, and metagenomic sequencing, was described in the Genogrouping section in Chapter 2.

It is necessary to note here an application of the electrospray ionization–ion trap mass spectrometry (ESI-MS) for the identification of the phage MS2 in the presence of a large excess of *E. coli* (Xiang et al. 2000). By this technique, the low ESI flow rate and precursor ion accumulation capability of the ion trap MS enabled high-sensitivity MS/MS analyses. Precursor ions were automatically selected and analyzed using tandem MS (MS/MS) to produce "global" MS/MS surveys and processed to yield two-dimensional MS/MS spectral displays. Thus, such global MS/MS surveys were demonstrated for *E. coli* lysates and enabled not only confident microorganism identification, but also sensitive detection of the phage MS2 (Xiang et al. 2000). Then, the phage MS2 was used as a model by the elaboration of a simple method for rapid identification of microorganisms by on-slide proteolytic digestion followed by MALDI-MS, or matrix-assisted laser desorption/ionization tandem mass spectrometry, without isolation or fractionation of microorganisms (Yao et al. 2002a,b). Finally, the phage MS2 was employed as a model by the elaboration of the universal Integrated Cell Culture-Mass Spectrometry, or ICC-MS, method which detected different infectious viruses by identifying viral proteins that appeared in cell cultures (Ye et al. 2019).

The phage enumeration was permanently developed and improved. Thus, the real-time fluorogenic RT-PCR assays were elaborated for detection of the phage MS2 (O'Connell et al. 2006). A combined method for the concentration and enumeration of the infectious FRNA phages, together with poliovirus and hepatitis A virus, from inoculated leaves of salad vegetables was elaborated (Dubois et al. 2006). By the involvement of the phage MS2 and poliovirus as models, the novel electropositive disposable capsule filter ViroCap was fabricated for concentration of viruses from deionized water and artificial seawater, as well as natural ground, surface, and seawater (Bennett et al. 2010).

The combination of the PCR and the pfu method for the enumeration led to the reliable analysis of mass balance of viruses, for example in a coagulation-ceramic microfiltration hybrid system, where the seeded phages MS2 and Qβ were employed (Matsushita et al. 2006). To selectively quantify the infective phages, the enzymatic treatment with proteinase K and ribonuclease, or ET (for enzymatic treatment) step, was introduced before the qPCR reaction and the ET-qPCR technique demonstrated accurate enumeration of the phage MS2 infectivity upon partial inactivation by three treatments: heating at 72°C, singlet oxygen, and UV radiation (Pecson et al. 2009). To maximally avoid false-positive results, a theoretical framework was elaborated to relate phage infectivity with genome damage measured by qPCR and tested with the UV-irradiation damaged phage MS2 (Pecson et al. 2011).

Further, with the phage MS2 as a model, combination of both ribonuclease A and proteinase K or ribonuclease A alone were employed for the enzymatic pretreatment of wastewater to minimize recovery of RNA from inactive phages (Unnithan et al. 2015).

However, Yang and Griffiths (2014b) showed that the ET RT-PCR method reduced, but did not eliminate completely, false-positive signals and proposed the two-step ET RT-PCR, in which the enzymes were added sequentially. The improved method was more effective at reducing false-positive signals than the one-step ET RT-PCR, which involved addition of both enzymes together (Yang and Griffiths 2014b). A novel culture-independent RT-PCR method for the source determination of fecal wastes in surface and storm waters was tested in samples that were collected in accordance with a standard operating procedure established by the USEPA New England Regional Laboratory (Paar et al. 2015).

A rapid and sensitive RT-qPCR test on the SYBR Green format was elaborated for the detection of the phage MS2 in wastewater (Gentilomi et al. 2008).

With the phage MS2 as a model, a novel online monitoring method for detection of viruses in flowing drinking water was elaborated by the integrated capture and spectroscopic detection of viruses (Vargas et al. 2009).

With the involvement of the phage MS2, a dual layer filtration system was designed for the simultaneous concentration of bacteria and phages from marine water (Abdelzaher et al. 2009). At the same time, the detection of the FRNA phages in fecal material was improved by extraction and polyethylene glycol precipitation, which assured highest recovery levels (Jones and Johns 2009). With the use of the phage MS2, extraction of RNA from stool suspensions was improved and standardized (Shulman et al. 2012).

Jones et al. (2009a) suggested also for the first time the use of ultrasonication as a post-treatment step to increase recovery of viruses from ultrafiltration devices. The phage MS2 was used as a model for the adaptation of the spot-titer culture-based method for phage enumeration (Beck et al. 2009). With the phage MS2 as a model, an adsorption-elution method together with RT-qPCR was fabricated, which was based on monolithic affinity filtration on columns consisting of a hydrolyzed macroporous epoxy-based polymer (Pei et al. 2012). Lanthanum-based flocculation method coupled with modified membrane filtration procedures was developed and evaluated with the phage MS2 as a model to detect viruses in large-volume, up to 40 L, water samples (Zhang Y et al. 2013). The phage MS2 was employed as a model by the concentration of viruses from seawater using granular activated carbon (Cormier et al. 2014). The nitrocellulose and ZetaPlus 60S membrane filters were compared for the simultaneous concentration of the FRNA phages and porcine teschoviruses and adenoviruses as indicators of fecal contamination of swine origin in river water (Jones et al. 2014a). Moreover, the phage MS2 contributed to the evaluation of a rapid and efficient filtration-based procedure for the separation and safe analysis of CBRN, or chemical, biological, radiological and nuclear agents, mixed samples that might be used as weapons in warfare and terrorism and contain both chemical and biological warfare agents (Bentahir et al. 2014). A new procedure based on concentration of the phage MS2 as a model on poly-L-lysine dendrigrafts, coupled with directed nucleic acid extraction and real-time PCR quantification, was proposed by Cadiere et al. (2013).

Generally, methods of the concentration and recovery of the FRNA phages, among other viruses, from water were summarized in a comprehensive review (Ikner et al. 2012). Later, a simple and cost-effective filter paper–based shipping and storage medium was developed for the environmental sampling of the FRNA phages when refrigeration was not possible (Pérez-Méndez et al. 2013). Then, ultracentrifugation was tried for the concentration of viruses from raw sludge, where the FRNA phages were involved as indicators (Jebri et al. 2014). A high-performance concentration method for viruses in drinking water was developed with the phage MS2 as a model (Kunze et al. 2015). A "nanofiltration" method was tested with the phage MS2, where superabsorbent poly(acrylamide-co-itaconic acid) polymer beads were to concentrate water samples (Xie et al. 2016). Concentration of the phage MS2 from river water was achieved by a combined ferric colloid adsorption and foam separation–based method, with MS2 phage leaching from ferric colloid (Suzuki et al. 2016, 2019). For the enumeration of airborne viruses, a personal electrostatic particle concentrator was fabricated with the phage MS2, as well as DNA phage T3, as models (Hong S et al. 2016). All studies on the concentration of the phage MS2 from water by membrane-based methods that were published since 2001 were reviewed exhaustively by Shi et al. (2017).

The phage MS2 was employed by the elaboration of the standoff detection of biological agents, which used laser-induced fluorescence (Farsund et al. 2012).

With the phage MS2 as a model, a simple sensor for viral particles was developed by measuring ionic conductivity through anodized alumina membranes (Chaturvedi et al. 2016). A porous silicon membrane-based electrochemical biosensor was proposed for label-free voltammetric detection of the phage MS2 (Reta et al. 2016).

With MS2 as the reference phage, Yang and Griffiths (2014a) developed the first fluorescence-based method for the detection of the FRNA phages in combination with the phage concentration. The method was based on testing the phage-mediated release of β-galactosidase and allowed detection of low numbers of the FRNA phages: when used alone, 1 log pfu/mL of the FRNA phages was detected within 3 h, while 0.01 pfu/mL was detected within 5 h when the method was combined with the concentration step.

To detect the aerosolized phage MS2, a novel method using electro-aerodynamic deposition and a field-effect transistor (FET) was developed, where MS2-antibody-bound particles were delivered to the FET during detection and neither a pretreatment antibody binding step on the FET channel nor washing process for MS2-antibody-binding were necessary (Park KT et al. 2015).

The phage MS2 was used as a model by the development of a high-sensitivity flow cytometer-based method, which ensured analysis of single viruses with a resolution comparable to that of electron microscopy and the throughput of flow cytometry (Ma L et al. 2016). It is noteworthy that Jebri et al. (2016) conducted removal of phages by transmission electron microscopy and published some pictures of the putative *Leviviridae* family members. The phage f2 was used as a model by the development of a novel and highly efficient method for isolating the FRNA phages from water and based on the electropositive silica gel particles (Liu W et al. 2017).

The phage MS2 was applied traditionally by the efficacy evaluation of commercial kits for viral DNA/RNA extraction (Saeidi et al. 2017).

It is necessary to consider that short MS2 RNA sequences, up to 100 bp, were not completely degraded by autoclaving and were recovered intact by molecular amplification (Unnithan et al. 2014a). Therefore, if short-indicator RNA sequences are used for virus identification and quantification, then post-autoclave RNA degradation methodology should be employed, which may include further autoclaving.

In order to distinguish between the viable and nonviable phage MS2, as well as norovirus and bacteria, a combined propidium monoazide-qPCR assay was proposed (Kim SY and Ko 2012). Differences in the enumeration of the phage MS2 with culture and RT-qPCR methods were analyzed by slow sand filtration at a pilot plant (Lodder et al. 2013). Propidium monoazide treatment coupled with long-amplicon qPCR assays were assessed for their ability to quantify infectious MS2 in pure cultures and following inactivation by a range of UV light exposures and chlorine doses (McLellan et al. 2016).

Recent methodological progress by measuring quantitative distribution of infectious FRNA phage genotypes allowed differentiation of infective strains from inactive strains in surface waters (Hata et al. 2016). In this study, the integrated culture RT-PCR coupled with the most probable number approach was applied to surface water samples, and an FRNA phage recovery method was developed to recover low FRNA phage concentrations without inactivation. The conclusion was that group I and group II genotypes tended to be predominant at locations impacted by treated and untreated municipal wastewater, respectively, and the numbers and proportions of the infectious RNA phages tended to be higher during the winter season when water temperature decreased (Hata et al. 2016).

Further improvement of the RNA phage monitoring was succeeded by the introduction of a novel concentration technique for their enumeration from 10-liter volumes of ambient surface waters such as lake, river, and marine and river water with varying turbidities (McMinn et al. 2017b). An anion-exchange resin was employed for the FRNA phage concentration in diverse water types (Chandler et al. 2017a). Application of the anion-exchange resin was elucidated before with the phages MS2, Qβ, and GA as models (Pérez-Méndez et al. 2014). An extensive review of the concentration techniques was published recently by Dincau et al. (2017).

A novel affinity system for the detection of the phage MS2 was fabricated using molecularly imprinted polymers (MIPs) (Altintas et al. 2015a). The MIP nanoparticles were artificial receptor ligands which could recognize and specifically bind to a target molecule and were more resistant to chemical and biological damage and inactivation than antibodies. The MIP targeting the phage MS2 as the template were designed and the regenerative MIP-based MS2 detection assay was developed using a new surface plasmon resonance biosensor which provided an alternative technology for the specific detection and removal of waterborne viruses. An extensive review was published at the same time on the rapid detection of viruses, predominantly, FRNA phages, using biosensors (Altintas et al. 2015b). High specificity of the novel method was confirmed by direct comparison of the MIPs targeted to MS2 or adenovirus (Altintas et al. 2015c). The principles of the molecularly imprinted polymers involving the RNA phages were reviewed recently (BelBruno 2019; Gast et al. 2019), while the use of the phage MS2 for the generation of the surface plasmon resonance sensors applicable in medical diagnosis was summarized by Saylan et al. (2018).

Interestingly, a vice versa approach by the phage-bacteria pair was invented and the MS2-covered long period fiber grating sensor, as one of the optic fiber sensors, was fabricated to bind and enumerate *E. coli* cells (Chiniforooshan et al. 2017). The phage MS2 amplification-coupled, bead-based sandwich-type immunoassay was developed to detect live *E. coli* cells (Mido et al. 2018). Earlier, the MS2 infection and detection of the MS2 coat protein was applied in the *E. coli* detection method that integrated immunomagnetic separation with the phage amplification prior to MALDI-MS analysis (Madonna et al. 2003).

It is surprising that the detection of a single MS2 capsid appeared possible. Thus, a single MS2 phage was detected with an extraordinary microcavity that married microcavity photonics with nanoplasmonic receptors (Arnold et al. 2015).

Meanwhile, the heterogeneous asymmetric recombinase polymerase amplification (haRPA) was applied for the phage MS2 detection in combination with the stepwise phage concentration from 1250 L drinking water into 1 mL (Elsäßer et al. 2018). Since then, digital PCR (dPCR) has become a promising technology for absolute quantification of nucleic acid without need of calibration curves. Following dPCR, various digital isothermal amplification methods were also developed which only required isothermal incubation. Among them, a loop-mediated isothermal amplification, or LAMP, became the most popular one and was adapted to the rapid enumeration of the phage MS2 as a model (Huang X et al. 2018; Lin et al. 2019).

The phage MS2 contributed to the generation of the handy devices for the rapid real-time laser-induced fluorescence detection and monitoring of microbial contaminants on solid surfaces before, during, and after decontamination (Babichenko et al. 2018). The phage MS2 was used as a model by the development of the QuantiPhage assay, a novel method for the rapid colorimetric detection of coliphages using cellulose pad materials (Rames and Macdonald 2019), a vertical flow-based paper immunosensor for rapid electrochemical and colorimetric detection of influenza virus (Bhardwaj et al. 2019), of the norovirus NanoZyme aptasensor (Weerathunge et al. 2019), and of a fully automated digital microfluidics platform for immunoassay of a wide range of pathogens (Coudron et al. 2019). Furthermore, the phage MS2 was involved in the proof of principle testing of a novel immunosensor based on hydrogen-terminated thermally hydrocarbonized porous silicon, or THCpSi, by means of electrochemical impedance spectroscopy, showing a detection limit of 4.9 pfu/mL (Guo et al. 2019), as well as by the generation of a biosensor for the detection of live/viable bacteria based on the response of the conductive polymer 4-(3-pyrrolyl) butyric acid to glucose-induced metabolites (Saucedo et al. 2018, 2019).

6 Ecology

There's only one degree of freshness—the first, which makes it also the last.

Mikhail Bulgakov

Water is life, and clean water means health.

Audrey Hepburn

A massive set of the RNA phage collections and a great deal of data about geographical distribution and natural sources of the RNA phages had been launched by the end of the 1970s, while the general molecular biology studies on the RNA phages reached a definite plateau at that time. Many molecular biologists turned their attention from the phages to the animal viruses. In contrast to molecular biology, the role of the RNA phages had started to grow, at the beginning of the 1980s, in ecological studies, where the RNA phages, among other bacteriophages, were recognized not only as a necessary element of the global ecological system, but also as a powerful surrogate of infectious viruses. As a consequence, a novel discipline known as phage ecology—in particular, as RNA phage ecology—emerged. Phage ecology, however, provided not only permanent control of the RNA phage appearance in the natural sources, but also contributed strongly to the steady development of the virus inactivation methodology.

The extensive geographical and natural source surveys performed by Itaru Watanabe, Kohsuke Furuse, their colleagues and many other researchers, the work of which was described in Chapter 5, paved the way to the birth of modern RNA phage ecology.

MEASURE OF POLLUTION

As we learned from the previous chapter, Furuse et al. (1981) were the first to track the main feature of the fecal pollution in water: the group I and IV phages were found mostly in animal sources, while the group II and III phages originated preferentially from human sources. This finding paved the way for the development of a simple method to source-track fecal contamination in water by use of the RNA coliphages, defined now mostly as the FRNA phages, as a model organism.

From the very beginning, the two distinctly different functions of a potential model organism were distinguished: the index function and the indicator function. The index function would inform about health risks and would appear in the estimations of fecal contamination, sewage contamination, type of fecal contamination, etc., while the indicator function should relate only to the effect of a performed treatment process (Havelaar and Pot-Hogeboom 1988). The criteria for the model organism were in fact the same in both index and indicator functions: the model organism must possess similar ecological characteristics as the pathogens to be modeled, greater resistance to the planned treatment than the pathogens in question, and simple propagation and purification methodology.

A bit surprisingly, the early studies did not detect the FRNA phages in feces of humans, cows, and pigs (Dhillon et al. 1976), or detected only occasionally some strains in fecal samples from a variety of mammals and birds (Osawa et al. 1981b). Further studies revealed rare FRNA-phage isolates from feces of humans, dogs, cows, and horses, but relatively high counts were obtained from pigs, sheep, and in particular, young broiler chickens, which was explained by the selection of IncF plasmids in animals raised in bioindustry (Havelaar et al. 1986; Havelaar and Pot-Hogeboom 1988; Havelaar et al. 1990b). Later, by the thorough investigation of the pork slaughter process, the FRNA phages were observed in wastewater, trailers, slaughter process water, and swine feces, and appeared to be the best suited for process control verification by wastewater treatment (Miller et al. 1998). Calci et al. (1998) found the male-specific phages (without any indication on the genome structure, DNA, or RNA) in fecal samples from cattle, chickens, dairy cows, dogs, ducks, geese, goats, hogs, horses, seagulls, sheep, and humans as well as in 64 sewerage samples. All animal species were found to harbor the male-specific phages, although the great majority excreted these viruses at very low levels. Nevertheless, the study demonstrated clearly that in areas affected by both human and animal wastes, the wastewater treatment plants are the principal contributors of the male-specific phages including the RNA phages to fresh, estuarine, and marine waters. Remarkably, the presence of the phage MS2 in the diet of Canada geese and pigeons increased greatly the presence of the phage in their feces, which were normally low (Ricca and Cooney 2000). Similarly, the RNA phage presence in animal feed indicated their appearance in feces (Maciorowski et al. 2001).

Not long ago, the FRNA phages were isolated from fecal samples of farms, wild animal habitats, and human wastewater plants, and phylogenetic analyses were performed on the partial nucleic acid sequences of 311 FRNA phages from various sources (Lee JE et al. 2009). The FRNA phages were most prevalent among geese (95%) and were least prevalent in cows (5%). Among the genogroups of the FRNA phages, the latter, which were isolated from animal fecal sources, belonged to either group I or group IV, and most from human wastewater sources were in groups II or III, in full accordance with the earlier observations. The FRNA phages isolated from various sources were divided into two main clusters. All FRNA phages isolated from human wastewater were grouped with the Qβ-like phages, while the FRNA phages isolated from most animal sources were grouped with the MS2-like phages. Statistical analysis revealed significant differences between human and animal phages, and the FRNA phages isolated from human waste were distinctively separated from those isolated from the animal sources. At the same time, the FDNA coliphages did not demonstrate significant difference in this sense. Therefore, this study demonstrated clearly that proper analysis of the FRNA phages could effectively distinguish fecal sources (Lee JE et al. 2009).

Although some difficulties might appear by the fecal sample analysis, the FRNA phages were isolated from sewage and were found consistently in a variety of wastewaters, the normal ranges being relatively independent of the type of wastewater. Data on the occurrence of the FRNA phages in a variety of wastewaters have been published in detail (Furuse et al. 1981; Havelaar and Hogeboom 1984; Furuse 1987; Havelaar and Pot-Hogeboom 1988; Havelaar et al. 1990b) but remained inconsistent with the great differences in counts between feces of e.g., humans and broiler chickens. In all types of wastewater examined, the FRNA phage counts were several orders of magnitude higher than those expected solely on the basis of the direct fecal input. It was concluded that the FRNA phages were able to multiply in the sewerage system, and their presence in a water sample appeared to be indicative of sewage pollution rather than direct fecal pollution (Havelaar and Pot-Hogeboom 1988).

However, the problem of the putative multiplication was complicated by the fact that the subunit synthesis and the assembly of F pili required minimum temperatures of 30°C and 25°C, respectively. Nevertheless, a special study showed that phage infection was possible at 20°C not only in the optimal growth medium, but also in natural waters, although less efficient, in the case when F pili were produced while growing at a temperature above 30°C (Havelaar and Pot-Hogeboom 1988). For example, the replicative cycle of the phage Qβ was prolonged when the temperature decreased, and the replication was ceased below 25°C (Woody and Cliver 1995). Therefore, the possibility of the FRNA phage multiplication in the environment was generally confirmed, but many questions remained unanswered: why do the FRNA phages not multiply in the intestines of warm-blooded animals? What is the frequency of cells bearing F pili in feces, and what factors govern its variation? What is the FRNA phage burst size under natural conditions? Concluding, the FRNA phages were declared as a *pseudo-index* of fecal contamination, since the fecal pollution of the environment did not directly lead to an increase in their number, but rather provided the ecological requirements necessary for their multiplication (Havelaar and Pot-Hogeboom 1988).

Concerning the above-postulated criteria for the model organisms, after simple laboratory methodology, the FRNA phages demonstrated great resistance to the possible treatment. The FRNA phages were capable of surviving well in water and during sewage treatment processes (Ayres 1977) and were relatively resistant to inactivation by disinfectants (Shah and McCamish 1972; Snead et al. 1980; Grabow et al. 1983, 1985), heat treatment (Burge et al. 1981), and sunlight (Kapuscinski and Mitchell 1983).

As for the FRNA phages in the role of *sewage indicators*, it was stated that in sewage, "RNA coliphage levels often

equal or exceed DNA virus titers, though the frequency of occurrence of RNA viruses is much lower than that of DNA viruses" (Poppell 1979, cited by Snowdon and Cliver 1989). Moreover, Poppell (1979) explained levels of the RNA coliphages encountered in sewage by a theoretical calculation based on the observed levels of the viruses in feces and concluded that the observed levels of the RNA coliphages in sewage fell within one order of magnitude of the expected levels.

WATER QUALITY

Although the FRNA phages were classified above as a *pseudo-index* model for the fecal, but rather sewage contamination of water, they have been suggested as one of the model organisms for the control of water quality, namely as ecological indicators of enteroviruses in various water systems. The coliphages in total were intended first to fulfill the role of the ecological indicators, in order to produce *virologically safe* water (Kott 1981; Šimková and Červenka 1981; IAWPRC 1983; Grabow et al. 1984; Funderberg and Sorber 1985; Borrego et al. 1987, 1990; Hernández-Delgado et al. 1991). A bit later, the next IAWPRC document (1991) acknowledged that the

F-specific RNA bacteriophages are a more homogeneous group than somatic coliphages and have generally a greater resistance. The value of somatic coliphages is further limited by their ability to multiply in unpolluted waters. Both groups are an index of sewage contamination rather than fecal contamination, hence their ecology is different from human enteric viruses. (Reprinted from IAWPRC Study Group on Health Related Water Microbiology. Bacteriophages as model viruses in water quality control. *Water Res.*, 25, 529–545, Copyright 1991, with permission from Elsevier.)

In their authoritative review at that early time, Snowdon and Cliver (1989) accepted that "small, round coliphages, especially those containing RNA, appear to offer some potential as indicators or indices of the presence of enteroviruses" and that the "RNA coliphages can and do replicate in host *E. coli* cells in the environment outside the intestines."

Nevertheless, after a thorough investigation of the concentration methods of a set of phages including the phage MS2, Karst et al. (1991) concluded that many difficulties made it probably impossible to use coliphages at all as "index" organisms or even as "indicators". Moriñigo et al. (1992) were also critical and considered total coliphages as more optimal indicators of the microbiological quality of the natural waters than the FRNA phages, since the latter "showed no direct relationship with the levels of fecal pollution, and this group was never detected in samples with a low level of enteroviruses."

At the same time, the FRNA phages, together with the *Bacteriodes fragilis* phages, were proposed again as a reliable future index-organism of drinking water pollution by sewage (Armon 1993; Armon and Kott 1993, 1995). The special role of the FRNA phages as indicators was reported by the assessment of the disinfection, chlorination, and UV irradiation efficacy (Tree et al. 1997).

INDICATORS OF FECAL POLLUTION

The employment of the FRNA phages in the role of clear indicators of fecal pollution was accepted generally by a large set of studies (Armon and Kott 1995, 1996; Araujo et al. 1997; Hellard et al. 1997; Lee JV et al. 1997; Calci et al. 1998; Cole et al. 2003; Gerba 2006; Stewart-Pullaro et al. 2006; Colford et al. 2007; Gino et al. 2007). However, it was necessary to keep in mind that no single indicator could fulfill all the needs of water quality monitoring, since each indicator had its own advantages and disadvantages, and the best results would be obtained by using combinations of indicators for different purposes (Grabow 1996).

Nevertheless, the FRNA phages were found highly useful for the tracking of the human and/or non-human origin of fecal pollution in environmental waters by the measure of groups II and III versus groups I and IV, respectively (Hsu et al. 1995; Sinton et al. 1998; Griffin et al. 2000; Grabow 2001; Brion et al. 2002; Schaper et al. 2002b; Sundram et al. 2002, 2005; Cole et al. 2003; Long et al. 2003, 2005; Noble et al. 2003; Vinjé et al. 2004; Stewart-Pullaro et al. 2006; Friedman et al. 2011).

Figure 6.1 demonstrates the distribution of the RNA phage group representatives that were isolated from fecal wastes of various animal species (Cole et al. 2003). It is seen that some group I phages were found also in human fecal wastes.

Furthermore, a model that could distinguish the human from the non-human sources and was suitable for recently and heavily polluted waters was elaborated by the machine learning and data mining technique (Belanche-Muñoz and Blanch 2008). The model searched for suitable tracers for microbial source tracking (MST), a term that included different methodological approaches to distinguish the origin of fecal pollution in water based on the use of microbial indicators. As a result, the model permitted determination of the source of fecal pollution, and the enumeration of the group II FRNA phages

appeared there as one of the key parameters. The group II FRNA phage marker figured further in an advanced set of markers for the detection of fecal contamination (Jeanneau et al. 2012). The usefulness of the group II marker was confirmed by a large interlaboratory study of the accuracy of the MST methods (Harwood et al. 2013).

Another predictive model was more empiric and built up by the analysis of an impressive set of data, including the FRNA phages (Costán-Longares et al. 2008).

The role of the FRNA phages as the fecal pollution indicators was addressed in large and mini reviews (Tyagi et al. 2006, 2009; Jofre 2007; Plummer and Long 2007; Stapleton et al. 2007; Stoeckel and Harwood 2007; Yates 2007; Campos 2008; Ehlers and Kock 2008; Templeton et al. 2008; Herzog et al. 2010; Lucena and Jofre 2010). Jofre et al. (2011) emphasized the greatest potential of the FRNA phages, including the group I/IV against II/III detection, together with the *Bacteroides* phages, for tracking sources of fecal pollution in water. Furthermore, novel extensive reviews were published on the subject (Goodridge and Steiner 2012; Lin and Ganesh 2013). It is noteworthy that recent reviews were rather critical, in light of novel data, about the importance of the FRNA phage grouping on the background of the plethora of new MST assays (Harwood et al. 2014). However, the role of at least additional indicators was never denied for the FRNA phages (Plummer et al. 2014). At the same time, the FRNA phages were respected in parallel with the *B. fragilis* phage GB-124, a novel and modern candidate for the detection of fecal pollution (McMinn et al. 2014).

The important role of the FRNA phages as the classical indicators of fecal pollution has been emphasized by the most recent reviews (Tran et al. 2015; Blanch et al. 2016; Jofre et al. 2016; Amarasiri et al. 2017; Jebri et al. 2017; Rodrigues and Cunha 2017; Dias et al. 2018; Haramoto et al. 2018; Hartard et al. 2018; García-Aljaro et al. 2019; Nappier et al. 2019). The high potential of the FRNA phages as viral indicators of fecal contamination when they were quantified from domestic and food-industrial effluents containing human, chicken, swine, or bovine wastes was confirmed strongly in recent large experimental studies (Barrios et al. 2018; Devane et al. 2019; Goh et al. 2019; Toribio-Avedillo et al. 2019; Worley-Morse et al. 2019). Remarkably, Barrios et al. (2018) emphasized high resistance of the FRNA phages to the primary and secondary treatments of the wastewater treatment plants as a clear benefit of the model. Auffret et al. (2019) evaluated the survival of the FRNA phages in swine and dairy manure, and did not recommend their use as an indicator of swine manure contamination by enteric viruses, since they might not be abundant in specific manures.

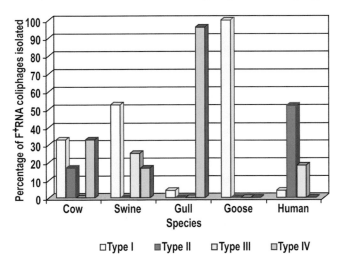

FIGURE 6.1 Percentages of F⁺ RNA coliphages isolated from fecal waste of various animal species. (Redrawn in color with permission of ASM from Cole D et al. *Appl Environ Microbiol.* 2003;69:6507–6514.)

SURVEILLANCE OF GROUND AND SURFACE WATERS

By the 1890s, it was decided by public health officials that general sanitation measures such as surveillance of ground and surface waters would be conducted to detect fecal pollution rather than individual pathogens, since there were too many pathogens, they were present in very small concentrations, and methods for their detection were not practical. *E. coli* became the primary indicator of fecal pollution at that time, but in the 1970s it was actualized that the fate of *E. coli* might differ substantially from the fate of pathogenic viruses in drinking water and especially in groundwater, since bacteria and viruses are differentially resistant to environmental conditions (Leclerc et al. 2000). However, the conclusion with regard to monitoring for enteric viruses was the same as that decided by public health officials in the 1890s: it is better to monitor for the indicators of fecal pollution than for specific pathogens, since they are too many, are present in low concentrations, and many of them could be still unknown. From the known pathogens, the most important and dangerous were the hepatitis A and E viruses, rotaviruses, astroviruses, adenoviruses, and caliciviruses including noroviruses. However, these groups of viruses did not correlate with each other by their occurrence, survival, and disinfection-resistant behavior (Leclerc et al. 2000).

As it has been brought out clearly above, the FRNA phages demonstrated real capability to perform a function of both the index and the indicator of the fecal pollution. Moreover, a reliable modern method that could assess the potential contamination of surface water with the FRNA phages, namely the plaque hybridization assay, was available since the 1990s (Hsu et al. 1995; Beekwilder et al. 1996a) and continued to develop rapidly, as explained in Chapter 5. Thus, a simple and inexpensive magnetite-organic flocculation method was developed (Bitton et al. 1981) for the concentration of coliphages from wastewater effluents and polluted lake water.

Despite many well-argued criticisms about the FRNA role in the potential groundwater surveillance, which were collected in the thorough review of Leclerc et al. (2000), mainly, because of their low concentration and status as an "indicator of indicators," namely, bacteria, the methodological approaches with the FRNA phages continued their development. The U.S. Environmental Protection Agency (USEPA) developed two methods: 1601 (USEPA 2001a) and 1602 (USEPA 2001b) for the detection of the FRNA phages, as well as somatic phages, in water. The exhaustive field testing of the "USEPA methods 1601 and 1602 for coliphage in groundwater" was performed successfully by Karim et al. (2004). Methods to detect coliphages in large volumes of water followed (Sobsey et al. 2004). A novel investigation on the methods 1601 and 1602 for groundwater was conducted in four different regional aquifers across the United States: the Southeast region, the Northeast region, the Southwest region, and the upper Midwest region (USEPA 2006). Ultimately, a modified USEPA method 1601 to indicate viral contamination of groundwater was published (Salter and Durbin 2012).

Recently, the phage MS2 was implemented as a surrogate to establish proficiency using a biosafety level 3 procedure for drinking water control (Mapp et al. 2016). By this study, the corresponding quality control criteria were elaborated for the phage MS2, a biosafety level 2 agent, to minimize safety hazards associated with biosafety 3 agents and to use the criteria to evaluate analytical proficiency during a demonstration exercise. The USEPA method 1602 was used recently by the enumeration of the FRNA phages for the estimation of the effect of rainfall on the microbial water quality of a tropical urban catchment in Singapore (Fang H et al. 2018).

WATER SURVEILLANCE

UNITED STATES

A massive multidisciplinary study, including the FRNA phages, was prompted by the discharge of primarily treated sewage by two offshore outfalls in Mamala Bay, Oahu (Paul et al. 1997). The impact of such activities on the water quality in the bay and at adjacent recreational beaches was revealed. As a part of this study, an RT-PCR test was developed to detect and enumerate the FRNA phages in polluted marine waters (Rose et al. 1997) and compared with the previously developed plaque hybridization assay (Hsu et al. 1995). Then, the water quality study of Homosassa Springs State Wildlife Park and surrounding areas was undertaken, in order to detect fecal contamination (Griffin et al. 2000).

The FRNA phages were accepted as reliable indicators of sewage contamination by the monitoring of streams in Hawaii (Luther and Fujioka 2002, 2004). The water surveillance of the Puget Sound Basin streams led to identification of sources with the clear preference of the group I or group II FRNA phages, depending on the major pollution source (Embrey 2001). The state of the FRNA phage source tracking, among other available indicators, at that time was exhaustively reviewed by Scott et al. (2002b).

The FRNA phages were tracked in environmental waters including storm samples as well as wastewaters, from North and South Carolina, Massachusetts, and Nebraska (Cole et al. 2003). The incidence of the FRNA phages together with enteric viruses was monitored four times over a year, once each season in Wisconsin household wells located near septage land application sites or in rural subdivisions served by septic systems (Borchardt et al. 2003).

To investigate human viral contamination in southern California urban rivers and its impact on coastal waters, the PCR detection of the F-specific phages, as well as of a set of human pathogenic viruses, was performed (Jiang and Chu 2004). The fecal contamination of agricultural soils before and after Hurricane Floyd-associated flooding in North Carolina was tested with the FRNA phages as one of the central indicators (Casteel et al. 2006).

The usefulness of the FRNA phages as an indicator of fecal contamination was then demonstrated by RT-PCR assay in Table Rock Lake on the Arkansas–Missouri border, in areas predisposed to fecal contamination (Dryden et al. 2006). Remarkably, by the identification of the fecal sources in the New River at the U.S.–Mexican border region, the group I and IV coliphages were predominantly identified, but no human-specific genotypes were detected, while the group III phages were only detected at the sampling site near the international boundary, indicating human fecal contamination

(Rahman et al. 2009). The FRNA phage groups were determined in an interesting study in South Carolina, where effects of the conversion of forested watersheds to suburban and urban ones occurred due to the population growth along the southeastern coast of the United States (DiDonato et al. 2009). The FRNA phages were helped by the detection of the cesspool discharge pollution in the state of Hawaii (Vithanage et al. 2011).

The FRNA phages were detected, including extensive grouping, in two different locations in the Salinas River and four locations in or near the city of Salinas on the central coast of California (Ravva et al. 2015). All groups were represented in this study: surprisingly, the "animal" groups I and IV were found in locations impacted by human activities. The seasonal prevalence of the FRNA during winter months was confirmed by this study.

Recently, the F⁺ and somatic coliphages were enumerated from recreational water samples collected in the three Great Lake Basin areas (Wanjugi et al. 2018). A quantitative microbial risk assessment (QMRA), which included detection of the phage MS2, was conducted to support renewal of the City of Vacaville, California, wastewater discharge permit (Seto et al. 2018). The group I phage MS2 and the group II phage GA were monitored, among other indicators, in impaired streams in the Prickett Creek watershed, West Virginia (Weidhaas et al. 2018).

Canada

Eighteen samples of treated drinking water and their source waters were tested in three Canadian cities in the province of Ontario for coliphages and they were all found to contain the FRNA phages (Palmateer et al. 1990). A unique set of 1155 water samples that were collected from the years 2008–2014 in a river basin of eastern Ontario was used for the FRNA phage grouping, as well as for the analysis for many other viruses (Jones et al. 2017). The FRNA phages were detected in 28% of the samples at maximum concentrations of 16,300 pfu per 100 mL.

Africa

The occurrence of the FRNA phages in polluted South African waters was first established by Kfir et al. (1991). The first molecular characterization of the FRNA phages in South Africa was performed by Uys (1999), a student of W.O.K. Grabow at the University of Pretoria. A special study on the application of the standard ISO (1995) method for the isolation of the FRNA phages combined with their further genotyping by original assays was published for Umgeni Water in South Africa (Sundram et al. 2006a,b). Remarkably, no differences were observed between the distribution of the FRNA phage genotypes by parallel source tracking in South Africa and Spain (Schaper et al. 2002b).

Later, the effectiveness of four wastewater treatment plants in the Gauteng Province in South Africa was evaluated systematically by the assessment of the FRNA phages, among other

indicators (Dungeni et al. 2010). Then, the water samples were collected from three critical points of the Temba, Klipdrift, and Wallmansthal water treatment plants: group II, which is of human fecal origin, was present in all the water samples, while the group II and III representatives were detected only in the raw water from Temba (Okeyo et al. 2013). The microbial monitoring of surface water in South Africa was reviewed by Luyt et al. (2012) and by Naidoo and Olaniran (2013). Complex surveys including the FRNA phage detection were performed on the Umgeni River in KwaZulu-Natal (Ganesh et al. 2014; Singh and Lin 2015). An unexpectedly low number of the FRNA phages was found during potable water examination from rural areas of the North West Province (Nkwe et al. 2015). Recently, the FRNA phages were monitored in the Umhlangane River, draining directly into the above-mentioned Umgeni River and making it a key component in KwaZulu-Natal's drinking water supply (Marie and Lin 2018).

From other African countries, the FRNA phages were tested in the three wastewater treatment plants located north and south of Tunisia (Yahya et al. 2015). The total coliphages were enumerated recently in drinking water sources in some regions of Ethiopia (Bedada et al. 2018).

New Zealand and Australia

The presence of the FRNA phages as the indicators of human viruses was checked first in the marine environment and rivers of New Zealand (Donnison and Ross 1995, 1999; Lewis 1995; Turner and Lewis 1995). Then, the transport of the FRNA phages was considered by the estimation of septic tank set-back distances between septic tank systems and the shorelines of Lake Okareka in New Zealand (Pang et al. 2004). After a series of large earthquakes struck the city of Christchurch, the distribution of microbiological indicators including the FRNA phages was studied in the Avon River of Christchurch for approximately 6 months (Devane et al. 2014). Remarkably, the strength of the correlation of microbial indicators with pathogen detection in water decreased in the following order: *E. coli* > FRNA phages > *C. perfringens*. The FRNA phages did not appear to accumulate in sediment and were recommended as an indicator of recent human sewage discharge in freshwater.

The relationship of a set of water quality indicators, including the FRNA phages, was established from six sites in an urban estuary in Sydney, Australia (Ferguson et al. 1996). The FRNA phages were included in an extensive report of a study funded by the Australian Water Recycling Centre of Excellence, in order to elaborate National Validation Guidelines for Water Recycling (Branch and Le-Clech 2015).

Israel

The presence of the FRNA phages, among other phages, was examined in more than 1000 samples collected between 1992 and 1994 and compared with that in Spain, and no significant differences were found (Armon et al. 1997). Later, the FRNA phage ecology was studied in a small agricultural community,

2–3 km from Haifa, in its sewer and oxidation pond system (Gino et al. 2007). Remarkably, the presence of the FRNA phages, as well as somatic phages, in sewer lines of the community was influenced by several factors, such as anionic detergents, nutrients, temperature, source (mainly infants), shedding, and survival capability of the host strain.

EUROPE

The relationships between bacteriological and viral indicators of sewage pollution, including the FRNA phages, and environmental variables in coastal water and weather were studied in the bathing waters of Gipuzkoa, the Basque Country (Serrano et al. 1998; Ibarluzea et al. 2007). The fates of the FRNA phages and bacterial pollution indicators were compared in the Moselle River, France (Skraber et al. 2002). Large microbial source tracking, which included the FRNA phage genotyping, was performed at three French estuaries in Brittany and Normandy (Gourmelon et al. 2007). The FRNA phages were detected in 71% of water samples, and group II and III representatives were present in 64% of these water samples, mainly in areas downstream of urban activities. Group I was detected only in three sampling sites with sufficient phage concentrations, and no phage from group IV was detected in any of the water samples. Remarkably, when present, group I represented a high proportion of the total phages, namely 50%–100% of hybridized phages (Gourmelon et al. 2007). Later, an extensive study including detection of the FRNA phage groups was completed in the large Daoulas catchment, Brittany, on the west coast of France (Mauffret et al. 2012). The phage MS2 was monitored not long ago by large-scale survey in the Seine River, together with enteric viruses, such as adenovirus, aichivirus, astrovirus, cosavirus, enterovirus, hepatitis A and E viruses, norovirus of genogroups I and II, rotavirus A, and salivirus, highlighting therefore the health status of the local population (Prevost et al. 2015).

The FRNA phages and human enteric viruses were monitored in domestic sewage from a small community in Finland during a period of 1 year (von Bonsdorff et al. 2002). The FRNA phages, among other indicators, were conducted recently in order to reveal the cause of the two waterborne norovirus outbreaks in southern and northern Finland (Kauppinen et al. 2018).

The levels of the FRNA phages, among other indicators, were established in bathing waters throughout Europe, where fresh and marine bathing waters were analyzed from north, south, east and west Europe: namely from Austria, Finland, France, Germany, Greece, Italy Spain, the Netherlands, and the United Kingdom (Contreras-Coll et al. 2002).

The microbiological quality of coastal waters of bathing sites, including the presence of the FRNA phages, was evaluated along the Achaia coastline in Greece (Vantarakis et al. 2005).

The FRNA phages in European surface waters were studied in frame of a broad European Union research program (Blanch AR et al. 2004, 2006). Furthermore, a clear relationship between the FRNA phage genogroups and human

adenoviruses was found in water samples from the Meurthe River, France, in an urbanized watershed with recognized anthropogenic influences (Ogorzaly et al. 2009).

In Germany, the FRNA phages were examined in water samples taken regularly at five sites in the upper reaches of the small river Swist located to the southwest of Cologne, which showed that river sections in intensively used areas turned out to be more contaminated than in the less intensively used regions (Franke et al. 2009).

The FRNA phages, among other indicators, were tested in sewage effluent and river water during the temporary interruption of a wastewater treatment plant in Norway (Grøndahl-Rosado et al. 2014).

A thorough investigation of the FRNA phages, together with their grouping, was achieved in water samples originating from municipal Rusałka Lake, near Szczecin, Poland (Śliwa-Dominiak et al. 2014).

CENTRAL AND SOUTH AMERICA

The occurrence and densities of the FRNA phages, among other phages proposed as indicators and bacterial indicators were measured in waters from 10 rivers in Argentina and Colombia, in parallel with rivers in France and Spain, which represented very different climatic and socio-economic conditions (Lucena et al. 2003, 2006). The numbers of the phage and bacterial indicators were nevertheless similar in these different geographical areas studied. A similar study was performed at the same time in the Bodocongó River in the semi-arid region of northeast Brazil (Ceballos et al. 2003). Recently, the FRNA phages were monitored in the São Bartolomeu Stream catchment, which serves as a source of drinking water supply in Viçosa, Minas Gerais State, Brazil (Andrade et al. 2019).

The FRNA phage groups as indicators of fecal contamination were quantified in Mexican tropical aquatic systems, with a special relationship to human adenoviruses (Arredondo-Hernandez et al. 2017). As a result, the group II FRNA phages demonstrated the strongest correlation for human adenovirus. Furthermore, the FRNA phages were tested, together with a novel indicator pepper mild mottle virus (PMMoV) in groundwater from a karst aquifer system in the Yucatan Peninsula, Mexico (Rosiles-González et al. 2017).

Recently, the total coliphages were enumerated in crop production systems irrigated with low-quality water in Bolivia (Perez-Mercado et al. 2018).

JAPAN

The presence of the FRNA phages showed a high correlation with that of noroviruses, enteroviruses, adenoviruses, and torque teno viruses during a year survey of surface water in the Tamagawa River, Japan (Haramoto et al. 2005). However, genotyping of the FRNA phages from the Tonegawa River basin did not show any correlation with the occurrence of human adenoviruses, suggesting that genotyping of the phages alone is inadequate for the evaluation of the occurrence of human viruses in aquatic environments (Haramoto et al. 2009). Next, FRNA phage genogrouping survey was performed in the Kofu basin (Haramoto et al. 2012c). Finally, a nationwide survey of indicator microorganisms including the FRNA phages was conducted in drinking water sources of Japan (Haramoto et al. 2012b). Total coliforms, *E. coli*, and the FRNA phages were detected in 98%, 52%, and 27% samples, respectively, and *E. coli* was judged to be the most suitable indicator of pathogen contamination of drinking water sources at the very beginning of the water control era. The animal group I and human groups II and III were found in 41%, 39%, and 3%, respectively, of 31 plaques isolated (Haramoto et al. 2012b). Remarkably, the FRNA phages and particularly the phage Qβ were used by the monitoring of the effluents of *johkasou*—small-scale onsite wastewater treatment systems that are very popular in Japanese rural areas (Setiyawan et al. 2013, 2014; Fajri et al. 2017). Recently, the FRNA phages were conducted by the water sampling in the coastal area of Tokyo following two rainfall events in October and November 2016 (Poopipattana et al. 2018).

CHINA

The concentrations of the FRNA phages and somatic coliphages were measured in effluent of the three wastewater treatment plants in the city of Beijing, while the FRNA phages were used for prediction of enterovirus concentrations (Li et al. 2006). The phage MS2 was used as an indicator by assessment of the drinking water treatment plants in Shanghai (Zhang et al. 2012). Then, the FRNA phages were tested in an extensive study on the presence of enteroviruses in recreational water in Wuhan, Hubei Province (Allmann et al. 2013).

INDIA

The FRNA phages were monitored in groundwater in the Ghaziabad District of Uttar Pradesh (Sharma et al. 2018).

SOUTHEAST ASIA COUNTRIES

The FRNA phages were identified by both hybridization and RT-PCR assays in tropical river waters of the Klang Valley, Malaysia (Yee et al. 2006). Moreover, the FRNA phages of the Klang Valley were subjected later to the extensive grouping (Ambu et al. 2014).

The F-specific RNA phages were tested, among other indicators, to assess the surface water and groundwater quality in Angkor, Cambodia, when 58 monitoring sites along and near the Siem Reap River, in Tole Sap Lake, and West Baray, the primary water resources in this region, were involved (Ki et al. 2009).

By a survey in the Kathmandu Valley, Nepal, among the four FRNA phage groups tested, only the group II phages were detected in three of the nine groundwater samples, while the river water sample was positive for the three groups I, II, and III representatives (Haramoto et al. 2011). It was concluded therefore that the genotyping demonstrated human fecal contamination of groundwater and both human and animal fecal contamination of river water.

The grouping of the FRNA phages followed by proper statistical analyses revealed fecal origins of either humans or animals by the analysis of the ground and surface waters in metropolitan Seoul and Gyeonggi Province in Korea (Lee JE et al. 2011). The FRNA phages were detected by the identification of human and animal fecal contamination after rainfall in the Han River (Kim et al. 2013). Recently, the FRNA phages were monitored at seven sites of Oncheon Stream, Suyeong River and Gwanganri Beach in Busan from January to November 2017, in order to investigate the occurrence of norovirus in rivers and beaches (Choi SH et al. 2018).

The FRNA phages, together with PMMoV as a novel indicator for drinking water quality, were tested in Hanoi, Vietnam (Sangsanont et al. 2016).

ANTARCTICA

The FRNA phages, among a set of other indicators, were detected in wastewater, water column samples, sediments, drinking water, and Weddell seal feces collected at McMurdo Station, Antarctica, before a wastewater treatment facility at McMurdo Station began operation (Lisle et al. 2004). It was emphasized that the contribution of seal feces to the indicator concentrations should be considered in this area.

DIFFERENT SURVIVAL IN WATER

The phage f2 was less stable in groundwater than bacteria and poliovirus (Bitton et al. 1983). The phages MS2 and f2 in dark survived better in freshwater than in seawater and at lower temperatures (5°–10°C) than at higher (20°–25°C) but MS2 survived much better than f2 and overall better than any bacteria tested (Evison 1988). Remarkably, the raw well waters were found at that time to contain less FRNA phages than the chlorine-treated waters (Dutka et al. 1989). Survival of the phage MS2 in soil (Hurst et al. 1980), water (Yates et al. 1986; Governal and Gerba 1997; Alvarez et al. 2000), artificial wetlands (Gersberg et al. 1987), and desert soils amended with anaerobically digested sewage sludge (Straub et al. 1992) was tested.

In an effort to find suitable indicators of enteric viruses in seawater, survival of the FRNA phages was compared with that of somatic *Salmonella* phages, hepatitis A virus, and poliovirus in coastal seawater from three geographic areas, southern California, Hawaii, and North Carolina, at 20°C (Callahan et al. 1995). The survival was greater for the *Salmonella* phages than for any of the other viruses; the FRNA phages and poliovirus were inactivated rapidly, while hepatitis A virus reductions were intermediate between the *Salmonella* and FRNA phages. The observed differences in virus survival suggested at that time the *Salmonella* phages but not the FRNA phages to be good indicators for enteric viruses in seawater (Callahan et al. 1995). However, the presence of the FRNA phages in seawater was recommended as a definite indication of the fecal pollution (Lucena et al. 1995).

The phage MS2 and the phage PRD1, a *Salmonella* double-stranded DNA phage from *Tectiviridae* family, were compared with hepatitis A virus and poliovirus for persistence in soil (Blanc and Nasser 1996). This time, the results indicated that the phage PRD1 could be more suitable than the phage MS2 to predict the persistence of pathogenic viruses, especially at ambient temperatures.

The elimination of the seeded phage MS2 and the DNA phage φX174 was compared during sewage treatment by natural lagooning or activated sludges (Benyahya et al. 1998). The phage MS2 degraded much faster in untreated coastal Santa Monica Bay seawater than the marine DNA phage H11/1 isolated from the North Sea (Noble and Fuhrman 1999).

The FRNA phages survived less successfully in different kinds of sludges than the somatic coliphages and the *B. fragilis* phages (Lasobras et al. 1999). The FRNA phages were compared with the somatic and *B. fragilis* phages during the treatment of solid and liquid sludge (Mignotte-Cadiergues et al. 2002). The phage MS2 persisted much less in seawater compared to the *B. fragilis* phages and showed more variation between the temperatures during investigation in Tampa Bay, Florida (McLaughlin and Rose 2006). Remarkably, the combination of high temperature and the presence of sand appeared to produce the greatest disruption to the phage MS2 (Anders and Chrysikopoulos 2006). Chandran et al. (2009) measured survival of the phage MS2 in pure human urine.

Concerning the FRNA phage group distribution, it was concluded from the corresponding extensive studies that the group I phages were the clear majority, about 90%, of those recovered from environmental surface waters, as well as the percentage of group I phages detected was greatest at background sites, while the percentage of group II phages was highest at human-impacted sites (Griffin et al. 2000; Brion et al. 2002; Cole et al. 2003; Vergara et al. 2015). However, this asymmetric distribution could be explained particularly by the differential survival of the various group representatives at the impacted surface water sites, which could impact the general distribution pattern of the recovered phage groups (Cole et al. 2003). Thus, it was found that the group I phages survived longer in lake water at both 4°C and 20°C than did the other subgroups (Long and Sobsey 2001: cited by Cole et al. 2003; Long and Sobsey 2004). Both Schaper et al. (2002) and Brion et al. (2002) found the increased survival of the environmentally derived group I and II coliphages at 25°C–37°C in comparison with the other FRNA groups. Earlier, the temperature effects on the FRNA phage survival, namely the preferential survival of the group II phages at lower temperatures in comparison with the other three groups, was acknowledged by Furuse (1987).

Nevertheless, although it seemed very likely that the presence of the FRNA phage groups in environmental surface waters could be influenced by differences in their environmental persistence, the group classification appeared to be useful by the identification of major sources of water quality impairment in the case when regional impacts and differential environmental survival of the FRNA phage groups would be considered (Cole et al. 2003).

The variability of the persistence of the FRNA phage groups under different conditions was checked in river water, where temperature and pH were characterized as the main factors of the persistence (Yang and Griffiths 2013).

Recently, the environmental persistence of all four FRNA phage groups was extensively studied in surface waters from the central coast of California (Ravva et al. 2016). Water temperature played a significant role in the persistence: all prototype and environmental strains survived significantly longer at 10°C compared to 25°C. Similarly, the availability of host bacterium was found to be critical in the FRNA phage survival. In the absence of *E. coli* F+, all prototypes of the FRNA phages disappeared rapidly, the longest surviving prototype was the phage SP. However, in the presence of the host, the order of persistence at 25°C was Qβ > MS2 > SP > GA and at 10°C it was Qβ = MS2 > GA > SP. Significant differences in survival were observed between prototypes and environmental isolates of the FRNA phages. While most environmental isolates disappeared rapidly at 25°C and in the absence of the host, the members of groups I and III persisted longer with the host compared to the members of groups II and IV (Ravva et al. 2016). Figure 6.2 presents the survival pattern of the FRNA phage prototype strains in surface waters, which was taken from the Ravva et al. (2016) paper.

Decay of the FRNA phages in sewage-contaminated freshwater of the San Diego Creek was quantified by Bayesian approach (Wu J et al. 2016). In contrast to the somatic coliphages, the decay rates of the FRNA phages were not significantly different between sunlight and shaded treatments. The decay rates of both FRNA and somatic phages in winter were considerably lower than those in summer.

The survival and partitioning behavior of two model non-enveloped phages MS2 and T3 were compared with those of two enveloped viruses: the phage φ6 and murine hepatitis virus in the untreated Ann Arbor wastewater (Ye et al. 2016). Decay rates of the phage MS2 and a large set of foodborne zoonotic pathogens were determined in clay and sandy soils amended with biosolids and manures (Roberts et al. 2016).

Recently, the decay rate of the phage MS2 and adenovirus was compared in seawater in accordance with the QMRA requirements (Eregno et al. 2018). The results showed that both the phage MS2 and adenovirus were inactivated relatively rapidly at 20°C in seawater collected from 1 m depth, while a slow inactivation was observed at 4°C in seawater collected from 60 m depth. The QMRA was applied in both pilot and full-scale waterworks by modeling the phage MS2 removal (Hokajärvi et al. 2018). Furthermore, the functional QMRA models were developed with a large set of indicators, including the FRNA phages (Dias et al. 2019). Boehm et al. (2019) performed a systematic review and meta-analysis of decay rates of waterborne coliphages and mammalian viruses in surface waters.

SURROGATES OF HUMAN VIRUSES

In contrast to the more disputable sources such as fresh water, the FRNA phages were found in very high numbers in all

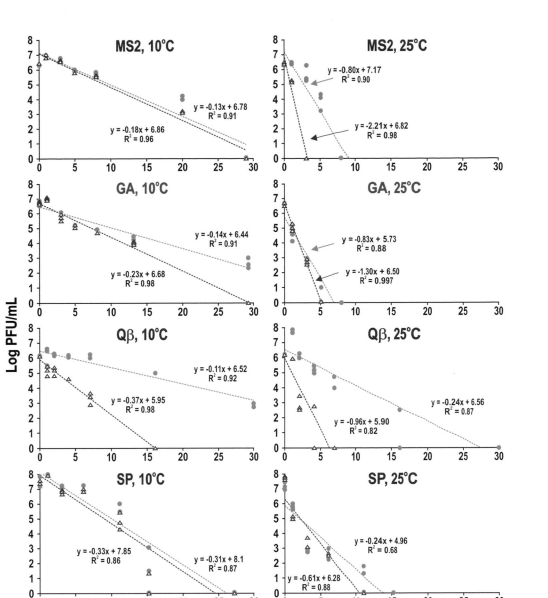

FIGURE 6.2 Survival of prototype strains of FRNA in surface waters. FRNA populations of four different genogroups were monitored at two temperatures and in the presence (filled blue circles) or absence of (open red triangles) the host *E. coli* F_{amp}. (Reprinted from Ravva SV, Sarreal CZ. *PLOS ONE*. 2016;11:e0146623.)

types of wastewater investigated and were suggested as indisputable indicator organisms, or nonpathogenic surrogates, for the human pathogenic virus presence and persistence, first of all, in the evaluation of disinfection processes for water treatment plants (Havelaar 1987a,b; Havelaar et al. 1987; Havelaar and Pot-Hogeboom 1988). Arie Hendrik Havelaar, one of the pioneers of the FRNA phage employment in the indicator role, acknowledged that the FRNA phages were seldom found in non-fecally contaminated wastewater, surprisingly low numbers of them were found in feces, and that the FRNA phages in wastewater effluent were highly resistant to chloramines and relatively resistant to UV inactivation (Havelaar and Nieuwstad 1985; Havelaar 1987b; Havelaar et al. 1990a). The phage MS2 was recommended as a surrogate for enteric

viruses after comparative inactivation of the phage MS2 with poliovirus 1 and echovirus 1 (Yates et al. 1985) or poliovirus 3 (Finch and Fairbairn 1991). The same idea was proposed in 1992 by Wilson et al. (1993). The FRNA phages were proposed as possible indicators for the presence and persistence of pathogenic viruses in water sources by Nasser et al. (1993) and by Kamiko and Ohgaki (1993). Finally, Havelaar et al. (1993) showed that the FRNA phages correlated highly with concentrations of pathogenic viruses in raw and treated wastewater, raw and partially treated drinking water, and surface and recreational waters. It is necessary to note, however, that some early studies were rather pessimistic about the FRNA phages as surrogates of enteric viruses during disinfection processes (Balluz et al. 1978; Balluz and Butler 1979).

The FRNA phages were known as highly resistant subjects to disinfection from the very beginning (Hajenian and Butler 1980a,b; Havelaar and Nieuwstad 1985; Keswick et al. 1985; Tyrrell et al. 1995).

The MS2 survival in river water under identical conditions over a period of at least 24 hours was found similar to that of the human pathogenic viruses: poliovirus type 1 (Sabin) and hepatitis A virus (Springthorpe et al. 1993). In seawater at 25°C, the F⁺ coliphages were inactivated faster than three enteric viruses: hepatitis A virus, poliovirus, and rotavirus, while in seawater-sediment mixtures at 5°C and 25°C or at 5°C in seawater, their survival was comparable to those of enteric viruses that died off rapidly (Chung and Sobsey 1993).

Nasser and Oman (1999) confirmed that the FRNA phages were more suitable than *E. coli* to predict the inactivation of pathogenic viruses in natural water sources, since the FRNA phages persisted for the longest time in the various water types, whereas *E. coli* inactivation was the fastest in groundwater at 4°C and 37°C.

The phage MS2 was inactivated by monochloramine (Berman et al. 1992). By monochloramine disinfection of water, the reduction of Norwalk virus by RT-PCR assay was compared with that of poliovirus 1 and MS2 that were assayed by RT-PCR, as well as by infectivity (Shin et al. 1998). Interestingly, norovirus reduction was about one $\log_{10}$ by RT-PCR assay, suggesting that norovirus was more susceptible to monochloramine than the other two viruses tested. Reductions of poliovirus 1 and MS2 by infectivity assays were about one $\log_{10}$ but there were no reductions of these viruses by RT-PCR assays. Hence, RT-PCR was found to underestimate virus inactivation by monochloramine (Shin et al. 1998). At that time, the FRNA phages were reviewed in detail as reliable surrogates of human enteroviruses in water (Li M and Hu 2004; Savichtcheva and Okabe 2006).

From the point of view of the disinfection efficacy, the phage MS2 was characterized as an organism that was clearly most resistant to disinfection by chlorination in comparison with other models including poliovirus (Tree et al. 2003). Neither *E. coli* nor enterococci or polio were adequate indicators of virus removal during sewage disinfection in this study. Thus, the phage MS2 was more resistant to inactivation than poliovirus under both laboratory and field conditions and showed similar behavior in both seeded and naturally contaminated samples (Tree et al. 2003). By UV irradiation, the phage MS2 was less sensitive than poliovirus type 1, but more sensitive than adenovirus (Meng and Gerba 1996) and less sensitive than feline calicivirus, but again more sensitive than adenovirus (Thurston-Enriquez et al. 2003).

In dechlorinated water, the survival rate of the phage MS2 was found higher than that of *E. coli* and feline calicivirus (Allwood et al. 2003, 2005). By bench-scale ozone disinfection, virus reductions were similar for the phage MS2, Norwalk virus and poliovirus type I (Shin and Sobsey 2003). Moreover, a systematic study on raw and anaerobically digested sewage sludge samples from two wastewater treatment plants in Athens showed good correlation of the FRNA phage presence with bacterial indicators over a 2-year period (Mandilara et al. 2006a,b). It must be emphasized that the phages and the FRNA phages among them but not fecal bacterial indicators were correlated with viral contamination in water (Baggi et al. 2001).

Therefore, the FRNA phages were intended as a conservative *fail-safe* model or *worst-case scenario* indicators of virus inactivation during sewage disinfection. When the phage MS2 and poliovirus type I were studied for persistence in water and inactivation efficiency, an advanced RT-PCR test was developed to reduce the risk of detection of inactivated viruses (Sobsey et al. 1998). The phages MS2 and β were tested as surrogates by the UV inactivation of protozoan *Cryptosporidium* (Fallon et al. 2007). The phage f2 was used as the best indicator for the survival of noroviruses (Armon et al. 2007). The phage MS2 also demonstrated good potential as a norovirus surrogate (Bae and Schwab 2008).

Nevertheless, some studies at that time denied capability of the fecal indicators to predict the presence of enteroviruses (Petrinca et al. 2009). Later, the extensive survey demonstrated clear statistical relationships between the FRNA phages and a suite of human enteric viruses: adenoviruses, astroviruses, noroviruses, and rotaviruses in the environment of Singapore (Rezaeinejad et al. 2014; Vergara et al. 2015).

The following FRNA phages were used as surrogates to evaluate the efficiency of various barrier materials that included membranes used in water treatment processes: MS2 (Jacangelo et al. 1995, 1997; Van Voorthuizen et al. 2001; Hu et al. 2003) and MS2, Qβ, and fr (Herath et al. 1999).

The presence of the FRNA phages and hepatitis E virus was studied for 6 months in waters proximal to swine concentrated animal feeding operations in North Carolina (Gentry-Shields et al. 2015). Although hepatitis E virus was detected in only one sample, the F⁺ coliphages were detected in 85% of samples, only 3% of which were FRNA phages (all of them belonged to group I); the near ubiquity of coliphages as well as the presence of hepatitis E virus suggested that current waste management practices may be associated with the dissemination of viruses of public health concern in waters proximal to concentrated animal feeding operation spray fields (Gentry-Shields et al. 2015). The role of the FRNA phages in the evaluation of the hepatitis E virus transmission by water was reviewed recently (Fenaux et al. 2019).

The FRNA phages, including f2, MS2, R17, and Qβ, were used routinely as surrogates for human pathogenic viruses in many soil and sand adsorption and transport studies (Goyal and Gerba 1979; Taylor et al. 1980; Bales et al. 1993; Penrod et al. 1996; Jin et al. 1997; Sakoda et al. 1997; Dowd et al. 1998; Thompson et al. 1998; Chattopadhyay and Puls 1999; Thompson and Yates 1999; Chu et al. 2000; Schijven and Hassanizadeh 2000; Ryan et al. 2002; Meschke and Sobsey 2003; Zhuang and Jin 2003a; Sinclair et al. 2009).

By the testing of the RNA phages as putative surrogates, the filtration in packed beds of quartz sand, which was strongly influenced by pore water pH over the environmentally important range of pH 5–7, was different in the case of the phage MS2 and the recombinant Norwalk virus (rNV) particles (Redman et al. 1997). The latter were morphologically and antigenically similar to live Norwalk strains but lacked nucleic acid and were therefore noninfectious. It was concluded that the phage MS2 might not mimic the subsurface filtration of Norwalk virus in natural systems due to differences in their electrostatic properties (Redman et al. 1997). Later, the influence of ionic strength on the electrostatic interaction of viruses with environmentally relevant surfaces was determined for the phages MS2 and Qβ, and Norwalk virus by Schaldach et al. (2006). In this study, the future prospect for the use of the RNA phages in the role of surrogates was much more optimistic.

It is interesting to note that the phage Qβ appeared as the surrogate typically used in Japan (Kamiko and Ohgaki 1989; Urase et al. 1996; Otaki et al. 1998), while the phage MS2 was the favorite surrogate in the United States and Europe. Further studies in Japan, however, exploited the phages MS2 and fr, together with the phage Qβ (Herath et al. 1999) or Qβ and MS2 (Herath 2000; Matsui et al. 2003; Matsushita et al. 2004). Two phages, MS2 and Qβ, were compared for their role of the surrogates in a coagulation-ceramic microfiltration system (Shirasaki et al. 2009a) and during the aluminum coagulation process (Shirasaki et al. 2009b). In the latter paper, it was concluded that the removal ratio of the infectious Qβ concentration was approximately two log higher than that of the infectious MS2 concentration at all coagulant doses tested. Thus, significant inactivation of the phage Qβ was observed during the coagulation process, whereas little or no inactivation of the phage MS2 occurred. In contrast, during the coagulation-ceramic microfiltration process (Shirasaki et al. 2009a), the removal of both infectious phages was similar under each coagulation condition: approximately six log reduction was achieved by the optimal coagulation conditions.

Furthermore, Shirasaki et al. (2010b) directly compared removal of the MS2 and Qβ surrogates with the removal of recombinant norovirus VLPs (rNV-VLPs) in a coagulation-rapid sand filtration process and arrived at rather pessimistic conclusions. The removal performance for the phage MS2 was somewhat larger than that for rNV-VLPs, meaning that MS2 could be not recommended as an appropriate surrogate for native norovirus because of inevitable overestimation of the removal performances. By comparison, the removal performance for the phage Qβ was similar to or smaller than that for rNV-VLPs. However, the removal performances for rNV-VLPs and Qβ differed between each unit process, namely, between the coagulation process and the following rapid sand filtration process. Therefore, the phage Qβ also was not recommended as an appropriate surrogate for native norovirus (Shirasaki et al. 2010b).

In a coagulation-ceramic microfiltration process, the direct comparison of rNV-VLPs with MS2 and Qβ surrogates led nevertheless to the conclusion that both phages have the distinct potential to become appropriate surrogates for native norovirus in the process in question, and, of the two, the phage Qβ could be the more conservative surrogate (Shirasaki et al. 2010c). In optimal conditions, more than four log removal was observed for the rNV-VLPs, whereas the removal ratios of Qβ and MS2 were approximately two log and one log, respectively, smaller than the ratio of rNV-VLPs.

The feasibility of in-line coagulation as a pretreatment for ceramic microfiltration was acknowledged (Shirasaki et al. 2010a), since the in-line coagulation-ceramic microfiltration system efficiently removed the phages Qβ and MS2, whereas infectious Qβ and MS2 were removed to similar levels by the two precoagulation methods tested, but the removal of the total MS2 particles was higher than that of the Qβ particles, possibly because of the selective interaction with the cake layer. Ceramic pot filters with different silver applications were tested with the phage MS2, together with *E. coli*, in a long-term experiment (van der Laan et al. 2014).

Gerba et al. (2015) acknowledged that the use of the FRNA phages MS2, Qβ, and fr, as well as the DNA phages PRD1 and φX174, as surrogates for testing of the drinking water treatment devices met the WHO and USEPA requirements. Furthermore, the phage GA was used as a reliable surrogate of norovirus in a study on the UV inactivation in the wastewater treatment facilities (Barrett et al. 2016).

By recent reviews (Mesquita and Emelko 2012; Sinclair et al. 2012; McMinn et al. 2017a), the usefulness of the FRNA phages as surrogates was acknowledged, first of all, because of the fact that the phage levels in all matrices examined were similar to those of viral pathogens. Furthermore, the FRNA phages closely mimicked viral pathogen persistence in wastewater treatment systems.

Nevertheless, some criticisms were turned, for example, against the phage MS2 that "is a valid surrogate for chemical disinfection processes but due to its *E. coli* source from various warm-blooded animals, it is not a good index of human enteric virus presence in environmental waters" (Ashbolt 2015). Moreover, meta-analysis of 12 international studies, which were conducted in 1992–2013 at 718 public drinking-water systems, nominated somatic coliphages and *E. coli* as better surrogates than the FRNA phages, although multiple indicators were also acknowledged (Fout et al. 2017). However, the FRNA phages, as an official monitoring instrument appeared at that time in the technical report of the European Food Safety Authority (Allende et al. 2017).

In the drinking water treatment plants employing coagulation-sedimentation, rapid sand filtration, ozonation, and biological activated carbon treatments in Japan, the classical phage MS2 and Qβ indicators behaved similarly to PMMoV, a novel prospective process indicator (Kato R et al. 2018). The data on the phage MS2 inactivation were taken into account in an extensive review of low-cost point-of-use water treatment systems for developing communities (Pooi and Ng 2018).

Recently, Lee S et al. (2019a) compared the reductions of the group II phages and norovirus GII during wastewater treatment and concluded that that the group II phages could be used as appropriate indicators of norovirus GII. Moreover, the group I phages were found to be appropriate indicators of viral reduction because of their high resistance to wastewater treatment compared with the other RNA phage groups and norovirus genotypes. As also pointed out above, the authors characterized them as a *worst-case scenario* organism.

Wyrzykowska-Ceradini et al. (2019) recognized the phage MS2 as a reliable surrogate of foot-and-mouth disease virus by development of a virucide decontamination test method for porous and heavily soiled surfaces.

INACTIVATION

The earliest characteristic of the RNA phage response to chemical agents was the resistance of the phage f2 against chloroform, since the latter, together with lysozyme, was added at the final infection stage to complete bacterial lysis (Loeb and Zinder 1961). Chloroform sensitivity remained a conventional test to differ the male-specific RNA phages, which were chloroform resistant, from the male-specific filamentous ssDNA phages, which were usually chloroform sensitive. Thus, all RNA phages that followed the phage f2 were resistant to chloroform, except for the phages M (Coetzee et al. 1983), pilHα (Coetzee et al. 1985a), and D (Coetzee et al. 1985b), which demonstrated chloroform sensitivity. The phages F_0lac h and SR were also sensitive to chloroform (Coetzee et al. 1986), although the phage F_0lac, the source of the phage F_0lac h, was chloroform resistant (Bradley et al. 1981a).

The pseudomonaphage PRR1 demonstrated unusual sensitivity to EDTA, in contrast to all other RNA phages including the pseudomonaphages 7s and PP7 (Olsen and Thomas 1973).

The susceptibility of the phages f2 and Qβ to the two popular ionic detergents, sodium dodecyl sulfate, an anionic detergent, and dodecyltrimethylammonium chloride, a cationic detergent, was found to be highly dependent on the pH value (Ward and Ashley 1979).

Concerning the general survival *in situ,* the "natural" inactivation of the phage MS2 was evaluated at that time by dialysis and storage (Nikovskaia et al. 1979).

The inactivation of the phage f2 was measured in nonaerated liquid and semiliquid animal wastes, in parallel with and depending on ambient temperature, pH, and type of animal waste (Pesaro et al. 1995). The effects of freezing and storage temperature on the MS2 viability were estimated by Olson et al. (2004). By natural inactivation in river, the FRNA phages were more resistant—with the exception, however, of the summer months—than fecal coliforms and enterococci in fresh waters (Durán et al. 2002). However, the *in situ* comparative microbial inactivation studies in aquifer created some doubts about the adequacy of the phage MS2 as a universal surrogate. Thus, the phage MS2 was found to decay at a much faster rate than poliovirus and coxsackievirus (Gordon and Toze 2003) or adenovirus (Ogorzaly et al. 2010; Sidhu and Toze 2012) in groundwater.

The monofunctional and bifunctional furocoumarin derivatives were studied by their dark- and light-mediated action on the phage MS2 (Toth et al. 1988).

A recent investigation examined how the inactivation of the phage MS2 in water was affected by ionic strength and dissolved organic carbon using static batch inactivation experiments at 4°C conducted over a period of 2 months (Mayotte et al. 2017). This study is especially interesting because of its mathematical approach, where experimental data were fit with constant and time-dependent inactivation models using traditional linear and nonlinear least-squares techniques and the generalized likelihood uncertainty estimation (GLUE).

An extensive review on the survival of phages, which included data on the RNA phages f2, MS2, Qβ, and PP7, under the influence of external factors, was published by Jończyk et al. (2011). The most recent review of the virus, including the FRNA phages' survival in water was published by Pinon and Vialette (2018).

To summarize the huge amount of literature on the disinfection processes, systematic annual reviews on the published literature are recommended (Munakata and Kuo 2014, 2015, 2016; Kuo 2017, 2018; Xue et al. 2018).

ULTRAVIOLET IRRADIATION

Ultraviolet (UV) irradiation of the first RNA phages has been proven in early publications. The phage f2 showed about one thirtieth the sensitivity to UV irradiation as the DNA coliphage T2 (Zinder 1963), a difference which was of the same order of magnitude as their relative nucleic acid contents. However, considering the single-stranded nature of the phage f2, perhaps a more valid comparison would be with the DNA phage φX174, which had a nucleic acid content of the same order of size: the phage f2 was seven times as resistant to UV as φX174, probably because f2 contained uracil rather than thymine. In addition, the phage f2 did not undergo either photoreactivation or host reactivation (Zinder 1965). As stated in the latter review, after UV irradiation of bacteria, their capacity to grow f2 was as sensitive as was the free phage (Zinder 1963). Zinder (1965) noted, however, that the dose to the bacteria was "so enormous that it would be difficult even to guess what is being inactivated in the host. Despite these heavy doses of UV, the bacteria were able to synthesize considerable amounts of phage RNA, if not infective particles (Fenwick et al. 1964)."

The early data on the UV irradiation effect on the RNA phages and their replication in bacteria were earned also for the RNA coliphages μ2 (Turri et al. 1964), β (Nonoyama and Ikeda 1964), fr (Winkler 1964), R17 (Fenwick et al. 1964), f2 (Neubauer and Závada 1965; Závada and Koutecká 1965; Závada et al. 1966; Werbin et al. 1967, 1968), and MS2 (Rappaport 1965; Sherban and Krivisky 1969, Karczag et al. 1972), as well as for the pseudomonaphage 7s (Feary et al. 1964). Moreover, the phage MS2 was chosen as a reference for the calibration of the powerful UV lamp that was used at that time for the studies of the radiosensitization mechanisms (Hotz and Walser 1970).

The action spectra of the phages MS2, R17, fr, and 7s in the range of 225–302 nm were first determined by Rauth (1965) using a large diffraction grating monochromator, which dispersed monochromatic UV light approximately 1.2 nm in bandwidth.

Uridine photohydrates, namely 6-hydroxy-5,6-dihydrouridine, were the major products formed upon irradiation of the phage R17 with UV light at 280 nm, while cyclobutane-type pyrimidine dimerization did not occur to an appreciable extent

(Cerutti et al. 1969; Remsen et al. 1970). The suppression of dimerization was the first indication that the RNA inside the phage is held in a rigid conformation in close contact with the phage protein. It was concluded therefore that uridine photohydrates represented a major part of the lethal damage caused by UV light in the phage R17 (Remsen et al. 1970). Only 5.3 uridine photohydrates in the total R17 genome were necessary and sufficient to decrease the translation of the R17 coat gene to 37% by the *in vitro* translation system (Remsen and Cerutti 1972). Generally, a unit length of phage RNA (in bases) was 2–3 times more sensitive than a unit length of DNA (in base pairs) with respect to the inactivation of messenger function (Ponta et al. 1979). Mattern et al. (1972) showed that cytidine photohydration occurred in the R17 RNA with 10–15 times lower efficiency than the uridine photohydration. The formation of uridine hydrates in the UV-irradiated MS2 phage was acknowledged by Yamada et al. (1973). Moreover, εN-(2-oxopyrimidyl-4)-lysine was found in the phage MS2 hydrolysates after UV irradiation at 254 nm or treatment with bisulfite because of the covalent cross-linkage of the RNA to the coat protein: the exocyclic amino group of the activated cytosine was substituted by the lysine ε-amino group of the protein (Budowsky et al. 1976).

Furthermore, kinetics of the UV-induced inactivation of the stirred and non-stirred solutions of the phage MS2 at 254 nm demonstrated the homogeneity of the phage population with respect to photosensitivity of infectivity (Budowsky et al. 1981). Such UV irradiation decreased the binding of the phage MS2 with anti-MS2 immunoglobulins, since the number of antigenic determinants exposed on the MS2 phage dropped (Budowsky and Kostiuk 1985). Occasionally, different resistance of the RNA coliphages to UV irradiation was determined in the order from more resistant to less resistant: MS2, GA, and Qβ (Furuse et al. 1967; Furuse and Watanabe 1971). The UV dose-response curve of the phage MS2 was characterized as following first-order kinetics without shoulders, as could be expected for a single stranded RNA virus (Harm 1980).

As noted in the entry about the RNA phages to the official *Virus Taxonomy*, "the inactivation by UV light and chemicals is comparable to that of other icosahedral viruses containing ssRNA" (van Duin and van den Worm 2005).

Concerning fine mechanisms of the UV-induced photochemical reactions, the modern investigation technique, quantitative MALDI-TOF mass spectrometry (MALDI-TOF-MS), detected significantly more RNA modifications than RT-qPCR after UV_{254} irradiation, suggesting that certain chemical modifications in the RNA were not detected by the reverse transcriptase enzyme (Qiao and Wigginton 2016). In contrast, by singlet oxygen $^1O_2$ treatment, the MALDI-TOF-MS tracked as much $^1O_2$-induced RNA damage as the RT-qPCR. After 5 hours of simulated sunlight exposure, neither MALDI-TOF-MS nor RT-qPCR detected significant damage. High-resolution ESI-Orbitrap MS analyses identified pyrimidine photohydrates as the major UV_{254} products, which likely contributed to the discrepancy between the MS- and RT-qPCR-based results (Qiao and Wigginton 2016). Recently,

the reactions that took place in single-stranded and double-stranded RNA and DNA viral genomes after the UV_{254} irradiation were checked in the phage MS2 genome, in parallel with the double-stranded RNA phage φ6, single-stranded DNA phage φX174, and double-stranded DNA phage T3 (Qiao et al. 2018).

The RNA coliphages appeared as the first reliable surrogates and monitoring controls by the preparation of blood products, where UV irradiation was involved in combination with other disinfectants in order to prevent transmission of viral diseases. By combined treatment with β-propiolactone and UV irradiation, the phage f2 was one of the four phages that have been used as a monitoring test for the inactivation of hepatitis viruses and HIV in plasma and plasma derivatives by the preparation of a virus-safe stabilized serum and coagulation factor concentrates during the period from 1981 to 1986 (Dichtelmüller and Stephan 1988; Stephan 1989). The phage R17 was used as one of the controls by the use of long-wavelength UV radiation at 320–400 nm combined with 8-methoxypsoralen treatment for the decontamination of platelet concentrates for transfusion medicine (Lin et 1989; Corash et al. 1992).

Further rapid development of the UV inactivation studies was connected to a large extent with the needs of wastewater disinfection. By UV irradiation of secondary effluent at the 200–300 nm range, the FRNA phages were positioned as better indicator organisms regarding viral inactivation by UV than classical indicators like *E. coli*, fecal streptococci, or somatic coliphages (Havelaar et al. 1987). Furthermore, the use of the FRNA phages as calibration models in the UV disinfection of wastewater was justified by Havelaar et al. (1991), where inactivation of the phage MS2 was thoroughly measured by UV radiation from a low-pressure mercury lamp. It appeared, however, that the pure MS2 culture was inactivated at a rate almost twice that of the naturally occurring FRNA phages. The latter were more resistant to UV irradiation than somatic coliphages, *E. coli*, and fecal streptococci (Havelaar et al. 1991). Nevertheless, the phage MS2 was used as a standard for the microbiological calibration of an UV reactor constructed for the UV disinfection of secondary wastewater effluent (Nieuwstad et al. 1991). The kinetics of batch UV inactivation of the phage MS2 and full microbiological calibration of an UV pilot plant was described by Nieuwstad and Havelaar (1994). The standard MS2 samples were prepared and distributed in an interlaboratory study (Schijven et al. 1995). Havelaar et al. (1995) reported the total process of the inactivation of viruses by drinking water treatment under full-scale conditions, where removal of the FRNA phages most closely followed that of viruses. A similar full-scale pilot plant study in a pretreated wastewater that contained about 70% secondary effluent from sewage treatment plants and 30% surface water showed that the phage f2 was more resistant than the tested bacteria (*E. coli*, *Salmonella*, and fecal streptococci) (Dizer et al. 1993a,b). In a large comparative investigation of the 254-nm UV inactivation of potential waterborne pathogens (hepatitis A virus, coxsackievirus, and rotavirus) the phage MS2 exhibited the greatest resistance (Battigelli et al. 1993). At the same time, the special role of UV radiation and ozonation as alternatives to chlorine and its derivatives by harmful disinfection of wastewater was emphasized (Rajala-Mustonen and Heinonen-Tanski 1995; Rajala-Mustonen et al. 1997). A novel male-specific RNA phage Raa$_4$ was used as an inactivation standard in these studies. As a result, the disinfection with ozone proved to inactivate the phage Raa$_4$ more rapidly than UV light or UV light together with hydrogen peroxide (Rajala-Mustonen and Heinonen-Tanski 1995). In parallel, the UV disinfection system was assessed in terms of the most stringent U.S. wastewater reuse standards, and the seeded phage MS2 as a standard demonstrated the greater resistance to the UV disinfection than the coliform group (Braunstein et al. 1996). The UV disinfection state of those times, with the pioneering role of the RNA phages, was reviewed exhaustively by Kuo and Smith (1996).

In line with the then European Directives on bathing water quality, the UV disinfection of wastewater was studied, and again the FRNA phages were found to be the most resistant of the tested pathogens (Moreno et al. 1997). In Austria, the time-dose reciprocity in UV disinfection of water was studied with the phage MS2 as one of the indicators (Sommer et al. 1998). At the same time in the United States, a solid study that included the phage MS2 as one of the standards, acknowledged the substitution of UV light for chlorine disinfection at a full-scale tertiary treatment facility (Oppenheimer et al. 1997). The usefulness of the phage MS2 as a standard was confirmed in the study that compared monochromatic low-pressure UV radiation, which was used by most wastewater treatment plants, with the polychromatic medium-pressure UV radiation source (Shin et al. 2000). Furthermore, the phage MS2 was classified as "a safe and easy-to-use biodosimeter for UV equipment qualification tests" by the evaluation of UV disinfection for wastewater recycling at West Basin, California (Savoye et al. 2000). In this study, the phage MS2 was three times more resistant to the UV irradiation than poliovirus. Furthermore, the phage MS2 was used as an indicator by the acceptance of the UV disinfection for reuse applications in the states of Florida and California (Sakamoto et al. 2001). Remarkably, storage and lag time between the preparation of the phage MS2 suspension and the UV irradiation test affected the MS2 susceptibility to the treatment (Jolis 2002).

The phage MS2 was employed as an indicator by the assessing UV reactor performance for treatment of finished water by the American Water Works Service Company (Bukhari and LeChevallier 2003). The bioassay validation of UV reactors in the United States was performed using the phage MS2 or *Bacillus subtilis* spores in accordance with the appropriate USEPA draft UV disinfection guidance; it was shown that the MS2 probes remained stable over the 24-h time suggested by the protocols (Fallon et al. 2004). In fact, it was highly important practically, because of the real laboratory capacities and numerous logistical challenges.

The phage MS2 and the DNA phage PRD-1 were more resistant than the DNA phages φX174 and T4 by both low pressure, low intensity and medium pressure, high intensity

UV irradiation (Asahina et al. 2002). The most resistant status of the phage MS2 was confirmed in more recent studies (Hu Xuexiang et al. 2012; Rodriguez et al. 2014), and the lack of the phage MS2 photoreactivation was also acknowledged (Rodriguez et al. 2014).

Overall, in the early 2000s, the MS2 phage was routinely used for UV reactor validation in the United States, while *B. subtilis* spores were often used for validation testing in Europe. Mamane-Gravetz et al. (2005) exposed these popular indicators to quasi-monochromatic UV irradiation across the microbicidal spectrum at wavelengths of 214, 230, 240, 254, 265, 280, and 293 nm. The phage MS2 was three times more sensitive to wavelengths near 214 nm compared to the 254-nm output of low-pressure lamps, while *B. subtilis* spores were most sensitive to wavelengths around 265 nm. Later, in an advanced study using a tunable laser that produced precise monochromatic light with single UV wavelengths at 10 nm intervals between 210 and 290 nm, the MS2 and Qβ action spectra exhibited a relative maximum near 260 nm and an increased sensitivity below 240 nm (Beck et al. 2015b). During exposition of the phage MS2 to the monochromatic UV irradiation from the tunable laser at wavelengths of between 210 and 290 nm, the RT-qPCR was performed to measure genomic damage across the germicidal UV spectrum for comparison with the genomic damage at 253.7 nm (Beck et al. 2015a). The results indicated that the rates of the RNA damage closely mirrored the loss of the viral infectivity across the germicidal UV spectrum. Therefore, the genomic damage appeared as the dominant cause of the MS2 inactivation from exposure to the UV irradiation, in contrast to adenovirus, for which the phage MS2 was used as a viral surrogate by validating polychromatic UV reactors (Beck et al. 2015a).

Above 240 nm, all bacteria and viruses tested exhibited a relative peak sensitivity between 260 and 270 nm. Of the coliphages, the phage MS2 exhibited the highest relative sensitivity below 240 nm, relative to its sensitivity at 254 nm, followed by Qβ and some DNA coliphages. The *B. pumilus* spores were more sensitive to the UV light at 220 nm than any of the coliphages. These spectra were required for calculating action spectra correction factors for medium pressure UV system validation, for matching appropriate challenge microorganisms to pathogens, and for improving UV dose monitoring. In addition, understanding the dose response of these organisms at multiple wavelengths could improve polychromatic UV dose calculations, and enabled prediction of pathogen inactivation from wavelength-specific disinfection technologies such as UV light-emitting diodes (LEDs), as described below.

A first putative disadvantage of the RNA phages as indicators by UV irradiation was noted when Bourrouet et al. (2001) found that bacteriophages of *Bacteroides fragilis* were more resistant to UV disinfection than somatic and FRNA coliphages. Later, too-high resistance of the phage MS2 was regarded as a disadvantage when the reduction equivalent fluence (REF) was considered with seeded model organisms (Hijnen and Medema 2005; Hijnen et al. 2006). For this reason, the phage Qβ, as well as the DNA phage T7, were suggested as more realistic surrogates, since their inactivation rate constants were more in the range of that for the more sensitive pathogens (Hijnen and Medema 2005). Nevertheless, the phage MS2 was accepted as the most suitable surrogate by the accreditation of a full-scale UV disinfection plant in Australia (Poon et al. 2006). In 2006, the USEPA published the UV Disinfection Guidance Manual, which replaced the 2003 draft Guidance, and confirmed the surrogate status of the phage MS2 for the UV reactor validation, and recommended the phage Qβ, as well as the DNA phages T1 and T7, as alternates for the phage MS2 for the validation of reactors for *Cryptosporidium* and *Giardia* credit (Wright et al. 2007). As explained above, the phage Qβ was more sensitive to UV light than the phage MS2, and hence reduce the reduction equivalent dose (RED) bias uncertainty associated with testing reactors for *Cryptosporidium* and *Giardia* credit. For UV reactors with relatively narrow UV dose distributions, the use of models such as Qβ instead of MS2 could result in significant cost savings while maintaining public health protection (Wright et al. 2007). In parallel, a novel innovative UV reactor was constructed and characterized by the MS2 and *E. coli* inactivation kinetics (Schmidt and Kauling 2007). Recently, the phage MS2 was acknowledged again as the most reliable surrogate for human pathogenic viruses for both low-pressure and medium-pressure UV disinfection in wastewater treatment processes and water reuse practice (Bae and Shin 2016).

Meanwhile, the putative protectors against the phage MS2 inactivation by the UV irradiation were identified. Thus, coagulation of the phage MS2 with kaolin clay particles or aggregation by alum into viral clumps prior to the UV exposure reduced the inactivation efficiency (Templeton et al. 2003, 2005). The humic acid–coated MS2 particles were, however, more resistant to the UV inactivation (Templeton et al. 2006b). The phage MS2 particles that were associated with iron oxide particles in groundwater were shielded from UV light (Templeton et al. 2006a). The water turbidity had a negative impact on UV disinfection due to particle association with the seeded phage MS2 (Hu et al. 2007). The phage MS2 was involved as an indicator in the large study on a potential of naturally occurring particles to protect indigenous coliform by the UV irradiation of the unfiltered surface water (Cantwell and Hofmann 2008).

Concerning mechanisms of the UV_{254} disinfection of viruses, aggregation of the phage MS2 appeared as responsible for the so-called tailing phenomenon, when treatment of aggregates permanently fused a fraction of viruses, which increased the likelihood of multiple infection of a host cell and ultimately enabled the production of infective viruses via recombination (Mattle and Kohn 2012).

Recently, the influence of the phage MS2 aggregation and adsorption onto particles on UV irradiation was studied with kaolinite and *Microcystis aeruginosa* as model inorganic and organic particles (Feng et al. 2016). Remarkably, in the absence of model particles, the MS2 aggregates which were formed in either 1 mM NaCl at pH = 3 or 50–200 mM ionic strength $CaCl_2$ solutions at pH = 7 led to a decrease in the MS2 inactivation efficacy because the virions located inside the aggregate were protected from the UV irradiation. In contrast, the presence of kaolinite and *M. aeruginosa* led to increase of the MS2 inactivation, since more MS2 virions were exposed due to adsorption on the particles.

After turbidity and microparticles, organic micropollutants, including soluble microbial products (SMPs), could affect the disinfection process of pathogens in drinking water treatment. It was demonstrated that sodium alginate, a natural polysaccharide and a model SMP, enhanced inactivation of the phage MS2 by the UV-C treatment (Song et al. 2015).

The UV irradiation of wastewater was combined with other prospective inactivators. Thus, a sublethal H_2O_2 concentration was shown to greatly enhance killing of the phages f2 and MS2 by near UV (300–400 nm) but not far UV (254 nm) (Eisenstark et al. 1986). The hydrogen peroxide alone was used rather for recovery of phages from natural environments than for disinfection, since it inactivated bacteria much more rapidly than phages, e.g., the model phage MS2 (Asghari et al. 1992).

The UV inactivation of the phage MS2 was enhanced by aqueous silver (Butkus et al. 2004, 2005; Kim et al. 2008). The synergistic effect of the UV irradiation with peracetic acid and H_2O_2 treatment against the phage MS2, as well as bacterial indicators, was demonstrated by Koivunen and Heinonen-Tanski (2005). The results of the UV_{295} nm/H_2O_2 advanced oxidation process led to the conclusion that the disinfection due to the presence of OH radicals was very small compared to the damage from the UV irradiation, although for the MS2 and the DNA coliphages T4 and T7 tested, there might be some oxidative enhancements that could assist disinfection efficacy (Mamane et al. 2007). The phage MS2 was used by the advanced UV/H_2O_2 oxidation study that compared efficiency of low-pressure and medium-pressure UV lamps for the treatment of natural waters containing micropollutants (IJpelaar et al. 2010). Then, sequential application of the UV as a primary disinfectant with and without H_2O_2 addition followed by free chlorine as secondary disinfectant was studied with the phage MS2 and *B. subtilis* spores as indicators (Cho et al. 2011). It was concluded that the efficiency of UV/free chlorine sequential disinfection processes, which were widely employed in drinking water treatment, could be significantly enhanced by adding H_2O_2 in the primary step and hence converting the UV process to an advanced oxidation process.

The advanced UV/H_2O_2 oxidation process was followed not only by the phage MS2 inactivation, but also by degradation of two fluorescent dyes, rhodamine B and fluorescein, that modeled the fate of organic pollutants (Timchak and Gitis 2012). It appeared that the inactivation of viruses was not affected by the presence of dyes but could be improved by the addition of hydrogen peroxide.

The efficacy of combinations of the UV/H_2O_2 and UV/peroxydisulfate treatments was compared by the inactivation of the phage MS2, *E. coli*, and *B. subtilis* spores as the most popular surrogates (Sun P et al. 2016). Addition of H_2O_2 and peroxydisulfate greatly enhanced the inactivation rate of MS2 by around 15-fold and 3-fold, respectively, whereas the inactivation of *B. subtilis* spores was only slightly enhanced. Based on the manipulation of solution conditions, reactive species responsible for the inactivation were identified and ranked by their disinfection efficacy: $\bullet OH > SO_4^{\bullet-} > CO_3^{\bullet-} \gg O_2^{\bullet-}/HO_2^{\bullet}$. The role of the phage MS2 by the development of the inactivation of pathogenic microorganisms by sulfate radical was reviewed recently by Xiao et al. (2019).

Comparison of cultural and molecular methodologies of the RNA phage enumeration by the advanced UV/H_2O_2 oxidation process revealed overestimation of the true number of infective virus by the qPCR detection (Sherchan et al. 2014).

The transmission of UV irradiation was affected slightly by the presence of chlorine and monochloramine, which resulted in some decrease of the MS2 inactivation (Örmeci et al. 2005). For this reason, further studies concentrated on the MS2 inactivation by sequential approach using UV as the primary disinfectant and chlorine or monochloramine as the secondary disinfectant (Shang et al. 2007). By the enumeration of the FRNA phages and many other indicators, it was proposed that the disinfection treatment in water reclamation plants should include at least one operating UV module and a minimal addition of chlorine (Montemayor et al. 2008). The phage MS2 was used as a benchmark for confirmation of the UV doses that were applied by the collimated beam in a study on the UV and chlorine disinfection of ampicillin-resistant and trimethoprim-resistant *E. coli* (Templeton et al. 2009b).

The combination of the UV and chlorination processes was especially attractive to reduce the risk of adenovirus infection (Rattanakul et al. 2014). The combination processes of UV and chlorine, either sequential or simultaneous application, seemed to be more effective than a standalone process in viral inactivation. However, when MS2 was more resistant against chlorine than adenovirus, the latter was more resistant to UV than MS2 (Rattanakul et al. 2014).

Importantly, the chlorine disinfection of the phage MS2 was mitigated to a different extent by oxide TiO_2, NiO, ZnO, SiO_2, and Al_2O_3 nanoparticles, where the shielding effect was probably caused by reduced free chlorine and free MS2 in the solution due to sorption onto the oxide nanoparticles (Zhang W and Zhang X 2015). In fact, all oxide nanoparticles exhibited strong adsorption capacity for MS2, except SiO_2, while TiO_2 and ZnO nanoparticles could enhance MS2 inactivation under solar irradiation but NiO and SiO_2 decreased MS2 inactivation.

Zyara et al. (2016a) found that the phage MS2 was intermediately resistant to chlorine, and thus it was not an optimal choice to indicate resistant viruses. A search for novel surrogates in this study resulted in a set of 18 chlorine-resistant coliphages (Zyara et al. 2016a,b), two of which, namely the strains 1 and 5, were characterized further as chlorine- and UV-resistant members of the RNA phage family (Zyara 2017).

The mechanism of the simultaneous action of UV and chlorine (UV/Cl_2) on the phage MS2 was unveiled by Rattanakul and Oguma (2017). The combined action led to production of OH$\bullet$ radicals, whereas the UV or chlorine alone did not produce OH$\bullet$ radicals. By the UV/Cl_2 treatment, synergistic effects on viral genome damage were observed, but were not directly due to OH$\bullet$ radicals. The ability of MS2 to penetrate the genome of the host bacteria was impaired, but its ability to attach to the host was not affected by the treatment. It was concluded that the major cause of virus inactivation in response to UV/Cl_2 was the damage to the viral genome caused by combined actions of chlorine species and OH$\bullet$ radicals (Rattanakul and Oguma 2017).

Combination of the UV irradiation with flocculation-chlorination sachets led to reduction of hepatitis E virus in water matrices where the phage MS2 was used as a control (Guerrero-Latorre et al. 2016).

Along with the development of the UV inactivation technologies, an interest in other possible inactivation surrogates grew. Thus, a thorough study was devoted to somatic coliphages (Lee and Sobsey 2011). Then, murine norovirus, feline calicivirus, and echovirus 12 were tested as potential surrogates for human norovirus, in parallel with the phage MS2, by UV_{254} light inactivation (Park GW et al. 2011).

The inactivation effect of the UV disinfection reactor was determined in the case of the high concentration of the phage MS2, 1×10^7 pfu/mL and 1×10^8 pfu/mL, and the relationship between UV dose and inactivation of the phage MS2 was developed using Bayesian theory (Wang Z et al. 2013).

The phage MS2 was used as a model in an advanced study on the UV sterilization of cell culture media in terms of chemical composition and the ability to grow cell cultures in the treated media (Yen et al. 2014). Recently, the efficacy of the UV-C irradiation for reducing foodborne pathogens in skim milk was tested with the use of the phage MS2 as a surrogate (Gunter-Ward et al. 2017). To prevent incidents of contamination in biopharmacy, a special review collected the available data on the UV sensitivity of adventitious agents in biopharmaceutical manufacturing (Meunier et al. 2017). This long list of viruses also included the phage MS2.

Recently, the UV-C irradiation at 254 nm was found to be an effective alternative for production of beverages: coconut water, a highly opaque liquid food, was irradiated using a novel flow-through UV system, where the phage MS2, among others, was used as a surrogate (Bhullar et al. 2018).

Because of the USEPA recommendation to use MS2 as a standard by the design and validation of UV reactors, the phage MS2 was chosen as a basic reference virus for the elaboration of a new concept of the Virus Sensitivity Index (VSI) of UV disinfection (Tang and Sillanpää 2015). The VSI was defined as the ratio between the first-order inactivation rate constant of a virus, k_i, and that of the phage MS2 during UV disinfection, k_r. Therefore, the VSI was recommended for ranking of the UV disinfection sensitivity of viruses in reference to the phage MS2.

The efficiency of UV irradiation was compared to that of the lagooning system, which was found less efficient in removing phages and viruses than bacterial indicators, except for the FRNA phages (Gomila et al. 2008).

Remarkably, UV irradiation at 222 nm wavelength was more effective compared to 254 nm for inactivation especially of germs with higher UV resistance as the phage MS2 and bacterial spores, but not for inactivation of the UV-sensitive bacteria (Clauss et al. 2009).

An interesting mathematical model that was based on base-counting of potential dimers in the virus genomes was presented for the prediction of UV susceptibility of RNA and DNA viruses (Kowalski et al. 2009). The results correlated well with available data on the UV inactivation rate constants, which included the well-studied MS2 inactivation parameters, among others.

It is important to note here that the term biodosimetry in respect to the phages MS2 and Qβ was explained in detail for the first time in the large study of Blatchley et al. (2008), which was devoted to the development of dyed microspheres as a new method for validation of the UV reactor systems. Biodosimetry was formulated as "a term used to describe validation approaches in which a (nonpathogenic) challenge microorganism is imposed on a UV reactor under a fixed set of operating conditions. The same organism is also subjected to a range of well-defined, single-valued UV doses using a bench-scale reactor…" (Blatchley et al. 2008). Thus, the MS2 biodosimetry was applied by the pilot-scale assessment of the impacts of transient particulate water quality events on the UV disinfection of indigenous total coliform bacteria in drinking water treatment (Templeton et al. 2009a).

The MS2 biodosimetry helped a lot for the validation of computational fluid dynamics (CFD), a novel method of the assessing UV reactor performance and understanding of the dominant processes occurring in the UV reactors (Wols et al. 2012). Furthermore, the MS2 biodosimetry was used to simulate the whole UV disinfection process (Li HY et al. 2017). A portable handheld UV light device for water disinfection was assessed with the phage MS2 (Abd-Elmaksoud et al. 2013).

Further validation process of the UV-reactors faced a problem of the resistance level. Thus, the most common and coincidentally most UV-resistant challenge microorganism MS2 ranged in resistance from 18 to 25 mJ/cm^2/log, while, for example, a low-pressure UV dose of 186 mJ/cm^2 was required for adenovirus, although not reported in any drinking water outbreaks. Therefore, more UV-resistant microorganism, namely spores of the fungus *Aspergillus niger*, was proposed for doing practical demonstrations of high UV resistances (Petri and Odegaard 2009).

Later, the greater UV-C resistance of double-stranded DNA viruses, namely human adenovirus 2 and polyomavirus JC, versus that of the phage MS2, was acknowledged by Calgua et al. (2014).

The phage Qβ was used to detect a viable but non-cultivable *E. coli* after UV irradiation (Ben Said et al. 2010; Myriam et al. 2010).

Despite all controversies, the phage MS2 was used recently in the investigation of potential synergies of viral disinfection from sequential exposure to multiple UV sources (Hull and Linden 2018) and of dose rate and temperature dependence by the UV-A and UV-B inactivation in optically clear water (Lian et al. 2018; Mbonimpa et al. 2018), as well as by the field evaluation of novel UV water disinfection systems (Younis

et al. 2018, 2019a,b). The UV-C light device was tested for the inactivation of the phage MS2, among other indicators (Wallace et al. 2019). The UV inactivation of the phage MS2 at 220 and 254 nm was studied by the influence of algal organic matter (Wang Yulin et al. 2019). Schijven et al. (2019) acknowledged also that the MS2 remains a good, conservative indicator for the adenovirus disinfection in drinking water by the UV irradiation.

PHOTOCATALYTIC DISINFECTION

Near-UV photocatalytic disinfection of the phage MS2 in TiO$_2$ aqueous suspensions was accomplished for the first time by Sjogren and Sierka (1994). A level of phage inactivation of 90% increased to 99.9% after 2 μM ferrous sulfate was added. The hydroxyl radical oxidation was suspected to be primarily responsible for the viral degradation observed (Sjogren and Sierka 1994). The disinfection efficacy of the UV light irradiation at 254 nm over a TiO$_2$ suspension was more effective than that of the UV alone, with the phage Qβ as an indicator (Lee S et al. 1997, 1998). The photocatalytic chemistry of TiO$_2$, including inactivation of the phages MS2 and Qβ, was reviewed at that time by Blake et al. (1999).

The kinetic of the MS2 photocatalytic inactivation was evaluated in different ionic species (Koizumi and Taya 2002b) and pH values (Koizumi and Taya 2002a). Different crystalline structures of TiO$_2$ particles, namely anatase and rutile types, were examined for photocatalytic deactivation of the phage MS2 (Sato and Taya 2006). Cho et al. (2005) studied the different inactivation behaviors of the MS2 phage and *E. coli* by TiO$_2$ photocatalytic disinfection and concluded that the phage MS2 was inactivated mainly by the free hydroxyl radical in the solution bulk but that *E. coli* was inactivated by both the free and the surface-bound hydroxyl radicals. The general principles of TiO$_2$ photocatalytic antimicrobial disinfection have been exhaustively reviewed (Foster et al. 2011; Markowska-Szczupak et al. 2011).

The UV/TiO$_2$ disinfection of murine norovirus, as a surrogate for human norovirus, was standardized with the help of the phage MS2 that was more resistant to 254-nm UV than murine norovirus (Lee J et al. 2008). Photocatalytic inactivation of the phage Qβ by the TiO$_2$-coated glass plates under low-intensity, long-wavelength UV irradiation was described (Ishiguro et al. 2011). Later, the inactivation kinetics of the phage MS2 and murine norovirus as a surrogate for human norovirus by the UV-A and UV-B irradiation were studied with and without TiO$_2$ (Lee and Ko 2013). Both MS2 and murine norovirus were highly resistant to UV-A, but the addition of TiO$_2$ enhanced the efficacy of the inactivation. In comparison, UV-B alone effectively inactivated both MS2 and murine norovirus and the addition of TiO$_2$ increased the inactivation of MS2, but not of murine norovirus (Lee and Ko 2013). The phage Qβ was used by the fabrication of the TiO$_2$ photocatalysts with different hollow and spherical morphologies (Yamaguchi et al. 2016).

The silver ion was capable of enhancing the TiO$_2$ photocatalysis on a set of indicators including the phage MS2 (Vohra et al. 2006). Furthermore, a special investigation on the phage MS2 acknowledged more than fivefold enhancement of the inactivation rate by nano-sized silver doping (Liga et al. 2011). The silver doping of TiO_2 and other photocatalysts was exhaustively reviewed by McEvoy and Zhang (2014).

The TiO_2 was also doped by other metals. Thus, Mn and Co and binary Mn/Co enhanced solar/TiO_2 activity, in the absence of UV light, for the MS2 phage inactivation (Venieri et al. 2015). Adsorption of the phage MS2 to the TiO_2 semiconductor film was enhanced by nitrogen doping (Li et al. 2007). Treatment of the phage MS2 with the palladium-modified nitrogen-doped TiO_2 photocatalyst illuminated by visible light was described by Li et al. (2008b).

At the end of the twentieth century and beginning of the nano era, it was discovered that metal oxides, after they have been powdered to the nanoparticle form, exhibited novel antibacterial and antiviral properties. Thus, MgO and CaO in nanocrystalline form, carrying active forms of halogens—for example, $MgO \cdot Cl_2$ and $MgO \cdot Br_2$—were able to decontaminate water of the phage MS2 in minutes (Koper et al. 2002). However, ZnO, as well as silver nanoparticles, did not inactivate the phage MS2 (You et al. 2011). Moreover, in a binary bacteria-phages system where the *E. coli* host was exposed to MS2 and silver or ZnO nanoparticles simultaneously, the latter increased the number of phages by two to six orders of magnitude and therefore facilitated the phage infection (You et al. 2011).

The situation changed cardinally when ZnO nanoparticles, by analogy with TiO_2, were found to exhibit photocatalytic properties, which remained non-displayed in the non-nano ZnO state and were used in the UV/ZnO combination. The fate of ZnO as a semiconductor and photocatalyst was reviewed by Bogdan et al. (2015).

Concerning the TiO_2 nanoparticles, the nano-TiO_2 was able to inactivate the phages MS2 and fr by UV irradiation (Gerrity et al. 2008) and the phage MS2 by UV irradiation as well as by the natural light radiation (Wu et al. 2010). The TiO_2(P25) nanoparticles, a standard commercially available material, were doped by silica for the MS2 inactivation (Jafry et al. 2011). The addition of silica increased the MS2 adsorption onto the catalyst and the hydroxide radical $HO\bullet$ was responsible for the antiviral action. The TiO_2(P25)-SiO_2 either produced more $HO\bullet$ than non-silica-doped material, or the enhanced adsorption of MS2 to the catalyst resulted in greater exposure to the $HO\bullet$, or both mechanisms might work in concert (Jafry et al. 2011). A simple method of modifying the TiO_2 photocatalysts with SiO_2 was developed so that the SiO_2 nanoparticles were simply mixed with TiO_2 in water under ambient conditions. The composite SiO_2-TiO_2 nanomaterials had markedly higher photocatalytic inactivation rates for the phage MS2: up to 270% compared to the unmodified TiO_2 (Liga et al. 2013).

The UV/TiO_2 photoelectrocatalytic system was designed wherein TiO_2(P25) nanoparticles were coated onto an indium tin oxide electrode and an electrical potential was applied under black light blue irradiation (Cho et al. 2011a). In this system, the MS2 inactivation was greatly enhanced by anodic potential, whereas cathodic potential completely inhibited inactivation, which was primarily caused by hydroxyl radicals, both in the bulk phase and on the TiO_2 surface.

The fluorinated TiO_2(P25) nanoparticle surface coatings demonstrated good inactivation effects on human norovirus and its surrogates, the phage MS2, as well as murine norovirus and feline calicivirus (Park GW et al. 2014). Recently, the phage MS2 was chosen to compare nano TiO_2, nano ZnO, nano Fe_3O_4, carbon nanotube, graphene, nano Ni, and nano TiO_2(P25), where the MS2 removal increased with the increase of the TiO_2(P25) concentration (Cheng R et al. 2018b).

Remarkably, no measurable MS2 inactivation was achieved with carbon-doped TiO_2 nanoparticles under visible light (Shim et al. 2016). Furthermore, the quartz sand protected the phage MS2 from inactivation by the TiO_2 nanoparticles in ambient light (Syngouna and Chrysikopoulos 2017). The heteroaggregation of the phage MS2 with the TiO_2 anatase nanoparticles was examined by Syngouna et al. (2019).

Solar photocatalytic inactivation of the phage MS2, as well as of the DNA phages φX174 and PR772 and *E. coli* by use of TiO_2, ZnO and ruthenium based complexes, namely Tris(2,2′-bipyridyl)dichlororuthenium(II) hexahydrate ($Ru(bpy)_3Cl_2$), was described by Mac Mahon et al. (2017). The phage MS2 was found to be the most susceptible phage target by the treatment with $Ru(bpy)_3Cl_2$, with the complete phage removal observed within the first 15 min of exposure.

The phage f2 as a model was inactivated under visible light irradiation on the one-dimensional Cu-TiO_2 nanofibers that were fabricated using the electrospinning method (Zheng et al. 2017, 2018; Cheng et al. 2019a). The introduction of Cu into TiO_2 effectively extended the spectral response of the TiO_2 nanoparticles to visible light.

In parallel, a new material with a novel structure, namely silver- and copper-loaded TiO_2 nanowire membrane (Cu-Ag-TiO_2), was prepared and found efficient by the inactivation of the phage MS2 evaluated for its efficiency to infect *E. coli* and the phage MS2 under both dark and UV light illumination (Rao et al. 2016).

The wide application of the RNA phages in the UV inactivation technologies renewed the interest to the photochemical pathways that are leading to virus inactivation by the UV_{254} irradiation and $^1O_2$ treatment (Wigginton et al. 2010). As a result, it was established by the MALDI-TOF-TOF and ESI-TOF techniques that the chemical modifications occurred in the MS2 coat protein with both treatments, although the majority of residues within the MS2 coat protein remained largely unaffected, even after eight log_{10} of the MS2 inactivation: one oxidation event was detected following $^1O_2$ treatment in an amino acid residue located on the capsid outer surface, while the UV_{254} treatment caused three chemical reactions in the coat protein, two of which were oxidation reactions with residues on the capsid outer surface. A site-specific cleavage also occurred with UV_{254} irradiation at a protein chain location on the inside face of the capsid shell, which could influence a contact of the affected residues with viral RNA and might play a role in virus inactivation (Wigginton et al. 2010).

This was the first example of the mass spectrometric tracking of the inactivation mechanisms. The ESI- and MALDI-based FT-ICR, Orbitrap, and TOF mass spectroscopy allowed unveiling of the smart mechanisms of the inactivation (Wigginton et al. 2012a). The cleavage site at the Cys46-Ser47 residues, a location of viral genome−protein interaction, was identified and the presence of viral RNA was found essential to induce this backbone cleavage. A cleavage mechanism was proposed by radical formation (Wigginton et al. 2012a). The phage GA did not exhibit, however, the site-specific protein cleavage.

Furthermore, Wigginton et al. (2012b) performed quantitative analysis of the total damage incurred by the phage MS2 upon inactivation induced by five common virucidal agents: UV, heat, hypochlorous acid, singlet oxygen, and chlorine dioxide. It was concluded in this study that each treatment targeted one or more virus functions to achieve inactivation: UV, singlet oxygen, and hypochlorous acid treatments generally rendered the genome nonreplicable, whereas chlorine dioxide and heat-inhibited host−cell recognition/binding. It is noteworthy that the Wigginton team later performed the study on the UV_{254} and free chlorine reactivity of genome, proteins, and lipids of the enveloped double-stranded RNA pseudomonaphage φ6, by full analogy and direct comparison with their previous pioneering studies on the phage MS2 (Ye et al. 2018).

Recently, the phages fr and MS2 were involved, among a long list of other viruses, in a complex study on the impact of capsid protein composition on virus inactivation by the UV irradiation and TiO_2 photocatalysis and removal of viruses by ferric chloride coagulation (Mayer et al. 2015). It was found that the oxidizing chemicals, including hydroxyl radicals, preferentially degraded amino acids over nucleotides, in accordance with the previous data (Wigginton et al. 2010), and the tyrosine residue appeared to strongly influence the virus inactivation. Inactivation of the phage MS2 was studied by using an N-doped TiO_2-coated photocatalytic membrane reactor (Horovitz et al. 2018) and by the binder-free immobilization of TiO_2 photocatalyst on steel mesh via electrospraying and hot-pressing (Ramasundaram et al. 2018).

The role of the phages f2 and MS2 by the development of the visible light-active photocatalysts for water disinfection was unveiled recently in exhaustive reviews (Uyguner Demirel et al. 2018; You et al. 2019).

UV-LEDS

The UV light-emitting diodes (UV-LEDs), were recommended to be used for disinfection purposes by Hamamoto et al. (2007) and Vilhunen et al. (2009). Since UV-LEDs do not contain toxic mercury, offer design flexibility due to their small size, and have a longer operational life than mercury lamps, they would be advantageous for the UV inactivation techniques. For a comparative inactivation of the phage MS2 and other bacterial and viral indicators, UV-LEDs were tested for the first time by Bowker et al. (2011). By inactivation of the RNA phages MS2 and Qβ using UV-LEDs operated at 255 and 280 nm, the latter was found more suitable for the practical disinfection of water, even though the inactivation efficiency of 280 nm was lower than that of 255 nm (Aoyagi et al. 2011). The UV-LED that produced the UV-A light with peak wavelength at 365 nm killed the phage MS2 and *Cryptosporidium parvum* oocyst which were, however, 7.7–7.8 times more resistant than MS2 (Hashimoto et al. 2013). A numerical computational fluid dynamics (CFD) model of a continuous-flow UV-LED water disinfection process was based on the inactivation data of the phages MS2 and Qβ and created a basis for the design of alternative point-of-use drinking water disinfection reactors in developing countries using UV-LEDs (Jenny et al. 2014). Furthermore, heuristic optimization of a continuous-flow point-of-use UV-LED disinfection reactor was elaborated by use of computational fluid dynamics, where the phage Qβ appeared as an experimental model (Jenny et al. 2015).

The phage Qβ, as well as goose parvovirus and avian influenza virus, were inactivated by a novel photocatalyst on polyethylene terephthalate film under UV-LED (Hasegawa et al. 2013).

A ring-shaped disinfection apparatus has been developed containing 20 UV-LEDs with emission at 285 nm and tested with *E. coli*, MS2, Qβ, and adenovirus, in order of growing resistance to the inactivation (Oguma et al. 2015). Then, Oguma et al. (2016) proposed a design concept on how to arrange UV-LEDs in a water disinfection apparatus: a ring-shaped UV-LED device composed of two units containing ten 285-nm UV-LEDs, each used to measure the inactivation efficiency of the phage Qβ and *E. coli*.

The UV-LEDs emitting at 260 nm were evaluated to determine the inactivation kinetics of a long set of indicators including the phage MS2, where the latter was the most resistant (Sholtes et al. 2016). The dual-wavelength UV-C LED unit, emitting at peaks of 260 nm, 280 nm, and the combination of 260 and 280 nm together was evaluated for the inactivation of the phage MS2 and adenovirus among other indicators, and the 260 nm UV-LED was found most effective for the phage MS2 (Beck et al. 2017). The UV-LEDs operating at a wavelength of 270 nm disinfected the water spiked with the phage MS2 and two novel chlorine-resistant RNA phages (Zyara et al. 2017) that have been selected before as prospective surrogates (Zyara et al. 2016a,b). Concluding, the tested wavelengths between 255 and 280 nm of UV-LEDs inactivated the RNA phages efficiently.

UV-LED continuous and pulsed irradiation with various pulse patterns under equivalent UV fluence at 265 nm was checked by the inactivation of the phage MS2 (Song et al. 2018). Song et al. (2019) evaluated the MS2 inactivation by various combinations of UV-C, UV-B, and UV-A LEDs.

The superior UV-LED inactivation of the phages MS2, Qβ, as well as of the DNA phage φX174, was achieved also by the UV-C treatment at 266 and 279 nm (Kim DK et al. 2017; Kim DK and Kang 2018). The UV-LEDs at peak emission wavelengths of 265, 280, and 300 nm were adopted to inactivate the phage Qβ, among bacterial indicators and pathogens *P. aeruginosa* and *Legionella pneumophila*, where the 265-nm UV-LED showed the most effective fluence (Rattanakul and Oguma 2018), as well as the phage MS2 in comparison with the inactivation of feline calicivirus in water (Oguma 2018). The UV-LED inactivation of the phages MS2 and Qβ, as well as of a set of bacterial indicators, was highest in the fluence-based inactivation rate constant k at 265 nm, followed by 280 nm and much lower at 300 nm (Oguma et al. 2019). According to this recent study, the k value at 280 nm was close to that at 265 nm for feline calicivirus and MS2, suggesting that 280 nm UV-LED could be a good option as 265 nm UV-LED to inactivate these viruses.

The remarkable role of the phages MS2 and Qβ in the studies relating to UV-LED disinfection was unveiled recently in two reviews with an emphasis on water treatment systems (Li X et al. 2019) or food applications (Hinds et al. 2019). Using data on RNA phage inactivation, Umar et al. (2019) presented an original view on the global role of UV irradiation, moving from the traditional paradigm of pathogen inactivation to controlling antibiotic resistance in water. The UVC-LED disinfection system was validated using the phage MS2 inactivation over a range of flow rates and water UV transmittances (Hull et al. 2019). Sholtes and Linden (2019) used the phage MS2 in a research exploring microbial inactivation of UV-LEDs at various wavelengths under continuous and pulsing operating conditions.

PHOTOSENSITIZING AGENTS

First, the phage MS2 was sensitized to visible light after treatment with acridine orange (Amati 1967). Later, a series of antiviral compounds consisting of an intercalating acridine–derived part were synthesized and tested on the phage MS2 (Csuk et al. 2004a,b,c).

The near-UV irradiation inactivated the phages MS2 and Qβ in the presence of chlorpromazine, a derivative of pheanothiazine, which induced phototoxic or photoallergic dermatitis as the side effect in patients treated with this drug (Matsuo et al. 1980, 1983).

Methylene blue and rose bengal dyes were able to photoinactivate the phages R17 and Qβ in visible light (Schneider et al. 1993, 1998, 1999; Floyd et al. 2004). The phage MS2 was also sensitive to methylene blue as a photosensitizer (Specht 1994). In parallel, methylene blue and other cationic dyes such as thionine and thiopyronine were tested for the Qβ photoinactivation (Jockusch et al. 1996), and then the methylene blue for the Qβ and f2 photoinactivation (Lee D et al. 1997) in UV-visible light.

The phage R17 was used in the study that selected the dye 1,9-dimethyl-3-dimethylamino-7-dimethylaminophenothiazine (1,9-dimethylmethylene blue) as the most efficient photosensitizer from a series of phenothiazine dyes (Wagner et al. 1998a; Hirayama et al. 2001) and adopted it to the preservation of red blood cells during storage (Wagner et al. 1998b).

The photoinactivation by synthetic porphyrin photosensitizers was tried on the phage MS2 and hepatitis A virus at the 365 nm UV light (Casteel et al. 2004).

Later, a tricationic porphyrin, 5,10,15-tris(1-methylpyridinium-4-yl)-20-(pentafluorophenyl)porphyrin tri-iodide (Tri-Py+-Me-PF) was tested on the phages MS2 and Qβ as a prospective photosensitizer, where the photodynamic efficiency varied with the phage type, the RNA phages being much more easily photoinactivated than the DNA ones (Costa et al. 2012b). Next, a Tetra-Py+-Me was tested with the phage Qβ (Costa et al. 2013). It is noteworthy that the photodynamic inactivation of phages, as well as of mammalian viruses, was reviewed extensively at that time by Costa et al. (2012a).

A photosensitizer containing aluminum tetrakis [bis (cholinyl)phenylthio)] phthalocyanine grafted onto silica was efficient by the inactivation of the phage MS2 and poliovirus and recommended for the water purification (Nedachin et al. 2011).

Furthermore, the photosensitized inactivation of the phage MS2 was achieved by periodate IO_4^- activation through dye-sensitization mimicking self-sensitized dye decolorization on TiO_2 (Yun et al. 2017). Simultaneous decolorization of dyes and production of oxidizing radicals occurred because of the visible-light-induced activation of periodate into reactive iodine radicals via sensitized electron transfer from an organic dye, Rhodamine B. The dye-sensitized IO_4^- activation process was highly effective in the MS2 inactivation (Yun et al. 2017).

The A protein was identified as a target by the photoinactivation of the phage MS2 with such photosensitizer as the 5, 10, 15, 20-tetrakis (1-methyl-4-pyridinio) porphyrintetra p-toluenesulfonate (TMPyP) (Majiya et al. 2018b). The studies were extended to the phage Qβ, bovine enterovirus, and murine norovirus and showed the rate of inactivation by the TMPyP treatment in the order MS2 > Qβ > norovirus > enterovirus (Majiya et al. 2018a). It was not possible to generate photodynamic inactivation–resistant MS2 in this study, although emergence of a mutation in the lysis protein was detected after serial exposure to the TMPyP.

PHOTOSENSITIZATION OF FULLEROL NANOPARTICLES

Significant inactivation of the phage MS2 exceeded UV-A irradiation at 315–400 nm alone when the polyhydroxylated C_{60} fullerene (fullerol), nanoparticles were present (Badireddy et al. 2007). The inactivation was initiated by the production of both singlet oxygen $^1O_2$ and superoxide by fullerol in the presence of UV light. Li et al. (2008a) published the first review on the potential application of the antimicrobial nanomaterials including the fullerol nanoparticles, as well as TiO_2, ZnO, silver, and other nanoparticles for water disinfection and microbial control. The phage MS2, among other nonenveloped viruses, was shown to be inactivated by singlet oxygen $^1O_2$ produced in the UV-A (315–400 nm) photosensitized aqueous suspensions of a 40 μM fullerol, a polyhydroxylated fullerene $C_{60}(OH)_{22-24}$ (Hotze et al. 2009).

Novel hexakis C_{60} derivatives with varying functionalities, i.e., $NH3^+$-, CO_2H-, or OH-terminated multifunctionalized C_{60}, inactivated the phage MS2 and *E. coli* as standard indicators more efficiently than commercial fullerol because of facile $^1O_2$ production (Lee I et al. 2009). Cationic aminofullerene hexakis, which likely exerted electrostatic attraction, exhibited exceptionally rapid MS2 inactivation even compared to the commercial nano-TiO_2 photocatalyst. Moreover, the cationic amine−functionalized C_{60} derivative was also photoactive in response to visible light from fluorescence lamps and sunlight by the MS2 inactivation (Cho et al. 2010).

The C_{60} aminofullerene was immobilized on silica gel to enable recycling and repeated use and demonstrated the efficient MS2 inactivation (Lee et al. 2010). The phage MS2 was employed in a technological study on the role of heterogeneities in fullerene nanoparticle aggregates affecting reactivity, bioactivity, and transport (Chae et al. 2010). Moor and Kim (2014) proposed a simple synthetic method for preparation of the solid-supported C_{60} visible light-activated photocatalysts that were tested successfully by the MS2 inactivation. Meanwhile, the most rapid and efficient MS2 inactivation was reported for the colloidal fullerene aggregates (Snow et al. 2014).

The destruction mechanism of the phage MS2 particles by inactivation with four UV-A illuminated fullerenes consisted of the loss of capsid structural integrity, localized deformation, and the reduced ability to eject genomic RNA into its bacterial host (Badireddy et al. 2012).

The novel grade of the antimicrobial nanomaterials as water disinfectants was reviewed by Hossain et al. (2014).

Finally, an unprecedented MS2 photoinactivation efficiency was achieved by the visible light photoactivity of supported fullerene photocatalysts using C_{70} fullerene (Moor et al. 2015).

GRAPHENE AND CARBON NANOMATERIALS

Photoinactivation of MS2 was achieved by graphene−tungsten oxide composite thin films with sheet-like surface morphology under visible light irradiation (Akhavan et al. 2012). The photocatalysis of the phage MS2 on surface of the graphene−tungsten oxide composite film resulted in a nearly complete destruction of the viral protein and a sharp increase in the RNA efflux after 3 h light irradiation at room temperature (Akhavan et al. 2012). The covalently synthesized graphene oxide−aptamer nanosheets damaged the MS2 capsid and nucleic acid under broad visible-light spectrum (Hu Xiangang et al. 2012). Magnetic Fe_3O_4/graphene composite removed efficiently the phage MS2, as well as bacterial indicators (Zhan et al. 2015).

Thorough reviews of the antimicrobial characters of single-component and functionalized graphene materials, as well as their applications for water disinfection and microbial control, were published recently (Lukowiak et al. 2016; Zeng et al. 2017).

The polymeric graphitic carbon nitride (g-C_3N_4), a metal-free robust photocatalyst, was able to inactivate the phage MS2 under visible light irradiation by distortion of the capsid shell (Li Y et al. 2016; Zhang C et al. 2018a,b). The photocatalytic inactivation of the phage f2 with Ag_3PO4/g-C_3N_4 composite under visible light irradiation was described by Cheng R et al. (2018a). The role of the phage MS2 by elaboration of the graphitic carbon nitride-based photocatalysts was reviewed recently by Zhang C et al. (2019a,b) and Murugesan et al. (2019).

The single-walled carbon nanotubes for removal of viral and bacterial pathogens were tested with the phage MS2 (Brady-Estévez et al. 2008, 2010a,b,c). The nanosponge electroporation, which used one-dimensional nanomaterials and allowed electroporation to occur at only several volts, was tested with the phage MS2 as one of the indicators (Liu C et al. 2013). Furthermore, the carbon-nanotube sponges enabled highly efficient MS2 inactivation by low-voltage electroporation (Huo et al. 2017). Recently, the heteroaggregation of acid-functionalized multiwalled carbon nanotubes with the phage MS2 was demonstrated under mono- and divalent cations and with Suwannee River humic acid (Merryman et al. 2019).

OTHER NANOPARTICLES

Nanoalumina fibers, 2 nm in diameter and approximately 0.3 μm long, were electroadhesively grafted to a microglass fiber and used for the retention of the phage MS2 (Tepper and Kaledin 2007).

Polycationic superparamagnetic core-shell nanoparticles, where cores consisted of magnetite clusters and shells of functional silica covalently bound to poly(hexamethylene biguanide), polyethyleneimine (PEI), or PEI terminated with aziridine moieties, inactivated the phage MS2 as well as a set of eukaryotic viruses (Bromberg et al. 2012). The electrospun poly(vinyl alcohol) nanofibers containing benzyl triethylammonium chloride inactivated the phage MS2 (Park JA and Kim 2015b).

Magnetic hybrid colloids (MHC), or a cluster of superparamagnetic Fe_3O_4 nanoparticles encapsulated with a silica shell, were decorated with silver and used to bite away the phage MS2 (Park et al. 2013; Park S et al. 2014). Shao (2014) showed that 100 mg/L silver nanoparticles were enough to completely deactivate the phage MS2. The latter was used also by disinfection performance of different nanosilver materials (Xu et al. 2014).

Silver nanowire-carbon fiber cloth nanocomposites were approved for the electrochemical point-of-use water disinfection with the phage MS2 and *E. coli* as models (Hong X et al. 2016). Silver nanoparticle-decorated silica hybrid composites were tested successfully with the phage MS2 and murine norovirus (Park et al. 2018). The synthesis, characterization, properties, applications, and therapeutic approaches of silver nanoparticles, including the role of the phage MS2, were reviewed extensively (Zhang XF et al. 2016; Deshmukh et al. 2019). The phages MS2 and PP7 were tested to address the antiviral properties of silver nanoparticles (Gokulan et al. 2018). The authors arrived at the interesting conclusion that the silver nanoparticles could lead to serious perturbations of the gut microbial ecosystem, leading to the inactivation of resident phages that play an important role in influencing gastrointestinal health.

The phage MS2 served as a test organism by the creation of nanopores in single-layer molybdenum disulfide nanosheets by the electrospray deposition of silver ions on a water suspension of the former for solar water disinfection (Sarkar et al. 2018).

Recently, the phage MS2 was employed by the evaluation of the ceramic nanofibers electrospinned with iron(III) oxide or copper(II) oxide (Schabikowski et al. 2019). The phage f2 was used by the characterization of the Fe/Ni nanoparticles, including the fine mechanisms of the phage f2 inactivation (Cheng et al. 2019b). The microbicidal effects of iron/copper Fe/Cu nanoparticles were explored on the phage MS2 and *E. coli* as standard models (Kim HE et al. 2019).

General reviews on the role of the RNA phages by the evaluation of the metal nanoparticles as a protective nano-shield against virus infection were published recently (Rai et al. 2016; Adegoke and Stenström 2019; Rikta 2019).

IONIZING RADIATION

The x-ray and γ-irradiation of the RNA phages, namely of the phage R17, was first evaluated by William Ginoza (1963) and reviewed in detail together with a set of the DNA phages (Ginoza 1967a,b). Concerning the two latter papers, it is curious to note that dual publication of the same paper was given in the 1967 volumes of *Annual Reviews: Annu. Rev. Microbiol.* and *Annu. Rev. Nucl. Sci.*, "to provide more coverage of radiobiology," as referenced by the editors. In addition, the action of radiation on viruses was reviewed by Dertinger and Jung (1970).

The phage μ2 was compared with the scrapie agent, herpes simplex and yellow fever viruses to determine the extent of protection afforded against ionizing radiation by anoxia, i.e., the absence of oxygen, when the freeze-dried preparations of viruses were irradiated by the electron beam from the MRC linear accelerator (Alper and Haig 1968). The gamma irradiation of the phages MS2 (Vinetskii 1966, 1967, 1969; Hotz 1968; Karczag et al. 1973; Fidy and Karczag 1974) and f2 (Petranović et al. 1971) was studied. At the same time, the killing of the highly ^{32}P-labelled phage MS2 by transmutation was found significantly lower than would be expected for a single-stranded phage (Lupker et al. 1973).

The phage f2, together with the DNA phage T4, was used as a model for the γ-radiation and heat destruction of viruses in serum (Ward 1979). The first systematic study on the inactivation behavior of the phage MS2, together with two representatives of the single-stranded and double-stranded DNA phages, toward γ-radiation was performed in order to evaluate their potential as viral indicators for the water disinfection by γ-irradiation (Sommer et al. 2001). The phage MS2 demonstrated unexpectedly high sensitivity to γ-radiation compared to the two DNA phages.

Because of this rather high sensitivity as compared to the pathogenic viruses, the phage MS2 was considered a nonsuitable indicator for the ionizing irradiation treatment. The phage MS2 was significantly more sensitive to ionizing radiation than *E. coli* (Gehringer et al. 2003). The inactivation rate of the phage MS2 was comparable but nevertheless different from those for feline and canine caliciviruses by ionizing, as

well as nonionizing, radiation treatment (De Roda Husman et al. 2004). Later, the phage MS2 appeared as the most resistant organism against ionizing radiation when it was compared with *E. coli*, feline calicivirus, and poliovirus (Tree et al. 2005).

A similar study on the γ-radiation of the phages, including the phages MS2 and f2, was completed by Chinese scientists to screen a suitable phage as a virus indicator by the γ-irradiation sterilization (Chen et al. 2008). As a result, the order of resistance of six microorganisms to γ-radiation from the biggest to the smallest was determined: *B. subtilis* var. *niger* sp. > phage MS2 > phage f2 > phage T4 > phage φX174 > *E. coli*. Next, the comparative γ-radiation effect on the naturally occurring somatic coliphages, FRNA phages, and *E. coli* was examined in raw sewage and sewage sludge (Jebri et al. 2013). These effects were compared with the effects of γ-radiation on the seeded phages MS2 and φX174 in distilled water, autoclaved raw sewage, and a peptone solution. It was remarkable that inactivation of the phages, in contrast to *E. coli*, was significantly greater in distilled water than in the other matrices. The somatic coliphages in raw sewage and sewage sludge were far more resistant than the FRNA phages. As a result, the phage φX174, but not MS2, was proposed as a suitable indicator for estimating virus inactivation by γ-irradiation, because of the greater resistance of the phage φX174 to the ionizing radiation.

PLASMA

The phage MS2 and bacterial indicators were inactivated by pulsed arc electrohydraulic discharge, where a plasma channel was created between a pair of iron electrodes set 0.5 mm apart in two bench-scale reactors, 0.7 and 3 L (Lee LH et al. 2008). The streamer corona discharge process inactivated the phage MS2 used as a surrogate of pathogenic viruses in water (Lee C et al. 2011).

Next, the sterilizing effect of atmospheric non-thermal plasma was tested with a set of indicators including the phage MS2 (Mizuno and Yasuda 2012). The cold plasma generated by electrical discharge was recommended to be employed for biodecontamination and sterilization of surfaces, medical instruments, water, air, food, even of living tissues without causing damage or side effects, was able to destroy the phage MS2 when it was in wet condition.

Furthermore, an in-house-designed atmospheric pressure, nonthermal plasma jet operated at varying helium/oxygen feed gas concentrations was able to inactivate the phage MS2 (Alshraiedeh et al. 2013). As a result, the atmospheric pressure, nonthermal plasmas were recommended for the rapid disinfection of virally contaminated surfaces with efficiency, which was superior to previously published inactivation rates for chemical disinfectants.

The phage MS2 was chosen as a model for the inactivation study by the atmospheric-pressure cold plasma using different gas carriers and power levels (Wu Yan et al. 2015).

The application of corona discharge-generated air ions was studied by the inactivation of the aerosolized phage MS2 and a susceptibility constant of the phage to different air ions was introduced (Hyun et al. 2017). The role of the phage MS2 by the possible sanitizing and disinfection applications of the corona discharge was reviewed by Hamade (2018) and by Šimončicová et al. (2019).

Use of the non-thermal plasmas for the inactivation of viruses, together with the role of the phage MS2, was reviewed by Pradeep and Chulkyoon (2016) and Mizuno (2017).

Guo et al. (2018) performed a systematic study on the phage MS2, as well as φX174 and T4, inactivation by surface plasma in argon mixed with 1% air and plasma-activated water. The inactivation of the phages was alleviated by the singlet oxygen scavengers, demonstrating that singlet oxygen played a primary role in this process. The reactive species generated by plasma damaged both nucleic acids and proteins, consistent with the morphological examination showing that plasma treatment caused the phage aggregation. Remarkably, prolonged storage had marginal effects on the antiviral activity of plasma-activated water (Guo et al. 2018).

SELECTIVE PHOTONIC DISINFECTION

The phage MS2 was involved as one of the targets into the radical selective photonic disinfection (SEPHODIS) idea (Tsen and Tsen 2016). The proposed SEPHODIS technology employed a single ultrashort pulsed laser beam to kill viruses, bacteria, and other pathogens through forced mechanical vibration. According to the authors, this could be a method of disinfection from which pathogens could not readily escape, while human cells are left intact.

METAL COMPOUNDS

At the very beginning of the RNA phage era, it appeared that the phages MS2 and f2 were sensitive to contact with an aluminum alloy surface or when they were diluted with fluids that had been in contact with aluminum, zinc, or magnesium (Yamamoto et al. 1964). The inactivation was believed to result from the simultaneous action of traces of Cu^{2+} and electrolytically formed H_2O_2 and were stimulated by addition of both, although neither alone was fully active when present in trace amounts, while the phages were protected by adding either catalase or EDTA (Yamamoto et al. 1964). A very low concentration of hydrogen peroxide (0.00015%) in the presence of 10 μM $CuSO_4$ inactivated the RNA phage MS2 (Yamamoto 1969). Incubation of the phage Qβ with a mixture of 100 mM ribose and 10 μM $CuSO_4$ resulted in a complete loss of viable phage after 20 min (Carubelli et al. 1995). The synergistic effect of cupric chloride with monochloramine was demonstrated by the MS2 inactivation (Straub et al. 1995). The phage MS2 was 10 times more sensitive than poliovirus to inactivation by electrolytically generated copper and silver ions, separately and in combination with free chlorine (Yahya et al. 1992).

The phage MS2 was found to be even more susceptible than *E. coli* to the oxidative stress induced by the Cu(II)/H_2O_2 system and caused by the hydroxyl radical ·OH, cupryl species Cu(III), and the superoxide radical $O_2^{·-}$, which were produced via the catalytic decomposition of H_2O_2 (Nguyen et al. 2013). The inactivation of the phage MS2, as well as *E. coli*, by Cu(II) was significantly enhanced, up to 5- to 100-fold, depending on the conditions, in the presence of a small amount of hydroxylamine (Kim HE et al. 2015).

A recent systematic study compared the effect of copper and silver compounds on the phage Qβ and influenza A virus H1N1 (Minoshima et al. 2016). The viruses were exposed to solid-state cuprous oxide Cu_2O, which efficiently inactivated both viruses, whereas solid-state cupric oxide CuO and silver sulfide Ag_2S showed little antiviral activity. Copper ions from copper chloride $CuCl_2$, as well as silver ions from silver nitrate $AgNO_3$ and silver oxide Ag_2O, showed little effect on the phage Qβ but inactivated influenza A virus.

Recently, a series of bench-scale experiments were conducted on ceramic filters with various amounts of copper and/or silver nanoparticles fired-in during the manufacturing process, where the phage MS2, together with *E. coli*, were employed as surrogates (Lucier et al. 2017). Given the potential health risks of inadequate filtration, it was recommended, however, that the quality control process should be examined and upgraded.

The inactivation kinetics of the phage MS2 as a surrogate for enteric viruses by dissolved ionic copper, which appeared during storage of water in copper containers, were measured (Armstrong et al. 2017). Copper inactivated MS2 at doses between 0.3 and 3 mg/L. It was concluded that copper could be a candidate for improving the safety of stored drinking water, although it required longer contact times than conventional disinfectants.

The phage MS2 was used as a control during iron electrocoagulation–microfiltration of surface water (Tanneru and Chellam 2012). However, higher reductions in the phage MS2 concentrations were obtained by aluminum electrocoagulation and electroflotation compared with conventional aluminum sulfate coagulation (Tanneru et al. 2013). The effect of electrochemical treatment using a sacrificial aluminum anode on the phage MS2 was evaluated further by Tanneru et al. (2014). It was established that the electrocoagulation generated only small amounts of free chlorine *in situ* but effectively destabilized the phage and incorporated it into $Al(OH)_3$ flocs during electrolysis. Therefore, the dominant phage control mechanism during aluminum electrocoagulation of saline waters was "physical" removal by uptake onto flocs rather than "chemical" inactivation by chlorine (Tanneru et al. 2014). It is necessary to note here that the direct action of electric field and current on the phage MS2 was not sufficient for efficient inactivation; the latter was induced by the electrochemically generated oxidants (Drees et al. 2003).

The phage MS2, together with φX174 and hepatitis A virus, was used in the study, where colloidal alumina particles were surface functionalized with amino, carboxyl, phosphate, chloropropyl, and sulfonate groups in different surface concentrations and characterized in terms of elemental composition, electrokinetic, hydrophobic properties, and morphology (Meder et al. 2013).

The phage R17 was inactivated by platinum II compounds: dichloro-ethylenediamine Pt II and cis- and trans-dichloro-diamine Pt II, when the mean lethal concentrations varied for R17 from 0.09 to 1.1 µg/mL, depending upon which compound was used and the manner in which the experiment was performed (Shooter et al. 1972).

The phage f2 was rapidly inactivated by iron (VI) ferrate, or FeO_4^{2-} ion, in buffered water and secondary treated sewage effluent, and was equally or less resistant to potassium ferrate than were most bacteria, including *E. coli* (Schink and Waite 1980). The ferrate inactivation of the phage Qβ was described by Kazama (1994). Furthermore, kinetics of the MS2 and GA inactivation by ferrate was examined in detail (Hu L et al. 2012). In this study, the mass spectrometry and qRT-PCR analyses demonstrated that both coat protein and genome damage increased with the extent of inactivation, suggesting that both may contribute to the phage inactivation, whereas coat protein damage was localized in the two regions containing oxidant-sensitive cysteine residues. The role of the phages MS2 and GA in the ferrate(VI)-based remediation of soil and groundwater was reviewed extensively by Kumar Rai et al. (2018).

The phage MS2 was inactivated successfully by the oxidants produced from Fenton's reagent, namely, $Fe[II]/H_2O_2$ (Kim JY et al. 2010). Further, the inactivation of the phage MS2 by iron- and copper-catalyzed Fenton systems was studied in detail (Nieto-Juarez et al. 2010; Nieto-Juarez and Kohn 2013) and principal parameters affecting the phage MS2 inactivation by the solar photo-Fenton process were characterized (Ortega-Gómez et al. 2015). Finally, the pathway describing the photo-Fenton-induced MS2 inactivation in wastewater was proposed (Giannakis et al. 2017a). The advanced role of the photo-Fenton process in wastewater treatment was acknowledged (Giannakis et al. 2017b,c) and extensively reviewed (Giannakis et al. 2016a,b, 2018; Giannakis 2018; Polo-López et al. 2018).

The phage MS2, together with φX174, was used in the first study on the removal and inactivation of waterborne viruses using zero-valent iron (You et al. 2005). Most of the viruses removed from solution were either inactivated or irreversibly adsorbed to iron. The MS2 inactivation was achieved further by zero-valent iron nanoparticles (Kim et al. 2011). Moreover, the latter caused more MS2 capsid damage than Fe(II). The nanoscale zero-valent iron also removed the phage f2 (Cheng et al. 2014, 2016).

The phage MS2 was used as a model in solid recent investigations that verified the phage inactivation as a function of ferrous iron oxidation (Heffron et al. 2019b,c).

The phage MS2 was also used by the successful assessment of ceramic microfilters that were coated with colloidal zirconia, namely, ZrO_2 nanopowder (Wegmann et al. 2008).

Brown and Sobsey (2009) used ceramic filter materials amended with iron oxides such as goethite, hematite (α-Fe_2O_3) and magnetite (Fe_3O_4) to remove the phage MS2 from drinking water. Furthermore, the phage MS2, as well as rotavirus, was removed using glass fiber coated with hematite nanoparticles (Gutierrez et al. 2009). Effects of solution pH, carbonate, and phosphate on the interaction of the phage MS2 with goethite during saturated flow was studied in detail (Zhuang and Jin 2008). Furthermore, iron oxides were impregnated on the surface of fiberglass to remove the phage MS2 from water, and the novel Fe-fiberglass material was characterized (Park JA et al. 2015b). The phage MS2 was used also as a model for the coagulation of viruses from water by polyferric chloride that was prepared by addition of NaOH to ferric chloride $FeCl_3$ (Shirasaki et al. 2016a,b).

The rhombohedral-like $CuFeO_2$ crystals showed promising efficiency in the inactivation of the phage Qβ (Qiu et al. 2012).

The arsenate As(V), which could be present in groundwater, had a mainly negative effect on MS2 removal by hematite through the competitive sorption, although As(V) itself was able to inactivate the phage MS2 (Park JA et al. 2015c).

Recently, the phage MS2 was used in the study that proposed zirconium (IV) oxychloride octahydrate and chitosan for the coagulation of waterborne pathogens in wastewater (Christensen et al. 2017).

NITROUS ACID

Inactivation by nitrous acid initiated the studies on the RNA phage genetics and resulted in numerous RNA phage mutants that are described in Chapter 9. Briefly, nitrous acid acted as a deamination agent on adenine, guanine, and cytosine, where adenine was converted to hypoxanthine, cytosine was converted to uracil, and guanine was converted to xanthine, which resulted in mispairing. The first nitrous acid inactivation, or mutagenesis, experiments were performed on the phages fr, or formerly ft5 (Kaudewitz and Knolle 1963), µ2 (Turri et al. 1964), and MS2 (Sherban 1969; Sherban and Krivisky 1969).

ALKYLATING AGENTS

A chemical mutagenic agent N-methyl,N-nitroso,N'-nitroguanidine was used for the generation of the phage fr mutants by Heisenberg and Blessing (1965). Later, the phage µ2 RNA was treated with the carcinogen N-methyl-N-nitrosourea and an evidence for O-methylation was gained (Lawley et al. 1971).

Bifunctional nitrogen mustard, or 2,2'-dichloro-N-methyldiethylamine, inactivated the phage MS2, while the monofunctional analog was either much less effective or ineffective (Yamamoto and Naito 1965). The bifunctional alkylating agents, analogs of di-(2-chloroethyl)-methylamine, inactivated the phage MS2, as well as single- and double-stranded DNA phages by intra-strand cross-linkage in DNA or RNA (Yamamoto et al. 1966).

However, the phage µ2 was inactivated by both mono- and difunctional sulfur mustards at relatively low extents of alkylation, while no degradation of alkylated RNA was detected (Shooter et al. 1971). The crosslinking of RNA to protein was observed with the difunctional agent, but this reaction was only a minor contribution to the inactivation. The inactivation resulted from the mono-alkylation of adenine or cytosine.

Furthermore, a thorough alkylation analysis was performed on the phage R17 with two groups of alkylating agents: (i) methyl methanesulfonate and dimethyl sulfate and (ii) N-methyl-N-nitrosourea and N-methyl-N'-nitro-N-nitrosoguanidine (Shooter 1974b). The extent of biological inactivation by these alkylating agents was explained by the breaks in the RNA chain, which resulted from hydrolysis of phosphotriesters formed in the alkylation reactions and the rate of hydrolysis increased rapidly as the pH was raised (Shooter et al. 1974a). The mechanism of the biological inactivation of the phage R17 by ethyl methanesulfonate and N-ethyl-N-nitrosourea was unveiled (Shooter and Howse 1975). Then, a group of eight alkylating agents was involved in the R17 studies (Shooter 1975) and acetoxy-dimethylnitrosamine was tested (Shooter and Wiessler 1976).

In parallel, two carcinogenic aralkylating agents 7-bromomethylbenz[a]anthracene and 7-bromomethyl-12-methyl benz[a]anthracene (Dipple and Shooter 1974) and 7,8-dihydrodiol-9, 10-oxides of benzo(a)pyrene (Shooter et al. 1977) were used for the studies on the phage R17.

BISULFITE

A limited number of covalent polynucleotide–protein crosslinks was induced in the phage MS2 by treatment with bisulfite and both A protein and coat protein were involved in this crosslinkage (Turchinsky et al. 1974). These data indicated the presence of some specific noncovalent interactions between RNA and proteins inside the virion.

HYDROXYLAMINE

High sensitivity of the phage fr to hydroxylamine in slightly alkaline solution was reported first by Hartmut Hoffmann-Berling et al. (1963a). Heisenberg and Blessing (1965) used hydroxylamine for the selection of the phage fr mutants. The high hydroxylamine sensitivity of the phages fr and f2, as well as of the single-stranded DNA phage φX174, but not of the double-stranded DNA phages of T series and λ, was confirmed by Vízdalová (1969).

The phage MS2 was inactivated with hydroxylamine (Budowsky and Pashneva 1971) and O-methylhydroxylamine (Budowsky et al. 1971, 1972, 1973, 1974a,b). In both cases, the rate of the MS2 phage inactivation was almost the same as that of the inactivation of the phage RNA. Hydroxylamine reacted only with uridine residues but O-methylhydroxylamine reacted also with cytosine residues, while reaction of both with adenine residues was negligible (Budowsky et al. 1971).

The correlation among chemical and functional changes induced by hydroxylamine in the MS2 genome was examined under conditions of predominant modification of either cytidine at pH 5.0 or uridine at pH 8.0 (Budowsky et al. 1974). In this context, Eduard Izrailevitch Budowsky (1971) later published an exhaustive review on the preparation of killed antiviral vaccines.

In connection with the hydroxylamine treatment, the question about the acid inactivation of the phage MS2 was brought up and the pH dependence of the inactivation rate showed it to be proportional to the hydrogen ion concentration (Dorsett et al. 1971).

CHLORINE, IODINE, AND BROMINE

For the first time, halogen inactivation studies were performed in 1964 and showed that both the phage f2 RNA and poliovirus RNA were resistant to iodination and that the inactivation of both the phage f2 and poliovirus were inhibited by increasing iodide ion concentrations (Hsu 1964; Hsu YC et al. 1965). The action of iodine on proteins was well known and included the oxidation of sulfhydryl groups or substitution into tyrosyl or histidyl components, while the iodide ion retarded the iodination of tyrosine.

The phage f2 was markedly more resistant to chlorine in water than either poliovirus or coliphage T2 (Shah and McCamish 1972). The comparative mode of action of chlorine, bromine, and iodine on the phage f2 was described first by Olivieri et al. (1975). The effect was dependent on the halogen. Chlorine inactivated naked f2 RNA at the same rate as the RNA within the phage, while the protein of the inactivated phage was still able to adsorb to the host. Bromine inactivated naked RNA at the same rate as it inactivated intact phage, but the RNA prepared from bromine-treated virus was significantly less inactivated than the intact virus and the protein moiety was regarded as the primary site of the bromine inactivation. Iodine acted by iodination of tyrosine and had almost no effect on the nucleic acid.

Further, the susceptibility of the phage f2 to chlorine and iodine in wastewater was compared with that of poliovirus (Cramer et al. 1976). It was found that the phage f2 was slightly more resistant to halogenation than poliovirus and appeared as an acceptable model for the enteric virus group by the wastewater disinfection. Resistance of the phage f2 to chlorine and five other popular disinfectants was compared to that of echovirus and coxsackievirus and found comparable or more expressed in the case of the phage f2 (Drulak et al. 1979). As for the mechanism of the phage f2 inactivation with chlorine, the rate of the inactivation was paralleled by the rate of ^{36}Cl incorporation into the phage, while the phage RNA demonstrated greater affinity for chlorine than did protein (Dennis et al. 1979). The clay-associated phage MS2 was more resistant to chlorine (Stagg et al. 1977).

The efficiency of the phage f2 inactivation by chlorine was studied further in parallel with that by bromine, chloride, and peracetic acid (Hajenian and Butler 1980a) and compared with the chlorine inactivation of poliovirus at various temperatures and pH (Hajenian and Butler 1980b). The phage f2 was found more resistant to chlorine than poliovirus also by Snead et al. (1980).

The phage MS2 was subjected to inactivation by chlorine, as well as iodine, hypochlorite, formaline, glutaraldehyde, quaternary ammonium salt (Lepage and Romond 1984), and a set of popular antiseptics and disinfectants (Garrigue 1984).

Nevertheless, the FRNA phages were found in the chlorinated waters that fulfilled the bacteriological criteria but still contained phages (Durán et al. 2003). The phage f2 was more resistant to chlorine than severe acute respiratory syndrome-associated coronavirus (SARS-CoV) (Wang et al. 2005). Inactivation of the phage MS2 by chlorine treatment was similar to that of norovirus but much faster than that of poliovirus (Shin and Sobsey 2008). The inactivation rates of Venezuelan equine encephalomyelitis virus and Sindbis virus were similar to each other and faster than that of the phage MS2 (Fitzgibbon and Sagripanti 2008).

Data on the indicator survival during chlorination at several municipal wastewater treatment plants in Rhode Island led to conclusion that the FRNA phages were a reliable indicator of viral contamination and die-off in estuarine waters (Armon et al. 2007). Further, it was confirmed that the phage MS2 was substantially more resilient to both free and combined chlorine and could be used as a conservative surrogate of poliovirus (Soroushian et al. 2010).

The phage MS2 was used for the assessment of the chlorine and bromine halogenated contact disinfection, where drinking water canisters contained N-halamine bromine or chlorine media (Coulliette et al. 2010).

Choe et al. (2015) observed effect of chlorine, bromine, and ozone treatment on the MS2 coat protein: the methionine residue was preferentially targeted, forming predominantly methionine sulfoxide (Choe et al. 2015). The kinetics of the phage MS2 chlorine disinfection were examined under the effect of pH, temperature, particulate matter, organic matter, and NH_3 (Cai et al. 2016). Recently, serious investigations of the parallel MS2 and human norovirus inactivation by the chlorine treatment of sewage (Kingsley et al. 2017) and postharvest leafy green wash water (Dunkin et al. 2017) were published.

The phage f2 was used as a model in the studies that were conducted to evaluate chloride dioxide as a less harmful disinfectant than chlorine (Noss and Olivieri 1985) and the reactivity of chlorine dioxide with RNA (Hauchman et al. 1986) and proteins (Noss et al. 1986).

The phage MS2, together with *B. subtilis* spores, was used by the evaluation of the water disinfection efficiency after chlorine dioxide treatment (Barbeau et al. 2005).

Furthermore, the FRNA phages, among other indicators, were controlled in an extensive comparative study on the efficiency of chlorine dioxide and peracetic acid at low doses in the disinfection of urban wastewaters (De Luca et al. 2008). The phage MS2 was used as a model in the investigation of the disinfection kinetics of very low concentrations of chlorine dioxide, which were applied in drinking water practice (Hornstra et al. 2011). The chlorine dioxide inactivation of enterovirus 71 in water was tested with the phage MS2 as a control, which appeared less resistant than enterovirus (Jin et al. 2013). Then, chlorine dioxide or UV resistance of echovirus was studied in parallel with the phage MS2 model (Zhong et al. 2017). Remarkably, the reaction of chlorine dioxide with the phage MS2 created products that deposited onto the phage particles and protected them from further disinfection (Sigstam et al. 2014). This protection took place on the coat protein, which was extensively but reversibly

modified during the disinfection process. Moreover, the chlorine dioxide−resistant MS2 mutants were identified and characterized by sequencing and electron cryomicroscopy (Zhong et al. 2016). Interestingly, the chlorine dioxide resistance was connected mostly with mutations in the A protein gene (Zhong et al. 2016).

The phage MS2 was used as one of the models by the successful inactivation of airborne bacteria and viruses using extremely low concentrations of chlorine dioxide gas in a hospital operating room (Ogata et al. 2016).

The phage MS2 was also employed as a model by the chloride-assisted electrochemical disinfection, where the electrochemical conversion of chloride to chlorine was acknowledged (Fang et al. 2006). In this context, it is notable that a direct electrolyzer was designed to investigate the electrochemical disinfection without the generation of chlorine and was tested by the MS2 inactivation (Kerwick et al. 2005). Then, the disinfection ability of two other electrolytic generation systems, ClorTec and MIOX, was compared with conventional hypochlorite disinfectant using the phage MS2 and three strains of *Bacillus subtilis* spores (Clevenger et al. 2007).

The role of the FRNA phages in the studies on chlorine, chlorine dioxide, and chloramine disinfection was extensively reviewed by Keegan et al. (2012).

As an alternative to chlorine and chlorine dioxide, chlorite (ClO_2^-) (Bichai and Barbeau 2006) and sodium hypochlorite (ClO^-) (Potgieter et al. 2009), solutions were regarded as prospective disinfectors, with employment of the phage MS2 as a model or FRNA phages as a testing target, respectively.

Concerning the iodine disinfection, the phage MS2 was a model in the successful NASA investigation within the space water reuse research program (Brion and Silverstein 1992, 1999; Marchin et al. 1997). It appeared that the mechanism of the iodine MS2 inactivation involved conformational changes to the protein coat, which could be partially reversed when the iodine residual disappeared (Brion and Silverstein 1999). Furthermore, the phages MS2 and GA, among other indicators, were used as models, where MS2 was the most susceptible to the iodine inactivation but the phage GA demonstrated the greatest iodine resistance (Brion et al. 2004). A comparative study was performed on the iodine inactivation with the phage f2, among other phages, as a model, where f2 appeared as the most sensitive to the iodine treatment (Chen et al. 2006b).

The phage MS2 served as a model on the evaluation of selected metal oxide nanoparticles, i.e., CeO_2, Al_2O_3, and TiO_2, that were capable of strongly adsorbing large amounts of chlorine, bromine, and iodine (Häggström et al. 2010). Overall, the halogen adducts of TiO_2 and Al_2O_3 were most effective.

The phage MS2 was used as a model by the assessment of the controlled iodine release from the iodinated polyurethane sponges for water decontamination (Aviv et al. 2013).

Furthermore, removal and inactivation of the phage MS2 was measured by drinking water treatment with ozone in the presence of bromide or iodide (Zhang Q et al. 2015).

The hydantoin-N-halamine derivatives conjugated on polystyrene beads demonstrated promising disinfectant abilities against the phage MS2 and other indicators (Farah et al. 2015). These compounds ensured the gradual release of oxidizing halogen in water at different rates.

Furthermore, the phage MS2 was used to assess two household water treatment devices that used monobrominated hydantoinylated polystyrene beads as a bromine deliverer (Enger et al. 2016).

Recently, the phage MS2 was used as a seeded model in a solid study on sequential chlorine and chloramine disinfection for wastewater reuse (Furst et al. 2018). The chlorine dioxide was found more effective than chlorine against the phage MS2, especially at pH values of >7.5 at which chlorine efficacies already declined (Grunert et al. 2018). However, free chlorine was found highly efficient by the phage MS2 inactivation in nitrified membrane bioreactor effluent at the City of Lathrop Consolidated Treatment Facility (Ikehata et al. 2018). The phage MS2 was used by the fabrication of antibacterial povidone-iodine-conjugated cross-linked polystyrene resins and investigation of their antimicrobial efficacy for water decontamination (Gao T et al. 2019).

AMMONIA

The kinetics of the phage f2 and poliovirus inactivation by NH_3 were compared in such a way that the effects of NH_3, NH_4^+, and OH^- were separated and the phage f2 was found 4.5 times more resistant to inactivation by NH_3 than was poliovirus, and that the effect of NH_4^+ was like that of Na^+, while any effects of OH^- were below the threshold of measurability (Cramer et al. 1983). The rate of the phage f2 and poliovirus inactivation by NH_3 was strongly influenced by temperature, where the change in the rate with increasing temperature in the range of approximately 10°C–40°C was greater for the phage f2 than for poliovirus (Burge et al. 1983). The group I RNA phages were more resistant to ammonia than other RNA phage groups (Schaper et al. 2002a). Furthermore, the phage MS2 was accepted as a reliable surrogate for the avian influenza virus H5N3 inactivation by the ammonia treatment of hatchery waste (Emmoth et al. 2007).

By inactivation with a commercial quaternary ammonium disinfectant, the response of both the phage MS2 and the DNA phage φX174 was similar to that of feline calicivirus and hepatitis A virus (Solomon et al. 2009).

Ammonia disinfection of hatchery waste, an animal byproduct of the poultry industry, was efficient against high-pathogenic avian influenza virus H7N1 and low-pathogenic avian influenza virus H5N3, as well as against bovine parainfluenza virus, feline coronavirus, and feline calicivirus used as models (Emmoth et al. 2011). However, the phage MS2 was more resistant than the listed viruses to all treatments (Emmoth et al. 2011) and proved too thermoresistant to be considered a valuable surrogate for avian influenza virus during thermal treatments (Elving et al. 2012).

An extensive study on the inactivation of the phage MS2 by ammonia under different temperature and pH conditions led to the conclusion that the loss in the MS2 infectivity was initiated by a loss in genome integrity, which was attributed to genome cleavage via alkaline transesterification (Decrey et al. 2015, 2016; Oishi et al. 2017). The phage MS2 was one of the indicators of a special study concerning the sanitization of toilets in developing countries with tropical and subtropical climates, where the use of intrinsic ammonia combined with high pH was found effective in producing a safe and highly valuable liquid that could be used as a fertilizer (Magri et al. 2015). However, the phage Qβ was found earlier to have the higher die-off in residentially operated bio-toilets in Japan (Nakagawa et al. 2006; Kazama et al. 2011; Kazama and Otaki 2011).

Quaternary ammonium silane, namely dimethyloctadecyl [3-(trimethoxysilyl) propyl] ammonium chloride-coated sand filter was implemented for the removal of the phage MS2 and other indicators from drinking water as a promising household device (Torkelson et al. 2012).

ALKALINE STABILIZATION

The phage MS2 was used as one of the indicators by the assessment of the class B alkaline, or lime, treatment, and the first study demonstrated that the MS2 monitoring could be reliable to the land application of biosolids while demonstrating no increased public health threat (Bean et al. 2007). The phage MS2 appeared as a conservative indicator of the efficacy of lime stabilization of adenovirus and rotavirus and was proposed as a useful indicator organism (Hansen et al. 2007). Furthermore, the utility of the phage MS2 as a surrogate was confirmed by comparative study with hepatitis A virus and reovirus (Katz and Margolin 2007).

The alkaline disinfection of compost with calcium lime and ash was investigated with the phage MS2 and *E. coli* as surrogates for enteric pathogens and showed the partial MS2 capsid damage and RNA exteriorization due to a raised pH, which was proportional to the amount of alkaline agents added (Hijikata et al. 2016). The phage MS2, among other microorganisms, was used recently by a hygiene assessment study of the alkaline dehydrated urine (Senecal et al. 2018).

PERACETIC ACID

The phage MS2 was sensitive to peracetic acid (PAA, or CH_3COOOH) (Maillard et al. 1994). With the phage MS2 as an indicator, peracetic acid was compared with UV irradiation and ozone after primary treatment of municipal wastewater in Montreal (Gehr et al. 2003). It was concluded that a single indicator organism might not be appropriate and the required dose of any of the disinfectants was unlikely to be economically viable. Further, inhibitory effect of peracetic acid on the phage MS2 was extensively reviewed by Kitis (2004). The inactivation of the FRNA phages by low doses of peracetic acid in comparison to the inactivation of the bacterial indicators was studied in a municipal wastewater plant (Zanetti et al. 2007).

The MS2 aggregation effect was studied by the phage inactivation with peracetic acid and the latter was recommended for the treatment of water containing viral aggregates (Mattle et al. 2011). The exposure of the phage MS2 that was deposited on different test materials to different formulations of a solid source of peracetic acid was studied by Buhr et al. (2014).

The phage MS2, together with the double-stranded DNA phage P001 infecting lactic bacteria, participated in a comparative study on the virucidal efficacy of peracetic acid, potassium monopersulphate, and sodium hypochlorite (Morin et al. 2015).

The comparative inactivation of the phage MS2 and murine norovirus by peracetic acid and monochloramine in municipal secondary wastewater effluent contributed sound reasoning for the employment of peracetic acid as an alternative to chlorine-based disinfection (Dunkin et al. 2017a,d). The efficacy of the human norovirus GI and GII inactivation by peracetic treatment was assessed in comparison with the inactivation of the phage MS2 and murine norovirus as surrogates (Dunkin et al. 2019). Moreover, the infectivity reduction of the phage MS2 and murine norovirus was examined by the combined peracetic acid–UV irradiation treatment in secondary wastewater effluent (Weng et al. 2018).

Park E et al. (2014) introduced the RT-qPCR technique, in addition to culture-based method, to study the inactivation of the FRNA phages by peracetic acid.

OZONE

Pavoni et al. (1972) reported first that the mechanism of destruction of the phage f2 and bacteria by ozone was oxidative. The RNA enclosed in the phage coat was inactivated less by the ozonation than were whole phage but inactivated more than the naked RNA, while the protein capsid was broken into subunits (Kim et al. 1980). The phage MS2 was found more sensitive to ozone than enteric viruses by the surface water disinfection (Helmer and Finch 1993).

Furthermore, the phage MS2 was inactivated by Peroxone, an advanced oxidation process generated by combining ozone and hydrogen peroxide (Wolfe et al. 1989).

Later, the phage MS2 was used as a model of the nebulization technique, which was used by the ozone inactivation in large volumes of body fluids, such as plasma, partial blood, and perhaps whole blood in a short time (Kekez and Sattar 1997).

Meanwhile, ozonation was found to be less efficient than mixed oxidants generated by the Purizer process by direct comparison of both methods for the inactivation of the phage MS2 (Holland et al. 2001, 2002). The phage MS2 was used as one of the indicators by the combined application of ozone and hydrogen peroxide (Sommer et al. 2004).

An extensive study was performed on the inactivation of the aerosolized phage MS2, among other popular phage indicators, by ozone, and the latter was found to be highly efficient (Tseng and Li 2006). Remarkably, the ozone dose for the 99% MS2 inactivation was two times higher than that for 90% inactivation for all tested phages. Further, the efficacy of ozone in the inactivation of the phage f2 in water was analyzed in detail (Ma and Wang 2009).

The phage MS2 was used as an indicator by the evaluation of the HiPOX reactor, an ozone-based pressurized in-vessel system that was installed in Dublin, California (Ishida et al. 2008).

The phage MS2 was employed as an indicator in a large study on the surrogate correlation models to predict trace organic contaminant oxidation and microbial inactivation during ozonation (Gerrity et al. 2011, 2012). Furthermore, the phage MS2 was used as a surrogate for the development of a refined laboratory system to study changes of biological aerosols in simulated atmospheric environments (Santarpia et al. 2012; Ratnesar-Shumate et al. 2015) and for the optimization of the micropollutant removal from the tertiary effluent (Schaar et al. 2013). A special ozone disinfection device was tested with the phage MS2 (Donofrio et al. 2013). The data on the MS2 inactivation were employed for the successful application of Bayesian analysis to wastewater ozonation (Carvajal et al. 2017a).

The phage Qβ was employed by the development of an ozone-assisted photocatalytic water-purification unit using a TiO_2 modified titanium mesh sheet (TMiP) as a filter (Ochiai et al. 2012).

Advanced kinetic studies of the phage MS2 inactivation by ozone disinfection were performed by Cai et al. (2014). Recently, Wolf C et al. (2018) presented the accurate and quantitative kinetic data regarding inactivation of the phages MS2 and Qβ, as well as the DNA phages φX174 and T4, by the ozone treatment, and it was concluded that ozone remains a highly effective disinfectant for virus control. Moreover, ozonation was combined efficiently with coagulation and ceramic membrane process for water reclamation, and the effects of ozonation on virus coagulation were presented where the phage MS2 was used as a test model (Im et al. 2018). The role of the phage MS2 by the introduction of the ozone treatment for pathogen removal from water was reviewed recently by Gomes J et al. (2019).

OTHER INACTIVATING AGENTS

During the early studies, the phage MS2 was inactivated by oxidized spermine, where the oxidation product, a dialdehyde, was shown as an affecting agent (Bachrach and Leibovici 1965). Later, the phage MS2 was inactivated by diepoxybutane (Budowsky et al. 1981) and β-propiolactone (Budowsky and Zalesskaya 1985, 1991).

The phage Qβ, as well as the isolated phage Qβ RNA, was inactivated by such electrophiles as methyl, ethyl and isopropyl methanesulfonates, diethyl pyrocarbonate, and autoclaved irradiated sucrose and glucose (Kondorosi et al. 1973). Of 20 tested amino acids, the phages MS2 and GA were sensitive to arginine alone, while the phages Qβ and SP were sensitive to arginine and lysine (Murata et al 1974). Some higher homologs of lysine derivatives consisting of L-lysine and dicarboxylic acids demonstrated strong inactivation of the phage MS2 (Kondo et al. 1984).

The inactivation of the phage MS2 by ascorbic acid was explained by free radical intermediates formed during the ascorbic acid oxidation and causing strand scissions in the MS2 RNA (Murata and Uike 1976).

The covalent photobinding of a single psoralen, namely 4′-(hydroxymethyl)-4,5′,8-trimethylpsoralen, molecule at 360-nm light was a lethal event to the phage MS2 genome (Karathanasis and Champney 1981).

The inactivation kinetics of the phage MS2 by oligoaziridines were presented (Tsvetkova and Nepomnyaschaya 2001). The inactivation of the phage f2 by glutaraldehyde was described by Chen et al. (2006a) and the MS2 inactivation by oxidized polyamines was reviewed by Bachrach (2007).

Dimethyldioxirane, generated *in situ* by adding acetone to an aqueous solution containing potassium peroxymonosulfate, or oxone, inactivated the phage MS2 (Wallace et al. 2005).

The polymeric N-halamine latex emulsions were synthesized for use in antimicrobial paints and demonstrated capability to inactivate many microbiological indicators including the phage MS2 (Cao and Sun 2009).

Performic acid was applied for advanced wastewater disinfection, where the phage MS2 showed the highest resistance among other tested viral and bacterial indicators (Karpova et al. 2013).

Chitosan, a natural nontoxic biodegradable polymer obtained by the deacetylation of chitin from the exoskeleton of crustaceans, decreased pfu of the phage MS2 and feline calicivirus, but not of murine norovirus (Su et al. 2009; Davis et al. 2012). Furthermore, the MS2-inhibiting effect of chitosan was confirmed, but no effect was found with such cationic compounds as cetyltrimethylammonium bromide, nisin, and lysozyme (Ly-Chatain et al. 2013). Davis et al. (2015) demonstrated a full picture of the antiviral chitosan activity, where two chitosans of 53 and 222 kDa were tested and chitosan treatments showed the greatest reduction of the phage MS2, followed by feline calicivirus, the phage φX174, and murine norovirus.

The phage MS2 was used by the chitosan coagulation as a pretreatment for ceramic water filters for household drinking water treatment (Abebe et al. 2016). A chitosan membrane functionalized with basic TMPyP photosensitizer, or 5,10,15,20-tetrakis (1-methyl-4-pyridinio) porphyrin tetra p-toluene sulfonate, demonstrated complete inactivation of the phage MS2 and was recommended for the efficient sunlight driven water disinfection (Majiya et al. 2019).

The antiviral activities of poly(phenylene ethynylene) (PPE)-based cationic conjugated polyelectrolytes (CPE) and oligo-phenylene ethynylenes (OPE) were investigated using two model viruses, the phage MS2 and the DNA phage T4, under UV/visible light irradiation; without irradiation, most of these compounds exhibited high inactivation activity against the phage MS2 and moderate inactivation ability against the phage T4 (Wang Y et al. 2011).

The inactivation of the phage MS2, among other human norovirus surrogates, was studied by such novel putative inactivators as benzalkonium chloride, potassium peroxymonosulfate, tannic acid, and gallic acid (Su and D'Souza 2012).

The phage MS2 was used by the evaluation of the N-bromodimethylhydantoin polystyrene beads as a potential agent for water microbial decontamination (Aviv et al. 2015).

A commercial phenolic-based disinfectant, Bi-OO-cyst, was more effective for the phage MS2 and porcine rotavirus inactivation than commercial iodophore-, peroxygen-, and glutaraldehyde-based disinfectants at all levels of organic matter concentrations in the farm environment (Chandler-Bostock and Mellits 2015).

Twenty biocidal agents based on monoaza-, diaza-, triaza-, and tetraazaadamantanes and their homo and dihomo analogs were tested successfully on the phage MS2 and other indicators, and the most active compounds were selected as prospective chlorine-free disinfectants (Zubairov et al. 2015).

Recently, a subset of novel synthetic cationic oligomeric conjugated polyelectrolytes (OPEs) was shown to display high antiviral activity against the phage MS2 (Martin TD et al. 2016). Concerning the mechanism of their action, the oligomers perturbed the structure of the MS2 capsid. By using a multiscale computational approach, including random sampling, molecular dynamics, and electronic structure calculations, an understanding was gained on the molecular-level interactions of the MS2 capsid with a series of OPEs that varied in length, charge, and functional groups. The binding was dominated by strong van der Waals interactions between the hydrophobic OPE backbone and the capsid surface, and strong electrostatic free energy contributions between the OPE charged moieties and the charged residues on the capsid surface. This knowledge provided important molecular-level insight into how to tailor the OPEs to optimize viral capsid disruption and increase the OPE efficacy to target amphiphilic protein coats of icosahedral viruses (Martin TD et al. 2016).

The phage MS2 was used as a surrogate to estimate the inactivation rates for enteric viruses by a hot 150°C air bubble column evaporator system that was suggested as a new energy-efficient treatment for water reuse applications (Garrido et al. 2017). However, the fact of the rapid MS2 inactivation by bubbling air or nitrogen gas through the suspension was observed much earlier (Trouwborst et al. 1974).

The efficient inactivation of the phage MS2 was achieved in water by hydrodynamic cavitation, which was the first proof that hydrodynamic cavitation might inactivate viruses (Kosel et al. 2017). The remarkable role of the phage MS2 by the elucidation of the effects of cavitation on different microorganisms was reviewed recently by Zupanc et al. (2019).

A systematic study on structure and inhibitory activity relationships was recently performed with the phage MS2 and the DNA phage P100 with 55 ionic liquids (Fister et al. 2017).

The potential of the natural bioactive compounds able to control the foodborne viral diseases was tested with the phage MS2, among other foodborne virus surrogates. Thus, crude aqueous extracts of 255 plant taxa growing in various areas of Greece were screened for antiviral activity against the phage MS2 and five DNA phages, and 38 extracts showed antiviral activity against one or more of them (Delitheos et al. 1997). Then, the ethanol extracts of 35 algae collected from several Greek shores were tested against the same phage collection and 5 samples were found to induce mild antiviral activity (Couladis et al. 1998).

The MS2 titer was decreased by cranberry juice and cranberry proanthocyanidins (Su et al. 2010a,b), pomegranate juice and pomegranate polyphenols (Su et al. 2010c, 2011), and grape seed extract (Su and D'Souza 2011). To determine the risk of the human norovirus transmission by contaminated blueberry juice, the MS2 survival, among other surrogates, was checked in the blueberry juice by high-pressure homogenization as a novel processing method (Horm et al. 2012). The perspectives of phytocompounds as natural alternatives to chemically-synthesized therapeutics for the control of human enteric viruses and the role of the phage MS2 in their development were summarized by D'Souza (2014). Furthermore, Kamimoto et al. (2014) found anti-MS2 effect of persimmon extract, where persimmon tannin appeared as an active ingredient. Extracts from five hydrophytes leaves, *Polygonum hydropiper*, *Polygonum orientale*, *Phragmites communis*, *Arundo donax*, and *Typha latifolia*, inhibited the phage f2, as well as the DNA phage T4, to different extents (Zhang N et al. 2009).

A special modified MS2 plaque reduction assay was elaborated for the rapid screening of antiviral plant extracts, when the MS2 titer was reduced by extracts of *Camellia sinensis* and *Scaevola spinescens* and *Aloe barbadensis* juice (Cock and Kalt 2010). Remarkably, the antiviral and antibacterial activities of the extract of *S. spinescens,* an endemic Australian native plant, were examined further in detail (Cock and Kukkonen 2011). As a result, the Australian Aboriginal usage of *S. spinescens* was validated and its medicinal potential was officially indicated. Some limited anti-MS2 activity was demonstrated by ethanolic extract from roots of *Scutellaria baicalensis* Georgi, commonly named Huangqin (Lu et al. 2011). Other components of pharmacopeias of multiple Australian Aboriginal tribal groupings, which traditionally inhabited the areas in which they grow, namely *Petalostigma pubescens* and *P. triloculare*, were subjected to gas chromatography-mass spectroscopy analysis to identify the bioactive petalostigma components and tested on the MS2 inhibition, among other indicators (Kalt and Cock 2014). Moreover, the stem, leaf root, and whole plant of two tomato cultivars, Pitenza and Floradade, appeared as natural sources of bioactive substances with antiviral activity and were able to inactivate the phage MS2 (Silva-Beltrán et al. 2015). The active compounds identified were gallic acid, chlorogenic acid, ferulic acid, cafeic acid, rutin, and quercetin. Furthermore, phenolic compounds of potato peel extracts were shown as inactivators of the phage MS2 and prospective natural protectors against human enteric viruses (Silva-Beltrán et al. 2017).

"Água-mel," a traditional Portuguese honey-based product, demonstrated antiviral properties by decrease of the phage Qβ infectivity (Miguel et al. 2013). Liao et al. (2020) reported remarkable antimicrobial, including anti-MS2, effects of Chinese rice wine in traditional wine-treated mud snails.

It is noteworthy that the *Tetrahymena thermophila* ciliate was found to remove the phage MS2 when the phage and ciliate were co-incubated in a simple salt solution (Pinheiro et al. 2008).

The MS2 particles were seen inside *Tetrahymena* within vesicles that had the shape and size of food vacuoles, and the engulfment of the phage into food vacuoles probably led to the MS2 inactivation. However, Akunyili et al. (2008) concluded that *T. thermophila* failed to inactivate the phages MS2 and φX174, although it had ingested and inactivated the phage T4.

The phage MS2 was found to be used as a minor carbon source for flagellates (Deng et al. 2014). Thus, the phage MS2 was actively ingested by the suspensions of feeder flagellates *Thaumatomonas coloniensis* and *Salpingoeca* sp. in contrast to the actively raptoriale grazer *Goniomonas truncata*.

Recently, a 1D nanostructure-assisted low-voltage electroporation disinfection cell (EDC) was shown to ensure complete inactivation of the phage MS2 and bacteria (Huo et al. 2018). The authors are aware that the EDC will change substantially the current production methods in the food and water industry.

HEAT

Concerning heat treatment in early studies, Feary et al. (1964) determined the temperature stability of the pseudomonaphage 7s at 60°C: over 90% of the plaque-forming ability of the phage 7s was inactivated during 60 min. At the same time, the heat inactivation of the phage R17 RNA, but not of the whole phage, was tested by Ginoza et al. (1964).

The thermal and alkaline degradation of the phage MS2 contributed to the important conclusion that the A protein is in contact with both RNA and coat protein within the phage particle (Oriel 1969).

Since the heat resistance of the phage f2 was highest among a long list of tested pathogens, the phage f2 was chosen as a standard organism for establishing time-by-temperature criteria for determining the level of destruction achieved by the wastewater treatment (Burge et al. 1981).

The phage f2 was used as a standard measure of the thermal inactivation of viruses during sludge treatment processes (Traub et al. 1986; Spillmann et al. 1987). Furthermore, the group I RNA phages were found to be relatively heat resistant, while the group III RNA phages were more susceptible to thermal inactivation and the latter were suggested as better indicators of thermal inactivation of pathogens (Schaper et al. 2002a; Feng et al. 2003; Nappier et al. 2006).

When the sludge was heated at 80°C and the sewage was heated at 60°C, the FRNA phages were more resistant to thermal inactivation than bacterial indicators, except for spores, but less resistant than the somatic coliphages and the *B. fragilis* phages (Mocé-Llivina et al. 2003).

The phage MS2, among other indicators, was involved in a study that demonstrated that the incorporation of the 60°C or 70°C pretreatment phase could dramatically increase pathogen inactivation during mesophilic and thermophilic anaerobic digestion of sewage sludge (Ziemba and Peccia 2011). By modern elucidation of the synergistic heat and solar UV effect on the phage MS2, the latter was found to be highly resistant to irradiation and heat, with a slightly synergistic effect observed only at 59°C and natural sun insolation of 5,580 kJ m^{-2} (Theitler et al. 2012). Remarkably, the MS2 inactivation rate by the mesophilic digestion of sludge could be influenced substantially by bacteria-produced proteases (Mondal et al. 2015). A pilot study of the auto-thermal aerobic digestion and ammonia demonstrated reliable removal of the FRNA phages, although the somatic phages and parasitic eggs remained more resistant (Nordin et al. 2018). Thwaites et al. (2018) examined the log$_{10}$ removal of the FRNA phages by the aerobic granular sludge and the conventional activated sludge systems, during the startup phase through to maturation. At the same time, the decay rates of the phage MS2, among other popular indicators, were measured in facultative pond sludge (Schwartz et al. 2019).

Comparative survival of the phage MS2, as well as of the lipid-containing double-stranded RNA phage φ6, was evaluated during thermophilic and mesophilic anaerobic digestion (Sassi et al. 2018). As a result, the phage φ6, but not MS2, was suggested as a surrogate for the Ebola virus.

The effect of temperature, in parallel with pH and NaCl, on the inactivation kinetics of the phage MS2 and murine norovirus as popular surrogates was determined by modern RT-PCR methods (Seo et al. 2012). Both MS2 and murine norovirus were rapidly inactivated at temperatures above 60°C. Overall, temperature had a greater effect on infectivity than salt or low pH, but MS2 was more resistant to high salinity than murine norovirus (Seo et al. 2012). The survival of the phage MS2 at different temperatures was compared on either lacquer coating rubber tree wood or stainless steel (Kim SJ et al. 2012).

The heat stability of the phage MS2, as well as that of norovirus, was determined in the presence of such food additives as NaCl, sucrose, and milk (Jarke et al. 2013).

Remarkably, the persistence studies of the MS2 and other viral genomes after autoclaving showed that the MS2 genome was the most vulnerable among those tested, with no amplification observed after 18 min of autoclaving (Choi et al. 2014).

Recently, Brié et al. (2018) described in detail the impact of chlorine and heat on the infectivity and physicochemical properties of the phage MS2. The phage MS2 contributed to an important recommendation for swine carcass pretreatment at high temperatures for sanitary security (Tápparo et al. 2018). The kinetics inactivation of the phage MS2, as indicator of pathogenic viruses, was determined during treatment at different temperatures (30°C, 40°C, and 50°C) with varying moisture contents (50%, 60%, and 70%), in order to model the sanitization of compost withdrawn from the composting toilet by setting post-treatment conditions (Darimani et al. 2018). The phage MS2 was chosen as the model virus to evaluate the efficiency of thermal inactivation of the process using a range of different hot gases to sterilize water at atmospheric pressure within a hot bubble column evaporator (HBCE) reactor (Garrido et al. 2018; Garrido Sanchis et al. 2018, 2019).

SUNLIGHT

The phage MS2 was employed as one of the standards by the evaluation of sunlight-induced mortality of viruses and *E. coli* in coastal seawater at relevant temperatures, where the phage MS2 appeared more resistant than *E. coli* (Kapuscinski and Mitchell 1983).

By the investigation of sunlight wavelengths inactivating fecal indicator microorganisms by "natural" disinfection of waste stabilization ponds in New Zealand, it appeared that the UV-B, UV-A, and blue-green visible radiation (<550 nm wavelength) all contributed appreciably to the FRNA phage inactivation (Davies-Colley et al. 1997, 1999, 2000, 2005). The FRNA phage inactivation was studied also in sewage-polluted winter and summer seawater of New Zealand shore (Sinton et al. 1999), as well as in seawater and freshwater in California (Noble et al. 2004).

Mechanisms of the sunlight inactivation of the phage MS2 were described (Kohn and Nelson 2007; Kohn et al. 2007). The sunlight inactivation of the FRNA phages, among a set of other indicators, was studied at Avalon Beach on Santa Catalina Island in southern California (Boehm et al. 2009; Rodríguez et al. 2012). Direct sunlight markedly stimulated the inactivation of the phage MS2 on grass surfaces irrigated with treated effluent (Sidhu et al. 2008).

The effect of sunlight on the phages MS2, Qβ, SP, and FI, as well as on the natural isolates and DNA phages, was measured at Avalon Bay, California, where MS2 appeared as the most resistant and potential indicator for the sunlight resistant human viruses in clear water when sunlight inactivation was the main removal mechanism (Love et al. 2010b). The action spectra for simulated sunlight were measured in clear water for the phage MS2 and the double-stranded DNA phage PRD1, and the sensitivity of both MS2 and PRD1 to photoinactivation was estimated in a frame from 285−345 nm (Fisher et al. 2011). The synergistic role of temperature and natural organic matter was studied by sunlight-mediated inactivation of the phage MS2 and porcine rotavirus (Romero et al. 2011). Furthermore, the photolytic and photocatalytic disinfection of the phage MS2, as well as of the DNA phages φX174 and PR772, was compared under both artificial UV irradiation and natural sunlight (Misstear et al. 2012).

The phages f2 (Wegelin et al. 1994) and MS2 (Walker et al. 2004; Kohn and Nelson 2007; Fisher et al. 2011, 2012), together with many other indicators, were at the background of the simple and easy solar disinfection system of water (SODIS), a system, which consisted of placing water into transparent plastic or glass containers, normally 2-L polyethylene terephthalate (PET) beverage bottles, and exposing them to sunlight. The SODIS germicidal effect combined therefore the solar light and UV radiation with thermal heating. Later, the phage MS2 and other surrogates were used by the search of enhancers to accelerate the SODIS protocol (Harding and Schwab 2012). For example, the photoinactivation of the phage MS2 on iron-oxide-coated sand, which enhanced virus inactivation in sunlit waters, was proposed (Pecson et al. 2012). An excellent review for more than 30 years of the SODIS, "from bench-top to roof-top," was published by McGuigan et al. (2012). An extensive review on the drinking water disinfection by solar radiation, or SODIS, as one of the simplest methods for providing acceptable quality drinking water, and the role of the phage MS2 in the establishment of the disinfection process was published by

Teksoy and Eleren (2017). The SODIS approach was enhanced later by moderate addition of Fe and sodium peroxydisulfate, under solar light, and demonstrated high efficiency by the MS2 inactivation at minute-range residence times (Marjanovic et al. 2018). Ryberg et al. (2018) presented an enhanced SODIS scheme that utilized erythrosine, a common edible food dye, as a photosensitizer to produce singlet oxygen for virus inactivation and to indicate the completion of water disinfection through photobleaching color change. The experimental results and predictions based on global solar irradiance data suggested that over 99.99% inactivation of the phage MS2 could be achieved within 5 min in the majority of developing countries, reducing the time for SODIS by two orders of magnitude. The edible dye−enhanced SODIS was proposed as an efficient water disinfection method that could potentially be used by governments and non-governmental organizations to improve drinking water quality in rural developing communities (Ryberg et al. 2018). Polo-López et al. (2019) used the phage MS2 for the evaluation of polypropylene buckets by the SODIS procedure.

The sunlight inactivation of the FRNA phages and norovirus in seawater was studied by RT-qPCR at summer and winter conditions in Ireland (Flannery et al. 2013b). The role of natural photosensitizers in coastal waters was emphasized by the sunlight inactivation of the phage MS2 and a set of human viruses (Silverman et al. 2013).

The pathogen and indicator inactivation in source-separated human urine heated by sun in sun-exposed cans was determined with use of the phage MS2 as an indicator (Nordin et al. 2013). The effect of temperature and sunlight on the stability of the phage MS2 and human adenoviruses was determined on lettuce and strawberry surfaces as representative fresh products (Carratalà et al. 2013).

The phage MS2 was inactivated by reactive oxygen species and triplet excited state of dissolved organic matter (DOM), which were produced by irradiation of natural and synthetic sensitizers collected in Singapore with simulated sunlight of wavelengths greater than 320 nm (Rosado-Lausell et al. 2013). The role of DOM as a natural photosensitizer that contributed to the inactivation of the phage MS2 was evaluated recently in detail (Wenk et al. 2019).

The FRNA phages were used to show that the natural solar radiation was efficient for of a real secondary effluent from a municipal wastewater treatment plant using added H_2O_2, TiO_2 and photo-Fenton in the compound parabolic collector photoreactors (Agulló-Barceló et al. 2013). In parallel, the sunlight inactivation of the phage MS2 was modeled in the absence of photosensitizers by using a photoaction spectrum (Nguyen et al. 2014). The phage MS2, in parallel with poliovirus, was used for the modeling of the sunlight endogenous and exogenous inactivation rates in wetlands (Silverman et al. 2015).

Mattle et al. (2015) showed that the MS2 inactivation occurred mostly by direct processes, namely by the absorption of sunlight by the virus, but not by indirect processes, i.e., the adsorption of sunlight by external chromophores, which subsequently generated reactive species, though indirect inactivation by $^1O_2$ also contributed to the MS2 inactivation.

Solar disinfection of the phage MS2, together with the DNA phage φX174 and human viruses, was studied in drinking water in PET bottles (Carratalà et al. 2015). Good inactivation of MS2 was achieved in Swiss tap water at 22°C; less-efficient inactivation was observed in Indian waters, due to their higher content of organic matter. High temperatures enhanced MS2 inactivation substantially.

The selective inactivation of the phage Qβ in the presence of bacteria was achieved by use of ground Rh-Doped $SrTiO_3$ photocatalyst and visible light (Yamaguchi et al. 2017). This is an interesting approach to support the bacterial fermentation industry, where phage infections are leading to the loss of fermented products such as alcohol and lactic acid.

The UV-based and sunlight-supported technologies in the frame of so-called advanced oxidation processes (AOPs) were recommended for the elimination of chemical and microbiological pollution of wastewaters by parallel investigation in Switzerland and in the Ivory Coast and Colombia (Giannakis et al. 2017c).

The sunlight-simulated inactivation of the phage MS2 was investigated to explain the potential shielding effects due to the MS2 aggregation and adsorption to particles in solutions, namely to kaolinite and *Microcystis aeruginosa* as model inorganic and organic particles, respectively (Wu Xueyin et al. 2018).

With the assistance of the phage MS2, the first application of solar-to-thermal converting nanomaterial was presented for the direct inactivation of bacteria and viruses in drinking water (Loeb et al. 2018). Gold nanorods, carbon black, and gold nanorod−carbon black composite materials were used as light absorbers and induced multiple scattering events, increasing photon absorption probability and concentrating the light within a small spatial domain, leading to localized, intense heating that inactivated microorganisms in close proximity (Loeb et al. 2018).

Finally, the phage MS2 was considered to be the best surrogate for studying sunlight disinfection in wastewater treatment ponds, or lagoons, that remain one of the most common types of technologies used for wastewater management worldwide, especially in small cities and towns (Verbyla and Mihelcic 2015).

Recently, the phage MS2 contributed to the improvement of pond walls through the addition of an inclined plane and the generation of a thin film, therefore increasing the exposure of pathogens to sunlight (Hawley and Fallowfield 2018). Zepp et al. (2018) used the phage MS2 to develop biological weighting functions (BWFs) for light-induced inactivation. The BWFs were used to model the inactivation of the phage MS2 over a range of conditions in aquatic environments that included two beach sites in Lake Michigan and one in Lake Erie. In parallel, four BWF models were compared to assess their ability to predict endogenous sunlight inactivation rates of the phage MS2 (Silverman et al. 2019a,b). Remarkably, the effect of the phage MS2 on the inactivation of its *E. coli* host was evaluated in an effort to recreate some of the main factors governing solar disinfection of wastewater, namely microbial growth, infection, and photonic flux (Voumard et al. 2019).

The role of the FRNA phages in the development of the sunlight-mediated inactivation of health-relevant microorganisms in water was reviewed exhaustively by Nelson et al. (2018). The role of the phage MS2 by the construction of wetlands for greywater recycle and reuse was reviewed by Arden and Ma (2018).

HIGH PRESSURE

Pressures up to 2.5 kbar promoted only 10% dissociation of the whole particles of the phage R17, while in the presence of urea between 1.0 and 5.0 M the pressure promoted complete reversible dissociation of the phage particles (Da Poian et al. 1993). The pressure-resistant and pressure-sensitive mutants of the phage MS2 were isolated (Lima et al. 2004). Then, the stability of the phages and virus-like particles was compared (Lima et al. 2006).

The effects of hydrostatic pressure on viruses, including RNA phages, were reviewed thoroughly by Oliveira et al. (2008); the authors described in detail how high pressure has been used to tackle basic and applied problems in virus biology.

Practically, the phage MS2 was used as an indicator by the development of the high and atmospheric pressure cycling technology for the inactivation of viruses in human plasma (Dusing et al. 2002). Then, the phage MS2 appeared in a study, which compared hydrostatic and hydrodynamic pressure by inactivation of foodborne viruses in a deli meat product (Sharma M et al. 2008). Furthermore, high-pressure homogenization was employed for viral inactivation in fluid foods, where the phage MS2, together with murine norovirus, were involved as human enteric virus surrogates (D'Souza et al. 2009).

The effectiveness of holding temperatures at the high-pressure processing was evaluated for inactivating the phage MS2 and murine norovirus as surrogates in pomegranate and strawberry juices and strawberry puree (Pan H et al. 2016).

ULTRASOUND

The phage MS2, together with two other surrogates, murine norovirus and feline calicivirus, was subjected to high-intensity ultrasound in buffer or orange juice (Su et al. 2010d). However, the ultrasound alone was not sufficient to inactivate virus in food, and combination with heat, pressure, and antimicrobials was recommended. Thus, the high-frequency ultrasound in combination with visible light was efficient in the inactivation of the phage MS2 and the DNA phage φX174 as model viruses (Chrysikopoulos et al. 2013).

CO$_2$

The phages MS2 and Qβ were inactivated by high levels of dissolved CO_2 (Cheng X et al. 2013). Under supporting high pressure, a high volume of CO_2 microbubbles was able to inactivate the phage MS2, as well as DNA phage T4 and *E. coli*, where the optimum conditions were found to be 0.7 MPa and an exposure time of 25 min (Vo et al. 2013). Furthermore,

inactivation of the phage Qβ, as well as of the DNA phage φX174, was achieved by the pressurized carbon dioxide (Vo et al. 2014). The role of the phage MS2 in the development of the CO_2 inactivation technology to the next standard sterilization technique was unveiled in a recent review (Soares et al. 2019).

DESICCATION, HUMIDITY, AND AEROSOLS

Interestingly, the intracellular phage f2 was extremely resistant to inactivation by desiccation, since it was not until the relative humidity was lowered to 30% that inactivation became apparent (Webb 1967). The desiccation increased resistance of the f2-infected cells to UV irradiation, and the general conclusion followed that the phage f2 RNA was almost completely resistant to desiccation, and therefore its biological integrity was assumed to be unaffected by loss of bound water (Webb and Walker 1968).

The studies of the airborne stability of the phage MS2 showed that a commercial MS2 preparation suspended in a buffered saline solution showed a rapid loss of viability at relative humidity above 30%, whereas a laboratory preparation containing 1.3% tryptone showed high recoveries at all relative humidity values studied (Dubovi and Akers 1970). The MS2 RNA was not damaged by aerosolization (Dubovi 1971). The inactivation of the phage MS2 was maximal in the aerosol particle in fluid phase, and became less at lower relative humidity where aerosol particles were expected to be in the solid state, and the effect was explained generally by surface inactivation at the air–water interface (Trouwborst and de Jong 1973). The inactivation of the aerosolized phage f2 was measured on a spray irrigation site, where f2, in contrast to bacteria, was reduced only 95.4% in the chlorinated effluent and was readily measured 137 m downwind (Bausum et al. 1982).

Later, the phage MS2 was employed for the evaluation of the virus retention by ventilator-circuit filters (Holton and Webb 1994). Foarde et al. (1999) chose the phage MS2 as a surrogate for various viruses, including influenza, as it had a similar shape and aerosol characteristics as human viruses despite the fact that it was slightly smaller and non-enveloped.

The charge distribution of the airborne phage MS2 was studied in detail, and a compact capture unit using soft X-ray irradiation and electrostatic precipitation was constructed for mounting in ventilation ducts or air supply systems (Hogan et al. 2004).

The phage MS2 was aerosolized within a biosolids spray application truck, and bioaerosols were collected at discrete downwind distances ranging from 2–70 m in order to develop an empirically derived transport model that could be used to predict downwind concentrations of viruses and bacteria during land application of liquid biosolids and subsequently assess microbial risk associated with this practice (Brooks et al. 2005). The seeded phage MS2 was employed by the evaluation of the bioaerosol emission rate and plume characteristics during land application of liquid biosolids to farmland near Tucson, Arizona (Tanner et al. 2005).

At the same time, the phage MS2 was employed as a model in the testing of collection efficiencies of aerosol samplers for virus-containing aerosols (Hogan et al. 2005; Tseng and Li 2005a) and comparative studies of the inactivation of the virus-containing aerosols by UV irradiation (Tseng and Li 2005b; Walker and Ko 2007) or ozone (Tseng and Li 2006). Barker and Jones (2005) used the phage MS2 to examine aerosol contamination caused by toilets. The role of the phage MS2 was detailed at that time in a large review on the methods for sampling of airborne viruses (Verreault et al. 2008).

The physical and viable bioaerosol penetrations through respirator filters were differentiated when the phage MS2 was used as an indicator for challenging respirator filters with bioaerosols and contributed to the differentiation of the physical and viable bioaerosol penetrations through popular respirator filters (Eninger et al. 2008a,c, 2009; Woo et al. 2010). The challenges of MS2 aerosols were used also for the assessment of iodine-treated filter media (Lee JH et al. 2009).

Furthermore, a PCR assay was developed to detect viruses and their fate in aerosol droplets, with the phage MS2 as a surrogate (Perrott and Hargreaves 2011). Moreover, the phage MS2 was chosen as one of the suitable models for the virus aerosolization (Gendron et al. 2010; Ge et al. 2014; Turgeon et al. 2014).

The phage MS2 aerosol was inactivated by hot air (Grinshpun et al. 2010). Then, the nebulized phage MS2 was exposed to various levels of relative humidity and temperature as well as to germicidal UV radiation, and remained the most resistant among other non-RNA phage indicators (Verreault et al. 2015). The main conclusion was that the aerosolized phages behaved differently under various environmental conditions and therefore the careful selection of viral simulants in bioaerosol studies would be necessary (Verreault et al. 2015). Other studies at that time also cast doubt on the MS2 suitability as a general surrogate for pathogenic viruses in aerosols (Appert et al. 2012).

The survival of the airborne MS2 phage, which was generated from human saliva, artificial saliva, and cell culture medium, was independent of particle size but was strongly affected by the type of nebulizer suspension, while human saliva was found to be much less protective than the cell culture medium and artificial saliva (Zuo et al. 2014).

The aerosolized phage MS2 was used by the fabrication of an antiviral air filter with SiO_2 nanoparticles surface-coated with silver nanoparticles (Joe et al. 2014, 2016). Jiang X et al. (2016) used the phage MS2 as a surrogate for the generation of an effective aerosol collector that utilized water-based condensation for collecting virus-containing aerosols.

An interesting study on the virus partitioning and aerosolization potential, with the phages MS2 and φ6 as models, showed that in a sludge from a wastewater treatment plant the viruses partitioned predominantly into the liquid fraction; no more than 0.8% of virions partitioned to the solids and no more than 6% to the material surface (Titcombe et al. 2016). Therefore, partitioning of viruses in wastewater did not appear to mitigate the potential for virus aerosolization, as most of the virus remained in the liquid phase. Furthermore,

aerosolization of the phage MS2, as well as φ6, from wastewater was described in connection with their status as Ebola virus surrogates (Lin K and Marr 2017).

The aerosolized phage MS2 was used as a competent human virus model by the *in situ* airborne virus inactivation by microwave irradiation (Wu and Yao 2014). Remarkably, the scanning electron images revealed visible damage to the viral surface after the exposure, and damage was observed to the viral RNA genes, among which the A protein gene was completely destroyed (Wu and Yao 2014). A set of novel halogen-containing nanomaterials was tested for their ability to inactivate the aerosolized phage MS2 (Grinshpun et al. 2012). The resistance of the aerosolized phage MS2, among other indicators, was measured against a set of popular germicidal products (Turgeon et al. 2016).

Furthermore, the phage MS2 appeared as a tool to provide an aerosol with precise particle sizes from 300 nm down to 45 nm (Walls et al. 2016). Then, a special method for the improved detection of the aerosolized phage MS2, in parallel with influenza viruses, was elaborated (Chandler et al. 2017b). The role of the FRNA phages by the decontamination of indoor air was reviewed by Duchaine (2016). Then, a thorough review on the technologies and photocatalytic materials for the air purification was published, where the phages MS2 and Qβ figured among other targets of the inactivation (Ren H et al. 2017).

Recently, the phage MS2 contributed to the generation of novel microorganism aerosol collectors (Lin XT et al. 2018; Pan M et al. 2018, 2019a,b; Yu et al. 2018). Schaeffer et al. (2018) used the phage MS2 to achieve the detection of viruses from bioaerosols using anion exchange resin. Concerning the inactivation of the aerosolized viruses, the phage MS2 served as a model for approaches using (i) vacuum UV photocatalysis for a flowthrough indoor air purifier with short irradiation time (Kim J and Jang 2018), (ii) noncontact ultrasonic transducer bound in high-efficiency particulate air (HEPA), filters (Versoza et al. 2018), and (iii) a packed bed nonthermal plasma reactor (Xia et al. 2019).

BIOFILMS

The surfaces within wastewater treatment plants are covered with biofilms offering adsorption sites to suspended solids and colloids, but also to enteric pathogens including viruses and protozoa. If the interaction of pathogens with biofilms would lead only to their attachment and inactivation, biofilms could be considered as a part of the water treatment process that would remove pathogens from the water phase. However, the biofilm-associated microorganisms may persist and further detach the entrapped infective pathogens. For the first time, the persistence of the spiked phage MS2 in biofilms was measured by Storey and Ashbolt (2001, 2003). It was concluded that virions have the potential to accumulate within biofilms, and problems could arise when clusters of the biofilm-associated enteric virions become detached from the substrata by hydrodynamic forces or sudden changes in disinfection regime. Further, the persistence of the FRNA phages, among other indicators, in the biofilms and in the corresponding wastewaters at 4°C and 20°C was characterized by Skraber et al. (2007): by both temperatures, the FRNA phages persisted longer in biofilms than in the corresponding wastewaters. The seeded phage MS2 attached to the drinking water biofilm within 1 h and persisted within the biofilm for up to 6 days after the inoculation, while viral genome was still detected at day 34, corresponding to the last day of the monitoring period (Helmi et al. 2008). Generally, the infectious FRNA phages, as well as also noroviruses, persisted longer in biofilm than in wastewater (Skraber et al. 2009a). Besides biofilms, the FRNA phages accumulated over 100 times more than the somatic coliphages in clayey sediments, and the phage inactivation in clayey and sandy sediments over a 1-month period at 15°C was negligible (Skraber et al. 2009b).

The detailed evaluation of the adsorption of viruses, including the phage MS2 to biofilms, showed the dependence on the isoelectric points, the pipe material, namely polyvinyl chloride, cement, or cast iron, and the hydrodynamic conditions (Helmi et al. 2010). When biofilms were formed by *E. coli* carrying a natural conjugative F-plasmid and generating F pili, the early biofilm formation was completely inhibited by addition of the f1 phage, but not by the phage MS2 (May et al. 2011).

Furthermore, the accumulation of enteric viruses on the surfaces within a drinking water distribution system was investigated in a reactor using the three FRNA phages: MS2, GA, and Qβ, where different adhesion profiles were obtained, and the phage MS2 was found to be the less adherent to drinking water biofilms (Pelleïeux et al. 2012). Remarkably, the phages Qβ and GA demonstrated different adsorption kinetics on drinking water biofilms under hydrostatic conditions (Hébrant et al. 2014). The phages Qβ and GA had a similar affinity for the biofilm, whereas the free diffusion in water was the rate limiting step for the adsorption of the phage GA, but not of the phage Qβ. The next study aimed to assess at pilot scale the effect of chlorination and water flushing on 2-month-old drinking water biofilms and above all, on the biofilm-associated FRNA phages MS2, GA, and Qβ (Pelleieux et al. 2016). The effect of chlorine on the biofilm-associated phages was limited to the upper layers of the biofilm and was not enhanced by an increase in hydrodynamic shear stress. A smaller decrease was observed for the phage MS2 than for the phages GA or Qβ after completion of the cleaning procedure. The role of the RNA phages in the biofilm applications was reviewed recently by Milho et al. (2019).

DYED PHAGE PARTICLES

The dyed phage MS2, together with gold nanoparticles, appeared as the two novel tools that facilitated the studies of virus transport, adsorption, and inactivation, especially by sand filtration, ultrafiltration, chlorination, and UV disinfection (Gitis et al. 2006; Gitis 2014). The generation of the dyed phages was established earlier by the conjugation of fluorescent dyes to the phage MS2 (Gitis et al. 2002a,b).

DIFFERENT RESISTANCE TO INACTIVATION

Although the phage MS2 demonstrated a clear advantage as a major subject of the RNA phage inactivation studies, some information for the comparative analysis of different RNA phages must be extracted. Thus, by the early UV irradiation studies, it was found that the phage MS2 was more resistant to the UV inactivation than the phage GA, but the latter was more resistant than the phage Qβ, where the phages MS2, GA, and Qβ represented the groups I, II, and III, respectively (Furuse et al. 1967; Furuse and Watanabe 1971). Furthermore, the greater resistance of the phages MS2 and GA phages and an intermediate sensitivity of the phage Qβ to the UV irradiation at 254 nm was detected, in comparison with the least-resistant poliovirus (Simonet and Gantzer 2006).

By the inactivation of the RNA phages by ascorbic acids and thiol reducing agents, namely dithiothreitol, 2-mercaptoethanol, L-cysteine, and thioglycollate, the sensitivities were arranged in the following order: Qβ, MS2, GA (Murata et al. 1972).

Furthermore, the effect of different inactivation pathways was evaluated on mixtures of the FRNA phage isolates belonging to the genotypes I, II, III, and IV (Schaper et al. 2002a). By natural inactivation, group I showed the highest persistence, which was significantly different from that of the group II, IV, or III phages. The pattern of resistance to extreme pHs, ammonia, temperature, salt concentration, and chlorination was similar, where the group I phages showed the highest persistence, followed by the group II, III, and IV phages. The group III and IV phages were the least resistant to all treatments, and resistance of the group III and IV phages to the treatments was similar, while the group II phages showed intermediate resistance (Schaper et al. 2002a).

The phages MS2 and Qβ differed by their survival at different pH and temperature conditions (Feng et al. 2003). The phage MS2 survived better in acidic conditions than in an alkaline environment, whereas the phage Qβ had a better survival rate in alkaline conditions than in an acidic environment. The inactivation rates of both phages were (i) lowest within the pH range 6–8 and the temperature range 5°C–35°C, (ii) increased when the pH was decreased to below 6 or increased to above 8, and (iii) increased with increasing temperature. The phage Qβ behaved peculiarly in extreme pH buffers, i.e., it was inactivated very rapidly initially when subjected to an extreme pH environment, although the inactivation rate subsequently decreased. Feng et al. (2003) concluded that the phage MS2 could be a better indicator than the phage Qβ, but either MS2 or Qβ could be used as a viral indicator within the pH range 6–9 and at temperatures not above 25°C.

By the iodine inactivation, the phage MS2 was more susceptible than the phage GA (Brion et al. 2004). By heat treatment at 53°C, the group I phages were more resistant than the group III phages (Nappier et al. 2006). By the high hydrostatic pressure treatment, the phages f2 and GA had much higher pressure resistances than the phages Qβ and SP (Guan et al. 2007).

The differential persistence of the FRNA phage groups was measured systematically in wastewaters, after inactivation in surface waters or after wastewater treatment and in mixtures of wastewater of human and animal origin (Muniesa et al. 2009). The group III and IV phages were the least resistant to all treatments.

Figure 6.3 demonstrates the proportion of the RNA phage groups after water treatment of municipal raw sewage, where the secondary effluents appeared after activated sludge treatment and the tertiary effluents appeared after disinfection of the secondary effluents with chlorine, UV irradiation, or both. The pasteurization included heating at 60°C for 30 min, a treatment commonly applied to treated sludges (Muniesa et al. 2009).

Figure 6.4, which is taken from a paper on the natural distribution of the RNA phage groups (Hartard et al. 2015), shows the group pattern in raw wastewater over time and intends to serve as a natural starting point for the data presented in Figure 6.3. These data are of utmost importance, since the genomes were both sequenced and tested for infectivity (Hartard et al. 2015).

The inactivation kinetics of the phages MS2, fr, and GA were compared by the UV_{254}, singlet oxygen $^1O_2$, free chlorine, and chlorine dioxide ClO_2 treatment (Sigstam et al. 2013). The genome damage was quantified by PCR, and protein damage was assessed by quantitative matrix-assisted laser desorption ionization mass spectrometry (MALDI-MS). The ClO_2 caused great variability in the inactivation kinetics between viruses and was the only treatment that did not induce genome damage. The inactivation kinetics were similar for all viruses when treated with disinfectants possessing a genome-damaging component, namely, free chlorine, $^1O_2$, and UV_{254}. On the protein level, the UV_{254} subtly damaged the MS2 and fr coat proteins, whereas the GA coat remained

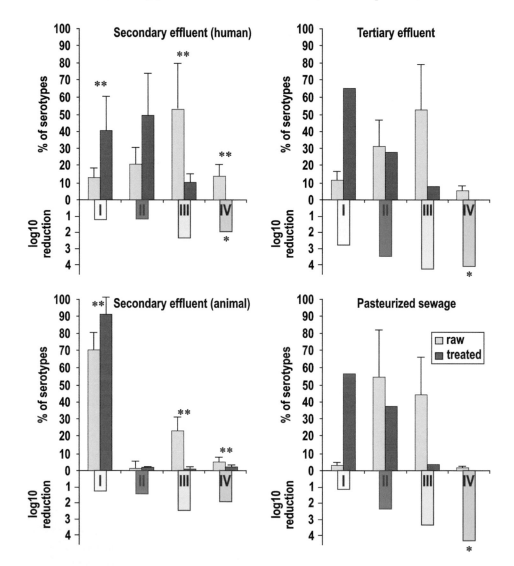

FIGURE 6.3 Distribution of the various subgroups of F-specific RNA bacteriophages after different wastewater treatments. Colored bars indicate the $\log_{10}$ reduction of the numbers of each subgroup observed when comparing the beginning (raw) and the end of the inactivation experiment (treated). Black and gray bars indicate the percentage of each subgroup in raw and treated samples. (*) Greater than. (**) Significantly different (P < 0.05 Student's t test). (Redrawn from *Water Res.*, 43, Muniesa M, Payan A, Moce-Llivina L, Blanch AR, Jofre J, Differential persistence of F-specific RNA phage subgroups hinders their use as single tracers for faecal source tracking in surface water, 1559–1564, Copyright, 2009, with permission from Elsevier.)

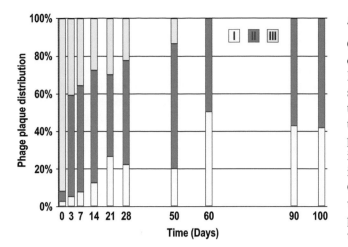

FIGURE 6.4 Infectious FRNAPH-I (yellow), FRNAPH-II (red), and FRNAPH-III (blue) distribution in urban wastewater. Between 16 and 24 plaques were isolated and typed for each analysis. (Redrawn in color with permission of ASM from Hartard C et al. *Appl Environ Microbiol.* 2015;81:6505–6515.)

intact. The $^1O_2$ oxidized a methionine residue in the MS2 coat but did not affect the other two phages. The molecular dynamics simulations indicated that the degradation was dictated by the solvent-accessible surface area of the individual amino acids. These data explained clearly why closely related phages could exhibit drastically different inactivation kinetics (Sigstam et al. 2013).

Behavior of the FRNA phage groups was examined in Japan at monthly intervals between March and December 2011 during wastewater treatment in raw sewage, aeration tank effluent, secondary-treated sewage, and return activated sludge (Haramoto et al. 2015). The group I phages were the most abundant in the secondary-treated sewage samples, while the group II phages were the most abundant in the other three sample types. The mean reduction ratio of the group I phages was much lower than that of the group II–IV phages (Haramoto et al. 2015).

Recently, Goetsch et al. (2018) found that the spiked phage Qβ was inactivated at a significantly higher rate in hydrolyzed urine than the spiked phage MS2. The authors concluded that the genome transesterification, but not disruption of the capsid structure, caused the phage inactivation.

Lee S et al. (2019b) performed a thorough evaluation of virus reduction at a large-scale wastewater reclamation plant by detection of the indigenous FRNA phage genotypes after ultrafiltration and UV treatment. The group I phages showed the highest UV resistance, followed by the group II, III, and IV. Remarkably, the resistance of the indigenous group FRNA phages was equivalent to that of the spiked phage MS2. The reduction of the total infectious FRNA phages determined by plaque assay was affected by the predominant FRNA phage genotype, presumably because of their different UV resistances. These results revealed that the indigenous group I FRNA phages could be a good alternative indicator to the spiked phage MS2 in view of virus reduction during water reclamation (Lee S et al. 2019b).

ADSORPTION AND COAGULATION

Generally, it was evident that viruses could be adsorbed onto solid surfaces and keep their infectivity for a long time. Moreover, it was quite true that the virus adsorption onto solid surfaces remained one of the major factors controlling their transport and survival in a water environment. Thus, the adsorbed phages ranged from 12%–30% of the total coliphages in the raw sewage, and over 97% of the coliphages in the activated sludge were associated with suspended solids, most of which were the FRNA phages (Ketratanakul and Ohgaki 1989). It was not surprising therefore that the stormwater treatment systems, where wetlands and ponds ensured physical sedimentation of particles of sand, silt, and clay to which pollutants adsorb, demonstrated high efficiency to remove the FRNA phages (Yousefi et al. 2001; Davies et al. 2003; Meuleman et al. 2003).

Historically, the RNA phage adsorption studies have been initiated by the attachment of the phage f2 to nitrocellulose filters (Lodish and Zinder 1965a), whereas the coagulation technique started with the coagulation and flocculation of the phage MS2 by aluminum sulfate (Chaudhuri and Engelbrecht 1970). Later, the kinetics of the phage MS2 and f2 adsorption to powdered nitrocellulose were determined as a function of temperature (Preston and Farrah 1988). Then, adsorption of the phage MS2 on nitrocellulose and cationic polysulfone membranes was studied in detail by Lytle and Routson (1995).

The phage f2 was one of the three phages compared to poliovirus for their adsorption to suspended soils (Moore et al. 1975). This study concluded that the phage f2 associated well with clays and did not retain infectivity by plaque assay in the adsorbed form, but nevertheless it could not be used as a surrogate to describe the general pattern of the poliovirus association. To estimate the competition between phage and organics for adsorption sites on soil and to determine the effect of natural organic matter on the adsorption, the influence of fulvic acid, a major fraction of natural soluble organic matter, on the phage MS2 adsorption was studied (Bixby and O'Brien 1979). In fact, the phage MS2 strongly adsorbed to soil in the absence and presence of fulvic acid, although the latter was capable of reversible phage inactivation.

Moreover, the concentrations of sodium dodecyl sulfate and alkylbenzyl sulfonate found generally in surface water (0.2–10 mg/L) had no effect on the adsorption of the phage f2 in soil, while concentration of 100 mg/L was the lowest intensity, which led to impaired phage adsorption and migration (Dizer 1990).

The phage MS2 demonstrated sorption to charge-modified silica (Zerda et al. 1985) and to montmorillonite, a very soft phyllosilicate mineral (Globa and Nykovskaya 1984). The montmorillonite was used for the modeling of Martian soils in a study devoted to the potential terrestrial microbiological contamination on Mars, where survival of the phage MS2, among other indicators, was measured in Martian environmental conditions of temperature and atmospheric pressure and composition (Moll and Vestal 1992). Remarkably, the phage MS2 survived the simulated Mars conditions better than the terrestrial environment, likely due to stabilization of the phage caused by the cold and dry conditions of the simulated Martian environment.

It is noteworthy that the attachment of the phage MS2 to flat bare silica surfaces, or especially to the Suwannee River Natural Organic Matter-coated silica surfaces, were 7 to 17 times higher in the presence of Ca^{2+} than in the presence of Mg^{2+} (Pham et al. 2009). The ghost cells, namely pig erythrocyte membranes, were used as a sorption matrix for the phage MS2 and f2 concentration from water (Armon et al. 1984).

Diatomaceous earth, when modified by *in situ* precipitation of metallic hydroxides, was efficient for the adsorption of the phage MS2 as well as of other tested viral indicators (Farrah and Preston 1991; Farrah et al. 1991). Later, diatomaceous earth was used in the ceramic depth filters that were, however, not able to remove the phage MS2 (Michen et al. 2012). Introduction of a novel virus adsorbent material, magnesium oxyhydroxide, into the diatomaceous earth filter matrix improved removal of the phage MS2 (Michen et al. 2013).

The deposition kinetics of the phage MS2 to bare and clay-coated silica surfaces was examined in both monovalent (NaCl) and divalent ($CaCl_2$ and $MgCl_2$) solutions under a wide range of environmentally relevant ionic strength and pH conditions, where the divalent ions greatly increased the phage deposition (Tong et al. 2012).

Magnetite (Fe_3O_4), a prospective component of the water treatment system, was found efficient for the MS2 adsorption, although some competition with clay was noticed (Atherton and Bell 1983a), with further degradation of the phage particles (Atherton and Bell 1983b).

The two types of activated carbon with different structural characteristics—granular carbon and an activated carbon fiber composite—were compared for the batch adsorption of the phage MS2 (Powell et al. 2000). Later, it was shown that the granular activated carbon filtration did not remove the phage MS2 by the drinking water treatment (Hijnen et al. 2010). The special super-powdered activated carbon with a median diameter of 0.7–2.8 µm was able to adsorb and remove the phage Qβ and, to a lesser extent, the phage MS2, while commercially available powdered activated carbons with a median diameter greater than 10 µm failed (Matsushita et al. 2013). It was concluded from these data that the hydrophobicity of the virus surface contributed greatly to virus removal: the more hydrophobic the surface, the greater the virus removal.

The coordinate and chemical bonding and adsorption (CCBA) process, which used clay as an adsorbent followed by alum flocculation and clarification, was tested for its ability to remove viruses by the seeded phage MS2 and poliovirus as indicators (Gersberg et al. 1988). Then, the coagulation with aluminum sulfate was performed to model the removal of hepatitis A virus and poliovirus, and the phage MS2 was used as a putative index of the pathogenic virus removal (Nasser et al. 1995). By batch coagulation treatments of water samples with different aluminum coagulants, the phages MS2 and Qβ were most efficiently inactivated with polyaluminum chloride, and the phage Qβ appeared as the most susceptible to the treatment

(Matsui et al. 2003; Matsushita et al. 2004). Polyaluminum chloride inactivated the phage f2, as well as poliovirus, in sewage water (Zhang X et al. 2009). As for the phage MS2, the aluminum Keggin polycation Al_{13} was identified as an effective tool for the MS2 coagulation (Stewart et al. 2009).

The adsorption of the four FRNA phages MS2, fr, GA, and Qβ to the five model surfaces with varying surface chemistries and to the three dissolved organic matter adlayers was studied systematically as a function of solution pH and ionic strength (Armanious et al. 2016a). The differences in the adsorption characteristics of the tested phages were linked to the differences in their capsid surface properties. Furthermore, competitive suppression of the FRNA phage adsorption by negatively charged dissolved organic matter was measured (Armanious et al. 2016b).

The attachment of the phage MS2, in parallel with that of the DNA phage φX174 and Aichi virus, was studied by atomic force microscopy (AFM) on sands of different surface properties: oxide-removed, goethite-coated, and aluminum oxide-coated (Attinti et al. 2010). This study was the first to employ AFM to the direct measure of the interaction forces between viruses and solid surfaces.

The phages MS2 and fr were acknowledged as suitable surrogates of adenovirus and feline calicivirus (Abbaszadegan et al. 2007, 2008) and echovirus and coxsackievirus (Mayer et al. 2008) by aluminum and ferric chloride coagulation. Furthermore, ferric chloride coagulation was combined with ceramic microfiltration for potable water production in Norway, and the phage MS2 was used as an indicator (Meyn et al. 2012).

The phage MS2 was removed by Mg/Al layered double hydroxide (LDH) in batch and column experiments, and the Mg/Al LDH-coated sand was recommended for water purification (Kim JH et al. 2012). The Ni/Al layered double hydroxide was also able to eliminate the phage MS2 in batch and flowthrough column experiments (Park JA et al. 2012).

The phages MS2 and Qβ were compared as models by the aluminum coagulation (Shirasaki et al. 2009b) and aluminum coagulation-ceramic microfiltration (Shirasaki et al. 2009a) processes by using river water spiked with these phages. As a result, the phage Qβ was found more sensitive to the virucidal activity of the aluminum coagulant. Furthermore, the mechanisms of the phage inactivation during coagulation with aluminum coagulants were revealed (Matsushita et al. 2011). Shirasaki et al. (2012) involved the phage f2, together with the DNA phage f1, in the studies on virus removal during the aluminum coagulation−rapid sand filtration and coagulation−microfiltration processes. The high-basicity polyaluminum coagulants improved removal of the phages MS2 and Qβ from river water, in comparison with commercially available aluminum-based coagulants (Shirasaki et al. 2014). The influence of flocculation parameters on the phages MS2 and Qβ was examined in a mechanistic study of flocculation with polyaluminum chloride (Kreißel et al. 2014).

The phages MS2 and fr were used as surrogates by the application of aluminum oxide−hydroxide nanofibers, 2 nm in diameter and approximately 250 nm long, which were electroadhesively grafted onto glass microfibers, thereby forming a macroscopic assembly of alumina nanofibers on the second solid in a highly organized matter (Kaledin et al. 2014).

The aggregation of the phage MS2 was induced by exposure to polyhexamethylene biguanide (Pinto et al. 2010a,b).

Dika et al. (2013b) performed a highly qualified modern study on adhesion of the phages MS2, GA, and Qβ to abiotic surfaces—1-dodecanethiol gold-coated surface, glass, polypropylene, and stainless steel—that differed significantly with regard to their surface roughness and hydrophobic/hydrophilic balance. Regardless of electrolyte concentration and surface roughness, the adhesion capacity of the phages systematically followed the hydrophobicity sequence MS2 < Qβ < GA. The capacity of each phage to adhere to the surfaces increased with increase in the degree of hydrophobicity and/or the roughness of the deposition substrate. Increasing the electrostatic interactions between phages and deposition surface by decreasing solution ionic strength led to a reduction in surface concentration of adhered phages, except in cases where the roughness of the deposition surface was significant.

The early experiments showed low adsorption of the phage MS2 to columns filled with quartz sand and aquifer material, in contrast to simian rotavirus (Herbold-Paschke et al. 1991). Later, the phage MS2 attachment onto quartz sand under batch experimental conditions was affected significantly by ambient temperature being greater at 20°C than at 4°C and by grain size, where the attachment decreased with increasing sand size (Chrysikopoulos and Aravantinou 2014). The coral sand from the Bonriki freshwater lens, South Tarawa, Kiribati, was tested for the removal of the phage MS2, but the ability to determine precise removal was compromised by the use of a plastic apparatus that adsorbed the phage (Burbery et al. 2015). The phage MS2 was employed by evaluation of the biochar-amended sand for virus retention (Sasidharan et al. 2016).

The transport and fate of the phage MS2 was studied in sediment and stormwater collected from a managed aquifer recharge site in Parafield, Australia (Sasidharan et al. 2017) and in a macro-tidal coastal basin, the Loughor estuary, located along the Bristol Channel in the southwest region of the United Kingdom (Abu-Bakar et al. 2017).

Attachment of the phage MS2 to the "used" and "new" sand during basin infiltration managed aquifer recharge was tested in Uppsala, Sweden (Mayotte et al. 2017b). The attachment was found relatively irreversible in the "new" sand case, while the attachment to the "used" sand was reversible and the MS2 inactivation was slowed.

The phage MS2 was removed by pyrophyllite clay in batch and column experiments, also in the presence of fluoride, a groundwater contaminant (Park JA et al. 2014a, 2017).

The phage MS2 demonstrated extensive adhesion to iron oxide particles in batch experiments with different types of iron oxide particles—maghemite with some goethite, hematite, and magnetite—where the MS2 adhesion efficiency decreased in the presented order of oxides (Park JA et al. 2014b; Park JA and Kim 2015a).

Electrocoagulation was reviewed as a future alternative of the coagulation technique for drinking water treatment (Ghernaout et al. 2011). In a large study on electrocoagulation, Symonds et al. (2015) involved the phage MS2 as one of the standard indicators.

The place of the phages MS2 and Qβ in the virus mitigation by coagulation was reviewed extensively by Heffron and Mayer (2016). Heffron et al. (2019a) suggested that overall treatment claims based on the phage mitigation for any iron-based technology should be critically considered due to higher susceptibility of the phages to inactivation *via* ferrous oxidation.

The surfactant-modified zeolite was recommended for removing viruses from groundwater in field experiments with the phage MS2 as a model, while iron oxide−coated sand was found to be inefficient (Schulze-Makuch et al. 2003a). Recently, the toxicity of a cationic surfactant, hexadecyltrimethylammonium bromide (HDTMA-Br), one of the most commonly used surfactants for zeolite modification, was determined for the MS2, among other standard indicators (Reeve and Fallowfield 2017). The applications of natural and surfactant-modified zeolites for water remediation and the role of the phage MS2 therein were reviewed recently by Reeve and Fallowfield (2018).

Natural interactions of the FRNA phages belonging to different groups with suspended matter and sediment in a river water system were presented in a recent advanced study of Fauvel et al. (2017b). The status of the phages was a function of the hydroclimatological conditions, where phage-solid association occurred during the peak of rainfall-runoff events and groups II, III, and IV were able to sorb to the riverbed sediment, where group II was more abundant in sediment than in surface water. Therefore, group distribution differed between water and sediment compartments (Fauvel et al. 2017b).

Recently, the phage MS2 was characterized as not an appropriate surrogate by the coagulation of the specific contaminant candidate list of viruses, which included coxsackievirus, echovirus, hepatitis A virus, recombinant norovirus virus-like particles, and murine norovirus (Shirasaki et al. 2017b). In contrast to PMMoV, a novel indicator, the phage MS2 did not appear as a reliable surrogate for the contaminant list viruses also by the micro- and ultrafiltration (Shirasaki et al. 2017a). Further evaluation of the suitability of PMMoV as a surrogate of human enteric viruses by coagulation–rapid sand filtration was performed in comparison with the phage MS2 and many other indicators (Shirasaki et al. 2018).

It is amusing that tethered honey bees, *Apis melifera*, were coaxed to fly in a miniature wind tunnel for a specific time interval, and adsorbed the phage MS2 from aerosol at a linear rate of 1% of the aerosol concentration for every −6.73 pC of electrostatic charge on the bee (Lighthart et al. 2005). Another intriguing study showed that the phage MS2 was removed efficiently by steam-exploded bark of the *Pinus radiata* pine tree (Lewis et al. 1995).

A special study was devoted to the possible adsorption of the phage MS2 to algae *Nannochloropsis salina*, a marine unicellular algal species with an average diameter of 2 μm and very simple ultrastructure, which could be a promising candidate for biodiesel production due to their ease of growth and ability to accumulate long unsaturated fatty acids (Unnithan et al. 2014b). It was concluded that the presence of algae and wastewater organics marginally enhanced viral viability, and the recovery of the viable phage MS2 was affected by the physiological state of algal cells as induced by age and algal culture media. Furthermore, high rate algal ponds were suggested as replacements for the waste stabilization ponds in South Australian towns and demonstrated reliable efficiency to remove the FRNA phages (Young et al. 2016; Fallowfield et al. 2018). Furthermore, Patchaiyappan et al. (2019) treated the phage MS2 with a natural coagulant, the seed extract of *Strychnos potatorum* L.f. (Loganiaceae), and showed 1.3 log reduction of MS2. The seeds of *S. potatorum* are known as nirmali seeds, or thethankottai in vernacular language. It is a commonly distributed deciduous species in southern and central parts of India, Sri Lanka, and Burma. The ancient Indian texts such as the *Sushrutha Samhita* report that the nirmali seeds can be used to clarify turbid water (Patchaiyappan et al. 2019).

FRACTURE FLOW AND TRANSPORT

The onset of the RNA phage transport studies can be related to the publication of Bales et al. (1989), who evaluated behavior of the phages MS2 and f2 in sandy soil and fractured tuff columns. Gross et al. (1991) used the seeded phage MS2 to look after virus movement in a soil treatment system.

McKay et al. (1993) added the phage MS2, as well as the DNA phage PRD1, to water in a trench-to-trench lateral groundwater flow, and the phages were used therefore for the first time as groundwater tracers over periods of several days. Both phages then contributed to further studies on the influence of flow rate on transport and retention of the phage tracers in a fractured shale saprolite, a highly weathered, fine-grained subsoil that retains much of the fabric of the parent bedrock (McKay et al. 2002). The seeded phage MS2, together with the DNA phages PRD1 and φX174, and attenuated poliovirus, were used by the control of viral transport in a sand-and-gravel aquifer under field pumping conditions (Woessner et al. 2001).

The phage MS2 contributed to the determination of a critical pH for the virus fate and transport in saturated porous medium, which was determined to be 0.5 unit below the highest isoelectric point of the virus and porous medium (Guan et al. 2003). According to this rule, when water pH was below the critical pH, the phage had an opposite charge to at least one component of the porous medium and was almost completely and irreversibly removed from the water.

The extent and pathways of water flow and virus transport in fractured, weathered granitic bedrock were investigated at a field site in southern California (Frazier et al. 2002). In the study, a suspension containing the phage MS2, sodium bromide, and blue dye was ponded at the soil–weathered bedrock interface and allowed to infiltrate for 9 h. The trench was then excavated and bedrock samples were collected and assayed for water, bromide, and MS2 content. These studies suggested that microbial contamination from sources such as septic fields and sewage ponds might pose a threat to the quality of groundwater and surface water in areas with saprolitic subsoils.

Penrod et al. (1996) investigated the filtration kinetics of the phage MS2 and the DNA phage λ in a model porous medium consisting of saturated beds of ultrapure quartz sand by pH of 5, where the phages and quartz possessed a net negative surface charge. Transport of not only seeded phage MS2, but also of the intrinsic FRNA phages into and through an alluvial gravel aquifer was evaluated by Sinton et al. (1997). The reduction of the FRNA phages by dune infiltration and estimation of their sticking efficiencies during transport was tested by Schijven et al. (1998). Furthermore, the phages MS2, Qβ, and fr, as model viruses, were employed for adsorption on cellulose and its derivatives, kaolin, carbon black, and river sediment as model solid surfaces (Sakoda et al. 1997). The phage MS2 was compared with Norwalk virus and poliovirus for the adsorption to soils suspended in treated wastewater (Meschke and Sobsey 1998).

The phages MS2 and Qβ, among other phages, were tested for the adsorption and transport through sandy soils in dependence of their isoelectric points (Dowd et al. 1998). The migration potential of the FRNA phages was assessed during a large groundwater recharge project (Yanko et al. 1999). The modeling removal of the seeded phages MS2 and PRD1 by dune recharge was described at Castricum, the Netherlands (Schijven et al. 1999).

The phage MS2, together with other indicators, was suggested for the role of a biotracer by monitoring effluent retention time in the constructed wetlands (Hodgson et al. 2003). The role of the tracer was performed by the phage MS2 to measure bacterial and virus movement in the unsaturated flow through sand filter (Sélas et al. 2003) or through the saturated dune sand (Schijven et al. 2003). The seasonal variations in effluent retention in a constructed wetland were traced by the phage MS2, among other viral and bacterial tracers (Hodgson et al. 2004). The presence of plants in wetlands significantly increased the inactivation of the phage MS2 and poliovirus (Karim et al. 2008).

The retention and transport of the phage MS2 in glass beads having hydrophilic or hydrophobic surface properties was tested under saturated and unsaturated flow conditions (Han et al. 2006).

The attachment and detachment models for the batch and column sandy soils were generated with the phages MS2 and Qβ, as well as with the DNA phages PRD1, φX174, and PM2 (Schijven et al. 2000a). The contribution of the FRNA phages to the studies on the transport of viruses and their removal by soil passage was reviewed extensively by Schijven and Hassanizadeh (2000).

The phage MS2, together with the DNA phage φX174, was used to study virus removal and transport in saturated and unsaturated sand columns in the Ottawa sand (Jin et al. 2000). The capability of the acquired model was validated further in the Taiwan lateric soils with the phage MS2 as an indicator (Hu et al. 2006).

The removal of the phage MS2, among other indicators, was studied by deep well injection into a sandy aquifer at a pilot field site in the Netherlands (Schijven et al. 2000b). The tracing experiments with the phage MS2 and other indicators were performed in an alluvial gravel aquifer at Burnham, near Christchurch, New Zealand (Sinton et al. 2000).

The adsorption and survival of the FRNA phages in comparison with fecal coliforms and somatic coliphages was measured in soil irrigated with wastewater, and the adsorption behavior of the FRNA phages was found intermediate between those of these two microorganisms (Gantzer et al. 2001). Remarkably, the rate of inactivation of all microorganisms was lower in soil than in wastewater and depended extensively on soil temperature and moisture content. However, by field investigations, only a small number of the FRNA phages was found in conventional, dewatered mesophilic anaerobically digested biosolids, and neither was detected in biosolids-amended agricultural soil (Lang et al. 2007).

The kinetic modeling of both phage MS2 and PRD1 transport through columns of saturated dune sand was performed, and their removal mechanism during transportation was presented (Schijven et al. 2002a,b; Schijven and Šimůnek 2002).

The fate and transport of the phages MS2, PRD1, and φX174 in UK aquifers was studied with a special emphasis to regard them as surrogates for pathogenic viruses (Collins et al. 2004, 2006). Similarly, the fate and transport of the phages MS2 and PRD1 was investigated during artificial recharge with recycled water in Los Angeles County (Anders and Chrysikopoulos 2005).

The MS2 transport was found sensitive to the groundwater pH and to the isoelectric points of minerals in the groundwater medium by the transport simulation in a 1-meter-long model aquifer (Schulze-Makuch et al. 2003b). The practical batch and column experiments were performed on the FRNA phage transport, among other indicators, in soil caused by surface and subsurface drip irrigation with treated wastewater (Kouznetsov et al. 2004). The data on the phage MS2 transport contributed strongly to the overall investigations of the transport of colloids in unsaturated porous media (Sirivithayapakorn and Keller 2003). The occurrence and levels of the FRNA phages, among other indicators, were measured in different sludges and biosolids (Guzmán et al. 2007b) with the help of the advanced isolation methods (Guzmán et al. 2007a). The role of the phage MS2 in the visualization techniques of the biocolloid transport processes at the pore scale under saturated and unsaturated conditions was reviewed at that time by Keller and Auset (2007).

The dissolved organic carbon had no influence on the retention and transport of the phage MS2 in a sandy soil (Cheng et al. 2007). However, the increasing concentrations of dissolved humic acid led to decrease of the MS2 retention on sand columns (Zhuang and Jin 2003b). Leaching of the phage MS2 from Class B biosolids and potential transport through soil was described by Chetochine et al. (2006).

By other data, the saturated pumice sand aquifer material assured high mass removal of the phage MS2, as well as of *E. coli*, as indicators. although additional dissolved organic carbon showed a progressive reduction in mass removal and retardation (Wall et al. 2008). This suggested that the organic matter competed with phages for the sorption sites, thus promoting phage transport, and the viral transport rates might be significantly greater in the contaminated compared with the uncontaminated pumice sand aquifers. In sandy soil column tests, the bonded and dissolved organic matters significantly reduced adsorption of the phage MS2 (Cao et al. 2010).

The influence of autochthonous microorganisms and water content on the MS2 adsorption and inactivation was studied in different soils by Zhao et al. (2008). The ecology of the RNA phages and other viruses in soils was reviewed exhaustively at this stage by Kimura et al. (2008) and Sidhu and Toze (2009).

By artificial recharge of groundwater, the efficient removal of the phage MS2 as a tracer was tested in a field study, where a 5-m-deep column of sand and an artificial esker, a 2-m-wide, 2-m-deep, 18-m-long bed of coarse sand, were constructed and river water was pumped at a rate of 40 L/h to the sand column (Niemi et al. 2004).

The phage MS2 and the DNA phage PRD1 were employed in the laboratory-scale virus transport experiments in columns packed with sand under saturated and unsaturated conditions (Anders and Chrysikopoulos 2009). The salinity was found critical for the phage MS2 sorption in saturated sand columns: the MS2 showed very weak adsorption in fresh water due to having the same negative charge as the sand, while increasing the salinity concentrations dramatically enhanced the MS2 adsorption (Cao H et al. 2009).

The phage MS2 was efficiently removed from drinking water in a 5 m long × 0.3 m diameter column of the saturated, heterogeneous gravel (Sinton et al. 2010, 2012) and in columns packed with the clean sand only and with the sand with a layer of zero-valent iron-sand mix (Shi et al. 2012). The iron oxide−amended biosand filters were evaluated with the phage MS2 (Bradley et al. 2011).

The phage MS2 was inactivated in the presence of quartz sand under static and dynamic batch conditions at different temperatures (Chrysikopoulos and Aravantinou 2012) and by the batch attachment to kaolinite and montmorillonite as model clay colloids (Chrysikopoulos and Syngouna 2012). The cotransport of clay colloids and viruses were continued in the laboratory-packed glass bead columns, where the phage MS2, as well as φX174, were used as model viruses but kaolinite and montmorillonite appeared as model clay colloids (Syngouna and Chrysikopoulos 2010, 2011, 2013, 2015, 2016). Further studies on the interaction of the phage MS2, φX174, and human adenoviruses with kaolinite and bentonite led to the conclusion that neither MS2 nor φX174 could be recommended as a suitable model for adenovirus in this respect (Bellou et al. 2015).

Remarkably, Wong et al. (2014) used for the first time in the transport experiments not only free MS2 phage, but also pilus-associated MS2 variants, when they compared transport of the phage MS2 and human adenovirus in saturated quartz sand columns. Although retention of such pilus-associated MS2 in the column was just slightly higher than that of the individual MS2 particles, the membrane filtration results indicated potentially important differences between removal of MS2 and pilus-associated MS2 by filtration with finer pore sizes. The effect of redox conditions on the removal of the phage MS2 during sand filtration was estimated by Frohnert et al. (2014). Later, the phage MS2 attachment under suboxic conditions was investigated at field scale in a dune area with artificial recharge of groundwater under suboxic conditions in a sandy aquifer (Hornstra et al. 2018).

He et al. (2014) found that kaolinite had no effect on the survival of the phage MS2. Nevertheless, further studies investigated the phage MS2 and Qβ removal in various clay minerals and clay-amended soils (Park JA et al. 2015a). The batch experiments in kaolinite, montmorillonite, and bentonite showed that kaolinite was far more effective by the MS2 removal than montmorillonite and bentonite, while the Qβ log removals were far lower than those of MS2 at all the kaolinite contents. In the presence of kaolinite colloids, breakthrough of the phage MS2 occurred concurrently with that of the colloidal particles, and the time taken to reach the peak virus concentration was therefore reduced (Walshe et al. 2010).

Furthermore, a laboratory-scale variable-aperture dolomite rock fracture was employed to compare the transport of the phage MS2, as well as the DNA phage PR772, with that of the virus-sized microspheres of 20 and 200 nm diameter (Mondal and Sleep 2013). It was concluded that the 20 nm microspheres were not suitable surrogates for the MS2 transport, whereas the 200 nm microspheres would be more suitable but nevertheless showed different responses to changes in solution chemistry compared to the phages.

The penetration of the phage MS2 through soils was described by a three-parameter model accounting for inactivation, reversible adsorption, and accumulation (Gitis et al. 2011). The numerical modeling was applied to the transport of the phage MS2 and other surrogates in segmented mesocosms packed with sand, sandy loam, or clay loam soil, in order to determine the effect of soil texture and depth in the soil treatment units (Morales et al. 2014). The kinetic transport model was applied for the field experiments with the rapid phage MS2 transport in an oxic sandy aquifer in the cold climate of central Norway (Kvitsand et al. 2015). The results demonstrated that the MS2 inactivation was negligible to the overall removal, and that the irreversible MS2 attachment to aquifer grains, coated with iron precipitates, played a dominant role in the MS2 removal. Although the total removal during 38-m travel distance and less than 2 days residence time was high, pathways capable of allowing virus migration at rapid velocities were present in the aquifer. Therefore, the risk of rapid transport of viable viruses should be recognized, particularly for water supplies without permanent disinfection (Kvitsand et al. 2015).

Microbial transport in a cold climate and coarse soil was investigated with the phage MS2 as a tracer after a norovirus outbreak at Lake Mývatn, Iceland (Gunnarsdottir et al. 2013). The tracer experiments with the phage MS2 were carried out also in a naturally discrete-fractured chalk core (Weisbrod et al. 2013).

The FRNA phage transport during and after rainfalls was described by Fauvel et al. (2016). Interestingly, the first arrival of phages was likely to be linked to the resuspension of riverbed sediments that was responsible for a high input of group II, but surface runoff contributed further to the second input of phages, and more particularly of group I. Fauvel et al. (2017a, 2019) accurately followed the propagation of the FRNA phages along 3 km of the Alzette River in Luxembourg and France. This work provided a new way to assess true *in situ* viral propagation along a small river.

The application of modern quantitative molecular tools (Wong and Molina 2017), the improved sampling methods (Turnage and Gibson 2017), and the modern resuspension technique from bed sediment in irrigation canals (Zhou K et al. 2017) were recommended insistently for the further elucidation of the transport behavior of enteric viruses in groundwater, with the reference on the special MS2 role in the previous and present studies. Moreover, the phage MS2 was used as a model by the first investigation of marine phages, the nonterrestrial viruses infecting marine host bacteria, for transport in sand-filled percolated columns (Ghanem et al. 2016).

It is remarkable that the phage MS2 stimulated generation of the pathogen-sized and protein-coated microspheres with modified surface charge, which could mimic different pathogens and be exploited as pathogen surrogates by transport studies in groundwater (Pang et al. 2009).

Recently, the phage MS2 served as one of the models in an interesting study that established the importance of streambank fencing from cattle by the input of fecal indicator organisms to watercourses (Kay et al. 2018). The unique role of the phage MS2 in the natural aquifer treatment was reviewed extensively (Donath et al. 2019).

FILTRATION BY SOIL AND SAND COLUMNS

The adsorptive capacity of different soils for the phage f2 and poliovirus were first detected by Schaub et al. (1982). To assess soil–aquifer treatment of sewage effluent for removal of viruses, the phage MS2, together with other tracers, was added to effluent applied to basins, which were constructed in the coarse sand alluvium near Tucson, Arizona (Powelson et al. 1991, 1993). Removal of the seeded phage MS2 from sewage effluents during saturated and unsaturated flow through soil columns was also studied (Powelson and Gerba 1994).

Miniature soil columns were used for the filtration of the seeded phage MS2, as well as hepatitis A virus, poliovirus, and echovirus (Sobsey et al. 1995). The soils studied were coarse sand, loamy sand, clay loam, and organic muck, where the latter, as well as coarse sand soils, were found unsuitable for land application of wastewater.

The phage MS2 as a possible surrogate for waterborne viral pathogens was used together with the recombinant Norwalk virus particles by the filtration in a quartz sand column (Redman et al. 1997). It was concluded, however, that the phage MS2 could not mimic the subsurface filtration of

Norwalk virus in natural systems because of significant differences of electrostatic properties of their particles.

The MS2 spiking was used as a quality control by the testing of the large USEPA-funded integrated membrane system (IMS) project based on the combination of biological activated carbon filtration, slow sand filtration, and reverse osmosis, and carried out by the Amsterdam Water Supply (Nederlof et al. 1998). The phage MS2 was used to evaluate intermittent soil infiltration of treated sewage for reuse in Chile (Castillo et al. 2001). At this stage, the role of the phage MS2 in the monitoring of the sand filtration was reviewed by Baveye et al. (2002). The phage MS2 appeared as a model by removal of microbes from municipal wastewater effluent by rapid sand filtration and subsequent UV irradiation in Finland (Rajala et al. 2003). The slow sand filtration was efficient for the elimination of the phage MS2 and of a set of bacterial indicators (Hijnen et al. 2004, 2005). The phage MS2, together with the DNA phage PRD1 and bacterial indicators, was used for the evaluation of a three-dimensional pilot-scale soil treatment system in the laboratory and during the testing of full-scale systems under field conditions (Van Cuyk et al. 2001, 2004).

A mathematical model to predict removal of pathogens by slow sand filtration as a function of variable operational conditions was established on the background of experimental data on the seeded phage MS2 (Schijven et al. 2013).

The phage MS2 participated in the study that led to an important conclusion that the presence of *in situ* metal oxides in soils was a significant factor responsible for virus sorption and inactivation (Chu et al. 2003).

The phage MS2, together with hepatitis A virus and poliovirus, when they were spiked into a simulated surface water and subjected to the flocculation with aluminum sulfate, were removed then efficiently by following high-rate filtration through long sand column, while the MS2 removal by filtration alone without the addition of aluminum was very poor (Nasser et al. 1995).

To enhance virus removal, the sand and granular activated carbon were coated with ferric and aluminum hydrous metal oxide Fe/Al hydroxides (Scott et al. 2002c). This coating changed the zeta potential of these filtration media from negative to positive at pH 6–9 and improved significantly removal of the phage MS2, as well as of the DNA phage PRD1 and poliovirus as models. The nanoporous aluminum oxide was tested for the coating of granular filter media, with the phage MS2 as a model (Lau et al. 2004). Furthermore, a carbonaceous nanofilter was fabricated, where the phage MS2 was used as the only indicator of virus removal (Mostafavi et al. 2009).

The phage MS2 appeared as a standard indicator among emerging waterborne pathogens in the evaluation of the effect of filtration conditions by the conventional continuous-flow pilot-scale treatment of raw water with aluminum coagulation, flocculation, sedimentation, and granular media, i.e., anthracite and sand gravel, filtration at the University of Wisconsin in Madison (Harrington et al. 2003).

The phage MS2 was found to adsorb on the sand coated with a novel material, namely, layered double hydroxides (LDH), a group of anionic clay-like nanocomposites with entrapped Zn^{2+} and Mg^{2+} together with Al^{3+} in an octahedral structure (You et al. 2003). Furthermore, the phage MS2 and the DNA phage φX174 were used as indicators by the elaboration of a novel LDH-based filtration method for the virus removal from raw river water (Jin et al. 2007).

The anthracite/sand column filtration was combined with subsequent UV disinfection and the phages MS2 and T4 were used as indicators (Templeton et al. 2007). The phage MS2 was used by the zeolite column filtration and application of quaternary ammonium chloride-treated zeolite in pilot filters (Abbaszadegan et al. 2006).

The removal of the spiked phage MS2 was controlled by the evaluation of design and operating conditions of the pilot-scale multistage slow sand filtration system (Anderson et al. 2009). Again, it was concluded that the sand filtration might not provide adequate virus removal and should be combined with a disinfection/inactivation step.

The rapid sand filtration process was characterized by the filtration of the spiked phage MS2, together with the DNA phages φX174 and T4, out of tap water and secondary effluents (Aronino et al. 2009). The removal efficiency for the phages MS2 and $Q\beta$ was detected by the RT-PCR technique (Langlet et al. 2009). Furthermore, RT-PCR was used in a detailed study on the removal of the phages $Q\beta$ and MS2 by slow sand filtration technology in a full-treatment water plant in Japan (Indah et al. 2012).

The promising removal of the FRNA phages as indicators of viruses was assessed in stormwater biofilters, among other indicators (Li et al. 2012).

The effect of sand composition on treatment performance in the popular biosand, or household-scale slow sand, filters was evaluated with the phage MS2 as an indicator (Elliott et al. 2015). The reduction of the phage MS2 as a surrogate was studied in parallel with the rotavirus reduction by the efficacy evaluation of the biosand filtration (Wang Hanting et al. 2016). This was the first study to determine the efficiency of rotavirus reduction by the biosand filtration in order to decrease diarrheal disease incidence.

Recently, the phage MS2 was used as an indicator by evaluation of biochar-modified stormwater biofilters (Afrooz et al. 2018). Katzourakis and Chrysikopoulos (2018) used the phage MS2 by the numerical elucidation of the biocolloid transport in geochemically heterogeneous porous formations with a three-dimensional mathematical model. Betancourt et al. (2019) studied removal of a set of viruses during transport through a saturated soil column and found that the phage MS2 removal was nonlinear and could be described well by a power law relation under saturated flow and predominantly anoxic redox conditions. Miao et al. (2019) explored for the first time a filter cartridge system with electropositive granule media for recovery of viruses from coastal water, where the phage MS2 was used as a traditional model. Ushijima et al. (2019) showed that only fine soil (1–4 mm) performed 5 $\log_{10}$ and 3 $\log_{10}$ reductions of *E. coli* and MS2, while coarse soil could not remove pathogens.

MEMBRANE FILTRATION

The application of membrane processes for virus retention seemed especially attractive to avoid the disadvantages of chemical disinfection by drinking water purification and wastewater reclamation. The phage MS2 was used to investigate the mechanisms of virus capture from aqueous suspension by a polypropylene microfilter (McGahey and Olivieri 1993). The phage Qβ, in parallel with the DNA phage T4, was employed as a model in the filtration studies where many types of membranes, including micro-, ultra-, and nanofiltration membranes of many different materials, were tested for virus retention (Urase et al. 1993, 1994, 1996). Moreover, the phage Qβ, together with the poliomyelitis virus vaccine, was involved into the pilot scale membrane micro- and ultrafiltration processes located in the eastern part of the Tokyo metropolitan area and surveyed for 6 months (Otaki et al. 1998). The role of microfiltration was studied with the seeded phage MS2 by the assessment of tertiary treatment technology for water reclamation in San Francisco (Jolis et al. 1996). Then, the seeded MS2 phage was used by the membrane microfiltration processes operated by the City of Los Angeles Bureau of Sanitation (Iranpour 1998) and again by the combined microfiltration and UV disinfection process for water reclamation in San Francisco (Jolis et al. 1999). The phage MS2 was used as one of the models in a thorough investigation on the influence of salts and pH on virus adsorption to different microporous filters including nitrocellulose, fiberglass, cellulose, and charged-modified fiber filters (Lukasik et al. 2000).

The phage MS2 was used as a biologic model for assessing virus removal by and integrity of high-pressure membrane systems in the pilot-scale water treatment process in two cities in California (Kitis et al. 2003). The phages MS2, Qβ, and fr contributed to the investigation of the virus rejection by *nuclepore* microfiltration membranes, a Whatman brand of polycarbonate membranes (Herath et al. 1998, 1999). Furthermore, a thorough theoretical analysis of the microfiltration process through Al_2O_3 membranes was performed with the phages MS2 and Qβ as models (Herath et al. 2000).

The retention of the phage MS2 on hydrophobic (GVHP) and hydrophilic (GVWP) 0.22-μm microfiltration membranes at different pH levels and with different salts was investigated by Van Voorthuizen et al. (2001). The FRNA phages were employed as one of the models to compare low-protein-binding polyvinylidene fluoride (PVDF) filters with more advanced polyether sulfone (PES) filters, which both avoided virus adsorption to membranes (Mocé-Llivina et al. 2003a). However, the practical evaluation of small-scale water purification devices showed that the latter, using only filtration through pores of 0.2–0.4 μm or larger, failed in removal of the FRNA phages, and employment of reverse osmosis was necessary to remove the FRNA phages at concentrations under the detection limit (Hörman et al. 2004). As a result, it was emphasized that the FRNA phage test should be obligatory for qualified testing of novel water purification devices.

To enhance the efficacy of the microfiltration process, electrocoagulation and chemical coagulation pretreatments

were employed, where the phage MS2 acted as a traditional indicator (Zhu et al. 2005a,b). The phage MS2 was used as a standard by combination of the filtration through ultra- and microfiltration membranes with aluminum precoagulation for virus removal from drinking water in a laboratory test unit (Fiksdal and Leiknes 2006). As before, the filtration only was inefficient for the virus removal.

Sano et al. (2006) were more optimistic about filtration feasibility and showed that the microfiltration of sewage sludge and treated wastewater on membranes with a pore size of 0.1 μm ensured removal of the RNA phages and indigenous noroviruses. Later, the MS2 removal testing led to the conclusion that ultrafiltration followed by UV irradiation was more efficient in terms of energy consumption than coagulation followed by ultrafiltration (Lee S et al. 2013). Further, ultrafiltration combined with coagulation-sedimentation was evaluated for reduction of the spiked phage MS2, among many other indicators, in a pilot-scale water reclamation plant in Okinawa, Japan (Lee S et al. 2017a,b). Finally, different combinations of ultrafiltration, nanofiltration, coagulation, UV irradiation, and reverse osmosis were studied over 2010–2014 in a pilot plant in Japan with the phage MS2 as an indicator (Yasui et al. 2018). It was concluded that reclaimed water could be considered acceptable for recreational enhancement by adding a UV, a nanofiltration membrane, or a reverse membrane treatment to the ultrafiltration membrane treatment process.

Fluorescent-dye-labeled MS2 phages, in parallel with gold particles, were used for the first time as nanoscale probes for the detection of breaches and evaluation of the integrity of ultrafiltration membranes (Gitis et al. 2006). Moreover, the aggregation and surface properties of the phages MS2, GA, Qβ, and SP were evaluated with implication for the membrane filtration processes, where the hydrophobic sequence SP ~ GA > Qβ > MS2 was established for the four group representatives (Langlet et al. 2008a). As a result, the phages SP and GA were not qualified as useful model systems for understanding membrane filtration processes because the particles might be easily eliminated by size exclusion simply by appropriate choice of the membrane pore size. Unlike SP and GA, the phages Qβ and MS2 exhibited significantly different electrokinetic and aggregation features with respect to pH and ionic strength conditions (Langlet et al. 2008a). Later, the phages MS2, Qβ, and GA were employed to assess the effectiveness of drinking water plants with respect to the removal of phages by a conventional pretreatment process (clarification coupled to a sand filtration) followed by an ultrafiltration membrane (Boudaud et al. 2012). The pore size was measured using aquasols of gold and silver nanoparticles and controlled by the retention of the phages MS2 and φX174 (Duek et al. 2012).

The phage MS2 was applied as an indicator for the efficient filtration of aerosolized bacteria and phages on the classical Whatman GF/A glass microfiber filters (Wang and Brion 2007). At that time, the phage MS2 played a crucial role in the validation of a novel material, namely nanoalumina fibers composed of the mineral boehmite AlOOH, as a revolutionary filtration approach that received the name Disruptor™

(Komlenic 2007; Tepper and Kaledin 2007; Li et al. 2009; Tepper et al. 2009; Kaledin et al. 1014, 2017). By combination of the Disruptor filtration with a fluorescence-based detection of the β-galactosidase release, Yang and Griffiths (2014a) proposed a novel method for rapid enumeration of low numbers of the infectious FRNA phages.

The effect of ozone on membrane fouling by microfiltration was evaluated with the phage MS2 as the only model by an attempt to integrate the ozone treatment with the microfiltration process (Oh et al. 2007).

Low-pressure membranes, which have been increasingly applied in the disinfection of wastewater effluent, were challenged with the phage MS2 at bench and pilot scales (Jacangelo et al. 2008). The phage MS2 and naturally-occurring FRNA phages were used by the assessment of the Filtration and Irrigated cropping for Land Treatment and Effluent Reuse (FILTER) scheme in Queensland, Australia (Chinivasagam et al. 2008).

The phage MS2 was used as an indicator by assessment of the ultrafiltration devices in water vending machines (Miles et al. 2009) and in water treatment devices designed for household use in low-income settings (Clasen et al. 2009).

The phage MS2 served as a seeded indicator by the combination of filtration and UV disinfection in a device for the treatment of harvested rainwater in Tucson (Jordan et al. 2008).

The elaboration of the RT-PCR method markedly simplified the application of the phages MS2 and Qβ as standards by membrane filtration in water treatment (Langlet et al. 2009). Interestingly, the Qβ phage was eliminated to a lower extent than the phage MS2, this being the case for all membranes considered in this study. The global role of the phages MS2 and Qβ, among other indicators, in ultrafiltration as a mechanism to supply drinking water in international development, was reviewed at that time by Davey and Schäfer (2009).

Pierre et al. (2010) quantified adsorption of the spiked phage MS2 not only on the filters, but also on the filtration equipment, and concluded that the most appropriate material to be used as a filtration test tank is Pyrex glass. Characterization of the ultrafiltration membranes by the MS2 retention was standardized by Causserand et al. (2010). Removal of the FRNA phages was used as an efficacy parameter by the evaluation of the membrane bioreactor technology in a full-scale municipal wastewater treatment plant in Bologna, Italy (Zanetti et al. 2010).

Apart from the direct disinfection needs, microporous filters were used for the concentration of the phages to be studied from large volumes of water and wastewater. Thus, positively (Goyal et al. 1980a,b; Singh and Gerba 1983) and negatively (Farrah 1982; Farrah and Preston 1985; Preston et al. 1988; Scott et al. 2002a) charged microporous filters were applied successfully for the adsorption with further elution of the phage MS2 and the DNA coliphages as models. The phage f2 survived on electropositive microporous filters for at least 4 weeks at 4°C (Keswick 1983; Keswick et al. 1983).

Méndez et al. (2004) developed a simple filtration-elution method to concentrate waterborne phages, including the FRNA phages from drinking water on the acetate-nitrate cellulose membrane filters, which was then validated according to ISO standards. Later, a protocol based on tangential flow filtration followed by ultracentrifugation was elaborated with the phage MS2 as one of the seeded indicators (Sylvain et al. 2009). This protocol was adapted to large environmental samples of surface, ground, and drinking waters from the Grand Duchy of Luxembourg in order to assess the occurrence of protozoan parasites, pathogenic viruses, somatic coliphages, and the FRNA phages in these samples. The filtration−elution procedures were improved further by Helmi et al. (2011).

The phage MS2, together with the double-stranded RNA phage φ6, was used for the evaluation of filters for the sampling and quantification of the RNA phage aerosols, where the qRT-PCR detection was strongly recommended (Gendron et al. 2010). A novel bioaerosol amplification unit for improved viral aerosol collection was constructed with the phage MS2 as the test agent (Oh S et al. 2010). The phage MS2 was used for the microwave (Woo et al. 2012b) and UV (Woo et al. 2012a) decontamination of filters loaded with viral aerosols.

The phage MS2 contributed to the elaboration of the knowledge-based protocol for the assessment of viral retention capability of cellulose acetate (Pierre et al. 2011) and dialdehyde cellulose (Woo et al. 2011) membranes, with specific attention to the effects of aggregation, adsorption, and inactivation of viruses during filtration. Furthermore, the phages MS2 and GA were used in the challenge tests of the membrane integrity evaluation for the popular ultrafiltration systems (Ferrer et al. 2013). The phage MS2 was used for monitoring of the integrity and ageing of the reverse osmosis membranes (Huang X et al. 2015; Antony et al. 2016).

The general theory of the filtration processes, which was dependent to a considerable degree on the data on the phages MS2 and Qβ, was presented by Antony et al. (2012).

The ultrafine cellulose and chitin nanofibrous membranes were verified by the MS2 testing for water purification (Ma H et al. 2011). The phage MS2 and E. coli were applied by the assessment of an anodic multiwalled carbon nanotube (MWNT) microfilter for drinking water treatment (Vecitis et al. 2011). A microporous filter composed of the microglass filaments coated with nanoalumina fibers was evaluated with the phage MS2 as a model (Ikner et al. 2011).

The phage MS2 was used by the assessment of the multilayered nanofibrous microfiltration membrane based on the cellulose nanowhiskers and characterized with high flux, low pressure drop, and high retention capability against both bacteria and phages (Ma H et al. 2012). The electrochemical multiwalled carbon nanotube (EC-MWNT) filter was fabricated and tested with the phage MS2 (Rahaman et al. 2012). Moreover, the MWNT filters were coated with copper(I) oxide, titanium(IV) oxide and iron(III) oxide nanoparticles and tested with the phage MS2, where the most efficient removal, greater than or equal to 99.99%, was obtained with the Cu$_2$O-coated MWNT membrane in the whole pH range (Németh et al. 2019).

The phage MS2, together with the DNA phage φX174, was used for the fabrication of a novel high virus retention membrane, namely sintered yttria-stabilized zirconia micro-tubes with tailored porosity (Kroll et al. 2012; Werner et al. 2014). Next, the ultrafiltration polyethersulfone membrane was graft-polymerized with zwitterionic SPP, or [3-(meth-acryloylamino) propyl] dimethyl (3-sulfopropyl) ammonium hydroxide, and the novel membranes were tested with the phage MS2 and human adenovirus and recommended for potable water reuse (Lu et al. 2017).

The phage MS2 was used as an indicator for the low-pressure membrane filtration of secondary wastewater effluent (Huang H et al. 2012b), including magnetic ion exchange pretreatment before low pressure membrane filtration (Huang H et al. 2012a). The phage MS2, together with φX174, was used in the evaluation of virus removal mechanisms by low pressure membranes (ElHadidy et al. 2013).

The aerobic membrane bioreactor (MBR) using a 0.4-μm hollow-fiber membrane module, was tested for the MS2 removal by Shang et al. (2005). The FRNA phages remained as reliable indicators by the pilot plant MBR use (Marti et al. 2011; De Luca et al. 2013). The nine different MBR systems were compared, using the phage MS2 as a model, by the municipal wastewater treatment (Hirani et al. 2012).

Meanwhile, the phage MS2 was used for assessment of the polysulfone ultrafiltration membranes impregnated with silver nanoparticles (Zodrow et al. 2009) and by assessment of bifunctional alumoxane/ferroxane nanoparticles for the removal of viral pathogens, where the importance of surface functionality to nanoparticle activity was highlighted (Maguire-Boyle et al. 2012). Further, the phages MS2 and GA were used for monitoring of membrane integrity during ultrafiltration performance (Ferrer et al. 2015).

The unique role of the phage MS2 as an ultrafiltration indicator on removal of pathogens by membrane bioreactors was demonstrated in an extensive review by Hai et al. (2014). Furthermore, the enhanced removal of the seeded phage MS2 (Fox and Stuckey 2015) and of the intrinsic FRNA phages (Peña et al. 2019) was demonstrated in anaerobic membrane bioreactors (AnMBRs). The phages MS2 and fr were compared for removal efficiency by MBR filtration (Chaudhry et al. 2015). The spiking of the phage MS2 into the MBR reactor to create a high viral load, which might be observed during an outbreak, was used by application of the USEPA-approved CANARY event detection software for real-time performance monitoring of decentralized water reuse systems (Leow et al. 2017).

The phage MS2 was used as an indicator by testing of the improved low-cost ceramic water filters for viral removal in the Haitian context (Guerrero-Latorre et al. 2015). Using the phage MS2 as a surrogate of human enteric viruses, the ceramic water filters were improved by doping with goethite (FeOOH) (Tsao et al. 2016), and a ceramic silicon carbide microfiltration membrane was fabricated for swimming pool water treatment (Skibinski et al. 2016). High-flow ceramic pot filters were fabricated and tested with the phage MS2 as an indicator (van Halem et al. 2017).

The phage MS2 was used for the evaluation of different ultrafiltration membranes during gravity-driven membrane filtration over 500 days (Lee D et al. 2019). The phage MS2 and poliovirus were used as survival markers by the introduction of preservative agents and antibiotics for increased virus survival on positively charged filters (Fagnant et al. 2017a,b).

The theoretical validation of a full-scale ultrafiltration plant was achieved through Bayesian modeling, with the data on the MS2 removal (Carvajal et al. 2017b). The highly-impressive role of the RNA phages MS2, Qβ, GA, and the FRNA phages in total by membrane-based water and wastewater treatment processes was highlighted in detail in a recent extensive review (Wu B et al. 2017).

The self-assembling bicontinuous cubic liquid crystals were presented for the first time as a novel method for the synthesis of membranes with a regular pore size, below 1 nm, and showed their efficiency by the phage Qβ rejection (Marets et al. 2017). The phages MS2 and φX174 were used for evaluation of antimicrobial electrospun poly(vinyl alcohol) nanofibers containing benzyl triethylammonium chloride (Park JA and Kim 2017).

Recently, the phage Qβ contributed to the elaboration of the efficient columnar liquid-crystalline nanostructured membranes that removed viruses efficiently and showed sufficient water permeation (Hamaguchi et al. 2018).

Liang et al. (2018) achieved the 6.74 log reduction of the phage MS2 by the reactive electrochemical membrane system, with titanium suboxide microfiltration membrane serving as the filter and the anode. The effect of divalent versus monovalent cations on the MS2 retention capacity of amino-functionalized ceramic filters was measured by Bartels et al. (2018, 2019). The phage MS2 was used to check novel microfiltration membranes that were modified with cationic polymer polyethyleneimine (PEI) for drinking water applications (Sinclair TR et al. 2018, 2019), photo-crosslinked PVA/PEI electrospun nanofiber membranes, and copper-coated cellulose-based water filters (Szekeres et al. 2018). Gallardo et al. (2019) used the phage MS2 by the generation of hollow fiber dialysis filters operated in axial flow mode for recovery of microorganisms in large volume water samples. Remarkably, the phage MS2 contributed markedly to the evaluation of ultrafiltration membrane breakage (Lee S et al. 2019c). Then, the phage MS2, among other indicators, was used in a study that demonstrated the importance of carefully designed sampling regimes when characterizing microorganism removal efficiencies of deep bed filters (Nilsen et al. 2019). The phage MS2 was used by detecting impact of repeated pressurization on virus removal by reverse osmosis membranes for household water treatment (Torii et al. 2019), as well as for the integrity monitoring of these membranes (Yoon 2019).

The detection of the intrinsic FRNA phages was involved in the great field study that evaluated the efficiency of a full-scale treatment facility in the UK, comprising of Trickling Filters and an experimental Aerated Constructed Wetland, where—for the first time—the fate of a set of bacterial indicators and enteric phages were simultaneously investigated under real operating conditions (Stefanakis et al. 2019). Dang and Tarabara (2019) presented the fine mechanisms of the MS2 deposition onto polyelectrolyte multilayers.

The role of the phage MS2 by the development of membranes for water purification applications was unveiled in exhaustive recent reviews (Govindan et al. 2018; Ma H and Hsiao 2018; Bodzek et al. 2019). Wickramasighe et al. (2019) reviewed the situation with the filtration of viruses by the preparation of human therapeutics, where the phages MS2 and PP7 were used as models.

OYSTERS AND MUSSELS

For the first time, the coliphages, without any indication however on the RNA phages, were found in shellfish and shellfish-raising estuarine waters and suggested for the role of the enteric virus surrogates by Vaughn and Metcalf (1975). Thus, the start was given for the replacement of *E. coli* as the pollution indicator in shellfish by more adequate virus surrogates. It was established soon that the FRNA phages were present in shellfish in higher concentrations than *E. coli*.

The first report on the isolation and enumeration of the phage tracers, including the phage MS2 tracer from the bivalve molluscan shellfish, was published by West and Wipat (1988). The FRNA phages were unable to replicate in hard-shelled clams *Mercenaria mercenaria* but the densities of the phages were stable for up to 7 days in shellfish held at ambient seawater temperatures (Burkhardt et al. 1992). The evidence of replication, although not observed in live shellfish, was found to occur in temperature-abused shellfish homogenates and supernatants, when the phage f2 was added and a suitable bacterial host was present. The further study led to a conclusion that the FRNA phages might not be a suitable indicator for virus removal from Pacific oysters and that the somatic phages might be better suited to this role because of rapid decline of the FRNA phage number in oyster tissue (Humphrey and Martin 1993).

Remarkably, *E. coli* and the FRNA phages were both bioaccumulated rapidly in oysters and mussels, with levels reaching an equilibrium after 24 h (Doré and Lees 1995). The contamination levels in oysters *Crassostrea gigas* ranged from 180,000−400,000 *E. coli* organisms and from 70,000−80,000 FRNA phages per 100 g of shellfish, while contamination levels in mussels *Mytilus edulis* were similar, ranging from 80,000−220,000 *E. coli* organisms and from 55,000−120,000 FRNA phages. During depuration, the time needed to reduce levels of the FRNA phage by 90% was considerably longer than that to reduce *E. coli*, while the FRNA phages were retained only in the digestive gland and were not sequestered into other internal tissues (Doré and Lees 1995).

The FRNA phages, among other phages and bacteria as indicators, were detected in oysters and their harvest waters in Calico Creek, near Morehead City, North Carolina (Chung et al. 1998). The indicator levels in oysters were highest for the FRNA phages and the group II phages predominated in the wastewater, receiving water, and oysters. As a result, the FRNA phages and their groups were proclaimed again as promising indicators of human enteric virus contamination in oysters and their harvest waters (Chung et al. 1998). Meanwhile, new tools were proposed for the monitoring of

virological quality, including detection of the FRNA phages as prospective indicators of norovirus contamination in oysters (Doré et al. 1998, 2000). However, Croci et al. (2000) did not detect the FRNA phages in Adriatic Sea mussels.

The preferential accumulation of the FRNA phages from estuarine waters was found by 1-week trials over a 1-year period in eastern oysters *Crassostrea virginica* (Burkhardt and Calci 2000). Remarkably, the greatest level of the FRNA accumulation continued from late November through January, with a concentration factor of up to 99-fold. This finding suggested that the seasonal occurrence of shellfish-related illnesses by enteric viruses was in part the result of the seasonal physiological changes undergone by the oysters that affected their ability to accumulate virus particles from estuarine waters (Burkhardt et al. 2000).

Next, the FRNA phages were detected in naturally polluted mussels *Mytilus galloprovincialis* that were collected at two sites: in the Delta of the Ebro River and at the outlet from a sewage treatment plant with ten times greater pollution (Muniain-Mujika et al. 2000). The results showed a different pattern in the proportions between the viral parameters when the source of the fecal pollution was close to or distant from the mussel growing area.

Nevertheless, the validity of the FRNA phages as enterovirus surrogates of shellfish contamination was not accepted by the corresponding study in France (Miossec et al. 2001). In Sweden, the FRNA phages were regarded as a reliable indicator with respect to enteroviruses but certainly not for adenoviruses in the blue mussel *Mytilus edulis* (Hernroth et al. 2002). However, the FRNA phages were significantly related to human adenoviruses and enteroviruses during depuration of naturally highly polluted mussels in Spain (Muniain-Mujika et al. 2002). It was concluded in this study that 5 days might be necessary to assess the sanitary quality of shellfish, and the FRNA phages could be used as a complementary parameter for evaluating the efficiency of the depuration treatment.

The broad evaluation of potential indicators of viral contamination in shellfish, before and after depuration treatments, was performed in order to establish their applicability to diverse geographical areas in the north and south of Europe: Greece, Spain, Sweden, and the United Kingdom (Formiga-Cruz et al. 2003). The FRNA phages were present in higher numbers in Northern Europe and seemed to be significantly related to the presence of viral contamination, with a very weak predictive value for hepatitis A virus, human adenovirus, and enterovirus and a stronger one for Norwalk-like virus. In Spain, further thorough comparative analysis of viral pathogens and potential indicators in oysters and mussels indicated that "the use of FRNA phages as potential indicators of viral contamination in shellfish deserves further analysis with higher number of samples" (Muniain-Mujika et al. 2003). In the United States, the seasonal trend for the levels of the FRNA phages in oysters was not observed (Shieh et al. 2003), although seasonal trends for depurated market-ready oysters in the UK had been reported by Doré et al. (2000).

Doré et al. (2003) confirmed the strong seasonal influence on the FRNA phage concentrations in shellfish, with a geometric mean count of 4503 pfu/100 g in the winter compared with 910 pfu/100 g in the summer, and proposed again the FRNA phages as an indicator of the viral risk associated with shellfish. In Norway, common blue mussels *Mytilus edulis*, horse mussels *Modiolus modiolus*, and flat oysters *Ostrea edulis*, which were obtained from various harvesting and commercial production sites along the Norwegian coast, were screened for the FRNA phages and for the presence of noroviruses, human adenoviruses, and human circoviruses (Myrmel et al. 2004, 2006). The seasonal variation of the FRNA phages was observed and the positive correlation was found between the FRNA phages and noroviruses. However, the FRNA phages were present in only 43% of the norovirus-positive samples (Myrmel et al. 2004). In Brittany, France, a study on the impact of wastewater input on shellfish quality was conducted over a period of 20 months and no marked seasonal differences were found for the FRNA phages and noroviruses (Pommepuy et al. 2004). The phage MS2 was found together with enteroviruses and adenoviruses in shellfish and cultured shrimp at the southwest coast of India, where some samples, however, were positive for enteric viruses but negative for MS2 (Umesha et al. 2008). Bioaccumulation, retention, and depuration of the phage MS2, as well as of hepatitis A virus, poliovirus, and murine and human noroviruses, were studied thoroughly in *Crassostrea virginica* and *C. ariakensis* oysters from the harvesting regions of the Chesapeake Bay (Nappier et al. 2008).

The proportions of the different FRNA phage groups in shellfish were checked for the first time in harvesting areas of Greece (Vantarakis et al. 2006). Again, the elevated numbers of the FRNA phages observed in the winter concurred with the known increased viral risk associated with shellfish harvested at that time of year in Greece. The majority of the FRNA phages detected in shellfish samples belonged to group IV, which indicated the possible presence of animal fecal material in the sample harvesting areas. The group II and III phages were present at low levels, but the popular group I represented 8% of the phages. Group IV showed clear seasonal distribution, more in winter, less in summer, whereas the other groups did not show any difference (Vantarakis et al. 2006). In New Zealand, the overall high concentration of the FRNA phages in the shellfish samples was confirmed, and group II, which has been associated commonly with human fecal pollution, was found at a high frequency and concentration in both the shellfish and river water samples (Wolf et al. 2008). However, all other groups were also identified, with frequent findings of the group I and III phages. At nine estuaries in the east, west, and Gulf coasts of the United States, more FRNA phages were typed from colder water than warmer waters, while the water salinity did not affect the FRNA phage levels (Love et al. 2008b). In this study, group I was found strongly prevalent in oysters, mussels, clams, and water. Group II followed, while the contribution of group III was minimal and that of group IV was negligible.

At this stage, extensive reviews on the enteric viruses of humans and animals in aquatic environments, and especially on viral contamination in shellfish and the role of the FRNA phages by the assessment of the contamination, were presented by Fong and Lipp (2005) and Le Guyader and Atmar (2007). Remarkably, Smiddy et al. (2006) described at that time the high pressure-induced inactivation of the phage $Q\beta$ in oysters.

The association between the levels of the FRNA phage and norovirus contamination was noted by Lowther et al. (2008) and the FRNA phages were proposed again to the role of a more reliable surrogate than *E. coli* for the detection of viral contamination. In fact, this study in the United Kingdom showed again the peaks of the norovirus and FRNA phage contamination during winter months, with average levels approximately 17 times higher in oysters sampled from October to March than during the remainder of the year, consistent with epidemiological data for the UK showing that the oyster-associated illness was confined to winter months. Furthermore, the use of the FRNA phages to predict the occurrence of norovirus in shellfish was supported strongly by the investigation of the phage contamination in Irish oysters *Crassostrea gigas*, which confirmed the seasonal trend described earlier (Flannery et al. 2009). However, further over a 1-year period study at the wastewater treatment plant providing secondary wastewater treatment (Flannery et al. 2012) confirmed the seasonal trend for noroviruses but not for the FRNA phages in oysters sampled adjacent to the wastewater treatment plant discharge. The phage GA was used for comparison of the RT-qPCR and infectious phage detection in wastewater and oysters (Flannery et al. 2013a). Remarkably, Flannery et al. (2014) showed later that the domestic cooking practices based on shell opening alone did not inactivate the infectious FRNA phages and norovirus in experimentally contaminated mussels; however, cooking mussels at temperatures above 90°C reduced infectious viruses to undetected levels within 3 min.

In France, Gourmelon et al. (2010a,b) applied an advanced microbial source tracking (MST) toolbox, which included the group II FRNA phage probes. Tracking with the MST, which was supplied additionally with the FRNA phage group I probes, of both oysters and estuarine waters in Brittany showed the higher occurrence of the human group II and allowed the identification of human fecal contamination as the predominant source of contamination in oysters (Mieszkin et al. 2013).

The phages MS2 and $Q\beta$, together with some DNA phages, were tried as pressure surrogates for enteric viruses, namely, hepatitis A and Aichi viruses, in artificial seawater or oyster slurry, where the phage T4 displayed similar pressure responses as hepatitis A virus, but the phage MS2 was the most pressure-resistant and similar to Aichi virus (Black et al. 2010).

In total, pros and cons of the FRNA phages as putative surrogates in shellfish were analyzed at that time in an exhaustive review written by Oliveira et al. (2011).

The reduction of the phage MS2 and other indicators was measured in artificially contaminated oysters *Crassostrea virginica* and hard-shell clams *Mercinaria mercinaria* over 5 days in flow-through depuration tanks, where depuration findings were compared to depuration regulations in the European Union and the United States (Love et al. 2010a). During commercial depuration of shellfish by chlorination, the FRNA phages were significantly reduced in 7 days but showed differences in the reduction rate between carpet shell clams *Venerupis pullastra* and mussels *Mytilus galloprovincialis*, with a faster depuration rate in mussels (Polo et al. 2014). Generally, clams showed slower depuration rates and higher contamination levels not only for the FRNA phages, but also for noroviruses and hepatitis A virus that were analyzed in this study. A mathematical model of the depuration process predicted the minimum depuration times for the FRNA phages and norovirus in shellfish (McMenemy et al. 2018). Excellent reviews on the depuration of shellfish from noroviruses and the role of the FRNA phages in the depuration process were published recently (McLeod et al. 2017a,b).

Remarkably, when Eastern oysters *Crassostrea virginica* were individually exposed to the phage MS2 for 48 h at 15°C followed by collective maintenance in continuously UV-sterilized seawater for 0–6 weeks at either 7, 15, or 24°C, it appeared that cooler temperatures dramatically enhanced the persistence of MS2 within oyster tissues (Kingsley et al. 2018). These data could serve as a guideline for regulatory agencies regarding the influence of water temperature on indicator phage after episodic sewage exposure.

Interest in the FRNA phages as putative surrogates in shellfish was raised again by Hartard et al. (2016). The authors confirmed the seasonal trend of the infectious FRNA phage concentrations that were higher than those found in water and approved therefore the phage bioaccumulation in shellfish. Moreover, the relationship between the presence of human norovirus genomes and those of the human-specific group II phages was underlined in shellfish collected throughout Europe (Hartard et al. 2016). Therefore, the specific detection of the infectious group II FRNA phage was proposed as an indication of the presence of infectious human noroviruses. Furthermore, the strong correlation between the presence of the group II FRNA phages and that of norovirus in shellfish impacted by fecal contamination was acknowledged when both viruses were detected using molecular approaches (Hartard et al. 2017b). This study also showed that the FRNA phages persisted at least as long as noroviruses did. In order to exploit the potential of the FRNA phages as indicators, the rapid and sensitive integrated cell culture RT-PCR method was elaborated to evaluate the FRNA phage group prevalence in oysters and applied to marketed products (Hartard et al. 2017a). Its application to marketed oysters over a 1-year period has made it possible to identify the winter peak classically described for the norovirus or FRNA phage accumulation. Recently, Lowther et al. (2018) acknowledged the positive association and correlation between the PCR-detectable levels of the group II FRNA phages and norovirus in oysters.

During controlled bioaccumulation from artificial seawater, the FRNA phages were bioaccumulated to the highest concentration in mussels *Mytilus edulis* after 6 hours of incubation but were below detection limit in oysters *Crassostrea gigas* throughout the exposure period (Olalemi et al. 2016a). This study demonstrated strong differences in the bioaccumulation rates among a set of pollution markers. A parallel study of these authors led to the conclusion that somatic coliphages could, however, be better pollution markers than the FRNA phages to predict the concentration of adenoviruses in mussels *M. edulis* and their overlying waters (Olalemi et al. 2016b).

Recently, Cho K et al. (2018) performed an exhaustive investigation in the shellfish-growing area in the Republic of Korea and recommended coliphages—in particular, the FRNA phages—as a useful tool for tracking fecal contamination and noroviruses in aquatic environments. Langlet et al. (2018) evaluated the effect of proteinase K on the MS2 and norovirus capsids by the recovery of viruses from shellfish, as well as the effect of shellfish digestive glands on the recovery. Based on the collected data including the FRNA phage accumulation, Gomes et al. (2018) proposed the invasive bivalves as a tool of the efficient biofiltration and bioaccumulation for water and wastewater decontamination.

Finally, recent reviews on the FRNA phages as enteric viral indicators in bivalve mollusk management could be recommended (Hodgson et al. 2017; ICMSF 2018).

FRUITS AND VEGETABLES

In connection with the fruits and vegetables, the FRNA phages, namely the phage f2, was mentioned as a destruction indicator, among other indicators, in a device for the anaerobic treatment of fruit, yard, and vegetable waste (Wellinger et al. 1993). In other words, 2 days of thermal treatment at 50°C–55°C were sufficient to destroy the phage f2.

The efficacy levels of different physical and chemical washing treatments were evaluated for the reduction of viral and bacterial pathogens, including the phage MS2 as a surrogate, from inoculated strawberries (Lukasik et al. 2003). The survival of the phage MS2 on lettuce and cabbage was detected together with that of *E. coli* and feline calicivirus by different temperatures and sanitizers (Allwood et al. 2004a). A stronger correlation of survival measures was observed between feline calicivirus and MS2 than between *E. coli* and either of the viral agents at 25°C and 37°C. Furthermore, the phage MS2 was applied as a surrogate for norovirus to study the virus survival when it was spiked into fresh produce items such as iceberg lettuce, baton carrot, cabbage, spring onion, curly leaf parsley, capsicum pepper, tomato, cucumber, raspberries, and strawberries (Dawson et al. 2005). It was concluded that MS2 could be used as an effective surrogate in similar studies, since it survived for prolonged periods, both in buffer and on fresh produce, at temperatures relevant to chilled foods, and it was not removed effectively by chlorine washing.

The variability of virus attachment patterns to butterhead lettuce was detected by comparison of the MS2 attachment with that of the DNA phage φX174, feline calicivirus, and echovirus, where the latter demonstrated the highest affinity to lettuce surface (Vega et al. 2005).

Both somatic and FRNA phages were consistently recovered from the roots of cucumber plants along with bacterial fecal indicators in hydroponic cucumber greenhouses (Xu and Warriner 2005). Nevertheless, despite the heavy contamination of plant roots, the cucumber fruits were within acceptable microbiological limits. It is noteworthy to mention that the model of the phage f2 uptake through plant roots was designed by Ward and Mahler (1982).

The attachment of the phage MS2 and other tested viruses to lettuce was mostly electrostatic: 1 M NaCl was the most effective treatment in desorbing viruses from the surface of lettuce at pH 7 and 8 (Vega et al. 2008). The phage MS2 was used as an indicator by the combined UV light and hydrogen peroxide treatment of lettuce, an alternative method for hypochlorite-based washes to reduce the carriage of viruses on fresh produce (Xie et al. 2008).

The phage MS2, together with hepatitis A virus, was used by the detailed assessment of chlorine disinfection of strawberries, cherry tomatoes, and head lettuce, where inactivation kinetics of MS2 and hepatitis A virus were similar, suggesting that MS2 could be useful as a process indicator and surrogate in actual practice (Casteel et al. 2008, 2009).

The advanced nested RT-PCR and TaqMan RT-PCR methods were introduced for detection of the seeded phage MS2, hepatitis A virus, and poliovirus in tomato sauce or blended strawberries (Love et al. 2008a). Further, duplex real-time qRT-PCR was elaborated for the detection of hepatitis A virus in raspberries using the phage MS2 as a process control (Blaise-Boisseau et al. 2010).

The role of the phage MS2 as an indicator of fecal contamination was reviewed by extensive assessment of microbial risks associated with cabbage, carrots, celery, onions, and deli salads made with these produce items in the US (Erickson 2010).

The phage MS2 participated in the study that found low-dose γ-radiation acceptable for the improving microbial quality and shelf life of fresh mint as one of the high-risk herbs, which is used without further cooking (Hsu et al. 2010).

The reduction in foodborne viral surrogates including the phage MS2 was achieved by high pressure homogenization, a method suitable for industrial application by decontamination of juices (D'Souza et al. 2009, 2011). The phage MS2 was used as one of the surrogates of human norovirus by treatment of pressurized steam combined with high-power ultrasound to decontaminate fresh raspberries and smooth plastic surfaces and utensils (Schultz et al. 2012). However, the steam-ultrasound treatment in its current format was not able to achieve sufficient decontamination of the contaminated raspberries. The phage GA was used as an artificial contaminator by the generation of a rapid, simple, and efficient method for detection of viral genomes on raspberries (Perrin et al. 2015). The phage MS2 was used as a process control by a complex work on the detection and typing of norovirus from frozen strawberries involved in a large-scale gastroenteritis outbreak in Germany, which affected about 11,000 people (Mäde et al. 2013). Then, the norovirus detection in frozen strawberries was improved, again with the phage MS2 as a permanent process control virus, after repeated identification of strawberries as vehicles for norovirus transmission causing large

gastroenteritis outbreaks (Bartsch et al. 2016, 2018, 2019). Furthermore, the phage MS2 was used as one of the long lists of indicators by the treatment of strawberries with a levulinic acid plus sodium dodecyl sulfate-based sanitizer, which demonstrated a good performance (Zhou Z et al. 2017). The phage MS2 was used as a process control by the evaluation of extraction methods for the norovirus detection in fruit and vegetable salads (Cheng D et al. 2018).

The phages MS2, GA, and Qβ, together with other indicators, were involved in the first large study on the adhesion of pathogenic viruses to food-contact surfaces, namely stainless steel and polypropylene, and food surfaces of lettuce leaves, strawberries, and raspberries (Deboosere et al. 2012). As a result, the phages MS2 and GA were defined as good models of viral adhesion on the inert and lettuce surfaces. Later, the interaction of the phage MS2 with common food processing and preparation surfaces of polyvinyl chloride and glass was assessed by atomic force microscopy and virus recovery assays (Shim et al. 2017).

The phage MS2 was used as an indicator by assessing potential of pulsed light to inactivate viruses on powdered black pepper, garlic, and chopped mint (Belliot et al. 2013). The inactivation functioned on glass beads but was only marginal at the food surface.

Holvoet et al. (2014) simulated a commercial fresh-cut lettuce wash process with the phage MS2 and other spiked indicators and concluded that potable water alone for washing contaminated crops was not adequate from a food safety perspective.

The efficacy of a chlorine-based sanitizer on fresh-cut romaine lettuce was detected with the phage MS2 as a surrogate for foodborne viruses during simulated commercial production using a small-scale processing line (Wengert et al. 2017). In fact, this study showed that the currently recommended commercial production practices were unable to effectively decrease viruses once they have attached to leafy greens during commercial processing.

The FRNA phages were recovered from the surface of parsley and leek in Neyshabour, Iran, in order to identify sources of fecal contamination, while the phage MS2 was used as a model for the recovery technique (Yavarmanesh et al. 2015). After grouping of the FRNA phages, it was concluded that the contamination of vegetables most likely originated from animal sources because of higher frequency of groups I and IV, absence of group III, and also the low frequency of group II. The phage MS2 was used also as a surrogate of enteric viruses for the source tracking of contamination on lettuce (Yazdi et al. 2017).

The phage MS2 inoculated into apple juice was employed as a challenge organism in a large study on the effect of UV-C irradiation as a prospective alternative disinfection technique (Islam et al. 2016). Finally, the phage MS2 as an accepted norovirus surrogate was used by the application of low-frequency-pulsed ohmic heating for inactivation of foodborne pathogens in tomato juice without causing electrode corrosion and quality degradation (Kim SS et al. 2017). The phage MS2 was inactivated more effectively at low frequency and was more sensitive to acidic conditions than pathogenic bacteria, while electrode corrosion and quality degradation of tomato juice were not observed regardless of frequency.

The phage MS2 was used as an indicator by the inactivation of foodborne pathogens on fresh and frozen strawberries using ozone (Zhou Z et al. 2018) or UV-C (Butot et al. 2018) and by the UV inactivation of bacteria and model viruses in coconut water using a collimated beam system (Bhullar et al. 2019). The phage MS2 appeared also as an indicator by the UV-C inactivation of cranberry-flavored water (Gopisetty et al. 2018, 2019). Then, the phage MS2 served also as a traditional surrogate for norovirus by evaluating the effects of washing and freezing strawberries on inactivation of the virus (Huang Licheng et al. 2019).

FOOD TOTAL

Generally, the application of the FRNA phages in food studies was connected directly with the enteric viruses, particularly human norovirus and hepatitis A virus, that remain nowadays the key foodborne pathogens. The general methodology for the enumeration of the FRNA phages, among other coliphages, in foods was elaborated first by Kennedy et al. (1986).

Since fish grown in contaminated ponds can be exposed to a variety of human pathogens, tilapia fish was seeded with the phage f2, and the maximum f2 concentration was detected in muscle tissue (Fattal et al. 1988, 1993).

Active microbial decontamination of tilapia fish was tried with the seeded phage MS2 and poliovirus as indicators, where the latter were tested in the skin, muscle, liver, spleen, and digestive tract, and faster decontamination of the fish was achieved by repeated changing of the water in the holding area (Zuaretz-Peled et al. 1996).

The recovery efficiencies of the FRNA phages MS2, GA, Qβ, FI, and SP were determined from the ground beef and chicken breast meat (Hsu et al. 2002). As a result, of the eight market food samples tested, the FRNA phages were detected in five and somatic coliphages in seven samples.

The phage MS2 was inactivated rapidly by accelerated natural lactic fermentation of infant food ingredients (Nout et al. 1989).

The FRNA phages were detected in a large number of various fresh market-ready produce from retail outlets throughout Minnesota, and in six samples that were collected at a restaurant during an investigation of a norovirus outbreak (Allwood et al. 2004b). Remarkably, the processed retail samples, i.e., cut, shredded, chopped, or peeled, appeared more likely to contain the FRNA phages than unprocessed samples. The authors concluded that the FRNA phages could be useful conservative indicators of fecal contamination in produce. The role of the phages and the FRNA phages among them in foods was reviewed thoroughly at that time by Hudson et al. (2005) and Pillai (2006).

The phage MS2, among other indicators, was used for the generation of an efficient hollow-fiber ultrafiltration device for the concentration and simultaneous recovery of multiple pathogens in contaminated foods (Kim et al. 2009).

In order to prevent norovirus transmission by food, the phage MS2, together with murine norovirus and feline calicivirus, was involved in the rather successful evaluation of trisodium phosphate as a chemical sanitizer instead of sodium hypochlorite (D'Souza and Su 2010).

The phage MS2 was used to compare different concentration methods for the detection of viruses present in bottled waters and those adsorbed to water bottle surfaces (Huguet et al. 2012).

The seeded phage MS2, together with murine norovirus and feline calicivirus, as surrogates of human norovirus, was tested for survival in milk, orange and pomegranate juice, and juice blends at 4°C refrigeration up to 21 days, since fresh fruits, juices, and beverages have been implicated in the human noroviral and hepatitis A virus outbreaks (Horm and D'Souza 2011). A full list of references on the norovirus surrogates, including the phage MS2, in food safety research was collected in a critical review by Richards (2012), who proclaimed a shift in the research priorities from surrogate research to volunteer studies. Bertrand et al. (2012) collected huge amount of data—a total of 658 data from 76 published studies—and performed qualified statistical analysis on the impact of temperature on the inactivation of enteric viruses and their surrogates, including the FRNA phages, in food and water. Further reviews were devoted to analytical methods for the detection of viruses in food, with a strong emphasize on the third Contaminant Candidate List (CCL-3) released by the USEPA in 2009 (Hartmann et al. 2012) and on the biosensors as novel platforms for the detection of food pathogens and allergens (Kumar et al. 2012). Both reviews presented the phage MS2 as a well-accepted and extensively studied surrogate of enteric viruses and a widely accepted model for the elaboration of novel techniques.

In order to prevent the hepatitis E virus distribution, which was common in pigs, the incidence and levels of contamination of hog carcasses with hepatitis E virus and with the FRNA phages, as well as other indicators, were measured at different stages of the dressing process (Jones and Johns 2012). As a result, the FRNA phages were assessed as a reliable indicator of the enteric virus contamination in a meat processing environment. An optimized method of recovery of the phage MS2 from meat was developed for this purpose (Jones et al. 2012). The survival of the phage MS2 and murine norovirus was measured on pork chops during storage and retail display, and both viruses were qualified as good surrogates of human norovirus (Brandsma et al. 2012).

The phage MS2 was used in the study on the recovery of viral RNA from oral fluid of pigs, a prospective source to determine herd health and documenting the circulation of viruses in commercial swine populations (Jones and Muehlhauser 2014). Furthermore, the study on hog carcasses was enriched by other indicators, such as porcine adenovirus and porcine teschovirus, and significant correlations were observed between the viable FRNA phages and porcine adenovirus and between the FRNA phages and porcine teschovirus but not between adenovirus and teschovirus at the various stages of pork processing (Jones and Muehlhauser 2017).

It was concluded that the FRNA phages could be a preferred indicator in the pork slaughter process, as they also provided an indication of infectivity. Remarkably, the viable group II and III phages were generally not detected at the earlier stages of the slaughter process, but they appeared after evisceration and in the retail pork samples.

Emmoth et al. (2017) presented the phage MS2 as a model for the inactivation of swine hepatitis E virus in pork products by high-pressure processing and by lactic acid and intense-light-pulse treatments. Furthermore, the phage MS2 was used as an internal process control by the screening of meat and meat products for hepatitis E virus in Switzerland (Moor et al. 2018) and Germany (Althof et al. 2019).

Jones et al. (2014b) determined the number of the FRNA phages, among other indicators, on the commercial vacuum packaged beef. Remarkably, the prevalence and mean log numbers of the FRNA phages were lower for the provincially registered plants than for the federally registered plants, while groups II and III associated with human origin were detected in 12% and 30% of samples that originated from provincially and federally registered plants, respectively.

The spiked phage MS2 was used as a surrogate of enteric viruses by the investigation of enteric virus survival on lamb and poultry meat under different conditions (Pezeshki et al. 2017a,b).

The phage MS2 deserved mention in a capital review on the further potential of nanoparticles and nanotechnology, including quantum dots and metal, silica, and magnetic nanoparticles, for sensing foodborne pathogens and their toxins in foods and crops (Koedrith et al. 2014). A following review, which mentioned the phage MS2, was devoted to recent developments in the use of viability dyes and quantitative PCR in the food microbiology (Elizaquível et al. 2014). The phage MS2 figured in reviews on high hydrostatic pressure processing as a promising nonthermal technology to inactivate viruses in high-risk foods (Lou et al. 2015) and on thermal inactivation of foodborne enteric viruses and their viral surrogates in foods (Bozkurt et al. 2015). The MS2 appeared also in a motivating review on the fate of foodborne viruses in the "farm-to-fork" chain of fresh produce (Li D et al. 2015) and in highly informative chapters in the recent book *Viruses in Foods* (Bright and Gilling 2016; Gibson and Borchardt 2016; Goyal and Aboubakr 2016).

The FRNA phages and human norovirus were tested in raw milk in the city of Mashhad, Iran (Yavarmanesh et al. 2015). Although statistics did not confirm the correlation between the occurrence of the FRNA phages and norovirus, the frequency distribution analysis indicated that three out of four raw milk samples containing high FRNA phage titers were also positive for noroviruses. To improve the extraction and detection of human norovirus from food matrices, the phage MS2, among a set of other process control models, and human norovirus were seeded onto romaine lettuce or sliced deli ham and tested for the efficiency of recovery (Gentry-Shields and Jaykus 2015). The phage MS2 contributed also to the evaluation of the enteric virus recovery from different parts of farmed salmon (Yavarmanesh et al. 2017).

The phage MS2 modeled survival of enteric viruses during the yogurt-making process and storage (Moradi Moghadam et al. 2017). It appeared that the heat treatment of milk at 85°C for 30 min markedly decreased the MS2 titers. In the same way, the spiked phage MS2 was used as a model in the production process of traditional Iranian butter and Doogh, an Iranian yogurt drink (Fatemizadeh et al. 2016), and by thermal treatment of cow milk (Hosseini et al. 2017). Pasteurization along with homogenization had the most marked effect on elimination of the phage MS2 from frozen ice cream (Ghadirzad et al. 2018). The phage MS2, among other indicators, was employed by the nonthermal processing technology for pasteurizing milk with a UV reactor based on the Taylor-Couette vortex flow (Rawa and Warriner 2018).

The UV fluence was quantified and verified using the phage MS2 by the UV-C irradiation of green tea as a non-thermal processing alternative to the heat pasteurization of tea beverages (Vergne et al. 2018).

The phage MS2 was used as an enteric virus surrogate by comparison of the commercial RNA extraction kits for the RT-PCR detection of the contamination on green onions (Xu et al. 2017).

The phage MS2, together with a set of viral indicators, participated in the extensive study of virus survival in food, namely, onto either the digestive gland of oysters or the surface of fresh peppers, over time under various temperature and relative humidity with further mathematical modeling (Lee et al. 2015).

The phage MS2 was used as one of the indicators by the studies on the putative contamination of the phyllosphere and stored grains of wheat *Triticum aestivum* (Schwarz et al. 2014a,b).

Although not directly related to food, it is noteworthy that the phage MS2 was used as one of the indicators by the evaluation of alkaline hydrolysis as an alternative method for treatment and disposal of infectious animal waste, namely, animal carcasses of pigs, sheep, rabbits, dogs, rats, mice, and guinea pigs (Kaye et al. 1998).

In order to prevent contamination of food contact surfaces with pathogens, the phage MS2, among other surrogates, was studied by the removal and transfer of viruses on stainless steel and nonporous solid surfaces by cleaning cloths, where the latter were of cellulose/cotton, microfiber, and nonwoven materials (Gibson et al. 2012). Moreover, Sinclair R et al. (2018) sprayed reusable grocery bags in a conventional grocery supermarket with the phage MS2 in order to trace the possible norovirus transmission pathway. The data showed that MS2 spread to all surfaces touched by the shopper; the highest concentration occurred on the shopper's hands, the checkout stand, and the clerk's hands.

With the phage MS2 as one of the models, Sommer et al. (2018) performed a thorough examination of virucidal potential of ionic liquids in the context of the potential food protection.

The outstanding role of the phage MS2, among other indicators, by the norovirus tracking and inactivation in foods was reviewed recently (Kamarasu et al. 2018; O'Sullivan et al. 2019; Sommer et al. 2019).

PATIENT CARE

Along with the general ecological importance described above, the FRNA phages have a direct impact by the cleaning praxis for patient safety, medical care, and medical applications.

The first studies to be mentioned in this section had, however, no direct relation to the cleaning praxis but were linked directly to human health. Thus, in connection with the growing resistance against beta-lactam antibiotics and aminoglycosides in hospitals, Schmid and Kayser (1976) used the phage MS2 by the characterization of the resistant Gram-negative bacteria that were isolated from patients in the University Hospital in Zurich. Wenger et al. (1978) tried modeling the behavior of phages that could be present in live virus vaccines. The phage MS2 together with the three DNA phages including one isolate from live virus vaccines were chosen for evaluation of their cytogenetic, proliferative, and viability effect on human lymphocytes. It appeared that the mitogen-stimulated lymphocytes and human embryonic kidney tissue cultures showed no increase in chromosomal abnormalities for high doses of the phage-infected versus control cultures, although some cells experienced inhibition of DNA synthesis but no cell death. Possibly, the phage attachment to the plasma membranes of lymphocytes might provoke the suppression of DNA synthesis (Wenger et al. 1978).

Attempts to involve the FRNA phages into the discrimination of the healthy subjects from the patients with internal and leukemic diseases (Furuse et al. 1983a) or with the gastrointestinal disturbances (Cornax et al. 1994) led to rather indefinite results. The FRNA phages, among many other bacterial and viral indicators, were tested by the correlation studies between the freshwater quality and health complaints among triathletes and run-bike-runners (Medema et al. 1995) and swimmers (Medema et al. 1997). Moreover, the value of the FRNA phages as an index of risk from recreational use of a freshwater environment was accepted by studies among users of a white-water course fed by the River Trent, UK (Lee JV et al. 1997). Recently, a systematic study was published on the transfer of the RNA phages MS2 and Qβ from liquid to skin at the skin–liquid interface that occurred during water-related activities (Pitol et al. 2017, 2018). Remarkably, up to 90% of the virus inoculated on the skin was transferred to the water when the skin remained wet, compared to 30 ± 17% when the skin was dry, and the transfer from skin to liquid was 41% higher when the recipient liquid was water as compared with saliva.

As to direct studies in the human gut, the RNA phages have been found in human feces at very low titers (Osawa et al. 1981; Havelaar et al. 1986). The gut could contain the RNA phages, but they seemed to be transient members that originated from dietary sources (Zhang et al. 2006). Furthermore, a recent metagenomic study revealed only a limited number of gut RNA phages in primates, while the majority of RNA viruses found in the gut were plant and human viruses (Krishnamurthy et al. 2016). The recent reviews confirmed the general statement that the RNA phages are infrequent visitors at the *ménage à trois* of host, bacteria, and phages in the human gut (Mirzaei and Maurice 2017; Manrique et al. 2017).

It is noteworthy that the phage f2 was employed as one of the indicators by studies on the stability of a group of selected viruses in cosmetics (von Rheinbaben and Heinzel 1992).

Most of the studies in connection with the RNA phages and human health were targeted to microbiological safety and disinfection control. Thus, at the early stage, the phage MS2 was involved as a model to evaluate virucidal hand (Jones et al. 1991; Sickbert-Bennett et al. 2004, 2005) and skin (Davies et al. 1993) disinfectants. The phage MS2 was also used by microbiological evaluation of the dental air-turbine handpiece (Ohsuka et al. 1994).

The phage R17 was used in studies on virus inactivation by photoactive phenothiazine dyes, such as methylene blue, in blood components (Wagner et al. 1998b; Wagner 2002). The phage f2 was used by the evaluation of the three glutaraldehyde-based disinfectants used in endoscopy (Jetté et al. 1995) and by the biological testing of a laboratory pathological waste incinerator (Le Blanc Smith et al. 2002).

The attachment of the FRNA phages, among other indicator microorganisms, was checked by the evaluation of surfaces potentially dangerous due the spread of pathogens, such as, for example, water containers (Momba and Kaleni 2002). Remarkably, the phage MS2 was found sensitive to contact with the Korean traditional bronze alloy Yugi, and therefore the bronze utensils were recommended to be used more widely in order to decrease the viral poisoning at food processing environments and hospitals (Hwang et al. 2009). The phage MS2 was inactivated within 1–2 minutes by microwave radiation (Park et al. 2006). Thereupon the authors suggested designing containers to be used in exposure of medical devices to microwave radiation.

The phage MS2 was used as a model to evaluate techniques for virus recovery from healthcare personal protective equipment, such as gowns, gloves, respirators, and splash-proof plastic goggles (Casanova et al. 2009) and from filtering face-piece respirators (Fisher et al. 2009). Thus, the phage MS2 contributed greatly to the evaluation of the certified N95 respirators and surgical masks for the protection efficacy against airborne viruses, where surgical masks showed a much higher particle penetration than the N95 respirators, but the latter nevertheless could not provide the expected protection level against small virions (Bałazy et al. 2006). Furthermore, the phage MS2 participated in the filter performance of the N95 and N99 facepiece respirators (Eninger et al. 2008b). The survival of the phage MS2 aerosols deposited on the N95 filtering face piece respirator samples was determined by Rengasamy et al. (2010). Moreover, the phage MS2 contributed to the development of a test system to evaluate procedures for decontamination of respirators containing viral droplets (Vo et al. 2009). The data of the MS2 survival on the filtering face-piece respirator coupons led to recommendations for reuse of the filtering facepiece respirators after decontamination as a strategy to conserve supplies during an influenza pandemic (Fisher and Shaffer 2010). The phage MS2 was used by the assessment of the flash infrared radiation disinfection of fibrous filters contaminated with bioaerosols (Damit et al. 2011). However, after outbreak of the 2009 influenza

A (H1N1) pandemic, the phage MS2 was not determined as a reliable surrogate for the influenza pandemic strain in the surface survival studies of the N95 respirators with respect to droplet persistence (Coulliette et al. 2014). The transfer of the phage MS2 from the N95 filtering facepiece respirators to hands was evaluated by Brady et al. (2017). At the same time, special antimicrobial agent-treated filtering facepiece respirators were tested with the phage MS2 and recommended as the first protection and mass prophylaxis candidates during a putative environmental catastrophe or bioterrorism attack (Ali 2016).

The transfer of the phages MS2 and fr, as well as of the DNA phage φX174, between fingerpads and glass was studied by Julian et al. (2009, 2010). It was remarkable that the transfer of the phage fr was significantly higher than of both MS2 and φX174, but the phage transfer between surfaces was reduced for recently washed hands. The hand–mouth transfer of the phage MS2, as well as *E. coli*, was measured after manipulation the beach sand, Chicago, Illinois (Whitman et al. 2009).

The transfer efficiency of the phage MS2, among other bacterial and viral indicators, from porous and nonporous fomites to fingers was measured under different relative humidity conditions (Lopez et al. 2013).

The effect of single- versus double-gloving on virus transfer to healthcare workers' skin and clothing during removal of personal protective equipment was measured with the phage MS2 as an indicator (Casanova et al. 2012). The MS2-charged cotton swatches were used as disinfection bioindicators by antiviral laundry processes in the healthcare sector (Gerhardts et al. 2009, 2012b).

The transmission of the phage MS2 from hands to surfaces in a public toilet scenario was studied among other microbiological indicators by Gerhardts et al. (2012a). The phage MS2 was employed by the evaluation of the potential for virus dispersal during hand drying by paper towels, a warm air dryer, and a jet air dryer, where the latter led to significantly greater and further dispersal of the seeded phage MS2 (Kimmitt and Redway 2016). The phage MS2 was a measure by the evaluation of an alcohol-based hand sanitizer intervention on the spread of viruses in homes (Tamimi et al. 2014, 2015a,b). The methods for the evaluation of handwashing efficacy including the MS2 data were reviewed by Conover and Gibson (2016).

The spiked phage MS2 was used as an indicator to examine the rapid dissemination of microorganisms from hospital floors to the hands of patients and to high-touch surfaces inside and outside of rooms (Koganti et al. 2016). Aerosols of the phage MS2 and *Mycobacterium vaccae* were used as surrogates by the performance of a duct-mounted air disinfection system in a hospital (Griffiths et al. 2005). The seeded phage MS2 was used by the investigation of the virus in a long-term 67-bed care facility using hygiene protocols in the United States (Sassi et al. 2015).

Remarkably, a special simulated vomiting device was constructed to demonstrate that virus aerosolization occurred in a simulated vomiting event: the phage MS2 appeared in this study as a surrogate of norovirus (Tung-Thompson et al. 2015).

The phage MS2, together with the DNA phage φX174, was used as a model to observe the spread of viruses in a hotel setting and to assess the effectiveness of a hygiene intervention in reducing their spread (Sifuentes et al. 2014). The phage MS2 and *P. aeruginosa* were used as markers for *Legionella pneumophila*, a cause of Legionnaires' disease, by the safety evaluation of domestic spa pools (Moore G et al. 2015).

The phage MS2 contributed as a model to the Healthy Workplace Project (HWP) that was targeted to the reduction of viral exposure in an office setting (Reynolds et al. 2016; Sifuentes et al. 2017). These studies assessed how implementation of the HWP providing hand sanitizers, disinfecting wipes, facial tissues, and use instructions had reduced viral loads in an office setting of approximately 80 employees after seeding fomites and the hands of volunteer participants with the phage MS2 tracer. The latter also played the central role by the evaluation of the so-called office wellness intervention (OWI) to reduce viral load in the workplace in an office building (Kurgat et al. 2019).

The phages MS2 and Qβ served as the spiked indicators during urine inactivation process in South Africa (Bischel et al. 2015a,b), subtropical Australia (Ahmed et al. 2017), and in Switzerland (Decrey and Kohn 2017).

After the 2014 West African Ebola virus disease outbreak, new challenges were posed for disinfection of personal protective equipment. Thus, a large number of glove and gown removal simulations were performed and fixated by the correlation between contamination of skin with fluorescent lotion and the phage MS2 (Tomas et al. 2015b). Then, an enclosed UV-C light booth was fabricated for disinfection of the contaminated personal protective equipment prior to removal and tested with the phage MS2 as an efficiency indicator (Tomas et al. 2015a). An extensive review of the personal protective equipment control in an Ebola outbreak (Fischer et al. 2015) substantiated the use of MS2 as a conservative surrogate, although filoviruses were huge enveloped single-stranded RNA viruses, and acknowledged its appropriateness to ensure the healthcare provider's safety.

Furthermore, the frequency of contamination of healthcare personnel during removal of the contaminated personal protective equipment was found similar for the phage MS2 and a novel reflective marker, a fluorescent lotion, visualized using flash photography, which allowed documentation of personnel contamination during the removal of personal protective equipment (Tomas et al. 2016). A pilot study was performed to assess the use of the fluorescent lotion, together with the phage MS2, in patient care simulations, and personnel were trained by dissemination of the phage MS2 and fluorescent lotion from a contaminated mannequin (Alhmidi et al. 2016).

The CDC Prevention Epicenters Program performed an extensive study on the assessment of self-contamination during removal of personal protective equipment for Ebola patient care with the phages MS2 and φ6 as surrogates (Casanova et al. 2016, 2018; Mumma et al. 2018) or the phage MS2 alone (Yarbrough et al. 2018). The FRNA phages were checked, among other indicators, by the assessment of recommended approaches for the safe handling, containment, and removal of excreta from Ebola and cholera treatment centers (da Silva et al. 2018).

During patient care simulations, the phage MS2 and cauliflower mosaic virus DNA performed similarly as surrogate markers of pathogen dissemination (Alhmidi et al. 2017b). An ethanol-based spray disinfectant significantly reduced contamination of the phage MS2 on material from gowns worn by personnel but was affected by the type of gown material and the correctness of fit (Koganti et al. 2017). Moreover, by the permanent MS2 control, the dissemination of viruses from computer touchscreens in patient waiting areas was reduced through patient hand hygiene and an automated UV-C touchscreen disinfection device (Alhmidi et al. 2017a, 2018). A recent tracking and controlling study of soft-surface contamination in healthcare settings was performed with the MS2 as a tracer (Sexton et al. 2018). The special aim of this study was to determine the efficiency of the USEPA-registered soft surface sanitizer in the health care environment. Alhmidi et al. (2019) used the MS2 tracking to evaluate contamination of healthcare personnel during removal of contaminated gloves. Furthermore, the MS2 tracer was employed to measure the resistance of fabric of the protective clothing to liquid and viral penetration (Li M et al. 2019). Wilson et al. (2019) developed a model to predict virus concentration on nurses' hands using data from the phage MS2 tracer study conducted in Tucson, Arizona, in an urgent care facility.

Further, the phage MS2 was tested as a possible surrogate of the Ebola virus by evaluation of the surface disinfection efficacy but was found too conservative against chlorine-based disinfectants, and double-stranded RNA phage φ6 was recommended finally as the appropriate surrogate, since it was slightly more resistant than the Ebola virus (Gallandat and Lantagne 2017). The Ebola virus outbreak initiated the study on the possible self-contamination of healthcare workers during donning and doffing of personal protective equipment, where the MS2 phage was used as a model (Kwon et al. 2017).

Remarkably, the phage MS2 as a virus surrogate was involved in the first study to document the survival characteristics of microbiological indicators in tattooing solutions (Charnock 2006).

The phage MS2 was chosen as an indicator by studies of recycled water pollution during machine clothes washing (O'Toole et al. 2009a) and by turf-grass irrigation (O'Toole et al. 2008, 2009b).

As to decontamination in public places, the efficacy of sodium hypochlorite to decontaminate fecally soiled stainless steel surfaces was evaluated simultaneously with the phage MS2 and feline calicivirus, murine and human norovirus, where MS2 together with murine norovirus appeared as the most resistant (Park GW and Sobsey 2011). The phage MS2, among a set of other indicators, contributed to the elaboration of the portable saturated steam vapor disinfection system with an aim to be used in schools, hospitals, and food-processing facilities and replace chemical disinfectants by the treatment of microbially contaminated surfaces (Tanner 2009). The exposition to hydrogen peroxide vapor and vapor hydrogen peroxide gaseous disinfection systems was tested with the MS2 suspensions, which were dried onto stainless steel carriers or in 10% and 50% of horse blood to simulate the virus being present in a spill of blood/bodily fluids in a hospital ward environment (Pottage et al. 2010). The inactivation of the phage MS2 on hard surfaces or in suspension by chemical disinfectants was reviewed and subjected to meta-analysis, in comparison with other norovirus surrogates, by Hoelzer et al. (2013).

To prevent transmission of nosocomial pathogens in hospitals, a light-activated antimicrobial surface, namely a medical grade silicone incorporating crystal violet, methylene blue, and 2 nm gold nanoparticles, was shown to significantly reduce the viable counts of the phage MS2 together with a set of bacteria and fungi (Walker et al. 2017). The addition of the phage MS2 to the toilet bowl before flushing allowed valuation of hospital-grade disinfectants on viral deposition on surfaces after toilet flushing (Sassi et al. 2018b). Recently, the phage MS2 was used by the generation of an UV-LED device for decontamination of shared pens in healthcare facilities (Emig et al. 2019).

Concerning the potential antiviral treatment of patients, Sankarakumar and Tong (2013) suggested an original method of preventing viral infections with polymeric virus catchers. This novel approach used high-affinity polymeric receptors prepared by a molecular imprinting technique to "catch" viruses. The virucidal action of the imprinted particles was rapid, dose dependent on virus and polymer concentration, and occurred due to the specific adsorption. The fabricated nanoparticles displayed remarkable positive antiviral results that significantly hindered viral infections as compared to the controls. This work was performed with the phage fr as a model (Sankarakumar and Tong 2013).

The unique role of the phage MS2 by the tracing of pathogens and reduction of microbial risks was acknowledged recently (Wilson et al. 2018).

PSEUDOMONAPHAGE PP7

All above sections in this chapter have been devoted to the participation of the F-specific, or FRNA, coliphages in the phage ecology. However, the pseudomonaphage PP7 appeared as the only non-coli phage that was involved in the ecological studies. The phage PP7 appeared first as a marker by the ultrafiltration studies (Oshima et al. 1994, 1995, 1996, 1998; Evans-Strickfaden et al. 1996; Aranha-Creado et al. 1997; Winona et al. 2001; Morales-Morales et al. 2003; Olszewski et al. 2005). As a logical result of the ultrafiltration studies, it was concluded that the phage PP7 could be used as a surrogate of mammalian viruses if the virus removal mechanism was based on size exclusion (McAlister et al. 2004).

A quantitative TaqMan RT-PCR assay was elaborated for the phage PP7, in order to quantify phage from the water samples (Rajal et al. 2007a,b). The phage PP7, together with the DNA phage φX174, was employed for the validation of a set of commercial small-virus-retentive filters (Lute et al. 2007; Sedillo et al. 2008). Finally, the phage PP7 was chosen by the Virus Filter Task Force organized by the Parenteral Drug Association (PDA) as a consensus standard to rate the virus filters (Lute et al. 2008; Brorson et al. 2008a). An internal virus polarization model was generated to explain virus retention by ultrafiltration (Jackson et al. 2014). Furthermore, a systematic statistical study was performed with the phage PP7 as a viral model and *S. typhimurium* as a bacterial model, in order to understand the effect of water chemistry on removal of the microorganisms by ultrafiltration (Cruz et al. 2017). The phage PP7 was used also as a well-accepted standard by the determination of pore size distributions of virus filtration membranes using gold nanoparticles (Kosiol et al. 2017) and by the retention of *Acholeplasma laidlawii*, a common contaminant of growth media for cell culture, by sterile filtration membranes (Helling et al. 2018).

By the purification studies, the technique of chromatofocusing was applied to the characterization and purification of the phage PP7 and two other model phages, φX174 and PR772, that were routinely used for the virus filters testing (Brorson et al. 2008b). The modern methods electrospray differential mobility analysis (ES-DMA) and quantitative amino acid analysis (AAA) were used to determine the PP7 particle concentration, as well as the concentration of the MS2 and φX174 particles (Cole et al. 2009). Later, the phage PP7 contributed to the PDA studies on the virus purification standards (Roush 2015). A stepwise approach was used to define binding mechanisms of surrogate phage particles, first of all, of the phage PP7, to multimodal anion exchange resin (Brown MR et al. 2017a,b). The phage PP7 bound to the multimodal resin via both anionic and hydrophobic moieties (Brown MR et al. 2017b) and the positively charged quaternary amine and the hydrophobic aromatic phenyl group were responsible for binding to the multimodal resin (Brown MR et al. 2017a).

By the PCR standards, the phage PP7 was used as an internal multiplex qPCR control for the concentration and detection of rotavirus (Fumian et al. 2010) and rotavirus, adenovirus, norovirus, and hepatitis A virus (Prado et al. 2013) in sewage and biosolid samples. The phages PP7 and MS2 that were labeled with either fluorescein-5-isothiocyanate (FITC) or Alexa Fluor®488 were used by confocal scanning laser microscopy to study virus retention during virus filtration (Bakhshayeshi et al. 2011). Furthermore, the phage PP7 contributed to the evaluation of the virus concentration efficiency in various water matrixes by different elution, concentration, and detection methods (Poma et al. 2013; Shapiro et al. 2013). The phage PP7 was used as an internal control by the assessment of gastroenteric viruses from wastewater in Uruguay River (Victoria et al. 2014b) and by rainfall events at urban beaches in Brazil

(Victoria et al. 2014a). Furthermore, the internal PP7 control was used for the detection of aichivirus (Burutarán et al. 2016) and variety of noroviruses (Victoria et al. 2016) in wastewater samples from Uruguay and of human sapoviruses in wastewater and stool samples from Rio de Janeiro (Fioretti et al. 2016) and gemycircularviruses in wastewater of Rio de Janeiro and in river water samples collected in Manaus, Amazon region (da Silva Assis et al. 2016). Then, the phage PP7 was involved as the internal control in the detection of noroviruses in Rio de Janeiro (Fioretti et al. 2018).

By phage mortality studies, the phage PP7 was subjected to inactivation by either metal copper or copper sulfate $CuSO_4$ (Li J and Dennehy 2011). Significant mortality was observed for the phage PP7 as well as for the double-stranded RNA phages φ6 and φ8, the single-stranded DNA phage φX174, and the double-stranded DNA phage PM2, while the double-stranded DNA phages PRD1, T4, and λ were relatively unaffected by copper. The seeded phage PP7 was used by optimization of the skimmed-milk flocculation method for recovery of adenovirus from sludge (Assis et al. 2017, 2018). It is necessary to note that the skimmed-milk flocculation method was used also with the phage MS2 as a model, among many other indicators, with an aim to simultaneously concentrate viruses, bacteria, and protozoa (Gonzales-Gustavson et al. 2017).

By ultrafiltration, the phage PP7 was employed recently in an advanced study of virus removal by polyethersulfone ultrafiltration membrane, with respect to size and surface charge under a broad range of relevant conditions of pH and ionic strength (Gentile et al. 2018). Moreover, the phage PP7 contributed to the differentiation between size exclusion and adsorptive effects during virus filtration, where the current understanding of mechanisms related to virus breakthroughs after temporary flow interruptions were broadened (Kosiol et al. 2019).

In food protection, the phage PP7 was used as an internal process control by the assessment of microbiological contamination of fresh, minimally processed, and ready-to-eat lettuces in Brazil (Brandão et al. 2014) and Argentina (Blanco Fernández et al. 2017).

In patient care, the recovery rate of the phage PP7 and murine norovirus was measured on both porous and nonporous formic as well as on rubberized surfaces of fomites in hospitals (Ganime et al. 2015).

By the combating of bacterial biofilms, the phage PP7 failed to inhibit planktonic growth of *P. aeruginosa* but led to a significant decrease in the level of the *P. aeruginosa* biofilm (Hosseinidoust et al. 2013a).

7 Genome

There is no greatness where there is not simplicity.

Lev Nikolaevich Tolstoy

The simplest things are often the truest.

Richard Bach

GENETIC MAKEUP

Full-length RNA phage genomes include the following major characteristics: size of the genome, number, size, and location of genes, as well as variations in the genomic organization among different phages. The exceptional plus-strand character of the RNA phage genome was proved experimentally *in vitro*: an RNA complementary to the phage R17 RNA, isolated from bacteria infected with the phage, failed to stimulate protein synthesis in the cell-free extracts of *E. coli* (Schwartz et al. 1969). A small amount of amino acid incorporation that was observed with the minus-strand correlated with the degree of contamination by the phage plus-strand RNA.

Table 7.1 presents the general characteristics of the known full-length RNA phage genomes that were compiled from the NCBI taxonomy browser by the *Leviviridae* family and arranged in accordance with the current taxonomical standards.

The small RNA phage genome involves four genes, where three of them—maturation, or A, protein, coat protein, and the phage-specific replicase subunit—constitute the three-gene minimum needed to make a phage. The fourth lysis gene appears as a specific addition to this minimum. The three basic genes are responsible for a set of regulatory functions that must give the RNA minus and plus strands in proper ratio, as well as the needed amounts of the four different proteins on time (Zinder 1980). The father of RNA phages, Norton D. Zinder (1980), characterized this genome orchestra as "a marvel of natural selection," where "almost each nucleotide performs several functions and in proper proportion." Chapters 16 and 17 of this book will show how this regulation is achieved, but it functions, essentially, through different affinity of the corresponding ribosome-binding sites, and next by the involvement of the two regulatory complexes formed on the RNA phage plus strand by coat protein and replicase.

Table 7.1 demonstrates first the strongest genomic difference of the *Allolevivirus* and *Levivirus* genera members that are present at the NCBI taxonomy browser. In addition to the three basic proteins—A and coat proteins and the replicase subunit—the *Allolevivirus* genome encodes a C-terminally extended coat protein known as the minor A1 protein, which appears as a result of ribosomal readthrough of a leaky opal UGA termination codon of the coat protein gene (Weiner and Weber 1971, 1973; Weiner et al. 1972) and is essential for the formation of viable Qβ particles *in vivo* (Hofstetter et al. 1974; Engelberg-Kulka et al. 1977, 1979a; Skamel et al. 2014). For the first time, the A1 protein was observed by the Qβ replication in spheroplasts exposed to low concentrations of actinomycin D or rifampicin in the presence of an amino acid mixture containing radioactive leucine or histidine: this protein, that was present also in the Qβ particles, was named IIb (Garwes et al. 1969). It was calculated that the 3–10 copies of the IIb protein incorporate into the virion. After 45 years, it was established precisely that the A1 protein, or the IIb protein of Garwes et al. (1969), is incorporated in 12 copies per Qβ virion (Skamel et al. 2014). It is required for infection, but its precise function is not yet fully deciphered. A recent electron microscopy visualization of foreign epitopes carried by the A1 protein within infectious Qβ particles showed that the A1 protein molecules are occupying corners of the Qβ icosahedron (Skamel et al. 2014). The A1 protein appeared as a special peculiarity of the group III and IV phages that are forming together the *Allolevivirus* genus. After the phage Qβ, the synthesis of the A1 protein was identified in the rifamycin-treated infected cells and its presence was documented within the particles of the phages TW19 and TW28, representatives of the group IV, in contrast to the phage GA of the group II, which was used in this study as a control (Aoi and Kaesberg 1976).

The lysis gene is not present in the *Allolevivirus* genus representatives. In this case, the A2, or maturation, or IIa, protein is playing the lysis function and replacing the lysis protein of the *Levivirus* genus (Karnik and Billeter 1983; Winter and Gold 1983; Bernhardt et al. 2001, 2002; Reed 2012; Reed et al. 2012, 2013; Cui et al. 2017).

FULL-LENGTH SEQUENCES

The phage MS2 genome was the first sequenced full-length genome in the world (Fiers et al. 1975a, 1976, 1977; Fiers 1979). This sequence allowed first unprecedented studies on the choice of code words (Grosjean et al. 1978; Fiers and Grosjean 1979). Furthermore, a plasmid containing a nearly full-size DNA copy of the phage MS2 genome was constructed (Devos et al. 1979b). Some novel MS2 sequences are represented now in GenBank (Domingo-Calap et al. 2009; Friedman et al. 2012; Oates et al. 2016; Arhancet et al. 2017, 2018). It is remarkable that Friedman et al. (2012) described an isolate DL52 as a strain of the phage MS2, a tendency that will develop further. The genome of the phage fr was sequenced by Adhin et al. (1990b). The phage fr was the second sequenced representative of group I and demonstrated overall sequence homology of 77% with the phage MS2. The greatest divergence between the fr and MS2 phages occurred in the 5′ terminal region of the A gene, while the lysis-replicase gene overlap, the coat gene, and the central region of the replicase gene were highly conserved.

TABLE 7.1

The Full-Length Genomes of the RNA Phages

RNA Phage	Genome, nt	Maturation Protein, aa	Coat Protein, aa	A1 Protein, aa	Replicase, aa	Lysis Protein, aa	Accession Number	References
				Allolevivirus				
FI strain 4184 b	4184	438	132	332	586		EF068134	Kirs and Smith (2007)
FI strain HB-P24	4243	441	132	330	576		FJ539133	Friedman et al. (2009)
FI strain HB-P22	4241	440	132	330	576		FJ539132	Friedman et al. (2009)
NL95	4248	442	132	330	576		AF059243	Beekwilder et al. (1995, 1996b)
SP	4276	450	132	331	576		NC_004301	Inokuchi et al. (1988)
SP ancestral	4276	450	132	331	576		GQ153935	Domingo-Calap et al. (2009)
SP mutant 2	4276	450	132	331	576		GQ153933	Domingo-Calap et al. (2009)
SP mutant 3	4276	450	132	331	576		GQ153934	Domingo-Calap et al. (2009)
Qβ	4215	421	133	329	586		NC_001890	Beekwilder et al. (1995, 1996b)
Qβ	4160	420	133	329	589		AY099114	Bacher et al. (2003)
Qβ strain BZ1	4219	420	133	329	589		FJ483844	Friedman et al. (2009)
Qβ strain VK	4218	420	133	329	589		FJ483843	Friedman et al. (2009)
Qβ strain BR12	4218	420	133	329	589		FJ483842	Friedman et al. (2009)
Qβ strain HL4-9	4221	420	133	330	589		FJ483841	Friedman et al. (2009)
Qβ strain TW18	4218	420	133	329	589		FJ483840	Friedman et al. (2009)
Qβ ancestral	4198	420	133	329	589		GQ153931	Domingo-Calap et al. (2009)
Qβ mutant 1	4154	420	133	329	589		GQ153928	Domingo-Calap et al. (2009)
Qβ mutant 2	4198	420	133	329	589		GQ153929	Domingo-Calap et al. (2009)
Qβ mutant 3	4153	420	133	329	589		GQ153930	Domingo-Calap et al. (2009)
Qβ isolate Anc(P1)	4217	420	133	329	589		AB971354	Kashiwagi et al. (2014)
M11	4217	421	133	329	588		AF052431	Beekwilder et al. (1995, 1996b)
MX1	4215	421	133	329	586		AF059242	Beekwilder et al. (1995, 1996b)
				Levivirus				
BZ13 strain DL20	3458	390	130		532	63	FJ483839	Friedman et al. (2009)
BZ13 strain T72	3393	390	130		532	64	FJ483838	Friedman et al. (2009)
BZ13 strain DL10	3412	390	130		532	63	FJ483837	Friedman et al. (2009)
GA	3466	390	130		532	63	X03869	Inokuchi et al. (1986)
KU1	3486	390	130		532	64	AF227250	Groeneveld et al. (1996)
fr	3575	393	130		545	71	X15031	Adhin et al. (1990)
MS2	3569	393	130		545	75	V00642	Fiers et al. (1976)
MS2	3569	393	130		545	75	NC_001417	Kastelein et al. (1982)
DL52 (JS group)	3525	393	130		545	75	JQ966307	Friedman et al. (2012)
MS2 Arhancet	3569	393	130		545	75	LQ281049 LP869662	Arhancet et al. (2018)
MS2 Oates 2016	3569	393	130		545	75	LF706680	Oates et al. (2016)
MS2 ancestral	3569	393	130		545	75	GQ153927	Domingo-Calap et al. (2009)
MS2 mutant 1	3569	393	130		545	75	GQ153924	Domingo-Calap et al. (2009)
MS2 mutant 2	3569	393	130		545	75	GQ153925	Domingo-Calap et al. (2009)
				Unclassified *Levivirus*				
C-1 INW-2012	3523	406	133		522	65	JX045649	Kannoly et al. (2012)
Hgal1	3562	407	132		523	65	JX045650	Kannoly et al. (2012)
M	3405	395	133		520	37	JX625144	Rumnieks and Tars (2012)
PRR1	3573	417	132		540	54	NC_008294 DQ836063	Ruokoranta et al. (2006)
				Unclassified *Leviviridae*				
AP205	4268	534	131		590	35	AF334111	Klovins et al. (2002)
φCB5	3762	372	123		655	136	HM066936	Kazaks et al. (2011)
LeviOr01	3669	463	129		544	72	LT821717	Pourcel et al. (2017)
PP7	3588	449	128		552	55	X80191	Olsthoorn et al. (1995)

(Continued)

TABLE 7.1 *(Continued)*

The Full-Length Genomes of the RNA Phages

RNA Phage	Genome, nt	Maturation Protein, aa	Coat Protein, aa	A1 Protein, aa	Replicase, aa	Lysis Protein, aa	Accession Number	References
AVE000 partial	4977	314	126 ? 168 ?		728		KT462694	Krishnamurthy et al. (2016)
AVE001 partial	5021	>805	203 ?		609		KT462695	Krishnamurthy et al. (2016)
AVE002 partial	4220	427	141 ? 210 ?		468		KT462696	Krishnamurthy et al. (2016)
Unclassified *Leviviridae* ?								
EC	>3180	430	138		>386		KF616862	Greninger and DeRisi (2015)
MB	3925	452	128		652		KF510034	Greninger and DeRisi (2015)

Note: The assembled genome EC and *Leviviridae sp.* representatives AVE000, AVE001, and AVE003 are included, although they are partial. The data are compiled from the NCBI taxonomy browser for the *Leviviridae* family (https://www.ncbi.nlm.nih.gov/Taxonomy/Browser/wwwtax.cgi?id=11989), with the exception of the assembled genomes EC and MB that still do no not appear under the *Leviviridae* family.

The first full-length Qβ sequence was achieved by Philipp Mekler (1981) in his PhD thesis under the guidance of the famous Martin A. Billeter in Zürich, but this genome never appeared as a journal paper and is not included in GenBank up to now. The Qβ genome, with amendments in nine single nucleotide positions, privately communicated by Billeter, was used by the analysis of the full-length SP genome, the first fully sequenced representative of the RNA phage group IV (Inokuchi et al. 1988). Later, a number of novel Qβ genomes were uploaded to GenBank (Beekwilder et al. 1995, 1996; Bacher et al. 2003; Friedman et al. 2009; Domingo-Calap et al. 2009; Kashiwagi et al. 2014). The phages M11 and MX1 from group III and the phage NL95 from group IV were sequenced by Beekwilder et al. (1995, 1996b).

The genome of the phage FI from group IV was sequenced by Kirs and Smith (2007) and by Friedman et al. (2009). Again, Friedman et al. (2009) regarded two independent phage isolates HB-P22 and HB-P24 as strains of the phage FI, by analogy to the above-mentioned isolate DL52 as a strain of the phage MS2. In a similar way, Friedman et al. (2009) characterized the phage isolates BZ1, VK, BR12, HL4-9, and TW18 as strains of the phage Qβ.

The first full-length genome of a representative of the RNA phage group II, namely of the phage GA, was published by Inokuchi et al. (1986). The phage KU1 genome, the next fully-sequenced representative of group II, was reported by Groeneveld et al. (1996). The KU1 genome had an insertion of 18 nucleotides in the start codon of its lysis gene as the most conspicuous difference in comparison with other group II members, such as GA and JP34; partial sequence of the latter was published earlier by Adhin et al. (1989). In the phages GA and JP34, the coat and lysis genes overlapped by one nucleotide in the configuration UAAUG. The 18-nt insertion in the KU1 genome was positioned between the A and the U of the start codon. It did not affect the coat reading frame, but it destroyed the AUG start codon and separated the previous overlapping of the genes

by 17 nucleotides. The insert created a UUG codon at its 3′ border, which served as the start site for the lysis protein synthesis in KU1.

Continuing the group II story, three isolates—DL10, DL20, and T72—were subjected to full-length sequencing and fixed as strains of the group II phage BZ13 (Friedman et al. 2009).

Two full-length genomes that are regarded now by the present taxonomy as unclassified *Levivirus* species, namely C-1 and Hgal1, were sequenced by Kannoly et al. (2012). The two phages were R-plasmid-dependent and required conjugative plasmids of the incompatibility groups IncC and IncH, respectively. The evident similarity of the C-1 and Hgal1 phages to the representatives of the *Levivirus* genus was declared clearly by the authors. Thus, the lysis function of the predicted lysis genes was confirmed experimentally. However, both phages were more related to the IncP-plasmid-dependent phage PRR1, which was routinely described as a pseudomonaphage, rather than the classical coliphage of the *Levivirus* genus. In full agreement with the previously postulated idea (Bollback and Huelsenbeck 2001), Kannoly et al. (2012) thought that the *Levivirus*-like genome organization was ancestral, and the *Allolevivirus*-like genomes were derived from the *Levivirus*-like genomes.

The IncM plasmid-dependent RNA phage M possessed the smallest known *Leviviridae* genome to date, namely 3405 nucleotides, but had the typical genome organization with the maturation, coat, and replicase genes in the 5′ to 3′ direction (Rumnieks and Tars 2012). The lysis gene was located in a different position than in other known *Leviviridae* phages and completely overlapped with the replicase gene in a different reading frame. Again, similarity to other known RNA phage sequences was rather low. Nevertheless, some similarity in protein sequences and RNA secondary structures at the 3′ untranslated region placed phage M together with phages specific for IncP, IncC, and IncH, but not IncF plasmid-encoded pili. It was suggested that the phage M represented a lineage that branched off early in the course of RNA phage specialization on different conjugative plasmids (Rumnieks and Tars 2012).

The full-length genome of the pseudomonaphage PRR1 was presented by Ruokoranta et al. (2006). In fact, the phage PRR1 had a broad host range due to the promiscuity of the receptor encoded by the IncP plasmid. By analogy to the *Levivirus* genus, the phage PRR1 genome demonstrated the same order of the classical three-gene combination: maturation, coat, and replicase with an overlapping lysis gene. Nevertheless, some characteristics of the phage PRR1 made it more similar to the members of the *Allolevivirus* genus. Thus, the PRR1 coat protein was closer to the Qβ and SP coats, while the replicase and lysis proteins were closest to the group II phages.

The group of the full-length genomes belonging to the unclassified *Leviviridae* species includes two phages that are regarded as pseudomonaphages. Thus, the line of the full-length pseudomonaphage genomes was started with the phage PP7 (Olsthoorn et al. 1995a). The obvious three open reading frames (ORFs) that coded for the apparent protein homologs of the RNA coliphages, i.e., maturation, coat proteins, and replicase, were identified. A fourth overlapping reading frame existed that probably encoded a lysis protein, similar to what has been found in the *Levivirus* genus. However, there was no significant nucleotide sequence identity between PP7 and the coliphages except for a few regions where homologous parts of proteins were encoded, most notably in the replicase gene. In these regions the nucleotide sequence similarity between PP7 and MS2 was no greater than between PP7 and the *Allolevivirus* phage Qβ. Moreover, Qβ and MS2 were no closer to each other than they were to PP7. Olsthoorn et al. (1995a) suggested that the phage PP7 branched off before the *Levivirus* and *Allolevivirus* coliphages diverged into separate groups. The genome of the pseudomonaphage LeviOr01 was published by Pourcel et al. (2017) and demonstrated the same four-gene structure as the phages PP7 and PRR1. This phage was distinguished by its unique property to establish the carrier state in the infected bacteria.

The genome of the only acinetophage AP205 was sequenced by Klovins et al. (2002) and demonstrated the characteristic three-gene combination for the maturation, coat, and replicase structure. However, it was distinct from both *Allolevivirus* and *Levivirus* genera by the appearance of the lysis gene at the 5′ terminus, preceding the maturation gene, but not at the usual position, by overlapping the coat and replicase proteins. After the putative lysis gene, one more ORF with unknown function appeared at the 5′ terminus of the AP205 genome. Other new features concerned the 3′-terminal sequence, where all RNA coliphages had three cytosine residues, but in AP205 there was only a single terminal cytosine. The AP205 genome did not have significant sequence similarity to coliphages, although important secondary structural features of the RNA were conserved—for example, 3′ untranslated region and the replicase-operator hairpin. Although AP205 had the genetic map of a levivirus, its 3′ untranslated region has the length and RNA secondary structure of an allolevivirus, and the phage AP205 was placed between Qβ and MS2 in an evolutionary sense (Klovins et al. 2002).

One of the caulophages, the φCB5, was subjected to full-length sequencing by Kazaks et al. (2011). The φCB5 revealed classical three-gene structure, but unlike other *Leviviridae* members, the lysis protein gene overlapped entirely with the replicase in a different reading frame. Moreover, the φCB5 lysis protein was about two times longer than that of the distantly related MS2 phage and presumably contained two transmembrane helices.

If all genomes described above belonged to the real phage isolates, the next and most recent group of the RNA phage sequences appeared as a result of the metagenomic search. First, Greninger and DeRisi (2015) assembled two draft genomes of the marine phages EC and MB from San Francisco wastewater. The EC and MB genomes were aligned by BLASTx to the phages Hgal1, M11, and Qβ and comprised a typical *Leviviridae* genome organization, consisting of the three-gene composition, with phage EC likely truncated by ~600 nucleotides at the 3′ end. However, both phages did not encode a readily apparent lysis or readthrough protein. The predicted replicase protein of the phage MB demonstrated 39% amino acid identity to that of the phage φCB5 over the entirety of the replicase protein, while the closest replicase to the phage EC demonstrated 42% amino acid identity to the phage Hgal1. Both replicases demonstrated <38% identity to members of the *Leviviridae* family already placed in a genus, which was equivalent to amino acid identity between *Allolevivirus* and *Levivirus* genera, consistent with the idea that these two viruses could form two novel genera within the *Leviviridae* family. The predicted maturation proteins were aligned with 32% amino acid identity to the phage BZ13 (EC) and 28% to the caulophage φCB5 (MB). The predicted phage EC coat protein aligned with 32% amino acid identity to the phage C-1, while the predicted phage MB coat protein failed to demonstrate significant amino acid alignments to any phage proteins (Greninger and DeRisi 2015).

By the metagenomic analysis of the samples collected from a range of ecological niches worldwide, including invertebrates and extreme microbial sediment, Krishnamurthy et al. (2016) have identified partial genome sequences of 122 RNA phage phylotypes that were highly divergent from each other and from the previously described RNA phages. Three partial genomes from this paper, namely AVE000, AVE001, and AVE002, are included in Table 7.1, since they are the longest and clearly demonstrate novel genome organization features. Thus, the genomes AVE000 and AVE001 are at least 4.95 and 5.02 kb long, respectively. The long AVE000 genome can be attributed to the presence of a novel >1.20 kb ORF of unknown function that is 5′ to and partially overlaps the maturation protein. The AVE001 genome is also expanded due to the presence of a strikingly large 2.39 kb ORF containing the maturation domain, which is larger than all the reference RNA phage maturation genes. The AVE002 is the first RNA phage to contain two nonoverlapping ORFs between the maturation and replicase genes, while neither of the two ORFs has discernable similarity to known proteins. While one of these ORFs likely represents the coat protein, the other ORF might represent a novel lysin or have homologous function to the Qβ read-through protein (Krishnamurthy et al. 2016).

PECULIARITIES OF THE GENOME

Figure 7.1 presents detailed graphical structures of the RNA phage genomes, but, in contrast to Table 7.1, the genomes are grouped by their host specificity but not by the current taxonomy principles, although the data are compiled from the same NCBI taxonomy browser.

The size of the microbiologically characterized RNA phage genomes varies from the smallest: 3405 nt of the phage M and 3458 of the phage BZ13 to the largest ones: 4268 nt of the phage AP205 and 4276 nt of the phage SP. The metagenomic data show that the RNA phage genomes may exceed 5000 nt by the strongly different genome organization. The three novel assembled genomes AVE000, AVE001, and AVE002 are intended to illustrate the possible broad deviations of the putative RNA phage genomes from the basic three-gene scheme.

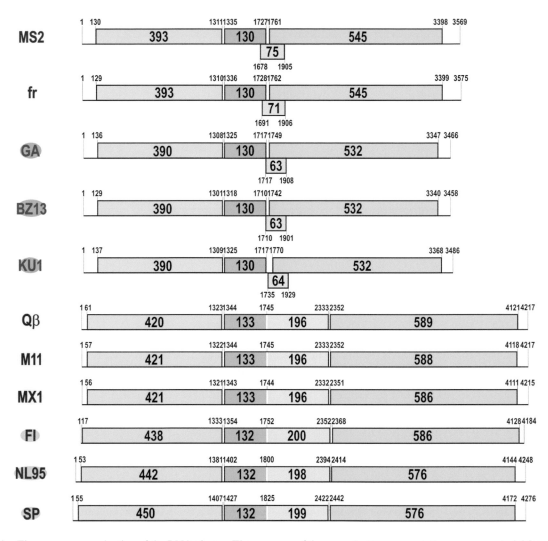

FIGURE 7.1 The genome organization of the RNA phages. The genomes of the group I–IV representatives are presented. Maturation, coat, replicase, and lysis genes are shown to scale and colored green, red, blue, and beige, respectively. Read-through part of the A1 protein by the *Allolevivirus* representatives is also colored beige, by analogy to the lysis gene. Length of the corresponding proteins is given by large numbers, but numbering of their location within the genome is indicated by small numbers. The specific colors of the RNA phage representative names are the same as in Figure 5.1 and indicate the group and/or the host. The genomes are collineated by the position of the coat protein gene. The data are compiled from the NCBI taxonomy browser. It is necessary to consider that the experimentally determined lengths of the phage coat proteins are always one aa residue shorter than the actual proteins in the database because the N-terminal methionine is cleaved off in infected *E. coli* cells. This explains some discrepancies in the coat protein numbering in different published works. In cases where more than one full-length genome was available, the following sequences were chosen to show here: BZ13 strain DL20 (Friedman et al. 2009), MS2 (Fiers et al. 1976), FI (Kirs and Smith 2007), SP (Inokuchi et al. 1988), and Qβ (Kashiwagi et al. 2014). The unassigned *Levivirus* (Hgal1 and C-1 INW-2012) and *Leviviridae* (all other) representatives are presented. The genomes acquired by the metagenomic sequencing have no color indication. The hypothetical genes are white. Three sequences of the *Leviviridae sp.* AVE000, AVE001, and AVE002 (Krishnamurthy et al. 2016) and the assembled genome EC are included, although they are not full-length. *(Continued)*

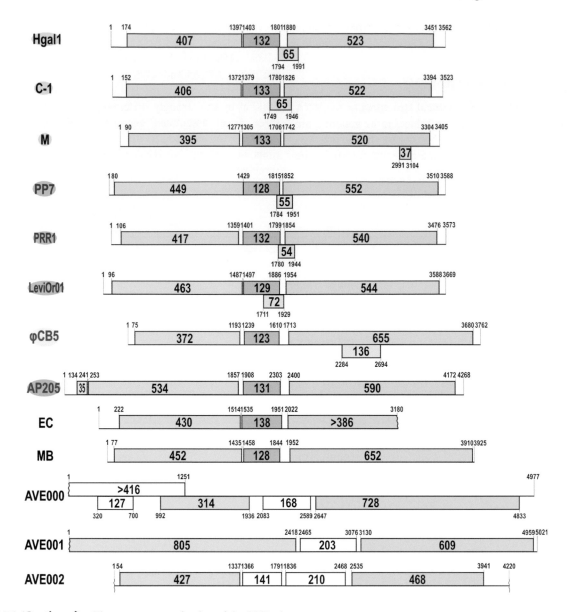

FIGURE 7.1 (Continued) The genome organization of the RNA phages.

There are noncoding regions at both ends of all fully-sequenced genomes. However, the length of these regions varies over rather wide limits. Thus, the 5′-untranslated regions could be from the shortest: 16 (the phage FI), 52 (the phage NL95), 54 (the phage SP), and 55 (the phage Qβ) nucleotides to the largest: 151 (C-1 INW-2012), 173 (Hgal1), and still 221 (the assembled genome EC) nucleotides. The 3′-untranslated regions vary by length from 15 (the assembled genome MB) to 144 (the assembled genome AVE000), 171 (the phage MS2), and 176 (the phage fr) nucleotides. However, many phages retain this length of about 100 nucleotides; for example, such numbers of the 3′-untranslated region length are popular as 104 (MX1, NL95, SP, Hgal1, C-1 INW-2012) or 118 (BZ13, KU1) nucleotides.

The order of genes by the well-attributed RNA phage genomes is always preserved as the maturation-coat-replicase in the 5′ to 3′ direction. It is not excluded that the same direction could be maintained by novel genomes such as AVE000,

AVE001, and AVE002. The position of the lysis gene is variable: mostly it overlaps the coat and/or replicase genes but it is located in-frame as the first 5′-terminal gene in the case of the phage AP205. In the *Allolevivirus* genus, no special lysis gene is present, but the lysis function can be performed by the A1 protein. No traces of the lysis gene or A1-like structure were found in the assembled genomes EC and MB.

Concerning the size of the RNA phage genes, the lysis gene varies from the smallest 35 (AP205) and 37 (M) to the largest 136 (φCB5) amino acid residues. The size of the maturation gene varies over a wide range: from minimal 314 (AVE000) to maximal 534 (AP205) or just more than 805 (AVE001) amino acid residues, while the classical representatives of the *Levivirus* and *Allolevivirus* genera possess maturation proteins of about 390 and 420–450 amino acid residues, respectively. The same is true for the replicase gene: it may vary from 520 (M), 522 (C-1 INW-2012), and 523 (Hgal1) to 652 (MB), 655 (φCB5) and still 728 (AVE000) amino acid residues. The classical *Levivirus* and *Allolevivirus* genomes have replicases of 532–545 and 576–589 amino acid residues, respectively. The coat genes are less variable. The smallest coat is 123 (φCB5) and the largest one is 138 (EC). Other coats are varying slightly in the range of 128–133 amino acid residues.

Two of the three RNA phage genes start with AUG (coat and replicase) and one with GUG (maturation), while they terminate with UAG or UAA, with the exception of the *Allolevivirus* coat, which is terminated by UGA and is subjected deliberately to the readthrough process. All the synonymous codons are used, and with frequencies that might be expected on a more or less random basis.

Each gene is separated from the next by intervals that are the first intergenic spaces that have been noted in molecular biology. These intervals vary from nothing (the maturation–coat space in PP7) to 107 (the coat–replicase space in MB) nucleotides. Most of the spaces are in the range of 20–40 nucleotides.

SIMILARITIES AND DISSIMILARITIES

The first statistical estimate of phylogeny for the *Leviviridae* family using maximum likelihood and Bayesian estimation was performed in a classical study by Bollback and Huelsenbeck (2001). At that time, the *Leviviridae* family appeared as a monophyletic group consisting of two clades representing the genera *Levivirus* and *Allolevivirus*, with the pseudomonaphage PP7 diverged from a common ancestor with the coliphage prior to the ancient split between these genera and their subsequent diversification. In fact, the phage PP7 was the only sequenced non-coliphage genome at that time. Bollback and Huelsenbeck (2001) regarded the *Levivirus* genus-like genomes as ancestors that catalyzed subsequent changes, which led to the current RNA phage genome organization and gene expression.

Figure 7.2 presents the standard Clustal W alignment result for the RNA phage genomes taken from Table 7.1. This simple Clustal W analysis cannot be regarded as a proper

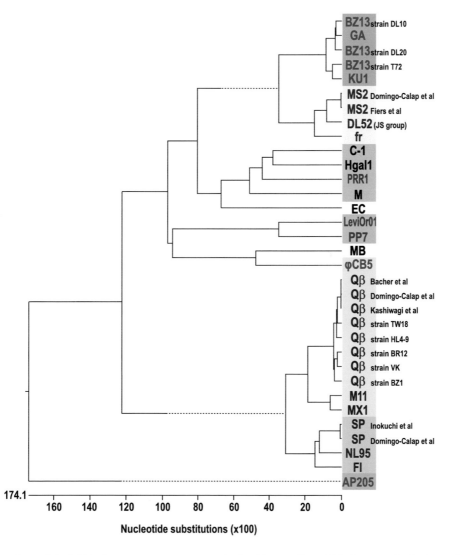

FIGURE 7.2 The comparison of the RNA phage genome sequences obtained by the Clustal W alignment in the MegAlign program from the DNASTAR Lasergene software suite. The designations and colors of the genomes correspond to those used in Table 7.1 and Figure 7.1.

phylogenetic reconstruction, but it illustrates nicely the idea that the phage AP205 is the most distant member of the family, when the most distant genomes AVE000, AVE001, and AVE002 are not involved into the alignment.

Thus, after the AP205 genome, other genomes demonstrate clear separation into *Allolevivirus* and *Levivirus*-like branches, where the four coliphage groups I–IV are clearly parted. The three pseudomonaphages and the caulophage φCB5, as well as the non-classified phages C-1, Hgal1, and M and the assembled genomes EC and MB are involved in sub-branches within the *Levivirus*-like branch. Moreover, when the assembled genome AVE001 was included in the alignment procedure, it appeared rather within the *Levivirus* branch as a neighbor of the pseudomonaphages. The assembled genomes AVE000 and AVE002 were, however, very far from all other genomes.

8 Genetics and Mutants

The difficulty in life is the choice.

George Moore
The Bending of the Bough, act IV, 1900

Real is what can be measured.

Max Planck

CONDITIONALLY LETHAL MUTATIONS

The concept of conditionally lethal mutations was established first by the work on the DNA phages λ (Campbell 1961) and T4 (Benzer and Champe 1962; Epstein et al. 1963) and alkaline phosphatase (Garen and Siddiqi 1962) in the early 1960s, just at the time when the RNA phages were discovered. The work on the RNA phage mutants was started very soon after their initial characterization and stimulated further rapid progress of the theory and praxis of the conditionally lethal mutations and chemical mutagenesis.

Until the 1960s, the genetic analysis of all bacteriophages rested upon a few types of mutants, those with an altered host range and those with a different plaque morphology, and inferences with regard to function were difficult to make: while mutants with altered host range were often related to phage attachment structures, those with altered plaque morphology might reflect any aspect of phage growth (Zinder 1965). Remarkably, the antiserum-resistant mutants of the phage f2 were isolated early (Valentine and Zinder 1964). Concerning the host range mutants of the RNA phages, the latter have never been known to mutate to infect F⁻ bacteria, although about 200 phage-resistant bacterial mutants from an *E. coli* HfrC strain were selected, but no phage mutants could be found that would attack these strains (Zinder 1965). As for the plaque morphology mutants, two types have been observed initially, those that produced a more turbid and those that produced a clearer plaque, but neither of these two types has proven to be stable and contained about 10^{-3} revertants (Valentine et al. 1964; Zinder 1965). Later, a morphologic mutant of the phage MS2, the so-called *rim* mutant, was isolated (Sherban 1969) and classified later as a coat protein missense mutant (Dishler et al. 1980b).

The situation changed cardinally with the introduction of two general classes of the conditionally lethal mutants: the temperature-sensitive (*ts*) mutants and the host-dependent (*amber*) mutants (Luria 1962; Epstein et al. 1963). By the classical review of Norton D. Zinder (1965) on the newly appearing conditionally lethal mutants with respect to the RNA phages,

the advantage of such mutants is that they should be obtained in all cistrons without necessarily any knowledge of cistron function. Any gene which is responsible for the production of protein might mutate to produce a protein with a lower

thermostability than that produced by the wild type. Host-dependent mutants depend on the presence in the bacteria of specific suppressor genes which relate to particular code words rather than to specific functions. These suppressor genes are assumed to modify some aspect of the protein synthesizing mechanism in such a way that a code word which is uninterpretable in the normal host (nonpermissive) can be interpreted in the mutant host (permissive).

Therefore, the conditionally lethal mutants acted as lethal ones under a set of prescribed conditions and appeared viable under another set of conditions.

From the very beginning, both *amber* and *ts* mutants of the RNA phages were obtained by chemical mutagenesis. The *ts* mutants of the phage f$_{can1}$ were isolated after mutagenesis with 5-fluorouracil: with phage grown in the presence of 10^{-4} M 5-fluorouracil, 25% of the yield were *ts* mutants (Davern 1964a). The nitrous acid treatment also induced *ts* mutants of the phage MS2 (Pfeifer et al. 1964) and f2 (Zinder 1965). Hydroxylamine was also used successfully as a chemical mutagen at that time (Zinder, unpublished, cited by Horiuchi 1975).

According to the above-mentioned studies, the conditionally lethal mutants were all relatively unstable and the stock cultures had about 10^{-3} revertants, since such mutants were rarely restored to full activity even when grown at permissive conditions and tended therefore to have a negative selective value (Zinder 1965). Some *ts* did not revert and were assumed to be either multiple or deletion mutants (Horiuchi 1975).

The first *amber* mutants, induced by nitrous acid treatment, were described by Zinder and Cooper (1964) for the phage f2. These mutants served to identify all the genetic functions of the phage. Furthermore, *opal* (Model et al. 1969) and *azure* mutants (Horiuchi and Zinder 1967) were obtained. However, no successful isolation of the *ochre* mutants has been reported (Horiuchi 1975).

The complementation and physiological analyses of the *amber* and *ts* mutants of the RNA phages have established three complementation groups and the existence of the three major RNA phage genes: maturation (or A), coat, and replicase by the phages f2 (Horiuchi et al. 1966), R17 (Gussin 1966; Gussin et al. 1966), and Qβ (Horiuchi and Matsuhashi 1970). This was possible due to the three important conclusions that were made at the early steps of the RNA phage genetic studies. First, the RNA molecules were not necessarily responsible for the synthesis of their individual coat proteins: the complementation might therefore also occur with the coat mutants (Valentine et al. 1964), and the rescued phage genomes could be enclosed in the protein shell encoded by the helper phage (Valentine and Zinder 1964). Second, the synthesis of the replicase subunit ceased long before much virus RNA, coat, or particles have been produced. The progeny RNA therefore did not synthesize the phage-specific replicase subunit in normal infection, but some of the f2 coat *amber* mutants caused hyperproduction of the replicase subunit (Lodish et al. 1964). Moreover, the tempo of the mutant incidence was inspired by very high mutation rates, about 10^{-3} bp/replication (Drake 1993), since the RNA phages did not have a proofreading mechanism, and the sequence diversity was tempered therefore only by the conservation of the most crucial structural and regulatory elements within the viral genome and specific proteins. Third, the three proteins were detected by the *in vivo* and *in vitro* synthesis and were associated, with the help of the phage mutants, with one of the three genes (Nathans et al. 1966, 1969; Oeschger and Nathans 1966; Viñuela et al. 1967, 1968; Fromageot and Zinder 1968; Viñuela 1968; Horiuchi et al. 1971). These proteins were characterized exhaustively by polyacrylamide gel electrophoresis.

The viability of the coat *amber* mutant-produced products was dependent on the type of *amber* suppression. While Su-1 (Weigert and Garen 1965) suppressor, which inserted serine residues at the *amber* codon position and led therefore to the glutamine → serine substitution, was valid, the Su-3 suppressor that inserted tyrosine residues and provoked glutamine → tyrosine substitution (Weigert et al. 1965) did not ensure the viable product in the coat protein mutants. Direct proof for the insertion of serine instead of glutamine by the f2 coat *amber* mutant in the Su-1 cells was published by Notani et al. (1965). The Su-2 suppressor did not change the amino acid residue within the *amber*-mutated proteins, since it inserted the same glutamine residue at the *amber* position, although the suppression efficiency was lower than in the case of the Su-1 and Su-3 suppressors (Weigert et al. 1965).

Later, a fourth complementation group was defined by a mutant of the phage f2 that grew only on the UGA suppressing strains but failed to lyse the non-permissive hosts while filling them with normal phage particles (Atkins et al. 1979b; Beremand and Blumenthal 1979; Model et al. 1979). This mutation occurred in the lysis gene whose reading frame was different from other genes and overlapped the known genes of the coat and/or replicase proteins.

POLAR MUTATIONS

The polar mutants that depressed the synthesis of gene products, which have been assumed to be controlled by a polycistronic message, were described in the histidine operon just at the time of the first molecular experiments on the RNA phages (Ames and Hartman 1963). By the RNA phage f2, a definite coat protein mutant hyperproduced replicase when it infected a strain with inacceptable permissive qualities, the nonsense triplet probably being read as nonacceptable missense, but this mutant did not hyperproduce replicase and/or RNA when it infected nonpermissive host (Lodish et al. 1964, Lodish and Zinder 1965b). Moreover, such mutants did not complement both coat and replicase mutants. Therefore, the coat protein was found to have the polar effect on the replicase synthesis, and some of the coat *amber* mutations appeared as being polar. The polar effect was further explained by the

necessity to translate the N-terminal coat part to expose the initiation region of the replicase gene.

GENE ORDER

At first, Ohtaka and Spiegelman (1963) interpreted the predominant synthesis of coat protein as resulting from a sequential translation of a polycistronic RNA, where the first gene at the 5′-end of the genome encodes the coat protein. This hypothesis was supported by the observation of a definite lag in the incorporation of histidine relative to that of other amino acids (histidine residue was not present in the coat proteins of the studied phages). In fact, this effect was a consequence of the polar effect of the coat onto the replicase synthesis. Nevertheless, the erroneous 5′-coat-maturation-replicase-3′ order of the RNA phage genes was accepted for some period (Engelhardt et al. 1968; Lodish 1968d; Robertson and Zinder 1969; Webster et al. 1969).

The problem was raised by the fact that the absence of genetic recombination, one of the major tools of genetic analysis, "has almost completely shut off the area of formal genetics of RNA phage" (Horiuchi 1975). According to the latter review, all attempts to detect genetic recombination between the RNA phage mutants have failed. The presence of high levels of revertants in the *amber* mutant stocks made the experiment difficult since the production of recombinants below the level of reversion was impossible to detect. Although many of the *ts* mutants did not revert ($<10^{-9}$) and were assumed to have either several *ts* mutations in a genome or a deletion, no recombinant was obtained even when these mutants were used. At last, it was concluded that the frequency of recombination, if any, in RNA phages was very low (Horiuchi 1975). Attempts to detect multiplicity reactivation of the phage f2 have also failed (Horiuchi, unpublished results, cited by Horiuchi 1975).

The correct order of the RNA phage genes was approved finally by the structural rather than the genetic methods as the 5′-maturation-coat-replicase-3′ in the case of the following phages: R17 (Gesteland and Spahr 1969; Jeppesen et al. 1970b), f2 (Lodish and Robertson 1969b; Robertson and Lodish 1970), M12 (Konings et al. 1970), and Qβ (Hindley et al. 1970; Staples et al. 1971). In doing this, the sequence analyses were used (Hindley et al. 1970; Jeppesen et al. 1970; Staples et al. 1971), while the positions of mutations within a gene were determined either by amino acid sequence analysis of a phage protein (Notani et al. 1965; Tooze and Weber 1967) or from the size of *amber* fragments produced in a cell-free, protein-synthesizing system (Webster et al. 1967; Lodish and Robertson 1969a; Model et al. 1969; Horiuchi et al. 1971).

FREQUENCIES OF MUTATIONS

Horiuchi (1975) presented an interesting summary of the appearance of nonsense mutations in each gene after usage of different chemical mutagens. This review is presented in Table 8.1. The distribution of mutants among genes was uneven. Thus, among *amber* mutants isolated from f2 or R17 after nitrous acid treatment, more than half the mutations were in

TABLE 8.1

Frequency of Appearance of Nonsense Mutations in Each RNA Phage Gene

Mutation	amber	amber	amber	opal	amber	
Phage	f2	R17	R17	f2	Qβ	
Mutagen	HNO₂	HNO₂	fluorouracil	fluorouracil	HNO₂	
Suppressor strain used	Su-1	Su-1	Su-1	Su⁺_UGA	Su-1	Su-3
			Number of Mutants			
Replicase	1	4	1	1	8	9
Maturation	6	22	25	6	1	2
Coat	5	9	6	0	6	0
Total	12	35	32	7*	15	11
Reference	Zinder and Cooper (1964)	Gussin 1966; Tooze and Weber (1966)	Tooze and Weber (1966)	Model et al. (1969)	Horiuchi and Matsuhashi (1970)	

*Op-3 mutant (discussed in text below) was omitted.

Source: Adapted with permission of *Cold Spring Harbor Laboratory Press* from Horiuchi K. In: *RNA Phages*. Zinder ND (Ed). Cold Spring Harbor Laboratory, Cold Spring Harbor, NY, USA, 1975, pp. 29–50.

the maturation protein gene, and less than 10% of them were in the replicase gene (Horiuchi 1975). Considering that the *amber* codon should derive from either a CAG or UGG codon, Horiuchi (1975) calculated that one would expect more than half of the *amber* mutations in the replicase gene and only 10% in the coat gene. However, this was clearly not the case for both the nitrous acid and fluorouracil treatment, although fluorouracil was assumed to cause mutations by base substitution during replication. A similar situation was observed for *opal*, or UGA, mutagenesis, where CGA codons were the targets of the mutagenic agents (Table 8.1).

Surprisingly, an entirely different bias was observed in the case of the phage Qβ, when more than half of the *amber* mutations obtained from Qβ with nitrous acid were in the replicase gene and less than 10% of them were in the maturation gene (Table 8.1). The uneven distribution of the induced mutations was explained by the existence of *hot spots* within the genome (Horiuchi 1975). Thus, there were 17 coat ambers of the phages R17 and f2 which were located at any of four sites (Tooze and Weber 1967; Webster et al. 1967). All four locations were sites where a glutamine residue was present in the wild type. Nine out of the seventeen mutants were at the position 6 (polar mutation), while other mutations were nonpolar and located in the following way: two of them were at the position 50, five were at the position 54, and one was at the position 70. The position 6 appeared therefore to be the *hot spot*. It is necessary to consider that the traditional numbering of the coat positions used here does not take into account the first methionine, which we considered by the numbering of the coat genes in Chapter 7.

COMPLEMENTATION

According to the valuable statements by Kensuke Horiuchi, one of the pioneers of RNA phage genetics, the complementation tests between *amber* mutants of the RNA phages have not been technically very easy (Horiuchi 1975). The results have been rather ambiguous because, first, the stocks of mutants

usually contained wild-type revertants at high frequencies, i.e., 10^{-3} or higher, sometimes 10^{-2}; second, the number of phages produced as the result of complementation was usually 5–300 per cell, much smaller than the enormous burst size (ca. 2,000) of wild-type phage; and third, in many cases the fraction of the cell population that was really infected by two phages was probably not very large. For instance, when permissive cells were mixedly infected with two *amber* mutants of Qβ at a multiplicity of 5 for each and were analyzed by the single burst technique, only 20% of the bursts were found to contain both mutants (Horiuchi and Matsuhashi 1970). Therefore, it was not easy to detect the production of progeny mutant phages or to demonstrate that they were not due to rescue by wild-type revertants (Horiuchi 1975). These difficulties were overcome by application of single-burst or quasi-single-burst techniques to the complementation tests (Valentine et al. 1964). Though this method was much more laborious than the usual complementation tests used for other phages, the complementation analyses of the RNA phage *amber* mutant were carried out using this technique, and the three genes were unambiguously demonstrated (Gussin 1966; Horiuchi and Matsuhashi 1970). It is noteworthy to add that Tooze and Weber (1967) examined 55 *amber* mutants of the phage R17 in total and did not find the fourth cistron.

No successful complementation between mutants of the phage Qβ and mutants of the phages f2 or R17 has been reported, whereas mutants of the phages f2 and MS2 do complement each other (Model et al. 1969).

FUNCTION OF MUTANT GENES

The phages with a mutation in the replicase gene, or gene 1 by the early classification (for more explanation see Horiuchi 1975), did not synthesize any phage RNA, including double-stranded RNA, in the infected cells under nonpermissive conditions (Gussin 1966; Horiuchi et al. 1966; Lodish and Zinder 1966b; Horiuchi and Matsuhashi 1970).

The replicase mutants did not convert the parental phage RNA into a double-stranded form under nonpermissive conditions (Lodish and Zinder 1966b). It was concluded therefore that the parental RNA was first translated inside the cell to produce phage RNA replicase, which then synthesized a complementary strand, thereby converting the parental RNA into the double-stranded form, and that the gene 1 coded for this RNA replicase. The replicase activity was observed in crude extracts of the wild-type phage-infected cells (August et al. 1963; Haruna et al. 1963; Weissmann et al. 1963a,b), but not in the nonpermissive cells infected with mutants of this group (Gussin 1966). The RNA extracted from *amber* or *opal* mutants of gene 1 failed to direct the synthesis of the replicase in cell-free extracts but produced a new, smaller peptide fragment, while addition to the extracts of suitable suppressor tRNA caused the synthesis of the complete replicase molecule (Model et al. 1969; Horiuchi et al. 1971).

The mutants of the maturation gene, or gene 2 by the early classification, produced, under nonpermissive conditions, defective phage particles that did not adsorb to host cells (Zinder and Cooper 1964; Heisenberg and Blessing 1965; Lodish et al. 1965). The defective particles demonstrated lighter density in CsCl gradient and contained smaller and noninfectious RNA, although such particles were recognized normally by anti-phage antibodies (Engelhardt and Zinder 1964). However, temperature shift experiments with *ts* mutants showed that the function of the gene 2 was a late one, and that the mutants synthesized a normal amount of infectious phage RNA (Horiuchi et al. 1966). Furthermore, it was found by careful purification from ribonuclease I−deficient bacteria that the particles had normal CsCl density and contained infectious RNA. When treated with ribonuclease, these particles became less dense, since their RNA was cleaved and some of it was lost from the particle (Argetsinger and Gussin 1966; Heisenberg 1966). It was concluded that the protein coded for by this gene was necessary for the proper assembly of intact phage particles, and that the improperly assembled particles in the absence of this protein were not only unable to adsorb to the cells but were also susceptible to the ribonuclease treatment. Thus, the gene 2 protein, which was a hypothetical entity at that time, was named maturation protein (Horiuchi 1975).

Furthermore, Nathans et al. (1966) identified the product of the gene 2 with the protein II in polyacrylamide gels and found it in purified phage particles as a minor component. Joan Argetsinger Steitz (1968b,c) isolated this protein from purified phage particles and demonstrated that the defective particles produced by the *amber* mutants of the gene 2 did not contain this protein. Moreover, she estimated the number of maturation protein monomers per phage particle as one. By addition of the isolated maturation protein to the reconstitution system of the functional phage from RNA and coat protein (Hohn 1967; Sugiyama et al. 1967), Roberts and Steitz (1967) were able to increase the yield of viable phage particles by more than two orders of magnitude, showing that the maturation protein was a necessary component of the phage particle. Remarkably, the maturation protein mutants of the

phages f2 and R17 lysed the nonpermissive cells as well as the wild-type phage did (Zinder and Cooper 1964; Gussin 1966).

According to Horiuchi (1975), the gene 2 mutants were generally leaky, in the sense that they produced one to two viable progeny phages per infected cell under nonpermissive conditions. Thus, when Su⁻ cells were infected with the gene 2 *amber* mutants and plated for infective centers on an Su⁺ indicator, almost all the infected cells formed plaques. This did not hold true for the gene 2 *amber* mutants of the phage Qβ, which did not form infective centers under the same conditions; neither did they lyse nonpermissive cells (Horiuchi and Matsuhashi 1970).

The mutants of the gene 3 were first identified as coat mutants by the amino acid analysis of bulk protein extracted from purified sus3 mutant of the phage f2 grown in Su-1 host, where one glutamine residue in wild-type coat had been replaced by a serine residue in the mutant (Notani et al. 1965). Zinder and Cooper (1964) found before that the *amber* mutants of this gene plated only on Su-1 hosts. Their plating efficiency on Su-2 host was about 0.1 of that on Su-1, and on Su-3 the number of plaques was almost equal to that of revertants, and this rule also held true for all six coat *amber* mutants isolated from Qβ (Horiuchi et al. 1975). Using a Qβ stock mutagenized with nitrous acid, 6 out of 15 *amber* mutants isolated on Su-1 host were coat *amber* mutants, while none of 11 *amber* mutants isolated on a Su-3 host were coat *amber* mutants. This plating property of the coat *amber* mutants has been used as a criterion for quick preliminary classification of newly isolated *amber* mutants (Horiuchi 1975). This phenomenon was explained by the incompatibility of the coat protein function with the replacement of a glutamine residue by a tyrosine, since the efficiency of suppression by Su-3 of 55% was at least as high as the Su-1 efficiency of 28%, while the poor growth on Su-2 host was attributed to the poor efficiency of the Su-2 suppression, namely 14% (Horiuchi 1975). The efficiency of suppression was determined by Garen (1968).

The mutation of the sixth position in the coat (in fact, seventh by the genome structure, taking the first methionine into account), as in the sus3 mutant of the phage f2, demonstrated a polarity effect on the replicase gene, or gene 1, and led to the absence of RNA replicase, double-stranded RNA, and infectious RNA (Lodish and Zinder 1966b). The conversion of parental RNA into the double-stranded RNA was also very poor (Roberts and Gussin 1967; Lodish 1968). Moreover, the complementation between the polar coat *amber* mutants and the gene 1 mutants was very weak (Gussin 1966). Remarkably, these effects of the sus3 mutation on the expression of the gene 1 disappeared not only in an Su-1 host but also in a nonpermissive Su⁺ host, i.e., Su-2 and Su-3, indicating the fact that the translation event of the coat gene was critical for prevention of the polarity effect. Figure 8.1 demonstrates the mechanism to overcome polarity in light of this observation.

Therefore, the ratio of phage-specific RNA synthesis in the infected Su⁻ host versus that in the Su⁺ host was a measure of the polarity of the coat *amber* mutants (Horiuchi 1975). By this criterion, the coat *amber* at the position 6 was the only polar mutant known in the phages f2 and R17, while the

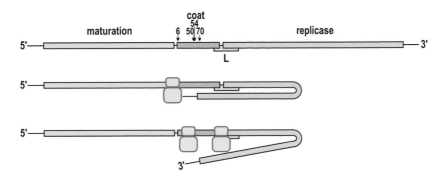

FIGURE 8.1 Mechanism of the polarity effect. The positions of the *amber* mutations in the coat gene are marked by arrows.

polarity was not detected with mutants at the positions 50, 54, and 70. The polar effect of the position 6 *amber* on the replicase synthesis was demonstrated also in the cell-free synthesis (Capecchi 1967b; Engelhardt et al. 1967; Capecchi and Webster 1975). In the case of Qβ, the existence of a gradient of polarity across the coat protein gene has been reported using coat *amber* mutants at positions 17, 37, and 86 (Ball and Kaesberg 1973a).

Later, the careful elucidation of the replication cycle of the polar mutant sus3 of the phage f2 and of the polar mutant am12 of the phage Qβ, which possessed an *amber* mutation at the position 37, was performed (Pumpen et al. 1978b). Figure 8.2

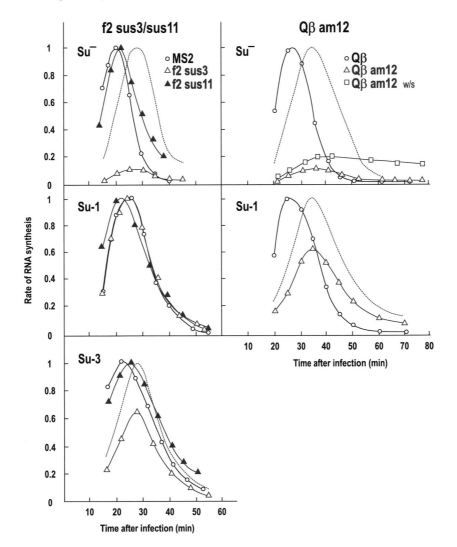

FIGURE 8.2 The normalized RNA synthesis rate of the RNA phage coat *amber* mutants in rifamycin-treated suppressor and non-suppressor *E. coli* cells. Rifamycin was added at 10 min (MS2, f2sus3, and f2sus11) or 15 min (Qβ and Qβam12) postinfection. Dashed red line represents normalization of the mutant RNA synthesis to 1.0; w/s, without synchronization by ribonuclease. (Based on data from Pumpen PP et al. *Genetika.* 1978b;14:1687–1695.)

shows that the synthesis of viral RNA under non-suppressor conditions in the presence of rifamycin produced the same gaussian pattern of rates as the synthesis of RNA by wild-type phage or the nonpolar-coat protein mutants.

However, the total amount of RNA was decreased approximately tenfold and the peak of RNA synthesis was displaced 7–10 min later. Most important was the obstacle that about a tenfold reduction number of infective centers was observed by direct detection of the infective centers with the ^{32}P-labeled parental phage RNA. This indicated that a certain time lapse was required to overcome the polarity of the parental RNA, this process being of single occurrence, exclusively on the parental RNA but not on the progeny strains. Therefore, it was concluded that the initiation of translation at the replicase gene starts on the nascent RNA chains within the replicative complexes and not on the fully synthesized templates with their complete secondary structure. The data obtained were not in contradiction with the hypothesis concerning the role of the repressor complex II, namely the replicase-RNA complex, which is described in Chapters 16 and 17.

The nonpolar coat mutants exerted a different type of effect on the phage replicase and RNA synthesis. If the rate of the replicase synthesis decreased after about 20 minutes in cells infected with wild-type phage, the nonpolar coat mutants did not stop the replicase synthesis and therefore induced hyperproduction of enzyme by 6–10 times that of wild type in nonpermissive Su⁻, Su-2, or Su-3 cells. This effect was demonstrated either by measurements of enzyme activity (Lodish et al. 1964) or by incorporation of radioactive amino acids into the replicase that was assayed on the SDS-gel electrophoresis (Viñuela et al. 1968; Nathans et al. 1969). These results indicated the repressor role of the coat protein by the replicase synthesis in the late stage of normal infection and led to conclusion that in the absence of functional coat protein, the infected cells could hyperproduce the active replicase enzyme (Lodish and Zinder 1966b). The repressor function of the coat protein on the synthesis of replicase has also been shown in the cell-free system (Sugiyama and Nakada 1967a; Robertson et al. 1968b; Eggen and Nathans 1969).

Remarkably, the behavior of the coat mutants supported to some extent the idea that the coat protein could be a lysis agent (Zinder and Lyons 1968). When a Su⁻ host was infected with nonpolar coat *amber* mutants, cell growth stopped at the time when the cell would start lysis by a permissive Su-1 host, but no lysis occurred. The nonpermissive Su-3 cells behaved in exactly the same way when they were infected with the coat *amber* mutants, while in Su-2 host, which translated the *amber* codon as glutamine, the wild-type amino acid, the coat *amber* mutants did cause cell lysis (Horiuchi 1975). These results therefore suggested the idea that the production of the functional coat subunit could be a prerequisite for cell lysis, and, moreover, that, specifically for the lysis event, such coat subunits need not be produced in amounts sufficient to give a large yield of particles (Zinder and Lyons 1968).

The Qβ readthrough protein was detected by the gel electrophoretic analysis of the infected cells and of the purified Qβ particles and was named IIb, by its near location to the IIa, or maturation, protein on the electrophoregram (Garwes

et al. 1969). The gel electrophoretic analysis of proteins synthesized by the Qβ *amber* mutants *in vivo* and by the *amber* mutant RNA *in vitro* showed that the IIb was the product of the coat gene, or the gene 3: when rifamycin-treated Su⁻ cells were infected with the gene 3 *amber* mutants, the peak IIb disappeared and purified defective particles produced by the Qβ gene 3 *amber* mutants did not contain the IIb protein but did contain the IIa protein (Horiuchi et al. 1971). The Qβ coat *amber* mutant am11 produced a fragment of coat protein in a cell-free extract instead of intact coat protein, and did not produce the IIb protein. Horiuchi et al. (1971) found that the IIb synthesis in a cell-free extract was stimulated by addition of tRNA from a UGA-suppressor strain when compared to the control with an equal amount of tRNA from the Su~ strain. This phenomenon is illustrated by Figure 8.3, which is taken from Fromageot and Zinder (1968) but also appeared in the classical review chapter written by Horiuchi (1975). Therefore, with the substantial help of mutants, it was concluded that the

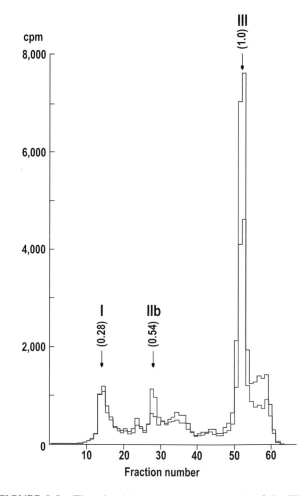

FIGURE 8.3 The stimulation of *in vitro* synthesis of the IIb, or A1, protein by addition of tRNA extracted from the UGA-suppressor strain. The proteins were synthesized in an S30 extract of Su⁻ cells using wild-type Qβ phage RNA as the messenger. (Red line) 450 μg/mL of tRNA extracted from K223 (Su-UGA) was added. (Black line) 450 μg/mL of tRNA from K56 (Su⁻) was added. Phage proteins I, IIb (A1) and III are indicated by arrows. (Redrawn with permission of Henri Pierre-Marcel Fromageot from Fromageot HPM, Zinder ND. *Proc Natl Acad Sci U S A.* 1968;61:184–191.)

IIb protein is a product of improper translation of the UGA chain punctuation signal of the Qβ coat protein not only *in vitro* but also *in vivo*, and that the *readthrough* product was incorporated into phage particles (Horiuchi et al. 1971).

AZURE MUTANTS

By heavy mutagenesis of the phage f2 with nitrous acid, the *azure* mutants were isolated, which were specifically restricted by *amber* suppressor genes in the host bacteria and appeared as the inverse of the *amber* mutants (Horiuchi and Zinder 1967). Thirteen *azure* mutants have been isolated and characterized and were found to be so similar to each other that they might have an identical defect (Horiuchi 1975). Among the three *amber* suppressor hosts shown, the restriction of *azure* mutants was most effective in Su-3 and least effective in Su-2. Thus, the relative order of the restriction qualitatively agreed with that of the efficiency of suppression, while the *ochre* suppressor strains did not restrict *azure* mutants, perhaps due to their low level of suppression. Similar MS2 mutants were isolated also by Marc Van Montagu (cited by Iserentant et al. 1980). It is remarkable that these mutants were obtained at high frequency, almost like *amber* mutants.

The *amber* suppressor genes restricted formation of infective centers, while each infected Su$^+$ host, once the infection was established, produced a normal phage yield. The number of infective centers formed by infection of the Su$^+$ cells by *azure* mutants depended on the multiplicity of infection. For infection of Su-3 cells with an *azure* mutant, the response was a linear function of multiplicity of infection up to multiplicities of about 10, suggesting that the infective centers formed by Su$^+$ strains were not due to heterogeneity in the population of the cells; i.e., when Su$^+$ cells were infected with *azure* mutants, there was a certain probability of forming an infective center per phage particle, not per cell, and this probability depended on the particular Su$^+$ gene present. Attempts to rescue or reverse the abortive infection by simultaneous infection with wild-type or *amber* mutants of f2 have failed.

Concerning to the author's review (Horiuchi 1975), the Su$^+$ cells that did not form infective centers formed colonies and did not produce any defective phage particles. Measurement of uracil incorporation into the UV-irradiated Su$^+$ cells showed that there was no detectable amount of phage RNA synthesis. The synthesis of double-stranded RNA did not occur either. Moreover, the conversion of the parental RNA into the double-stranded form did not occur in Su$^+$ cells in all the 13 *azure* mutants isolated.

The authors excluded the possibility that the *azure* mutants might only poorly adsorb to or inject their RNA into Su$^+$ cells, since, first, the adsorption of the *azure* mutants to Su$^+$ cells occurred at a normal rate. Second, when spheroplasts prepared from Su$^+$ or Su$^-$ cells were infected with the RNA samples extracted from the *azure* mutants and compared with RNA from wild-type phage, restriction of the *azure* mutants by the Su$^+$ host was observed. Third, when Su$^-$ and Su-3 cells infected for 10 minutes with the ^{32}P-labeled *azure* mutants were washed well in the cold, the same amount of radioactivity was recovered from either cell. Analysis of lysates of these washed cells,

obtained by treatment with lysozyme and EDTA followed by freezing and thawing, showed that 80% of the radioactive material in either lysate was sensitive to ribonuclease.

Since ^{32}P-labeled RNA inside the phage particle was completely resistant to ribonuclease, this result strongly suggested the idea that the phage RNA was normally injected into the cells when the Su-3 host was abortively infected with the *azure* mutants. Therefore, the step at which the growth of the *azure* mutants was blocked by suppressor genes was probably either in the translation of parental RNA strand or synthesis of the first complementary RNA strand on the parental RNA.

The authors proposed two types of mechanisms, depending on whether the *amber* codon suppressed was in the bacterial genome or in the phage genome. In the former instance, one would assume that the *E. coli* strains have an *amber* codon in some gene, the product of which, when suppressed, recognizes the *azure* mutation in such a way as to prevent RNA replication.

In the other instance, the authors assumed that the natural chain terminating codon of the replicase (e.g., UAA) has mutated into the *amber* codon UAG in the *azure* mutants. The suppression of the latter should result then in the addition of some amino acids to the C-terminus of the gene product, which would cause malfunction of the phage replicase (Horiuchi and Zinder 1967).

Remarkably, the efficiency of the restriction of the *azure* mutants was higher than the efficiency of suppression. In the case of Su-3, 97% of Su$^+$ cells infected with the *azure* mutants gave abortive infection, while the reported efficiency of suppression was 55% (Horiuchi 1975). To explain this amplification of the effect of suppression, one would have to assume some mechanism like a "lethal" effect of the suppressed product of replicase protein due to the multimeric structure of the replicase (Horiuchi and Zinder 1967; Horiuchi 1975).

It is remarkable that the phage FIC appeared as a spontaneous *azure* mutant of the group IV phage FI (Hirashima et al. 1977).

Finally, the mystery of the *azure*, or reverse *amber*, mutants was solved (Iserentant et al. 1980) due to the full-length sequencing of the MS2 genome (Fiers et al. 1976), which revealed the UAG termination of the maturation (Fiers et al. 1975b) and replicase (Vandenberghe et al. 1975) genes. Meanwhile, a certain degree of readthrough of both the maturation gene (Remaut and Fiers 1972a,b) and replicase (Atkins and Gesteland 1975) genes was shown to occur in *amber* suppressor strains. The MS2 replicase gene ended with a UAG codon, followed seven triplets further by an in-phase UAA triplet. When the three different MS2 *azure* mutants obtained by the nitrous acid treatment were characterized at the nucleotide sequence level, all three *azure* mutants contained an A → G transition in this UAA second stop codon of the replicase gene, resulting in a second suppressible UAG codon (Iserentant et al. 1980). Analysis of revertants demonstrated that the *azure* mutation can be counteracted either by a true site reversion at the second stop or by the creation of a new stop signal for the replicase gene, either UAA or UGA, before or at the first stop, or beyond the second stop. Figure 8.4,

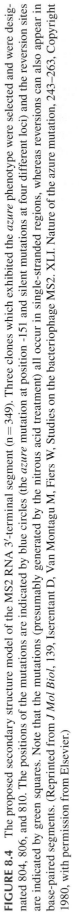

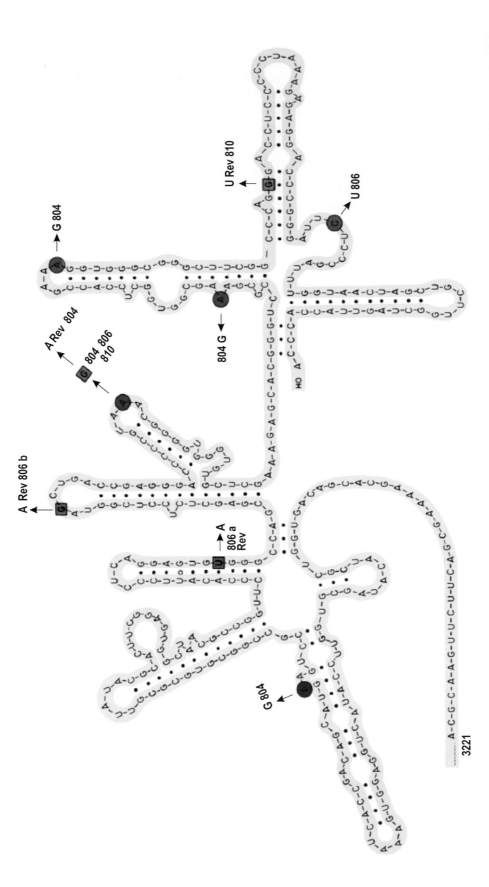

FIGURE 8.4 The proposed secondary structure model of the MS2 RNA 3′-terminal segment (n = 349). Three clones which exhibited the *azure* phenotype were selected and were designated 804, 806, and 810. The positions of the mutations are indicated by blue circles (the *azure* mutation at position -151 and silent mutations at four different loci) and the reversion sites are indicated by green squares. Note that the mutations (presumably generated by the nitrous acid treatment) all occur in single-stranded regions, whereas reversions can also appear in base-paired segments. (Reprinted from *J Mol Biol*, 139, Iserentant D, Van Montagu M, Fiers W, Studies on the bacteriophage MS2. XLI. Nature of the azure mutation, 243–263, Copyright 1980, with permission from Elsevier.)

which is taken from Iserentant et al. (1980), demonstrates clearly the appearance of the *azure* mutants.

Op-3 MUTANT

By the search for *opal* mutants of the phage f2, a mutant, op-3, was isolated which did not lyse Su$^-$ cells but lysed any of the three independently isolated UGA suppressor strains. In the Su$^-$ host, the op-3 mutant made a normal burst of phage, which could be released by artificial lysis with lysozyme and EDTA. No differences in infectivity and physical properties were found among the particles made in the Su$^-$ and Su$^+$ hosts. The mutant complemented mutants in each of the other three complementation groups.

Therefore, it was logical to place the op-3 mutant in a novel fourth complementation group, which could be different from the three recognized before. The paper of Model et al. (1979), which described the unique properties of the op-3 mutant, appeared after some lag period and the existence of the fourth complementation group based on the op-3 data was first postulated much earlier and was provided with a set of reliable arguments (Horiuchi 1975). It is noteworthy that Model et al. (1979) indicated that the op-3 mutant "was isolated by L. Lyons, using methods previously described (Model et al. 1969)," therefore, by the 5-fluorouracil mutagenesis. At the same time the paper of Model et al. (1979) was published, Beremand and Blumenthal (1979) found a 75 amino acid long lysis protein *in vivo* and located the lysis gene on the known sequence of the phage MS2 (Fiers et al. 1976). Moreover, Atkins et al. (1979b) translated RNA of the phage MS2 in a system composed of rabbit reticulocyte ribosomes and ascites cells supernatant and found the same peptide. In both instances this lysis peptide disappeared when the op-3 RNA was used. As a result, the lysis gene was located in a plus one phase starting at an AUG codon, 47 codons from the distal end of the coat gene continuing through 36 codons of intergenic space and on into 142 codons of the replicase gene. The op-3 mutation was localized on the MS2 sequence as the C → U exchange in the second position of the second UCG codon of the replicase and within its ribosome-binding site, as well as at the position 30 of the overlapping lysis gene in the arginine codon CGA (Atkins et al. 1979b). According to the review paper written by Zinder (1980),

> several puzzling features of the lysis mutant became clear. Its delayed growth was probably due to slow replicase production. However, its overproduction of replicase in time was probably the result of poor coat protein binding, hence poor repression. The position of the lysis gene also explains why mutations affecting coat gene translation lyse cells poorly: the ribosome-binding site of the lysis gene probably never becomes available.

The story of the unique op-3 mutant remains in the history of virology as a brilliant example of the scrutinizing approach of both geneticists and molecular biologists to unexpected and nontrivial problems.

STREPTOMYCIN SUPPRESSION

The *amber* mutations were found to be suppressed by streptomycin (Valentine and Zinder 1964b). Then, it was specified that only polar coat *amber* mutants of the phages f2 and Qβ, in contrast to nonpolar ones, were suppressed by streptomycin (Kuwano and Endo 1969). Later, a *ts* streptomycin-resistant *E. coli* mutant was found (Engelberg-Kulka et al. 1977), where ribosomes demonstrated temperature sensitivity in the suppression of the *amber* codon (Zeevi et al. 1979). The restriction of *amber* suppression at 42°C was related to an alteration in ribosomal protein S12 of the streptomycin-resistant mutant. Since the phage Qβ required the readthrough protein, its growth was inhibited at the nonpermissive temperature in contrast to that of the phages f2, MS2, and R17 (Engelberg-Kulka et al. 1979b). The Qβ particles, which were produced in the streptomycin-resistant *E. coli* mutant at 41°C, were defective (Engelberg-Kulka et al. 1979a). Such particles did not adsorb to F-piliated bacteria, and they were deficient in both minor capsid proteins of Qβ: maturation, or IIa, or A2, and readthrough, or IIb, or A1.

MUTANTS OF THE PHAGE f2

As stated above, one of the first systematic works on the RNA phage genetics was performed by Norton D. Zinder and his colleagues with the phage f2. In this early period, a thorough analysis of a series of *amber* mutants was presented (Engelhardt and Zinder 1964; Lodish et al. 1964; Valentine and Zinder 1964a,b; Valentine et al. 1964; Zinder and Cooper 1964; Engelhardt et al. 1965; Lodish and Zinder 1965; Lodish et al. 1965; Notani et al. 1965; Valentine and Strand 1965). The mutants were generally obtained by nitrous acid treatment (Zinder and Cooper 1964) and divided by complementation into two groups: the sus1 mutant-like group and the sus3 mutant-like group (Valentine et al. 1964). The sus3 mutation has been shown to be in the gene directing the structure of the coat protein, or gene 3 by the early classification (Notani et al. 1965). Moreover, the sus3 mutant itself demonstrated the polarity effect, while the popular sus3-like mutant sus11 did not, as it appeared later. At the same time, it was shown that the sus11 mutant strongly overproduced replicase in nonpermissive host, in contrast to the sus3 mutant (Lodish et al. 1964; Lodish and Zinder 1965). Then, Valentine and Zinder (1964b) showed that the sus3 mutation could be phenotypically repaired by streptomycin, while Engelhardt and Zinder (1964) determined the RNA infectivity of the sus1 and sus3 mutants on spheroplasts of the permissive host. Later, the phenotypic suppression of the *amber* mutants was shown to work with polar coat mutants only: the f2 mutant sus3 and the R17 mutant amB2, but not with sus11 and other nonpolar f2 and R17 coat *amber* mutants (Kuwano and Endo 1969).

The substitution of the glutamine residue by the serine one within the f2 coat protein was shown for the *amber* mutant sus4A and a set of other members of the sus3-like mutant class, which differed from the sus11 mutant (Notani et al.

1965). Direct suppressing action of tRNA from permissive host on the translation of the coat *amber* mutant RNA *in vitro* was demonstrated by Engelhardt et al. (1965). The site of the serine insertion, instead of glutamine, into the coat protein was also located in this paper.

When the sus1 mutants infected the nonpermissive host, a normal number of defective phage particles was produced (Zinder and Cooper 1964). The first explanation described these *dead* particles as lacking a basic protein (Zinder 1965). The sus1-like *ts* mutants were isolated after nitrous acid treatment or growth of the phage f2 in medium containing 5-fluorouracil, and the appropriate defective *dead* particles were characterized in detail (Lodish et al. 1965). Valentine and Strand (1965) showed inability of the sus1 mutant to adsorb on F pili. Further, the sus1-like mutants were identified as the maturation, or gene 2, mutants (Horiuchi et al. 1966).

Horiuchi et al. (1966) continued the study of Lodish et al. (1965), isolated a full three-gene set of *ts* mutants by nitrous acid or 5-fluorouracil treatment, and clearly formulated the existence of the three mutant groups. Thus, direct homology of the previously described host-dependent *amber* and novel *ts* mutants in frame of the three-gene theory was established in this study. By homologization, the previously obtained sus10 *amber* mutant was identified as the replicase mutant (Horiuchi et al. 1966). Furthermore, the f2 *ts* mutants were employed by the work that proposed the semi-conservative replication mechanism of the phage f2 RNA (Lodish and Zinder 1966c).

The behavior of the polar sus3 coat protein *amber* mutant, in contrast to the nonpolar sus11 mutant, was interpreted in terms of a model coupling genetic translation and transcription (Lodish and Zinder 1966a) that played a further crucial role in the understanding of the regulatory mechanisms of phage replication (Lodish and Zinder 1966b). Finally, the *amber* mutations of the sus3 and sus11 were located to the f2 coat positions 6 (7 by the gene structure) and 70 (71 by the gene structure), respectively, when the according suppressible peptide fragments were found by the *in vitro* translation of the corresponding RNAs (Webster et al. 1967). Moreover, the polarity of the sus3 mutation, in contrast to the sus11 mutant, was demonstrated by the *in vitro* translation: when polar coat mutant RNA was translated in Su⁻ extracts, only monosomes were formed, while when wild-type RNA or RNA containing the nonpolar coat mutation were translated under these conditions, polysomes as well as monosomes were observed (Engelhardt et al. 1967). Then, the sus3 mutant contributed to the identification of so-called *suppressor A* allele that facilitated the survival of the untranslated mRNA distal to nonsense codons, which was associated with the lack of an endonuclease thought to degrade the exposed mRNA (Morse and Primakoff 1970).

The f2 *amber* mutants played a crucial role at the end of the 1960s by the establishment of the specific tRNAs as suppressing entities, rather than fragments of ribosomal RNA or a structural gene for a ribosomal component as possible suppressor candidates. Thus, the *in vitro* translation directed by the sus3 RNA was used by the identification and purification of the serine-specific *amber* suppressor tRNA from *E. coli* cells carrying the Su-1 gene (Söll 1968). The f2 sus13 *amber* mutant, an analog of the sus3 mutant, was employed in the studies of the *ochre* suppressor *E. coli* strain and contributed to the identification of tRNA as the molecules which cause suppression (Gartner et al. 1969).

The *opal* mutants of the phage f2 were induced by 5-fluorouracil mutagenesis and the replicase mutant op9 was studied as the first, in parallel with the *amber* sus10 replicase mutant (Model et al. 1969). It was concluded that the UGA codon, like the UAG codon, caused premature termination of the replicase chain, and that the termination could be suppressed by the appropriate tRNA *in vitro*. Remarkably, after one replicase mutant, namely, sus10, six of the remainder *opal* mutants were defective in the maturation gene but no coat protein *opal* mutants were found. Again, the addition of su$_{UGA}$ tRNA to op-9 RNA-stimulated su⁻ extracts *in vitro* resulted in an increased formation of intact replicase and a decreased synthesis of the replicase fragment.

The phage f2 appeared as a source of unique and unusual types of mutations. Thus, the novel type of mutants, namely the *azure* mutants that were described above, was isolated for the phage f2 (Horiuchi and Zinder 1967). Among a set of the f2 *opal* mutants (Model et al. 1969), the lysis-defective op-3 mutant was isolated (Model et al. 1979). As described above, it paved the way to the identification and detailed characterization of the fourth, namely, lysis, gene of the RNA phages (Atkins et al. 1979b; Beremand and Blumenthal 1979).

The phage f2 sus1 and sus3 mutants were used by the characterization of a novel efficient *E. coli amber* suppressor strain KO1 that did suppress the sus1, but not the sus3 mutant (Aoi and Watanabe 1975). Then, the f2 sus3 mutant RNA was used to approve suppressor activity of the appropriate tRNA from *Pneumococcus* (Gasc et al. 1979).

Finally, the polar mutant sus3 of the phage f2, together with the polar mutant am12 of the phage Qβ (the *amber* mutation in the position 37) contributed markedly (Pumpen et al. 1978b) to the general regulation scheme of the phage replication (Pumpen et al. 1978a). In the latter study, the f2 sus11 mutant was used also.

The first antibody-resistant mutant, namely f24, was selected by plating the phage f2 in the presence of a high concentration of anti-f2 antiserum (Valentine and Zinder 1964a). The f2 and f24 phages demonstrated phenotypic mixing by the simultaneous infection.

MUTANTS OF THE PHAGE MS2

The first MS2 mutants were isolated by Robert L. Sinsheimer's team. Thus, by addition of 5-fluorouracil to infected cells 15 min after infection, Pfeifer et al. (1964) isolated seven *ts* mutants, some of which were blocked at an early stage of infection, while others were defective at a near-terminal step, and achieved complementation of their function. These mutants were supposed to have a missense mutation in the replicase or coat genes. Remarkably, in one combination which was analyzed in detail, the complementation was asymmetric, resulting in enhanced synthesis of infective RNA and mature phage of only one type.

The action of 5-fluorouracil on the growth and the properties of the 5-fluorouracil-containing phage MS2 were studied in detail (Shimura and Nathans 1964; Shimura et al. 1965). The 5-fluorouracil-induced defective MS2 particles (Shimura et al. 1967) closely resembled the *dead* particles produced by certain *amber* mutants of the phage f2 (Lodish et al. 1965), as well the similar defective particles of the phages fr (Heisenberg 1966) and R17 (Argetsinger and Gussin 1966). The fluorouracil-RNA fragments from such particles represented about two-thirds of the 5'-end of the phage MS2 RNA molecule and directed the synthesis of phage coat protein but not the RNA replicase in the cell-free system *in vitro* (Shimura et al. 1968).

Marc Van Montagu (1966, 1968) isolated a set of the *amber* and *opal* suppressor sensitive mutants of the phage MS2 by nitrous acid treatment and performed the appropriate complementation tests. These mutants were used for the identification of suppressors (Van Montagu et al. 1967) and overproduction of replicase (Gillis et al. 1967). Then, differences between the nitrous acid–induced and the hydroxylamine-induced *amber* mutants of the phage MS2 were studied (Vandekerckhove et al. 1971). The *amber* mutations of the coat were localized on the amino acid sequence of the phage MS2 (Vandekerckhove et al. 1969; Van Assche et al. 1974) and the maturation mutants, which were obtained by nitrous acid or hydroxylamine treatment, were characterized by polyacrylamide gel electrophoresis (Vandamme et al. 1972). Moreover, antiserum-resistant mutants (Van Assche et al. 1972) and acidic coat mutants (Van Assche and Van Montagu 1974) of the phage MS2 were isolated. The MS2 mutants were used in the studies on the regulation of the phage replication (Remaut et al. 1974). The nonpolar coat protein mutants am601 and am623 carrying mutations in positions 70 and 50, respectively, as well as the maturation mutant am302, synthesizing 90% A protein fragment in nonpermissive cells (Vandamme et al. 1972), were used by the studies on regulatory replication control (Pumpen et al. 1978a).

A set of 12 MS2 *amber* mutants was generated by nitrous acid treatment by Cedric I. Davern (cited by Viñuela et al. 1968). The mutant MU9 from this collection appeared to have a nonpolar *amber* mutation at the position 70 of the coat gene, while the mutants MU3 (as well as similar MU6, MU7, MU10, and MU12 mutants) and MU11 carried mutation in the maturation gene (Viñuela et al. 1968). The protein synthesis induced by the MU9 mutant, in comparison with the sus3 mutant of the phage f2, was characterized by polyacrylamide gel electrophoresis on the background of the dropping host protein synthesis (Sugiyama and Stone 1968b). The suppression of the MU9 mutant was examined by analyzing the production of intact coat protein under Su-1 and Su-3 suppression both *in vivo* and *in vitro*, where Su-3 appeared as a very poor suppressor *in vivo* and a reasonably good suppressor *in vitro* (Sugiyama et al. 1969). The 70th position (71st by the genome structure) of the MU9 *amber* mutation was determined by Katze and Konigsberg (1969).

The full three-gene set of the MS2 *amber* mutants was generated by nitrous acid mutagenesis and used for elaboration of the regulation scheme for the RNA phage replication (Nathans et al. 1969). The overproduction of the replicase by coat protein mutants in nonpermissive strain was confirmed in this study.

In the former Soviet Union, the first conditionally lethal mutants of the phage f2 were described by Alikhanyan and Piruzian (1968). Then, the mutagenesis of the phage MS2 was performed readily by Aleksandr Samsonovitch Krivisky and Eduard Izraelivitch Budowsky with their colleagues. First, the mutagenic effects of nitrous acid and UV-irradiation on the phage MS2 was studied (Krivisky and Sherban 1969, 1970; Sherban 1969; Sherban and Krivisky 1969). Then the mutagenic action of O-methylhydroxylamine (Budowsky et al. 1971, 1972, 1973, 1974a) and hydroxylamine (Budowsky and Pashneva 1971; Budowsky et al. 1974b) on the phage MS2 was elucidated. Moreover, the reversion of two MS2 *amber* mutants of the maturation gene, namely am309 and am606 (Vandamme et al. 1972) by hydroxylamines was demonstrated (Budowsky et al. 1975, 1978).

During these studies, the morphologic MS2 mutants having a white dense ring around negative colonies, so-called *rim* mutants, were isolated (Sherban 1969). One of them, a typical mutant 40 (Bespalova et al. 1976), was defined later as a *ts* coat mutant carrying a missense mutation in the coat gene (Dishler et al. 1980b).

A novel class of the *ts* mutants with disrupted repressor activity of the replicase, so-called repressor complex II mutants, was isolated by Andris Dišlers from the Elmārs Grēns team after nitrous acid treatment of the MS2 coat *amber* mutant am623 with mutation at the position 50 of the coat (Dishler et al. 1980a). The typical mutant ts130 from this unique class demonstrated the 10-minute displacement of the RNA synthesis peak and 20-times overproduction of the replicase subunit in comparison to the original phage MS2am623.

The MS2 mutant ts13 was selected with a special aim to elaborate procedures for the detection of small differences in ribonucleic acids of large molecular mass (Robinson et al. 1969b). It is worth mentioning here the experiments on the perpetuation of the *ts* mutants in the surviving bacteria (Tsuchida et al. 1966), where the phage β, a close relative of the phage MS2 (Nonoyama et al. 1963), was employed. The *ts* mutants were obtained by nitrous acid treatment, and one of the mutants, ts-9, with a lesion in the early replication step, demonstrated the perpetuation of its genome for at least 7 hours in about 10% of growing *E. coli* cells at nonpermissive 43°C temperature and in the presence of anti-β serum. The other early step mutant, ts-2, revealed incomplete perpetuation, while the late step mutant ts-5 and the wild-type phage β did not perpetuate at all (Tsuchida et al. 1966).

A set of MS2 mutants was shown to have an additional silent mutation Met → Ile at position 108 of the coat protein (Van Assche et al. 1974). As transitions were more frequent than transversions, one would have expected an AUA codon in this position in the mutant RNAs. This problem was of special interest at that time, since the AUA codon was one of the best candidates for a modulation role in the control of translation in *E. coli*. The presence of this AUA in the gene for the protein made in major amounts upon viral infection would impose serious doubt on the theory of modulation. Min Jou et al. (1976) have directly proven that, in fact, the isoleucine residue at position 108 of the coat protein gene is specified by the non-rate−limiting AUU codon, in agreement with a modulation type of control of protein synthesis.

MUTANTS OF THE PHAGE fr

Knolle and Kaudewitz (1962) reported that they have treated the phage ft5, an early designation for the phage fr, by nitrous acid and have therefore isolated the first artificially-induced RNA phage mutants. They were plaque-size mutants that formed stable small-plaque clones (Kaudewitz and Knolle 1963). The plaque-size fr-A92, fr-A102, and mostly fr-A105 mutants from the latter paper were used by the studies on the interference and mutual exclusion with the wild-type fr, due to their easily distinguishable phenotypes (Knolle 1966a), and with the non-infectious f2 mutant sus3 in nonpermissive conditions (Knolle 1966b). Moreover, the fr-A105 mutant demonstrated ability to rescue the f2 mutants sus1 and sus3 (Knolle 1966c).

After hydroxylamine treatment, Heisenberg and Blessing (1965) isolated the same two groups of the complementing *amber* mutants of the phage fr, as Zinder and Cooper (1964) previously isolated for the phage f2, and described in detail the potential maturation gene mutant su$_m$, which formed incomplete virus particles with only 70% of the normal amount of RNA. Formation of the defective particles was studied further in detail (Heisenberg 1966), including employment of an anti-RNA antiserum (Heisenberg 1967). By the *in vitro* self-assembly of the defective fr particles from RNA and coat protein, Hohn (1967) observed their similarity to the defective particles formed *in vivo* by the su$_m$ mutant.

MUTANTS OF THE PHAGE R17

The first published *amber* mutant of the phage R17, namely the coat mutant am11B, has played a central role in the breakthrough paper that identified a serine tRNA as the *nonsense* suppressor in the cell-free synthesis system *in vitro* (Capecchi and Gussin 1965) by James D. Watson's laboratory. Three complementation groups, A, B, and C, in the phage R17 *amber* mutants that were obtained after nitrous acid mutagenesis were presented a bit later, and the functions of the three genes were described exhaustively (Gussin 1966; Gussin et al. 1966). The amB2, one of the coat mutants, was identified as polar, while the am11B that appeared earlier in the Capecchi and Gussin (1965) paper was classified as nonpolar. The am11B RNA was used by the studies on the initiation of viral protein synthesis in *E. coli* extracts (Kolakofsky and Nakamoto 1966). Further, the amC13 *amber* mutant was used to identify the phage-specific replicase subunit in the cell-free protein synthesis system (Capecchi 1966a).

More *amber* mutants of the phage R17, especially of the B, or coat, gene, in order to look for the potential polarity gradient,

were isolated after nitrous acid and 5-fluorouracil mutagenesis (Tooze and Weber 1967). Moreover, as mentioned above, the authors tried to find evidence for the existence of more than three cistrons in the R17 genome. Since the coat protein amino acid sequence had just been established at that time for the closely related phage f2 (Konigsberg 1966; Konigsberg et al. 1966; Weber et al. 1966; Weber and Konigsberg 1967) and for the R17 itself (Weber 1967), where potentially susceptible glutamine residues were located at positions 6, 40, 50, 54, 70, and 109, and tryptophan residues at positions 32 and 82 (always add one for the position at the gene structure), coat gene was of special interest. Among 13 coat protein mutants, seven had the mutation at the position 6, one had its mutation at the position 50, and the other five had the mutation at the position 54. The general conclusion was that the position 50 and 54 mutants were not strongly polar, whereas the position 6 mutants were strongly polar (Tooze and Weber 1967). Another important observation indicated that each RNA genome is preferentially but not exclusively replicated by its own replicase.

As stated above, the polar coat mutant amB2 (position 6), together with the f2 mutant sus3 and in contrast to the nonpolar amB11 (position 50) and amB21 (position 54) coat mutants, contributed to the phenotypic suppression of the polar *amber* mutants by streptomycin (Kuwano and Endo 1969).

The amB2 RNA, together with the wild-type R17 RNA, was used by the sequencing of oligonucleotide coding for the first six amino acid residues of the R17 coat protein (Robinson et al. 1969a). A series of the R17 *amber* mutants was tested for the inhibition of the synthesis of bacterial proteins and ribosomal RNA, where the latter was found inhibited with phage mutants retaining the ability to produce replicase (Spangler and Iglewski 1972). It was found that the inhibition of 16S ribosomal RNA synthesis occurred solely at the transcriptional level. Following the approaches to the maturation mutants of other phages, the changes in the capsid structure and stability of defective particles of the R17 amA31 mutant were investigated thoroughly by serological and chemical methods (Iglewski 1977).

The full spectrum of the R17 amber mutants was used by the investigation of polyribosome patterns in infected cells (Phillips et al. 1969). Moreover, the amB2 mutant was used by studies on the translation mechanisms by the characterization of ribosome-releasing factors (Ryoji et al. 1981b), re-initiation of translation from the triplet next to the *amber* termination codon in the absence of ribosome-releasing factor (Ryoji et al. 1981a), and measuring the *amber* codon readthrough translation efficiency *in vitro* (Ryoji et al. 1985).

A number of *ts* mutants of the phage R17 was isolated after the 5-fluorouracil mutagenesis, and one of them, ts24, was specified as the replicase mutant (Igarashi et al. 1970a) and used by the studies of the host cell metabolism during phage infection (Igarashi et al. 1970b).

MUTANTS OF THE PHAGE GA

The *amber* (Aoi 1973) and *ts* (Aoi et al. 1973) mutants of the phage GA were isolated and characterized. The maturation mutant amH8 was used by the characterization of the above-mentioned *E. coli* suppressor strain KO1 (Aoi and Watanabe 1975).

An interesting intergroup serological mutant, GAsus5H, was isolated from the phage GA maturation *amber* mutant GAsus5 (Harigai et al. 1986). The mutant was spontaneous but appeared after sequential electrophoretic selection for the particle charge. As a result, the serological character of the GAsus5H mutant was changed from the phage GA of group II to that of the phage MS2 of group I. The authors insisted that such serological change from the group II phage GA type to the group I phage MS2 type occurred at quite high frequencies in the natural population, and minor changes of the amino acid in the coat protein might cause drastic changes in the serological and electrophoretic properties of the phage particles. Thus, the mutant has now become sensitive to anti-MS2 serum and resistant to anti-GA serum. The deduced amino acid sequence showed that five amino acids were substituted in the mutant, and three of the five became identical to the corresponding amino acids of the MS2 coat. However, the mutant did not complement MS2 (Harigai et al. 1986).

MUTANTS OF THE PHAGES Qβ, SP, AND FI

A number of *amber* mutants of the phage Qβ were isolated after nitrous acid treatment and classified into the three groups which appeared to be related to the three basic genes of the phages f2 and R17 (Horiuchi and Matsuhashi 1970). A bit later, 12 amber mutants of the phage SP belonging to group IV were isolated with the same nitrous acid mutagenesis and located into the three complementation groups (Ando et al. 1976). From intergroup complementation tests, the phage SP demonstrated a fairly close relationship to Qβ, while f2 had no such relationships to Qβ or SP. The phage FIC was mentioned above as a spontaneous *azure* mutant of the group IV phage FI (Hirashima et al. 1977).

The polarity gradient in the expression of the replicase gene of the phage Qβ was established by Ball and Kaesberg (1973a) with nine Qβ *amber* coat mutants isolated by Horiuchi and Matsuhashi (1970). To do this, the N-terminal coat protein fragments were synthesized *in vitro* by a non-suppressing *E. coli* cell extract directed by the mutant RNAs and characterized by polyacrylamide gel electrophoresis, when the mutant codons were identified: in three cases the amber mutation was at position 17; in five cases, at position 37, and in one case at position 86 (add one to get the amino acid position by the genome structure). The polar effect was measured by RNA synthesis and Qβ replicase activities in infected, non-suppressing cells and their amounts were related for the position of the *amber* mutation. From a practical point of view, it was greatly important that the *amber* mutant amB86 carrying *amber* mutation at the position 86 demonstrated five to eight times overproduction of replicase than the wild-type Qβ phage (Palmenberg and Kaesberg 1973). Moreover, it was found that the Su⁻ bacteria infected with the amB86 mutant carried the viral RNA in a plasmid-like state for many bacterial generations.

The electrophoretic mutant QβE1 was isolated among a number of species that moved more rapidly than the principal component of Qβ (Strauss and Kaesberg 1970). The coat protein of QβEl differed from that of Qβ only by a single amino acid substitution of a lysine residue in Qβ by either glutamic acid or, more probably, by glutamine in QβEl. This resulted in the fact that the QβEl coat was more negatively charged by one or two units than the Qβ coat.

Another electrophoretic mutant, 27-2, was discovered upon examination of the electrophoretic characteristics of several Qβ *ts* mutants (Radloff and Kaesberg 1973). The 27-2 mutant was especially interesting since it differed by a set of narrow, well-defined bands in electrophoresis instead of a single, major, anomalously wide band of Qβ. This behavior resulted from the packing mode of the readthrough A1 protein into phage particles. In the mutant, the A1 proteins were found in the virions only in multiples of three and formed partially resolved bands, whereas wild-type virions differed by only a single A1 protein and formed therefore a wide single band by electrophoresis. The most rapidly migrating band of the mutant contained defective virions that did not possess the A1 protein at all.

From the point of view of the replication regulation, very interesting *ts* mutants of the phage Qβ with lesions in the replicase gene were isolated by Gupta et al. (1975). The three groups of such mutant were identified. A detailed analysis showed that the defective replicase could be involved in the inhibition of both RNA and protein synthesis.

From the global theoretical point of view, the phage Qβ and its replicase *amber* mutant am1 contributed markedly by the evidence that the yeast suppressors of UAA and UAG nonsense codons work efficiently *in vitro* via tRNA in a eukaryotic cell-free protein-synthesizing system (Gesteland et al. 1976).

The maturation amber mutant Qβam205 was used for the "pseudolysogenization" by the phage Qβ (Watanabe et al. 1979). As a result, fairly stable lysogenic-like bacteria were established from the mutant and its nonpermissive host. These bacteria contained few mature phages intracellularly, continued to grow with ability to produce the phage spontaneously, and showed an immunity-like response against homologous phage infection. Since these characteristics were maintained by growth in liquid medium containing anti-Qβ serum, the authors regarded these cells as pseudolysogenic bacteria.

Finally, the polar mutant am12 of the phage Qβ (the *amber* mutation in the position 37), together with the polar mutant sus3 of the phage f2, contributed markedly to the general regulation scheme of the phage replication (Pumpen et al. 1978a,b).

The novel capabilities of mutagenesis were opened by the unique properties of the Qβ RNA synthesis *in vitro* by Qβ replicase and introduced systematically by Charles Weissmann's team. Thus, the high abilities of the Qβ replicase system allowed site-directed mutagenesis during *in vitro* propagation of the phage Qβ RNA (Flavell et al. 1974, 1975; Domingo et al. 1975). A point mutation was generated in the highly-conserved extracistronic region, namely at the 16th position from the 3′-end of Qβ RNA. When a mixture of about 60% wild-type and 40% mutant RNA was repeatedly replicated, the mutant RNA was enriched to 80%, showing that at least this point mutation in the terminal sequence of Qβ RNA did not impair its *in vitro* replication, but in fact slightly accelerated it. This mutation led, however, to loss of phage infectivity (Sabo et al. 1975, 1977). Nevertheless, an infectious extracistronic mutant of the phage Qβ was prepared by the site-directed insertion of an A → G transition at position 40 from the 3′-end (Domingo et al. 1976). Thus, it was demonstrated that the viability of the phage Qβ was not fully dependent on the unique nucleotide sequence in the 3′-extracistronic RNA segment.

The site-directed mutagenesis was applied also to the initiator region of the bacteriophage Qβ coat gene (Taniguchi et al. 1975; Taniguchi and Weissmann 1978b). For this aim, the Qβ plus strands with a 70 S ribosome bound to the coat gene initiation site were used as template for Qβ replicase. The minus strand synthesis proceeded until the replicase reached the ribosome. The ribosome was removed, and elongation was continued in a substrate-controlled, stepwise fashion. The nucleotide analog N^4-hydroxyCMP was introduced into the positions complementary to the third and fourth nucleotides of the coat gene. The minus strands were elongated to completion, purified, and used as a template for Qβ replicase. The final plus strand preparation consisted of four species, with the sequences -A-U-G-G- (wild type), and mutant: -A-U-A-G-, -A-U-G-A-, and -A-U-A-A- at the coat initiation site. The ribosome-binding capacity of the mutant RNAs was measured. These ingenious approaches of the site-directed mutagenesis of the Qβ RNA were reviewed by Taniguchi (1978) and Weissmann et al. (1979a,b).

MUTANTS OF THE PHAGE PP7

The reliable methodology of the reverse genetics was elaborated recently for the pseudomonaphage PP7 (Lee JY et al. 2019). This protocol includes three fundamental steps: (i) creation of a promoter-fused cDNA, (ii) generation of a clone into mini-Tn7-based vector, and (iii) introduction of the clone into nonsusceptible hosts. The PP7 cDNA was fused to the T7 promoter, which was cloned in mini-Tn7 plasmid. This construct was introduced into *P. aeruginosa* PAK and *E. coli* HB101. While the functional phage assembly was observed in *E. coli*, the phage titer from the overnight culture of an *E. coli* clone was significantly lower than that of a PAK clone in the absence of T7 polymerase expression. The *E. coli* strain expressing T7 polymerase showed ∼10³-fold increase in phage production compared to the *E. coli* HB101 strain. Using this protocol, the authors created the reverse genetic system for the phage MS2 as well. These reverse genetic systems could be exploited to introduce desired mutations into the phage genes.

9 Particles

A mighty flame followeth a tiny spark.

Dante Alighieri

As the builders say, the larger stones do not lie well without the lesser.

Plato

MOLECULAR MASS

The RNA phages were discovered at a time when many novel biophysical methods were developed and introduced into structural studies of natural biopolymers, first of all, of viruses and their genomes. Thus, the RNA phages were used as a special test area for numerous novel approaches. The first 10 years of the structural phage history were reviewed systematically by Helga Boedtker and Raymond F. Gesteland (1975).

Table 9.1 presents a short summary of the biophysical determination of the RNA phage molecular mass. In addition, the molecular mass of the phage μ2 was provisionally estimated as 4×10^6 daltons (Ceppellini et al 1963).

The first parameters discovered were traditionally the sedimentation constant $S_{20,w}$ and the diffusion constants $D_{20,w}$. The RNA phage particles sedimented as a single sharp component, the sedimentation constants are given in Table 9.1. The sedimentation constants $S_{20,w}$ of the group I phages were practically identical, 77-81S, within experimental error, while the phage Qβ had a sedimentation constant of about 84S.

The molecular mass measurements of both R17 and MS2 have been carried out independently in two different laboratories for each, and the results were in excellent agreement. Thus, the phage MS2 was first studied by Strauss and Sinsheimer (1963) and later by Overby et al. (1966a), and both reported a molecular mass of 3.6×10^6 daltons. A molecular mass of 3.6×10^6 was also reported for the phage R17 (Gesteland and Boedtker 1964), which agreed well with the more recent value of 3.8×10^6 (Camerini-Otero et al. 1974a). In general, the accuracy of the light-scattering measurements was about ±5% due to errors in the absolute calibration of the light-scattering instrument and cells, the concentration determinations, and extrapolations to zero angle and zero concentrations. The phage, being a small spherical subject, displayed no concentration dependence and very little angular dependence of scattered light (Boedtker 1968b; Boedtker and Gesteland 1975).

Therefore, the early biophysical estimations demonstrated very good agreement with the parameters that were obtained later when precise molecular mass calculations appeared available. Thus, the average of all biophysically obtained values for the phage R17 led to the molecular mass of 3.85×10^6, which was in agreement with the molecular mass calculated from the viral RNA and protein composition (Camerini-Otero et al. 1974a).

The estimates of the RNA content within the particle varied from 28%–29% for f2 (Zinder 1966–1967) and 30% for fr (Marvin and Hoffman-Berling 1963b) to 31.5% for MS2 (Strauss and Sinsheimer 1963) and 31.7% for R17 (Enger et al. 1963). Using the average value of 30.4%, the molecular mass of the RNA was predicted as 1.2×10^6 for R17 and 1.3×10^6 for Qβ, assuming both phage types have the same RNA content (Boedtker and Gesteland 1975). Fischbach and Anderegg (1976) estimated the RNA content within the phage R17 as 30.6%.

Concerning water content, there were 1–1.2 grams of water per gram of phage (Boedtker and Gesteland 1975). These estimates were obtained from a comparison of the volume calculated from the size of the particle with that of the dry volume calculated from the molecular mass timed the partial specific volume (Fischbach et al. 1965), and from the frictional ratio (Marvin and Hoffmann-Berling 1963b).

The isoelectric points of the phages MS2, Qβ and fr were compiled by Boedtker and Gesteland (1975). In fact, they were found to differ very much: the phage fr had an isoelectric point of 9 (Marvin and Hoffmann-Berling 1963b), Qβ had 5.3, and that of MS2 was 3.9 (Overby et al. 1966a), and the large difference between fr and MS2 seemed surprising. As was expected by Boedtker and Gesteland (1975), further structural data explained the isoelectric point differences by the obviously large differences in the surface charge due to differences in the coat protein primary sequence and in the way the coat protein is arranged in the particle. Later, the isoelectric point of the phage MS2 appeared important by the elaboration of the whole particle microelectrophoresis technique (Penrod et al. 1995).

The CsCl equilibrium density gradient centrifugation demonstrated that the buoyant densities of the phages MS2, GA, and Qβ as the representatives of groups I, II, and III, respectively, differed, and appeared as 1.46, 1.44, and 1.47 g/cm³ (Nishihara and Watanabe 1969). The same buoyant density of 1.46 g/cm³ for MS2 was reported earlier by Strauss and Sinsheimer (1963). Therefore, the buoyant densities of the phages were different and specific for each group but were not different within the same group, since a set of representatives of each group was tested: the phages MS2, ZR, MY, GR, R17, and f2 for the group I, the phages GA, SD, SW, KJ, and EI for group II, and the phages Qβ, VK, ST, NM, and NH for group III (Nishihara and Watanabe 1969).

Remarkably, since the molecular mass of the phage MS2 was estimated at about 3–4 megadaltons, it would be expected that the phage would not pass through filters of various sizes with low molecular mass cutoff values of less

TABLE 9.1

Molecular Mass of the RNA Phages

Phage	S$_{20,w}$	Molecular Mass, $\times 10^6$	Method	References
R17	77.4	4.19	Sedimentation-diffusion	Enger et al. (1963)
	79–80	3.60	Light scattering	Gesteland and Boedtker (1964)
		3.81	Light scattering	Pusey et al. (1972)
	78.9	4.02	Sedimentation-diffusion	Camerini-Otero et al. (1974a)
		3.80	Light scattering	Camerini-Otero et al. (1974a)
		4.02	Light scattering	Camerini-Otero et al. (1974b)
		3.83	Light scattering	Loewenstein and Birnboim (1975)
		3.90	Turbidity	Camerini-Otero et al. (1974a)
		3.65	Sedimentation equilibrium	Camerini-Otero et al. (1974a)
		3.98	Sedimentation-viscosity	Camerini-Otero et al. (1974a)
		3.62	X-ray scattering	Zipper et al. (1971)
		3.60	X-ray scattering	Fischbach and Anderegg (1976)
MS2	81	3.60	Light scattering	Strauss and Sinsheimer (1963)
	81	5.30	Light absorption	Möller (1964)
	78.5	3.87	Sedimentation-diffusion	Overby et al. (1966a)
		3.60	Light scattering	Overby et al. (1966a)
fr	79	4.30	Sedimentation-diffusion	Marvin and Hoffmann-Berling (1963b)
		4.10	Sedimentation-viscosity	Marvin and Hoffmann-Berling (1963b)
		3.62	X-ray scattering	Zipper et al. (1971)
μ2	76	3.8 ± 0.2	Calculated from the RNA percentage and molecular mass	Piffaretti and Pitton (1976)
Qβ	84.3	4.29	Sedimentation-diffusion	Overby et al. (1966a)
		4.20	Light scattering	Overby et al. (1966a)
		4.55	Light scattering	Camerini-Otero et al. (1974b)
φCB5	70.6 ± 2.0		Sucrose gradient centrifugation	Bendis and Shapiro (1970)

Source: The older data are taken from Boedtker H, Gesteland RF. Physical properties of RNA bacteriophages and their RNA. In: RNA Phages. Zinder ND (Ed). Cold Spring Harbor Laboratory, Cold Spring Harbor, NY, USA, 1975, pp. 1–28 with more recent additions.

than 1 megadaltons. However, it was discovered that the phage MS2 passed through filters with 750 K-, 500 K-, and 300 K-dalton values but was retained on the 100 K-dalton filter (Wick and Patrick 1999a).

It is interesting that the phage fr was used as a reference by determination of the biophysical parameters for so-called Australia-SH-antigen, or 22-nm hepatitis B surface antigen, or HBsAg, particles from the HBsAg-carrier blood (Schober et al. 1971). As a result, the average diameter of the HBsAg particles was calculated to 21.2 ± 2 nm, which corresponded well to the size of the HBsAg 22-nm particles estimated by electron microscopy.

SIZE

The first reported diameters ranging from 20 nm for f2 (Loeb and Zinder 1961) to 27 nm for M12 (Hofschneider 1963) agreed reasonably well with the diameter of the dry particle calculated from the molecular mass and the partial specific volume, assuming the particle is a sphere, so that diameters of 20.6 nm for MS2 and 21.2 nm for Qβ were calculated on the basis of the biophysical data (Boedtker and Gesteland 1975). When the diffusion coefficients D of the RNA phages were determined by the method of intensity fluctuation spectroscopy

and the appropriate molecular mass values were determined, the calculated hydrodynamic radii appeared as 140 ± 2 Å for R17 and 151 ± 2 Å for Qβ (Camerini-Otero 1974b). Therefore, a reasonable agreement of the latter with the electron microscopical observations was achieved. It is noteworthy that the phage R17 was employed by the development of the light-scattering technique (Pusey et al. 1974).

Table 9.2 presents the electron microscopy data on the size of particles. In fact, all RNA phages appeared very similar in electron microscope photographs. The obtained diameters of the particles are given in comparison with the high-resolution x-ray crystallography measurements that appeared since 1990 when the first high resolution structure of the RNA phages, namely, of the phage MS2, was published (Liljas and Valegård 1990; Valegård et al. 1990). The actual particle size data in Table 9.2 were compiled from the VIPERdb (http://viperdb. scripps.edu) database (Carrillo-Tripp et al. 2009). This highly informative database is designated and maintained by Vijay S. Reddy.

Remarkably, the x-ray scattering measurements resulted in an estimate of the mean outer diameter 26.6 nm for the phage R17 (Fischbach et al. 1965; Fischbach and Anderegg 1976). Later, this diameter was found to be 26.34 nm in the case of both R17 and fr (Zipper et al. 1971a). These diameters

TABLE 9.2

Diameters of the RNA Phages Determined by Electron Microscopy in Comparison with the High-Resolution 3D Data

Phage	Diameter, nm	References	Diameter, X-ray Resolution, nm		References
			Outer	Average	
f2	20	Loeb and Zinder (1961)			
MS2	26	Strauss and Sinsheimer (1963)	28.8	27.6	Valegård et al. (1986, 1990, 1991); Liljas and Valegård (1990); Golmohammadi et al. (1993)
	25	Overby et al. (1966a)			
M12	27	Hofschneider (1963a)			
fr	21	Marvin and Hoffmann-Berling (1963b)	28.6	27.6	Bundule et al. (1993); Liljas et al. (1994)
	26.34	Zipper et al. (1971)			
R17	20	Crawford and Gesteland (1964)			
	26.6	Fischbach et al. (1965); Fischbach and Anderegg (1976)			
	26.34	Zipper et al. (1971)			
ZIK/l, ZJ/1, ZG/l	22.5	Bradley (1964a,b)			
7s	25	Feary et al. (1964)			
φCB5	21–23	Schmidt and Stanier (1965)	28.6	28.0	Plevka et al. (2009a)
	23	Bendis and Shapiro (1970)			
PP7	25	Bradley (1965b)	30.2	29.2	Tars et al. (2000a,b)
Qβ	25	Overby et al. (1966a)	29.4	28.6	Valegård et al. (1994a); Golmohammadi et al. (1996)
PRR1	25	Olsen and Thomas (1973)	28.8	27.8	Persson et al. (2008)
φCp2	29.2	Miyakawa (1976)			
φCp18	28.5	Miyakawa (1976)			
$F_0 lac$	28	Bradley et al. (1981a)			
t	25	Bradley et al. (1981b)			
C-1	27 ± 2	Sirgel et al. (1982)			
Iα	24	Coetzee et al. (1982)			
M	27	Coetzee et al. (1983)			
pilHα	25	Coetzee et al. (1985a)			
D	27	Coetzee et al. (1985b)			
$F_0 lac$ h	28	Coetzee et al. (1986)			
SR	28	Coetzee et al. (1986)			
AP205	27–30	Klovins et al. (2002)	27.6	27.2	Shishovs et al. (2016)
GA			28.8	27.8	Tars et al. (1997)

are closer to the recent measurements by high-resolution techniques than the electron microscopy-based estimations.

Interestingly, the first laser beat frequency spectroscopy measurement gave rather unexpected 41.1 ± 1 nm value for the MS2 diameter that was explained by large hydration (French et al. 1969). However, a bit later, photon-correlation spectroscopy of the phage R17 yielded a value of a hydrodynamic radius of only 140 Å, or 28 nm diameter (Camerini-Otero et al. 1974b). The further light scattering data led to 27 nm diameter for the phage R17 determined by Newman et al. (1974) and 27.8 nm determined by Pusey et al. (1972). This apparent discrepancy among closely-related MS2 and R17 phages was solved by Nieuwenhuysen and Clauwaert (1978), who reported full experimental identity of the hydrodynamic radius and volume, hydration, and molecular mass of both R17 and MS2. At that time, the 27 nm diameter for the

phage MS2 was obtained also by small-angle neutron scattering methodology (Jacrot et al. 1977).

COMPOSITION

The RNA phage particles were originally regarded to consist of the two components: the single-stranded RNA molecule and the coat protein (Zinder 1965) until the A, or maturation, protein was identified as a single copy complexed with the phage RNA (Steitz 1968a,b,c). By the *Allolevivirus* representatives, this maturation protein was designated A2, or IIa in early studies. The maturation protein appeared as a mandatory component to ensure the infectivity of the phage particles (Roberts and Steitz 1967). By the *Allolevivirus*, but not by the *Levivirus* representatives, the A1, or IIb in early studies, was identified first in the Qβ particles as a third component of the particle (Weiner and Weber 1971). As explained before, the A1 protein was nothing else as the C-terminally extended coat protein after ribosomal readthrough of the leaky *opal* termination codon of the coat gene (Weiner and Weber 1971). The A1 protein was found essential for formation of the viable Qβ particles (Hofstetter et al. 1974; Engelberg-Kulka et al. 1977, 1979a). According to more recent studies, 12 copies of the A1 protein are present in the Qβ particle (Skamel et al. 2014).

SYMMETRY

The spherical, or isometric, RNA phage particles displayed icosahedral symmetry. The name comes from the Greek είκοσι, meaning "twenty," and έδρα, meaning "seat." After the hypothesis proposed by Crick and Watson (1956, 1957) to explain the structure of the spherical virus particles, the icosahedral nature of the RNA phage symmetry was acknowledged generally after the mid-1960s (Bradley 1964b, 1965b; Gussin et al. 1966; Vazquez et al. 1966; Hohn and Hohn 1970; Dunker 1974; Crowther et al. 1975; Dunker and Paranchych 1975). The more detailed story of the general phage symmetry, together with the appropriate illustrations, is expounded in Chapter 1.

The early models of the particle structure of the phages R17 (Finch 1968, cited by Crowther et al. 1975) and fII (Hohn and Hohn 1970) have suggested that the 180 protein subunits were arranged so that they avoided the fivefold and threefold (quasi-sixfold) positions of a T = 3 surface lattice. A novel interpretation of the electron microscopy data was suggested by Crowther et al. (1975). Thus, the RNA phages contributed strongly to the idea of the structural analysis of macromolecular assemblies by image reconstruction from electron micrographs (Crowther et al. 1975), which was exhaustively reviewed at that time by R. Anthony Crowther and Aaron Klug (1975) and a bit later by Crowther (1982). The long way from these first applications to a recent impressing electron cryomicroscopy of the RNA phage particles (Koning et al. 2016; Dai et al. 2017; Twarock et al. 2018a,b), which enabled the study of the phage particles in unprecedented detail, will be tracked in Chapter 21.

MORPHOLOGY

Concerning the early electron microscopy resolutions, David E. Bradley, a guru of phage electron microscopy, revealed 92 capsomers within the icosahedral RNA coliphage particle, although a picture with 42 possible subunits was also presented (Bradley 1964b, 1965b). Based on morphology, Bradley (1965b) supposed some difference of coliphages from the pseudomonaphages, since the phage 7s, the first representative of the latter, demonstrated greater detail in the capsid morphology (Feary et al. 1964); the capsomers appeared to be larger than those of the coliphage ZIK/l and their appearance was more consistent with an icosahedron with the 42 capsomers. However, the pseudomonaphage PP7 isolate, a close analog of the phage 7s, which was obtained by Bradley (1965b), looked the same as the RNA coliphages

with no marked capsomer structure. Finally, Bradley (1965b) concluded that the difference was more likely to be due to the negative staining technique rather than to the real difference of the phage structure. However, Bradley (1966b) suggested an opinion that the RNA coliphages were not structurally identical with the pseudomonaphages, which was mostly based again on the same nice micrographs of Feary et al. (1964). In the next large review, Bradley (1967) repeated the idea that the coli- and pseudomonaphages are morphologically distinct because of the differences in the capsomer size.

The excellent micrographs of the phage R17, including two-dimensional crystalline arrays, were published by Vasquez et al. (1966). The authors indicated the icosahedral symmetry with 32 morphological units, probably situated on the lattice points of a T = 3 icosahedral surface lattice, although they admitted that the particle morphology cannot be easily distinguished by the electron microscope.

As clearly explained in an excellent review of Hohn and Hohn (1970), the RNA phages could be regarded as one of the first examples to confirm the theory of the *quasiequivalent arrangement of subunits*, which was intended to overcome the fact that most isometric viruses, including the RNA phages, had more than 60 identical subunits (Caspar and Klug 1962). According to this theory, the postulate of strictly mathematical equivalence of the subunits in viral shells could be dropped without violating the physical principle that in the formation of the capsid the same contacts between subunits were used over and over again. According to the theory, more than 60 identical subunits could be arranged on the surface of an icosahedron in similar binding of all subunits with their neighbors if a certain variability in the angle of the bonds and/or a certain deformation of the subunits could be permitted (Hohn and Hohn 1970).

As a result, Hohn and Hohn (1970) proposed alternate models for the icosahedral arrangement of 180 identical subunits into the triangulation number T = 3 particles with no clustering or with clustering in trimers and in hexamers and pentamers (see Figure 1.5 in Chapter 1).

It is noteworthy that the phage R17 helped at that time to evaluate the damage of the biological samples by the electron beam during electron microscopy, although it remained unclear how the results of this study could be applied to the case of negatively stained particles (Thach and Thach 1971).

The next ingenious spatial model was proposed not long after (O'Callaghan et al. 1973a; Dunker and Paranchych 1975; Paranchych 1975), which was based on the electron microscope and ultracentrifugation studies of the phage R17 preparations that had been subjected to controlled degradation by guanidine hydrochloride. It was proposed that the R17 protein coat was not arranged into 32 morphological capsomers as postulated by Vasquez et al. (1966), but rather formed a network of protein arising from 20 "donut-shaped" nonagons, or capsomers containing nine coat protein subunits, arranged in icosahedral symmetry. The A protein was envisaged as being attached to the surface of the RNA core and protruding through the coat protein network at one of the empty spaces existing at each of the 12 vertices of the icosahedron (Paranchych 1975).

COAT-RNA CONTACT

The Raman spectroscopy technique was applied to answer questions about the cooperation of RNA and protein components in the RNA phages. The first laser-excited Raman spectrum of the RNA phage was reported for the phage R17 (Hartman et al. 1973). It contained a large number of Raman lines assignable to scattering by vibrations of the nucleotide residues of RNA and the amino acid residues of protein capsomers. The Raman lines from specific nucleotide vibrations in the phage were compared with their counterparts in the spectrum of protein-free RNA to suggest many similarities of RNA structure in the phage and protein-free states. However, the average configuration of guanine residues in the phage was apparently very different from that of protein-free RNA, suggesting that guanine played an important role in the RNA-protein interactions (Hartman et al. 1973). Furthermore, the systematic laser–Raman study was performed to compare the Raman spectra of the native MS2 phage, heat-degraded MS2 phage, MS2 capsids free of RNA, and MS2 RNA free of protein (Thomas et al. 1976). The results of this study provided new information on the structures and interactions of the MS2 and its component molecules. For example, it was concluded, before the further high-resolution structures, that the coat molecules in the native phage maintained a conformation determined largely by regions of β-sheet (~60%) and random chain (~40%) structures. No disulfide bridges were found. The protein–protein interactions within the phage particle were stable up to 50°C. The vibrational Raman optical activity study was undertaken later by the phage MS2, in parallel with a range of different structural types of virus exemplified by filamentous DNA phage fd, tobacco mosaic virus, satellite tobacco mosaic virus, and cowpea mosaic virus (Blanch et al. 2002; McColl et al. 2003).

The circular dichroism of the phage μ2 was close to the sum of the native RNA and protein components and thermal absorbance profiles of the phage, and its RNA showed that there was no energetically important interaction between the double-helical parts of the RNA and the protein (Isenberg et al. 1971). Then, the circular dichroism was used to compare the isolated and the natural *in situ* phage R17 RNAs (Bobst et al. 1976). The thermal circular dichroism melting profiles exhibited the presence of free energy of interaction between RNA and coat protein. It was suggested that the heat and pH induced conformational alterations of the R17 RNA *in situ* coinciding with loss of infectivity, which occurred after an *in situ* alteration of nucleic acid–coat protein interaction (Bobst et al. 1976).

In another study, the circular dichroism, combined with hydrodynamic measurements of gross size and shape, of the empty f2 shells, which have been obtained by short alkaline treatment, did not demonstrate any major alterations, while combined circular dichroism spectra of the isolated protein shell and RNA components were very similar to the circular dichroism of the intact phage f2 (Henkens and Middlebrook 1973). These results indicated that the protein–nucleate interactions in the phage structure did not result in major conformational alteration of the macromolecular components.

The RNA–protein interactions in the phages MS2 and Qβ were compared with those in messenger ribonucleoproteins and ribosomes via phosphorus-31 NMR relaxation (Bolton et al. 1982). The NMR parameters were measured for the RNA-coat complexes and for free RNA, where the results were dependent on the particular complex examined. The ^{31}P NMR results for the phages MS2 and Qβ suggested that the RNA phosphodiester moiety in the particles had an altered mobility compared to that of free RNA.

The presence of specific noncovalent interactions between the RNA and proteins within the MS2 particles was proven by bisulfite-induced formation of polynucleotide–protein crosslinks, where both coat and maturation proteins were involved (Turchinsky et al. 1974).

DEFECTIVE PARTICLES

The history of the defective particles is narrated clearly by Boedtker and Gesteland (1975). Thus, the first recognized type of the defective particles was connected with the mutation in the maturation, or A protein, gene. The infection of nonpermissive host with phage carrying an *amber* mutation in the A protein gene led to production of noninfectious, defective particles in the case of the phages f2 (Lodish et al. 1965), R17 (Argetsinger and Gussin 1966), and fr (Heisenberg and Blessing 1965). The lack of the A protein in such particles was demonstrated by Steitz (1968b).

Kaerner (1969) made the defective A protein-deficient particles of the wild-type nonmutant phage fr by using the fact that the coat protein did not contain histidine. In his experiment, infected histidine-auxotrophic cells were starved for histidine at a time when the replicase protein was synthesized. Therefore, the A protein was not synthesized, while the synthesis of coat protein continued. The defective particles without A protein contained RNA, but, unlike normal phage particles, the RNA in them was partially sensitive to ribonuclease so that under normal growth and purification conditions only RNA-deficient particles were found, as demonstrated earlier for the maturation gene mutants of the phages f2 (Engelhardt and Zinder 1964; Lodish et al. 1965) and fr (Heisenberg and Blessing 1965). However, when the particles were grown in an RNase I⁻ host, or the ribonuclease effect was minimized maximally during purification, the particles of the phages R17 (Argetsinger and Gussin 1966) and fr (Heisenberg 1966) were obtained, which possessed non-degraded infectious RNA, although they were not infectious themselves.

According to the above-cited review (Boedtker and Gesteland 1975), the defective particles looked normal in the electron microscope and had normal density, but sedimented at only 70S as compared to 80S for the normal phage. Upon treatment with ribonuclease, either during preparation or by addition of pancreatic ribonuclease, about 30% of the phage RNA was degraded, resulting in a rather heterogeneous population of particles with a lower density in CsCl and a sedimentation constant of about 74S. It was concluded that the phage particles that have been assembled without A protein seemed to have a portion of their RNA exposed to the exterior. The evidence from the experiments using the defective RNA as a messenger for *in vitro* protein synthesis suggested that the 5′ end of the RNA might be preferentially cleaved off.

The direct serological and chemical comparison of the phage R17 and R17amA31 defective particles was performed by Iglewski (1977). The immunodiffusion analysis demonstrated identity between intact R17 and amA31 capsids and between dissociated subunits of both R17 and amA31 and purified coat protein. Remarkably, the radioimmunoassays detected an antibody in R17 antisera that bound intact R17 but could not be absorbed from R17 antisera with amA31, while the R17 antibody remaining in amA31-absorbed sera did not neutralize infectivity of R17 phage. The differences between the surface composition of R17 and amA31 capsids were also detected by iodination, where capsids of R17 bound approximately four times more ^{125}I than amA31. Finally, the amA31 capsids dissociated under milder conditions of the sodium dodecyl sulfate treatment than R17 capsids.

The second type of defective particles resulted from the substitution of uracil by 5-fluorouracil in the RNA of the phages MS2 (Shimura et al. 1965) and f2 (Lodish et al. 1965). If the concentration of 5-fluorouracil in the growth medium was chosen so that 80% of the uracil residues were replaced by the analog, the majority of the progeny phage particles were defective in that they were noninfectious, did not adsorb to bacteria, were heterogeneous in buoyant density, and were lighter than normal phage (Shimura et al. 1967). Nevertheless, these defective particles appeared normal in the electron microscope and had a normal protein complement, but about 30% of the RNA was missing. The 5-fluorouracil-substituted RNA of the phage MS2 appeared as a 65% fragment that had 1.5 times the normal amount of pppGp after alkaline hydrolysis, showing that the normal 5′ end was present and that the 3′ end was defective (Shimura et al. 1968). The 5-fluorouracil-substituted RNA was effectively translated *in vitro*, yielding coat protein but no replicase, indicating that the replicase gene was located at the 3′ part of the MS2 RNA (Shimura et al. 1968). As noted properly by Boedtker and Gesteland (1975), the origin of the defective 5-fluorouracil particles remained unclear. Shimura et al. (1967) noticed that the defective particles showed serological differences from normal phage and hypothesized that the phage proteins made in the presence of 5-fluorouracil were altered because of coding errors, which resulted in the assembly of imperfect virions which did not properly protect the RNA from nuclease attack. Other possible mechanisms for the formation of the 5-fluorouracil particles, such as altered A protein, altered RNA folding, or altered interaction between the 5-fluorouracil-substituted RNA and normal proteins, have not been ruled out.

In a similar way, the f2 particles formed in the presence of 5-azacytidine had nearly normal sedimentation coefficient and buoyant density, but their infectivity with respect to populations of normal phage was reduced, while RNA extracted from these particles was partly degraded to small fragments and had very low template activity (Doskočil and Šorm 1971).

The third type of defective particles was represented by abnormal structures that were found in the normal wild-type populations as naturally-occurring components. Thus, an abnormal minor structure was found for the phage MS2 in normal lysates (Rohrmann and Krueger 1970a). These particles were recognized by their lower buoyant density in CsCl density gradients and were thus called *light* (L) particles. The infectivity of the L particles was low, but they possessed normal RNA that was infectious in spheroplasts and had normal amounts of coat protein and A protein.

Regīna Renhofa, from Elmārs Grēns' team, found a defective phage fraction with different buoyant density within the wild-type MS2 population, which was not minor in fact and reached up to 50% of the total phage yield (Renhof and Gren 1972). This fraction appeared as a specific, a bit diffuse band by the CsCl density gradient centrifugation, demonstrated low infectivity, and possessed specifically fragmented RNA, although the $S_{20,w}$ value for both the defective particles and MS2 was the same and determined as 80S. The RNA:protein and A-protein:coat protein ratios did not differ in the defective and native MS2 particles, while RNA of the defective particles was composed mainly of three specific fragments, their length being about 60%, 50%, and 40% of that of the native MS2 RNA (Renhof et al. 1974). The major break occurred in the 3′-terminal half of MS2 RNA, which led to the formation of short 3′-terminal and long 5′-terminal fragments with 40% and 60% of the whole RNA length (Renhof et al. 1974). The molecular mass of the specific RNA fragments within the defective particles was estimated further as 6.5×10^5, 5.5×10^5, and 4.4×10^5 with molar ratio 5:4:9, respectively (Renhof et al. 1975).

Propst Ricciuti and Haywood (1974) detected production of the defective MS2 particles in sulfur-starved *E. coli*, when a 100-fold decrease in the yield of plaque-forming units was explained by the formation of noninfective phage particles lacking A protein.

In the case of the *Allolevivirus* representatives, the Qβ particle variants avoided of the A1 protein, formed noninfectious particles (Radloff and Kaesberg 1973; Engelberg-Kulka et al. 1977, 1979a).

After the naturally produced defective particles, the structural changes were induced in the wild-type phages by such actions as salt, pH, or temperature treatment, which resulted in artificially modified particles, similar in some respects to natural defective particles. Steitz (1968b) first showed that heating the phage R17 at 46°C in relatively low ionic strength (0.15 M NaCl) buffer converted the phage from its normal 80S form to a 45–50S form. The latter form had a normal set of RNA and protein (including the A protein), so that the observed decrease in the sedimentation rate must be due to a change in the shape. The similar effect of heat was observed with the phage MS2 (Verbraeken and Fiers 1971). These particles retained 30%–50% infectivity (Steitz 1968a; Verbraeken and Fiers 1971, 1972b). The RNA within these altered particles was partially sensitive to ribonuclease: Verbraeken and Fiers (1972b) found that MS2 heated in 0.005 M Tris has lost 60% of its RNA, while Steitz (1968a) reported the nearly total (about 90%) sensitivity of the R17 RNA to the action of ribonucleases.

Such treatments as freezing and thawing (Hohn 1967) and heating at higher salt concentrations (Kaesberg 1967; Oriel and Koenig 1968) led not only to the ribonuclease sensitivity, but also to the total loss of infectivity, in contrast to the low-salt treatment described above.

RECONSTITUTION

The history and general principles of the RNA phage reconstitution have been touched on briefly in the Reconstitution section of Chapter 1. By going back in the history, the first two-component reconstitutions were performed for the phages fr (Hohn 1967) and MS2 (Sugiyama et al. 1967). In both cases, the reconstitution was performed from phenol-extracted phage RNA and coat protein isolated by the acetic acid method. Although the reconstitution products resembled the authentic phage in electron microscopy, they were noninfectious due to lack of the A protein.

Roberts and Argetsinger-Steitz (1967) reconstituted the phage R17-like particles from three components, by addition of the purified A protein to the RNA and coat protein. The same was done later in the case of the phage fr (Herrmann R. Thesis, Univ. of Tubingen, 1970, cited by Hohn and Hohn 1970). Addition of the A protein to the system resulted in the improved infectivity of the reconstituted particles, although the yield of infectious particles still remained low because of the problems with the A protein purification. The A protein was highly insoluble in aqueous buffers and could be obtained at high concentrations only in denaturing solvents (Roberts and Argetsinger-Steitz 1967).

The particles with the same buoyant density and sedimentation behavior as authentic phage particles were obtained by *in vitro* stimulation with fr RNA of (i) S30 extract from the phage fr-infected *E. coli* (but not from uninfected) or (ii) a purified amino acid incorporating system derived from uninfected cells (Knolle 1969). In the case of the S30 extracts from the fr-infected cells, the stimulating effect of the fr and MS2 RNAs on the outcome of the virus-like particles (VLPs) was similar, but Qβ RNA was ineffective.

The infective particles were reconstituted *in vitro* from purified RNA, coat protein, and a minor protein of the phage Qβ, where the precise nature of the minor protein was not defined, but the infectivity was dependent clearly on the presence of this protein (Hung and Overby 1969). Later, Hofstetter et al. (1974) reported a new method for the purification of all Qβ phage components and showed that each of the three purified capsid proteins, namely coat, A1, and A2, was required absolutely for the formation of the infectious phage particles.

The four kinds of particles were reconstituted with RNA and coat protein from the genetically unrelated phages MS2 and Qβ, namely two homologous and two heterologous, with respect to RNA and protein (Hung et al. 1969b). However, once Qβ RNA (or MS2 RNA) reacted with a few molecules of either Qβ or MS2 protein to form a nucleoprotein complex, or initiation complex, by authors' terminology, it formed a VLP only with subsequent addition of the same protein. In this study, phenol-extracted RNA was used, but proteins were obtained by the guanidinium-HCl method. Under conditions where MS2 and Qβ phage RNAs competed for one phage protein, the formation of homologous particles prevailed (Ling et al. 1969). However, when two proteins competed for one RNA, Qβ protein was predominantly incorporated into particles with either RNA. Remarkably, the particles with mixed protein coat were never detected. Thus, the interaction between protein subunits appeared to have absolute species specificity (Ling et al. 1969). The *in vitro* preferences by the phage reassembly were confirmed by the *in vivo* experiments. Thus, in the case of double Qβ and MS2 infection, the two phages interfered with each other and resulted in reduced infection by each phage, but the produced phage progenies were authentic Qβ and MS2 (Ling et al. 1970). No phenotypic mixing, i.e., mixed coat particles, and genomic masking, i.e., hybrid particles, were detected. The mixed coat protein particles were obtained later with the two related representatives of group IV, namely SP and FI (Miyake and Shiba 1971). More recently, mixed spherical and rod-like particles were obtained from the group I and II phages fr and GA by *in vitro* reassembly from a mixture of the coat protein dimers together with *E. coli* ribosomal RNA or by the simultaneous *in vivo* expression of the fr and GA coat genes (Rumnieks et al. 2009). However, coinfection by the two phages led to the authentic fr and GA virions only.

A complex consisting of one RNA strand and six protein subunits, so-called complex I, was regarded as a possible precursor during the self-assembly process (Hohn 1969a). However, complex I appeared as very specific in respect to the RNA partner, and its formation rendered the RNA neither more nor less advantageous in forming VLPs with additional protein. Moreover, it was concluded that complex I could be a type of RNA not to be used in phage formation, in order to preserve its modulated messenger function (Hohn 1969a).

Rohrmann and Krueger (1970c) identified at least two intermediates present during the self-assembly of the phage MS2, which were antigenically unrelated to the capsid.

Continuing his work on the above-mentioned noninfectious VLPs, which appeared during histidine starvation (Kaerner 1969), Kaerner (1970) presumed that the association of viral RNA with a histidine-containing protein, presumably maturation protein, was an early step in the process of phage maturation, while the complexes of viral RNA and coat protein formed in absence of the maturation protein were generally excluded from the maturation process.

The ability of different nonspecific RNAs to participate in the VLP formation was confirmed further by the involvement of not only fragments of the phage RNAs or genetically distal RNAs, but also of plant virus TMV and TYMV RNA and even polyuridylic acid and polyvinyl sulfate, while only RNAs smaller than 20 nucleotides remained inactive (Hohn 1969c; Hohn and Hohn 1970). The reverse was also true, since the plant virus cowpea chlorotic mottle virus (CCMV) protein was self-assembled around the phage f2 RNA (Hiebert et al. 1968).

Moreover, the assembly of VLPs, or the RNA phage VLPs, by modern terminology, was observed in the absence of any

RNA in the case of the phages fr (Herrmann et al. 1968; Hohn 1969b; Schubert and Frank 1970a,b, 1971; Zipper et al. 1971b, 1973), f2 (Matthews 1970; Zelazo and Haschemeyer 1970; Matthews and Cole 1972a,b,d), R17 (Samuelson and Kaesberg 1970), and MS2 (Oriel and Cleveland 1970; Rohrmann and Krueger 1970c).

Thus, for the first time, the self-assembly of the empty fr particles was achieved from the fr coat protein that was obtained after the phage RNA was precipitated with acetic acid (Herrmann et al. 1968). By measuring the circular dichroism, the reversible depolymerization of the phage fr coat by acetic acid was found to be coupled with a reversible denaturation, occurring in several stages (Schubert 1969). The native conformation of the coat protein subunit was replaced by a more helical one when the phage coat was split with 11 M acetic acid, but after lowering the acetic acid concentration below 0.3 M the peptide chain began to refold into the native conformation. Furthermore, Schubert and Frank (1971) showed that the fr VLPs could be re-associated from the coat subunits prepared by the glacial acetic acid treatment. The two kinds of aggregates were observed: VLPs shown by electron microscopy to be hollow spheres, and particles that consisted of a virus-like center and additional protein layers surrounding the nucleus, or *multi-shell* particles. The type of the observed aggregate was mainly determined by salt and protein concentrations. The high ionic strength (>0.5) favored the formation of VLPs, whereas at low ionic strength and high protein concentration (>0.5 mg/mL), the multi-shell particles were assembled. At the same time, Schubert and Frank (1970b) used also 2-chloroethanol for the reversible depolymerization of the protein shell of the phage fr: after removal of RNA and replacement of the organic solvent by water, the VLPs were obtained by dialysis of the protein against neutral buffers of high ionic strength, whereas multi-shell particles were formed in buffers of low ionic strength. The action of 2-chloroethanol was very similar to that of acetic acid. The clear evidence that the reaggregated VLPs of the phage fr were indeed empty shells, which were very similar to the protein shell of the native phage, was gained not only by the electron microscopy but also by the small angle x-ray scattering measurements (Zipper et al. 1971b).

After the acetic acid treatment, hollow protein shells similar in structure to the protein portion of the phage R17 were obtained when the phage R17 was exposed to pH 10.6 to yield a protein and products of alkaline hydrolysis of the viral RNA (Samuelson and Kaesberg 1970).

Zipper et al. (1973) performed an excellent comparative small-angle x-ray scattering study of the VLPs derived from the phage fr by alkaline degradation of the phage RNA, as well as by self-assembly of the coat protein subunits in the absence of nucleic acid. The radial dimensions of both types of particles were practically identical to those of the protein shell of the native phage, with the inner radius being 10.6 nm and the outer radius being 13.2 nm. The $S_{20,w}$ values of the empty particles were also similar: the alkaline-treated particles sedimented with 42S, while the reaggregated protein particles were found to sediment on the average with 45.4S.

Oriel and Cleveland (1970) presented a 9-M urea method for the reassembly of the phage MS2 protein subunits without RNA. The gained particles resembled intact MS2 in size and antigenicity. Moreover, using optical rotatory dispersion and circular dichroism, Oriel et al. (1971) fixed conformational changes of the coat during reassembly from guanidine hydrochloride. According to the obtained data, the structure of the coat resembled β-structure with little or no α-structure.

The search for the putative precursors of the coat shell led to the discovery of the so-called 11S component that appeared by the partial disaggregation of the phage f2 with guanidine hydrochloride (Zelazo and Haschemeyer 1969). The molecular mass of this stable degradative intermediate indicated the presence of 19 ± 1 polypeptide chains, while electron microscopy revealed a predominant ellipsoidal structure with dimensions of about 103×50 Å. The appearance of the 11S structures was compatible with the 180 subunit-60 capsomer model, but not with the 32 or 90 capsomer models of 180 subunits, which were discussed at that time. Furthermore, Zelazo and Haschemeyer (1970) succeeded in the reassembly of the 11S subunits into the 37S particles that closely resembled those expected for the RNA-free virus-like capsids. The latter appeared during dialysis of the 11S subunits to pH 4 at 24 to 34°C, while the same dialysis at low temperatures provoked further dissociation of the 11S subunits into 5.5S subunits.

The 11S forms were detected also by the sequential study of alterations in the phage R17 structure induced by guanidine hydrochloride (O'Callaghan et al. 1973a). Dialysis of the phage R17 against 3.0 M guanidine at 4°C resulted in a loss of infectivity and subtle changes in the sedimentation properties of the resulting noninfectious particles. Upon increasing the temperature of the reaction from 4 to 37°C, the three particle forms were found to have S values of 78, 58, and 44. Moreover, the protein fragments directly released from the 44S particles were found to be morphologically similar to the 11S subunits produced during the phage breakdown in higher concentrations of guanidine. The similar 11S structures were obtained from the phages R17, MS2, and f2 after treatment with 1 M sodium thiocyanate (MacColl et al. 1972).

The effect of pH, temperature, ionic strength, ion species, and protein concentration were defined accurately for the self-assembly of the empty f2 shells, while the presence of the phage RNA promoted shell formation under conditions where shells normally would not form spontaneously (Matthews and Cole 1972d). The self-assembly of the f2 shells was inhibited completely and reversibly by 0.3 M urea, as well as by other compounds containing the group N-C=N, such as purines, pyrimidines, pyridines, guanidinium chloride, imidazole, formamide, semicarbazide, citrulline, and arginine (Matthews and Cole 1972b). Concerning the effect of chemical modifications on the shell-forming ability of the f2 coat, the cleavage by cyanogen bromide resulted in the loss of the latter, although the fragments produced were active in inhibiting shell formation by unmodified protein (Matthews and Cole 1972a). The reactions aimed at sulfhydryl and methionine, as well as at arginine, led to loss of the ability to form shells.

The cleavage of the C-terminal tyrosine by carboxypeptidase produced a modified coat protein which was incapable of forming shells, but which prevented intact monomers from achieving the conformation necessary for shell formation. Remarkably, both shells and phage treated with carboxypeptidase did not release the C-terminal tyrosine residue (Matthews and Cole 1972a).

In the context of the empty particles, the first fluorescence study of the phage f2 was performed (Kitchell et al. 1977). The fluorescence properties of the phage f2 and its empty shell predicted location of tryptophan residues between subunits and an unusual tyrosine environment in both cases. Later, Kitchell and Henkens (1979) reported a single, low molecular mass protein, apparently a monomer, but not the 11S structure, as the result of the urea or guanidine hydrochloride treatment of the empty f2 capsids. The authors explained the appearance of the previously described 11S structure by a stabilizing role of the phage RNA.

The intermediates of the phage MS2 assembly *in vivo* were examined by Bonner (1974). The labeled MS2 protein appeared first as a low-molecular-mass peak at the tops of gradients, then as a peak at 40S and as a large number of almost inseparable structures between 40 and 80S, and finally as the 80S mature phage particles. During the chase of a short labeling period, radioactive phage protein was found to disappear from gradients in the same temporal order as it appeared; the soluble peak disappeared first, followed by the 40–70S region. The chased label appeared quantitatively in the 80S phage peak. The labeled phage RNA was found to appear first in the 40S peak, then in the structures between 40 and 70S, and finally in 80S phage particles. The order of disappearance of labeled phage RNA during a chase was the same as its appearance.

The first report on the dissociation of the phage Qβ and its discrete substructure was presented by Takamatsu and Iso (1982). After the controlled dissociation by SDS, the pentamers and hexamers were detected as the chemical phage entities of the phage Qβ. Their numbers per particle were about 12 for pentamers and about 20 for hexamers, consistent therefore with the theoretical expectation from the quasi-equivalent packing of 180 identical subunits in the coat protein shell. Both pentamers and hexamers were stabilized by intermolecular disulfide bonds, in contrast to the components of the group I phages described above. It was shown also in this breakthrough study for the first time that the A1 protein molecule could be substituted for a coat protein in a pentamer and a hexamer.

After the first high-resolution structure of the phage MS2 was published (Liljas and Valegård 1990; Valegård et al. 1990), Da Poian et al. (1993) examined the reversible pressure dissociation of the phage R17. In the absence of urea, pressures up to 2.5 kbar promoted only 10% dissociation, while in the presence of urea between 1.0 and 5.0 M, the pressure caused complete, reversible dissociation of the phage particles. At the lower urea concentrations, the reversible dissociation of the phage R17 particles showed no dependence on protein concentration, indicating a high degree of heterogeneity

of the particles, but higher urea concentrations, 2.5–5.0 M, resulted in the progressive restoration of the protein concentration dependence of the pressure dissociation. At still higher urea concentrations, 5.0–8.0 M, the irreversible phage dissociation took place at atmospheric pressure. In contrast, the dissociation of the isolated dimers of the capsid protein was dependent on protein concentration to the extent predicted for a stochastic equilibrium, and the dimers were much less stable than the whole virus both to dissociation by pressure or by urea. Remarkably, the experiments demonstrated that the *thermodynamic individuality* of the phage particles arose in conformational differences in the assembled phages, and that there was a direct relation between the stability of the particles and their heterogeneity.

Moreover, a theoretical model was presented that accounted for the facilitation of the pressure dissociation of the phage R17, and for the partial restoration of the concentration dependence of the dissociation, by the presence of sub-denaturing concentrations of urea (Weber et al. 1996).

Later, the dissociation process of the phage MS2 was studied by NMR relaxation measurements and, for the first time, the presence of intermediates was shown clearly (Anobom et al. 2003).

Recently, the disassembly of the Qβ VLPs was monitored by ¹⁹F-NMR (Leung et al. 2017).

It is necessary to add that the reconstitution-like experiments have been performed with the phage RNAs and non-phage coat proteins. Thus, typical plant virus rods were reconstituted with the MS2 RNA and coat protein subunits of tobacco mosaic virus (TMV) or turnip yellow mosaic virus (TYMV) (Matthews and Hardie 1966). The authors found that neither MS2 RNA nor MS2 RNA reconstituted with TMV protein showed any infectivity for *E. coli*. Similar "mixed reconstitution" between MS2 RNA and TMV protein, which resulted in the formation of TMV-like rods, was reported also by Sugiyama (1966). The spherical particles of cowpea chlorotic mottle virus (CCMV) particles were assembled *in vitro* around the phage f2 RNA (Hiebert et al. 1968). The MS2 RNA stimulated *in vitro* assembly of purified retrovirus, namely, Mason-Pfizer monkey virus, CANC proteins, which consisted of a fusion protein comprised of the capsid and nucleocapsid domains of Gag (CANC) and its N-terminally modified mutant (DeltaProCANC) (Ulbrich et al. 2006; Kuznetsov et al. 2007; Voráčková et al. 2014; Füzik et al. 2016; Píchalová et al. 2018).

Concerning foreign non-coat proteins, Dunker and Anderson (1975) tried the binding of the gene-5 protein from the filamentous phage fd to the R17 RNA and found it several hundred-fold weaker than the binding to the fd DNA. As to other phage proteins, Suau et al. (1980) studied the interaction of the MS2 RNA with the phage T4 coded 32-protein (P32), the first of the group of proteins which bound tightly to single-stranded DNA. Concerning plant viruses, the coat proteins of tobacco streak virus, brome mosaic virus, cucumber mosaic virus, and southern bean mosaic virus recognized specific sites on alfalfa mosaic virus RNA 1, but not on the MS2 RNA (Zuidema and Jaspars 1985).

MONITORING

A set of biophysical methods was proven to pave the way to the highly sensitive detection of microorganisms. As one of the first approaches to be mentioned, the phage MS2 was labeled by tritium to a high specific radioactivity of 20–50 Ci/mmole and retained infectivity, while the radioactivity was distributed by 1:3 ratio between the phage RNA and protein (Neiman et al. 1986).

Since the 1980s, mass spectrometry started to be involved more and more in the rapid and sensitive identification of microorganisms. The matrix-assisted laser desorption/ionization (MALDI) spectrum of an unfractionated aliquot of growth medium containing the phage MS2 and the *E. coli* host cells was presented by Fenselau et al. (1997) in the Proceedings of the First Joint Services Workshop on Biological Mass Spectrometry. Remarkably, the coat protein of the phage MS2 provided the dominant signal in the spectrum with the 13,787 mass, while this signal represented approximately 1 fmol of protein in the mixture. Furthermore, this rapid method for MS2 detection was developed and applied to TMV and Venezuelan equine encephalitis virus, and the method's sensitivity in the low-femtomole range, as estimated by titering plaque-forming units of MS2, was confirmed (Thomas et al. 1998). In another experiment, the MS2 coat protein was captured from solution on an antibody column, the protein was released and digested, and the proteolytic peptide fragments were analyzed by MALDI time-of-flight (TOF) mass spectrometry (Ashton et al. 1999). The role of the phage MS2 by the early development of the mass spectrometry analyses, together with instrumentation, sample collection, sample preparation, and algorithms for data analysis, was reviewed by Fenselau and Demirev (2001). The most impressive conclusion of this review was that these analyses should be carried out optimally in less than 5 minutes. Furthermore, Yao et al. (2002a,b) presented the MALDI-based method for microorganism identification, where no isolation or fractionation of microorganisms and no special search algorithm or database were needed but samples of the unfractionated intact microorganisms were subjected to brief on-slide proteolytic digestion. Next, the phage MS2 was used as a standard when detection was simplified by rapid acid digestions, which were carried out using a microwave system and provided an efficient alternative to the enzymatic digestion-based methods for virus identification (Swatkoski et al. 2007). The specificity of the proposed MS2 identification allowed for the discrimination between MS2 and other closely related RNA phages, namely BO1, M12, JP501, and fr. Although the phage MS2 coat and the R17 coat differed in intact molecular mass by a single dalton, the high-sequence coverage and specificity provided by this chemical digestion method enabled such discrimination. In the case of the phage BO1, a single amino acid change was sufficient to disrupt six of the peptide mass matches (Swatkoski et al. 2007).

Using electrospray ionization mass spectrometry (ESI-MS), Tito et al. (2000) obtained a mass spectrum for the phage MS2 with a measured molecular mass of 2,484,700 ± 25,200 Da,

which was to date the largest complex observed by the ESI-MS TOF methodology. Then, Cargile et al. (2001) employed electrospray quadrupole ion trap mass spectrometry to confirm the presence of the phage MS2 in *E. coli* lysates based on the isolation and fragmentation of the multiply-charged molecular ions of the MS2 coat. It is necessary to note here that the MS2 sample was analyzed earlier using the integrated virus detection system (IVDS) instrument, or more directly by the gas-phase electrophoretic mobility molecular analyzer (GEMMA) detector, which was one stage of the IVDS instrument, while the GEMMA detector consisted of an electrospray unit to inject samples into the detector (Wick and McCubbin 1999). This approach made it possible to overcome the mass spectrometry limitations to low resolution mass information. Thus, the electrospray ion mobility spectrometry detected the MS2 modal diameter as 23.6 nm (full width at half maximum [FWHM], 22.9–24.3 nm), therefore, with an accuracy that was sufficient to differentiate the phage MS2 from rice yellow mottle virus (RYMV) with modal diameter 27.9 nm and FWHM of 27.0–28.6 nm (Thomas JJ et al. 2004). Furthermore, the GEMMA analyzer was used for the detailed analysis of the common cold virus HRV and its subviral particles, where data on both phage MS2 and RYMV were employed as standards (Weiss et al. 2015). The effect of the MS2 sample matrix composition on its characterization by the IVDS and electrospray ionization mass spectrometry (ESI-MS) was described by Wick et al. (2007).

The phage MS2 was used further for the lowering the detection limit of the MALDI mass spectrometry assays for bacteria when a biological amplification procedure based on the intrinsic lytic infection cycle of the phages for their host cells was implemented (Madonna et al. 2003). Thus, the presence of *E. coli* was detected by the increased concentration of the MS2 coat. In a similar manner, MALDI-TOF mass spectrometry was used for simultaneous detection of the two target bacterial pathogens: *E. coli* infected with the phage MS2 and *Salmonella* infected with the DNA phage MPSS1 (Rees and Voorhees 2005).

In the early 2000s, the further development of the mass spectrometry detection techniques was stimulated by the anthrax attacks of 2001 that demonstrated the inability to rapidly detect and identify the threat posed by weapons of mass destruction (Griffin and McLuckey 2004). This study elaborated a linear ion trap for biological agent detection and identification, where the phage MS2 was used, as always, as a traditional model. Furthermore, pyrolysis-gas chromatography-ion mobility spectrometry (Py-GC-IMS) was employed to detect and classify deliberately the released bioaerosols in outdoor field scenarios, where the phage MS2 was one of the models (Snyder et al. 2004). The phage MS2 assisted by elaboration of the arrayed TOF mass spectrometer for time-critical detection of hazardous agents (Cornish et al. 2005) and by improvement of the bioaerosol mass spectrometry in real time (Steele et al. 2005).

The application of the charge reduced electrospray size spectrometry led to the resolution of whole viruses, including the phage MS2, which demonstrated a mobility diameter of

24.13 ± 0.06 nm and remained highly viable after the electrospray process (Hogan et al. 2006).

Desorption electrospray ionization-mass spectrometry (DESI-MS) was evaluated as an alternative approach to MALDI-TOF-MS, where a matrix compound was required, and succeeded in the detection of the MS2 coat protein from crude samples with minimal sample preparation (Shin et al. 2007).

At that time, the phage MS2 was used by Monte Carlo simulations using a coarse-grained model to investigate the location of RNA as a function of the degree of dodecahedral distribution of capsid charges ranging from a spherical symmetric to a complete dodecahedral distribution (Angelescu and Linse 2007, 2008). Later, Angelescu and Caragheorgheopol (2015) tested the influence of the shell thickness and charge distribution on the effective interaction between the two like-charged hollow spheres.

The electrospray technique was used to investigate the characteristics of the airborne phage MS2 particles (Jung et al. 2009). The electrosprayed airborne MS2 demonstrated non-agglomerated particles with the geometric mean diameter of 23.8 ± 0.49 nm and maintained their monodisperse size distribution with good stability and uniformity for more than 1 hour. The concentrations of the phages MS2 and PP7 were determined with the electrospray differential mobility analysis and quantitative amino acid analysis (Cole et al. 2009). To interpret the electrospray ionization mass spectra of bioparticles, the phage MS2 was used as a model by the elaboration of an algorithm that automatically minimized the standard deviation in a series of related ion peaks with varying numbers of charges, and allowed the determination of the correct charge state in a peak series (Tseng et al. 2011).

To optimize MALDI-TOF MS-based bacteria detection, Cox et al. (2012) performed a modeling that predicted the time during a given phage infection when a detectable signal would occur. The modified model used a series of three differential equations composed of predetermined experimental parameters, including the phage burst size and burst time to predict progeny phage concentrations as a function of time, while significant agreement between the mathematically calculated phage growth curves and those experimentally obtained by the MALDI-TOF MS was observed.

Generally, the historical role of the phage MS2 in the development of the MALDI-MS approach was thoroughly reviewed in the context of rapid profiling of the: (i) recombinant protein expression (Russell 2011), (ii) investigation of virus structure, dynamics, assembly, and interaction with their hosts (Uetrecht and Heck 2011), (iii) methodological problems of the mass measurements (Peng WP et al. 2014), and (iv) intrinsic problems of the single-molecule mass spectrometry (Keifer and Jarrold 2017). Remarkably, the latter review also highlighted the role of the hepatitis B core antigen, another famous VLP carrier, together with the RNA phage coats, in the development of mass spectrometry applications.

The phage PP7 was used as one of the standards by the evaluation of electrospray differential mobility analysis (ES-DMA) (Pease et al. 2009, 2011; Guha et al. 2011; Pease

2012). The latter was proposed as a potential potency assay for the routine virus particle analysis in biomanufacturing environments, e.g., by the evaluation of vaccines and gene delivery products for lot release. In fact, the ES-DMA appeared as a rapid particle sizing method and distinguished intact particles from degraded ones.

The phage Qβ was deposited on a nano-assisted laser desorption/ionization (NALDI) plate, in order to characterize the particles with nanoprojectile secondary ion mass spectrometry (SIMS) (Liang et al. 2013). The scanning electron microscope images verified that the integrity of the phages was preserved on the NALDI substrate.

The Raman spectroscopy and the surface-enhanced Raman spectroscopy (SERS) were employed for the discrimination of the phage MS2, the double-stranded DNA phage PRD1, and *E. coli*, which produced differentiable Raman spectra (Goeller and Riley 2007).

The size and aggregation level of the phage MS2 were measured at that time also by the dynamic light scattering under different ionic strength and pH values, as student work at the University of Illinois (Diaz 2008). Later, time-resolved dynamic light scattering was used to measure the aggregation kinetics of the phage MS2 across a range of solution chemistries to determine what factors might destabilize viruses in aquatic systems (Mylon et al. 2010). The authors showed that the MS2 aggregation could not be induced, and the phage MS2 was stable even at salt concentrations greater than 1.0 M in monovalent electrolytes LiCl, NaCl, and KCl, while the MS2 aggregation could be induced by divalent Ca^{2+} electrolytes. These results were confirmed by small-angle x-ray scattering experiments (Mylon et al. 2010). Remarkably, Mylon et al. (2010) thought that viruses like MS2 have adopted similar strategies for stability against aggregation, including a net negative charge under natural water conditions and using polypeptides that formed loops extending from the surface of the protein capsid for stabilization.

The concerns of chemical and biological warfare agent proliferation stimulated interest in the collection of signature data in the terahertz region of the phage MS2 RNA, among other bio-simulants (Majewski et al. 2005). Recently, Park SJ et al. (2017) demonstrated highly sensitive detection of the phage MS2, as well as the double-stranded DNA phage PRD1, on the metamaterial surface using the terahertz split-ring resonators with various capacitive gap widths. The dielectric constants of the virus layers in the THz frequency range were first measured using thick films, and the large values found identified them as efficient target substances for dielectric sensing.

Since solubilization of biological samples in organic solvents would be of high importance for the application of biophysical methods, a systematic study on such solubilization and stabilization of the phage MS2 was performed by Johnson et al. (2007). Although direct extraction of the MS2 from an aqueous phase into isooctane containing 2 mM aerosol OT (AOT)/water/isooctane, a proven approach for the organic solubilization of many proteins, was not successful, the pre-dried samples of MS2 were solubilized through the direct addition of organic solvents containing 500 mM AOT. The solubilized phage MS2 was derivatized with stearic acid in chloroform, illustrating that bioconjugation reactions could be performed on organic-solubilized capsids using reagents that were completely insoluble in water. Furthermore, the organic-solubilized phage remained infectious after heating at 90°C for 20 min, whereas the phages in aqueous buffer or dried with nitrogen were nonviable following the heat treatment protocol.

The first detection of fluorescently labeled Qβ phage on a planar integrated optofluidic chip was presented by Rudenko et al. (2007, 2008, 2009). The detection sensitivities were on the order of 1000 viruses within an 85 femtoliter excitation volume. These studies suggested considerable potential for an inexpensive and portable sensor capable of discrimination between viruses of different sizes.

Concerning capillary electrophoresis of viruses, the behavior of the phage MS2 and related antibodies was described first by the development of the capillary isoelectric focusing with liquid-core waveguide laser-induced fluorescence whole-column imaging detection (CIEF-LCW-LIF-WCID) approach (Liu and Pawliszyn 2005a,b) and reviewed in detail by Wu XZ et al. (2005) and later by Kremser et al. (2007). This successful study was reviewed again in the context of the recent advances in the analysis of biological particles by capillary electrophoresis (Kostal and Arriaga 2008; Silvertand et al. 2008; Kremser et al. 2009). Then, microfluidic chip gel electrophoresis (CGE) was developed for protein profiling and virus identification, where the phage MS2 was used as one of the profile models (Fruetel et al. 2008, 2013). Recently, capillary zone electrophoresis of the phages MS2 and Qβ was studied in the context of their interaction with sodium dodecyl sulfate (SDS) (Sautrey et al. 2018). Although SDS has been commonly employed in virus electrophoresis, the way by which SDS interacted with the surface of viruses remained poorly known. The authors showed that the surface hydrophobicity of phages is a key factor influencing their mobility and that the SDS−virus association was driven by hydrophobic interactions at the surface of virions.

Concerning fluorescence measurements, the single-particle fluorescence spectrometer (SPFS) system capable of measuring two UV-laser excited fluorescence spectra from a single particle on-the-fly was applied to the phage MS2 (Pan et al. 2008). Furthermore, this approach was applied in a BSL-3 laboratory by the spectrally-resolved fluorescence cross-sections of aerosolized biological live agents and simulants, including the phage MS2, using five excitation wavelengths (Pan et al. 2014). The laser-induced fluorescence of the phage MS2, among other suspended biological agents, was then investigated by Manninen et al. (2009). The detection and characterization of biological aerosol particles in atmosphere using laser-induced fluorescence, including the role of the phage MS2, was reviewed by Pan (2015).

The surface plasmon resonance was used to examine the adsorption kinetics of the phage PP7, among other biological models, to anion exchange surfaces (Riordan et al. 2009).

Concerning the label-free detection of single protein using the nanoplasmonic−photonic hybrid microcavity, the detection and sizing of the phage MS2 with a mass of 6 ag from the resonance frequency shift of the whispering gallery mode−nanoshell hybrid (WGM-h) resonator was first reported (Dantham et al. 2012) and used for the further development of the approach for a set of small proteins (Dantham et al. 2013).

Concerning novel chromatography approaches, the phage MS2 was used by the elaboration of non-woven electrostatic media for the chromatographic separation of biological particles, such as the nanoalumina fibers electroadhesively grafted to a microglass fiber (Tepper et al. 2009). Optimized size-exclusion chromatography on novel carriers was applied to the MS2 isolation and quantitation by the RT-qPCR detection (Farkas et al. 2015a).

It is necessary to mention here the exhaustive review on the isoelectric points of viruses (Michen and Graule 2010), where data on viruses that replicate in hosts of kingdom plantae, animalia, and bacteria were collected and all available isoelectric points of the RNA phages GA, SP, f2, MS2, Qβ, μ2, and PP7 were listed. Remarkably, the phage MS2 was measured 10 times and showed a mean isoelectric point value of 3.5 with a standard deviation of 0.6. It is noteworthy that the two more recent studies by Langlet et al. (2008a) and Yuan et al. (2008) determined the alteration of the isoelectric point by a change in water chemistry, e.g., ionic strength or ionic composition (Michen and Graule 2010).

Concerning the recent studies on the physical properties of the phage MS2, Furiga et al. (2011) showed that in high-ionic-strength solutions, in contrast to low-ionic-strength solutions, the phage kept its biological activity under static conditions but it quickly lost its infectivity during the ultrafiltration process. Remarkably, increasing the ionic strength decreased both the inactivation and the capsid breakup in the feed suspension and increased the loss of infectivity in the filtration retentate. The hydrophobicity of the phage MS2 was measured in comparison to other popular models, such as rotavirus and some artificial particles (Farkas et al. 2015b) or DNA phages φX174 and PRD1 (Dika et al. 2015).

The effect of the RNA core on interfacial interactions of the phage MS2 was investigated first by Nguyen et al. (2011). After removal of the RNA core, the empty intact capsids were characterized and compared to untreated MS2, while electron density of untreated MS2 and RNA-free MS2 were characterized by transmission electron microscopy and synchrotron-based small angle spectroscopy (SAXS). Remarkably, the suspensions of both particles exhibited similar electrophoretic mobility across a range of pH values. Similar effects were observed at pH 5.9 across a range of NaCl or $CaCl_2$ concentrations. Both suspensions showed insignificant aggregation over 4 h in 600 mM NaCl solutions. In the presence of Ca^{2+} ions, the aggregation of both types of particles was consistent with the earlier aggregation studies and was characterized by both reaction-limited and diffusion-limited regimes occurring at similar Ca^{2+} concentrations. Despite some differences in the kinetics of adsorption to the air−water interface, the changes in surface tension, which resulted from particle adsorption, showed no difference between the untreated MS2 and

RNA-free MS2. The surface elasticity and surface viscosity at the interface were low for both the untreated phage and the RNA-free capsid. Further evaluation showed that the charged core heavily influenced the local potential within the soft particle, while the potential distribution outside the particle in the salt solution was found to be weakly dependent on the core features (Phan et al. 2013). These findings were consistent with previous experiments showing the minor impact of the nucleate core of the phage MS2 on its overall electrical properties. These conclusions were commented on and improved to some extent by McDaniel et al. (2015).

In parallel, Dika et al. (2011) performed a comparative study on the impact of internal RNA on the aggregation and electrokinetics of the phage MS2 and its VLPs. The essence of the VLPs will be outlined in detail in Chapters 19, 20, and 21. Thus, the MS2 VLPs exhibited electrophoretic mobility larger in magnitude than that of MS2, and both had similar isoelectric point of $\sim$4. The electrokinetic results reflected a greater permeability of the MS2 VLPs to electroosmotic flow, developed within/around these soft particles during their migration under the action of the applied electrical field. The presence of some remaining negatively charged component within the VLPs was acknowledged. In addition, the phage MS2 systematically formed aggregates at pH values below the isoelectric point, regardless of the magnitude of the solution ionic strength, whereas the MS2 VLPs aggregated under the strict condition where the pH was relatively equal to the isoelectric point at sufficiently low salt concentrations, $<$10 mM (Dika et al. 2011). These conclusions contradict the above-described findings of Nguyen et al. (2011) who reported identical electrokinetic and aggregation characteristics for the MS2 and its VLPs. Dika et al. (2013a) later compared three purification schemes, and explained this contradiction as originating from the different purification methods adopted prior to measurements. Then, the effect of heat on the physicochemical properties of the phage MS2, infectious and noninfectious, was evaluated (Brié et al. 2016). The heat-induced inactivation process of the infectious phages caused hydrophobic domains to be transiently exposed and their charge to become less negative. The particles also became progressively permeable to small molecules. When exposed to a temperature higher than their critical temperature (72°C), the particles were disrupted, and the genome became available for ribonucleases (Brié et al. 2016).

Meanwhile, Nap et al. (2014) investigated the role of solution conditions in the phage PP7 capsid charge regulation. The overall charge of the virus capsid arose as a consequence of a complicated balance with the chemical dissociation equilibrium of the amino acids and the electrostatic interaction between them, and the translational entropy of the mobile solution ions, i.e., counterion release. By investigation of the impact of reducing and oxidizing agents on the structure and infectivity of the phage Qβ, Loison et al. (2016) found that the physiological redox conditions modified disulfide bonds of the capsid of Qβ phage without decreasing infectivity, whereas the total reduction disrupted the capsid, while chlorine oxidation disrupted especially minor proteins.

Furthermore, experimental studies with the phage MS2 that was modeled as a soft particle with rigid core of radius 10.3 nm were performed (De et al. 2016; Gopmandal et al. 2016; Gopmandal and Ohshima 2017). As a result, the effect of core charge density, electrolyte pH, ionic concentration, and the degree of inhomogeneity of the polymer layer on the soft particle electrophoresis was established.

Then, the phage MS2 was used by the elaboration of the Quenching of Unincorporated Amplification Signal Reporters (QUASR; read as "quasar") method that improved application of the reverse-transcription loop-mediated isothermal amplification (RT-LAMP) for the simplified diagnostic tests of RNA viruses (Ball et al. 2016).

Moreover, the MS2 particle was used as a model of the soft negatively-charged polyelectrolyte layer by the physical evaluation of the interaction between the two charged-hard-core soft particles (Bui et al. 2017). First, the expression of interaction energy between two identical soft particles was derived. Then, the numerical calculations were performed by investigating the dependence of the interaction energy on the distance between the soft particles and the concentration of salt solution.

Recently, the phage MS2 was used as a model by the elaboration of a dielectrophoresis method for the rapid and selective concentration of bacteria, viruses, and proteins using alternating current signal superimposition on two coplanar electrodes (Han et al. 2018).

All monitoring methodologies that have involved immunological applications are presented in Chapter 15.

AGGREGATION

The phenomenon of virus aggregation, including data on the phage MS2, was recently summarized systematically by Gerba and Betancourt (2017). They noted first that the aggregates may have a significant impact on the quantification and behavior of viruses in the environment, while aggregates may be formed in numerous ways. For example, the RNA phages may form crystal-like structures in the host cell during replication, as described in the Crystalline inclusions section of Chapter 4, or may form aggregates due to changes in environmental conditions, as mentioned in the Adsorption and coagulation section of Chapter 6. Generally, the aggregates tended to form near the isoelectric point of the virus, under the influence of certain salts and salt concentrations in solution, cationic polymers, and suspended organic matter. Nevertheless, both aggregation and disaggregation processes were greatly influenced by virus type. For this reason, Gerba and Betancourt (2017) collected data on aggregate types, methods used to reduce and eliminate aggregation, size of aggregates, and detection and quantification of aggregates. The impact of aggregation on survival in water, disinfection outcome, and filtration effectivity was also evaluated. Briefly, the phage MS2, among many other viruses, was susceptible to aggregation in the presence of Ca^{2+} rather than Mg^{2+} (Pham et al. 2009), while anions in solution might also play a role in the aggregate formation. Thus, the phage MS2 was found to form aggregates of 300–400 nm in size when stored in phosphate-buffered saline, but not when bicarbonate was used as a buffer (Yuan et al. 2008). It was theorized that this was due to the phosphate linking of the amino acid lysine in the proteins of the phage coat, resulting in the formation of aggregates (Gerba and Betancourt 2017).

The phage MS2 could be made to aggregate by adjustment of the pH or salt concentration and type of cationic salt in suspension, for example, by lowering the pH to near the phage's isoelectric point pI = 3.9 (Mattle et al. 2011). The rate of aggregation was clearly dependent upon the pH. After 1 h at pH 3.0, the radius of the MS2 aggregates was 1000 nm, whereas at pH 3.6 the radius was 750 nm.

The cationic organic molecules, such as the polyhexanethylene biguanide cationic disinfectant, also induced the formation of MS2 aggregates, which changed the disinfectant inactivation kinetics, thus overestimating the effectiveness of the disinfectant (Pinto et al. 2010b). Using light scattering, it appeared that aggregates of the phage MS2 up to 500 nm in diameter were formed, greatly decreasing the number of phage plaque-forming units. Such popular disinfectants as chlorine dioxide induced the MS2 aggregation that caused an average 21-fold increase in resistance against disinfection. Chlorine dioxide was found to promote the formation of MS2 aggregates through particle destabilization (Barbeau et al. 2004, 2005).

When Langlet et al. (2007) compared the four RNA phages during removal by membrane filtration, it appeared that the aggregate formation was dependent upon the nature of the virus, pH, and ionic strength. The Qβ aggregation occurred when the pH decreased in addition to increasing ionic strength, while the phages SP and GA aggregated over the entire range of pH-ionic strength conditions tested. It was suggested that the phages MS2 and Qβ at neutral pH and low ionic strength met the right criteria for the particles to express a negative charge. The aggregation of the RNA phage particles, and especially of the recombinant RNA phage VLPs, remains an urgent problem nowadays and needs further attention in the context of broad nanotechnological applications.

STANDARDS AND INTERNAL CONTROLS

The RNA phages have often been used by standardization because of their high stability and remarkable level of knowledge about their structure. Thus, Doel and Mowat (1985) used the phage MS2 to standardize quantification of foot-and-mouth disease virus during the sucrose gradient procedure. The MS2 particles, among other surrogates, were used by the calibration of an early PCR microchip instrument, the quenching of unincorporated amplification signal reporters (ANAA) (Belgrader et al. 1998). The MS2 phage was used as a standard by the fabrication of an array biosensors for simultaneous identification of bacterial and viral analytes (Rowe et al. 1999) and as a simulant by the generation of assays for detection of biological warfare agents, including poxvirus (McBride et al. 2003; Thomas JH et al. 2004). However, a novel simulant of poxvirus, the *Cydia pomonella* granulovirus, replaced MS2 much later (Garnier et al. 2009).

The phage MS2 was used frequently as an "incorrect" antigen, or control; for example, by the elaboration of a microfluidic chip-compatible bioassay based on single-molecule detection with high sensitivity and multiplexing, adopted particularly for the detection of botulinum toxoid A (Burton et al. 2010).

As a favorite viral model, the phage MS2 was used by the elaboration of clinical assays; for example, for the detection of respiratory viruses in aerosols (Perrott et al. 2009).

The phage MS2 particles were employed frequently as internal controls due to their ease of propagation and since they represented well-characterized encapsidated RNA. The phage MS2 was introduced first as the internal control by the specific detection of hepatitis C virus (Dreier et al. 2005), human norovirus (Dreier et al. 2006; Rolfe et al. 2007; Brooks et al. 2018; Leone et al. 2018; Lun et al. 2018a,b; Wong RS et al. 2018; Cannon et al. 2019), human immunodeficiency virus (Agarwal et al 2007), respiratory syncytial virus and various influenza viruses (Beck et al. 2010; Chidlow et al. 2010; Ellis and Curran 2011; Wu LT et al. 2013; Zimmerman et al. 2014; Stumpf et al. 2016), hepatitis E virus (Slot et al. 2013; Wang B et al. 2018), rotavirus (Gautam R et al. 2016b), Ross River virus (Faddy et al. 2018), Zika virus (Boujnan et al. 2018; Fernandes da Costa et al. 2018; Lee AHF et al. 2018), Japanese encephalitis virus (Bharucha et al. 2018a,b, 2019), vesivirus (Renshaw et al. 2018); human mastadenoviruses (Lun et al. 2019), and herpes simplex virus 1 and 2 (Navidad et al. 2019). The phage MS2 internal control was also employed in modern kits for the surveillance of emerging pathogens, such as Mayaro and Oropouche virus, especially in areas where differential diagnosis of Dengue, Zika, and Chikungunya viruses should be performed (Naveca et al. 2017). Moreover, the phage MS2 was used as the internal control by the evaluation of transmission of viral vaccine strains, first of all, of Sabin oral polio vaccine (Altamirano et al. 2018a,b, Jarvis et al. 2018; van Hoorebeke et al. 2018) and Rift Valley fever phlebovirus vaccine (Balkema-Buschmann et al. 2018), as well as by the estimation of the impact of rotavirus vaccines in the post-rotavirus vaccine era (Hassan et al. 2019; Praharaj et al. 2019).

The internal phage MS2 control played an important role in the evaluation of respiratory infections in human. Thus, it was used by the generation of the xTAG respiratory virus panel (RVP) test that was capable of detecting 20 viruses and subtypes simultaneously in a single patient sample (Krunic et al. 2007; Mahony et al. 2007). Then, the MS2 internal control assisted in the generation of highly sensitive detection of respiratory syncytial virus (Eboigbodin et al. 2017; Yun et al. 2018). The xTAG RVP approach was used subsequently in many circumstantial studies (Fillatre et al. 2018; Gosert et al. 2018; Huang SH et al. 2018; Zhao et al. 2018; Gonsalves et al. 2019; Kellner et al. 2019; Xie et al. 2019a,b; Yoo et al. 2019).

The phage MS2 served as an internal control by the elaboration of novel extraction methods (Nanassy et al. 2011) and by the detection and quantitation of viral agents of gastroenteritis, namely noroviruses GI and GII, rotavirus, astrovirus, sapovirus, and adenovirus (Liu et al. 2011; Vocale et al. 2015), or adenovirus, enterovirus, and noroviruses GI and GII (Viau et al. 2011).

The phage MS2 was used as an internal control in FDA-approved tests for the molecular detection of respiratory pathogens, including influenza (Ginocchio 2011), or single-tube multiple RT-PCR for the detection of 13 common virus types/subtypes associated with acute respiratory infection (Zhang Dan et al. 2016).

Ninove et al. (2011) demonstrated suitability of the phage MS2 as the internal control, in parallel with the phage T4, on a large number of clinical samples for a standard laboratory of medical virology. These results were obtained in a routine-based evaluation including 8950 clinical specimens, representing 36 types of samples, submitted for PCR detection of selected viruses including DNA viruses (herpes viruses, JC and BK polyoma viruses, parvovirus B19, adenoviruses) and RNA viruses (enterovirus, influenza virus, respiratory syncytial virus, human metapneumovirus, rhinovirus, Toscana virus, West Nile virus, lymphocytic choriomeningitidis virus, Dengue virus, Chikungunya virus). This technique was used further successfully by the evaluation of respiratory and gastrointestinal infections (Hoang et al. 2019; Ly et al. 2019).

The internal phage MS2 control was employed by combined human parechovirus−enterovirus real-time RT-PCR detection (Selvaraju et al. 2013) and simultaneous detection of 19 enteropathogens (Liu J et al. 2013).

Birger et al. (2018) used the internal phage MS2 control for the detection of adenovirus, coronavirus, human metapneumovirus, rhinovirus, influenza virus, respiratory syncytial virus, and parainfluenza virus. Chasqueira et al. (2018) detected influenza A/B, human parainfluenza virus 1–4, adenovirus, human metapneumovirus, respiratory syncytial virus, rhinovirus, enterovirus, human coronavirus, and human bocavirus. Miao et al. (2018) used the MS2 control for enterovirus, rotavirus, astrovirus, norovirus, and adenovirus detection.

The phage MS2, along with the phages T4 and M13, was applied by the simultaneous identification of DNA and RNA viruses present in pig feces, using process-controlled deep sequencing (Sachsenröder et al. 2012).

The phage MS2 contributed as the internal control to the ecologically important studies that established the absence of coronaviruses, paramyxoviruses, and influenza A viruses in seabirds in the southwestern Indian Ocean (Lebarbenchon et al. 2013) and controlled circulation of wild poliovirus type 1 (Hindiyeh et al. 2014). Moreover, the MS2 phage particles appeared as the most suitable internal control in a study dealing with rat brain development (Fedoseeva et al. 2014).

After an outbreak of Ebola virus in western Africa in March 2014, the phage MS2 was among the first proposed surrogates (Bibby et al. 2014). Then, the phage MS2 was used as the internal control of the extraction/amplification efficiency in modern RT-PCR kits for Ebola virus surveillance (Bailey et al. 2016; Liu J et al. 2016a,b; Poliquin et al. 2016; Weller et al. 2016; Wonderly et al. 2019).

The sequencing of the first hepatitis D virus RNA WHO International Standard was performed with the phage MS2 as the internal standard (Pyne et al. 2017). The improved assays to detect six genotypes of hepatitis C virus (Neto et al. 2017) and to differentiate between RNA and DNA of HIV-1-based lentiviral vectors (Pavlovic et al. 2017) also involved the MS2 internal standard.

Surmounting the borders of the predominantly viral diagnostic, the phage MS2 was employed as the internal control by the detection of *Cryptosporidium* as a leading cause of diarrhea in children (Graef et al. 2018; Schnee et al. 2018), analysis of sputa microbiota of chronic obstructive pulmonary disease (COPD) patients (Jubinville et al. 2018), and elaboration of a clinical metagenomic sequencing assay for the pathogen detection in cerebrospinal fluid (Miller S et al. 2019). The phage MS2 was used as the internal control in a huge study on the etiology, burden, and clinical characteristics of diarrhea in children in low-resource settings (Platts-Mills et al. 2018).

Concerning the methodological progress, the phage MS2 particles contributed strongly, as the well-established internal control, to the re-evaluation of the concentration techniques for such virus particles as human immunodeficiency virus 1, hepatitis B virus, and hepatitis C virus from low-titer specimens (Sundarrajan et al. 2018) and to the performance and workflow assessment of six nucleic acid extraction technologies for use in resource-limited settings (Beall et al. 2019).

The unquestioned authority of the phage MS2 as the popular internal control for detection of numerous disease biomarkers was acknowledged in a recent review (Kaushik et al. 2018).

After the favorite phage MS2, the phage PP7 was also employed as the internal control, at first, by comparison of different commercial RNA extraction kits (Poma et al. 2012) and by the estimation of the quantitative microbial risk assessment (Poma et al. 2019). As the phage MS2, the phage PP7 was employed for the detection of hepatitis E virus (Pisano et al. 2018), enteric viruses (Lizasoain et al. 2018a,b; Masachessi et al. 2018), noroviruses (Fumian et al. 2019; Prado et al. 2019), and JC polyomavirus (Levican et al. 2019).

The phage Qβ was used as the internal control in a study dealing with the fine mechanisms of simian immunodeficiency virus infection in rhesus macaques (Obregon-Perko et al. 2018).

10 RNA

Positive anything is better than negative nothing.

Elbert Hubbard

There is no excellent beauty that hath not some strangeness in the proportion.

Francis Bacon

BIOPHYSICS

The data on the molecular mass measurements of the genomic RNA of the RNA phages are presented in Table 10.1. Figure 10.1 is intended to illustrate the early biophysical methods for the RNA characterization.

The first direct detection of the molecular mass of the phage RNAs was performed by light scattering. According to the special structural review by Boedtker and Gesteland (1975) for the classical *RNA Phages* book, there were three light scattering studies of the phage RNAs, and in each case a molecular mass of between 1,000,000 and 1,100,000 daltons was reported. The first light scattering study was performed by Strauss and Sinsheimer (1963), who reported a molecular mass of $1,050,000 \pm 100,000$ for the MS2 RNA in 0.2 M NaCl, 0.01 M Tris buffer, pH 7.0. Although they observed higher molecular mass in lower salt concentrations, this was probably the result of aggregation due to contamination by traces of heavy metal ions. The molecular mass of the R17 RNA detected further by light scattering was independent of counterion concentration, since solutions were extensively dialyzed against EDTA (Gesteland and Boedtker 1964). A final value of 1,100,100 daltons was obtained after correcting for concentration and polydispersity (Boedtker and Gesteland 1975). The light scattering studies by Overby et al. (1966b) gave values of 1,000,000 and 900,000 daltons for the MS2 and Qβ RNAs, respectively. However, their light scattering data did not have a linear dependence on the scattering angle, they used relatively high concentrations of RNA, and their solutions contained Mg^{2+}, which might induce aggregation. These obstacles could be the reason for the discrepancy of the Qβ RNA molecular mass estimations with the real value.

In parallel with the light scattering, the direct RNA molecular mass calculations were performed with measurements of sedimentation and viscosity values. Boedtker and Gesteland (1975) recalculated the molecular mass using the following values of critical parameters: a value of 2.28×10^6 for β, an experimental parameter, and 0.55 for $\bar{v}$, the partial specific volume, namely the values that have been derived from an average of all the reported determinations (Boedtker 1968b), and the R17 RNA molecular mass was corrected to that of the Na^+ salt rather than the K^+ salt which was measured. These results are given in bold in Table 10.1. In fact, the R17 RNA

molecular mass determinations by the sedimentation-viscosity were in quite good agreement with the light scattering results on this RNA.

Concerning the indirect molecular mass measurements, the equations relating the sedimentation constant to molecular mass date back to 1958, when Hall and Doty found that for calf liver ribosomal RNA $S_{25,w} = 0.021\ M^{0.49}$ (Boedtker and Gesteland 1975). This equation was only applicable to this RNA when the measurements were made at 25°C in the same buffer (0.01 M K^+ phosphate buffer, 1:1). If, however, the measurements were made under denaturing conditions, such as after reaction with formaldehyde, a general equation could be derived valid for all RNA samples provided the measurements were made in the same buffer and at the same temperature. When the R17 RNA was reacted with 1.1 M formaldehyde and its sedimentation constant measured, it was found to be identical to that obtained for 23S *E. coli* rRNA, and based on the molecular mass of the latter, the molecular mass of R17 RNA was again found to be 1,100,000 (Boedtker 1968a). Remarkably, under identical conditions, the Qβ RNA was found to have a molecular weight of 1,300,000, or about 10% larger than that of R17 RNA (Boedtker and Gesteland 1975). The reaction of the R17 RNA with formaldehyde or formamide was described in detail by Boedtker (1967, 1968a,b, 1976), Fenwick (1968), and Boedtker and Lehrach (1976). Denaturation of the phage MS2 was also achieved with dimethyl sulfoxide (Strauss et al. 1968).

Concerning estimates based on the percentage of the phage RNA within the phage particles (Enger et al. 1963; Marvin and Hoffman-Berling 1963b; Strauss and Sinsheimer 1963; Möller 1964; Zinder 1966–1967), the average value of 30.4% led to the molecular mass 1.2×10^6 for the R17 RNA and 1.3×10^6 for the Qβ RNA (Boedtker and Gesteland 1975).

Concerning RNA size measurements, the radii of gyration of R17 and MS2 RNA have been measured by either light scattering or viscosity at several salt concentrations, and the corresponding results are listed in Table 10.2, taken with permission from Boedtker and Gesteland (1975). The size increased dramatically as the counterion concentration was reduced tenfold. When the sedimentation constant decreased, the size increased. Thus, the traditional identification of phage RNA as a 26S species was only valid if the sedimentation rate was measured in 0.1 M salt at 20°C (Boedtker and Gesteland 1975).

The circular dichroism was used to follow changes in the R17 RNA conformation accompanying its thermal denaturation and to gain insight into the forces stabilizing the RNA structure (Phillips and Bobst 1972; Bobst et al. 1974). The double-stranded f2 RNA replicative forms accounted in fact for the appearance of an intense positive band in the circular dichroism spectra (Evdokimov et al. 1976). Later, the circular

TABLE 10.1

Molecular Mass of the Genomic Phage RNAs

Group	Phage	$S_{20,w}$	Method	Molecular Mass, Da	References	Calculated by Sequence, Da
I	MS2		Light scattering	1,050,000 ± 100,000	Strauss and Sinsheimer (1963)	1,147,398.37
	MS2	25.8	Light scattering	1,000,000	Overby et al. (1966b)	
	MS2		pppGp number	1,207,500		
	MS2		Ribonuclease digestion	1,210,000 ± 70,000	Haegeman et al. (1971)	
	MS2		Gel electrophoresis	1,230,000	Reijnders et al. (1973)	
	MS2		Gel electrophoresis	1,040,000	Kaper and Waterworth (1973)	
	MS2		Electron microscopy	1,200,000	Jacobson and Bromley (1975)	
	R17		Calculation: 31.7% RNA within the phage	1.300,000	Enger et al. (1963)	–
	R17	26.2 (0.1 M KCl; 0.01 K⁺ phosphate)	Sedimentation-viscosity	1,080,000–1,370,000 **1,270,000**	Mitra et al. (1963)	–
	R17	18.4 (0.01 K⁺ phosphate)	Sedimentation-viscosity	950,000–1,210,000 **1,100,000**	Mitra et al. (1963)	–
	R17		X-ray irradiation	1,000,000	Ginoza (1963)	–
	R17		γ-irradiation	800,000 ± 20,000	Ginoza (1963)	–
	R17		Light scattering	1,100,100	Gesteland and Boedtker (1964)	–
	R17		Ribonuclease digestion	1,100,000	Sinha et al. (1965b)	–
	R17		Gel electrophoresis	1,300,000	Boedtker (1971)	–
	R17		Gel electrophoresis	1,100,000	Benike et al. (1975)	–
	R17		Electron microscopy	1,300,000	Benike et al. (1975)	–
	R17		X-ray scattering	1,100,000	Fischbach and Anderegg (1976)	–
	M12			1,200,000	Delius and Hofschneider, unpublished, cited by Ammann et al. (1964a)	–
	M12		Ribonuclease digestion	~1,100,000	Sinha et al. (1965a)	–
	μ2		Gel electrophoresis	1,300,000	Staynov et al. (1972)	–
	fr	24.6 (0.02 M citrate)	Sedimentation-viscosity	1,450,000 **1,570,000**	Marvin and Hoffmann-Berling (1963b)	1,151,100.20
	f2		Electron microscopy	1,300,000	Chi and Bassel (1974)	–
	f2		Electron microscopy	1,230,000 ± 150,000	Edlind and Bassel (1977)	–
	μ2		Gel electrophoresis	1,100,000 ± 100,000	Piffaretti and Pitton (1976)	–
	MS2, f2, R17, JP501	24	Gel electrophoresis	1,210,000	Furuse et al. (1979b)	
II	GA, SD, TH1, BZ13, KU1, JP34	23	Gel electrophoresis	1,200,000	Furuse et al. (1979b)	
	GA					1,112,025.69
	BZ13					1,109,320.88
	KU1					1,118,293.26
III	Qβ	28.9	Light scattering	900,000	Overby et al. (1966b)	1,352,212.33
	Qβ		Gel electrophoresis	1,500,000	Boedtker (1971)	
	Qβ		Gel electrophoresis	1,400,000	Reijnders et al. (1973)	
	Qβ		Sedimentation	1,300,000	Boedtker and Gesteland (1975)	
	Qβ		End-group labeling	1,210,000	Hindley et al. (1970)	
	Qβ	32 (0.1 M NaCl)			Bendis and Shapiro (1970)	

(Continued)

TABLE 10.1 (*Continued*)

Molecular Mass of the Genomic Phage RNAs

Group	Phage	$S_{20,w}$	Method	Molecular Mass, Da	References	Calculated by Sequence, Da
	Qβ		Electron microscopy	1,550,000	Chi and Bassel (1974)	
	Qβ		Electron microscopy	1,530,000 ± 140,000	Edlind and Bassel (1977)	
	Qβ		Electron microscopy	1,640,000 ± 60,000	Vollenweider et al. (1978)	
	Qβ, VK, ST, TW18	27	Gel electrophoresis	1,390,000	Furuse et al. (1979b)	
	M11					1,353,167.14
	MX1					1,353,491.96
IV	SP, FIC, TW19, TW28, MX1, ID2	28	Gel electrophoresis	1,420,000	Furuse et al. (1979b)	
	FI					1,345,326.04
	NL95					1,366,437.68
	SP					1,376,223.20
?	ZIK/1	28 ± 3			Bishop (1965)	
Levivirus-like	Hgal1					1,144,198.68
	C-1					1,131,780.17
	M					1,092,375.79
Pseudomona-phages	PP7		Gel electrophoresis	1,030,000	Benike et al. (1975)	1,153,615.50
	PP7		Electron microscopy	1,240,000	Benike et al. (1975)	
	PP7		Electron microscopy	1,260,000 ± 160,000	Edlind and Bassel (1977)	
	PRR1					1,148,017.31
	LeviOr01					1,180,559.13
Caulophage	φCB5	30 (0.1 M NaCl)			Bendis and Shapiro (1970)	1,208,250.31
Acinetophage	AP205					1,368,379.53
Assembled genomes	EC					1,022,802.58
	MB					1,261,389.64

Note: The order of the RNA presentation corresponds to that of the phage genome order in Figure 7.1. The molecular mass values corrected by Boedtker and Gesteland (1975) are given in bold.

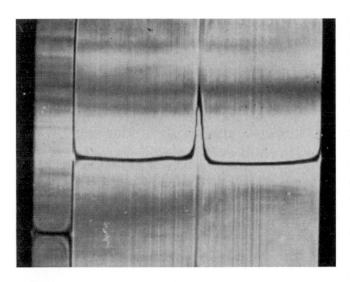

FIGURE 10.1 The Schlieren sedimentation diagram of R17 RNA (approximately 1.2 mg/mL) in a KCl-phosphate buffer in the 30-mm cell taken 19 min after reaching top speed of 50,750 rpm. (Reprinted with permission of Sankar Mitra from Mitra S et al. *Proc Natl Acad Sci U S A.* 1963;50:68–75.)

dichroism was used by the interaction of MS2 RNA with a synthetic Gly/Arg-rich peptide corresponding to residues 676–692 of human nucleolin in order to establish the role of N$^\omega$-arginine dimethylation in the perturbance of the helical structure in nucleic acids (Raman et al. 2001). Recently, the circular dichroism spectra were used to measure transitions in the MS2 RNA secondary structure (Borodavka et al. 2015).

The Raman spectra of the R17 RNA were tried to indicate the frequencies and intensities characteristic of the P-O stretching vibrations, which could provide a basis for the quantitative determination of the RNA secondary structure (Hartman et al. 1973; Thomas and Hartman 1973). Paleček and Doskočil (1974) applied differential pulse-polarographic analysis to the double-stranded f2 RNA. Paleček and Fojta (1994) used the f2sus11 RNA to elaborate differential pulsed voltammetric determination of RNA at the picomole level in the presence of DNA.

The fluorescent staining of the phage RNAs with acridine orange, in the case of the phages ZIK, ZJ, and 7s, was elaborated by Bradley (1966a). This approach, which was enriched by possible post-staining treatment color changes, allowed

TABLE 10.2
Relation of Size to Sedimentation Constant

Phage	Buffer	Temperature, °C	$S_{20,w}$	R_G, Å	Method	References
MS2	0.2 NaCl, 0.01 M Tris pH 7	6	31	160	Light scattering	Strauss and Sinsheimer (1963)
R17	0.1 M KCl, 0.01 M K$^+$ phosphate	20	26.2	225	Viscosity	Mitra et al. (1963)
R17	0.1 M NaCl, 0.01 M Na acetate	20	25.7	192	Light scattering	Gesteland and Boedtker (1964)
R17	0.01 M K$^+$ phosphate	20	18.4	360	Viscosity	Mitra et al. (1963)
R17	0.01 M NaCl, 0.001 M EDTA	20	16.5	485	Light scattering	Gesteland and Boedtker (1964)

Source: Reprinted with permission of Cold Spring Harbor Laboratory Press from Boedtker H, Gesteland RF. In: *RNA Phages.* Zinder ND (Ed). Cold Spring Harbor Laboratory, Cold Spring Harbor, NY, USA, 1975, pp. 1–28.

the determination of the type and strandedness of the phage RNAs by examination under the fluorescence microscope.

GEL ELECTROPHORESIS

Although the empirical relation between the log of the molecular mass and the mobility on polyacrylamide gels has been applied to estimation of the RNA molecular mass, the validity of this relation depended on there being a direct proportionality between the molecular mass and the molecular volume for all RNA species, and this was shown to be incorrect (Gesteland and Boedtker 1964). However, it was shown further that it is possible to measure RNA mobilities under denaturing conditions, either after reaction with formaldehyde (Boedtker 1971) or in formamide (Staynov et al. 1972). Using TMV RNA and *E. coli* 23S and 16S rRNA as standards, the molecular mass of the R17 RNA was found to be 1,300,000, and that of Oβ 1,500,000 if the RNA samples were reacted with either 1.1 or 2.2 M formaldehyde and then run on gels polymerized in the presence of formaldehyde (Boedtker 1971).

The mobility of the RNA of the phage μ2, a close analog of the phages MS2 and R17, on formamide gels was exactly as predicted for its presumed molecular mass of 1,300,000 when compared to a semilog plot of molecular mass versus mobility for the standards defined above and rabbit reticulocyte RNA, yeast RNA, and *E. coli* tRNA in addition (Staynov et al. 1972). The R17 RNA has also been run on formamide gels, and the value of 1,300,000 was confirmed for the molecular mass of this phage RNA (Boedtker and Gesteland 1975). Thus, when measured under denaturing conditions on polyacrylamide gels, the phage RNAs demonstrated molecular mass appreciably higher than those found in direct molecular mass determinations on undenatured RNA. Comparing biophysical measurements and polyacrylamide electrophoresis data, Boedtker and Gesteland (1975) concluded that the best estimate of the molecular mass of the R17 RNA and related MS2 and μ2 RNAs was 1,300,000 and that of the Qβ RNA was 1,500,000. These values are a bit greater than the authentic figures that can be calculated from the full genomic sequences. The latter are posted for comparison with the historical data in Table 10.1.

To follow directly to the electrophoretic separation of RNAs in the gel, toluidine blue complexes with the MS2 RNA, among other RNA models, were applied (Popa and Bosch 1969).

GENERAL PROPERTIES

The ultraviolet absorption spectrum of the phage RNAs was typical of that of high molecular mass RNAs. For example, in the case of the R17 RNA, it exhibited a maximum at 258 nm and a minimum at 230 nm. The ratio of the absorbance at 260 nm to that at 230 nm, in KCl-phosphate buffer at pH 7.0, was 2.28; the 260/280 ratio was 2.23 (Mitra et al. 1963). The secondary structure of the phage RNAs was explored mainly with absorbancy–temperature profile measurements. The midpoint T_m of the thermal transition for the R17 RNA was 58°C. The R17 RNA was almost entirely randomly coiled in the presence of 1% formaldehyde at 70°C (Mitra et al. 1963). It was reasonably assumed that the phage RNA was completely in a random coiled configuration at high temperature and that at lower temperature it was, in part, in a helical configuration. Because the helical configuration had a lower absorbancy at 260 nm, the depression of the absorbancy below its value for a random coil was taken as a measure of the fraction of residues that participated in the helix.

As a result of these early calculations, 82% of helical form was predicted for the R17 RNA within the particle (Mitra et al. 1963). Boedtker (1967) estimated the involvement into base pairing as $73 \pm 5\%$ of the nucleotides, while Isenberg et al. (1971) predicted $63 \pm 5\%$ of base pairing in the μ2 RNA. The experimentally determined base compositions presented in Table 10.3 demonstrate uniform distribution of nucleotides.

The actual calculated ratios for the fully sequenced genomes are more representative, since they cover maximally different genomes. These ratios are presented in Table 10.4. From the latter it becomes evident that the base ratios of the phage RNAs are quite similar but not identical. Although at first glance the amounts of the four nucleotides seem quite equimolar, some nuances are evident. Thus, the group II and III phages are relatively rich in uracil. The "true" pseudomonaphages PP7 and LeviOr01 are poor in adenine and in the A+U pair, while the acinetophage AP205 is rich in the A+U pair.

Recently, Almpanis et al. (2018) demonstrated the definite correlation between the bacterial G+C content and the genome

TABLE 10.3

Experimentally Determined Ratio of Nucleotides in the Phage Genomic RNAs

Group	Phage	A	G	C	U	References
I	f2	0.221	0.259	0.268	0.251	Loeb and Zinder (1961)
		0.229	0.286	0.250	0.235	Bautz and Heding (1964)
	MS2	0.224	0.275	0.245	0.256	Strauss and Sinsheimer (1963)
		0.232	0.267	0.253	0.248	Strauss and Sinsheimer (1963)
	R17	0.248	0.225	0.271	0.255	Paranchych and Graham (1962)
		0.231	0.263	0.249	0.257	Mitra et al. (1963)
	fr	0.243	0.271	0.249	0.237	Hoffmann-Berling et al. (1963)
	β	0.233	0.216	0.272	0.280	Nonoyama et al. (1963)
		0.232	0.270	0.240	0.256	Nonoyama and Ikeda (1964b)
	μ2	0.238	0.254	0.237	0.271	Ceppellini et al. (1963)
III	Qβ	0.229	0.241	0.238	0.292	Feix et al. (1968)
?	ZIK/1	0.236	0.239	0.242	0.283	Bishop and Bradley (1965)
	ZJ/1	0.243	0.238	0.237	0.282	Bishop and Bradley (1965)
	ZG	0.251	0.272	0.240	0.237	Bishop and Bradley (1965)
	ZS/3	0.248	0.280	0.238	0.234	Bishop and Bradley (1965)
	ZL/3	0.248	0.269	0.249	0.284	Bishop and Bradley (1965)
	α15	0.247	0.263	0.250	0.240	Bishop and Bradley (1965)
–	PP7	0.209	0.274	0.259	0.258	Edlind and Bassel (1977)

size and G+C content of associated plasmids and phages, where genomes of the phages M, KU1, C-1 INW-2012, and MS2 isolate DL52 were included among many other models.

The actual value of the buoyant density of the MS2 RNA at 40°C was determined as 1.901 g cm^{-3} by ultracentrifugation in a CsCl density gradient (Bruner and Vinograd 1965). Later,

Daniel and Banin (1970) got 1.902 g cm^{-3} for this value and confirmed therefore the buoyant density of the MS2 RNA.

By the small-angle x-ray scattering data, the shape of the MS2 RNA was determined as a both flat and elongate coil of about 620 Å diameter and a ratio of about 2:1:0.5 for the mean radii of gyration in the three directions of space (Zipper and

TABLE 10.4

Calculated Ratio of Nucleotides in the Phage Genomic RNAs

Group	Phage	A	G	C	U	A+U	G+C
I	MS2	0.2340	0.2597	0.2614	0.2449	0.4788	0.5212
	fr	0.2448	0.2657	0.2481	0.2414	0.4862	0.5138
II	GA	0.2435	0.2369	0.2418	0.2778	0.5214	0.4786
	BZ13	0.2394	0.2383	0.2435	0.2788	0.5182	0.4818
	KU1	0.2504	0.2315	0.2338	0.2843	0.5347	0.4653
III	Qβ	0.2279	0.2416	0.2397	0.2907	0.5186	0.4814
	M11	0.2345	0.2454	0.2419	0.2779	0.5124	0.4873
	MX1	0.2410	0.2456	0.2399	0.2735	0.5146	0.4854
IV	FI	0.2318	0.2622	0.2502	0.2557	0.4876	0.5124
	NL95	0.2354	0.2632	0.2451	0.2564	0.4918	0.5082
	SP	0.2449	0.2622	0.2404	0.2526	0.4974	0.5026
Levivirus-like	Hgal1	0.2524	0.2414	0.2378	0.2684	0.5208	0.4792
	C-1	0.2478	0.2450	0.2393	0.2680	0.5158	0.4842
	M	0.2402	0.2382	0.2414	0.2802	0.5204	0.4796
Pseudomonaphages	PP7	0.2152	0.2740	0.2681	0.2425	0.4576	0.5421
	PRR1	0.2480	0.2463	0.2457	0.2600	0.5080	0.4920
	LeviOr01	0.2178	0.2766	0.2715	0.2341	0.4519	0.5481
Caulophage	φCB5	0.2326	0.2523	0.2531	0.2621	0.4947	0.5053
Acinetophage	AP205	0.2683	0.2163	0.2224	0.2931	0.5614	0.4386
Assembled genomes	EC	0.2421	0.2607	0.2509	0.2459	0.4881	0.5116
	MB	0.2211	0.2662	0.2545	0.2578	0.4790	0.5208

Folkhard 1975). Furthermore, the small-angle x-ray scattering measurements revealed high dependence of the MS2 RNA pattern on the Mg^{2+} ion concentration as a result of changes in the secondary and tertiary structure of RNA (Ribitsch et al. 1985). Therefore, the ellipsoidal dimensions of 632:306:140 Å were estimated for the MS2 RNA. Then, the small-angle x-ray scattering profile of the MS2 RNA was reanalyzed using multiple reconstruction methods (Zipper and Durschlag 2007a,b).

It is noteworthy that the MS2 RNA was purified efficiently by chromatography on benzoylated-naphthoylated DEAE cellulose (Sedat et al. 1967). Later, Hung (1969) elaborated a simple DEAE-cellulose method for isolating biologically active RNA from the phages Qβ and MS2. A chromatographic procedure at different temperatures with columns of cellulose CF-11 was able to distinguish the extent of the RNA secondary structure as tested with the phage f2 RNA, among other RNA models (Engelhardt 1972). The adsorption chromatography of nucleic acids, and the Qβ RNA among them, on siliconized porous glass was elaborated as a useful method for the separation of tRNA, rRNA, and mRNA (Mizutani 1983). The MS2 RNA was employed as a model by the separation of single- and double-stranded RNA forms, which was based on their differential binding to silica particles at high concentration of the chaotropic agent guanidinium thiocyanate (Beld et al. 1996). The presence of divalent Ca^{2+} cations in solutions greatly enhanced the deposition kinetics of the MS2 RNA on silica surfaces (Shen et al. 2011). Furthermore, the MS2 RNA was employed by the elaboration of the nucleic acid separations using superficially porous silica particles (Close et al. 2016).

HYBRIDIZATION

The story of nucleic acid hybridization was reviewed brilliantly by Sol Spiegelman (1974), including his first experiments on the hybridization of the MS2 RNA with bacterial DNA (Doi and Spiegelman 1962). Later, the history of the nucleic acid hybridization was narrated by Giacomoni (1993).

The perfect RNA–RNA hybridization methodology was generated with plus and minus strands of the MS2 or Qβ RNAs by introduction of 40% formamide that lowered the melting temperature of double-stranded polynucleotides and allowed hybridization at lower temperatures, thus reducing the danger of chain scission and depurination during prolonged incubations (Friedrich and Feix 1972).

In the context of the hybridization and polynucleotide hybrids, the discovery (Stein and Hausen 1969; Hausen and Stein 1970) of ribonuclease H (H for *hybrid*), an enzyme degrading the RNA moiety of DNA–RNA hybrids, in calf thymus extracts played an enormous role for the further development of molecular biology and gene engineering. In these papers, the double-stranded M12 RNA was used as a representative of an RNA-RNA hybrid that was not a target for the enzyme.

PREPARATIONS FOR SEQUENCING

The digestion of the phage RNAs to smaller oligonucleotides by ribonucleases was the first step for nucleotide sequencing.

However, it helped also for the detection of the molecular mass when the latter was calculated from the fragments obtained after digestion with pancreatic ribonuclease. The total number of nucleotides was calculated from the total radioactivity of all the fragments and the number and specific activity of a particular oligonucleotide. Thus, the R17 RNA was found to be 3342 nucleotide residues long (Sinha et al. 1965b) and the length of the MS2 RNA was estimated as 3500 ± 200 nucleotides (Haegeman et al. 1971). Taking the mass of the average Na^+ nucleotide as 345, the molecular mass of the MS2 RNA resulted in approximately $1,210,000 \pm 70,000$. Meanwhile, the molecular mass of the Qβ RNA was measured by synthesizing the RNA *in vitro*, using Qβ replicase and ^{32}P-GTP and ^{14}C-GTP in a defined ratio, which resulted in a value of 3500 nucleotides (Hindley et al. 1970). The calculated mass values are included in Table 10.1.

The kinetics of the MS2 RNA degradation by ribonuclease A, heat, and alkali and the presence of configurational restraints in this RNA were studied systematically by Strauss et al. (1968).

The specific fragments of the phage RNAs were of natural origin or obtained by limited digestion with different ribonucleases. Thus, as described in Chapter 9, the specific RNA fragments with molecular mass of 6.5×10^5, 5.5×10^5 and 4.4×10^5 in molar ratio 5:4:9 were detected in the defective particles of the wild-type phage MS2 (Renhof et al. 1972, 1974, 1975).

When the Qβ RNA was digested with 5 ng of pancreatic ribonuclease per mg RNA in 0.2 M Tris buffer pH 8.5 at 0°C for 30 min, most of the parent molecule was cut at predominantly one site, producing two fragments corresponding to 32% and 68% of the intact RNA and carrying its 5′ and 3′ ends, respectively (Bassel and Spiegelman 1967). The R17 RNA was cleaved at one site with ribonuclease IV and 5′-one-third and 3′-two-thirds fragments were obtained (Spahr and Gesteland 1968). Digestion of the R17 RNA with pancreatic ribonuclease under conditions identical to those used with the Qβ RNA led to production of as many as 10 high molecular mass RNA fragments (Thach and Boedtker 1969). The 3′ end-labeling made it possible to locate three largest fragments at the 3′ end of the molecule. Remarkably, the difference in the number of fragments produced was not a function of the extent of digestion because the 10 fragments could be detected even when most of the parent molecule remained intact (Boedtker and Gesteland 1975). Boedtker and Stumpp (1968) also tried ribonuclease T4 to produce the R17 RNA fragments. The native R17 RNA was insensitive, however, to attack by ribonuclease V, an exonucleolytic activity associated with ribosome movement on the substrate RNA, but it was attacked if it was either heated or fragmented (Kuwano et al. 1970).

The controlled cleavage of the R17 RNA was achieved by treatment with ascorbate and Cu^{2+} ions (Wong et al. 1974).

The highly specific fragmentation pattern that has been observed after limited ribonuclease digestion was believed to be the result of the secondary structure of the phage RNA, when the folding of the polynucleotide strand into helical regions made those regions less susceptible to enzymatic attack than the rest of the molecule. In fact, the R17 and MS2 RNA fragments that have been produced by T1 ribonuclease digestion and sequenced were puzzled into RNA structures in which most, but not all, of the guanines after which scission had occurred were located in regions that were not hydrogen-bonded regions (Adams et al. 1969a,b; Jeppesen et al. 1970a,b; Min Jou et al. 1972a,b). However, the factors other than secondary structure, possibly tertiary structure, were suspected for their role in the specificity of the phage RNA fragmentation (Boedtker and Gesteland 1975).

A bit later, large fragments of the R17, MS2, and Qβ RNAs were obtained by limited digestion with ribonuclease T1 (Fuke 1974). In the case of the R17 and MS2 RNAs, two pairs of large fragments were obtained that probably resulted from specific cleavages of the whole molecules. The cleavages of the MS2 RNA were produced at points 36% and 47% from the 5′ end. Later, 15 ribonuclease T1-resistant large oligonucleotides of the MS2 RNA were allocated on the genome (Fuke 1976).

MS2 SEQUENCING: WALTER FIERS' TEAM

The history of full-length MS2 sequencing was narrated thoroughly by Walter Fiers (1975), and more recently by Pierrel (2012). The work was started only 2 years after the discovery of the phage f2 by Loeb and Zinder (1961). Following the Fiers' (1975) story, interest in the phage RNA was stimulated at that time by its potentially decisive role in the studies of translation mechanisms and genetic code, replication and its regulation, virus–host interactions, as well as in the studies of evolutionary aspects of life, since a set of similar phages was discovered during 1961–1963. Fiers (1975) mentioned in his history the role of previous studies on ribonucleases when Markham and Smith (1952) were the first to successfully fractionate oligonucleotides and showed that pancreatic ribonuclease, later referred to as ribonuclease A, was, in fact, a very specific enzyme which cleaved only after pyrimidine nucleotides, leaving a 3′-phosphate end group. Then, ribonuclease T1 of *Aspergillus oryzae* was characterized, which cleaved only after G residues (Sato-Asano 1959). Next, the 1968 Nobelist Har Gobind Khorana and his coworkers, also Walter Fiers, elucidated the mechanism of action of 3′- and 5′-exonucleases, showing how, by stepwise removal of terminal nucleotides, these enzymes could be used to determine the nucleotide sequence of oligonucleotides (Razzell and Khorana 1959a,b; 1961; Fiers and Khorana 1963a,b,c). This progress was supported at that time by the development of column chromatographic separation of oligonucleotides (Tomlinson and Tener 1962). At that time, the 1968 Nobelist Robert William Holley and coworkers completed the primary structure determination of the 77-nucleotide-long alanine-tRNA (Holley et al. 1965).

The two important preliminary decisions must be mentioned. First, Fiers decided to work with ^{32}P-labeled viral RNA, as this provided a level of sensitivity at least 1000-fold better than detection by optical density. Moreover, this label has the advantage that the specific activity of the four mononucleotides was absolutely the same and could easily be determined either by Geiger counter, liquid scintillation counter, or autoradiography. Second, the problem of genetic stability of the phage RNA was comprehended. In order to minimize this problem, the main MS2 virus stock was prepared in 1963 from the twice-purified plaque, and all later studies were done on cultures grown from this stock with as few intermediary subcultures as possible.

Concerning specific enzymatic digestion of the MS2 RNA, polypurine tracts were obtained first after total digestion with ribonuclease A (De Wachter et al. 1965; Fiers et al. 1965a,b; Lepoutre and Fiers 1965; Vandenbussche and Fiers 1966). As this enzyme cleaved only after U and C residues, all oligonucleotides present in the final digest should have the general structure $(Pup)_nPyp$, i.e., a series of n purine nucleotides followed by a 3′-terminal pyrimidine nucleotide.

Subsequently, all longer polypurine tracts were individually isolated and sequenced (Min Jou and Fiers 1969b; Haegeman et al. 1971) and allocated to larger RNA fragments (Min Jou and Fiers 1969a; Min Jou et al. 1969). When the amino acid sequence of the f2 coat protein was the first to be determined (Weber and Konigsberg 1967), followed by that of R17 (Weber 1967), one could write possible nucleotide sequences for the coat gene. Unfortunately, none of the longer ribonuclease A products could possibly be derived from this part of the viral genome. As it later turned out, only one heptanucleotide was actually part of the coat gene.

Concerning the 3′-terminal sequence, Sugiyama (1965) showed that alkaline hydrolysis of MS2 RNA releases a single nucleoside, namely adenosine. Thus, the genome was recognized as a linear, not a circular polynucleotide. Labeling of the 3′-end with tritiated borohydride allowed isolation of an oligonucleotide containing the labeled 3′ after hydrolysis with RNase T1. The viral RNA was uniformly labeled with ^{32}P and terminally labeled with ^{3}H, and only a single ^{3}H-labeled peak of 8 to 10 residues was obtained after enzymatic digestion. The last step consisted of the actual nucleotide sequence determination that was achieved by partial hydrolysis with venom exonuclease, separation of the partial products, and determination of the nucleotide composition $GUUACCACCCA_{OH}$ -3′ (De Wachter and Fiers 1967). Weith and Gilham (1967) independently came to the same sequence for the 3′-end of the f2 RNA.

As Fiers (1975) noticed, the most important conclusion that could be drawn from the 3′-terminal sequence was that it contained none of the termination codons for polypeptide synthesis: UAA, UAG, or UGA. It was the first indication that the 3′-end of a polynucleotide was not sufficient for the polypeptide chain release by the ribosome and that the terminal non-translated region must serve another important function. Another interesting finding was that the last three nucleotides, namely CCA, were the same as found in all tRNAs. This terminal sequence had already been identified as the 3′-end of tobacco mosaic virus (TMV) RNA by Steinschneider and Fraenkel-Conrat (1966) and was later also found at the 3′-end of the Qβ RNA (Dahlberg 1968) and of brome grass mosaic virus RNA (Glitz and Eichler 1971). Although R17 RNA-pC and R17 RNA-pCpC lacking 1 or 2 cytidine residues, respectively, were unable to accept nucleoside monophosphates in the enzymic reaction catalyzed by tRNA 3'-terminal nucleotidyltransferase, it was found that they were able to form complexes with the enzyme (Igarashi and McCalla 1971).

The presence of the 3'-terminal adenosine was not necessary for viral replication, since it was removed from R17 RNA without loss of infectivity as assayed in a spheroplast system, while removal of the penultimate C residue was lethal (Kamen 1969). Similar findings were reported for the Oβ RNA (Rensing and August 1969).

Concerning the 5'-terminus, Takanami (1966) found 5'-triphosphoryl 3'(2')-monophosphoryl nucleoside (pppNp) at the 5'-end of the f2 RNA. This residue was identified as pppGp by Roblin (1968), who worked with the R17 RNA. Similar results were soon obtained for the genomes of MS2 (De Wachter et al. 1968b,c), f2 (Dahlberg 1968), Oβ (Dahlberg 1968; Watanabe M and August 1968a), and R23 (Watanabe M and August 1968a). In the case of the MS2 RNA, identification was improved by direct comparison with the tetraphosphate derivatives of all four common nucleosides, which have been synthesized for this purpose by Messens and Van Montagu (1968). Then, the tetranucleotide pppGGGU that contained the 5'-terminal pppGp was identified (De Wachter et al. 1968a,d). The structure of the 5'-terminal oligonucleotide was confirmed later by the complete characterization of the heptanucleotides of the ribonuclease A digest (Haegeman et al. 1971). By the isolation of the 5'-terminal oligonucleotide of the Qβ RNA, the two sequences were always present in approximately equal amounts, namely pppGGGGAAC and pppGGGGGAAC, even in independent preparations derived from single plaques (De Wachter and Fiers 1969, 1970; De Wachter et al. 1969). Remarkably, both MS2 and Qβ RNAs did not contain an initiation codon for protein synthesis at their 5'-ends. This was the first clear evidence that also at the 5'-end an untranslated sequence was present. It seemed that the 5'-end was closely connected to the mechanism of viral RNA replication. When RNAs of MS2, f2, R17, and Qβ started at their 5'-end with a series of three or more G residues, the complementary sequence of three or more C residues was present at the 3' end. The presence of the pppGp at the 5'-end allowed estimation of the MS2 RNA length as approximately 3500 nucleotides (Fiers et al. 1968; Vandenberghe et al. 1969). The pppGp was found also at the 5'-end of the MS2 RNA minus-strand. Moreover, on the basis of the relative pppGp content, the number of growing chains per minus strand template was estimated as the 2.5 nascent strands per molecule (Van Styvendaele et al. 1968; Fiers et al. 1969; Vandenberghe et al. 1969).

The general organization of the MS2 RNA was solved due to partial digestions. Thus, RNA components sedimenting at 21S and at 15S, in addition to the intact 27S form, were characterized, where the 21S component corresponded to approximately a 3'-terminal two-thirds fragment and the 15S to a 5'-terminal one-third fragment of the molecule (Fiers 1967; De Wachter and Fiers 1971). In fact, this selective degradation was the first indication of a specific three-dimensional folding of the viral RNA. Similar fragments were obtained for the Qβ (Bassel and Spiegelman 1967) and R17 (Spahr and Gesteland 1968) RNAs.

This cleavage into the two fragments made it possible to assign unambiguously all previously identified polypurine sequences to either the 5' fragment or the 3' fragment (Min Jou and Fiers 1969a; Min Jou et al. 1969).

The partial digestion with the ribonuclease T1, together with the sequencing method developed at that time by the double 1958 and 1980 Nobelist Frederick Sanger and coworkers (Sanger et al. 1965; Brownlee et al. 1967, 1968), provided a real breakthrough for the nucleotide sequence analysis. Thus, the treatment of the phage MS2 RNA with a low concentration of ribonuclease T1, at 0°C and analysis of the resulting digest by polyacrylamide electrophoresis led to a reproducible discrete number of bands (Min Jou et al. 1968). Then, the advanced technique for isolation and sequencing of partial fragments was adopted (De Wachter and Fiers 1971), two-dimensional polyacrylamide electrophoresis (De Wachter and Fiers 1972), optimized partial ribonuclease splitting (Contreras and Fiers 1971), and gradient chromatography on DEAE paper (Fiers et al. 1971) were introduced. First, the untranslated 5'-terminal 129 nucleotides long sequence preceding the first gene to the A-protein initiation codon at the position 130 was determined (De Wachter et al. 1971a,b,c). In parallel, 104 nucleotides were sequenced at the 3'-end (Min Jou et al. 1970; Contreras et al. 1971). The 3'-sequence was further extended (Vandenberghe and Fiers 1973) until it reached 361 nucleotides and indicated UAG as the replicase termination signal followed by a 174 nucleotide-long untranslated segment (Vandenberghe et al. 1975).

Generally, the classical procedure of sequencing was used, which consisted of the combination of partial digestions by two ribonucleases, when, for example, one part of the fragment preparation was digested with ribonuclease T1 and another part with ribonuclease A, or other ribonucleases, e.g., ε-carboxymethyllysine-41-pancreatic ribonuclease, or CM-ribonuclease, which split only PypPu links under appropriate conditions (Contreras and Fiers 1971), were involved. This allowed ordering of the oligonucleotides, which was the last step in the sequencing.

Thus, the nucleotide sequence of the gene coding for the bacteriophage MS2 coat protein was determined (Min Jou et al. 1971a,b, 1972a,b; Haegeman et al. 1972). Remarkably, by the sequencing, numerous indications were noticed for an ordered structure in many of the pure fragments, when the two strands remained together and behaved as a single molecule until their denaturation in the first dimension of the two-dimensional gel separation method (Fiers et al. 1971; Min Jou et al. 1971a,b, 1972a,b). The fragments showed a tendency to renature in the form of a dimer after denaturation.

In parallel with the achievements of the Fiers' team at the beginning of the 1970s (De Wachter and Fiers 1970), the nucleotide sequences around the initiation codons of the three viral genes of the phage R17 were known from the work of Joan Argetsinger Steitz (1969b). It made it possible not only to establish the 5'-terminal nucleotide sequence up to position

125 (De Wachter et al. 1971a,b,c), but also to find that the last 16 nucleotides of the fragment were identical to the 5′-terminal segment of the ribosome-binding region of the R17 A protein gene. Considering the length of the overlap, and moreover since it contained the heptanucleotide AGGAGGU, which occurred only once in the MS2 RNA (Haegeman et al. 1971), it was concluded that the obtained 5′-terminal nucleotide sequence spanned the entire region between the physical 5′ end and the first gene. This finding also confirmed the fact that the A protein, or maturation, gene was 5′-proximal. Its initiation codon, AUG in the R17 RNA and GUG in the MS2 RNA, started at nucleotide 130 and was preceded by a 129-nucleotide-long untranslated leader sequence. The heptanucleotide sequences present in the pancreatic ribonuclease digest of the MS2 RNA were characterized earlier (Haegeman et al. 1970).

Figure 10.2 presents the 5′-terminal leader sequence of the MS2 RNA. According to Fiers (1975), the 129-nucleotide-long untranslated segment could specify a particular complementary sequence at the 3′ end of the minus strand and recognition by replicase.

Then, there was some similarity between the proposed secondary structure for the 5′ leader sequence and the cloverleaf structure of tRNA. This similarity was even more pronounced when the complement, the 3′ end of the minus strand, was compared to tRNA. It should be emphasized that the minus strand of the MS2 RNA also ended with …CCA (Fiers 1975), as did the MS2 RNA plus strand and the Qβ RNA plus and minus strands (Weber and Weissmann 1970). In this connection with the tRNA-like structure, it is necessary to mention here that the elongation factors Tu-Ts, which functioned in the protein synthesis machinery as carriers of charged tRNA, have been identified as components of the Qβ replicase complex (Blumenthal et al. 1972) and could be a part of the replicase complex of the group I phages (Fedoroff and Zinder 1971). This aspect of the RNA phage function will be unveiled systematically in Chapter 13. Moreover, tRNA nucleotidyltransferase added back CMP and AMP to genome fragments of the RNA phages MS2, R17, and Qβ, which have been obtained by incubation with snake venom phosphodiesterase, in conditions in which the enzyme remained highly specific of CCA-deprived tRNAs (Prochiantz et al. 1975).

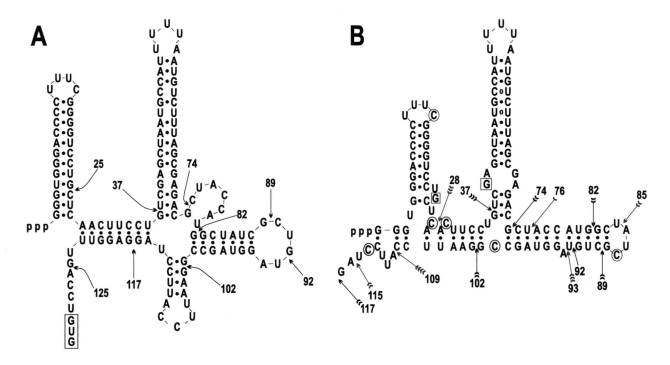

FIGURE 10.2 Sequencing of MS2 RNA. (A) The 5′-terminal leader sequence of MS2 RNA. The sequence of the first 132 nucleotides is shown in the form of a secondary structure model. The arrows point to preferential splitting points by ribonuclease T1. The main fragment in the partial digests extends from 0 to 125. The G-U-G triplet at position 130–132 is the initiation codon of the A protein cistron. The preceding segment remains untranslated (De Wachter et al. 1971a; Volckaert and Fiers 1973). (Volckaert G, Fiers W: Studies on the bacteriophage MS2. G-U-G as the initiation codon of the A-protein cistron. *FEBS Lett.* 1973. 35. 91–96. Copyright Wiley-VCH Verlag GmbH & Co. KGaA. Reproduced with permission.) (B) Proposed secondary structure of the 5′ end of the bacteriophage MS2 RNA genome. The nucleotide sequence is taken from De Wachter et al. (1971a). A red circle around a cytidine indicates partial reactivity (<50%) with methoxyamine and black circles indicate extensive reactivity (>50%). Guanosines which react with kethoxal are boxed. G⁷⁴ modification is not indicated, because it was observed in the fragment of 82 nucleotides, where it occurs in a single-stranded region. Black arrows are directed at T1 RNase cutting points under conditions of partial digestion (De Wachter et al. 1971a); red arrows point to sites cleaved by Cm-RNase (Contreras and Fiers 1971). The number of feathers on each arrow indicates an increasing susceptibility of the site to cleavage (Fiers et al. 1976). The symbol (o) between two bases indicates that the union does not add to the stability of the secondary structure (i.e., $\Delta G = 0$) (Iserentant D, Fiers W. Secondary structure of the 5′ end of bacteriophage MS2 RNA: Methoxyamine and kethoxal modification. *Eur J Biochem.* 1979. 102. 595–604. Copyright Wiley-VCH Verlag GmbH & Co. KGaA. Reproduced with permission.)

These observations might indicate that the phage RNAs contained certain features probably present in all tRNAs and recognized by the transferase.

The 5′-terminal nucleotide sequence was solved up to position 117 for the R17 RNA (Adams and Cory 1970; Adams et al. 1972a,b) and up to position 74 for the f2 RNA (Ling 1971). Both sequences were identical to the corresponding MS2 RNA sequence. As Fiers (1975) noted, this was remarkable in view of the 3%–4% divergence in RNA sequence between the translated regions of these phages (Fiers et al. 1971; Robertson and Jeppesen 1972). This suggested again a highly specific three-dimensional folding for the untranslated leader sequence and/or of its complement which could play an important biological role.

The 3′-terminal region of the R17 RNA was known up to position -51 and was identical to its MS2 counterpart (Cory et al. 1972). This illustrated again the evolutionary stability of the untranslated regions.

Concerning the coat protein gene, a series of four hairpins likewise derived from the coat gene were identified by working with partial ribonuclease T1 digests (Min Jou et al. 1971a,b). By using milder digestion conditions, these hairpins were prolonged at one or both ends (Fiers et al. 1971). On the basis of this primary sequence data, a model for the secondary structure, called by authors the "flower" model, was proposed. This classical breakthrough model is presented in Figure 10.3. As Fiers (1975) commented, the main hairpins were rather well supported by the experimental data, but other aspects of the model were often only based on the stability estimates according to Tinoco et al. (1971) and should be regarded as very tentative. In the flower model, approximately 67% of the nucleotides were involved in base pairing.

The first nucleotide fragments were identified as being derived from the coat gene because their genetic information matched with a segment of the amino acid sequence which was known for the R17 coat polypeptide (Weber 1967; Weber and Konigsberg 1967). As the MS2 coat sequence was thought to be identical to the R17 one, it was not surprising to find almost complete agreement over the entire gene between the nucleotide sequence and the amino acid sequence. The only exceptions were at positions 11, 12, and 17, where nucleotide sequence indicated aspartic acid, asparagine, and aspartic acid instead of Asn11, Asp12, and Asn17, respectively. Vandekerckhove and Van Montagu (1974b) have independently determined the amino acid sequence of the MS2 coat protein and have directly confirmed these amino acid changes.

The agreement between the nucleotide sequence of the coat gene and the amino acid sequence confirmed in a most direct way the genetic code-word dictionary, compiled by Marshall Warren Nirenberg, Severo Ochoa, and Har Gobind Khorana to the early 1970s.

Concerning the sequence of the intercistronic regions, Nichols (1970) has sequenced a region, the left part of which corresponded to the end of the R17 coat gene. An UAA codon was identified as the termination signal; the immediately following UAG codon was interpreted as a kind of safety device.

As the last seven nucleotides at the 3′ end of the sequence overlapped with the 5′ end beginning of the ribosome-binding region of the replicase gene (Steitz 1969a,b), Nichols (1970) concluded that this untranslated intercistronic region was 36 nucleotides long. The same sequence was established for the corresponding region of the MS2 RNA (Fiers et al. 1971; Min Jou et al. 1972a,b). The only difference was a single C to U transition. As Fiers (1975) noted, it was of some interest that the intercistronic region in R17 could possibly code for a hexapeptide, although this has never been found to occur in vitro or in vivo. Due to the single transition, such a possibility was not actual in MS2.

The other intercistronic region was located between the A protein gene and the coat gene. As the amino acid sequence at the carboxyl end of the A protein was not known at that time, it was not possible to identify directly the nucleotide sequence containing the termination signal of the A protein gene. This termination codon, however, must be a single UAG codon because it was suppressible in various Su+ strains with a normal efficiency (Remaut and Fiers 1972a,b). Such readthrough resulted in a prolonged A protein containing some 30 additional amino acids, which did not incorporate into phage particles. The sequence of 160 nucleotides preceding the initiating AUG of the coat gene was identified (Contreras et al. 1973). As Fiers (1975) noted, inspection of this sequence for a UAG codon, which would not be preceded by another nonsense codon in the same reading frame and which would only be followed by another nonsense codon in the same phase some 90 nucleotides farther on, made it possible to identify the termination signal of the A protein unambiguously (see Figure 10.3). This assignment was confirmed by Vandekerckhove et al. (1973a,b), who isolated and sequenced some peptides derived from the carboxyl end of the A protein. It was concluded that the untranslated intercistronic region was 26 nucleotides long.

The nucleotide sequences preceding the initiating AUG of the coat protein have been reported for R17 (Steitz 1969a,b; Adams et al. 1972a) and, although less complete, for f2 (Gupta et al. 1969, 1970). Except for an A to G transition which occurred in some R17 stocks, all intercistronic regions were apparently identical.

Concerning the A protein gene, the 5′-terminal leader sequence until nucleotide 132 was extended and the initiation codon for the A protein gene was identified as GUG in MS2 (Volckaert and Fiers 1973) and not AUG as in R17 (Steitz 1969a,b). No other base changes were observed in this region, either before or after the GUG, and also the N-terminal part of the amino acid sequence was the same (Fiers 1975). Since the R17 and MS2 were nearly identical phages in all molecular biological aspects, including the level of the in vivo A protein synthesis, it led to the conclusion that AUG and GUG could be equally efficient initiator codons. As Fiers (1975) commented later, the identification of GUG as the initiator codon of the A protein gene marked the first time that this codon has been found in this function in a natural messenger RNA.

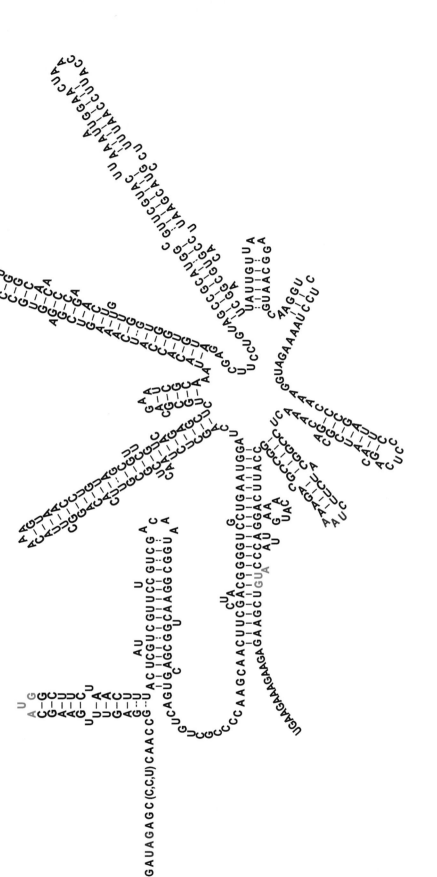

FIGURE 10.3 The "flower" model. The nucleotide sequence of the last part of the A protein gene, the first intercistronic region, the coat gene, the second intercistronic region, and the first part of the polymerase gene is given in the form of a secondary structure. The initiation codons and termination signals are shown green and red, respectively. The original arrows showing the splitting points with RNase T1 are omitted. (With kind permission from Springer Science+Business Media: *Arch Intern Physiol Biochim*, The nucleotide sequence of the coat protein cistron of bacteriophage MS2 RNA: derivation of a secondary structure model, 80, 1972b, 401–403, Min Jou W, Haegeman G, Ysebaert M, Fiers W.)

Many other segments of the A protein gene have been identified, either because they contained reference polypurine sequences or on the basis of other evidence (Contreras et al. 1973; Haegeman and Fiers 1973).

The structure of the A protein gene was first reported at the joint meeting 1974 organized by the Biochemical Societies of Belgium, the Federal Republic of Germany, and the Netherlands in Düsseldorf (Min Jou et al. 1974), included in the short progress report on the status of the MS2 genome (Fiers et al. 1974), and finally published in 1975 (Fiers et al. 1975b).

Concerning the replicase gene, the first series of nucleotide segments derived from the replicase gene were reported in 1972 (Contreras et al. 1972). Finally, the full-length 3569 nucleotide-long MS2 sequence was presented first in September 1975 at the 3rd International Congress for Virology in Madrid (Fiers et al. 1975a) and published in the *Nature* April 8, 1976 issue (Fiers et al. 1976). At once, Walter Fiers published quite popular, but deep and detailed stories of the MS2 full genome (Fiers 1976, 1977). A rapid commentary and interpretation of this historical event was published by Piechocki (1976). The full impact of the complexity of the first genomic structure and the enormity of the work required in the sequence determination was vividly portrayed by commercial educational models constructed at that time by Leach (1977).

MS2 SEQUENCING: OTHER GROUPS

The adenosine was identified after alkaline hydrolysis as the 3′-terminal nucleoside of both f2 and MS2 RNAs (Lee and Gilham 1965).

The 5′ sequence of MS2 RNA was identified as pppGp-GpUp after the ribonuclease A treatment (Glitz 1968). Then, the 5′-linked end of the MS2 RNA, in parallel with the f2 RNA and TMV RNAs, was determined by labeling with periodate oxidation followed by reduction with tritiated borohydride (Glitz et al. 1968). The 5′-sequence was determined as …GUUACCACCCA, whereas …GCCCA was confirmed as the 5′-terminal sequence of TMV RNA.

The sequence differences in the RNAs of the MS2 and its nitrous acid *ts* mutant were identified after ribonuclease T1 digestion and fractionation of the obtained oligonucleotides on two consecutive DEAE-Sephadex columns (Robinson et al. 1969b).

After isolation and sequence analysis of a series of oligonucleotides from complete ribonuclease T1 digests of the R17 RNA (Jeppesen 1971), the homologous sets of large T1-resistant oligonucleotides were isolated from the MS2 and f2 RNAs (Robertson and Jeppesen 1972). The direct comparison of nucleotide sequences from the sets of RNA fragments from the R17, MS2, and f2 genomes allowed an estimate of the overall degree of variation among the three RNAs as 2.8%–3.4%, and yielded additional evidence that noncoding regions showed significantly lower variation. Robertson and Jeppesen (1972) hypothesized that the phages R17, MS2, and f2 were apparently not linear descendants but could have diverged from a common ancestor.

Concerning fragmentation of natural RNAs, the remarkable activities of George W. Rushizky's team must be emphasized. Thus, the separation of higher MS2 oligonucleotides of equal chain length from ribonuclease T1 digests was described first (Rushizky et al. 1965b). Then, preparation of large oligonucleotides was achieved from the MS2 RNA adsorbed on DEAE-cellulose and hydrolyzed *in situ* with one of four different ribonucleases, including ribonuclease A and *Bacillus subtilis* ribonuclease, where the latter demonstrated significant differences between the base ratios of the oligomers and those of the MS2 RNA (Rushizky et al. 1966). Polylysine was complexed to the MS2 RNA in order to obtain protected fragments of suitable length (Sober et al. 1966). The phenol-denatured MS2 coat protein was tried as a potential ion exchanger by column chromatography of oligonucleotides, which could complement the properties of such popular adsorbents as DEAE-cellulose, BD-cellulose, or MAK (Rushizky 1969a). Then, a tricky method was published by which the MS2 RNA could be hydrolyzed preferentially to yield large oligomer fractions (Rushizky 1969b). By this approach, the MS2 RNA was first adsorbed on DEAE-cellulose and then hydrolyzed with ribonuclease from *Bacillus amyloliquefaciens*. The ribonuclease U2 from *Ustilago sphaerogena*, which exhibited a preference for the hydrolysis of purine phosphodiester linkages, was tested with the MS2 RNA (Rushizky and Mozejko 1971, 1973a). Then, micrococcal nuclease was tried with the MS2 RNA (Rushizky and Mozejko 1973a). Moreover, the intact MS2 particles were reacted with succinic anhydride to modify the protein coat and then treated with ribonuclease T1 to obtain controlled hydrolysis of the viral RNA (Rushizky and Mozejko 1973b). Finally, S1 nuclease from *Aspergillus oryzae*, one of the most useful enzymes of the coming gene engineering era, was purified, characterized, and used for removal of single-stranded portions from MS2 RNA (Rushizky et al. 1975).

By the introduction of other novel ribonucleases, the MS2 RNA was used as a model for treatment with bull semen ribonuclease, an enzyme similar to pancreatic ribonuclease A in regard to its specific activity toward single-stranded RNA, but able also to degrade the double-stranded MS2 RNA regions (Libonati and Floridi 1968, 1969).

As pointed out above, the large MS2 RNA fragments, together with the R17 and Qβ oligonucleotides, were obtained by limited digestion with ribonuclease T1 (Fuke 1974), and 15 oligonucleotides were allocated later on the MS2 genome (Fuke 1976).

A large, complex study was performed on the addressed fragmentation of the MS2 RNA by efficient collaboration of laboratories from Moscow and Riga (Metelev et al. 1978, 1980a,b; Stepanova et al. 1979). In this study, ribonuclease H from *E. coli* was used for selective cleavage of MS2 and R17 RNAs in the region of a heteroduplex formed in RNA with an oligodeoxyribonucleotide complementary to a certain part of the RNA. It was shown that ribonuclease H split the molecule of RNA in the position corresponding to the 3′-end of the heteroduplex. Later, this approach was developed further by crosslinking of 15-mer DNA fragments to the *Thermus thermophilus* ribonuclease HI at different enzyme positions (Chon et al. 2002). As a result, site-specific cleavage of the MS2 RNA was achieved successfully by this thermostable DNA-linked ribonuclease H.

The spatial organization of the MS2 RNA was studied at that time in Moscow. Thus, the exposed sites for single-strand- and double-strand-specific nucleases were identified by limited hydrolysis with nuclease S1 and double-strand-specific snake venom ribonuclease (Grechko et al. 1982). Next, this study was expanded with the double-strand-specific ribonuclease III treatment (Grechko et al. 1985b). The ribonuclease III was an *E. coli* nuclease specific for double-stranded RNA (Robertson et al. 1967). The restoration of the three-dimensional structure of the MS2 RNA was evaluated after heating above the melting point (Grechko et al. 1985a) and the accessibility to the S1 and snake venom nucleases was combined by evaluation with thermal stability of the MS2 RNA (Grechko et al. 1987). Fluorescent dyes were involved in these studies on the secondary structure of the MS2 RNA, where the interaction of ethidium bromide with double-stranded, and acridine orange with single-stranded fragments was evaluated in a wide range of ionic strength, ion compositions, and at various pH (Borisova et al. 1984a,b, 1987).

BY-PRODUCTS AND DIRECT CONSEQUENCES OF THE MS2 SEQUENCING

First, the fully solved MS2 sequence allowed extensive comparison of all known sequence fragments from the related RNA phages belonging to the group I (Min Jou and Fiers 1976b). The sequenced segments of the closely related phages R17 and f2 reached 23.9% and 11.50% of the genome, respectively. According to the comparative sequence analysis, the overall degree of variation was estimated as 3.9% (MS2 - R17), 3.4% (MS2 - f2), and 3.7% (R17 - f2). All the differences observed were accounted for by single base substitutions: transitions were highly predominant (86%). No change was found in the untranslated terminal regions and only two mutations occurred in the intercistronic regions. From a total of 34 observed variable sites located in translated regions, 25 were neutral point mutations and only 9 led to an amino acid change, mostly rather conservative. The amount of variation in double-stranded regions (16 out of 36 cases) was much lower than the relative degree of secondary structure of the RNA (roughly two-thirds) would predict. Min Jou and Fiers (1976b) concluded therefore that the selective pressure preserved at least certain aspects of the three-dimensional conformation of the RNA phage genome. The MS2 RNA sequence was analyzed for the number of repeats, and the expected number of repeats was calculated for random nucleic acid sequences (de Wachter 1981).

In parallel with the MS2 sequencing, many technical and theoretical problems were solved by Fiers' team. Thus, reaction of MS2 RNA with formaldehyde was studied in detail (Slegers and Fiers 1970a,b,c). This reaction was usually used to stabilize the denatured forms of nucleic acids as conformation-independent structures. If the MS2 RNA sedimented at 27S at neutral pH and high ionic strength, upon unfolding by reaction with formaldehyde its sedimentation rate decreased to 13.4S. Slegers and Fiers (1970a,b,c) found, however, that if the reaction with formaldehyde was carried out at acidic pH and at low ionic strength, a fast-sedimenting, homogeneous species was formed. The sedimentation coefficient of this species varied between 32S and 41S, depending on the reaction conditions, and it was regarded to be an MS2 RNA dimer. Moreover, both the ribosomal RNA species from *E. coli* were found to form similar dimers, and the conditions for their formation were exactly the same as for the MS2 RNA. Most important, the process was highly sequence specific and no hybrid interaction between 16S and 23S RNA was observed. It was thought that the putative dimers, which appeared under defined conditions, were stabilized by formaldehyde-induced crosslinks. However, further studies did not support the dimer assumption and the 37.5S structure of the MS2 RNA, as well as similar ribosomal RNA species, were qualified by refined molecular mass measurements as fast-sedimenting, contracted monomers (Slegers and Fiers 1972). This fast-sedimenting species was homogeneous and stable at neutral pH (Slegers and Fiers 1973b). The most compact form sedimented at 46S and was obtained after short reaction times with formaldehyde at high temperature or after long reaction times at 35°C. Removal of more than half of the bound formaldehyde had no effect on the compactness of the molecule, although most of the original secondary structure had not yet re-formed (Slegers and Fiers 1973b).

These findings stimulated further investigation of the compact MS2 RNA forms. Thus, the MS2 RNA, which sedimented at 27S in a neutral buffer, was converted to a compact 57S conformation at pH 3.8 in the absence of formaldehyde (Slegers and Fiers 1973a). Requirements for this conversion, besides protonation, were small concentrations of Mg^{2+} ions and a low ionic strength. On the other hand, after heating in the presence of EDTA and at low ionic strength, the MS2 RNA was unfolded to an 11.7S form at pH 6.8 and to 10.5S at pH 3.8. The compact 57S form had lost at least 50% of its secondary structure, as determined by its hypochromicity, and corresponded to a monomer species by light-scattering analysis (Slegers et al. 1973).

Many technical novelties were applied during the MS2 sequencing and characterization, such as the mini-fingerprinting procedure by analysis of ^{32}P-labeled MS2 RNA (Volckaert et al. 1976), and modification of hairpin loops of the MS2 RNA with methoxyamine (Iserentant and Fiers 1976) or kethoxal (Min Jou and Fiers 1976a). Both methoxyamine and kethoxal modifications contributed to the refinement of the 5′-end of the MS2 RNA (Iserentant and Fiers 1979). This refined structure is shown in Figure 10.2 to compare with the first version of the secondary structure. A novel approach toward the determination of the tertiary structure of the MS2 coat protein was attempted (Vingerhoed and Van Montagu 1977).

It is especially noticeable that the results of sequencing at that time were always accompanied with the secondary structure estimations. The main reason was the fact that the sequencing process was tightly connected with the secondary structure of RNA because of the use of ribonucleases as the major sequencing instruments. Therefore, it was not surprising that the full sequence of the MS2 coat gene evoked theoretical studies on the mutual influence of the secondary structure of the gene and its coding potential (Ball 1973a,b,c). In fact, considerations on the role of the secondary structure in messenger RNAs were encouraged earlier (Ball 1972) by the sequenced coat gene fragments of the R17 RNA (Adams et al. 1969a,b; Steitz 1969b; Jeppesen et al. 1970a; Nichols 1970). It was evident that two conflicting requirements pressed on the MS2 coat protein gene: that for the protein structure of the gene product, and that for the RNA secondary structure of the gene itself. Neither requirement could be fulfilled independently of the other, so the evolved structures must represent a compromise. With the whole MS2 coat gene, Ball (1973b) concluded that those amino acid residues which, when represented by base-paired codons, impose the most severe limitations on the overall protein sequence, showed a clear tendency to avoid being encoded in base-paired regions of the messenger RNA. This was the first deciphered example of mutual influence of the secondary structure and the information content of a messenger RNA. The MS2 genome possessing evident secondary structure was compared with the φX174 genome, with scarce self-complementarity (Figueroa et al. 1977; Kovàcs et al. 1981; Shepherd 1981).

The MS2 sequence opened possibilities for the doublet frequency count calculations, i.e., set of frequencies of the sixteen possible two-base sequences (Elton 1975). Remarkably, in the MS2 RNA, the doublet frequencies of the translated regions of the genome resembled those in the host *E. coli*, whereas those in the intercistronic regions differed substantially. These findings were discussed in relation to the hypothesis that the translation rate in *E. coli* system is dependent on the doublet frequency and codon usage patterns (Elton et al. 1976). Furthermore, Elton and Fiers (1976) examined rhythmic variations in purine run frequencies in the phage RNA and found that a strong rhythmic effect was localized in two subregions of the MS2 genome, in part of the Qβ RNA, and in the three RNAs transcribed from the phage λ genome. The nearest-neighbor doublets in the protein-coding regions of the MS2 RNA were calculated by Jukes (1977).

From the point of view of the fine translation mechanisms, the full-length MS2 genome ensured extensive analysis of the codon usage and the choice of code words (Grosjean et al. 1978; Fiers and Grosjean 1979; Grosjean and Fiers 1982). Erickson and Altman (1979) examined the MS2 genome by computer for internal patterns, where a nucleotide sequence was analyzed as a Markov chain. One of the more surprising results of this analysis was the discovery that the noncoding sequences in the genome were as highly ordered, although in a different sense, as the genes themselves. It was also discovered that the codon frequency distributions for the three genes were similar.

Among amino acid codons that required a third-position pyrimidine, a significant bias favoring the use of cytidine over uracil was detected in the MS2 RNA, while there was a bias favoring pyrimidines over purines among amino acid codons with fourfold degeneracy (Fitch 1976). It was hypothesized that this could arise from selection against wobble pairing in the interaction of transfer RNA and messenger RNA.

Based on the secondary structural model of MS2 RNA, Hasegawa et al. (1979) found that, in base-pairing regions of the phage RNA, there was a bias in the use of synonymous codons which favored C and/or G over U and/or A in the third codon positions, and that in non-pairing regions, there was an opposite bias which favored U and/or A over C and/or G. This finding was interpreted as a result of selective constraint, which stabilized the secondary structure of the MS2 RNA (Hasegawa et al. 1979). A potential of the MS2 RNA to form an unusually large number of stable hairpins was evaluated by Nussinov et al. (1980). The RNA folding algorithm was elaborated and tested with the MS2 genome as a model (Nussinov and Pieczenik 1984a,b). Then, it was calculated that the paired MS2 RNA regions had a higher G+C content than unpaired regions, which could reflect selection for high G+C content to encourage pairing, but a re-analysis of the data together with computer simulation suggested that it was an automatic consequence in any RNA sequence of the way it folded up to minimize its free energy (Bulmer 1989).

The secondary structure of MS2 RNA was analyzed in the context of so-called self-induced structural switches, i.e., switches in their secondary structure, which were accompanied by changes in their function, e.g., by RNA chaperones (Nagel and Pleij 2002). The phage RNAs did not require transacting factors for the structural switch, which was therefore indicated as a "self-induced switch." The latter started from a metastable structure, which was maintained for some time allowing or blocking a particular function of the RNA. The folding dynamics of the RNA secondary structure were studied further with the MS2 RNA as a classical model (Voss et al. 2004; Wolfinger et al. 2004). The MS2 RNA structure was involved by the molecular dynamics' free energy simulation study on the conformational transitions in the RNA single uridine and adenosine bulge structures (Barthel and Zacharias 2006).

The sequencing of the MS2 RNA, as well as of the appropriate regions of f2, R17, and Qβ RNAs, led to the breakthrough discovery of the Shine-Dalgarno sequences. Thus, the Australian scientists John Shine and Lynn Dalgarno (1974) postulated that the 3′-terminal sequence of *E. coli* 16S ribosomal RNA, namely PyACCUCCUUA$_{OH}$, interacts with mRNA and that the 3′-terminal UUA$_{OH}$ is involved in the termination of protein synthesis through base pairing with terminator codons. The sequence ACCUCC was proclaimed in turn to recognize a conserved sequence found in the ribosome binding sites of various phage RNAs, and the full set of these conserved sequences within prokaryotic mRNAs derived its name from the authors. Furthermore, the involvement of the 3′-terminal region of the 16S ribosomal RNA was extended also to the elongation phase and mechanisms of the messenger translocation, where MS2 RNA was used as a central model (Sedláček et al. 1978, 1979).

Engelberg and Schoulaker (1976) proposed a hypothesis of so-called RNA phage parasitism, which was based on the extensive analysis of the putative sequence homologies between 3′-end of the 16S ribosomal RNA and the corresponding segments of the MS2 and Qβ RNAs.

The knowledge gained on the MS2 genome contributed strongly to the rapid progress in gene engineering, which was started at that time. The first success related to the covalent addition of polyA segments to the 3′-end of RNA with ATP:RNA adenyltransferase from *E. coli*, where MS2 RNA was used as a natural model (Devos et al. 1974, 1976b). The size of the polyA tails and the specific binding of the treated material to an oligo(dT)-cellulose column (Devos et al. 1976b), as well as removal of polyA segments from RNA (Gillis et al. 1977) and addition of oligoC with adenylyltransferase (Gillis et al. 1976) were described. The 3′ oligoA segment was added also to the Qβ RNA (Devos et al. 1976c). Finally, the synthesis of discrete reverse transcripts of polyadenylated phage RNA was achieved by avian myeloblastosis virus RNA-dependent DNA polymerase in the presence of oligo(dT) as a primer (Devos et al. 1976a, 1977). The results of these studies led to conclusions that were very important for the growing field of gene engineering, namely that it was not the secondary structure of the template but its primary sequence that was the major determinant of the discrete partial transcripts. Then, the successive incorporation of the different dNTPs added to the primer, using MS2 RNA-polyA or Qβ RNA-polyA as a template, was in complete agreement with the known 3′-terminal sequence. Under the conditions used and with the MS2 RNA-polyA or polyA as template, a ribosyl primer could not be substituted for the deoxyribosyl primer. Similarly, with the MS2 RNA-oligoC only (dG)$_{10}$ was an efficient primer for the reverse transcription. The ribonucleoside triphosphates were very poor substrates (Devos et al. 1977). At last, the polyadenylation allowed construction and characterization of a plasmid containing a nearly full-size DNA copy of MS2 RNA (Devos et al. 1979b). The copy was synthesized again with the avian myeloblastosis virus RNA-dependent DNA polymerase. After the MS2 RNA template was removed from the complementary DNA strand with T1 and pancreatic ribonuclease digestion, the complementary DNA became a good template for the synthesis of double-stranded MS2 DNA with *E. coli* DNA polymerase I. The double-stranded MS2 DNA was inserted into the *Pst*I restriction endonuclease cleavage site of the *E. coli* plasmid pRR322 by means of the poly(dA) • poly(dT) tailing procedure. The entire MS2 RNA was faithfully copied, since the restriction cleavage site map of the cloned MS2 DNA corresponded to the cleavage map predicted from the primary structure of the initial RNA, and nucleotide sequence of the 5′ and 3′-end regions of the MS2 DNA was correct. Therefore, the full-length DNA copy of the viral genome, except for 14 nucleotides corresponding to the 5′-terminal sequence of MS2 RNA, was inserted into the plasmid. This approach predetermined the further traditional scheme for the cloning and expression of messenger RNAs by the gene engineering methodology.

Symbolically, the first full-copy cloning of MS2 RNA was accompanied with the insertion of the translocatable element IS1 (Devos et al. 1979a), a constant and imminent threat by cloning procedures in future. The IS1 element was inserted between the N-proximal part of the ampicillin gene and the poly(dA)•poly(dT) linker, and a repetitious sequence of nine base pairs in length was generated by the translocation process.

It is noteworthy to add here that, from the very beginning of the reverse transcription era, the f2, MS2, and Qβ RNAs were used as matrices for testing of the RNA-dependent DNA polymerase activity of RNA tumor viruses, first, of the enzyme from avian myeloblastosis virus (AMV) (Spiegelman et al. 1971; Leis and Hurwitz 1972). The Qβ RNA was employed by the studies of the virion-associated RNA-dependent DNA polymerase of avian sarcoma virus (ASV) (Novak et al 1979). Moreover, the MS2 genome fragments were used traditionally to test novel gene engineering methodology, such as a popular method for the direct cloning of PCR-amplified nucleic acid (Mead et al. 1991).

Remarkably, the MS2 genome was checked for the presence of the recognition sites of the host restriction enzymes and was found not to underrepresent them, in contrast to DNA phages that avoided the 6-base palindromes (Sharp 1986).

The next direct vital consequences of the MS2 sequencing resulted in the development of promising methodologies of high-level protein synthesis in bacteria. Thus, the MS2 knowledge contributed strongly to the pioneering cloning and expression of the human fibroblast interferon gene by Fiers' team, in collaboration with Biogen Company (Derynck et al. 1980). Then, a way was paved to the birth of the popular virus-like particle (VLP) technology (Kastelein et al. 1983a). These gene engineering achievements will be discussed in the nanotechnology chapters of this book, starting from Chapter 19.

The impressing 1963–1976 history of Fiers' *RNA Phage Lab* in Ghent, from genetic code to gene and genome, was published not long ago (Pierrel 2012). According to this story, the success of the *RNA Phage Lab* illuminated two decisive shifts in post-war biology: the emergence of molecular biology as a discipline in the 1960s and of genomics in the 1990s.

Meanwhile, in the former USSR, the joined scientist group from Lev Kisselev's team in Moscow and Elmārs Grēns' team in Riga performed DNA synthesis catalyzed by reverse transcriptase of avian myeloblastosis virus on the MS2 RNA template with synthetic octa- and nonadeoxyribonucleotide primers complementary to the coat protein binding site and the region of the coat protein gene, respectively (Frolova et al. 1977). The synthesis product consisted of discrete DNA fractions of different length, including transcripts longer than 1000 nucleotides. The coat protein inhibited RNA synthesis if it was initiated at its binding site but had no effect on DNA synthesis initiated at the coat protein gene. The MS2 coat protein binding site, namely an RNA fragment of 59 nucleotides, served as a template for polydeoxyribonucleotide synthesis in the presence of octanucleotide primer and reverse transcriptase, where the product of synthesis was homogenous and corresponded to the length of the template.

R17 SEQUENCING

Although the full-length genome did not succeed in the case of the R17 RNA, 23.9% of which was sequenced, the R17 sequencing contributed markedly to the history of the phage

RNA sequencing. In fact, this sequencing was initiated by the famous Frederick Sanger and dealt as a place of the development of Sanger's finterprinting approach (Sanger 1971). Moreover, the R17 sequencing contributed by a novel approach for the isolation and determination of the ribosome-protected ribonuclease-resistant initiation regions. Finally, the R17 sequencing played a crucial role in the correct ordering of the RNA phage genes (Jeppesen et al. 1970b).

The Cold Spring Harbor Symposium in 1969 was really the stage where the full-length genome plans from Ghent, Cambridge, and Zürich were intended. Among the nine sessions there, one was dedicated to the "sequences of mRNA" (cited here and further on the Symposium from Pierrel 2012). At this session, Joan Argetsinger Steitz (1969a) proclaimed the R17 ribosome binding sites. After earning a PhD at the James Watson's Harvard Laboratory, she had gone to Cambridge and began to work on the phage sequence. The other paper on the R17 sequence was reported by the post-doctoral student Jerry Adams, PhD student Peter Jeppesen, Frederick Sanger, and his technician Bart Barrell (Adams et al. 1969b). After development of the fingerprint and homochromatography techniques by the 5S RNA sequencing, a noncoding RNA, Sanger was looking for the genetic code confirmation and turned to the R17 RNA. The sequence of 57 nucleotides in the R17 coat protein gene confirmed directly the genetic code, showed its degenerative character, and suggested that highly ordered base-paired structures in the R17 RNA could be involved in the regulation of gene expression and packing of the RNA into the phage particle (Adams et al. 1969a). As reported at the Cold Spring Harbor Symposium in 1969,

> since this is the first time that a sequence from a mRNA has been determined by chemical means and shown to correspond to the sequence of amino acids in the protein for which it codes, the results can be regarded as one of the most direct confirmations of the correctness of the genetic code. (With kind permission from Springer Science+Business Media: Adams et al. 1969b, cited by Pierrel 2012.)

In April 1971, Sanger reported in his Hopkins Lecture that about 50% of the R17 coat protein gene were sequenced (Sanger 1971). Although Sanger was rather reserved about the completion of the gene in his lecture, the full sequence of the MS2 coat protein gene was resolved soon by Fiers' team (Min Jou et al. 1972a,b). It is necessary to note that collaboration of both Cambridge and Ghent labs was at very high level, since Willy Min Jou, a PhD student from Fiers' lab and one of the lab's leaders, learned the fingerprint technique in 1968 at Cambridge. Not long after, Frederick Sanger shifted completely to DNA sequencing, although the R17 sequencing did continue to be carried out in his lab (cited by Pierrel 2012).

As mentioned before, the sequence of the R17 ribosome binding sites (Adams et al. 1969a,b; Steitz 1969a,b) constituted a considerable help when it became possible to relate unique MS2 oligonucleotides to defined genetic regions. Moreover, Jerry M. Adams and coworkers (Adams and Cory 1970; Cory et al. 1970) were the first to report the complete sequence of a 74-nucleotide-long fragment derived from the 5'-terminus

of the R17 RNA. Subsequently, several other regions of the R17 coat gene became known (Nichols 1970; Jeppesen 1971; Jeppesen et al. 1972). Then, Robertson and Jeppesen (1972) compared a series of T1 oligonucleotides isolated from the f2, R17, and MS2 RNAs and estimated that the divergence was about one nucleotide in 30 to 35 for each combination.

The sequence of 51 nucleotides at the 3'-end of the R17 RNA was determined from partial T1 ribonuclease and ribonuclease IV digests (Cory et al. 1972). Then the 117-nucleotide sequence from the 5' end to the maturation protein gene (Adams et al. 1972b) and the 23 nucleotides of the region immediately preceding the coat protein gene (Adams et al. 1972a) were determined.

Next, a historical breakthrough in gene regulation mechanisms occurred when Bernardi and Spahr (1972a,b) isolated and sequenced the 59 nucleotide fragment after ribonuclease T1 digest of the R17 RNA and coat protein complex. This fragment contained the punctuation signal between the coat protein and replicase genes and represented the site on the RNA where the coat protein acted as a translational repressor. A procedure for the preparation of highly labeled intact ^{32}P-RNA from the phage R17 (Gesteland and Spahr 1970) helped greatly for this success. This method was further improved by Muller and Noll (1976) with the same phage R17.

Steitz (1974) showed later that the R17 coat protein was able to recognize the isolated R17 replicase initiator region of 29 nucleotides.

Novel sequences from the R17 RNA were resolved later at Cambridge: 50 nucleotides (Rensing and Schoenmakers 1973), 73 nucleotides (Rensing 1973), 33 nucleotides (Rensing et al. 1973), and 54 nucleotides (Rensing et al. 1974).

At last, Atkins and Gesteland (1975) showed that the replicase gene of the RNA phages R17, MS2, and f2 had a single UAG terminator codon by translation of the RNA from the wild-type phages in bacterial cell-free extracts containing an *amber* suppressor, where 30%–40% of the replicase appeared with an approximate molecular mass of 63,500, slightly larger than the major replicase product of 63,000 daltons.

In parallel, the isolation, characterization, and cataloging of the R17 and M12 oligonucleotides obtained by ribonuclease A digestion were carried out by Paul Kaesberg's team (Sinha et al. 1965a,b; Thirion and Kaesberg 1968a,b, 1970). The catalogs of products of limited hydrolysis of the R17 and Qβ RNAs by ribonuclease T1 were arranged by Samuelson (1969). An original flow cell packed with anthracene was used for the analysis of column eluents by mixing the eluent with a dioxane scintillator (Thirion 1970). By this technique, the number of isomeric tri- and tetranucleotides were determined in the ribonuclease A digests of the R17 and M12 RNAs after a ribonuclease T1 digestion, and similarities and differences of these phages were identified. Further, the elution positions of the R17 RNA T1 digest pattern were mapped by marker oligonucleotides of known composition and chain length by DEAE-cellulose and DEAE-Sephadex chromatography (Kelley et al. 1971). The sequence of the first six amino acid residues of the coat protein gene was determined after ribonuclease T1 digestion of RNAs from the phage R17 and from its polar coat amber mutant amB2 (Robinson et al. 1969a). Of the products isolated, only one that coded for these six amino acids was different.

Richard Roblin (1968b) found that the pppGp was located at the 5'-terminus of the R17 RNA and suggested using it as a natural label of the 5'-terminus of the phage RNAs. Further, Roblin (1968a) found that the 5'-terminal sequence must be pppGpXpYp---, where X was a purine and Y a pyrimidine, and proclaimed therefore that an N-formylmethionine initiator codon cannot be present directly at the 5'-terminus of the R17 RNA.

The sequenced R17 RNA fragments were used as a background by the elaboration of a simple and highly popular method to estimate the secondary structure in ribonucleic acids (Tinoco et al. 1971). Remarkably, a theoretical analysis on the compatibility of hairpin loops with protein sequences suggested an additional unobserved loop in the R17 RNA (White et al. 1972). Moreover, direct physical evidence, namely melting behavior, of secondary structure (Gralla et al. 1974) was demonstrated for the 59 nucleotides long R17 RNA fragment recognized by coat protein when functioning as a translational repressor (Bernardi and Spahr 1972a,b). The secondary structure of this fragment was confirmed also by high-resolution proton NMR evaluation (Hilbers et al. 1974).

As described for the MS2 RNA above, the R17 sequencing stimulated the growing interest in potential deoxynucleotide primers for reverse copying of messenger RNAs. Thus, a ^{3}H-labeled complementary hexadeoxynucleotide was synthesized (Metelev et al. 1974a), complexed with the R17 RNA (Metelev et al. 1974b), and used as a specific primer for the reverse synthesis of the R17 RNA (Metelev et al. 1974c).

f2 SEQUENCING

According to the above-mentioned Min Jou and Fiers (1976b) calculations, 11.50% of the f2 genome was sequenced in total.

The tetraphosphate pppNp was found at the 5'-end of the f2 RNA by Takanami (1966). This tetraphosphate was identified further as pppGp by Dahlberg (1968). The 5'-terminal pppGp was identified also in the nascent f2 RNA tails (Robertson and Zinder 1968, 1969; Webster et al. 1969). Next, the 5'-linked end of the f2 RNA, in parallel with the MS2 RNA, and TMV RNA was determined (Glitz et al. 1968). Since the labeled oligonucleotide from ribonuclease T1 digests of the f2 RNA was chromatographically indistinguishable from that of the MS2 RNA, it was supposed that the f2 5'-terminal sequence was the same as the ...GUUACCACCCA sequence determined for the MS2 RNA. Finally, the 5'-terminal nucleotide sequence was solved up to position 74 for the f2 RNA and found identical to the corresponding R17 sequence (Ling 1971).

Concerning the 3'-terminal sequence, the adenosine was found as the 3'-terminal nucleoside of both f2 and MS2 RNAs (Lee and Gilham 1965) and the content of the 3'-terminal ribonuclease T1 fragment was determined as -Gp-(3Cp,2Up,2Ap)-CpA for the f2 RNA (Lee and Gilham 1966). Then, the 3'-terminal undecanucleotide of f2 RNA was sequenced as GUUACCACCCA$_{OH}$ by the periodate oxidation-β-elimination approach (Weith and Gilham (1967). The isolation and characterization of the terminal polynucleotide fragments from the f2 and Qβ RNAs was described by Lee et al. (1970). The obtained 3'-terminal f2 sequence was confirmed by Dahlberg (1968), who also determined the same 3'-terminal sequence for the R17 RNA. The reaction of dimedone, or 5,5-dimethyl-1,3-cyclohexanedione, with dialdehydes resulting from periodate oxidation of the 3'-hydroxyl terminus of RNA, was suggested for the terminal labeling of high molecular mass RNAs, and its efficiency was demonstrated on the f2 RNA (Glitz and Sigman 1970a,b). The f2 RNA, together with the Qβ and GA RNAs as representatives of three unrelated RNA phage groups, served as a model for the elaboration of a novel approach for the isolation of 3'-terminal polynucleotides from RNA molecules (Rosenberg and Gilham 1971; Rosenberg et al. 1971). By this method, the 3'-terminal polynucleotides obtained by nuclease digestion of large RNAs were selectively bound to columns of cellulose derivatives containing covalently bound dihydroxyboryl groups.

The 61-nucleotide-long ribosome binding site of the coat gene initiation region within the f2 RNA was isolated after the ribonuclease T1 treatment and sequenced as a fragment specifying the first 7 amino acids of the coat protein and extending 41 nucleotides toward the 5' end from the AUG initiator codon (Gupta et al. 1969, 1970). This sequence was quite identical to the corresponding R17 sequence (Argetsinger-Steitz 1969; Adams, Cory and Spahr 1972). Remarkably, Gupta et al. (1970) called attention to the UUUGA sequence which was separated from the AUG codon by 2 or 3 nucleotides in the known initiation sites, and that this sequence was complementary to the TΨC containing loop of the initiator tRNA$_F^{Met}$.

It should be noted here in the ribosome binding connection that the ribosome-f2 RNA interaction studies were initiated by Takanami et al. (1965). The ribosome-unattached portion of the ^{32}P-labeled f2 RNA was removed by digestion with pancreatic ribonuclease, and the attached fragment was isolated from the ribosome as a homogeneous fragment with a sedimentation coefficient averaging 2.1S. The nucleotide composition of the fragment was determined as about 21% A, 25% C, 35% G, and 20% U, and compared with 22.4% A, 26.5% C, 25.9% G, and 25.2% U for the total f2 RNA. Therefore, it was concluded for the first time that the ribosome was attached to a specific locus of the phage RNA. Later, Takanami (1968) studied attachment of ribosomes to the 5'-terminally ^{32}P-labeled fragments of the f2 RNA of the four different lengths that were obtained after treatment with low concentrations of pancreatic ribonuclease. In parallel, Hindley (1968) reported the specific fragmentation of the ^{32}P-labeled f2 RNA when bound to ribosomes after the ribonuclease T1 treatment.

According to the above-mentioned study of Robertson and Jeppesen (1972) on the sets of large MS2 and f2 T1-resistant oligonucleotides that were homologous to the corresponding R17 fragments, it appeared probable that a total of 15–20 base replacements in the phage coat protein gene might accompany the single amino-acid difference between the f2 coat protein and that of either R17 or MS2.

The two coat protein gene fragments, 42 and 57 nucleotides in length, from a partial ribonuclease T1 digest of the f2 RNA, which differed from their R17 counterparts, were sequenced, and 4 nucleotide substitutions were found (Nichols and Robertson 1971). Three of these substitutions occurred within the coat gene, including one that accounted for the only amino acid difference between the coat proteins of the two phages; the other was located immediately after the termination triplets of the coat gene.

The UAG codon was identified as a single replicase gene terminator in the f2, R17, and MS2 genomes by translation of their RNAs in *E. coli* cell-free extracts containing an *amber* suppressor (Atkins and Gesteland 1975).

It is noteworthy that the f2 RNA, together with human 5S RNA, was used as a model for the application of the fingerprinting techniques to iodinated nucleic acids that appeared ^{125}I-labeled principally at cytidine residues (Robertson et al. 1973). The iodination approach showed little sensitivity to potential structure in the single-stranded RNA molecules and did not change the specificity of ribonucleases. Later, the f2sus11 RNA served as a model to compare activity of monomers and cross-linked dimers of bovine ribonuclease A (Palmieri and Libonati 1977).

Qβ SEQUENCING

In contrast to the MS2, R17, and f2 sequencing described above, the Qβ sequencing was performed after another scenario. The latter was announced by Martin Billeter from Charles Weissmann's team in Zürich at the same Cold Spring Harbor Symposia in 1969 (Billeter et al. 1969b). Whereas RNA sequencing had been done until then by degradation procedures, Weissmann's team introduced an effective use of the Qβ replicase by incorporating *in vitro* a ^{32}P radioactive precursor. The idea was brilliant. Later, Sanger (1988) quoted this method as a root for his DNA sequencing approach using DNA polymerase:

> As an alternative to partial hydrolysis, some work had been done on the possibility of using copying techniques for sequencing, particularly by Charles Weissmann and his colleagues (Billeter et al. 1969c). The RNA bacteriophage Qβ contains an RNA polymerase that copies its own RNA, and they had developed elegant pulse-labeling techniques for the RNA and for deducing sequences. An obvious enzyme to choose for copying DNA was DNA polymerase … (Cited with permission of Annual Reviews from Sanger F. Sequences, sequences, and sequences. *Annu Rev Biochem*. 1988;57:1–29.)

The details of the novel approach were published by Billeter et al. (1969a). Using the novel technique, the team managed to obtain a 175-nucleotide-long sequence that remained, as Pierrel (2012) noted, the longest sequence that had been made for 2 years. However, the team did not get as far as publishing the full sequence of the Qβ genome using this method. According to the science history comments (Pierrel 2012), "the precise reason why the team gave up RNA sequencing remains to be studied." The idea and its realization were

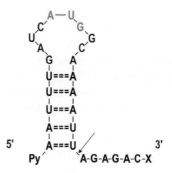

FIGURE 10.4 Possible looped structure of the Qβ coat protein gene sequence. (With kind permission from Springer Science+Business Media: *Nature*, Sequence of a ribosome binding site in bacteriophage Qβ-RNA, 224, 1969, 964–967, Hindley J, Staples DH.)

described in the excellent review articles (Weissmann et al. 1973a,b). Remarkably, King (1973) published a computer algorithm for the Qβ RNA sequencing, which used data on pancreatic and T1 ribonuclease digested fragments which were obtained in Weissmann's team by synchronized *in vitro* synthesis of the Qβ RNA.

In parallel, a sequence of 26 nucleotides, which included the initiation region of the Qβ coat protein gene, was isolated and determined by scientists from the University of Bristol (Hindley and Staples 1969). Figure 10.4 presents the possible looped structure of this sequence. The comparison of the obtained sequence with the three ribosome binding sites of the R17 RNA determined by Steitz (1969a,b) showed in each case a common pentanucleotide sequence UUUGA preceding the AUG initiation codon, an indication to the future Shine-Dalgarno sequence. In fact, Hindley and Staples (1969) supposed that this sequence must be involved in the ribosome recognition of some genes, or cistrons, in a polycistronic messenger.

This study contributed strongly to the correct location of the coat protein gene within the phage Qβ genome, which has been achieved by joint effort of the Zürich and Bristol groups (Hindley et al. 1970). The coat protein gene was located in the middle of the genome and found to begin between the 1100th and 1400th nucleotide from the 5′-terminus of the Qβ RNA.

The collective Zürich and Bristol data on the first 5′-terminal 175 nucleotides and correct location of the Qβ coat protein gene were summarized and presented at the 59th General Meeting of The Society for General Microbiology at Oxford on 24–25 September 1970 (Billeter et al. 1970).

At the same time, the 3′-end groups of Qβ plus and minus strands were determined in Zürich by the terminal labeling with KB[3]H₄ (or NaB[3]H₄) following oxidation with sodium periodate (Weber and Weissmann 1970). The Qβ plus strands, both extracted from the viral particle or from infected bacteria, as well as minus strands, were terminated predominantly by adenosine, and to a lesser degree by cytidine. The Qβ RNA synthesized *in vitro* by the purified Qβ replicase also had mostly adenosine at the 3′-terminus.

It is necessary to highlight here the contribution of other groups to the determination of the terminal Qβ sequences. Thus, the determination of the 3′-terminal oligonucleotides

(Dahlberg 1968; Weith et al. 1968; Weith and Gilham 1969; Lee et al. 1970) and 5′-terminal tetraphosphate pppGp (Dahlberg 1968; Watanabe M and August 1968a) was achieved. The 3′-terminus was determined as (G)CCCUCCUCUCUCCCA$_{OH}$ (Dahlberg 1968). The two 5′-terminal Qβ RNA sequences, pppGGGGAAC… and pppGGGGGAAC…, were identified in nearly equal amounts by Fiers' team (De Wachter and Fiers 1969; De Wachter et al. 1969). The 5′- and 3′-termini of the Qβ RNA were described also by Young and Fraenkel-Conrat (1971), who found unique 5′-terminal sequence ppp-GGGGAC…, in contrast to the previous study. The Qβ RNA, together with the f2 and GA RNAs, served further as a model for a novel approach to isolate 3′-terminal polynucleotides from RNA molecules (Rosenberg and Gilham 1971; Rosenberg et al. 1971). Then, large fragments of Qβ RNA, together with fragments of R17 and MS2 RNAs, were obtained by limited digestion with ribonuclease T1 (Fuke 1974).

The longer 3′-terminal Qβ sequence of 52 nucleotides was deduced by the Zürich and Bristol groups from the 5′-terminus of the Qβ minus strand that was determined by the Qβ replicase technique, as described above (Goodman et al. 1970). The authors emphasized that no termination signals were detected within the last 32 nucleotides, and a notable homology was noticed between the 3′-ends of the Qβ and R17 RNAs.

The situation in the Qβ sequencing was reported regularly by progress reports (Billeter 1971; Billeter et al. 1972; Sastry et al. 1974; Billeter et al. 1975; Escarmis et al. 1975) or in the classical reviews devoted to the RNA phage story in total (Weissmann et al. 1973a,b). Remarkably, Weissmann et al. (1973a,b) have presented some unpublished Qβ sequences and boxed the potential future Shine-Dalgarno sequences.

The ribosome binding site of the Qβ replicase gene was isolated and sequenced when initiation at this site was allowed by use of ^{32}P-labeled Qβ RNA that was fragmented by treatment of the phage with alkali at high ionic strength (Staples and Hindley 1971). The 31-nucleotide sequence contained three possible initiator codons but the AUG nearest the 3′-end was implicated as the initiator codon since the derived N-terminal tripeptide corresponded to the predicted Qβ replicase sequence (Skogerson et al. 1971). The penta-nucleotide AGGAU, common with the pentanucleotide in the appropriate R17 site, was found on the 5′-side of the initiator codon. The ribosome binding site of the Qβ maturation gene was localized (Staples et al. 1971) due to the obstacle that the 5′-terminal Qβ sequence was available via de novo replicase synthesis used, as described above, for the Qβ sequencing. The 330 nucleotides at the Qβ 5′-terminus were known at that time, 175 of which were published (Billeter et al. 1969c) but others remained unpublished (cited by Staples et al. 1971). Therefore, the newly synthesized 5′-terminal Qβ fragments of 170 and 500 nucleotides were used to be complexed with ribosomes. The 70S complexes were treated with pancreatic ribonuclease and the fragment was isolated, which contained the initiation codon at position 62. The latter was followed by nucleotide sequence that corresponded to the expected N-terminal sequence Pro-Lys-Leu-Pro- of the Qβ maturation, or A2, or IIa protein (Weiner and Weber 1971).

The coat and replicase gene ribosome binding sites were extended further by isolation of several ribonuclease T1-resistant fragments (Porter and Hindley 1972). One of them was 66 nucleotides in length, contained neither an AUG nor a GUG codon, and preceded coat protein gene, and other fragment, 82–83 nucleotides in length, contained the replicase gene ribosome binding site. Porter et al. (1974) published the sequence of the 83-nucleotide-long ribosome binding site of the Qβ replicase gene.

The sequence of 97 nucleotides preceding the coat gene was known thanks to the Qβ replicase-driven protection of the Qβ replicase binding site on the Qβ RNA against ribonuclease T1 degradation (Weber et al. 1972). The 3′-terminal sequence of this RNA segment was identical to the 5′-terminal half of the ribosome binding site of the coat protein gene. It was one of the mechanisms of the replication regulation when Qβ replicase competed directly with ribosomes for the coat initiation site, which accounted for its repressor activity. For further details, see Chapters 16 and 17.

Since the lack of the translational repression between Qβ and the group I phages remained intriguing, statistical calculations were performed at that time on possible sequence homologies of the Qβ and R17 RNAs (Köhler and Rohloff 1974). It was concluded that both RNAs were quite different in their properties, although they could be derived from a common precursor.

By analogy with the MS2 and R17 RNA properties mentioned earlier, the Qβ RNA could receive back CMP and AMP at the 3′-end by treatment with snake venom phosphodiesterase (Prochiantz et al. 1975).

It is noteworthy, although not directly connected with the Qβ sequencing, that at that time Eikhom (1975) succeeded in isolation of free Qβ minus strands, which were not involved in a firm duplex structure. These minus strands were extracted from the cells under conditions of mild lysis and low salt concentrations and purified by electrophoresis on polyacrylamide gels. Later, the Qβ, as well as MS2, minus strands, which have been prepared by in vitro transcription, were each separated into two forms on nondenaturing agarose gels (Shaklee 1991). However, only a single form of RNA was observed for minus strands denatured with glyoxal prior to agarose analysis, demonstrating that the two forms were conformers in each case. The truncated and elongated versions of these RNAs, analyzed under nondenaturing conditions, demonstrated that a region of RNA complementary to the maturation gene was at least partially responsible for the presence of these conformers in both Qβ and MS2 minus strands.

Further development of the Qβ sequencing occurred in the obstacles of the gene engineering approaches. As in the case of the MS2 RNA, the Qβ RNA was extended at the 3′-end with the polyA segment by terminal riboadenylate transferase purified from calf thymus (Gilvarg et al. 1974, 1975a) The polyadenylated Qβ RNA retained full infectivity in a spheroplast assay system. However, the progeny viruses did not contain polyA termini, indicating an in vivo rectification of the in vitro alteration (Gilvarg et al. 1975a). While the polyA-Qβ RNA functioned normally as messenger for the synthesis of virus-specific proteins, it had lost its capacity to serve as template for the Qβ replicase. The template function was restored, however, by phosphorolysis with polynucleotide phosphorylase. It was concluded that a host enzyme, perhaps polynucleotide phosphorylase, removed part or all of the adenylate residues prior to replication of the RNA *in vivo* (Gilvarg et al. 1975b). The Qβ RNA was elongated also with a 3'-terminal oligoC tract (Mekler and Billeter 1975). Meanwhile, polyA sequences were added to the 3'-terminus of the Qβ RNA by ATP:RNA adenylyltransferase from *E. coli* by Fiers' team (Devos et al. 1976c). By tail lengths not exceeding 200 nucleotide residues, the physical properties of Qβ-RNA-polyA were only slightly different from those of the original RNA, but almost complete abolishment of template activity, even by short oligoA stretches, was found, in agreement with the conclusions of Weissmann' team.

The next Qβ RNA step in gene engineering consisted of the preparation of Qβ DNA-containing plasmid by use of poly(dA)•poly(dT) tail technique (Mekler et al. 1977). Now, further Qβ sequencing could continue by the methodology of the DNA sequencing, namely, by the Sanger and/or Maxam and Gilbert methodology. Moreover, the hybrid pCR1 plasmids carrying the complete Qβ DNA copy inserted in either orientation gave rise to the Qβ phage progeny in the bacterial host (Taniguchi et al. 1978a,b).

However, in parallel, the ribonuclease T1 fragmentation of the Qβ RNA was successfully applied, when 29 oligonucleotides, 11–26 nucleotides in length, were isolated, sequenced, and located within the genome (Billeter 1978). This led to the determination of the first half of the Qβ coat protein gene, namely 239 nucleotides located immediately adjacent to the initiation triplet (Escarmis et al. 1978). Remarkably, the primary structure and the secondary structure model derived from it did not provide any evidence of homology with the corresponding RNA region of the MS2 RNA. Shapira and Billeter (1978) published a hybridization procedure that was used for the isolation of specific RNA segments applied to the analysis of the Qβ RNA. This tricky method included addition of AMP residues using terminal adenylate transferase, hybridization to the ^{32}P-labeled Qβ RNA minus strands synthesized *in vitro*, partial digestion with ribonuclease T1, and recovery by chromatography on poly(U)-Sephadex.

Remarkably, at that time, a short segment of the Qβ coat protein gene, namely the A-A-A-C-U-U-U-Gp fragment, was synthesized at a milligram scale by a combination of enzymatic methods using the phage T4 RNA ligase and the thermophilic polynucleotide phosphorylase (Kikuchi and Sakaguchi 1978).

Meanwhile, a rapid method for the RNA sequencing was described (Kramer and Mills 1978). By analogy with Sanger's dideoxy chain-termination method of DNA sequencing (Sanger and Coulson 1975), this method employed the 3'-deoxy analogs of the ribonucleoside triphosphates as specific chain terminators during RNA synthesis. This method was tested by sequencing of the MDV-1 minus RNA, a molecule that was synthesized *in vitro* by the Qβ replicase. The story of these short Qβ variants will be presented in Chapter 18.

At last, thanks to efforts of Martin A. Billeter and Philipp Mekler, the full-length Qβ sequence was achieved. It appeared in the PhD thesis (Mekler 1981) but was never published as a paper or submitted to the GenBank.

The Qβ RNA sequencing stimulated interest in the role of secondary structure in the RNA resistance to ribonucleases. Libonati and Palmieri (1978) studied the resistance of the Qβ and f2sus11 RNAs to different nucleases, including ribonuclease A and S1 nuclease, at different conditions. The authors concluded that the secondary structure was neither the only nor the most important variable in determining the susceptibility of double-stranded RNA to ribonucleases. Other factors, such as the effect of ionic strength on the enzyme and/or the binding of enzyme to nucleic acids, played an important role in the process of double-stranded RNA degradation by ribonucleases specific for single-stranded RNA. Then, a two-dimensional model was constructed for the central hairpin of the Qβ RNA and tested with a set of ribonucleases and chemical probes (Skripkin and Jacobson 1993). The model was supported by the SP RNA data. As a result, it was shown that the readthrough region of the readthrough protein A1 formed a separate structural domain and suggested that it functions as a nucleation site that participates in the folding and refolding of the molecule during replication and translation. Moreover, the secondary structure of the entire 4217 nucleotide sequence of the Qβ RNA was predicted in one computer run using the computer program MFOLD that computed RNA structures within any prescribed increment of the computed minimum free energy (Jacobson and Zuker 1993). The final "energy dot plot" revealed five large, well-determined, independent structural domains that covered approximately 50% of the viral genome. The predicted structural domains were consistent with and provided support for five large structural domains identified previously by quantitative electron microscopy in the Qβ RNA. Furthermore, the model was confirmed by experimental data on the mutated Qβ RNAs (Arora et al. 1996; Jacobson et al. 1998).

SEQUENCING OF OTHER PHAGE RNAs

The global sequencing of many RNA phages was performed further by the rules of the classical gene engineering approaches, namely, by cloning and Sanger's dideoxynucleotide DNA sequencing.

Thus, Yoshio Inokuchi and colleagues contributed strongly at that time by the search of homology among RNA coliphages in their 3'-terminal regions (Inokuchi et al. 1979, 1982; Inokuchi 1981). Eight cDNAs of the 3'-terminal RNA regions from the four groups of RNA coliphages, namely, MS2, JP34, GA, Qβ, VK, SP, TW18, and FI, up to 200–300 nucleotides in length, were synthesized by AMV reverse transcriptase and used for the phage RNA-cDNA hybridization tests (Inokuchi et al. 1979). The RNAs from one serological group were annealed in extremely high frequency with a cDNA complementary to an RNA from the same group. On the other hand, when annealed with RNAs from other groups, the frequency of hybrid formation was as low as that annealed with tRNA.

It was concluded that the RNA sequences in the 3′-terminal untranslated regions were similar within the serological group. Group I was represented in this study by the phages MS2, R17, JP501, FR1 (a local isolate, do not confuse with the phage fr), and BO-1; group II by the phages JP34, GA, TH1, SD, KU1, and BZ13; group III by the phages Qβ, VK, ST, and TW18; and group IV by the phages SP, ID2, TW19, TW28, and FI.

In order to go more deeply into the genealogical relationships among four groups I–IV, the 3′-terminal 200–260 nucleotides were sequenced for 14 phage RNAs, namely of the phages MS2, JP501, FR1, and BO-1 from group I, JP34, GA, TH1, KU1, and BZ13 from group II, Qβ, VK, and ST from group III, and SP and TW28 from group IV (Inokuchi et al. 1982). The sequences of the phage RNAs within the same group were extremely homologous, about 90%. On the other hand, when the sequences were compared with those from other groups, they were seen to be only about 50%–60% homologous between group I and group II, and about 50% homologous between group III and group IV. In other combinations, such as groups I (or II) and III, and groups I (or II) and IV, however, the extent of homology was small. The sequences up to 30 residues from the 3′-end were found to be about 90% homologous between groups I and II, and between groups III and IV.

The full-length 3466 base sequence of the group II phage GA was revealed by Inokuchi et al. (1986). The sequence was related to that of the MS2 RNA, including the coat protein binding site which was involved in translational repression of the replicase gene, and a hairpin in a region proximal to the lysis protein gene, which have been especially emphasized by authors. A bit later, Inokuchi et al. (1988) deciphered the complete 4276-base-long nucleotide sequence of the group IV phage SP. It was related to the Qβ sequence: the sequences for the coat and central region of the replicase were strongly conserved, as well as the replicase binding sites. Remarkably, the base composition of the SP and Qβ RNAs differed significantly from one another, and most of the differences were accounted for by a strong preponderance of U in the third position of each codon of Qβ relative to SP.

In parallel, the first 1392-base-long sequence from the unusual and intriguing group I phage fr genome was cloned by Alexander Tsimanis from Elmārs Grēns' team in Riga, and the appropriate cDNA sequence was published (Berzin et al. 1986, 1987). Then, the complete fr sequence of 3575 nucleotides was determined in collaboration with Jan van Duin's team in Leiden (Adhin et al. 1990b). Overall sequence homology between fr and MS2 reached 77%. The greatest divergence between these phages occurred in the 5′-terminal region of the A gene, while the lysis-replicase gene overlap, the coat gene, and the central region of the replicase gene were highly conserved. It is noteworthy that the fr RNA sequence was used immediately by the elaboration of a new approach for the synthesis of oligoribonucleotides with the E. coli RNA polymerase and immobilized synthetic DNA-templates (Denisova et al. 1989).

Meanwhile, van Duin's team continued successful decoding of the most unusual and rare RNA phage genomes. Thus, the nucleotide sequence of the coat and lysis genes of the phage JP34 genome was determined and the discrepancy between serological and biophysical classification of the RNA phages was resolved (Adhin et al. 1989). The phage JP34 was classified as a member of group II, since the nucleotide similarity was more than 95% for group II, while with group I it was less than 45%. The altered serotype of JP34 was due to the change of three critical amino acids of the coat protein to residues present in the group I phage MS2 at the homologous positions.

Next, the complete nucleotide sequence of the pseudomonaphage PP7 was published (Olsthoorn et al. 1995). It made it possible to classify the phage PP7 as a levivirus, although no significant nucleotide sequence identity between PP7 and the group I and II coliphages was observed, except for a few regions where homologous parts of proteins were encoded, most notably in the replicase gene. However, in these regions the nucleotide sequence similarity between the PP7 and MS2 genomes was no greater than between PP7 and the group III and IV coliphages. Moreover, the Qβ and MS2 were no closer to each other than they were to PP7.

Then, the nucleotide sequence of the group II phage KU1 was determined (Groeneveld et al. 1996). The most conspicuous difference in the comparison with other group II members such as GA and JP34 was the presence of an insertion at the start codon of the lysis gene. In GA and JP34, the coat and lysis genes overlap by one nucleotide in the configuration UAAUG. The 18-nt insertion in KU1 was positioned between the A and the U of the start codon. It did not affect the coat reading frame, but it destroyed the AUG start codon and separated the previously overlapping genes by 17 nucleotides. The insert created a UUG codon at its 3′ border which served as the start site for the KU1 lysis protein. The phages M11 and MX1 from group III and the phage NL95 from group IV were sequenced by Beekwilder et al. (1995, 1996b).

At last, the complete nucleotide sequence was achieved for the most mysterious RNA phage, namely the acinetophage AP205, no analogs or relatives of which have been found up to now (Klovins et al. 2002). The phage AP205 gained special popularity due to its prominent role in viral nanotechnology, as described in the chapters following Chapter 19.

Klovins et al. (2002) showed that the content and order of the three major AP205 genes, as well as the lack of the readthrough protein, made it homologous to the *Levivirus* genus. A major digression from the known *Levivirus* members was the apparent absence of the AP205 lysis gene at the usual position, overlapping the coat and replicase proteins. Instead, two small open reading frames (ORFs), were present at the 5′-terminus, preceding the maturation gene. One of these ORFs might encode a lysis protein. Other new features concerned the 3′-terminal sequence. In all RNA coliphages, there were always three cytosine residues at the 3′-end, but in AP205 there was only a single terminal cytosine. Generally, the AP205 RNA did not demonstrate significant sequence identity with the coliphage RNAs.

The appearance of the novel RNA phage genomes stimulated studies oriented on the similarities and dissimilarities of overall phage, DNA and RNA, genomes (Blaisdell et al. 1996). Moreover, this appearance generated the phylogenetic and evolutionary studies directed specifically to the *Leviviridae* family (Bollback and Huelsenbeck 2001).

The rapid progress in the methodology, which was supported also by the urgent needs of the ecological evaluation, i.e., environmental pollution and water quality, as described in Chapter 6, led to the novel sequenced RNA phage genomes, mostly coliphages. Thus, Vinjé et al. (2004) validated their novel RT-PCR assay by using 190 field and prototype of FRNA and FDNA strains and presented a panel of 351 enriched FRNA and FDNA phage samples. They presented a potentially new subgroup of coliphages, which was called JS and included strains having more than 40% nucleotide sequence diversity with the known levivirus subgroups represented by MS2 and GA.

The next unusual phage, the pseudomonaphage PRR1, but rather the broad host range phage, was sequenced by Dennis H. Bamford's team in Helsinki (Ruokoranta et al. 2006). The broad host range of the phage PRR1 was a result of the promiscuity of the receptor encoded by the IncP plasmid and used by the phage for infection. The comparison of the phage PRR1 with other members of the *Leviviridae* family placed it rather into the *Levivirus* genus but with some characteristics more similar to those of members of the *Allolevivirus* genus.

The next mysterious phage, namely caulophage φCB5, was subjected to full-length sequencing by Kaspars Tārs' team in Riga (Kazaks et al. 2011). All four characteristic levivirus genes were found but, unlike other leviviruses, the lysis protein gene of φCB5 entirely overlapped the replicase gene in a different reading frame. Then, Rumnieks and Tars (2012) determined the full-length nucleotide structure of the IncM-dependent phage M. The latter possessed the smallest known *Leviviridae* genome and had the typical genome organization with maturation, coat, and replicase genes in the 5' to 3' direction. The lysis gene was located, however, in a different position than in other known *Leviviridae* phages and completely overlapped with the replicase gene in a different reading frame. The sequence identities of the phage M proteins to those of other RNA phages did not exceed 25% for maturation protein, 51% for coat protein, and 41% for replicase.

The IncC and IncH group-dependent RNA phages, C-1 and Hgal1, respectively, were sequenced by Kannoly et al. (2012). Again, they repeated the characteristic organization of the levivirus, but not allolevivirus genome and were most closely related to the IncP-plasmid-dependent RNA phage PRR1.

Nowadays, the metagenomic sequencing is going to contribute more and more novel information on the RNA phages. Thus, the marine levivirus EC and MB genomes were assembled by the metagenomic sequencing of organisms in San Francisco wastewater (Greninger and DeRisi 2015). At last, Krishnamurthy et al. (2016) presented the metagenomic sequencing of 122 partial genomes of the RNA phage phylotypes that were highly divergent from each other and from the previously described RNA phages.

Coming back after the future metagenomic sequencing to the early RNA phage history, the studies must be mentioned that have been devoted to the RNA phages other than those already described. Thus, total digestion with ribonuclease A was undertaken on the M12 RNA, in parallel with R17 RNA (Sinha et al. 1965a; Thirion and Kaesberg 1968a,b, 1970; Thirion 1970).

For the phage R23 RNA, the guanosine triphosphate was identified at the 5'-terminus by Watanabe and August (1968a).

Concerning the μ2 RNA, a simple spectrophotometric method was elaborated for the determination of the nucleotide composition of the μ2 RNA hydrolysates (Richards 1968). The method depended on the measurement of a difference spectrum between an aliquot of the hydrolysate at pH 7 and one at pH 11.8, where this difference spectrum depended only on the concentrations of UMP and GMP present, while the quantities of CMP and AMP were obtained from a spectrum at pH 7 after allowance for the contribution of the UMP and GMP. Then, the distribution of purine nucleotides in the μ2 RNA was determined by complete ribonuclease A digestion (Matthews 1968c). Some oligonucleotides from the μ2 RNA ribonuclease A digests were fractioned on DEAE-Sephadex columns (Matthews 1968a). It is noteworthy that Matthews (1968b) applied Cerenkov counting to column chromatography of ^{32}P-labeled substances. Next, Gould et al. (1969) fractionated low molecular mass fragments of the μ2 RNA by polyacrylamide gel electrophoresis.

The phage GA RNA, together with the f2 and Qβ RNAs, served as a model for the elaboration of a novel approach for the isolation of 3'-terminal polynucleotides from RNA molecules (Rosenberg and Gilham 1971; Rosenberg et al. 1971). The GA RNA was employed by the generation of a novel method consisting of the separation of oligonucleotides, nucleotides, and nucleosides on columns of polystyrene anion-exchangers with solvent systems containing ethanol (Asteriadis et al. 1976).

For the caulophage φCp2, the RNA termini were identified as guanosine tetraphosphate pppGp at the 5'-terminus and adenosine A_{OH} at the 3'-terminus (Fujiki et al. 1978).

The pioneering place of the RNA phages, as the first sequenced gene and the first sequenced genome, was localized recently on the sequencing time-scale in a short sequencing history (Crowgey and Mahajan 2019).

ELECTRON MICROSCOPY

The detailed electron microscopic evaluation of the R17 RNA, as well as of its replicative form and replicative intermediate, was performed for the first time by Granboulan and Franklin (1966a). The size of a set of RNA molecules from animal cells was evaluated on the basis of the calculated length to molecular mass relation of the R17 RNA and the idea was proposed that the electron microscope might be employed for the determination of the molecular mass of RNAs (Granboulan and Scherrer 1969).

In fact, a bit later, electron microscopy was used directly to the determination of molecular mass of the phage RNAs. Thus, Chi and Bassel (1974) presented a corresponding method for the preparation of single-stranded RNA, where treatment with formaldehyde at elevated temperatures removed secondary structure, and molecules were spread in a protein monolayer from 50% formamide onto a 50% formamide hypophase. Jacobson and Bromley (1975) assumed the molecular mass of the MS2 RNA to be 1.2×10^6 by their electron microscopic molecular mass measurements of Rous sarcoma virus RNA. The molecular mass of the f2, Qβ, and PP7 RNAs was determined by the formaldehyde–formamide method in the electron microscopic study of Edlind and Bassel (1977). Remarkably, the modal length of the f2, Qβ, and PP7 RNAs was determined in this study as 1.0, 1.2, and 1.0 μm, respectively. Kolakofsky et al. (1974) determined the mean length of the R17 RNA as 1.2 μm. Benike et al. (1975) estimated the length of the R17 and PP7 RNAs as 1.22 and 1.16 μm, respectively. Vollenweider et al. (1978) refined molecular mass of the Qβ RNA: the latter was found to be 1.30–1.38 times longer than MS2 RNA and its length was calculated as 4652–4936 nucleotides with an average of 4792, based on the known MS2 RNA length of 3569 nucleotides by Fiers et al. (1976). The phage RNA molecular mass values obtained by electron microscopy are included in Table 10.1.

Furthermore, Jacobson (1976) developed a method allowing electron microscopic demonstration of the secondary structure of viral RNA. Depending on $MgCl_2$ concentration, the MS2 RNA displayed one to three large open loops which ranged in size from 10%–20% of the total RNA length, and smaller closed loops which were approximately 3%–5% of the total RNA length. Within one spreading, the conformation of the molecules was variable. However, the average complexity of the molecules increased with increasing salt, and individual loops that were infrequent at low salt increased in frequency with increasing salt. It was concluded that all RNA molecules could be described by one basic pattern of the secondary structure formation. Furthermore, Jacobson and Spahr (1977) subjected two unequal MS2 fragments produced by the ribonuclease IV treatment to the electron microscopic evaluation (Spahr and Gesteland 1968). On the one hand, the location of the cleavage site was predicted with respect to the secondary and tertiary structure of the RNA and the RNA conformation around this site. On the other hand, the approximate location of individual viral genes was obtained within the secondary structure map by comparing the map of native RNA with known sequence data.

In parallel, Edlind and Bassel (1977) compared by electron microscopy the secondary structures of the f2, Qβ, and PP7 RNAs. While the f2 RNA demonstrated a central open loop and four symmetrically placed hairpins, similar to the data of Jacobson (1976) for the closely related phage MS2, the Qβ RNA had a central open loop and a smaller terminal loop and the PP7 RNA had two large closed secondary structures, one of which was nearly central. With increased denaturing conditions, the central closed structure of the PP7 RNA was converted into an open loop. The central structures of all three phages included about 700 nucleotides.

Remarkably, Naora and Fry (1977) confirmed the presence of the secondary structures in the MS2 RNA but did not find any traces of the secondary structure in eukaryote messenger RNAs, such as rat liver mRNA and hen and rabbit globin mRNA.

Then, the electron microscopic spreading of the MS2 RNA was performed in the presence of the polyamine spermidine under conditions where the molecules were extensively base-paired (Jacobson et al. 1985). In this study, computer modeling was used to identify some of the nucleotide sequences that might be associated with the structures visualized by electron microscopy. Later, the secondary structure of the Qβ RNA was examined by electron microscopy in the presence of varying concentrations of spermidine (Jacobson 1991). As a result, the size and position of large structural features that ranged in size from 170–1600 nucleotides and covered 70% of the viral genome were mapped. Remarkably, a loop containing approximately 450 nucleotides was located at the 5′-end of the RNA and included the initiation region for the viral maturation protein. A large hairpin containing approximately 1600 nucleotides was located in the center of the molecule. It was multibranched and included most of the coat gene, the readthrough region of the A1 gene, and approximately one-third of the viral replicase gene. It was concluded that the secondary structures of the Qβ RNA differed significantly from structures that were described previously for the MS2 RNA but were similar to structures observed by electron microscopy in the SP RNA (Jacobson 1991). It is noteworthy that the MS2 RNA was capable to form tight complexes with another polyamine, namely spermine (Ikemura 1969).

SECONDARY STRUCTURE: JAN VAN DUIN'S TEAM

The activity of Jan van Duin's team by the sequencing of novel rare RNA phage genomes was associated with the interest in the RNA secondary structure. Thus, a thorough evaluation of the secondary structure of the central region of the MS2 RNA of 900 nucleotides, at positions 1297–2190, was performed first (Skripkin et al. 1990). It was huge work involving enzymatic and chemical sensitivity of specific nucleotides, phylogenetic sequence comparison, and the phenotypes of constructed mutants. The specific regions were compared with the appropriate sites in the RNAs of the phages M12, R17, GA, Qβ, SP, and just newly sequenced phages JP34 and fr. The substantial conservation of helices between the related groups I and II was found. The authors put to the forefront the idea that the secondary structure was needed not only to confer protection against ribonuclease, but also to prevent annealing of plus and minus strands, besides the clear needs required to regulate translation and replication. The translation and replication needs were emphasized especially by the authors, since the central MS2 RNA section contained three translational start sites, as well as the binding sites for the coat protein and the replicase enzyme.

In parallel, the secondary structure at the 3′-terminal region of the four groups of the RNA coliphages, represented by MS2, GA, Qβ, and SP, was compared with that of tRNA (Adhin et al. 1990a). The secondary structure models for the 3′ noncoding region of the coliphage RNAs were proposed and based on comparative sequence analysis and on previously published data on the sensitivity of nucleotides in the MS2 RNA to chemical modification and enzymes. It was concluded that, in contrast to the coding regions, the structure at the 3′-terminus was characterized by stable regular helices. Next, comparison of the homologous helices indicated that only those base pair substitutions were allowed that maintained the thermodynamic stability of the genome. Then, although the overall structure similarity of the phage RNAs with tRNA was low, one common element was detected, which consisted of a quasi-continuous helix of 12 base pairs that could be the equivalent of the 12 base pair long coaxially stacked helix, formed by the TΨC arm and the aminoacyl acceptor arm in tRNA. As in tRNA, this structure element started after the fourth nucleotide from the 3′-end. Finally, the phage RNAs contained a large variable region of about 35 nucleotides bulging out from the quasi-continuous helix. It was speculated that the variable loop in the present-day tRNA could be the remnant of the variable region found in the phage RNAs. The variable region contained overlapping binding sites for the replicase enzyme and for the maturation protein, and could serve as a switch from replication to packaging (Adhin et al. 1990a).

The PP7 RNA sequencing opened up fresh opportunities for the search for the conserved and regulatory sequences in the RNA phages (Olsthoorn et al. 1995a). Thus, several regulatory RNA secondary structure features that were present in the coliphages were identified also in the PP7 RNA, although the sequences involved cannot be aligned. Among these were the coat protein binding helix at the start of the replicase gene, structures at the 5′- and 3′-terminus of the RNA, a replicase binding site, and the structure of the coat protein gene start. Remarkably, some of these features resembled the MS2 type coliphages but others the Qβ type (Olsthoorn et al. 1995a). The KU1 RNA sequencing (Groeneveld et al. 1996) made it possible to compare the KU1 and JP34 RNAs and construct a secondary structure model that accounted for the insertion of a separate stem-loop in the KU1 structure at the start codon of its lysis gene.

The highly impressive secondary structure models were generated for the Qβ RNA. First, for the last two domains, positions 2980–4123 (Beekwilder et al. 1995), and then, for the first three domains of Qβ RNA, positions 1–860 (Beekwilder et al. 1996b). The Qβ RNA models were constructed by the use of information on the newly sequenced RNAs of the three phages MX1, M11, and NL95, the full-length sequences of which appeared for the first time in these two papers and were intended to provide the necessary data for comparative analysis, as well as on the previously known SP RNA sequence (Inokuchi et al. 1988). The models were supported by phylogenetic comparison, nuclease S1 structure probing, and computer prediction using energy minimization and a Monte Carlo approach.

Figure 10.5 is intended to give the impression about the last two domains of the Qβ RNA. The 3′-terminal domain, which was about 100 nucleotides long, contained most of the 3′ untranslated region and folded into four short, regular hairpins.

The adjacent 3′ replicase domain contained about 1100 nucleotides. The hairpins in this protein-coding domain were much longer and more irregular than in the 3′ untranslated region. Both domains were defined by long-distance interactions. The secondary structure was not a collection of hairpin structures connected by single-stranded regions. Rather, the RNA stretches between the stem–loop structures were all involved in an extensive array of long-distance interactions that contracted the molecule to a rigid structure in which all hairpins were predicted to have a fixed position with respect to each other. A general feature of the model was that the helices tend to be organized in four-way junctions with little or no unpaired nucleotides between them. As a result, there was a potential for coaxial stacking of adjacent stems (Beekwilder et al. 1995).

Figure 10.6 illustrates the secondary structure of the first three domains of the Qβ RNA (Beekwilder et al. 1996b).

This part of the Qβ RNA contained the 60 nucleotide 5′ untranslated region and the first 800 nucleotides of the maturation protein gene. The RNA adopted a highly ordered structure in which all hairpins were held in place by a network of long-distance interactions, which formed three-way and four-way junctions. Only the 5′-terminal hairpin was unrestrained, while connected by a few single-stranded nucleotides to the body of the RNA. The start region of the A-protein gene, which was a part of the network of long-distance interactions, was base paired to the three noncontiguous downstream sequences.

The fidelity of the secondary structure models was compared by the generation of the site-directed mutants. Thus, a fully double-stranded hairpin of 17 bp, which was the ribonuclease III target, was inserted into a noncoding region of the MS2 RNA genome and the stem underwent Darwinian evolution to a structure that was no longer a substrate for ribonuclease III (Klovins et al. 1997b). The cloverleaf model of the 3′-terminal sequence of the MS2 RNA was confirmed directly by the *in vitro* experiments, where ribosome binding to the maturation gene was faster than refolding of the denatured cloverleaf (Poot et al. 1997). When the three stem–loop structures were replaced by a single five-nucleotide loop and folding delay had virtually disappeared, i.e., the RNA folded faster than ribosomes could bind, the perturbation of the cloverleaf by an insertion made the maturation start permanently accessible.

A long-range interaction in the Qβ RNA that bridged the thousand nucleotides between the M-site, or major binding site for Qβ replicase, and the 3′ end was confirmed by the mutations that destabilized the long-range interaction and were virtually lethal to the phage, whereas base pair substitutions had little effect (Klovins et al. 1998). The existence of a long-range pseudoknot, base pairing eight nucleotides in the loop of the 3′-terminal hairpin to a single-stranded interdomain sequence located about 1200 nucleotides upstream, close to the internal replicase binding site, was identified (Klovins and van Duin 1999). This pseudoknot explained how replicase binds at an internal segment, the M-site, some 1450 nucleotides away from the 3′-end. The introduction of a

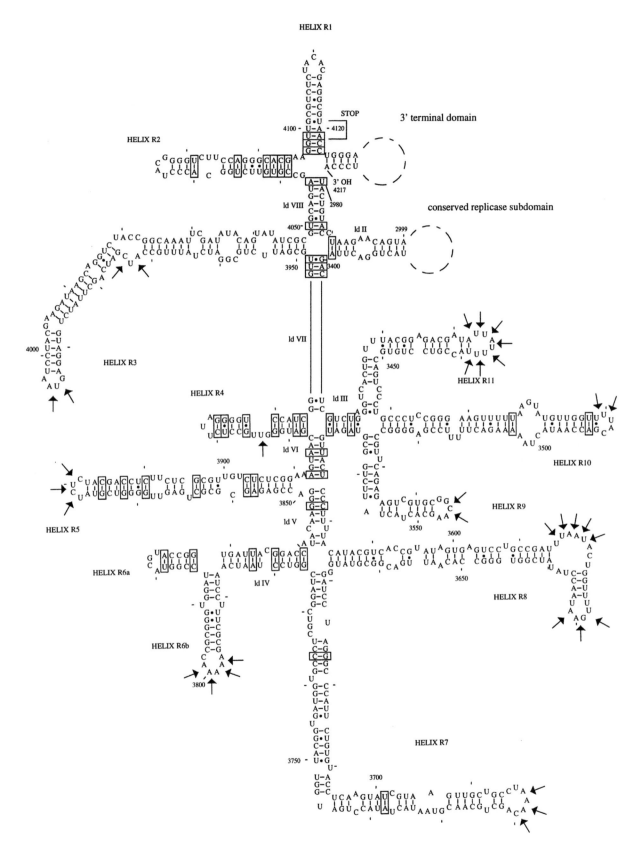

FIGURE 10.5 Proposed secondary structure for the 3′ replicase domain of Qβ RNA (2980 to 4123). The 3′-terminal domain and the conserved replicase subdomain (3000 to 3390) are presented schematically here as broken circles. Covariations with other group III and IV phages are indicated by a box. S1 reactivities are indicated by arrows. Continuous lines connect adjacent bases and are introduced for practical reasons, as are the angles in the helices. Long-distance interactions (ld) have roman numerals. (Reprinted from *J Mol Biol.*, 247, Beekwilder MJ, Nieuwenhuizen R, van Duin J, Secondary structure model of the last two domains of single-stranded RNA phage Qβ, 903–917, Copyright 1995, with permission from Elsevier.)

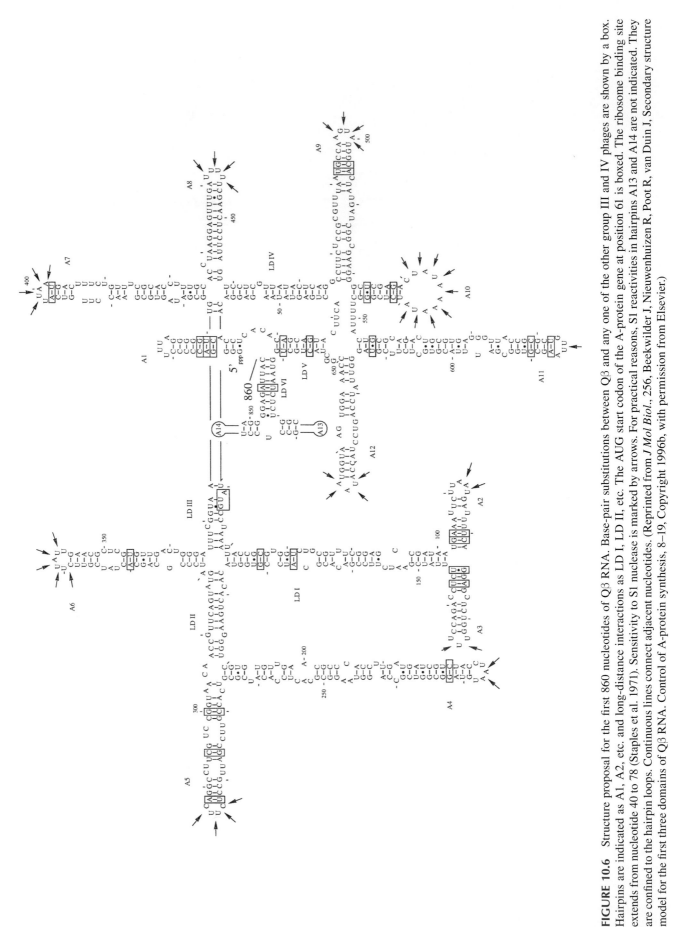

FIGURE 10.6 Structure proposal for the first 860 nucleotides of Qβ RNA. Base-pair substitutions between Qβ and any one of the other group III and IV phages are shown by a box. Hairpins are indicated as A1, A2, etc. and long-distance interactions as LD I, LD II, etc. The AUG start codon of the A-protein gene at position 61 is boxed. The ribosome binding site extends from nucleotide 40 to 78 (Staples et al. 1971). Sensitivity to S1 nuclease is marked by arrows. For practical reasons, S1 reactivities in hairpins A13 and A14 are not indicated. They are confined to the hairpin loops. Continuous lines connect adjacent nucleotides. (Reprinted from *J Mol Biol.*, 256, Beekwilder J, Nieuwenhuizen R, Poot R, van Duin J, Secondary structure model for the first three domains of Qβ RNA. Control of A-protein synthesis, 8–19, Copyright 1996b, with permission from Elsevier.)

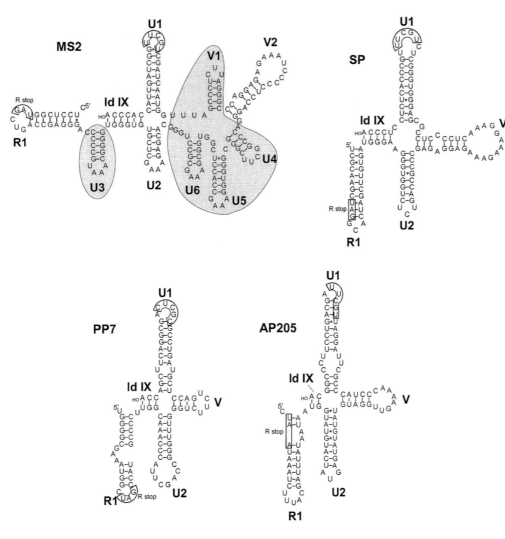

FIGURE 10.7 Comparison of RNA secondary structures in the 3′ UTR in four different RNA phages. The difference between Qβ and SP is that Qβ has a small stem–loop of 12 nt between hairpins V and U1, whereas SP only has 3 nt here (CGC). SP is therefore even more similar to AP205 and PP7 than Qβ is. The boxed sequence at the top of hairpin U1 is conserved in all RNA phages. In Qβ this sequence was shown to form a pseudoknot with its complement about 1200 nt upstream. The stop codon of the replicase is boxed. ld is long-distance interaction. Structures for the coliphages were derived by phylogenetic comparison (Adhin et al. 1990; Beekwilder et al. 1995; Olsthoorn et al. 1995) and are usually predicted by computer programs. End structures for PP7 and AP205 are predicted by RNA MFOLD, except ld IX. Comparison with the folding of the coliphages, especially species IV, strongly argues for its presence, however. The shadowed areas in MS2 are the structure elements absent from SP. (Reprinted with permission of Microbiology Society from Klovins J et al. *J Gen Virol.* 2002;83:1523–1533.)

single mismatch into this pseudoknot was sufficient to abolish replication, but the inhibition was fully reversed by a second-site substitution that restored the pairing. The pseudoknot was part of an elaborate structure that seemed to hold the 3′-end in a fixed position *vis-a-vis* the replicase binding site (Klovins and van Duin 1999).

The sequencing of the most distant RNA phage genome, namely AP205, opened novel opportunities for the secondary structure evaluations (Klovins et al. 2002). As expected, the phage AP205 and the coliphages did not have significant sequence identity; yet, important secondary structural features of the RNA were conserved. Figure 10.7 shows this resemblance in the case of the 3′ untranslated region and of the replicase–operator hairpin. Interestingly, although AP205 had the genetic map of a levivirus, its 3′ untranslated region had the length and RNA secondary structure of an allolevivirus. Sharing features with both MS2 and Qβ suggested that, in an evolutionary sense, AP205 should be placed between Qβ and MS2 (Klovins et al. 2002).

SECONDARY STRUCTURE: FURTHER INQUIRY

After the classical secondary structure evaluation by van Duin's team, the MS2 and KU1 genome fragments were used to test a Kinwalker, a heuristic program that calculated a folding trajectory for an RNA sequence, i.e., a sequence of secondary structures connecting the unfolded state with the thermodynamic ground state (Geis et al. 2008). The algorithm allowed co-transcriptional folding and was used to fold sequences of up to about 1500 nucleotides in length. In parallel, the MS2 fragments were used to validate another computational approach to simulate RNA folding kinetics (Tang et al. 2008).

Later, the problem of the secondary and higher MS2 RNA structures was touched by the extensive studies on the self-assembly of the phage MS2 (Borodavka et al. 2012, 2013, 2015, 2016). The novel approach consisted of single-molecule fluorescence assay to monitor the RNA conformation and virus assembly in real time, where the MS2 RNA was labeled with Alexa-fluor-488 (Borodavka et al. 2012). This technique made it possible to monitor the fate of the MS2 RNA in the absence and presence of the coat protein (Borodavka et al. 2013). The specific process of the RNA phage self-assembly is described in Chapter 21.

Recently, Borodavka et al. (2016) started with an assumption that long RNAs with nonrepeating sequences usually adopt highly ramified secondary structures and could be better described as branched polymers. To test whether a branched polymer model could estimate the overall sizes of large RNAs, the authors employed fluorescence correlation spectroscopy to examine the hydrodynamic radii of a broad spectrum of biologically important RNAs, ranging from viral genomes, including the MS2 RNA, to long noncoding regulatory RNAs. In fact, the relative sizes of long RNAs measured at low ionic strength corresponded well to those predicted by theoretical approaches that treated the effective branching associated with the secondary structure formation.

MS2 RNA AND PROGRAMMED CELL DEATH

The MS2 RNA was used as a specific substrate by the studies of the prokaryotic toxin-antitoxin systems, which are responsible for the growth regulation of bacteria under stress conditions. In the *E. coli* MazE-MazF system, which belongs to the type II toxin-antitoxin systems, the MazF toxin is a sequence-specific mRNA endoribonuclease, or interferase, that initiates a programmed cell death pathway in response to various stresses (Engelberg-Kułka et al. 2005). Thus, the MazF toxin was found to function as a specific mRNA interferase and to cleave mRNAs at ACA sequences, in order to inhibit protein synthesis and induce cell growth arrest.

When MS2 RNA was used as a substrate for the MazF interferase from *Myxococcus xanthus*, it was cleaved into two major bands of approximately 2.8 and 0.8 kb with many minor bands between them, suggesting that the MS2 RNA might contain a preferential cleavage site for the MazF, which was mapped (Nariya and Inouye 2008). The MazF activity to cleave the MS2 RNA was completely inhibited when it was preincubated with purified MazE, the antidote for the MazF (Yamaguchi et al. 2009). Working on the MazF homologs from *Mycobacterium tuberculosis*, Zhu et al. (2008) developed a new general method for the determination of the recognition sequences longer than three bases for the mRNA interferases with the use of the MS2 RNA as a substrate and CspA, an RNA chaperone, which prevented the formation of secondary structures in the RNA substrate. Using this method, it was found that pentad sequences, such as UU↓CCU, CU↓CCU, or U↓CGCU, were targets of the MazF homologs. Recently, tRNA was identified as a new target of one of the MazF homologs from *M. tuberculosis* (Schifano et al. 2016).

The same MS2 RNA and CspA chaperone-based assay was used to identify a pentad sequence U↓ACAU as a specific target of the MazF from *Staphylococcus aureus* (Zhu et al. 2009).

Furthermore, with the use of the MS2 RNA as a substrate, the MqsR gene, a crucial regulator for quorum sensing and biofilm formation in *E. coli*, was identified as a GCU-specific mRNA interferase (Yamaguchi et al. 2009).

With the MS2 RNA and CspA chaperone, the *Bacillus subtilis* MazF was found to cleave a pentad U↓ACAU, in contrast to the *E. coli* MazF cleaving specifically the ACA sequences (Park JH et al. 2011).

Using the MS2 RNA as a substrate, a straightforward approach was elaborated for preventive and therapeutic treatments of infection by HIV-1, HCV, and other single-stranded RNA viruses by the intramolecular regulation of the MazF toxicity (Park JH et al. 2012a,b). Thus, the MazE-MazF toxin-antitoxin system of *E. coli* was engineered by fusion of a C-terminal 41-residue fragment of antitoxin MazE to the N-terminal end of toxin MazF with a linker having a specific protease cleavage site for either HIV-1 protease, NS3 protease of HCV, or factor Xa. These fusion proteins formed a stable dimer to inactivate the ACA-specific mRNA interferase activity of the MazF. When the fusion proteins were incubated with the corresponding proteases, the MazE fragment was cleaved from the fusion proteins, releasing active MazF, which then acted as an ACA-specific mRNA interferase cleaving the MS2 RNA.

The MS2 RNA cleavage assay was used to characterize the MazE-MazF system of the emerging nosocomial pathogen *Clostridium difficile*, which exhibited selective, not global, mRNA cleavage (Rothenbacher et al. 2012). This was especially important, since this pathogen was able to survive antimicrobial therapy and switch from inert colonization to active infection. Again, the MS2 RNA was employed to characterize the MazEF toxin-antitoxin homolog from *Staphylococcus equorum* (Schuster et al. 2013). The system was shown to cleave at the UACAU motifs, mainly after and occasionally before the first uracil.

Remarkably, the MS2 RNA test contributed to the highly important study, where the replacement of all arginine residues with canavanine, a toxic arginine analog, was achieved for the first time in the MazF interferase from *Bacillus subtilis* (Ishida et al. 2013). To do this, the single-protein production system in *E. coli* was used. The resulting MazF variant changed its specificity and gained a 6-base recognition sequence, UACAUA, for RNA cleavage instead of the 5-base sequence, UACAU.

It is noteworthy that the crystal structures of the *B. subtilis* MazF were presented in complex with mRNA substrate, thereby directly providing the structural basis underlying the sequence-specific recognition of the RNA substrate by the MazF (Simanshu et al. 2013). Again, the MS2 RNA was used as a substrate for the characterization, including crystallographic and NMR solutions, of the SaMazF toxin from *Staphylococcus aureus* (Zorzini et al. 2014) and for the crystallographic solution of the *E. coli* MazF toxin in complex with a 7-nucleotide substrate (Zorzini et al. 2016) and of the *B. subtilis* MazF in complex with its RNA substrate (Ishida et al. 2017). Moreover, the role of the $V_{16}V_{17}$ sequence was studied by replacing it with the TK, GG, AA, or LL amino acid residues. The substitution mutants thus constructed showed significant differences in the cleavage specificity of the MS2 RNA. The primer extension analysis of the cleavage sites revealed that the $V_{16}V_{17}$ sequence played an important role in the recognition of the 3′-end base of the RNA substrate.

The MS2 RNA substrate made it possible to identify the MoxT toxin of *Bacillus anthracis*, a ribonuclease, which exhibited U↓ACAU sequence specificity but was distinct from the other MazF family toxins by its ability to cleave RNA in DNA-RNA hybrids (Verma and Bhatnagar 2014).

Then, the MS2 RNA was employed by the characterization of the translation-dependent mRNA cleavage by YhaV, one of the toxins in the type II toxin-antitoxin systems in *E. coli* (Choi et al. 2017). Recently, the MS2 RNA contributed to the identification of MazF homolog in *Legionella pneumophila* with the RNA cleavage site AACU (Shaku et al. 2019).

The largest family of the type II toxin-antitoxin systems, virulence-associate proteins (VapBC), was described first in several *Rickettsia* genomes and found functional in bacteria and eukaryotic cells, as reviewed by Audoly et al. (2011). The rickettsial toxin VapC cleaved the MS2 RNA, while VapB antitoxins had no detectable effect on RNA but counteracted the toxins ribonuclease activity when forming a complex (Audoly et al. 2011). Remarkably, the toxic effect of VapC in mammalian cells, through its ribonuclease activity, was prevented by dexamethasone. The authors concluded in this context that early mortality following antibiotic treatment of some bacterial infections could be prevented by administration of dexamethasone. Winther and Gerdes (2011) found, however, that VapC exhibited only weak ribonuclease activity toward the MS2 RNA *in vitro* but attacked preferentially initiator fMet-tRNA in the anticodon stem−loop between nucleotides +38 and +39 *in vivo* and *in vitro*. Consistently, VapC inhibited translation *in vivo* and *in vitro* but the translation reactions could be reactivated by the addition of VapB and extra charged tRNA$_{fMet}$.

Over half of the *Mycobacterium tuberculosis* toxin-antitoxin modules were VapBC family members (Sharp et al. 2012). This study presented detailed data on the MS2 RNA cleavage. Thus, the MS2 RNA was cleaved at ACGC and AC(A/U)GC sequences. The cleavage activity of the VapC was comparatively weak relative to the MazF toxin from *M. tuberculosis*. In contrast to other toxin-antitoxin system toxins, translation inhibition and growth arrest preceded mRNA cleavage, suggesting that the RNA binding property of VapC, not RNA cleavage, initiates toxicity. The VapC exhibited stable, sequence-specific MS2 RNA binding in an electrophoretic mobility shift assay. Therefore, the activity of VapC was mechanistically distinct from other toxin-antitoxin system toxins because it appeared to primarily inhibit translation through selective, stable binding to the MS2 RNA (Sharp et al. 2012). McKenzie et al. (2012) were the first to replace the MS2 RNA by computer-generated *Pentaprobes* to determine ribonuclease sequence-specificity of the VapC cleavage. Nevertheless, a huge study on the VapC toxins from the

extremely thermoacidophilic archaeon *Metallosphaera prunae*, originally isolated from an abandoned uranium mine, was performed with the MS2 RNA as a substrate (Mukherjee et al. 2017). The consensus VapC cleavage sites were mapped on the MS2 RNA. In fact, *M. prunae* utilized VapC toxins for post-transcriptional regulation under uranium stress to enter a cellular dormant state, thereby providing an adaptive response to what would otherwise be a deleterious environmental perturbation.

Recently, the MS2 RNA contributed to the identification of 11 toxin-antitoxin systems in *Aggregatibacter actinomycetemcomitans*, which represented the MazEF, RelBE, and HipAB families of the type II toxin-antitoxin systems (Schneider et al. 2018).

MS2 RNA AS A SUBSTRATE AND INTERACTION PROBE

MS2 RNA was used as a substrate by the characterization of many enzymes, such as, for example, the tRNA isopentenyltransferase from maize root tips and kernels (Holtz and Klämbt 1978).

The binding of the intermediate filament protein vimentin that was able to discriminate between single-stranded RNA and DNA was studied with a variety of naturally-occurring DNAs and RNAs, including MS2 RNA (Traub et al. 1983). The MS2 RNA contributed then to the characterization of the neurofilament triplet proteins from porcine spinal cord by their specificity to bind single-stranded RNA or DNA (Traub et al. 1985).

The MS2 RNA contributed to the studies of microbial proteases. Thus, the proteolysis of calmodulin by fungal protease was greatly enhanced in the presence of dGTP and MS2 RNA, while only moderate proteolytic activation of bacterial proteases was observed in the presence of the MS2 RNA (Kuo et al. 1990b). Moreover, the MS2 RNA effect was studied on different substrates of both fungal and bacterial proteases (Kuo et al. 1990a, 1991).

The MS2 RNA, which lacks introns, was used as an RNA component by the studies of heterogeneous nuclear ribonucleoprotein particles (hrRNPs). Thus, the MS2 RNA was employed by the interaction with the single-stranded nucleic acid-binding protein HD40, the major protein component of the 30S hrRNP from the brine shrimp *Artemia salina*. The HD40 induced extensive unfolding of the MS2 RNA by destabilization of helices and disruption of the secondary structure, while conformations of double-stranded DNA and double- or triple-stranded synthetic polynucleotides were not affected by the HD40 (Nowak et al. 1980). Structure of the HD40 complexes with polyribonucleotides including MS2 RNA was studied by Thomas et al. (1983).

The MS2 RNA secondary structure was disrupted also by the C-type hnRNP protein from HeLa cells (Kumar et al. 1987). Moreover, the MS2 RNA was used as an RNA component by the elucidation of the interactions of the HeLa cell hnRNP (Wilk et al. 1983, 1991). The specific interaction with the MS2 RNA was used by the methylation studies of the recombinant hnRNP protein A1 (Rajpurohit et al. 1994).

Further studies on the enzymatic methylation of the recombinant hnRNP protein A1 employed again the MS2 RNA as a system component (Kim et al. 1998).

The MS2 RNA was complexed with recombinant *Chlamydia trachomatis* histone H1-like proteins (Pedersen et al. 1994, 1996).

The MS2 RNA was tried as an activator of recA protein, but did not demonstrate the good recA activator properties because of its strong secondary structure (DiCapua et al. 1992).

The MS2 RNA contributed as a substrate to the evaluation of the pokeweed antiviral protein (PAP), that was isolated from the leaves of the pokeweed plant *Phytolacca americana* and characterized initially as a ribosome-inactivating protein (Rajamohan et al. 1999). The PAP acted as the site-specific RNA N-glycosidase and removed catalytically a single adenine base from a highly conserved loop of the large rRNA species in eukaryotic and prokaryotic ribosomes. In contrast to ricin A chains, the PAP-I, PAP-II, and PAP-III isoforms demonstrated ability to depurinate not only MS2, but also HIV-1 RNA, and appeared as potent anti-HIV agents (Rajamohan et al. 1999).

Finally, concerning pioneering studies of the reparation of alkylation damage to RNA, the MS2 RNA was shown to be repaired by human hABH3 and bacterial AlkB oxidative demethylases (Aas et al. 2003). Moreover, the AlkB and hABH3 expressed in *E. coli* reactivated the methylated MS2 RNA *in vivo*, illustrating the biological relevance of this repair activity and establishing RNA repair as a potentially important defense mechanism in living cells. The MS2 RNA, among other substrates, was used also in the study on the human AlkB homolog 1, which appeared in fact as a mitochondrial protein (Westbye et al. 2008). With the participation of the MS2 RNA, the first functional characterization of AlkB proteins was achieved from three plant viruses: grapevine virus A, blueberry scorch virus, and blackberry virus Y, representing *Flexiviridae* and *Potyviridae* families (van den Born et al. 2008). The viral AlkB proteins efficiently reactivated the methylated MS2 genome when expressed in *E. coli*. They preferred RNA over DNA substrates, and thus represented the first AlkBs with such substrate specificity. In contrast, most bacterial AlkBs had low activity on RNA, and the AlkB-mediated RNA repair was regarded as rare, if not absent, in bacteria (van den Born et al. 2009). This was also supported by the observation that the activity of AlkB of *E. coli* on RNA substrates was ~tenfold lower than on DNA (Aas et al. 2003).

Recently, the MS2 RNA was used as a matrix by the engineering of a thermostable viral polymerase using metagenome-derived diversity for highly sensitive and specific RT-PCR, which resulted in the highly promising PyroPhage polymerase (Heller et al. 2019).

It is noteworthy that, besides the popular MS2 RNA, the phage PRR1 RNA was employed as a model of single-stranded RNA by the generation of a novel method for separation and purification of RNA molecules (Levanova and Poranen 2018).

MS2 RNA AS A CONTROL

Starting from the sucrose gradient times, the MS2 RNA was used as a favorite molecular mass marker (Martin and Byrne

1970). For example, the MS2 RNA contributed as a PAGE marker by the classical elucidation of the hepatitis B core and e antigen synthesis in rodent cells transformed with hepatitis B virus DNA (Gough 1983).

The MS2 RNA appeared as a favorite matrix by the evaluation of the reverse transcriptase reactions (Nakane et al. 1991a,b; Froussard 1992; Hart et al. 1992; Orr et al. 1992; Pyra et al. 1994; ten Haaft et al. 1995; Hackett et al. 1996; Chang et al. 1997; Maudru and Peden 1997; Kaushik et al. 2000; Mu et al. 2000; Brorson et al. 2001; Müller and Wirth 2002; Kothapalli et al. 2003; Pittoggi et al. 2003; Sciamanna et al. 2003; Beraldi et al. 2006; Lee YJ et al. 2012; Moser et al. 2012; Vermeire et al. 2012; Xie et al. 2012; Blatter et al. 2013; Fang and Beland 2013; Gualtieri et al. 2013; Uphoff et al. 2015; Miranda and Steward 2017). Nowadays, the MS2 RNA remains a traditional matrix by screening for the novel reverse transcriptase activities (Schwertz et al. 2018; Hagen et al. 2019; Raghunathan and Marx 2019; Stoltz et al. 2019).

The MS2 RNA was used as a carrier by the RNA isolation and construction of cDNA libraries, as, for example, by the pioneering identification of the hepatitis C virus, at that times still known as so-called parenteral non-A, non-B hepatitis agent, using a recombinant cDNA approach (Choo et al. 1992), and in other studies (Leemhuis et al. 1993; Cornetta et al. 1994; Kenner et al. 1999; Smith et al. 2007).

The MS2 RNA was used by the generation of a novel method for the 3′-end-labeling of various types of RNA molecules with digoxigenin, or biotin-modified nucleotides, using terminal deoxynucleotidyl transferase (Rosemeyer et al. 1995). With the MS2 RNA as a prototype, Syrchin and Mendzhul (2002) applied the terminal transferase for the generation of the ^{32}P-labeled hybridization probes by their studies on a cyanophage.

The MS2 RNA contributed to the evaluation of cell-to-cell transport of the infectious RNA in plants, when it was injected into transgenic *Nicotiana benthamiana* plants (Lough et al. 1998).

The RT-PCR assays of viruses frequently involved the MS2 RNA as a carrier and/or control, as, for example, by the detection of rotavirus (Lovmar et al. 2003; Katz et al. 2017; Japhet et al. 2018), respiratory syncytial virus (Kabir et al. 2015), hepatitis E virus (Girón-Callejas et al. 2015; Germer et al. 2017; Marziali et al. 2019), influenza viruses (Lee HK et al. 2016), and foot-and-mouth disease virus (Goller et al. 2018). The internal MS2 RNA control was employed in a large arbovirus surveillance in Australia (Kurucz et al. 2019).

After the viral RT-PCR reactions, the MS2 RNA was used as an internal standard for the precise quantification of ribosomal RNA (Hardiman et al. 2008; Angénieux et al. 2016) and as a negative control in the studies of the SIV/macaque model of the HIV CNS disease (Knight et al. 2018; Mangus et al. 2019). The known amount of the spiked MS2 RNA enabled normalization of absolute molecule counts from RT-PCR reactions, such as in the recent remarkable studies on the alternative splicing in cancer (Ludlow et al. 2018) and the roles of structural domains of the RNA polymerase-binding protein A in mycobacteria (Prusa et al. 2018).

It is noteworthy that the internal MS2 RNA control was employed by the safety/efficacy trial of the first clinical pig islet xenotransplantation in New Zealand (Wynyard et al. 2014). Moreover, the MS2 RNA was used as a stabilizing carrier in the important study on the autologous osteochondral transplantation of autografts for the treatment of small to medium cartilage defects occurring in knee and ankle joints (Bauer C et al. 2019).

The MS2 RNA was used also as a competitor in the RNA affinity capture studies (Dannull et al. 1994; Tsukahara et al. 1996; Campbell and Rein 1999; Harris et al. 2006).

Djikeng et al. (2008) applied MS2 RNA by the generation of a reliable methodology for the genome sequencing of RNA and DNA viruses. The MS2 RNA was used by the elaboration of the degenerate oligonucleotide primer-PCR approaches for the massive parallel sequencing of viral genomes in cell lines (McClenahan et al. 2014). Furthermore, the MS2 RNA contributed markedly to the generation of the quantitative miRNA expression platforms in numerous microRNA studies (Mestdagh et al. 2014; Ameling et al. 2015; Graham et al. 2015; Hirota et al. 2015; Hou et al. 2015; Kuosmanen et al. 2015; Panach et al. 2015; Starikova et al. 2015; Gautam A et al. 2016; Thorsen et al. 2017; Ramón-Núñez et al. 2017; Wang WX et al. 2017; Åkerman et al. 2018; Caserta et al. 2018; Denk et al. 2018; Doumatey et al. 2018; Hsieh et al. 2018; Ji et al. 2018; Kopkova et al. 2018; Maciejak et al. 2018; Martínez-Hernández et al. 2018; Romeo et al. 2018; Shirshova et al. 2018; Takamura et al. 2018; Zeka et al. 2018; Androvic et al. 2019; Atkin et al. 2019; Grasso et al. 2019; Ishinaga et al. 2019; Mayr et al. 2019; Mello-Grand et al. 2019).

11 Maturation Protein

Everything has beauty, but not everyone sees it.

Confucius

It is easier for a camel to pass through the eye of a needle if it is lightly greased.

Kehlog Albran

The mysterious story of the maturation protein, or maturase, or A protein, or RNA-protecting protein, or IIa protein, or A2 protein, is an intriguing but typical example how the parallel development and combination of different novel methodologies led in the revolutionary 1960s to the brilliant final solution, in the case when the strategies were targeted systematically to both structural and functional features of the gene/protein in question.

MUTANT ALLUSIONS

The first indications on the existence of a specific maturation protein have been received from genetics. As Norton D. Zinder (1965) wrote in his first classical review, "Combining the results so far would give the RNA phages three genes: an early function gene directing the synthesis of a viral RNA polymerase and two late functioning genes, the coat protein determinant, and the still uncertain su-1 determinant." The su-1, or sus1, from *su*ppressor-*s*ensitive, which was later decided to name these mutants, phenomenon was found for the conditionally lethal mutants of the phage f2 (Engelhardt and Zinder 1964; Zinder and Cooper 1964; Lodish et al. 1965), as described in Chapter 8. The sus1 mutants produced *dead* noninfectious particles in the nonpermissive conditions. These sus1 mutant-produced phage particles were unable to adsorb on F pili (Valentine and Strand 1965) but lysed the nonpermissive cells as well as wild-type phage did in the case of the phages f2 (Zinder and Cooper 1964) and R17 (Gussin 1966). The similar mutants were found also for the phage fr (Heisenberg and Blessing 1965) and the corresponding defective particles carrying noninfectious degraded RNA were described (Heisenberg 1966, 1967). As a result, Heisenberg (1966) assumed the existence of a phage-specific protein that must be essential for establishing or stabilizing the conformation of the phage RNA. The idea of an unknown crucial component of the phage particles was confirmed indirectly by the reconstitution attempts of the phages MS2 (Sugiyama et al. 1967) and fr (Hohn 1967) when the self-assembly *in vitro* of the phage RNA and coat protein led to formation of the defective phage particles like those formed *in vivo* by the corresponding *dead* particle mutant.

The sus1 phenomenon appeared for the first time under the name of a *maturation* function when the temperature-sensitive f2 mutants were divided into three groups and the hypothetical gene 2 was proclaimed to be responsible for the sus1 phenomenon, as well as for the homologous function of the corresponding *ts* mutants (Horiuchi et al. 1966). By analogy with the phages f2 and fr, the existence of the three complementation groups and the necessity of the maturation protein was established also for the phages R17 (Gussin 1966; Gussin et al. 1966), and Qβ (Horiuchi and Matsuhashi 1970). Gussin (1966) was the first to introduce designation of the *gene A* to encode a protein "necessary for the synthesis of infective phage particles" within the phage R17 genome. The existence of the A protein as a structural unit present in very small amounts but obligatory for the proper encapsulation of the phage RNA into the phage coat was proposed by Gussin et al. (1966). In this paper, the A protein was estimated to contain 150–300 amino acid residues. The non-degraded phage RNA from mutant particles grown in ribonuclease-deficient *E. coli* host was susceptible to nuclease attack, and such particles were still noninfectious (Argetsinger and Gussin 1966). Nevertheless, the A protein, or maturation protein, remained to the end of 1960s as a bit more than a pure genetic abstraction that could be in some way required for the infectivity of the phage.

GEL ELECTROPHORESIS

The first biochemical evidence of the existence of the three or four phage-specific proteins was obtained by the sucrose gradient technique, although the resolution level of the latter was rather low (Haywood and Sinsheimer 1965; Gussin et al. 1966). The strong progress in the characterization of the phage-specific proteins was achieved due to the introduction of sodium dodecyl sulfate gel electrophoresis. Thus, Nathans et al. (1966) discerned the three phage MS2-specific proteins *in vivo*, one of which was identified as an *RNA-protecting protein*. This minor protein was present also within the phage MS2 particles and appeared by the *in vitro* translation of the phage RNA (Nathans et al. 1966). The lack of the corresponding *maturation* or *RNA-protecting* protein by the *in vivo* and *in vitro* synthesis in the phage MS2 maturation mutants was briefly mentioned by Nathans et al. (1966) but described in detail a bit later (Eggen et al. 1967; Viñuela et al. 1967a,b,c).

The SDS-gel electrophoretic pattern of the *in vivo* synthesized Qβ proteins was published by Garwes et al. (1969). Here, the first approximation of the size of the hypothetical maturation protein, or peak IIa in this paper, was made as 39,000 Da, or ~339 amino acid residues. Further comparison of the *in vivo* and *in vitro* synthesized IIa proteins of the phage Qβ confirmed this estimation (Jockusch et al. 1970).

The mobilities of the maturation proteins in gel electrophoresis were used, together with the mobilities of RNAs and coat proteins, for the grouping of the RNA coliphages

(Furuse et al. 1979b). The average molecular mass of the maturation proteins from the groups I, II, III, and IV were estimated as 4.48, 4.45, 4.50, and 4.8×10^4, respectively. Table 11.1 presents a compilation of the data on the estimated and actual molecular mass of the maturation proteins.

ISOLATION

The A protein, in precise accordance with the author's designation, was isolated from the phage R17 and identified as a protein of molecular mass in the range of 35,000–45,000 Da, but the value of 38,000 was used to calculate the number of histidine residues, namely, five, in the molecule (Steitz 1968c). The fact that the R17 coat protein did not contain histidine was exploited by the A protein identification, where the A protein within the phage was labeled specifically with the ^{3}H-histidine. The incorporation of radioactive histidine into highly purified wild-type R17 revealed the presence of a protein which behaved as a single polypeptide chain upon gel electrophoresis, ion exchange chromatography, and gel filtration (Steitz 1968b,c). The isolation problem was complicated by the extreme insolubility of the phage structural proteins in aqueous buffers, when all manipulations had to be performed in the presence of denaturing solvents. The phage was first dissociated in 6 M guanidine-HCl, then dialyzed into 8 M urea, and subjected to ion exchange chromatography on phosphocellulose that was found to best satisfy the isolation criteria, where the A protein remained bound to the column. The A protein was then eluted with a linear NaCl gradient in urea-Tris, pH 8.7. The next purification step, chromatography on agarose, was performed in the presence of SDS (Steitz 1968c). Based on thorough calculations, the existence of approximately one A protein molecule per phage particle was estimated (Steitz 1968b,c). The A protein was definitely not present in the defective particles of the corresponding gene 2 mutants of the phage R17 (Steitz 1968b). The conclusion that the maturation protein constituted a structural component of the RNA phage particle was confirmed at the same time for the phage MS2, where this polypeptide was absent from defective particles produced by the growth of two MS2 *amber* mutants in nonpermissive cells (Viñuela et al. 1968).

The isolation of the A protein has also made possible the *in vitro* reconstitution of the infective R17 phage, where the addition of the isolated A protein to an assembly mixture containing the R17 RNA and coat protein resulted in the reconstitution of several hundred-fold more infective particles than was observed when the A protein was not included (Roberts and Steitz 1967). Remarkably, as highlighted by the author, the optimum infectivity was obtained when the A protein and the RNA were present at a molar ratio of one to one, lending support to the above suggestion that each R17 particle contained only one A protein molecule (Steitz 1968c). Finally, Steitz (1968c) suggested that the A protein might be an attachment organelle, or it could be wrapped inside the coat protein of the phage, where it would confer infectivity by determining the overall conformation of the particle. Again, it was emphasized that the extreme insolubility of the A protein and

its tendency to adhere to many types of surfaces have made experiments to test these hypotheses very difficult to perform.

A more rapid and simpler procedure for the isolation of the maturation protein was developed by Osborn et al. (1970b), where treatment of the phage R17 with 66% glacial acetic acid disrupted the particle into soluble coat protein and an insoluble complex of RNA and maturation protein. Similar results were also obtained with the phage f2 (Osborn et al. 1970b). For the phage Qβ, the presence of a second minor capsid protein, the readthrough protein IIb, complicated the purification of the maturation, or IIa protein, but all three capsid proteins of Qβ were obtained in pure form if the RNA-free, carboxymethylated virions were subjected to ion exchange chromatography on diethylaminoethyl cellulose DEAE52 in the presence of 6 M urea (Weiner and Weber 1973). If only small quantities of maturation protein were required, preparative SDS-gel electrophoresis allowed the isolation of the three capsid proteins (Weiner and Weber 1971). The first isolation and purification steps of the maturation steps were thoroughly reviewed in the classical *RNA Phages* book by Weber and Konigsberg (1975).

CHEMICAL STUDIES

Therefore, the amino acid composition of the maturation protein of the phage R17 has been determined (Steitz 1968b,c). Moreover, its N-terminal sequence was found to be Met-Arg-Ala-Phe-Ser-Ala-Leu-Asx (Weiner et al. 1972). At that time, the sequence of the 43 C-terminal residues was determined for the MS2 maturation protein and provided direct evidence for the nucleotide sequence and reading frame proposed by the MS2 RNA sequencing (Vandekerckhove et al. 1973a). The RNA sequence preceding the coat protein initiation region of the phage R17 (Cory et al. 1970) differed only at one position from that found in MS2 (Contreras et al. 1973). Therefore, it seemed that the maturation proteins of both phages must be very similar in amino acid sequence. Nevertheless, the R17 maturation protein had a C-terminal lysine residue (Weber and Konigsberg 1975) rather than the arginine residue found in MS2 (Contreras et al. 1973).

The Qβ maturation protein, or IIa protein, showed a higher molecular mass, 41,000 Da, than the R17 maturation protein of approximately 38,000 Da. The N-terminal sequence of the Qβ maturation, or IIa, protein was determined as Pro-Lys-Leu-Pro (Weiner and Weber 1971; Weiner et al. 1972). The comparison of this sequence with the known RNA sequence at the known at that time 5′ end of Qβ RNA (Billeter et al. 1969a,b,c) showed that the maturation protein was initiated at the AUG codon starting in nucleotide 61 of the RNA sequence (Weiner and Weber 1971), a conclusion independently verified by the isolation of the corresponding ribosome initiation site (Staples et al. 1971). The GUG codon was identified later as the initiation codon of the MS2 maturation gene (Volckaert and Fiers 1973).

FATE OF THE MATURATION PROTEIN

The definite proof of the maturation protein copy within the phage particle stimulated hypotheses for the possible spatial

TABLE 11.1
Molecular Mass of Maturation Proteins

Group	Phage	Method	Estimated Molecular Mass, Da	Expected Length, aa	Reference	Molecular Mass, Calculated by Genome Sequence, Da	Actual Length, by Genome Sequence, aa	Charge	pI
I	MS2	Gel electrophoresis	40,000		Richelson and Nathans (1967)	43,983.19	393	+13.67	9.69
	MS2	Gel electrophoresis		240	Viñuela et al. (1968)				
	MS2	Gel electrophoresis	42,000		Vandamme et al. (1972)				
	MS2	Gel electrophoresis	42,500 ± 500		Vandekerckhove and Van Montagu (1974)				
	MS2	Gel electrophoresis	44,500		Furuse et al. (1979b)				
	R17	Sucrose gradient	35,000–40,000	150–300	Gussin et al. (1966)				
	R17	Gel filtration	38,600	350	Steitz (1968b)				
	R17	Gel electrophoresis, amino acid composition			Steitz (1968c)				
	R17	Gel electrophoresis	38,000		Osborn et al. (1970b)				
	R17	Gel electrophoresis	44,500		Furuse et al. (1979b)				
	fr				Piffaretti and Pitton (1976)	43,956.39	393	+15.67	9.77
	M12					43,940.18	393	+12.64	9.54
	μ2	Gel electrophoresis	40,500						
	JP501	Gel electrophoresis	44,500		Furuse et al. (1979b)	44,187.61	393	+14.67	9.71
	ZR	Gel electrophoresis	44,500		Furuse et al. (1979b)				
	BO1, FR1	Gel electrophoresis	45,000		Furuse et al. (1979b)				
II	GA	Gel electrophoresis	44,000		Furuse et al. (1979b)	44,384.91	390	+21.71	9.92
	SD, TH1	Gel electrophoresis	44,000		Furuse et al. (1979b)				
	BZ13, DL20	Gel electrophoresis	45,000		Furuse et al. (1979b)	44336.74	390	+18.55	9.82
	KU1	Gel electrophoresis	45,000		Furuse et al. (1979b)	44,369.66	390	+16.08	9.73
	JP34	Gel electrophoresis	44,000		Furuse et al. (1979b)				
III	Qβ	Gel electrophoresis	39,000	339	Garwes et al. (1969)	48,547.30	420	+7.94	9.30
	Qβ	Gel electrophoresis	39,000		Jockusch et al. (1970)				
	Qβ	Gel electrophoresis	42,000		Horiuchi et al. (1971)				
	Qβ	Gel electrophoresis	41,000		Weiner and Weber (1971)				
	Qβ	Gel electrophoresis	45,000		Furuse et al. (1979b)				
	VK, ST, TW18	Gel electrophoresis	45,000		Furuse et al. (1979b)				
	M11					48,619.37	421	+12.74	9.61
	MX1	Gel electrophoresis	48,000		Furuse et al. (1979b)	48,486.26	421	+11.54	9.44

(Continued)

TABLE 11.1 (Continued)
Molecular Mass of Maturation Proteins

Group	Phage	Method	Estimated Molecular Mass, Da	Expected Length, aa	Reference	Molecular Mass, Calculated by Genome Sequence, Da	Actual Length, by Genome Sequence, aa	Charge	pI
IV	FI	Gel electrophoresis	49,000		Furuse et al. (1979b)	50,042.22	438	+16.24	9.78
	SP	Gel electrophoresis	48,000		Furuse et al. (1979b)	48,547.30	450	+7.94	9.30
	ID2	Gel electrophoresis	48,000		Furuse et al. (1979b)				
	TW19, TW28	Gel electrophoresis	47,000		Furuse et al. (1979b)				
	NL95					50,499.74	442	+20.57	10.11
?	ZIK/1								
Levivirus-like	Hgal1					45,441.95	407	+22.53	10.02
	C-1					45,471.30	406	+27.17	10.19
	M					43,436.48	395	+14.10	9.55
Pseudomonaphages	PP7	Gel electrophoresis	25,000 ?		Dhaese et al. (1977a)	50,822.12	449	+11.57	9.09
	PRR1	Gel electrophoresis	40,000		Dhaese et al. (1977b)	47,327.24	417	+18.90	9.90
	LeviOr01					52,300.15	463	+13.04	9.50
Caulophage	φCB5					40,676.49	372	+16.14	10.02
Acinetophage	AP205					60,981.02	534	+26.73	9.92
Assembled genomes	EC					47,265.86	430	+24.98	9.97
	MB					50,664.57	452	+8.17	8.95

Note: The order of the maturation protein presentation corresponds to that of the phage genome order in Figure 7.1 and Table 10.1.

models is how this copy could be integrated into the coat-formed capsid and exposed on its surface to be accessible by the F pili (O'Callaghan et al. 1973a; Dunker and Paranchych 1975; Paranchych 1975). If the earlier model proposed by O'Callaghan et al. (1973a) suggested that the maturation protein might reside at one of the vertices of the icosahedron, Dunker and Paranchych (1975) decided that the maturation protein could protrude through the openings produced by the subunits which circumscribed either the threefold axis or the fivefold axis. They thought that either of these open spaces between the coat protein subunits was probably large enough to accommodate one molecule of the maturation protein.

The maturation protein appeared to be attached to the phage RNA, since the maturation protein–RNA complexes have been isolated from infected cells during the late stages of the growth cycle (Richelson and Nathans 1967) and have been indicated also by the phage maturation during analysis of the sequential steps in the *in vivo* assembly of the phage fr (Kaerner 1970). It was assumed, therefore, that the maturation protein stabilized the highly coiled RNA molecules, when without the maturation protein the poorly packaged RNA could be made sensitive to ribonuclease attack, possibly as a result of extrusion of part of the RNA from the particle.

Such strong interaction with the phage RNA was also evidenced by the data on the maturation protein penetration into infected cells together with RNA in the form of an RNA-maturation protein complex by the phages MS2 (Kozak and Nathans 1971) and R17 (Paranchych et al. 1971; Krahn et al. 1972).

When the phage MS2 was converted into expanded, slower-sedimenting but still infective particles by moderate heat treatment in low ionic strength buffer, digestion with ribonuclease removed two-thirds of the RNA molecule, but the A protein was not released by the enzymatic digestion and remained associated with the converted particles both during expansion and after ribonuclease digestion (Verbraeken and Fiers 1972b). In high salt, the A protein was preferentially lost from the viral particles (Verbraeken and Fiers 1972a). The A protein-less particles remained physically intact, unlike "defective particles," but were not infective. This confirmed again the idea that the A protein was not only essential for the proper assembly, but also for the subsequent infection (Verbraeken and Fiers 1972a).

The interaction of the phage R17 particles with F piliated bacteria at 37°C led to the cleavage of the maturation protein molecule into two smaller components of molecular mass about 24,000 and 15,000, and these polypeptide fragments were transferred into the cell along with the phage RNA (Krahn et al. 1972).

Novel arguments to the external location of the maturation protein were gained when the intact phage MS2 conjugates with dinitrophenol (DNP) were prepared and purified (Curtiss and Krueger 1973). The maturation protein isolated from these DNP-MS2-conjugated particles contained covalently attached DNP (Curtiss and Krueger 1974). In the latter paper, the external location of the maturation protein was confirmed by extensive lactoperoxidase iodination of the maturation protein within the intact unconjugated MS2 particles.

A tight RNA-maturation-protein complex was isolated from the acetic acid precipitate of the highly purified phage M12 (Leipold 1975; Leipold and Hofschneider 1975). This complex was an important factor in controlling the relative plating efficiency of a phage preparation and was infectious itself for intact *E. coli* cells. The molar ratio of RNA to protein in the complex was found to be 1:1. In contrast to the M12 RNA alone, which could infect spheroplasts only, the RNA-maturation-protein complex infected the intact *E. coli* cells via F pili and produced infectious progeny particles like the original phage (Leipold and Hofschneider 1975, 1976). All of the infectivity disappeared if the ionic bonds of RNA to maturation protein in the complex were dissociated in 0.5 M sodium chloride buffer at 37°C. Therefore, it was concluded that the binding of the maturation protein to the RNA was a prerequisite for the infectivity on intact host cells, and that both components of the complex were held together by ionic bonding (Leipold and Hofschneider 1975, 1976). It is noteworthy that the infectivity of the complex on intact *E. coli* cells was enhanced by incubation with the polyamine spermidine (Leipold 1977).

Similar results were obtained in parallel by Shiba and Miyake (1975), who reconstituted the RNA-maturation-protein complexes by mixing the appropriate purified components of the phages MS2 and GA. Such complex was much more sensitive to ribonuclease, trypsin, and pronase than were intact phage particles. The extreme sensitivity to ribonuclease suggested that a naked RNA was a component of the complex. Later, Shiba and Suzuki (1981) localized the maturation protein in the MS2 complex by ribonuclease digestion, and two types of RNA fragments were obtained: one was 5'-terminal and originated from the maturation gene, and the other corresponded to the untranslated 3'-terminal region.

The infectious RNA-maturation-protein complex of the phage R17 was isolated by the coprecipitation with acetic acid, as before in the case of M12 (Reynolds and Paranchych 1976). It was also noted that a significant fraction of the phage RNA within the complex was resistant to ribonuclease digestion (Wong and Paranchych 1976a). Iglewski (1976) got the analogous R17 RNA-maturation-protein complex by the reconstitution from the purified RNA and the maturation protein purified by DEAE cellulose chromatography from disrupted phage.

Finally, the RNA-maturation-protein complex was obtained for the phage f2 by the acetic acid precipitation (Cody and Conway 1981b).

SEQUENCING

In parallel with the nucleotide sequencing of the MS2 maturation gene that was completed in 1975 (Fiers et al. 1974; Min Jou et al. 1974; Fiers et al. 1975b), Marc Van Montagu's team worked on the amino acid sequencing of the maturation protein of the phage MS2. First, the 43 aminoacid sequence at the C-terminus of the maturation protein was determined (Vandekerckhove et al. 1973a) and the use of fluorescamine was adopted to the sequence analysis of the MS2 structural proteins, purified at the nanomole scale (Vandekerckhove and

Van Montagu 1973b). Then, the complete solubilization of the poorly soluble maturation protein was achieved by chymotrypsin cleavage, and 53 chymotryptic peptides were obtained (Vandekerckhove and Van Montagu 1974a). The chymotryptic peptides were localized with fluorescamine under circumstances enabling complete amino acid analysis and/or sequence determination (Vandekerckhove and Van Montagu 1974b).

The tryptic peptides were obtained, and a preliminary reconstruction of the maturation protein primary sequence was performed by overlapping of the tryptic peptides with the chymotryptic ones and in correlation with the nucleotide sequence of the maturation gene (Nolf et al. 1974).

Finally, a series of three papers covered the global amino acid sequencing story of the MS2 maturation protein. Thus, Nolf et al. (1977) published the isolation of the maturation protein, determination of the N- and C-terminal sequences, and isolation and amino acid sequence of the tryptic peptides. Vandekerckhove et al. (1977) presented the isolation and sequence determination of the chymotryptic peptides, while Vandekerckhove and Van Montagu (1977) reported the

isolation and sequencing of thermolytic peptides and soluble cyanogen bromide fragments, as well as final alignment of 363 amino acid residues of a total of 393.

HOMOLOGY

Figure 11.1 presents the Clustal V alignment of the maturation protein sequences. All sequences are translated from the corresponding nucleotide sequences. It is quite obvious that the general similarity pattern repeats that obtained for the full-length RNA phage genomes. In this case, however, the maturation protein sequences from partially sequenced genomes, such as JP501 and M12, are also included.

SPATIAL STRUCTURE

A first attempt, namely, a polyconfiguron model, was tried by Loyd Y. Quinn (1978) to imagine the tertiary structure of the MS2 maturation protein. He presented an algorithm for decoding mRNA by stepwise "translation" of each codon into the

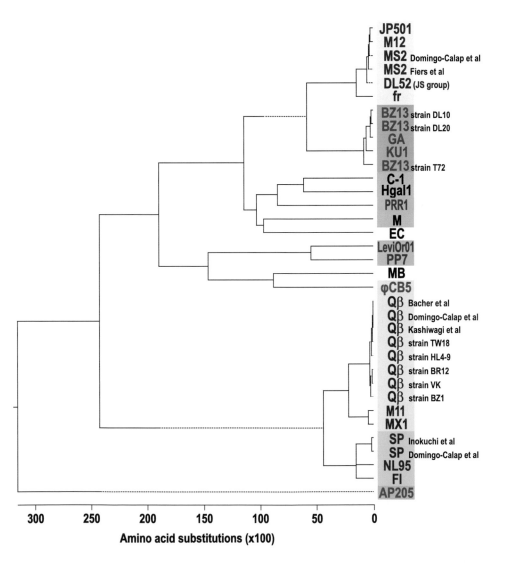

FIGURE 11.1   The comparison of the maturation protein sequences obtained by the Clustal V alignment in the MegAlign program from the DNASTAR Lasergene software suite. The designations and colors of the genomes correspond to that used in Table 7.1 and Figure 7.1.

corresponding amino acid configuron, so that a polyconfiguron composed of the 393 amino acid residues of the maturation protein was generated. This polyconfiguron had the topography of the native maturation protein, as it has been predicted by a variety of functional and physicochemical procedures.

A role of the maturation protein in the global arrangement of the phage protein shell was studied by small-angle neutron scattering, when analysis of the coat protein shell of recombinant versions of MS2 that lacked the A protein revealed dramatic differences compared to the wild-type MS2 (Kuzmanovic et al. 2006b). Moreover, it was shown by electron cryomicroscopy that the maturation protein replaced a single coat protein dimer and formed the phage particle together with remaining 178, but not 180, coat protein monomers (Dent et al. 2013; Koning et al. 2016). The asymmetric reconstruction of the MS2 particle with the exposed maturation protein molecule was performed by electron cryomicroscopy at 3.6 Å (Dai et al. 2017). The external location of the maturation protein made it possible to explain why the maturation protein appeared as a target by photodynamic inactivation of the phage MS2 (Majiya et al. 2018a,b), which was described earlier in Chapter 6. All these three-dimensional phenomena will be discussed in more detail in Chapter 21.

Finally, crystal structure of the Qβ maturation protein, or protein A2 in the case of Qβ, was resolved at 3.3 Å (Rumnieks and Tars 2017). The overall structure of the Qβ A2 protein is presented in Figure 11.2.

The protein has a bent, highly asymmetric shape and spans 110 Å in length. Apart from small local substructures, the overall fold of the maturation protein does not resemble that of other known proteins. The protein is organized in two distinct regions, an α-helical part with a four-helix core, and a β-stranded part that contains a seven-stranded sheet in the central part and a five-stranded sheet at the tip of the protein. The Qβ maturation protein has two distinct, positively charged areas at opposite sides of the α-helical part, which are involved in genomic RNA binding. The genomic RNA binding of the Qβ A2 protein is presented in Figure 11.3.

The coat protein- or RNA-binding residues are not preserved among different ssRNA phage maturation proteins; instead, the distal end of the α-helical part is the most evolutionarily conserved, suggesting the importance of this region for maintaining the functionality of the protein. Figure 11.4 presents the conserved residues of the RNA phage maturation proteins in the context of the three-dimensional structure of the Qβ A2 protein. All these images are taken from the breakthrough study on the maturation protein structure, which was performed recently by Rumnieks and Tars (2017).

LYSIS

Unexpectedly, the Qβ maturation, or IIb, or A2, protein was exposed by triggering the cell lysis (Karnik and Billeter 1983). The overproduction of the A2 protein after cloning and expression of its gene in *E. coli* (see Chapter 19) led to definite lethality (Winter and Gold 1983). The basis of this lethality was identified as cell lysis that correlated with the A2 protein synthesis. The plasmid-derived A2 protein specifically complemented the Qβ A2 amber mutants. The lysis activity was abolished in the clones that synthesized truncated or internally deleted A2 polypeptides ranging from 10% to 95% of the length of wild-type protein. It was concluded therefore that the host lysis was promoted by the Qβ maturation protein itself, rather than by a separate lysis protein (Winter and Gold 1983). The cloned A2 gene of the phage SP initiated an efficient lysis of host cells (Nishihara 2002), as will be described in more detail in Chapter 14.

The A2 protein was demonstrated to be necessary and sufficient for the host cell lysis by targeting the MurA, or UDP-N-acetylglucosamine enolpyruvyltransferase, an enzyme involved in bacterial peptidoglycan biosynthesis from the pathway of cell wall construction (Bernhardt et al. 2001). The process of the A2-induced cell lysis when the resulting peptidoglycan was weakened in structure, thereby allowing the phage-infected bacterium to lyse, releasing progeny phage, was reviewed in further detail by Bernhardt et al. (2002). The N-terminal 179 residues of the A2 protein still retained lytic function but the next largest fragment, 1-171, did not (Langlais 2007, cited by Reed 2012). It is necessary to remember here that the inhibition of bacterial cell wall mucopeptide synthesis as a function of the Qβ infection was asserted first by Ozaki and Valentine (1973).

The finding of Bernhardt et al. (2001) stimulated in turn a comeback of the phage antibacterials as an alternative strategy for combating bacterial infections (Fischetti 2001).

The A2 protein did not possess any muralytic activity but blocked synthesis of murein precursors *in vivo* by inhibiting

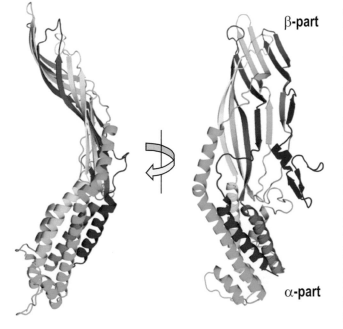

β-part

α-part

FIGURE 11.2 The overall structure of the Qβ A2 protein. The protein is shown in two orientations rotated by 90° and rainbow-colored blue to red from the N to the C terminus. (Reprinted from *J Mol. Biol.*, 429, Rumnieks J, Tars K, Crystal structure of the maturation protein from bacteriophage Qβ, 688–696, Copyright 2017, with permission from Elsevier.)

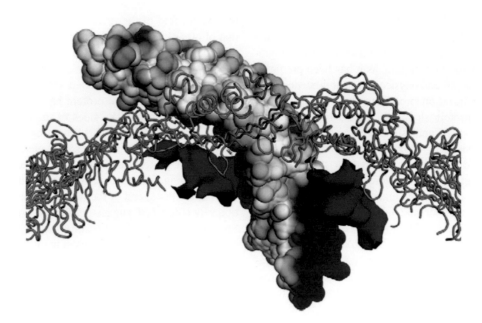

FIGURE 11.3 Genomic RNA binding of the Qβ A2 protein. The surface of the A2 protein is colored according to its electrostatic potential. Two positively charged regions of the protein are involved in interactions with genomic RNA hairpins, represented as red surfaces. The surrounding coat protein molecules are shown in gray as a ribbon model. The genomic RNA hairpins that make the most pronounced contacts with the A2 protein are represented as red surfaces cut from the cryo-EM volume (EMDB accession number EMD-8254) (Gorzelnik et al. 2016) and contoured at 2.5σ. (Reprinted from *J Mol. Biol.*, 429, Rumnieks J, Tars K, Crystal structure of the maturation protein from bacteriophage Qβ, 688–696, Copyright 2017, with permission from Elsevier.)

MurA, the catalyst of the committed step of murein biosynthesis. The A2-resistance mutation was mapped to an exposed surface near the substrate-binding cleft of MurA. This mutant, designated as *rat1* (*r*esistant to *A-t*wo), was shown to be a Leu-to-Gln change at position 138 of MurA. The purified Qβ virions inhibited wild-type MurA, but not the mutant MurA, *in vitro*. Moreover, the authors isolated the Qβ mutants that were able to overcome the A2-resistance mutation. These mutants were postulated to be compensatory for the L138Q mutation in the A2−MurA interface.

However, the true inhibitory mechanism of the A2 protein was deciphered later (Reed 2012; Reed et al. 2012). It was shown with a two-hybrid study that the A2 protein and

MurA interacted directly. The interaction between A2 and MurA was dependent on a substrate-induced conformational change. Thus, the A2 protein preferentially recognized MurA liganded with uridine 5′-diphosphate-N-acetylglucosamine (UDP-NAG), thereby preventing catalysis by occluding phosphoenolpyruvate (PEP) from accessing the active site. It is necessary to comment here that MurA catalyzes the transfer of the enolpyruvyl (EP) moiety from PEP to UDP-NAG, yielding two products, UDP-NAG-EP and inorganic phosphate.

The MurA inactivated purified Qβ particles, casting doubt on the notion that the A2 must assemble into particles prior to MurA inhibition (Reed et al. 2013). Furthermore, the quantification of the A2 protein induced from a plasmid indicated

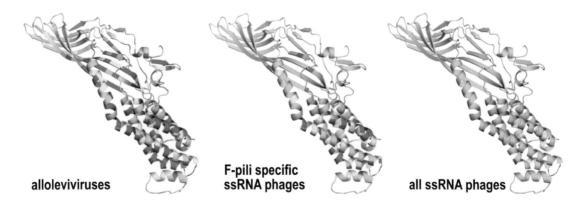

FIGURE 11.4 Conserved regions of the RNA phage maturation proteins. The conserved residues are represented in the context of the three-dimensional structure of the Qβ A2 protein. The completely conserved residues are colored orange and similar residues in yellow. (Reprinted from *J Mol. Biol.*, 429, Rumnieks J, Tars K, Crystal structure of the maturation protein from bacteriophage Qβ, 688–696, Copyright 2017, with permission from Elsevier.)

that lysis was entrained when the amount of the lysis protein was approximately equimolar to that of the cellular MurA. Qβ *por* mutants, isolated as suppressors that overcame the *murA*ʳᵃᵗ mutation that reduced the affinity of MurA for A2, were shown to be missense mutations in the A2 protein that increased the translation of the latter. Because of the increased production of the A2, the *por* mutants had an attenuated infection cycle and reduced burst size, indicating that a delicate balance between assembled and unassembled A2 levels could regulate lysis timing (Reed et al. 2013).

Finally, the structure of the Qβ-MurA complex was resolved by single-particle cryo-electron microscopy, at 6.1 Å

resolution, where the outer surface of the β-region in A2 was identified as the MurA-binding interface (Cui et al. 2017). Moreover, the pattern of MurA mutations that blocked Qβ lysis and the conformational changes of MurA that facilitated A2 binding were found to be due to the intimate fit between A2 and the region encompassing the closed catalytic cleft of substrate-liganded MurA. Figure 11.5 reveals the structure of Qβ bound with MurA and the structural mechanism of A2-mediated host lysis.

The general principles of the RNA phage-induced lysis by the special lysis protein of the *Levivirus* genus are presented in Chapter 14.

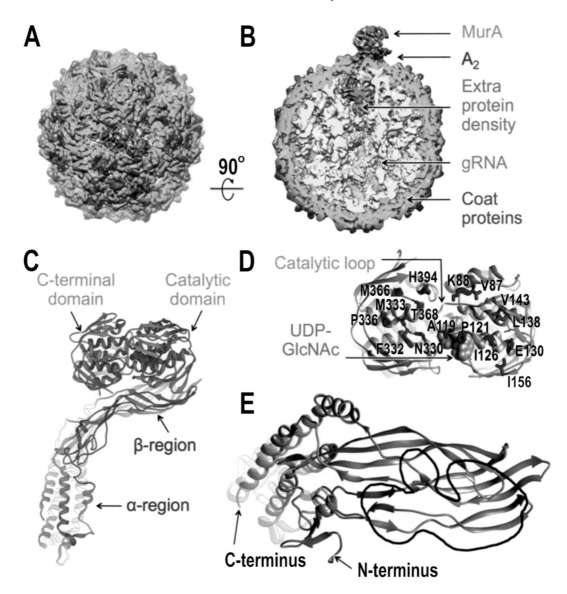

FIGURE 11.5 The structure of Qβ bound with MurA reveals the structural mechanism of A2-mediated host lysis. (A) The structure of Qβ bound to MurA with MurA colored orange. (B) A 90° turn and cutaway view of Qβ shows MurA bound to the maturation protein. (C) The ribbon model of A2 bound to MurA with uridine diphosphate N-acetylglucosamine (UDP-GlcNAc) in the active site (cornflower blue). (D) The ribbon model of MurA viewed from the MurA-A2 interface. The point mutations that render MurA resistant to A2 are labeled and shown as red stick models. Locations of the catalytic loop and the UDP-GlcNAc are indicated by black arrows. (E) Ribbon model of A2 as viewed from the MurA–A2 interface. The region interacting with MurA, encompassing the N-terminal β-sheet region of residues 30–120, is outlined by a black lasso. The N and C termini are indicated by black arrows. (Cui Z et al. Structures of Qβ virions, virus-like particles, and the Qβ-MurA complex reveal internal coat proteins and the mechanism of host lysis. *Proc Natl Acad Sci U S A.* 114:11697–11702, Copyright 2017 National Academy of Sciences, U.S.A.)

12 Coat Protein

Clothes make the man. Naked people have little or no influence on society.

Mark Twain

Too much of a good thing is wonderful.

Mae West

COMPOSITION AND SEQUENCING

The coat protein was the first RNA phage protein to be faced directly. As the most accessible in quantity and the shortest one, it was first to be subjected to the amino acid content determination and then to the amino acid sequencing. In 1972, the coat protein gene became the first-ever gene sequenced (Min Jou et al. 1972a,b), as described earlier in Chapter 10.

From the earliest studies, it was clear that each phage RNA molecule is enclosed in a protein shell constructed from a large number of identical subunits. Zinder (1965) stated in his first classical review that this bulk coat protein has a molecular mass of 15,000 for the monomer consisting of some 133 amino acids in the case of the phage f2. Moreover, 12 peptides have been resolved after trypsin digestion and Dowex chromatography. The lack of the amino acid histidine was fixed as a striking compositional feature of the f2 coat protein. The histidine residues were also lacking in the coat protein product obtained by the *in vitro* translation of the f2 RNA (Nathans et al. 1962). Zinder (1965) stated also that the f2 coat protein had tyrosine at its C-terminus, although the N-terminal amino acid was found to be masked. None of the coat proteins of related RNA phages that have been studied to that date contained histidine: both R17 and M12 (Enger and Kaesberg 1965), as well as MS2 (Ohtaka and Spiegelman 1963). This feature helped later by the separate labeling and identification of two other general phage proteins: maturation protein and replicase.

Enger and Kaesberg (1965) suggested that the molecular mass of the coat protein subunits of R17 and M12 could be about 14,200 Da. This was the minimum estimate consistent with the observed amino acid composition, while the phage R17 contained one more mole of lysine residues and one less mole of glutamic acid or glutamine residues than M12 per 14.2×10^3 g protein. According to Gussin et al. (1966), the same value for the molecular mass of the coat protein subunit was derived by dividing the total molecular mass of the phage protein (2.5×10^8) by 180, the number of subunits consistent with the phage symmetry properties, as suggested by John T. Finch and Aaron Klug.

Concerning rapid progress in the amino acid sequencing of the f2 coat, Konigsberg et al. (1966) isolated and characterized the tryptic peptides from the f2 coat, determined N- and C-terminal amino acid residues as alanine and tyrosine,

respectively, and estimated length of the coat as 129 residues. Konigsberg (1966) arranged the tryptic peptides in the f2 coat and determined its N-terminus as an Ala-Ser-Asn-Phe-Thr-Gln-Phe sequence. Then, Weber et al. (1966) published the full-length f2 coat protein sequence of 129 amino acid residues. Finally, Weber and Konigsberg (1967) revised this preliminary sequence a bit and reported that the coat protein subunits of MS2 and R17 consist like f2 of 129 amino acids. There was only one difference between the sequences of MS2 and R17 as compared to f2, namely, Leu88 in f2 was replaced by methionine in R17 and MS2. The full-length amino acid sequence of the R17 coat was published by Weber (1967). The MS2 coat protein was sequenced at the same time by Lin et al. (1967) and a bit later by Vandekerckhove et al. (1969). Moreover, the *in vivo* and *in vitro* synthesized MS2 coats were compared (Lin and Fraenkel-Conrat 1967). Table 12.1 presents the most important data on the biochemical structure of the coat protein subunits.

In parallel, the coat of the phage fr was sequenced and showed the same 129 amino acid length as the f2 coat but contained 19 amino acid replacements. First, the amino acid composition, tryptic peptides, and partial sequences of the fr coat were reported (Wittmann-Liebold 1965; Wittmann and Wittmann-Liebold 1966). The full-length fr coat sequence was published by Wittmann-Liebold (1966). Next, Wittmann-Liebold and Wittmann (1967) compared the complete amino acid sequences of the phages fr and f2 and were the first to propose the grouping of RNA phages according to their coat proteins into three groups: (I) MS2, f2, M12 and R17, (II) fr and (III) Qβ. The MS2, M12, R17, and f2 were very strongly related to each other: MS2 and M12 seemed to be identical, and f2 and R17 differed from these by only one amino acid replacement. Further, the ZR coat was sequenced and no difference between amino acid sequence of the ZR coat and those from the phage R17 seemed to be present, but the ZR coat differed from that of the phage f2 by a single amino acid substitution (Nishihara et al. 1970).

The remarkable difference between MS2 and Qβ coat proteins was noted first by Overby et al. (1966b). The coat protein subunits were of molecular mass 15,500 and 14,000 for Qβ and MS2, respectively. The MS2 protein, like other group I representatives, lacked histidine, while the Qβ coat protein lacked both histidine and tryptophan, as well as methionine.

The amino acid sequence of the Qβ coat of 131 amino acid residues was published in comparison with that of the f2 phage by Konigsberg et al. (1970). The exhaustive details of the Qβ coat sequencing appeared the next year (Maita and Konigsberg 1971). The Qβ coat sequence was revised later by the nucleotide sequencing (Stoll et al. 1977). The 22nd amino acid was asparagine rather than aspartic acid, and an additional amino acid, serine, was found between proline

TABLE 12.1

Molecular Mass of Coat Proteins

Group	Phage	Molecular Mass by Amino Acid Sequencing, Da	Length, aa	References	Molecular Mass, Calculated by Genome Sequence, Da	Actual Length, by Genome Sequence, aa	Charge	pI
I	f2	13,709.53	129	Weber and Konigsberg (1967)	–	–	+1.85	8.48
	R17	13727.57	129	Weber (1967)	–	–	+1.85	8.48
	fr	13,735.50	129	Wittmann-Liebold (1966)	13,866.70	130	+0.86	7.95
	MS2	13,728.56	129	Lin et al. (1967)	13,859.76	130	+0.85	7.96
	ZR	13,727.57	129	Nishihara et al. (1970)	13,858.77	130	+1.85	8.48
	M12	–	–	–	13,858.77	130	+1.85	8.48
	JP501	–	–	–	13,816.73	130	+0.85	7.96
	ZR	–	–	–	13,859.76	130	+0.85	7.96
	BO1	–	–	–	13,843.71	130	+0.85	7.96
II	GA	–	–	–	13,682.54	130	+2.91	9.35
	SD	–	–	–	13,756.62	130	+2.91	9.35
	TH1	–	–	–	13,740.62	130	+2.91	9.35
	TL2	–	–	–	13,740.62	130	+2.91	9.35
	JP500	–	–	–	13,741.61	130	+1.91	9.04
	KU1	–	–	–	13,726.60	130	+2.91	9.39
	JP34	–	–	–	13,741.61	130	+1.91	9.04
III	Qβ	14,122.94		Maita and Konigsberg (1971); Stoll et al. (1977)	14,254.14	133	+2.85	8.87
	M11	–	–	–	14,198.14	133	+2.85	8.87
	MX1	–	–	–	14,198.10	133	+1.85	8.47
IV	FI	–	–	–	14,164.20	132	+2.86	8.86
	SP	–	–	–	14,129.06	132	+1.86	8.49
	NL95	–	–	–	14,143.12	132	+5.85	9.58
?	ZIK/1	12,000 ± 100	102 ± 2	Robinson (1972)	–	–	–	–
Levivirus-like	Hgal1	–	–	–	14,276.31		+0.06	7.09
	C-1	–	–	–	14,633.62	133	+0.09	7.16
	M	–	–	–	14,432.50	133	+1.92	9.06
Pseudomona- phages	PP7	13,889.84	127	Dhaese et al. (1977a, 1980)	14,021.04	128	+2.03	8.48
	PRR1	14,535.53	131	Dhaese et al. (1977b, 1979)	14,695.76	132	+1.25	8.45
	LeviOr01	–	–	–	13,422.29	129	+2.08	9.43
Caulophage	φCB5	–	–	–	13,614.21	123	−0.58	6.62
Acinetophage	AP205	–	–	–	14,008.90	131	+2.02	8.49
Assembled genomes	EC	–	–	–	14,618.78	138	+1.09	8.59
	MB	–	–	–	14,216.98	128	−1.42	6.33

Note: The order of the coat protein presentation corresponds to that of the phage genome order in Figure 7.1 and Table 10.1.

in position 55 and arginine in position 56. The revised Qβ coat structure reached the 132 amino acid residue length and agreed with the nucleotide sequence (Escarmis et al. 1978).

Nishihara et al. (1969) compared the amino acid content of the coats of six phages, by two from each of three groups: MS2 and ZR from group I, GA and ZD from group II, and Qβ and VK from group III. They found that the group II phages possessed C-terminal sequence (Tyr, Phe)-Ala, in contrast to four others with the Tyr as the C-terminal residue. The N-terminal residue was always Ala. The group II phage coats had no histidine, cysteine, and methionine residues.

Modak and Notani (1969) turned at that time to the phage f4, a species from the first series of the RNA phages (Loeb and Zinder 1961). By the coat amino acid composition, the phage f4 differed from f2 in having an additional methionine residue per molecule. Possibly that was why the phage f4 cross-reacted with f2 antisera. Next, the phage μ2 was classified as a representative of group I, due to the coat amino acid composition as well as its antigenicity (Piffaretti and Pitton 1976). The molecular mass of the μ2 coat was estimated as 14,400 Da.

A breakthrough in the coat structures of pseudomona-phages was achieved by Marc C. Van Montagu's team in

Ghent. Preliminary data on the first coat protein sequence from pseudomonaphages, namely of the broad range phage PRR1, were published by Dhaese et al. (1977b). The complete amino acid sequence of the coat protein of the phage PRR1 was presented a bit later (Dhaese et al. 1979). A complete set of different protease and chemical cleavages was undertaken in order to overlap all peptides. The first report on the coat sequencing of the next pseudomonaphage, PP7 (Dhaese et al. 1977a), was followed also by the complete amino acid sequence (Dhaese et al. 1980). Like the coat proteins of the coliphages MS2 and Qβ and of the broad host range RNA phage PRR1, the PP7 coat protein was highly hydrophobic and contained a cluster of basic residues between positions 40–60. In contrast to the most RNA phages including the phage PRR1, N- and C-termini of the PP7 coat were serine and arginine residues, respectively. Hirashima et al. (1982) determined the classical alanine and tyrosine as the N- and C-termini also in the case of the three representatives of the group IV phages: SP, FI, and ID2.

The next studies on the full-length RNA phage coat structures occurred via nucleotide sequencing, including mostly full-length genome sequencing. The respective data are included in Table 12.1. From partial genome sequences, six coat protein gene structures were published: ZR and BO1 from group I, and SD, TH1, JP500 and TL2 from group II (Nishihara et al. 2006). The length of the coats, namely, 130 amino acid residues, corresponded to the expected ones for groups I and II. It is necessary to mention also that the coat sequence of one of the oldest classical RNA phages, namely, M12, was submitted to the Genbank by Groeneveld et al. from van Duin's team in 1999.

The phage ZIK/1 remained as an unresolved mystery. This phage was described as similar to the phage Qβ by its relatively low G+C composition of the phage genome (48.4% and 48.1% by Qβ and ZIK/1, respectively, and more than 52% for MS2, f2, and fr) (Bishop and Bradley 1965). Robinson (1972) found that the ZIK/1 coat did not contain histidine, methionine, and cysteine but was only about 102 amino acid residues in length. Therefore, the phage ZIK/1 seemed different from the known phages by the size and composition of the coat protein.

N-TERMINAL MODIFICATION

As it appears from Table 12.1, the amino acid-sequenced coats were always shorter for one amino acid residue than their counterparts deduced from the nucleotide sequence. This discrepancy was explained finally when Housman et al. (1972) found that the N-terminal formylmethionine was removed when the nascent chain of the phage f2 coat protein reached 40–60 amino acids in length. It was known before that all protein synthesized *in vitro* in the cell-free *E. coli* extracts retained an N-terminal formylmethionine or methionine residue (Lengyel and Soll 1969). By contrast, no completed protein isolated from growing *E. coli* cells contained formylmethionine as its N-terminal residue, and only 30% had an N-terminal methionine (Waller 1963). Since growing

E. coli cells initiated protein synthesis with formyl-methionine tRNA (Eisenstadt and Lengyel 1966), these results suggested that the cell-free extracts were defective in the enzyme or enzymes which normally removed the amino-terminal formylmethionine or formyl residues (Adams and Capecchi 1966; Capecchi 1966b; Webster et al. 1966). The emotional moment of this important point was touched on in Chapter 1. Once and for all, Housman et al. (1972) showed clearly that the f2 coat protein synthesized by an infected cell was initiated with the N-terminal sequence fMet-Ala-Ser..., but the N-terminal fMet residue was removed while the polypeptide chain was still nascent.

Furthermore, Ball and Kaesberg (1973b) investigated the fate of the formylmethionine residue during synthesis of the Qβ coat protein under various conditions *in vitro*. The Qβ coat protein provided an ideal system for such studies, since it contained no internal methionine residues. As a result, the formylmethionine residue was retained by all the nascent chains but by only about 50% of the completed coat protein molecules under standard conditions of the cell-free protein synthesis. If 2-mercaptoethanol was omitted from the cell-free system, the formylmethionine residue was cleaved during the course of peptide chain elongation. All nascent peptides which contained fewer than 40 ± 5 amino acids retained the formylmethionine residue. Thereafter, the proportion of the nascent peptides lacking the residue increased with peptide length to about 70% for nearly full-length nascent peptides and complete released coat protein molecules (Ball and Kaesberg 1973b).

ISOLATION

The appropriate isolation techniques were reviewed thoroughly by Weber and Konigsberg (1975). According to this review, the choice of procedure depended on whether it was desired to have the RNA or the protein in native form. One method involved treating the phage solution with urea and LiCl when the RNA precipitated and was removed by centrifugation (Nishihara et al. 1970). The coat protein remained in solution and was used when the native protein was required. Another method, which destroyed the RNA but left the coat protein in solution in a relatively native state, was to treat the phage solution with 50% acetic acid (Fraenkel-Conrat 1957). The RNA precipitated, while the coat protein remained in the supernatant and was acceptable for the phage or capsid reconstitution. The undegraded RNA could be obtained from the phage by phenol extraction when the RNA remained in the aqueous phase and the coat protein precipitated at the interface between the aqueous and phenol layers. The protein could then be useful for structural studies after washing the precipitate with ethanol and ether to remove phenol. Finally, the protein could also be prepared by heating the phage to 100°C for 10 minutes in 2 M KCl (Konigsberg et al. 1966).

AGGREGATION

One of the most characteristic features of the RNA phage coat proteins consisted of their tendency to aggregate even in moderately dissociating solvents such as 50% acetic acid (Weber and Konigsberg 1975). The f2 coat protein first formed hexamers (Konigsberg et al. 1966) which, after standing for some time, aggregated further and eventually precipitated from solution. The coat proteins could be kept in the monomeric form in 1% SDS or in 6 M guanidine-HC1. As discussed in Chapters 1 and 9, the capsid-like structures were formed when the coat proteins were allowed to aggregate under carefully controlled conditions (Matthews and Cole 1972d). The so-called 11S structure as a putative precursor of the coat shell formation was discussed in the Reconstitution section in Chapter 9.

If a large molar excess of RNA was mixed with native coat protein from the same group of phages, a Complex I was formed where from one to six equivalents of coat protein were bound to the RNA (Sugiyama et al. 1967; Spahr et al. 1969). The importance and striking regulatory role of the Complex I will be discussed in Chapters 16 and 17.

PHYSICAL PROPERTIES

When the optical rotatory dispersion of the phage MS2, its RNA and coat protein subunit were measured, it was concluded that the unusual rotatory dispersion of the coat protein probably reflected tertiary structure, which persisted upon assembly (Oriel and Koenig 1968). The optical rotatory dispersion and circular dichroism data of the reassembled MS2 particles were interpreted as those of β structure within the protein (Oriel et al. 1971). By measuring the circular dichroism, the conformational changes were detected, which occurred during reversible depolymerization of the fr coat protein with acetic acid (Schubert 1969).

An ingenious procedure was invented for the radioactive labeling of the Qβ and f2 coats *in vitro*, in which ³H- or ¹⁴C-methyl groups were attached to the protein amino groups by reductive alkylation (Rice and Means 1971).

As described in the Monitoring section in Chapter 9, the direct analysis of the MS2 coat protein was performed by MALDI (Thomas et al. 1998) and ESI mass spectrometry (Tito et al. 2000; Xiang et al. 2000). Moreover, by analysis of the ions formed by dissociation of the intact protein ions, the MS2 coat was identified when expressed in *E. coli* (Cargile et al. 2001). Next, the charge-state-dependent fragmentation behavior of the MS2 coat protein was summarized over a wide range of charge states, and a novel approach was invented to obtain fragmentation behavior from multiple parent ion charge states after concentration and purification of the protein ions in the gas phase (He et al. 2002). The putative use of the Raman optical activity to characterize the secondary and tertiary structure of the polypeptide backbone including the MS2 coat was reviewed by Blanch et al. (2004).

HOMOLOGY

Figure 12.1 demonstrates the Clustal V alignment of the traditional RNA phage coat protein sequences. The sequences were translated from the corresponding nucleotide sequences including some partial genome sequences. The amino acid sequences were used only in the two cases when the corresponding nucleotide sequences were still not available, namely, for the f2 and R17 coats. To make the homologous pattern reliable, the most distant coats of the phages AP205, φCB5, and MB were excluded from the analysis. The general similarity pattern remains those obtained for the full-length RNA phage genomes and maturation proteins.

As discussed in the Metaviromics section in Chapter 2, Kaspars Tārs' team performed an unprecedented similarity analysis of the coat proteins of all known levi-like RNA genomes, obtained mostly from the metagenomic data (Liekniņa et al. 2019). The results of this great similarity analysis are presented in Figure 12.2.

Liekniņa et al. (2019) started from the observation that the majority of the partial levi-like genomes possessed an open reading frame (ORF) between the conserved maturation and replicase genes. Although the ORFs showed great variation in length and sequence, and in numerous cases no similarity to the known Leviviridae phage coats or any other proteins, these ORFs could be identified as putatively encoding the coat protein. Due to the often very weak sequence similarity and broadly variant protein length that ranged from 105−208 residues compared to only 122–132 in the previously known phages, the authors employed a BLAST similarity analysis followed by UPGMA clustering based on the BLAST bit score ratios (hit score/self-score). The resulting clustering analysis, while not a proper phylogenetic reconstruction, provided useful information regarding the diversity of the novel coat proteins and their relatedness to the previously known phages. Thus, almost a half, or over 80, of the known levi-like phage coat sequences clustered into an MS2-like supergroup with approximately ten recognizable subgroups. Two of those represented the currently recognized *Levivirus* and *Allolevivirus* F pili-specific phage genera, with an adjacent cluster containing the other conjugative pili-specific phages M, C-1, PRR1, and Hgal1. The two additional neighboring coat clusters were comprised solely of metagenomic sequences, with increasing coat length and decreasing coat similarity to the known phages. The supergroup included some additional smaller clusters with low similarity to MS2 or Qβ, one of which contained the previously known pseudomonaphage PP7. A φCB5-like supergroup emerged as the second largest, that included the caulophage φCB5 and more than 30 related coat sequences from the metagenomic data. These further clustered into five subgroups, all formed by short and rather uniform-length proteins (most around 115–120 residues). One more supergroup with some 30 sequences emerged which was comprised entirely of metagenomic sequences and contained two major recognizable subgroups. The coats in this AVE015-like supergroup, named after a representative phage from the group, were considerably bigger than those of MS2- or φCB5-like phages, with an average length around 165 residues, with some of them spanning 180 residues. The three supergroups comprised approximately 80% of all the known levi-like phage coat sequences (Liekniņa et al. 2019). The coat genes were expressed by the authors of this study in *E. coli* and

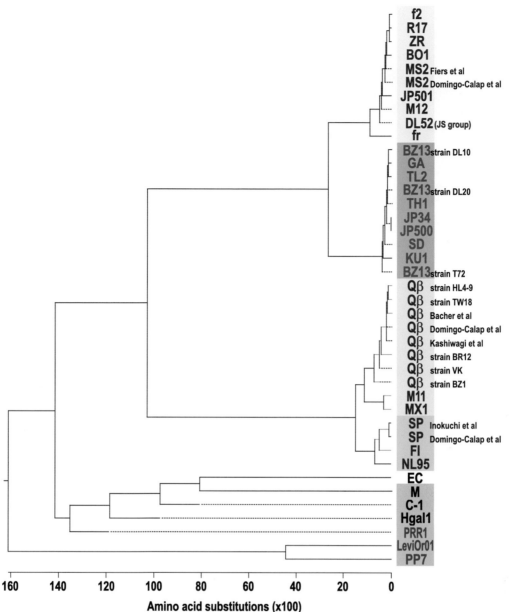

FIGURE 12.1 The comparison of the coat protein sequences obtained by the Clustal V alignment in the MegAlign program from the DNASTAR Lasergene software suite. The designations and colors of the genomes correspond to that used in Table 7.1 and Figure 7.1. The very distal AP205, φCB5, and MB coats are excluded from the alignment.

contributed to the unprecedent expansion of the RNA phage VLP carriers, as will be described in the Novel VLP models section in Chapter 25.

SPATIAL STRUCTURE

The three-dimensional structures of the coat proteins, assembled virions, RNA phage-derived virus-like particles, as well as of their chimeras, will be presented in Chapter 21.

LYSIS

The coat protein seemed also to be associated with cell lysis. Zinder and Lyons (1968) studied the cell lysis after infection

with the phage f2 and its coat *amber* mutants. When an Su⁻ strain of *E. coli* was infected with the polar *amber* coat protein mutant with mutation at position 6, there was no difference between infected and uninfected cells. With the nonpolar *amber* mutant with mutation at position 70, the Su⁻ cells stopped growing but did not lyse. When *E. coli* Su-1 or Su-2 cells were infected with the same polar and nonpolar *amber* coat protein mutants, the suppression and lysis of the cells were observed. With the *E. coli* Su-3 cells, which ensured nonfunctional insertion of tyrosine, the cells stopped growing but without lysis. It was concluded that the functional and fully completed coat proteins were required for cells to lyse. Only after 14–15 years, were the clear experimental proofs reported that the coat protein did not have an obligatory role

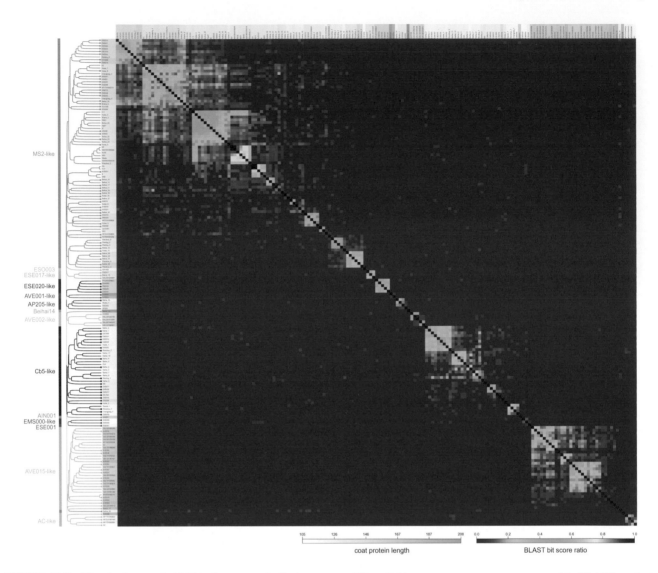

coat protein length   BLAST bit score ratio

FIGURE 12.2 The single-stranded RNA phage coat similarity groups. All available coat sequences were compared in BLAST analysis followed by UPGMA clustering based on BLAST bit score ratios (hit score/self score). The resulting dendrogram and heat map representation of coat diversity are presented. The orange shading of coat labels corresponds to their length distribution. The coat proteins that were studied experimentally in this paper are indicated with dots at branch tips. (Reprinted from Liekniņa I et al. *J Nanobiotechnology.* 2019;17:61.)

in the lytic process itself (Kastelein et al. 1982; Coleman et al. 1983). The actual lysis mechanisms of the RNA phages are described in Chapters 11 and 14.

READTHROUGH PROTEIN

As stated in Chapters 7 and 8, the analysis of *amber* mutants indicated the three genes in the Qβ genome (Horiuchi and Matsuhashi 1970) which encoded the maturation protein, coat protein, and replicase (Horiuchi et al. 1971). Meanwhile, the studies of the Qβ-infected cells revealed the four phage-specific proteins, three of which were incorporated into the virion (Garwes et al. 1969). Following the nomenclature of the latter paper, the fourth protein was called IIb, in contrast to the neighboring maturation protein IIa. The molecular mass of the IIb protein was estimated as 36,000. The three proteins:

maturation, or IIa, protein with the molecular mass 41,000, IIb protein with the molecular mass 36,000, and coat protein with the molecular mass 14,000, were present in a molar ratio of approximately 1.3:5:95 within the phage virion (Weber and Konigsberg 1975). This fourth Qβ protein was termed the A1 protein, in contrast to the A2 maturation protein, by Kondo et al. (1970). This parallel nomenclature, A1 and IIb, was also retained later. The presence of the fourth protein was also evident in the studies of other groups, as by Kamen (1970) and by Strauss and Kaesberg (1970) in their exhaustive papers on the gel electrophoresis of the phage Qβ and its protein components. The production of the IIb protein was dependent on the synthesis of normal coat protein, since the coat *amber* mutants did not synthesize the IIb protein (Kamen 1970; Horiuchi et al. 1971).

The natural translational readthrough at the leaky UGA termination signal of the coat protein gene was declared soon as the possible origin of the IIb, or A1, protein, and the latter got the trivial name of the *readthrough protein* (Moore et al. 1971; Weiner and Weber 1971). Moreover, Weiner and Weber (1971) found that the N-terminal sequence of the IIb protein was identical for at least eight residues with the N-terminal sequence of the Qβ coat protein. This result was possible due to characterization of the four Qβ-coded proteins by their N-terminal sequences using a micro-dansyl Edman degradation that was published a bit later (Weiner et al. 1972). The identity of the N-terminal portions of IIb protein and coat protein was shown also by Moore et al. (1971) who isolated an N-terminal pentapeptide of the IIb protein. In parallel, Hirsh (1970) found that a mutant tryptophan tRNA mediated high-level UGA suppression in a UGA suppressor strain, while tryptophan tRNA from the non-suppressing parental strain and no other tRNA was also active, albeit at only 3% the level of Su+ tRNA (Hirsh 1970; Hirsh and Gold 1971).

Finally, Weiner and Weber (1973) gave clear evidence of the natural readthrough at the UGA termination signal and published a detailed story of the readthrough protein. In fact, the molar fraction of this readthrough, or IIb, or A1, protein relative to normal coat protein in the viral capsid increased from 2.2% to 7.2% when a UGA suppressor strain was used as a host for the Qβ infection. Moreover, the partial amino acid sequence, which included the suppressed termination signal, was obtained for the IIb protein. This sequence proved that the single UGA codon was used alone as the natural translational termination signal of the phage Qβ coat cistron. The evidence was also presented that in both the Su− and Su+UGA host, the ratio of the readthrough protein to the normally terminated coat protein was 1.5 to threefold higher *in vivo* than in the purified virus. Thus, in the process of self-assembly, the viral capsid preferred to incorporate normally terminated coat protein rather than the readthrough product. Although the virus discriminated against the IIb protein and preferred to incorporate the normal coat protein, the Qβ phage preparations with up to 15% of the protein appearing as the IIb protein were routinely obtained (Weber and Konigsberg 1975). The molecular mass

of the IIb readthrough protein exceeded that of the Qβ coat protein by 22,000, and this corresponded to about 200 amino acids, which must be coded (Weiner and Weber 1973). At the same time, the intercistronic region between coat and replicase genes was estimated as approximately 600 nucleotides long (Weiner and Weber 1971, 1973).

As described in the Mutants of the phages Qβ, SP, and FI section in Chapter 8, Radloff and Kaesberg (1973) isolated the so-called electrophoretic Qβ mutant 27-2 that differed by its behavior to pack the readthrough A1 protein into phage particles. In the mutant particles, the A1 protein was found in the virions only in multiples of three, while wild-type virions differed by only a single A1 protein and therefore formed a wide single band of the phage by electrophoresis. Radloff and Kaesberg (1973) were the first to postulate that the Qβ particle variants avoided of the A1 protein are noninfectious. The urgent need of the A1 protein for the formation of the viable Qβ particles was clearly approved by the reconstitution of infectious phage particles that was stringently dependent on the addition of the A1 protein (Hofstetter et al. 1974). Later, the noninfectious character of the Qβ particles avoided of the A1 protein and unable therefore to recognize F pili was confirmed by a series of physiological studies (Engelberg-Kulka et al. 1977, 1979a,b; Zeevi et al. 1979).

Before long, the readthrough protein was found in the group IV phages TW19 and TW28, by direct comparison with the group II phage GA which did not possess the protein in question (Aoi and Kaesberg 1976). Thus, the readthrough protein was acknowledged as a special landmark of the *Allolevivirus* genus. The general properties of the readthrough proteins and the readthrough extensions are collected in Table 12.2. Figure 12.3 demonstrates the Clustal V alignment of the full readthrough proteins in comparison with the analogous tree of the readthrough extensions.

All sequences that appear in Table 12.2 and Figure 12.3 were translated from the corresponding nucleotide sequences. The general phylogeny pattern repeats those obtained for the full-length RNA phage genomes and maturation and coat proteins.

TABLE 12.2

Molecular Mass of the A1 Proteins and Readthrough Extensions of the *Allolevivirus* Coat Proteins

Group	Phage	Molecular Mass of the A1 Protein, Da	Length of the A1 Protein, aa	Charge	pI	Molecular Mass of the Extension, Da	Length of the Extension, aa	Charge	pI	References
III	Qβ	36,134.02	329	+1.91	7.94	21,711.66	195	−1.04	5.94	Kashiwagi et al. (2014)
	M11	35,893.41	329	−3.09	5.29	21,527.06	195	−6.03	4.61	Beekwilder et al. (1995, 1996)
	MX1	35,880.26	329	−5.89	4.98	21,513.95	195	−7.83	4.51	Beekwilder et al. (1995, 1996)
IV	FI	36,183.99	332	−3.26	5.13	21,851.57	199	−6.20	4.47	Kirs and Smith (2007)
	SP	36,203.76	331	−5.25	4.87	21,906.49	198	−7.20	4.41	Inokuchi et al. (1988)
	NL95	36,174.95	330	−0.81	6.59	21,863.61	197	−6.76	4.81	Beekwilder et al. (1995, 1996)

Note: The length and properties for the extensions are given without the readthrough tryptophan. Molecular mass is calculated by the translated nucleotide sequence of the corresponding genome.

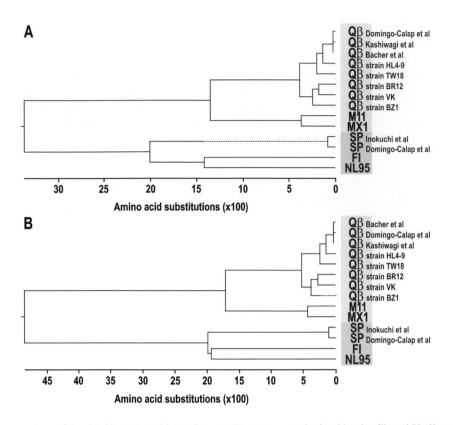

FIGURE 12.3 The comparison of the A1 (A) and readthrough parts (B) sequences obtained by the Clustal V alignment in the MegAlign program from the DNASTAR Lasergene software suite. The designations and colors of the genomes correspond to that used in Table 7.1 and Figure 7.1.

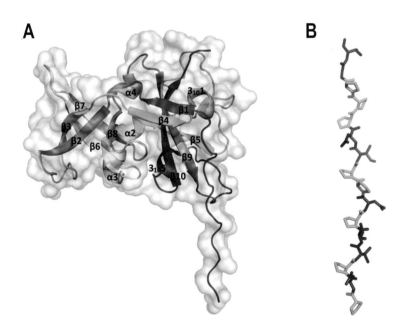

FIGURE 12.4 Structure of the readthrough domain. (A) Overall structure of the domain. The protein is represented as a cartoon model rainbow-colored blue (N-terminus) to red (C-terminus) and overlaid with a surface representation of the domain (light grey). (B) A detailed view of the polyproline helix. In the first 16 residues of the model, prolines are represented in cyan and other residues in deep blue. (Rumnieks J and Tars K. *Protein Sci.* 2011;20:1707–1712. Copyright Wiley-VCH Verlag GmbH & Co. KGaA. Reproduced with permission.)

A

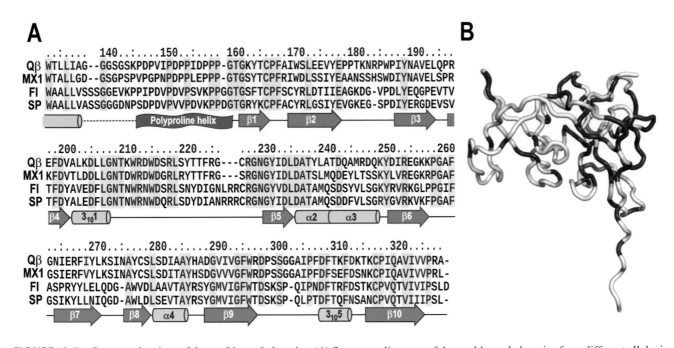

B

```
        ..:..... 140..:....150..:..... 160..:....170..:...180..:....190..:..
QB   WTLLIAG--GGSGSKPDPVIPDPPIDPPP-GTGKYTCPFAIWSLEEVYEPPTKNRPWPIYNAVELQPR
MX1  WTALLGD--GSGPSPVPGPNPDPPLEPPP-GTGSYTCPFRIWDLSSIYEAANSSHSWDIYNAVELSPR
FI   WAALLVSSSGGEVKPPIPDVPDVPSVKPPGGTGSFTCPFSCYRLDTIIEAGKDG-VPDLYEQGPEVTV
SP   WAALLVASSGGGDNPSDPDVPVVPDVKPPDGTGRYKCPFACYRLGSIYEVGKEG-SPDIYERGDEVSV
```

```
       ..200..:....210..:....220..:.    ...230..:....240..:....250..:....260
QB   EFDVALKDLLGNTKWRDWDSRLSYTTFRG---CRGNGYIDLDATYLATDQAMRDQKYDIREGKKPGAF
MX1  KFDVTLDDLLGNTDWRDWDGRLRYTTFRG---SRGNGYIDLDATSLMQDEYLTSSKYLVREGKRPGAF
FI   TFDYAVEDFLGNTNWRNWDSRLSNYDIGNLRRCRGNGYVDLDATAMQSDSYVLSGKYRVRKGLPPGIF
SP   TFDYALEDFLGNTNWRNWDQRLSDYDIANRRRCRGNGYIDLDATAMQSDDFVLSGRYGVRKVKFPGAF
```

```
       ..:....270..:....280..:....290..:....300..:....310..:....320..:...
QB   GNIERFIYLKSINAYCSLSDIAAYHADGVIVGFWRDPSSGGAIPFDFTKFDKTKCPIQAVIVVPRA-
MX1  GSIERFVYLKSINAYCSLSDITAYHSDGVVVGFWRDPSSGGAIPFDFSEFDSNKCPIQAVIVVPRL-
FI   ASPRYYLELQDG-AWVDLAAVTAYRSYGMVIGFWTDSKSP-QIPNDFTRFDSTKCPVQTVIVIPSLD
SP   GSIKYLLNIQGD-AWLDLSEVTAYRSYGMVIGFWTDSKSP-QLPTDFTQFNSANCPVQTVIIIPSL-
```

FIGURE 12.5 Conserved regions of the readthrough domains. (A) Sequence alignment of the readthrough domains from different allolevi-viruses. Conserved residues are colored red; of these, identical residues are shaded yellow and nonidentical light yellow. Assigned secondary structure elements are presented below the alignment. A dashed line represents the portion for which no experimental data are available; the a-helix from secondary structure prediction is drawn as a pale blue cylinder. (B) Mapping of the conserved regions on the three-dimensional structure of the readthrough domain. Identical and nonidentical but conserved residues are colored red and yellow-orange, respectively. (Rumnieks J and Tars K. *Protein Sci.* 2011;20:1707–1712. Copyright Wiley-VCH Verlag GmbH & Co. KGaA. Reproduced with permission.)

The crystal structure of the readthrough domain from the phage Qβ A1 protein was resolved at a resolution of 1.8 Å by Riga scientists Jānis Rūmnieks and Kaspars Tārs (Rumnieks and Tars 2011). The domain consists of a heavily deformed five-stranded β-barrel on one side of the protein, and a β-hairpin and a three-stranded β-sheet on the other. Several short helices and well-ordered loops are also present throughout the protein. The N-terminal part of the readthrough domain contains a prominent polyproline type II helix. The overall fold of the domain is not similar to any published

structure in the Protein Data Bank. Figure 12.4 presents the crystal structure in question.

Figure 12.5 shows homologically conserved regions of the readthrough domains and their localization on the three-dimensional structure.

Concerning the localization of the A1 protein within the phage particles, a recent electron microscopy visualization of foreign epitopes carried by A1 protein within infectious Qβ particles showed that the A1 protein molecules are occupying corners of the Qβ icosahedron (Skamel et al. 2014).

13 Replicase

Nothing worse could happen to one than to be completely understood.

Carl Gustav Jung

Tis but a part we see, and not a whole.

Alexander Pope

ACTIVITY

According to the first classical review, a novel enzyme appeared after infection by the phage f2 (Zinder 1965). The first indirect evidence for the early appearance of the enzyme came from many sources. First, histidine was not required late in infection, but was essential during the first 15 minutes of infection. Then, chloramphenicol added at the moment of infection inhibited the synthesis of viral RNA, as measured by several techniques. If added within 10–15 minutes, it did allow normal synthesis of viral RNA even though protein synthesis was inhibited in the case of the phages f2 (Cooper and Zinder 1963) or R17 (Paranchych and Ellis 1964). The parental phage RNA molecules were converted to a double-stranded form and the process of replication started, as shown for the first time in the case of the phage MS2 (Kelly and Sinsheimer 1964). The replication process on the whole will be described in Chapter 17. Here, it is important to stress that the conversion of the parental phage RNA into the double-stranded RNA was inhibited when chloramphenicol was added at the time of infection, indicating therefore the appearance of a new phage-induced enzyme which was involved in the synthesis of viral RNA (Kelly and Sinsheimer 1964). As Norton D. Zinder (1965) noted, it was the first case when phage RNA made phage RNA.

The three laboratories Zinder, Spiegelman, and Weissmann's have reported the detection of such an enzymic activity after infection by the phages f2 (August et al. 1963; Kaye et al. 1963) and MS2 (Haruna et al. 1963; Weissmann and Borst 1963; Weissmann et al. 1963a,b). Spiegelman and Hayashi (1963) were the first to introduce the term *replicase* as a shortened form of the official term RNA-dependent RNA polymerase. Weissmann et al. (1963a,b) used a term RNA synthetase, while August et al. (1963) preferred the term RNA polymerase. This enzyme seemed at the beginning rather unstable, and its characterization had raised several novel problems. It polymerized ribonucleoside triphosphates into trichloroacetic acid-insoluble material. The activity was sensitive to ribonuclease yet insensitive to deoxyribonuclease or actinomycin D. It required all four triphosphates, Mg^{2+}, and a triphosphate-generating system for maximum activity. Thus, it appeared possible to distinguish the phage-generated replicase from all of the host's normal nucleoside polymerizing enzymes including the DNA-dependent RNA polymerase.

However, the problem of the enzyme's template for the replicase activity remained unclear. Weissman et al. (1963a,b) were unable to separate the enzyme from RNA sufficiently to permit analysis of the template. Nevertheless, they purified a double-stranded RNA from the infected cell, together with the enzyme, and considered this RNA to be the template. Haruna et al. (1963) purified the phage-induced replicase 1000-fold and found it to be stimulated only by phage RNA. August et al. (1965) were the first to purify the replicase enzyme from cells infected with the coat mutant f2 sus11 that overproduced replicase by 20 times against the normal amount of enzyme. The authors purified the enzyme 200-fold and found that addition of any single-stranded RNA was stimulatory. Further, the product of the reaction was studied by two groups. Weissman et al. (1964a,b) concentrated on the properties of the replicative forms of the MS2 RNA that will be the main subject of Chapter 17. Shapiro and August (1965a,b) analyzed preferentially the properties of the f2-induced enzyme, including the dependence of the product on the nature of the added template RNA. They were the first to conclude from nearest-neighbor analysis that the actual template was the strand complementary to the added RNA and that the RNA synthesis proceeded in an antiparallel fashion, by analogy with the *E. coli* DNA polymerase. Exhaustive studies on the MS2 replicase were performed also by Sol Spiegelman's team (Haruna and Spiegelman 1965a,b,c).

The appearance of the *in vivo* replicase activity occurred 5–10 min after the phage f2 infection, and the activity increased for 20 or 25 min (Lodish et al. 1964). As Zinder (1965) noted, there were 15–30 min before half of the viral RNA, and 20–30 min before half of the progeny f2 phage were made (Cooper and Zinder 1963). Lodish et al. (1964) found that the enzyme was synthesized in normal amounts even if RNA synthesis was inhibited markedly by uracil or adenine starvation of host bacteria prior to infection. Lodish et al. (1964) were the first to postulate that, in general, only input parental RNA molecules served as templates for enzyme synthesis.

Davis et al. (1964) supposed that the enzyme, in the MS2 studies, was irreversibly bound to, and replicated therefore, only that RNA molecule which directed its synthesis in infected cells. Harvey F. Lodish (cited by Zinder 1965) found that it was not the case and the enzyme, in the f2 studies, synthesized on one RNA was able to convert another RNA into the double-stranded form, even in the presence of chloramphenicol.

Later, in his next classical review, Zinder (1980) concluded that the replicase protein was made in large amounts, about 10% of coat protein, when the coat protein control was released, but normally *in vivo* it could be detectable only enzymatically. Moreover, the replicase subunit itself was found to have a regulatory function (Kolakofsky and Weissmann 1971a,b; Weber et al. 1972), but this is the main subject of Chapters 16 and 17.

REPLICASE SUBUNIT

The phage-specific replicase subunit of the enzyme was first identified clearly as a separate band on gel electrophoresis in the case of the phage MS2 by the *in vivo* and *in vitro* syntheses (Nathans et al. 1966; Eggen et al. 1967; Viñuela et al. 1967a; Vandamme et al. 1972). It is necessary to add that, in contrast to other phage-specific proteins, the replicase subunit was never sequenced by the amino acid sequencing technologies, although N-terminal sequences of the Qβ and R17 replicase subunits were determined by the modified Edman degradation of proteins purified on a nanomole scale by gel electrophoresis (Weiner et al. 1972). The N-terminus of the Qβ replicase subunit from the purified Qβ enzyme was determined as Ser-Lys-Thr-Ala-Ser (Weiner and Weber 1971).

The full-length Qβ replicase subunit of approximately 550 amino acid residues was obtained as a product of the Qβ RNA-stimulated cell-free protein synthesis (Jockusch and Hindennach 1972). It was identical in size to the corresponding chain of the purified replicase. Interestingly, the authors deduced a net charge of −27 and a cysteine content of >15 residues from its electrophoretic behavior at pH 8.5. The *in vitro* synthesized Qβ replicase subunit was able to bind Qβ RNA and some constituents of the *E. coli* extract to form a 20 S complex (Jockusch and Hindennach 1972). Moreover, the *in vitro* synthesized Qβ replicase was found further capable to the polyC-dependent polyG synthesis after some partial purification (Happe and Jockusch 1973).

Finally, Happe and Jockusch (1975) showed that the *in vitro* synthesized Qβ replicase subunit was able to reassemble with other three components (see below) to form active enzyme.

Table 13.1 presents estimations of the molecular mass of the phage-induced replicase subunit, as well as actual parameters of the replicase polypeptide, which were deduced later from the nucleotide sequences of the corresponding genomes.

HOMOLOGY

Figure 13.1 demonstrates the Clustal V alignment of the replicase subunits. All sequences were translated from the corresponding nucleotide sequences of full-length genome sequences. The partial replicase sequences that have been deduced for some genomes were not considered. The general pattern is similar to those obtained for the full-length RNA phage genomes and maturation proteins. Ohnishi (1985, 1988) found a considerable homology of the MS2 replicase with the C-terminal regions of *E. coli* DNA polymerase I and T7 phage DNA polymerase. The homologous stretches were also found among the MS2 replicase, picornavirus RNA-dependent RNA-polymerase, and certain polypeptides involved in the replication of RNA genomes of alphaviruses, tobamoviruses, and tricornaviruses, where the common

sequences were located in the C-terminal regions of the viral proteins (Morozov and Rupasov 1985).

It was noted in the Relatedness to other taxa section in Chapter 2 that the RNA phage replicases were chosen to look for the possible homology of the *Leviviridae* representatives with other members of the virus kingdom. This problem is touched in more detail in Chapter 18. An excellent and highly informative alignment of the RNA phage Qβ, NL95, SP, FI, PRR1, fr, MS2, and GA replicases with localization of the temperature-sensitive and lethal mutations was published by Kidmose et al. (2010).

Qβ REPLICASE

After the first hard trials on the replicases induced by the group I RNA phages f2 (August et al. 1963) and MS2 (Weissmann et al. 1963), the activities were turned generally to the Qβ replicase (Haruna and Spiegelman 1965c) and met with great success and a brilliant future. The Qβ replicase enzyme demonstrated unexpected organization and mysterious properties that led to numerous discoveries, including evolutionary ones, which are presented in Chapter 18. The true history of the Qβ replicase was written clearly by Robert I. Kamen, one of the main founders of the Qβ replicase story (Kamen 1975).

Haruna and Spiegelman (1965) partially purified the Qβ replicase and obtained for the first time the enzyme preparations that were strictly dependent on the addition of the exogenous Qβ RNA as a template for the RNA synthesis. Next, Spiegelman et al. (1965) accomplished the first *in vitro* synthesis of the fully infectious viral genome. As Kamen (1975) noted in his classical review, the exciting demonstration that the Qβ replicase synthesized the biologically active progeny RNA molecules *in vitro* stimulated the intensive use of the enzyme in studies of the mechanism of phage RNA replication.

The early studies concentrated first on the general biochemical properties of the RNA replication including so-called autocatalytic synthesis of the viral RNA *in vitro* (Haruna and Spiegelman 1965a,b, 1966a,b; Mills et al. 1966; Pace and Spiegelman 1966a,b; Weissmann and Feix 1966; August and Eoyang 1967; Banerjee 1967; Banerjee et al. 1967; Bishop et al. 1967a,b,c; Eikhom and Spiegelman 1967; Hori 1967; Hori et al. 1967; Pace et al. 1967a,b; Spiegelman 1967; Spiegelman et al. 1967).

It is necessary to emphasize here the clear demonstration of the RNA minus strand synthesis *in vitro* by Charles Weissmann's team (Weissmann and Feix 1966; Feix and Weissmann 1967; Feix et al. 1967a,b; Weissmann 1967), functioning of the single-stranded RNA minus strands *in vitro* as templates for the synthesis of viral RNA plus strands (Weissmann et al. 1967), and purification and characterization of the Qβ minus strands (Pollet et al. 1967). Weissmann's team showed that, on incubation with Qβ replicase and nucleoside triphosphates, noninfectious minus strands directed a rapid synthesis of the infectious viral RNA in substantial excess over the added template (Feix et al. 1968). To distinguish the radioactive viral plus and minus strands, an isotope

TABLE 13.1

Molecular Mass of the Phage-Specific Replicase Subunits

Group	Phage	Method	Estimated Molecular Mass, Da	Expected Length, aa	References	Molecular Mass, Calculated by Genome Sequence, Da	Actual Length, by Genome Sequence, aa	Charge	pI
I	MS2	Gel electrophoresis		330	Viñuela et al. (1968)	60,837.52	545	+13.39	9.16
	MS2	Gel electrophoresis	55,000		Remaut et al. (1968)				
	MS2	Gel electrophoresis	60,000		Furuse et al. (1979b)				
	MS2	Gel electrophoresis	63,000		Atkins and Gesteland (1975)				
	R17	Sucrose gradient; Sephadex chromatography	50,000	450	Capecchi (1966)				
	R17	Gel electrophoresis	63,000		Atkins and Gesteland (1975)				
	f2	Gel electrophoresis	63,000		Fedoroff and Zinder (1971b)				
	f2	Gel electrophoresis	63,000		Atkins and Gesteland (1975)				
	f2	Gel electrophoresis	61,000		Stallcup et al. (1976)				
	fr					61,340.06		+14.86	9.14
II	GA	Gel electrophoresis	57,000		Furuse et al. (1979b)	59,928.75	532	+5.35	7.99
	GA	Gel electrophoresis	60,000		Yonesaki and Haruna (1981)				
	BZ13 DL20					59,734.55	532	+7.34	8.21
	KU1					60,041.02	532	+11.34	8.59
III	Qβ	Gel electrophoresis	49,000	426	Garwes et al. (1969)	65,530.71	589	+2.16	7.48
	Qβ	Gel electrophoresis	52,000		Jockusch et al. (1970)				
	Qβ	Gel electrophoresis	66,000		Kamen (1970)				
	Qβ	Gel electrophoresis	69,000		Kondo et al. (1970)				
	Qβ	Gel filtration	55,000		Horiuchi et al. (1971)				
	Qβ	Gel electrophoresis	64,000		Weiner and Weber (1971)				
	Qβ	Gel electrophoresis	60,000	550	Jockusch and Hindennach (1972)				
	Qβ	Gel electrophoresis	62,000						
	VK	Gel electrophoresis	45,000						
IV	M11					66,041.81	588	+4.59	8.04
	MX1					65,953.67	586	+6.87	8.21
	FI	Gel electrophoresis	62,000		Furuse et al. (1979b)	66,374.92	586	+10.69	8.75
	SP	Gel electrophoresis	62,000		Furuse et al. (1979b)	65,397.93	576	+3.16	7.79
	NL95					65,012.79	576	+11.81	8.77
Levivirus-like	Hgal1					58,694.01	523	+11.50	8.87
	C-1					59,548.94	522	+6.56	8.39
	M					58,934.18	520	+15.97	9.54
Pseudomonaphages	PP7					63,318.74	552	+11.56	8.91
	PRR1					60,983.69	540	+14.01	9.11
	LeviOr01					61,913.83	544	+2.95	7.73
Caulophage	φCB5					74,105.60	655	+20.27	9.24
Acinetophage	AP205					66,543.44	590	+3.54	7.59
Assembled genomes	MB					72,845.13	652	+7.43	8.19

Note: The order of the replicase subunit presentation corresponds to that of the phage group order in Figure 7.1 and Table 10.1.

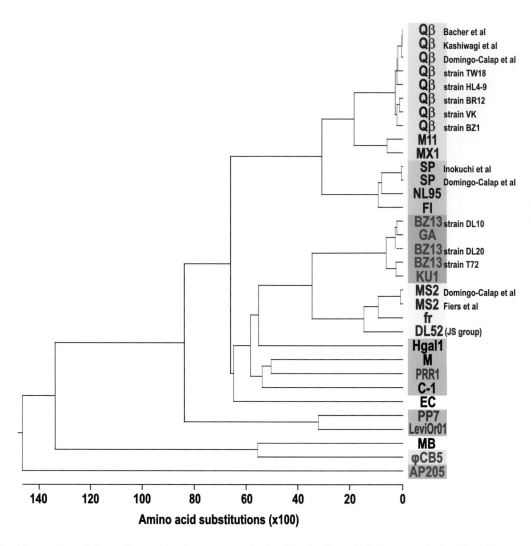

FIGURE 13.1 Comparison of the replicase subunit sequences obtained by the Clustal V alignment in the MegAlign program from the DNASTAR Lasergene software suite. The designations and colors of the genomes correspond to that used in Table 7.1 and Figure 7.1.

dilution method was proposed and used to show that the Qβ replicase, when primed with Qβ plus strands, first synthesized minus strands and later predominantly plus strands, but when primed with Qβ minus strands, produced plus strands from the very outset of the reaction (Weissmann et al. 1968a).

In parallel, J. Thomas August's team found that synthetic ribopolymers can function as templates in the Qβ replicase reaction (Hori et al. 1967). Moreover, this team concentrated on the special issue of so-called host factors. Franze de Fernandez et al. (1968) were the first to proclaim clearly that a factor fraction essential for the *in vitro* synthesis of Qβ RNA could be isolated from both infected and uninfected *E. coli* cells. In fact, August's team noted in the early reports that a factor fraction was required for maximal activity in the reaction directed by the Qβ RNA plus strands (August and Eoyang 1967; Banerjee et al. 1967; Hori et al. 1967). The story of the host factors required for the replication of phage Qβ RNA *in vitro* was summarized exhaustively by Kuo et al. (1975), the central persons who developed this field.

The next strong contribution of August's team consisted of the discovery of another specific and natural RNA template

for the Qβ replicase, 6S RNA (Banerjee and August 1969; Banerjee et al. 1969b). This small RNA was first found during an investigation of the endogenous template present in crude extracts of Qβ-infected *E. coli*. This story is, however, more related to the molecular evolutionary studies and will be discussed therefore in Chapter 18.

A bit later, Hori (1970) studied the structure of the Qβ replicative complex by glycerol-density gradient centrifugation and found that the two types of the enzyme-RNA complex, namely, 45S and 34S, were functional as active replicating sites of Qβ RNA *in vitro*.

As for Spiegelman's team, Pace et al. (1967a) introduced examination of the Qβ replicase reaction by gel electrophoresis, which markedly improved the resolution of the newly synthesized RNA and allowed analogies with the replicative complexes found by the *in vivo* replication of the phages M12 (Francke and Hofschneider 1966a,b) and R17 (Franklin 1966). The electrophoretic separation of viral nucleic acids on polyacrylamide gels was later described in detail by Bishop et al. (1967b). In fact, Mills et al. (1966) showed that the Hofschneider's structures (HS), namely, a noninfectious structure possessing a low sedimentation value 15S which became infectious on heat denaturation, appeared also in the latent period that preceded the appearance of newly synthesized infectious units in the Qβ *in vitro* system.

Spiegelman's team at this stage raised the problem of a novel polarity by the *in vitro* Qβ replication (Bishop et al. 1967c). The authors hypothesized that, although the synthesis of all polynucleotides has been shown to occur from 5′–3′, there were no theoretical reasons for excluding the possibility of 3′–5′ polymerizations, and tried to find support for the idea and resolve the so-called *DNA dilemma* of a single growing point for the two antiparallel chains. This dilemma of the bacterial chromosome and its manner of replication appeared at the same time when extensive studies of the RNA phages had been started (Bonhoeffer and Gierer 1963; Cairns 1963). Therefore, Bishop et al. (1967c) thought that the RNA synthesis started with 5′–3′ polymerization of the negative complementary strand and was followed by a reversal of the chemical polarity to yield a novel 3′–5′ polymerization of new plus strands. The "pulse-chase" experiments indicated the so-called Franklin's structure (FS) (Franklin 1966) *in vivo*— namely, the replicative form with growing plus strands—as an immediate precursor of the viral RNA plus strands in the Qβ replicase reaction (Pace et al. 1968b). Mills et al. (1968) rapidly confirmed the finding of Feix et al. (1968) on the negative strands as excellent templates for the Qβ replicase and acknowledged the 5′–3′ direction of synthesis of plus strands when the enzyme was employing free negative variant strands as templates.

The two alternative mechanisms of the chain elongation, either from the 5′- to the 3′-terminus or from the 3′- to the 5′-terminus, were also addressed in the exhaustive study from August's team (Banerjee et al. 1969a). It was concluded clearly that the direction of growth of both Qβ RNA and the strand complementary to Qβ RNA was from the 5′-triphosphate to the 3′-hydroxyl terminus. The observation that guanosine triphosphate was the 5′-terminus of the complementary strand indicated that the Qβ replicase reaction *in vitro* did not give rise to a complete complementary copy of Qβ RNA, since the hydroxyl terminus of the phage RNA was adenosine, not cytidine. The problem of the 3′-terminal nucleotide was touched on in Chapter 10.

At this stage, the excellent idea of the Darwinian experiments was triggered (Levisohn and Spiegelman 1968, Mills et al. 1968). This special subject of the use of Qβ replicase will be described in detail in Chapter 18.

Meanwhile, Haruna et al. (1967) isolated the replicase enzyme of the group III phage VK, a closest relative of the phage Qβ, and did not find any difference in the specific template requirements of both Qβ and VK replicases, which used RNAs of the group III, but not of the group I and II phages. Then, Haruna et al. (1971) isolated the replicase enzyme of the group IV phages SP and FI, which showed a template specificity different from that of the Qβ replicase. The response of these replicases to RNA was in accord with that reported for the Qβ replicase, the preference being clearly for its own and intact RNA. The Qβ RNA had a template activity corresponding to 20% of that observed with homologous RNA for the SP and FI replicases, and *vice versa*. Thanks to the excellent quality of the replicases induced by the group III and IV phages, Miyake et al. (1971b) attempted to apply a new criterion, the template specificity of RNA replicases, for the grouping of these phages, as presented in the *template specificity of replicases* section in Chapter 2. According to this grouping, the phages Qβ, VK, and ST belonged to one group III, and the phages SP and FI to another group IV, while some similarity existed between the groups III and IV. The subgroups a, b, and c were distinguished in group III, and subgroups a and b in group IV (see Chapter 2).

Watanabe et al. (1968a) looked for the specific inhibition of the Qβ replicase, in comparison with the DNA-dependent RNA polymerase from *E. coli*. Some chemical compounds, as ethine derivatives, demonstrated strong inhibitory effect on Qβ replicase, while no such specific agent was found among 13 kinds of well-known antibiotics, including rifamycin and chromomycin A_3, although pluramycin had some inhibitory effect. This study was continued further with about 1000 synthetic chemical compounds, and the inhibitory effect of the ethine derivatives was acknowledged (Haruna et al. 1970). Furthermore, tobacco mosaic virus (TMV) RNA was found as a specific natural inhibitor of Qβ replicase, while other tested RNAs, including phage, animal and plant viruses, *E. coli*, and yeast RNAs, were inert (Okada et al. 1969). The binding of the TMV RNA to the Qβ enzyme, but not to the Qβ RNA, was directly demonstrated by retention of the RNA-enzyme complex on a nitrocellulose filter, whereas the relative affinity of the Qβ enzyme to TMV RNA was comparable to that for Qβ RNA (Okada et al. 1971). Once the Qβ synthesis had started, however, the reaction was not inhibited by TMV RNA during the chain growth reaction. Moreover, the Qβ replicase reaction was inhibited by fragmented Qβ RNA when Qβ RNA was heated at a high temperature before reaction (Yamane et al. 1973). Both Qβ and SP replicases were inhibited specifically by the tomato aspermy virus particles (Okada and Haruna 1969).

The Qβ replicase reaction was inhibited by four sulfhydryl-blocking reagents: 4-chloromercuribenzoic acid (PCMB), $HgCl_2$, N-ethylmaleimide, and iodoacetic acid (Ohki and Hori 1972). The inhibition, caused by a mercaptide-forming reagent such as PCMB and $HgCl_2$, was reversed by incubating the inhibited enzyme with an excess amount of dithiothreitol and 2-mercaptoethanol, suggesting that the replicase contains SH groups which participate, in some fashion, in the function of the Qβ enzyme. The PCMB-inhibited enzyme still maintained the template-binding ability (Ohki and Hori 1972).

The finding on the Qβ replicase inhibition by polyethylene sulfonate was of a special importance. The polyethylene sulfonate at low concentrations inhibited the initiation, but not elongation, of RNA synthesis by Qβ replicase (Kondo and Weissmann 1972b). It allowed in turn the controlled, stepwise synthesis of RNA by the Qβ enzyme (Bandle and Weissmann 1972). The polyethylene sulfonate affected a step subsequent to the primary binding of Qβ RNA to the replicase and prior to the formation of the first few internucleotide bonds.

Therefore, if addition of polyethylene sulfonate to the Qβ RNA-directed Qβ replicase reaction after initiation has taken place, it led to the exclusive formation of free, single-stranded minus strands. Moreover, the inhibition by polyethylene sulfonate could be reversed by the addition of an appropriate quantity of protamine sulfate.

The initiation of the Qβ replicase-directed synthesis, but not elongation of the initiated polynucleotide chains, was inhibited also by low concentrations of the aurintricarboxylic acid (ATA) dye (Blumenthal and Landers 1973).

Kapuler and Spiegelman (1970) compared the Qβ replicase with the *E. coli* DNA-dependent RNA polymerase by the requirements for substrate selection among base analogs. In the Qβ replicase reaction, the synthesis in the presence of substrate analogs displayed abortive kinetics, implying that the replicase reaction can be uniquely curtailed by the substrate analogs.

The work on the general elucidation *in vitro* of the Qβ replicase enzymatic function, first of all, on the initiation of the RNA synthesis and on the RNA intermediates involved, was essentially completed by 1970. This stage of the Qβ replicase studies was extensively reviewed by the main participants of the story (Haruna 1967; August and Shapiro 1968; August et al. 1968, 1969; Eoyang and August 1968; Pace et al. 1968a,c; Spiegelman 1968–1969; Spiegelman et al. 1968; Weissmann 1968; Weissmann et al. 1968b; August 1969; Stavis and August 1970; August et al. 1970, 1973; Bishop and Levintow 1971; Hori 1972, 1973b; Ito and Haruna 1972).

Remarkably, at the end of this early stage, a simple method for producing large quantities of cells of *E. coli* Q13, rich in the Qβ replicase, together with a reliable purification method, was published by Sargeant et al. (1971).

SUBUNITS OF THE Qβ REPLICASE

After the first stage of studies that were directed to the general elucidation of the Qβ replicase as a whole enzyme, attention was redirected to the protein components involved within the enzyme. Eikhom and Spiegelman (1967) were the first to dissociate the Qβ replicase into a *heavy* and *light* component by successive centrifugations in sucrose gradients. While neither component responded to added Qβ RNA, their combination reconstituted the fully active enzyme system. Meanwhile, the heavier component was associated with a remarkable ability to synthesize polyG, an activity which was completely dependent on the addition of polyC as a template. The relative positions of these components in the gradient indicated that the *heavy* component was approximately 130,000 and the *light* one about 80,000 in molecular mass. The simple calculation led to the conclusion that the phage genome was too small to encode the total Qβ replicase size in the case when the latter could be constructed from non-identical subunits. Therefore, it was assumed for the first time that one of the replicase components was a host-specified protein (Eikhom et al. 1968). If the *heavy* component was regarded as being programmed by the viral RNA, the *light* component was found in uninfected cells (Eikhom et al. 1968). Later, the polyC-dependent

polyG-synthesizing activity of the *heavy* component was characterized in detail (Hori 1973b; Mitsunari and Hori 1973).

Finally, Robert I. Kamen (1970) and Masatoshi Kondo et al. (1970) revealed the fine structure of the Qβ replicase that appeared as a surprisingly complex enzyme. The highly purified Qβ replicase contained four different polypeptide chains, designated I, II, III, and IV, in order of decreasing molecular mass. Figure 13.2 presents the classical pattern of the purified

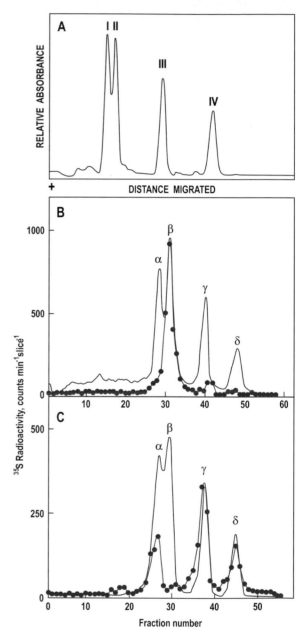

FIGURE 13.2 (A) SDS-polyacrylamide gel pattern of Oβ RNA replicase. The enzyme was purified, and electrophoresis conditions were as described (Kamen et al. 1972) except that fast green was used for staining. Adapted with permission of *Cold Spring Harbor Laboratory Press* from Kamen (1975). (B) Qβ replicase purified from a mixture of ^{35}S-labeled Qβ-infected cells and unlabeled Qβ-infected cells. (C) Qβ replicase purified from a mixture of ^{35}S-labeled uninfected cells and unlabeled Qβ-infected cells. ^{35}S-label is marked red. (Adapted with permission of Taylor & Francis Group from Kondo M. *Arch Int Physiol Biochim.* 1975;83:909–948.)

TABLE 13.2
Properties of the Subunits of the Qβ Replicase Enzyme

Subunit	Molecular Mass	Origin	Identification	Function
I (α)	70,000	Host	Ribosomal protein S1 (translational interference factor i)	Qβ RNA binding
II (β)	65,000	Phage	–	RNA polymerase
III (γ)	45,000	Host	EF-Tu (ψr)	RNA chain initiation
IV (δ)	35,000	Host	EF-Ts	RNA chain initiation

Qβ replicase by gel electrophoresis, which is taken from Kamen's chapter (Kamen 1975) in the classical *RNA Phages* book. Table 13.2, which is adopted from the same chapter, gives identification of the polypeptides that are forming the Qβ replicase enzyme. The quality and purity of the Qβ replicase was improved markedly by the introduction of a new method for the complete purification of the enzyme from cells infected with the Qβ am12 coat mutant, which strongly overproduced the enzyme and did not lyse cells (Kamen 1972).

Three of the enzyme polypeptides, I, III, and IV by Kamen (1970) or α, β, and δ by Kondo et al. (1970), were *E. coli* gene products. Only one, subunit II, or β, was encoded by the phage genome. Surprisingly, the three host subunits came from the bacterial protein synthetic system. As Kamen (1975) noted rightly, the RNA phage replicases constituted a novel class of enzymes containing protein chains encoded by two distinct genomes and readapting preexisting proteins to a viral function apparently different from their original role in the uninfected cell.

Remarkably, the replicase preparations which resolved into four polypeptides by SDS gel electrophoresis behaved as a single protein species when analyzed in nondenaturing solvents by agarose gel column chromatography (Eoyang and August 1971), Sephadex G-200 thin-layer chromatography, sucrose, or glycerol gradient centrifugation at high ionic strength (Kamen 1970; Kondo et al. 1970), or by electrophoresis on Cellogel cellulose acetate strips (Kamen 1975). No resolution of the single band into multiple components was seen within the pH range 7.0–9.5 but examination of the protein eluted from the single band showed it to contain all four subunits, in approximately unimolar stoichiometry (Kamen et al. 1972b, Kamen 1975). It was possible to reconstitute enzyme missing subunit I (Kamen et al. 1972a,b) or dissociate the subunit III + IV complex from replicase by low-salt centrifugation (Kamen 1970). These observations led to the clear conclusion (Kamen 1975) that the coelectrophoresis of the four polypeptide chains coincident with the activity of intact replicase was due to their association to form a single molecule, and not to a chance identity in electrophoretic mobility. This conclusion was further supported by the observed resolution of replicase into four distinct bands when it was electrophoresed in the presence of 6 M urea (Kamen et al. 1972b).

A provocative observation concerning the content of the RNA-replicating enzymes was made at that time by Rosenberg

et al. (1972). The authors isolated polyC-dependent RNA replicase of encephalomyocarditis virus and noticed that the molecular mass of its four out of five polypeptides, namely, 72,000, 65,000, 45,000, and 35,000, was almost identical to the reported molecular mass of the four subunits of the Qβ replicase. Further studies, however, revealed the 56,000-dalton polypeptide as a predominant component of the enzyme that was active also with Qβ RNA as a template, among other RNAs (Traub et al. 1976).

The subunit I of the Qβ replicase was initially identified with the so-called translational interference factor *i*. The translational repressor activity of the Qβ replicase (which will be described in detail by the repressor complexes in Chapter 16 and by the regulation of phage replication in Chapter 17) suggested that the components involved in the initiation of protein synthesis could be involved in the Qβ enzyme. Groner et al. (1972a,b) discovered a factor that was involved in translation and inhibited ribosome binding *in vitro* at the RNA phage coat gene ribosome binding site. This factor was originally called translational interference factor *i*. As reviewed by Kamen (1975), the activity of the factor *i* was qualitatively similar to the translational repressor function of the Oβ replicase (Kolakofsky and Weissmann 1971a,b), but differed in that the factor *i* worked equally well in inhibiting ribosome binding to MS2 or Oβ RNA, while the Oβ replicase showed at least 10 times more efficient inhibition of its homologous RNA. Furthermore, many molecules of the factor *i* were required per phage RNA molecule, whereas only one molecule of the Oβ replicase was sufficient for complete inhibition. Nevertheless, antiserum prepared against pure factor *i* gave a precipitation reaction of identity in double-immunodiffusion tests between factor *i* and either intact Oβ replicase or the subunit I + II enzyme fragment, but did not cross-react with the subunits III + IV (Groner et al. 1972c). Since host subunit I and factor *i* further proved to have identical mobilities by SDS gel electrophoresis, Groner et al. (1972c) concluded that they are the same protein.

The Oβ replicase lacking subunit I was unable to utilize the Qβ plus strand RNA as template at the ionic strength optimum of the complete enzyme but the normal plus strand–dependent activity was regenerated by addition of one molar equivalent of the factor *i* purified from uninfected cells or of authentic replicase-derived subunit I (Kamen et al. 1972b). Therefore, the two proteins were functionally and structurally indistinguishable.

Meanwhile, the protein S1, the largest protein of the 30S *E. coli* ribosomal subunit, was implicated in the binding of both synthetic and natural mRNAs to the ribosome (van Duin and Kurland 1970; van Duin and van Knippenberg 1974; van Knippenberg et al. 1974). In parallel, the RNA binding has been invoked as the mechanism of action of translational interference factor *i* (Jay and Kaempfer 1974a,b; Miller and Wahba 1974). Finally, the factor *i*, the replicase host subunit I, and the protein S1 were identified to be the same protein by polypeptide sequence, as well as by immunological and functional criteria (Hermoso and Szer 1974; Inouye et al. 1974; Wahba et al. 1974).

As Kamen (1975) noted in his *RNA Phages* chapter, the identification of subunit I and factor *i* with the ribosomal protein S1 clarified a point of confusion in the replicase literature. The interference factor *i* was thought to inhibit ribosome binding at the Qβ coat gene (Groner et al. 1972a), while the factor *i* was implicated in the promotion of replicase binding to the same region of the viral RNA. The supposition that the interference caused by the factor *i* was actually the effect of the S1 competing with itself allowed one to understand both results in terms of the hypothesis that the protein S1 acted through recognition of specific RNA sequences, whether it was functioning as a ribosomal protein or as a replicase subunit. Therefore, the host protein S1 was required for the specific binding of the Qβ replicase to the Qβ plus strand RNA. This binding led to formation of the repressor complex II, one of two repressor complexes that ensured regulation of the phage replication. Both repressor complexes I and II are the central subjects of Chapters 16 and 17.

The interactions of Qβ replicase with Qβ RNA were investigated directly by treating replicase-Qβ RNA complexes with ribonuclease T1, and by characterizing enzyme-bound RNA fragments recovered by a filter binding technique (Weber et al. 1972, 1974; Meyer et al. 1975, 1981; Weber 1975). As a result, two internal binding sites were determined within the Qβ RNA. One region, or S-site, which was located at about 1250–1350 nucleotides from the 5′-end, overlapped with the initiation site for coat protein synthesis; this interaction was thought to be inessential for template activity but rather to be involved in the regulation of the protein synthesis. The other region, or M-site, which was located at about 2550 and 2870 nucleotides from the 5′-end, was essential for the template activity. The binding to this site was dependent on the presence of magnesium ions. More detail on the structure of both sites will be presented by the analysis of the repressor complex II in Chapter 16. The role of the repressor complex II by the regulation of the whole replication cycle will be described in Chapter 17. The nucleotide sequences of the RNA fragments from the two sites were determined and found to have no common features. Meanwhile, the replicase binding at the 3′-end of Qβ RNA could not be demonstrated, except when initiation of RNA synthesis was made it possible to occur in the presence of GTP and host factor. If instead of intact Qβ RNA, a complete ribonuclease T1 digest of Qβ RNA was allowed to bind to replicase, the oligonucleotides from the S-site and the M-site, as well as oligonucleotides from a region close to the 3′-end, were found to have the highest affinity to the enzyme (Meyer et al. 1981). The M-site was regarded as the specific template recognition site since Schwyzer et al. (1972) showed that 3′-terminal fragments of Qβ RNA were competent as replicase templates, provided they were longer than 50% of the total RNA length—namely, they retained the M-site.

The finding that the Qβ replicase recognized two regions within the central RNA portion was consistent with the flowerlike secondary and potential complex tertiary structure of the phage RNAs, as described in Chapter 10. Thus, the beginning of the coat gene could be in very close spatial proximity to the very beginning of the replicase gene. It would thus be possible for a single replicase molecule to simultaneously bind sequences from both sections of the putative RNA binding sites.

It was established clearly that the function of the subunit I was consistent with the stabilization of the binding complex between the replicase and its Qβ plus strand template, since the Qβ replicase lacking subunit I retained capability to the phosphodiester bond formation. The subunit I-deficient enzyme functioned normally with all templates except Qβ plus strand RNA and, even with the latter template, chain elongation and termination were unaltered (Kamen et al. 1972; Kamen 1975). The replicase subfragment comprising only subunits I and II (Kamen 1970) retained the binding specificity of the complete enzyme (Weber et al. 1972). It was concluded that the second subunit involved in specific Qβ RNA template binding was therefore the phage-specific polypeptide II, whereas subunits III and IV were not essential for viral RNA template binding and they did not detectably bind to RNA (Kondo et al. 1970; Landers et al. 1974; Kamen 1975). Since templates other than the Qβ plus strand, e.g., the Qβ minus strand, could be utilized by subunit I-deficient replicase, it appeared therefore that the subunit II alone could be responsible for specific binding with these RNAs.

The Qβ replicase subunits III and IV were identified as the protein synthesis elongation factors EF-Tu and EF-Ts (Blumenthal et al. 1972). The subunit IV and EF-Ts showed identical mobilities on SDS- or pH 4.5 urea-polyacrylamide gels, had a common N-terminal amino acid sequence, were precipitated by antibody to EF-Ts, and functioned in the GDP exchange assay for EF-Ts. The subunit III and EF-Tu had identical molecular mass and common antigenic determinants, shared extremely high affinity for guanosine nucleotides, and bound Phe-tRNA. The equimolar quantities of the complete Qβ replicase or replicase-derived subunits III and IV were able to replace the EF-TuTs complex in stimulating polyU-dependent polyphenylalanine synthesis (Landers et al. 1974). The functional equivalent of EF-TuTs and the subunits III and IV in RNA synthesis was established (Blumenthal et al. 1972) using a refinement of the reconstitution assay developed by Kamen (1970). The recovery of the phage Qβ replicase, EF-Tu and EF-Ts activities was achieved after denaturation in 8 M urea (Blumenthal and Landers 1976). The affinity of the subunits III and IV to the subunits I and II was measured by Blumenthal et al. (1976).

Both subunits III and IV were shown to be necessary for RNA chain initiation, and EF-Tu and EF-Ts purified from uninfected cells were able to replace the subunits III and IV by the reconstitution of the Qβ replicase's polyC-dependent activity.

As known, the function of EF-TuTs in protein synthesis consisted of the binding of aminoacyl-tRNA to ribosomes by protein elongation (Lucas-Lenard and Lipmann 1971). Surprisingly, within the Qβ enzyme they participated in the initiation process. It was suggested that EF-Tu bound tightly to GTP but not to other nucleoside triphosphates, since the Qβ replicase could only initiate product chains with GTP. The EF-Tu therefore supplied the initiating nucleotide as a "primer" whose 3'-OH was extended by the replicase subunit II. The EF-Ts might then catalyze the release of the new 5'-end of the product molecule from the replicase, allowing chain elongation to proceed (Kamen 1975).

Generally, the subunits III + IV were released from the Qβ replicase complex, while the subunits I + II were complexed with the Qβ RNA (Kondo et al. 1970; Landers et al. 1974). Thus, the Qβ replicase, consisting primarily of subunit II, was still able to elongate previously initiated copoly(U,G) chains, at the normal elongation rate but could not initiate new chains (Landers et al. 1974). Therefore, the EF-TuTs complex was not essential for the RNA chain elongation with at least synthetic polymer templates and the guanosine nucleoside, but not aa-tRNA, while the EF-Tu binding site was probably required for the RNA chain initiation by the Qβ replicase. The EF-Tu was the only subunit of the replicase, which was able to bind ppGpp, or guanosine 5'-diphosphate, 2' (3') diphosphate (Blumenthal et al. 1972). The subunits III and IV were qualified definitively as the nucleotide binding site of the Qβ replicase by Hori (1974).

It is noteworthy that prior to the identification of replicase subunits III + IV as the EF-TuTs, Travers et al. (1970a,b) had reported that at least one of these subunits was a protein that was called ψr, which stimulated the transcription of ribosomal RNA by the *E. coli* DNA-dependent RNA polymerase. The involvement of the ψr in the rRNA synthesis has since been challenged (Haseltine 1972). The interest in this question from the RNA phage side was initiated by the fact that the phage-specific RNA synthesis competed in infected cells with the synthesis of ribosomal RNA, as explained in the Phage effect on host section in Chapter 4. However, Travers (1973) subsequently reaffirmed that the EF-Tu did indeed play a critical role in ribosomal RNA synthesis. Then, Travers and Buckland (1973) reported the detection of the RNA polymerase-EF-Tu complexes *in vivo*. As Kamen (1975) noted, as one of the direct players in this story, the distinct possibility remained that the EF-TuTs complex was normally part of both the transcriptional and translational systems in uninfected bacteria.

Participation of the elongation factor EF-Ts was confirmed by the presence of the *ts*-mutated EF-Ts within the Qβ replicase purified from the mutant *E. coli* strain HAK88 carrying an altered thermolabile EF-Ts version (Kuwano et al. 1973, 1974; Hori et al. 1974).

The RNA replicase of the group IV phage SP had the same structure as the Qβ replicase (Mori et al. 1978). The enzyme was found to be composed of four non-identical polypeptides, i.e., subunits I, II, III, and IV, with molecular mass of 74,000, 69,000, 47,000, and 36,000 daltons, where the largest polypeptide was identical with the ribosomal protein S1, and subunits III and IV with polypeptide chain elongation factors EF-Tu and EF-Ts, respectively. Intriguingly, while the binding activity of the S1 to polyU was retained, the GDP binding activity of the EF-Tu was masked in the SP replicase. It was concluded that the participation of the S1 in the SP replicase was required functionally, but the EF-Tu and EF-Ts factors played the structural role only.

EF Tu-Ts

The subunit relationships within the Qβ replicase were determined first by intramolecular crosslinking (Young and Blumenthal 1975). The treatment of the Qβ enzyme, consisting of the four subunits of molecular mass 70,000, 65,000, 45,000, and 35,000 with protein crosslinking reagents, resulted in the formation of the three covalently bound complexes of molecular mass 215,000, 135,000, and 80,000. The 215,000 molecular mass complex was composed of one each of the replicase subunits, while the 135,000 complex was composed of the two larger subunits and the 80,000 complex was made up of the two smaller subunits by crosslinking these two subunits in the absence of the larger pair. The increasing ionic strength stabilized the large complex at the expense of the two smaller complexes. The presence of the stoichiometric amounts of Qβ RNA during crosslinking dramatically reduced formation of the large complex, but other natural and synthetic RNAs reduced the formation of this complex to a lesser extent.

The elongation factor Tu-Ts complex, when covalently crosslinked with dimethyl suberimidate, was able to reconstitute functional Qβ replicase, although it lacked the ability to catalyze the known host functions catalyzed by the individual elongation factors (Brown and Blumenthal 1976b). The sample of the Qβ replicase with crosslinked EF-TuTs replacing the individual elongation factors lacked EF-Tu and EF-Ts activities but could initiate transcription of both polyC and Qβ RNA normally and had approximately the same specific activity as the control enzyme. Remarkably, the denatured Qβ replicase formed with the crosslinked EF-TuTs was found to renature much more rapidly than untreated enzyme and, in contrast to the normal replicase, its renaturation was not inhibited by GDP. Moreover, the aminoacyl-tRNA binding site of EF-Tu was not required for the Qβ replicase activity, since the latter was not affected by inhibitors of the aminoacyl-tRNA binding activity of EF-Tu (Brown and Blumenthal 1976a). Thus, when Qβ replicase was treated with kirromycin, an antibiotic which modified EF-Tu activity, the Qβ RNA replication activity was only slightly affected, while the protein synthetic activity of the EF-Tu in the replicase complex was eliminated. The treatment of the pure EF-Tu with kirromycin, however, prevented it from functioning in the renaturation of the Qβ replicase (Brown and Blumenthal 1976a). Remarkably, kirromycin induced a similar conformational change in EF-Tu as the EF-Ts did, thereby opening the guanine nucleotide binding site, and increased tenfold the trypsin cleavage rate of the EF-Tu (Blumenthal et al. 1977). The trypsin-cleaved EF-Tu still could bind GDP and EF-Ts and could function in the Qβ replicase, but it was no longer spontaneously renatured following denaturation in urea.

The guanine nucleotide binding site of the EF-Tu was blocked in the "cross-linked" Qβ replicase that contained the covalently linked EF-TuTs complex (Blumenthal 1977). The initiation of transcription of polyC by both native and "cross-linked" replicase was competitively inhibited by high concentrations of GDP and ppGpp, while elongation of pre-initiated polynucleotide chains was not. When Qβ replicase was denatured with 8 M urea, the rate of its renaturation following dilution in a high-salt buffer was inhibited by relatively low concentrations of GDP, ppGpp, and GTP, but in this assay the "cross-linked" replicase was not inhibited by guanine nucleotides. Similarly, inactivation of the Qβ replicase, but not of the "cross-linked" replicase, upon incubation at 30°C was accelerated by low concentrations of GDP, ppGpp, and GTP. Thus, the guanine nucleotides inhibited renaturation of denatured replicase and caused inactivation of the Qβ replicase by interacting at the binding site on the EF-Tu. On the other hand, the inhibition of the transcription initiation by the guanine nucleotides involved a different site, presumably the site at which GTP binds to initiate transcription (Blumenthal 1977).

Therefore, the results gained by Thomas Blumenthal's team demonstrated clearly that the elongation factors EF-Tu and EF-Ts functioned in the Qβ replicase as a complex, rather than the individual polypeptides, and did not perform their known protein biosynthetic function in the RNA synthetic reaction. This work was reviewed exhaustively by the authors (Blumenthal et al. 1978; Blumenthal and Carmichael 1979), and the essential methodological details of the approach were presented (Blumenthal 1979).

This technique made it possible to replace the endogenous EF-Tu in the purified Qβ replicase with the EF-Tu from a variety of sources with wild-type and mutant *tufA* and *tufB* genes that both code for the EF-Tu polypeptides which appeared to be identical, or nearly so (Blumenthal et al. 1980). Thus, the EF-Tu was purified from strains containing (i) a wild-type *tufA* gene only, (ii) a kirromycin-resistant mutant *tufA* gene only, and (iii) a kirromycin-resistant mutant *tufA* gene and a mutant *tufB* gene which coded for EF-Tu that did not bind ribosomes. When each of these EF-Tu preparations was inserted into the Qβ enzyme, the wild-type *tufA* gene product and the *tufB* gene product functioned apparently normally, but the kirromycin-resistant *tufA* gene product caused the formation of an altered enzyme. The Qβ replicase containing kirromycin-resistant EF-Tu was unstable and was rapidly inactivated in the reaction mixture, but this property resulted in an apparent increase in template specificity: while the wild-type Qβ replicase transcribed polyC and other synthetic RNA species, the mutant enzyme did so only in the presence of Mn^{2+}. These data made it possible to acknowledge that the EF-Tu was involved in maintenance of the enzyme structure, which, in turn, was implicated in the template specificity (Blumenthal et al. 1980; Van der Meide et al. 1980). Moreover, the elongation factors from *Caulobacter crescentus* were able to substitute the *E. coli* elongation factors in the Qβ replicase and ensure its apparently normal RNA synthetic activity, although significant differences were found in the molecular mass of EF-Ts and relative affinities of guanine nucleotides, sensitivity to trypsin cleavage, and rate of heat denaturation of EF-Tu (Stringfellow et al. 1980). Furthermore, the Qβ replicase containing the *Bacillus stearothermophilus* elongation factor EF-Ts, tightly complexed with *E. coli* EF-Tu, was fully active in the transcription of a variety of templates, although the homologous *B. stearothermophilus*

EF-TuTs complex was not effective in the reconstitution of the Qβ enzyme (Stringfellow and Blumenthal 1983). Later, the *E. coli* elongation factor EF-Ts within the Qβ replicase molecule was replaced by its homolog from the thermophilic bacterium *Thermus thermophilus* (Vasiliev et al. 2010).

To determine more directly whether the Qβ replicase activity required the tRNA binding site of the EF-Tu, the enzyme was reconstituted with EF-Tu • GTP covalently cross-linked to aminoacyl-tRNA (Guerrier-Takada et al. 1981). The crosslinked EF-Tu was incorporated into the Qβ replicase as extensively as EF-Tu and restored the Qβ enzyme activity at least as well as an equivalent amount of EF-Tu. These results showed clearly that the aminoacyl-tRNA binding site on the EF-Tu was not required for the assembly or activity of the Qβ enzyme.

The gene engineering methodology offered novel strategies to elucidate the Qβ replicase complex. Thus, Fukano et al. (2002) aimed to construct an autocatalytic mRNA amplification system (this will be described exhaustively in Chapter 18) in the rabbit reticulocyte cell-free system using α-less replicase, which could replicate some small RNAs, despite lacking the S1 protein. In this system, the autocatalytically amplifiable mRNA could be constructed by inserting the replicase gene into a small RNA. To do this, the three mRNAs, encoding the phage-specific replicase subunit, EF-Tu, and EF-Ts, were co-translated in the rabbit reticulocyte cell-free system, but no replicase activity was detected in the reaction solution although all of the components were expressed. However, when EF-Tu and EF-Ts were fused to give an EF-Ts-EF-Tu chimera, the active α-less replicase was successfully obtained (Fukano et al. 2002).

Then, a wild-type *E. coli* strain with two plasmids, one encoding EF-Tu with a His-tag and another encoding the Qβ viral replicase subunit, was generated (Mathu et al. 2003). After confirmation that this system also yielded active Qβ replicase complexes, a set of mutations interfering with the tRNA binding properties of EF-Tu was introduced. A successful separation of the two possible types of complexes, wild-type and mutant, was achieved by affinity chromatography and a direct link between the vital tRNA binding capacities of these mutant EF-Tu species and the *in vitro* activities of the corresponding Qβ replicase complexes was established. This delicate approach made it possible to conclude that the tRNA binding properties of EF-Tu were of vital importance for the Qβ replicase activity with a direct correlation between the affinity for tRNA of the EF-Tu species used and the *in vitro* activity of the resulting Qβ replicase complex (Mathu et al. 2003).

Considering that the four subunits of Qβ replicase holoenzyme must be in equivalent molar ratio of the subunits, and that the synthesis of the S1 and EF-Ts in cells is relatively low, Wang (2004) performed coexpression of the phage-specific β-subunit with the other three subunits to determine whether the availability of the host subunits was the contributing factor for the low solubility of the overproduced Qβ replicase when it was expressed alone. Among the different combinations of coexpression experiments, the solubility was found to increase slightly when the β-subunit was coexpressed together with the EF-TuTs. Facing the same problem of the insolubility of the overexpressed β-subunit, Kita et al. (2006) elaborated two procedures to overcome it. First, both EF-Tu and EF-Ts subunits were expressed together with the β subunit, as described above. Second, the β subunit was genetically fused with EF-Tu and EF-Ts and the fused protein, a single-chain α-less Qβ replicase, was found mostly in the soluble fraction and could be readily purified (Kita et al. 2006).

Karring et al. (2004) presented the first EF-TuTs pair ever described that was completely inactive in the Qβ polymerase complex despite its almost full activity in protein synthesis. The phage resistance correlated strongly with an observed instability of the mutant EF-TuTs complex that was inspired by the lack of the coiled-coil motif in the mutant EF-Ts. It was proposed that the role of EF-Ts in the Qβ replicase was to control and trap EF-Tu in a stable conformation with affinity for the RNA templates while unable to bind aminoacyl-tRNA.

It is noteworthy that the presence of the *E. coli* elongation factors within the Qβ replicase inspired the search for the homologous elongation factor EF1 of plant origin within the turnip yellow mosaic virus RNA replicase, but the fine hypothesis was not confirmed (Pulikowska et al. 1981). Later, it was found that the RNA-dependent RNA polymerases (RdRps) from plant viruses, such as brome mosaic virus (Quadt et al. 1993) and tobacco mosaic virus (Osaman and Buck 1997; Yamaji et al. 2006) formed replicative complexes with the eukaryotic translation factor eIF-3. The RdRps from animal viruses, such as poliovirus (Harris et al. 1994), vesicular stomatitis virus (Das et al. 1998), and bovine viral diarrhea virus (Johnson et al. 2001) formed replicative complexes with eEF-1A, a eukaryotic counterpart of EF-Tu, or eEF-1A and 1B, a eukaryotic counterpart of EF-Ts. The participation of the components of the host cell translation machineries in the transcription and replication of viral RNA genomes was reviewed at that time by Lai (1997). Furthermore, it was found that the eEF-1A binding to the aminoacylated viral RNA repressed minus strand synthesis by the RNA-dependent RNA polymerase of turnip yellow mosaic virus and therefore coordinated the competing translation and replication functions of the genomic RNA (Matsuda et al. 2004).

More recently, the functions of eEF-1A as a protein chaperone and a virus replication cofactor were reviewed exhaustively and clearly highlighted the role of eEF-1A in the viral transcription, translation, assembly, and pathogenesis of a diverse group of viruses, from human immunodeficiency type 1 and West Nile virus to tomato bushy stunt virus (Li et al. 2013).

RIBOSOMAL PROTEIN S1

The protein S1 appeared as an obligate participant of the translation process including translation of the RNA phage messenger. Thus, the translation of the phage MS2 RNA *in vitro* depended strictly on the presence of S1 (van Dieijen et al. 1975). The S1 protein was uniquely involved in the initiation of protein synthesis on the native MS2 RNA, while

mildly unfolded phage RNA functioned as the template in the complete absence of S1 (van Dieijen et al. 1976). It was proposed at this stage of studies that the S1 was a part of a site of the 30S ribosome, also including the 3'-end of the 16S RNA and the initiation factor IF3, which functioned exclusively in the recognition of the native MS2 and Qβ RNAs, among other native messenger RNAs (van Dieijen et al. 1976, 1977). The involvement of the S1 in the assembly of the initiation complex was fully acknowledged when anti-S1 antibodies were involved in the study (van Dieijen et al. 1978). The breakthrough methodology of the approach used was presented in detail by van Duin et al. (1979).

Meanwhile, the two polynucleotide binding sites were detected on the S1 protein: the site I recognized either single-stranded DNA or RNA and did not discriminate between adenine- and cytidine-containing polynucleotides, while the site II binding was highly specific for RNA over DNA and showed a marked preference for cytidine polynucleotides over the corresponding adenine-containing species (Draper et al. 1977). Every S1 molecule carried both the site I and the site II. The authors hypothesized that one or both of the S1 binding sites could therefore be involved in facilitating recognition of the appropriate RNA sequences, including those involved into the regulatory function of the Qβ replicase within the repressor complex II (see Chapters 16 and 17). Remarkably, the S1 protein, when associated with the ribosome 30S subunit, did not interact with 16S ribosomal RNA, but its binding was determined mostly by protein–protein interactions, while it crosslinked efficiently to the Qβ RNA (Boni et al. 1982).

In turn, Goelz and Steitz (1977) found that the S1 protein was able to recognize two sites on Qβ RNA. A bit earlier, Senear and Steitz (1976) detected an oligonucleotide spanning residues −38 to −63 from the 3' terminus of the Qβ RNA as an S1 binding site, but Goelz and Steitz (1977) reported the strong interaction of the S1 with a second oligonucleotide in Qβ RNA, which was derived from the region recognized by Qβ replicase just 5' to the Qβ coat protein cistron.

The modified S1 species, whose properties in nucleic acid binding and protein synthesis had been previously established, were used for the reconstitution of the Qβ replicase lacking S1 (Chu et al. 1982; Cole et al. 1982). The S1 protein derivatized with N-ethylmaleimide reconstituted the modified replicase that was 81% as active as the replicase reconstituted with unmodified S1. These results indicated that the helix-unwinding properties of the S1, which were known to be inactivated by N-ethylmaleimide modification, were not required for the Qβ RNA replication and that the sulfhydryl-derivatized region of the S1 was not utilized in the replicase subunit contacts. In contrast with its established ability to replace *E. coli* S1 on the ribosome, the *Caulobacter crescentus* S1 did not reconstitute the Qβ RNA transcription activity.

Guerrier-Takada et al. (1983) looked for the activity of the discrete S1 fragments in the Qβ replicase function. By the reconstitution of the enzyme, the N-terminal region of S1 was required for the S1 subunit interactions within the replicase, since a trypsin-resistant fragment lacking the N-terminal 31%

of S1 was functionally inactive, but the C-terminal region of S1 was dispensable for the Qβ replicase function because a mutant of S1 lacking 21% of the C-terminal portion of the chain was as active as the wild-type S1 within the Qβ replicase. The structure and functions of the S1 protein, including its role within the Qβ replicase, had been reviewed in detail by that time (Subramanian 1983).

When RNA-protein interactions between the Qβ plus strand RNA and the components of the Qβ replicase system were studied by deletion analysis and electron microscopy, the formation of looped complex structures resulting from simultaneous interactions with the Qβ replicase at the S-site and the M-site was also observed when the purified protein S1 only was used (Miranda et al. 1997). It was suggested therefore that these internal interactions of the Qβ replicase with the Qβ RNA were mediated by the S1 protein.

It appeared later that the S1 protein, together with the host factor HF, or Hfq (see below), copurified in molar ratios with DNA-dependent RNA polymerase of *E. coli* (Sukhodolets and Garges 2003). The purified S1 associated independently with the RNA polymerase, and the Hfq binding to polymerase occurred in the presence of S1. The protein S1 was capable of significant stimulation of the RNA polymerase transcriptional activity from a number of promoters; the stimulatory effect was observed on linear as well as supercoiled DNA templates. Therefore, the S1 protein functioned at the border of transcription and translation in uninfected cells. Furthermore, the S1 protein was identified as a potent activator of transcriptional cycling *in vitro* (Sukhodolets et al. 2006). The authors supposed that *in vivo*, the cooperative interaction of multiple RNA-binding modules in the S1 might enhance the transcript release from the RNA replicase, alleviating its inhibitory effect and enabling the core enzyme for continuous re-initiation of replication.

Although it was commonly believed that the S1 protein binds to a template RNA at the initiation step in both protein and RNA syntheses and is not involved in later events, Vasilyev et al. (2013) showed that in the Qβ replicase-mediated RNA synthesis, the S1 protein functions at the termination step by promoting release of the product strand in a single-stranded form. This function of a terminator is fulfilled by the N-terminal fragment comprising the first two S1 domains. In parallel, Duval et al. (2013) found that the S1 protein endowed the ribosomal 30S subunit with an RNA chaperone activity that was essential for the docking and the unfolding of structured mRNAs, and for the correct positioning of the initiation codon inside the decoding channel. The first three OB, or oligonucleotide/oligosaccharide-binding, fold domains of the S1 protein retained all its activities on the 30S subunit. Remarkably, the S1 was not required for all mRNAs and acted differently on mRNAs according to the signals present at their 5′-ends. The taxonomic distribution, repeats, and functions of the S1 domain-containing proteins as members of the OB-fold family were analyzed by Deryusheva et al. (2017). A short but definitive review on the contribution of the S1 protein to the structure and function of the Qβ replicase was published recently by Kutlubaeva et al. (2017).

FINE STRUCTURE OF THE Qβ REPLICASE

The antibodies were used as the first specific probes to elucidate the roles of the host polypeptides by the *in vitro* replication of Qβ RNA by the Qβ replicase as well as by the polyC-directed synthesis of polyG by this enzyme. Thus, Carmichael et al. (1976) performed the first immunochemical analysis of the Qβ replicase complex by rabbit gamma-globulins specifically directed against each of the bacterial host proteins required for the *in vitro* replication of Qβ RNA: elongation factors EF-Tu and EF-Ts, the 30S ribosomal protein S1, and the host factor HFI, or Hfq. Interestingly, the polyC-directed reaction was much less sensitive to the antibody inhibition than was the Qβ RNA-directed reaction.

The first three-dimensional model of the Qβ replicase was constructed by electron microscopy studies using a negative staining technique (Berestowskaya et al. 1988). The Qβ replicase complex appeared as a compact structure having a size of 100 ± 10 Å. which was subdivided into the two unequal bipartite subparticles. The important conclusion was made that all the constituent subunits, including the protein Sl, acquired a globular conformation when associated in the replicase complex. It was especially intriguing, since hydrodynamic and small-angle x-ray scattering studies of the isolated Sl protein had led to the conclusion that it was a highly elongated molecule with an axial ratio of 1:10, its length of 210–280 Å being comparable with the maximal ribosome dimension (Subramanian 1983). It was thought that the S1 protein retained its elongated shape upon binding to the ribosome.

The crystal structure of the Qβ replicase complex formed by the subunit II and EF-TuTs was resolved at 2.5 Å resolution (Kidmose et al. 2010). Whereas the basic catalytic machinery in the viral subunit appeared similar to other RNA-dependent RNA polymerases, a unique C-terminal region of the subunit I was engaged in extensive interactions with EF-Tu and might contribute to the separation of the transient duplex formed between the template and the nascent product to allow exponential amplification of the phage genome. It was noted by the authors that the evolution of resistance by the host appeared to be impaired because of the interactions of the subunit II with parts of EF-Tu essential in recognition of aminoacyl-tRNA. Generally, this was the first structure of an RNA-dependent RNA polymerase (RdRp) complex consisting of virus- and host-encoded proteins. The structure revealed the nice mechanism by which the virus elegantly takes advantage of the normal function of EF-Tu in translation to replicate its own genome.

The three-dimensional structures are reprinted here from Kidmose et al. (2010). Figure 13.3 presents the Qβ replicase core complex.

Figure 13.4 shows intermolecular contacts within the Qβ replicase, which are established by hydrogen bonds and electrostatic interactions.

Figure 13.5 presents the RNA binding and surface properties of the subunit II.

In addition, the chimeric Qβ replicase was produced in *E. coli* cells by coexpression of the genes encoding the subunit

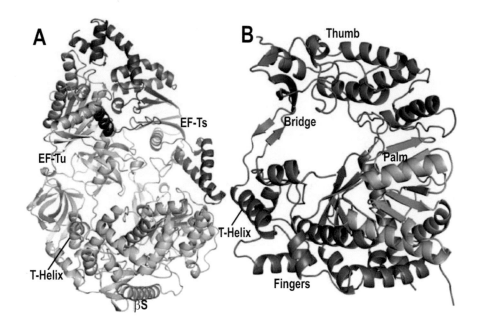

FIGURE 13.3 Structure of the Qβ replicase core complex. (A) Cartoon representation of the monomeric core replicase containing β-subunit (green, labeled "βS") in complex with EF-Tu (yellow) and EF-Ts (dark red). (B) The domains of the β-subunit: the thumb domain is colored blue, the palm domain green, the fingers domain magenta, and the bridge region gray. (Kidmose RT et al. Structure of the Qβ replicase, an RNA-dependent RNA polymerase consisting of viral and host proteins. *Proc Natl Acad Sci U S A*. 107:10884–10889, Copyright 2010 National Academy of Sciences, USA.)

II and thermophilic EF-Ts from *Thermus thermophilus*, isolated and crystallized for further three-dimensional resolution (Vasiliev et al. 2010).

In parallel, the crystal structure of the core Qβ replicase, comprising the subunit II and EF-TuTs, was determined at 2.8 Å resolution for the calcium ion-bound form (Ca-form), and at 3.2 Å resolution for the magnesium ion-bound form (Mg-form) (Takeshita and Tomita 2010).

Figure 13.6 presents the overall structure of the Qβ replicase by Takeshita and Tomita (2010).

The subunit II has a right-handed structure, and the EF-TuTs binary complex maintains the structure of the catalytic core crevasse of the subunit II through hydrophobic interactions, between the finger and thumb domains of the subunit II and domain-2 of EF-Tu and the coiled-coil motif of EF-Ts, respectively. These hydrophobic interactions are required for the expression and assembly of the Qβ replicase complex. Thus, EF-Tu and EF-Ts have chaperone-like functions in the maintenance of the structure of the active Qβ replicase. Figure 13.7 shows the structural differences between EF-Tu and EF-Ts within the Qβ replicase and in the EF-TuTs binary complex, as was resolved earlier at 2.5 Å by Kawashima et al. (1996).

Figure 13.8 demonstrates the structure of the subunit II alone. The structure of the core Qβ replicase revealed two distinct tunnels leading to the catalytic site of the subunit II. One tunnel allows the access of an incoming nucleotide, and the other is for the access of the template RNA to the catalytic site.

Figure 13.9 shows the RNA elongation model by the Qβ replicase. The modeling of the template RNA and the growing RNA in the catalytic site of the Qbeta replicase structure

also suggested that the structural changes of the RNAs and EF-TuTs should accompany processive RNA polymerization and that the EF-TuTs within the Qβ replicase could function to modulate the RNA folding and structure. Concerning the location of the S1 protein, the latter might bind to the bottom of the boat-like structure of the core Qβ replicase and recruit the 3'-part of the genomic RNA to the positively charged area near the template tunnel entrance. The binding of the S1 protein to specific internal sites of the viral genomic RNA would facilitate the initiation of the replication and transcription of the viral genomic RNA (Takeshita and Tomita 2010).

Next, Takeshita and Tomita (2012) presented a pattern for the RNA polymerization by the Qβ replicase. It was revealed that at the initiation stage, the 3'-adenine of the template RNA provides a stable platform for the *de novo* initiation. The EF-Tu within the Qβ replicase forms a template exit channel with the subunit II. At the elongation stages, the C-terminal region of the subunit II, assisted by the EF-Tu, splits the temporarily double-stranded RNA between the template and nascent RNAs before translocation of the single-stranded template RNA into the exit channel. Therefore, the EF-Tu within the Qβ replicase modulates RNA elongation processes in a distinct manner from its established function in protein synthesis. Figure 13.10 demonstrates the structure of the initiation stage.

The next in-depth illustrations unveiled the fine structures of the primary elongation stages, the separation and translocation of RNAs, and template RNA channels (Takeshita and Tomita 2012).

Moreover, Takeshita et al. (2012) solved the problem of the template-independent 3'-adenylation activity, which was

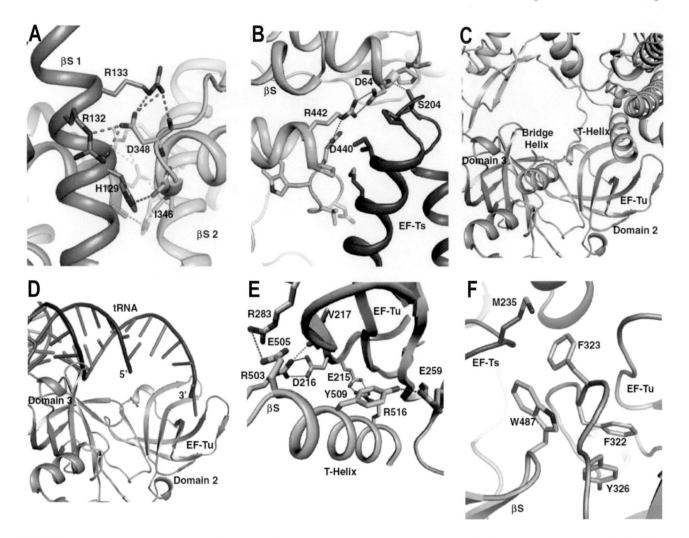

FIGURE 13.4 Intermolecular contacts within the Qβ replicase. Hydrogen bonds and electrostatic interactions are shown as dashed lines. (A) Details of the β-subunit (βS): β-subunit interface. (B) Contacts between the EF-Ts (dark red) coiled-coil motif and the β-subunit (green) thumb domain. (C) The interface between the β-subunit and EF-Tu (yellow) involving the T helix and bridge helix from the β-subunit and domains 1 and 3 from EF-Tu. (D) The complex of EF-Tu and aa-tRNA in the same orientation as in c. (E) Close-up of the β-subunit T-helix interaction with EF-Tu domain 2. (F) Insertion of the β-subunit thumb domain between EF-Tu domain 3 and EF-Ts. (Kidmose RT et al. Structure of the Qβ replicase, an RNA-dependent RNA polymerase consisting of viral and host proteins. *Proc Natl Acad Sci U S A.* 107:10884–10889, Copyright 2010 National Academy of Sciences, USA.)

required for the efficient viral RNA amplification in the host cells. They determined the structure of a complex that included Qβ replicase, a template RNA, a growing RNA complementary to the template RNA, and ATP. The structure represented the terminal stage of RNA polymerization and revealed that the shape and size of the nucleotide-binding pocket became available for ATP accommodation after the 3′-penultimate template-dependent C-addition (Takeshita et al. 2012).

Then, Takeshita et al. (2014) analyzed the crystal structure of the Qβ replicase, consisting of the virus-encoded subunit II, EF-Tu, EF-Ts, and the N-terminal half of the S1 protein, which was still capable of initiating Qβ RNA replication. This N-terminal half preserved three of six contiguous OB folds of the S1 protein. The structural and biochemical studies revealed that the two N-terminal OB-folds of S1 anchored the S1 onto the subunit II, and the

third OB-fold was mobile and protruded beyond the surface of the subunit II. The third OB fold interacted mainly with a specific RNA fragment derived from the internal region of Qβ RNA, and its RNA-binding ability was required for the replication initiation of the Qβ RNA. Thus, the third mobile OB-fold of S1, which was spatially anchored near the surface of the β-subunit, primarily recruited the Qβ RNA toward the subunit II, leading to the specific and efficient replication initiation of the Qβ RNA, and the S1 functioned as a replication initiation factor, beyond its established function in protein synthesis (Takeshita et al. 2014). Figure 13.11 presents the full mechanism of the initiation of the negative strand Qβ RNA synthesis with the direct participation of the S1 protein.

The breakthrough achievements on the Qβ replicase structure and function were reviewed patently by the authors (Tomita 2014; Tomita and Takeshita 2014). It was emphasized

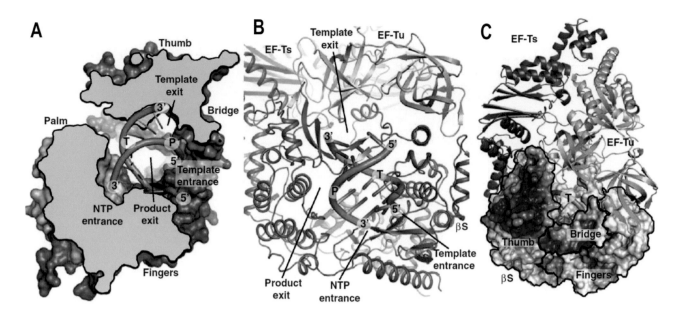

FIGURE 13.5 RNA binding and surface properties of the β-subunit. The double-stranded RNA in the internal cavity of the β-subunit (colored as in Figure 13.3b) was docked by comparison with the structure of the Norwalk virus RdRp, PDB ID code 3BSN (Zamyatkin et al. 2008). Template and product strand are labeled "T" and "P," respectively, and 3′ and 5′ ends are labeled as well. (A) Cross-section through the β-subunit displaying the suggested template and NTP entrance channels together with the template exit channel. (B) Cartoon representation of the core replicase indicating the four putative channels. (C) Surface representation displaying the electrostatic potential of the β-subunit colored according to charge (*Blue* 10 kt/e, *White* 0, *Red* −10 kt/e) with EF-Tu and EF-Ts shown in cartoon mode. The positively charged blue surface patch containing elements of the bridge and thumb domains of the β-subunit is a possible binding site for the S1 protein, template, and product. The thumb, bridge, and finger domains of the β-subunit are indicated on the surface representation with a black outline. (Kidmose RT et al. Structure of the Qβ replicase, an RNA-dependent RNA polymerase consisting of viral and host proteins. *Proc Natl Acad Sci U S A.* 107:10884–10889, Copyright 2010 National Academy of Sciences, USA.)

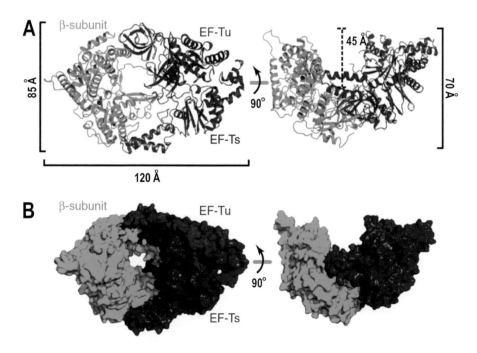

FIGURE 13.6 Overall structure of the Qβ replicase. (A) Ribbon and (B) surface models of Qβ replicase. The Qβ replicase adopts a boat-like structure. The β-subunit, EF-Tu and EF-Ts are colored green, red, and blue, respectively. (Takeshita D and Tomita K. Assembly of Qβ viral RNA polymerase with host translational elongation factors EF-Tu and -Ts. *Proc Natl Acad Sci U S A.* 07:15733–15738, Copyright 2010 National Academy of Sciences, USA.)

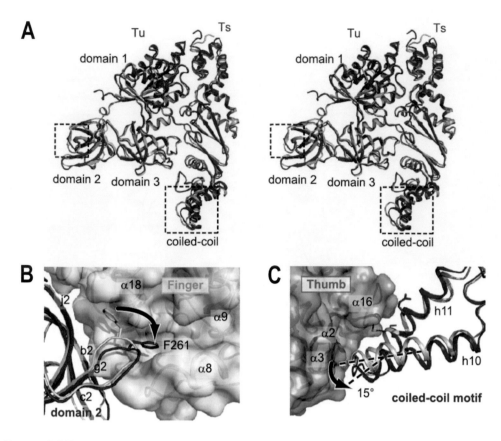

FIGURE 13.7 Structural differences between EF-Tu and -Ts. (A) A stereo-view comparison of the structures of EF-Tu:Ts in Qβ replicase (Tu and Ts are colored red and blue, respectively) and in the EF-Tu:Ts binary complex (Kawashima et al. 1996) (Tu and Ts are colored gray). The structural differences are observed in domain 2 of EF-Tu and a coiled-coil region in EF-Ts and are enclosed by boxes. (B and C) Detailed views of structural differences of EF-Tu and EF-Ts between the EF-Tu:Ts complex in Qβ replicase and the EF-Tu:Ts binary complex. The structures of EF-Tu and -Ts in the binary complex are colored gray. (Takeshita D and Tomita K. Assembly of Qβ viral RNA polymerase with host translational elongation factors EF-Tu and -Ts. *Proc Natl Acad Sci U S A*. 07:15733–15738, Copyright 2010 National Academy of Sciences, USA.)

specially that the most important questions concerning the mechanism of RNA polymerization by the Qβ replicase and the detailed functions of the EF-Tu, EF-Ts, and S1 had remained enigmatic for almost 50 years, since the first isolation of the Qβ replicase, due to the absence of structural information about the Qβ enzyme (Tomita 2014). Finally, in the last 5 years, the crystal structures of the core Qβ replicase, consisting of the β-subunit, EF-Tu and EF-Ts, and the structure of Qβ replicase comprising the β-subunit, EF-Tu, EF-Ts and the N-terminal half of S1, were reported.

A bit later, the roles of the subunit II and of the first two OB folds of the S1 protein in the recognition of the Qβ RNA plus strands by the Qβ replicase complex were deciphered independently by the use of x-ray crystallography, NMR spectroscopy, as well as sequence conservation, surface electrostatic potential, and mutational analyses (Gytz et al. 2015). Figure 13.12 presents the overall structure of the monomeric form of the Qβ replicase core complex bound to the first two OB folds of S1. Taken together, these data acknowledged that the subunit II and the protein S1 cooperatively bind the Qβ RNA plus-strands by replication initiation, and ensure template discrimination during replication initiation.

The construction of the gene-engineered Qβ enzyme structures that were used by the crystal structure investigations are presented in the Replicase gene section of Chapter 19.

Recently, Loveland and Korostelev (2018) presented the excellent electron cryomicroscopy data that explained how the S1 protein cooperates with other ribosomal proteins to form a dynamic mesh near the mRNA exit and entrance channels to modulate the binding, folding, and movement of mRNA.

HOST FACTORS

As stated briefly above, the maximal activity of the Qβ enzyme needed the presence of a host factor fraction in the case when the Qβ plus strands were used as a template (August and Eoyang 1967; Banerjee et al. 1967; Hori et al. 1967). The requirement for this fraction seemed to bear a quantitative relationship to the Qβ RNA but not to the Qβ enzyme (Franze de Fernandez et al. 1968; Franze de Fernandez and August 1968). The factor fraction was not required for enzyme activity when synthetic polymers (Hori et al. 1967), minus strands, or other RNA molecules (August et al. 1968) were used as templates. From the very beginning, two macromolecular

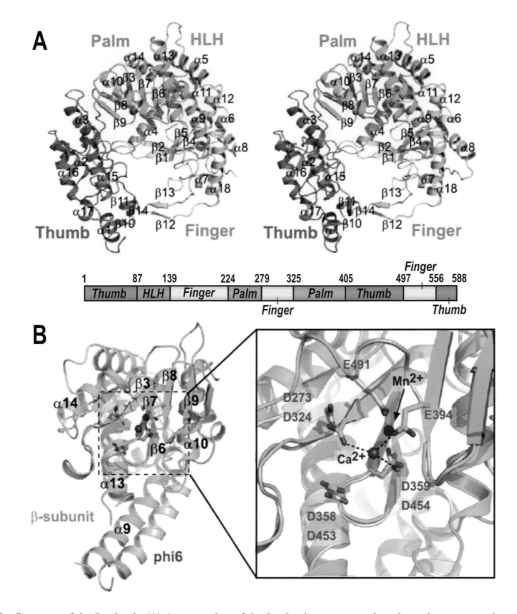

FIGURE 13.8 Structure of the β-subunit. (A) A stereo view of the β-subunit structure and a schematic representation of the β-subunit. The thumb, palm, and finger domains and the HLH, or long helix-loop-helix α4–loop–α5 region, are presented. (B) Superposition of the palm domain of the β-subunit (Ca-form, colored cyan) onto that of phi-6 RdRp (colored beige, PDB code: 1HHS (Butcher et al. 2001) (Left). A close-up view of the superposed catalytic core regions (Right). The carboxylates in the catalytic sites are depicted by stick models, and the corresponding amino acid residues are labeled in blue for the β-subunit and orange for phi-6 RdRp. The calcium ion in the structure of the β-subunit is colored brown. Only one metal ion was observed in the structure of the β-subunit of Qβ replicase. The manganese ion in the structure of phi-6 RdRp is colored blue. (Takeshita D and Tomita K. Assembly of Qβ viral RNA polymerase with host translational elongation factors EF-Tu and -Ts. *Proc Natl Acad Sci U S A.* 07:15733–15738, Copyright 2010 National Academy of Sciences, USA.)

factors, designated host factor I (HFI) and host factor II (HFII), were resolved (Shapiro et al. 1968). The HFI was found to be highly resistant to heat and to the treatment by proteases, while the HFII was heat-sensitive and completely inactivated when treated with pepsin (Shapiro et al. 1968). In the sense of the additional factors, Walter Fiers' team (Gillis et al. 1969) contributed with the sound argument that the necessity for the host factors might reflect the used purification procedure either by Eoyang and August (1968) or by Haruna and Spiegelman (1965c).

As summarized by Kuo et al. (1975), the HFI was purified finally from uninfected *E. coli* Q13 cells by liquid polymer phase fractionation, heating at 82°–85°C, DEAE-cellulose chromatography, QAE-Sephadex chromatography, and zone sedimentation to the approximately 3000-fold enrichment (Franze de Fernandez et al. 1972). Because the HFI was heat stable and retained substantial activity even after heating to 100°C, it was possible to remove contaminating proteins by heat treatment at an early step in the procedure without significant loss of activity. However, to determine whether heat

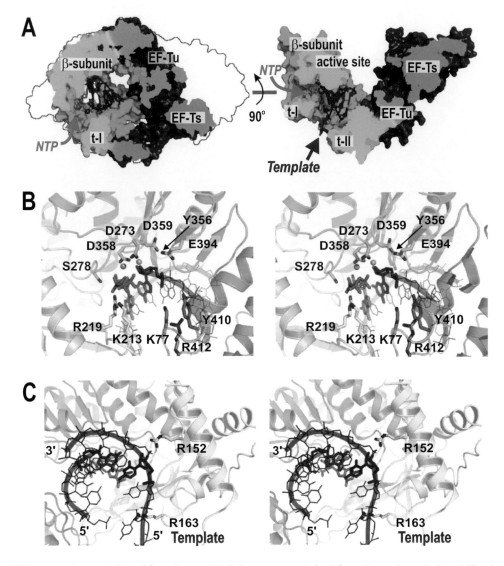

FIGURE 13.9 RNA elongation model by Qβ replicase. (A) Substrate tunnels in Qβ replicase (tunnels I and II), shown in cross sections. Tunnel-I (t-I) and tunnel-II (t-II) are for the access of incoming nucleotides and a single-stranded template RNA, respectively. The template RNA, the growing RNA, and the incoming nucleotide are colored blue, red, and orange, respectively, and are shown in stick models. (B and C) Detailed view of the catalytic site of Qβ replicase, representing the elongation stage of RNA polymerization, in A. The incoming nucleotide and the 3′-nucleoside of the growing RNA are shown as stick models in B. Two magnesium ions are modeled in B using the structure of Norwalk virus RdRp (colored purple, PDB code: 3BSO), representing the RNA elongation stage (Zamyatkin et al. 2008). The nucleoside at the 3′-end of the growing RNA and the nucleoside of the template RNA that base pairs with the nucleoside at the 3′-end of the growing RNA are shown in stick models in C. (Takeshita D and Tomita K. Assembly of Qβ viral RNA polymerase with host translational elongation factors EF-Tu and -Ts. *Proc Natl Acad Sci U S A.* 07:15733–15738, Copyright 2010 National Academy of Sciences, USA.)

treatment modified the physical properties of the factor, an alternate purification procedure, phosphocellulose chromatography, was substituted for the heat treatment (Hayward and Franze de Fernandez 1971). The properties of the HFI purified by these two different procedures appeared to be the same, and the specific activities of the final purified fractions were comparable. If the apparent molecular mass in the initial studies was measured by gel filtration (Hayward and Franze de Fernandez 1971) or by zone sedimentation (Franze de Fernandez et al. 1972) as 70,000–80,000, the molecular mass of the single polypeptide was resolved by SDS-polyacrylamide gel electrophoresis as 12,000–15,000 daltons.

In parallel, the HFI was identified by gel electrophoresis as a molecule of 12,500 daltons (Gillis et al. 1971; Van Emmelo et al. 1973). Therefore, the HFI appeared as a complex structure consisting of four to six polypeptide chains of the same molecular mass.

The Qβ RNA replication was completely dependent on the HFI. A small amount of RNA synthesis was observed after prolonged incubation in the absence of the HFI, but in this case the reaction product was 6S RNA (see Chapter 18), which could be replicated in the absence of factors (Banerjee et al. 1969b). The synthesis of this RNA was presumably due to contamination of the enzyme preparation with a small

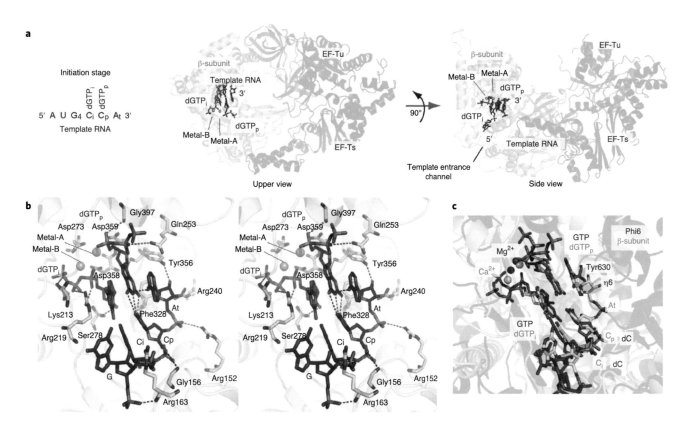

FIGURE 13.10 Structure of the initiation stage. (a) Overall structure of the initiation complex of core Qβ replicase, showing the β-subunit (yellow), EF-Tu (red), EF-Ts (blue), the template RNA (blue sticks) and 3′-dGTPs (red sticks). The sequence of the template RNA is shown. (b) Stereo view of the catalytic core of the β-subunit of the initiation complex, showing calcium ions (gray, metals A and B), 3′-dGTPs (red sticks) and the template RNA (blue sticks). (c) Superposition of the catalytic site of the initiation complex of Qβ replicase (yellow) onto that of φ6 RdRp (magenta; Butcher et al. 2001). RNAs and nucleotides are depicted by stick models. (With kind permission from Springer+Business Media: *Nat Struct Mol Biol.*, Molecular basis for RNA polymerization by Qβ replicase, 19, 2012, 229–237, Takeshita D, Tomita K.)

amount of 6S RNA since it also occurred when the Oβ RNA template was omitted from the reaction mixture (Kuo et al. 1975). As mentioned above, the HFI concentration required for the maximum rate of synthesis varied according to the amount of the Qβ RNA added to the reaction, but was independent of the enzyme concentration. In fact, one molecule of the HFI of the 75,000 molecular mass was required per molecule of the Qβ RNA and binding of the HFI to RNA was demonstrated, while this binding did not require the presence of other components required for the RNA synthesis (Kuo et al. 1975). Remarkably, the binding was not limited to the template RNA, since the HFI was able to bind to RNA of the phages R23 and f2, as well as to ribosomal RNA of *E. coli*, but no binding was observed with the double-stranded reovirus RNA or with the double-stranded or single-stranded DNA (Kuo et al. 1975).

The HFII was obtained by extraction of the crude cell extract with guanidine hydrochloride and perchloric acid, and the acid-soluble fraction then was purified by carboxymethyl-cellulose column chromatography and Sephadex gel filtration (Kuo 1971). The apparent molecular mass of the active HFII fractions ranged from 5000 to approximately 30,000. The HFII protein was basic and contained 25% arginine and lysine, and the average molar ratio of basic to acidic residues

was about 2. The basic proteins with the HFII activity were a major constituent of *E. coli*, comprising approximately 3% of the total proteins (Kuo and August 1971, 1972). It was concluded that many of the basic proteins of *E. coli* that were stimulating in the replication of the Qβ RNA *in vitro* were structural proteins of ribosomes.

The effect of changes in the HFII concentration on the Qβ RNA replication was complex. At low levels, the rate of synthesis was not proportional to the HFII concentration, suggesting the possibility of a cooperative effect. At higher concentrations, the HFII proteins were inhibitory (Kuo et al. 1975). Anyway, the requirement for the HFII in the reaction was related to the RNA concentration (Franze de Fernandez et al. 1968).

Because many different basic proteins of *E. coli* were active as the HFII, the effect of histones, protamine, and other basic proteins was also tested in the reaction and it was found that lysine-rich and arginine-rich calf thymus histones and also salmon sperm protamine sulfate completely substituted for the HFII required for the Qβ RNA replication (Kuo and August 1972). Nevertheless, the histones or protamine did not substitute for the HFI. Remarkably, the Qβ coat protein satisfied the requirement for the HFII in the Qβ replication, the maximal rate of synthesis was achieved with 16–20 molecules of coat protein per molecule of RNA (Kuo et al. 1975).

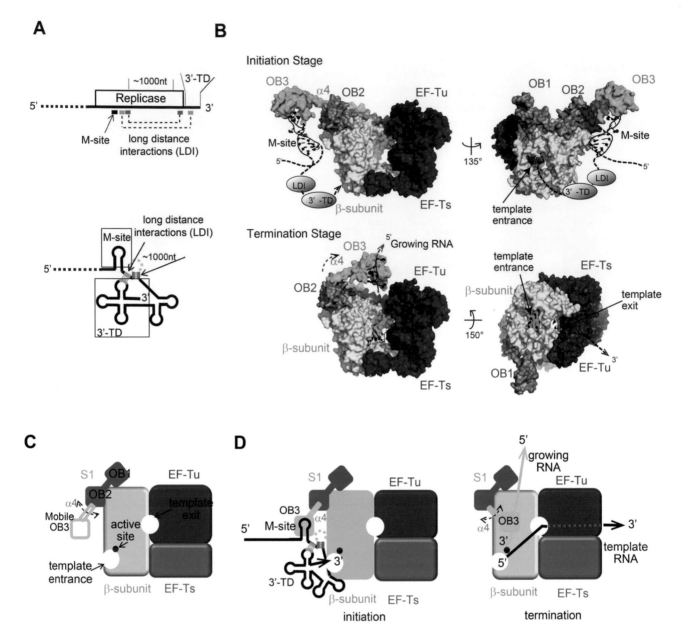

FIGURE 13.11 Mechanism of initiation of negative strand Qβ RNA synthesis. (A) One-dimensional simplified view of the 3′-half of Qβ RNA (above) and two-dimensional view of the 3′-half of Qβ RNA. The long-distance interactions (LDI) in the RNA (Klovins et al. 1998; Klovins and van Duin 1999) are depicted by dotted lines, and the regions used for interactions in the Qβ RNA are shown. 3′-TD stands for 3′-terminal domain. (B) Structure of Qβ replicase containing R1–3 domains of the S1 protein. The mobile R3 was modeled. The R3 is capable of pivotal rotation, using the α helix (α4) between OB2 and OB3 as a swing arm for both initiation and termination of stages. The template Qβ RNA at the replication initiation stage (upper) and the template and growing RNAs at the replication termination stage (lower) are modeled. LDI and 3′-TD in Qβ RNA were simplified. (C) Simplified cartoon of Qβ replicase containing the N-terminal half of S1. The β-subunit, EF-Tu, EF-Ts, and S1 (OB1 and OB2) and the mobile OB3 of S1 are depicted. (D) A simplified model of the initiation of negative strand RNA synthesis (left). R3 interacts with the M-site of the Qβ RNA. A simplified model of RNA synthesis termination (right). The interaction between the growing RNA and the mobile R3 of S1 at the termination stage triggers the release of the RNA product from the complex (Vasilyev et al. 2013). (Takeshita D, Yamashita S, Tomita K. *Nucleic Acids Res.* 2014, by permission of Oxford University Press.)

Moreover, the coat proteins from the phages f2 and R23 were also active in the role of the HFII, but at concentrations tenfold greater than that required for the Qβ coat protein (Kuo et al. 1975).

Concerning the specific role of the host factors in the Qβ replication, it was clear that it related to the Qβ RNA binding. The mechanism of the Qβ RNA binding during replication belonged to priorities of August's team (Silverman and August 1970; Silverman 1971). Thus, it was clarified first that the HFI was involved at some very early steps in the reaction (Franze de Fernandez et al. 1972). The problem of the Qβ RNA binding was further studied by Silverman (1973a,b). In these studies, the number of RNA sites bound by the Qβ replicase and the affinity of the enzyme for different RNA sites was

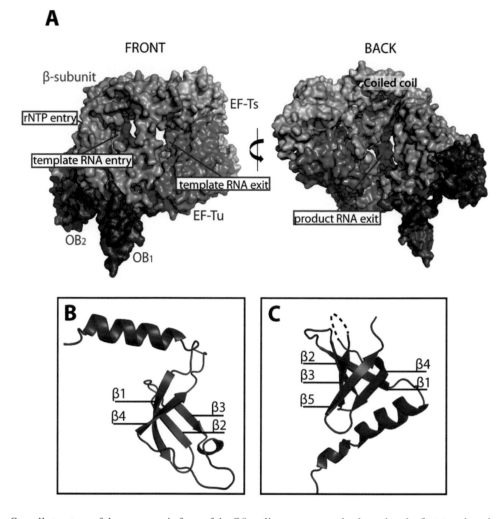

FIGURE 13.12 Overall structure of the monomeric form of the Qβ replicase core complex bound to the first two domains, OB1–2, of ribosomal protein S1. The OB1–2 domains bind at distant sites on the β-subunit and are connected by an extended 15 amino-acid linker region, which is only partly visible in the electron density. (A) Surface representation of the monomeric Qβ replicase complex. The following color code is applied: β-subunit (orange), EF-Tu (blue), EF-Ts (green), and OB1–2 (magenta). The front and back of the complex are related by an approximately 180° vertical rotation. (B) Structure of the OB1 domain illustrating the presence of a flexible N-terminal helix packing against the β-subunit. The orientation of the domain matches closely the back view of the complex. The four β-strands are denoted β1–4. (C) Structure of the OB2 domain having the classical OB fold and a helix, which is part of the connecting linker between OB1 and OB2. The five β-strands are denoted β1–5. The missing density for loop L34 between β3 and β4 is indicated by a dotted line. (Gytz H et al. *Nucleic Acids Res.* 2015, by permission of Oxford University Press.)

measured by equilibrium partition of enzyme and enzyme-RNA complexes between two liquid polymer phases. This technique has a major advantage over the zone centrifugation, gel filtration, and nitrocellulose filter techniques commonly employed in such studies, in that both products and reactants were at chemical equilibrium (Silverman 1973a).

The association constant for the Qβ enzyme binding to the Qβ RNA was more than tenfold greater than that for binding to the f2 RNA over a wide range of salt concentrations. The specific *in vitro* binding of the Qβ replicase to the Qβ RNA was detected by the formation of an enzyme-Qβ RNA complex that did not exchange bound RNA molecules and was not dissociated by 0.8 M NaCl. The formation of this non-dissociating complex required both host factors and GTP, but not ATP, CTP, UTP, or Mg^{2+} ions. In the presence of the HF and GTP, a maximum of one active enzyme molecule was bound per molecule of the Qβ RNA template. The complex formation at 0°C was not observed, and the rates of the reaction were maximal at 37°C. The reaction occurred with the intact Qβ RNA, but not with bacterial or other phage RNAs. The complex formation and the initiation of synthesis were separate, however, because in the absence of Mg^{2+} ions, when complex formation occurred readily, no RNA synthesis could be detected (Silverman 1973b).

The limiting concentration of the host factors was supposed by Kuo et al. (1975) as a cause of the inhibition of the Qβ replication by an excess of Qβ replicase that has been observed by Kondo and Weissmann (1972a). The original explanation suggested that a second molecule of the Qβ RNA might interact with the enzyme in a nonspecific manner and thus prevent the 3′-terminus of the RNA strand correctly bound to the enzyme from entering the initiation site.

It is necessary to mention here another independently described host factor, so-called factor B, a component of *E. coli* B cells (Nishihara et al. 1972). It was found by the studies on the group IV phage SP replicase. The *heavy* component of the SP replicase, which was obtained by analogy with the Qβ replicase (Eikhom et 1968), did have the SP replicase activity for the SP RNA as template by itself, in contrast to that of the Qβ replicase. Moreover, the template specificity of the SP replicase was changed by the factor B to the recognition of Qβ RNA and some reduction of the SP RNA recognition (Nishihara et al. 1972). No indications on the relatedness of the factor B to the HFI and HFII did appear.

The overall feeling against the role of the traditional host factors HFI and HFII was nevertheless far from full unambiguity. Thus, Kamen et al. (1974) acknowledged that the HFI was a specific agent but offered evidence that the requirement for the HFII is due to an inhibitor present in certain Qβ RNA preparations and that this inhibitor was counteracted by basic proteins. It was suggested that the source of this inhibitor is Bentonite, used by some laboratories in the extraction of Qβ RNA from virions (Kamen 1975). Palmenberg and Kaesberg (1974) demonstrated the synthesis of complementary strands of the Qβ RNA plus strands, as well as of different heterologous RNAs, with the Qβ replicase in the absence of host factors, but in the presence of Mn^{2+} ions. So, the RNAs complementary to R17 RNA and brome mosaic virus RNAs were synthesized, but replication of these templates was not achieved. It is necessary to add here that Haruna and Spiegelman (1965c) noticed earlier that the addition of Mn^{2+} ions could relax the template specificity of the Qβ replicase.

HOST FACTOR I, OR HF, OR Hfq

Therefore, the HFII faded away and the studies were concentrated on the HFI that will figurate further under the simple *host factor*, or HF, or, later, Hfq, name. Gradually, the Hfq protein that was discovered originally as an accessory factor of the phage Qβ replicase became one of the central players in cellular physiology through its active participation in the bacterial translation and transcription regulation mechanisms.

But first, the HF was purified extensively by polyadenylate-cellulose chromatography (Carmichael 1975) and found to be associated primarily with ribosomes (Carmichael et al. 1975). This location was established both by complement fixation assays with highly specific antiserum directed against the HF, and by the *in vitro* stimulation of Qβ RNA replication by the Qβ replicase. The complement fixation assay has provided the estimate that there were approximately 2500 copies of the HF polypeptide per cell. It was confirmed by this exhaustive study that the HF had a monomer molecular mass of 12,000 by sodium dodecyl sulfate-polyacrylamide gels, and a native hexamer molecular mass of 72,000 as judged by the stoichiometric interaction of the HF with the Qβ RNA, by sedimentation in sucrose velocity gradients, and by sodium dodecyl sulfate gel mobility when incompletely disaggregated. The HF was a potent inhibitor of the *in vitro* polyA-directed polylysine protein-synthesizing system but had less effect on the

in vitro translation of polyU, R17 RNA, late T7 mRNA, or endogenous *E. coli* mRNA. The amino acid composition and N-terminal sequence ruled out the possibility that the HF was one of the known 30S or 50S *E. coli* ribosomal proteins. The finding that the HF was associated with ribosomes *in vivo* completed the demonstration that all the host-supplied proteins required for the Qβ RNA replication *in vitro* were either associated with ribosomes or were involved in the protein-synthetic machinery of the cell (Carmichael et al. 1975).

Meanwhile, a direct analog of the *E. coli* HF was identified in *Pseudomonas putida* (DuBow and Blumenthal 1975). It was detected in crude extracts by both functional activity in the Qβ RNA replication assay and by the immunodiffusion with antibody made against *E. coli* HF. Both HFs had similar but not identical molecular mass as judged by SDS-polyacrylamide gel electrophoresis. Like *E. coli* HF, *P. putida* HF was found to be associated with ribosomes and to bind tightly to polyA. The pure *P. putida* HF had full functional activity in the *in vitro* Qβ RNA replication assay (DuBow and Blumenthal 1975). Furthermore, the HF homologs were detected by both immunological and functional tests in *Acinetobacter calcoaceticus*, *Klebsiella pneumoniae*, and *Pseudomonas aeruginosa*, but it was not detectable by these criteria in *Bacillus stearothermophilus*, *Bacillus subtilis*, *Caulobacter crescentus*, *Micrococcus lysodeikticus*, *Rhodopseudomonas capsulata*, or *Saccharomyces cerevisiae* (DuBow et al. 1977). Therefore, the HF activity and antigenicity were conserved among certain Gram-negative bacterial species. Remarkably, the HF played a role of an aid in elucidating phylogenetic relationships within the *Pseudomonas* genus (Dubow and Ryan 1977). Thus, by the polypeptide's antigenicity and by the Qβ RNA replication test, the HFs were not detected in *Pseudomonas diminuta* or *P. vesicularis*, while the other 11 tested species contained cross-reacting material to the *E. coli* HF.

Hori and Yanazaki (1974) studied the interaction of the HF with the Qβ RNA by competition with synthetic homopolymers, and the uridine-rich RNA regions were found responsible for this interaction. Next, it was found that the HF had a higher affinity than the S1 protein to Qβ and R17 RNAs (Senear and Steitz 1976). The HF recognized a single site in the R17 RNA located in the replicase cistron, and two sites of the Qβ RNA, one of which was located approximately 60 nucleotides from the 3'-end of Qβ RNA. All three HF binding sites had portions rich in adenylate residues, and all were bound by HF when contained in oligonucleotides which were predicted to exist only in single-stranded form. It was suggested therefore that the HF, as well as the S1 protein, were directly involved in the recognition of the 3'-end of the Qβ RNA by the Qβ replicase through their specific interaction with the 3'-terminal region of the Qβ RNA (Senear and Steitz 1976).

The HF interaction was studied with oligoriboadenylates (de Haseth and Uhlenbeck 1980a) and Qβ RNA, where an equilibrium competition assay with $(pA)_{15}$ or $(pA)_{27}$ probes was used (de Haseth and Uhlenbeck 1980b). Of the homopolymers tested, only polyriboadenylate recognized the HF with a high affinity. At low ionic strength of 0.1 M NaCl, the binding to the Qβ RNA was much stronger than to the oligoadenylates, but the situation was reversed upon fragmentation of the Qβ RNA with ribonuclease T1. Increasing the ionic strength resulted in a drastic reduction of the affinity of the HF for the Qβ RNA over a relatively narrow salt range (0.1–0.3 M NaCl). The tight binding of the HF to the Qβ RNA was proposed to result from the binding of an aggregate which could interact with several low-affinity sites on the RNA simultaneously.

Blumenthal and Hill (1980) found that the HF, as well as the S1 protein, allowed initiation of the Qβ replicase reaction on the favored templates at a reduced GTP concentration.

In contrast to Mn^{2+} ions that produced this effect with all templates, the HF was specific for the Qβ RNA only. When both HF and Mn^{2+} were present, transcription of the Qβ RNA occurred at a much lower GTP concentration. Thus, both HF and Mn^{2+} appeared to reduce the GTP concentration requirement, but by different mechanisms (Blumenthal and Hill 1980).

Finally, Hans Weber's team demonstrated two-looped structures by ribonuclease degradation and electron microscopic studies for complexes consisting of the Qβ RNA plus strands with the HF but without the Qβ enzyme (Barrera et al. 1993). This study suggested that the role of the HF on the Qβ plus strand template was to bring the 3′-end into the proximity of the S-site/M-site domain, where Qβ replicase could initiate on it. In contrast, the 3′-end of the Qβ RNA minus strand appeared to be directly available to the enzyme. Later, it was defined more exactly with the deletion mutant RNAs that the looped complexes observed with the HF might not involve the S- and M-sites themselves but adjacent downstream sites (Miranda et al. 1997). Moreover, an additional internal HF interaction was mapped at position 2300 with several mutant RNAs. The Qβ RNA molecules with 3′-truncations formed 3′-terminal loops with similar efficiency as the wild-type RNA, indicating that recognition of the 3′-end by the HF was not dependent on a specific 3′-terminal base sequence (Miranda et al. 1997). The capability of the HF to mediate the access of the subunit II to the 3′ end of the Qβ plus strand was acknowledged by an evolutionary experiment in which the phage Qβ was adapted to an *E. coli* Q13 host strain with the inactivated *hfq* gene encoding the HF (Schuppli 1997). This strain initially produced phage at a titer approximately 10,000-fold lower than the wild-type strain and with minute plaque morphology, but after 12 growth cycles, the phage titer and plaque size had evolved to levels near those of the wild-type host. The RNAs isolated from the adapted Qβ mutants were efficient templates *in vitro* for the Qβ replicase without the HF. Electron microscopy showed that the mutant RNAs, in contrast to the wild-type RNA, efficiently interacted with the Qβ replicase at the 3′-end in the absence of the HF. The same set of the four mutations in the 3′-terminal third of the genome was found in several independently evolved phage clones. One mutation disrupted the base pairing of the 3′-terminal CCC_{OH} sequence, suggesting that the HF stimulated activity of the wild-type RNA template by melting out its 3′-end. Furthermore, it was found that each single substitution, of

total five that were present in the mutants, produced only small effects, but that in combination the four previously described mutations synergistically accounted for most of the observed adaptation of the evolved phages (Schuppli et al. 2000).

THE *hfq* GENE

Ten years after Thomas Blumenthal's team's last activity and concurrently with Hans Weber's team activity on the HF function, the *hfq* gene encoding HFI, or simply HF, or Hfq by more recent classification, was cloned using synthetic probes designed by the partial amino acid sequence of the Hfq and mapped at 94.8 min on the *E. coli* chromosome downstream of the *miaA* gene involved in the 2-methylthio-N^6-(isopentyl)-adenosine (ms^2i^6A) tRNA modification (Kajitani and Ishihama 1991). The sequence of the cloned *hfq* gene indicated that the HF, or Hfq, is a small protein of molecular mass 11,166 and consisting of 102 amino acid residues.

The synthesis rate of the Hfq at the exponential-growth phase of *E. coli* was higher than at the stationary phase, and it increased concomitantly with the increase in cell growth rate, while its intracellular level was about 30,000–60,000 molecules per cell in the pentameric, but not hexameric, form, the majority being associated with ribosomes as one of the salt wash proteins (Kajitani et al. 1994). A significant amount of the Hfq, about 10%–20%, was found in the cellular fraction containing the bacterial nucleoid. It was suggested finally that the Hfq is one of the growth-related proteins.

The first *hfq* gene mutants were constructed and characterized by a chromosomal insertion mutation (Tsui et al. 1994). The *hfq* mutation exhibited a wide range of growth defects and phenotypes that resembled those reported for bacteria containing mutant histone-like proteins. These results confirmed the critical physiological role of the Hfq in *E. coli*. A functional complementation of the *hfq* gene product to the *E. coli hns* gene product was found especially interesting because of the frequent copurification of the *hns* gene product during the Hfq purification process (Shi and Bennett 1994).

A *nrfA* gene, a direct functional analog of the *hfq* gene, was identified in *Azorhizobium caulinodans* (Kaminski et al. 1994). The deduced amino acid sequence of the *nrfA* gene product was similar to that of the Hfq. The similar structure and function were attributed to the *yrp* gene of *Yersinia enterocolitica* (Nakao et al. 1995). The *brg* gene of avirulent *Erwinia carotovora* was shown to function as an analog of the *hfq* gene (Chuang et al. 1999).

Sonnleitner et al. (2002) reported the functional replacement of the *E. coli hfq* gene by the homolog of *P. aeruginosa*, namely, hfq_{Pa} gene that encoded a product of only 82 amino acid residues in length. The 68 N-terminal amino acids of the Hfq_{Pa} showed 92% identity with the Hfq_{Ec} product and the Hfq_{Pa} was shown to functionally replace Hfq_{Ec} for the phage Qβ replication. Therefore, the functional domain of Hfq was localized within its N-terminal domain. A C-terminally truncated Hfq_{Ec} lacking the last 27 aa was capable for the functional replacement of the full-length Hfq protein (Sonnleitner et al. 2002, 2004). Moreover, Sonnleitner et al. (2004) mapped the Hfq amino acid positions influencing the Qβ replication. Meanwhile, the Hfq was qualified as a global regulator of *P. aeruginosa* O1 virulence and stress response (Sonnleitner et al. 2003). The Hfq was essential also for the virulence of *Vibrio cholerae* (Ding et al. 2004) and *S. typhimurium* (Sittka et al. 2007). The global role of the Hfq in bacterial pathogens was reviewed at that time by Chao and Vogel (2010). Later, the intrinsic role of the Hfq by the *E. coli* infection of *D. melanogaster* flies was demonstrated, where the Hfq inhibited the killing of bacteria by *Drosophila* phagocytes after engulfment (Shiratsuchi et al. 2016). In parallel, the Hfq was found essential also for the *Yersinia enterocolitica* virulence in mice, where it influenced production of surface pathogenicity factors, in particular, of the lipopolysaccharide and adhesins mediating interaction with host tissue (Kakoschke et al. 2016). It was concluded by the authors that the Hfq might constitute an attractive target for the developing of new antimicrobial strategies. The critical role of the Hfq in the bacterial virulence was addressed in recent reviews (Feliciano et al. 2016; Heroven et al. 2017).

Muffler et al. (1996) found that the Hfq participated in the translation of the *E. coli rpoS* gene that encodes σ^S subunit of RNA polymerase, a global regulatory factor involved in several stress responses including the osmotic control of the gene expression (Hengge-Aronis 1996). The Hfq was required also for the efficient expression of the *rpoS* gene in *S. typhimurium* (Brown and Elliott 1996, 1997). Vytvytska et al. (1998) reported that the Hfq bound mRNA of the *E. coli ompA* gene that was coding for the major outer membrane protein A, or OmpA, in a growth rate-dependent fashion, and regulated its stability by interfering with ribosome binding, which in turn resulted in rapid degradation of the transcript.

Muffler et al. (1997) were the first to conclude that the synthesis of more than 30 proteins was altered in the *hfq* null mutants and that the Hfq was a global regulator involved in the regulation of many σ^S-dependent and σ^S-independent genes. They were also the first to use the term *RNA chaperone* for the Hfq when they discussed the concept of an RNA-binding protein acting in a way on the RNA structure that was reminiscent of the way chaperones acted on proteins.

Then, small RNAs appeared to be targets for the Hfq (Wassarman et al. 2001). The Hfq not only facilitated the interaction between small RNAs and their mRNA targets, but also protected small RNAs from endonucleolytic attack by ribonuclease E (Moll et al. 2003).

Meanwhile, the Hfq protein was found to bind both supercoiled and linear plasmid DNA, but the binding seemed to be sequence-nonspecific (Takada et al. 1997). Among other *E. coli* DNA-binding proteins, the Hfq appeared as a curved DNA-binding protein with the highest preference for curved DNA (Ali Azam and Ishihama 1999; Ali Azam et al. 1999). It led to an attractive hypothesis that the Hfq with binding activities to both DNA and RNA could play a role in the functional coordination between DNA and RNA such as transcription-translation coupling (Ali Azam and Ishihama 1999). Moreover, in tight connection with the RNA phage activity, the Hfq was qualified as a regulator of the F-plasmid TraJ and

TraM synthesis in *E. coli* (Will and Frost 2006), encoded by the genes that were described generally in the Genetics of F pili section in Chapter 3.

The Hfq protein was found in the *E. coli* nucleoid (Talukder and Ishihama 2015). Recently, a novel critical role was established for the Hfq. It was found that the Hfq serves as a critical switch that modulates bacteria from high-fidelity DNA replication to stress-induced mutagenesis (Chen and Gottesman 2017).

Intriguingly, the Hfq was acknowledged recently as one of the central elements by the RNA-mediated crosstalk between bacterial core genome and foreign genetic elements (Miyakoshi 2019).

FINE STRUCTURE OF THE Hfq

The ring-like donut-shaped hexameric Hfq structure with a sixfold symmetry was demonstrated by electron microscopy (Møller et al. 2002; Zhang et al. 2002). Both papers also demonstrated the possible evolutionary connection between the Hfq and spliceosomal Sm and Sm-like, or Lsm, proteins that have been found in eukaryotes and in *Archaea*. These proteins formed ring-like hetero-heptamers in eukaryotes which were the main components of the spliceosomal small nuclear ribonucleoproteins (snRNPs). As such, they took part in the RNA splicing but also participated in many other RNA processing events (for review see Will and Lührmann 2001).

The crystal structure of the essential N-terminal 72 amino acid residues of the Hfq protein was resolved at a resolution of 2.15 Å by Sauter (2003). This structure is demonstrated in Figure 13.13 and shows the Hfq monomer displaying a characteristic fold of the Sm proteins and forming a homohexamer, in agreement with older biochemical data. Overall, the structure of the *E. coli* Hfq ring was similar to the one recently described for *S. aureus* (Schumacher et al. 2002). This confirmed that bacteria contained a hexameric Sm-like protein which was likely to be an ancient and less specialized form characterized by a relaxed RNA binding specificity. A detailed structural comparison showed that the Sm-fold was remarkably well conserved in the bacteria *Archaea* and *Eukarya*, and represented a universal and modular building unit for the oligomeric RNA binding proteins.

Arluison et al. (2006) showed further how the Hfq assembled into toroidal homo-oligomers that bind singlestranded RNA. This substantial review accented the similarities between the structures, functions, and evolution of Sm/Lsm proteins and Hfq, first of all, the property to polymerize into well-ordered fibers, whose morphologies closely resembled those found for Sm-like archaeal proteins (Arluison et al. 2006). Figure 13.14 shows the structural model of the fibrillar Hfq from the electron microscopy and image analysis.

The history and important role of the bacterial Sm-like protein Hfq was reviewed at that time to present it as a key player in RNA transactions, first of all, as a pleiotropic regulator that modulates the stability or the translation of an increasing number of mRNAs (Valentin-Hansen et al. 2004).

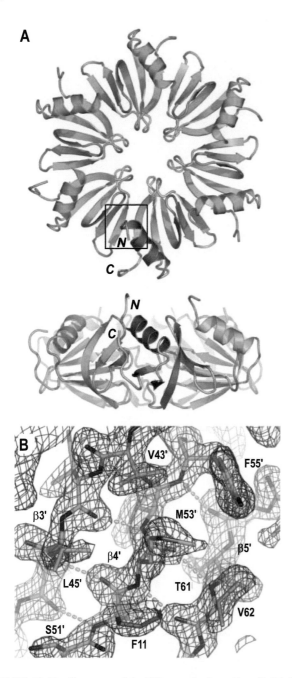

FIGURE 13.13 Structure of the Hfq protein from *E. coli*. (A) Top and side views of the Hfq hexameric donut. Secondary structure elements are highlighted in one monomer with the N-terminal α-helix in pink and the five β-strands in blue. N- and C-termini pointing toward the top of the hexamer are indicated. (B) The dimer interface and H-bond interactions between strands β4′ and β5 of adjacent subunits. The 2Fo-Fc composite omit map (level 1.6σ) is shown in the region indicated by a square in (A). This figure was prepared using PyMol (Delano Scientific, San Carlos, CA). (Sauter C, Basquin J, Suck D. *Nucleic Acids Res.* 2003, by permission of Oxford University Press.)

The Hfq protein purified to homogeneity showed ATPase activity (Sukhodolets and Garges 2003). This finding was not surprising, since some representatives of the extended family of the eukaryotic Sm-like proteins that shared sequence homology with Hfq (Møller et al. 2002; Zhang et al. 2002)

were also ATPases. The plausible ATP-binding site within the Hfq was mapped (Arluison et al. 2007) and studied using molecular dynamics simulations (Lazar et al. 2010).

Finally, Beich-Frandsen et al. (2011b) published the first crystal structure of the full-length 102-amino-acid Hfq molecule of *E. coli* at 2.85 Å resolution. These data revealed that the presence of the C-terminus changed the crystal packing of *E. coli* Hfq. In parallel with the crystallization, the structure of the *E. coli* Hfq was determined in solution by small-angle x-ray scattering (Beich-Frandsen et al. 2011a). The structural work on the Hfq homologs continued and there were more than 20 Hfq crystal structures collected in the Protein Data Bank to year 2012. Moreover, the idea to compare crystal and solution structures prevailed. Thus, Kadowaki et al. (2012) showed that the Hfq from the nitrogen-fixing bacterium *Herbaspirillum seropedicae* had the same folding structure in crystals and solution. It was of special interest that the hitherto puzzling C-terminal domain of the Hfq was found responsible *in vitro* for self-assembly into long amyloid-like fibrillar structures, and the normal localization of Hfq within membrane-associated coiled structures *in vivo* required this C-terminal domain (Fortas et al. 2015). Then, Santiago-Frangos et al. (2016) showed that the C-terminal domain of the *E. coli* Hfq was not needed to accelerate RNA base pairing but was required for the release of double-stranded RNA, and also mediated competition between small RNAs, offering a kinetic advantage to sRNAs that contacted both the proximal and distal faces of the Hfq hexamer.

Concerning hexamer to monomer equilibrium of the *E. coli* Hfq, Panja and Woodson (2012) found that the Hfq underwent a cooperative conformational change from monomer to hexamer around 1 μM protein, which was comparable to the *in vivo* concentration of the Hfq and above the dissociation constant of the Hfq hexamer from many RNA substrates. Above 2 μM protein, the Hfq hexamers associated in high-molecular-mass complexes. It was shown clearly that the Hfq was most active in RNA annealing when the hexamer was present. Selective 2′-hydroxyl acylation and primer extension, small-angle x-ray scattering, and Monte Carlo molecular dynamics simulations were applied to build up a low-resolution model of *E. coli* Hfq bound to the *rpoS* mRNA, a bacterial stress response gene that is targeted by three different sRNAs (Peng Y et al. 2014). Next, the ESI-MS analyses were used to characterize the native *E. coli* Hfq obtained through methods that preserved its posttranslational modifications (Obregon et al. 2015). This approach showed that the majority of the cellular Hfq cannot be extracted without detergents, and that purified Hfq can be retained on hydrophobic matrices. Analyses of the purified Hfq and the native Hfq complexes observed in whole-cell *E. coli* extracts indicated the existence of dodecameric assemblies likely stabilized by interlocking C-terminal polypeptides originating from separate Hfq hexamers and/or accessory nucleic acid. The cellular Hfq was redistributed between transcription complexes and an insoluble fraction that included protein complexes harboring polynucleotide phosphorylase. This distribution pattern was consistent with a function at the interface of the apparatuses responsible for

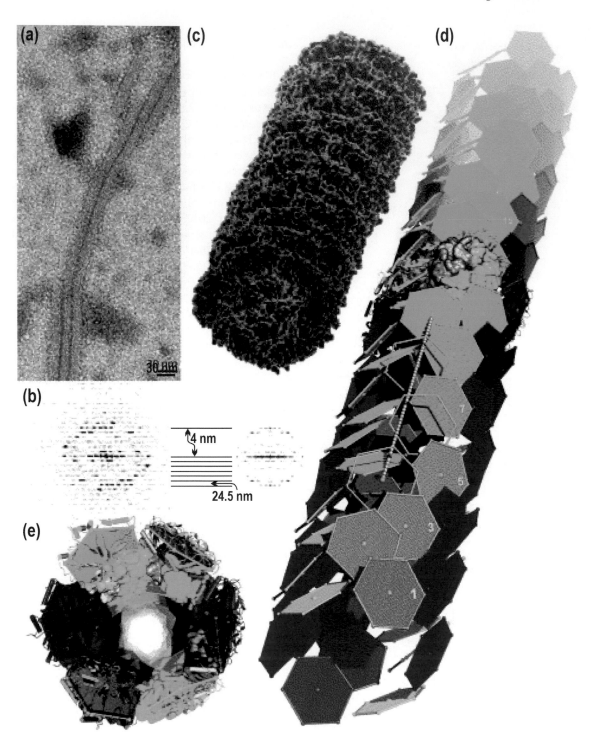

FIGURE 13.14 Structural model of fibrillar Hfq from electron microscopy (EM) and image analysis. (a) Transmission EMs of *E. coli* Hfq reveal bundles of regular, unbranched fibers. (b) Fiber images were analyzed by alignment and clustering, and averaged power spectra of a single windowed sub-image from the major class of observed fibers (<S(*f*)Obs>) are shown in the left panel. Note the good agreement between this <S(*f*)Obs> and the transform derived from resolution-filtered back-projections of the best candidate 3D structural model ((b), right panel). Putative models such as those shown in (c) were based upon layer-line spacings in these spectra, as indicated in (b). Two views of a refined, atomically detailed 3D model for Hfq fibers are shown in (d) and (e). Hfq hexamers are illustrated as hexagonal plates, with the color ramped from blue (nearest) → green (furthest). In order to convey the overall architecture of individual layers and inter-layer packing within each fiber, various hexamers from the 12th and 13th layers are rendered as solvent-accessible surface areas or as ribbon cartoons. The dotted yellow line in (e) spans a distance of ≈245 Å, and presumably corresponds to the strong layer-line spacing in (b). (Reprinted from *J Mol Biol.*, 356, Arluison V et al. Three-dimensional structures of fibrillar Sm proteins: Hfq and other Sm-like proteins, 86–96, Copyright 2006, with permission from Elsevier.)

the RNA synthesis and degradation. It was concluded that the Hfq functions as an anchor/coupling factor responsible for de-solubilization of RNA, and it is tethered to the degradosome complex (Obregon et al. 2015). Zheng et al. (2016) compared the RNA binding and RNA annealing activities of the Hfq from *E. coli*, *P. aeruginosa*, *Listeria monocytogenes*, *B. subtilis*, and *S. aureus* using minimal RNAs and fluorescence spectroscopy, and concluded that the RNA annealing activity increased with the number of arginine residues in a semi-conserved patch on the rim of the Hfq hexamer, and correlated with the previously reported requirement for the Hfq in the sRNA regulation.

Jackson and Sukhodolets (2017) genetically engineered chimeric Hfq6 complexes, in which C-termini of the Hfq subunits were substituted with a sequence derived from human histone H2B that included multiple functionally significant amino acids whose modifications have been linked to carcinogenesis. This substitution resulted in an enhanced formation of dodecameric assemblies by the chimeric Hfq, a result pointing to the possibility of a functional homology between these motifs in proteins from distant kingdoms. It was hypothesized that these putative Palindromic Self-recognition motifs could act as proteins' *cohesive ends* that could allow the protein complexes carrying such motifs to interact dynamically and dissociate-reassociate in response to stress and/or growth phase-specific changes (Jackson and Sukhodolets 2017). In parallel, the C-terminal domain of the Hfq was found responsible for the compaction into a condensed form of double stranded DNA by using various experimental methodologies, including fluorescence microscopy imaging of single DNA molecules confined inside nanofluidic channels, atomic force microscopy, isothermal titration microcalorimetry, and electrophoretic mobility assays (Malabirade et al. 2017). In parallel, the acidic C-terminal tips were found to mimic nucleic acid to auto-regulate RNA binding to the Sm ring (Santiago-Frangos et al. 2017).

During the high-resolution period in the Hfq studies, exhaustive reviews on the structural and functional peculiarities of the Hfq (Vogel and Luisi 2011; Sobrero and Valverde 2012; De Lay et al. 2013; Wagner 2013; Gottesman and Storz 2015; Becker 2016) and both Hfq and S1 (Hajnsdorf and Boni 2012) as pleiotropic RNA-binding proteins were published. Cech et al. (2016) reviewed in turn the role of the Hfq in the regulation of the DNA-related processes.

It is intriguing that the Qβ replication was chosen to test the hypothesis that the Hfq synthesis in *E. coli* could be down-regulated under simulated microgravity (Cabrera et al. 2018). The aim of the authors was to study the virulence of bacteria and viruses in connection with the wellbeing of human health during space travel. Furthermore, the Hfq-assisted RNA annealing strategy for the specific microRNA analysis, and possibly to the RNA related tumorigenicity and disease diagnosis, was detailed recently by Zhang K et al. (2019). Generally, the novel molecular interactions that broadened the functions of the RNA chaperone Hfq were reviewed exhaustively by dos Santos et al. (2019).

SOME SPECIAL FEATURES OF THE Qβ REPLICASE ENZYME

As briefly noted above, the intrinsic and most intriguing property of the Qβ replicase, after its unusual content, consisted of the ability to replicate a lot of shortened Qβ RNA variants. This unique feature will be presented and discussed in detail in Chapter 18, in the context of the famous Darwinian experiments.

As for the enzymatic activities, one of the first sound reviews that presented fresh knowledge on the Qβ replicase subunit structure with possible enzymatic role of each subunit was published by Masatoshi Kondo (1975).

At the same time, the pure Qβ replicase was isolated by Elmārs Grēns' team, who concentrated on the template specificity of the Qβ replicase (Berzin and Gren 1975). Thus, the effect of synthetic polynucleotides and phage RNAs on the polyC-dependent synthesis of polyG was studied. It appeared that single-stranded polyU and poly(dT) were strong inhibitors, whereas structured polynucleotides polyA, MS2 RNA, as well as double-stranded complexes polyA-polyU and polyA-poly(dT) did not affect the polyG synthesis. It was suggested that the contact region of template with the enzyme had single-stranded structure, and the affinity of polynucleotides to the Qβ replicase was determined by the degree of the secondary structure. In this context, Kondo (1976) showed that polyA, among the homopolymers examined, inhibited preferentially the synthesis of the Qβ minus strand in the presence of host factor, while polyC and polyU inhibited efficiently the syntheses of both Qβ plus and minus strands. As a result, a specific interaction of polyA with the host factor was suggested.

Meanwhile, it was found that any oligo- or polynucleotide that was able to offer a CCC-sequence at the 3′-terminus and a second CCC-sequence in a defined steric position to the Qβ replicase was an efficient template, and the corresponding chemical modifications could convert non-template RNAs to template RNAs (Küppers and Sumper 1975; Küppers 1976). Similar conclusions were made by Feix and Sano (1975), who studied the initiation specificity of the RNA synthesis catalyzed by the polyC-dependent Qβ replicase activity. The initiation efficiency of a series of oligoC with various chain lengths was proportional to the template size. The synthetic riboheteropolymers containing cytidylic acid were accepted as templates only if they contained at their 3′-end a cytidylic acid sequence of more than 5 nucleotides. Such an oligo-cytidylate sequence served also as an initiator sequence for copying the non-cytidylic-acid-containing part of the het-erotemplate, while the RNA synthesis always began with the incorporation of GTP, even if the 3′-terminating nucleotide of the template was not cytidylic acid. Therefore, the requirement for the polyC sequence could be most simply fulfilled by addition of an oligoC stretch containing at least 5–20 residues to the 3′-end of the template candidate (Feix and Sano 1975). Moreover, poly(dC) was recognized as a template by the Qβ replicase and served as an initiator sequence for further RNA synthesis (Feix and Sano 1975).

Yoshinari et al. (2000) stated that the CCA box served as the RNA initiation place by the Qβ replicase, by analogy with the RNA-dependent RNA polymerases of turnip yellow mosaic and turnip crinkle viruses. Each enzyme was able to initiate transcription from several CCA boxes within these RNAs, and no special reaction conditions were required for these activities.

The most CCA repeats in an RNA, consisting of 12 CCAs, functioned as independent sites of the *de novo* RNA initiation by the Qβ replicase, with initiation occurring opposite the 3′-C residue of each CCA (Yoshinari and Dreher 2000). The Qβ replicase was capable of internal initiation remote from the 3′-end, although predominant initiation occurred close to the 3′-end. The precise 3′-terminal sequence in (CCA)n-containing RNAs influenced the number and position of active initiation sites near the 3′-terminus. The C residues were required at the initiation site, whereas the position of purines, especially A residues, influenced the selection of initiation sites. The template activity of (CCA)n RNAs was positively correlated with the number of CCA repeats. The three CCA repeats added to the 3′-end of a non-template 83-nucleotide RNA were sufficient to activate transcription. The CCA boxes therefore acted as strong initiation sites in the absence of specific cis-acting signals derived from Qβ RNA, although the efficiency of initiation was modulated by surrounding sequence (Yoshinari and Dreher 2000).

Then, the special role of the 3′-terminal CCCA in directing transcription of RNAs by the Qβ replicase was confirmed by Tretheway et al. (2001) who used ca. 30-nucleotide-long templates that were expected to be free of secondary structure and permitting unambiguous analysis of the role of template sequence in directing transcription. In fact, the 3′-terminal CC<u>C</u>A directed transcriptional initiation to opposite the underlined C; the amount of transcription was comparable between RNAs possessing upstream (CCA)n tracts, A-rich sequences, or a highly folded domain. The removal of the 3′-A from the 3′-CCCA resulted in five- to tenfold-lower transcription, emphasizing the importance of the non-templated addition of 3′-A by the Qβ replicase during termination (Tretheway et al. 2001).

However, a survey of the sequences of the natural Qβ replicase templates from the 6S RNA-derived group indicated that the replicase-template recognition model of Küppers and Sumper (1975) cannot be generally correct (Schaffner et al. 1977). The exciting problem of the natural shortened Qβ templates such as Darwinian variants, midi-. mini-, micro, and nanovariants will be described in detail in Chapter 18 as an essential part of the precellular evolutionary studies.

The Qβ replicase reaction was adapted to a set of practical purposes. First, due to the capability of Mn²⁺ ions to relax the template specificity of the Qβ replicase, the latter was used to synthesize RNA probes for non-polyadenylated RNAs such as 5S, 18S, 28S ribosomal RNAs, and histone mRNA (Obinata et al. 1975). Remarkably, these copies appeared to be the same size as the parental RNA strand. Second, in addition to the well-established primer-independent initiation, the new type of initiation by primers, or oligonucleotides complementary to the template, was tested with synthetic polyribonucleotides as templates (Feix and Hake 1975). This allowed the copying of RNAs otherwise not utilized as a template by the Qβ replicase. Thus, rabbit globin mRNA containing 3′-terminating polyA sequence was copied with the Qβ replicase in the presence of oligoU$_6$ as a primer and Mg²⁺ ions (Feix 1976). In parallel, the RNA complementary to rabbit globin mRNA was synthesized by the Qβ replicase in a reaction primed by either oligoU or oligo(dT) (Vournakis et al. 1976).

Owens and Diener (1977) compared the two of the above-described approaches by the synthesis of RNA complementary to potato spindle tuber viroid (PSTV). First, in the presence of Mn²⁺ ions, the synthesis led to a mixture of heterogeneous RNAs, the largest products of which appeared to be nearly complete PSTV copies. However, when a polyC sequence was added to the 3′-terminus of PSTV, the Qβ replicase synthesized a high-molecular-mass RNA in the presence of Mg²⁺ and without any primer, in accordance with the oligoC rules postulated earlier (Feix and Sano 1975; Küppers and Sumper 1975).

Remarkably, the polyadenylation of the Qβ RNA led to almost complete abolishment of template activity, even by short oligoA stretches. Furthermore, the polyadenylated Qβ RNA inhibited the normal replication reaction of the Qβ RNA by removal of the host factor HFI, in the same way as did free polyA (Devos et al. 1976c).

The template specificity of the Qβ replicase was reviewed at that time by Fukami and Yonesaki (1978). A bit later, Blumenthal (1980b) showed in the context of the template specificity that different templates required different GTP concentrations for initiation. He showed for the first time that the Qβ replicase was able to transcribe heterologous natural RNA species in the absence of Mn²⁺ if sufficient GTP was present, while each RNA tested required a different GTP concentration for initiation. It was concluded therefore that the Mn²⁺ ions reduced the concentration of GTP required for initiation (Blumenthal 1980b).

Concerning specific inhibitors of the Qβ replicase reaction, the substrate, metal, and template effects were studied on the Qβ enzyme inhibition by ortho- and pyrophosphate (Brooks and Andersen 1978), as well as by a large set of guanosine triphosphate analogs (Brooks 1979).

Most intriguing, the highly purified Qβ replicase, when incubated without any added template, synthesized self-replicating RNA species in an autocatalytic reaction (Sumper and Luce 1975). Strong evidence was offered that the production of this RNA was directed by templates generated *de novo* during the lag phase. The enzyme contamination by traces of RNA templates was ruled out by strong experimental arguments. The products of this reaction were established by Biebricher et al. (1981a,b, 1982). Nevertheless, Hill and Blumenthal (1983) demonstrated clearly that, in the *de novo* reaction, the Qβ replicase was replicating undetected contaminating RNA molecules. These conclusions were challenged in turn by Biebricher et al. (1986). The problem of the template-free synthesis will be discussed in detail in Chapter 18, by the evolutionary experiments.

Meanwhile, Thomas Blumenthal published two exhaustive reviews on the structure and function of the Qβ replicase (Blumenthal 1980a, 1982).

Further intriguing activities on the Qβ replicase function related to the so-called YGDD structural motif. The conserved Tyr-X-Asp-Asp motif was identified after extensive alignment of replicases from plant and animal viruses, such as tobacco mosaic, brome mosaic, alfalfa mosaic, Sindbis, foot-and-mouth disease, polio, encephalomyocarditis, and cowpea mosaic viruses (Kamer and Argos 1984). Since in poliovirus the aligned sequence was identified as an RNA-dependent polymerase, the sequence was used further for the identification of the polymerases in the other viruses. As a result, the conserved fourteen amino acid residue segment consisting of an Asp-Asp sequence flanked by hydrophobic residues was found in retroviral reverse transcriptases, influenza virus, cauliflower mosaic virus, hepatitis B virus, as well as in the phage MS2, suggesting this span as a possible active site or nucleic acid recognition region for the polymerases (Kamer and Argos 1984; Argos 1988). Inokuchi and Hirashima (1987) changed the Gly residue at position 357 of the Qβ replicase within the conserved sequence Y_{356}-G_{357}-D_{358}-D_{359} to Ala, Ser, Pro, Met, or Val and examined the replicase activity *in vivo*. The cells carrying the variant plasmids lost the replicase activity and severely inhibited the proliferation of the phage Qβ or the group IV phage SP by suppressing synthesis of the phage RNA. In contrast, substitution of the Gly residue at position 390 showed only a slight inhibitory effect, although the replicase activity was also lost. It was concluded that the cells harboring an altered replicase at the conserved segment interfered specifically with the wild-type phage and different but related phage infections. When such YGDD-defective replicases were tested by the *in vitro* reaction, they exhibited only 2%–6% of the enzyme activity of the lysate from those expressing the wild-type replicase (Inokuchi et al. 1994a). However, the defective replicases, especially one with the Ala substituted for the Gly, recovered enzyme activity when Mn^{2+} was added to the reaction mixture.

Then, Inokuchi and Hirashima (1990) studied the similar interference phenomenon but with the Qβ replicase deleted at the C-terminal region. Thus, the cells expressing the Qβ replicase gene with up to 17% deletion from the C-terminus of the protein prevented the proliferation of Qβ phage. However, in the case when the deletion extended beyond 25% from the C-terminus, the cells showed no interference. Remarkably, when the interference took place, the phage coat protein synthesis was inhibited. The authors concluded that the region between amino acids 440 and 487 of the Qβ replicase subunit was involved in the interference and suggested that the defective replicase inhibited the phage coat protein synthesis by competing with the ribosomes at the initiation site of the coat gene (for detail on the repressor complexes see Chapter 16). It is noteworthy that similar interference by dysfunctional replicase was found at that time by the replication of potato virus X in tobacco plants and was compared with the Qβ interference phenomenon (Longstaff et al. 1993).

Further studies revealed that the C-terminal part of the phage-specific subunit II, or β, participated in the RNA recognition by the Qβ replicase (Inokuchi and Kajitani 1997). Thus, the terminal 18 amino acid residues at positions 571–588 were dispensable for the replicase reaction. Subsequent deletions up to the Ala_{565} residue reduced the RNA polymerizing activity of the Qβ replicase *in vivo* but increased it *in vitro*. The mutant replicases with the enhanced *in vitro* RNA polymerizing activity were found to have relaxed template specificity for ribosomal RNAs and cellular RNAs, as well as Qβ RNA. The deletions beyond the Ile_{564} residue abolished both the RNA polymerizing activity and the binding ability to the plus strand of the MDV-1, a midivariant of the Qβ-related RNAs that will be described in detail in Chapter 18.

The mapping of the functional domains within the Qβ replicase was performed by tricky gene engineering approaches (Mills 1988; Mills et al. 1989). This story will be described in Chapter 18, since it was connected logically to the evolutionary studies using the Qβ replicase and a set of its self-replicating RNAs.

The next important feature of the Qβ replicase appeared by the identification of two classes of RNA ligands that bound Qβ replicase with nanomolar equilibrium dissociation constants (Brown and Gold 1995a,b). The RNA ligands to the two sites, referred to as site I and site II, were used further to investigate the molecular mechanism of the RNA replication employed by the Qβ replicase (Brown and Gold 1996). The replication inhibition by site I- and site II-specific ligands defined two subsets of the replicable RNAs. When provided with appropriate 3′-ends, ligands to either site served as replication templates. The UV crosslinking experiments revealed that site I was associated with the S1 subunit, while site II was connected with the EF-Tu and with the polymerization activity of the phage-specific subunit of the holoenzyme. The practical methodology for the generation of the replicase inhibitors was elaborated by Chen H et al. (1996).

The kinetics of the Qβ replication was evaluated by means of analytical and computer simulation methods (Biebricher et al. 1983). Despite the complexity of the reaction mechanism, the conventional concepts of steady-state and dynamic enzyme catalysis and plausible values of the rate and stability constants for the elementary reactions sufficed to provide detailed understanding of the RNA replication kinetics. It appeared that the main features could be described with simple formulas that were analogous to traditional descriptions of enzyme kinetics. In the twenty-first century, the kinetic properties of the Qβ replicase were investigated experimentally and compared to those of the *T. thermophilus* DNA polymerase (Nakaishi et al. 2002). Then, the exhaustive kinetic analysis of the entire RNA amplification process by the Qβ replicase was undertaken again by Hosoda et al. (2007).

The problem of thermal stability of the Qβ replicase was solved by the identification of the two enzyme forms with different thermal stabilities but identical RNA replication activity (Ichihashi et al. 2010). The authors found that the replicase underwent conversion between these forms due to oxidation, and the Cys_{533} residue in the catalytic subunit II, or β, and

Cys$_{82}$ residue in the EF-Tu subunit were essential prerequisites for this conversion to occur. The thermostable replicase contained therefore the intersubunit disulfide bond between these cysteines.

The influence of double-stranded RNA regions on the Qβ replicase polymerizing abilities was evaluated recently by Tomita et al. (2015), since the effect of the size of such double-stranded regions remained unknown. The authors prepared RNA templates hybridized with complementary RNA or DNA strands of various sizes and analyzed their replication by the Qβ replicase. The latter synthesized the complementary strand as long as the template RNA was hybridized with no more than 200 nucleotide fragments, although the replication amounts were decreased.

It is noteworthy that the Qβ replicase appeared as a tool for RNA sequencing. As just mentioned in the Qβ sequencing section in Chapter 10, a rapid method for the RNA sequencing was developed by Kramer and Mills (1978). By analogy with Sanger's dideoxy chain-termination method of DNA sequencing (Sanger and Coulson 1975), it employed the 3′-deoxy analogs of the ribonucleoside triphosphates as specific chain terminators during RNA synthesis. Later, Axelrod and Kramer (1985) compared the sensitivity of the Qβ replicase to the presence of the 3′-deoxyribonucleoside 5′-triphosphate chain terminators with that of the T7 and SP6 RNA polymerases and found that the latter were 100–1000 times more sensitive than the Qβ enzyme.

NON-*ALLOLEVIVIRUS* REPLICASES

The successful isolation of the replicase from the phage f2-infected cells (Fedoroff and Zinder 1971a,b, 1972a,b, 1973; Fedoroff 1972, 1975) confirmed the previous assumption that the replication mechanisms of different RNA phages must be similar. The three host proteins were in place, indicating not only the same mechanism of replication as by the group III and IV phages, but also, at least for those parts of the replicase polypeptide chains that interacted with the EF-TuTs and S1 proteins, a rather similar amino acid sequence. In this context and to that time, Weber and Konigsberg (1975) noted that the replicase of the phages R17 (Steitz 1969b), MS2 (Min Jou et al. 1972b), and Oβ (Weiner and Weber 1971) showed an identical N-terminal sequence Ser-Lys-Thr-. Nevertheless, the purification and characterization of the group I and II replicases faced enormous difficulties and never reached the level of success by the enormous applications, which fell to the Qβ replicase lot.

It is noteworthy that one of the first reports concerning the non-*Allolevivirus* replicases was devoted to the replicase of the phage FH5 that was immunologically related to f2, MS2, and M12 and therefore could be regarded as a representative of group I (Manifacier and Cornuet 1968; Manifacier et al. 1968). The authors observed an interesting phenomenon that the FH5 replicase synthesized predominantly RNA plus strands at 34°C, while at 42°C only the RNA minus strands were synthesized. Similar temperature behavior was also found later for the R17 replicase (Astier-Manifacier and Cornuet 1968). Moreover, participation of a host factor from noninfected cells, possibly phosphatase, was proposed in this paper as a necessary component of the *in vitro* RNA replication system. Unfortunately, this interesting study with the phage FH5 was never continued.

f2 REPLICASE

The presence of an RNA-dependent RNA polymerase in the lysates of the phage f2-infected *E. coli* cells was observed for the first time by Zinder's team (August et al. 1963). It was highly important that the pioneers of the replicase search introduced the classical replicase assay that contained (i) ribonucleoside triphosphates as substrates, (ii) phosphoenol pyruvate and phosphoenol pyruvate kinase as a nucleoside triphosphate generating system and a high concentration of potassium phosphate buffer to inhibit polynucleotide phosphorylase activity, and (iii) deoxyribonuclease to inhibit the DNA dependent-RNA polymerase activity. With this assay there was virtually no incorporation of ribonucleotides into acid-insoluble material when extracts were prepared from uninfected cells. Moreover, August et al. (1963) observed for the first time the overproduction of the replicase activity by nonpolar coat protein *amber* mutant f2 sus11 and the absence of this activity by the polar coat protein *amber* mutant f2 sus3 infection in nonpermissive cells. The activity represented the existence of the template−enzyme complexes in the infected cells, but it appeared highly difficult to purify the stable template-dependent f2 enzyme (August et al. 1965; Shapiro and August 1965a,b), as it was managed later by the group III and IV phages.

The method for obtaining reproducible, template-free preparations of the f2 replicase was developed almost 10 years after the first activities by Fedoroff and Zinder (1971a,b). The properties of the f2 replicase were studied in detail (Fedoroff 1972; Fedoroff and Zinder 1972a,b). Finally, this highly successful story was reviewed exhaustively by Nina Fedoroff (1975).

One of the first steps to the f2 replicase success consisted of the introduction of a safe and simple RNA replication assay based on the polyC-dependent polyG synthesis in the presence of rifamycin that inhibited the DNA-dependent RNA polymerase of the host. The assay obviated the need for the bacterial protein factors necessary for the complete Oβ replicase reaction, since GTP polymerization by the Qβ replicase was factor-independent (August et al. 1968). As Fedoroff (1975) acknowledged, this simple "polyG polymerase" assay proved applicable and extremely useful in the purification of the f2 replicase. When extracts of uninfected cells and cells infected with the replicase-hyperproducing f2 sus11 mutant were freed from nucleic acids by liquid−polymer phase partitioning, the infected cell extracts displayed substantial rifampicin-insensitive polyG polymerase activity.

The full process of the valiant f2 replicase purification and characterization process was published by Fedoroff and Zinder (1971b, 1972a,b), but the difficulties of purification and storage were exclusively commented on by the corresponding review (Fedoroff 1975). Thus, the template-dependent enzyme preparations obtained after early purification steps had a half-life of approximately 12 hours at 0°C. Such preparations were also quite fragile to freezing and thawing, losing about half of their initial activity in a single freeze-thaw cycle. The storage of the f2 enzyme at −13°C in a buffer containing 50% glycerol at high ionic strength prolonged the half-life to about 1 month (Fedoroff and Zinder 1971). However, once frozen, the enzyme was stored in liquid nitrogen with no loss of activity (Fedoroff 1975). The low stability of the f2 enzyme necessitated rapid, gentle purification procedures, first of which was DEAE-cellulose that ensured substantial purification with minimal loss of activity, good separation from the bacterial DNA-dependent RNA polymerase, and about 20-fold purification extent. The subsequent purification by affinity chromatography on cellulose containing bound f2 RNA and by glycerol gradient centrifugation at high ionic strength achieved an additional 20- to 40-fold purification (Fedoroff 1975).

The template-specific nucleotide polymerization by the f2 replicase was assayed in the presence of various bacterial and viral RNAs (Fedoroff and Zinder 1972a). When assays were carried out under the standard polyG polymerase assay conditions, the f2 enzyme preparation showed substantial nucleotide-polymerizing activity only in the presence of free f2 minus strands. Very little activity was observed either with the pure f2 plus strand RNA or with the native f2 replicative ensemble (RE), in which the minus strands occurred primarily in a double-stranded configuration (Robertson and Zinder 1969), but the heat-denatured RE, which comprised a mixture of free plus and minus strands, was the active template. The phage Oβ, reovirus, and E. coli RNAs did not support nucleotide polymerization by the f2 replicase. The activity with the f2 plus strand RNA became detectable under somewhat different assay conditions, which included a brief preincubation of the f2 enzyme at high ionic strength and subsequent incubation at an ionic strength of 0.1 (Fedoroff and Zinder 1972a). The modified assay conditions enhanced the activity of the enzyme in the presence of both f2 plus strands and denatured f2 RE as template, but did not reduce the ability of the enzyme to discriminate between homologous and heterologous template RNAs. Therefore, the f2 replicase demonstrated a higher affinity for minus, rather than plus, strand RNA (Fedoroff 1975). Generally, the duration of the *in vitro* synthetic reaction was fairly brief, and little incorporation of nucleotides into acid-insoluble material was observed beyond 40 minutes of incubation at 30°C, but the amount of product synthesized rarely exceeded 50% of the amount present as template. The product of the reaction was found to consist primarily of extensively ribonuclease-resistant structures sedimenting at 13-15S (Fedoroff 1975), similar to the double-stranded replicative forms of the phage RNA observed *in vivo* (Erikson and Franklin 1966), while only a small fraction of the reaction product showed the sedimentation characteristics and ribonuclease sensitivity of intact, single-stranded viral 27S RNA.

By analogy with the Qβ replicase (Kamen 1970; Kondo et al. 1970), the f2 replicase consisted of the four subunits with the estimated molecular mass 75,000, 63,000, 46,000, and 33,000 daltons (Fedoroff and Zinder 1971b), where the subunit II corresponded to the phage-induced replicase protein, but the subunits I, III, and IV were host proteins S1, EF-Tu, and EF-Ts, respectively.

Fedoroff and Zinder (1972b) studied specific properties of the f2 replicase and compared both polyC-dependent, or polyG polymerase, and f2 minus strand-dependent, or f2 replicase, activities with respect to a number of parameters. For both activities, the inactivation was extremely rapid at 30°C in buffers of low ionic strength, when half of the initial activity was lost within 10 minutes, with a more gradual inactivation thereafter. The optimal incubation temperature, 25°–30°C, was rather low. The overall kinetics differed for the two reactions. The reaction rate remained fairly constant during the first 10 minutes of the replicase reaction, decreasing gradually thereafter. The polyG polymerase reaction, on the other hand, was virtually complete within 1 minute. No inhibition was observed in the presence of either E. coli ribosomal RNA or Qβ RNA. The polyU depressed both the polyG polymerase and the replicase reactions, although it did not support polyA synthesis (Fedoroff 1975).

The f2 replicase appeared therefore to be fundamentally similar to the Qβ replicase, since the same host proteins occurred in both enzymes. Nevertheless, despite their structural similarity, the Qβ and f2 enzymes differed functionally. While in the absence of host protein factors, the Qβ replicase utilized Qβ minus strands, oligoC-containing polymers, and small variant RNAs as templates, as reviewed at that time by Stavis and August (1970), the f2 enzyme, while active with cytidylic acid as template, showed no activity with either f2 plus or minus strands (Fedoroff and Zinder 1971). Meanwhile, the Oβ replicase used the f2 minus strands as a template with an efficiency of 10%–20% relative to the Qβ minus strands, although no activity was detected with the f2 plus strand RNA (Fedoroff 1972, 1975).

The differential factor requirements of the Qβ and f2 enzymes were found (Fedoroff and Zinder 1973). Thus, the highly purified polyG polymerase, unlike its Qβ counterpart, required the presence of an easily dissociable factor for activity with the f2 minus strand template as well as with the f2 plus strand template, but the chemical identity of the stimulating factor was not established. When added to the highly purified f2 polyG polymerase, the Qβ host factors HFI and HFII had no effect on its f2 replicase activity with either f2 plus or f2 minus strand template, either alone or in combination with the f2 stimulating factor recovered by high salt elution from RNA cellulose. These results indicated that the f2 stimulating factor was either different from the Qβ factors or contained some additional component not required by the Qβ replicase (Fedoroff and Zinder 1973).

Therefore, also after this thorough study, the f2 replicase appeared considerably less efficient than the Qβ replicase that generated a vast excess of product single strands over input template strands. The *in vitro* f2 replicase reaction ceased, however, before net synthesis of RNA was achieved.

MS2 REPLICASE

Together with the f2 replicase, the MS2 replicase belonged to the first studied RNA replicases described by Sol Spiegelman's (Haruna et al. 1963) and Charles Weissmann's (Weissmann and Borst 1963; Weissmann et al. 1963a,b, 1964b; Borst and Weissmann 1965; Weissmann 1965; Weissmann et al. 1966; Feix et al. 1967a) teams. These preparations, while unresponsive to exogenous template RNA, were capable of synthesizing both complementary strands of viral RNA in the presence of ribonucleotide triphosphate precursors, suggesting that the complex contained both the viral replicase and endogenous viral RNA template. By the presentation of the methodology for the Qβ replicase isolation, Eoyang and August (1968) noted that the isolation of replicases from cells infected with the phages f2, MS2, or R17 phages was impeded in large measure by the lability of the enzymes. Although Haruna et al. (1963) succeeded by obtaining the template-dependent MS2 replicase, the results were like the early results on the f2 replicase and both teams turned rapidly to the more prospective Qβ replicase.

The template-dependent MS2 replicase preparations were described further by Walter Fiers' team (Fiers, Verplancke and Van Styvendaele 1966; Lunteren Symposium on Regulatory Mechanisms in Nucleic Acids and Protein Biosynthesis, cited by Steitz 1968c; Fiers et al. 1967). This enzyme preparation, like the f2 replicase, had a high degree of template specificity, while only homologous MS2 RNA was found to support efficient ribonucleotide polymerization by the MS2 enzyme in vitro. The template-dependent MS2 replicase, which displayed autocatalytic kinetics of RNA synthesis, behaved as a large complex, having a sedimentation coefficient of about 40S (Fiers et al. 1967). Gillis et al. (1967) found that the nonpolar MS2 coat protein amber mutants, MS2 am305 and am601 among them, overproduced the replicase enzyme activity. Then, Gillis et al. (1970) compared the effect of salt concentration on the polyribonucleotide synthesizing activities in extracts of E. coli cells infected with phages MS2, its nonpolar coat mutant am305, and Qβ, as well as uninfected cells that represented DNA-dependent RNA polymerase activity.

The enzyme similar to that of Fiers' team was purified partially later by Elmārs Grēns' team, who used the wild-type MS2-infected E. coli cells as a source of the enzyme (Pumpen et al. 1972a,b, 1974). The enzyme was rather unstable and only partially dependent on the added template. During these studies, ^{14}C-uracil was found for the first time to substitute for ^{14}C-UTP in the unpurified cell extracts (Pumpen et al. 1972b), which led further to the elaboration of an original method for the simple and easy preparation of ^{14}C-UTP from ^{14}C-uracil by the complex of E. coli enzymes (Pumpen and Gren 1972).

Some interesting circumstantial evidence on the MS2 replicase was published by Abdel-Hady et al. (1972). They found that the fluorophenylalanine treatment of the MS2-infected cells led to the reduced infectivity of the produced RNA, and speculated that the fluorophenylalanine incorporation into the replicase might produce an enzyme more likely to make errors in the fidelity of RNA production.

Additional circumstantial evidence, this time on the presence of the EF-Tu within the MS2 replicase, came from Lupker et al. (1974a,b). They found a thermosensitive EF-Tu mutant of E. coli that was unable to replicate the bacteriophage MS2 at 42°C but permitted phage production at 37°C.

Finally, a strong argument for the clear differences in the host factor requirements for the MS2 and Qβ replicases was presented by Su et al. (1997). The authors found that the inactivation of the hfq gene led to the 5000-fold reduction of the phage Qβ production but had no effect on the MS2 replication.

GA REPLICASE

Yonesaki and Haruna (1981) reported the successful isolation and purification to a homogeneous state of the group II phage GA replicase. The purified GA replicase was found to contain four different subunits, numbered I, II, III, and IV, the molecular mass of which were 74,000, 60,000, 47,000, and 36,000, respectively. Three of them, I, III, and IV, proved to be host-coded proteins, ribosomal protein S1, EF-Tu, and EF-Ts, respectively, while subunit II was phage-specific, in accordance with the structure of the Qβ and f2 replicases described above. As expected, the RNA replicase was separated into two components on a phosphocellulose column, one composed of subunits I and II and the other composed of subunits III and IV. By combination of both components at 0°C, 60% of the replicase activity was restored within 10 minutes. Although the purified GA replicase catalyzed the GA phage RNA-directed synthesis of template-size RNA, the maximum level of the product RNA synthesized was less than 20% of the amount of template RNA added. Moreover, the product RNA included only the RNA minus strands and not the plus strands (Yonesaki and Haruna 1981).

A certain E. coli factor was found to extensively stimulate RNA synthesis by the GA replicase (Yonesaki and Aoyama 1981). This factor, named GA-HF, was partially purified from an uninfected cell extract and characterized. In the presence of the GA-HF, the GA replicase synthesized 50–100 times more RNA than was synthesized in its absence, and was capable of synthesizing both the minus and plus strands. Nevertheless, this factor could not be replaced by the host factor HFI, or Hfq, of the Qβ story. In the presence of the GA-HF, the GA RNA replication system had a characteristic template specificity, where the group I and II phage RNAs, but none of the group III and IV phage RNAs, showed template activity (Yonesaki and Aoyama 1981).

Furthermore, the template activity of various phage RNAs was compared thoroughly for the template activity for the minus strand synthesis with the GA replicase (Yonesaki et al. 1982). Again, the group I (MS2, R17, JP501, and f2) and II (SD, JP34, and JP500) phage RNAs showed almost equal template activities with the GA replicase, while those of group III (Qβ, ST, and MX1) revealed no template activity, but those of group IV (SP, FI, TW28, and TW19) showed lower but distinct template activity with the GA replicase. On this basis, some affinity similarities between the group II and IV phage RNAs was proposed (Yonesaki et al. 1982).

R17 REPLICASE

Unexpectedly, Satomi J. Igarashi and her colleagues proclaimed principally different structure of the R17 replicase that anticipated the involvement of the *E. coli* DNA-dependent RNA polymerase. Thus, the R17 replicase was found to be copurified with the DNA-dependent RNA polymerase during DEAE-cellulose chromatography and to be inhibited by rifamycin, a specific inhibitor of the DNA-dependent RNA polymerase (Igarashi and Bissonette 1971a). Then, the R17 RNA replicase required optimum salt concentrations very close to those of the RNA polymerase (Igarashi and Bissonette 1971a). Moreover, upon high salt treatment and chromatography on the second DEAE-cellulose column, the R17 replicase was dissociated into two components, one of which, so-called φ factor, was found only in phage-infected cells, while another one, DR, continued to co-chromatograph with the host RNA polymerase activity and could be replaced in the R17 replicase reaction by the RNA polymerase from uninfected cells (Igarashi and Bissonette 1971b). The fraction DR could be derived even from the female *E. coli* strain and catalyzed the R17 RNA-directed RNA polymerization in the presence of the φ-factor. It was concluded therefore that the phage-specific factor φ had a close relationship to the host DNA-dependent RNA polymerase or its subunit(s) (Igarashi and Bissonette 1971b).

To support the idea of the host RNA polymerase as a component of the R17 enzyme, Igarashi (1972) isolated rifamycin-resistant mutants of *E. coli* that allowed the unhampered R17 replication. The R17 replicase from the infected rifamycin-resistant or rifamycin-sensitive *E. coli* was then dissociated into the φ and DR components, where the latter was found to carry the site for sensitivity to rifamycin (Igarashi 1973). It is necessary to add here that Igarashi (1972, 1973) regarded the RNA phage replication as the rifamycin-sensitive process. Appropriate discussion on the subject is given in the Rifamycin section in Chapter 4. Igarashi and Elliott (1975) published a special study that localized the rifamycin-sensitive stage for the R17 growth to the first 20 min of its infectious cycle when the synthesis of phage components, but not the assembly, took place. The authors concluded that the RNA synthesis, predominantly progeny plus strand synthesis, was definitely sensitive to rifamycin in contrast to the other RNA phage systems. It is noteworthy that during these studies Igarashi (1975) elaborated an efficient method for the large-scale preparation of the R17 template RNA in gram quantities using a 300-liter fermenter.

Unfortunately, nobody made an attempt to continue the R17 replicase investigation in order to explain the unusual data obtained in this piece of hard work. Moreover, finishing this marvelous story of the RNA phage replicases, it is regrettable to note that nobody has tried replicases of the RNA caulophages, pseudomonaphages, and the only acinetophage, to say nothing about the recently found novel members of the *Leviviridae* family. This field remains a blank spot on the RNA phage map.

14 Lysis Protein

Absence of proof is not proof of absence.

Michael Crichton

There is nothing insignificant in the world. It all depends on the point of view.

Johann Wolfgang von Goethe

DISCOVERY

In contrast to the three classical *necessary-and-sufficient* phage genes described in the previous three chapters, the lysis gene has no permanent size and position within the phage genome. As presented in the full-length sequences and the peculiarities of the Genome sections of Chapter 7, the lysis genes differ not only by their location within the genome and by the different mode of the reading frame overlapping with the coat and/or replicase genes (let alone the phage AP205, where lysis gene does not overlap but precedes the other genes in the same frame), but also by their length and amino acid composition. Moreover, the members of the *Allolevivirus* genus do not possess the lysis gene at all, and fulfill their lysis function by the specific action of the maturation, or A2, or IIb protein, as described in the Lysis section in Chapter 11. Furthermore, the assembled EC and MB genomes did not disclose the presence of the lysis gene, although otherwise they are very similar to the *Levivirus* genomes. The lysis gene—like open reading frames were also not localized within the recent *Leviviridae* genomes, such as AVE000, AVE001, and AVE0002 (see Figure 7.1).

The detective story of lysis protein discovery has been narrated earlier in the Op-3 mutant section of Chapter 8. Briefly, the only mutant, namely f2 op-3, that was identified by Model et al. (1969), constituted the fourth complementation group of the RNA coliphages, which was elucidated after 10 years lag period as performing the lysis function (Model et al. 1979). In parallel with the genetics, Beremand and Blumenthal (1979) observed the f2 phage-specified 75 amino acid peptide *in vivo* and allocated this product within the phage MS2 genome that was sequenced to that time by Fiers et al. (1976). At the same time, Atkins et al. (1979b) discovered the same 75 amino acid peptide when translated MS2 RNA *in vitro* in the ascites cell supernatant with the rabbit reticulocyte ribosomes.

The similar ribosomal frameshift events that led to the appearance of the lysis protein were found in fact ordinary in the phage life cycle. Thus, the so-called protein 5, a product of the plus one frameshift fusing the coat and lysis genes, was detected both *in vivo* (Beremand and Blumenthal 1979) and *in vitro* (Atkins et al. 1979a,b). Dayhuff et al. (1986) later characterized the nature of the ribosomal frameshifting with regard to the MS2 RNA template.

As Zinder noted in his classical review (Zinder 1980), "the mechanism by which the lysis protein lyses the cells is obscure.

It is not in any way a lysozyme." He mentioned in this context the single-stranded DNA phage φX174 that also possessed a small lysis gene that was identified in a second reading frame within another gene (Barrell et al. 1976) and speculated that

perhaps the lysis function of these phages was only needed after the size of the genome had been fixed by the particle size. Therefore, they had to enlist a protein produced by occasional misreading from within an existent gene. In time the advantage of being able to leave efficiently infected cells would select for more and more gene-like properties of the primitive lysis gene. (Zinder 1980).

It was attempted nevertheless to reveal parallels by the *Levivirus*- and *Allolevivirus*-employed lysis modes. Thus, a cryptic lysis gene, so-called Rg-lysis gene, was found near the start of the Qβ replicase gene in the +1 frame (Nishihara et al. 2004). The authors found a frameshift mutation in the Qβ replicase gene, which caused cell lysis. The mutant had a single base deletion 73 nucleotides 3′ from the start of the replicase gene with consequent translation termination at a stop codon 129–131 nucleotides further in the 3′ direction. This 67-aa protein mediated cell lysis by a different mechanism and via a different target than that caused by the A2 protein that was described in the Lysis section in Chapter 11. The 43-aa C-terminal part of the 67-amino acid product encoded by what in the wild-type genome was the +1 frame, was rich in basic amino acids. The synthesis of a counterpart of this lysis product in the wild-type phage infection would require a hypothetical ribosomal frameshifting event. Furthermore, location of the novel product at the start of the replicase gene was similar to that of the *Levivirus* lysis gene, raising significant evolutionary implications for the finding (Nishihara et al. 2004). According to the evolutionary implications, the Rg-lysis gene, or an Rg-like-lysis gene, was regarded as an ancestral form of the RNA phage lysis genes that might have occurred during the early days of the phage Qβ evolution (Nishihara et al. 2006). This speculation was consistent with that of Bollback and Huelsenbeck (2001), who proposed that the ancestral RNA phage genome was small, and its lysis gene could be located around its coat protein and replicase genes.

It is interesting to note that the first gene overlap in bacteria was found only in 2003, when Feltens et al. (2003) described the out-of-frame overlapping *rnpA* and *rpmH* genes in the thermophilic bacterium *Thermus thermophilus*.

HOMOLOGY

As Table 14.1 demonstrates, the lysis proteins differ strongly by their size, effective charge, and pI values among the different groups of the RNA phages. The amino acid composition of the lysis proteins is also very different. Figure 14.1 shows

TABLE 14.1

Molecular Mass of Lysins

Group	Phage	Method of Identification, Da	Reference	Molecular Mass, Calculated by Genome Sequence, Da	Actual Length, by Genome Sequence, aa	Charge	pI
I	f2	Synthesis *in vivo*, gel electrophoresis	Beremand and Blumenthal (1979)				
	MS2	Synthesis *in vitro*, gel electrophoresis	Atkins et al. (1979b)	8870.31	75	+ 6.05	10.91
		Expression *in vivo*	Nishihara (2003)	8870.31	75	+ 6.05	10.91
		Mutational analysis	Chamakura et al. (2017a,c)	8870.31	75	+ 6.05	10.91
	BO1	Expression *in vivo*	Nishihara et al. (2006)	8870.31	75	+ 6.05	10.91
	ZR	Expression *in vivo*	Nishihara et al. (2006)	8870.31	75	+ 6.05	10.91
	M12			8913.33	75	+ 5.05	10.21
	JP501			8845.20	75	+ 3.05	9.51
	fr	Expression *in vivo*	Adhin and van Duin (1989)	8204.61	71	+ 2.02	8.56
II	GA	Expression *in vivo*	Nishihara (2002)	7357.77	63	+ 1.21	8.19
	SD	Expression *in vivo*	Nishihara et al. (2006)	7334.83	63	+ 3.21	9.48
	TH1	Expression *in vivo*	Nishihara et al. (2006)	7334.83	63	+ 3.21	9.48
	TL2	Expression *in vivo*	Nishihara et al. (2006)	7334.83	63	+ 3.21	9.48
	JP500	Expression *in vivo*	Nishihara et al. (2006)	7334.83	63	+ 3.21	9.48
	BZ13			7334.83	63	+ 3.21	9.48
	DL20						
	KU1	Expression *in vivo*	Groeneveld et al. (1996)	7496.98	64	+ 3.21	9.48
	JP34	Expression *in vivo*	Adhin et al. (1989)	7334.83	63	+ 3.21	9.48
Levivirus-like	Hgal1	Expression *in vivo*	Kannoly et al. (2012)	7570.17	65	+ 6.04	10.58
	C-1	Expression *in vivo*	Kannoly et al. (2012)	7296.67	65	+ 4.07	10.45
	M	Expression *in vivo*	Rumnieks and Tars (2012)	4150.99	37	+ 0.88	8.19
Pseudomonaphages	PP7			6346.61	55	+ 4.84	9.70
	PRR1	Expression *in vivo*	Ruokoranta et al. (2006)	5976.17	54	+ 3.04	9.50
	LeviOr01			7841.07	72	+ 9.01	11.98
Caulophage	φCB5	Expression *in vivo*	Kazaks et al. (2011)	14688.37	136	+ 10.15	10.65
Acinetophage	AP205	Expression *in vivo*	Klovins et al. (2002)	4247.20	35	+ 4.07	9.93

Note: The order of the lysin presentation corresponds to that of the phage genome order in Figure 7.1 and Table 10.1.

the only weak traces of homology among lysis proteins from the various phage groups. It is necessary to emphasize that the phylogenetic tree presented in Figure 14.1 is rather formal and intended to show general relation tendencies but not to measure the real phylogenetic distances among the lysin proteins of the *Leviviridae* members.

Some local structural relatedness was detected for the lysis proteins of the RNA phages and colicinogenic plasmids (Lau et al. 1987). Despite their dissimilar sizes and origins, it was suggested that the colicinogenic lysis proteins could be functionally analogous and evolutionarily related to those of the RNA, as well as of the icosahedral single-stranded DNA phages.

After sequencing of the AP205 genome, Klovins et al. (2002) noted impartially that the RNA phage lysis proteins were remarkably variable in size and, even among the four closely related group I and II coliphages, namely, MS2, fr, GA, and KU1, the amino acid sequence identity of this protein

was no more than 20%. When the PP7 and AP205 were included in the comparison, the sequence identity became insignificant. The authors concluded that there were apparently no precise amino acid requirements for these proteins, which simply dissipated the protonmotive force by inserting the inner membrane. These relaxed requirements would make it not very difficult for a phage to evolve a lysis gene, either by exploiting a second reading frame in an already dedicated sequence, as the coliphages and the PP7, or by using a vacant sequence, as the AP205 did (Klovins et al. 2002).

Nishihara et al. (2006) were the first to perform thorough phylogenetic analysis of a large set of the lysis genes comprising the group I phages MS2, M12, fr, BO1, JP501, and ZR, the group II phages GA, SD, JP34, JP500, KU1, TH1, and TL2, the pseudomonaphage PP7, as well as the Qβ Rg-lysis gene. This analysis showed that groups I and II were clearly dispersed into two clusters, while the PP7 and Qβ genes were separated phylogenetically from the group I cluster.

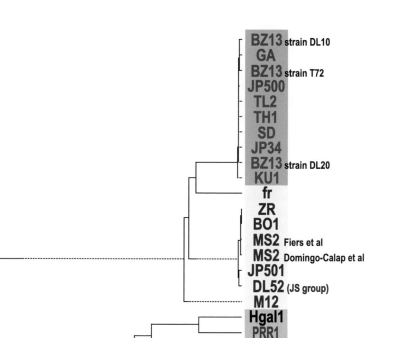

FIGURE 14.1 Comparison of the lysine sequences obtained by the Clustal V alignment in the MegAlign program from the DNASTAR Lasergene software suite. The designations and colors of the genomes correspond to that used in Table 7.1 and Figure 7.1.

NATURAL EXPRESSION OF THE LYSIS GENE

The actual activity of the lysis gene was confirmed by a study on the defective lysis of streptomycin-resistant *E. coli* cells infected with the phage f2 (Cody and Conway 1981a). The cell-free extracts from both the parental and mutant strains synthesized the potential lysis protein in considerable amounts in response to the formaldehyde-treated f2 RNA but not in response to the untreated f2 RNA. It was concluded therefore that the initiation at the lysis protein gene might be favored by translation errors that expose the normally masked initiation site, but the streptomycin-resistant ribosomes, known to have more faithful translation properties, might be unable to efficiently synthesize the lysis protein (Cody and Conway 1981a). Furthermore, Cody and Conway (1981b) observed that the complexes of the f2 RNA and maturation protein transfected *E. coli* cells much better than the protein-free RNA did. This time, it was concluded that the complexes, either native or noninfectious when treated with formaldehyde, introduced the lysis gene into the cells and caused leakage of large macromolecules. This observation was consistent with an earlier finding (Cody and Conway 1981a) that the formaldehyde-treated f2 RNA stimulated the *in vitro* synthesis of the lysis protein. The previously described streptomycin-resistant mutant (Cody and

Conway 1981a) remained lysis defective. The rate of leakage was increased, however, in a double *E. coli* mutant resistant to both streptomycin and rifampin, which was lysed normally by the phage f2 (Cody and Conway 1981b).

It was established soon that the translation of the MS2 lysis gene RNA required a frameshift of ribosomes reading the upstream coat protein mRNA, when the out-of-phase fraction of ribosomes terminated at either one of two nonsense codons just preceding the lysis gene and the termination was followed by the initiation at the start codon of the lysis gene (Kastelein et al. 1982).

The lysis gene therefore offered an interesting example of a nontranslatable message, while the start of the gene was not directly accessible to ribosomes. For this reason, translation of the lysis gene was strictly coupled to the passage of ribosomes over the preceding coat gene. However, small deletions, introduced about 40 nucleotides 5′ to the start codon of the MS2 lysis gene, removed the initiation barrier and opened the cistron to independent translation (Kastelein et al. 1983b). The authors presented an RNA secondary structure accounting for the closed state of the ribosome binding site.

Berkhout et al. (1985a) demonstrated by recombinant DNA procedures that the first 40 amino acids of the lysis protein were dispensable for its function. Thus, all the genetic information essential to the synthesis of the active C-terminal peptide lay within the overlap with the replicase gene, whereas all dispensable residues were encoded in the overlap with the coat protein gene and in the intercistronic region. This suggested that the overlap with the coat protein gene was not required for extra coding capacity but served to regulate the expression of the lysis gene (Berkhout et al. 1985a). Moreover, a synthetic peptide, covering the C-terminal 25 amino acids of the lysis protein, was capable to dissipate the electrochemical potential, and the possible connection between the dissipation of the proton-motive force and bacteriolysis was proposed (Goessens et al. 1988).

Berkhout et al. (1985b) showed further that the amount of lysis protein synthesized in such non-coupled MS2 coat/lysis gene constructs was inversely related to the translation frequency of the coat gene, i.e., expression of the lysis gene was switched off when coat protein synthesis was switched on. At the second overlap, replicase synthesis was not affected by the normal, low-level translation of the lysis gene but could be suppressed if the efficiency of translation over the lysis gene was raised to a high level. Thus, with the assistance of the lysis gene, it was concluded that the passage of elongating ribosomes over an initiation site interfered with translational starts at that site (Berkhout et al. 1985b).

Furthermore, the RNA secondary structure regulating the MS2 lysis gene expression was determined (Schmidt et al. 1987). The existence of such hairpin between nucleotides 1636 and 1707 agreed with the structural mapping data and also with the conservation of base pairing in the related phage M12. After identification of the hairpin structure, Berkhout et al. (1987) examined how translation of the coat gene activated the start site of the lysis gene. The authors concluded that the movement of ribosomes through the hairpin was not sufficient to expose the lysis gene, but the activation was ensured by the translation termination at the overlapping coat gene. Therefore, the endpoint of the translation event was critical for the initiation at the lysis gene and the termination at the natural end of the coat gene triggered the lysis response, but further downstream terminations did not. The activation of the lysis gene was suppressed when the stability of the lysis initiator hairpin was increased by mutations that created additional base pairs. Berkhout et al. (1987) assumed that the ribosome, terminating at the coat reading frame, covered part of the lysis hairpin, thereby destabilizing the secondary structure, and this was sufficient to promote the binding of a vacant ribosome to the lysis gene start. Therefore, instead of the ribosome frameshift proposed by Kastelein et al. (1982), another mechanism was suggested where a proportion of ribosomes terminating coat protein synthesis scanned backwards the template and initiated translation of the lysis gene (Adhin and van Duin 1990). Later, the effects of the secondary structure on ribosome binding was confirmed by mutations of the MS2 coat protein, which were clustered ahead of the overlapping lysis gene without changing the amino acid sequence of the coat protein but disrupting a local hairpin that controlled the lysis gene (Klovins et al. 1997a).

Concerning other group I phages, Adhin and van Duin (1989) found that the translational regulation of the lysis gene required a UUG initiation codon of the lysis gene in the group I phage fr. It was highly important, since the UUG initiation codon was nonfunctional in the *de novo* initiation but was activated by the translational termination at the overlapping coat gene. Therefore, in the case of the phage fr, the UUG initiation codon was crucial for gene regulation, since it excluded uncontrolled access of ribosomes to the start of the lysis gene. The replacement of UUG by either GUG or AUG resulted in the loss of genetic control of the lysis gene (Adhin and van Duin 1989). In spite of distinguishing feature, the full-length sequencing of the phage fr genome revealed more than 85% identity with the phage MS2 in the central part of the replicase gene and the lysis-replicase gene overlap, although the average homology between MS2 and fr was 77% (Adhin et al. 1990b). Moreover, the homology among MS2 and fr reached 86% in the essential C-terminal region of the lysis gene, while the similarity in the first 33 amino acids that were found dispensable for the lytic activity (Berkhout et al. 1985a) was merely 48% (Adhin et al. 1990b).

Concerning the other coliphage groups, the translation of the lysis gene was coupled to the coat gene translation also by the group II phage JP34, by analogy to the regulation described above and found originally in the group I phages (Adhin et al. 1989). In contrast to the JP34 and GA phages, another group II phage KU1 had an insertion of 18 nucleotides in the start codon of its lysis gene (Groeneveld et al. 1996). If in GA and JP34 the coat and lysis genes overlapped by one nucleotide in the configuration UAAUG, the 18-nucleotide insertion in KU1 was positioned between the A and the U of the start codon. This insertion did not affect the coat reading frame, but it destroyed the AUG start codon and separated the previously overlapping genes by 17 nucleotides. The insertion created a UUG codon at its 3′ border which served as the start site for the KU1 lysis protein synthesis. Again, by analogy to the group I phages, such as MS2 and fr, expression of the lysis gene in KU1 was coupled to termination of the coat gene translation (Groeneveld et al. 1996).

Concerning non-coliphages, although the pseudomonaphage PP7 did not demonstrate any significant nucleotide sequence identity with the group I and II coliphages, it possessed the putative lysis gene in the same place as its MS2 counterpart, i.e., starting near the end of the coat protein gene in the +1 frame with respect to the coat gene, extending through the intercistronic region, and overlapping the 5′ region of the replicase gene (Olsthoorn et al. 1995a). Like its MS2 counterpart (Atkins et al. 1979b; Beremand and Blumenthal 1979; Coleman et al. 1983), the putative PP7 lysis protein had a positively charged N-terminal part followed by 30 mainly hydrophobic residues at the C-terminus. This composition was characteristic of the lysis proteins of the phages, and the hydrophobic domain was thought to cross the cytoplasmic membrane and to short circuit its electric potential, as reviewed at that time by Young (1992). Therefore, Olsthoorn et al. (1995a) thought that the interaction with the membrane could apparently be achieved by a variety of hydrophobic amino acids, since strong homologies have not turned up when comparing the sequences of all known RNA phage lysis proteins. Although the PP7 lysis protein with its 54 amino acids seemed small compared to the 75 amino acids for the MS2 counterpart, it was known that truncated versions of the MS2 lysis protein were still functional as long as they retained their 30 C-terminal residues (Berkhout et al. 1985a; Goessens et al. 1988).

Again, as in the case of the phages fr (Adhin and van Duin 1989) and KU1 (Groeneveld et al. 1996), the start codon of the PP7 lysis gene was UUG.

In the acinetophage AP205 (Klovins et al. 2002), the major digression from the other *Leviviridae* family members consisted in the apparent absence of the lysis gene at the usual position, overlapping the coat and replicase genes. In contrast to other leviviruses, the lysis protein of the phage AP205 was encoded at the 5′-end of the genome in a non-overlapping fashion. Thus, two small open reading frames were present at the 5′ terminus, preceding the maturation gene. One of them, nucleotides 134–238, was only 35 amino acid residues long, contained a pronounced Shine-Dalgarno sequence at the optimal distance upstream of the AUG start codon, and encoded the lysis protein. There was a clustering of the positively charged residues at the N-terminus, while hydrophobic amino acids were concentrated near the C-terminus in accordance with the proposed distribution for the phage lysis proteins (Young 1992). When this putative lysis gene was cloned into an expression vector behind the strong inducible P_L promoter, its induction in *E. coli* halted cell growth, while expression of the AP205 cDNA covering the nucleotides 2000–2700, where the coliphage lysis protein was located, did not show any effect on the cell growth (Klovins et al. 2002).

In the pseudomonaphage PRR1, there were two successive unusual start codons, AUU and GUG, with six nucleotides in between, in the putative lysis gene (Ruokoranta et al. 2006), which supported the proposed "scanning model" for the initiation of the lysis gene translation (Adhin and van Duin 1990). The lysis protein of PRR1 has one predicted transmembrane helix at the 24 to 46 residues. To confirm the lysis gene function, the putative gene at nucleotide region 1780 to 1944 was inserted into an expression vector, with the start codon changed to AUG, and the cell lysis occurred upon promoter induction (Ruokoranta et al. 2006).

In the caulophage φCB5 (Kazaks et al. 2011), the lysis gene entirely overlapped with the replicase in a different reading frame, unlike other *Leviviridae* members. The lysis protein of the phage φCB5 was 135 amino acid residues long and therefore exceeded about two times the length of the group I and II lysis proteins. The lysis gene had a strong Shine-Dalgarno sequence, an unusual start codon, UUG, and contained two transmembrane helices. Upon its cloning and expression in *E. coli* (Kazaks et al. 2011), the optical density decreased in a manner similar to that described earlier for the phage AP205 (Klovins et al. 2002) and PRR1 lysis proteins (Ruokoranta et al. 2006). When both potential transmembrane helices were cloned and expressed separately, the growth of *E. coli* was halted, but no decrease in the optical density value was observed (Kazaks et al. 2011).

The lysis function was also confirmed empirically for the putative lysis genes of the phages C-1 and HgaI1 (Kannoly et al. 2012). The assignment of the lysis genes was first

supported by the presence of a single predicted transmembrane domain in both sequences. Then, the corresponding nucleotide sequences 1749 to 1946 for C-1, with exchange of the start codon from GUG to AUG, and 1794 to 1991 to HgaI1 were cloned and expressed in *E. coli* cells in a medium-copy-number expression vector. The culture decline appeared visible 38 and 54 min after induction for the expressed C-1 and HgaI1 lysis genes, respectively (Kannoly et al. 2012).

Finally, Rumnieks and Tars (2012) presented the lysis gene of the IncM pili-dependent phage M. The M lysis gene completely overlapped with the replicase gene but was located more distal than in the phage φCB5. It encoded a 37 amino acid residue-long polypeptide that contained a transmembrane helix like the other known lysis proteins of leviviruses. After cloning and expression under a T7 promoter, a clone containing an approximately 1000 nucleotide-long fragment spanning nucleotides 2098 to 3129 of the phage genome led to a clear cell lysis after induction of the promoter. Examination of this sequence located a likely candidate for the lysis gene between nucleotides 2991 and 3104. This was based on the following criteria: (i) the only open reading frame in the fragment with a significant length, since the shortest known *Leviviridae* lysis protein was that of phage AP205 with 34 amino acids; (ii) the presence of a transmembrane helix with over 95% probability; (iii) the presence of the initiation codon UUG with a rather strong Shine-Dalgarno sequence GAGG nine nucleotides upstream; (iv) the RNA secondary structure prediction showing the initiation codon of the putative gene on top of an AU-rich stem-loop that would presumably have sufficiently low thermodynamic stability to promote the initiation of translation. To verify the lytic function of the gene, the selected stretch together with the original SD sequence and UUG initiation codon was cloned in an inducible protein expression vector, and the induction resulted in almost complete cell lysis some 45 minutes after, thus demonstrating that the approximately 150 nucleotide-long stretch was sufficient to encode the functional lysis protein (Rumnieks and Tars 2012). Figure 14.2 presents the most important data concerning the M lysis gene.

LYSIS MECHANISM

In contrast to the Qβ A2-driven lysis mechanism that was described in the Lysis section in Chapter 11, the effect of the *Levivirus* lysis proteins did not yet result in the brilliant structural patterns, but the lysis mechanism was deciphered clearly enough.

The early attempts to explain the lysis mechanism used by the phage MS2 led to an observation that the fluorophenylalanine resistance of *E. coli* K12 changed the MS2 plaque morphology and increased efficiency of the phage propagation in liquid cultures (Jenkins et al. 1974). The authors explained this phenomenon by the assumption that the mutation triggered inhibition of early lysis in infected cells and that the progeny phages could be released from infected *E. coli* cells by a process which involved the production of a lysozyme. The latter was demonstrated in the MS2-infected, but not in

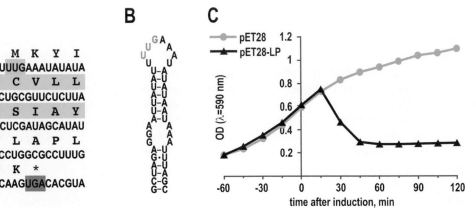

FIGURE 14.2 Lysis protein of phage M. (A) The lysis gene. The Shine-Dalgarno sequence is underlined and initiation and termination codons are indicated by green and pink shading, respectively. The translated amino acid sequence is given above the RNA sequence and the putative transmembrane helix is shaded gray. (B) An RNA hairpin around the initiation codon of the lysis gene. The initiation codon and the Shine-Dalgarno sequence are indicated. (C) Verification of the lysis gene. Growth of *E. coli* cells harboring either empty vector (pET28) or a plasmid with the cloned lysis gene (pET28-LP) before and after the induction of protein synthesis is shown. (Reprinted from Rumnieks J, Tars K. *BMC Microbiol.* 2012;12:277.)

the uninfected cells by a lysozyme test following the decrease in turbidity at 420 nm of a suspension of *Micrococcus lysodeikticus*.

The discovery of the lysis gene actualized the old hypothesis about the coat protein as a lysis trigger (Zinder and Lyons 1968). This study showed, as explained above in the Lysis section in Chapter 12, that when two different coat gene *amber* mutants, which were derived from glutamine codons, were suppressed by a tyrosine-inserting suppressor, lysis did not ensue, whereas suppression by a less efficient, but functional glutamine-inserting suppressor did lead to lysis. Remarkably, despite the partial gene overlap, neither of these mutants affected the lysis gene sequence. It could be assumed that either coat protein, alone or as a part of the virion, acted as a cofactor in the lytic process or a regulatory role of functional coat protein in the phage development affected the lysis gene expression.

Kastelein et al. (1982) got independent evidence that the translation of the coat and lysis genes was coupled, when clones producing large amounts of coat protein also lysed quickly. In contrast, the clones synthesizing small amounts of coat protein showed delayed lysis. This phenomenon resulted inevitably from the translational coupling of both genes. The genetic engineering approaches made it possible to separate both genes (Kastelein et al. 1982; Coleman et al. 1983). It led to the clear conclusion that the coat protein did not participate as a physical moiety in the lytic process, but its role did not go beyond ribosomes traversing the coat gene.

Harkness and Lubitz (1987) engineered a novel lysis protein E-L by combining the N-terminal 32 amino acid sequence of the E protein, a lysis protein of the DNA phage φX174, and the C-terminal 51 amino acid sequence of the *Leviviridae* lysis (L) protein. The approach was based on the observation that the hydrophobic domains of the E and L proteins occurred in reverse orientation at the N- and C-terminus, respectively. As a result, the chimeric gene product had very

high lysis-inducing activity. Later, an entire family of chimeric E-L genes was tested for the *E. coli* lysis (Witte et al. 1998).

Going into the fine mechanisms of the lysis triggered by the *Leviviridae* lysis proteins, Goessens et al. (1988) employed a synthetic peptide corresponding to the 25 C-terminal residues of the MS2 lysis protein, as mentioned briefly above. The effects of the synthetic peptide were examined on the electrochemical potential generated in *E. coli* membrane vesicles by generating hydrophilic pores and in liposomes reconstituted with cytochrome C oxidase. In all cases, the peptide dissipated the electrochemical potential and also induced the release of carboxyfluorescein, but not of inuline, from protein-free liposomes. Thus, the evident connection between the dissipation of the proton-motive force and bacteriolysis was found (Goessens et al. 1988).

The expression of the cloned MS2 lysis protein, which was sufficient to lyse the wild-type *E. coli*, did not cause lysis of the *mdoA* mutants lacking the osmoregulatory membrane-derived oligosaccharides (Höltje et al. 1988). The authors supposed that the lysis induction by the *Leviviridae* members would therefore be dependent on the presence of osmoregulatory membrane−derived oligosaccharides, since the lysis protein normally found in the membrane fraction was not stably inserted into the membranes of the corresponding *mdoA* mutant; rather, degradation and release from the membrane occurred. It was concluded that the membrane-derived oligosaccharides played an important role in maintaining a proper arrangement of inner and outer *E. coli* membrane, a prerequisite for a functional insertion of the MS2 lysis protein (Höltje et al. 1988).

Furthermore, Walderich et al. (1988) found that the bacterial lysis induced by the expression of the cloned MS2 lysis gene in *E. coli* was under the same regulatory control mechanisms as penicillin-induced lysis. The changes in the fine structure of the murein were found to be the earliest physiological changes in the cell, taking place 10 minutes before the onset of cellular lysis and inhibition of murein synthesis. The lysis protein was present predominantly in the cytoplasmic membrane and in a fraction of intermediate density and, to a lesser degree, in the outer membrane, irrespective of the conditions of growth. However, only under lysis-permissive conditions could a 17% increase in the number of adhesion sites between the inner and outer membranes be observed. Thus, a causal relationship between lysis and the formation of lysis protein-induced adhesion sites seemed to exist (Walderich et al. 1988).

However, the mode of action of the phage MS2 lysis protein did not involve a direct interaction with the murein synthesis machinery as is the case for lysis induced by β-lactam antibiotics (Ursinus-Wössner and Höltje 1989). The *E. coli* mutants with defects in various penicillin-binding proteins, which were involved in murein synthesis, were found to show normal lysis sensitivity toward the cloned MS2 lysis protein. No change in the capacity of the binding proteins to bind ^{14}C-labeled penicillin G was observed in the presence of the MS2 lysis gene product (Ursinus-Wössner and Höltje 1989). The phage MS2 lysis protein was not able to induce lysis of the *E. coli* minicells that were refractory toward standard procedures known to induce bacteriolysis of DNA-containing *E. coli* cells (Markiewicz and Höltje 1992).

The specific localization of the MS2 lysis protein in the membrane adhesion sites of the *E. coli* cell wall was determined by immunoelectron microscopy (Walderich and Höltje 1989). Thus, the lysis protein was present mostly in clusters of the adhesion sites visible after plasmolysis. In contrast, a quite different distribution of the lysis protein was found in the cells grown under conditions of penicillin tolerance, i.e., at pH 5, a condition that protected cells from the lysis protein−induced lysis. It was concluded that lysis of the *E. coli* host was a result of the formation of specific lysis protein−mediated membrane adhesion sites (Walderich and Höltje 1989).

The intracellular *E. coli* lytic systems and their mode for disruption of bacterial cells, including those of the phages MS2 and Qβ, were reviewed at that time by Dabora and Cooney (1990) and by Kellenberger (1990).

In a large review on lysis mechanisms and regulation, the presence of a hydrophobic transmembrane helical domain within the protein was proclaimed as the most typical feature of so-called holins, a class of proteins that may create a lesion in the cytoplasmic membrane through which the murein hydrolase passes to gain access to the murein layer (Young 1992). The available evidence suggested that the holins oligomerize to form nonspecific holes, and that this hole-forming step was the regulated step in the phage-induced lysis, while the correct scheduling of the lysis event is as much an essential feature of the holin function as is the hole formation itself. However, with the lysis strategy used by the phage MS2, as well as by the single-stranded DNA phage φX174, no murein hydrolase activity was synthesized. Instead, there was a single species of small membrane protein, unlike the holins in primary structure, which caused disruption of the cell envelope by activation of cellular autolysins. The presence of the

hydrophobic transmembrane helix was defined as obligatory also for the RNA phage lysis proteins (Young 1992).

In the exhaustive review on murein hydrolases in *E. coli*, Höltje (1995) confirmed the principles formulated by Young (1992) and emphasized that MS2, as well as φX174, encoded hydrophobic lysis proteins that, being devoid of hydrolytic activity themselves, seemed to trigger the autolytic system of *E. coli*, where the MS2 lysis protein induced the formation of contact sites between the cytoplasmic membrane and the outer membrane of the envelope (Walderich and Höltje 1989). Finally, Ry Young et al. (2000) published an exhaustive review on cell lysis mechanisms, including those used by the RNA phages.

It is remarkable to note in this context an early observation that the phage f2 lysed late less-active log-phase or early stationary-phase *E. coli* cells more readily than rapidly growing log-phase cells, whereas φX174 lysed cells in both phases of growth (Groman and Suzuki 1966).

The excellent transmission electron micrographs were obtained for *E. coli* cells that were lysed by cloned lysis genes of the phages GA from group II and SP from group IV (Nishihara 2002) and by the phage MS2 from group I (Nishihara 2003). The pictures revealed various morphological aspects of the intermediates of lysing cells. Thus, the cells induced by the SP lysis, or, in fact, A2, or maturation, gene became stretched and tapered in shape, and fragmentation of the parts of the cells had also occurred. The cells induced by the GA lysis gene showed many ballooning structures on the cell surfaces, and others leaking material through the cell wall. However, some balloon-like structures also appeared on the surfaces of cells induced by the SP lysis gene (Nishihara 2002). By the phage MS2 infection, the transmission electron micrographs showed the ballooning structures that appeared on the cell surfaces and the others that were leaking materials through the cell wall. The scanning electron micrographs revealed many extruded structures from the cells. Therefore, both transmission and scanning electron micrographs showed various morphological aspects of the intermediates in the lysing process (Nishihara 2003).

Since the MS2 lysis protein acted without inducing bacteriolytic activity or inhibiting net peptidoglycan synthesis, it was intriguing to find host genes required for the lysis gene-mediated lysis (Chamakura et al. 2017c). Thus, spontaneous insensitivity to L lysis (Ill) mutants were selected as survivors of the lysis gene expression and shown to have a missense change of the highly conserved proline (P330Q) in the C-terminal domain of DnaJ. In the $dnaJ_{P330Q}$ mutant host, the lysis gene-mediated lysis was completely blocked at 30°C without affecting the intracellular levels of the lysis protein. At higher temperatures (37°C and 42°C), both lysis and the lysis protein accumulation were delayed. The lysis block at 30°C in the $dnaJ_{P330Q}$ mutant was recessive and could be suppressed by the L overcomes *dnaJ* (*Lodj*) alleles selected for restoration of the lysis phenomenon. All three *Lodj* alleles lacked the highly basic N-terminal half of the lysis protein and caused lysis ∼20 minutes earlier than the full-length lysis gene. The DnaJ protein was found to form a complex with the full-length lysis protein. This complex was abrogated by the P330Q mutation and was absent with the Lodj truncations. Therefore, in the absence of interaction with DnaJ, the N-terminal domain of the lysis protein interfered with its ability to bind to its unknown target. The lysis retardation and DnaJ chaperone dependency conferred by the nonessential, highly basic N-terminal domain of the lysis protein resembled the SlyD chaperone dependency conferred by the highly basic C-terminal domain of the E lysis protein of the phage φX174, suggesting a common theme where single-gene lysis can be modulated by host factors influenced by physiological conditions (Chamakura et al. 2017c).

PROTEIN ANTIBIOTICS

Although solid data have been collected on the RNA phage lysis proteins to the Millennium time, Bernhardt et al. (2002) were forced to confess in their great review that the lysis mechanism by the RNA phage lysis proteins remained a mystery. Nevertheless, this review extended the concept of the *protein antibiotics* initiated by the A2 protein of the phage Qβ (Bernhardt et al. 2001) described in the Lysis section of Chapter 11 over the RNA phage lysis proteins. The term *amurin* was proposed for the lysis proteins that lacked muralytic activity but somehow subverted peptidoglycan integrity (Bernhardt et al. 2002). At that time, the A2 protein and the E protein of the phage φX174 as the first representatives of the protein antibiotics were numbered together with the putative phage therapy tools (Inal 2003). In fact, the concept of phage therapy to treat bacterial infections was born with the discovery of the bacteriophage almost a century ago but seemed barely acceptable to the RNA phages, whereas their lysis proteins gave some confidence to the possible application. Generally, the current renaissance of phage therapy was fueled by the dangerous appearance of antibiotic-resistant bacteria on a global scale, especially for human use.

If the action of the MS2 lysis protein still remained obscure, the mechanisms to overcome peptidoglycan barrier of the cell wall were clarified clearly for the two putative protein antibiotics. Thus, the A2 protein of the phage Qβ and the E protein of the single-stranded DNA phage φX174 inhibited the MurA and MraY steps of the peptidoglycan synthesis pathway

(Bernhardt et al. 2001, 2002). Unexpectedly, the third reliable protein antibiotic was found in the genome of the smallest RNA phage M (Chamakura et al. 2017b). The M lysis protein inhibited the translocation of the final lipid-linked peptidoglycan precursor, called lipid II, across the cytoplasmic membrane by interfering with the activity of MurJ. The novel finding that the M lysis protein inhibited a distinct step in the peptidoglycan synthesis pathway from the A2 and E proteins indicated that small phages, particularly the *Leviviridae* members, have a previously unappreciated capacity for evolving novel inhibitors of the cell wall biogenesis despite their limited coding potential and presenting new protein antibiotics (Chamakura et al. 2017b; Rubino et al. 2018).

To further characterize amurins, a thorough mutational analysis of the MS2 lysis protein was performed (Chamakura et al. 2017a). In contrast to the three amurins described above, the MS2 lysis protein acted by a still unknown mechanism. To identify residues important for the lytic function, randomly mutagenized alleles of the MS2 lysis gene were generated, cloned into an inducible plasmid, and the transformants were selected on agar containing the inducer. From a total of 396 clones, 67 were unique single base-pair changes that rendered the lysis gene nonfunctional, 44 of which were missense mutants and 23 were nonsense mutants. Most of the nonfunctional missense alleles that accumulated in levels comparable to the wild-type allele were localized in the C-terminal half of the lysis gene, clustered in and around an LS dipeptide sequence. The LS motif was used to align the lysis genes from the RNA phages lacking any sequence similarity to MS2 or to each other. This alignment revealed a conserved domain structure, in terms of charge, hydrophobic character, and predicted helical content. None of the missense mutants affected membrane-association of the lysis protein. Several of the lysis gene mutations in the central domains were highly conservative and recessive, suggesting a defect in a heterotypic protein–protein interaction, rather than in direct disruption of the bilayer structure, as had been previously proposed for the lysis protein (Chamakura et al. 2017a). Figure 14.3 presents the unique pattern of the mutational analysis of the MS2 lysis gene.

Recently, Streff and Piefer (2019) raised an objection to the conclusion that the DnaJ chaperone was a host factor involved

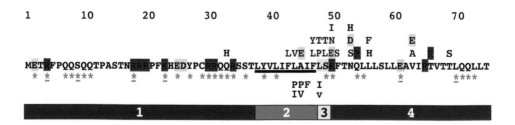

FIGURE 14.3 Mutational analysis of the MS2 lysis gene. The four domains of lysin are represented by numbered boxes. Missense alleles with lysis defects but without a defect in protein accumulation are indicated above the lysin sequence. Missense changes that do not affect lytic function are indicated below the lysin sequence. Green asterisks indicate all possible codon positions where a nonsense mutation could be accessed by a single nucleotide change; underlined asterisks indicate positions where no nonsense mutants were obtained in the mutagenesis. Basic and acidic residues are highlighted in red and blue, respectively. (Reprinted with permission of the Microbiology Society from Chamakura KR, Edwards GB, Young R. *Microbiology.* 2017a;163:961–969.)

in the MS2 L-mediated cell lysis (Chamakura et al. 2017c). The authors thought that this study did not describe a direct protein–protein interaction between L and DnaJ. Thus, the question remained whether DnaJ along with MS2 L was necessary and sufficient for lysis, or if there were additional host proteins involved. In order to answer this question and investigate host protein interactions using a biochemical approach, a hemagglutinin epitope tag was inserted into a plasmid containing the MS2 L gene and an arabinose-inducible promoter. Once transformed into *E. coli*, the tagged L protein was therefore detectable using a Western blot and could be purified by immunoprecipitation. Using these methods, Streff and Piefer (2019) expected that any host protein that interacts with the L protein would be co-immunoprecipitated.

Substantial reviews on the phage lysis machinery were published recently (Cahill and Young 2019; Chamakura and Young 2019).

OTHER APPLICATIONS

The MS2 lysis gene was used in the bacterial ghost technology. The latter were empty envelopes of Gram-negative bacteria, which were produced by controlled expression of the cloned lysis gene E of the single-stranded DNA phage φX174 (Szostak et al. 1996). Generally, the bacterial ghosts retained the structural integrity of native cell envelopes, therefore representing excellent vaccine candidates endowed with intrinsic adjuvant properties. In addition to proteins, this system could be used for the packaging of drugs, nucleic acids, or other compounds for numerous applications.

A family of vectors was constructed based on the above-mentioned chimeric E-L gene (Harkness and Lubitz 1987). The novel vectors were composed of cassette systems coding for antibiotic selection genes, membrane targeting sequences comprised of the E-L chimera under translational control of the *lac* system, and a fully functional φX174 gene E under the transcriptional control of the λ promoters P_L or P_R and cI857 λ repressor, while the E-L sequence was reengineered to allow insertion of foreign sequence into three multiple cloning sites flanked by T3 and T7 promoter sequences for *in vitro* transcription and translation (Szostak et al. 1990).

Later, a technique was elaborated to immobilize plasmid DNA to the inner membrane of bacterial ghosts (Mayrhofer et al. 2005). The self-immobilizing plasmid DNA (pSIP) carried a tandem repeat of a modified lactose operator *lacO* sequence, which was recognized by a fusion protein composed of the repressor of the lactose operon LacI and a hydrophobic membrane anchoring sequence LV derived from the MS2 lysis protein. The LacI-LV fusion protein produced from pSIP was immobilized in the cytoplasmic membrane of *E. coli* via the hydrophobic sequences of the truncated lysis protein LV, whereby the LacI repressor domain simultaneously bound to the *lacO* elements on the pSIP. During the lysis process, most of the cytoplasmic proteins and nucleic acids were expelled through the E-specific lysis tunnel, but the anchored plasmid DNA was retained in the bacterial cell envelopes.

Such combination of the plasmid immobilization procedure based on the MS2 lysis protein with the protein E–mediated lysis technology represented an efficient *in vivo* technique to produce non-living DNA carrier vehicles (Mayrhofer et al. 2005). It is noteworthy that the *E. coli* ghosts were applied to display the shortened variant 1–149 of the hepatitis B virus core (HBc) on the inner and outer membrane of the ghosts, where the C-terminally shortened HBc(1–149) sequence was fused to the amino acids 21–75 of the MS2 lysis protein, among other constructs (Jechlinger et al. 2005).

The *scanning model*, or the *ribosome delivery mechanism*, that was used to conjugate translation of the coat and lysis genes in the leviviruses (Adhin and van Duin 1990) was applied in the artificial genetic selection protocol, which made it possible to find for a given target gene an individual efficient ribosome binding site from a random pool (Zhelyabovskaya et al. 2004). In the proposed vector, the downstream gene possessed the weak ribosome binding site from the MS2 lysis gene that was virtually incapable of independent translation initiation. The latter was accomplished when ribosomes were delivered to this site from the upstream gene. This protocol had two advantages: (i) the efficiency of translational conjugation was not dependent on the primary structure of the upstream gene, which made the construct versatile with respect to the target gene, and (ii) the efficiency of the translation initiation of the downstream gene was largely reduced compared to the upstream gene, which would mean limited expression of the selection marker under overexpression of the target gene. Therefore, the microorganism survival in the presence of antibiotic depended on expression of the target gene, while putting no special requirements on this gene (Zhelyabovskaya et al. 2004).

Later, the MS2 lysis gene was used for improved cell lysis and DNA transfer from *E. coli* to *B. subtilis* (Juhas et al. 2016). Thus, the lysis efficiency was improved by introducing different combinations of the MS2 lysis gene with the lysis genes of the DNA phages φX174 and λ under the control of the thermosensitive λ repressor into the *E. coli* chromosome. The lysis of the engineered autolytic cells was therefore inducible by a simple temperature shift. Such autolytic cells were efficient by transferring plasmid and bacterial artificial chromosome (BAC)-borne genetic circuits from *E. coli* to *B. subtilis* (Juhas et al. 2016). Furthermore, the lysis genes of the phage MS2, together with its counterparts from the phages φX174 and λ, within the engineered genetic circuit, were located downstream of the pT7 promoter, which was regulated by the T7 RNA polymerase, within the *E. coli* chromosome (Juhas and Ajioka 2017). In this system, the T7 RNA polymerase regulated by the thermosensitive repressor drove the expression of the phage lysis genes and significantly increased the efficiency of the cell lysis and the transfer of the plasmid and BAC-encoded DNA from the lysed *E. coli* into the *B. subtilis* target.

Finally, the MS2 lysis gene was fused C-terminally with GFP to track protein localization within the cell and provide a "handle" for future pull-down experiments (Rasefske and Piefer 2017).

15 Immunology

Not everything that can be counted counts, and not everything that counts can be counted.

Albert Einstein

All theory, dear friend, is gray, but the golden tree of life springs ever green.

Johann Wolfgang von Goethe

SEROLOGICAL CRITERIA

When sequencing did not dominate by the maintenance of the strain purity, the serological criteria appeared as the safest and the easiest way to the clear identification of the individual RNA phage species, as demonstrated in the Coliphages and antibodies sections of Chapter 1, as well as with the long list of the serologically identified phages in Table 1.1. The first paper on the RNA phages (Loeb and Zinder 1961) showed this, and neutralization by a specific antiserum became for a long time the most critical indicator of phage identity, as well as of the relationships among different RNA phages (Zinder 1965). From the very beginning, the specific anti-phage sera were prepared by immunization of rabbits. Such phage-specific antibodies served not only for the identification and grouping of an RNA phage in question, but also for the easy prevention of contamination with non-related phages.

For the first time, the known immunological characteristics of the phage MS2, namely, anti-MS2 antiserum, were used in Japan for the isolation of 13 novel strains of the RNA phages, one of which, the well-known phage β, was studied further in detail (Nonoyama 1963). At the same time, the serological relatedness of the phage M12 to the phage fr was determined (Hofschneider 1963a). A year later, the first antiserum-resistant mutants of the phage f2 were isolated (Valentine and Zinder 1964a). The neutralization to a similar extent of the phages R17 and μ2 with anti-f$_{Can1}$ antiserum was described in this early period (Davern 1964b).

Watanabe and August (1967a) described in detail how to use antiserum for phage selection. To test whether a phage was inactivated by antiserum, it was advised to add an appropriate concentration of antiserum to the top agar. The identical aliquots of an appropriate dilution of the phage were then plated, one with and one without antiserum. The replica-plating techniques were also recommended, the method being to stab successively through a sensitive organism in the presence and absence of antiserum in the top agar. To select immunologically unrelated phages, plating of sewage samples in the presence of antiserum or by incubation of sewage samples with antiserum, was employed. It was emphasized especially that each phage isolated in this manner must be tested subsequently with the antiserum to confirm its immunologic unrelatedness (Watanabe and August 1967a).

The serological properties and relationships among different RNA phages remained a long-lasting top priority by further studies on the RNA phages. Thus, the serological grouping of the RNA phages was employed as the first systematic taxonomical approach of RNA phage classification. The serogroup affiliation of the RNA phages was indicated in Table 1.1 and, when possible, specific comments were provided on the serological peculiarities of a phage in question.

SEROGROUPING

The full history and principles of the RNA phage serogrouping, as well as the content of the four coliphage serogroups I–IV, were described in the Serogrouping section of Chapter 2. It is necessary to accept from the very beginning that the current official taxonomy of the RNA phages described in Chapter 2 arose from the immunological characteristics and has been structured up to now in full accordance with the historically longstanding serogrouping.

A first attempt to discover the immunological differences among the set of recently found RNA phages was performed by David William Scott. Rabbit antisera against the three RNA phages MS2, M12, and f2 were tested with these phages and with other RNA phages known at that time: f4, fr, ft5 (new name fr), FH5, R17, and β. All tested RNA phages cross-reacted well with each antiserum and were therefore neutralized by each antiserum. Although neutralization tests demonstrated close serological relatedness of the tested phages, the Ouchterlony double diffusion tests showed that they were not fully identical and contained serologically similar but distinct antigens. It was finally concluded that "the RNA phages may all be serological mutants of the same phage" (Scott 1965). Figure 15.1 demonstrates an excellent example of the classical Ouchterlony double diffusion test in the context of RNA phage characterization.

At almost the same time, two distinct serological groups were found for the six new RNA coliphages: ZIK/1 and ZJ/1 were similar to each other but were serologically dissimilar to the other four coliphages: ZG, ZS/3, ZL/3, and α15, which were serologically related but fell into two subgroups (ZG + ZS/3 and α15 + ZL/3) (Bishop and Bradley 1965). Although all six phages showed a definite level of cross-reactivity, it was apparent that none of the six coliphages was serologically identical.

Later, the phage β and two novel phages B1 and I were found serologically related but not identical to the phage MS2 (Yuki and Ikeda 1966).

A real taxonomic breakthrough happened when Overby et al. (1966a) showed that the Qβ phage was serologically different from the MS2 and therefore from all other immunologically related phages studied before. Next, Itaru Watanabe's

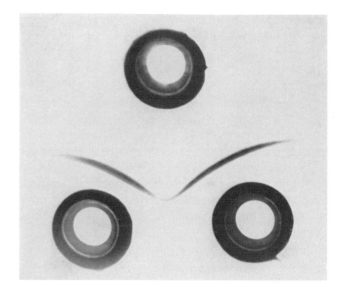

FIGURE 15.1 Serological relationship between R17 and M12 bacteriophages identified by Ouchterlony's double radial diffusion. Precipitin lines were formed by co-diffusion of M12 antiserum (upper well) and 0.3% solutions of R17 (lower left well) and M12 (lower right well) through agar gel (0.8 mL of 0.75% Oxoid agar no. 3, 0.9% in sodium chloride on a 6 cm² square area of a microscope slide). The diffusion proceeded for 24 hr at 37°C in a water-saturated atmosphere. (Reprinted from *J Mol Biol.* 13, Enger MD, Kaesberg P, Comparative studies of the coat proteins of R-17 and M-12 bacteriophages, 260–268, Copyright 1965, with permission from Elsevier.)

team succeeded in the isolation of 30 novel RNA phage strains and identified the three serological groups of the RNA phages (Sakurai et al. 1967; Watanabe et al. 1967a). In these studies, a weak cross-reaction was observed between the group I phages MS2, f2, and R17 (and many others) and the group II phages GA and SD (and many others), but the group III phages Qβ and VK (and again many others) showed virtually no cross-reaction to the group I and group II members (Watanabe et al. 1967a). Figure 2.6, which was taken from Watanabe et al. (1967a), was intended to demonstrate the cross-reactive inactivation of the RNA phages representing the three serological groups by the appropriate three antisera.

The serological group IV appeared when the phages SP (Sakurai et al. 1968), FI (Miyake et al. 1969), and TW19 and TW28 (Miyake et al. 1971a; Aoi and Kaesberg 1976) were discovered and characterized by the synthesis of "readthrough" proteins typical for the group III and IV phages. The detailed serological analysis led to identification of the subgroups within the basic serological groups I–IV (Miyake et al. 1968; Furuse et al. 1978), as described in the Serogrouping section of Chapter 2.

The resolution of the putative serological intermediates, namely the phages JP34 and JP500 between groups I and II (Furuse et al. 1973), the phage MX1 between groups III and IV (Furuse et al. 1975), and the phage ID2 between all four groups (Furuse et al. 1978) was also discussed in the Serogrouping section of Chapter 2. The list of the group members was presented there in Table 2.2. It was suggested at that time to assemble groups I and II into the large group A, and

groups III and IV into the large group B (Ando et al. 1976; Furuse et al. 1978). The large A and B groups were therefore logical progenitors of the *Levivirus* and *Allolevivirus* genera, respectively, in the nearest future. The phages MS2, GA, Qβ, and SP became the traditional reference strains for the serogroups, later genogroups, I, II, III, and IV, respectively. The first exhaustive review on RNA phage serogrouping was published by Kohsuke Furuse (1987).

Concerning the non-coli phages, their serological characterization was rather poor. The first discovered pseudomonaphages phages 7s (Feary et al. 1963, 1964) and PP7 (Bradley 1965b, 1966) were shown to be serologically related to each other but not identical. Feary et al. (1964) presented an unusual pattern of the phage 7s inactivation by the specific antiserum, which appeared to occur at different rates with respect to the reaction time, and determined a K value of 350 during the earlier stage of the reaction (0–5 min) and a K value of 200 during the later stages of the reaction (5–20 min).

The phage PRR1 (Olsen and Thomas 1973; Olsen and Shipley 1973) was not compared serologically to the pseudomonaphages 7s and PP7.

More serology has been done for the caulophages. Thus, the latter formed the three different groups: IV (φCB8r, φCB 9), V (φCB2, φCB4, φCB5, φCB12r, φCB15), and VI (φCB23r) (Schmidt and Stanier 1965). The caulophages φCp2 and φCpl8 were different from each other serologically and different from the previous RNA phage φCB5 (Miyakawa et al. 1976).

No serological characteristics were published for the numerous RNA phages that were specific for the *E. coli* hosts carrying pili-encoding plasmids of different incompatibility groups, such as the phage M (Coetzee et al. 1983) and many others, while Maher et al. (1991) published exhaustive serological investigation on the phages pilHα and Hgal specific for the incompatibility group HI and HII plasmids (see Chapters 2 and 3 for more detail).

IMMUNOGLOBULINS

The RNA phages contributed greatly to the early elucidation of the general structure of antibodies—in particular, to the interaction of the heavy H and light L chains and to the reconstitution of the complex formed by the H and L chains. These studies were undertaken by the Nobelist 1972 Gerald Maurice Edelman and his colleagues, in tight collaboration with Norton D. Zinder. The anti-f2 antibodies, as well as anti-f1 antibodies, from guinea pigs were selected for studies, since the phage neutralization assay measured activity over a wide range with high sensitivity (Edelman et al. 1963). Thus, after separation of the H and L chains of the specifically purified guinea pig anti-phage antibodies, the neutralizing activity of the chains declined to low levels, while mixing the H and L chains resulted in partial restoration of activity. It was concluded that both H and L chains contributed to the immunological specificity, and both the activity and specificity of antibodies were ensured by a complex function of both chains. Moreover, the first hypothetical models of the H and L chain interactions were proposed (Edelman et al. 1963). Then, by using anti-f2 antibodies as one of the models, Olins and Edelman (1964) described in detail the reconstitution conditions of the 7S γ-globulins from the H and L polypeptide chains with greater than 30% yield. The molecular mass of 160,000 was determined for the complex. Using ^{131}I and ^{125}I labels on the different types of chains, combined with ultracentrifugation of chain mixtures in sucrose density gradients, the 7S product was found to contain both isotopes in ratios consistent with the presence of two L and two H chains (Olins and Edelman 1964).

Fougereau et al. (1964) reconstituted the H and L chains of purified antibodies from sheep or guinea pigs and directed against the phages f2 and f1. It appeared that hybrid mixtures of complementary chains of anti-phage antibodies and γ-globulins of the same animal showed less activity than the homologous mixtures of the H and L chains from antibodies of the same specificity. It was remarkable that the 7S interspecies molecular hybrids could be formed between H or L chains of sheep γ-globulins and the complementary chains of guinea pig γ-globulins. In contrast to the results obtained within a single animal species, mixtures of H or L chains of anti-phage antibodies from one species with the complementary chains of antibodies to the same phage from the other species showed slight or negligible potentiation of neutralizing activity (Fougereau et al. 1964).

The role of the bivalent binding of rabbit antibodies in the phage R17 neutralization was investigated, and the 30-fold enhanced affinity of the bivalent antibody as compared to the monovalent one was demonstrated (Klinman et al. 1967). The important role of bivalence was attributed to the increased affinity of the antibody for the phage arising from both the bivalence of the antibody and the presence of multiple identical antigenic determinants on the phage. The authors suggested that the bivalence was the result of an evolutionary development whose selection was favored because of the increased protection it provided against infectious disease.

The RNA phages also played an important role in the progress of the antibody synthesis *in vitro*. Thus, Robert G. Krueger (1965) found that streptomycin altered the *in vitro* synthesis of antibody to the phage MS2 in spleen and lymph node cells from immunized rabbits in a manner that appeared to be similar to the effect of the antibiotic on bacterial systems, although streptomycin has not been reported to affect protein synthesis in mammalian cells. The further studies by Krueger (1966a) showed that the antibody molecules synthesized in the presence of streptomycin had altered reactivities to the phage MS2 in that they did not measurably combine with complete infectious phage but still combined with incomplete noninfectious phage. The author suggested designating the two antibodies synthesized *in vitro* in the presence or absence of streptomycin as N-antibody and SM-antibody, respectively, to differentiate them from serum antibody produced *in vivo*. Next, Krueger (1966b) found that the tritiated dihydrostreptomycin was taken up by intact splenic cells cultured from rabbits immunized with the phage MS2, and was bound to the 40S ribosomal subunit of the 76S splenic ribosome. Later,

it was found that the antibodies synthesized in cell cultures in the absence of streptomycin neutralized phage MS2 to a high titer, whereas antibodies synthesized in the presence of streptomycin had reduced neutralizing capacities (Watkins and Krueger 1970). The difference in the neutralizing activity was explained by the assumption that the antibody preparations synthesized in the presence of the drug had either lower avidity and/or affinity for the phage MS2 than did antibody preparations synthesized in the absence of the drug.

Meanwhile, the primary *in vitro* synthesis of antibody has been achieved with a mouse spleen-thymus organ culture system 54 hours after it was incubated for 18 hours with the phage R17 (Saunders and King 1966).

David T. Rowlands Jr (1967) turned to the comparison of the precipitation and neutralization activities of the phage f2 by the rabbit antibodies, where γG-immunoglobulin, 5S pepsin fragments, and Fab′ fragments were used to test the hypothesis that steric hindrance was important in the virus neutralization. There was no difference in the capacity of intact γG-immunoglobulin and 5S pepsin fragments to precipitate the phage, while the Fab′ fragments did not precipitate it. In all cases the kinetics of neutralization were of the first order, where the γG-immunoglobulin was most active and the Fab′ fragment was least active, and the 5S pepsin fragments were intermediate in activity. It was concluded that the virus neutralization was not entirely explicable on the basis of steric hindrance (Rowlands 1967).

Furthermore, Dudley et al. (1970) estimated the kinetics of neutralization of the phage f2 by the rabbit γG-antibodies or their fragments. In all cases, the neutralization was first-order with respect to both phage and antibody. The minimum values for the rate constants were of the order of $10^7 \ M^{-1} \ sec^{-1}$. The kinetic observations were consistent with the view that the binding of a single antibody molecule could bring about phage neutralization. There are two ways in which a single antibody molecule can affect neutralization: (i) binding at or near some critical site on the phage may block its function, and (ii) binding may disturb the general architectural design of the protein shell of the phage. Although the rate of neutralization varied directly with antibody size, the consideration of the activation parameters spoke against the dependence of neutralization on simple steric factors. The addition of antibodies directed against rabbit γG-immunoglobulin or the 5S pepsin fragment caused approximately a threefold neutralization enhancement (Dudley et al. 1970).

The phage MS2 immunization was used by the recovery of allotypic expression in the allotype-suppressed rabbits (Seto et al. 1975). It is noteworthy that the x-ray irradiation followed by the immunization procedures promoted the recovery of the suppressed allotype expression in anti-bovine serum albumin antibody and in bulk immunoglobulins, but not in the anti-MS2 phage antibody.

The antibody response in mice against the phage f2 was studied in connection with the early clonal expression of IgG antibody and the possible frequent mutations in the immunoglobulin genes (Obel 1980). Thus, the spleen cells from donor mice, exhibiting a clonally very restricted early antibody response to the phage f2, were transferred by limiting cell dose into irradiated syngeneic recipients. The antibodies produced in such recipients were analyzed with regard to isoelectric points. The most antibodies in the recipients were found to differ from the donor, but the pI range into which they fall was frequently fairly narrow and related to the pI of the donor. It was suggested therefore that the most recipient clones might be derived from a common ancestral precursor cell by a somatic mutation in immunoglobulin genes, conceivably subsequent to antigenic stimulation. If such a process should indeed account for the observed diversity, it would imply an unusually high mutation rate (Obel 1980).

Although proper epitopes of the MS2 coat have never been mapped precisely, as will be demonstrated below, the epitope-recognizing complementarity-determining region (CDR) peptides forming the antibody combining site of the monoclonal antibody against the RNA phage fr were found to neutralize the phage activity by Alexander Tsimanis' group in Riga (Steinbergs et al. 1996). Thus, a mouse hybridoma FR52 secreting neutralizing monoclonal antibody was constructed, and demonstrated specificity not only for the RNA phage fr, but also for the phages MS2 and GA. Then, the genes encoding the variable domains of the H and L chains were cloned and sequenced, and the corresponding CDR peptides were chemically synthesized. The CDR peptides were tested for their ability to neutralize the activity of the phage fr and related phages MS2 and GA. The CDR-derived peptides H2, L2, and L3 interacted with the fr particles and neutralized the phage fr activity. Two of these peptides, H2 and L3, partly neutralized the phage MS2, while the peptide L1, and especially the peptide L3, neutralized the activity of the phage GA. These results provided an excellent system for further antibody−antigen interaction studies and raised the possibility that the simple CDR peptides could serve as a new class of antiviral molecules (Steinbergs et al. 1996).

More recently, the CDRs from thermostable single-domain anti-MS2 antibodies (sdAbs) were sequenced and characterized for their binding affinity (Liu JL et al. 2013a). To achieve this, the first immune sdAb library was prepared from llamas immunized with the phage MS2 and purified MS2 coat protein. The best antibody from the library was used for the development of the highly sensitive MS2 test. Furthermore, the four sdAbs with specificity for the MS2 coat protein were expressed as fusions with alkaline phosphatase, and demonstrated permanent thermal stability and increased binding affinity as compared to the original sdAbs (Liu JL et al. 2013b).

It is noteworthy that these studies were performed by the most advanced modern methods. Thus, the melting temperature of the sdAb and sdAb−alkaline phosphatase fusion proteins was measured by a combination of circular dichroism, differential scanning calorimetry, and fluorescence-based thermal shift assay, while binding kinetics were assessed using surface plasmon resonance.

Miklos et al. (2012) developed a protocol using the Rosetta macromolecular modeling suite and designed a supercharged, highly thermoresistant anti-MS2 single-chain variable antibody fragment scF$_V$, where substitution of up to 14 residues with arginine or lysine resulted in marked resistance to thermal denaturation while retaining or improving antigen-binding activity. This scF$_v$ fragment was used further by the development of a micro-emulsion technology for the directed evolution of antibodies (Buhr et al. 2012).

It is also noteworthy that the phage MS2 RNA was used by the detection for the anti-RNA antibodies in the sera of patients with systemic lupus erythematosus, but not in sera of patients with other connective tissue diseases (Eilat et al. 1978). The competition experiments showed that the anti-RNA antibodies preferentially bound native single-stranded RNA as compared with synthetic single- and double-stranded polyribonucleotides, where the anti-MS2 RNA population was most heterogeneous.

PHYLOGENY OF ANTIBODIES

Generally, the RNA phages contributed strongly to the recognition of the peculiarities in the immune system of amphibians, jawless and bony fish, and marsupials, as well as the fundamental differences existing between the serological responses of invertebrates and vertebrates.

The immunization with the phage f2 led to discovery of the immunoglobulins in the primary immune response of the bullfrog *Rana catesbiana*, which paved the way for the further studies on the phylogenetic origins and evolution of the antibody structure (Marchalonis and Edelman 1966). In this study, the molecular mass and amino acid composition of the amphibian γM and γG type immunoglobulins were determined. Furthermore, the sea lamprey *Petromyzon marinus* was found to produce specific antibodies after immunization with the phage f2 (Marchalonis and Edelman 1968). It was shown therefore that the amphibian and jawless fish antibodies revealed the same chain structures as those of their counterparts in higher animal species. However, it is noteworthy that the results on lamprey still remained unexplained in the absence of immunoglobulins in these species of fish (Matsunaga and Rahman 1998). The latter review proposed that these results were attributable to the binding activities of activated C3 and C4 component for foreign cells and molecules. Later, the dual nature of the adaptive immune system in lampreys was discovered (Guo et al. 2009). It appeared that the jawless vertebrates use so-called variable lymphocyte receptors (VLRs), comprised of leucine-rich-repeat segments as counterparts of the immunoglobulin-based receptors that jawed vertebrates use for the antigen recognition. The highly diverse VLR genes were somatically assembled by the insertion of variable leucine-rich-repeat sequences into the incomplete germline VLRA and VLRB genes. The VLRA(+) and VLRB(+) lymphocytes resembled those of the mammalian T and B cells, and the VLRB lymphocytes bound native antigens and differentiated into the VLR antibody-secreting cells. Therefore, the early work of Marchalonis and Edelman (1968), together with further finding of the T-like and B-like lymphocytes in lampreys, offered new insight into the evolution of the adaptive immunity.

The phage f2 was employed further to study the primary and secondary immune responses of the South American marine toad *Bufo marinus* (Lin et al. 1971). The anti-f2 antibodies were identified in the serum 2 weeks after primary immunization, and peak antibody levels were reached at 6 weeks. Although both IgM and IgG antibodies could be found in toad serum, most of the antibody activity in the animals persisted in the IgM fraction until 8 weeks after immunization. When a second injection of antigen was given 4 weeks after primary immunization, there was a marked increase in the total serum antibody activity, and IgG antibodies were found as early as 4 weeks after the second injection of antigen. These findings suggested that the toads were capable of a true secondary response to the phage f2 (Lin et al. 1971). Furthermore, Lin and Rowlands (1973) studied the thermal regulation of the immune response in the *B. marinus* toads after single injections of the phage f2. The immune response was markedly inhibited in animals kept at 15°C as compared to the controls at 25°C. The appearance of serum antibodies was delayed in animals kept at 15°C for the first post-immunization week, but their peak antibody levels were similar to those in toads maintained at 25°C throughout. The transfer of animals from 25°C to 15°C 2 weeks after immunization only temporarily depressed the serum antibody levels but caused a marked delay in conversion from heavy to light antibodies.

At the same time, Rowlands and Dudley (1968) reported isolation of immunoglobulins of a marsupial, namely the adult opossum *Didelphys virginiana*, after immunization with the phage f2. Analysis of the serum and the immunoglobulins suggested that these animals occupied a position intermediate between eutherian mammals and lower vertebrates in terms of their immune system. Next, the development of humoral immunity to the phage f2 was evaluated in opossum "embryos" (Rowlands and Dudley 1969). Those "embryos," which were 5 days or older at the time of immunization, had an immune response to the phage f2, where γM antibodies were most prominent but γG antibodies could also be detected. Furthermore, Rowlands (1970) compared the humoral immunity to the phage f2 in adult opossums with that in New Zealand White rabbits. Antibodies were found in the serum of opossums 7 days after the subcutaneous injection of the antigen, and peak antibody responses were observed between 10 and 21 days after immunization. The second injection of the phage antigen resulted in the increased antibody activity. In either case, the level of serum antibody reached in opossums was less than that in rabbits. More striking, however, was the relatively slow conversion from γM to γG antibodies in opossums. Remarkably, the course of the immune response in adult opossums was more nearly like that of cold-blooded vertebrates than that of eutherian mammals (Rowlands 1970). Further studies of the early immune response and immunoglobulins of opossum embryos after immunization with the phage f2 led to the conclusion that the earliest antibodies appeared as 13S "embryonic" antibodies localized between 19S and 7S positions in the sucrose density gradients, while the IgM antibodies appeared later as the embryonic antibodies decreased, but the IgG antibodies were never prominent and were the last to appear in measurable amounts during development (Rowlands et al. 1972). Then,

Rowlands et al. (1974) defined the order of appearance of the responsiveness in opossum embryos to different phages, proteins, and haptens. A distinct hierarchy was observed when the levels of antibodies were compared with the age of the pouch young. Remarkably, the phage f2 was the first in the rank of immunogens, followed by the DNA phages φX174 and T4, haptens, and proteins. The pioneering f2-employing work on the immune system of such "not proper mammals" as marsupials was reviewed later in detail by Jurd (1994).

The RNA phages also played a historical role in the study of the fish immune system. The live phage MS2 was selected for these studies as a highly immunogenic species, while nonpathogenic in fishes. Moreover, the antibody response could be quantified in the case of the RNA phages by the most sensitive and reproducible technique to that time. Thus, O'Neill (1979) measured the humoral antibody response to single intraperitoneal inoculations of the phage MS2 in brown trout *Salmo trutta*. The immune memory was observed with enhancement of both the time and level of antibody production, though an initial suppression was observed 7 days after secondary inoculation. The antibody titers were related to the MS2 concentration, and adjuvants were found to enhance the response, where an increased enhancement of antibody production by Freund's complete adjuvant over that of incomplete adjuvant was observed. The blood clearance of the phage MS2 in *S. trutta* showed that the primary challenge of MS2 was cleared from the blood within 2.5 days, while the clearance of the second challenge required an extra 3.5 days and the induction of the secondary antibody response was delayed, though immune memory was observed later (O'Neill 1980a). Then, the primary and secondary immune responses of three teleosts, *S. trutta*, *Cyprinus carpio*, and *Notothenia rossii*, to the phage MS2 were presented (O'Neill 1980b).

Next, O'Neill (1981a) turned to pollutants as environmental stressors and measured the effects of intraperitoneal lead and cadmium on the humoral immune response of *S. trutta* to the phage MS2. It was apparent that single intraperitoneal doses of lead and cadmium resulted in a substantial reduction of antibody titer in the MS2-immunized trout and that within the time limits of the experiment there was no recovery. In the case of the two lethal concentrations, the antibody was totally eliminated from the sera and death ensued. Remarkably, the cadmium exposure suppressed those responses mediated by T-lymphocytes and macrophages. A reduction in the number of B-like cells, as well as the loss of helper and memory cell activity, could have been responsible for the reducing antibody titer in the lead- and cadmium-dosed trout. The reduction in the number and activity of immune effector cells would also account for the cadmium-dose-dependent suppression of the antibody response after a re-challenge with the phage MS2. Moreover, the *S. trutta* and *C. carpio* were exposed to low levels of waterborne heavy metals including Ni, Zn, Cu, and Cr, and the humoral antibody response to the phage MS2 was followed during a 38-week exposure period (O'Neill 1981b). Suppression of the immune response was observed in fish exposed to all four heavy metals, while the total suppression of the humoral antibody response was found only in *C. carpio* exposed to Cu or Cr, and these fish exhibited symptoms of acute toxicosis. The time for the primary blood clearance of the live phage MS2 was increased in *S. trutta* exposed to the heavy metals, with the exception of Zn-exposed fish, and in *C. carpio* exposed to Cu. Following the suppressed primary responses, the Ni-exposed *S. trutta* and Zn-exposed *C. carpio* exhibited an adjuvant-like response to the second bacteriophage challenge. It is noteworthy that the immune response of the Antarctic teleost, marbled rockcod *Notothenia rossii* to the phage MS2 was also studied in detail (O'Neill 1981c).

The role of the pioneering J.G. O'Neill's studies in the revelation of the immunosuppression as a link in the etiology of pollutant-associated disease among marine and estuarine fishes was reviewed extensively by Zelikoff (1993).

Later, the phage MS2 assisted by the investigation of furunculosis caused by bacterium *Aeromonas salmonicida*, a serious infectious disease of fish which has caused losses in both wild populations of salmon and in aquaculture (Hussain et al. 2000). Thus, bacteria-free supernatants of broth cultures of *A. salmonicida* inhibited the humoral immune response, but not the cell-mediated immune response, of Atlantic salmon to the phage MS2. It was found that the immunosuppressive factor was the 64 kDa serine protease secreted by *A. salmonicida*, while the principal lethal toxin of *A. salmonicida*, the glycerophospholipid:cholesterol acyltransferase, did not inhibit the immune response of salmon.

The "immune" response in insects compared with vertebrates was studied by Ziprin and Hartman (1971). They followed to the fate of the phage MS2 after injection of the intact particles into larvae of the darkling beetle *Tenebrio molitor*. By the patterns of the phage clearance from hemolymph, the authors concluded that the insects did not produce agents capable of inactivating the phage particles.

TARGET OF ANTIBODIES

After identification by the specific antisera, as described above, the RNA phages were recognized very early as a suitable model system to study antigenic properties of viruses. Thus, Lacy R. Overby (1967) presented the phages Qβ and MS2 as such models, and proposed numerous sites on the phage surface that were recognized by antibodies. Moreover, it was observed that the neutralization reaction of the phages Qβ and MS2 with specific antiserum did not follow first-order kinetics. To explain this phenomenon, the idea was proposed that antibody or antibodies interacted with a phage particle in two ways, one resulting in inactivation and the other in protecting the particle from inactivation. This necessarily presupposed that there are two or more combining sites on the phage particle. The simplest model would be that, in order to inactivate the phage, the antibody must combine with a specific crucial site on the phage and that the interaction of antibodies with other sites would not inactivate. The combination of antibodies with the noncrucial site, however, would decrease the chances of "hits" at the crucial site. Both reactions could follow the kinetics of a first-order reaction, but with differing rate constants. The observed rate of neutralization as measured by

the plaque test would thus be a superposition of the two reactions (Overby 1967). The deviation from first-order kinetics was approximated mathematically in the Qβ neutralization experiments (Hale et al. 1969). This paper supported the idea of the development of resistant phages by interaction with antiserum components. Hung et al. (1969a) were among the pioneers who correlated the serological reactivity of the phage Qβ with the spatial structure of the coat protein. Thus, the relative affinities of neutralizing antibodies to either intact phage, its isolated structural protein and RNA, and reconstituted nucleoprotein particles were compared (Hung et al. 1969a). It appeared that all antigens tested except Qβ RNA interacted with the same active sites of anti-Qβ serum in competition against the phage Qβ. The general conclusion was that the quaternary structures played a vital role in interaction between the phage coat and neutralizing antibody. Moreover, Hung et al. (1969a) mentioned an unpublished observation of A.A. Hirata, who found that immunization with intact Qβ induced no antibody specific for the Qβ RNA or for the isolated coat protein. A bit later, the paper in question was published, where the phage Qβ, reconstituted phage, and a coat protein monomer-succinylated lipopolysaccharide complex were shown to be highly immunogenic in rabbits but the coat monomer was slightly immunogenic, whereas the coat protein aggregate and the RNA were nonimmunogenic (Hirata et al. 1972). The peak antibody response to the reconstituted phage and to the coat protein monomer-succinylated lipopolysaccharide complex was slightly lower than that of the whole phage. It was concluded therefore that the immunogenicity of the antigen was dependent upon a specific quaternary structure imparted to the coat monomer, according to the authors' opinion, by the RNA or succinylated lipopolysaccharide.

It is worthy of mention here that the phage fr was used by the elaboration of a special serological screening method for the detection of free C-terminal amino acids in virus coat proteins (Anderer et al. 1967). However, it was the fr coat where none of the antisera to C-terminal amino acids showed significant precipitation.

At the same time, MacColl et al. (1972) demonstrated that the 11S subunits, potential substructures of the phage particles, which have been described in the Reconstitution section of Chapter 9, were recognized by the anti-phage antibodies. This finding supported again the idea formulated above that the immunization of rabbits with the RNA phages led to induction of antibodies that recognized rather conformational than linear epitopes. The corresponding epitopes could be present in the 11S substructure.

Robert G. Krueger (1969a) was the first to correlate the serological cross-reactivity with the peculiarities of the primary structure of the coat proteins, after Scott (1965) demonstrated that the rabbit antisera generated against the phages MS2, M12, and f2, cross-reacted with all of the phages, and he suggested that these phages, together with f4, R17, FH5, β, and fr, were closely related but different phages that were serological mutants of the same phage. In parallel, it was detected that the MS2 and M12 coat proteins were identical in their amino acid sequence (Enger and Kaesberg 1965, Konigsberg et al. 1966) and that they differed by two different amino acid exchanges from the coat proteins of phages R17 and f2 (Enger and Kaesberg 1965, Konigsberg et al. 1966, Lin et al. 1966; Vasquez et al. 1966) and by many more exchanges from the coat protein of the phages fr (Weber and Konigsberg 1967) and Qβ (Overby et al. 1966b). Krueger (1969a) concluded that the single amino acid differences in the coat proteins of these phages showed a correlation with the serological cross-reactions observed with these phages. The serological differences among these phages was demonstrated by antiserum specific for the phages MS2, R17, f2, fr, and Qβ. Remarkably, the adsorption of either hyperimmune anti-MS2 or anti-fr sera with appropriate concentrations of any of the RNA phages not only removed activity for the phage used for adsorption but uniformly reduced the neutralizing activity of the serum for all of the RNA phages, demonstrating that the antigenic determinant on the coat proteins of these phages were similar. Krueger (1969a) emphasized the role of the tertiary configuration of the coat protein by the antigen−antibody reaction. Moreover, he noted that the antiserum used in the studies could readily distinguish antigenic differences between MS2 and M12, which apparently had identical primary structure in their coat proteins, as well as R17 and f2, which appeared to differ only by one amino acid in their coat proteins. Thus, it was acknowledged clearly that the specific antibodies should show the strong correlation with the amino acid sequence and, moreover, should be able to distinguish small amino acid differences.

Furthermore, Krueger (1969b) showed the effect of homologous and heterologous antigenic stimulation on the antibody synthesis *in vitro*. The spleen and lymph node cells from rabbits immunized with the phages MS2 and fr synthesized neutralizing and complement-fixing antibody after *in vitro* antigenic stimulation with the related heterologous RNA phages. The phages MS2, M12, R17, f2, f4, β, and fr, but not Qβ were used in this study. In the presence of more distantly related heterologous phages, a relatively small amount of 7S antibody specific for the phage used to immunize the intact rabbit was synthesized. However, in the presence of very closely related heterologous phages, relatively large amounts of 7S antibody highly cross-reactive for these heterologous phages were produced. The unrelated phages, such as T2 or T4, did not elicit any immune response at all.

Furthermore, Krueger (1970) reported that the repeated immunization of rabbits with the phage MS2 resulted in the appearance of an increasing number of neutralizing antibodies showing increased affinity with the homologous phage antigen. In parallel, there was a progressive narrowing of the specificity of the antibodies produced after each antigenic stimulation as evidenced by a decrease in the extent of cross-reactions observed with the related heterologous RNA phages.

At the same time, the first systematic quantitative analysis was made by the inactivation of the phage MS2 (Rappaport 1970). Thus, the data were obtained relating the fraction of MS2 surviving to the number of rabbit antibody molecules bound, from the minimum of one antibody molecule adsorbed per virus particle to a maximum of 80–85 per virus particle.

It was calculated that between 20 and 40 binding sites on the phage particle were critical, the covering of any one of which resulted in inactivation. In extreme antibody excess, approximately 10^{-4} of the phage population survived. In this antibody range, a maximum of 80–85 antibody molecules were bound to a phage particle, leaving 95–100 antigenic sites free, explaining the small but constant probability of survival in extreme antibody excess, so-called persistent fraction that was found also in the case of animal viruses (Rappaport 1970). Remarkably, Parren and Burton (2001) in their extensive review, and later Klasse and Burton (2007) by their calculations on the number of antibody molecules for the West Nile virus neutralization regarded an average of four immunoglobulins as the necessary and sufficient MS2 neutralizing dose, with a reference to the pioneering paper by Irving Rappaport (1970).

Rohrmann and Krueger (1970b) compared the neutralizing and precipitating activities of 10 hyperimmune rabbit antisera to the phage MS2. The neutralization constant of these antisera was directly proportional to their concentration of precipitating antibodies. In the quantitative precipitin analysis, the phage MS2 had a minimal antigenic valence of 90 in extreme antibody excess, and at equivalence the ratio of antibody molecules to phage particles was 32. Again, the isolated MS2 protein subunits were antigenically unrelated to intact or empty viral capsid particles (Rohrmann and Krueger 1970b). Moreover, Rohrmann and Krueger (1970a) acknowledged that the normal and so-called *light* phage particles (see the Defective particles section in Chapter 9) were antigenically related, whereas the coat protein monomers and dimers were unrelated to the higher-molecular-mass particles.

Next, Curtiss and Krueger (1973) purified and characterized the phage MS2 hapten conjugates to either dinitrophenol or picolinimidate in order to employ these conjugates as antigens in the generation, detection, and characterization of antibody generated against the hapten−protein conjugates. In fact, the hapten conjugates demonstrated decreased ability to be neutralized by the anti-MS2 serum (Curtiss and Krueger 1973). When rabbits were immunized with highly conjugated dinitrophenylated phage MS2 containing 150 dinitrophenol groups per phage particle, the dominant reactivities were carrier or hapten-carrier specific rather than hapten specific (Curtiss and Krueger 1975). Thus, the dinitrophenol hapten did not appear to play an immunodominant role. Nevertheless, Chapter 22 will present successful vaccines where similar principles of conjugation have been used.

Moreover, as described in the Fate of the maturation protein section in Chapter 11, the purification of the phage MS2 dinitrophenol conjugates provided a system for localization of the single molecule of the A protein in the capsid of the MS2 phage particle (Curtiss and Krueger 1974).

Meanwhile, Witte and Slobin (1972) examined the neutralization of the trinitrophenyl haptenated phage f2 by anti-hapten rabbit antibodies. First it was found that whole antibody was much more effective in the neutralization reaction than were the corresponding Fab fragments. Second, the relative effectiveness of antibody in neutralizing the f2 conjugates was

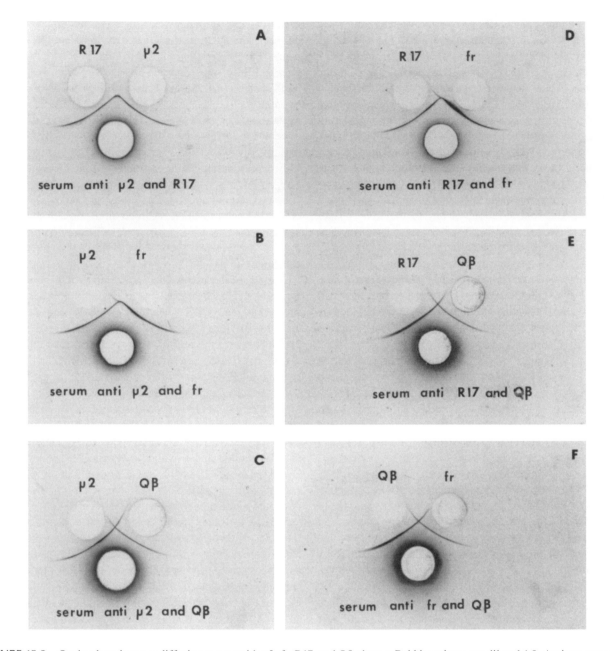

FIGURE 15.2 Ouchterlony immunodiffusion assays with μ2, fr, R17, and Qβ phages. Rabbit antisera are diluted 1:8. Antigen concentration, 1 mg/mL. Reprinted with permission of ASM from Piffaretti and Pitton (1976).

controlled by the degree of phage modification. A quantitative study of the infectivity of the lightly substituted f2 conjugate led to a model of f2 neutralization, in which a single antibody molecule must directly interact with a single critical site on the surface of f2 in order to inactivate the phage, namely by attachment at or in the vicinity of the maturation protein.

In contrast to these data, the standard anti-phage antibodies were targeted exclusively against coat protein. Thus, the infectious reconstituted MS2 RNA-maturation-protein complex that has been described in the Fate of the maturation protein section of Chapter 11 was resistant to the neutralizing anti-MS2 antibodies (Shiba and Miyake 1975).

Concerning the immunological relatedness of the RNA phages, highly convincing visual results were obtained by Piffaretti and Pitton (1976) when they compared the phage μ2 with the phages R17, fr, and Qβ. Figure 15.2 is intended to acknowledge the high demonstrating potential of the Ouchterlony (1965) reaction. Thus, it is clear from this pattern that the phages μ2 and R17 are immunologically identical; there is also a relationship between the phages μ2 and fr, but the latter seems to have some uncommon antigenic determinants with the former.

The same pattern was obtained for the phages R17 and fr. In contrast, the phage Qβ has no detectable immunological relationship with μ2, fr, or R17. The neutralization assays to the RNA phages with the homologous and heterologous antisera confirmed the excellent image data of the Ouchterlony immunodiffusion technique (Piffaretti and Pitton 1976).

Finally, a theoretically highly important finding was published by Snippe et al. (1976). They found that in thymectomized, irradiated, and bone marrow—reconstituted (TXBM) mice, the response was normal when a high dose of the phage MS2 was used for immunization, while with a 50-fold lower dose of MS2, the response was impaired. This finding indicated a T-cell involvement in the anti-MS2 antibody formation. More evidence for the T-cell connection with the MS2 immunization was obtained from experiments with anti-thymocyte serum or cyclophosphamide-treated mice. The ³H-thymidine incorporation experiments demonstrated that both B and T cells were active upon the *in vitro* stimulation with the phage MS2. Therefore, it was concluded that the phage MS2 is a thymus-dependent antigen that could be only thymus independent in high doses.

MAPPING

Surprisingly, the RNA phages that became one of the most favorite virus-like particle (VLP) carriers and served as a reliable background of the foreign epitope mapping, as Chapters 20 through 23 will present, were never subjected to modern in-depth epitope mapping. This phenomenon can be explained partially by the fact that the RNA phage-neutralizing antibodies recognized conformational, or discontinuous, B cell epitopes, as demonstrated above. This conclusion also followed from the numerous published and unpublished observations of Elmārs Grēns' team.

Harry Langbeheim, Ruth Arnon, and Michael Sela were the first to perform mapping of the RNA phage coat epitopes (Langbeheim et al. 1976). Thus, the MS2 coat protein was cleaved with cyanogen bromide to yield three fragments, possessing the sequences 1–88, 89–108, and 109–129. The peptides corresponding to the sequences 89–108 and 109–129 were synthesized chemically.

It was found that only the peptide Glu-Leu-Thr-Ile-Pro-Ile-Phe-Ala-Thr-Asn-Ser-Asp-Cys-Glu-Leu-Ile-Val-Lys-Ala-Met spanning from positions 89–108 was capable of inhibiting phage neutralization by specific antiserum, while the synthetic peptide 109–129 had no capacity to interfere with neutralization of the phage MS2. The synthetic antigen prepared by attachment of the peptide 89–108 to multichain poly(DL-alanine) induced antiserum in rabbits, which was capable of neutralizing MS2 activity almost as efficiently as the antiserum prepared against the intact coat protein. This inactivation was specific, because it could in turn be totally inhibited by the 89–108 peptide. The peptide 109–129 conjugated with multichain poly(DL-alanine) was unable to induce the production of phage-neutralizing antibodies in rabbits (Langbeheim et al. 1976). This paper was of a special importance, since it provided the first clear evidence that antibodies raised against a totally synthetic molecule, an epitope of the natural protein, could show antiviral activity; namely, they successfully inactivated the viable phage.

Next, the synthetic antigen comprising the peptide 89–108 was attached to a carrier multi-poly-DL-alanyl-poly-L-lysine and ensured with a small-molecular-mass water-soluble peptidoglycan prepared from *Bacillus megaterium* for its adjuvant effect by the immunization of guinea pigs (Langbeheim

et al. 1978a). When injected in phosphate-buffered saline or Freund's incomplete adjuvant, the synthetic antigen did not elicit any measurable anti-phage activity, and addition of the peptidoglycan by simple mixing did not bring about a significant increase in antibody production. However, when the peptidoglycan was chemically linked to the synthetic antigen conjugate, it had a marked adjuvant effect when the material was administered in Freund's incomplete adjuvant, almost identical to the extent of the effect of Freund's complete adjuvant. The peptidoglycan was therefore capable of replacing mycobacteria (Langbeheim et al. 1978a).

Then, the synthetic antigen consisting of the 89–108 peptide attached to the multi-poly-DL-alanyl-poly-L-lysine carrier was compared with the intact phage MS2 by the cellular immune response in guinea pigs, both *in vivo* by the delayed-type hypersensitivity skin test and *in vitro* by lymphocyte transformation assay (Langbeheim et al. 1978b). Both the synthetic conjugate and the phage MS2 elicited high homologous cellular reaction, but no significant cross-reaction was observed between them.

Furthermore, the chemical conjugation of the N-acetylmuramyl-L-alanyl-D-isoglutamine (MDP) adjuvant to the synthetic antigen markedly increased the induction of the MS2-neutralizing antibodies in rabbits, while immunization with a mixture of the synthetic antigen and MDP brought about only slight enhancement in the titer of neutralizing antibodies (Arnon et al. 1980). This was a real example how a completely synthetic antigen evoked neutralizing antiviral antibodies when administered in aqueous solutions.

The story of how the MS2 coat was mapped for antigenic epitopes was narrated in the context of chemically defined antiviral vaccines by Ruth Arnon (1980). Later, an exhaustive review of how peptide vaccines work was published by David J. Rowlands (1992). He emphasized in particular the pioneering role of the synthetic MS2 peptides, together with the tobacco mosaic virus—neutralizing experiments, in this field. The impressive MS2 story was also mentioned by Michael Sela (1996) and Michael J. Francis (1996) in their reviews on synthetic vaccines for infectious and autoimmune diseases.

Budowsky and Kostiuk (1985) observed that the ultraviolet 254-nm irradiation of the phage MS2 resulted in the decrease of the number of antigenic determinants exposed on the phage surface by the irradiation-induced decrease of the anti-phage antibody binding. It was explained by the shielding of the antigenic determinants, which could be caused by rearrangement of coat protein molecules and/or of the capsid induced by photomodification of non-antigenic fragments of coat protein and/or of intraphage RNA.

After the complete nucleotide sequence of the group II phage GA was obtained, Inokuchi et al. (1986) compared the hydrophobicity profiles of the GA and MS2 coats in the hope of identification of the antigenicity determining sites which distinguished GA from MS2. It was finally concluded that a few amino acid residues of the coat protein could affect the antigenicities of GA and MS2. This assumption was supported by the properties of the GA*sus*5H mutant that appeared spontaneously from the phage GA maturation *amber* mutant

GAsus5 (Harigai et al. 1986) and was described earlier in Chapter 8. The serological and electrophoretic properties of the GAsus5H mutant became similar to those of the group I phage MS2. The deduced amino acid sequence showed that five amino acids were substituted in the mutant, and three of the five became identical to MS2, resulting in increased molecular mass of the coat protein. Nevertheless, it did not complement MS2 *in vivo* (Harigai et al. 1986). The critical role of only a limited number of amino acid residues in the antigenic pattern of the phage coats was confirmed further by the sequence of the JP34 coat (Adhin et al. 1989).

IMMUNODETECTION

The three classical types of ELISAs—antibody capture, antigen capture, and two-antibody sandwich assays—were adjusted for the detection of the phage MS2 with monoclonal anti-MS2 antibodies 5A4-1 and 3A2-12 and rabbit polyclonal serum (Menking et al. 1994). The same antibodies were used further for the quantitative immunoligand assay on the light-addressable potentiometric sensor (LAPS). The latter used a specially coated biotinylated nitrocellulose membrane onto which fluoresceinated immunocomplexes were immobilized, followed by reaction with an anti-fluorescein urease-conjugated antibody. Analyte quantitation was accomplished by placing the urease-labeled immunocomplex into a reader containing a 100-mM solution of urea. Hydrolysis of the urea produced a potentiometric shift proportional to the amount of urease present in solution, and hence sensitive and reliable detection of the target protein. The optimal variant with polyclonal sera, but not with monoclonal antibodies, which experienced steric hindrance after conjugation to biotin or fluorescein, led to the reproducible detection limit of 10^7 phage particles by 5-min incubation (Menking et al. 1994).

The phage MS2 was detected with the same 10^7 pfu/mL sensitivity limit by the first array biosensor capable for simultaneous identification of bacterial, viral, and protein analytes in 14 minutes (Rowe et al. 1999). The sensor utilized a standard sandwich immunoassay format where antigen-specific "capture" antibodies were immobilized in a patterned array on the surface of a planar waveguide, and bound analyte was subsequently detected using fluorescent tracer antibodies. The polyclonal rabbit anti-MS2 antibodies were employed by the assay.

The phage MS2 was used further by comparison of multiplexed techniques for detection of bacterial and viral proteins, where the immobilized antibody microarrays were compared to the Luminex flow cytometry system that utilized suspensions of polystyrene microbeads covalently coupled with capture antibodies (Rao et al. 2004). Again, the rabbit polyclonal anti-MS2 antibodies were used. The limit of detection was 4.38×10^7 pfu/mL in the protein microarray and 3.51×10^6 pfu/mL in the Luminex system.

The phage MS2 was tested by a paramagnetic bead-based electrochemical immunoassay (Thomas JH et al. 2004). The immunoassay sandwich was made by attaching a biotinylated rabbit anti-MS2 immunoglobulin to a streptavidin-coated bead, capturing the phage, and then attaching a rabbit anti-MS2 IgG-β-galactosidase conjugate to another site on the phage. The β-galactosidase converted p-aminophenyl galactopyranoside (PAPG) to p-aminophenol (PAP). The PAPG was electroinactive at the potential at which PAP was oxidized to p-quinone imine (PQI), so the current resulting from the oxidation of PAP to PQI was directly proportional to the concentration of antigen in the sample. The immunoassay was detected with rotating disk electrode amperometry and an interdigitated array electrode. The most optimal conditions resulted in the 90 ng/mL, or 1.5×10^{10} MS2 particles/mL limit of detection (Thomas JH et al. 2004; Bange et al. 2005, 2007). Furthermore, a fully automated fluidic system for a bead-based immunoassay with electrochemical detection was developed for the phage MS2 with the 990 ng/mL, or 1.6×10^{11} particles/mL, limit of detection (Kuramitz et al. 2006). These immunodetection approaches were reviewed extensively at that time by Mao et al. (2009).

The affinity-purified goat anti-MS2 immunoglobulin was studied by the capillary isoelectric focusing with liquid-core waveguide laser-induced fluorescence whole-column imaging detection (CIEF-LCW-LIF-WCID) approach that was mentioned in the Monitoring section in Chapter 9 (Liu and Pawliszyn 2005a,b; Wu XZ et al. 2005).

Tok et al. (2006) demonstrated a novel biosensing platform using engineered nanowires as an alternative substrate for sandwich immunoassays. The nanowires were built through submicrometer layering of different metals by electrodeposition within a porous alumina template. Although variety of metals could be deposited, this study employed stripes of gold, silver, and nickel. Owing to the permutations in which the metals could be deposited, a large number of unique yet easily identifiable encoded nanowires could be included in a multiplex array format. As a result, the 1×10^5 pfu/mL detection limit was reached for the phage MS2 with the use of polyclonal anti-MS2 antibodies for the nanowire coating.

Next, a new method, using thin films of nanoporous silicon, was suggested for improving the sensitivity of the phage MS2 detection (Rossi et al. 2007). Such porous silicon surface–based biosensor allowed the detection limit of 2×10^7 pfu/mL. Therefore, the system exhibited sensitivity and dynamic range similar to the Luminex liquid array–based assay, while it outperformed the micro-array methods. Again, the polyclonal rabbit anti-MS2 antibodies were used by this approach.

Bhatta et al. (2010) presented the universal multipurpose optical microchip biosensors for real-time label-free detection of the phage MS2 among a wide range of biological agents, such as bacteria, viruses, and toxins. The SpectroSens™ chips, containing high-precision planar Bragg gratings, were exploited as low-cost, robust refractive index sensors, while the sensitivity to biological agents was conferred by functionalizing the sensing surface with antibodies selected against targets of interest—the sheep polyclonal antibody in the case of the phage MS2. The further development of the SpectroSens consumables enabled simultaneous 16-channel data acquisition, including the phage MS2 detection (Bhatta et al. 2011).

Grego et al. (2011) adopted Tunable Wavelength Interrogated Sensor Technology (TWIST) for the point-of-care diagnostics of infectious diseases, where the phage MS2 was used as one of the numerous models. Generally, the TWIST platform was an optical evanescent wave sensor which enabled a label-free immunoassay-based portable instrument. The approach was based on input grating coupler sensors serving as functionalized sensing devices. The binding of the target analyte to the receptor-coated grating was detected by wavelength interrogation in the telecom spectral range. It was demonstrated that high-performance volumetric sensing could be achieved using a compact, low-cost telecom laser as light source. Given the height of the reaction chamber of 0.12 mm and the assumption that all of the phage antigens in the sample were captured by the antibody, the 2×10^{10} particle/mL was calculated as necessary phage concentration (Grego et al. 2011).

An immunochromatographic lateral flow assay detecting the phage MS2 as a model for viral pathogens was developed by using functionalized M13 phage-based nanoparticles as ultrasensitive reporters (Adhikari et al. 2013).

The surface plasmon resonance–based immunosensor was elaborated for the detection of the phage MS2 as a frequently used model organism in bioaerosol studies (Usachev et al. 2012, 2013). For the first time, a murine monoclonal anti-MS2 antibody was used in this study. Generally, this technology could be regarded as the first in the world real-time bioaerosol monitor.

Electrochemical biosensors featuring oriented antibody immobilization via electrografted and self-assembled hydrazide chemistry were applied for the MS2 detection by Prieto-Simón et al. (2014). This study presented two new functionalization strategies that preserved proper folding and binding potential of antibodies by forcing their oriented immobilization. Both strategies were based on the formation of hydrazone bonds between aldehyde groups on the Fc moieties of periodate-oxidized antibodies and hydrazide groups on functionalized gold electrodes. Those hydrazide groups were introduced by electrografting of diazonium salts or by self-assembly of mono- and dithiolated hydrazide linkers, resulting in films with tailored functional groups, and thus antibody distribution and spacing. The MS2 detection was performed through either a direct assay using electrochemical impedance spectroscopy or through a sandwich assay using differential pulse voltammetry. The best results of 15 pfu/mL were achieved for immunosensors based on mixed monolayers of hydrazide and hydroxyl diothiolated linkers (Prieto-Simón et al. 2014).

The work of Prieto-Simón et al. (2014) was employed for the use of carbon nanotubes (CNTs) as building blocks in the design of electrochemical biosensors (Prieto-Simón et al. 2015). The two approaches, based on tailored single-walled CNTs (SWCNTs) architectures, were presented to develop immunosensors for the phage MS2. In the first approach, the SWCNTs were used in the bottom-up design of sensors as antibody immobilization support. The carboxy-functionalized SWCNTs were covalently tethered onto gold electrodes via carbodiimide coupling to cysteamine-modified gold electrodes. These SWCNTs were hydrazide-functionalized by electrochemical grafting of diazonium salts. The site-oriented immobilization of antibodies was then carried out through hydrazone bond formation. In the second approach, the SWCNTs were decorated with iron oxide nanoparticles. The diazonium salts were electrochemically grafted on the iron-oxide-nanoparticle-decorated SWCNTs to functionalize them with hydrazide groups that facilitated the site-directed immobilization of antibodies via hydrazone coupling. These magnetic immunocarriers facilitated MS2 separation and concentration on an electrode surface. This approach minimized nonspecific adsorptions and matrix effects and allowed low limits of detection (12 and 39 pfu/mL in buffer and in river water, respectively) that could be further decreased by incubating the magnetic immunocarriers with larger volumes of sample (Prieto-Simón et al. 2015). The polyclonal rabbit and monoclonal murine anti-MS2 antibodies were used in both studies.

The phage MS2 contributed to the construction of an electrical immunosensor based on dielectrophoretically-deposited single-walled carbon nanotubes (SWCNTs) for the detection of influenza virus H1N1 (Singh et al. 2014). The immunosensors with a 2-μm channel length detected 1 pfu/mL of the influenza virus selectively from the phage MS2. The MS2 chip was used as a control for the H1N1 detection using a reduced graphene oxide (RGO)-based electrochemical immunosensor integrated with a microfluidic platform (Singh et al. 2017). The RGO has recently gained considerable attention for use in electrochemical biosensing applications due to its outstanding conducting properties and large surface area.

Garvey et al. (2014) introduced a diagnostic platform utilizing lithographically fabricated micron-scale forms of cubic retroreflectors, arguably one of the most optically detectable human artifacts, as reporter labels for the use in sensitive immunoassays. In fact, such retroreflectors were widely used on bicycles, traffic signs, and safety vests because of their high detectability. They returned incident light in a narrow beam directly back to the illumination source and therefore were significantly brighter than surfaces that scattered light. The authors demonstrated the applicability of this novel optical label in a simple assay format in which retroreflector cubes were first mixed with the sample. The cubes were then allowed to settle onto an immunocapture surface, followed by inversion for gravity-driven removal of nonspecifically bound cubes. The cubes bridged to the capture surface by the analyte were detected using inexpensive, low-numerical-aperture optics. The sensitivity of the method reached 10^4 MS2 particles/mL, using suspended, microfabricated retroreflector cubes as optical labels conjugated to the rabbit polyclonal anti-MS2 antibodies (Garvey et al. 2014).

An amperometric immune biosensor kit for MS2 detection was elaborated by using flow technology (Braustein and Braustein 2014).

A lateral flow immunoassay (LFA) was used for MS2 detection using the fluorescently labeled M13 phage as reporter and single-reporter counting as the readout (Kim J et al. 2015). The AviTag-biotinylated M13 phage was functionalized with antibodies using avidin-biotin conjugation and fluorescently labeled with AlexaFluor 555. The individual phage M13 bound to the phage MS2 as a target captured on an LFA membrane strip was imaged using epifluorescence microscopy. Using automated image processing, the number of the bound phage in micrographs was counted as a function of target concentration. Biotinylated rabbit anti-MS2 antibodies were used in this study (Kim J et al. 2015).

It is remarkable that the molecularly imprinted polymers (MIPs), or plastic antibodies, that have been mentioned in the Enumeration section of Chapter 5, appeared as analogs of natural antibodies and were employed for specific and sensitive recognition of viruses, including the phage MS2, with the combination of biosensors (Altintas et al. 2015a,b,c).

Meanwhile, a simple ionic conductivity sensor for the phage MS2 was based on anodized alumina membranes and demonstrated the lowest detection concentration $\sim$7 pfu/mL with pure MS2 and $\sim$30 pfu/mL in the 1:1 mixtures of MS2:Qβ (Chaturvedi et al. 2016). This study used polyclonal anti-MS2 antibodies.

The recent advances in electrochemical immunosensors, including the remarkable MS2 contribution, were exhaustively reviewed by Piro and Reisberg (2017).

Finally, the photodynamic inactivation of the phages MS2 (Majiya et al. 2018b) and Qβ (Majiya 2018a) that was described earlier in the Photosensitizing agents section in Chapter 6 should be remembered. As it followed from the latter, as well as from the Spatial structure section in Chapter 11, the maturation protein appeared as a target by the MS2 photodynamic inactivation (Majiya et al. 2018a,b). In fact, the four polyclonal rabbit antibodies were raised against the four sites on the maturation protein. Using these antibodies, it was demonstrated that the rate of photodynamic inactivation was relative to loss of antigenicity of the two sites on the maturation protein. Although photodynamic inactivation caused aggregation of the MS2 particles and crosslinking of the MS2 coat, these inter- and intracapsid changes did not correlate to the rate of the photodynamic inactivation. Therefore, the deliberate involvement of the anti-maturation antibodies contributed to solution of the problem.

16 Translation

When you look long into an abyss, the abyss looks into you.

Friedrich Nietzsche

The pure and simple truth is rarely pure and never simple.

Oscar Wilde

THE ESSENTIALS

As noted in the Translation section of Chapter 1, it is necessary to understand from the outset that the RNA phage messengers played a crucial role in the decryption of the translation mechanisms and genetic code, and contributed greatly to the central subject of molecular biology. First, the discovery of the RNA phages offered a ready source of homogeneous messenger RNA. The RNA phage messengers initiated the classical *in vitro* studies of the *necessary-and-sufficient* components of the translational machinery. Since the 1970s, the R17, MS2, and Qβ genes were the sources of the the first messenger RNAs with known sequence. In 1976, the whole MS2 genomic structure was made available for the studies of translation principles. This made it possible to decipher the genetic code and establish the frequency of codon usage. Moreover, it subsequently emerged that all possible nucleotide triplets were employed by the genetic code. Thus, the obtained nucleotide sequences of the RNA phage genes have confirmed the validity of the genetic code, while the versatility of the latter was established by the translation of the RNA phage messengers in eukaryotic cell-free systems.

Knowledge of the terminal and intergenic noncoding regions was acquired for the first time. The principles and instruments of protein initiation and termination were disclosed. Thus, the prevalent role of AUG as a start codon was ascertained, but the GUG, UUG, and AUU triplets were appreciated as the minor start codons. The usage principles, general contribution, and functional difference of the termination codons UAG, UAA, and UGA as the protein punctuation signals were established. Moreover, the RNA phage messengers contributed to the determination of the mechanism of nonsense suppression.

The studies on the RNA phage messengers outlined an approach to the ribosome structure and function mechanisms, including the factors involved into the initiation, elongation, and termination processes, as well as the role and function of transport RNAs. The formylated methionine tRNA was found directly because of the RNA phage studies. It led further to the conclusion that all *E. coli* proteins were started by N-formyl methionine. The studies of ribosome binding to phage RNAs led to discovery of the ribosome recognition sites by correct polypeptide initiation on mRNAs, including the famous Shine-Dalgarno (SD) sequences. Remarkably, the phage f2 RNA was used by the first examination of the apparent changes in ribosome conformation during protein synthesis by transition to the pre- and post-translocation complexes from the initiation one (Waterson et al. 1972).

The phage RNA-directed *in vitro* protein synthesis was used to estimate the effect that various compounds exerted at the level of protein synthesis. The phage RNAs contributed to the clarification of the action mode of a long list of antibiotics, such as bottromycin, chloramphenicol, edeine, erythromycin, evernimicin, gentamicin, hygromycin, josamycin, kasugamycin, minosaminomycin, negamycin, neomycin, pactamycin, purpuromycin, siomycin, spectinomycin, streptomycin, tetracycline, tuberactinomycin, viomycin, and virginiamycin (Schwartz 1965; Anderson P et al. 1967; Luzzatto et al. 1968, 1969a,b; Modolell and Davis 1968, 1969; Gurgo et al. 1969; Mizuno et al. 1970; Cocito and Kaji 1971; Modolell et al. 1971; Okuyama et al. 1971, 1972; Celma et al. 1972; Kozak and Nathans 1972a; Okuyama and Tanaka 1972; Szer and Kurylo-Borowska 1972; Legault-Demare et al. 1973; Stewart and Goldberg 1973; Tai et al. 1973a; Wallace et al. 1973; Pinkett and Brownstein 1974; Szer and Leffler 1974; Uehara et al. 1974; Otaka and Kaji 1976; Reusser 1976; Suzukake and Hori 1977; Yamada and Bierhaus 1978; Klita and Szafrański 1980; Yamada et al. 1980; van Buul et al. 1984; Leclerc et al. 1991; Landini et al. 1992; McNicholas et al. 2000; Ganoza and Kiel 2001; Lovmar et al. 2004). In his excellent review on inhibitors of the ribosome function, Sidney Pestka (1971), the future "father of interferon," has clearly pointed out the great value that the phage RNA-directed *in vitro* protein synthesis has achieved in studying the mechanism of action of various antibiotics. Later, the MS2-direct *in vitro* protein synthesis was used by the generation of a designer antibiotic on the basis of paromomycin and the x-ray resolution of its complex with a model aminoacyl site of the 30S ribosomal subunit (Russell et al. 2003). In parallel, the MS2 system contributed to the NMR resolution of the ribosome-binding sites of thiazole antibiotics; namely, thiostrepton and micrococcin (Lentzen et al. 2003).

The MS2 RNA-programmed *in vitro* protein synthesis was employed by examination of small peptides that inhibited translation by targeting the specific sites of ribosomal RNA within the ribosome and not only accelerated production of new antibiotics but also provided new tools for ribosomal research (Llano-Sotelo et al. 2009). Such peptides were selected by a principally novel approach when a heptapeptide phage-display library was screened for the affinity to the h18 pseudoknot of 16S rRNA, which played a central role in functions of the ribosome decoding center and was directly involved in controlling the accuracy of the aminoacyl-tRNA selection (Ogle et al. 2001). As a result, two of the selected peptides could interfere efficiently with both bacterial and eukaryotic translation.

The MS2 RNA was used as a model by the characterization of the depurination activity of saporin (Barbieri et al. 1994), a ribosome-inactivating protein that appeared as a representative of the multigene family of proteins in various parts of the *Saponaria officinalis* plant. Interest in the ribosome-inactivating proteins (RIPs), which were produced in fact not only by plants, but also by bacteria and fungi, grew recently in connection with their potent anti-HIV activity (Yadav and Batra 2015).

The binding of the phage f2 RNA to *E. coli* ribosomes was used as a model system to study interactions of triphenylmethane dyes, including aurintricarboxylate (ATA), which appeared as novel inhibitors of mRNA translation (Grollman and Stewart 1968). Siegelman and Apirion (1971) found that ATA preferentially inhibited initiation, while elongation of the polypeptide chain was not appreciably affected by the *in vitro* translation of the R17 RNA.

Then, the f2 RNA-directed cell-free protein synthesis system made it possible to establish that trimethoprim, an inhibitor of dihydrofolate reductase, presumably blocked protein synthesis by making N-formylmethionyl-tRNA unavailable (Shih et al. 1966; Klein et al. 1970).

Furthermore, using the f2 RNA-directed system, Boon (1971) first showed that purified colicin E3 would inactivate ribosomes *in vitro*. The subsequent investigation showed that the colicin cleaved the 16S ribosomal RNA near the 3' end to inactivate the 30S ribosome (Bowman et al. 1971; Boon 1972). By analogy, the ribosomes from *Enterobacter cloacae* cells that have been treated with bacteriocin cloacin DF13 produced by the *E. cloacae* strain DF13 were unable to support the MS2 RNA-directed protein synthesis (De Graaf et al. 1971). The cloacin-treated ribosomes fully lost their ability to sustain polypeptide synthesis due to specific cleavage of 16S rRNA but were still able to form initiation complexes with the MS2 RNA (Baan et al. 1976). Remarkably, the cloacin-cleaved ribosomes retained their fidelity and specificity, and started translation only on the coat gene of the MS2 RNA (Twilt et al. 1979).

Small amounts of diphtheria toxin inhibited the R17 RNA-directed amino acid incorporation in the *E. coli* cell-free system (Tsugawa et al. 1970). The authors suggested that diphtheria toxin might prevent the binding of mRNAs by successfully competing with the AUG for the ribosomal binding sites.

The action mechanism of such a popular translation inhibitor as fusidic acid was clarified finally in the R17 RNA-directed *in vitro* protein synthesis system (San Millan et al. 1975). The latter was also used by the studies of tiamulin, a semisynthetic pleuromutilin derivative, on the bacterial polypeptide chain initiation (Dornhelm and Högenauer 1978). The cell-free extracts of tiamulin resistant mutants of *E. coli* exhibited an increased *in vitro* resistance to tiamulin in the R17 RNA-dependent protein synthesis systems, and it was a property of the 50S subunit (Böck et al. 1982).

The MS2 RNA-directed *in vitro* protein synthesis was employed by the evaluation of the inhibitory mechanisms of oxazolidinones, a new class of synthetic antibiotics with good activity against Gram-positive pathogenic bacteria (Shinabarger et al. 1997). Furthermore, MS2 RNA-directed

in vitro protein synthesis was used by the characterization of mutations in the ribosomal RNA recognition domain for the ribotoxins ricin and sarcin (Macbeth and Wool 1999).

The studies of the tobacco mosaic virus RNA-programmed protein synthesis in the cell-free *E. coli* extracts were developed in tight connection with the models directed by the phage R17 (Boedtker and Stumpp 1966) and f2 (Schwartz 1967) RNAs. MS2 RNA was used as a permanent control by the *in vitro* translation in the *E. coli* extracts programmed by plant virus, namely, tobacco mosaic virus, alfalfa mosaic virus, satellite tobacco necrosis virus, and turnip yellow mosaic virus RNAs (Hoogendam et al. 1968; Reinecke et al. 1968; Albrecht et al. 1969a,b; Castel et al. 1979).

Furthermore, the experience in the f2 RNA-directed cell-free synthesis contributed substantially to the successful implementation of the synthesis of specific proteins in *E. coli* extracts programmed by poliovirus RNA which were performed by Nobelist 1975 David Baltimore's team (Rekosh et al. 1969, 1970). The same was true for the studies on hemoglobin synthesis in extracts of rabbit reticulocytes, as a direct continuation of the f2 RNA-programmed syntheses (Lodish et al. 1971). The Qβ RNA-directed *in vitro* protein synthesis was shown to be increased when a 0.5-M KCl wash fraction from rabbit reticulocyte ribosomes was added to the Krebs ascites cell lysate (Metafora et al. 1972). The fidelity of the MS2 RNA translation initiation in rabbit reticulocyte lysates was studied later by Jackson (1991).

The f2 RNA-programmed protein synthesis system was also employed in the studies that led to discovery of novel tRNAs, such as tRNA^{Gly} from *Staphylococcus epidermidis*, which participated in peptidoglycan synthesis but not in protein synthesis (Stewart TS et al. 1971).

Next, the R17 RNA-programmed *in vitro* system was employed by the discovery of *magic spots*, the nucleotides ppGpp and pppGpp, which were synthesized on ribosomes when wild-type, but not rel⁻, strains of *E. coli* were starved for the required amino acids (Haseltine and Block 1973). The signal for making the *magic spots* was the presence of an uncharged tRNA in the ribosomal acceptor site, and these compounds did not accumulate if the ribosomes were actively engaged in protein synthesis. However, the ppGpp and pppGpp exerted no effect on the messenger-directed synthesis of the MS2 coat protein (Yang et al. 1974).

The order and different efficacy of the phage protein synthesis *in vivo* and *in vitro* promoted for the first time the general idea of translational regulation. Moreover, the idea of the global regulation of the phage replication through translational mechanisms was generated. A set of such mechanisms was established. First, regulation by the accessibility of the initiation sites to ribosomes, i.e., regulation by the secondary and tertiary RNA structure, where polarity of the coat protein *amber* mutations appeared as the first example, and mechanisms of the initiation of the maturation and lysis genes followed. Second, the structure of the three initiation regions revealed the role of the SD sequences determining the initiation strength. Third, a mechanism of the strong translational control was confirmed, when the clear examples of the two repressor complexes formed by coat protein (repressor complex I) and replicase (repressor complex II) were acknowledged. Therefore, the translation process appeared to be coupled strongly to the replication of the phage RNA thru the two repressor complexes *pari passu* with the simultaneous implementation of both processes that were exploiting common components, such as S1 protein and factors EF-Tu, EF-Ts, and Hfq, as described in detail for the RNA phage replicase enzyme in Chapter 13. The biochemical principles of the repressor complexes I and II are described in the present chapter, but the complete pattern of the regulation of the phage replication is the major subject of Chapter 17.

The *in vitro* translation system offered a unique opportunity to obtain functionally complete proteins and measure their sizes by their mobility in a polyacrylamide gel. First, it was clearly established that the three RNA phage proteins produced *in vitro* did not differ from their analogs that appeared *in vivo*. Second, the fidelity of the *in vitro* translation system was verified, and the successive step-by-step elucidation of the translation process was announced.

It must be admitted finally that the first 10 years of the RNA phage studies proceeded with a tremendous emphasis on the RNA-directed protein synthesis. A comprehensive survey of the role of the RNA phages by decrypting of the general translation mechanisms was performed in the two first books devoted completely to the RNA phages in the 1970s (Gren 1974; Zinder 1975). In the classical *RNA Phages* book edited by Norton D. Zinder (1975), the Nobelist 2007 Mario R. Capecchi and Robert E. Webster (1975) presented the RNA phage RNAs as templates for *in vitro* protein synthesis, while Harvey F. Lodish (1975) described the regulation of RNA-directed *in vitro* protein synthesis by RNA tertiary structure, and Joan Argetsinger Steitz (1975) summarized her revolutionary studies by the ribosome recognition of the initiator regions in the RNA phage genome. The excellent review of Marilyn Kozak and Daniel Nathans (1972b) is also highly recommended, as well as more general reviews of that time on translation mechanisms (Lengyel and Söll 1969; Lucas-Lenard and Lipmann 1971; Haselkorn and Rothman-Denes 1973). In the 1980s, the exhaustive review of Elmārs Grēns (Gren 1984) is worthy of mention. This large review described exhaustively the recognition of messenger RNAs during translational initiation in *E. coli* and included virtually all known examples from the RNA phages.

In the historical context, it is paradoxical that the first revolutionary results of Daniel Nathans et al. (1962a,b) on the *in vitro* synthesis of a phage protein were used as an argument by one the last attempts to disprove the validity of the messenger concept (Hendler 1963). The historical role of the RNA phages in the unveiling of the fine mechanisms of protein synthesis and translational control was outlined recently (Tahmasebi et al. 2019).

TRANSLATION *IN VIVO*

Stephen Cooper and Norton D. Zinder (1963) were the first to obtain clear evidence that an early protein synthesis step is required in the phage f2 replication. This step was related to the synthesis of a phage-specific subunit of the RNA replicase enzyme, and several groups have each described such an enzyme in extracts of the phage f2 (Kaye et al. 1963) and MS2 (Haruna et al. 1963; Weissmann and Borst 1963; Weissmann et al. 1963a,b) infected cells and just partially purified it, as narrated in the Non-allolevivirus replicases section of Chapter 13.

At the end of the 1960s–beginning of the 1970s, the three phage-specific proteins were characterized by the their mobility in a gel electrophoresis: the coat protein of approximately 14,000 daltons, a maturation protein of approximately 40,000 daltons, and a replicase of approximately 65,000 daltons by the phage MS2 (Viñuela et al. 1967a,b; Remaut et al. 1968; Sugiyama and Stone 1968a,b; Nathans et al. 1969), f2 (Fromageot and Zinder 1968), and Qβ (Horiuchi et al. 1971) infection. An additional protein A1 of approximately 36,000 daltons was synthesized in the Qβ-infected bacteria and consisted of approximately 200 amino acids attached to the C-terminus of the complete coat protein (Horiuchi et al. 1971; Moore et al. 1971; Weiner and Weber 1971). From the RNA sequence analysis of partial nuclease digests of the phage RNA, a 5′ to 3′ gene order of maturation protein-coat protein-replicase was established at that time for the phages R17 (Jeppesen et al. 1970a,b) and Qβ (Hindley et al. 1970; Staples et al. 1971). The fourth protein, lysis protein, was found later, as described in detail in Chapter 14.

The *in vivo* studies revealed the definite order of the appearance of the phage proteins. The replicase subunit appeared first in a relatively low amount, while the highly efficient coat protein synthesis occurred late in the infection. The moderate synthesis of the maturation protein and low synthesis of the lysis protein were observed late in the infection. The synthesis of the lysis protein was coupled tightly to the coat synthesis. Remarkably, the order of the protein synthesis did not correspond to the gene order. The efficiency of each protein synthesis was definitely different. One may conclude from the *in vivo* data on the protein synthesis that the translation of the phage RNA was subjected to circumstantial regulation. The existence of such regulation was confirmed *in vivo* by the fact that the natural ratios of the phage proteins could be surmounted first by mutations in the coat and replicase, but not of the maturation and lysis genes. As we now know, these mutations affected functioning of the repressor complexes I and II. Second, the natural ratio of the phage protein synthesis *in vivo* was disrupted by the inhibition of the protein synthesis by chloramphenicol, which again led to the failure of the repressor complex function. The true mechanisms of the translation regulation by the repressor complexes I and II was deciphered *in vitro* and will be presented below, but the general pattern of the replication mechanisms will be summarized in Chapter 17.

Most of the phage-specific RNA was localized *in vivo* in ribosomal fractions of the infected cells. Thus, in a lysate of MS2-infected cells, 70% of the phage progeny single-stranded RNA was present in polyribosomes that contained 2 to 5 ribosomes, while no pool of free RNA molecules appeared during infection (Godson 1968). Fenwick (1971b) measured the density of ribosomes bearing messenger RNA in the R17-infected and uninfected bacteria. The R17 RNA that was associated with groups of 1, 2, 3, or 4 ribosomes was detected. It was obvious that the distributions of mRNA were quite different in infected and in uninfected bacteria, where the latter demonstrated only few or no polygenic mRNA molecules comparable in size to the R17 RNA. In parallel, Iglewski (1971) came to the same conclusion that the labeled RNA was associated with larger polysomes from the R17-infected cells as compared to control cells. Truden and Franklin (1972) appraised the R17 RNA-specific polysomes as containing four or more ribosomes. The 30S ribosomal subunits through trimer-size polysomes, which were associated with all of the R17-specific proteins and were predominant in the infected cell, synthesized only coat protein. These structures accumulated as products derived from larger polysomes. The appreciable amounts of viral coat protein remained attached to ribosomes and polysomes during R17 replication, supporting the idea of the repressor role of this protein, when the coat was preferentially synthesized, but the synthesis of non-coat proteins was suppressed (Truden and Franklin 1972).

TRANSLATION *IN VITRO*

The f2 RNA exhibited characteristics of a messenger RNA (Nathans et al. 1962a,b) in the cell-free protein synthesizing system of Nirenberg and Matthaei (1961) and stimulated the incorporation of amino acids into a product which has been shown to be similar to the phage coat protein. The f2 RNA also stimulated the incorporation of histidine into an acid-insoluble product (Nathans et al. 1962a; Valentine and Zinder 1963) which definitely was not coat protein. It is noteworthy that the *in vitro* synthesis of the coat protein directed by the f2 RNA (Nathans et al. 1962a,b) was, together with hemoglobin (Bishop et al. 1961), among the two first products of the cell-free *in vitro* system, which have been identified chemically as specific proteins by peptide analysis. At this early period, the RNAs of the phages M12 (Doerfler et al. 1963) and MS2 (Möller and von Ehrenstein 1963) were also found to be capable for programming amino acid incorporation in the *E. coli* cell-free system. Moreover, Möller and von Ehrenstein (1963) were the first to demonstrate that synthetic polynucleotides may act as specific inhibitors of protein synthesis, which was induced by the MS2 RNA. The ZIK/1 RNA was also an efficient template in the *E. coli* cell-free system and produced coat protein as the major product, as well as a number of non-coat polypeptides (Robinson 1972).

Daniel Nathans (1965) was the first to present tryptic peptides of the MS2 RNA-directed product, which corresponded to the tryptic peptides of the phage coat protein. Moreover, the formation of whole-coat protein molecules as the predominant product *in vitro* was shown by co-chromatography on Sephadex with an authentic phage coat subunit. It was confirmed also that the MS2 RNA directed the synthesis of histidine-containing polypeptides, other than the coat, in addition to the coat protein. Immunological relatedness of the

in vitro-synthesized MS2 coat to the authentic protein was presented by Wade (1967). Knolle (1969) observed the incorporation of the labeled amino acids into the phage fr- and MS2-like particles in the *E. coli* cell-free protein synthesis system. Remarkably, the MS2 RNA-programmed amino acid incorporation was stimulated by addition of ribosomal 5S RNA (Kirtikar and Kaji 1968). The whole-protein synthesis mechanisms relying mostly on the phage RNA-programmed experiments were described at that time in the exhaustive review of Heinrich Matthaei et al. (1968). The detailed methodology of the MS2 RNA-programmed *in vitro* synthesis in the *E. coli* cell-free extracts was described by Nathans (1968).

Then, the fidelity of the *in vitro* translation of the R17 RNA in the *E. coli* cell-free system was confirmed chemically by the peptide mapping of the synthesized coat protein (Burns and Bergquist 1970). The ^{14}C-labeled N-terminal peptides of the Qβ, f2, and R17 proteins synthesized in the *E. coli in vitro* system were identified, isolated, and sequenced by Osborn et al. (1970a). The *in vitro* translation of the MS2 RNA was described at that time by Elmārs Grēns' team (Rozentāls and Grēns 1971).

The Qβ proteins, namely, coat, A1, and replicase subunit, were synthesized *in vitro* in sufficient amounts and mapped by the appropriate fingerprint patterns after tryptic digestion of the isolated products (Hindennach and Jockusch 1974).

The complementary chain, or RNA minus strand, did not demonstrate the messenger activity, as demonstrated with the highly purified R17 minus strand under conditions optimal for the *in vitro* protein synthesis in the *E. coli* cell-free extracts (Schwartz et al. 1969). Remarkably, although incapable of associating with ribosomes and of initiating protein synthesis on its own, the minus strand RNA enhanced the incorporation of amino acids directed by the R17 plus RNA. As a result, an association of the two strands under the incubation conditions for cell-free protein synthesis was suggested (Schwartz et al. 1969).

Zinder (1980) noted in his classical review portraying the RNA phage that the basic f2 translation studies were done in parallel in his laboratory and in that of James D. Watson, and no single discovery was made in one laboratory alone. Zinder wrote that "although the procedures were often different, the results were always the same. This competition was hard on the nerves, particularly those of the students. However, since most results had been replicated in two laboratories, they were quickly and generally accepted.Thus, 70% of the protein synthesized on the f2 RNA was its coat protein; replicase or its fragments amounted to 30% of the protein synthesized, according to Lodish (1968d), and no maturation protein could be found" (Zinder 1980).

This unique period in the *in vitro* translation studies deserved the special reviews written by Harvey F. Lodish (1975) and Mario R. Capecchi and Robert E. Webster (1975) for the *RNA Phages* book. A special emphasis on the translation studies was also made in the first RNA phage book written by Elmārs Grēns (Grēns 1974). Lodish (1975) emphasized at the very beginning of his review the considerable efficiency differences between the *in vivo* and *in vitro* translations. He calculated from the most optimal data that, in even the most efficient bacterial cell-free systems available, such as preincubated crude extracts of *E. coli* (Webster et al. 1967), the amount of the protein produced, expressed as micromoles of coat protein synthesized per micromole of RNA added, was about four, whereas in the infected cell this number was probably at least 10- to 40-fold greater, as calculated from Fromageot and Zinder (1968) for the phage f2 and from Nathans et al. (1969) for the phage MS2.

Capecchi and Webster (1975) also turned to the moment that the amount of protein synthesized *in vitro* was much less than that expected from the *in vivo* experiments. Thus, whereas one expected at least 40 coat protein molecules per RNA molecule in the infected cell (Nathans et al. 1969), only 4 phage coat proteins per RNA molecule were obtained *in vitro* (Engelhardt et al. 1967; Webster et al. 1967).

Furthermore, Capecchi and Webster (1975) calculated that the rate of protein synthesis was much slower *in vitro* than expected. The normal rate of the chain elongation expected in *E. coli* from *in vivo* studies has been estimated at approximately 1000 amino acids per minute (Forchhammer and Lindahl 1971). The phage RNA directed the *in vitro* polymerization of amino acids into protein at a rate of 30–80 amino acids per minute, less than 10% of that found *in vivo* (Webster and Zinder 1969b; Tai et al. 1973a,b).

Therefore, at best only about 20% of the f2 RNA added to cell lysates was ever attached to ribosomes and presumably directed the protein synthesis. One could never be sure that the general properties of the added mRNA population were representative of the active fraction (Webster and Zinder 1969b; Lodish 1975). According to Lodish's (1975) qualified statement, in most fractionated cell-free systems, containing washed ribosomes, purified initiation factors, and so forth, the efficiency of protein synthesis was much less, since some factors or components were limiting in these systems relative to the whole cell. Moreover, the phage RNAs used were generally isolated by phenol extraction, followed by precipitation with ethanol, and it was not at all clear that the 3D structure of the resultant RNA in solution was similar to what occurred either inside the virion or inside the bacterial cell. Next, the ionic conditions were adjusted so as to give the optimum *in vitro* synthesis, irrespective of whether this is the *natural* concentration. Nonetheless, the *in vitro* studies in cell-free systems confirmed the data concerning regulation of phage protein synthesis in the phage-infected cells *in vivo*, first, by the use of *amber* mutants in the phage proteins and by deduction of a putative regulatory function of the protein in question (Lodish 1975). It is noteworthy in this context that the MS2 and Qβ RNAs that were later purified on reverse-phase-chromatography-5 columns retained their messenger activity and directed the initiation of their respective gene products in approximately the proper relative proportions (Campbell et al. 1980).

Generally, when the purified f2 RNA was used as a messenger in the *E. coli* cell-free system, the predominant product was the coat protein (Nathans et al. 1962a,b). The major non-coat protein formed was replicase produced as about one-third

of the coat protein in the case of RNAs from the phages R17 (Capecchi 1966a), MS2 (Eggen et al. 1967; Viñuela et al. 1967a,b), and f2 (Lodish 1968d). Yamazaki and Kaesberg (1966) observed the *in vitro* synthesis of a product similar to R17 coat protein, but the major product was characterized as a very basic polypeptide resistant to trypsin and similar to lysine-rich histones. Igarashi and Paranchych (1967) found that the stimulation of amino acid incorporation by the R17 RNA occurred at low Mg^{2+} concentrations of 8–9 mM, and only when fMet-tRNA was present to initiate protein synthesis by virtue of its ability to bind to the ribosome-phage RNA complex. At that time, William Paranchych's team had undertaken some remarkable attempts to improve the efficiency of the R17 RNA-directed *in vitro* protein synthesis (Igarashi 1969; Igarashi and Paranchych 1969, 1970a,b; Krahn and Paranchych 1970).

Takeda (1969) showed the stimulatory effect of polyamines, such as spermine and spermidine, on the MS2 RNA-programmed *in vitro* protein synthesis and their ability to replace Mg^{2+} ions and to thereby cause a shift in the optimal Mg^{2+} concentration. Later, Watanabe et al. (1981) found that spermidine ensured differential stimulation of the MS2 and Qβ RNA-directed protein syntheses. Thus, the degree of stimulation was in the order replicase greater than A protein, while the synthesis of coat protein was not stimulated significantly by spermidine. Then, Praisler et al. (1984) found that the MS2 RNA translation was supported by spermidine, while putrescine had no further stimulatory effect, but diaminoheptane enhanced the rate of translation. The spermidine stimulation of the MS2 RNA-directed replicase synthesis in the *E. coli* cell-free system disappeared with the addition of methylglyoxal bis(guanylhydrazone) (MGBG), a structural analog of polyamines (Ohnishi et al. 1985). Then, aminopropylcadaverine, a compensatory polyamine, stimulated the *in vitro* synthesis of the MS2 replicase (Igarashi et al. 1986). It is remarkable in this context that the MS2 RNA-directed *in vitro* replicase synthesis was stimulated at least twofold in the presence of 1 mM spermidine by so-called polyamine-induced (PI) protein that was induced in the polyamine-requiring mutant of *E. coli* following polyamine supplementation of polyamine-starved cells (Mitsui et al. 1984). To relate the *in vitro* translation data to live bacteria, Miyamoto et al. (1993) demonstrated the stimulatory effect of polyamines on the synthesis of the MS2 replicase *in vivo*.

Remarkably, the translational *in vitro* capacity of *E. coli* with the R17 and Qβ RNAs as messengers was several times higher if the extracts were prepared from cells harvested in early exponential phase or grown under conditions of good aeration compared to extracts prepared from cells harvested in a later growth phase or grown under semi-aerobic conditions (Isaksson et al. 1980). In low-activity extracts, the replicase production was preferentially affected.

Since the coats of the group I phages did not contain histidine, Ohtaka and Spiegelman (1963) succeeded in showing that the MS2 RNA-directed synthesis *in vitro* of the majority of non-coat protein, measured by incorporation of histidine into protein, lagged behind that of the coat protein. This result was confirmed in the case of R17 (Capecchi 1967b), f2 (Engelhardt et al. 1967), and MS2 (Eggen et al. 1967). Using N-formyl-^{35}S-methionyl-tRNA to label specifically the amino termini of the proteins synthesized in the cell-free system, it was shown that the f2 RNA initiated the synthesis of coat protein and only two additional proteins, that one of these non-coat proteins, synthesized at about 20%–30% the amount of the coat protein, was replicase, and that the initiation of the replicase synthesis followed initiation of the coat synthesis (Lodish 1968d).

When binding of aminoacyl-tRNAs to ribosomes programmed with the MS2 RNA was studied, alanyl-tRNA was also extremely well bound to the initiation complex, besides fMet-tRNA, while seryl- and lysyl-tRNAs were also bound, although to a far lesser extent as compared with the alanyl-tRNA reaction (Voorma et al. 1969, 1971).

The qualified analysis of the *in vitro*-synthesized proteins was made possible by the broad introduction of the polyacrylamide gel electrophoresis technique, as referenced above and also earlier in Chapters 8 and 11–14. Further insight into the relationship between initiation of the coat and replicase proteins came from the *in vitro* studies on the coat *amber* mutants, as it will be described below in the Polarity section.

HETEROLOGOUS TRANSLATION *IN VITRO*

Stunningly, Schwartz et al. (1965) showed that the f2 RNA could direct the synthesis of the characteristic coat protein peptides in extracts from the chloroplasts of the single cell flagellate eukaryote *Euglena gracilis*, an organism quite unlike *E. coli*. Then, Schwartz et al. (1967) isolated N-formylmethionine from the product synthesized in extracts of *E. gracilis* and showed that the N-terminal chymotryptic peptide from the *E. gracilis* product was identical to the same peptide from the product made in extracts of *E. coli*. Later, the purified chloroplast initiation factor IF-3 of *E. gracilis* stimulated formation of the initiation complex on *E. coli* ribosomes with the MS2 RNA (Kraus and Spremulli 1986). In fact, this was the first report of an organellar ribosome dissociation factor.

Then, the RNAs of Qβ (Aviv et al. 1972; Morrison and Lodish 1973; Schreier et al. 1973), MS2 (Gordon et al. 1972), f2 (Morrison and Lodish 1973), and R17 (Schreier et al. 1973) were shown to function as messengers in the cell-free amino acid–incorporating systems derived from mammalian cells and tissues. In all cases, one of the main products synthesized was a polypeptide resembling the phage coat protein. Moreover, the RNA from the Qβ coat *amber* mutant am11 containing a termination codon at the coat position 37 directed the synthesis of the corresponding coat fragment (Morrison and Lodish 1973). Next, the cell-free extracts of the Krebs II mouse ascites cells directed by the Qβ RNA synthesized the expected Qβ replicase and, in addition, a polypeptide which resulted from initiation of protein synthesis in the interior of the replicase gene (Morrison and Lodish 1974). Moreover, the formyl-^{35}S-methionine was incorporated into several initiator sequences not found in authentic f2 proteins. These incorrect

initiations on the phage RNAs by mammalian ribosomes were in marked contrast to the correct initiation by the mammalian ribosomes on the mammalian viral and cellular RNAs in the corresponding cell-free systems (Morrison and Lodish 1974). The methodology of the Qβ RNA translation in cell-free extracts of mammalian cells was presented in detail by Villa-Komaroff et al. (1974). Furthermore, Atkins et al. (1975) stimulated translation of the R17, Qβ, and MS2 RNAs in a mammalian cell-free system by addition of polyamines spermidine, spermine, or putrescine.

The Qβ RNA was translated successfully by cell-free extracts from wheat embryos (Davies and Kaesberg 1973). The three products were synthesized which co-electrophoresed with the Qβ proteins synthesized in the *E. coli* extracts, and the smallest one was clearly identified as coat protein. Furthermore, optimal conditions were selected for the efficient Qβ RNA translation, among other synthetic and natural messages, by the wheat germ protein-synthesizing system (Rychlik and Zagórski 1978). The Qβ RNA translation in wheat germ, as well as in the cell-free extracts of reticulocytes, contributed further strongly by the elucidation of the role of the messenger capping, namely, of the 5′-terminal 7-methylguanosine residue, in translation of mammalian mRNAs (Bergmann and Lodish 1979).

A bit surprisingly, the f2 RNA did not bind to ribosomes of *Clostridium pasteurianum*, a Gram-positive anaerobic soil bacterium, and demonstrated no significant stimulation of protein synthesis in a homologous system or with *E. coli* initiation factors (Himes et al. 1972). Further detailed studies showed that the salt-washed clostridial ribosomes, and not the initiation factors, controlled the specificity of the f2 RNA translation (Stallcup and Rabinowitz 1973a,b,c). By further broadening the scope of tested bacteria, Stallcup et al. (1976) established that ribosomes from the two Gram-negative bacteria, *E. coli* and *Pseudomonas fluorescens*, were able to translate the f2 RNA, while ribosomes from the Gram-positive bacteria *C. pasteurianum*, *Streptococcus faecalis*, and *Bacillus subtilis* failed. However, the crude initiation factor preparations from both vegetative and sporulating *B. subtilis* cells were active in promoting the initiation factor-dependent translation of the Qβ RNA by washed *E. coli* ribosomes (Chambliss and Legault-Demare 1977).

Looking for the specificity of the *Enterobacteriaceae* translation systems, Howk et al. (1973) found that the Qβ and MS2 RNAs were translated by cell-free extracts of *Shigella alkalescens* and to a lesser extent *Citrobacter freundii*, but not of *Salmonella typhimurium*, *S. pomona*, and *Proteus morganii*. The ribosomal rather than the soluble components appeared responsible for the ineffectiveness in translation. In parallel, Bassel and Curry (1973) found that the *S. typhimurium* extracts translated the f2 and Qβ RNAs only 10%–15% as fast as the *E. coli* extracts did. However, similar amounts of phage RNA-ribosome complexes were observed in both the *S. typhimurium* and *E. coli* systems, indicating that the different activities observed might be attributed to different rates of peptide elongation or to the formation of complexes at different sites on the RNA strand.

The response of the MS2 RNA-directed *in vitro* protein synthesis to pressure was measured in the barotolerant deep-sea *Pseudomonas bathycetes* cell-free extracts, in comparison with the *E. coli* and *P. fluorescens* extracts (Landau et al. 1977). In all cases, the responses to pressure were parallel to those obtained when unwashed ribosomes were utilized. It led to the conclusion that the initiation factors were interchangeable among these organisms, and that these factors did not play a critical role in determining the pressure responses of the protein-synthesizing systems.

Furthermore, the effects of low temperature on the MS2 RNA-programmed *in vitro* protein synthesis were compared in mesophilic *E. coli* with psychrotrophic *P. fluorescens* cell-free extracts (Broeze et al. 1978). The *P. fluorescens* cell extracts synthesized protein at linear rates for up to 2 h at 5°C, whereas *E. coli* cell extracts synthesized protein for only 25 min, although the rates of polypeptide elongation remained identical for both organisms over the range of 25°–0°C.

In the cell-free extracts of the rickettsia-like, obligate intracellular bacterium *Coxiella burnetti*, the Qβ RNA translation proceeded at a rate and to an extent equal to that obtained in the conventional *E. coli* system and revealed the major product that co-electrophoresed with the Qβ coat protein (Donahue and Thompson 1980, 1981). When translation of the Qβ RNA by *C. burnetti* extracts was conducted at temperatures elevated above 37°C, both polypeptide elongation and frequency of initiation were increased, and the ratios of products changed (Donahue and Thompson 1987).

The ribosomes that were isolated from an enterotoxigenic strain of *Clostridium perfringens* cells in vegetative or in sporulation phase could equally translate the R17 RNA with either vegetative or sporulation initiation factors (Smith and McDonel 1980).

The cell-free extracts of the moderate halophile *Vibrio costicola* translated the R17 RNA when fMet-tRNA was present and produced a protein of the same electrophoretic mobility as the R17 coat protein (Choquet and Kushner 1990).

The highly efficient cell-free protein synthesis system of an extremely thermophilic eubacterium *Thermus thermophilus* HB27 allowed production of the MS2 RNA translation products at a rate of more than 5 μg per hour per 1.9 mg of ribosomes at 65°C when the production continued linearly for at least 340 min (Uzawa et al. 1993b). However, when no polyamine was added, the system did not produce the proteins. The results revealed the physiological importance of a branched quaternary polyamine, tetrakis(3-aminopropyl) ammonium, in thermophile protein biosynthesis (Uzawa et al. 1993a). Later, the MS2 system was used as a control in the cell-free DNA-directed polypeptide synthesis by the extreme thermophiles *T. thermophilus* HB27 and *Sulfolobus tokodaii* (Uzawa et al. 2002, 2003).

The R17 RNA was able to stimulate the cell-free protein synthesizing system that was obtained from *Saccharomyces cerevisiae* yeast mitochondria, in contrast to the yeast cytoplasmic fraction, and demonstrated sensitivity to inhibitors of bacterial protein synthesizing systems (Scragg et al. 1971). Later, when the cell-free system from *S. cerevisiae*

was optimized for the translation of both homologous yeast mRNA and for a number of heterologous eukaryotic mRNAs, only very low translation efficiency of the Qβ and MS2 RNAs was detected (Tuite et al. 1980).

POLARITY

The polarity effect was described in the Function of the mutant genes section of Chapter 8 and illustrated in Figure 8.1. Thus, the *amber* mutations near the beginning of the coat, as in the f2sus3 polar coat *amber* mutant, exhibited the polar effect that resulted in a decreased rate of the replicase synthesis in the case of the phages f2 (Engelhardt et al. 1967; Lodish 1968a,b,c,d) and R17 (Capecchi 1967b), but *amber* mutations at sites 50, 54, and 70 of the coat gene did not have this effect in f2 (Engelhardt et al. 1967) and R17 (Roberts and Gussin 1967; Tooze and Weber 1967).

In fact, when an *amber* mutation was present at the site in the coat gene that determined the sixth amino acid residue from the amino-terminal end of the coat protein, only a small hexapeptide was synthesized in the case of R17 (Adams and Capecchi 1966) and f2 (Webster et al. 1966), and little, if any, replicase was formed (Zinder et al. 1966; Capecchi 1967b; Engelhardt et al. 1967). As Capecchi and Webster (1975) noted, it was estimated that approximately 40 amino acids of the coat protein must be translated before the secondary structure has been altered enough to unmask the replicase gene for translation (Stewart ML et al. 1971).

Similar studies were performed with *amber* mutants from either MS2 (Fiers et al. 1969) or Qβ (Ball and Kaesberg 1973). As mentioned earlier in the Function of the mutant genes section of Chapter 8, the classical polarity gradient was found in the Qβ genome by the *in vivo* investigations of the coat *amber* mutants at positions 17, 37, or 86 (Ball and Kaesberg 1973).

Thus, it was concluded that the translation initiation of the replicase gene normally was dependent on the translation of the first part of the coat gene, but not necessarily on the synthesis of the C-terminal region of the coat protein. In other words, the ribosomes in an *E. coli* cell-free system initiated directly the coat synthesis, but the initiation of the replicase synthesis required normally some event concerned with synthesis of the beginning of coat protein (Lodish 1975).

This conclusion was strengthened first by the dipeptide assay when the translocation reaction was completely blocked, either by addition of antibiotics, such as fusidic acid, or by washing the ribosomes to remove EF-G factor (Lodish 1969a; Roufa and Leder 1971). Such systems formed only the first peptide bond of the viral proteins, since synthesis stopped after formation of the formyl-methionyl-aminoacyl-tRNA complex. Using the intact f2 RNA, the assay did show the first peptide bond fMet-Ala of the coat but not fMet-Ser of the replicase (Lodish 1969a). Second, using ^{32}P-labeled phage RNA, the RNA sequences were determined, which were bound to the initiation complex, consisting of both ribosomal subunits, initiation factors, fMet-tRNA and phage RNA, and protected from the ribonuclease digestion (Steitz 1969a,b). Again, when the intact unfragmented R17 (Steitz 1969a,b) or f2 (Gupta et al. 1970) RNA was used, the site for initiation of coat synthesis, but not of replicase synthesis, was protected.

As it followed from the RNA sequencing results presented in Chapter 10, the tertiary structure of the phage RNA could be responsible for the polarity effect when some unfolding of the phage RNA occurred during translation of the first part of the coat gene, so that a ribosome could initiate synthesis of the replicase subunit.

The role of the RNA spatial structure by the recognition of the initiation sites was confirmed by the elimination of the polarity effect when the f2sus3 RNA was fragmented (Lodish 1968c,d, 1975). In this case, the *in vitro* initiation of replicase occurred immediately, without the lag period characteristic of the replicase synthesis by wild-type f2 RNA. This was confirmed also by the dipeptide assay (Lodish 1969a,c) and by the equal recognition of the three initiation sites on the ^{32}P-labeled R17 RNA, which was fragmented by radioactive decay (Steitz 1969a,b).

The elimination of the polarity effect occurred also in the cases when wild-type RNA was subjected to the treatments that changed the tertiary structure. Thus, Fukami and Imahori (1971) did find that when the R17 RNA was heated to 60° or 70°C in the presence of Mg^{2+} ion and then cooled to 37°C, the ratio of the *in vitro*-produced phage proteins was changed on gel electrophoresis. One other way to irreversibly denature phage RNA without markedly inhibiting ability to direct cell-free protein synthesis was to react the RNA under mild conditions with formaldehyde (Boedtker 1967; Lodish 1970a), when the formaldehyde was crosslinking the amino groups of two nucleotide bases by forming an -NH-CH$_2$-NH- group and about ten formaldehyde molecules remained bound to each RNA after repetitive precipitations of the latter from ethanol (Lodish 1975). The treated RNA synthesized four to five times as much replicase as did normal f2 RNA. Moreover, when the f2sus3 RNA was treated with formaldehyde, the polar effect of the coat mutation on the replicase synthesis was abolished, and both treated f2 and f2sus3 RNAs synthesized essentially the same amount of replicase (Lodish 1970a). This finding was also confirmed convincingly by the dipeptide assay (Lodish 1970a, 1975).

It is worth mentioning here that another substance, namely, O-methylhydroxylamine, or methoxyamine, was employed by the *in vitro* translation of the f2 RNA (Filipowicz et al. 1972a,b). In contrast to hydroxylamine, methoxyamine was not reactive toward uracil but modified all exposed unpaired cytosines (Filipowicz et al. 1972a; Wodnar-Filipowicz et al. 1975). The modified f2 RNA retained an unchanged ordered structure and showed the same capacity to direct fMet-tRNA binding to *E. coli* ribosomes as unmodified f2 RNA. Since this binding corresponded mainly to the coat protein initiation site, it was suggested that unpaired cytosines presented in the coat protein initiation site were not involved in the specific recognition by ribosomes. However, when f2 RNA was treated with methoxyamine in the presence of guanidine-HCl, the resulting modified f2 RNA became irreversibly unfolded and stimulated the fMet-tRNA binding about 50 times more efficiently than native f2 RNA (Filipowicz

et al. 1972b, 1976; Szkopińska et al. 1975). These studies of Przemysław Szafrański's team in Warsaw were reviewed thoroughly at that time (Zagórska and Szafrański 1973; Szafrański et al. 1974a,b). The parallel modification of the f2 and brome mosaic virus RNAs was undertaken in order to directly compare the effect of the methoxyamine modification on the prokaryotic and eukaryotic messengers (Zagórska et al. 1982). Remarkably, when 16S rRNA was treated with methoxyamine, the modified 30S subunits lost their ability to associate with 50S ribosomes and to bind the native phage f2 RNA, but not polyU (Wrede et al. 1977).

The nature of the polarity effect was understood finally by the RNA sequencing and modeling of its tertiary structure, as described in Chapter 10. Thus, the determination of the entire nucleotide sequence of the MS2 coat gene (Min Jou et al. 1972a,b), together with Joan Aegetsinger Steitz's classical sequence studies of the replicase initiation site (Steitz 1969), provided strong support for the notion that the replicase initiation site in intact RNAs was masked by the RNA three-dimensional structure. A sequence of 11 bases, including the AUG at the beginning of the replicase gene, was a close complement of a sequence of nucleotides coding for amino acids 24–32 of the coat gene. Figure 16.1 presents this structure under the (h) sign, as well as the possible secondary structures for the RNA phage initiator regions that were taken from Steitz's pioneering review on initiator sequences (1975). This arrangement was substantiated by the fact that these two RNA sequences were isolated together from partial nuclease digests (Steitz 1975).

When ribosomes translate the region of the coat gene corresponding to amino acids 24–32, they have to open up this double-stranded structure; this may well be sufficient for other ribosomes to bind at the replicase initiation site. As Steitz (1975) noted, the 3′ fragments of R17 RNA which were clipped near the coat initiator loop but did include codons 24–32, the remainder of the coat gene and the entire replicase gene lacked polarity, suggesting that portions of the molecule 5′ to the coat gene were required to stabilize the (h) structure (Jeppesen et al. 1970b).

An alternative configuration, namely, the classical replicase initiation site, which is shown on Figure 16.1g, can also be proposed for the MS2 (or R17) coat-replicase intergenic region (Min Jou et al. 1972b). In this form, the replicase initiator codon is buried in a helical region of significant theoretical stability and would thereby be presumed to be non-initiating. Steitz (1975) thought that the binding of a few molecules of coat protein, which has been shown to occur specifically in this region of the R17 RNA within the repressor complex I (Bernardi and Spahr 1972a), simply stabilized the (g) loop and thereby efficiently blocked translation of the replicase gene. An RNA conformation that did not exhibit the polarity effect has also been reported (Fukami and Imahori 1971). Perhaps this reflected yet another structural configuration of the region surrounding the replicase initiator in R17 RNA (Steitz 1975).

The critical role of the secondary and tertiary structure of phage RNA in the translation process was confirmed

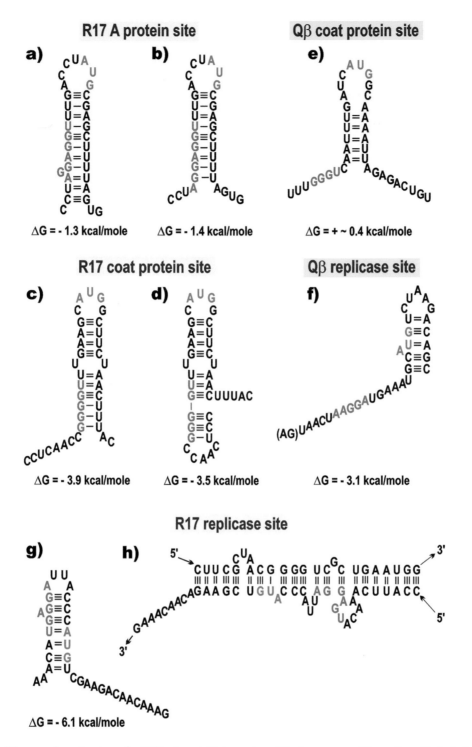

FIGURE 16.1 Possible secondary structures for RNA phage initiator regions. The free energy of formation of hairpin loops was calculated according to Gralla and Crothers (1973). ΔG values for structures containing many internal GU base pairs are considered less reliable than those for structures which do not. Structure (h), including the R17 replicase ribosome binding site, is taken from Min Jou et al. (1972a). Initiator AUG triplets are indicated green. The Shine-Dalgarno sequences of the corresponding initiation sites are depicted grass green. (Redrawn with permission of Cold Spring Harbor Laboratory Press from Steitz JA. Ribosome recognition of initiator regions in the RNA bacteriophage genome. In: *RNA phages*. Zinder ND (Ed). Cold Spring Harbor Laboratory, Cold Spring Harbor, NY, USA, 1975, pp. 319–352.)

indirectly by the inhibition of the f2 RNA-directed protein synthesis *in vitro* by poly(rI-rC), which occurred just due to an interaction between the poly(rI-rC) and f2 RNA (Chao et al. 1971). Such synthetic polynucleotides as poly(AUG), poly(UG), poly(AUC), and poly(AG) were observed to inhibit

the translation initiation of the f2 maturation and coat genes (Okuyama and Tanaka 1973). The f2 maturation gene was affected more significantly than the fr coat by poly(AUG) and poly(UG), whereas the f2 coat by poly(AUC), but poly(AG) markedly blocked initiation of both genes at the same level. In all cases, the synthetic messengers interfered with the binding of the f2 RNA to ribosomes.

Grubman and Nakada (1974) showed clearly that the coat and replicase genes were translated noncontinuously in the MS2 RNA-directed *in vitro* protein-synthesizing system: the ribosomes that completed the synthesis of the coat dissociated from the MS2 RNA and did not read through the intergenic gap. The translation of the replicase gene required ribosomes other than those translating the coat gene.

Remarkably, the polarity of translation, together with the general template activity and termination of the nascent peptide chains from the MS2 replicase gene, was not affected by limited exonucleolytic action of snake venom phosphodiesterase when 5–6 3′-terminal nucleotides of the MS2 RNA were removed (Jansone et al. 1975). It appeared therefore that the ultimate 3′-terminus was unessential by the translation of the native MS2 genome.

It is also noteworthy that 32.5% of the MS2 RNA-specified products by the *in vitro* cell-free protein synthesis were precipitated with antibodies directed against the phage MS2 (Moorman et al. 1978).

Berkhout and van Duin (1985) approved directly the molecular mechanism of the polarity phenomenon. They performed the deletion mapping on the cloned MS2 cDNA and showed that the ribosomal binding site of the replicase gene was masked really by a long-distance base pairing to an internal coat gene. The removal of the internal coat gene region led to uncoupled replicase synthesis. Therefore, the gene engineering approach confirmed structurally the original polarity model.

Furthermore, van Himbergen et al. (1993) narrowed by deletion studies the region in the coat gene over which ribosomes should pass to activate the replicase start from initial 132 nucleotides to the only 6 base pairs as the basis of the translational polarity. The 3′ side of the complementarity region was in the coat-replicase intergenic region, some 20 nucleotides before the start codon of the replicase. The 5′ side encoded amino acids 31 and 32 of the coat protein. Mutations that disrupted the long-range interaction abolished the translational coupling, while repair of the base pairing by second-site base substitutions restored translational coupling (van Himbergen et al. 1993).

NASCENT CHAINS

The tertiary structure of the full-length phage RNA also prevented the initiation of the maturation protein, but in a different way than in the replicase case. The problem became more complicated due to the low level of the maturation protein production in the *E. coli* cell-free system. Indeed, only when peptide mapping of the f2-specific proteins labeled with f[^{35}S]Met-tRNA$_f$ was used to identify the proteins produced *in vitro* (Lodish and Robertson 1969a,b) was it possible to show unambiguously that the maturation protein was indeed produced (Lodish 1975). The synthesis of the maturation protein was independent of that of the coat or replicase. The addition of coat protein to the phage RNA, which almost

totally prevented synthesis of replicase by the formation of the repressor complex I, as described below, had no effect on production of the maturation protein (Lodish and Robertson 1969a).

Surprisingly, ribosomes from different bacteria, either *Bacillus stearothermophilus* or *B. megaterium*, initiated at any temperature, exclusively the f2 maturation protein; at 37°C they made the same amount of the maturation protein per μg of added RNA as did *E. coli* extracts (Lodish 1969b). In experiments mixing components from *Escherichia* and *Bacillus* cells, the effect was determined only by the origin of the 30S ribosomal subunits, while the origin of the cell supernatant or initiation factors remained indifferent (Lodish 1970b).

Furthermore, the examination of heterologously reconstituted *E. coli* and *B. stearothermophilus* ribosome binding to the ^{32}P-labeled R17 or Qβ RNA, as well as of the phage RNA directed ^{35}S-fMet dipeptide synthesis, indicated that the gene specificity was conferred primarily by the protein fraction of the 30S subunit (Goldberg and Steitz 1974).

The approximately tenfold stimulation of the coat and replicase synthesis was achieved when purified S1 from *E. coli* was added to *B. stearothermophilus* ribosomes at roughly one-to-one molar ratio (Isono and Isono 1975). This finding was explained by the simple fact that *B. stearothermophilus* lacked a protein equivalent to the *E. coli* S1 (Isono and Isono 1976).

It is noteworthy in this context that ribosomes from two psychrophilic organisms, *Pseudomonas* sp. 421 and *Micrococcus cryophilus*, translated MS2 RNA at 37°C, yielding RNA replicase and coat protein of the same electrophoretic mobility and in the same relative amounts as the *E. coli* ribosomes (Szer and Brenowitz 1970).

However, cell-free extracts of *Rhodopseudomonas sphaeroides* recognized and translated the R17 coat but not the replicase gene (Chory and Kaplan 1982). Moreover, the formaldehyde-treated R17 RNA was totally inactive in the *R. sphaeroides* system.

The role of the three-dimensional structure of f2 RNA in the restriction of the maturation protein initiation was revealed conclusively when the structure of the f2 RNA was partially denatured by formaldehyde and the ability of the RNA to direct synthesis of the maturation protein in *E. coli* extracts increased 20-fold, and essentially equal amounts of coat, maturation, and replicase proteins were made (Lodish 1970a, 1975). Then, although the f2 RNA in extracts of *B. stearothermophilus* directed the synthesis exclusively of the maturation protein, the amount of this protein made increased 20-fold as the temperature changed from 37° to 65°C (Lodish 1971).

The increase of the R17 RNA messenger activity in the *B. stearothermophilus* cell-free system with increasing temperatures also reflected a loss of ordered structure in the RNA at higher temperatures (Dunlap and Rottman 1972). Ota et al. (1972) supposed existence of three different conformations of the R17 RNA which could be responsible for the different protein-synthesizing activity *in vitro*.

Steitz (1969a,b) showed that when fragmented R17 RNA was used as a messenger, the initiation sequence for the maturation initiation was bound specifically to *E. coli* ribosomes. Therefore, it was concluded finally that the RNA tertiary structure restricted the maturation protein initiation site to about 5% that of the coat, but when the RNA structure was partially unfolded, the initiation at this site increased 20-fold, to a level approximately that of the coat (Lodish 1975).

It was supposed logically that the predominant template for the maturation protein synthesis in the infected cell is not native single-stranded RNA, but some other structure, probably the nascent RNA chains within the replicative intermediate (RI), which is described in more detail in the next chapter. According to the statement by Lodish (1975), this would be attractive for two reasons: first, the single-stranded nascent chains of the RI are enriched in sequences corresponding to the maturation protein, so they would be expected to synthesize maturation protein, relative to coat, in a higher ratio than would mature RNA (Fiers et al. 1969; Robertson and Zinder 1969). Second, one would imagine that in the infected cell the ratio of replicative intermediate to the mature single-stranded RNA would decrease during the course of infection, and this could account for the decreased rate of the maturation protein synthesis, relative to coat, during the course of infection.

Nevertheless, when the intact RI was translated in the *E. coli* cell-free system, the major product was the coat protein (Engelhardt et al. 1968). However, the relative amount of the maturation protein was considerably higher than with the single-stranded RNA, and the maturation:coat ratio approached that of the 1:6 found in the infected ceil (Robertson and Lodish 1970). Moreover, according to the comprehensive analysis performed by Lodish (1975), it was not clear whether the increased synthesis of the maturation protein by the RI was solely due to the increased gene frequency; it could be due, in part, to an increased rate of initiation per site, possibly because of a more open conformation on the RI. Thus, the RI also directed synthesis of proportionately more replicase than did the single-stranded f2 RNA, and the polar effect of the f2sus3 coat mutation on replicase synthesis was reduced when the f2sus3 RI was used as the template (Robertson and Lodish 1970; Lodish 1975). In this connection, Lodish (1975) suggested that certain constraints on replicase synthesis were relaxed in the nascent RNA template, possibly by a more relaxed tertiary structure of RNA.

INITIATION

At the very beginning, it was shown that a distinct N-formyl-methionine tRNA (fMet-tRNA$_f$), discovered by Marcker and Sanger (1964), was necessary to initiate the synthesis of the phage coat protein *in vitro* (Adams and Capecchi 1966; Kolakofsky and Nakamoto 1966; Webster et al. 1966; Clark 1967). This was subsequently shown to be the case for all the phage-specific proteins (Viñuela et al. 1967c; Lodish and Robertson 1969b) and probably all proteins synthesized in *E. coli*, while the AUG or GUG were the codons which bound fMet-tRNA$_f$ (Capecchi 1966b).

As mentioned in the Translation section of Chapter 1, this was simply the inactivity of the *E. coli* deformylase under conditions of the *in vitro* protein synthesis that allowed the formylmethionine to remain attached to each protein and thereby facilitated the discovery of its role in the protein initiation, thanks to the RNA phages (Adams 1968). As Capecchi and Webster (1975) noted later, this lability of the deformylase also allowed large amounts of the N-formyl-methionine-containing N-terminal hexapeptide of the coat protein to be isolated from the *in vitro* reaction programmed by phage RNA containing the *amber* mutant at the appropriate position. The hexapeptide obtained in this way was used as a substrate in the investigations on the nature of the deformylase and peptidase required for the removal of the N-formyl-methionyl group *in vivo* (Adams 1968; Takeda and Webster 1968; Livingston and Leder 1969).

Later, during *in vivo* studies in *E. coli*, Housman et al. (1972) established that removal of formyl-methionine from the f2 nascent chains took place normally on the ribosome, when the nascent chain was about 50 amino acid residues long. Since the growing chain of such length was protected by the ribosome from proteolytic digestion with trypsin or pronase *in vitro*, they suggested that similar protection of the fMet residue of nascent chains occurred by ribosomes.

The binding of f2 RNA to ribosomes was independent of GTP, initiation factors, and amino acyl-tRNAs, while the f2 RNA-directed binding of fMet-tRNA to ribosomes was dependent upon GTP, initiation factors, and the presence of the formyl group (Anderson JS et al. 1967). In parallel, it was found that the fMet-tRNA$_f$ binding directed by the phage f2 RNA messenger took place on 30S in the absence of the 50S subunits (Nomura and Lowry 1967). Moreover, the 50S subunits inhibited this specific binding of fMet-tRNA$_f$, although they stimulated the binding of other aminoacyl-tRNA directed by their respective codons. Therefore, due to the phage f2 RNA, the first step in protein synthesis was identified as the formation of an initiation complex consisting of the 30S particle, mRNA, and fMet-tRNA$_f$, and a reasonable explanation was found for the existence of two basic subunits in the ribosomes (Nomura et al. 1967). The critical role of the 30S subunit, but not of the whole 70S ribosome, was also observed by the initiation of the MS2 RNA-programmed protein synthesis (Grubman and Nakada 1969).

The early kinetics of the MS2 and R17 RNA-programmed synthesis in *E. coli* extracts indicated that methionyl-tRNA was formylated before it was bound to the ribosome (Economou and Nakamoto 1967). It is noteworthy that the fixation of fMet-tRNA$_f$ on *E. coli* ribosomes was also studied with the phage β RNA (Temmerman and Lebleu 1968). The initiation process by the phage fr RNA-programmed *in vitro* synthesis was studied exhaustively by Heinrich Matthaei's team (Swan et al. 1969).

The f2 RNA-fMet-tRNA initiation system was used by the studies on the fine structure and function of *E. coli* ribosomes when translational fidelity and efficiency of protein-deficient subribosomal particles were evaluated (Traub et al. 1968).

When *E. coli* ribosomes were tested for activity in initiation with the R17 RNA as a messenger, the ability of 30S and 50S subunits to form a 70S couple at Mg^{2+} concentrations above 4 mM was discovered as the most critical criterium for the ribosome activity (Noll et al. 1973). Therefore, the nativity of topological elements necessary for subunit association were conclusive by the correct ribosome function.

The f2 RNA was used by the bifunctional imidoester, dimethylsuberimidate, crosslinking studies of ribosomes, which confirmed the conclusion that 70S ribosomes must dissociate into subunits to initiate protein synthesis with natural mRNAs, but not with polyU (Slobin 1972). The efficiency of the f2 RNA translation was diminished selectively, in contrast to that of late T4 mRNA, after alkylation of *E. coli* ribosomes containing bound initiation factors with N-ethylmaleimide or iodoacetate (Singer and Conway 1973b). At the same time, the degree of the alkylation of the MS2 RNA by dihydroxyepoxy-tetrahydrobenzo[a]pyrene correlated with the impaired template capacity, where translation of the replicase gene was much more sensitive to modification than either coat or maturation protein synthesis (Sagher et al. 1979).

The phage RNAs contributed to the discovery of the protein initiation factors. Thus, the wash from the ribosomes was found to contain protein factors which, when added back to the more purified system, allowed translation of the phage RNAs to a degree similar to that in the S30 system (Brawerman and Eisenstadt 1966; Eisenstadt and Brawerman 1966, 1967; Revel and Gros 1966; Stanley et al. 1966; Revel et al. 1968a,b; Brawerman 1969; Mangiarotti 1969b; Wahba et al. 1969a,b). The initiation factors were purified and characterized biochemically into the three distinct proteins, IF-1, IF-2, and IF-3, by using f2 RNA as a messenger (Dubnoff et al. 1972a). These components were enumerated and their connection with phage RNAs was characterized at that time in the exhaustive reviews of Mangiarotti (1969a), Lucas-Lenard and Lipmann (1971), and Haselkorn and Rothman-Denes (1973).

All three of the initiation factors were required for the normal initiation of translation in a phage RNA-directed *in vitro* protein-synthesizing system. According to the brief statement given by Capecchi and Webster (1975), the initiation factor IF-3 has bound to the 30S ribosomal subunit and helped to mediate the association of the 30S subunit to the available specific initiator regions on the phage RNA (Sabol et al. 1970, 1973; Berissi et al. 1971; Sabol and Ochoa 1971; Benne et al. 1973; Suttle et al. 1973). When IF-3 was attached to the 30S ribosomal subunit, it also appeared to prevent association of the 30S subunit with a 50S subunit to form the 70S ribosome (Kaempfer 1971; Lee-Huang and Ochoa 1971; Subramanian and Davis 1971). At least two distinct IF-3 have been isolated, each specific in recognizing distinct messenger RNAs (Pollack et al. 1970; Revel et al. 1970a,b: Grunberg-Manago et al. 1971; Lee-Huang and Ochoa 1971, 1972, 1973; Vermeer et al. 1971, 1973a,b,c; Yoshida and Rudland 1972). The specific binding of the IF-3 to the MS2 RNA was reported at a sequence in the untranslated region near the 3′-terminus of the molecule (Johnson and Szekely 1977; Fowler and Szekely 1979). The obligatory presence of IF-3 on the 30S subunit was confirmed further by van der Hofstad et al. (1978), who showed that the MS2 RNA binding preceded fMet-tRNA binding in the initiation complex, but the IF-3 and the IF-2-fMet-tRNA complex

were mutually exclusive on the 30S ribosome. Nevertheless, the stepwise assembly process of the initiation system allowed the MS2 RNA-directed synthesis of the complete coat protein in the cell-free translational system, deprived of the IF-3 (Zipori et al. 1978). The methodological aspects of the IF-3 binding to the MS2 RNA were described further by Johnson and Szekely (1979).

The initiation factor IF-2 was responsible for the transfer of the fMet-tRNA$_f$ to the messenger RNA-30S ribosomal subunit. A ternary complex composed of IF-2, GTP, and fMet-tRNA$_f$ served as an intermediate in this process (Lockwood et al. 1971; Rudland et al. 1971). Using the R17 and MS2 RNAs as messengers, two species of IF-2 proteins also have been isolated, but no real difference in their activities has been detected (Mazumder 1971; Miller and Wahba 1972, 1973; Miller et al. 1973).

The initiation factor IF-1 has bound to the 30S subunit in the presence of IF-2, GTP, fMet-tRNA$_f$, and messenger RNA (Thach et al. 1969), and was required for the proper translation initiation using phage RNA. If the normal GTP hydrolysis associated with initiation was allowed to occur, the presence of IF-1 allowed IF-2 to act catalytically (Chae et al. 1969; Dubnoff et al. 1972b; Benne et al. 1972; Kay and Grunberg-Manago 1972). As Capecchi and Webster (1975) wrote, it was suggested that one of the functions of IF-1 might be to promote the dissociation of IF-2 from the 30S initiation complex and GDP, thereby allowing the IF-2 to recycle. The combination of the 30S ribosome, fMet-tRNA$_f$, phage RNA, GTP, and the three initiating factors was termed the 30S initiating complex. The addition of the 50S ribosome resulted in the release of the factors, hydrolysis of GTP, and formation of the 70S ribosome-phage RNA- fMet-tRNA$_f$ complex (70S complex), which functioned further in the elongation process.

Using the MS2 RNA as a messenger, the IF-1 did not influence the binding of the IF-2 and IF-3 factors to the initiation complex in the absence of fMet-tRNA (van der Hofstad et al. 1976). However, in the presence of fMet-tRNA, it was found that the enhancement of the fMet-tRNA binding by IF-1 was accompanied with an equimolar increase in binding of IF-2. Therefore, the functions of the IF-1 in dissociation and recycling of the IF-2, described above, could be explained as a pleiotropic effect of the IF-1 basic action, i.e., by a conformational change of 30S subunits.

The formation of the initiation complex by the R17 RNA together with fMet-tRNA, initiation factors, and the 30S and 50S ribosomal subunits was shown to be dependent on temperature (Lu et al. 1971).

The ribosomes from chloramphenicol-treated *E. coli* cells showed a marked decrease in the MS2 RNA-specific binding of fMet-tRNA and MS2 RNA-directed protein synthesis when compared to normal ribosomes, while no decrease in polyU-directed polyphenylalanine synthesis was observed (Young and Nakada 1971). This phenomenon appeared due to reduction of the IF-2 and IF-3 initiation factors, since addition of the latter derived from normal ribosomes restored the ability of the ribosomes from drug-treated cells to carry out the MS2 RNA-directed protein synthesis.

The discovery of the special interference factor *i* (Revel et al. 1973a,b) must be mentioned here. The role of the factor *i* was elucidated in detail by the initiation of the R17 RNA translation (Jay and Kaempfer 1975a,b). This *i* factor was identified soon as the ribosomal protein S1 (Hermoso and Szer 1974; Inouye et al. 1974; Wahba et al. 1974), one of components of the Qβ replicase enzyme, as described in detail in the Subunits of the Qβ replicase section of Chapter 13. It was established further that only the *E. coli* 30S subunits that contained the protein S1 were able to interact with the MS2 RNA without any need for the initiation factors (Szer and Leffler 1974; Szer et al. 1975). Within the initiation complex, the S1 protein was found in the immediate neighborhood of the 3′-terminus of the 16S RNA (Czernilofsky et al. 1975; Van Dieijen et al. 1975, 1976, 1977, 1978), together with the initiation factor IF-3 (van Duin et al. 1975). The fine mechanisms of the protein S1 interactions were studied further in the R17 RNA-directed *in vitro* protein synthesis (Sobura et al. 1977). In the MS2 RNA-programmed protein synthesis, the presence of the S1 in 30S subunits was indispensable for the formation of 30S initiation complex, where helix-unwinding capacities of the S1 were discovered (Kolb et al. 1977). The methodological aspects of the S1 functioning were described at that time by Van Duin et al. (1979).

Further studies showed that the 30S ribosomal subunits of *E. coli* were forming a stable complex with the MS2 RNA or Qβ RNA at 37°C in the absence of initiator tRNA (Van Dieijen et al. 1978; Van Duin et al. 1980). Such preformed complex functioned as a precursor of initiation since, and entered the ribosome cycle in the presence of inhibitors of the *de novo* initiation.

It was, however, rather surprising that the initiation complex of *E. coli* ribosomes, R17 RNA, and fMet-tRNA was still able to translate the N-terminal portion of the R17 coat gene and to release the polypeptide chain at an *amber* termination codon after extensive ribonuclease digestion when no intact macromolecular rRNA was retained (Kuechler et al. 1972). Thus, it was proved experimentally that the intact rRNA was not required for proper ribosomal function.

The R17 RNA-directed *in vitro* synthesis was performed successfully also after total reconstitution of functionally active *E. coli* 50S ribosomal subunits from 23S RNA, 5S RNA, and the total proteins from 50S subunits, when no requirement for 30S subunits or other components was found (Nierhaus and Dohme 1974). However, treatment of *E. coli* ribosomes by bisulfite that converted 4.5% of the uracil residues of the rRNA into 5.6-dihydrouracil-6-sulfonate markedly impaired the formation of the initiation complex on the MS2 RNA as well as the total MS2-directed protein synthesis (Braverman et al. 1975).

When *E. coli* 16S rRNA was cleaved at nine regions belonging to different RNA domains with ribonuclease H or subjected to deletions, it retained the ability to assemble with total ribosomal protein into 30S-like particles and maintained the MS2 RNA-directed protein synthesis (Afonina et al. 1991). A similar approach for the probing the function of conserved RNA structures in the 30S subunit of *E. coli* ribosomes was used by Almehdi et al. (1991) with the same *in vitro* protein synthesis system primed with the MS2 RNA.

A first general model of the protein initiation based on experiments with the R17 RNA as a messenger was presented by Markus Noll and Hans Noll (1972, 1974a,b, 1976). This model proposed a two-stage mechanism, the first stage requiring dissociation into subunits and initiation factors and the second involving direct attachment of ribosomes.

The R17 RNA-directed protein synthesis system was used by the identification of the individual functions of the *E. coli* 30S ribosomal proteins as determined by *in situ* immunospecific neutralization (Lelong et al. 1979).

The fate of the initiation factors after phage infection was studied. First, ribosomes isolated from *E. coli* infected with the phage T4 were found to interact only weakly with the R17 RNA to form initiation complexes *in vitro* (Dube and Rudland 1970). Such ribosomes discriminated against the coat protein and replicase initiation sites and bound appreciably only to the maturation gene initiation site, and this change was clearly connected with the initiation factor fraction (Steitz et al. 1970), possibly IF-3 (Yoshida 1972). Similar observations that a factor was required for the translation of the MS2 (Hsu and Weiss 1969; Klem et al. 1970; Hsu 1973) or f2 (Schedl et al. 1970; Singer and Conway 1973a) RNAs in extracts from the T4-infected *E. coli* cells were reported. However, continuing their studies on the inhibition of the f2 replication by superinfection with phage T4 (Goldman and Lodish 1971) and using the f2 and Qβ RNAs, as well as T4 and T7 mRNAs as messengers, Goldman and Lodish (1972) found that there was no change in the specificity of initiation factors after T4 infection, but alterations in the amounts of factors were probably not essential either for shutoff of host protein synthesis or for the transition from early to late T4 protein synthesis. Similar conclusions were also made by studies of the translational initiation defects arising in *E. coli* infected with the phages Qβ, T7, and λ (Leder et al. 1972). The mechanism by which the T4 infection actually restricted the f2 RNA translation was presented at last by Goldman and Lodish (1975). In fact, the T4 late mRNA outcompeted f2 RNA by the *in vitro* cell-free synthesis, since it had a higher rate constant for polypeptide chain initiation than did phage the f2 RNA.

Furthermore, the efficiency of the MS2 RNA-programmed cell-free protein synthesis was compared in F⁺ and F⁻ *E. coli* cells after the phage T7 infection (Yamada and Nakada 1975). The authors concluded that the abortive infection of F⁺ cells by T7 resulted from a general reduction of the translational activity, but not of specific translational inhibition of the late T7 mRNA.

A novel low-molecular-mass factor was isolated from *E. coli* cells, which specifically inhibited the initiation of the R17 RNA-directed *in vitro* protein synthesis by impairing the binding of fMet-tRNA$_f$ to form the 70S initiation complex (Clark 1980).

Remarkably, the DEAD-box RNA helicases (for a recent review, see Redder et al. 2015) were contributing to the initiation, presumably, by unwinding of the ribosome binding sites. Thus, whereas artificial RNAs devoid of secondary structures were translated efficiently *in vitro* in the absence of CsdA, or DeaD by original name, one of such helicases, the MS2 RNA, required CsdA for efficient translation (Lu et al. 1999).

The kinetics of the 30S initiation complex formed by the MS2 RNA and fMet-tRNA were studied with limiting concentrations of *E. coli* ribosomes to be an obligatory intermediate in the formation of the 70S initiation complex (Blumberg et al. 1979). The formation of the 70S initiation complex began with an induction period and was proportional to the concentration of the 30S complex, which rapidly rose to a peak. The entire time course of the sequential pseudo-first-order, second-order reaction was reproduced accurately by the overall rate expression. By using limiting concentrations of messengers, it appeared that the MS2 RNA contained no specific signal that enhanced its rate of 30S complex formation with *E. coli* ribosomes and initiation factors; the pseudo-first-order rate constants obtained with synthetic polynucleotides were 12–45 times higher than that with the MS2 RNA. Then, the kinetics of the MS2 RNA-directed protein synthesis were measured at seven temperatures between 30° and 47°C by using ribosomes isolated from a wild-type strain and seven temperature-sensitive mutants of *E. coli* (Rahi et al. 1979).

Furthermore, the analysis of initial velocity kinetic data was used to examine the order in which fMet-tRNA and the coat gene of the R17 or Qβ RNA were bound to the 30S ribosome subunit (Ellis and Conway 1984). These data were obtained using a quantitative assay for protein synthesis in *E. coli* extracts, where the rate of accumulation of protein product was dependent on the concentration of the messenger and was partially dependent on fMet-tRNA. The results from the initial velocity and alternative substrate experiments were consistent with a rapid equilibrium ordered mechanism, as opposed to a rapid equilibrium random mechanism. The analysis of the rate of coat protein synthesis at varied concentrations of mRNA and fixed concentrations of fMet-tRNA indicated that fMet-tRNA was the first substrate to bind to the 30 S subunit when either coat gene was used as a messenger. This scheme assumed the existence of a relatively slow step in protein synthesis that occurred after both the initiating tRNA and messenger were bound to the ribosome and which allowed substrate addition to reach thermodynamic equilibrium (Ellis and Conway 1984).

More recently, a mathematical modeling of the translation initiation was performed for the estimation of the translation efficiency to computationally design mRNA sequences with desired expression levels in prokaryotes (Na et al. 2010). The problem was regarded as a critical element of the emerging field of synthetic biology, and the authors specifically emphasized the fact that even a few nucleotide modifications around the start codon might alter translational efficiency up to 250-fold and dramatically change protein expression. Finally, they confirmed the feasibility of the generated model using previously reported expression data on the MS2 coat protein.

INITIATION SITES

The structural clue to the initiation puzzle was found by the pioneering studies of Joan Argetsinger Steitz on the initiation sites of the R17 RNA. This study was summarized in her classical review in the *RNA Phages* book (Steitz 1975). Importantly, up to the year 1975, almost all knowledge about initiator regions for protein synthesis came from the RNA phages, and RNAs of the latter were the best understood natural messenger RNAs available. As Steitz (1975) noted neatly, the ease with which the single-stranded RNA genomes of these viruses could be labeled to high specific activity and purified from the phage particles had made them ideal subjects for RNA sequence analysis. Moreover, the structural information on the initiator regions from the RNA phage genomes was the first meaningful structural information correlated with the highly important biological function.

Steitz argued experimentally that the initiation of polypeptide chains in bacteria is a highly specific and accurate process from the point of view of the messenger. Thus, she sequenced the actual initiation sites, where ribosomes attached and started protein synthesis on the phage RNA. As stated above, the initiation complex contained, in addition to the 30S ribosomal subunit and the messenger RNA containing the initiator AUG or GUG codon, the fMet-tRNA$_f$, GTP, and the initiation factors IF-1, IF-2, and IF-3.

Therefore, the isolation, identification, and characterization of the initiation sites from the three major genes of the group I phage R17 and the group III phage Qβ, as well as the coat initiation site of the group I phage f2, were achieved. It was clear at that time that the natural messengers did not begin with AUG or GUG and that ribosomes did not find initiators by simply threading onto the 5′ end of an mRNA and sliding along until the first AUG or GUG triplet was reached.

Takanami et al. (1965) were the first to show that the specific regions of the phage RNA could be protected from ribonuclease digestion by association with ribosomes. When sequence studies of radioactive phage RNAs first became feasible, three groups—Steitz (1969) with R17, Hindley and Staples (1969) with Qβ, and Gupta et al. (1969, 1970) with f2—undertook the isolation and analysis of the initiator fragments from the ^{32}P-labeled viral genomes. As Steitz (1975) described in her classical review, the complexes with ribosomes were formed under conditions of polypeptide chain initiation *in vitro* when ribosomes did not move beyond the initiation sites on an mRNA in the absence of elongation factors. T1 or pancreatic ribonuclease was used to trim away the ends of the messenger RNA not involved in the initiation complex. The radioactive RNA fragments extracted from the ribosomes were subjected to sequence analysis, using techniques developed by Sanger and his colleagues (Barrell 1971). In each case, it was shown that the amount of ^{32}P-labeled RNA remaining bound to ribosomes after fractionation away from the bulk of the degraded RNA depended on the inclusion of fMet-tRNA$_f$ in the original initiation reaction. This provided assurance that the interaction of the ribosome with the phage messenger was occurring at sites actually specifying the initiation of protein synthesis. Moreover, the fraction of total radioactive RNA remaining complexed with the ribosome (approximately 1%) was compatible with protection of about 30–40 nucleotides (Steitz 1975).

In the original work with the f2 and Qβ RNAs, special precautions were taken to ensure that intact RNA alone was added to the binding reaction. Thus, only the initiator region at the beginning of the coat protein gene was isolated from these two phage RNAs. With the R17 RNA, the use of RNA that had undergone radiolytic autodegradation permitted ribosome binding and analysis of all three initiator regions. The state of fragmentation of the RNA was therefore shown to influence the relative affinity of ribosomes for several binding sites. One of the three R17 RNA sites that corresponded to the beginning of the maturation gene was obtained from *E. coli* ribosomes in very low yield. Extending an observation made at that time by Harvey F. Lodish's team (Lodish 1969b; Lodish and Robertson 1969b), Steitz (1969a,b) used ribosomes from *B. stearothermophilus* to protect this region selectively and with high enough efficiency to permit completion of its sequence.

Concerning the Qβ initiation sites, Weiner and Weber (1971) observed that the oligonucleotide sequence starting at position 62 from the 5′ terminus of Qβ RNA (Billeter et al. 1969c) matched their determined N-terminal amino acid sequence for the Qβ maturation protein. Furthermore, Staples et al. (1971) demonstrated that this site was easily accessible to *E. coli* ribosomes only in the nascent RNA molecule, which they synthesized *in vitro*. The protection by ribosomes from *B. stearothermophilus* in this case had not yielded the maturation protein initiator region as it did with the R17 RNA (Steitz 1973b).

The areas adjacent to the initiator codons were next examined for the presence of any base sequence or local secondary structure homology that might allow ribosomes to distinguish these as true initiator regions. In accordance with the classical review (Steitz 1975), the initiation sites with their putative secondary structures are shown in Figure 16.1.

The values for the free energy (ΔG) of formation of each structure (at 25°C in 1 M NaCl) was calculated according to the formulation of Gralla and Crothers (1973). As Steitz (1975) noted, the structure (e), previously proposed for the Qβ coat protein initiator (Hindley and Staples 1969), and (a) and (b) for the R17 maturation gene site were thus of questionable stability, while the alternative configurations for the R17 RNA coat site, (c) and (d), had a theoretically higher probability of existence. The remaining phage initiator sequences could not be folded into hairpin forms which have the initiator triplet exposed and give ΔG values of less than 0, according to Gralla and Crothers (1973), or positive stability numbers calculated by the earlier scheme of Tinoco et al. (1971). The two replicase sites (f) and (g), however, could be drawn to contain favorable secondary structures, in which the initiator AUG codons were sequestered in helical regions.

The next survey step included rebinding of the initiator fragments to ribosomes. As described by Steitz (1975), to establish the relative ribosome affinity of the isolated R17 initiator regions, the mixtures of the three sites was prepared by binding ribosomes to ^{32}P-labeled phage RNA in various states of degradation. After isolation, the resulting fragments were on the average 30 residues long, somewhat shorter than the regions shown in Figure 16.1.

By rebinding to *B. stearothermophilus* ribosomes, the exclusive recognition of the maturation protein initiation site was totally retained by the fragments. By rebinding to *E. coli* ribosomes, unexpected results were obtained. The coat initiation site did not demonstrate the highest affinity for *E. coli* ribosomes. Moreover, the R17 RNA initiation sites were not recognized in ratios reflecting their relative concentrations in the binding reaction. Instead, the maturation site attached with high efficiency, while rebinding of the coat and replicase sites was barely detectable (Steitz 1973a, 1975).

In a similar way, to isolate the beginning of the Qβ replicase gene, the fragmented RNA, produced by mild alkaline hydrolysis or by autoradiolysis, was used by Staples and Hindley (1971) and by Steitz (1972) in ribosome binding experiments. When the binding of the isolated RNA fragments to ribosomes was tested, only the longest one of the Qβ RNA fragments from the coat protein gene ribosome binding site, which ranged from 30 to 120 nucleotides, was able to form initiation complexes (Porter and Hindley 1973).

Since no single recognition element could be convincingly identified, Steitz (1975) proposed an idea that the three initiation sites on the phage RNA could be recognized by different components of the protein-synthesizing machinery. Thus, Steitz's work paved the way to the SD sequences—in fact, to the direct RNA-RNA interaction involving the messenger on the one hand, and the ribosomal RNA on the other, which provided the basis for selectivity in ribosome binding and will be described in the next section. It is remarkable in this context that even when the sites were exposed, the initiation of synthesis at the maturation and replicase genes did not reach the level of that for the coat genes, thus indicating a type of control other than secondary structure (Capecchi and Webster 1975).

In contrast to the *E. coli* ribosomes, the wheat germ ribosomes bound preferentially at the beginning of the lysis peptide and replicase gene of the fragmented R17 RNA, and at a third site which might be derived from the C-terminal region of the A protein gene (Kozak 1980), similarly to that reported previously in a mammalian translational system (Atkins et al. 1979b). Remarkably, more extensive fragmentation of the R17 RNA permitted wheat germ ribosomes to bind and protect a great many additional sites. Thus, it was concluded that the presence of an exposed 5′-terminus on an RNA molecule could be *necessary and sufficient* for attachment of eukaryotic ribosomes.

A global review on the genetic signals and nucleotide sequences in messenger RNAs was published by Steitz (1979).

SHINE-DALGARNO

The RNA phage messengers helped to conclude that the secondary and tertiary structures of the intact genome served to regulate protein initiation. However, as Steitz (1975) noted in her classical review, no universal sequence or structural homology within the ribosome-protected regions were found. Moreover, when the RNA conformation was disrupted by fragmentation or formaldehyde treatment, no novel initiation sites did appear, and ribosomes preferred to bind to the real initiation sites. The question arose as to what information directed them to reside in the immediate vicinity of the initiator codon, and how might portions of the RNA beyond the AUG or GUG triplets drive the initiation activity.

Because of the basic knowledge collected on the RNA phage messengers, this specific sequence came to light in 1974 and was given the famous Shine-Dalgarno, or SD, sequence name. In fact, Australian scientists John Shine and Lynn Dalgarno (1974, 1975), and then Joan Argetsinger Steitz and Karen Jakes (1975), presented evidence that ribosome binding might be mediated by the complementary base pairing between sequences in the initiation sites on mRNAs and sequences at the $3'$ end of 16 s rRNA. Thus, the $3'$-terminal sequence of *E. coli* 16S ribosomal RNA, namely, $PyACCUCCUUA_{OH}$, interacted with mRNA where the $3'$-terminal UUA_{OH} was involved in the termination of protein synthesis through base pairing with terminator codons, but the ACCUCC sequence in turn recognized a conserved sequence found in the ribosome binding sites of various phage RNAs, the full set of which derived its name from the authors.

Furthermore, it was suggested that this interaction was promoted by the S1 protein (Dahlberg and Dahlberg 1975), the same 30S ribosomal protein that was required for both the mRNA binding and for the replication of the phage genome, as presented in Chapter 13. Remarkably, the same S1 protein inhibited formation of the initiation complex with the R17 RNA (Miller et al. 1974). The extent of the complementary base pairing was limited to 3–7 base pairs, so that binding must also be influenced by initiation factor–mediated binding of fMet-tRNA, by other interactions of mRNA with the ribosome, and by possible secondary or tertiary folding of the phage RNA, as noted earlier. *Exempli gratia*, in the RNA phages, seven base pairs of the SD sequence were formed by *E. coli* 16S rRNA with the maturation initiation site, compared with three or four base pairs with the coat initiation site. This explained the "unexpected" results on Steitz's rebinding survey, which were presented at the end of the previous section. The SD sequences of the corresponding initiation sites are depicted in Figure 16.1.

It was highly remarkable in this context that, in the absence of fMet-tRNA, ribosomes of *E. coli* were able to bind non-initiator sites of Qβ RNA (Taniguchi and Weissmann 1979). While the coat initiation site was a dominant recognized fragment in the presence of fMet-tRNA, three oligonucleotides were recognized in the absence of fMet-tRNA, none of which was close to a protein initiation site, but all three fragments showed striking complementarity to the $3'$-terminal region of *E. coli* 16 S RNA. These experiments showed that the interaction of 30S ribosomes with the SD region preceding the initiator codon of the Qβ coat gene was insufficient to direct correct placement of the ribosome on the viral RNA, and that an additional contribution from the interaction of fMet-tRNA with the initiator triplet was required for the ribosome binding to the initiator region.

In direct support of the SD accuracy, Tadatsugu Taniguchi and Charles Weissmann (1978a) showed that the oligonucleotide $AGAGGAGGU_{OH}$, eight bases of which were complementary to the sequence $ACCUCCUUA_{OH}$ at the $3'$-terminal region of 16S rRNA, was bound tightly to the 30S ribosome and was released as a complex with the $3'$-terminal 49-mer of 16S rRNA after digestion with cloacin DF13. This oligonucleotide either strongly inhibited the Qβ RNA binding to 70S ribosomes or decreased the stability of the initiation complex formed in its presence. In parallel, Taniguchi and Weissmann (1978b) performed site-directed mutagenesis in the initiator region of the Qβ coat gene, as described in the Mutants of the phages Qβ, SP, and FI section in Chapter 8. In this context, it should be remembered that Tadatsugu Taniguchi, a PhD student of Charles Weissmann in Zürich, is first known nowadays for his pioneer research on interferons and interferon regulatory factors.

Following the Taniguchi and Weissmann (1978a) study, Trudel et al. (1981) prepared this nonanucleotide in larger quantities and confirmed its effect. Furthermore, no selectivity was observed in the inhibition of the coat protein and replicase protein synthesis.

Proceeding with the SD accuracy, the initiation of the R17 RNA-directed *in vitro* protein synthesis was inhibited by a synthetic pentanucleotide G-A-dG-dG-U complementary to the $3'$ end of *E. coli* 16S rRNA (Eckhardt and Lührmann 1979). The oligonucleotides such as G-A-G-G, which have been complementary to the C-U-C-C region at the $3'$-end of 16S rRNA, inhibited the R17 RNA-dependent binding of the initiator fMet-tRNA to 30S ribosomal subunits (Schmitt et al. 1980). Remarkably, if phage RNA was replaced by A-U-G, the same oligonucleotides stimulated the binding of fMet-tRNA to the 30S subunits. Then, the deoxyoctanuclotide dAAGGAGGT, which was complementary to the $3'$-end of 16S RNA, inhibited the formation of the complex between the 30S subunit and the MS2 RNA (Backendorf et al. 1980). However, when the complex was preformed, the octanucleotide could not prevent entry of the complex into the ribosome cycle upon supplementation with the components for protein synthesis. Moreover, the site-specific binding of the octanucleotide was shown by labeling 16S RNA *in situ* at its $3'$-end with ^{32}P-pCp and T4 RNA ligase (Backendorf et al. 1981). The ribosomal protein S21 ensured the potential of the 30S subunit to bind the octanucleotide and to accept the MS2 RNA as a messenger, while the presence or absence of S1 on 30S subunits had no effect on their octanucleotide-binding property (Backendorf et al. 1981). The defect of the S21-deficient 30S ribosomes was traced back to their inability to bind the MS2 RNA at the initiation step of protein synthesis (Van Duin and Wijnands 1981). When S21 was labeled at its single cysteine group with a fluorescent probe, it was found to bind to the 50S ribosomal subunits, as well as to the 30S and 70S particles, in the MS2 RNA-dependent reactions (Odom et al. 1984). At this time, one of the first theoretical analyses of the SD role was published (Bahramian 1980).

Next, a series of nonionic oligonucleotide analogues, the deoxyribooligonucleoside methylphosphonates, complementary to the $3'$-end of 16S rRNA, were found to inhibit effectively the MS2 RNA-directed protein synthesis but not that directed by synthetic polynucleotides as either polyU or polyA, or by globin mRNA in the reticulocyte system (Jayaraman et al. 1981).

The sequence AGGACAUAUGp, which was identical to the initiation site of the Qβ maturation protein, was bound to the E. coli 70S ribosomes substantially better than related probes in which the 16S complementary region, AGCA, was replaced by the anticomplementary region, UCCU, or was altogether absent (Neilson et al. 1980).

Steitz et al. (1977) found that the three R17 RNA initiator regions did not exhibit equal requirements for the initiation factors and S1 protein during formation of the initiation complex. Thus, both initiation factors and S1 stimulated ribosome binding to the beginnings of the coat and replicase genes to a greater extent than they promoted recognition of the A protein initiation site. The differential effects were therefore inversely correlated with the degree of the mRNA-16S rRNA complementarity exhibited by the three initiator regions. Remarkably, the S1 protein suppressed ribosome binding to spurious sites in the R17 RNA.

The mutant S1 protein containing 77% of the polypeptide chain length was almost as effective as the whole S1 in the MS2 RNA binding and unfolding, while a large trypsin-resistant S1 fragment of 66% did not bind MS2 RNA and did not support the protein synthesis in vitro (Thomas et al. 1979). The translation of the MS2 RNA contributed to the further mapping of the repeating homologous sequences in nucleic acid binding domain of the S1 protein (Suryanarayana and Subramanian 1984), as touched on also in Chapter 13. Thus, the S1 protein contained four repeating, homologous stretches of sequences in its RNA binding domain, but the functionally active mutant form contained only three repeating stretches. When an S1 fragment with two of the homologous stretches was employed, it was functionally active in translating polyU and polyA but totally inactive in translating the MS2 RNA.

Later, the sites were localized within the Qβ, MS2, and fr RNAs, which have been involved in the interaction with the S1 protein during formation of complexes with 30S ribosomal subunit (Boni et al. 1986, 1991; Boni and Isaeva 1988). Thus, the authors used a specially developed procedure based on efficient UV-induced crosslinking of the S1 protein to the messenger RNA (Boni et al. 1991). The targets for the S1 were localized on the Qβ and fr RNAs upstream from the coat protein gene and contained oligoU-sequences. In the case of the Qβ RNA, this S1 binding site overlapped the S-site for the Qβ replicase and the site for the S1 binding within a binary complex. It seemed reasonable that similar U-rich sequences might represent the S1 binding sites on bacterial mRNAs.

An unexpected opinion was published by Laughrea and Tam (1989), who tried a set of synthetic mDNAs and mRNAs, about 19 nucleotides in length, which had 4−8 nucleotides long SD sequences that were similar to the R17 coat and maturation protein initiation sites. They found that the S1 protein and initiation factors had no effect on the rate and extent of the ribosomal binding of these mDNAs and mRNAs. The authors concluded that the main function of the S1 protein or initiation factors included neither the direct recognition or binding

of short messengers to the 30S subunits nor their rejection as "false" for possessing the wrong sugars or an insufficient length.

Then, short RNAs of 25–36 nucleotides in length with sequences of the R17 maturation protein initiation site and labeled at their 5′ or 3′ ends with fluorescent probes were employed to the generation of the first topological map of the corresponding mRNA region on the 30S subunit by fluorescence spectroscopic techniques (Czworkowski et al. 1991).

Balakin et al. (1992, 1993) further confirmed the point that the cell-free translation of the MS2 or fr RNAs was not possible when ribosomal 30S subunits were S1-depleted or treated with anti-S1 antibodies, while an artificial mini-mRNA containing an extended 9-nucleotide-long SD sequence appeared as the S1-independent in a cell-free system.

Remarkably, a structural and functional homolog of the E. coli S1 protein was isolated from ribosomes of Gram-positive Micrococcus luteus (Muralikrishna and Suryanarayana 1987).

The ML-S1 protein reacted with the E. coli anti-S1 serum with an immunological partial-identity reaction and stimulated the MS2 RNA-dependent translation, although to a much lesser extent than the E. coli S1 protein.

The MS2 RNA was used as a standard model for the in vitro translation by a systematic study on the selection of the mRNA translation initiation region by E. coli ribosomes, when genes specifying model mRNAs of minimal size and coding capacity, with or without the SD sequence, were assembled, cloned, and transcribed in high yields (Calogero et al. 1988). This study led to the logical suggestion that the function of the SD interaction was to ensure a high concentration of the initiation triplet near the ribosomal peptidyl-tRNA binding site, whereas the selection of the translational start was achieved kinetically, under the influence of the initiation factors, during decoding of the initiator tRNA.

With the use of the MS2 RNA-programmed in vitro protein synthesis system, La Teana et al. (1993) studied the influence of the rare initiation triplet AUU on the mRNA translation by comparing the activity of two pairs of model mRNAs that differed in the length of SD and spacer sequences. The model mRNAs were evaluated by their competition with the MS2 RNA-directed amino acid incorporation.

In conclusion, as Li (2015) noted in his recent review, the lack of understanding of the general strategy used by bacteria to tune translation efficiency remains the urgent problem. This urgency was strengthened by forthcoming differences between translation of natural and synthetic mRNAs when the latter are used to express recombinant proteins in bacteria. The problem is going to be solved by the development of array-based oligonucleotide synthesis and ribosome profiling that can provide new approaches to address this issue. At the present day, these high-throughput studies pointed out a statistically significant but mild contribution from the mRNA secondary structure around the start codon, although the major determinant for translation efficiency remains unknown. Li (2015) came to a paradoxical conclusion that bacteria do not use the SD sequence to tune the translation efficiency. In this context, he pointed to the work of Shine and Dalgarno (1974), who noted at that time that the MS2 maturation gene with the strongest SD site was in fact most weakly translated. Therefore, 40 years after their discovery, the major determinant of the translation efficiency is still largely unknown (Li 2015).

ELONGATION

As Capecchi and Webster (1975) noted, the usefulness of *in vitro* cell-free protein synthesis programmed with phage RNAs for studying the mechanism of polypeptide chain elongation arose from the ability of these messengers to direct the synthesis of polypeptide products with known sequence, first of all, of the phage coat protein. The structure of the latter was determined for the first time both on the amino acid (Weber et al. 1966; Weber and Konigsberg 1967) and on the nucleotide (Min Jou et al. 1972a,b) levels.

The elongation, or translocation, or stepwise addition of each aminoacyl residue to the growing polypeptide chain was the third step after the (i) codon-directed binding of the aminoacyl-tRNA to the 70S initiation complex consisting of 70S ribosome-messenger RNA-fMet-tRNA$_f$ and (ii) peptide bond formation. In addition to the ribosome-messenger RNA-aminoacyl-tRNA complex, the elongation step required GTP and three protein factors, designated as elongation factors EF-Tu, EF-Ts, and EF-G. The content and functioning of the complex were reviewed exhaustively at that time by Lucas-Lenard and Lipmann (1971) and by Haselkorn and Rothman-Denes (1973).

Thus, the major advantage of the phage RNA-directed protein synthesis by the elongation studies was the ability to pinpoint with great precision the stage of the elongation cycle at which the system rests after any given operation. As Capecchi and Webster (1975) noted, this was possible because (i) following the stepwise addition of any given amino acid, the polypeptide products were readily separable and characterized (Capecchi 1967a; Kuechler and Rich 1970), and (ii) the position of the messenger RNA relative to the ribosome could be determined (Gupta et al. 1971; Thach SS and Thach RE 1971). The addition of the ternary complexes EF-Tu-GTP-aa-tRNA to the 70S initiation complex in the absence of EF-G resulted in the formation of the dipeptide fMet-Ala and the release of EF-Tu-GDP in the case of the coat protein initiation complex. This dipeptide was bound to the ribosomal complex through tRNA$_{ala}$. Furthermore, under these conditions the peptidyl-tRNA will not react with puromycin, thereby operationally positioning the peptidyl-tRNA in a non-puromycin-reactive site or state. If the above experiment was repeated substituting for GTP the nonhydrolyzable analog guanylyl-5'-methylene-diphosphonate (GMP-PCP), one observed that alanyl-tRNA$_{ala}$ still bound to the initiation complex, but EF-Tu was not released from the ribosome, and peptide bond formation did not occur (Kuechler 1971). By this operation, the binding of the incoming aminoacyl-tRNA could be separated from subsequent peptide bond formation, as reviewed by Capecchi and Webster (1975). Therefore, the result, if not

the function, of GTP hydrolysis was the release of EF-Tu. Concomitant with this release of EF-Tu-GDP, the aminoacyl-tRNA was positioned for the peptidyl acceptance.

The ribosome involved in this 70S ribosomal complex protected the mRNA from ribonuclease digestion up to four codons beyond the AUG initiator codon. However, the codons beyond the alanyl codon, although in the protected environment of the ribosome, were not positioned for aminoacyl-tRNA binding prior to translocation in the Qβ messenger (Roufa et al. 1970a,b). These studies clearly indicated that only two tRNAs were bound to the ribosome prior to translocation and made it possible to distinguish between multi-site models of ribosomal function. As a result, the two-site model of the ribosome (Watson 1964) was accepted. However, it is noteworthy that Swan and Matthaei (1971) presented evidence for the four-site model of ribosomal function by their studies on the aminoacyl-tRNA binding directed by the phage fr RNA. Remarkably, a detailed assessment of Matthaei's conception was given in Riga within the scope of the VII International Symposium on the Chemistry of Natural Products, the first international symposium on the subject in the former Soviet Union (Kisselev and Venkstern 1970).

As Capecchi and Webster (1975) emphasized, the determination of the composition of the mRNA fragment, which was protected by the ribosome, could be used to define the position of the mRNA relative to the ribosome in the complex. Thus, after a translocation, one additional codon in the 3' direction of the mRNA should become resistant to nuclease digestion, but the translocation did not occur until EF-G and GTP were added to the reaction mixture (Gupta et al. 1971). The addition of these components resulted in the conversion of the fMet-alanyl-tRNA$_{ala}$ from a puromycin-unreactive state to a puromycin-reactive state, movement of the ribosome one coding unit in the 3' direction, release of tRNA$_{fmet}$, and release of EF-G-GDP from the complex. GTP hydrolysis was required for the above reactions. The elongation factor G participated primarily in the release of deacylated tRNA, but the translocation *per se* was catalyzed by the 50S ribosomal subunit (Tanaka et al. 1971). The EF-Ts was bound to neither the ribosome nor the ternary complex EF-Tu-GTP-aa-tRNA, but it was involved in the release of GDP from the EF-Tu-GDP complex, resulting in an EF-TuTs complex. In the absence of EF-Ts, the action of EF-Tu was stoichiometric rather than catalytic, because EF-Tu became trapped in the EF-Tu-GDP complex released from the ribosome.

Using the one-step polypeptide chain elongation technique, Hsu (1971) estimated the time required for translation of the MS2 coat gene as approximately 12 minutes at 31°C under the conditions described; the rate of translation appeared to be nonuniform. Remarkably, the consequences of limiting the rate of elongation of protein synthesis *in vitro* was examined later also with the MS2 RNA as a messenger (Goldman 1982). Thus, the concentration of Trp-tRNA$_{trp}$ was manipulated by varying the amount of exogenously added tryptophan in extracts from an *E. coli* mutant in which the tryptophanyl-tRNA-synthetase had a higher K_M for tryptophan. The author concluded that the variation of the elongation rate could be a means of regulating gene expression, both directly, by slowing or accelerating the rate of protein synthesis, and indirectly, by leading to varying three-dimensional structures of the messenger RNA when progress of the ribosomes was perturbed.

Concerning the temperature control of the elongation process, the latter continued when an exponentially growing culture of *E. coli* was cooled to below 8°C while initiation of protein synthesis was blocked (Friedman et al. 1971). These studies were assisted by the phage R17 and f2 RNAs. A sharp decrease in the elongation rate occurred, which led to incomplete translation of the longer, namely, replicase, gene, by lowering of Mg^{2+} concentration from the optimal value of 12.6–18.6 mM to the suboptimal one of 10.6 mM (Zagórski et al. 1972).

Remarkably, the RNA of the coat *amber* mutant R17amB2, which directed the synthesis of the N-terminal hexapeptide fMet-Ala-Ser-Asn-Phe-Tyr in the *in vitro* system, was used to show that peptidyl transferase, the ribosomal enzyme that synthesized peptide bonds, was able to catalyze the formation of ester linkages between α-hydroxyacyl-tRNA and aminoacyl-tRNA (Fahnestock and Rich 1971).

The intrinsic complexity of the elongation step was supported by the fact that the cell-free incorporation of amino acids from tRNA charged with 20 amino acids stimulated by the phage fr RNA was competitively inhibited by Phe-tRNA^{Phe}, while deacylated Phe-tRNA had no significant influence (Scheulen et al. 1973).

The nonuniform size distribution of the MS2 coat nascent chains led to an assumption that the rate of chain elongation during coat synthesis was impeded by the MS2 secondary structure and served as the origin for the accumulation of nascent chains of discrete sizes (Chaney and Morris 1979).

By studies on conformations of the nascent MS2 coat peptides on ribosomes, Tsalkova et al. (1998) established that only a small but significant portion of these peptides of 4.5 kDa, i.e., approximately 44 amino acid residues in length or larger, appeared to be able to react with antibodies while they were still bound to the ribosomes. Remarkably, the corresponding size for the 4.5 kDa peptide could be 66 Å if the latter was α-helical or a diameter of about 25 Å if it had a spherical conformation. In contrast, the nascent MS2 coat peptides smaller than about 4.5 kDa could not react with antibodies while bound to ribosomes.

The ribosomes from temperature-sensitive mutants of *E. coli* were examined for defects in the MS2 RNA-directed cell-free protein synthesis (Champney 1980). Thus, the ribosomes with alterations in ribosomal proteins S10, S15, or L22 showed reduced activity in the MS2 RNA translation at 44°C and were more rapidly inactivated by heating at this temperature compared to the control ribosomes. Moreover, a good correlation between the *in vitro* and *in vivo* translational activity of the *E. coli* mutants was observed (Armstrong-Major and Champney 1985). Furthermore, the MS2 RNA translation system was used by the functional evaluation of ribosomes lacking protein L1 from an appropriate *E. coli* mutant (Subramanian and Dabbs 1980). The data indicated that the L1 protein exerted its effect at the elongation step.

The MS2 RNA-programmed assay contributed to the characterization of two novel factors: ribosome modulation factor (RMF) and hibernation promoting factor (HPF) (Ueta et al. 2008). The RMF and HPF dimerized most 70S ribosomes to form 100S ribosomes during the stationary phase of growth in *E. coli*, where the process of 100S formation was termed "ribosomal hibernation." In the MS2 RNA-dependent leucine incorporation assay, the bound HPF was removed and hardly inhibited normal translation.

TERMINATION

The fidelity of the initiation and elongation of the *in vitro* translation was quite good, since all three phage-specific proteins were synthesized, each beginning with a formyl-methionyl residue and directed in accordance with the genomic instructions, as it followed from the previous sections. The same was true for the termination process, since each polypeptide was released free of the final tRNA and thus from the ribosomal complex, when the chain was completed, and ribosome reached a chain termination signal (Capecchi 1967c; Webster et al. 1967). The precise termination was also observed when mutant phage RNAs were used that contained UAG or UGA codons internally in any of the three genes. As described above, a fragment peptide was synthesized containing the N-terminal portion of the particular phage protein prematurely terminated at the site of the UAG or UGA codon in these cases (Capecchi 1967b; Webster et al. 1967; Model et al. 1969; Horiuchi et al. 1971; Ball and Kaesberg 1973a).

As stated earlier in the Conditionally lethal mutations section of Chapter 8, the mechanism of suppression was deciphered due to the application of the RNA phage messengers. Thus, addition of tRNAs isolated from the appropriate suppressing strains of bacteria to the *in vitro* reactions led to identification of amino acids, which might be inserted specifically at the site of the UAG (Capecchi and Gussin 1965; Engelhardt et al. 1965; Notani et al. 1965; Weber et al. 1966; Zinder et al. 1966) or UGA (Chan et al. 1971) codon, thus allowing synthesis of the whole phage protein.

Remarkably, the leaky UGA termination played an intrinsic role in the replication cycle of the group III and IV phages by the controlled production of the A1 protein, as described in Chapters 8 and 12. It is noteworthy that the synthesis of the A1 protein in a cell-free protein synthesizing system from *E. coli* could be regulated by the Mg^{2+} concentration (Hirashima et al. 1979). Thus, at 6 mM of magnesium acetate, the major product of the synthesis was coat protein, while at 12 mM of magnesium, it was replaced by readthrough protein. This enhanced synthesis was substituted by the addition of 0.25 mM of spermine or 1 mM of spermidine to 6 mM of magnesium. Therefore, magnesium ion or combination of magnesium and polyamines caused leaky UGA termination *in vitro*.

The function of mediating polypeptide chain termination in prokaryotic organisms has been ascribed to two proteins, the release factors RF-1 and RF-2 (Capecchi and Klein 1969;

Caskey et al. 1969; Klein and Capecchi 1971). The RF-1 activated hydrolysis of the peptidyl-tRNA bond in response to the termination codons UAA and UAG, and the RF-2 did it in response to UAA and UGA (Scolnick et al. 1968; Beaudet and Caskey 1970; Beaudet et al. 1970). To study the mechanism of polypeptide chain termination in cell-free extracts of *E. coli*, Ganoza and Nakamoto (1966) developed a reliable method for the separation of polypeptides from peptidyl-tRNAs by column chromatography.

As Capecchi and Webster (1975) noted, the f2 or R17 RNA-directed protein-synthesizing system was instrumental in developing an assay to detect the release factors. The assay employed RNA from a mutant of R17 or f2, in which the seventh codon in the viral coat protein gene has mutated from CAG, which coded for glutamine, to UAG, which coded for release. In the unfractionated *E. coli* cell-free amino acid incorporating system, RNA from this mutant directed the synthesis of a small amino-terminal coat protein fragment, the hexapeptide fMet-Ala-Ser-Asn-Phe-Thr, which was released free of tRNA (Webster et al. 1967; Bretscher 1968). The cell-free system contained whatever factors were required for release. In order to control the release of the polypeptide chain, the synthesis had to be performed in a stepwise fashion using a partially fractionated cell-free amino acid incorporating system. By this method, a substrate for examining the mechanism of polypeptide chain termination was generated. This substrate, the 70S ribosomal-mRNA complex containing the coat protein N-terminal hexapeptide, was used to look for the *E. coli* supernatant factors which would mediate the release of the hexapeptide. Surprisingly, the search revealed not the expected hypothetical chain-terminating tRNA, but rather a protein which mediated release of the polypeptide chain (Capecchi 1967c). A more convenient assay for polypeptide chain termination was subsequently developed by Caskey et al. (1968). This assay followed the release of fMet from a ribosomal-AUG-fMet-tRNA$_f$ complex in response to added terminator trinucleotides.

Furthermore, the requirement of the release factors RF-1 and RF-2 was examined for the termination of the complete R17 proteins. This was done by employing anti-RF1 and anti-RF2 antisera as specific inhibitors to control the *in vitro* release of complete proteins (Capecchi and Klein 1970). These studies have shown that either RF-1 or RF-2 sufficed to release completed molecules of the R17 coat and replicase. Further, the codon specificities of the release factors were confirmed using intact messenger RNAs.

As reviewed exhaustively by Capecchi and Webster (1975), the experiments in which a release factor was specifically inactivated with antisera showed that the endogenous release factors followed the predicted pattern of response to the terminator codons in the RNAs from R17 or related mutants. Thus, only RF-1 mediated release of the hexapeptide in response to UAG within the coat protein gene, while only RF-2 mediated release of an R17 replicase fragment in response to a UGA codon within the replicase gene, and either RF-1 or RF-2 mediated release of intact coat protein in response to a UAA codon at the end of the coat gene. It

remained unclear why UAA was followed by a second terminator codon, UAG, at the end of the coat protein gene. Since such tandem signals were not a general rule, it was speculated that the second codon acts as a safety device which can signal release in case the first codon is misread (Capecchi and Webster 1975).

The R17 RNA-programmed *in vitro* synthesis allowed the identification of a non-ribosomal factor Z that was distinct from G and T translocation factors, as well as release factors, but was essential for the completion of messenger RNA translation during or subsequent to recognition of the signal for the polypeptide chain termination (Phillips 1971).

A detailed study of the Mg^{2+}-dependent dissociation of 70S ribosomes before and after MS2 RNA-programmed protein synthesis *in vitro* was performed by Jan van Duin et al. (1970). By further monitoring of the fate of the complex containing one 70S ribosome bound to the f2 RNA at the coat initiation site, it was found that the polypeptide termination resulted in the release of ribosome as subunits (Martin and Webster 1975). A transient 46S complex composed of the f2 RNA and 30S subunit, presumably the result of the initial release of the 50S subunit, served as an intermediate in this process. This 46S complex subsequently broke down to a free f2 RNA and 30S ribosomal subunit.

The ribosome releasing or recycling factor (RRF) was found by the translation of the R17 amB2 RNA having an UAG codon at the seventh triplet of the coat gene (Ogawa and Kaji 1975a,b). The RRF was purified to homogeneity, demonstrated a molecular mass of 23,500, and inhibited amino acid incorporation into proteins programmed by a mutant R17 amB2 RNA (Ryoji et al. 1981b). In this system, the major polypeptide formed in the absence of this factor had a molecular mass very close to the authentic R17 coat protein, suggesting that ribosomes may read through the *amber* codon in the absence of the RRF. It was shown further that, in the absence of the RRF, the ribosome, which has released the N-terminal hexapeptide at the *amber* codon, stood on the mRNA and subsequently reinitiated translation "in phase" immediately after the *amber* codon without formylmethionine (Ryoji et al. 1981a). However, the readthrough translation of the R17 amB2 coat gene occurred to some extent also in the presence of the RRF, as well as other termination factors (Ryoji et al. 1985). It was confirmed by peptide fingerprinting and amino acid sequencing that this coat-like protein was produced as a result of reinitiation of translation from the eighth codon, but not by amino acid misinsertion to the *amber* mutation codon. The readthrough by amino acid misinsertion in this system became predominant only when the Mg^{2+} concentration was higher than 16 mM (Ryoji et al. 1985).

Later, by studies on the phage GA coat and lysis protein synthesis, where both genes were translationally coupled through an overlapping termination and initiation codon UAAUG, Inokuchi et al. (2000) found that the RRF might not release ribosomes from the junction UAAUG, but it was essential for correct ribosomal recognition of the AUG codon at the initiation site for the lysis gene.

Then, the *in vitro* translation system of the MS2 RNA was used by the exhaustive study on the *in vivo* effect of inactivation of the RRF by the unscheduled translation in a temperature-sensitive RRF strain (Hirokawa et al. 2004).

Another protein rescue factor that did not coincide with the factors RF-1, RF-2, and RF-3 was identified with the f2 RNA as a messenger, acted at the level of coat protein termination, but also affected the synthesis of non-coat protein products (Ganoza et al. 1973). However, it was established later, after purification and characterization of the rescue factor, that the latter had rather early effect on the f2-directed assay for translation, but not on the initiation step (van der Meer and Ganoza 1975) The rescue factor regulated rather conformational events during the transition from initiation to elongation that were required for proper reinitiation and accurate propagation and termination of translation (Ganoza et al. 1989).

Finally, the requirement for the factor called protein W was demonstrated by the full *in vitro* reconstruction of the translation process from purified components and with the MS2 or f2sus3 RNAs as messengers (Ganoza et al. 1985; Green et al. 1985). The complex of ribosomes, precharged aminoacyl-tRNAs, and the sum of the purified proteins involved in the initiation (initiation factors; IF-1, IF-2, and IF-3), propagation (elongation factors; EF-Tu, EF-Ts, and EF-G), and termination (release factors; RF-1 or RF-2) of protein synthesis did not work without the protein W. The latter was purified free of all translation factors, activating enzymes, and other proteins such as the rescue factor implicated in translation. It was proposed that the factor W prevented the spurious entrance of deacylated tRNA into the particles. Further detailed study revealed that the factor W functioned by ejecting tRNAs from ribosomes in a step that preceded the movement of mRNA during translocation (Ganoza et al. 1995).

As described in Chapters 7, 8, and 12, the termination signal UGA at the end of the coat protein gene of the group III and IV representatives led to appearance of a novel readthrough protein A1, which arose from the misreading of the UGA codon as a tryptophan codon (Horiuchi et al. 1971; Moore et al. 1971; Weiner and Weber 1971).

INTERGENIC SEQUENCES

In his response to the classical *RNA Phages* book, Gary Gussin (1975) emphasized the outstanding role of the intergenic sequences of the RNA genome, including ribosome binding sites, because of their high conservation level. Thus, the MS2 and R17 RNAs differed by only 1 base in 35 in the region between the coat and replicase genes, and by only 1 base in 26 nucleotides between the maturation and coat genes. According to his comment, this was an extent of conservation at least as great and perhaps greater than that seen in the coding regions. Such striking conservation of the intergenic regions was necessitated by the essential functions they served, including regulatory ones, which provided for the specific recognition of the RNA fragments, including the intergenic sequences by phage proteins. The most crucial regulatory functions in the RNA phages were performed by repressor complexes I and II formed by coat protein and replicase, respectively.

RNA FRAGMENTS AS MESSENGERS

Bassel (1968) was the first to show that not only full-length Qβ RNA, but also its large fragments, directed the *in vitro* protein synthesis, where the large fragment, derived from the 3′-end of Qβ RNA, directed the synthesis of coat protein. Moreover, the coat protein was the first protein synthesized *in vitro* both by complete Qβ RNA strands and the large Qβ RNA fragment.

Shimura and Nathans (1968) described the *in vitro* translation of the MS2 RNA fragments extracted from RNA-deficient particles, which resulted from growth of MS2 in the presence of 5-fluorouracil (Shimura et al. 1967). Such fluorouracil-RNA fragments represented about two-thirds of the 5′-end of the phage RNA molecule and directed the synthesis of phage coat but not of replicase. A bit later, the fragments of R17 RNA were used for the *in vitro* protein synthesis (Gesteland and Spahr 1969).

Later, the translational activity of the phage RNA fragments was studied by Grēns' team in Riga. Thus, the phage fr RNA fragments covering the complete or part of the replicase gene initiation region were used as templates in the formation of a ribosomal initiation complex *in vitro* (Berzin et al. 1982; Tsielens et al. 1982).

These studies showed that, besides the minimal initiation region including the SD sequence and initiator AUG, the flanking regions were also involved in the process and were responsible for additional interactions with the ribosome. Such flanking regions possibly contributed to the stability of specific contact between the ribosome and the phage RNA template, realized by the minimal initiation region length (Berzin et al. 1982).

The knowledge of the structure of the initiation sites and SD sequences initiated successful experiments on their synthetic analogs, including mutational changes in the synthetic initiation sites. The global work on the synthetic analogs was performed at that time by Grēns' team. Thus, Regīna Renhofa's group synthesized enzymatically three 20-base polyribonucleotides corresponding to the minimal initiation region for the replicase gene of the phages MS2 and fr or having destroyed SD sequence (Renhof et al. 1985). The SD-containing polynucleotides demonstrated the corresponding template activity, while the SD-missing one did not. A next set of hepta- and decaribonucleotides was prepared enzymatically as components to construct more 20-base polyribonucleotide templates of the MS2 and fr replicase initiation sites (Nikitina et al. 1986).

Then, Stankevich et al. (1995) chemically synthesized a set of oligoribonucleotides, 15−24 members in length, containing native and modified MS2 and fr replicase initiation sites, and tested them for the interaction with ribosomes. The 15–19-mer oligoribonucleotides were synthesized then as a model of the translation initiation site of the phage fr replicase (Kumpin'sh et al. 1995).

MISCELLANEOUS BY THE *IN VITRO* TRANSLATION

The undermethylated tRNAs did not support the Qβ and MS2 RNA-directed *in vitro* protein synthesis (Stulberg et al. 1976). Therefore, a sufficient level of the tRNA base modifications was necessary to function in protein synthesis.

Nakamoto and Vogl (1978) studied the accessibility and selection of the initiator site of mRNA, with the MS2 and R17 RNAs among models, by the *in vitro* initiation and polypeptide synthesis. They concluded that the ribosomes did not have any special affinity for the viral RNA and that the selection of the initiator site in protein synthesis might be critically determined more by the accessibility of the initiator codon than by ribosomal recognition of the site.

Without going into specific details, it is necessary to mention that the MS2 RNA was used as a standard criterium for the initiation factor dependence in the exhaustive study of how nucleotides contiguous to AUG affected translational initiation (Ganoza et al. 1978).

Remarkably, ribosomal proteins S3, S4, S5, S9, S13, L2, and L17, in addition to the S1 protein, formed complex with the MS2 RNA (Sarapuu and Villems 1982). By ultraviolet irradiation at 254 nm of the initiation complex, the MS2 RNA crosslinked to the IF-3 and ribosomal proteins S3, S4, S5, S7, S9, and S18, manifesting the direct interaction with the MS2 RNA (Broude et al. 1983). At the same time, the fMet-tRNA interacted with proteins S4, S5, S9, S11, S14, and S15-S17 inside the preinitiation complex formed by fMet-tRNA with 30S subunit and MS2 RNA in the presence of initiation factors (Broude et al. 1985).

The initiation complex formation with the MS2 RNA was affected by the chemical modification *in situ* of *E. coli* 30S ribosomal proteins by the site-specific reagent pyridoxal phosphate, where the protein S3 was the primary cause of the inhibition but a role was also played by the ribosomal proteins S1, S2, and S21 (Ohsawa and Gualerzi 1983).

By using photoreactive derivatives of fMet-tRNA$_{fMet}$ bearing arylazido groups scattered statistically over guanosine residues, the irradiation of the 70S ribosome-MS2 RNA-fMet-azido-tRNAfMet complexes resulted in covalent linking of the tRNA derivative to the ribosomes, where 30S proteins were preferential targets of photoaffinity labeling (Babkina et al. 1987).

It is noteworthy that bisulfite up to 10 mM did not affect the MS2 RNA-directed protein synthesis in the *E. coli* cell-free system (Robakis et al. 1983).

The first immune electron microscopy experiments on ribosomes led to the localization of the hapten-labeled N-terminal methionine of the MS2 coat grown as far as the 42nd amino acid residue (Ryabova et al. 1988). This study allowed direct visualization of a channel for the nascent peptide inside the ribosome.

SECONDARY AND TERTIARY STRUCTURE

The intrinsic value of the higher RNA structures, secondary and tertiary, by the phage gene translation was demonstrated

from 1985–2009 in the exhaustive studies originally performed by Jan van Duin's team. Thus, the full account on the secondary structures of almost all prokaryotic mRNA initiation sites known at that time were collected in a comprehensive review (de Smit and van Duin 1990a). The detailed analysis showed that it would be quite surprising to find a simple linear relationship between the efficiency of a ribosome binding site and the fraction of unfolded mRNA molecules. The authors proposed that a non-sequence-specific interaction of ribosomes with single-stranded RNA constituted the first step in the process of initiation.

A quantitative analysis of the relationship between translational efficiency and the mRNA secondary structure in the initiation region was presented for variants of the MS2 coat protein initiation site (de Smit and van Duin 1990b). The stability of a defined hairpin structure containing a ribosome binding site was estimated, and the results revealed a strict correlation between translational efficiency and the stability of the helix. No evidence was found to confirm the assumption that exposure of only the SD region or the start codon preferentially favored recognition. The translational efficiency was strictly correlated with the fraction of mRNA molecules in which the ribosome binding site was unfolded, indicating that the initiation was completely dependent on spontaneous unfolding of the entire initiation region. At the same time, ribosomes did not appear to recognize nucleotides outside the SD region and the initiation codon. Figure 16.2 presents the MS2 coat initiation site with the mutations introduced in the coat initiator hairpin.

Therefore, the information about the secondary structures of the messengers used, before and after mutagenesis within the SD regions, was found especially crucial, since intrastrand base pairing of a ribosome binding site could have a profound influence on its translational efficiency (de Smit and van Duin 1994b). Varying by site-directed mutagenesis the extent of the SD complementarity in the MS2 coat gene, the

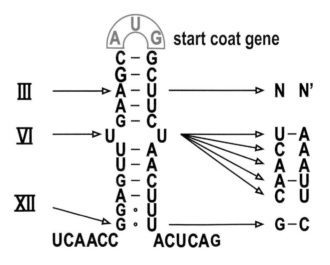

FIGURE 16.2 Mutations introduced in the coat initiator hairpin. Mutated base pairs are denoted by roman numerals. -, Watson-Crick base pair; •, G•U pair. (Reprinted with permission of Maarten H. de Smit from de Smit MH, van Duin J. *Proc Natl Acad Sci U S A.* 1990b;87:7668–7672.)

authors found that mutations reducing the SD complementarity by one or two nucleotides diminished translational efficiency only if ribosome binding was impaired by the structure of the messenger. Surprisingly, in the absence of an inhibitory structure, these mutations had no effect. In other words, a strong SD interaction was compensated for a structured initiation region. The authors considered therefore the translational initiation on a structured ribosome binding site as a competition between intramolecular base pairing of the messenger and binding to a 30S ribosomal subunit. The good SD complementarity provided the ribosome with an increased affinity for its binding site and thereby enhanced its ability to compete against the secondary structure. Remarkably, this function of the SD interaction closely paralleled the RNA-unfolding capacity of the ribosomal S1 protein (de Smit and van Duin 1994b). The quantitative analysis of literature data on the ribosome binding sites was collected in order to support this idea (de Smit and van Duin 1994a). This compilation showed that the efficiency of translation was determined by the overall stability of the structure at the ribosome binding site, whether the initiation codon itself was base paired or not. The structures weaker than −6 kcal/mol usually did not reduce translational efficiency. Below this threshold, all systems showed a tenfold decrease in expression for every −1.4 kcal/mol (de Smit and van Duin 1994a).

Olsthoorn et al. (1995b) demonstrated coevolution of the RNA helix stability and SD complementarity in the MS2 coat translational start region. Thus, the changes were made in the SD that shortened or extended its complementarity to the 3′ end of 16S rRNA, and their evolution to a stable pseudorevertant species was monitored *in vivo* in live MS2 phage. The results showed that the phages in which the SD complementarity was decreased evolved an initiator hairpin of lower stability than wild-type, while those in which the complementarity was extended evolved a hairpin with increased stability. It was concluded that weaker SD sequences still allowed maximal translation if the secondary structure of the ribosome-landing site was destabilized accordingly. Alternatively, translation-initiation regions with a stronger secondary structure still allowed maximal expression if the SD complementarity was extended. These findings supported the idea that the SD interaction helped the ribosome to melt the structure in the translation-initiation region (Olsthoorn et al. 1995b).

Next, a simple solution was found to relieve the initiation at the coat gene from the dependence of the upstream sequences (de Smit and van Duin 1993). The maximal translation of the MS2 coat gene required a contiguous stretch of native MS2 RNA that extended hundreds of nucleotides upstream from the translational start site, since deletion of these upstream sequences from MS2 cDNA plasmids resulted in a 30-fold reduction of translational efficiency. The authors showed by site-directed mutagenesis that this low level of expression was caused by a hairpin structure centered around the initiation codon. When this hairpin was destabilized by the introduction of mismatches, the expression from the truncated messenger increased 20-fold to almost the level of the full-length construct. Thus, the translational effect of hundreds of

upstream nucleotides was mimicked by a single substitution that destabilized the hairpin structure. It was speculated that the upstream RNA within the full-length MS2 RNA somehow reduced the inhibitory effect of the hairpin structure on translational initiation in a way that the hairpin did not impair the ribosome binding (de Smit and van Duin 1993).

Furthermore, the translational control of the MS2 maturation protein synthesis was explained by RNA folding (Groeneveld et al. 1995). The secondary structure of the untranslated 130-nt leader preceding the maturation gene was deduced by phylogenetic comparison and by probing with enzymes and chemicals. It was established that the RNA folded into a cloverleaf, i.e., a complex of the three stem-loop structures enclosed by a long-distance interaction, which was essential for the translational control. Its 3′-moiety contained the SD region of the maturation protein gene, whereas its complement was located 80 nt upstream, i.e., about 30 nucleotides from the 5′-terminus of the RNA chain. Figure 16.3 presents the proposed RNA structure of the 5′-untranslated leader of the MS2 RNA together with base differences with the group I phages fr and M12, and of the group II phage KU1 showing some differences with the group II phage GA.

The mutational analysis showed that the predicted base pairing repressed expression of the maturation protein gene. The proposed translational model was based on the observation that reducing the length of the intervening sequence reduced expression, whereas increasing the length had the opposite effect. Remarkably, this model did not explain the finding of Robertson and Lodish (1970) that the replicative intermediates were more active than full-sized phage RNAs in directing ribosomes to the maturation gene start. It was incompatible also with the structure designed by Fiers et al. (1975b), in which the maturation gene translation was controlled by a long-distance interaction between the gene start and a sequence some 700 nucleotides downstream. This interaction did not stand up in the comparative analysis with the other known sequences of the group I and II phages (Groeneveld et al. 1995).

Direct evidence for the regulation of the MS2 maturation gene translation by the RNA folding kinetics was presented by Poot et al. (1997). Thus, the *in vitro* experiments showed that ribosome binding to the maturation gene was faster than refolding of the denatured cloverleaf, and this folding delay appeared to be related to special properties of the leader sequence. When the three stem-loop structures of the 130-nt leader were replaced by a single five-nucleotide loop, the folding delay had virtually disappeared, suggesting that now the RNA folded faster than ribosomes could bind. This perturbation of the cloverleaf by an insertion made the maturation start permanently accessible (Poot et al. 1997).

At the same time, van Duin's team generated the secondary structure model for the Qβ RNA, as demonstrated in the Secondary structure: van Duin's team section of Chapter 10. First, the secondary structure model of the last two domains of Qβ RNA (see Figure 10.5) was published (Beekwilder et al. 1995). Then, the secondary structure for the first three domains of Qβ RNA, 860 nucleotides in total (see Figure 10.6), appeared (Beekwilder et al. 1996b). This structure was

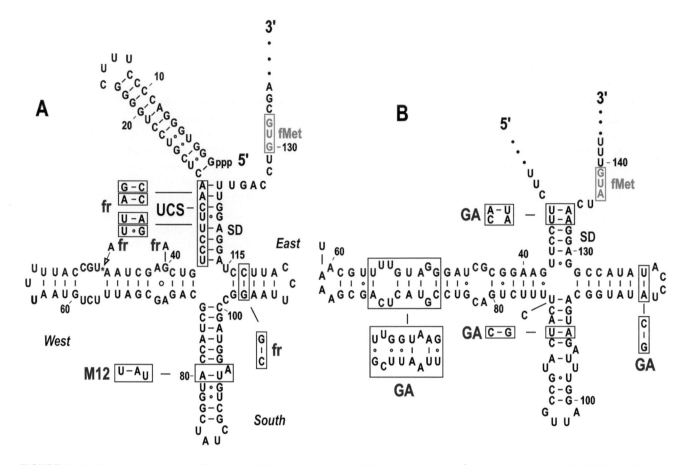

FIGURE 16.3 Secondary structure of the phage RNAs. (A) Proposed RNA structure of the 5′ untranslated leader of MS2 RNA. The start codon at position 130, the upstream complementary sequence at position 29–36, and the base differences with the group I phages fr and MS2 are boxed. (B) Proposed RNA structure of the 5′ untranslated leader of group II phage KU1, showing differences with group II phage GA. Additional substitutions in GA are found at positions 61, 83, and 86. (Redrawn with permission of *RNA* Journal and RNA Society from Groeneveld H et al. *RNA*. 1995;1:79–88.)

supported by phylogenetic comparison, nuclease S1 structure probing, and computer prediction using energy minimization and a Monte Carlo approach. To provide the necessary data for the comparative analysis, the MX1, M11, and NL95 RNAs were sequenced. Together with the known sequences of the Qβ and SP RNAs, this yielded five sequences with sufficient sequence diversity to be useful for the comparative analysis. The constructed secondary structure model explained convincingly the control mechanism of the maturation protein synthesis. Generally, the RNA adopted a highly ordered structure, in which all hairpins were held in place by a network of long-distance interactions which formed three-way and four-way junctions. Only the 5′-terminal hairpin of the 60-nucleotide untranslated region was unrestrained, while connected by a few single-stranded nucleotides to the body of the RNA. The start region of the maturation gene, which was part of the network of long-distance interactions, was base paired to three noncontiguous downstream sequences. As a result, the translation was expected to be progressively quenched when the length of the nascent chains increased. This feature explained the previous observation that the Qβ maturation protein synthesis was started only on short nascent strands. As narrated above, the translational control of the maturation protein in

the MS2 RNA was controlled by the kinetics of the RNA folding. Beekwilder et al. (1996b) explained this basic difference by the different RNA folding pathways in the Qβ and MS2 genomes.

Furthermore, the initiation control of the MS2 maturation gene was explained finally by van Duin's team (van Meerten et al. 2001a). Thus, when the MS2 RNA folding was at equilibrium, the maturation gene was untranslatable because the 5′ 130-nucleotide leader adopted a well-defined cloverleaf structure in which the SD sequence of the maturation gene was taken up into long-distance base pairing with an upstream complementary sequence. The synthesis of the maturation protein took place transiently while the RNA was synthesized from the minus strand. This required that formation of the inhibitory cloverleaf was slow. *In vitro*, the folding delay was on the order of minutes. The clear evidence was presented that this postponed folding was caused by the formation of a metastable intermediate. This intermediate was a small local hairpin that contained the upstream complementary sequence in its loop, thereby preventing or slowing down its pairing with the SD sequence. The mutants in which the small hairpin could not be formed made no detectable amounts of the maturation protein and were barely viable. Apparently, here the cloverleaf formed quicker than ribosomes could bind. On the other hand, the mutants in which the small intermediary hairpin was stabilized produced more maturation protein than the wild type and were still viable (van Meerten et al. 2001a).

Then, de Smit and van Duin (2003a,b) described how ribosomes may deal with the rapid folding kinetics of mRNA, when the stable base pairing at a translational initiation site might inhibit translation by competing with the binding of ribosomes. The novel standby model proposed that a 30S subunit must already be in contact with the mRNA while this was still folded, to shift into place as soon as the structure was opened. Thus, the single-stranded regions flanking the structure might constitute a standby site to which the 30S subunit could attach nonspecifically. As a result, a steady-state kinetic model for the early steps of translational initiation was proposed. The kinetic model provided an explanation of why the earlier equilibrium competition model predicted implausibly high 30S-mRNA affinities. Because all RNAs were structured to some degree, the standby binding was regarded as a general feature of the translational initiation (de Smit and van Duin 2003a,b).

The standby mechanism proposed by de Smit and van Duin (2003a,b), which postulated in short that the 30S ribosome binds nonspecifically to a standby site, namely, an accessible site on the mRNA, and waits for a transient opening of a stable ribosome binding site hairpin, was supported by experimental data that were collected later and illustrated nicely by Unoson and Wagner (2007).

The role of the self-induced structural switches in the regulation of the MS2 RNA translation was analyzed, among other similar examples, by Nagel and Pleij (2002) and by Wolfinger et al. (2004).

Priano et al. (1997) demonstrated on the Qβ RNA that the repression of one gene could activate translation of another one. Thus, the translation of the Qβ maturation gene was mediated by the presence of the Qβ replicase. This activation did not require RNA replication, translation of a second gene, or any direct protein-RNA binding at the maturation gene initiation site. According to the proposed model, the Qβ maturation gene remained translationally "off" by two means: (i) the thermodynamic stability of an RNA structure that greatly discouraged, but did not eliminate, ribosome access at the maturation start site; and (ii) the presence of the stronger, proximal coat gene ribosome binding site. Moreover, the maturation gene expression was switched "on" when ribosome entry at the coat initiation site was repressed by the Qβ replicase, thereby allowing ribosomes to compete for the weaker, upstream maturation start site.

Meanwhile, Walter Fiers' team proposed a hypothesis that the folding of the MS2 coat protein could be modulated by translational pauses resulting from the MS2 secondary structure and codon usage (Guisez et al. 1993). Thus, the possible translational pauses within the MS2 coat were located on the basis of a distribution plot of rare codons and RNA secondary structure (see Figure 10.3). Moreover, it appeared that the position of certain codon pauses corresponded with the size of some nascent polypeptide intermediates which have been isolated from the MS2-infected cells. Other accumulated polypeptide intermediates seemed to be related to the RNA regions where double-stranded secondary structures occurred, which probably impeded the movement of ribosomes during chain elongation. Later, Komar (2009) continued this approach and performed an attempt to generate a unifying concept that would combine the basic features governing self-organization of proteins into complex three-dimensional structures. He emphasized the idea that messenger RNAs contain an additional layer of information beyond the amino acid sequence that fine-tunes the *in vivo* protein folding, which is largely believed to start as a co-translational process. It was concluded that the translation kinetics might direct the co-translational folding pathway and that translational pausing at rare codons might provide a time delay to enable independent and sequential folding of the defined portions of the nascent polypeptide emerging from the ribosome. The conclusion was based on, among other data, the studies of the co-translational folding of the MS2 coat protein (Hardesty et al. 1993).

More recently, Jayant et al. (2010) established that the single-strandedness of the Qβ RNA at one ribosome binding site directly affected translational initiations at a distal upstream gene. Thus, in the Qβ RNA, the sequestering of the coat gene ribosome binding site in a putatively strong hairpin stem structure eliminated synthesis of the coat protein and activated protein synthesis from the much weaker maturation gene initiation site, located 1300 nucleotides upstream. As the stability of a hairpin stem comprising the coat gene SD site was incrementally increased there was a corresponding increase in translation of the maturation protein. The effect of the downstream coat gene ribosome binding sequence on the maturation gene expression appeared to have occurred only in *cis* and did not require an AUG start codon or initiation of coat protein synthesis. In all cases, no structural reorganization was predicted to occur within the Qβ RNA. As a result, it was suggested that the protein synthesis from a relatively weak translational initiation site was greatly influenced by the presence or absence of a stronger ribosome binding site located elsewhere on the same RNA molecule. The data were consistent with a mechanism in which multiple ribosome binding sites competed in *cis* for translational initiations as a means of regulating protein synthesis on a polygenic messenger RNAs (Jayant et al. 2010).

At the same time, the gene expression rate was estimated by integrating the opening probability of the SD sequence over the entire folding process by the expression of the MS2 maturation gene (Tapia et al. 2010). This study demonstrated, first of all, the capabilities of a computational technique based on algorithms for the robot motion planning that could study both protein and RNA motion. This technique composed an approximate map, or model, of the molecule's energy landscape. With this model, the folding pathways and study landscape properties were extracted, which included relative folding rates and population kinetics (Tapia et al. 2010).

RIBOSOMAL FRAMESHIFTING

The addition of Ser^{AGU}_{AGC} tRNA to the *E. coli* cell-free protein synthesizing system resulted in up to 100% of the ribosomes translating the MS2 coat gene shifting into the −1 reading frame (Atkins et al. 1979a). An analogous phenomenon was seen at a much lower level without the tRNA addition, where a shift into the +1 frame could also be detected. Thus, the translation with the endogenous tRNA levels yielded proteins which had the N-terminus of the coat protein, but which were substantially larger than the coat protein and comprised about 5% of the coat translation. Since the lysis gene overlapped the 3′-end of the coat gene in the +1 frame, it was concluded that the reading frame shift into the +1 frame yielded a hybrid protein. In parallel, ribosomes translating the replicase gene shifted into the −1 frame near the distal end of the gene. This frameshifting was promoted by Thr^{ACU}_{ACC} tRNA (Atkins et al. 1979a). The coat-lysis fusion protein was discovered in the f2-infected cells *in vivo* by Beremand and Blumenthal (1979).

Then, the frameshifted translation was detected by the translation of the MS2 RNA with mammalian ribosomes and led to the localization of the f2 op3 mutation and discovery of the lysis gene (Atkins et al. 1979b), as described in the Op-3 mutant section of Chapter 8. In short, the main binding site for mammalian ribosomes on the MS2 RNA was located nine-tenths of the way through the coat protein gene and translation initiated at an AUG triplet in the +1 frame yielded a 75 amino acid lysis polypeptide which terminated within the replicase gene at a UAA codon, also in the +1 frame. The same protein was made in an *E. coli* cell-free system, but only in very small amounts.

The coat protein shifted into the −1 reading frame, called protein 7 and having molecular mass of 18,000–19,000 daltons, was studied by Laughrea (1981). Thus, during *in vitro* translation of the R17 and MS2 RNAs, the protein 7 synthesis was stimulated by ethanol and streptomycin, while mutation in the ribosomal protein S12 had no detectable effect.

Spanjaard and van Duin (1988) found that the translation of the sequence AGG-AGG yielded 50% ribosomal frameshift. Thus, when the sequence 5′-AAG-GAG-GU-3′, which was complementary to the 3′-terminus of *E. coli* 16S rRNA, was inserted in a reading frame into the fused MS2 coat-interferon gene (for details, see Chapter 19), just before the interferon part, the translation over the sequence yielded a 50% ribosomal frameshift, if the reading phase was A-AGG-AGG-U. The other two possible frames did not give shifts. The introduction of a UAA stop codon before (UAA-AGG-AGG-U) but not after (A-AGG-AGG-UAA) the AGG codons abolished the frameshift. The change in the reading phase occurred exclusively to the +1 direction. The efficient frameshifting was induced also by the sequence A-AGA-AGA-U. The arginine codons AGG and AGA were read by a minor tRNA and suppression of frameshifting took place when a gene for the minor tRNA^{Arg} was introduced on a multicopy plasmid. It was suggested that frameshifting during translation of the A-AGG-AGG-U sequence was due to the erroneous decoding of the tandem AGG codons and arose by depletion

of the tRNA^{Arg}. Therefore, the complementary of tandem AGG codons to the 3′-terminus of 16S rRNA was regarded as a coincidence and apparently not related to the shift. However, replacing the AGG-AGG sequence by the optimal arginine codons CGU-CGU did not increase the overall rate of translation (Spanjaard and van Duin 1988).

Garcia et al. (1993) broadened the frameshift studies to the genomic RNA of beet western yellows virus (BWYV), which contained a potential translational frameshift signal in the overlap region of open reading frames ORF2 and ORF3. The frameshifting was assayed both *in vivo* and *in vitro* using plasmids containing the wild-type and modified versions of the putative BWYV shift signal placed between the MS2 coat gene and the *lacZ* gene. The signal, composed of a hepta-nucleotide slippery sequence and a downstream pseudoknot, was similar in appearance to those identified in retroviral RNAs. The efficiency of the signal in the eukaryotic system was low but significant, as it responded strongly to changes in either the slip sequence or the pseudoknot. In contrast, in *E. coli* there was hardly any response to the same changes. Replacing the slip sequence to the typical prokaryotic signal AAAAAAG yielded more than 5% frameshift in *E. coli*. In this organism, the frameshifting was highly sensitive to changes in the slip sequence but only slightly to disruption of the pseudoknot. In contrast, the eukaryotic assay systems were barely sensitive to changes in either AAAAAAG or in the pseudoknot structure in this construct. It was therefore concluded that eukaryotic frameshift signals were not recognized by prokaryotes. On the other hand, the typical prokaryotic slip sequence AAAAAAG did not lead to significant frameshifting in the eukaryotes (Garcia et al. 1993).

Finally, a new site of ribosomal frameshifting in *E. coli*, namely, CCC.UGA, was characterized and called "+1 shifty stop" (de Smit et al. 1994). To activate expression of a human transferrin-encoding cDNA in *E. coli* by translational coupling, it was placed in an expression plasmid downstream from a 5′-terminal fragment from the MS2 replicase gene. The resulting construct was found to produce, besides the desired transferrin, a protein with the mobility of a fusion product of the N-terminal replicase fragment and transferrin. Analysis of available mutants showed that this fusion resulted from +1 ribosomal frameshifting at the end of the MS2 replicase reading frame. This region contained the sequence CCC.UGA, suggesting that before termination occurred, the tRNA^{Pro} might dislodge from the CCC codon and reassociate with the +1 triplet CCU. By further site-directed mutagenesis, it was demonstrated that both the CCC codon and the termination codon were indeed required for the observed 2%–4% frameshifting. When either triplet was changed, the frequency of frameshifting dropped to 0.3% or less (de Smit et al. 1994).

REINITIATION

The process of the translational reinitiation in *E. coli* was studied in a two-gene system, the upstream MS2 coat gene and the downstream rat interferon gene, where expression of the downstream reporter gene was dependent on the translation of an upstream reading frame (Spanjaard and van Duin 1989). The dependence was almost absolute. The upstream translation increased expression of the downstream gene by two to three orders of magnitude. In the absence of the SD region, the reinitiation occurred but its efficiency was about 10% of that found in the SD carrying counterpart. This finding contributed strongly to the scanning model for translational reinitiation in eubacteria (Adhin and van Duin 1990). According to this model, the terminated but not released ribosome reached neighboring initiation codons by lateral diffusion along the mRNA. The model was based on the finding that introduction of an additional start codon between the termination and the reinitiation site consistently obstructed ribosomes to reach the authentic restart site. Instead, the ribosome now began protein synthesis at this newly introduced AUG codon. This ribosomal scanning-like movement was bidirectional, had a radius of action of more than 40 nucleotides in the model system used, and activated the first encountered restart site. The ribosomal reach in the upstream direction was less than in the downstream one, probably due to dislodging by elongating ribosomes. The proposed model had parallels with the scanning mechanism postulated for eukaryotic translational initiation and reinitiation (Adhin and van Duin 1990).

LARGE-SCALE TRANSLATION

As described in the Cell-free synthesis section of Chapter 4, the concept of the high-yield *in vitro* translation was generated by Alexander Sergeevich Spirin's team (Spirin et al. 1988). As mentioned there, the cell-free translation system was efficient enough to produce infectious MS2 phage particles (Katanaev et al. 1996), and the MS2 RNA translation process was maximally optimized (Ryabova et al. 1995).

Katanaev et al. (1995) used the Qβ RNA as a vector for the *E. coli* cell-free translation, where dihydrofolate reductase mRNA was inserted into the Qβ RNA instead of its coat protein gene. The translation of this recombinant mRNA resulted in the synthesis of dihydrofolate reductase, which was two orders of magnitude higher than that in the case of translation of the control dihydrofolate reductase mRNA. In additional, it resulted in a significantly enhanced synthesis of the Qβ replicase as compared with its synthesis when the original Qβ RNA was used.

The capabilities of the cell-free system for the RNA replication, expression, and cloning was reviewed at that time by Chetverin (1996) and later by Spirin (2001). The methodology of the cell-free synthesis was described in detail by Spirin's team (Ryabova et al. 1998; Alimov et al. 2000).

Furthermore, a microfabricated reactor was constructed for the cell-free protein synthesis, which employed the MS2 RNA as a messenger (Nojima et al. 1998, 2000). The main emphasis was placed on the microtechnological importance of the subject as a new system for medical and biochemical applications. The authors highlighted the fact that the MS2 coat was synthesized in a 226 nL microfabricated space.

As just described in the Cell-free synthesis section of Chapter 4, cell-free synthesis was used for the quite impressive

production of the MS2 virus-like particles (Bundy et al. 2008). Moreover, the *E. coli*-based cell-free transcription–translation system was able to synthesize infectious phages such as MS2, φX174, and T7 (Garamella et al. 2016; Rustad et al. 2017).

GENETIC CODE

As noted clearly in the classical Zinder (1965, 1980) reviews, the simple addition of the f2 RNA to the *E. coli* extract not only led to an understanding of the mechanisms for protein synthesis but contributed strongly to the full deciphering of the genetic code, especially genetic code degeneracy. If the genetic code was discovered by the addition of polyU to the cell-free extract, the usage of the code words was clarified after long RNA phage stretches, and finally full-length MS2 genome was brought to the translation studies. Walter Fiers (1975) noted that 49 different nucleotide triplets were found to code for the 129 amino acids of the MS2 coat protein. For a few amino acids, the choice between the degenerate codons seemed to be nonrandom. Perhaps most noteworthy was that some codons were apparently avoided, like AUA for isoleucine, UAU for tyrosine, ACA for threonine, and AGU for serine. Indeed, all codons which previously had not been encountered in the coat gene were found in the maturation protein gene. This means that all 61 possible sense code words and two nonsense codons were normally used in the phage genome. The two of the three MS2 genes started with AUG and one with GUG, while they terminated with UAG or UAA. The UGA termination was found later in the Qβ genome. The advantages of the code degeneracy markedly enhanced the base-pairing possibilities, especially if the phasing of opposite strands was such that third letters did not face each other (Fiers 1975). Finally, it was accepted that in the phage genome "all of the synonymous codons are used and with frequencies that might be expected on a more or less random basis. *E. coli*'s vocabulary is equivalent to the dictionary of genetic words" (Zinder 1980). Generally, the MS2 genome presented 1068 codons assigned. Figure 16.4 shows the distribution of the codons in the four MS2 genes.

Historically, the first observations on the fidelity of translation, or in other words, on the punctuation in the genetic code, in such natural messages as phage RNAs were presented by Zinder et al. (1966). The authors proposed that there existed properties of RNA primary and/or secondary structure that maintained the fidelity of translation, and such translational fidelity was retained in bacterial extracts by the f2 RNA translation when coat protein of proper composition was obtained. In parallel, Fiers (1966) indicated the rhythmic character of the genetic code of the phage RNAs by quantitative analysis of the distribution of purine nucleotide sequences that were found in the MS2 (Fiers et al. 1965a,b) and R17 and M12 (Sinha et al. 1965a,b) RNAs.

The growing primary structure data on the RNA phage genomes opened a door for the first exhaustive studies on the codon distribution (Fitch 1972, 1974, 1976; Grantham 1972; Ball 1973a; Fiers 1973; Grantham et al. 1980). By analysis

Maturation

1st	U	C	A	G	3rd
U	Phe 6	Ser 5	Tyr 4	Cys 0	U
U	Phe 10	Ser 6	Tyr 12	Cys 3	C
U	Leu 8	Ser 8	Ochre 0	Opal 0	A
U	Leu 6	Ser 10	Amber 1	Trp 12	G
C	Leu 7	Pro 5	His 2	Arg 7	U
C	Leu 8	Pro 5	His 3	Arg 6	C
C	Leu 5	Pro 4	Gln 9	Arg 6	A
C	Leu 2	Pro 3	Gln 9	Arg 3	G
A	Ile 1	Thr 10	Asn 2	Ser 4	U
A	Ile 8	Thr 6	Asn 15	Ser 3	C
A	Ile 7	Thr 5	Lys 5	Arg 3	A
A	Met 1+6	Thr 6	Lys 9	Arg 4	G
G	Val 8	Ala 6	Asp 8	Gly 15	U
G	Val 7	Ala 12	Asp 5	Gly 6	C
G	Val 9	Ala 7	Glu 5	Gly 2	A
G	Val 9	Ala 10	Glu 12	Gly 5	G

Coat

1st	U	C	A	G	3rd
U	Phe 1	Ser 3	Tyr 0	Cys 1	U
U	Phe 3	Ser 2	Tyr 4	Cys 1	C
U	Leu 1	Ser 2	Ochre 1	Opal 0	A
U	Leu 0	Ser 2	Amber 1	Trp 2	G
C	Leu 2	Pro 2	His 0	Arg 3	U
C	Leu 2	Pro 1	His 0	Arg 1	C
C	Leu 2	Pro 2	Gln 1	Arg 0	A
C	Leu 0	Pro 1	Gln 5	Arg 0	G
A	Ile 4	Thr 4	Asn 4	Ser 0	U
A	Ile 4	Thr 4	Asn 6	Ser 4	C
A	Ile 0	Thr 0	Lys 5	Arg 0	A
A	Met 1+2	Thr 1	Lys 1	Arg 0	G
G	Val 4	Ala 5	Asp 1	Gly 3	U
G	Val 4	Ala 2	Asp 3	Gly 3	C
G	Val 3	Ala 6	Glu 2	Gly 2	A
G	Val 3	Ala 1	Glu 3	Gly 1	G

Replicase

1st	U	C	A	G	3rd
U	Phe 12	Ser 7	Tyr 5	Cys 5	U
U	Phe 16	Ser 12	Tyr 16	Cys 2	C
U	Leu 8	Ser 6	Ochre 0	Opal 0	A
U	Leu 5	Ser 10	Amber 1	Trp 9	G
C	Leu 7	Pro 9	His 4	Arg 11	U
C	Leu 15	Pro 4	His 6	Arg 13	C
C	Leu 8	Pro 4	Gln 7	Arg 4	A
C	Leu 7	Pro 10	Gln 8	Arg 8	G
A	Ile 7	Thr 4	Asn 11	Ser 4	U
A	Ile 13	Thr 12	Asn 7	Ser 9	C
A	Ile 12	Thr 8	Lys 9	Arg 4	A
A	Met 1+9	Thr 7	Lys 16	Arg 2	G
G	Val 9	Ala 15	Asp 19	Gly 19	U
G	Val 10	Ala 6	Asp 14	Gly 7	C
G	Val 6	Ala 8	Glu 9	Gly 8	A
G	Val 7	Ala 12	Glu 13	Gly 10	G

Lysis

1st	U	C	A	G	3rd
U	Phe 2	Ser 1	Tyr 1	Cys 1	U
U	Phe 3	Ser 0	Tyr 1	Cys 0	C
U	Leu 1	Ser 1	Ochre 1	Opal 0	A
U	Leu 3	Ser 3	Amber 0	Trp 0	G
C	Leu 3	Pro 1	His 1	Arg 0	U
C	Leu 2	Pro 0	His 0	Arg 2	C
C	Leu 1	Pro 2	Gln 6	Arg 2	A
C	Leu 2	Pro 1	Gln 3	Arg 1	G
A	Ile 0	Thr 5	Asn 2	Ser 1	U
A	Ile 3	Thr 2	Asn 0	Ser 0	C
A	Ile 0	Thr 1	Lys 2	Arg 3	A
A	Met 1	Thr 1	Lys 0	Arg 0	G
G	Val 0	Ala 0	Asp 1	Gly 0	U
G	Val 0	Ala 0	Asp 0	Gly 0	C
G	Val 1	Ala 1	Glu 2	Gly 0	A
G	Val 2	Ala 2	Glu 1	Gly 0	G

FIGURE 16.4 Codons found in the MS2 coat, A protein, replicase, and lysis genes.

of the pattern and chance in the use of the genetic code by the MS2 genome in parallel with the known data on *E. coli*, φX174, and rabbit globin, Berger (1978) noted a significant deficiency of purines in the third position of fourfold degenerate codons. According to his analysis, there was no consistent selection against uracil in pyrimidine-restricted codons. For many amino acids, the choice between code words appeared random, while for arginine, isoleucine, and probably glycine, the distinct biases existed, which could be explained in terms of the tRNA availability (Berger 1978). At the same time, the MS2 RNA was studied for the correlation between the stability of the codon-anticodon interaction and the choice of code words (Grosjean et al. 1978). The nonrandom distribution of the degenerate code words in the MS2 RNA was partially explained by considerations of the stability of the codon-anticodon complex in prokaryotic systems. The authors noted that wobble codons were positively selected in codons having G and/or C in the first two positions. In contrast, the wobble codons were statistically less likely in codons composed of A and U in the first two positions. The analyses of nucleotides adjacent to 5'- and 3'-ends of codons indicated a nonrandom distribution as well. It was thus likely that some elements of the RNA evolution were independent of the structural needs of the RNA itself and of the translated protein product.

Hasegawa et al. (1979) connected the bias in the code word usage with the MS2 secondary structure. Based on the MS2 secondary structural model, they showed that in the base-paired regions of the MS2 RNA there was a bias in the use of synonymous codons which favored C and/or G over U and/or A in the third codon positions, and that in the non-paired regions there was an opposite bias which favored U and/or A over C and/or G. This finding was interpreted as a result of selective constraint, which stabilized the MS2 secondary structure.

Remarkably, Walter Fiers and Henri Grosjean (1979) were the first to comment on the total nucleotide sequence of the *lac* repressor gene *I* of 361 codons (Farabaugh 1978), where the codon usage was quite different from that of the MS2 and φX174 genomes. It was emphasized in this comment that the MS2 genome had optimized translational efficiency, especially of the coat gene, which needed the proper choice of degenerate code words, in contrast to the *lacI* gene, where *lac* repressor protein was made at very low levels. Thus, the possible relation of selective pattern of the codon usage to a difference in translational efficiency was proposed. Continuing this idea, general principles of the preferential codon usage in highly-expressed prokaryotic genes were postulated (Grosjean and Fiers 1982). The authors hypothesized that both the optimization of codon-anticodon interaction energy and the adaptation of the population to codon frequency, or *vice versa*, in highly expressed genes were part of a strategy that optimized the efficiency of translation. Conversely, codon usage in weakly expressed genes, such as repressor genes, followed exactly the opposite rules. It was concluded therefore that, in addition to the need for coding an amino acid sequence, the energetic consideration for the codon-anticodon pairing, as well as the adaptation of codons to the tRNA population, might have been important evolutionary constraints on the selection of the optimal nucleotide sequence.

The codon context of the MS2 genes, together with known enterobacteria and plasmid genes, was used in a search for correlations between the presence of one base type at codon position III and the presence of another base type at some other position in adjacent codons, in comparison with eukaryotic sequences (Gouy 1987). In enterobacterial genes, the base usage at codon position III was correlated with the third position of the upstream adjacent codon and with all three positions of the downstream codon, while plasmid genes were free of context biases. Remarkably, the MS2 codons had no biased context, whereas the phage λ genes partly followed the trends of the host bacterium, and the phage T7 genes had biased codon contexts that differed from those of the host.

By computer simulation of the codon usage in the context of the MS2 secondary structure, two remarkable conclusions were made (Bulmer 1989). First, concerning the suggestion that the higher G+C content in paired regions of the MS2 genome than in unpaired ones reflected selection for high G+C content to encourage pairing, it was concluded that it was an automatic consequence in any RNA sequence of the way it folded up to minimize its free energy. Second, concerning the idea that the three registers in which pairing could occur in a coding region (Fitch 1974) were used differentially to optimize the use of the redundancy of the genetic code, the examination of the data showed only weak statistical support for this hypothesis.

Menninger (1983) attempted a computer simulation of ribosome editing. In this study, a stochastic model of protein synthesis was modified by including the process of dissociating peptidyl-tRNA from ribosomes. As a result, the model mimicked in detail the experiments by Goldman (1982) on the cell-free synthesis of the MS2 coat protein during tryptophan starvation.

The RNA phages contributed remarkably to the studies of the transport RNA diversity in the context of the organization and evolution of the genetic code. Thus, the R17 RNA-directed *in vitro* synthesis system that was dependent on added tRNAs contributed to the characterization of the redundant tRNAs from yeast (Bergquist et al. 1968). As a result, the four glycine tRNAs were isolated from brewer's yeast and shown to recognize the same code word, being structurally different. The same R17 RNA-dependent *in vitro* system was used to evaluate the effect of 5-fluorouracil incorporation into tRNAs from *E. coli* (Lowrie and Bergquist 1968).

The strongest contribution of the RNA phage messengers was achieved, however, to the two-out-of-three reading hypothesis (Lagerkvist 1978). According to this hypothesis, a codon might be read by relying mainly on the Watson-Crick base pairs formed with the first two codon positions, while the mispaired nucleotides in the third codon and anticodon wobble positions make a comparatively small contribution to the total stability of the reading interaction. Thus, the MS2 RNA-dependent *in vitro* protein synthesis was used by the study on the differential utilization of leucyl-tRNAs in *E. coli* (Holmes et al. 1977). In parallel, the same system was employed by the codon-anticodon recognition studies in the valine codon family (Mitra et al. 1977). It was established that the three anticodons each recognized all four valine codons and concluded that the genetic code, as far as the valine codons were concerned, was operationally a two-letter code, i.e., the third codon nucleotide had no absolute discriminating function. With the MS2 RNA-programmed system, the relative efficiency of anticodons in reading the valine codons was investigated (Mitra et al. 1979). The MS2 RNA-programmed system was also employed to evaluate aberrations of the classical codon reading scheme during protein synthesis *in vitro* with alanine tRNAs (Samuelsson et al. 1980). Again, each of the anticodons was able to read all four alanine codons, but under conditions of no competition.

Mitra (1978) used the MS2 RNA-programmed system to investigate the ability of $tRNA_1^{Lys}$ (CUU) to read the codon AAA, the predominant lysine codon in the MS2 coat protein gene, which contained five AAA codons and only one AAG. The conclusion that the $tRNA_1^{Lys}$ recognized the codon AAA was challenged by Elias et al. (1979), who found no detectable formation of coat protein in the system and concluded that the $tRNA_1^{Lys}$ cannot read the codon AAA. Furthermore, the MS2 RNA-programed *in vitro* system was used to evaluate the translational errors that occurred by the reading of lysine and glutamine codons in conditions when one of the isoacceptor tRNAs was the only source of the appropriate amino acid for protein synthesis (Lustig et al. 1981).

The MS2 RNA-programmed *in vitro* synthesis system was employed by the investigation of the ability of glycyl-tRNAs with different anticodons to read the glycine codons (Samuelsson et al. 1983). Again, each of the isoacceptors tested could read all of the four glycine codons in the MS2 coat protein gene, and the two-out-of-three hypothesis was confirmed. Concerning the codon discrimination and anticodon structural context, Lustig et al. (1989) used the appropriately modified MS2 RNA-directed system for the evaluation of the consequences when a glycyl-tRNA was changed by site-directed mutagenesis in the wobble position.

The phage MS2 presented another unique opportunity for the codon usage by mistranslation studies. It was observed that the MS2 coat protein showed some charge heterogeneity *in vivo*, since in most strains there was a basic satellite of the native protein (Parker et al. 1983). This basic satellite was greatly diminished or absent in strains with the streptomycin-resistant allele *rpsL*, a mutation which led to increased translational accuracy. Sequencing of the N-terminal 19 amino acids of the satellite protein showed that the asparagine codon AAU at amino acid 12 was misread approximately eight times more frequently than the AAC at amino acid 3. It was concluded that the satellite species was the result of basal level lysine for asparagine substitution. These substitutions were caused by preferential misreading of AAU codons at a frequency of approximately 5×10^{-3}, tenfold higher than the average error frequency (Parker et al. 1983). The asparagine starvation increased the error frequency in the MS2 coat protein to over 0.3 mistake per asparagine codon, while no such increase was found when the host was starved for arginine, histidine, isoleucine, or proline (Parker et al. 1980). The peptide analysis and amino acid sequencing showed that there was a sixfold greater frequency of errors at AAU codons than at AAC codons when the MS2-infected cells were starved for asparagine (Johnston et al. 1984). Remarkably, this ratio was the same as that found in unstarved cells where the overall error frequency was 100-fold less.

The basal-level codon misreading of asparagine codons was examined further in a number of *E. coli* strains (Parker and Holtz 1984). In most strains, the heterogeneity was consistent with misreading of AAU codons at a frequency of $3–6 \times 10^{-3}$, while strains with streptomycin resistance mutations in the *rpsL* gene had reduced levels of misreading.

The synthesis of the full-sized MS2 coat protein, encoded by the phage RNA or by its DNA copy, was greatly reduced by starvation for phenylalanine, which led to leucine misincorporation frequencies of 0.1 and 0.6 at UUC codons in the *argI* transcript of *E. coli*, but no detectable misincorporation at an UUU codon (Parker and Precup 1986). This reduction in the MS2 coat amount was unaffected by the *rpsL* mutation.

To identify the missense misreading of asparagine codons as a function of codon identity and context, Precup and Parker (1987) constructed a series of the MS2 coat gene derivatives by site-directed mutagenesis. The mutant set constructed had either AAU or AAC as codon number three in the gene with each possible adjoining 3′ base. The lysine incorporation in the coat protein encoded by these genes showed that AAU was misread from four- to ninefold more frequently than AAC with any 3′ context. Therefore, the experiment did not support the hypothesis of the context effect on the misreading phenomenon.

This ingenious system (Parker and Precup 1986; Precup and Parker 1987), where the MS2 coat synthesis was induced from a cDNA copy of the MS2 coat gene carried on a multicopy plasmid, was used further successfully in a study on the reduced misreading of asparagine codons (Hagervall et al. 1998).

An interesting study was undertaken in order to compare the accuracy of the Qβ coat protein translation in whole cells and by extracts (Khazaie et al. 1984a,b). The most important conclusion was that even the best systems *in vitro* available at that time did not match accuracy *in vivo*. However, the differences varied markedly with the nature of the error being measured. Thus, after infection, the host protein synthesis was eliminated by adding rifamycin (as described in Chapter 4), and the radioactive phage-specified proteins were separated by one- or two-dimensional gel electrophoresis (Khazaie et al. 1984a). The errors leading to a change in the pI of the coat protein occurred at a rate of 0.05 per molecule, while the coat protein UGA stop codon was misread 6.5% of the time. In fact, these error rates were similar to data in some recent publications, but much higher than the canonical $3–4 \times 10^{-4}$. These errors further provided a reference point *in vivo* to which the translation of the same message by *E. coli* extracts could be compared (Khazaie et al. 1984b). Even the lowest Qβ error rates *in vitro* were still an order of magnitude greater than those for polyU or poly(U-G) translation. Comparing the Qβ translational errors made *in vitro* to those found in the whole cells, the histidine misinsertions were almost twice as frequent, the errors leading to the coat protein charge change were six times more frequent, and the tryptophan misinsertions were at least 15 times more frequent *in vitro*.

It remained intriguing that the RNA extracted from the MS2 phage particles was able to accept radioactive leucine and serine in the presence of tRNA-activating enzymes, since leucine tRNA was bound very tightly to the phage particles (Di Natale and Eilat 1976). Moreover, the pattern of the phage-associated leucyl-tRNA isoacceptors was different from that of normal *E. coli* leucyl-tRNA. It was also different from the pattern of host leucyl-tRNA isoacceptors found in *E. coli* lysate following MS2 phage infection. The authors suggested that these tRNAs were some modified forms of the normal leucine tRNA.

The MS2 RNA-directed *in vitro* synthesis was employed in the study that acknowledged that the conserved 5S rRNA complement to tRNA was not required for translation of natural mRNA (Zagorska et al. 1984). The same system also contributed to the study of the modification-dependent structural alterations in the anticodon loop of *E. coli* tRNA^{Arg} (Baumann et al. 1985) and to the evaluation of the generated mutants of the *E. coli* fMet-tRNA (Seong and RajBhandary 1987a,b).

As stated in the Conditionally lethal mutations section of Chapter 8, the mechanism of suppression by specific tRNAs was discovered. In addition to the general data presented there, the yeast super-suppressors could be mentioned here (Capecchi et al. 1975). Such yeast super-suppressors were altered tRNAs capable of translating a nonsense codon *in vitro*. The suppression was assayed by translation of the Qβ coat *amber* mutant RNA in the cell-free protein-synthesizing system derived from mouse tissue culture cells, or L-cells.

Remarkably, the L-cell protein-synthesizing system also responded to *E. coli* suppressor tRNA. This indicated that the biochemical mechanism for nonsense suppression was very similar in yeast and *E. coli*. Moreover, this finding provided additional evidence that the *amber* codon functions as one of the mammalian chain-terminating codons.

The properties of the Qβ readthrough protein A1 led to the conclusion that normal tRNA^Trp in *E. coli* was able to suppress UGA codon in the case when adenine residue was adjacent to the 3′ side of UGA in mRNAs which coded for the readthrough proteins (Engelberg-Kulka 1981). It was suggested that the nature of the nucleotide following a UGA codon determined whether the UGA signaled efficiently or inefficiently the termination of the polypeptide chain synthesis.

The Qβ RNA-programmed *in vitro* system was used in a large investigation of how point mutations affected the *E. coli* tryptophan tRNA by the codon-anticodon interactions and UGA suppression (Vacher et al. 1984). Among other things, this study explained why the hypermodification was necessary for the efficient suppression of the UGA terminator of the Qβ coat *in vitro*.

It is noteworthy that the fidelity of the MS2 RNA translation *in vitro*, namely, the specificity of codon-anticodon interactions, was stimulated by polyamines (Abraham 1981). Furthermore, the effect of polyamines on the translation fidelity was studied *in vivo* in the MS2-infected *E. coli* strain auxotrophic for polyamine synthesis (McMurry and Algranati 1986; Nastri et al. 1993). The error was measured by the incorporation of ^3H-histidine into MS2 coat, the gene of which did not code for histidine, and the synthesis of a basic variant of MS2 coat protein, in which a lysine erroneously replaced an asparagine, causing a change in isoelectric point. It was found that when cell cultures were supplemented with polyamines there was no effect on the first type of the error, and the second type decreased twofold. The average erroneous incorporation per mol coat protein in the presence of polyamines was 1.43 ± 0.59 mmol histidine and 25–34 mmol lysine/asparagine substitution. The reason for the different effects of polyamines on the two types of error remained unknown (McMurry and Algranati 1986).

Concerning the recent studies on the codon usage, it is noteworthy that the phage MS2 RNA was employed by testing of the special green fluorescent protein gene *lgfp* that was constructed with the lowest-usage codons to monopolize various minor tRNAs and therefore inhibit translation indirectly (Kobayashi 2015). The induction of such artificial gene led to the downregulation of the network genes, and the idea was proposed for the development of novel nonspecific virus defense systems in *E. coli* and human cells. Kobayashi (2015) assumed that the artificial gene networks could control gene expression and make organisms as controllable as robots.

Ma NJ and Isaacs (2016) used the phage MS2, together with the phages M13, P1, and λ, in the study where they assessed the ability of viruses as horizontally transferred genetic elements to propagate on the genomically recoded *E. coli* strain lacking UAG stop codons. The alternative genetic code conferred resistance to the phages and impaired conjugative plasmids F

and RK2. By recoding UAG codons to UAA in phages and plasmids, the viral infectivity and conjugative function was restored. Propagating viruses on a mixed community of cells with standard and alternative genetic codes reduced viral titer, and over time viruses adapted to the alternative genetic code. Remarkably, the authors suggested the genomic recoding as a strategy to stabilize engineered biological systems, for example, by generation of production and deployment of safe GMOs into open systems in clinical medicine or environmental bioremediation.

NON-COLIPHAGES

It should be clear from the previous sections that the lion's share of the translation data was obtained with the RNA coliphages. Concerning the non-coliphages, only few investigations touched on them in the translation context. Thus, the counterpart of the *E. coli* initiation factor IF-3 was isolated from *C. crescentus*, purified to homogeneity, and used in comparative studies on *in vitro* translation of the caulophage φCB5 RNA, in parallel with the MS2 RNA (Leffler and Szer 1973a,b). Such C-IF-3 factor substituted for the *E. coli* IF-3 and promoted correct translation of the MS2 RNA by *E. coli* ribosomes. Conversely, the *E. coli* IF-3 substituted for the C-IF-3 in the translation of the φCB5 RNA by the *C. crescentus* ribosomes. However, each phage RNA could be translated only by host ribosomes, or by mixed ribosomes containing the host 30S subunit.

The pseudomonaphage PP7 RNA was translated by the *E. coli* cell-free amino acid incorporating system (Bassel et al. 1974). In this study, the PP7 messenger activity was compared with that of the Qβ and f2 RNAs. The rate of the ^14C-leucine incorporation directed by the PP7 RNA was greater than by either Qβ or f2 RNA. The response to changes in the phage RNA concentrations was similar in all the systems, reaching a saturation level at 0.75–1.0 mg of RNA per mL of reaction mixture. The analysis of complete reaction mixtures of the PP7 RNA and of the Qβ RNA systems by sucrose gradient centrifugation showed generally similar patterns for both RNAs. The principal differences were that in the PP7 system a slightly higher percentage of RNA formed ribosome complexes and that the polysomes were somewhat smaller. The PP7 RNA was also degraded more extensively during the reaction than was the Qβ RNA. The analysis of the products of the reactions by acrylamide gel electrophoresis showed that the PP7 coat protein was the only identifiable product of the PP7 RNA-directed system, suggesting that only the coat protein gene was translated by *E. coli* ribosomes. Furthermore, the PP7 RNA fragments protected by both *P. aeruginosa* and *E. coli* ribosomes were sequenced (Bassel and Mills 1979). Only one binding site appeared available on the PP7 RNA, and this site was recognized by ribosomes of both species. The PP7 RNA binding site was approximately 38 nucleotides long and contained two AUG sequences and a purine-rich segment near the 5′-end that was complementary to segments near the 3′-ends of the 16S ribosomal RNAs of both *P. aeruginosa* and *E. coli*. In order to establish which of the AUG codons acted as the initiator, the N-terminal amino acid sequence of the PP7 coat protein was determined. This sequence was compatible with the codon sequence following the second AUG codon. Remarkably, the extent of the reaction of PP7 RNA with *E. coli* ribosomes was greater than with *P. aeruginosa* ribosomes, but no qualitative differences in the initial interaction between intact PP7 RNA and the ribosomes of either species were indicated (Bassel and Mills 1979).

In parallel, Davies and Benike (1974) translated the PP7 RNA in the *P. aeruginosa* cell-free extracts. The S30 extracts translated the PP7 RNA quite efficiently, but the isolated ribosomes incubated with an *E. coli* S100 supernatant preparation were even more active. The three distinct products were made which corresponded to the *in vivo* labeled PP7 phage polypeptides judged by SDS-polyacrylamide gels. This study made it possible to conclude clearly that the PP7 genome had at least three genes. By analogy to coliphages, the major product of the *in vitro* synthesis, synthesized in great excess of the other products, coelectrophoresed with authentic coat protein prepared from labeled phage. The tryptic peptide analysis of this product was identical to that of the PP7 coat protein. It was concluded that this *in vitro* polypeptide was the product of the coat protein gene. The PP7 coat was also efficiently synthesized when the PP7 RNA was translated by *E. coli* ribosomes, but the other PP7 genome products were not made, except occasionally in very small amounts. Moreover, the *P. aeruginosa* ribosomes translated the PP7 RNA more efficiently than the R17 and Qβ RNAs. The authors emphasized these facts as an indication of some differences by the gene recognition between *P. aeruginosa* and *E. coli* ribosomes or in their initiation factors.

The functional PP7 homology to the coliphages was further confirmed by the structural alignments (Olsthoorn et al. 1995a). The regulatory RNA secondary structure features that were present in the coliphages were also identified in the PP7 RNA, although the sequences were hard to align. Among such intrinsic elements were the coat protein binding helix at the start of the replicase gene, the replicase binding site, and the structure of the coat protein gene start. Some of these features resembled the MS2-type coliphages but others the Qβ type. These structures will be analyzed in more detail below in the sections devoted to the repressor complexes I and II. The similar functional homology to coliphages was also revealed when the full-length genome of the caulophage φCB5 was sequenced (Kazaks et al. 2011). By analogy to the coliphages, the phage PRR1 coat recognized its RNA and formed the structure similar to the complex I (Persson et al. 2013).

REPRESSOR COMPLEX I

The early observations with the *in vitro* MS2 RNA translation in the *E. coli* cell-free extract denoted a possible regulation of the order and efficiency of the protein synthesis, and the idea was born that a definite translational control mechanism must exist to ensure preferential expression of the coat gene (Ohtaka and Spiegelman 1963). The first evidence suggesting that the phage coat proteins were involved in regulating

the translation of the replicase gene followed from the properties of polar and nonpolar *amber* mutants of the coat gene, as described in detail in Chapter 8, and the role of a translational repressor was proposed first for the coat proteins of the phages f2 (Lodish et al. 1964; Lodish and Zinder 1965, 1966a) and R17 (Gussin 1966). At the same time, the capability of a few coat molecules to cosediment with R17 RNA was first observed by sucrose gradient sedimentation of the *in vitro* protein synthesis products (Capecchi and Gussin 1965). The idea was developed further by the exhaustive studies on the f2 replicative intermediates (Robertson et al. 1968b) and by the localization of the R17 replicative intermediate in the polysome fraction of the infected bacteria (Hotham-Iglewski et al. 1968).

Tsutomu Sugiyama et al. (1967) were the first to identify the existence of the specific coat protein complexes I and II with the MS2 RNA. Thus, the MS2 RNA and coat interacted *in vitro* and formed the complex I at low ratios of coat to RNA. The complex I was composed of one molecule of RNA and a small number, probably not more than six, of the coat protein molecules. It had a sedimentation coefficient very close to that of free RNA. The formation of the complex I was highly specific, and the MS2 RNA could not be substituted by other RNAs such as *E. coli* rRNA and tRNA, as well oligonucleotide homopolymers (Sugiyama et al. 1967). The complex II was identified as noninfectious phage-like particles, as described in Chapter 9. Moreover, Tsutomu Sugiyama and Daisuke Nakada (1967a,b, 1968, 1969, 1970) demonstrated that the complex I represented the regulatory mechanism that ensured the preferential synthesis of the coat in the MS2 RNA-directed *in vitro* protein synthesis system. When the MS2 RNA was incubated with a small amount of the MS2 coat protein before being added to the *in vitro* protein-synthesizing system, the synthesis of histidine-containing proteins was greatly reduced, while the synthesis of the coat protein was little affected. These findings were presented later in more detail (Sugiyama 1969).

At the same time, Richard Ward et al. (1967, 1968) and Kathleen Eggen and Daniel Nathans (1967, 1968, 1969) showed the same translational repression of the replicase synthesis by the coat protein in the *in vitro* protein synthesis programmed by the f2 or MS2 RNAs, respectively. The *in vivo* properties of the appropriate MS2 mutants led to the conclusion that the MS2 coat served as a repressor of the synthesis of replicase and maturation protein (Nathans et al. 1969). In fact, the coat did not suppress the synthesis of the maturation protein. The repressor role of the coat protein was confirmed by the *in vivo* and *in vitro* studies of the MS2 nonpolar coat *amber* mutants (Remaut et al. 1974).

The specificity of the complex I was shown clearly by Hugh Robertson et al. (1968b). Thus, the interaction of the coat protein with the RNA was quite specific, since the f2 coat protein did not bind to the Qβ RNA or repress the synthesis of the Qβ-specific replicase, and the Oβ coat protein had no effect on the f2 RNA.

Ward et al. (1970) demonstrated in a detailed study the binding and repression of the native M12 RNA by several molecules of its homologous coat protein that caused a shift in the size of polysomes formed under the direction of this RNA species during cell-free protein synthesis. This shift in the distribution of phage polysomes closely resembled that which occurred in the cell between early and late times after phage infection. This finding indicated clearly that translational control, as it has been described from experiments *in vitro*, could be also applied to reactions taking place in the M12-infected cell *in vivo*.

Thomas Hohn (1969a,c) and H.C. Kaerner (1970) regarded the complex I of the phage fr, which consisted of the fr RNA and six coat subunits, as a precursor of the phage self-assembly. The specific repressor function of the Qβ coat was also revealed (Robertson et al. 1968) and used to predict the first three amino acids of the Qβ replicase (Skogerson et al. 1971).

The full account of the complex I-performed translational control of the RNA phage replication was given at that time by the example of the phages f2 (Lodish and Robertson 1969b) and MS2 (Sugiyama 1969).

The coat protein within the complex I, or repressor complex I, blocked the initiation step in the translation of replicase, as shown directly in the case of the phages R17 (Spahr et al. 1969), f2 (Lodish 1969a; Roufa and Leder 1971), and Qβ (Skogerson et al. 1971). First, the R17 coat binding site was mapped to roughly a 60% fragment of R17 RNA from the 3′-terminus, and binding of one molecule of the coat per RNA molecule was postulated (Spahr et al. 1969). Next, Alberto Bernardi and Pierre-François Spahr (1972a,b) mapped precisely the binding site and showed that the R17 coat was bound directly to the initiation site of the R17 replicase and that ribosomes were unable to initiate the translation of the replicase gene. The R17 repressor complex I was stable enough to protect a portion of the RNA from nuclease digestion. Thus, Bernardi and Spahr (1972a,b) were able to isolate a nucleotide fragment of 59 bases that included six last codons of the coat protein, the intergenic region, and the first two codons of the replicase. The precise genomic position of the isolated fragment was established due to alignments with the nucleotide sequence of a fragment of the R17 RNA that corresponded to the last six amino acids of the coat and extended on the 3′-side of the coat gene for 26 nucleotides (Nichols 1970), and with the sequence of the replicase initiation site (Steitz 1969b). Thus, the coat binding site included the ribosome binding site for the replicase gene (Steitz 1969b), giving a molecular basis for the repression observed. Figure 16.5 presents the region of the MS2 genome containing regions protected by coat protein or ribosome from ribonuclease digestion, with the famous hairpin loop originally proposed by Bernardi and Spahr (1972a).

In fact, the size of the isolated RNA fragment was larger than would be expected on the basis of the relative dimensions of the coat protein and the RNA fragment. Bernardi and Spahr (1972a) proposed that part of the isolated RNA fragment extending from its 5′-terminus into the first part of the intergenic region was not protected directly by the coat protein but could form a hydrogen-bonded loop that was protected from the T1 ribonuclease cleavage. It was supported by the earlier finding that intact coat protein was synthesized at the same rate *in vitro* when either free R17 phage RNA or the repressor complex I served as the messenger. This complex was formed with RNA from all the phages in the group I and any of the proteins in the group I, but neither the complex I formation nor repression was found in the heterologous binding attempts with the Qβ coat protein or Qβ RNA. Joan Argetsinger Steitz (1974) showed later that the R17 coat protein was able to recognize the isolated R17 replicase initiation site of 29 nucleotides.

The binding site for the Qβ coat protein on the Qβ RNA was mapped by Hans Weber (1976). Again, the complexes of the coat protein and ^{32}P-RNA were subjected to the ribonuclease T1 degradation. The three main fragments obtained, 88, 71, and 27 nucleotides in length, all consisted of sequences extending from the intergenic region to the beginning of the replicase gene. This finding confirmed the universal character of the translational regulation in the RNA phages: in the Qβ

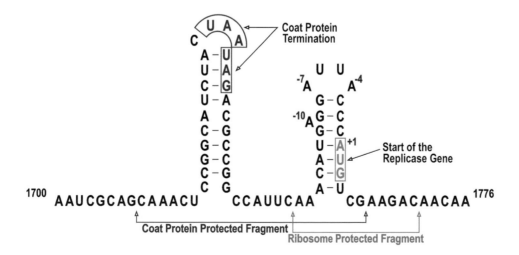

FIGURE 16.5 Sequence of a section of MS2 genome containing the translational repression site. R17 RNA has an identical sequence except that residue 1732 is a U. (Redrawn with permission of Taylor & Francis Group from Uhlenbeck OC et al. *J Biomol Struct Dyn.* 1983;1:539–552.)

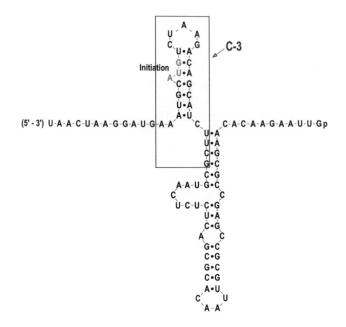

FIGURE 16.6 Possible secondary structure of Qβ RNA near the coat protein binding site. On the basis of the ΔG values given by Tinoco et al. (1973), the free energies of helix formation for the small and the large loop are about −6.6 and −23.6 kcal (−27.6 and −98.7 J), respectively. (Reprinted from *Biochim Biophys Acta*, 418, Weber H, The binding site for coat protein on bacteriophage Qβ RNA, 175–183, Copyright 1976, with permission from Elsevier.)

replication, as in the case of the R17 and other group I phages, the coat protein acted as a translational repressor by binding to the ribosomal initiation site of the replicase gene. Figure 16.6 presents the original picture from the paper (Weber 1976). By analogy with the R17 coat, the Qβ coat protein was bound to the replicase initiation site of the Qβ RNA (Weber 1975).

After this general introduction, the data on the repressor complexes I will be divided in accordance with the specific phages used by the authors. However, it is important to remember that the nucleotide sequences recognized by the coat proteins were nearly identical but differed markedly from the corresponding site in the Qβ RNA, as stated above. Thus, the 59-nucleotide operator was fully identical in the MS2 and ZR sequences and differed from them by one nucleotide exchange in R17, by two nucleotides in M12, by three nucleotides in JP501, and by 11 nucleotide exchanges in the phage fr. Accordingly, the group I phage coat proteins were very similar except for the fr coat protein, which differed in 17 positions from the MS2, R17, ZR, and f2 coat proteins, as demonstrated in Chapter 12. Remarkably, the regions of identity for all the coat proteins in the group I covered residues 20–54. Since all of the group I coat proteins prevented translation of the replicase genes in all the group I RNAs, Bernardi and Spahr (1972) suggested that the amino acids included into the 35-residue motif 20–54 were important in the protein-nucleic acid interactions that led to the formation of the complex I.

Besides the Elmārs Grēns (1974) and *RNA Phages* (1975) books, the translational control of the RNA phage protein

synthesis was reviewed excellently at that time by Harvey F. Lodish (1976).

REPRESSOR COMPLEX I: f2

The specificity and stoichiometry of the f2 coat-RNA complexes were studied in 1970s in Warsaw. First, up to 8 mol of coat protein per 1 mol of RNA were found within the complex I (Chroboczek et al. 1973). Its formation proceeded equally well at temperatures from 0°−45°C in medium with and without magnesium ions, as well as in the presence of 4 mM EDTA. The next step of the coat-RNA association, namely, the phage-like complex, contained up to 200 mol of coat protein per 1 mol of RNA, and its formation was inhibited by the presence of magnesium ions. Formaldehyde- or methoxyamine-treated f2 RNA, in which only exposed bases were modified, showed a normal pattern of the phage-like complex formation, whereas formation of the complex I was inhibited or abolished. Next, the molar ratio of the coat versus RNA within the complex I was reduced to six, as determined by sucrose gradient sedimentation (Shafranski et al. 1975). The ribonuclease digestion of the glutaraldehyde-fixed complex resulted in a mixture of products in which the hexamers of coat protein molecules were predominant. The authors emphasized for the first time that under conditions required for the phage protein synthesis, the coat protein occurred in solution as a dimer. Therefore, the translational repression of the RNA replicase gene was due to the cooperative attachment of three coat dimers to phage template, forming a hexameric cluster on the RNA strand. It was especially emphasized that the first portions of coat in the replication process were unable to affect the translation of the replicase gene because of the low coat concentration. The translational repressor complex I was formed later when coat reached sufficiently high concentration (Shafranski et al. 1975). When complexes formed at different coat-RNA input molar ratios were isolated and tested for template activity in the *in vitro* protein synthesis system, the strong inhibition of the replicase gene translation was observed when each of the RNA strands was associated with six coat molecules (Zagórska et al. 1975). The formation of the complex I was prevented by sulfhydryl group-protecting agents.

Generally, it was concluded that the translational repression of the replicase gene was due to the cooperative attachment of three coat dimers to the phage template and formation of the hexameric cluster on the RNA strand (Chroboczek and Zagórski 1975).

REPRESSOR COMPLEX I: R17

The R17 complex I was studied exhaustively by Olke C. Uhlenbeck's team in Urbana, Illinois, and further in Boulder, Colorado. They were the first to conduct the entire enzymatic synthesis of 21-nucleotide fragment corresponding to the positions from −17 to +4 of the replicase gene (Krug et al. 1982). By the synthesis, the T4 RNA ligase was used to join shorter oligomers. This sequence was identical with the R17 replicase initiator region and also encompassed the binding domain of the R17 coat protein. The resulting fragment had a secondary structure with the expected thermal stability, and demonstrated the same affinity by the coat binding as the 59-nucleotide fragment isolated from the R17 RNA and described above (Krug et al. 1982). The kinetic and equilibrium properties of the interaction between the 21-nucleotide RNA operator and the R17 coat were studied by the filter retention assay (Carey and Uhlenbeck 1983). The kinetics of the reaction were consistent with the equilibrium association constant and indicated a diffusion-controlled reaction. The temperature dependence of K_a gave $\Delta H = -19$ kcal/mol. This large favorable ΔH was partially offset by a $\Delta S = -30$ cal mol^{-1} deg^{-1} to give a $\Delta G = -11$ kcal/mol at 2°C in 0.19 M salt. The binding reaction had a pH optimum centered around pH 8.5, but pH had no effect on the ΔH. While the interaction was insensitive to the type of monovalent cation, the affinity decreased with the lyotropic series among monovalent anions. The ionic strength dependence of K_a revealed that ionic contacts contributed to the interaction. Most of the binding free energy, however, was regarded as a result of nonelectrostatic interactions (Carey and Uhlenbeck 1983).

Furthermore, the nuclease protection and selection experiments defined the binding site to 20 nucleotides which formed a typical hairpin (Carey et al. 1983a). The synthetic 21-nucleotide operator competed effectively for coat protein with the intact R17 RNA. The binding between the coat protein and the synthetic 21-nucleotide operator behaved as a simple bimolecular reaction. In contrast to the above-mentioned f2 data, the complex stoichiometry was defined as one coat monomer per RNA operator. It was especially emphasized that all the coat protein existed in a monomer form when protein concentration was below about 10 nM. The dissociation constant K_d of about 1 nM was obtained at 2°C in buffer containing 0.19 M salt. The interaction was highly sequence specific since a variety of RNAs failed to compete with the 21-nucleotide fragment for the coat protein binding (Carey et al. 1983a).

The 23-nt sequence variants of the R17 RNA operator were synthesized and their affinity to the R17 coat was measured by the nitrocellulose filter binding assay (Carey et al. 1983b). The experiments using oligomers truncated on the 3′- and 5′-termini allowed precise determination of the edges of the binding domain to the −17 and +1 of the replicase gene. The several oligomers, which disrupted one or more of the base pairs in the binding site, failed to bind the coat protein, establishing the importance of the RNA secondary structure for the interaction. The upper portion of the hairpin loop, encompassing residues −12 to +1, was essential for protein binding (Carey et al. 1983b).

The full story of the equimolar R17 operator and coat interactions was reviewed at that time by Uhlenbeck et al. (1983). This review mentioned more than 30 different synthesized variants of the R17 operator. Generally, out of the five single-stranded residues examined, four were found essential for protein binding, whereas the fifth could be replaced by any nucleotide. One variant was found to bind better than the wild-type sequence. The substitution of nucleotides that

disrupted the secondary structure of the binding fragment resulted in very poor binding to the protein.

Then, two of the seven single-stranded residues were found essential for the RNA-protein interaction after substitution by other nucleotides (Romaniuk et al. 1987). In contrast, three other single-stranded residues were allowed for the substitution without altering the binding. When several of the base-paired residues in the binding site were altered in such a way that pairing was maintained, little change in the affinity was observed. However, when the base pairs were disrupted, the coat protein did not bind. These data suggested that while the hairpin loop structure was essential for protein binding, the base-paired residues did not contact the protein directly. As a result, a model for the structural requirements of the R17 coat protein binding site was generated. The model was successfully tested by demonstrating that oligomers with sequences quite different from the replicase initiator were able to bind the coat protein (Romaniuk et al. 1987).

Remarkably, the introduction of a cytidine in place of one of the two single-stranded uridines in the R17 replicase translational operator resulted in a much tighter binding to R17 coat protein (Lowary and Uhlenbeck 1987). The complex containing the variant RNA had a binding constant about 50 times greater than the one with the wild-type RNA.

To examine the role of the bulged A residue (see Figure 16.5), 14 different nucleotides were introduced to its position within 3 different coat protein−binding fragments, and the bulged A residue was found essential for the coat binding (Wu and Uhlenbeck 1987).

The binding of the 21-nucleotide RNA operator with the R17 coat was inhibited by prior incubation of the latter with 5-bromouridine, but not with 5-bromodeoxyuridine (Romaniuk and Uhlenbeck 1985). The RNA binding activity of the bromouridine-inactivated protein was restored upon incubation with dithiothreitol. Surprisingly, unmodified cytidine nucleotides also inactivated coat protein, with a specificity similar to the modified analogs. The inactivation of the coat protein by N-ethylmaleimide or p-(chloromercuri)-benzenesulfonate indicated that a cysteine residue was located near the RNA binding site. As a result, a transient covalent interaction between an essential pyrimidine and a protein sulfhydryl group was proposed (Romaniuk and Uhlenbeck 1985).

Meanwhile, Carey et al. (1984) synthesized RNA fragments corresponding to the site of translational repression for both the wild-type f2 and the f2 op3 mutant, and showed that the affinity of the closely-related R17 coat for the mutant and wild-type RNA operators was the same. Moreover, the f2 op3 and R17 coats were bound to the wild-type R17 operator with essentially identical dissociation constants. Thus, it was proven clearly that the altered regulation of replicase protein synthesis in the f2 op3 mutant did not appear to be due simply to a reduced affinity of the translational repressor for its target site (for more detail, see Chapters 8 and 14).

Logically, the translation repression and coat self-assembly functions seemed to be related, and the translational RNA operator was tried as a trigger of the capsid assembly (Beckett and Uhlenbeck 1988). It actually appeared that the translational repression complex served as a nucleation seed for the subsequent assembly. Indeed, the binding of the translational operator to the coat dimer triggered polymerization of the protein into the T = 3 capsids, where about 90 RNA operator molecules were encapsulated. To understand the role of the operator and non-operator RNA sequences in the R17 capsid assembly *in vitro*, a variety of RNAs was tried for the capsid assembly (Beckett et al. 1988). For a series of RNA oligomers of the same chain length, the sequences that bound the coat dimer with a lower affinity required higher concentrations of RNA and protein for assembly. Among a series of nonspecific RNA molecules of differing lengths, lower protein and RNA concentrations were required for assembly of capsids containing longer RNAs. For RNA molecules of any length, the presence of a single high-affinity translational operator sequence lowered the concentration requirements for capsid assembly. However, the advantage for encapsidation provided by the operator sequence was small for large RNA molecules. The experiments indicated that in the overall assembly process the interaction of coat protein with nonspecific sequences was at least as important as its interaction with the specific translational operator sequence (Beckett et al. 1988). However, we need to remember at this point that the coat self-assembly was shown to proceed without any triggering RNAs, as presented in the Reconstitution section of Chapter 9.

Furthermore, the thiophosphate substitutions within the operator were employed to determinate the RNA-protein contacts (Milligan and Uhlenbeck 1989). The 17 sequence variants of the R17 operator were synthesized with different NTP(αS)s and tested for the coat binding and the thiophosphate positions were deduced, which altered the binding affinity. Of the 21 phosphate positions in the molecule, two were found to decrease the K_a threefold when substituted with a thiophosphate, one position decreased the K_a 10-fold, and one position increased the K_a 10-fold, while substitution of any of the other 17 positions with thiophosphates did not alter the K_a. The four positions that altered the K_a were located in a uniquely structured region of the RNA, and it was postulated that these thiophosphates affect binding because they contact coat protein directly (Milligan and Uhlenbeck 1989).

When synthetic RNAs contained two coat binding sites, they bound coat protein in a cooperative manner, resulting in a higher affinity and reduced sensitivity to pH, ionic strength, and temperature when compared with RNAs containing only a single site (Witherell et al. 1990). The cooperativity contributed up to −5 kcal/mol to the overall binding affinity, with the greatest cooperativity found at low pH, high ionic strength, and high temperatures. Similar solution properties for the encapsidation of the related phages fr and f2 suggested that the cooperativity was due to favorable interactions between the two coat proteins bound to the RNA. This system therefore resembled an intermediate state of phage assembly. No cooperative binding was observed for RNAs containing a single site and a 5′ or 3′ extension of nonspecific sequence, indicating that the R17 coat had a very low nonspecific binding affinity (Witherell et al. 1990).

The well-characterized RNA operator was used to investigate the crosslinking of the coat protein to 5-bromouridine-substituted RNA using medium-wavelength UV light (Gott et al. 1991b). The formation of the covalent complex was dependent upon the presence of 5-bromouridine at position −5 of the RNA and specific binding of the RNA by the coat dimer. The amount of crosslinking increased with time and depended on the light source and conditions used. The 5-bromouridine-substituted RNA at position −5 complexed with a dimer of the coat protein and photocrosslinked to the coat protein in high yield upon excitation at 308 nm with a xenon chloride excimer laser (Willis et al. 1994). The tryptic digestion of the crosslinked nucleoprotein complex followed by Edman degradation of the tryptic fragment bearing the RNA indicated crosslinking to tyrosine 85 of the coat protein. A control experiment with a Tyr85 to Ser85 variant coat protein showed binding, but no photocrosslinking at saturating protein concentration.

The exhaustive review on the performed complex I studies was published at that time by Uhlenbeck's team (Witherell et al. 1991).

The circular permutation analysis was employed further to examine binding effects caused by a single break in the phosphodiester backbone (Gott et al. 1993). This method revealed that breakage of all but one phosphodiester bond within a well-defined binding site substantially reduced the binding affinity. This was probably due to the destabilization of the hairpin structure upon breaking the ribose phosphates at these positions. One circularly permuted isomer with the 5′ and 3′ ends at the bulged nucleotide bound with wild-type affinity. However, extending the 5′ end of this coat isomer greatly reduced binding, making it unlikely that this circularly permuted binding site would be active when embedded in a larger RNA. The circular permutation analysis located the 5′ and 3′ boundaries of protein binding sites on the RNA: the 5′ boundary of the R17 coat protein site was defined two nucleotides shorter (nucleotides −15 to +2) than the previously determined site (−17 to +2). It is necessary to note here that Carey et al. (1983b) mapped this region by −17 to +1, as stated above. The smaller binding site was verified by terminal truncation experiments. Then, a minimal binding fragment (−14 to +2) was synthesized and was found to bind tightly to the coat protein. The site size determined by the 3-ethyl-1-nitrosourea-modification interference was larger at the 5′ end (−16 to +1), probably due, however, to steric effects of ethylation of phosphate oxygens. These impressive permutation data were reviewed by Pan and Uhlenbeck (1993).

Finally, the three-dimensional conformation of a 24-nucleotide variant of the R17 RNA operator was resolved by proton NMR, molecular dynamics, and energy minimization (Borer et al. 1995). Generally, the imino proton spectrum was consistent with the base-pairing requirements for the coat protein binding known from biochemical studies. The SD ribosome binding site was exposed, and only two apparently weak base pairs would have to break for the 16S rRNA to bind and the ribosome to initiate translation of the replicase gene. Although the loop form must be regarded as tentative, the known interaction sites with the coat protein were easily accessible from the major groove side of the loop (Borer et al. 1995). The hairpin structure was refined further by Kerwood and Borer (1996) and Nooren et al. (1998). The results of these pioneering investigations showed that the NMR structure determinations of 20–30-mer RNAs were practical without isotope labeling.

Finally, Bardwell and Wickens (1990) generated the highly prospective affinity chromatography approach for the isolation of the R17 RNA-coat complexes formed *in vivo* or *in vitro*. The method exploited the highly selective binding of the R17 coat to the short hairpin. The RNA containing that hairpin was bound to the coat protein that has been covalently bound to a solid support, and the bound RNA-protein complexes could be eluted with excess R17 operators. The binding to immobilized coat protein was highly specific and enabled one to separate an RNA of interest from a large excess of other RNAs in a single step. Surprisingly, the binding of an RNA containing non-R17 sequences to the support required two recognition sites in tandem; a single site was insufficient. The optimal conditions were identified for purification of the specific RNAs by comparing specific binding, i.e., retention of RNAs with recognition sites, to nonspecific binding, i.e., retention of RNAs without recognition sites, over a range of experimental conditions. These results suggest that binding of immobilized coat protein to RNAs containing two sites was cooperative. This study led to the real revolution in the gene tethering and imaging methodologies, as will be described in Chapter 24.

It is noteworthy that just the R17 coat dimers were chosen as a prototype by the study on the formation of the dimers from the pool of the hepatitis B virus core antigen (HBc) molecules as precursors of the virus core particles and on the concentration requirements to their further cooperative self-assembly into the HBc capsids (Seifer et al. 1993; Seifer and Standring 1995).

REPRESSOR COMPLEX I: MS2

The direct structural studies of the MS2 repressor complex I were performed by Grēns' team in Riga. Thus, Berzin et al. (1976) found that the MS2 coat protected two major fragments of 59 and 103 nucleotides in length from the ribonuclease T1 digestion, where the 59-nucleotide fragment was an analog of the R17 fragment isolated earlier by Bernardi and Spahr (1972a,b) and differed from the latter by two (one, according to the present GenBank data) nucleotide exchange (Berzin 1978a). The structure is presented in Figure 16.5. Both MS2 RNA fragments contained all the necessary information for the specific interaction with the MS2 coat protein and provided the normal polypeptide chain initiation at the replicase gene. The true initiator codon only was recognized by a ribosome despite the presence of a few additional AUG triplets within the polynucleotides. The removal of the beginning of the replicase gene in the 59-nucleotide fragment covering the −53 to +6 region by limited T1 hydrolysis led to a complete

loss of its ability to bind both the coat and ribosome. After partial ribonuclease T1 digestion and chemical modification of the 59-nucleotide fragment by kethoxal, it was found that the hairpin (b), which was more labile and existed in equilibrium with its open form, played the essential role in the complex I formation (Berzin et al. 1978b). Further mapping by nuclease S1 reduced the coat binding site to −35/33 to +6 and finally to −17 to +6 regions (Jansone et al. 1979).

Starting from the 59-nucleotide fragment that retained full activity in the ribosome binding, the S1 nuclease shortening from the 5′-end resulted in the finding that the ribosome binding signal comprised nucleotides between positions 21 and 33 before the initiation codon (Borisova et al. 1979). The fragments protected by ribosomes but starting 17 or 21 bases before the AUG were unable to rebind the ribosomes.

In parallel with the MS2 operator, the corresponding phage fr operator was isolated (Tsielens et al. 1982) and checked for the initiation activity (Berzin et al. 1982). The copies of both MS2 and fr RNA operators of 20 nucleotides were synthesized by Regīna Renhofa's group, who used RNA ligase and reached record-size level of the enzymatic RNA synthesis at that time (Renhof et al. 1985; Nikitina et al. 1986). The synthetic fragments were studied in parallel for the coat binding and the initiation of translation. Remarkably, an operator with the destroyed SD sequence did not manifest any functional activity either as a template or in binding to the coat protein (Renhof et al. 1985).

The weighty contribution to the mechanism of translational repression was made by David S. Peabody in Albuquerque. First, he verified the idea of Romaniuk and Uhlenbeck (1985), who proposed a transient covalent interaction between the essential pyrimidine and protein sulfhydryl group within the R17 complex I. Peabody (1989) performed the codon-directed mutagenesis to determine the importance of each of the two coat protein cysteines for the repressor function *in vivo*. The results indicated that Cys46 could be replaced by a variety of amino acids without loss of repressor function. Cys101, on the other hand, was more sensitive to substitution. Most position 101 substitutions inactivated the repressor, but one, namely, arginine, resulted in normal repressor activity. Although the possibility of a transient covalent contact between Cys101 and RNA was not categorically ruled out, the construction of double mutants demonstrated that cysteines were not absolutely required for translational repression by the coat protein (Peabody 1989).

Next, Peabody (1990) constructed a genetic two-plasmid system that placed the synthesis of a hybrid replicase-β-galactosidase enzyme under translational control of the coat protein. This pioneering system permitted the straightforward isolation of mutations that affected the repressor function of the coat protein. Moreover, this approach allowed genetic dissection of the translational repression from the ability of the coat protein to self-assemble. The two-plasmid system expressed the MS2 coat protein and the replicase-β-galactosidase fusion protein from different, compatible plasmids containing different antibiotic-resistant determinants, where the coat protein expressed from the first plasmid repressed synthesis of the replicase-β-galactosidase fusion protein encoded by the other plasmid. The mutations in the translational operator or in the coat protein resulted in constitutive synthesis of the enzyme. First, mutations were constructed that introduced serines for the cysteine 46 and 101 residues. Both of these mutations resulted in defects not only in the translational repression, but also in the capsid-forming ability (Peabody 1990). The loss of translational repression, as well as the ability to form capsids, was shown with the two-plasmid system by the deletion of nucleotides encoding the C-terminal two amino acids of the MS2 coat (Peabody and Ely 1992). The corresponding mutant was completely defective for repressor activity, probably because of an inability to form dimers. In fact, this finding pointed out the interdependence of the structural and regulatory functions of coat protein. However, some translational-repression mutants displayed a super-repressor phenotype, repressing translation from the wild-type and a variety of mutant operators better than did the wild-type coat protein. At least one mutant probably bound RNA more tightly than wild-type. The other mutants, however, were defective for assembly of virus-like particles, and self-associated predominantly as dimers. Peabody and Ely (1992) proposed that this assembly defect accounted for their super-repressor characteristics, since failure to assemble into virus-like particles elevated the effective concentration of repressor dimers. This hypothesis was supported by the observation that deletion of 13 amino acids from the FG loop (positions 68–80) known to be important for assembly of dimers into capsids also resulted in the same assembly defect and in the super-repressor activity. Then, Peabody (1993) isolated a number of repressor-defective coat mutants and constructed the first spatial picture of the RNA binding site by relating the specific amino acid substitutions to the timely emerged three-dimensional structure of the MS2 coat (Liljas and Valegård 1990; Valegård et al. 1990). This picture will be discussed further in Chapter 21. To ensure that the repressor defects were due to substitution of binding site residues, Peabody (1993) screened the mutants for retention of the ability to form virus-like particles, eliminating mutants whose repressor defects were secondary consequences of protein folding or stability defects. Therefore, each of the variant coat proteins was purified and its ability to bind operator RNA *in vitro* was measured. Finally, the substituted sites were localized on the three adjacent strands of the coat protein β-sheet, and the sidechains of the affected residues formed a contiguous patch on the interior surface of the MS2 coat.

Wittily, Lim et al. (1994) attempted to convert the MS2 coat to the RNA-binding specificity of GA through the introduction of GA-like amino acid substitutions into the MS2 coat protein RNA-binding site. The effects of the mutations were determined by measuring the affinity of the coat protein variants for RNA *in vitro* and by measuring translational repression *in vivo*. The five substitutions were found that affected RNA binding, where one dramatically reduced binding of MS2 coat protein to both operators but three others compensated for this defect by nonspecifically strengthening the interaction. Another substitution accounted for the ability

to recognize the differences in the RNA loop sequence (Lim et al. 1994). Later, Spingola and Peabody (1997) sought to confer on MS2 coat protein the RNA-binding specificity of a more distant relative, namely, the Qβ coat, which showed only ~23% amino acid sequence identity to the MS2 coat. Nevertheless, the authors showed that as few as one or two amino acid substitutions were sufficient to do so. The genetic selection for the ability to repress translation from the Qβ translational operator led to the isolation of several MS2 mutants that acquired binding activity for the Qβ RNA. Some of these mutants also had reduced abilities to repress translation from the MS2 translational operator. These changes in RNA-binding specificity were the results of substitutions of amino acid residues 87 and 89. The additional codon-directed mutagenesis experiments confirmed earlier results, showing that the identity of Asn87 was important for the specific binding of the MS2 RNA. The Glu89 residue, on the other hand, was not required for recognition of the MS2 RNA, but prevented binding of the Qβ RNA. In the close connection with the "recognition mutants," Hirao et al. (1998–1999) selected RNA molecules that bound to the substitution variants of the MS2 coat and assayed the aptamers for their ability to discriminate between closely related protein targets. Despite the fact that the optimal binding species were identified for each target, many of the aptamers could readily cross-recognize related, non-cognate targets. The search for such limits of specificity was important, especially for possible use in the putative aptamer pharmaceuticals.

Lim and Peabody (1994) continued the evaluation of mutations that increased the affinity of a translational repressor for RNA and isolated additional suppressors of the operator-constitutive mutation. This time they used the FG-deleted mutant as the starting point for further mutagenesis so as to avoid isolating more assembly-defective mutations. As before, each of the mutants isolated by this approach was a super-repressor *in vivo*, more tightly repressing both the mutant and wild-type operators *in vitro* than did the wild-type protein. So, each protein bound the operators from 3 to 7.5-fold more tightly than the normal MS2 coat. The positions of the amino acid substitutions in the MS2 coat appeared consistent with the idea that they extended the usual RNA-binding site by introducing new interactions with RNA.

In order to test the importance of the strength of the coat protein–RNA interaction, Peabody (1997a) introduced one of these so-called super-repressor mutations into the viral genome. The resulting recombinant was defective for plaque formation, presumably because replicase synthesis was prematurely repressed. Nevertheless, a small number of plaques was obtained. These resulted from second-site reversion mutations that allowed the phage to escape super-repression and fell into two categories. Those of the first type contained nucleotide substitutions within the translational operator that reduced or destroyed its ability to bind coat protein, showing that this interaction was not necessary for genome encapsidation. The revertants of the second type were double mutants, in which one substitution converted the coat initiator AUG to AUA and the other substitutes an A for the G normally present two nucleotides upstream of the coat start codon. The mutation of the coat protein gene initiation from AUG to AUA, by itself, reduced the coat protein synthesis to a few percent of the wild-type level. The second substitution destabilized the coat initiator stem-loop and restored coat protein synthesis to within a few-fold of the wild-type levels.

Going further, Peabody and Lim (1996) realized a brilliant idea of a covalent heterodimer. Since the C-terminus of each monomer closely approached the N-terminus of the other monomer within the dimer (see Chapter 21), this proximity of the N- and C-termini suggested the possibility of duplicating the coat sequence in such a way as to create a translational fusion between the identical subunits. The authors speculated that the existence of such a "two-domain" monomer would permit the construction of recombinants in which different halves of the sequence could contain different amino acid substitutions, thus creating the functional equivalent of obligatory heterodimers. As a result, all possible pairwise combinations of the repressor-defective mutants were introduced into heterodimers and tested for translational repression. Based on the resulting complementation pattern, a model of the RNA binding was proposed. Since the x-ray structure of the MS2 phage-operator RNA complex was reported at that time (Valegård et al. 1994b), the genetically identified RNA binding site was colocalized with the structurally defined one.

Furthermore, to test the significance of Thr45 that appeared in both adenine-binding pockets for a bulged adenine A_{-10} and a loop adenine A_{-4}, all 19 amino acid substitutions were introduced (Peabody and Chakerian 1999). It was difficult to determine the effects of the mutations on the RNA binding because every substitution compromised the ability of coat protein to fold correctly. However, the genetic fusion of coat protein subunits reverted these protein structural defects, making it possible to show that the RNA binding activity of coat protein tolerated substitution of Thr45, but only on one or the other subunit of the dimer. The single-chain heterodimer complementation experiments suggested that the primary site of Thr45 interaction with RNA was with A_{-4} in the translational operator. Either contact of Thr45 with A_{-10} made little contribution to the stability of the RNA-protein complex, or the effects of Thr45 substitution were offset by conformational adjustments that introduced new, favorable contacts at nearby sites (Peabody and Chakerian 1999). After the Thr45 residue, the study of the adenosine-binding pockets was extended to four other amino acid residues utilized by the pockets, namely Val29, Ser47, Thr59, and Lys61 (Powell and Peabody 2001). Summarizing briefly, the Val29 and Lys61 residues formed important stabilizing interactions with both A_{-4} and A_{-10}. Meanwhile, the Ser47 and Thr59 residues interacted primarily with A_{-10}, while the important interactions with Thr45 were restricted to A_{-4}.

Finally, Peabody and Al-Bitar (2001) developed a clever method to isolate the MS2 mutants with altered assembly and solubility properties. Thus, the colonies expressing MS2 coat from a plasmid were covered with an agarose overlay under conditions that caused the lysis of some of the cells in each colony. The proteins thus liberated diffused through the overlay at rates depending on their molecular sizes. After transfer of the proteins to a nitrocellulose membrane, probing with the coat protein–specific antiserum revealed spots whose sizes and intensities were related to the aggregation state of the coat protein. As a result, the three classes of mutants were identified: (i) defective in inter-subunit interactions and failing to form virus-like particles without disrupting the ability to synthesize large quantities of soluble coat protein, (ii) defective in folding and forming insoluble aggregates, and (iii) soluble second-site revertants of insoluble mutants.

The revolutionary contribution to the protein-RNA interactions within the repressor complex I was made by Peter G. Stockley's team in Leeds, in tight cooperation with Lars Liljas' team in Uppsala. First, Haneef et al. (1989) used a novel modeling technique, based on combining information from several preexisting structures to generate a three-dimensional model for the RNA stem-loop responsible for the translational repression of the MS2 replicase. As a result, the stem-loop structure of 17 out of 20 nucleotides was found as the minimum sequence information required for the specific coat binding, with a special emphasis on the adenines: bulged A_{-10}, and A_{-7} and A_{-4} within the loop. The specific features of the model were tested experimentally by chemical and enzymatic structural probes, where the model and chemical modification data were in part consistent. Talbot et al. (1990) used synthetic oligoribonucleotides to probe RNA-protein interactions in the MS2 complex I. Among other things, this paper developed the idea (Romaniuk and Uhlenbeck 1985) of a transient covalent bond between the single-stranded uridine residue at position −5 and a cysteine side chain of the coat protein, which was doubted impartially, as stated above, by Peabody (1989). The obtained results were discussed in the light of the new data in support of the Michael adduct formation by the contact of the cysteine side chain and the base in question (Talbot et al. 1990). The use of the synthetic RNA in the study of the MS2 RNA-protein interactions was reviewed at that time by Goodman et al. (1991).

Supporting the general model of the stem-loop, the crucial adenines A_{-10}, A_{-7}, and A_{-4} were found apparently hyper-reactive by diethylpyrocarbonate modification (Talbot et al. 1991). The MS2 stem-loop containing 4-thiouridine was synthesized, but no UV crosslinking was observed between the MS2 coat and the stem-loop with 4-thiouridine substitutions at either position (McGregor et al. 1996).

The actual structural studies of Stockley's team (Stockley et al. 1993, 1995; Stonehouse et al. 1996a,b) involved maximally the three-dimensional knowledge on the MS2 coat, and were performed in tight cooperation with Liljas' team in Uppsala, who determined in 1990 the first high-resolution structure of the RNA phages, namely, of the phage MS2 (for more detail, see Chapter 21). The novel performance potential for the complex I was approached by the unusual finding that the RNA molecules encompassing the minimal translational operator recognition elements could be soaked into crystals of the RNA-free coat protein shells, allowing the RNA to access the interior of the capsids and make contact with the operator binding sites (Valegård et al. 1994b). The soaking technique allowed the determination of the x-ray structure of the RNA-protein complexes formed. Thus, the crystal structure at 3.0 Å resolution was reported for the complex between recombinant MS2 capsids and the 19-nucleotide operator (Valegård et al. 1994b). This was the first example of a structure at this resolution for a sequence-specific protein-RNA complex on the virus level. The structure showed for the first time the real sequence-specific interactions between conserved residues on the protein and RNA essential for the complex I formation. The exclusive importance of this study was operatively reviewed (Nagai 1996; De Guzman et al. 1998; Hermann and Patel 2000; Feig et al. 2001; Babitzke et al. 2009; Thapar et al. 2014; Gelinas et al. 2016).

By using chemically synthesized operators carrying modified functional groups at defined nucleotide positions, it was confirmed clearly that the complexes formed between the operator stem-loop and RNA-free phage capsids were identical to those formed in solution between the operator and the single coat protein dimer (Stockley et al. 1995).

Furthermore, the crystal structures of two complexes between recombinant MS2 capsids and RNA operator fragments have been determined at the 2.7 Å resolution (Valegård et al. 1997). The three-dimensional studies clearly confirmed the intrinsic role of the adenines A_{-10}, A_{-7}, A_{-4} and a pyrimidine at position -5. The RNA stem-loop, as bound to the protein, formed a crescent-like structure and interacted with the surface of the β-sheet of a coat protein dimer (for more detail, see Chapter 21). It made protein contacts with seven phosphate groups on the 5′ side of the stem-loop, with a pyrimidine base at position -5, which stacked onto a tyrosine, and with two exposed adenine bases, one in the loop and one at a bulge in the stem. The replacement of the wild-type uridine with a cytosine at position -5 increased the affinity of the RNA to the dimer significantly. The complex with RNA stem-loop having cytosine at this position differed from that of the wild-type complex mainly by having one extra intramolecular RNA interaction and one extra water-mediated hydrogen bond (Valegård et al. 1997). Figure 16.7 presents the actual secondary structures of the two RNA operators within the complex I.

The novel soaking technique paved the way to massive study of the modified specifically targeted oligoribonucleotides, or, in other words, aptamers, by the complex I formation. Grahn et al. (1999) checked four RNA hairpin variants with longer or shorter stems than normal RNA hairpin. It was concluded that the four loop nucleotides and a two-base-pair stem but without the unpaired nucleotide were sufficient for binding to the coat protein shell, while an aptamer containing a stem with the unpaired nucleotide but missing the loop nucleotides did not bind to the protein shell.

Continuing the crystallographic studies, three different RNA aptamers were selected to bind the MS2 coat (Hirao et al. 1998–1999). Despite their different secondary structures, the 2.8 Å resolution of the capsid-RNA complexes was achieved, revealing that non-Watson-Crick base pairs in the stem and a 3-nucleotide loop could all be accommodated in such complexes (Convery et al. 1998; Rowsell et al. 1998).

The three-dimensional structures were also determined for the complexes of the stem-loop with the MS2 capsids bearing mutations in the coat protein (van den Worm et al. 1998). Thus, the universally conserved amino acid residue Thr45, as well as Thr59, which were found part of the specific RNA binding pocket and interacted directly with the RNA, the former through a hydrogen bond, the latter through hydrophobic contacts, were studied. The crystal structures of the MS2 capsids formed by the Thr45Ala and Thr59Ser mutants were resolved both with and without the 19 nt wild-type operator hairpin bound. Surprisingly, despite the deleted contacts, the RNA hairpin bound to the proteins and both resulting complexes were similar to the wild-type one (Valegård et al. 1994, 1997). In fact, the Thr59Ser mutant was isolated earlier by screening for mutants deficient in the repression of replicase synthesis (Peabody 1993; Stockley et al. 1993). Both the Thr59Ser and Thr45Ala mutants were unable to prevent phage infection in an *in vivo* repression assay, but only the Thr45Ala mutant had a significantly higher equilibrium dissociation constant *in vitro* than the wild-type protein (Lago et al. 1998).

After the Thr59Ser and Thr45Ala substitutions, Lago et al. (1998) also studied single Glu76Asp and Pro78Asn mutants by *in vivo* and *in vitro* functional assays. Therefore, the substitutions were inserted at amino acid residues which either interacted with the operator RNA or were involved in stabilizing the conformation of the FG loop, the site of the major quasi-equivalent conformational change (for more detail, see Chapter 21). The results with the variant RNA operators revealed the robustness of the operator-coat protein interaction and the requirement for both halves of a protein dimer to contact RNA in order to achieve tight binding. It was suggested that there might be a direct link between the conformation of the FG loop and RNA binding. This point was strengthened by the recent finding of Stockley's team that the Pro78 from the FG loop was essential for viral infectivity (Hill et al. 1997).

Grahn et al. (2000) presented the crystal structure at 2.85 Å resolution of the complex, where an RNA hairpin modified at position -5, which has been shown previously to be pyrimidine-specific, was soaked into the MS2 coat protein crystals. In this hairpin, the wild-type uracil was replaced by a pyridin-4-one derivative. This base was missing the exocyclic 2-oxygen, which formed the hydrogen bond to the Asn87 side chain. The modified complex structure showed an unprecedented major conformational change in the loop region of the RNA, whereas there was almost no change in the conformation of the protein. In fact, this complex showed a different mode of RNA binding than all other MS2 RNA-coat complex structures determined so far. The uracil -6 base, not previously believed to have any role in contacting the protein, made the stacking contact with TyrA85 instead of uracil -5. The hydrogen bond to AsnA87 was not made, despite evidence that this interaction must be critical for sequence specificity (Lim et al. 1994; Johansson et al. 1998).

Altogether, these results showed that the consensus could easily be broken, challenging the traditional understanding of the sequence specificity. To understand the classical

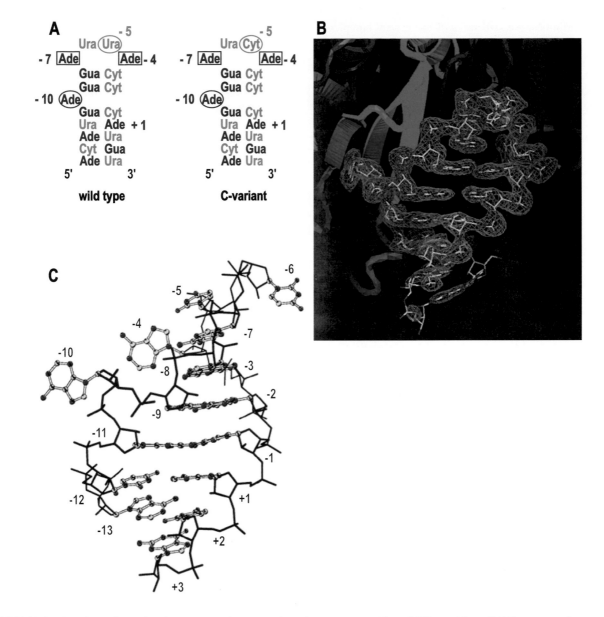

FIGURE 16.7 The three-dimensional structures of two complexes between recombinant MS2 capsids and RNA operator fragments. (A) Sequence, numbering and predicted secondary structure of the two RNA fragments. The numbering is relative to the AUG start codon for the replicase gene at position +1. (B) Drawing of the repressor-RNA complex. The electron density for the wildtype RNA is included. (C) The wild-type RNA stem-loop, as bound to the A/B dimer. The RNA is composed of a base-paired stem, containing a single-bulged nucleotide and a four-nucleotide loop. (Reprinted from *J Mol Biol.*, 270, Valegård K et al., The three-dimensional structures of two complexes between recombinant MS2 capsids and RNA operator fragments reveal sequence-specific protein-RNA interactions, 724–738, Copyright 1997, with permission from Elsevier.)

pyrimidine requirement at the position −5, Grahn et al. (2001) determined the x-ray structures of six MS2 RNA hairpin-coat complexes having different substitutions at the position −5, namely, 5-bromouracil, pyrimidin-2-one, 2-thiouracil, adenine, and guanine. Remarkably, the structure of the 5-bromouracil complex was determined at 2.2 Å resolution, which was the highest for any MS2 RNA-protein complex. All the complexes showed very similar conformations, despite variation in affinity in solution. The stacking of the −5 base onto the tyrosine side chain was found the most important driving force for the complex formation. A number of hydrogen bonds that were present in the wild-type complex were not crucial

for the binding, as they were missing in one or more of the complexes.

Continuing the nucleotide specificity problem of the 19-nucleotide hairpin, Helgstrand et al. (2002) resolved the structures of complexes between the MS2 capsid and operators with guanosine, cytidine, and two non-native bases, 2′-deoxy-2-aminopurine and inosine, at position −10 instead of the bulged A$_{−10}$, as well with guanosine at −10, but with multiple potential alternative conformations of the stem-loop. At the position −7, the native adenine was substituted with a cytidine. Of these, only the G$_{−10}$, C$_{−10}$, and C$_{−7}$ variants showed interpretable density for the RNA hairpin. In spite of large differences in binding affinities, the structures of the variant complexes were very similar to the wild-type operator complex. For the G_{-10} substitutions in hairpin variants that could form bulges at alternative places in the stem, the binding affinity was low, and a partly disordered conformation was seen in the electron density maps. The affinity was similar to that of the wild type when the base pairs adjacent to the bulged nucleotide were selected to avoid alternative conformations. Both purines bound in a very similar way in a pocket of the protein. In the C_{-10} variant, which had very low affinity, the cytidine was partly inserted in the protein pocket rather than intercalated in the RNA stem. The substitution of the wild-type adenosine at the position −7 by pyrimidines gave strongly reduced affinities, but the structure of the C_{-7} complex showed that the base occupied the same position as the A_{-7} in the wild-type RNA. It was stacked in the RNA and made no direct contact with the protein (Helgstrand et al. 2002). The crystal structure of the MS2 capsid complex with the 2′-deoxy-2-aminopurine at the position −10 was resolved by the 2.8 Å resolution (Horn et al. 2004). Substituting 2′-deoxy-2-aminopurine for the A_{-10} in the aptamer yielded an RNA with the highest reported affinity for the MS2 coat protein. The refined x-ray structure showed that the 2′-deoxy-2-aminopurine made an additional hydrogen bond to the protein compared to A_{-10} that was presumably the principal origin of the increased affinity.

Finally, the structural basis of the RNA binding discrimination between the phages MS2 and Qβ was tested (Horn et al. 2006). Thus, the x-ray crystal structures were resolved for eight complexes of MS2 mutants that have been shown to relax this discrimination *in vitro*, namely, Asn87Ser and Glu89Lys (Lim et al. 1996; Spingola and Peabody 1997) with both the MS2 and Qβ stem-loops. The results were consistent with the hypothesis that the side chains of residues Asn87 and Glu89, together with the differing secondary and hence tertiary structures of the RNA operators, regulated specificity *via* a combination of steric clashes and electrostatic repulsion.

In parallel, the conformational probing was simplified by the introduction of the fluorescent methods. In order to probe the dynamics of both the wild-type operator and the aptamer RNAs, the fluorescent nucleotide 2′-deoxy-2-aminopurine was incorporated at key adenosine positions (Parrott et al. 2000). This appeared in fact as a sensitive probe for the conformational changes. The recognition by coat protein was enhanced, unaffected, or decreased, depending on the site of substitution, consistent with the known protein-RNA contacts seen in crystal structures of the complexes. It was concluded that the detailed conformational dynamics of aptamers and wild-type RNA ligands for the same protein target were remarkably similar. The fluorescent technique was used to measure kinetics of the complex I formation between the MS2 coat dimer and members of three distinct RNA aptamer families which were known to bind to the same site on the protein (Lago et al. 2001). Remarkably, the complex I consisted of a single-coat dimer and one RNA molecule. The authors concluded that the conformational changes in the protein ligand during formation of the complex I might play a role in the triggering of the capsid self-assembly.

Furthermore, single-molecule fluorescence resonance energy transfer (SM-FRET) and fluorescence correlation spectroscopy (FCS) were used to probe the details of the thermal stem-loop unfolding reaction in solution (Gell et al. 2008). The stem-loops were derivatized with the appropriate donor and acceptor dyes on the 5′ and 3′ ends, respectively. In addition to the wild-type sequence, a number of sequence variants was studied in order to be able to relate sequence/structure to the folding behavior. The single-molecule fluorescence assays suggested that these RNAs existed in solution as ensembles of differentially base-paired/base-stacked states at equilibrium. The wild-type RNA samples demonstrated two distinct ensembles, implying the existence of a significant free energy barrier between folded, as a stem-loop, and unfolded ensembles. The experiments with sequence variants were consistent with an unfolding mechanism, in which interruptions to base-paired duplexes, in this example by the single-stranded loop and a single-base bulge in the base-paired stem, as well as the free ends, acted as nucleation points for unfolding. Strikingly, the U-to-C replacement of the position −5, which created the high-affinity form of the operator for the coat protein binding, resulted in dramatically different (un)folding behavior, revealing distinct subpopulations that were either stabilized or destabilized with respect to the wild-type sequence. This result suggested additional reasons for the selection against the C-variant stem-loop *in vivo* and provided an explanation for the increased affinity (Gell et al. 2008).

After the successful R17 story described above, Uhlenbeck's team turned to the MS2 complex I because of the availability of the MS2 crystal structure at that time. The MS2 coat-operator binding was evaluated by nitrocellulose binding assay, in contrast to the MS2 studies described above. First, the effects of mutations in the MS2 coat expected to disrupt capsid assembly were studied (LeCuyer et al. 1995). The six different mutations in the FG loop structure were selected, in which hydrophobic residues were replaced with charged residues. Most of these proteins formed capsids in *E. coli*, when coat gene was expressed from a recombinant plasmid, but not in an *in vitro* assembly assay, suggesting that interdimer interactions were weaker than in the wild-type MS2. These mutant proteins reduced the free energy of cooperative protein binding to a double-hairpin RNA from its wild-type value of −1.9 kcal/mol. Several of the variants that have large effects on cooperativity have no effect on RNA affinity, suggesting that protein-RNA interactions could be affected independently of dimer-dimer interactions. The V75E+A81G protein, which showed no measurable cooperativity, bound operator RNA equally well as the wild-type protein. It allowed studies of the RNA binding properties independent of capsid assembly. Then, the role of 2′-hydroxyl groups in an RNA-protein interaction was studied (Baidya and Uhlenbeck 1995). When ribose was replaced by deoxyribose in the MS2 operator, only one, namely, U_{-5}, out of 15 positions tested, reduced protein affinity by 1.6 kcal/mol. This supported the crystallographic finding that the 2′-hydroxyl of U_{-5} interacted with the carboxylate group of the Glu63 residue (Valegård et al. 1994). The complementary experiments introducing ribose into a DNA hairpin confirmed the putative protein contact. Several arguments suggested these riboses were required to maintain an A-form helix in this region of the binding site. Finally, a minimum requirement of four 2'-hydroxyl groups for the wild-type coat protein binding was determined, one of which was at the −5 position, and other three in the upper stem in any combination (Baidya and Uhlenbeck 1995).

Next, the thermodynamic contribution of a stacking interaction between Tyr85 in the MS2 coat and a single-stranded pyrimidine C_{-5} in its RNA binding site was examined (LeCuyer et al. 1996). The mutation of Tyr85 to Phe, His, Cys, Ser, or Ala decreased the RNA affinity by 1–3 kcal/mol. Since the Phe85, His85, and Cys85 proteins formed UV photocrosslinks with iodouracil-containing RNA at the same rate as the wild-type protein, the mutant proteins interacted with the RNA in a similar manner. The experiments with specifically substituted phosphorothioate RNAs confirmed a hydrogen bond between the hydroxyl group of tyrosine and a phosphate predicted by the crystal structure.

At that time, summarizing reviews were published on the fine structure-supported RNA recognition by the MS2 coat (Johansson et al. 1997) and on the new aptamer co-crystal technique (Uhlenbeck 1998) as a reaction to the Convery et al. (1998) paper.

Further, Johansson et al. (1998) presented a thermodynamic analysis of the unique U_{-5} change to C_{-5} in the loop of the MS2 hairpin that increased the binding affinity to the protein by nearly 100-fold. A series of protein mutations and RNA modifications were employed to evaluate the thermodynamic basis for the improved affinity. The obtained data were in good agreement with a comparison of the co-crystal structures of the two complexes, where small differences in the two structures were seen at the thermodynamically important sites (Valegård et al. 1997).

The phosphorothioate-substituted RNA was used to investigate the thermodynamic role of phosphates in the MS2 coat complex with the RNA operator (Dertinger et al. 2000). A single phosphorothioate linkage was introduced at 13 different positions. In each case, the R_P and S_P stereoisomers were separated and their affinities to the MS2 coat protein were determined. The comparison of these biochemical data with the crystal structure of the protein-hairpin complex indicated that introduction of a phosphorothioate only affected the binding at sites where a protein-phosphate contact was observed in the crystal structure. It was concluded that the phosphorothioate-containing oligoribonucleotides should also be useful for mapping phosphate contacts in the RNA-protein complexes for which no crystal structure was available.

Next, methylphosphonates were involved as analogs for detecting phosphate contacts in the MS2 coat-operator complex (Dertinger and Uhlenbeck 2001). Thus, single methylphosphonate stereoisomers were introduced at 13 different positions in the RNA hairpin. By comparing these data to the available crystal structure of the complex, it appeared that all phosphates that were in proximity to the protein showed a weaker binding affinity when substituted with a phosphorothioate. However, in two cases, a methylphosphonate isomer

either increased or decreased the binding affinity, where no interaction could be detected in the crystal structure. It seemed possible that the methylphosphonate substitutions at these positions affected the structure or flexibility of the hairpin. The utility of the methylphosphonate substitution was compared to the phosphate ethylation (Gott et al. 1993) and phosphorothioate substitution (Dertinger et al. 2000) experiments performed previously on the same system.

Further, Dertinger et al. (2001) studied the contribution of the contact between the 2'-hydroxyl group of ribose at the position −5 and Glu63 in one subunit of the covalently fused MS2 coat dimer that was prepared in accordance with the idea of Peabody (1997b). When the Glu63 residue was substituted with glutamine, aspartate, or alanine, the binding affinity of the hairpin for the protein was weakened by 12- to 100-fold, similar to that observed with deoxyribose at the position −5. However, the specificity of the three mutant proteins for RNAs with various modifications at the 2'-position of ribose −5 differed dramatically. While the Glu63Asp protein resembled the wild-type protein in preferring the 2'-hydroxyl group over a proton or a bulky 2'-substituent, both the Glu63Ala and Glu63Gln proteins preferred bulky 2'-substituents over the 2'-hydroxyl group by more than 100-fold. These experiments emphasized the ease with which the specificity of a protein-nucleic acid interaction could be changed at thermodynamically important sites (Dertinger et al. 2001).

Finally, Hobson and Uhlenbeck (2006) performed alanine scanning of the MS2 coat and revealed protein-phosphate contacts involved in thermodynamic hot spots. Since the co-crystal structure of the MS2 coat dimer with the RNA operator found eight amino acid side chains contacting seven of the RNA phosphates, these eight amino acids and five nearby control positions were individually changed to an alanine residue, and the binding affinities of the mutant proteins to the RNA were determined. In general, the data agreed well with the crystal structure and previous RNA modification data. Interestingly, the amino acid residues that were energetically most important for complex formation clustered in the middle of the RNA binding interface, forming thermodynamic hot spots, and were surrounded by energetically less relevant amino acids. In order to evaluate whether or not a given alanine mutation caused a global change in the RNA-protein interface, the affinities of the mutant proteins to RNAs containing one of 14 backbone modifications spanning the entire interface were determined. In three of the six protein mutations tested, the thermodynamic coupling between the site of the mutation and the RNA groups that could be even more than 16 Å away was detected.

At the same time, Uhlenbeck's team turned to the genomic systematic evolution of ligands by exponential (SELEX) enrichment, which was an experimental procedure that allowed extraction from an initially random pool of oligonucleotides of the oligomers with a desired binding affinity for a given molecular target. The library consisting of *E. coli* DNA that was transcribed *in vitro* was used, and it was found that the MS2 coat bound several *E. coli* mRNA fragments, *rffG* (involved in formation of lipopolysaccharide in the bacterial outer membrane) and *ebgR* (lactose utilization repressor) among them, more tightly than it bound the complex I operator (Shtatland et al. 2000). Remarkably, this was the first RNA genomic SELEX, as commented by the appropriate review article (Djordjevic 2007). Later, the MS2 aptamers were selected by the newly developed aptamer selection express (ASExp) system (Fan et al. 2008).

In parallel with the long-term systematic studies on the structure of the complex I that were reviewed above, the contribution of other groups to the subject must be mentioned. Thus, the MS2 complex I was employed for the demonstration of the photo-crosslinking with the 5-iodocytidine (Meisenheimer et al. 1996) or 5-bromouridine (Norris et al. 1997) substituted RNAs to the associated protein and to the detailed methodology of the nucleoprotein photo-crosslinking using halopyrimidine-substituted RNAs (Meisenheimer et al. 2000). Later, the UV-induced photochemical crosslinking was achieved with the MS2 coat and its operator, in which adenosine residues were replaced by 8-azidoadenosine (Gopalakrishna et al. 2004).

van Duin's team contributed by the evolutionary reconstruction of the MS2 hairpin deleted from the phage genome (Olsthoorn and van Duin 1996a; Klovins et al. 1997; Licis et al. 1998).

Arnold and Fisher (2000) incorporated 5-fluorouridine into several sites within a 19-mer MS2 RNA operator. The ^{19}F NMR spectra demonstrated the different chemical shifts of helical and loop fluorouridines of the hairpin secondary structure. This study demonstrated, first of all, that the overall ensemble average hairpin conformation was strongly salt-dependent. Therefore, the electrostatic considerations were suspected to be involved in the balance between different hairpin conformers as well as the duplex-hairpin equilibrium. Second, the obtained data demonstrated that the ^{19}F NMR could be a powerful tool for the study of conformational heterogeneity in RNA, especially for probing the effects of metal ions on the RNA structure.

Moreover, using NMR spectroscopy, Smith and Nikonowicz (2000) determined the structural effect of stereospecific phosphorothioate substitution at five positions in the MS2 RNA hairpin. At most sites, the substitution had little or no effect, causing minor perturbations in the phosphate backbone and increasing the stacking among nucleotides in the hairpin loop. At one site, however, the phosphorothioate substitution caused an unpaired adenine A_{-10} to loop out of the RNA helix into the major groove. These results indicated that the phosphorothioate substitution could substantially alter the conformation of the RNA at positions of irregular secondary structure, complicating the use of substitution-interference experiments to study RNA structure and function.

Remarkably, the recognition of the RNA encapsidation signal by the yeast L-A double-stranded RNA virus demonstrated clear analogies with the MS2 complex I (Fujimura and Esteban 2000). Thus, the L-A encapsidation signal contained a 24-nucleotide stem-loop structure with a 5-nucleotide loop and an A bulged at the $5'$ side of the stem. Interestingly, the MS2 coat recognized and promoted the catalytic activity of *Candida* group I ribozyme (Xie et al. 2009).

The MS2 coat-operator interactions were used as an experimental basis by the elaboration of a distance-dependent statistical potential function which could predict the specificity and relative binding energy of the RNA-binding proteins (Zheng et al. 2007). At the same time, these interactions were employed as a model by the elaboration of an infrared fluorescence imaging system by labeling of oligonucleotides with a fluorescent dye FAM (Wang K et al. 2010).

By quantitative analysis of the MS2 coat and RNA interactions on a massively parallel array, the remarkable biophysical and evolutionary landscapes were revealed (Buenrostro et al. 2014). By analyzing the biophysical constraints and modeling mutational paths describing the molecular evolution of the MS2 operator from low- to high-affinity hairpins, the widespread molecular epistasis and a long-hypothesized, structure-dependent preference for the G:U base pairs over the C:A intermediates were quantified in evolutionary trajectories.

Recently, the MS2 coat-operator interaction was evaluated by biolayer interferometry for changes in affinity as a result of phosphorodithioate replacing phosphate (Yang et al. 2017). A structure-based analysis of the data provided insights into the origins of the enhanced affinity of the RNA-protein interactions triggered by the phosphorodithioate moiety.

Finally, Poudel et al. (2017) approached the MS2 coat-operator interaction by using large-scale *ab initio* computation centered on critical aspects of the consensus protein-RNA interactions recognition motif. The density functional theory (DFT) calculations were carried out on two nucleoprotein complexes, a wild type and a mutated type corresponding to the Protein Data Bank entries 1ZDH (Valegård et al. 1997) and 5MSF (Rowsell et al. 1998). The calculated partial charge distribution of residues and the strength of hydrogen bonding between them made it possible to locate the exact binding sites with strongest hydrogen bondings identified to be Lys43-A_{-4}, Arg49-C_{-13}, Tyr85-C_{-5}, and Lys61-C_{-5}, due to the change in the sequence of the mutated RNA. Because of computational limitations, these calculations were restricted to a single subunit of an asymmetrical unit of the virus, including the MS2 coat monomer and operator. Remarkably, this appeared to be the largest *ab initio* quantum computation performed on a complex biomolecular system to date (Poudel et al. 2017). Figure 16.8 presents the final outline of the interactions.

The MS2 operator was chosen as a model in the recent study on the machine-learning of the fitness landscape by the virus assembly (Dechant and He 2019).

REPRESSOR COMPLEX I: fr

The fr coat-RNA complex I was studied in Riga by Grēns' team, in parallel with the MS2 complex I. As cited above, the fr RNA operator was isolated and sequenced from the complex I and complexed with both fr and MS2 coats (Tsielens et al. 1982). The operator length was shortened to 27 nucleotides and the RNA hairpin loop corresponding to the -23 to $+4$ positions of the fr replicase gene was found sufficient to bind the fr coat protein. The minimal nucleotide sequence of the fr

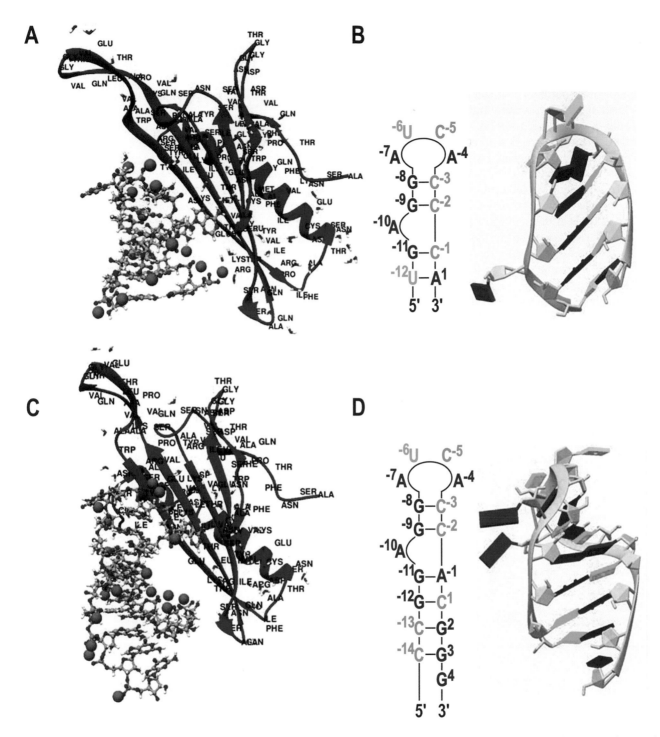

FIGURE 16.8 Relaxed structure of the MS2 CP-RNA complex with protein subunit A and RNA subunit R for the two nucleoprotein complexes: wild-type (PDB ID 1ZDH) and mutated (PDB ID 5MSF). (A) Protein-RNA complex in 1ZDH. (B) Structure and sequence of RNA in 1ZDH. (C) Protein-RNA complex in 5MSF. (D) Structure and sequence of RNA in 5MSF. In (A) and (B), the red ribbon denotes the MS2 CP, the ball and stick denote nucleotides of RNA, purple spheres denote Na, and stick represents the water molecules. (Reprinted with permission from Poudel L et al. Impact of hydrogen bonding in the binding site between capsid protein and MS2 bacteriophage ssRNA. *J Phys Chem B.* 121:6321–6330. Copyright 2017 American Chemical Society.)

and MS2 RNA binding sites differed therefore by 2 of 20 minimal length nucleotides, while the two coat proteins differed, as we remember from the previous sections, by 17 of 129 amino acids. The isolated operator RNA fragments were used as templates in the formation of the ribosomal initiation complex *in vitro* and made it possible to pinpoint the boundaries of

the initiation region of the replicase gene (Berzin et al. 1982). As a result, it was emphasized that the flanking regions of the initiation site were responsible for additional interactions with the ribosome, in addition to the initiator AUG and the SD sequence, where the flanking regions contributed possibly to the contact stability of the ribosome with the template.

As described in the previous section, the copies of the fr operator, in parallel with the MS2 operator, were synthesized enzymatically (Renhof et al. 1985; Nikitina et al. 1986). The operator with the destroyed SD sequence was inactivated by the recognition of both coat and ribosome (Renhof et al. 1985). Later, the fr RNA operator fragments were synthesized chemically (Kumpin'sh et al. 1995).

Furthermore, the interaction of the fr coat with its translational operator site was compared with the previously studied R17 interaction by Uhlenbeck's team (Wu et al. 1988). The RNA operator length was reduced to 21 nucleotides and covered the sequence of the −17 to +4 positions of the replicase gene. An analysis of the binding of the fr coat protein to the 24 operator variants revealed that the two proteins recognized operator sequences in virtually the same way. However, the fr coat was bound to nearly every RNA 6- to 14-fold tighter than the R17 coat. Since the fr operator was a weaker binding variant and the fr coat protein showed a different temperature dependence of binding, it was unlikely that the two systems have different K_a values *in vivo*. The RNA fragments containing the operator sequences initiated the capsid assembly with both fr and R17 coats. Moreover, the two coat proteins formed a mixed capsid *in vitro* (Wu et al. 1988). At that time,

the Bachelor thesis at the University of Illinois (Kastelic 1985) was devoted to the interaction of the fr coat with native and synthetic fr RNA operator variants.

REPRESSOR COMPLEX I: GA

The GA coat protein differs from the MS2 coat at 51 positions out of 130 amino acid residues by generally satisfying alignment. Despite the large number of amino acid changes that are scattered throughout the protein, these proteins are clearly homologous and can be aligned along their entire length. The primary structure of the appropriate GA 59-nucleotide, by analogy to the original coat binding site of the group I phages, differed from that of MS2 at 27 alignment positions, although both minimal sites were fairly similar by secondary structure. Figure 16.9 presents the corresponding operator hairpins of different phages including that of GA.

By analogy with the MS2 system, Peabody (1990) constructed the genetic two-plasmid system that placed the synthesis of the hybrid replicase-β-galactosidase enzyme under translational control of coat protein also for the phage GA. The binding of the GA coat protein to its operator of 23 nucleotides in length was determined with a K_a of 71 μM^{-1} (Gott et al.

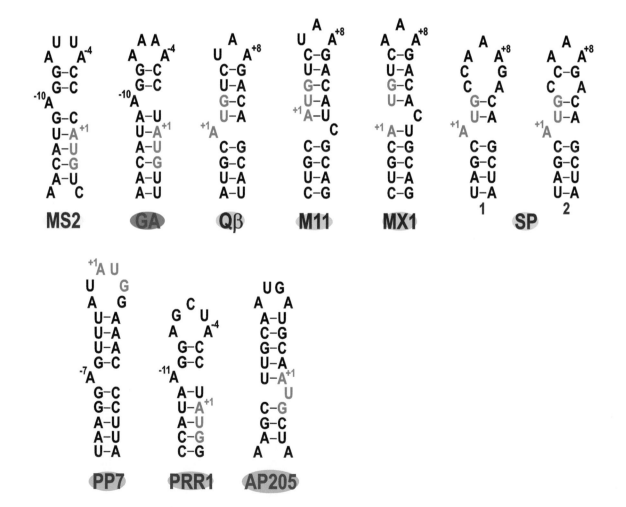

FIGURE 16.9 Replicase-operator hairpins in different RNA phages. Start codons are denoted green. (The data are compiled from the three original papers: Klovins et al. 2002; Spingola et al. 2002; Persson et al. 2013.)

1991a). This interaction differed kinetically from the R17 coat-operator interaction. Structurally, the GA hairpin differed from its MS2 or R17 counterpart by having adenines at positions −5 and −6 and possibly by the position of the unpaired A in the stem, since either A_{-10} or A_{-11} could pair with U_{-1} in the GA sequence. These differences were particularly interesting, since the MS2 and R17 coats had a strong preference for pyrimidines at loop residue −5 and required the precise location of an unpaired purine at −10 for binding. Unlike the group I phage proteins, the GA coat did not distinguish between two alternate positions for the unpaired purine and did not show high specificity for a pyrimidine at position −5 of the loop.

The specific potential of the GA operator to form one of two different A•U base pairs, while leaving the other adenine unpaired or bulged, was studied by heteronuclear NMR spectroscopy (Smith and Nikonowicz 1998). This RNA motif was studied in a 21-nucleotide hairpin containing the GA operator whose 4-nucleotide loop was replaced by a more stable MS2 loop. The structure of this hairpin was resolved to an overall precision of 2.0 Å, and the base pair involving the 5′-proximal adenine appeared to be the major conformation. This specific feature of the GA hairpin was of particular interest because of the sequence similarity it shared with the U2 snRNA-intron branch-site pairing within the spliceosome (Smith and Nikonowicz 1998; Newby and Greenbaum 2001).

Recently, the three-dimensional structure of the GA operator hairpin was resolved by heteronuclear NMR spectroscopy (Chang et al. 2017). In contrast to the loop sequence 5′-$A_{-7}U_{-6}U_{-5}A_{-4}$-3′ of the MS2 operator hairpin, in which the 5′-proximal adenine base stacks on the loop-closing base pair, the A_{-7} base of the GA hairpin exhibited motions on the intermediate timescale and appeared disordered. The residues A_{-6}, A_{-5}, and A_{-4} of the loop formed a well-ordered 3′ stack that was contiguous with the closing base pair of the loop. The secondary structure of the (A-A)-U bulge in the stem was constrained by NOE data and consistent with relaxation data that indicated the 3′-proximal adenine of the bulge was flexible and remained unpaired. Although the 5′-AAAA-3′ of the phage GA operator increased the conformational disorder and dynamics of the (A-A)-U motif relative to the 5′-AUUA-3′ loop, according to Smith and Nikonowicz (1998), the secondary structure of the bulge favored 5′-proximal adenine-uridine pairing (Chang et al. 2017).

REPRESSOR COMPLEX I: Qβ

The coat of the group III phage Qβ is about 25% identical to the MS2 coat by amino acid sequence but possesses highly similar tertiary structures. It was no doubt that both phages utilize a common structural framework to perform their repressor functions. However, the Qβ operator structure appeared quite different. Figure 16.9 demonstrates the Qβ operator, among other repressor complex I operators.

The direct demonstration of the Qβ coat repressor complex I *in vivo* was achieved by Grumet et al. (1987a,b). This study showed that the Qβ coat did confer resistance to the Qβ infection when the Qβ coat gene was expressed in susceptible *E. coli* strain. The Qβ coat expression had no obvious detrimental effect on the host, but even low-level, constitutive synthesis of the Qβ coat ensured very high levels of resistance to the Qβ infection. This low-level expression of the Qβ coat also produced an intermediate level of resistance to the closely related phage SP but failed to protect against the group I phage f2.

Witherell and Uhlenbeck (1989) characterized by the nitrocellulose filter binding assay the interaction between the Qβ coat and its operator hairpin of 29 nucleotides in length covering the −9 to +20 region according to the replicase start. The association constant K_a of $400\,\mu M^{-1}$ was determined at 4°C, 0.2 M ionic strength, pH 6.0. The salt dependence of K_a revealed that four to five ion pairs might be formed in the complex, although approximately 80% of the free energy of complex formation was contributed by nonelectrostatic interactions. The truncation experiments revealed that the Qβ coat binding required only the presence of a hairpin with an eight-base-pair stem and a three-base loop, while the sequence of the stem was not important for the Qβ coat recognition and only one of the three loop residues was essential. Remarkably, a bulged adenosine was not required for the Qβ coat binding. It was concluded that the Qβ coat binding was achieved primarily by the secondary structure and not by the sequence of the operator.

Lim et al. (1996) constructed the two-plasmid system for the Qβ coat protein as an equivalent to the previously generated and above-described two-plasmid system for the MS2 coat (Peabody 1990). In this system, the Qβ coat expressed from one plasmid translationally repressed synthesis of the replicase-β-galactosidase fusion protein expressed from the second plasmid. As a result, the Qβ mutants were isolated, which defined the RNA-binding site of the Qβ coat protein and showed that, as with MS2, it resided on the surface of a large β-sheet. The mutations were also described that converted the Qβ coat protein to the RNA-binding specificity of the MS2 coat. This study led to rather unexpected results that the binding sites of the two coat proteins could be easily converted by mutation to the RNA-binding specificity of the other (Lim et al. 1996). Figure 16.10 presents a schematic view of the β-sheet surface of the Qβ, MS2, and PP7 coats where the RNA-binding sites were resided.

Furthermore, as mentioned above in the MS2 repressor complex I section, Spingola and Peabody (1997) isolated the two MS2 coat mutants, Asn87Ser and Glu89Lys, that bound the Qβ RNA operator. The x-ray crystal structures were resolved for these specificity-relaxing mutants with both the MS2 and Qβ stem-loops (Horn et al. 2006).

Next, Spingola et al. (2002) succeeded to understand the RNA-binding specificity of yet another coat protein, that of the group IV phage SP, a fairly close relative of the phage Qβ. The amino acid sequences of the Qβ and SP coat proteins were 80% identical. This high degree of similarity extended to the surfaces of their β-sheets, where their RNA-binding sites were located. Remarkably, nearly all the amino acids so far implicated in RNA binding by the Qβ coat were conserved in the SP coat, the only difference being the Tyr89Phe substitution. This suggested that the SP coat protein ought to have an RNA-binding specificity similar, if not identical, to that of the Qβ

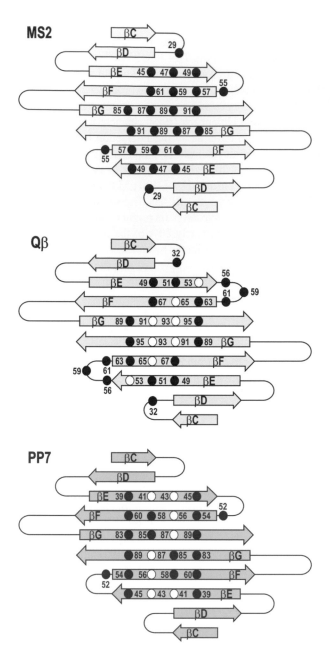

REPRESSOR COMPLEX I: PP7

The PP7 coat showed only 13%–15% amino acid sequence identity to that of the MS2 coat. As demonstrated in the preceding Non-Coliphages section, Bassel and Mills (1979) were the first to sequence the PP7 RNA fragments protected by both *P. aeruginosa* and *E. coli* ribosomes. The PP7 operator is presented in Figure 16.9. By analogy with the phages MS2, GA, and Qβ, Lim et al. (2001) generated the genetic two-plasmid system in which coat protein repressed translation of the replicase-β-galactosidase fusion protein. As a result, David S. Peabody's team was the first to demonstrate directly that (i) the PP7 coat is a translational repressor, (ii) an RNA hairpin containing the PP7 replicase translation initiation site is specifically bound by the PP7 coat both *in vivo* and *in vitro*, indicating that this structure represents the translational operator, and (iii) the RNA binding site resides on the PP7 coat β-sheet. The mapping of the PP7 coat RNA binding site is presented in Figure 16.10. The PP7 coat RNA binding activity was specific since it did not repress the translational operators of the MS2 or Qβ phages. Moreover, the reliable conditions were elaborated for the purification of the PP7 coat and for the reconstitution of its RNA binding activity from disaggregated virus-like particles. The dissociation constant K_a for the PP7 complex I *in vitro* was determined to be about 1 nM (Lim et al. (2001).

Figure 16.11 compares the different binding modes of the coat dimer-operator complexes of the phages MS2, Qβ, and PP7.

Furthermore, Lim and Peabody (2002) described the specific structural RNA requirements for the PP7 coat-operator recognition. The recognition specificity differed substantially from those of the MS2 and Qβ coats. Using designed variants of the wild-type RNA and selection of the binding-competent sequences from random RNA sequence libraries, i.e., SELEX, it was found that tight binding to the PP7 coat was favored by the existence of an eight-base-pair hairpin with a bulged purine on its 5′ side separated by four base pairs from a six-nucleotide loop having the sequence Pu-U-A-G/U-G-Pu. However, another structural class possessing only some of these features was capable of binding almost as tightly (Lim and Peabody 2002).

To investigate the structural basis of the RNA recognition by the PP7 coat, the crystal structures were determined for the truncated version of the PP7 coat protein that was deficient in capsid assembly, so-called PP7 ΔFG, both in the unbound form, to 1.6 Å resolution, or in complex with the 25-nt RNA hairpin, to 2.4 Å resolution (Chao et al. 2008). The deletion of the residues 67–75 of the central FG loop in the PP7 ΔFG version left the coat without any further cysteines and also kept it from forming higher oligomeric structures. This study showed that the PP7 and MS2 coat proteins represented two distinct solutions to the problem of the sequence-specific recognition of an asymmetric RNA hairpin by a symmetric binding surface. The coevolution of the PP7 coat and its translational operator had resulted in the formation of adenine recognition pockets whose orientation relative to the dimer axis differed considerably from those of the MS2 coat protein.

FIGURE 16.10 A schematic representation of the structural locations of amino acids substituted in repressor-defective, assembly competent mutants of MS2, Qβ, and PP7. The β-sheet surface of the protein which, in the assembled particle, is oriented toward the interior of the capsid. Positions which contain identical amino acids at structurally homologous positions are indicated as filled circles. (Redrawn with permission of American Society for Biochemistry and Molecular Biology from Lim F et al. *J Biol Chem.* 2001;276:22507–22513.)

coat. However, inspection of the structure of the translational operator predicted from the SP genome sequence suggested that the SP coat bound a rather different RNA structure. As Figure 16.9 illustrates, the SP operator was expected to form a six-nucleotide stem with a bulged A and a seven-nucleotide loop. However, despite these significant structural differences, the Qβ and SP proteins shared the same RNA-binding specificity. In fact, each protein bound the other's operator just as well as it bound its own (Spingola et al. 2002).

This dramatic rearrangement explained the inability of the two coat proteins to bind the other's RNA target. Moreover, the extended β-sheet surface revealed by the PP7ΔFG RNA structure was a flexible scaffold capable of recognizing diverse RNAs with high specificity (Chao et al. 2008).

Remarkably, the PP7 coat-operator binding data were used by the elaboration of the optimal protein-RNA area (OPRA) algorithm, a propensity-based method to identify the RNA binding sites on proteins (Pérez-Cano and Fernández-Recio 2010a,b). This approach postulated the protein-RNA complexes as the most essential structures in living organisms, understanding of which at molecular level could be a key goal not only from a basic biological point of view, but also for putative biotechnological and therapeutic purposes.

Recently, the PP7 complex I was employed by the elaboration of a novel methodology using a specific RNA-protein interaction to quench the fluorescent RNA spinach (Roszyk et al. 2017). Thus, for Förster resonance energy transfer (FRET), which is a common tool to study RNA-protein interaction networks, the two interacting molecules have to be fluorescently labeled. The "spinach" is a genetically encodable RNA aptamer that starts to fluoresce upon binding of

an organic molecule. Therefore, it is a biological fluorophore tag for RNAs. However, the spinach has never been used in a FRET assembly before. Roszyk et al. (2017) utilized the spinach quenching as a fully genetically encodable system and investigated the PP7 complex I quantitatively by quenching the spinach aptamer. To do this, they designed three cysteine mutants, A22C, A62C, and N93C, for chemical labeling via thioether formation with the red acceptor fluorophore Tide Fluor 3, as well as fusion proteins of PP7ΔFG with the red fluorescent protein mCherry. This study showed for the first time the usage of the spinach aptamer in RNA-protein interaction measurements, even under lysate conditions.

REPRESSOR COMPLEX I: PRR1

The crystal structure of the PRR1 complex I was resolved by Persson et al. (2013). The PRR1 operator is presented in Figure 16.9. In contrast to the PP7 complex, the overall conformation of the RNA was found to be similar to that of the MS2 phage, and the residues that were involved in the PRR1 RNA binding were the same as in the MS2 coat. Nevertheless, the arrangement of the nucleotide bases in the loop of the

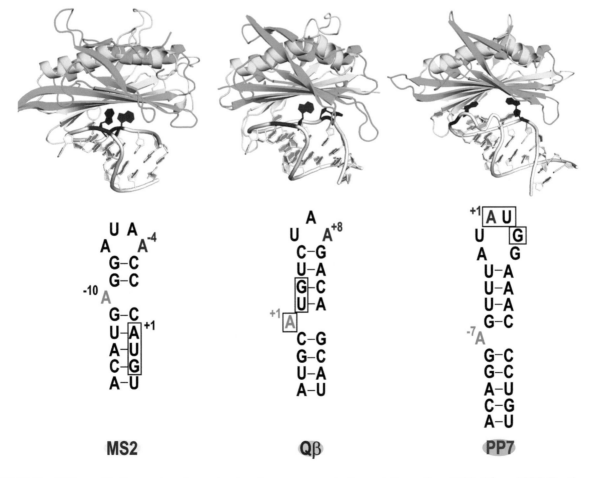

FIGURE 16.11 Different binding modes of operator stem-loop sequences to the coat dimers from MS2, Qβ, and PP7. For the operator stem-loop sequences, adenosine, which is required for the binding of loop nucleotides, is shown in red, while bulged adenosine is shown in blue. The replicase gene initiation codon is boxed. (Reprinted with permission of S. Karger AG, Basel from Pumpens P et al. *Intervirology.* 2016;59:74–110.)

stem-loop was different, leading to a difference in the stacking at the conserved Tyr86, which was equivalent to the Tyr85 residue in the MS2 coat.

REPRESSOR COMPLEX II

The repressor complex II is intended to switch the phage RNA from translation to replication. The model of replicase-controlled initiation of both coat protein and replicase synthesis was suggested by Daniel Kolakofsky and Charles Weissmann (1971a,b). Figure 16.12 presents the original Kolakofsky-Weissmann model that was proposed for the Qβ replicase because of the availability of the highly purified but still functional Qβ replicase for the *in vitro* evaluation, as described in detail in Chapter 13. The replicase in the repressor complex II does bind to the coat protein ribosome binding site, as shown *in vitro* for Qβ replicase and Qβ RNA (Weber et al. 1972; Meyer et al. 1975; Vollenweider et al. 1976).

As described in Chapter 13, the Qβ replicase subunit I, or ribosomal protein S1, alone demonstrated ability to form complexes similar to the Qβ RNA-replicase complex and to inhibit ribosome binding *in vitro* at the RNA phage coat gene ribosome binding site. This phenomenon was discovered

originally by Groner et al. (1972a,b). However, the protein S1 worked equally well in inhibiting ribosome binding to the MS2 or Qβ RNA, while the Qβ replicase showed at least 10 times more efficient inhibition of its homologous RNA. Moreover, more than one molecule of the S1 was required per phage RNA molecule, whereas only one molecule of the Qβ replicase was sufficient for complete inhibition. Finally, it was generally accepted that the S1 protein realized the specific binding to Qβ plus strand RNA by recognition of the specific RNA sequences whether it was functioning as a ribosomal protein or as the Qβ replicase subunit I.

As described in Chapter 13, the two internal binding sites of the Qβ replicase were determined within the Qβ RNA after ribonuclease T1 digestion of the Qβ RNA-replicase complexes (Weber et al. 1972, 1974; Meyer et al. 1975, 1981; Weber 1975). The genomic map of these sites is presented in Figure 16.13.

The S-site covers the initiation site of the coat gene at positions 1247–1346, while the M-site is located at about 2545–2867 nucleotides from the 5′ end and is essential for template activity. The electron microscopy of the Qβ RNA-replicase complexes after chemical crosslinking revealed the frequent formation of looped structures, in which the RNA fragment

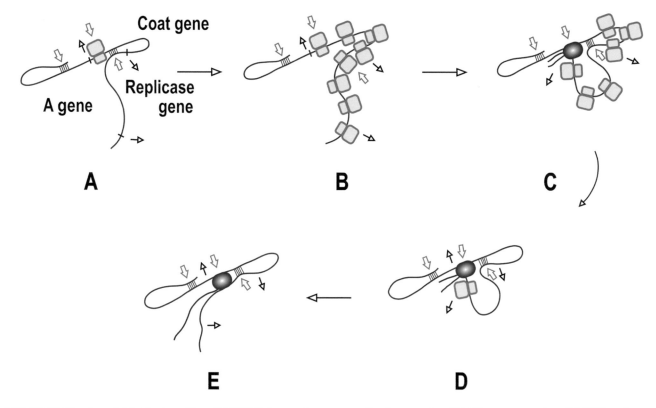

FIGURE 16.12 A model for the transition of Qβ RNA from polysome to replicating complex. (A) Ribosomes attach to Qβ RNA at the coat initiation site. The initiation site of the replicase gene is unavailable because of the secondary structure of the RNA. (B) Translation of the coat gene ensues, and the initiation site of the replicase gene is exposed. The replicase gene is translated. (C) When Qβ replicase becomes available, it attaches to the initiation site of the coat protein and blocks attachment of ribosomes in this position. The RNA can again refold preventing initiation at the replicase gene. (D) The RNA is cleared of ribosomes. (E) Qβ replicase can now attach to the 3′ terminus and initiate synthesis of the minus strand. The model supposes that the A cistron initiation site of complete Qβ RNA is always unavailable to ribosomes because of the secondary structure of the mature RNA; it is believed to be translatable only on nascent RNA strands. (Redrawn in color by permission from Springer Nature. *Nat New Biol.*, Kolakofsky and Weissmann, copyright 1971a,b and the legend from Biochim Biophys Acta, 246, Kolakofsky D and Weissmann C, Qβ replicase as repressor of Qβ RNA-directed protein synthesis, 596–599, Copyright 1971, with permission from Elsevier.)

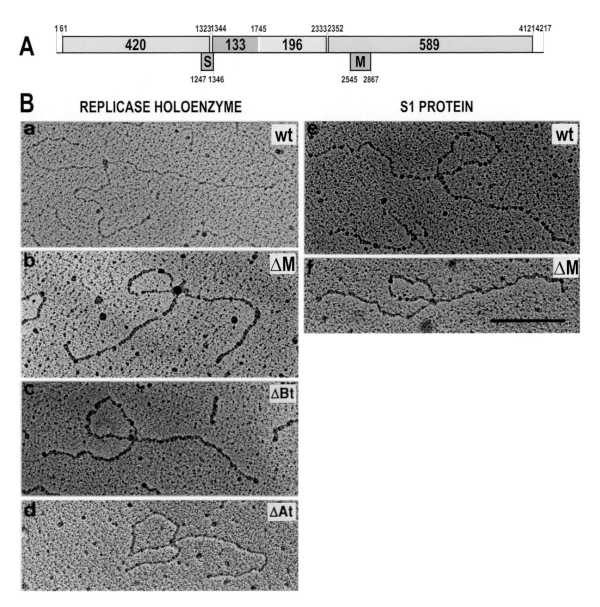

FIGURE 16.13 Complexes of Qβ RNA with Qβ replicase. (A) Map of the Qβ genome with the positions of S-site and M-site sequences. (B) Electron micrographs of looped complexes of wild-type and deletion mutant Qβ RNAs with replicase holoenzyme or isolated S1 protein: (a) wild-type RNA with replicase; (b) ΔM RNA with replicase; (c) ΔBt RNA with replicase; (d) ΔAt RNA with replicase; (e) wild-type RNA with protein S1; (f) ΔM RNA with protein S1. The ΔBt and ΔAt are truncated RNA variants prepared by run-off transcription from the wild-type genomic plasmid cleaved by two different restriction sites located several hundred nucleotides from the 3'-end. The scale bar represents 700 nucleotides. (Reprinted from *J Mol Biol.*, 267, Miranda G et al., Recognition of bacteriophage Qβ plus strand RNA as a template by Qβ replicase: role of RNA interactions mediated by ribosomal proteins S1 and host factor, 1089–1103, Copyright 1997, with permission from Elsevier.)

was linked to the enzyme at the S- and the M-sites simultaneously (Vollenweider and Koller 1975, 1976). Therefore, by the tertiary structure of the phage RNA, the start of the coat gene must be situated close to the region immediately following the start of the replicase gene. The protein S1 was regarded as the component of the Qβ replicase, which was responsible for this binding and further stabilization of the binding complex, or the repressor complex II.

This point was confirmed by the observations that the subunit I-deficient Qβ enzyme retained capability to the phosphodiester bond formation and functioned normally with all templates except the Qβ plus strand RNA and, even with the latter template, chain elongation and termination were

unaltered (Kamen et al. 1972; Kamen 1975). At the same time, the replicase subfragment comprising only subunits I and II retained the binding specificity of the complete enzyme (Kamen 1970; Weber et al. 1972). Therefore, the phage specific Qβ replicase subunit II was involved in the specific template binding within the repressor complex II (Kondo et al. 1970; Landers et al. 1974; Kamen 1975).

Figure 16.13 presents the looped structure that was observed in electron microscopy for the repressor II complex by Miranda et al. (1997). To map the complex, the internal, 50-terminal, and 30-terminal deletions were assayed for the template activity with the Qβ replicase *in vitro*. The formation of looped complex structures, previously reported and explained as simultaneous interactions with replicase at the S and the M-site (Vollenweider and Koller 1976), was abolished by deleting the S-site but, surprisingly, not by deleting the M-site. Remarkably, the same types of complexes observed with replicase were also formed with the purified protein S1, as seen in Figure 16.13. Therefore, the S1-mediated binding of the Qβ replicase complex to Qβ RNA was confirmed. In fact, the protein S1 possessed two polynucleotide binding sites: site I bound either single-stranded DNA or RNA, while site II was highly specific for RNA over DNA and showed a marked preference for cytidine polynucleotides over the corresponding adenine-containing species (Draper et al. 1977).

As mentioned in Chapter 13, the host factor I, or Hfq factor, was also suspected to form similar complexes by binding to the S- and M-sites or adjacent sites and in addition to the 3′-end, resulting in double-looped structures. The pattern of looped complexes observed by Miranda et al. (1997) with the deletion mutant RNAs suggested that the binding of host factor might not involve the S- and M-sites themselves but adjacent downstream sites.

Remarkably, the ability of the protein S1 to interact simultaneously with the two different regions of the phage RNA, namely, the ribosome-binding site of the coat protein gene and the internal region of the replicase gene, was confirmed also in the case of the phage MS2 (Boni et al. 1986), where direct binding experiments remained impossible because of the problems with the isolation of the group I phage replicases, as described in the Non-allolevivirus replicases section of Chapter 13. Later, using the original crosslinking technique, the S1-binding sites were localized also on the fr RNA, in parallel with the Qβ and MS2 RNAs (Boni and Isaeva 1988; Boni et al. 1991). As mentioned above in the Shine-Dalgarno section of the present chapter, the targets for S1 on the MS2, fr, and Qβ RNAs were localized upstream from the coat protein gene and contained oligoU-sequences. The S1-protein binding site that was found in these studies overlapped the known S-site for the Qβ replicase, suggesting the analogous mechanism of the repressor complex II formation in the case of other than Qβ RNA phages, at least in the group I RNA phages.

Therefore, the M-site region as originally defined from replicase-bound RNA fragments comprised a genome segment of over 300 nucleotides that, at least in its linear extension, appeared to be far too large for a protein-recognition domain. Further mutational analysis was performed to map the M-site on a finer scale. Thus, the fine mutational analysis

of the M-site showed that the essential elements within the 2545–2867 sequence consisted of the two successive stem-loop structures followed by a bulge loop of unpaired purines, located at nucleotides 2696–2754 on the tip of a long, imperfectly base-paired stalk (Schuppli et al. 1998). The mutational changes affecting the sequences of paired or unpaired nucleotides in this segment reduced the template efficiency only mildly. The only severe effects were observed when one of the helical stems or the unpaired bulge was completely deleted or substantially shortened. It was concluded that the three-dimensional backbone arrangement of these three elements constituted the feature recognized by replicase. The role of the long stalk remained undetermined, because mutations that either stabilized or disrupted its base pairing barely affected template activity, and even deletion of a major portion of one of its strands did not cause complete inactivation. Remarkably, the lack of an active M-site on the Qβ RNA template had the same quantitative and qualitative effects on template recognition as the absence of the S1 protein from replicase in the presence of the wild-type RNA. Therefore, the M-site interaction explained most of the role of the S1 protein in the recognition of Qβ RNA by the Qβ enzyme. The long-range interaction in the Qβ RNA that bridged the thousand nucleotides between the M-site and the 3′ end was mapped by van Duin's team (Klovins et al. 1998; Klovins and van Duin 1999), as described in Chapter 10.

Further structural analysis of the Qβ genome showed that the Δ159 deletion located within the readthrough domain 1549–2419, a 850-nucleotide hairpin, led to the uncoupling of the co-translational regulation of the viral coat and replicase genes *in vivo*, but the coupling was restored by the Δ549 deletion within the same domain by naturally evolved pseudorevertant (Jacobson et al. 1998). The structural analysis by electron microscopy of the genomic RNAs showed that several long-range helices at the base of the readthrough domain, that suppressed translational initiation of the viral replicase gene in the wild-type genome, was destabilized in the Δ159 RNA. In addition, the structure of local hairpins within the readthrough region was more variable in the Δ159 RNA than in the wild-type RNA. The stable RNA secondary structure was restored in the readthrough domain of the Δ549 RNA. These data suggested that the structure throughout the Qβ readthrough domain affected the regulation of the replicase expression by altering the likelihood that long-range interactions at the base of the domain would form.

17 Replication

The most incomprehensible thing about the world is that it is comprehensible.

Albert Einstein

To put everything in balance is good, to put everything in harmony is better.

Victor Hugo

THE ESSENTIALS

As it follows from Chapters 13 and 16, the specific RNA synthesis process is tightly coupled with the phage RNA translation, and the whole replication cycle is fully subjected to the translational control. First, both replication and translation share common interchangeable components, namely, the ribosomal protein S1, the elongation factors EF-Tu and EF-Ts, and the Hfq chaperone. Second, RNA replication occurs simultaneously with RNA translation on the nascent RNA plus strand chains. Third, the parameters of the whole replication cycle are determined by special regulatory mechanisms, namely, the polarity phenomenon and the controlling action of the two repressor complexes I and II that were described in Chapter 16.

The first review that addressed the replication intermediates in the course of RNA phage infection was presented by Norton D. Zinder (1965). The first exhaustive review devoted exclusively to the RNA replication process was written for the classical *RNA Phages* book by one of the pioneers in the field, Hugh D. Robertson (1975). He was the first to emphasize that the serious problems in the RNA replication studies *in vivo* arose from the lack of a neat separation of the transcription, replication, and translation processes, in contrast to the situation with the organisms possessing DNA genomes. The first computer simulation of the RNA phage replication was generated by Knijnenburg and Kreischer (1975), who based the regulation mechanism generally on the polarity effect and capability to reach necessary replicase concentrations.

Experimentally, the structures undergoing viral RNA replication were found *in vivo* in the polysome region of infected cells and as stated above, used the interchangeable host macromolecules at the same time and at the same place. Thus, the parental phage RNA, which could be as little as one molecule per cell, was found in the polysomes in the first minutes of infection (Godson and Sinsheimer 1967; Hotham-Iglewski and Franklin 1967), acting as messenger for the synthesis of the phage-specific subunit of the viral RNA replicase and starting the synthesis of the RNA minus strand. This total interdependency of the translation and replication processes, which functioned as a global network in infected cells, was probably the reason why the fully functional replicases of the

RNA phages other than the group III and IV representatives were not isolated, as described in Chapter 13.

It followed from the translation studies that the replicative intermediates of RNA replication, so-called nascent strains, were used as translation messengers, at least, for the synthesis of the maturation protein. The phage RNAs were deliberated from ribosomes by the repression complex II. The replicase synthesis with the following RNA synthesis was blocked late in infection by the repressor complex I. The parental RNA phage molecules were rapidly converted by the phage-specific replicase enzyme into a "double-stranded" replicative form, while the complementary RNA minus strand, which did not encode proteins, served as an obligatory intermediate of the RNA replication. The double-stranded RNA was not detectable in the uninfected *E. coli* cells. Further studies were needed to appreciate that the "double-stranded" replicative intermediates were not in fact double-stranded, and to reveal the actual structure of the replicative intermediate.

As stated above in the Qβ replicase section of Chapter 13, the problem of the direction of the RNA phage growth was solved in favor of the 5′-triphosphate to the 3′-hydroxyl terminus for both RNA plus and minus strands (August et al. 1968; Feix et al. 1968; Mills et al. 1968; Pace et al. 1968; Banerjee et al. 1969a). It helped to reach once and for all the conclusion that the RNA plus strands served as templates for the RNA minus strands, and that these in turn served as templates for the progeny plus strand production always in the 5′ → 3′ direction (Fiers et al. 1968; Robertson and Zinder 1968, 1969; Spiegelman et al. 1968; Vandenberghe et al. 1969).

The 5′-end of the phage RNA was found to fulfill functions connected with the RNA replication processes. Both the group I (MS2, R17, f2) and III (Qβ) phages started at their 5′-end with a series of three or more G residues, while a complementary sequence of three or more C residues was present at the 3′ end, disregarding the terminal A residue. The addition of an A at the 3′-end of the product chain occurred both *in vivo* and *in vitro* by the chain termination (Rensing and August 1969; Weber and Weissmann 1970), as described in Chapter 13. A similar series of C residues was also found at that time at the end of several plant viral RNAs, such as tobacco mosaic virus, brome grass mosaic virus, satellite virus of tobacco necrosis virus, and tobacco rattle virus (for references, see Fiers 1975). As Walter Fiers (1975) noted, these similarities were undoubtedly not due to a coincidence, but most likely were closely connected to the general mechanism of viral RNA replication.

Weissmann et al. (1964b) postulated the asymmetric character of the RNA phage replication when the synthesis of plus strands exceeded that of minus strands. This asymmetry was confirmed by a special study on the phage β (Nonoyama and Ikeda 1966). In total, between 5 and 10 plus strands

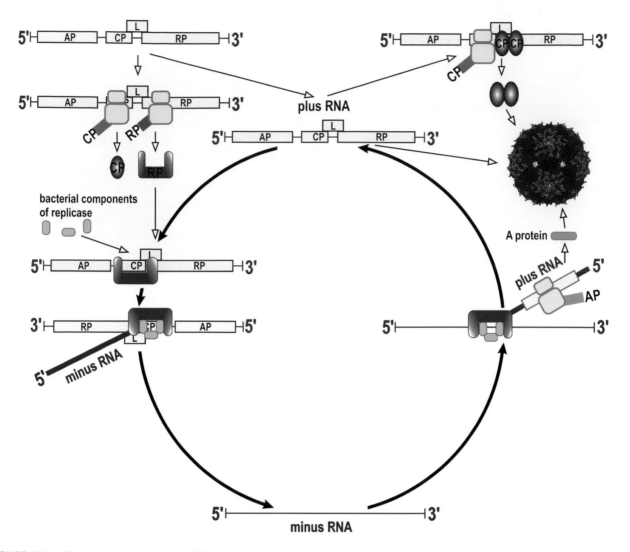

FIGURE 17.1 The general scheme of the RNA phage replication.

were made per minus strand by the phage RNA synthesis (Weissmann et al. 1968a). At that time, the first reliable RNA replication scheme was proposed (Weissmann et al. (1968c). This scheme is redrawn from the original paper and presented in Figure 1.10. It demonstrated for the first time a scenario showing how the replication intermediate could exist as open single-stranded structures but may be supposed to collapse, spontaneously or under the influence of external agents, to yield double-stranded structures.

The general flowchart of the RNA phage replication has been presented in Figure 1.11. Figure 17.1 shows the detailed cycle of the replication events, which takes into account all experimental information that was described previously in Chapters 13 and 16.

EARLY STUDIES

During early studies of the RNA phages, one of the most thoroughly investigated problems was the appearance and state of the phage-specific RNA in the infected cells. After the three groups reported the presence of the RNA-dependent replicase in the f2 (August et al. 1963) and MS2 (Weissman et al. 1963a,b;

Spiegelman and Doi 1963) infected cells, as described in the Activity section of Chapter 13, the search for the replication intermediates was started. The above-mentioned papers registered the appearance of the double-stranded RNA together with the detection of the specific replicase activity.

Weissman et al. (1964b) have characterized the MS2 RNA replication product on the basis of its sensitivity to pancreatic ribonuclease and found that half of the product was resistant to ribonuclease and had a density in CsCl less than that of a single-stranded RNA, as would be expected for the double-stranded RNA. The authors showed that the ribonuclease-resistant RNA was converted to nuclease sensitivity by heating and reannealing in the presence of excess viral RNA. This indicated that the part of the product was homologous to the viral plus strand. They were the first to suggest that the replication occurred in a semi-conservative fashion, with the double-stranded template giving rise primarily to the viral strand and not its complement. The sharp melting temperature also supported the double-stranded nature of the product.

Thus, Weissman et al. (1964b) formulated the concept of the double-stranded replicative form (RF) by analogy to the φX174 RF, in the RNA phage-infected cell. The RF contained material that annealed specifically with the phage RNA and had the density as that presumed for the double-stranded RNA, although it was too small to have been a whole RNA molecule in a duplex structure. The x-ray diffraction (Langridge et al. 1964) showed the RF to be analogous to reovirus RNA, known to be double-stranded, and to double-stranded DNA in the A configuration. The possible RNA replication intermediates, including the double-stranded RF structures, were studied extensively by other teams for the phages R17 (Ellis and Paranchych 1963; Erikson et al. 1964, 1965, 1966; Fenwick et al. 1964; Erikson 1966; Erikson and Gordon 1966; Erikson and Erikson 1967; Franklin 1967a,b; Iglewski and Franklin 1967a,b,c; Fenwick 1969, 1971a,b; Knolle 1972b), β (Nonoyama et al. 1963; Nonoyama and Ikeda 1964b, 1966), MS2 (Kelly and Sinsheimer 1964, 1967a,b,c; Ochoa et al. 1964; Kelly et al. 1965; Billeter et al. 1966a,b,c; Cramer and Sinsheimer 1971), fr (Kaerner and Hoffmann-Berling 1964a,b; Knolle 1967d), M12 (Ammann et al. 1964a,b; Francke and Hofschneider 1966a,b, 1969; Hofschneider et al. 1967a,b; Francke et al. 1970; Hehlmann and Hofschneider 1970), ZIK/1 (Bishop 1965, 1966a,b), and f2 (Lodish and Zinder 1966a,b,c; Engelhardt et al. 1968; Robertson and Zinder 1968, 1969). The RNA phage replication was reviewed exhaustively at that time by Erikson and Franklin (1966), Shapiro and August (1966), Spiegelman and Haruna (1966a,b), Weissmann and Ochoa (1967), August and Shapiro (1968), Erikson (1968), Lodish (1968b), Spiegelman (1968–1969), Spiegelman et al. (1968), Valentine et al. (1969b), Stavis and August (1970), and Bishop and Levintow (1971). Remarkably, as mentioned briefly above, the double-stranded RNA was not detectable in the uninfected cells or in the infected cells within the first 20 minutes after infection. The amount of the RF increased during the next 20–40 min, reaching a level of about 1% of the total cellular RNA or about 1000 molecules per cell (Weissmann et al. 1964a). It is noteworthy that the procedure used by the isolation of the MS2 RF, including digestion with pancreatic ribonuclease, was used successfully at that time by the isolation of the replicative form of tobacco mosaic virus (Weissmann et al. 1965).

The first isolation of the double-stranded RF involved treatment with ribonuclease to remove other RNA and yielded material with a sedimentation constant of 7–9S, much smaller than the ca 16S expected of a double-stranded nucleic acid of molecular mass of about 2×10^6 (Nonoyama et al. 1963; Kaerner and Hoffmann-Berling 1964a,b; Langridge et al. 1964). Peter H. Hofschneider's team was the first to avoid nuclease treatment and isolate the double-stranded RNA, which was 12,000 Å long when regarded in the electron microscope and therefore corresponded to the complete duplex (Ammann et al. 1964a,b). The preparations containing these structures became infective for protoplasts after heat denaturation. The double-stranded RNA failed to inhibit the formation of translation initiation complexes on R17 RNA, to perform overall synthesis of R17 proteins, or to support the ability of bacterial initiation factor IF-3 preventing the association of 30S and 50S ribosomal subunits into single ribosomes (Jay et al. 1974). However, the IF-3 was bound to

double-stranded RNA with lower apparent affinity than to either R17 RNA or 30S ribosomal subunits. In fact, these findings explained the resistance of bacterial protein synthesis to the double-stranded RNA.

It is noteworthy that that the first studies of those times on the replicative intermediates of eukaryotic viruses, such as Semliki Forest virus from the *Togaviridae* family (Friedman 1968), referred to the replicative intermediates of the RNA phages and looked for the possible analogies with them.

REPLICATIVE INTERMEDIATE

To the end of the 1960s, two classes of the replicative complexes containing both the initiating templates and early RNA product were discussed. The first type of such intermediates, called Hofschneider's complexes (HS), corresponded to the double-stranded structures found *in vivo* by Ammann et al. (1964a,b) and Francke and Hofschneider (1966a,b). The second type of intermediates, called Franklin's complexes (FS), was described as a double-stranded RNA template with nascent, partially completed, single-stranded viral RNA chains (Fenwick et al. 1964; Erikson and Franklin 1965, 1966; Franklin 1966, 1967a,b).

Pace et al. (1968) suggested the temporal order of the events when the HS appeared first in the infected cells and then were followed 1 minute later by the FS. Altogether, these structures were referred as replicative intermediates (RI), the name they were originally assigned by Erikson et al. (1964). Van Styvendaele and Fiers (1967) avoided the ribonuclease treatment by the replicative intermediate isolation and found two distinct, partially ribonuclease-resistant components in the soluble fraction extracted from the MS2-infected *E. coli* cells, which sedimented at 17–18S and 13–14S in a sucrose gradient. The 17–18S component was probably the first phage-specific structure after infection and was not detected in cells infected for 30 minutes. Similar structures were found also after R17 infection (Iglewski and Franklin 1967a,b). These studies, which formulated the replicative form RF as a complete double-stranded structure consisting of the RNA plus and minus strands along its entire length and the replicative intermediate RI as the double-stranded RNA containing nascent single strands, led finally to the purification of the full-length R17 RNA minus strand (Iglewski and Franklin 1967c).

Strong evidence against the presence of double-stranded RNA in replicating structures arose from a clearer understanding of nucleases in *E. coli*. Thus, the double-stranded RNA-specific ribonuclease III was discovered (Libonati et al. 1967; Robertson et al. 1967, 1968a), which would make impossible the RNA phage growth without being destroyed, if the RNA plus and minus strands would form the double-stranded structures.

Thus, for the growth in the presence of the ribonuclease III, the RNA phages were obliged to use sequences and structures which avoided signals to activate such ribonuclease activities. It was believed therefore that *in vivo* both the plus and minus strand templates were themselves single strands with nascent molecules as single-stranded tails (Zinder 1980). The double-stranded material was found following the usual deproteinization procedures. Apparently, the strands hybridized to each other during extraction (Erikson and Franklin 1966; Robertson and Zinder 1969) and were either artefacts of the extraction generated by the collapse of the native replicative intermediates or structures no longer involved in the RNA replication (Bishop and Levintow 1971). The electron microscopic study of the non-denatured R17 replicative intermediates (Thach and Thach 1973) and the biochemical characterization of the nascent M12 and Qβ RNA in extracts from both RNase I⁻ and RNase III⁻ E. coli strain AB105 (Keil and Hofschneider 1973; Kindler et al. 1973) were consistent with this conclusion.

It is noteworthy that the ribonuclease III and the double-stranded f2 RNA were used by the elaboration of sensitive methods for the detection and characterization of double helical RNAs (Robertson and Hunter 1975).

As Norton D. Zinder (1980) brought out clearly, the fact that the replicative form was never double stranded, solved another problem, particularly for a nucleic acid whose major product was plus strands:

Consider a viral strand acting as a template for the minus strand. Enzymes and nucleotides flow from its 3′ end to its 5′ end, giving a double-stranded molecule with the enzyme hanging on by its teeth at a now blunt-ended template. The enzyme might turn around and by a displacement synthesis make a new plus strand. It might once again repeat the process and synthesize a new minus strand. Shuttling back and forth, plus and minus strands would be made in equal numbers. However, in order to synthesize the requisite excess of plus strands, the enzyme would have to fall off one end of the double strand, return to the other end, and repeat the displacement synthesis. The molecular complications of such models caused most investigators to believe that there must be two different enzymes, one to make plus strands and the other to make minus strands. It is far simpler to synthesize single strands. (Reprinted with permission of S. Karger AG, Basel from Zinder N. Intervirology 1980;13:257–270.)

Concerning the number of the nascent plus RNA chains per minus strand template within the RI, it was estimated on the basis of the relative pppGp content to the average 2.5 nascent strands per molecule (Vandenberghe et al. 1969). This calculation was based on the reasonable assumption that the speed of the RNA synthesis was uniform along the template (Fiers 1975). Other authors came to a lower estimate, approximately one per molecule (Granboulan and Franklin 1968; Robertson and Zinder 1969). As Fiers (1975) commented, the fragile nascent strands could be easily detached from the template during various manipulations.

It is highly important that the replicative intermediates contained templates for the phage protein synthesis, where properties of the nascent plus strands differed from those of intact plus strands, as it appeared in vitro (Engelhardt et al. 1968; Robertson and Lodish 1970). Moreover, the phage coat

protein was capable to interact with isolated replicating structures (Robertson et al. 1968b), probably influencing the specific pattern of the phage RNA synthesis.

According to the in vitro data, the replicative intermediate consisting of the template, enzyme, and products behaved as a large associate of 40–50S (Feix et al. 1967b; Hori 1970). The RNA was fully sensitive to the single-strand-specific ribonuclease, pancreatic ribonuclease A, and resistant to the double-strand-specific ribonuclease III (Robertson, Webster and Zinder 1968; Libonati et al. 1967). The replicative intermediate was chromatographed on cellulose ethanol columns (Franklin 1966) as fully single stranded (Weissmann et al. 1968b). However, the treatment of the native replicative intermediate with protein denaturants converted the RNA into a form which was largely double-stranded by the same criteria (Weissmann et al. 1968a).

It is notable that similar treatment of mixtures of purified Qβ plus and minus strands did not induce the double-strand formation. Weissmann et al. (1968b) proposed that the replicase itself held the template and product RNA in a configuration which was hydrogen-bonded over a short distance near the chain growth point but was otherwise single-stranded. If it was the case, any treatment which denatured the replicase allowed collapse of the intermediate into the double-stranded form.

When the MS2-infected E. coli cells were opened under very mild conditions, the native replicative intermediate RI was estimated by size to the 35S sedimentation constant and by number to the 300 of such 35S ensembles per cell at 15 min after infection, while the number of complete virions was only 24 (van de Voorde et al. 1968). At 30 min, the number of the 35S ensembles has increased to 7000 per cell, and the complete viruses to 5000–20,000. Similar estimations of the size of the replicative intermediate followed also from another MS2 study (Cramer and Sinsheimer 1971).

The replicating phage RNA was found in polysomes. After 4−5 minutes of the MS2 infection, approximately 30% of the total parental ³²P-RNA was associated with the polysomes, 38% was associated with the 30S ribosome subunit, 15% was free RNA, and 16% was residual whole phage (Godson and Sinsheimer 1967). According to this pioneering study, at minutes four to ten of infection, when the portion of the ribonuclease-resistant parental RNA increased, a new component containing the parental ³²P-RNA appeared, sedimenting just in front of the 50S ribosome subunit. The ribonuclease-resistant 12S RNA, or core RNA by Fenwick et al. (1964), was present in both this 50S region and in the polysomes but none was present in the 30S or free RNA regions. The ribonuclease-resistant 12S RNA probably appeared first in the 50S region and later in the polysome region. The parental ³²P-RNA of cells infected for 8 minutes could be dissociated, in low Mg²⁺ concentration, from the polysomes as a fast-sedimenting component of 40S, or the replicative intermediate described above. Remarkably, the addition of 50 μg/mL of chloramphenicol to E. coli cells caused the disappearance of most of the host-cell polysomes (Godson and Sinsheimer 1967). Further, 70% of the progeny single-stranded MS2 RNA was found in polyribosomes that contained two to five ribosomes (Godson 1968). The double-stranded phage RNA was mostly associated with a structure that sedimented with the largest polyribosomes and contained little phage RNA that was infective to protoplasts. Remarkably, no pool of free RNA molecules appeared during infection, but the viral RNA strands were always associated with other molecules. After 30 minutes of infection, there were enough viral RNA strands to change the cellular polyribosomes to predominantly small phage polyribosomes. At lysis, there were still several thousand phage RNA molecules that were not encapsulated in the whole phage (Godson 1968).

In parallel, the R17 double-stranded RNA was also found in the polysome fraction, when particularly prominent concentrations of the double-stranded RNA appeared in polysomes of size 11–14 and in the pentamer region (Hotham-Iglewski and Franklin 1967). As in the previous study, some double-stranded RNA sedimented close to the 50S subunit and belonged possibly to the replicative intermediate. These findings were developed further and led to the strong belief that the phage coat could play its regulating role in the phage replication cycle at the level of transcription by the preferable binding to the replicative intermediate (Hotham-Iglewski et al. 1968), the idea that was made public a little earlier (Robertson et al. 1968b). According to this idea, the binding of the coat to the replicative intermediate created a differential rate of RNA synthesis between the end of the nascent single strand containing the coat gene and the remaining portions of this strand containing the other genes (by the then assumption that the coat gene was 5'-terminally located in the phage genome). This could lead to a temporary interruption in RNA synthesis which would take time to overcome. In other words, Robertson et al. (1968b) suggested that the coat might act as a repressor in the true sense, that is, at the level of transcription, as opposed to a modulating factor acting at the level of translation as the repressor complex I, as described in the corresponding sections of the previous chapter. Knolle and Hohn (1970) found that the R17 coat reacted with preparations of the R17 replicative intermediates RI in the form of the multi-stranded RNAs and yielded complexes containing phage-like particles at high protein inputs. At the same time, the RNA complexes containing predominantly double-stranded RNA, or the replicative form RF, were considerably less efficient in promoting particle formation than those rich in single strands. Similarly, the binding of coat protein at low-input multiplicities was less efficient with RF than with the RI preparations.

Further studies of the phage-specific polysomes showed that the synthesis of replicase, maturation, and coat proteins in *E. coli* infected with the phage R17 occurred mainly on polysomes containing four or more ribosomes (Truden and Franklin 1969, 1972). The 30S ribosomal subunits through trimer-size polysomes, which were associated with all of the R17-specific proteins and were predominant in the infected cell, synthesized only coat protein.

These structures could accumulate as products derived from larger polysomes as a result of failure in the release of nascent polypeptides after termination of chain growth. The appreciable amounts of viral coat protein remained attached to ribosomes and polysomes during R17 replication, supporting the repressor role of the coat. Again, the time course of synthesis of the phage-specific proteins, obtained from the polysomes of infected cells, demonstrated the regulated R17 messenger RNA translation consistent with the idea that the coat protein was preferentially synthesized whereas the synthesis of non-coat proteins was suppressed (Truden and Franklin 1969, 1972).

In order to see if the coat protein could in fact bring about such a halt to phage RNA synthesis in a specific manner, the effect of the isolated Qβ coat was tested when the Qβ plus strand RNA was copied by the Qβ replicase (Robertson 1975). Thus, the addition of the Qβ coat to the Qβ replicase reaction caused inhibition of the RNA synthesis. In particular, it was evident that the Qβ coat addition at the beginning of the reaction caused shutoff of RNA synthesis at about 12 minutes. If the reaction was allowed to go on for 12 minutes and then the coat protein was added along with radioactive RNA precursors, little or no incorporation was observed in the presence of the Qβ coat. It would appear therefore that by 12 minutes an RNA species was present whose ability to act as a template for continued synthesis was abolished by the Qβ coat protein. The inhibition of the RNA synthesis appeared to be specific, since f2 coat protein did not give such an effect in the Qβ replicase reaction. However, the molar ratios of the coat to RNA required for the inhibition effect were somewhat higher than those required to obtain the specific effect on the *in vitro* translation: about 100:1 versus 6:1 (Robertson et al. 1968b; Sugiyama and Nakada 1968). Since several reports have suggested that phage-like particles began to form at such ratios of coat to RNA, it was important to rule out this effect as an explanation for the specific inhibition of the Qβ RNA synthesis observed here. No peak of material sedimenting at 80S as the expected position of phage-like particles was observed, and the majority of the radioactive RNA product sedimented in a fairly broad region between about 20 and 40S (Robertson 1975).

According to data of Haywood et al. (1969) and Haywood (1971), the newly synthesized MS2 RNA appeared first in fractions that were bound probably to bacterial membranes. In parallel, Van de Voorde and Fiers (1971) acknowledged the presence of the MS2 replicative intermediates in the membrane fraction of bacteria and concluded that the newly synthesized phage RNA was not specifically associated with the bacterial membrane in the same way as bacterial DNA, but they did not exclude a special function of the membrane-associated MS2 RNA fraction.

Later, Haywood (1973) found the replicative intermediate—in other words, all the RNA species containing MS2 complementary RNA—in the membrane eluate and the membrane itself. Thus, two classes of the membrane binding were detected. After *E. coli* membranes were isolated in the presence of 6 mM Mg^{2+}, they were washed with buffer containing no Mg^{2+} to yield a fraction containing material bound only in the presence of divalent cations, or *membrane eluate*, and that bound in the absence of divalent cations, or *membrane*.

The amount of ^{14}C-uracil in the replicative RNA found in the *membrane eluate* increased with time of labeling, whereas that in the replicative RNA in the *membrane* reached a plateau in 1–2 minutes. This finding was consistent with the precursor–product relationship. Most of the label entering single-stranded viral RNA came from the replicative RNA in the *membrane* eluate. This result suggested that the replicase components, or factors required for the complementary-strand synthesis, were bound to membrane even in the absence of divalent cations, and that the replicase was no longer bound to these factors when it was making the bulk of the progeny single-stranded RNAs.

As Robertson (1975) noted further in his review, once assays had been established for the initiation and completion of all three f2 RNA phage proteins (Lodish and Robertson 1969) and the gene order of the three phage genes was determined (Jeppesen et al. 1970b), it became possible to determine the amounts of each protein initiated and completed by nascent strands and intact f2 RNA single strands. Thus, Robertson and Lodish (1970) carried out such investigations and found that the nascent-phage RNAs allowed efficient initiation and completion of coat synthesis, whereas the replicase synthesis was initiated but never completed. Moreover, the initiation of the maturation protein synthesis occurred at rates up to 10 times that with intact single strands.

Figure 17.2 presents a model that was proposed for the phage RNA synthesis *in vivo* by Robertson (1975). The model considered the following major requirements. First, it was clear that some form of functional circularity in the synthesis of viral RNA must exist, while the synthesis begins at the 3′ end of the template strand in both steps of the reaction plus and minus strand synthesis, as described above. In the second step, the production of plus strands, the minus strand must repeatedly reinitiate synthesis at its 3′ end, although it completes synthesis at its opposite 5′ end. Each minus strand must serve many times as template. Thus, the enzyme must be able to return to the 3′ end of the template efficiently. One way for this to happen would be for the 3′ end of the template to remain near the enzyme active site throughout RNA synthesis. The resulting circular structure could be held together by the enzyme or one of its subunits. The studies on the R17 or Qβ replicase mutants did in fact show that their genetic complementation *in vivo* was highly asymmetric, suggesting that the enzyme molecule remained preferentially bound to a single replicating structure (Tooze and Weber 1967; Horiuchi and Matsuhashi 1970).

Another property of the reaction which must be accounted for was the inherent strand selection of the enzyme, since both *in vitro* and *in vivo* the predominant product was plus strand RNA. This point was touched on in the Qβ replicase section of Chapter 13, when the roles of the Qβ enzyme components were analyzed. Finally, the RNA replication model must also provide a reasonable explanation of why, when the viral component of the replicase was overproduced *in vivo*, the proportion of viral plus strands that was doubled up was greatly increased, while the rate of plus strand synthesis was

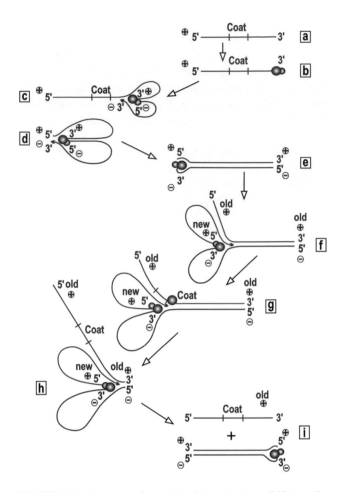

FIGURE 17.2 The butterfly model of bacteriophage RNA synthesis. The minus strand synthesis occurs in sections (a–e) and plus strand synthesis occurs in sections (e–i). The coat protein gene is placed at about the midpoint of the plus strand for reference. Its presence on one of the two plus strands in sections (e–i) is not necessarily meant to indicate which of the two hypothetical strands is acting as mRNA. (Redrawn with permission of Cold Spring Harbor Laboratory Press from Robertson HD. In: *RNA Phages*. Zinder ND (Ed). Cold Spring Harbor Laboratory, Cold Spring Harbor, NY, USA, 1975, pp. 113–145.)

no higher than in the wild-type infected cells (Lodish and Zinder 1966a; Weissmann et al. 1968a).

According to Robertson's (1975) comments to the postulated model presented in Figure 17.2, the first step (a,b) was the binding of an enzyme complex onto the 3′ end of the viral plus strand. The drawing shows two types of subunits, although as we know, four polypeptides plus several additional factors were really involved. The postulated functions of the enzyme considered in this model were (i) a single polymerizing activity, probably residing in the phage-coded subunit of the replicase complex, as discussed in Chapter 13, (ii) an activity binding the 3′ end of the template strand (whether it be plus or minus), and (iii) an activity binding at or near the 5′ end of the growing nascent strand (again, regardless of whether it is plus or minus). As the RNA synthesis begun, two subunits of the several involved remained attached to the 3′ end of the plus strand template and the 5′ end of the new minus strand, respectively (c). As RNA synthesis continued (d), two large loops were formed as the enzyme was attached to each end of the newly synthesized RNA, as well as to one end and the middle of the template strand. Upon phenol extraction or other deproteinizing treatment, the enzyme would be removed and the loops would collapse into linear double-stranded structures. The RNA synthesis was postulated to require the formation of at least a transient double-stranded region near the site of synthesis. When RNA synthesis reached the 5′ end of the plus strand template (e), it terminated. Now a switching reaction must occur, so that the 3′ end of the plus strand was replaced at the 3′ binding site of the appropriate subunit by the newly made 3′ end of the minus strand. This switchover was crucial for the synthesis of the plus strands. It was postulated therefore that the 3′ binding site had a substantially higher affinity for the 3′ end of the minus strand than for the 3′ end of the plus strand. This would also explain why factors must be added for successful initiation of the minus strand synthesis, whereas plus strand synthesis *in vitro* required no additional factors. The role of the functional subunits in the switchover reaction was twofold. First, once RNA synthesis had terminated, neither the 3′ end of the new minus strand nor the 5′ end of the old plus strand was attached to the enzyme complex, and the 3′ end of the old plus strand must be detached. During this process, the attachment of the enzyme to the 5′ end of the minus strand was the only link between the enzyme and its template. Second, although the two strands might renature over significant parts of their length, continued attachment of the 5′ end of the minus strand would prevent complete renaturation. Once the switch in 3′-end binding had been accomplished (e), the attachment of the 5′ end of the minus strand to an enzyme subunit ceased, and the enzyme was built into a complex with two full-sized strands. What happened next was one of two things: if the "free minus strand" school was correct, the plus strand in (e) would spontaneously float away, and plus strand synthesis would resemble a repeat of the (a–e) steps with the signs of the strands reversed. The steps (f–i) showed the synthesis of a new plus strand in the presence of the old one. The minus strand was now the template, and its 3′ end was bound to the appropriate enzyme subunit. This attachment was permanent and would survive through multiple rounds of the plus strand synthesis. As the plus strand synthesis commenced, one of the two 5′ ends (here it was depicted as the new one) was bound by a subunit of the enzyme, while the other one was a free tail. If there were no old plus strands involved, then this nascent plus strand would have its 5′ end bound by the enzyme.

Although the overall process *in vivo* was semiconservative, either new or old plus strands could be displaced in any single round of synthesis. An added complication of having the old plus strand present was that it encoded a potentially active replicase gene, which could be postulated to be kept covered by sufficient interaction with the minus strand template so that its ability to act as mRNA was low. This would also be expected because, as shown in (g), when this region near the end of the coat gene and the beginning of the replicase gene was exposed, it should bind one or a few monomers of the coat protein. Thus, the step (g) showed the late mRNA complex. The translation would occur upon the exposed coat gene of whichever of the two plus strands remained free. Then the block of RNA synthesis was overcome, the RNA synthesis was completed (h), and we were left with a free plus strand and a template structure, still bound by the 3′ end of the minus strand to the enzyme complex (i). No further switching would occur. The termination of the RNA synthesis was followed only by the release of the plus strand from the 5′ binding site. In general, this model had the advantage of being usable regardless of the number of plus strands present during the reaction and regardless of the exact time when a completed plus strand was released, be it before, during, or after the synthesis of the next one (Robertson 1975).

It is noteworthy that the elucidation of the replicative intermediates of the RNA phages was used at that time as an experimental prototype by the studies of the replicative intermediates of poliovirus (Bishop and Koch 1969).

REGULATION OF THE REPLICATION CYCLE

After the repressor complexes I and II were found to regulate translation of the replicase gene, it remained only to establish the possible link between the cessation of the replicase gene translation and the discontinuation of the phage RNA synthesis. In other words, it was necessary to directly link the regulation of the RNA synthesis to the translational regulation. This goal required first the identification of a basic replication cycle, without superposing of many elementary cycles due to lack of the synchronization of infection and continuing growth of the host cells, as well as thorough quantification of the phage-specific syntheses. On the whole, this work was finished in the late 1970s by Elmārs Grēns' team (Pumpen and Gren 1975, 1977; Bauman et al. 1978; Pumpen et al. 1978a,b, 1982; Dishler et al. 1980a,b). According to their approach, rifamycin was selected as the most suitable agent to avoid contamination of bacterial macromolecular synthesis and to isolate, together with the ribonuclease treatment, the single replication cycle, with the most minimal distortion of the phage-specific syntheses. The unique character of the rifamycin action by the separation of the phage-directed syntheses was discussed in the Rifamycin section of Chapter 4.

The examples of the bell-shaped Gaussian-like single replication cycles for the phages MS2 and Qβ are presented in Figure 17.3. The time parameters of these cycles appeared dependent, first of all, on the presence of ribonucleases in the bacterial host (Bauman et al. 1978), as presented in the Rifamycin section of Chapter 4. Figure 4.8 illustrated the prolongation of the replication cycle in the *E. coli* strain Q13 lacking ribonuclease I and just more prominent prolongation in the *E. coli* strain AB105 lacking both ribonucleases I and III, in contrast to the ribonuclease-positive *E. coli* strain AB259. The maximum RNA synthesis in the ribonuclease-deficient strains was reached later, and the rate of RNA accumulation declined more slowly. Some other *E. coli* strains, including Su-1 and Su-3 suppressors, possessing a complete set of ribonucleases repeated the same pattern of the replication cycle

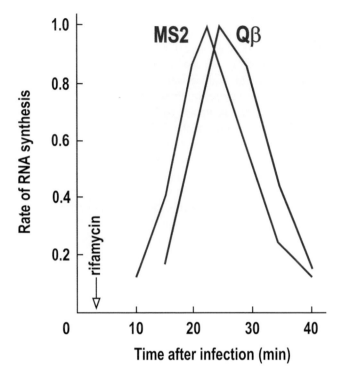

FIGURE 17.3 The normalized cycles of MS2 and Qβ RNA synthesis in *E. coli* AB259 cells in the presence of rifamycin since 3 min postinfection.

as that in the *E. coli* strain AB259. One might be tempted to assume that the observed differences in the RNA synthesis pattern were due to the lower ribonuclease content in the strains Q13 and AB105, which had as its consequence a longer lifetime of the phage RNA.

These characteristics were strain-specific and, moreover, independent of the cell growth rate, which defined only the time constraints of the phage release, as demonstrated earlier in Figure 4.9. Similarly, when the Qβ-infected cells were shifted from glucose-minimal to succinate-minimal medium 20 min after infection in the presence in rifamycin, the shift-down did not affect the rate of phage RNA replication, although production of progeny phage was about fourfold slower in down-shifted cultures than in the cultures in glucose medium (Leschine and Jacobson 1979). However, rifamycin inhibited the cell division machinery, which appeared important for the phage release (the lysis protein was not known at that time). As mentioned in Chapter 4, these experiments were performed with the group I phages MS2, f2, fr, and R17 and the group III phage Qβ, and always led to the same conclusions (Bauman et al. 1978).

The single RNA replication cycle was extended strongly by the replication of the polar coat *amber* mutants, such as Qβ am12 or f2 sus3, in the non-suppressive *E. coli* strains (Pumpen et al. 1978b). To a lesser degree, this effect was observed by the nonpolar coat *amber* mutants in the non-suppressive host strains. These typical extensions of the replication cycle by the coat mutants were described earlier in the Function of the mutant genes section of Chapter 8 and depicted in Figure 8.2.

Figure 17.4 demonstrates how the rate of protein synthesis was detected for either wild-type RNA phages or their coat *amber* mutants in the *amber* suppressor and non-suppressor strains of *E. coli* in the presence of rifamycin (Pumpen et al. 1978a).

Summarizing, it was demonstrated that the rates of synthesis of the phage-specific replicase and RNA minus strands dropped off concurrently in both wild-type and coat protein mutant-infected Su⁻ and Su⁺ cells after 10 and 15 min postinfection, respectively. The rate curves for the replicase and minus-strand RNA syntheses became practically superimposed. The rate of the RNA plus strand synthesis started to decline 5 to 10 min later in both cases. The excessive synthesis of the replicase subunit in the coat protein mutant-infected cells was accompanied by a similar overproduction of the RNA minus strands, but not of the plus strands. In all cases, the decline in the replicase synthesis was found to be invariably followed by a reduction in the total phage RNA synthesis after a certain period of time. This period was rather prolonged in the ribonuclease-deficient strains. This was true for all tested phages including both MS2 and its nonpolar coat protein mutant MS2 am623, which are shown here as prototypes in Figure. 17.4, as well as for other similar pairs tested, except that the decrease of the replicase and minus strand synthesis was observed in MS2 am623 and other nonpolar coat mutants, such as f2 sus11, a double mutant for both A protein and coat protein, 3 to 5 min later, and the observed delay resulted in a two- to fourfold replicase overproduction. This excessive synthesis of the replicase by the nonpolar coat mutants was accompanied by a similar overproduction of the minus strand, but not of the RNA plus strands (Pumpen et al. 1978a).

The presented data showed clearly that the syntheses of the phage-specific replicase subunit and RNA were somehow coupled. A hypothesis was put forward to explain the molecular mechanism of such coupling between the syntheses of the replicase and RNA minus strands. According to this hypothesis, the initiation of the RNA minus strands required unfolding of the 3′-terminal part of the RNA plus template, which was probably affected by the ribosome during translation of the replicase gene. The initiation of the plus-strand synthesis at the same time appeared to be independent of the replicase gene translation. If the RNA minus strands were functionally unstable and possessed a definite half-life, the rate of plus-strand synthesis must decrease after a period equal to the lifetime of an RNA minus strand in the absence of the replicase synthesis. The regulation of the replicase synthesis was sufficient in such a case to exert effective control over viral RNA replication (Pumpen et al. 1978a).

The idea was supported by the earlier data demonstrating the breakdown of the phage fr RNA in the host cells (Sarkar and Dürwald 1966), as well as the degradation of the parental MS2 RNA to nucleotides as soon as 10 to 15 min after infection (Engelberg and Artman 1970; Cramer and Sinsheimer 1971). Then, the elongation of the lifetime of the RNA minus strands, which possibly took place in the ribonuclease-deficient strains such as Q13 and AB105, might be able to increase the number of the RNA plus-strand synthesized by one minus-strand template. This would result in the protracted replication cycles in these strains that are seen in Figure 17.3. If this assumption was valid, the mechanism of coupling of minus-strand synthesis with the replicase synthesis should consist in making accessible the 3'-end and/or some other parts of the plus-strand templates for the initiation of RNA minus-strand synthesis. This accessibility was presumably secured by the ribosome movement along the replicase gene during its translation, which unfolded the RNA macrostructure and exposed the replicase binding sites (Pumpen et al. 1978a). This probably explained why the partially purified f2 replicase synthesized almost exclusively RNA plus strands on the f2 replicative ensemble as template (Fedoroff and Zinder 1973; Fedoroff 1975), as described in the Qβ replicase section of Chapter 13.

This hypothesis was supported not only by the *amber* mutant data, as shown in Chapter 8, but also when the partial suppression of protein synthesis was applied to the single replication cycle of the wild-type phages. If such treatments as addition of chloramphenicol, tetracycline or streptomycin, or valine-isoleucine restriction were applied at definite times after infection and provoked lowering of the bulk coat synthesis, the coat control over the replicase synthesis was abolished. As it follows from Figure 17.5, this led to the strong prolongation of the RNA synthesis cycle, synchronously with the replicase synthesis.

The clear correlation between the pattern of the RNA synthesis and the level of residual protein synthesis was found to be the most essential feature of the prolongation effect. At low inhibition values, the RNA synthesis declined more rapidly, although with a definite delay. When the inhibition of protein synthesis reached 60%−90%, which could be achieved by applying 10, 25, and 50 μg of chloramphenicol per mL, the maximum prolongation of the RNA synthesis was observed. The inhibition of the protein synthesis by more than 99%, caused for example by 200 μg of tetracycline per mL, resulted in an immediate decrease in the RNA synthesis. For example, when chloramphenicol inhibited protein synthesis by 92%, as estimated from the synthesis of both maturation and coat proteins, the rate of the replicase synthesis was reduced only twofold in the MS2-infected cells or fourfold in the MS2am623-infected cells. Moreover, the chloramphenicol treatment maintained the synthesis of replicase at an appreciable rate for a long period of time in both MS2- and MS2am623-infected cells. This resulted in the strong overproduction of the replicase subunit in the chloramphenicol-treated cells, in comparison with the untreated control. Again, the rate curve of the RNA minus strand synthesis invariably followed that of the replicase synthesis after the chloramphenicol treatment. Besides temporal coincidence of their rate curves, the replicase and RNA minus strands were synthesized in nearly equimolar amounts in the presence of chloramphenicol.

A similar effect of the protein synthesis inhibitors on the viral RNA accumulation was observed in all phages under study, including maturation protein, coat protein, and double

maturation and coat *amber* mutants, where the replicase was the only synthesized phage protein in non-suppressor cells. This indicated that the excess of the replicase itself may be capable of suppression of the replicase and then RNA synthesis in the absence of coat protein. The prolongation of the replicase synthesis was always followed by the prolonged synthesis of RNA minus strands in all cases. Moreover, both replicase and RNA minus strands were formed in nearly equal amounts when protein synthesis was partially inhibited. Therefore, the observed coupling of both replicase and minus-strand RNA synthesis offered a possibility for full control over viral RNA replication by means of control of the replicase synthesis on the translational level, assuming the functional instability and relatively short half-life of the phage RNA plus and minus strands (Pumpen et al. 1978a, 1982).

If the RNA replication could be controlled by switching off the replicase synthesis, both repressor complexes I and II were responsible for this control. The decrease in the total RNA synthesis in the nonpermissive cells infected with the coat protein mutant indicated that transformation of the repressor complex II into the initiation complex for minus-strand synthesis postulated by Kolakofsky and Weissmann (1971a,b) and studied further by Kolakofsky et al. (1973) might be limited under certain conditions, viz., in the case when the replicase gene was not occupied by ribosomes. Early in the wild-type phage replication, when the molar ratio of ribosomes to RNA was relatively high and the absolute amounts of replicase appeared to be low, the repressor complex II transformed readily into the initiation complex. On the contrary, in the cells infected with the coat protein mutants having excessive replicase synthesis, the ratio of replicase to RNA was increased at any stage of replication. This increase was accompanied by a permanent drop in the ribosome-to-RNA ratio, since accumulation of the RNA plus strands failed to produce phage particles. Therefore, the number of ribosome-free RNA plus strands involved in the repressor complex II and unable to initiate the RNA minus-strand synthesis rose, especially at late stages of replication, thereby shutting off any further RNA synthesis.

The binding of the replicase to the initiation site of the coat protein within the repressor complex II was confirmed indirectly by the marked lowering of the readthrough maturation protein A_s in the suppressor strains Su-1 and Su-3 infected with nonpolar *amber* mutants of the group I phages (Pumpen et al. 1978a). The readthrough A_s protein was first observed by Remaut and Fiers (1972a,b) due to the prolongation of the maturation protein by the addition of some 30 ammo acids in Su⁺ cells, since the maturation gene was terminated by the *amber* UAG codon. Unlike the normal maturation protein, the longer polypeptide was probably not incorporated into mature phage particles (Remaut and Fiers 1972a,b). The lowering of the A_s synthesis in the Su-3 cells, in contrast to the Su⁻ cells, is clearly seen in the continuation of Figure 17.4. This strong decrease clearly indicated that the intergenic region, where the A_s protein could be prolonged, was occupied by the repressor complex II.

Therefore, the mechanism-controlling translation of the phage templates appeared at the same time to be responsible

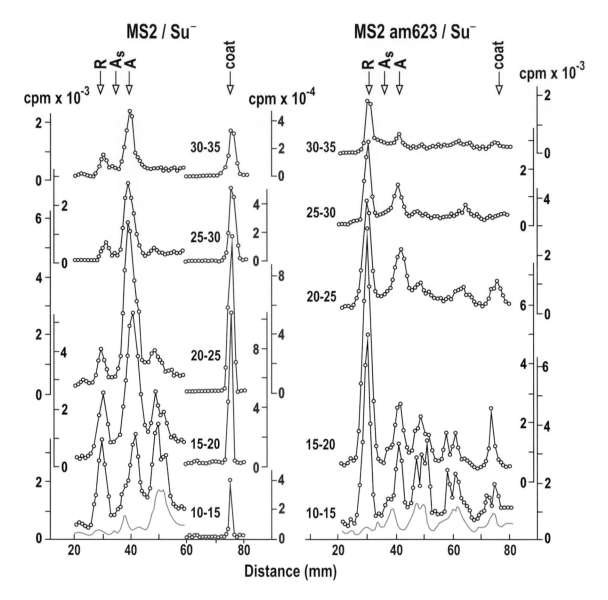

FIGURE 17.4 The typical polyacrylamide gel electrophoresis of ^{14}C-labeled proteins from MS2- and nonpolar coat protein mutant MS2am623-infected non-suppressor *E. coli* AB259 (Su$^-$) cells. Rifamycin was added at 3 min postinfection. Arrows indicate the peak positions of replicase (R), A_s protein (A_s), A protein (A), and coat protein (C). The numbers on the electropherograms signify the pulse time (minutes postinfection). The protein synthesis in rifamycin-treated uninfected cells (from 15 to 20 min relative to the time of infection) is shown by the green line. (Redrawn with permission of ASM from Pumpen P, Bauman V, Dishler A, Gren EJ. *J Virol.* 1978;28:725–735.) *(Continued)*

for the regulation of the viral RNA replication, since the RNA synthesis and the translation of replicase gene were found to be coupled. Such a regulatory mechanism seemed reasonable with respect to the final efficiency of the total phage production.

To confirm the proposed hypothesis, special regulatory *ts* mutants of the phage MS2 with decreased repressor activity of the replicase were obtained by Andris Dišlers (Dishler et al. 1980a). To do this, Dišlers from the Grēns team, subjected the phage MS2am623, the nonpolar coat *amber* mutant at position 50, to the nitrous acid mutagenesis and selected temperature-sensitive mutants with strongly prolonged RNA synthesis at nonpermissive 42°C temperature in the nonpermissive to the *amber* mutation Su$^-$ cells. A special trick was

used by Dišlers in the preliminary selection of the mutants, when the negative plaques were allowed to form for a definite time at 37°C and then the plates were moved to 42°C. The plaques that did not enlarge after the transfer were chosen for the further work, while the plaques that grew larger were discarded. Figure 17.6 presents the unique pattern of the RNA and replicase synthesis by this group of mutants. The most studied mutant, ts130, demonstrated the peak of the RNA synthesis displaced for 10 minutes to the right and strong overproduction of the replicase.

At 42°C, the replicase/RNA ratio in the ts130 mutant was 20 times higher than that of the original phage MS2 am623. It was assumed that there was a decrease in the activity of the repressor complex II in the ts130 mutant, resulting in the loss

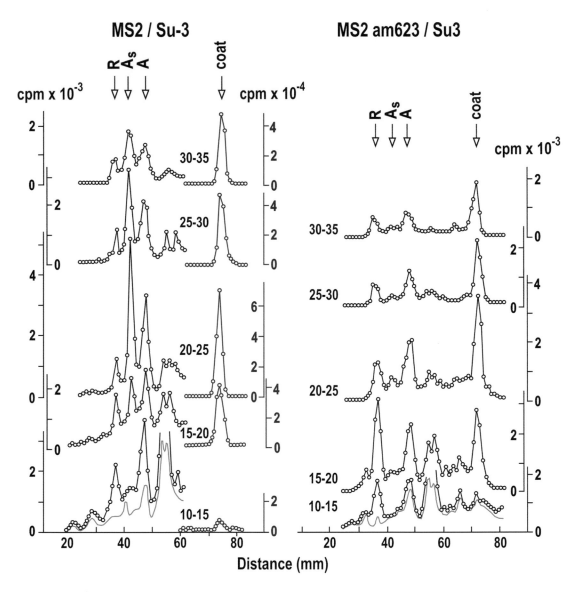

FIGURE 17.4 (Continued) The typical polyacrylamide gel electrophoresis of ^{14}C-labeled proteins from MS2- and nonpolar coat protein mutant MS2am623-infected Su-3 suppressor *E. coli* H12R8a (Su-3) cells.

of the replicase control over its own synthesis, while the control of the coat protein synthesis remained unaffected (Dishler et al. 1980a).

Figure 17.7 presents a flowchart that describes the proposed regulation scheme and is a simplified version of Figure 1.11 in the regulation section of Chapter 1. This flowchart was described by a system of differential equations that was generated by Grēns' team together with their good friend, physicist Boris Isaakovich Kaplan, in 1975, as mentioned in the Replication cycle section of Chapter 4. Varying the constants, the mathematical model was able to simulate the full cycle of the RNA phage replication.

The known mechanisms of interaction of the Qβ RNA and replicase were described in detail in Chapter 13, while the true structures of the repressor complex II were analyzed in the corresponding sections of Chapter 16.

It is noteworthy that the theoretical interest in the regulation of replication of single-stranded RNA viruses is not exhausted today. Thus, a stochastically perturbed model of the single-stranded RNA virus replication was proposed recently (Shaikhet et al. 2019). Moreover, the Qβ replication was chosen as an RNA replication prototype, and many improvements were made to achieve target gene expression enhancement directly by increasing mRNA abundance (Yao et al 2019).

INDUCTION OF INTERFERON

After the first experiments with synthetic complexed polynucleotide poly(I:C), the famous Maurice Ralph Hilleman's team published data that the double-stranded RNA extracted from the *E. coli* cells infected with the phage MS2 was highly active as an inducer of interferon in mice and rabbits with following resistance to viral infections *in vivo* and *in vitro* (Field et al. 1967). The authors stated that this substance, called MS2 RF RNA, could be of special promise as a practical source of double-stranded RNA for clinical purposes,

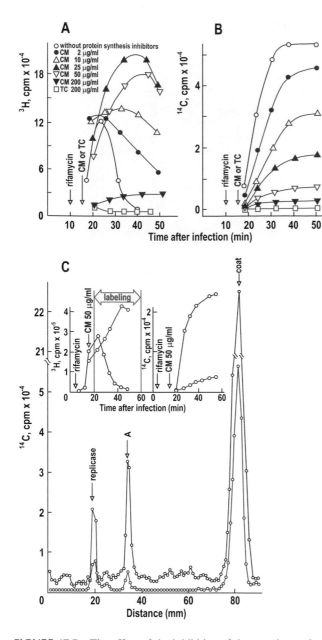

FIGURE 17.5 The effect of the inhibition of the protein synthesis on the rate of the phage RNA synthesis–prolonged RNA replication. (A) Rate of MS2 RNA synthesis and (B) accumulation of viral proteins in *E. coli* AB259 at different concentrations of protein synthesis inhibitors. Rifamycin was added at 10 min, chloramphenicol or tetracycline was added at 15 min, and ^{14}C-phenylalanine was added at 17 min postinfection. (C) Scaling up of the MS2 replicase synthesis by the protein synthesis inhibition. The experimental conditions are presented in the inset. (Redrawn with permission of ASM from Pumpen P, Bauman V, Dishler A, Gren EJ. *J Virol.* 1978a;28:725–735.)

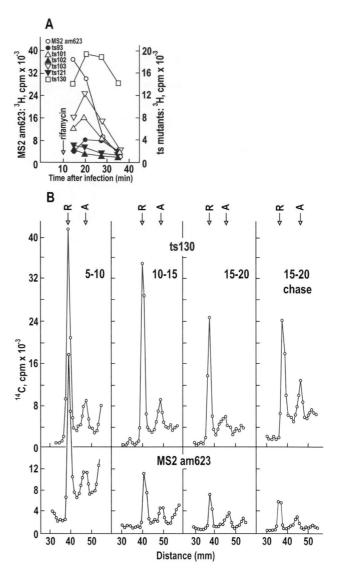

FIGURE 17.6 The RNA (A) and protein (B) synthesis directed by the regulatory *ts* mutants at 42°C. The numbers on the electropherograms signify the pulse time (minutes postinfection). (Based on data from Dishler AV, Pumpen PP, Gren EJ. *Genetika.* 1980a;16:1719–1728.)

while the single-stranded RNA derived from the MS2 virion was inactive. A bit earlier, Lodish and Zinder (1965b) were the first to notice the marked overproduction of double-stranded 7S RNA by nonpolar f2 coat protein mutants in nonpermissive conditions. As a result, due to these two occasions, double-stranded phage RNA aroused considerable practical interest by the induction of interferon. Next, Hilleman's team confirmed in a discussion that the factors considered

influential in making a polynucleotide, including the double-stranded MS2 RNA, an efficient inducer of interferon, were double-strandedness, sugar moiety, and molecular size of the polynucleotides, as well as their thermal stability and resistance to enzymatic degradation (Field et al. 1970).

Shortly thereafter, Maurice R. Hilleman, a microbiologist who developed over 40 vaccines, contributed to the discovery of hepatitis viruses and was characterized by Robert Gallo as "the most successful vaccinologist in history," presented the first preclinical studies in animal models with the double-stranded RNAs of the phage MS2 and its nonpolar coat protein mutant MU9, among other double-stranded RNAs (Hilleman 1970). The major conclusion was that all the active RNAs were double stranded, while the single-stranded RNAs were not.

Due to high levels of the double-stranded RNA, the non-suppressor *E. coli* cells infected with the MS2 *amber* mutant

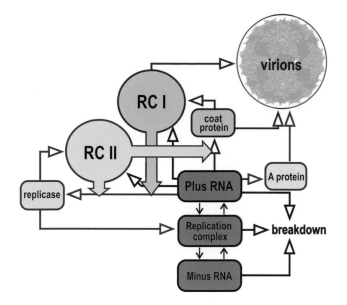

FIGURE 17.7 The general flowchart of the RNA phage replication.

MU9 were employed as a tool by the development of an efficient process for the large-scale double-stranded RNA production in fermentors (Lago et al. 1972). Over 300 μg of dsRNA/mL were accumulated after MU9 infection of cultures grown to high density, and this was approximately 300 times the amount of dsRNA made by the wild-type MS2 infection of cells.

Meanwhile, Kawade and Ujihara (1969) reported the interferon-inducing activity of the single-stranded MS2 RNA. It was, however, not highly surprising, taking into account the high level of the double-helicity in the phage RNA, as described in detail in Chapter 10. Nevertheless, the commercial phage MS2 RNA was found inefficient by the protection of mice from the *Pseudomonas aeruginosa* infection (Gonggrijp et al. 1983). Later, Kawade (1987) admitted that his results were negative by the treatment of animal cells with infectious MS2 RNA. It is also noteworthy that virus-like particles, but not, however, of the RNA phages, were tested at that time as potential interferon inducers and were regarded as a potential source of the double-stranded RNA (Planterose et al. 1970).

At the same time, Leonard et al. (1969) used the double-stranded RNA of the pseudomonaphage PP7 and confirmed its high efficiency by interferon induction, but found that the double-stranded RNA was intrinsically toxic in mice and rats and that thymic atrophy, lymphocyte transformation, and interferon production appeared to be inseparably linked and might be manifestations of the same activity.

A bit paradoxically, Kalmakoff et al. (1977) tried to get an antiviral response in insects, namely, in the cultured *Aedes aegypti* cells, with the double-stranded RNA of their original RNA phage Hd (Bilimoria and Kalmakoff 1971). Although *A. aegypti* cells could incorporate a comparable amount of dsRNA, no antiviral response was observed.

A fertile field for the study of the double-stranded RNA of the phage f2 was found in the then Czechoslovakia and the former Soviet Union. Jiří Doskočil from Praha was one of the first enthusiasts by these studies (Doskočil et al. 1971;

Fuchsberger et al. 1972). This interest led to the further serious structural and functional studies of the double-stranded f2 RNA as an efficient interferon inducer (Paleček and Doskočil 1974; Rovenský et al. 1975; Lackovic and Borecký 1976; Mayer et al. 1976; Sołtyk et al. 1976; Táborský and Dolník 1977; Táborský et al. 1977; Lackovic et al. 1979; Pekárek et al. 1979). Borecký et al. (1978) reported the results of a 5-year study of the curative effect of the double-stranded f2 RNA in viral dermatoses and eye diseases. The study was performed in five hospitals in viral dermatoses such as herpes simplex recidivism, herpes zoster, and male genikeratonconjunctivitis herpetica and conjunctivitis lignosa. The results of clinical tests indicated that the preparation of the double-stranded f2 RNA applied topically was harmless for man, and in the majority of cases had a beneficial effect in the disease.

The thorough structural studies of the double-stranded f2 RNA, first of all, on its compactness in solution and melting curve parameters, were initiated also in Moscow (Evdokimov et al. 1976, 1977; Permogorov et al. 1980). In parallel, the interferonogenic and antiviral activity of the double-stranded f2 RNA was investigated in tissue culture and on animals (Kadyrova et al. 1978). Thus, the intraperitoneal administration of the drug in a dose of 10 gamma per a mouse protected the infected animals from death after infection with Semliki Forest virus or tickborne encephalitis virus. The treatment was accompanied by production of up to 1280 units/mL of interferon in the blood serum of the animals. Remarkably, the combination of the drug with rimantadine, a popular anti-influence drug at that time, increased the protective effect both in the tissue culture and *in vivo*. The same positive effect of the double-stranded f2 RNA was observed in the cultures of lymphoblastoid and somatic human cells (Novokhatskii et al. 1979). Furthermore, the use of vaccine in combination with the double-stranded f2 RNA in tickborne encephalitis gave up to 75% survival of the infected mice (Barinskii et al. 1979). As a result, 4 hours after administration, interferon could be detected in mouse sera in a concentration up to 640 units/mL, which accumulated more rapidly in immune mice and remained at a higher level 18 h after inoculation of the inducer. The *in vitro* action of the double-stranded f2 RNA was compared with that of the poly(I:C) (Bukata et al. 1979; Labzo et al. 1980). The aerosol administration of the double-stranded f2 RNA, which employed an ultrasonic aerosol generator, was effective in inducing interferon production in mice and monkeys (Nosik et al. 1980). Thus, by prophylactic inhalation of the interferon inducer 4 and 24 hours before inoculation protected 50%–75% of mice against a fatal influenza infection. Interestingly, the use of individual masks for monkeys was presented. Furthermore, a high degree of protection of mice inoculated intracerebrally with street rabies virus was demonstrated after intracerebral inoculation of the double-stranded f2 RNA both 24 hours before and 4 hours after virus infection, with a considerable increase of the average life span of the animals, while intramuscular and intraperitoneal inoculation of the inducer was ineffective (Gribencha et al. 1983). The double-stranded f2 RNA was effective also in the murine model of experimental tickborne encephalitis (Knoroz et al. 1985). Moreover, an antiproliferative effect of the double-stranded f2 RNA was demonstrated in Raji lymphoblastoid cells (Lavrukhina and Ershov 1985). Then, the double-stranded f2 RNA was used as a genocorrector and exhibited a protective effect on both microorganisms and mammals, reducing the strength of the experimental mutagenic effects (Zolotareva et al. 1993a,b).

The Czechoslovak and Soviet studies on interferon induction by the double-stranded f2 RNA were commented on in the huge review on interferon inducers (Stewart 1981). At the same time, the general role of the double-stranded RNAs, including those of the phages f2 and MS2 with their interferon-inducing capabilities, was reviewed by Libonati et al. (1980).

By their evaluation of double-stranded RNA analogs as interferon inductors, Johnston and Stollar (1978) found that the double-stranded f2 RNA did not induce anti-nucleic antibody response in rabbits, in contrast to some synthetic double-stranded polynucleotides.

Later, the double-stranded RNA of the phage f2sus3 was found able to induce the interferon-induced protein kinase dsRNA-activated inhibitor (DAI) of translation, which plays a key role in regulating protein synthesis in higher cells (Manche et al. 1992). Once activated, in a process that involved autophosphorylation, the DAI phosphorylated the initiation factor eIF-2, leading to inhibition of polypeptide chain initiation. The activity of DAI was controlled by RNA regulators, including dsRNA activators and highly structured single-stranded RNAs, which blocked activation by dsRNA. This mechanism finally explained the effect of the double-stranded phage RNAs.

Finally, it is noteworthy that the single-stranded MS2 RNA was studied as a potential interferon inductor in the complex with tilorone, an orally active interferon inducer, in Kyiv (Karpov 1997a,b). Thus, the complex was formed due to the intercalation of tilorone into the two-stranded regions of the single-stranded RNA. As a result, tilorone stabilized the spontaneously forming double-stranded RNA regions. The Ukrainian efforts on interferon induction were reviewed substantially by Spivak et al. (2007). It is noteworthy that the effect of the RNA-tilorone combinations *in vitro* was compared with that of the drug Larifan (Karpov and Zholobak 1995, 1996; Karpov et al. 2001).

Remarkably, the early efforts on interferon induction by double-stranded RNA preparations were mentioned recently in the large authoritative review on the 50-year history of the search for selective antiviral drugs (De Clercq 2019).

LARIFAN

In Riga, Guna Feldmane and Valts Loža were the true pioneers of the double-stranded phage RNA as an interferon inductor. They started the appropriate activities in the 1970s and used the double-stranded RNA of the nonpolar f2sus11 *amber* mutant. As a result, they carried the studies to the real registered pharmaceutical named Larifan. The latter consisted of purified double-stranded f2 RNA. The history of

Larifan can be followed by journal papers (Feldmane et al. 1984; Brūvere et al. 1991; Loža et al. 1991), but more informatively by numerous meeting abstracts and conference papers that are listed now at the internet site of their company.

Larifan was declared an efficient interferon inductor, exerting a therapeutic effect on herpes, papilloma, influenza, and other viral infections (Loža and Feldmane 1996; Loža et al. 1996). The effect of Larifan on the expression of melanoma-associated antigens was studied (Heisele et al. 1998). Recently, the Larifan effect was evaluated by the production of a wide spectrum of cytokines and chemokines in *ex vivo* cultivated peripheral blood mononuclear cells (Veinalde et al. 2014; Pjanova et al. 2019).

In the 1990s, Larifan, in parallel with Reaferon, a recombinant interferon produced in the former Soviet Union, was subjected to clinical trials on volunteers (Sokolova et al. 1991c). It was shown that the preparations increased the activity of $2',5'$-oligoadenylate synthetase in the lymphocytes and protein kinase in the plasma of 60%–70% of the volunteers, while the effect of both Reaferon and Larifan on lymphocyte protein kinase activity was determined for the first time.

The fine mechanism of the Larifan action on cells was studied by Sokolova et al. (1991a).

Larifan inhibited multiplication of *Chlamydiae* in the lungs and lymph nodes of mice and was regarded as promising in therapy of chlamydiosis (Vinograd et al. 1990).

Concerning virus infections, Larifan inhibited the reproduction of cytomegalovirus in cell culture (Sokolova et al. 1991b) as well as HIV-1 replication in cell cultures when administered simultaneously with the inoculum, while its effect was comparable with that of azidothymidine in concentrations providing cell protection (Nosik et al. 1992). Larifan was sufficiently effective against flavivirus and alphavirus infections (Barinskii et al. 1996). The drug worked in immunocompromised mice inoculated intraperitoneally with HSV-1 and HSV-2 3 days after a single radiation exposure (Alimbarova and Barinskii 1996). Then, Larifan appreciably prolonged the life span of monkeys infected with smallpox (Loginova et al. 1997). The drug also demonstrated anticarcinogenic, antitumor, and antimetastatic properties (Gasparian et al. 1991; Zolotareva et al. 1991; Barinskii et al. 1994b; Loginova et al. 1996).

Concerning vaccine applications, Larifan significantly enhanced action of a commercial herpes vaccine in mice (Barinskii et al. 1993), enhanced specific antibody production, and at least doubled cell-mediated immune response by immunization of mice with recombinant yeast hepatitis B vaccine (Barinskii et al. 1994a). Moreover, Larifan used parenterally was combined with herpetic vaccine to treat severe recurrent herpes in 32 patients and led to amelioration of the clinical symptoms of recurrence (Potekaev et al. 1992). Ultrasonic spraying of Larifan solution was employed for treating acute herpetic stomatitis in 100 children (Mamedova et al. 1991). Interferon was found more effective by ultrasonic spraying, but Larifan was still effective enough. Furthermore, the drug demonstrated high antiviral efficacy against Omsk hemorrhagic fever virus strain *Ondatra* in experiments with laboratory animals. The drug prevented the death of 65% infected mice and significantly decreased infection severity in rabbits (Loginova et al. 2002). However, virus reproduction on cell culture was suppressed mildly, while human adenovirus serotype 2 was not suppressed by Larifan *in vitro* at all (Nosach et al. 1998). Immunomodulatory therapy with Larifan was studied in women with genital papillomavirus infection (Rogovskaya et al. 2002).

Larifan was characterized in detail among other interferon inducers in the thorough review of the Russian experience in screening, analysis, and clinical application of novel interferon inducers (Tazulakhova et al. 2001). The next exhaustive review on commercial interferon inducers (Silin et al. 2009)

described Larifan as a suitable drug against influenza, acute and chronic respiratory viral infections, Aujeszky's disease, foot-and-mouth disease, canine distemper, rabies, Omsk hemorrhagic fever, and herpes virus after systemic or local administration. The drug was presented as a product of the "Pharm" manufacturer in Riga.

According to the current web information, numerous Larifan products in different formulations are produced by Larifans Ltd in Riga since 1997. The long list of products includes prescription medicines for injections, cosmetic products, such as the Larifan lip balm, ointment, toothpaste, spray, foot cream, lotion, and hygienic gel. Special products including Larifan candies, pessaries, and suppositories are also available for sale.

18 Evolution

It is not the strongest of the species that survive, nor the most intelligent, but the one most responsive to change.

Charles Darwin

The measure of intelligence is the ability to change.

Albert Einstein

DARWINIAN EXPERIMENTS

The RNA phage evolution story started with the Qβ replicase, Spiegelman's monster, and the extracellular Darwinian experiments. At the very beginning, Haruna and Spiegelman (1965a) showed that when the Qβ replicase reactions were started at template concentrations below those required to saturate the enzyme, the RNA synthesis followed an autocatalytic curve. When the saturation concentration level was reached, the kinetics became linear. The autocatalytic behavior below saturation of the enzyme implied that the newly synthesized product could in turn serve as templates for the reaction. Moreover, when RNA from a mutant phage Qβ was used to start the reaction with wild-type Qβ enzyme, the mutant RNA was produced, proving that the RNA was the only instructing agent by the synthesis (Pace and Spiegelman 1966a). These findings encouraged Spiegelman's team to enter the next phase of the investigation and examine the infectivity of the synthesized material. The data established that the autocatalytic Qβ replicase reaction was indeed generating self-propagating replicas of the input RNA (Spiegelman et al. 1965). In this study, a serial dilution test was employed for the first time, in order to get direct proof that the newly synthesized RNA was infectious. A series of tubes was prepared, each containing 0.25 mL of the standard reaction mixture with the Qβ replicase, free nucleotides, and some salts, but no added template. The first tube was seeded with 0.2 μg of Qβ RNA and incubated for a period adequate for the synthesis of several micrograms of radioactive RNA. An aliquot was then transferred to the second tube, which was in turn permitted to synthesize about the same amount of RNA, a portion of which was again transferred to a third tube, and so on for 15 times. In fact, it was enough to ensure that the last tube contained less than one strand of the input template. A serial dilution experiment established therefore that the newly synthesized RNA was fully as competent as the original viral RNA to program the synthesis of viral particles in spheroplasts and to serve as templates for the generation of more copies.

The serial dilution experiments led to an idea for studying the evolution of a self-replicating nucleic acid molecule outside of a living cell (Mills et al. 1967). Spiegelman's team believed that the serial dilution situation mimicked an early precellular evolutionary event, when environmental selection presumably operated directly on the genetic material. By their words, in the universe provided to them in the test tube, the RNA molecules were liberated from many of the restrictions derived from the requirements of a complete viral life cycle. A restraint imposed was that they retained whatever sequences were involved in the recognition mechanism employed by the Qβ replicase. Thus, the sequences that coded for the coat proteins and replicase components might now be dispensable. Under these circumstances, it was of great interest to design an experiment that attempted an answer to the following question: "What will happen to the RNA molecules if the only demand made on them is the Biblical injunction, multiply, with the biological proviso that they do so as rapidly as possible?" (Mills et al. 1967). The conditions required were readily attained by a serial transfer experiment (Spiegelman et al. 1965), in which the intervals of synthesis between transfers were adjusted to select the first molecules completed. The authors assumed that the elimination of sequences, which would be dispensable under the conditions of the experiment, could confer a selective advantage. In accordance with this expectation, it was found that as the experiment progressed, the multiplication rate increased, and the product became smaller. In total, 75 transfers were performed. The synthesis of the biologically competent RNA ceased between the fourth and fifth transfers. A dramatic increase in the RNA synthesis rate occurred between transfers 8 and 9. By the 74th transfer, 83% of the original genome had been eliminated and its base composition was changed substantially.

According to the traditional sucrose gradient analysis, there was no material synthesized corresponding to the original 28S viral RNA by the 9th transfer, and a major component at about 20S and a minor one at about 15S were seen in its place. This pattern was essentially maintained through the 15th transfer, but by the 30th transfer the major component has decreased to 15S and the minor one to about 14S. The product of the 38th transfer showed variant RNA, which no longer was split into two peaks and was about 12S by the 74th transfer. It was very helpful that the fine analysis of the synthesized products became available due to the concurrent introduction of the gel electrophoresis technique (Bishop et al. 1967b) after a previous era of sucrose gradients. As a result, the molecular mass of about 1.7×10^5 daltons, corresponding to approximately 550 nucleotides, was determined for the final variant RNA. Mills et al. (1968) demonstrated that, by analogy to the full-length Qβ RNA, the negative strand of the little variant served as an excellent template for the Qβ replicase. The same sort of Replicative complexes, namely, HS and FS, as described in the replicative intermediate section of Chapter 17, appeared and the direction of synthesis of plus strands was 5′ to 3′. The molecule was named *little monster* by authors (Spiegelman et al. 1967) and *Spiegelman's monster* by other colleagues, but officially it was termed the V-1 variant. The full account of the story was published also by the *American Scientist* as the 80th Jubilee lecture 1967 (Spiegelman 1967). The result caused a sensation and the future prospects were suggestive, since such abbreviated molecules could open up a novel pathway for a highly selective interference with the replication of the complete viral genome. In his impressive lecture presented at the Nobel Workshop "Chemical Origin and Early Evolution of Life" held in Stockholm in December 1970, Sol Spiegelman developed an idea of precellular evolution, compared the Qβ replication with Xerox machines and von Neumann's *self-duplicating automata* and noted that the Darwinian experiment had its own built-in paleontology, for each tube could be frozen and its contents expanded whenever one decided to examine what happened at that particular period in the evolutionary process (Spiegelman 1971).

After the first Darwinian experiment that led to the V-1 variant that was isolated in fact under selection pressures designed to encourage rapid completion of replication, it was attempted to generate clones descended from individual RNA strands (Levisohn and Spiegelman 1968). As a first step, a new mutant, characterized by an ability to initiate synthesis at a low level of the template input and named V-2, was isolated. The selection of this type of mutant was accomplished by modifying the earlier, serial-selection procedures, when the time intervals between transfers were decreased but the dilution remained constant. In novel experiments, the time was kept constant and the dilution was increased as the transfers were continued. The variant RNA that evolved by the 17th transfer was called Variant 2, or V-2. The V-2 was clearly superior over the V-1 in growth at low template level and initiated synthesis with 0.29×10^{-18} g of RNA, which corresponded to the mass of one strand. Kinetic analysis revealed that, during the exponential growth phase, the V-2 had a doubling time of 0.403 minute as compared with 0.456 min for the V-1. In a 15 min period of logarithmic growth, the V-2 experienced more doublings (i.e., a 16-fold increase) than the V-1. The V-1 and V-2 had a similar electrophoretic mobility on polyacrylamide gels and the same base composition. They did, however, differ in sequence, as determined by the oligonucleotide fingerprint patterns. The molecular mass of the V-2 was estimated as 1.74×10^5 daltons.

The shortened RNA variants possessed evident advantages not only for the detailed examination of the replicative process, but also for the greater accessibility to the sequence determination at that time. Thus, Bishop et al. (1968) demonstrated that the 5′ end of the V-2 plus strand begun with ppp-GpGpGpGpApApCpCp, while the minus strand begun also with pppGp at the 5′ end and was followed by a long sequence rich in purines.

As narrated exhaustively in the above-mentioned original lecture (Spiegelman 1971), after the isolation of the V-1 and V-2 variants possessing increased growth rates under standard conditions, Spiegelman's team turned to selection of "nutritional mutants," when the syntheses were run under less-than-optimal conditions with respect to a component or parameter of the reaction (Levisohn and Spiegelman 1969). It was assumed that if a variant arose that coped with the imposed suboptimal condition, the continued transfers should lead to its selection over the wild type. This was done with variations in the level of the ribonucleoside triphosphates. Thus, the rate of the V-2 synthesis began to decrease sharply as the low level of CTP. A serial transfer experiment at 2 nM of CTP per reaction was initiated with Qβ RNA, culminating after 10 transfers in the appearance of V-4. A second series of transfers at 1 nM of CTP per reaction was then started with the V-4 and after 40 transfers led to the isolation of the V-6. The variants V-4 and V-6 replicated 28% and 56% better, respectively, than did the V-2 at low levels of CTP, but they did not show significant difference in chain length on polyacrylamide gel. Moreover, there was no significant difference in base composition. Another variant, V-3, was isolated with limiting CTP, starting with the V-2 instead of the Qβ RNA. Remarkably, the V-3 possessed phenotypic properties indistinguishable from those of the V-4. Thus, one could arrive at the V-4 phenotype either from the Qβ RNA or from the V-2.

The next selection was performed for a variant resistant to tubercidin triphosphate (TuTP), an inhibitor of the Qβ RNA synthesis *in vitro*, where tubercidin was an analog of adenosine, in which the nitrogen atom in position 7 was replaced by a carbon atom. The variant V-6 was chosen to start a series of transfers in a reaction mixture that contained limiting concentration of ATP of 1–5 nM and the variant V-8 was isolated. The doubling time of the V-8 in the reaction mixture with 1.5 nM of ATP was 2.8 min as compared with 8.4 min for the starting V-6 and the replication rate of the V-8 in a reaction mixture with 5 nM of ATP was inhibited fourfold with the addition of 30 nM of TuTP. A serial transfer was initiated with the V-8 in the inhibitory medium and led to the isolation of the V-9. The doubling time of the V-9 in the presence of TuTP was 2.0 min as compared with 4.1 min for the V-8. Therefore, the V-9 exhibited a specifically increased resistance to the TuTP (Levisohn and Spiegelman 1969).

Then, Saffhill et al. (1970) selected variants resistant to ethidium bromide, or 3,8-diamino-5-ethyl-6-phenylphenanthridinium bromide, an agent that could exert the inhibitory influence by direct interaction with the RNA molecule. The transfers were initiated with the V-2, continued in the presence of increasing levels of ethidium bromide, and the resistant variants V20γ, V40γ, and V50γ were selected. Remarkably, both V-2 and V40γ variants demonstrated similar size by the gel electrophoresis. Thus, ethidium bromide–resistant RNAs probably arose through changes in the base sequence, but not in the length. The fact that the V40γ enhanced the fluorescence of ethidium bromide less than the V-2 variant indicated that the V40γ bound with ethidium bromide less strongly than did the V2. The adaptation to ethidium bromide engendered some cross-resistance to proflavine sulfate, although the resistance to the latter was not as great as to ethidium bromide. Therefore, it was apparent from even the limited set of successful examples that the number of identifiable mutant molecules possessing prespecified phenotypes was restricted only by the ingenuity exercised in devising the appropriate selective conditions. In total, the selection experiments revealed an unexpected wealth of phenotypic differences that a replicating nucleic acid molecule can exhibit (Spiegelman 1971).

It soon became obvious that no matter how amusing, the continuing Darwinian selection experiments would not bring the scientific community any closer to the kind of understanding of either the evolutionary process being observed or of the replication mechanism without knowledge of the primary structure. It was therefore decided to get a self-replicating molecule with completely known sequence (Kacian et al. 1972). The above-described variants from the Darwinian experiments were all about 550 nucleotides long, and it was intriguing to find smaller replicating molecules. By this decision, Spiegelman's team (Kacian et al. 1972) was encouraged by the report of the 6S replicating RNA molecules in the cells infected with the phage Qβ (Banerjee et al. 1969b).

6S RNA

J. Thomas August's team (Banerjee and August 1969; Banerjee et al. 1969b) discovered a small RNA in the phage Qβ infected *E. coli*. This RNA was named 6S RNA after its sedimentation coefficient and demonstrated capability to be replicated *in vitro* by the purified Qβ replicase enzyme. Although it represented a small fraction of the total RNA isolated directly from the infected cells, it was the predominant nucleic acid component in a partially purified fraction of the Qβ replicase. The 6S RNA was not detected in the uninfected *E. coli*. The general requirements of the *in vitro* reaction with the 6S RNA as template were similar to those with Qβ RNA except that neither of the two host cell macromolecular factors was required. The RNA product of the reaction was synthesized manyfold in excess of the added template and was indistinguishable from the template in its sedimentation and template properties. It had a chain length of approximately 110–130 nucleotides. Considering the specificity of the Qβ replicase for template, as described in Chapter 13, the 6S RNA was expected to contain at least some nucleotide sequences in common with Qβ RNA, such as those recognized by the enzyme for binding or initiation of synthesis. In fact, it contained pppGp at the 5′-terminus. The RNA product was highly resistant to ribonuclease, although nearest-neighbor analysis did not indicate that a completely double-stranded structure was being replicated. The 6S RNA was found different in size and base composition from the V-1 RNA reported by Mills et al. (1967) and described above. Moreover, the V-1 RNA was presented as having evolved *in vitro* from the Qβ RNA template, whereas the 6S RNA was found in the infected cells and contaminated the partially purified fractions of the Qβ RNA polymerase (Banerjee et al. 1969b).

In fact, the presence of RNA with a sedimentation coefficient of 6–7S was well documented earlier by Robert L. Sinsheimer's team in cells infected with the phage MS2 (Kelly et al. 1965; Lunt and Sinsheimer 1966). This RNA was present in greater amounts when the infected bacteria were treated with actinomycin D or ultraviolet light (Kelly et al. 1965). These investigators suggested that the 6S RNA might

represent a device for replication of a specific portion of the viral genome (Kelly et al. 1965), or a degradation product which accumulated under conditions causing a poor yield of phage RNA (Lunt and Sinsheimer 1966). Lodish and Zinder (1966a) found that almost all of the ribonuclease-resistant RNA synthesized after infection with the f2sus11 mutant was 7S material and that a large proportion of the parental RNA was converted into this form. The 7S RNA was interpreted as representing fragments of double-stranded RNA analogous to the molecules described as replicative forms in Chapter 17. Later, the accumulation of the ribonuclease-resistant 6-7S RNA was observed in the presence of rifamycin and chloramphenicol by the replication cycle experiments described in the preceding chapter (Bauman et al. 1978; Pumpen et al. 1978a). Thus, the 6-7S RNA accumulated when rifamycin was added at 3 min, but the phenomenon was not observed if the drug was given at 12 min postinfection. The significant amounts of 6-7S RNA were also accumulated after chloramphenicol treatment at 50 μg/mL, when protein synthesis was inhibited about 90% and strong production of both replicase and minus RNA was provoked, but this time irrespective of the time of rifamycin addition (Bauman et al. 1978). The authors speculated that the accumulation of 6-7S RNA, which probably arose from the replicative complex degradation, was due to a certain defect in the late RNA degradation stages dependent on host protein synthesis. It seems now, however, that this phenomenon was tightly connected with the enhanced production of the RNA minus strands as the direct consequence of the greater replicase production in light of the tight coupling of both replicase and minus RNA syntheses.

The *in vitro* synthesis of RNA catalyzed by the Qβ replicase was studied using a single-stranded 6S RNA template (Prives 1970, 1971; Prives and Silverman 1972). Whereas the Qβ RNA replication resulted in the synthesis predominantly of the single-stranded Qβ RNA, the predominant reaction product of the 6S RNA replication was found to be double stranded. When treated with formaldehyde to dissociate complementary base pairs, the 6S RNA exhibited a decrease in molecular mass, as indicated by its slower sedimentation rate and faster electrophoretic mobility. However, a small amount of the 6S RNA was in a single-stranded configuration. The frequencies of complementary dinucleotides in purified 6S RNA were not equal, and 10%–20% of the RNA was hydrolyzed by ribonuclease. The electrophoretic mobilities of the native and nuclease-treated 6S RNA indicated that the single-stranded RNA might be present as "tails" at one or both termini of the molecule. The template activity of the 6S RNA was observed only with heat-denatured RNA and was lost upon renaturation (Prives and Silverman 1972).

Meanwhile, the major 3′- (Trown and Meyer 1973) and 5′-terminal (Trown 1973) sequences of the Qβ 6S RNA were determined. The predominant 3′ sequence was $GpCpCpA_{OH}$ with lesser amounts of $GpCpC_{OH}$ and $GpCpCpG_{OH}$. Since the sole 5′-terminal base of the Qβ RNA was G, these results provided another example of the ability of the Qβ replicase to add a noncomplementary adenosine to the 3′ end and the first example of an ability to add a guanosine. Thus, all major sequences found were considered derivatives of the sequence GpCpC$_{OH}$. This sequence differed significantly from those of other Qβ replicase templates studied thus far, and thus reaffirmed the requirement for the additional internal structural features by which Qβ replicase recognized its templates (Trown and Meyer 1973). As for the 5′-terminal sequence of the Qβ 6S RNA, approximately 95% of the molecules terminated with the same sequence, ppp-GpGpCp. This sequence was the complement of the only major 3′-sequence found in this RNA. Both strands of the 6S RNA therefore appeared to have identical 3′- and 5′-terminal trinucleotide sequences (Trown 1973).

The structure and template properties of the short 6S RNA led Walter Fiers' team to an idea of the improved synthetic templates for the Qβ replicase, which should start with an oligoG sequence, so that the complement would contain a 3′-oligoC stretch (Gillis and Opsomer 1974).

MDV-1: A MIDI-VARIANT

Turning back to the activities of Spiegelman's team, they discovered that the 6S species was a mixture of similar, but not identical, molecules. One of these "minivariants" was cloned, named MV-1, and examined in detail with respect to its 5′-terminal sequence and replicative features (Kacian et al. 1971). Thus, the nucleotide sequence at the 5′ terminus of the cloned MV-1 was determined as pppGpGpGpApUp. Unfortunately, these molecules possessed structural characteristics that made it difficult to recognize replicative stages or to separate plus from minus strands. Kacian et al. (1972) continued the search for a suitable candidate and succeeded in isolating the molecule of 218 nucleotides long in the course of examining products of the Qβ replicase reactions run without added RNA templates, since the 6S RNA was regarded as a contaminant of the Qβ replicase preparations. The novel molecule possessed the features desired for a definitive analysis of the replicating mechanism including separation of antiparallel complementary strands. Despite its small size, this molecule was able to mutate to the previously determined phenotypes. It therefore permitted the precise identification of the base changes required to mutate from one phenotype to another in the course of the extracellular Darwinian selection experiments. This variant of the 6S RNA was termed midivariant, or MDV-1, to indicate that it was smaller than the 550 nucleotide V-class of variants and was some 90 nucleotides longer than the group of the 6S "minivariants" as the MV-1. Thus, the length of the MDV-1 RNA molecule was only 6% that of the phage Qβ RNA, and since the initiating AUG triplet was absent, it was untranslatable into protein. In other respects, the two replicating molecules had much in common. Both were accepted by the Qβ replicase, implying that the MDV-1 had retained the structure used by the Qβ replicase to recognize and replicate the Qβ-like RNA. Both RNAs directed the synthesis of complements and both gave rise to similar multistranded, double-stranded, and single-stranded structures in the protein-free product (Kacian et al. 1972).

The complete nucleotide sequence of the MDV-1 RNA plus and minus strands of 218 nucleotides each was published soon in a great *Science* paper (Mills et al. 1973). The primary sequence contained a surprising number of intrastrand antiparallel complements, a peculiarity generating the potentiality for the extensive secondary and tertiary structures containing antiparallel stems and loops. The authors were highly optimistic in the implications of the primary structure for the extracellular Darwinian experiments. They wrote with pride: "For the first time, we can now ask, and answer, the following question: "Precisely what base changes have occurred in mutating from one phenotype to another?" (Mills et al. 1973).

To do this, Kramer et al. (1974) selected, by a serial transfer experiment in the presence of ethidium bromide, a mutant RNA that was more resistant to ethidium inhibition than was the wild-type MDV-1 RNA. The complete nucleotide sequence of the mutant RNA was determined, and the three nucleotides in the mutant sequence were found to be different from those in the wild type. The mixture of the mutant and wild-type RNAs present in successive transfers was also sequenced. It appeared that each of the three-point mutations occurred at a different time. These results showed that the mutant RNA did not arise from a preexisting strand present in the wild-type population, but rather occurred *de novo* in the course of the experiment. The chemical basis of the resistance consisted of the elimination of ethidium binding sites due to the specific alterations in the nucleotide sequence, since the mutant RNA was found to bind less ethidium than the wild-type molecules (Kramer et al. 1974).

As an Albert Lasker Basic Medical Research Award winner, Sol Spiegelman (1974) published an impressive self-report on two sets of highly promising investigations: the Qβ replicase together with the Darwinian experiments and molecular hybridization that became one of the most widely used molecular biologic techniques for a long time. It is noteworthy that at practically the same time Spiegelman's team worked on the reverse transcriptase from avian myeloblastosis virus and published an account for the synthesis of a DNA complement of the Qβ RNA, together with other genetically unrelated templates (Spiegelman et al. 1971).

The evolutionary studies of Spiegelman's team were interpreted at that time by evolutionary theorists who demonstrated that the fundamental theorem of natural selection published in 1930 by Sir Ronald Aylmer Fisher (Fisher 2017) can define time variations in the mean rate of synthesis for a heterogeneous population of replicating polymers (Davis 1978).

The established MDV-1 sequence allowed exploration of many intrinsic replication phenomena. Besides theoretical problems, the MDV-1 sequence contributed first to the elaboration of the RNA sequencing with radioactive chain-terminating ribonucleotides, by analogy with Sanger's dideoxy chain-termination method of DNA sequencing, as mentioned earlier in the R17 sequencing section of Chapter 10. The MDV-1 minus strand RNA was sequenced, and the resulting sequence agreed completely with the known sequence of 220 nucleotides in length with an added A at the 3′-terminus; therefore, 221 nucleotides altogether (Kramer and Mills 1978). Because the secondary structure of the MDV-1 minus RNA was known, it helped to contribute to the problem of the band compression that was due to the persistence of secondary structures during electrophoresis by Sanger's sequencing.

As for the theoretical replication problems, Mills et al. (1978) noticed the template-determined, variable rate of RNA chain elongation by the MDV-1 replication. The electrophoretic analyses of partially synthesized strands showed that a few of the elongation intermediates were much more abundant than others, reflecting a variable rate of chain elongation. Thus, the progress of the Qβ replicase reaction was temporarily interrupted, and then resumed spontaneously, with a finite probability at a relatively small number of specific sites in the sequence of the MDV-1 RNA. Since the time spent between these pause sites was negligible compared with the time spent at pause sites, the mean time of chain elongation was well approximated by the sum of the mean times spent at each pause site. The nucleotide sequence analysis of the most prominent elongation intermediates indicated that they all have the potential to form a 3'-terminal hairpin structure. This suggested that the marked variability in the rate of chain elongation was due to the formation of terminal hairpins in the product strand, or the reformation of hairpins in the template strand (Mills et al. 1978). Later, LaFlamme et al. (1985) compared the pausing during the MDV-1 replication by the Qβ replicase with the pausing during the MDV-1 cDNA transcription by the *E. coli* RNA polymerase. Although the transcripts by the latter were virtually identical to the MDV-1 RNA, the locations at which the RNA polymerase paused were different and apparently were not related to the sequences that formed hairpins. The authors concluded that the hairpin stability *per se* cannot be used to predict the occurrence of pausing during transcription.

Developing further the theoretical replication problems, Dobkin et al. (1979) turned with the MDV-1 variant to the unsolved problems of the Qβ enzyme-driven replication. They intended to answer the following intrinsic questions: Are the multistranded structures required intermediates in the replicative process? Can a single replicase molecule complete the entire replicative cycle, and in particular, can it dissociate from the complex? To determine whether a multienzyme complex was a necessary intermediate, they performed experiments under conditions of the template excess, in which only monoenzyme complexes were present. The results demonstrated that a single replicase molecule was competent to carry out a complete cycle of the MDV-1 replication and that the multistranded structures were not required intermediates in RNA replication. These experiments also suggested that, at the end of each replicative cycle after the completion of chain elongation, the complex released the product RNA before the replicase dissociated from the template (Dobkin et al. 1979).

An in-depth report on the phage Qβ replication and Darwinian experiments was given at that time by Berndt Küppers (1979).

Using sodium bisulfite that converted specifically the single-stranded cytidines to uridines, Mills et al. (1980) observed a pattern of chemical modification that strongly supported the existence of the secondary structure in the MDV-1 plus strands. The reactivity of 45 of the 76 cytidines in the MDV-1 plus strand RNA was determined by the nucleotide sequence analysis, where only 14 of these 45 cytidines were converted to uridine. The treatment of the RNA with methoxyamine, another single-strand-specific cytidine modification reagent, gave results in good agreement with the bisulfite data. The secondary structure consistent with the chemical modification data was proposed. Finally, the published sequence contained corrections to the original primary structure (Mills et al. 1973) and explained the final 221 nucleotide length of the MDV-1, where two additional cytidines occurred at nucleotide positions 90 and 173, a terminal adenosine occurred at position 221, and an adenosine was present at position 105. The modification of the MDV-1 plus strand by bisulfite rendered it inactive as a template for the RNA replication. This inactivation and the modification of the cytidines at the 3' end of the molecule occurred at very similar rates. It appeared that one or more of the cytidines in the 3'-terminal sequence were required for the template activity and that changes within this sequence had lethal consequences.

The well-characterized MDV-1 variant provided novel opportunities for the ultraviolet photosensitivity studies of an RNA with respect to its activity as a template by the replication process (Ryan et al. 1977, 1979). Finally, O'Hara and Gordon (1980) located the sites of the replication inhibition caused by the irradiation of the MDV-1 plus and minus strands with 254-nm light. Presumably, this inhibition was caused by uridine hydrates and pyrimidine cyclobutane dimers, which were previously shown to be formed at levels of 3.5% and 0.3% per the MDV-1 plus strand, respectively. Each of the ten inhibition sites corresponded to regions in the template, which contained two or more pyrimidines, including at least one uridine. At each site, the replication inhibition occurred at the two or three adjacent bases. The sites of the photolesions were analyzed with respect to the known primary sequence and the proposed secondary structure of the MDV-1 RNA (O'Hara and Gordon 1980).

The relatively short MDV-1 molecule made it possible to follow the secondary structure formation during RNA synthesis (Kramer and Mills 1981). Some of the secondary structures seen in nascent chains were observed to form, then to dissociate in favor of an alternative structure, and then to re-form, as chain growth continued. Thus, Kramer and Mills (1981) showed experimentally for the first time that the secondary structures in an RNA molecule were in a state of dynamic equilibrium, and that the extension of a sequence by chain growth, or the reduction of a sequence by processing, might result in significant changes in the secondary structures that were present.

Nishihara et al. (1983) localized the Qβ replicase recognition site in the MDV-1 RNA. The fragments of the latter, which missed nucleotides at either their 5' end or their 3' end, were still able to form a complex with the Qβ enzyme. By assaying the binding ability of fragments of different lengths, it was established that the binding site was determined by nucleotide sequences that were located near the middle of the MDV-1 RNA. The fragments missing nucleotides at their 5' end were able to serve as templates for the synthesis of complementary strands, but the fragments missing nucleotides at their 3′ end were inactive, indicating that the 3′-terminal region of the template was required for the initiation of the RNA synthesis. The nucleotide sequences of both the 3′ terminus and the central binding region of the MDV-1 plus strand were almost identical to the sequences at the 3′-terminus and at an internal region of the Qβ minus strand. Thus, 30 of 35 nucleotides, that occurred at the 3′ end of the MDV-1 plus strand between nucleotides 187 to 221, were identical with nucleotides 4186–4220 at the 3′ end of the Qβ minus strand, and 40–46 nucleotides that occurred near the center of the MDV-1 plus strand between nucleotides 81–126 were identical with the nucleotides 84–129 of the Qβ minus strand. The homologous internal regions of the MDV-1 plus strand and Qβ minus strand were remarkably similar in their location with respect to the 5′ end of each molecule. It was concluded that the MDV-1 RNA was recognized and replicated by the Qβ replicase because it shared these homologous sequence regions with the Qβ RNA (Nishihara et al. 1983).

Bausch et al. (1983) investigated with the MDV-1 the apparent requirement that the Qβ replicase must add a non-templated adenosine to the 3′ end of newly synthesized RNA strands. When the authors used the abbreviated MDV-1 plus RNA templates that lacked either 62 or 63 nucleotides at their 5′ end, the MDV-1 minus strands were released from the replication complex, yet they did not possess a non-templated 3′-terminal adenosine. It was concluded that the addition of an extra adenosine was not an obligate step for the release of completed strands. Since the abbreviated templates lacked a normal 5′ end, it seemed probable that a particular sequence at the 5′ end of the template was required for terminal adenylation to occur.

The role of the secondary structures by the evolution of the RNA phages was examined closely by Priano et al. (1987).

The kinetic and thermodynamic characteristics of the interaction between the MDV-1 RNA and Qβ replicase were determined under various conditions by using the gel-retardation assay as well as the filter retention (Werner 1991). The accurate determination of the equilibrium constants for both the MDV-1 plus and minus strands showed that the MDV-1 binding to the Qβ replicase was three orders of magnitude stronger than the binding of the full-length Qβ RNA to the enzyme. The association rate of the Qβ replicase with the plus MDV-1 was significantly larger than with the minus MDV-1, whereas the dissociation of the Qβ replicase from the minus MDV-1 strand proceeded approximately twice as fast as the dissociation from the plus MDV-1. The distinct differences in the association and dissociation rates of the plus and minus strands of the MDV-1 were observed under nearly all tested conditions. It is noteworthy that the interaction of the MDV-1 and Qβ replicase was compared with that of the R17 coat protein and RNA, which was determined earlier by the same methods (Carey et al. 1983a; Carey and Uhlenbeck 1983; Lowary and Uhlenbeck 1987), as described in the Repressor complex I: R17 section of Chapter 16.

The highly-conserved stem-loop structure in the MDV-1 RNA contributed to the localization of the two recognition elements on the Qβ RNA minus strand that were essential for the template activity (Schuppli et al. 1994). To do this, a series of minus strand RNAs with internal or external deletions were prepared by *in vitro* transcription from suitable expression plasmids, and the template activities of the deletion mutants were determined by single-round replication assays. The first element was a segment in the 5′-terminal region at positions 4078–4132 and contained the stem-loop, whose sequence was recognized to be highly conserved in the MDV-1 RNA. The second element was defined by two partially complementary sequence segments in the 3′-terminal region at positions 557–576 and 24–35, that appeared to be engaged in the long-range base pairing and might form the stem of a large secondary structure domain whose branches were not necessary for the template recognition. Remarkably, the results obtained with the Qβ replicase holoenzyme and core enzyme were identical. They confirmed the absence of any role of the S1 protein in the interaction of the Qβ replicase with the minus strand RNA, and further emphasized the profound difference in the interactions of Qβ replicase with the plus and minus strand, as described in Chapter 13.

MICRO- AND NANOVARIANTS

In parallel with the MDV-1 studies, the sequence of the second small RNA that was connected with the 6S Qβ RNA story and named WSI was determined by Charles Weissmann's team (Schaffner et al. 1974; Schaffner and Weissmann 1975). It was curious that the two sequences had little in common and that neither bore an obvious relationship to known sequences of the Qβ plus or minus strands. As the MDV-1, the WSI variant appeared in the non-templated Qβ replicase reactions. It is necessary to stress in this connection that the species similar to the 6S RNA were frequently obtained when even highly purified Oβ replicase was incubated without added template at low ionic strength, after a lag period of 2–20 min (Banerjee et al. 1969; Kacian et al. 1971; Kamen et al. 1972; Prives and Silverman 1972). According to Kamen (1972) and his authoritative review with many personal communications (Kamen 1975), some replicase preparations, particularly those derived from cells infected with plaque-purified wild-type Qβ or Qβ nonpolar coat mutants, were reportedly free of such endogenous activity.

Meanwhile, Spiegelman's team sequenced another small replicating molecule named microvariant RNA (Mills et al. 1975). In the course of examining the endogenous Qβ replicase reactions in the absence of exogenous template, which gave rise to the MDV-1 RNA, a smaller and apparently unrelated variant was also noted. This microvariant RNA was smaller, 114 nucleotides only, than the preceding MDV-1 of 220 nucleotides, the length given in this paper, but not 218, as was stated before (Mills et al. 1973). The microvariant and MDV-1 RNAs revealed no significant sequence similarity. Since the Qβ replicase mediated the synthesis of both of these disparate RNA molecules, the primary sequence seemed not to be the sole determining factor in the processes of the enzyme recognition and replication, and the key was to be found in the secondary or tertiary structures, as described in Chapters 13 and 16.

Furthermore, Weissmann's team published nucleotide sequences of the two minor nanovariant RNAs, namely, WSII and WSIII, in addition to the previously sequenced major nanovariant WSI (Schaffner et al. 1977). All these variants were derived from the molecules that were synthesized in the absence of added template. With a length of 91 nucleotides per single strand, the nanovariant RNAs were the smallest molecules known to be replicated. The sequences of oligonucleotide fragments belonging to further related species were analyzed as well. Within the family, the 21 minor variations were due to some substitutions and one apparently to a deletion of four nucleotides. The nanovariant RNA was bound to the Qβ replicase. Analysis of the protected RNA segments after ribonuclease T1 digestion showed that, as in the case of the Qβ RNA, the enzyme preferentially bound to the internal regions of the RNA, although replication started at the 3′ end. A comparison of the structures of the WSI RNA with that of the Qβ RNA, and the MDV-1 and microvariant RNA, two unrelated clones of the 6S RNA sequenced by Mills et al. (1973, 1975), showed some limited homology of certain regions of the midivariant and possibly of the nanovariant WSI to the terminal segments of the Qβ RNA. However, the only feature common to all Qβ replicase templates was the 3′-terminal -C-C-C-(A)$_{OH}$ which apparently did not participate in any secondary structure. Later, the variant WS-s(+) was synthesized enzymatically and tested for the site-directed cleavage with *E. coli* ribonuclease H, using complementary chimeric oligonucleotides containing deoxyribonucleotides and 2′-O-methylribonucleotides (Shibahara et al. 1987). As a result, it was cleaved at a single site, either at a hairpin loop or at a stem region. Although the name of this 90-nucleotide variant contained the (+) sign, it corresponded by its sequence to the WSI minus strand.

Remarkably, the model primitive tRNA with the nucleotide sequence GGCCAAAAAAAGGCCp, which was synthesized using T4 RNA ligase, was recognized by the Qβ replicase as a template (Kinjo et al. 1986).

DE NOVO SYNTHESIS?

As mentioned briefly in the Some special features of the Qβ replicase enzyme section of Chapter 13, Sumper and Luce (1975) from Manfred Eigen's team offered strong evidence for a new type of the template-free *de novo* RNA synthesis, catalyzed by the Qβ replicase, in which truly self-replicating RNA structures as long as 200 nucleotides were produced. These sequences were not homopolymeric or strictly alternating, and they were adapted to the environmental conditions applied during their generation. Qβ enzyme contamination by traces of RNA templates was ruled out by the following arguments: (i) additional purification steps did not eliminate this phenomenon, (ii) the lag phase was lengthened to several hours by lowering substrate or enzyme concentration, (iii) different enzyme concentrations led to RNA species of completely

different primary structure, (iv) addition of oligonucleotides or preincubation with only three nucleoside triphosphates affected the final RNA sequence, and (v) manipulation of conditions during the lag phase resulted in the production of RNA structures that were adapted to the particular incubation conditions applied, e.g., RNA resistant to nuclease attack or resistant to inhibitors or even RNAs "addicted to the drug," in the sense that they only replicate in the presence of a drug like acridine orange (Sumper and Luce 1975).

In contrast to other RNA and DNA polymerases that were capable of *de novo* synthesis in the absence of template but gave only highly ordered sequences, the template-free synthesis of RNA by the Qβ replicase led to the truly self-replicating RNA molecules with defined and nonrepetitive structures. A puzzling fact was the observation that under optimal conditions such as high substrate and enzyme concentration—but only at these—the products formed in separate but otherwise identical template-free RNA synthesis experiments gave very similar fingerprint patterns, suggesting the operation of an instruction mechanism.

Studying the sequence of the template-free synthesized RNA published by Mills et al. (1973), Sumper and Luce (1975) noticed the high abundance of the nucleotide sequence UUCG and its complement CGAA. The UUCG appeared as often as seven times in this RNA and was located in the unpaired regions of the molecule. Since the UUCG was common to most tRNAs as TψCG in the "TψC loop," it was speculated that the Qβ replicase would probably make a preferential use of certain oligonucleotide sequences for the final assembly of RNA molecules. Such a discrimination would limit the number of possible sequences. Moreover, during the replication phase, the sequences produced were subjected to selection for the self-replication capabilities, when both plus and minus strands could be recognized by the Qβ enzyme. Then, from several self-replicating RNAs produced simultaneously during the lag phase, the fastest replicating species would outgrow its competitors. This might explain the similarity of the RNAs produced under the optimal conditions (Sumper and Luce 1975).

It is noteworthy that the revolutionary paper of Sumper and Luce (1975) was supported and communicated by Manfred Eigen, the 1967 Nobelist, one of the grounders of the theory of quasispecies and the chemical hypercycle, namely, the cyclic linkage of reaction cycles as an explanation for the self-organization of prebiotic systems.

Further work with the direct participation of Eigen revealed intrinsic details of the *de novo* RNA synthesis by the Qβ replicase (Biebricher et al. 1981a,b). First, the products of the template-free RNA synthesis were investigated by gel electrophoresis and fingerprinting techniques (Biebricher et al. 1981b). A multitude of self-replicating RNA species appeared in the early phases of reaction with variable lengths and sequences. The template-free synthesis in different samples under completely identical conditions yielded RNA products with very different and unrelated fingerprints. The early products rapidly underwent an evolution process that altered the phenotypic properties of the self-replicating RNA and led to a concomitant increase of the replication efficiency. The fingerprints and electrophoretic mobilities of the self-replicating RNA species were altered discontinuously during the evolution process. The evolution process ended with the selection of the optimized self-replicating RNA species whose phenotypes were conserved even after many serial transfers. Some optimized RNA species, which were termed minivariants (MNV), were shorter than the MDV-1 and apparently had related sequences, since they contained many identical spots in their fingerprints. Among the MNV variants, the MNV-11 sequence of 86 nucleotides represented an optimized variant that could be easily cloned and amplified.

Next, the kinetics of the template-free and template-instructed RNA synthesis were compared (Biebricher et al. 1981a). The template-instructed RNA synthesis had different growth rates in the exponential (excess enzyme) and the linear (excess template) phase of growth. In the absence of exogenous template, the Qβ replicase synthesized self-replicating RNA after an initial lag phase. The lag time was determined by extrapolating the growth curve to the time of appearance of the first self-replicating strand. The growth rates in the exponential and linear phase, and especially the times of the lag phase for nucleotide incorporations under identical template-free conditions, showed considerable scattering in contrast to the deterministic behavior of the template-instructed synthesis. Evaluation of the kinetic data revealed that the time lag of the template-free synthesis was strongly dependent on the concentration of the nucleoside triphosphate and the Qβ enzyme. The lag time was approximately inversely proportional to the powers 2.75 of the nucleotide and 2.5 of the enzyme concentrations, respectively, both being lower limit values. The rate of the template-instructed RNA synthesis was linearly proportional to the Qβ enzyme concentration and less than linearly proportional to the triphosphate concentration, in accordance with a substrate dependence of a Michaelis-Menten type of mechanism. It was concluded that the kinetic data could not be reconciled with the proposition that the template-free synthesis was due to low concentrations of templates present, as impurities in the incorporation mixture and giving rise to autocatalytic RNA synthesis by a template-instructed mechanism. Moreover, a cyclic model was presented that described all essential steps of the replication of the self-replicating RNAs (Biebricher et al. 1981a). The results obtained were therefore in full agreement with the finding of Sumper and Luce (1975), who presented evidence that the Qβ replicase synthesized RNA de novo in the absence of exogenous template. What was difficult to understand, however, was the fact that under their standard conditions only one RNA species was reproducibly synthesized. It was 220 nucleotides long and closely related, if not identical, to the MDV-1 RNA characterized originally by Mills et al. (1973). Biebricher et al. (1981a,b) were not able to reproduce this particular result. In contrast, the template-free synthesis always resulted in the emergence of completely different RNA species each time, even if conditions were held as constant as possible, and the species were almost all considerably shorter than the MDV-1. In fact, this finding served as an additional argument in favor of the de novo synthesis (Biebricher et al. 1981b). The species closely resembled in size, heterogeneity, and sensitivity to ionic strength the 6S RNA found in the Qβ-infected cells by Banerjee et al. (1969). Moreover, Biebricher et al. (1981b) proposed that the 6S was indeed a product of the de novo synthesis by the Qβ replicase in vivo.

By self-replication of the above-mentioned MNV-11 variant in the presence of salt, the more salt-resistant variants SV-11 of 113 nucleotides and the MDV-1 of 220 nucleotides were selected, and the secondary structures of these and two other unrelated salt-resistant variants were determined (Biebricher et al. 1982). Moreover, the double-stranded duplices of the plus and minus strands of the RNAs were compared with the well-defined double-stranded DNA segments of comparable length. It is remarkable in the context of the previous information that the replication of the MDV-1 was highly resistant to and even stimulated by the addition of ammonium salt.

The properties of the three related variants MNV-11, SV-11, and MDV-1 were compared with the two stable variants SV-5 and SV-7 that were obtained by the de novo RNA synthesis at relatively high concentrations of ammonium sulfate, followed by further selection for salt resistance. These variants did not seem to be related to the first three variants. Their molecular mass was higher than those of the RNA products synthesized de novo under normal conditions (Biebricher et al. 1982). Finally, the experimental results on the replication of various templates by the Qβ replicase served as a basis to the deep and general theoretical analysis of the reaction mechanisms of the RNA replication performed by Eigen's team by means of analytical and computer simulation methods (Biebricher et al. 1983).

Remarkably, despite the complexity of this mechanism, the conventional concepts of the steady-state and dynamic enzyme catalysis and plausible values of the rate and stability constants for the elementary reactions sufficed to provide detailed understanding of the RNA replication kinetics. The main features were described with simple formulas that were analogous to the traditional descriptions of enzyme kinetics (Biebricher et al. 1983). Furthermore, Eigen's team subjected the whole process of Darwinian selection to investigation by analytical and computer-simulation methods (Biebricher et al. 1985). For this system, the relative population change of the competing species was found to be a useful definition of the selection value, calculable from measurable kinetic parameters and concentrations of each species. The critical differences in the criteria for selection were shown to pertain for the replicase/RNA ratios greater than or less than 1, for the case when formation of double-stranded RNA occurred and when comparisons were made of closed with open systems. At a large excess of enzyme, the RNA species grew exponentially without interfering with each other, and the selection depended only on the fecundity of the species, i.e., their overall replication rates. For the RNA concentrations greater than the replicase concentration, the selection of species was governed by their abilities to compete for enzyme. Under conditions where formation of double strands occurred, the competition led to a coexistence of the species; the selection values vanished, and the concentration ratios depended only on the template binding and double-strand formation rates. The approach to coexistence was rapid, because when its competitors were in a steady state, a species present in trace amount was amplified exponentially. When formation of the hybrid double strands occurred at a substantial rate, the coexistence of hybridizing species was essentially limited to cases where the formation rate of the heterologous double strands was smaller than the geometric mean of the formation rates of the homologous double strands. The authors found simple analytical expressions for the main aspects of the selection process at limiting cases, e.g., in the steady states (Biebricher et al. 1985).

Nevertheless, after the series of persuasive kinetic experiments, the eternal question "Does Qβ replicase synthesize RNA in the absence of template?" was asked again (Hill and Blumenthal 1983), as mentioned briefly in the Some special features of the Qβ replicase enzyme section of Chapter 13. The study suggested that, by the *de novo* reaction, the Qβ replicase was replicating still previously undetected contaminating RNA molecules. The Qβ replicase produced by the phosphocellulose column chromatography in the presence of a low concentration of urea and lacking the S1 protein did not synthesize detectable RNA in the absence of added template over 4 h. However, the ability to perform a reaction kinetically indistinguishable from the *de novo* synthesis reaction was restored to the highly purified enzyme by adding a heat-stable, alkali-labile component of the Qβ replicase preparations. In all fairness, one would admit that these clever combination experiments did not entirely exclude the possible ability of the Qβ replicase itself for the self-dependent synthesis of the "contaminating" seeds of the 6S RNA.

It is no wonder that Biebricher et al. (1986) published strong arguments in support of the template-free RNA synthesis by the Qβ replicase. The incorporation products seen by Hill and Blumenthal (1983) after "restoration of the *de novo* synthesis" were regarded as resulting from the template-instructed amplification of RNA impurities present in their standard enzyme preparations. Moreover, clear evidence was presented that, under the conditions required for *de novo* synthesis, the Qβ replicase prepared according to Hill and Blumenthal (1983) was also capable of *de novo* synthesis (Biebricher 1986; Biebricher et al. 1986). Furthermore, it was shown that the Qβ replicase condensed nucleoside triphosphates to more-or-less random oligonucleotides. It was conclusive that highly purified preparations of the Qβ replicase isolated from uninfected *E. coli* cells containing plasmids carrying the replicase gene showed the same *de novo* synthesis characteristics as those isolated from infected bacteria.

At that time, a solid account on the template-free synthesis with the Qβ replicase producing self-replicating RNA species of 70–250 nucleotides in length and on the role of this phenomenon for the study of the *in vitro* evolution processes was given by Christof K. Biebricher (1987). A bit later, Biebricher (1992) presented a quantitative analysis of mutation and selection in self-replicating RNAs and argued that the behavior of the latter mimics in many aspects the complex Darwinian behavior of living organisms. Such analysis was possible, since the RNA synthesis profiles were described by compact equations. The selection behavior of competing RNA species could be precisely predicted, using these equations, from kinetic parameters of the species. At low concentrations, the RNA species were selected for overall growth rate (fecundity), at higher concentrations, for rapid binding of replicase (selection for competition), and at still higher concentrations, for minimizing losses caused by formation of inactive double strands. Finally, an ecosystem might be established where the different species coexist, their relative concentrations being functions of their kinetic parameters. The analysis of competition and selection was extended to mutants of a species. The model made it possible to choose experimental conditions where quantitative measurement of mutation rates and selective values of mutants was possible. The interplay of mutation and selection resulted in establishing a *quasispecies* distribution, where mutants were represented according to their rates of mutational formation and their selective values. The replicating RNA clones, when amplified, rapidly build up quasispecies distributions containing pronounced "hot spots," produced predominantly by error propagation of nearly neutral mutants. Therefore, the primitive model system showed the same complex Darwinian behavior as observed in evolution of biological systems (Biebricher 1992).

It is noteworthy that the classical review of Biebricher and Eigen (1988) on the Qβ replicase was reprinted by CRC Press after 30 years (Biebricher and Eigen 2018).

The experimental and theoretical studies of Eigen's team were used as a background for the theoretical calculations of Ariel Fernández (1988a,b) on the onset and stochastic dynamical constraints in the *de novo* RNA replication. Furthermore, the autocatalytic MDV-1 replication was used for a set of the productive predictions and simulations (Fernández 1989a,b,c, 1991). Earlier, Peter Schuster (1984a,b) used the obtained data for his work on the kinetic theory of molecular evolution. At the same time, Klaus Dose (1984) interpreted the data in his study on the origin of cellular life with self-replicating polynucleotides that had to be formed by abiotic processes on the prebiotic Earth.

It should be stressed that the group IV phage SP replicase demonstrated the same template-free synthesis, by analogy with the Qβ replicase (Fukami and Haruna 1979). Moreover, the crossed template specificity of both SP and Qβ replicases was examined using the variant RNAs as templates. The three variant RNAs, one 8S generated by the Qβ replicase and two, 6S and 5.2S, generated by the SP replicase, were isolated from the reaction mixtures incubated in the absence of exogenous template RNA. All these RNAs were found to be active as templates for both SP and Qβ replicases, though homologous RNA exhibited activities about three times higher than heterologous RNA with either enzyme, in agreement with the results obtained in the phage RNA-dependent reactions. In these reactions, faithful replication of variant RNA was observed, and the amount of the RNA synthesized was in a manyfold excess over the template RNA added.

Later, Biebricher et al. (1993) turned back to the general problems of the template-directed and template-free RNA synthesis. They insisted on the fundamentally different nature of both the template-instructed and template-free synthesis and were concerned that reports of problems with contamination in other laboratories should not prevent the scientific community from investigating a primary model system for evolution at the molecular level. According to their opinion, the *de novo* reaction gave clear answers as to how information was formed by evolutionary events. Therefore, the authors outlined as clearly as possible the evidence for the template-free synthesis of self-replicating RNA and disproved the explanation of the emerging RNA species in the template-free reactions by residual RNA contaminants in the incubation mixture.

MOLECULAR PARASITES

In contrast to Biebricher's strong position, Chetverin et al. (1991) from Alexander Spirin's team thought that the spontaneous RNA synthesis by the Qβ replicase was after all template directed. The authors regarded the laboratory air as an immediate source of the template RNA and showed the ways to eliminate, or at least substantially reduce, the harmful effects of the spontaneous synthesis. The solitary RNA molecules were detected in a thin layer of agarose gel containing Qβ replicase, where they grew to form colonies that became visible upon staining with ethidium bromide. This result provided a powerful tool for the RNA cloning and selection *in vitro*. Moreover, the authors showed that the replicating RNAs, similar to those growing spontaneously, were incorporated into the Qβ phage particles and propagated *in vivo* for a number of the phage Qβ generations. Therefore, Chetverin et al. (1991) concluded that these RNAs were the smallest known molecular parasites, and in many aspects they resembled both the defective interfering genomes of animal and plant viruses and plant virus satellite RNAs.

Moody et al. (1994) elaborated a special method for the purification of gram quantities of highly purified Qβ replicase, where the enzyme was produced from a tightly regulated expression plasmid and no phage or phage RNA was introduced into the system, except for the RNA encoding the replicase message. The expression was induced late in the growth cycle from seed stocks of transformed *E. coli* cells that were themselves not passaged, thus minimizing the time available for the replicase holoenzyme to evolve the replicable RNA *in vivo*. Finally, the Qβ enzyme was purified under rigorously aseptic conditions, eliminating contamination with exogenous replicable RNAs from the laboratory environment. A number of modifications were elaborated to standard reaction conditions that suppressed or significantly delayed the appearance of spontaneous replication products by hours or days, while permitting the replication and detection of single molecules of exogenously added template molecules in 15 min or less. Nevertheless, under relaxed reaction conditions or during very long incubations, the enzyme preparation was still capable of spontaneously generating a variety of the replicable RNAs. In

an effort to understand the origin of these RNAs, Moody et al. (1994) developed a method of cloning the products of such reactions without *a priori* knowledge of their sequence. The analysis of a number of such spontaneous replication products showed that they were related to *E. coli* nucleic acids. It was concluded that the spontaneous replication products were derived from low levels of *E. coli* nucleic acids that contaminated the Qβ replicase preparation. These poorly replicating RNAs of the tRNA origin then evolved into the replicable species under relaxed conditions or long incubations of untemplated reactions. There appeared therefore to be no necessity to invoke the more exotic hypothesis that any amplifiable RNAs arose from the random untemplated polymerization of nucleotides in the reaction (Moody et al. 1994).

It is remarkable in this context that tRNA[His] was the only stable *E. coli* RNA with 3′-terminal CCCA sequence that was shown to be critical for Qβ amplification (see Chapter 13). In fact, *in vitro*-generated transcripts corresponding to tRNA[His] served as poor templates for the Qβ replicase, due to the inaccessibility of the partially base-paired CCCA (Tretheway et al. 2001).

MOLECULAR COLONIES

The work of Chetverin et al. (1991) gave onset to the highly prospective molecular colony concept. To clone RNA molecules *in vitro*, Chetverina and Chetverin (1993) reported an advanced RNA colony procedure in an immobilized medium. The medium contained a complete set of nucleotide substrates and purified Qβ replicase. RNA amplification in the immobilized medium resulted in the formation of separate colonies, each comprising a clone or the progeny of a single RNA molecule. The colonies were visualized by staining with ethidium bromide, by utilizing radioactive substrates, and by hybridization with sequence-specific labeled probes. The number and identity of the RNA colonies corresponded to that of the RNAs seeded. When a mixture of different RNA species was seeded, these species were found in different colonies. The possible implementations of this technique included a search for recombinant RNAs, very sensitive nucleic acid diagnostics, and gene cloning *in vitro*.

The molecular colony technique allowed discovery of the nonhomologous RNA recombination in a cell-free system by spontaneous rearrangements of RNA sequences in the absence of any protein or ribozyme (Chetverin et al. 1997; Chetverin 1999; Chetverina et al. 1999), as will be presented below in the corresponding Recombination section in this chapter.

An attempt was made to employ the Qβ molecular colony technique by diagnostics of virus infections, but the target-specific inserts, even short ones, often suppressed replication of natural replicase templates (Chetverin and Chetverina 2002). Hence the authors preferred to use the molecular colony technique in the PCR version, where one DNA molecule or seven RNA molecules were sufficient for the direct estimation of the titer of a target by counting DNA colonies (Chetverina et al. 2002). The technique was accelerated and simplified significantly by transfer of the molecular colonies onto a nylon membrane followed by membrane hybridization with fluorescent oligonucleotide probes (Chetverina et al. 2007).

The history and progress of the scientific and practical applications of the molecular colonies was analyzed in detail by Chetverin and Chetverina (2007). In particular, they noticed that Mitra and Church (1999) published a PCR version of the molecular colony technique under the name of polony (polymerase *colony*) technology. Next, they noticed that Szostak (1999) was the first to indicate that molecular colonies, analogous to those formed when RNA was amplified in agarose with Qβ replicase, could have represented a precellular variant of compartmentalization in the RNA World. In this case, the compartmentalization was provided not by an envelope but by a relatively low diffusion rate of macromolecules in a porous matrix as compared with the diffusion rate of low molecular mass substances. According to Szostak, the RNA colonies could be formed in moist clay and other porous minerals (cited by Chetverin and Chetverina 2007). This idea was further developed by Alexander Spirin (2002, 2005a,b), who postulated that the mixed colonies comprising the following three species of macromolecules could be the first individuals in the RNA World: (i) ligand-binding RNA molecules for selective adsorption and accumulation of necessary substances from the environment; (ii) a set of ribozymes catalyzing the metabolic reactions of nucleotide synthesis; and (iii) a ribozyme catalyzing complementary replication of all RNA molecules of the colony.

As Chetverin and Chetverina (2007) concluded, the RNA colonies could easily exchange their genetic material in the absence of envelopes, i.e., perform horizontal gene transfer within the population, including the air exchange, as the authors demonstrated earlier (Chetverin et al. 1991).

Later, the full story of the mysterious Qβ replicase enzyme, including the molecular colony technique, was narrated by the actual actors of this part of the Qβ history (Chetverin and Chetverina 2008). The current protocols for the detection of RNA molecules by growing RNA colonies in gels containing Qβ replicase or cDNA and visualizing them by hybridization with fluorescent probes directly in the gel or by transfer to a nylon membrane were published recently by Chetverina and Chetverin (2015).

AUTOCATALYTIC SYNTHESIS OF HETEROLOGOUS RNAS

Finally, the time was ripe for the onset of the gene engineering approach to the RNA vectors. Thus, the MDV-1 plus strands were used as a special vector for the cloning of heterologous RNA (Miele et al. 1983). Donald R. Mills and Fred Russel Kramer came forward among the most advanced enthusiasts of the idea. A heterologous RNA sequence, in this case decaadenylic acid, was copied *in vitro* by the Qβ replicase, when it was inserted between nucleotides 63 and 64 of the MDV-1 plus RNA, using phage T4 RNA ligase. The insert was located away from regions of the template known to be required for the enzyme binding and for the initiation of complementary strand synthesis, namely, into a hairpin loop on the exterior of

the molecule. The reaction was autocatalytic, and the amount of recombinant RNA increased exponentially, achieving a 300-fold amplification within 9 minutes (Miele et al. 1983). The birth of the recombinant RNA technology was welcomed warmly in the detailed *Science* response, saying "If the replication system works, Qβ replicase will be elevated … to being a molecular workhorse in a major industry" (Lewin 1983).

It would be fair to mention here that the polyA segments were inserted earlier between two Qβ RNA molecules which have been partially degraded, one from the 3′-end and another from the 5′-end, and a few recombinant phages were obtained which carried about 40-nucleotide-long polyA stretch (Shen and Jiang 1982).

The attempts to implement the MDV-1 vector for diagnostic purposes followed. First, Chu et al. (1986) attached biotin to the 5′-terminus of the MDV-1 plus strand via a disulfide linker. This biotinylated RNA was combined with avidin to give a product that was readily purified by gel electrophoresis. The RNA-biotin-avidin adduct, and the RNA released from it by reductive cleavage of the linker arm, replicated normally and was suggested to be a suitable reporter for a variety of replication-assisted bioassays involving biotinylated antibodies or biotinylated nucleic acid probes. The power of the exponential amplification mentioned in this paper deserves special merit. In fact, the RNA reporter was amplified exponentially, with a population doubling time of 36 sec, resulting in the synthesis of 10^6 copies of each molecule in 12 min. Figure 18.1 presents the famous MDV-1 plus strand molecule taken from this paper, with the secondary structure improved after the appropriate calculations according to Zuker and Stiegler (1981).

After addition of the biotin to the MDV-1, Chu and Orgel (1988) developed general methods for joining together, via

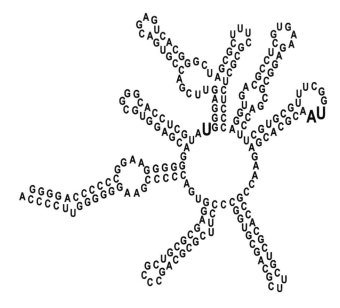

FIGURE 18.1 Nucleotide sequence of mutant MDV-1 (+) RNA. The RNA is folded into the secondary structure predicted by computer to be most stable (Zuker and Stiegler 1981). Large letters identify the three nucleotide substitutions that distinguish this sequence from wild-type MDV-1 (+) RNA. (Chu BC, Kramer FR, Orgel LE, *Nucleic Acids Res*, 1986, by permission of Oxford University Press.)

cleavable disulfide bonds, either two unprotected polynucle-otides or a polynucleotide and a peptide or protein. To join two oligonucleotides, each was first converted to an adduct in which cystamine was joined to the 5′-terminal phosphate of the oligonucleotide by a phosphoramidate bond. The adducts were mixed and reduced with dithiothreitol. Oxidation by atmospheric oxygen occurred to yield the required dimer. To join an oligonucleotide to a cysteine-containing peptide or protein, the 5′-cystamine oligomer was first converted to a 2′-pyridyldisulfide adduct and then reacted with an excess of the peptide or protein. This procedure was used first to join the MDV-1 RNA with a 16mer deoxynucleotide probe. This adduct hybridized with a complementary target DNA. Then, the MDV-1 RNA was joined to human IgG and the MDV-1 RNA-IgG adduct bound to a complementary anti-IgG.

Furthermore, a sequence specific for the protozoan para-site *Plasmodium falciparum* was cloned within the MDV-1 vector (Lizardi et al. 1988). The probe sequence was inserted within the hairpin loop on the exterior of the MDV-1 RNA. The recombinant RNAs hybridized specifically to comple-mentary DNA, were replicated at the same rate as the MDV-1 RNA, despite their additional length, and were able to serve as templates for the synthesis of a large number of RNA cop-ies. The Qβ replicase initiated with only 0.14 femtograms of recombinant RNA, or about 1000 copies, could produce 129 nanograms of recombinant product in 30 min, which repre-sented a one-billion-fold amplification (Lizardi et al. 1988). This paper served as a background for the bioassays described below. Figure 18.2 shows the nucleotide sequences of the recombinant transcripts folded into the secondary structures predicted according to Zuker and Stiegler (1981).

The simple construction of the recombinant RNAs was achieved using the recombinant DNA technique. After gen-eration of the appropriate plasmid with a synthetic polylinker within the MDV-cDNA sequence, the modified RNAs were synthesized by transcription with T7 polymerase from recom-binant pT7-MDV-poly plasmids, where the recombinant MDV-1 derivatives were placed under the T7 promoter. The recombinant RNAs were therefore generated by inserting any heterologous DNA sequence into one of the unique restriction sites in the polylinker of pT7-MDV-poly, and then utilizing the resulting plasmid as a template for transcription by T7 RNA polymerase (Lizardi et al. 1988). Therefore, amplifica-tion of the RNA probes was suggested as a logical alternative to the DNA polymerase chain reaction (PCR).

The first quantitative assay based on such replicable hybridization probes was constructed by Lomeli et al. (1989). The construction scheme followed precisely to that previ-ously described by Lizardi et al. (1988). A 30-nucleotide-long probe complementary to a conserved region of the *pol* gene in HIV-1 genome was inserted into the MDV-1 vector. After hybridization, the probe-target hybrids were isolated by reversible target capture on paramagnetic particles. The probes were then released from their targets and amplified by incubation with the Qβ replicase. The amount of RNA syn-thesized in each reaction, approximately 50 ng, was sufficient to measure without using radioisotopes. The kinetic analysis

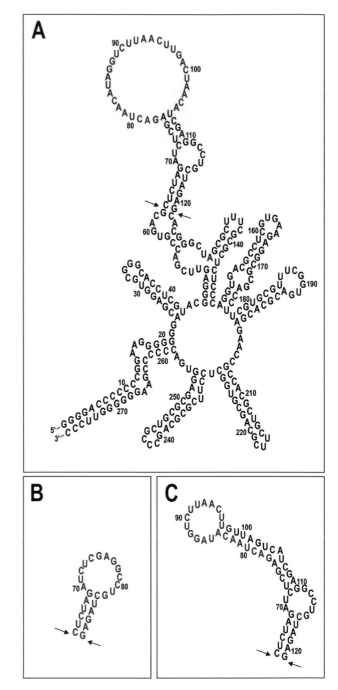

FIGURE 18.2 The nucleotide sequences of the recombinant tran-scripts folded into the secondary structures predicted according to Zuker and Stiegler (1981). (A) The MDV-fal-un (+) RNA contains a 58-nucleotide insert (shown between the arrows) in place of the 3-nucleotide segment, AGU, that occurs in natural MDV-1 (+) RNA (Mills et al. 1973). The secondary structures that the computer pro-gram predicts in the region of the recombinant outside the insert are identical to the secondary structures that were experimentally iden-tified in MDV-1 RNA (Mills et al. 1980; Kramer and Mills 1981), suggesting that the probe sequence would have little effect on the topology of the MDV-1 domain. Red letters indicate the nucleotides that are complementary to *P. falciparum* DNA. The lower panels show the predicted secondary structure of the inserts present in MDV-poly(+) RNA (B) and MDV-fal-st (+) RNA (C). (Reprinted by permission from Springer Nature. *Bio/Technology*, Lizardi et al., Copyright 1988.)

of the reactions demonstrated that the number of HIV-1 targets originally present in each sample could be determined by measuring the time it took to synthesize a particular amount of RNA. The authors suggested that clinical assays involving replicable hybridization probes will be simple, accurate, sensitive, and automatable (Lomeli et al. 1989).

Continuing the theoretical studies on the expression mode of the heterologous inserts, Axelrod et al. (1991) generated 14 recombinant RNA templates for the $Q\beta$ replicase, each having either an exogenous inverted repeat sequence or a sequence with no repeat. The replication rates for the RNAs that putatively contained secondary structures in the recombinant sequences ranged from 33%–69% that of the wild-type MDV-1 RNA control, regardless of the size of the inserted hairpin. Moreover, most of the newly synthesized RNA was present as single strands. Alternatively, the RNAs that contained exogenous sequences not expected to form secondary structures exhibited replication rates less than 25% that of the MDV-1. In each case, the reaction rate was correlated with the length of the insertion, and the majority of product RNA consisted of duplexed molecules (complementary plus and minus strands hybridized together). When the same recombinant RNAs were used in reactions, in which the molar amount of RNA template was 10^6–10^7 times lower than that of the replicase, only those that putatively contained secondary structures survived in the replication reaction. Therefore, these results were fully consistent with the theory that the hairpin structure formation during RNA synthesis directly influenced the regeneration of the single-stranded RNA products.

RQ VARIANTS

The opportunity to use the $Q\beta$ replicase templates as vectors provoked interest in the novel replicable sequences and a class of so-called related to $Q\beta$ (RQ) sequences was sequenced, in addition to the three previously sequenced structures: MDV-1 of 221 nucleotides (Mills et al. 1973), microvariant of 114 nucleotides (Mills et al. 1975), and nanovariant WSI of 91 nucleotides (Schaffner et al. 1977). In fact, some of the RQ sequences could be more efficient $Q\beta$ replicase templates than the known templates, including the $Q\beta$ genomic RNA, and could provide the identification of novel sites within the vector RNA, where foreign RNA sequences could be inserted without affecting replication. In the search for novel vectors, Munishkin et al. (1988), from Spirin's team, sequenced a new RQ RNA, 120 nucleotides long, which was isolated from the products of spontaneous synthesis by the nominally RNA-free $Q\beta$ replicase preparation. The minus strand of this RNA appeared to be a recombinant RNA, composed of the internal fragment of $Q\beta$ RNA, approximately 80 nucleotides long, and the 33-nucleotide-long 3′-terminal fragment of E. coli tRNA$_1$^{Asp}. This seemed to be the first strong indication of the RNA recombination in bacterial cells. The entire recombination phenomenon will be described below in the Recombination section of this chapter.

The RQ sequences were labeled further with numbers that corresponded to the length of the sequence. Thus, Munishkin et al. (1991) isolated the RQ135 variant from the products of spontaneous RNA synthesis in the template-free in vitro $Q\beta$ replicase reaction and showed its high efficiency as the $Q\beta$ replicase template. Surprisingly, the RQ135 sequence consisted entirely of segments that were homologous to ribosomal 23S RNA and the phage λ origin of replication but were unrelated to the $Q\beta$ genomic RNA. Avota et al. (1991), from Elmārs Grēns' team, cloned and transcribed the DNA copies of the autocatalytic RNAs RQ135 and RQ134.

The structural consensus of the RNA templates replicated by the $Q\beta$ replicase was summarized at that time by Voronin (1992). This sequence analysis has revealed that all these RNAs were potentially capable of forming a consensus secondary structure element that was represented by stalk which was formed by the 5′-GGG … and … CCCA-3′ complementary stretches at the termini of the replicating RNA molecules and adjacent 5′- and 3′-hairpins, which might form a stacking with the stalk. The structure found was rather similar to the analogous structure in the tRNA molecule. The genomic RNAs of the phage $Q\beta$ and other related phages demonstrated a similar structural element. The replicable RNA vectors, together with the prospects for cell-free gene amplification, expression, and cloning, were described in the large review by Chetverin and Spirin (1995).

The DNA cloning possibilities opened a novel horizon for the deep analysis of the RNA species synthesized by the $Q\beta$ replicase without a template. Thus, Biebricher and Luce (1993) sequenced several independent template-free products of 25 to about 50 nucleotides by cloning their cDNAs into plasmids. While their primary sequences were unrelated except for the invariant 5′-terminal G and 3′-terminal C clusters, their tentative secondary structures showed a common principle: both their plus and minus strands had a stem at the 5′ terminus, while the 3′ terminus was unpaired. The DNA cloning was extremely helpful for this aim, since direct accumulation of sufficient quantities of the early template-free synthesis products was prevented by the inherent irreproducibility of the synthesis process and by the rapid change of the products during amplification by evolution processes, but large amounts of such RNAs was synthesized in vitro by transcription from the cDNA clones with T7 RNA polymerase.

Zamora et al. (1995) designed artificial short-chained RNA species by transcription from synthetic oligodeoxynucleotides with T7 RNA polymerase, starting from the basic assumption that the leader stem structure must be at the 5′ terminus, while the 3′ terminus must be unpaired. Thus, a synthetic short RNA species known to be replicated was amplified, forming a stable quasispecies; i.e., its sequence was conserved during hundreds of replication rounds. A synthetic mutant of this sequence that stabilized the leader in one strand but favored a 3′-terminal stem in the other one led to the complete loss of template activity. When new RNA sequences with the described structural requirements were designed and synthesized, their template activity was too low to be directly measurable; however, incubation with the $Q\beta$ replicase produced replicating RNA whose sequence was closely related to the synthesized RNA species. The most likely interpretation was that the designed sequences were in a low mountainous region in the replication fitness landscape and were optimized during amplification by the Qβ replicase to a nearby fitness peak. Therefore, the structural features postulated to be required for replication were not only conserved but even improved in the outgrowing mutants (Zamora et al. 1995).

Furthermore, Rohde et al. (1995) determined the mutant spectrum of the above-mentioned minivariant MNV-11 replication by retrotranscribing the RNA to DNA and cloning it into plasmids. A surprisingly broad mutant distribution was found: the consensus sequence never made up more than 40% of the total population and was accompanied by many mutants. Most mutants had several base exchanges, insertions, and/or deletions; up to 9 of the total 86 nucleotides were changed. The mutants found had replication rates comparable to those of the wild-type and were thus enriched in the population by selection forces. When the growth conditions were changed, the mutant distribution center was shifted. Remarkably, the published consensus sequence of the MNV-11 did not have the highest growth rate of the mutants but was rather the best adapted to the various selection forces governing the growth phases the replicating RNA went through, i.e., it had found an optimal compromise between the rates of overall replication, enzyme binding, and double strand formation (Rohde et al. 1995).

The MNV-11 was used as a model by the examination of a novel vector for generating RNAs with defined 3′ ends and its use in antiviral strategies (Schwienhorst and Lindemann 1998). This transcription system was constructed in a way that allowed trimming of 3′ termini of RNA transcripts in *E. coli* by endogenous ribonuclease P and directed posttranscriptional cleavage 3′ of the target sequence. The RNA released proved to function as an active template for the Qβ replicase. Moreover, *E. coli* cells producing these short-chain replicator molecules no longer supported multiplication of the phage Qβ upon infection.

The MNV-11 was used as the Qβ replicase template by the testing of the unique automated multichannel PCR and serial transfer machine that was designed by Eigen's team as a future tool in evolutionary biotechnology (Schober et al. 1995). The idea consisted of the evolutionary *in vitro* optimization of macromolecules, and the automated machine processed up to 960 samples in parallel, with special sealed plastic reaction vessels and appropriate handling devices.

Rather suddenly, Biebricher and Luce (1996) discovered the template-free generation of RNA species that were replicated with the phage T7 RNA polymerase by incubating high concentrations of this enzyme with substrate for extended time periods. The products differed from sample to sample in sequence and length ranging from 60–120 nucleotides. The authors regarded the mechanism of autocatalytic amplification of RNA by T7 RNA polymerase to be analogous to that observed with the Qβ replicases. With enzyme in excess, exponential growth was observed; linear growth resulted when the enzyme was saturated by RNA template. The secondary structures of all species sequenced turned out to be hairpins. The RNA species were not accepted as templates by the Qβ replicase, or by the RNA polymerases from *E. coli* or phage SP6, while T3 RNA

polymerase was partially active. The template-free RNA production was completely suppressed by addition of DNA to the incubation mixture. No replicating RNA species were detected *in vivo* in cells expressing the T7 RNA polymerase (Biebricher and Luce 1996). This finding indicated a wider occurrence of the RNA replication than previously assumed. Biebricher and Luce (1996) pointed to viroids, the smallest agents of infectious diseases known, as the remarkable natural example of a parasitic RNA directing its own replication by a host RNA polymerase (Diener et al. 1993).

Later, Wettich and Biebricher (2001) detected the template-free replication also in the case of the RNA polymerase holoenzyme from *E. coli*. The replication products were heterogeneous in length; the different lengths appeared to be different replication intermediates. No double-stranded RNA was found, even though base pairing was certainly the underlying basis of the replication process. The reaction was highly sensitive: a few RNA strands were sufficient to trigger an amplification avalanche.

Christoph K. Biebricher's team combined their efforts with Grēns' team in order to clarify the problem of the *in vivo* appearing 6S RNA (Avota et al. 1998). Thus, the RNA of *E. coli* infected with the phage Qβ was isolated and screened for replicable short-chained RNA. In contrast to earlier assumptions, it appeared that the short-chained replicable RNA was a very minor part of the RNA synthesized within the infection cycle. The replicable RNA isolated from infected cells was derived from cellular RNA, in particular 23S rRNA and 10Sa RNA, and from the Qβ RNA itself. Surprisingly, none of the many RNA species known from *in vitro* experiments was found. The RNA species isolated were all inefficient templates. No replicable RNA could be isolated from non-infected cells. Even in the cells expressing high amounts of the Qβ replicase, very few RNA species could be isolated. It was concluded that the RNA generated *in vitro* in the template-free synthesis was not derived from the RNA species found *in vivo*, and the replicable RNA found *in vitro* was generated by a mechanism fundamentally different from the one operating *in vivo* (Avota et al. 1998).

Under strong influence of the Qβ self-replication, other examples of defective or "selfish" RNA molecules optimized for replication efficiency were discovered. Thus, Zhang and Brown (1993) turned to a family of single- and double-stranded RNA molecules, termed RNA plasmids, that replicated in a DNA-independent manner in the mitochondria of maize plants. The 719-nucleotide sequence of the so-called RNA b lacked open reading frames of significant length, and shared certain structural similarities with the Qβ RNA variants. Moreover, the small double-stranded RNA element that was isolated from a moderately hypovirulent strain of the chestnut blight fungus *Cryphonectria parasitica* demonstrated more relatedness to the Qβ RNA than to other hypovirulence-associated double-stranded RNAs (Polashock and Hillman 1994).

By further studies on the RQ135 variant, Ugarov et al. (2003) examined the template properties of the 5′ and 3′ fragments obtained after cleavage of the RQ135 RNA at an internal site.

Whereas the 3′ fragment inherited the initiation oligoC cluster, the 5′ fragment inherited the 5′ terminal oligoG. Because the two clusters were in separate molecules, the exponential amplification was not expected but the ability of the fragments to direct the synthesis of their complementary copies was checked. Unexpectedly, the authors found that, although only the 3′ fragment inherited the initiator oligoC cluster, the 5′ fragment was also capable of template activity allowing the Qβ replicase to initiate and elongate. However, whereas the enzyme recognized the 3′ fragment in the same way as it did with the intact RQ135 RNA, a quite different mechanism was employed for the initiation of the RNA synthesis on the 5′ fragment and a number of its derivatives. The most striking features of the new mechanism were that the initiation occurred without regard to the oligoC clusters and in the absence of GTP, and did not lead to the formation of a stable replicative complex that was characteristic of the typical Qβ replicase templates. Thus, by the different initiation mechanisms, the Qβ replicase discriminated between the legitimate and illegitimate templates. It was concluded that triggering the GTP-dependent conformational transition at the initiation step could serve as a discriminative feature of legitimate templates, providing for the high template specificity of the Qβ replicase (Ugarov et al. 2003).

Furthermore, by using the 139-nucleotide-long derivative of the RQ135^{-1} RNA as one of the most efficient Qβ replicase templates, and its 75-nucleotide-long 5′-terminal and 109-nucleotide-long 3′-terminal fragments that supplemented each other to the entire RNA sequence and contained foreign extensions at the truncated ends, Ugarov and Chetverin (2008) demonstrated that, in contrast to the illegitimate templates, the 5′ terminus of the legitimate template cooperated with the 3′ terminus during initiation and contributed to the stability of the replicative complex after initiation. The dissociation of the molecule into 5′-terminal and 3′-terminal fragments and even point mutations at the 5′ terminus reduced the rate of the initiation step and the overall rate of template copying, increased the GTP requirement, and destabilized the post-initiation complex. The results suggested that the cooperation between the opposite 5′ and 3′ ends was an important distinguishing feature of the legitimate Qβ replicase templates (Ugarov and Chetverin 2008).

In a felicitous review, Chetverin (2011) numbered five great paradoxes of the Qβ replicase-driven amplification including its uneven template specificities. Thus, this review presented the secondary structure models of broad RQ families coming from different laboratories. Moreover, the author suggested a hypothetical structure of the Qβ replicating complex in which several complementary RNA strands were synthesized at a time on the same template which was secured in a circle by the terminal helix. The terminal helix formed by every replicating RNA might first be a structural element distinguishing the legitimate templates from illegitimate ones, and, second, by securing the legitimate template in the circular configuration, prevent its annealing with nascent strand during elongation. This hypothetical complex is presented in Figure 18.3.

Recently, Alexander Chetverin (2018) published an exhaustive review dedicated to his 30 years of studies on the Qβ replicase.

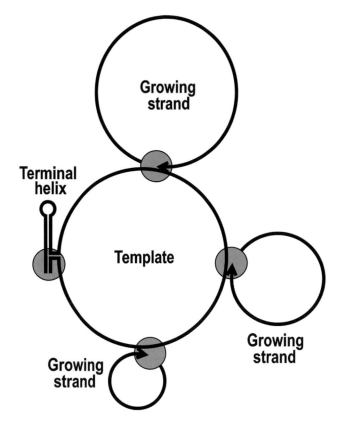

FIGURE 18.3 Hypothetical structure of a replicating complex in which several complementary RNA strands are synthesized at a time on the same template which is secured in a circle by the terminal helix. (Reprinted by permission from Springer Nature. *Mol Biol (Mosk).* Chetverin, Copyright 2011.)

MAPPING OF Qβ REPLICASE

As for the purely scientific exploitation of the MDV-1 vector capabilities, Donald R. Mills (1988) turned to the engineering of the recombinant messenger RNAs that could be replicated and expressed inside bacterial cells by the Qβ replicase, where the latter was tricked into recognizing, binding to, and replicating of the host-cell message RNAs. Therefore, the first successful replication of an active messenger RNA was achieved by a phage replicase *in vivo*. In this system, the Qβ replicase gene was constitutively expressed from a plasmid vector present in an *E. coli* host, while a second plasmid directed the transcription of a replication-competent, phage-like template RNA. This 680-nucleotide transcript, named N-RNA, contained the specific sequences required for the Qβ replicase function and a 360-nucleotide sequence that was complementary to the messenger RNA for the DNA phage λ N protein. Therefore, the active messenger RNA for the λ N protein was expressed only upon replication of the N-RNA by the Qβ replicase. The persistence of the active messenger RNA was ensured by applying the stringent selective pressure, namely, by employing of an *E. coli* host strain that required the λ N protein for growth under specific selective conditions. This new replication system provided a potential means of amplifying other recombinant messenger RNAs and their protein products *in vivo*. Moreover, it was regarded as helpful for gaining insight into how the viral RNA genomes promiscuously acquired additional information from host RNA sources and further evolved this information to their own advantage (Mills 1988).

This excellent approach was developed further for the fine mapping of the functional domains of an RNA-dependent RNA polymerase within the Qβ replicase enzyme (Mills et al. 1989). In the replication system generated after Mills (1988), the first plasmid actively expressed the Qβ replicase, while the second plasmid simultaneously transcribed an RNA template for the Qβ replicase. Contained within this RNA template were sequences complementary to a precursor of the *amber* tyrosine tRNA suppressor, supF, and utilizing the Qβ replicase, this RNA template directed the synthesis of the supF precursor tRNA sequence in the positive orientation. The bacterial tRNA maturation enzymes then functioned to generate mature supF tRNA from the precursor. This entire process was carried out in a non-suppressor *E. coli* strain under selective conditions that required the presence of supF tRNA for survival. In this way, the bacterial selection was fully dependent upon the presence of the mature supF tRNA. Since the supF tRNA was only generated if the transcribed RNA templates were replicated, the cell growth was a direct assay for the presence of the functional Qβ replicase. In such a way, a functional region of the Qβ replicase was localized after an extensive mutational analysis by an oligonucleotide linker-insertion mutagenesis that introduced small amino acid insertions. The analysis of 37 different mutant clones indicated that the Qβ replicase can accept amino acid substitutions and insertions at several sites at the N- and C-termini without abolishing functional activity *in vivo* or *in vitro*. However, the disruption within the internal amino acid sequences resulted almost exclusively in a nonfunctional enzyme. The results suggested that the central region of the Qβ replicase subunit contained a rigid amino acid composition that was required for the replicase function, whereas the N- and C-termini were much more receptive to small amino acid insertions and substitutions (Mills et al. 1989). Chapter 13 presents the current three-dimensional structures of the Qβ enzyme to be compared with these early considerations.

Using the same method, Mills et al. (1990) probed for the functional cis-acting RNA sequence elements that were contained within the coding structure of the Qβ replicase. Thus, for the first time, the regions of the cis-acting RNA elements within the bacteriophage Qβ replicase were mapped by analyzing the infectivities of the 76 replicase gene mutant phages in the presence of a helper replicase. Two separate classes of the mutant Qβ phage genomes (35 different insertion mutants, each containing an insertion of 3–15 nucleotides within the replicase gene and 41 deletion genomes, each having from 15–935 nucleotides deleted from different regions of the gene) were constructed, and their corresponding RNAs were tested for the ability to direct the formation of the progeny phage particles. Each mutant phage was tested for the plaque formation ability in an *E. coli* F⁺ host strain that supplied the helper Qβ replicase *in trans* from a plasmid DNA. Of the 76 mutant genomes, 34% were able to direct the phage production at or close to the wild-type levels, another 36% also produced the phage particles, but at much lower levels than those of the wild-type phage, and the remaining 30% produced no phage at all. This made it possible to map the regions that contained the functional cis-acting RNA elements and to correlate them further with the regions of RNA that were solely required to code for the functional RNA polymerase activity (Mills et al. 1990).

In parallel, the spatial neighborhood of the active center of the Qβ replicase was selectively modified by a method of self-catalyzed affinity labeling, where the MDV-1 RNA was used as the template (Hartmann et al. 1988; Lindner et al. 1991). By this method, in the template-directed, mainly intramolecular enzymatic catalysis, the product ^{32}P-GpG became specifically attached to the β subunit of the Qβ replicase. Using limited digestion of the radioactively labeled polypeptide by cyanogen bromide or N-chlorosuccinimide, Lindner et al. (1991) mapped the attachment site to the region of the β subunit between Trp93 and Met130, where Lys95 was the amino acid most likely to be modified, suggesting that the Lys95 was located near the nucleotide binding site in the active center.

The phage Qβ showed the highest mutation rate reported among viruses (Bradwell et al. 2013) and therefore should be particularly sensitive to nucleoside analog mutagenesis. In fact, it was shown that the experimental populations of Qβ exposed to the nucleoside analog 5-azacytidine (AZC) experienced strong decreases in the phage yield, which were concomitant with increases in the mutation frequency (Cases-González et al. 2008). The serial transfers in the presence of AZC led to the emergence of two amino acid substitutions in the Qβ replicase, Thr210Ala and Tyr410His, which reduced sensitivity to AZC but had strong fitness costs in the absence of the drug (Arribas et al. 2011). Despite being located outside the catalytic site, both mutants reduced the mutation frequency in the presence of the drug (Cabanillas et al. 2014). However, they did not modify the type of the AZC-induced substitutions, which were mediated mainly by ambiguous base pairing of the analog with purines. Furthermore, the Thr210Ala and Tyr410His substitutions had little or no effect on the replication fidelity in untreated viruses. Also, both substitutions were costly in the absence of AZC or when the action of the drug was suppressed by adding an excess of natural pyrimidines, uridine, or cytosine. Overall, the phenotypic properties of these two mutants were highly convergent, despite the mutations being located in different domains of the Qβ replicase (Cabanillas et al. 2014).

DIAGNOSTICS

After the first successful diagnostic attempts mentioned above (Lizardi et al. 1988; Lomeli et al. 1989), the Qβ replicase amplification approach was reviewed at that time with sympathy and compared with the growing PCR method and recommended strongly for further development as a reasonable alternative to PCR (Fox et al. 1989; Persing and Landry 1989; Kwoh and Kwoh 1990; Pritchard 1990; Van Brunt 1990; de la Maza 1991; Birkenmeyer and Mushahwar 1991; Cahill et al. 1991; Pallen and Butcher 1991; Abramson and Myers 1993; Quirós and González 1993; Berche 1994; Hagen-Mann and Mann 1995; Pfeffer et al. 1995). Remarkably, Jean-Michel Pawlotsky (1995) referred at that time to the Qβ amplification as one of the possible assays for the hepatitis C virus diagnostics.

Kramer and Lizardi (1990) presented the essentials of the Qβ replicase-based methodology to clinicians and ensured them that the amplifiable hybridization probes might enable the development of extremely sensitive clinical assays. The authors explained that the novel molecules consisted of a probe sequence embedded within the sequences of the replicable RNA. The molecules were first hybridized to target sequences in a conventional manner. The probe-target complexes were then isolated, and the probes were released from their targets. The released probes were then amplified by incubation with the Qβ replicase. The Qβ replicase copied the probes in a geometrically increasing manner. The doubling process was very rapid, resulting in as many as one billion copies of each molecule in 30 min. The amount of RNA that was made was large enough to be measured without using radioisotopes. Theoretically, these assays should be extraordinarily sensitive, since only one probe molecule was required to start the amplification process. In practice, the sensitivity of the assays was limited by the presence of nonhybridized probes that persisted despite extensive washing of the probe-target hybrids. Generally, the limit of detection was about 10,000 molecules of the target. However, replicable probes were now being prepared that included a "molecular switch," which was a region of the RNA that had undergone a conformational change when the probe sequence hybridized to its target. The protocols were being developed that linked the signal generation to the state of this switch. The simplicity and speed of the enzymatic steps that were required facilitated automation of the assays (Kramer and Lizardi 1990; Lizardi and Kramer 1991). Remarkably, Cahill et al. (1991) were the first to point clearly to the dramatic difference between the Qβ amplification and PCR by the mechanisms of action and modes of application. Thus, the PCR method amplified the target sequences between two priming oligonucleotides, and in essence amplified a portion of the analyte. The Qβ replicase, on the other hand, amplified a specific template molecule hybridized to target sequences and therefore amplified a signal component of the system. For this reason, the Qβ replicase amplification could have applications in areas other than for the detection of the nucleic acid sequences. The earliest priority of the patent application for *Amplification of midivariant DNA templates* was dated October 16, 1990.

Pritchard and Stefano (1990) presented their version of the amplification assay for HIV-1, which was a more complicated variant in comparison with the original test described above (Lomeli et al. 1989). In fact, so-called reversible target capture was employed in order to reduce the background by decreasing levels of the nonspecifically-bound reporter probe. To do this, a synthetic target RNA was hybridized in cell lysates prepared with guanidine thiocyanate with an RNA reporter probe and four deoxyoligonucleotide "capture"

probes. The RNA reporter probe was a recombinant MDV-1 RNA molecule carrying a 40-nucleotide-long fragment of the HIV-1 *pol* gene and generated by transcription from a cloned cDNA template. The capture probes were synthetic oligonucleotides that were complementary to the target nucleic acid and that bored 3' poly(dA) tails of 150 nucleotides. The ternary hybrids of target RNA with capture probe and reporter probe were captured on oligo(dT)-derivatized paramagnetic particles by hybridization with the dA tails of the capture probes. The nonhybridized reporter probes were removed by washing and successively eluting and recapturing the ternary hybrids on fresh particles. After three cycles of elution and capture, the hybrids were eluted in a low ionic strength buffer and the MDV-1 RNA reporter probes were amplified directly by the Qβ replicase. The amplified product RNA was detected by fluorescence using propidium iodide. The assay detected one femtogram, or 600 molecules of the synthetic target RNA containing the *pol* region of HIV-1, and the complete assay took about 2.5 hours. This methodology led to highly satisfying results by the practical HIV-1 diagnostics, demonstrating a 1000-fold greater sensitivity than the isotopic method (Pritchard and Stefano 1991).

Qβ amplification gave rise to the ultrasensitive assay for the detection of *Chlamydia trachomatis* (Shah et al. 1994). The assay used the principles of sandwich hybridization, reversible target capture, and Qβ replicase amplification. The 35-nucleotide probe complementary to the *Chlamydia* 16S rRNA was inserted at the unique restriction site located between nucleotides 63 and 64 of the MDV-1 vector plus strand by the DNA engineering technique. The recombinant replicable detector was transcribed into RNA with T7 RNA polymerase. Following reversible target capture, detection of the signal was accomplished by replication of the detector molecule by the Qβ replicase in the presence of propidium iodide. A specific assay signal was detected from as few as 1000 molecules above the background. Furthermore, the Qβ amplification test for *C. trachomatis* was compared directly with the semiquantitative competitive PCR assay that was also directed against the 16S rRNA gene (An et al. 1995b). The Qβ replicase and PCR assays were quantitative over 10,000- and 1000-fold ranges of organisms, respectively. In contrast to the PCR test, the Qβ assay showed no evidence of the sample inhibition. It was concluded that both the Qβ and PCR assays should allow quantitative investigation of infections due to *C. trachomatis*. The advantage of the Qβ replicase test was the fact that it was targeted against highly labile RNA and facilitated investigations into the role of active persisting infection in culture-negative inflammatory conditions (An et al. 1995b). This test was included in the detailed methodological presentation of the popular *C. trachomatis* assays (Chernesky and Mahony 1999). Ngan (1997) commented in an update on the *C. trachomatis* detection techniques that the Qβ amplification test was less subjected to inhibitory substances in urine than the PCR test, although the cell culture remained the gold standard for the legal *C. trachomatis* diagnosis.

In a novel *C. trachomatis* test, Stefano et al. (1997) employed a set of specially selected 23S rRNA probes within the MDV-1 vector. The optimal probe set showed a sensitivity of 10^3 molecules of 23S rRNA and could detect a single elementary body of *C. trachomatis* or 1–10 elementary bodies added to a clinical matrix of pooled negative human cervical swab samples. The assay showed approximately 10^9-fold discrimination over *C. pneumonae* rRNA. The high levels of cultured *C. albicans*, *E. coli*, *S. aureus*, or *N. gonorrhoeae* had no detectable effect on assay background or the ability to detect a single elementary body (Stefano et al. 1997).

After the clinical *C. trachomatis* test, the next Qβ replicase assay was elaborated for the detection of *Mycobacterium tuberculosis* directly from spiked human sputum (Shah et al. 1995b). As in the case of the *Chlamydia* test, the insertion of the specific probe into the same place of the MDV-1 vector was used, but the 35-nucleotide probe was targeted this time against 23S rRNA of *M. tuberculosis*. The assay sensitivity was 10^3 purified rRNA targets or 1 CFU of *M. tuberculosis* spiked into *M. tuberculosis*-negative human sputum. Remarkably high levels of other mycobacterial rRNA, approximately 10^7 organisms, including rRNAs of *M. avium* and *M. gordonae*, did not interfere with the sensitivity of the assay (Shah et al. 1995b). Then, the Qβ replicase test was compared directly with the appropriate PCR assay (An et al. 1995a). The sensitivities of the Qβ replicase and PCR assays of spiked sputum samples were 0.5 and 5.0 CFU per assay reaction, respectively. In contrast to the PCR test, the results of the Qβ assay were not inhibited by sputum. It was concluded that the Qβ replicase assay performed well in comparison with both culture and PCR and should offer a rapid means for detecting and controlling infection due to *M. tuberculosis*. Thus, Shah et al. (1995a) used the Qβ assay for detection of *M. tuberculosis* directly from clinical specimens, namely, from digested sputum pellets. A total of 261 specimens submitted to the three tuberculosis testing laboratories were analyzed. In comparison with the culture test, the overall assay sensitivity and specificity were 97.1 and 96.5%, respectively. After analysis of discrepant results, the positive predictive value and the negative predictive value were determined as 87.8% and 99.5%, respectively. Therefore, the Qβ replicase assay was characterized as a rapid, sensitive, semiquantitative, and specific test for the direct detection of *M. tuberculosis* from clinical specimens. Smith et al. (1997a) adapted a prototype automated instrument platform in which probes were amplified with the Qβ replicase. The assay was based on the amplification of the specific detector probe following four cycles of the background reduction by reversible target capture in a closed disposable pack. The assay sensitivity, specificity, and positive and negative predictive values in this pilot study were 79%, 98%, 97%, and 85%, respectively. Next, Smith et al. (1997b) presented results of the automated Qβ replicase amplification assay *Galileo* for *M. tuberculosis* in the clinical trial. It is noteworthy that 10 test packs at a time were automatically processed by the *Galileo* analyzer without operator intervention following loading of samples. In fact, this was the first report of a clinical study with a fully automated amplification probe hybridization assay for the detection of pathogens directly from a clinical specimen.

Using the model automated Qβ amplification instrument, the ribosomal RNA targets were detected from the four respiratory pathogens, namely, *Mycobacterium avium* complex (23S), *Mycoplasma pneumoniae* (16S), *Pneumocystis carinii* (18S), and *Legionella pneumophila* (16S) (Stone et al. 1996). The sandwich hybridization, reversible target capture, detector probe amplification, and fluorescent signal detection occurred in the closed disposable packs. The limits of detection were found at the 10^5–10^4 targets. The high level of discrimination against competitor RNAs was achieved. Therefore, the automated system demonstrated high utility and versatility of the direct-from-sample testing of respiratory pathogens.

It is remarkable that at this active time of the Qβ replicase invention, the RT-PCR method but not the Qβ amplification assay was employed for detection of the phage Qβ itself in environmental water samples, where primers were derived from the Qβ replicase gene (Sunun et al. 1995).

In parallel with the first clinical success, progress continued in the Qβ assay technique. Thus, Burg et al. (1995) published an improved system that measured the real-time fluorescence of the amplified MDV-1 RNA with propidium iodide, quantified the amounts of RNA, and was supposed for the clinical diagnostics. Then, Burg et al. (1996) discovered that the replication characteristics of the MDV-1 RNA molecules with internally placed probe sequences were dramatically affected by short spacer elements at the 5′ and 3′ ends of the probe sequence. For a given probe sequence, the sequences of the flanking spacer elements affected replication sensitivity by six orders of magnitude and replication rate by threefold. By taking advantage of spacer elements, internal MDV-1 probes were developed that permitted reproducible real-time fluorescence detection of a single RNA molecule in less than 25 min through amplification with the Qβ replicase. The RNA structural analysis of such probes suggested that the spacer elements functioned by allowing the RNA to fold in a way that substantially maintained the tertiary structure of the MDV-1 domain. The MDV-1 reporter probes with suitable replication properties were obtained from libraries of the RNA molecules in which the probe sequence was flanked by many different spacer elements generated by random nucleotide synthesis (Burg et al. 1996).

Tyagi et al. (1996) described a new assay strategy that solved the problem posed by the persistence of nonhybridized amplifiable probes. The probe molecules were redesigned so that they could not be amplified unless they hybridized to their target. The authors divided the recombinant RNA probes into the two separate molecules, neither of which could be amplified by itself because neither contained all the elements of sequence and structure that were required for replication. The division site was located in the middle of the embedded probe sequence. When these "binary probes" were hybridized to adjacent positions on their target, they could be joined to each other by incubation with an appropriate ligase, generating an amplifiable reporter RNA. The nonhybridized probes, on the other hand, because they were not aligned on a target, had a very low probability of being ligated. By combining this target-dependent ligation step with a new and simpler hybrid-isolation step, the signal generation was strictly dependent on the presence of the target molecules, no background signals were generated, and the resulting assays were extraordinarily sensitive. The strategy was based on the MDV-1 vector and used the HIV-1 detection as a prototype. The assay detected fewer than 100 nucleic acid molecules and provided quantitative results over a wide range of target concentrations. The authors claimed that this assay can detect any infectious agent (Tyagi et al. 1996). The experimental details of HIV-1 detection by the hybridization probes containing a molecular switch, including an attempt to use ribonuclease III to digest the nonhybridized probes, was published a bit later by Blok and Kramer (1997).

Preuß et al. (1997) employed the Qβ replicase amplification to develop the probing of RNA-protein interactions using pyrene-labeled oligodeoxynucleotides. Thus, the Qβ replicase bound efficiently small RNAs by recognizing pyrimidine residues, such as a probe of the sequence 5′-d(TTTTTCC) that was 5′-end-labeled with pyrene. In this construct, the proximal thymine residues efficiently quenched the fluorophore emission in solution. Upon stoichiometric binding of one probe per polymerase molecule, the pyrene steady-state fluorescence increased by two orders of magnitude, the fluorescence anisotropy increased, and a long fluorescence lifetime component of 140 ns appeared. With addition of the replicable RNA, the steady-state fluorescence decreased in a concentration-dependent manner and the long lifetime component was lost. The sensitivity of the described fluorometric assay allowed in principle the real-time detection of the Qβ replicase-driven RNA amplification (Preuß et al. 1977).

MUTAGENESIS SYSTEM

A novel unexpected application of the Qβ-related RNA together with the Qβ amplification was found by Kopsidas et al. (2006). They employed a novel mutagenesis system that utilized the error-prone Qβ replicase to create a diverse library of the single-domain antibody fragments based on the shark IgNAR antibody isotype. The coupling of these randomly mutated mRNA templates directly to the translating ribosome allowed the *in vitro* selection of affinity matured variants showing enhanced binding to target, the malarial apical membrane antigen 1 (AMA1) from *Plasmodium falciparum* (Kopsidas et al. 2006). The unique capability of the Qβ replicase to generate highly diverse mRNA libraries for *in vitro* protein evolution was confirmed further by Kopsidas et al. (2007). As the authors claimed, the mutational spectrum of the Qβ replicase was close to the ideal. Thus, the Qβ replicase generated all possible base substitutions with an equivalent preference for mutating A/T or G/C bases and with no significant bias for transitions over transversions. The high diversity of the Qβ replicase-generated mRNA library was demonstrated, as mentioned above, by evolving the binding affinity of a single-domain V_{NAR} shark antibody fragment (12Y-2) against AMA-1 via ribosome display. The binding constant of the 12Y-2 was increased by 22-fold following two consecutive but discrete rounds of mutagenesis and selection. The mutagenesis method was also used to alter the substrate specificity of β-lactamase which did not significantly hydrolyze the antibiotic cefotaxime. Two cycles of RNA mutagenesis and selection on increasing concentrations of cefotaxime resulted in mutants with a minimum 10,000-fold increase in resistance, an outcome achieved faster and with fewer overall mutations than in comparable studies using other mutagenesis strategies (Kopsidas et al. 2007).

COUPLED REPLICATION-TRANSLATION

In parallel with the diagnostic constructions, it was attempted to couple the outstanding replication capacities of the Qβ-related vectors with the translation of the desirable products. Thus, the 804-nucleotide-long messenger sequence encoding the chloramphenicol acetyltransferase was embedded within the MDV-poly vector carrying the appropriate polylinker, as mentioned above (Wu et al. 1992). The resulting 1023-nucleotide-long recombinant RNA was exponentially replicated by the Qβ replicase, and the product RNA served as a template for the cell-free translation of the biologically active chloramphenicol acetyltransferase enzyme. The chloramphenicol acetyltransferase production was prolonged markedly (Ryabova et al. 1994) when the coupled replication-translation reactions were carried out in a continuous-flow format (Spirin et al. 1988), as described in the Cell-free synthesis section of Chapter 4. The results suggested that the mechanism of replication and translation in the coupled reactions mimicked the mechanism by which Qβ RNA was simultaneously replicated and translated in the Qβ-infected *E. coli*, where protein synthesis occurred on nascent RNA strands, as described in Chapter 16.

In parallel, Spirin's team used another vector, namely, RQ135, in order to combine the Qβ amplification with the *E. coli* cell-free translation system. Thus, they inserted the messenger of the dihydrofolate reductase enzyme into the RQ135 RNA supplied with the appropriate restriction sites (Morozov et al. 1993). The enhancement of the enzyme production was associated first with the replication asymmetry of the Qβ replicase products, where the plus strands were produced in large excess over the minus strands. As a result, the coupled replication-translation of the amplifiable RNA recombinants was suggested by the authors as a useful model for studying the fundamental aspects of virus amplification, as well as a practical solution to the large-scale protein synthesis *in vitro*, as described earlier in the Cell-free synthesis section of Chapter 4. Furthermore, the high-level synthesis of the dihydrofolate reductase and chloramphenicol acetyltransferase was achieved in the cell-free synthesis system with the recombinant RNAs, where the appropriate mRNAs of the enzymes were inserted into the RQ135 vector (Ugarov et al. 1994). The use of the cell-free system for RNA replication, expression, and cloning was reviewed at that time by Chetverin (1996). The basic methodology of the coupled replication-translation was described in detail by Ryabova et al. (1998).

Furthermore, the RQ135-based vectors were designed to produce a functionally active protein carrying the Strep-tag oligopeptide at its C-terminus in a standard low-cost cell-free system (Alimov et al. 2000). The presence of the Strep-tag allowed the synthesized protein to be easily isolated on a streptavidin-agarose column under mild conditions, and the entire procedure could be completed in one working day.

RNA WORLD

The first approach to a synthetic minimal cell in which the reproduction of the membrane and the replication of the internalized RNA molecules proceeded simultaneously was achieved by Biebricher's team (Oberholzer et al. 1995). The enzymatic RNA replication by the Qβ replicase was performed in self-reproducing vesicles, namely, in oleic acid/oleate vesicles simultaneously with the self-reproduction of the vesicles themselves. The Qβ replicase together with the RNA template and the four ribonucleotides were entrapped inside the vesicles. The water-insoluble oleic anhydride was then added externally. It was bound to the vesicle bilayer where it was catalytically hydrolyzed, yielding the carboxylate surfactant *in situ*, which then brought about growth and reproduction of the vesicles themselves. These experiments showed for the first time the possibility of realizing a compartmentalized RNA replication system that was able to self-reproduce simultaneously under normal laboratory conditions. Since the reproduction of the boundary was due to a reaction that took place within the boundary and was catalytically induced by the boundary itself, the process was defined as autopoietic. This might suggest that also in the evolution of the first minimal life forms the boundaries might have performed much more than a simple protective shell function. By combining RNA replication with the principles of autopoiesis, Oberholzer et al. (1995) operated a bridge between the two more accepted views on the theory of minimal life, the one based on the cellular autopoietic view and the other based on the RNA World (Celander and Cech 1991).

Furthermore, the experimental studies of the RNA evolution *in vitro* were reviewed by Biebricher and Gardiner (1997) in the context of the Manfred Eigen's (1971) theory on the origin of life and its subsequent extensions. The vesicle approach of Oberholzer et al. (1995) was analyzed critically by Pier Luigi Luisi (2002) in his aspiration toward the engineering of minimal living cells. Finally, Oberholzer and Luisi (2002) included the Qβ experiments in the exhaustive survey of the use of liposomes for constructing cell models.

In the context of the Eigen's ideas and the general RNA World concept, the traveling waves of *in vitro* evolving RNA must be mentioned (Bauer et al. 1989). Thus, the populations of the short self-replicating RNA variants were confined to one side of a reaction-diffusion traveling wave front propagating along thin capillary tubes containing the MNV-11 vector and the Qβ replicase. The propagation speed was accurately measurable with a magnitude of about 1 micron/sec, and the wave persisted for hundreds of generations (of duration less than 1 min). The evolution of RNA occurred in the wavefront, as established by front velocity changes and gel electrophoresis of samples drawn from along the capillary.

The high population numbers of approximately equal to 10^{11}, their well-characterized biochemistry, their short generation time, and the constant conditions made the system ideal for the evolution experiments. Moreover, McCaskill and Bauer (1993) established that the traveling waves arose spontaneously without added RNA and provided a model system for major evolutionary change. The mathematical modeling of the Qβ amplification was performed at that time also by Bull and Pease (1995), in order to demonstrate why the PCR was more resistant to the *in vitro* evolution than the Qβ-driven RNA amplification.

The theoretical calculations of Brian K. Davis on the Qβ self-replication contributed markedly to the general idea of the RNA World. He analyzed quantitatively the MDV-1 evolution *in vitro* (Davis 1991a), the population fitness during competitive replication at elevated mutation rates (Davis 1991b), and the production of more complexity than entropy in the replication (Davis 1994). Then, Davis (1995a) calculated that the replicate-template annealing reaction during transcription significantly retarded replication *in vitro*, while annealing between complementary RNA strands free in solution had far less significance. By comparison of 17 self-replicating RNA species of 77–370 nucleotides in length, Davis (1995b) demonstrated that the stability in the replicative form of RNA molecules transcribed by the Qβ replicase provided a sequence-dependent indicator of their fitness. In light of these calculations, Darwin's fundamental precept, "survival of the fittest," could be appraised as an experimentally testable hypothesis (Davis 1995b). Generally, the Qβ amplification system was proven to clearly display the intrinsic features of the RNA World and, interestingly, it revealed a rapid path for evolution of the first self-replicating molecule on Earth (Davis 1995a). This appreciable piece of work was finished with a theory of evolution that included prebiotic self-organization and episodic species formation (Davis 1996). The calculations simulating *de novo* formation of self-replicating RNA molecules in the Qβ replicase system explained how spontaneous changes in the strand secondary structure promoted the transition from random copolymerization to the template-directed polymerization. The natural selection among competing self-propagators gave way to a principle of wider scope, stating that evolution optimally damped the physicochemical forces causing change within an evolving system (Davis 1996).

A marked contribution to the RNA World was brought by novel calculations on the autocatalytic reaction networks based on the template-dependent replication and specific catalysis, in the traditions of the previous ideas of Manfred Eigen, Peter Schuster, and Peter Stadler (Hecht et al. 1997).

Next, Strunk and Ederhof (1997) constructed two machines for automated evolution experiments *in vitro*. In contrast to the early dilution experiments, both machines enabled the monitoring of growing populations of RNA or DNA molecules in real time, using high-sensitivity glass fiber laser fluorimeters and an automated sample handling facility for volumes in the microliter range. The growth conditions were kept constant by means of the serial-transfer technique, that is, the successive transfer of a small fraction of a growing population into a fresh solution containing no individuals prior to the transfer. Therefore, the serial transfer technique was modified to work with large populations and constant growth conditions. Thus, in the single-channel evolution machine isothermal amplification reactions of the Qβ system and other models were monitored successively in single test tubes. This machine was particularly well suited for the investigation of optimal adaptation to altered environmental conditions, as experimentally demonstrated in the evolution of an RNA quasispecies using ribonuclease A as the selection pressure. The new variant of RNA appeared very rapidly, within approximately 80 generations, without stable intermediates, and it was selected by steadily increasing the ribonuclease concentration during the serial-transfer experiment. The other machine was a consequent extension of the single-channel machine and was designed to allow the multichannel detection of up to 960 samples simultaneously. Thus, the high-throughput screening was applied to evolution experiments (Strunk and Ederhof 1997).

After 30 years of Sol Spiegelman and Leslie Orgel's *in vitro* evolutionary studies, Oehlenschläger and Eigen (1997) published an article that was dedicated to Leslie Orgel on the occasion of his 70th birthday. By applying the principle of serial transfer pioneered in the laboratories of Spiegelman and Orgel, the authors used the fast growth selection pressure for the *in vitro* evolution of the self-sustained sequence replication (3SR) system. In the first step of the 3SR reaction, the RNA template was reversely transcribed by HIV-1 reverse transcriptase, followed by a second strand synthesis, and the transcription of the resulting dsDNA by T7 RNA polymerase. At the end of the exponential growth phase of the 3SR reaction, an aliquot of the reaction mixture was transferred into a new sample containing only buffer, nucleotides, and enzymes, while RNA template molecules were provided by the transfer. The conditions in the exponential growth phase allowed the RNA molecules to be amplified in a constant environment. Thirty years after Spiegelman's experiment, this study was another answer to the question he posed: "How do molecules evolve if the only demand is the Biblical injunction: multiply?" The answer derived from the 3SR amplification system that was mimicking a part of the HIV-1 replication cycle *in vitro* was the same as 30 years ago: the RNA molecules adapted to the new conditions by throwing away any ballast not needed for fast replication (Oehlenschläger and Eigen 1997).

The 40 years of Spiegelman's team's experiments and of the *in vitro* evolution were celebrated by the circumstantial review that was written by Gerald F. Joyce (2007) and dedicated to Leslie Orgel on the occasion of his 80th birthday. Figure 18.4 presents the cover of the review in question. The review carefully summarized the concepts and methods for the directed evolution of the RNA molecules *in vitro*. Moreover, Joyce emphasized that the *in vitro* evolution established the Darwinian evolution as a chemical process for the first time, and paved the way for the many directed evolution experiments that followed.

A year later, Kita et al. (2008) constructed a simplified life model involving encapsulated macromolecules termed

FIGURE 18.4 Sol Spiegelman on the cover. (Joyce GF: Forty years of *in vitro* evolution. Angew Chem Int Ed Engl. 2007. 46. 6420–6436. Copyright Wiley-VCH Verlag GmbH & Co. KGaA. Reproduced with permission.)

a "self-encoding system" in which the genetic information was replicated by the self-encoded Qβ replicase in liposomes. In this context, it is noteworthy that the concept of the autocatalytic amplification of mRNA encoding the β subunit in an *in vitro* system, namely, in the rabbit reticulocyte cell-free system, was suggested by Fukano et al. (2002). The authors realized the first stage of the aim and constructed the genetically fused EF-Tu and EF-Ts subunit that was able to support the enzyme activity, described earlier in the EF Tu-Ts section of Chapter 13.

The model of Kita et al. (2008) contained one template RNA sequence and an *in vitro* translation system that was reconstructed from purified translation factors as the machinery for decoding the information that was encoded on the RNA. The liposomes provided a biologically relevant environment. In the *in vitro* translation system, the β subunit of the Qβ replicase was synthesized from the template RNA that encoded the protein. The Qβ replicase then replicated the template RNA used for its production. Thus, in this reaction, the Qβ replicase was generated through translation of the self-encoded gene. This mimicked current living systems. The ability of a new functional gene to be incorporated into the RNA sequence was foreseen to allow the self-encoding system to exhibit additional phenotypes or functions. This system was one of the simplest artificial multicomponent self-encoding systems and was composed of only 144 gene products in which the information unit, or genotype, and the functional unit that replicated the information, or phenotype, were encoded on different molecules, and importantly had the potential to evolve by being compartmentalized in liposomes (Kita et al. 2008). Ichihashi et al. (2008) turned to the careful analysis of the translation-replication balance in this minimal self-encoded system. As followed from the classical Kolakofsky and Weissmann (1971a,b) model (for details, see the Repressor complex II section in Chapter 16), both the ribosomes and the Qβ replicase intended to bind the same RNA for translation and replication, respectively. Thus, there could be a dilemma existing in the course of the phage replication: the effective RNA replication required high levels of the replicase translation, but excessive translation in turn inhibited replication. The competition between the ribosomes and replicase was evaluated by constructing a kinetic model that quantitatively explained the behavior of the self-encoding system. Both the experimental and theoretical results consistently indicated that the balance between translation and replication was critical for an efficient self-encoded system, and the optimum balance was determined (Ichihashi et al. 2008). This approach further developed the investigation of the phage regulation processes, namely, repressor complexes, that were described in detail in Chapter 16.

It is remarkable that the compartmentalization in the water-in-oil emulsion repressed the spontaneous amplification of RNA by the Qβ replicase and avoided the risk of the contamination by the short RQ variants, especially by the kinetic analysis (Urabe et al. 2010). This approach allowed the exponential Qβ RNA amplification without detectable amplification of the RQ RNAs.

To further develop the concept of the artificial cell, Bansho et al. (2012) established a sophisticated mathematical model in order to gain a quantitative understanding of the complex reaction network. The sensitivity analysis predicted that the limiting factor for the present replication reaction was the appearance of parasitic replicators. It was confirmed experimentally that the repression of such parasitic replicators by compartmentalization of the reaction in water-in-oil emulsions improved the duration of the self-replication. Remarkably, the main source of the parasite was genomic RNA, probably by nonhomologous recombination. The latter will be described in the corresponding Recombination section of this chapter. Therefore, Bansho et al. (2012) overcame the unpredictability of the parasites and found conditions that satisfied the achievement of the long-lasting translation-coupled replication. Figure 18.5 presents a schematic drawing of the translation-coupled RNA self-replication system.

This impressive study was previewed by Lehman (2012), who emphasized the important moment that the compartmentalization into the water-in-oil droplets was able to ameliorate the problem of the parasitic sequences, but only if the droplets were small. In the opinion of the expert, this work helped to both recapitulate abiogenesis and optimize synthetic biology.

Ichihashi et al. (2013) constructed an evolvable artificial cell model from an assembly of biochemical molecules. The artificial cell model contained the artificial genomic RNA that replicated through the translation of the encoded Qβ replicase. A long-term replication experiment of 600 generations was performed. The mutations were spontaneously introduced into the RNA by replication error, and highly replicable mutants dominated the population according to

the Darwinian principles. During evolution, the genomic RNA gradually reinforced its interaction with the translated replicase, thereby acquiring competitiveness against selfish parasitic RNAs. This study provided the first experimental evidence that replicating systems could be developed through the Darwinian evolution in a cell-like compartment, even in the presence of parasitic replicators. Then, Mizuuchi et al. (2014) tested the adaptive evolutionary ability of the artificial RNA genome replication system to a reduced-ribosome environment. It was observed that the mutant genome compensated for the reduced ribosome concentration by the introduction of several mutations around the ribosome-binding site in order to increase the translation efficiency.

However, the projects to reconstruct the cellular system from a set of defined molecules were endangered by the formation of double-stranded RNAs. To resolve the double-stranded RNA problems, Usui et al. (2013) elaborated the first kinetic model of the double-stranded RNA formation during replication of long RNAs by the Qβ replicase. Besides the genomic Qβ RNA, the two recombinant derivatives of the MDV-1 vector were used in this study: one carrying the β subunit encoding sequence and other encoding the β subunit and a part of the Qβ readthrough protein, 2125 and 3035 nucleotides in length, respectively. The authors showed that it was possible to suppress the double-stranded RNA formation by modification of the RNA sequence. Thus, they showed that the dsRNA formation could be explained quantitatively by two distinct pathways, i.e., (i) dsRNA formation during the replication process and (ii) dsRNA formation by hybridization through collision between the newly synthesized negative-strand RNA and positive-strand RNAs. Nevertheless, the levels of dsRNA formation by both reactions were reduced substantially by insertion of the readthrough part of the Qβ genome (Usui et al. 2013).

Furthermore, Usui et al. (2014) expanded the previous kinetic model by inclusion of the effects of ribosome concentration on the RNA replication. The expanded model quantitatively explained the single- and double-strand formation kinetics during replication with various ribosome concentrations for two artificial long RNAs described above. This expanded model provided useful frameworks to understand the precise replication mechanism of the Qβ replicase with ribosomes and to design amplifiable RNA genomes in the translation-coupling systems.

Finally, to solve once and for all the problem of the inactive double-stranded RNAs, Usui et al. (2015) introduced 28 various predicted structures into the MDV-1 vector. As a result, a simple rule regarding the single-stranded RNA genome was extracted, namely, replication with less double-stranded RNA formation or less GC number in loops. Then, the authors designed an artificial RNA that encoded the α domain of the β-galactosidase gene based on this rule. Moreover, the evidence was obtained that this rule governed the natural genomes of all bacterial and most fungal viruses presently known. Therefore, this study revealed one of the structural design principles of a single-stranded RNA genome that replicated continuously with less double-stranded RNA formation (Usui et al. 2015).

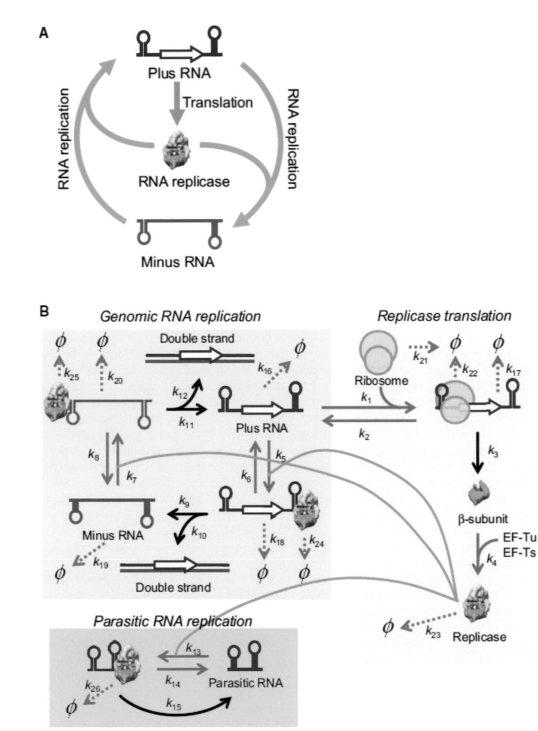

FIGURE 18.5 Schematic drawing of the translation-coupled RNA self-replication system. (A) Simple scheme. The system contained genomic plus RNA and a modified cell-free translation system (PURE system). The plus RNA carried the gene encoding the coliphage Qβ catalytic subunit of the RNA replicase. The subunit is translated from genomic plus RNA and forms an active replicase with the other subunits, EF-Tu and Ts, which are components of the cell-free translation system. The heterotrimer replicase synthesizes the complementary strand, the minus RNA, using the plus RNA as a template. Then, using the minus RNA as a template, the replicase synthesizes complementary plus RNA. (B) Detailed mathematical model. This reaction model is composed mainly of three parts: genomic RNA replication, replicase translation, and parasitic RNA replication. Black arrows represent the process in which RNA or replicase was synthesized. The gray arrows represent binding or dissociation processes. Dotted arrows and the symbol φ represent processes in which components are degraded or inactivated irreversibly. For simplicity, some of the pathways are not shown here, such as returning of the replicase from the RNA-replicase complex after complementary strand synthesis. In this model, the complementary strand, double-stranded RNA formation, and degradation of parasitic RNA were neglected. (Reprinted from *Chem Biol*, 19, Bansho Y, Ichihashi N, Kazuta Y, Matsuura T, Suzuki H, Yomo T, Importance of parasite RNA species repression for prolonged translation-coupled RNA self-replication, 478–487, Copyright 2012, with permission from Elsevier.)

Meanwhile, Tomita et al. (2015) prepared for the first time the long fully double-stranded RNA of 2041 nucleotides, which included recognition sequences for the Qβ replicase and the β subunit. The replication of this RNA by the Qβ replicase was minimal. The authors showed that the Qβ replicase synthesized the complementary strand as long as the template RNA was hybridized with no more than 200 nucleotide fragments, although the replication amounts were decreased.

Sunami et al. (2016) studied the effect of liposome size on the coupled replication-translation system. As before, the RNA was replicated by the Qβ replicase, which was translated from the RNA in giant liposomes encapsulating the cell-free translation system. A reporter RNA encoding the antisense strand of β-glucuronidase within the MDV-poly vector was introduced into the system to yield a system readout as green fluorescence. The coupled replication-translation was hardly detectable in larger liposomes of 230 fL but was more effective in smaller liposomes of 7.7 fL. Therefore, the smaller microcompartments considerably enhanced the coupled replication-translation process because the synthesized molecules, such as RNA and replicases, were more concentrated in smaller liposomes.

Mizuuchi et al. (2016) investigated whether and how the RNA genome adapted to four different conditions when different sets of translation factors, or initiation factors, or methionyl-tRNA formyltransferase, or release factors, or ribosome recycling factor were omitted. The Qβ amplification was started in this case with an RNA genome obtained after 128 rounds of *in vitro* evolution in the previous study (Ichihashi et al. 2013). It was found that the replication efficiency increased with the number of rounds of replication under all the tested minimized conditions. The types of dominant mutations differed depending on the condition, thus indicating that this simple system adapted to different environments in different ways. The authors supposed that even a primitive self-replication system composed of a small number of genes on the early Earth could have had the ability to adapt to various environments. The novel constructions of longer RNA retaining replication capability were generated recently (Mizuuchi and Ichihashi 2018; Ichihashi 2019; Ueda et al. 2019).

Yoshiyama et al. (2016) continued the translation-coupled RNA replication in the water-in-oil emulsion in the dynamic micro-sized compartments under continuous stirring and flow. To do this, a special continuous stirred-tank reactor was constructed, and 4800-rpm stirring was used to induce the droplet dynamics that enabled the maintenance of the whole process. The fresh substrate for the reaction and a cell-free translation system was supplied into the stirring tank in the form of droplets. The stirring was not just for mixing but for inducing the continuous coalescence and breakage dynamics of the droplets. The authors showed also that, by tuning the activity of the process, the outbreak of the parasitic amplified RNA could be suppressed. Therefore, it became possible to maintain the replication reaction in the micro-sized compartments as the stirring-tank reactors did for cell cultures.

The next step in these studies was the design of a new combinatorial selection method for a replicable RNA by the Qβ replicase to find an RNA sequence with the secondary structures and functional amino acid sequences of the encoded gene (Yumura et al. 2017). Thus, the RNA sequences based on their *in vitro* replication and *in vivo* gene functions were selected. First, the α-domain gene of β-galactosidase was used as a model-encoding gene, with functional selection based on blue-white screening. The selected sequence improved the affinity between the minus strand RNA and the Qβ replicase. Then, Yumura et al. (2017) established an *in vivo* selection method applicable to a broader range of genes by using an *E. coli* strain with one of the essential genes complemented with a plasmid. As a result, the more replicable RNAs encoding functional *serS* gene encoding serine-tRNA ligase enzyme were obtained.

These impressive approaches for understanding the origin of the self-replication and evolution with Qβ amplification as a model were reviewed exhaustively by Ichihashi and Yomo (2016). The employment of the giant vesicles for the compartmentalization of such evolution processes as the Qβ amplification was reviewed recently by Sugiyama and Toyota (2018).

It is noteworthy that norovirus RNA replicase was involved at that time in the RNA amplification and autonomous evolution studies *in vitro*, in parallel with the traditionally used Qβ replicase (Arai et al. 2013). The norovirus enzyme was chosen because of its simplicity, since it consisted of a single polypeptide chain and its molecular mass, 56 kDa, was not so large. Moreover, the enzyme was preferable to Qβ replicase by the amplification of the double-stranded RNA (Arai et al. 2014).

The possible impact of both point mutations and premature termination errors in the RNA World models was considered by Tupper and Higgs (2017). The authors used Qβ replication, where the full-length RNA was rapidly replaced by very short nonfunctional sequences, as a major argument. If the same thing were to occur when the polymerase was a ribozyme, this would mean that termination errors could potentially destroy the RNA World. Tupper and Higgs (2017) showed that it was not the case, at least in their RNA World model. The phenomenon of the Qβ amplification was discussed in the context of the origin of life and the synthesis of the *de novo* life (Duim and Otto 2017). The main conclusion of the review was that the full understanding of the requirements for the open-ended evolution would provide a better understanding of how life could have emerged from the molecular building blocks and what is needed now to create a minimal form of life in the laboratory.

Looking recently for the RNA origin of life, Muying Zhou (2018) proclaimed the RNA phages, such as MS2 and Qβ, as the first of the most primitive *chickens* of the RNA World and their replication as the essential *egg-chicken* loop. Remarkably, the A, A1, lysis, and coat proteins, which do not participate in the formation of the *egg-chicken* loop were classified by this theory as *passengers*. Zhou M (2018) found it amazing that the most primitive *chickens* are still living in the existing world.

PHYLOGENY

As Zinder (1980) noted in his classical review, "consideration of the origin of phages is at best a difficult exercise."

He thought at that time that the speculation on the origin of the RNA phages had even less substance than that on the origin of the DNA phages but acknowledged that the first organized could have an RNA genome. Then, the RNA phages could be derived from one of two sources. It could be a legacy of the evolving primitive RNA that survived by becoming a parasite, or alternatively, it could reflect an initial effort at sexuality by the early RNA genomes much like that described for DNA. Thus, the RNA phages might be old, a remembrance of the time when genetic material was RNA. However, the RNA phages could also have arrived in a DNA world, possibly derived from messenger RNA. Therefore, "as we find them today, as long as there are bacteria, there will be RNA phage" (Zinder 1980).

Blaisdell et al. (1996) were the first to look for the similarities and dissimilarities of the phage genomes in a collection of 23 phages with temperate, lytic, and parasitic life histories, with varied sequence organizations and with different hosts and different morphologies. The list also included the RNA phages GA and MS2. They noticed among other things that the single-stranded RNA phages had the dinucleotide relative abundance values closest to those for random sequences, presumably attributable to the mutation rates of the RNA phages being much greater than those of the DNA phages. In fact, the phages of the family *Leviviridae* had among the highest known mutation rates, namely, 10^{-3} bp/replication (Drake 1993) and some of the smallest RNA genomes known of 3500–4200 nucleotides.

Concerning the possible evolutionary routes of the Qβ and the whole RNA phage replication system, the finding of Rodríguez-Cousiño et al. (1991, 1998) was of very high interest at that time. They discovered that the yeast *S. cerevisiae* 20S and 23S RNA replicons possessed the 5′- and 3′-terminal nucleotide sequences resembling those of the RNA phages and were amplified by the replicon-encoded RNA-dependent RNA polymerases similar to the Qβ replicase. The similar putative RNA-dependent RNA polymerases were encoded also by a mitochondrial virus-like RNA in the Dutch Elm Disease ascomycete fungus, *Ophiostoma novo-ulmi*, by other viruses and virus-like RNAs, and by the *Arabidopsis* mitochondrial genome (Hong et al. 1998).

As presented in Chapters 2 and 7, Jonathan P. Bollback and John P. Huelsenbeck (2001) were the first to publish a systematic study on the phylogeny of the *Leviviridae* family. Using nucleotide sequences from the coat and replicase genes, they presented the first statistical estimate of phylogeny for the family *Leviviridae* using maximum-likelihood and Bayesian estimation. The study revealed that the RNA coliphage species were a monophyletic group consisting of two clades representing the genera *Levivirus* and *Allolevivirus*. The authors thought that the pseudomonaphage PP7 diverged from its common ancestor with the coliphage prior to the ancient split between these genera and their subsequent diversification. The differences in the genome size, gene composition, and gene expression were shown with a high probability to have changed along the lineage, leading to the *Allolevivirus* through gene expansion. The change in the genome size of

the *Allolevivirus* ancestor might have catalyzed subsequent changes that led to their current genome organization and gene expression (Bollback and Huelsenbeck 2001). In contrast, Furuse (1987) thought that the Qβ-like ancestor gave rise to an MS2-like genome through a major deletion, according to the "gene contraction" hypothesis. An indirect prediction of this hypothesis was that the ancestral protophage contained a readthrough protein and was therefore more similar in genomic organization to the *Allolevivirus* phage. A number of experiments, including the Darwinian ones (Mills et al. 1967, 1975; Schaffner et al. 1977) and others (Klovins et al. 1997b), have demonstrated that the deletions can occur in the Qβ genome. Moreover, at that time there were no known duplication or recombination and repair systems in RNA viruses, and the "deletion" or "gene contraction" hypothesis had more biochemical support than an "insertion/duplication" or "gene expansion" hypothesis. Bollback and Huelsenbeck (2001) noticed that the gene expansion hypothesis required some mechanism by which the ancestral phage could give rise to a larger daughter species. An increase in the genome size could be caused by either a duplication or recombination event. Philipp Mekler (1981) observed a number of repeats and "quasi-repeats" of homologous nucleotide sequence tracts arranged in the same order within the Qβ coat readthrough gene. This suggested some tentative support for a duplication mechanism in the origin of the group III and IV coliphage. In addition, high sequence similarity within the two different repeat groups suggested two separate and distinct duplication events (Mekler 1981). The alternative mechanism, recombination, has also become a plausible explanation, as the experimental work has demonstrated the homologous *in vivo* recombination in the Qβ genome (Chetverin et al. 1988, 1991; Biebricher and Luce 1992; Palasingam and Shaklee 1992), as well as other occasions including nonhomologous recombination, which are described in the appropriate Recombination section.

Thus, according to Bollback and Huelsenbeck (2001), the ancestral phage genome increased in size on the order of 600 nucleotides after the divergence of the *Allolevivirus* and *Levivirus* lineages, but before the diversification of the *Allolevivirus* lineage, by either recombination or intramolecular duplication. A number of additional changes appeared to have been coupled with this genome expansion: notably, the loss of the lysis coding region, the evolution of a novel readthrough protein, and an increase in translation of the maturation protein. An increase in the genome size might be implicated in catalyzing all of these changes via changing the secondary structure of the RNA molecule. The lack of the lysis gene in the *Allolevivirus* was compensated for by the higher levels of the maturation protein which functioned in the phage release. Although the genome evolution in viruses was typically believed to be characterized by an economization of the genome, Bollback and Huelsenbeck (2001) indicated that increases in genome size might have played an important role in viral evolution.

At the time of the Bollback and Huelsenbeck's (2001) paper, the pseudomonaphage PP7 was the only sequenced non-coliphage genome (Olsthoorn et al. 1995a). The novel sequences of the non-coliphages presented a lot of novel information. Thus, the acinetophage AP205 was characterized by the absence of the lysis gene at the usual position and the single cytosine residue at the 3′ end instead of the typical three cytosine residues but nevertheless demonstrated the conserved secondary structure features (Klovins et al. 2002). Although the AP205 genome possessed the genetic map of a levivirus, its 3′ UTR has the length and RNA secondary structure of an allolevivirus. Sharing features with both MS2 and Qβ suggested that, in an evolutionary sense, the AP205 should be placed between the Qβ and MS2 (Klovins et al. 2002). By analogy, Ruokoranta et al. (2006) placed the PRR1 genome in the *Levivirus* genus with characteristics shared with alloleviviruses. The genome of the caulophage φCB5 revealed different length and genomic location of the lysis gene (Kazaks et al. 2011). The IncM plasmid-dependent RNA phage M was smaller than other family members and demonstrated again the diversity by the lysis gene location (Rumnieks and Tars 2012). The authors thought that the phage M represented a lineage that branched off early in the course of the RNA phage specialization on different conjugative plasmids.

After the sequencing of the IncC and IncH plasmid-dependent phages C-1 and Hgal1 genomes, Kannoly et al. (2012) confirmed the general idea that the *Levivirus*-like genome organization was ancestral, and the *Allolevivirus*-like genome organization appeared derived.

As it follows from the given information on full-length genomes and as noted previously in Chapter 14, the diversity of the lysis genes was of special interest to the RNA phage evolution. Thus, the mutational appearance of the Rg lysis gene of the phage Qβ, which originally did not demonstrate the special lysis gene, could be regarded as a possible evolution event, together with any other hypothetical equivalent ribosomal frameshifting-generated product (Nishihara 2004). The Rg lysis product lysed cells very quickly, perhaps quicker than optimal for high phage yield. Taking into account the variable position of this small gene, it is conceivable that the lysis gene had arisen more than once in the RNA phages and that in fact the lysis genes in the MS2, PP7, and AP205 genomes were functional analogs rather than homologs (Klovins et al. 2002).

As Krishnamurthy et al. (2016) noted rightly, the role of the RNA phages by evolution and modulation of microbial populations in ocean, soil, and animal ecosystems is still poorly understood, in part because of the still limited number of the known RNA phage species. As touched on in Chapters 2, 5, 6, 7, and 10, the identification of the highly divergent 122 RNA phage phylotype sequences, but not the phages as such, paved the way for further progress in evolutionary studies.

RNA-DEPENDENT POLYMERASES

Turning to the data on the RNA phage protein evolution, it is worth noting at the beginning that their coats were employed in the early phage history as subjects by the elucidation of the controversy over the Darwinian versus non-Darwinian evolution, or selectionist versus neutralist (Holmquist 1975).

However, the replicases, as the most conservative proteins, but not the coats, have come under the scrutiny in evolutionary studies, as stated earlier in the Relatedness to other taxa section of Chapter 2. In fact, Eugene V. Koonin grounded the phylogeny of the RNA-dependent RNA polymerases of the positive-strand RNA viruses in the late 1980s (Kunin et al. 1987, 1988; Koonin 1991a,b; Koonin and Dolja 1993) and the MS2, GA, Qβ, and SP replicases were the first *Leviviridae* members, representing the four groups of the coliphages which have been involved in the phylogenetic trees of evolutionary conserved fragments of the RNA-dependent RNA polymerases. This resulted in delineation of the three large supergroups, where the sequences of segments of approximately 300 amino acid residues originating from the central and/or C-terminal portions of the polymerases could be aligned with statistically significant scores within each of the supergroups (Koonin et al. 1991b). The specific consensus patterns of conserved amino acid residues were derived for each of the supergroups. The RNA phages appeared unexpectedly within the supergroup II together with carmo-, tombus-, dianthoviruses, a subset of luteoviruses (barley yellow dwarf virus), pestiviruses, hepatitis C virus, and flaviviruses. The three supergroups got the names *Picornavirata*, *Flavivirata*, and *Rubivirata* that were considered as a class, with subsequent divisions into orders, families, and genera or groups, as discussed in Chapter 2. However, Koonin and Dolja (1993) noted that the phylogenetic position of the RNA phages is very uncertain, and so is the pathway of their evolution. Nevertheless, the similarity between the phage replicases and the polymerases of eukaryotic viruses of the supergroup II, as well as the closer relationship with polymerases of yeast dsRNA genetic elements (Rodríguez-Cousiño et al. 1991, 1998), made it possible to speculate that the RNA phages might have evolved from eukaryotic viruses by horizontal transfer (Koonin and Dolja 1993).

At the same time, another partially resolved tree was presented, where *Leviviridae* did not belong to the supergroup II of Koonin and Dolja (1993) but were classified as an outgroup of the plus strand RNA viruses (Bruenn 1991). A bit earlier, comparing the RNA-dependent RNA-polymerases including the RNA phage replicases with the RNA-dependent DNA polymerases, or reverse transcriptases, Poch et al. (1989) have found that the polymerases and reverse transcriptases were related proteins, although this was only based on the collinearity and conservation of four sequence motifs shared between them. This made it possible to speak about the large general group of the RNA-dependent polymerases.

Furthermore, Zanotto et al. (1996) reevaluated the phylogenies of the RNA-dependent RNA and DNA polymerase sequences by a Monte Carlo randomization procedure, bootstrap resampling, and phylogenetic signal analysis. Although clear relationships between some viral taxa were identified, overall the sequence similarities and phylogenetic signals were insufficient to support the proposed evolutionary groupings of RNA viruses. Likewise, no support for the common ancestry of the RNA-dependent RNA polymerases and reverse transcriptases was found.

Much later, the evolution of tertiary structure of a long list of the viral RNA-dependent polymerases including the Qβ replicase was studied by Černý et al. (2014). The authors stated that the sequence similarity of the enzymes was too low for the phylogenetic studies, although general protein structures were remarkably conserved. The major strength of this work consisted of the unification of the sequence and structural data into a single quantitative phylogenetic analysis, using the powerful Bayesian approach. The resulting phylogram of the enzymes demonstrated that the RNA-dependent DNA polymerases of viruses within the *Retroviridae* family clustered in a clearly separated group, while the RNA-dependent RNA polymerases of double-stranded and single-stranded RNA viruses were mixed together. This evidence supported the hypothesis that the enzymes replicating the single-stranded RNA viruses evolved multiple times from the enzymes replicating the double-stranded RNA viruses, and vice versa. The authors recommended their phylogram as a scheme for the general RNA virus evolution. This phylogenetic tree of the evolution of the RNA-dependent polymerases and possibly of the RNA viruses is shown in Figure 18.6.

As noted in Chapter 2, metagenomics opened new horizons for evolutionary studies. In this context, Valerian V. Dolja and Eugene V. Koonin (2013, 2018) demonstrated clearly how the virus metagenomics transformed our understanding of virus diversity and evolution, and the extensive horizontal virus transfer was revealed. The major expansion of many groups of the RNA viruses through metagenomics allowed the

construction of substantially improved phylogenetic trees for the conserved virus genes; primarily for the RNA-dependent RNA polymerases.

For the *Leviviridae* family, the metagenomic challenged the status quo by the broad survey of the RNA phages in diverse environments that dramatically expanded their known host range and spectrum of genome architectures (Krishnamurthy et al. 2016). For the first time, the single-stranded RNA phages were detected in viral marine metagenomes and in a Gram-positive bacterium. In addition to the four proteins encoded by most members of the *Leviviridae*, several proteins with no recognizable homologs in databases were identified.

As Dolja and Koonin (2018) noticed, although the *Leviviridae* phages might not have sired most of the eukaryotic plus strand RNA viruses, there was a direct line of descent from the *Leviviridae* family to several eukaryotic virus families. This evolution pattern is presented in Figure 18.7, which is taken from the great Dolja and Koonin (2018) review.

The first such eukaryotic family is *Narnaviridae*, which includes capsid-less, noninfectious, vertically transmitted RNA elements that reproduce either inside the mitochondria (genus *Mitovirus*) or in the cytosol (genus *Narnavirus*) of the host cells. The *Narnaviridae* hosted by fungi remained the only known descendants of the *Leviviridae* phages until the unexpected discovery of the genus *Ourmiavirus* that includes several plant viruses related to *Narnaviridae*. The recent massive metagenomics study has identified numerous viruses related to *Ourmiavirus* and narna-like viruses in holobionts of diverse invertebrates. Although some of these viruses could originate from host-associated fungi, the sheer genomic diversity of these

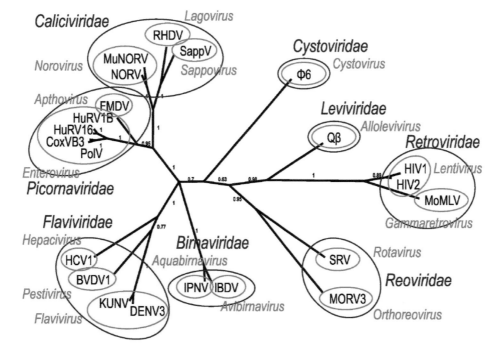

FIGURE 18.6 Phylogenetic tree of the evolution of viral RNA-dependent polymerases. Phylogenetic tree was calculated by an analysis unifying sequence and structure information. Only names of virus species coding the viral RNA-dependent polymerases are listed in the tree. Individual virus species are grouped in genera (blue) and families (red) according the actual ICTV virus taxonomy. (Reprinted from Černý J, Černá Bolfíková B, Valdés JJ, Grubhoffer L, Růžek D. *PLOS ONE*. 2014;9:e96070.)

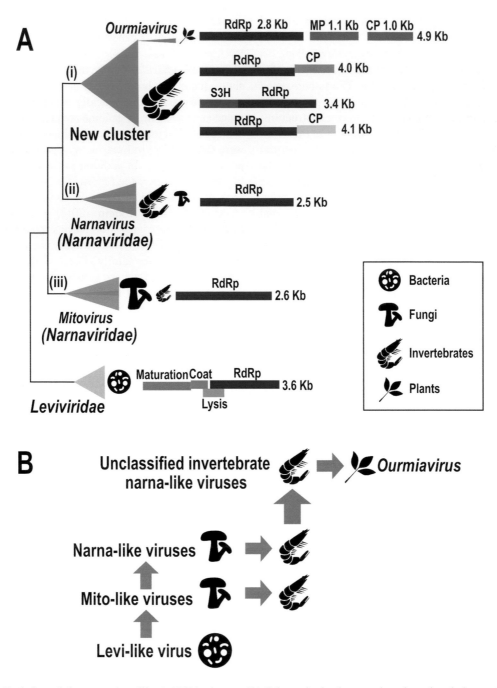

FIGURE 18.7 Evolution of the narnavirus-like (+)RNA viruses. (A) Schematic dendrogram based on the phylogenetic tree for RNA-dependent RNA polymerases (RdRp) of Narna-Levi clade from Shi et al. (2016a). Major clusters of related viruses are shown as triangles colored in accord with virus host ranges: gray, bacteria; olive, fungi; blue, invertebrates; green, plants. Rough diagrams of typical virus genomes for each cluster showing encoded proteins (rectangles; homologous proteins are in the same color), their functions and the genome size in kilobases (Kb), are at the right. CP, capsid protein; MP, movement protein; S3H, superfamily 3 helicase. (B) Hypothetical scenario for the evolution of narnavirus-like viruses. Vertical arrows denote virus transmission that accompanies host evolution, whereas horizontal arrows show presumed horizontal virus transfer (HVT) between distinct host organisms. (Reprinted from *Virus Res*, 244, Dolja VV, Koonin EV, Metagenomics reshapes the concepts of RNA virus evolution by revealing extensive horizontal virus transfer, 36–52, Copyright 2018, with permission from Elsevier.)

capsid-less or encapsidated viruses implies a broad host range, most likely among invertebrates. The currently defined, limited host range of the narna-like viruses that includes fungi, invertebrates, and plants might seem to be poorly compatible with the apparent origin of the RNA-dependent RNA polymerases of these viruses from the *Leviviridae* phages. Given the bacterial

origin of mitochondria, it can be assumed that a levivirus-like virus of the α-proteobacterial ancestor of the mitochondrion was brought along during eukaryogenesis, and subsequently, upon the transition of the endosymbiont to the organelle life style, lost its coat and other non-replicational genes, becoming a selfish RNA element of the genus *Mitovirus*. The *Narnavirus* elements could originate from either a *Mitovirus* that escaped to the cytosol or (less parsimoniously) from another levi-like virus that adopted to reproduction in the cytosol. This scenario implies that the mitoviruses were ancestral in eukaryotes even though these viruses have so far not been discovered in any hosts other than the unikont and archaeplastida supergroups. Although the descendants of the *Leviviridae* phages might have been lost in the rest of the eukaryotic lineages, it was suspected that the major reason for their apparently narrow spread is the limited scope of protist metagenomics. Whereas mitoviruses and narnaviruses are most likely to be of ancestral, protoeukaryotic origin, a different evolutionary scenario appears likely for the ourmia-like virus branch. Given that the plant ourmiaviruses represent but a small twig within a huge and diverse lineage (i) of the levi-like viruses (Figure 18.7A), most of which so far have been detected in invertebrates, horizontal virus transfer from herbivorous invertebrates (e.g., nematodes or arthropods) to plants emerges as the preferred evolutionary scenario (Figure 18.7B). From an evolutionary taxonomy standpoint, the three major lineages of phage-derived plus strand RNA viruses of eukaryotes, the mito-like, narna-like,

and ourmia-like viruses, each can be expected to eventually attain the family status. Furthermore, although the criteria for higher taxa of viruses are not currently well established, it appears likely that an umbrella taxon for these viruses, probably at the order level, eventually would be created (Dolja and Koonin 2018).

According to Dolja and Koonin (2018), the RNA phages of the family *Leviviridae* apparently gave rise to a distinct lineage of eukaryotic plus strand RNA viruses: mitoviruses, narnaviruses, ourmiaviruses, and their numerous invertebrate relatives via horizontal virus transfer spanning two domains of life by way of mitochondrial symbiosis (Figure 18.7). Following the original horizontal virus transfer, presumably from the proto-mitochondrion to the proto-eukaryote, the Levi-Narna clade was firmly established in fungi, being represented by the *Mitovirus* and *Narnavirus* genera. As to plant *Ourmiavirus*, which clusters with the invertebrate narna-like viruses, there is little doubt about their origin through horizontal virus transfer from invertebrates (Figure 18.7). Finally, Figure 18.8 presents the evolutionary relationships between viruses and their host organisms.

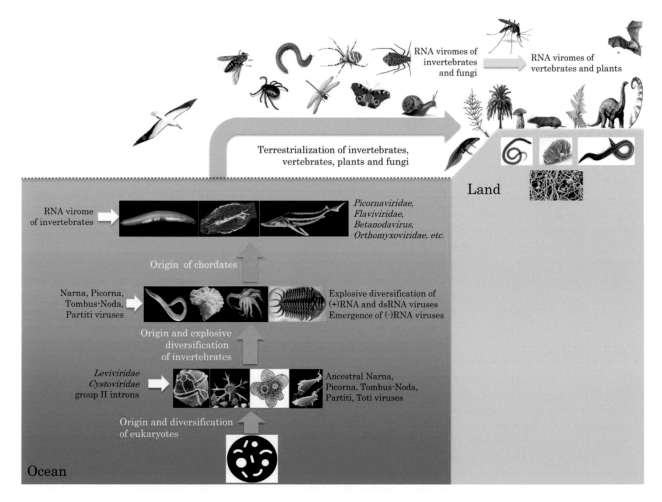

FIGURE 18.8 Evolutionary relationships between viruses and their host organisms. The vertical and bended arrows denote major transitions in the evolution of the cellular organisms, whereas horizontal arrows show hypothetical events of horizontal virus transfer (HVT) from the viromes of the ancestral organisms to the newly evolving groups of organisms. (Reprinted from *Virus Res*, 244, Dolja VV, Koonin EV, Metagenomics reshapes the concepts of RNA virus evolution by revealing extensive horizontal virus transfer, 36–52, Copyright 2018, with permission from Elsevier.)

The most recent phylogeny of the RNA-dependent RNA polymerases and reverse transcriptases included the *Leviviridae* family into the branch 1 out of 5, together with mitoviruses, ourmiaviruses, and narnaviruses (Wolf YI et al. 2018). It is noteworthy that the resulting tree encompassed 4617 RNA virus enzymes. Wolf YI et al. (2018) proposed the following evolution scenario for the branch 1 and their descendants. A levivirus-like ancestor that, like the extant members of the *Leviviridae*, possessed a capsid protein unrelated to single jellyroll capsid proteins (SJR-CPs), gave rise to naked eukaryotic RNA replicons known as mitoviruses and narnaviruses. These replicons consisted of a single RNA-dependent RNA polymerase gene and replicate in mitochondria and in the cytosol of the host cells of fungal and invertebrate hosts, respectively. The narnavirus RNA-dependent RNA polymerase is the ancestor of that of the expanding group of ourmiaviruses. Thus, the evolution of this branch apparently involved the loss of the structural module of leviviruses, which yielded naked RNA replicons that reproduced in the mitochondria of early eukaryotes. A group of these replicons subsequently escaped to the cytosol, which was followed by the reacquisition of unrelated structural modules from distinct lineages of eukaryotic viruses inhabiting the same environment. This complex evolutionary scenario emphasizes the key role of modular gene exchange in the evolution of RNA viruses (Wolf YI et al. 2018).

Recently, the existence of replicating mitochondrial virus in plants was demonstrated for the first time (Nerva et al. 2019). The 2.7 Kb contig was assembled from RNAseq data of infected *Chenopodium quinoa*, a plant species commonly used as a test plant in virus host-range experiments. This contig had highest similarity to mitoviruses found in plant genomes. The Northern blot analyses confirmed the existence of plus and minus strand RNA corresponding to the mitovirus genome, while no DNA corresponding to the genomic RNA was detected. The evolutionary links of the mitoviruses to the *Leviviridae* family were reviewed exhaustively by Marilyn J. Roossinck (2019).

A special alignment of the Qβ replicase subunit with the *Flavivirus* RNA-dependent RNA polymerases was performed recently by Selisko et al. (2018).

RECOMBINATION

The first *in vivo* RNA phage recombination event was described by Shen and Jiang (1982), when polyA molecule was inserted between two partially degraded Qβ RNA, one from the 3'-end and another from the 5'-end, as mentioned briefly above in the Autocatalytic synthesis of heterologous RNAs section. After infection with these mosaic RNA molecules, a few recombinant phages carrying polyA were obtained, one of which demonstrated the presence of the polyA fragment of 49 nucleotides. Then, as mentioned above in the RQ variants section, the 120-nucleotide-long recombinant RNA, named RQ120 and capable of the autocatalytic *in vitro* synthesis by the Qβ replicase, was discovered within the products of spontaneous synthesis by the nominally RNA-free Qβ replicase

preparation (Munishkin et al. 1988). Unexpectedly, the minus strand of the RQ120 was a recombinant RNA consisting of the internal Qβ RNA fragment of 80 nucleotides connected with the 33-nucleotide-long 3' terminal fragment of *E. coli* tRNA$_1^{Asp}$. This conclusion was totally unexpected, as it had long been believed that the recombination should not occur between RNA phage genomes, but failure to detect the recombinants by genetic analysis might have been due to the high mutation rate in the RNA genomes (Horiuchi 1975; Holland et al. 1982). Munishkin et al. (1988) pointed out in this connection that the large fragment of eukaryotic tRNA^{Asp} was found at the 5' terminus of the Sindbis virus defective interfering particle RNAs (Monroe and Schlesinger 1983).

Furthermore, Munishkin et al. (1991) isolated the natural recombinant RQ135 RNA from the products of spontaneous RNA synthesis in an *in vitro* Qβ replicase reaction that was incubated in the absence of added RNA. The RQ135 was unrelated to the genomic Qβ RNA variant but explained segments that were homologous to ribosomal 23S rRNA and the phage λ origin of replication. The authors concluded that the Qβ replicase recognition was not confined to viral RNA but could appear as a result of recombination among other RNAs that usually occurred in cells. In parallel, Biebricher and Luce (1992) determined that the above-mentioned variant SV-11, this time characterized as 115, but not 113 nucleotides as before, contained an inverse duplication of the high-melting domain of the MNV-11 variant. Thus, the SV-11 was a recombinant between the plus and minus strands of the MNV-11 resulting in a nearly palindromic sequence. During chain elongation in the replication, the chain folded consecutively to a metastable secondary structure of the RNA, which could rearrange spontaneously to a more stable hairpin-form RNA. While the metastable form was an excellent template for the Qβ replicase, the stable RNA was unable to serve as template. Remarkably, when initiation of a new chain was suppressed by replacing GTP in the replication mixture by ITP, the Qβ replicase added nucleotides to the 3' terminus of RNA. The replicase used parts of the RNA sequence, preferentially the 3'-terminal part for copying, thereby creating an interior duplication. The reaction also added nucleotides to the 3' terminus of some RNA molecules that were unable to serve as templates for the Qβ replicase (Biebricher and Luce 1992).

The recombination as a mechanism involved in the reversion of the phage Qβ mutants *in vivo* was disclosed by Palasingam and Shaklee (1992). Thus, the Qβ phage RNAs with inactivating mutations, namely, 8-base insertion or 17-base deletion, within their replicase genes, were prepared from the modified Qβ cDNAs and transfected into *E. coli* spheroplasts containing the Qβ replicase provided *in trans* by a resident plasmid. The replicase-defective Qβ phage particles that were produced by these spheroplasts were detected as normal-sized plaques on lawns of cells containing the plasmid-derived Qβ replicase but were unable to form plaques on cells lacking this plasmid. When individual replicase-defective phage was isolated and grown to high titer in cells containing the plasmid-derived Qβ replicase, the revertant Qβ phage was obtained at a frequency of ca. 10^{-8}. To investigate the mechanism of this reversion, a point mutation was placed into the plasmid-derived Qβ replicase gene by site-directed mutagenesis. The Qβ mutants amplified on cells containing the resultant plasmid also yielded the replication-competent revertants. The sequencing showed that the original mutation insertion or deletion was no longer present in the phage revertants but that the marker mutation placed into the plasmid was now present in the genomic RNAs, indicating that the recombination occurred by the reversion process. Further experiments demonstrated that the 3′ noncoding region of the plasmid-derived replicase gene was necessary for the reversion-recombination of the deletion mutant, whereas this region was not required for the reversion or recombination of the insertion mutant. The rates of recombination were estimated to be of the order of 10^{-8} per 1500-nucleotide RNA segment (Palasingam and Shaklee 1992).

Jan van Duin's team constructed an infectious cDNA clone of MS2 to monitor the response of MS2 to mutations in its genome (Olsthoorn et al. 1994). As the wild-type MS2 cDNA clone produced ~10^{12} phages per mL culture after overnight growth in E. coli, it was possible to reveal extremely rare evolutionary events like the repair of a 19-nucleotide deletion (Olsthoorn and van Duin 1996a). Thus, Olsthoorn and van Duin (1996a) have monitored the evolution of insertions in the two MS2 RNA regions of known secondary structure, where coding pressure was negligible or absent. The base changes and shortening of the inserts proceeded until the excessive nucleotides could be accommodated in the original structure. The stems of hairpins could be dramatically extended but the loops could not, revealing natural selection against the single-stranded RNA. The 3′ end of the MS2 maturation protein gene formed a small hairpin with an XbaI sequence in the loop. This site was used to insert XbaI fragments of various sizes. The phages produced by these MS2 cDNA clones were not wild type, nor had they retained the full insert. Instead, every revertant phage had trimmed the insert in a different way to leave a four- to seven-membered loop to the now extended stem. Similar results were obtained with inserts in the 5′ untranslated region. The great number of different revertants obtained from a single starting mutant as well as sequence inspection of the crossover points suggested that the removal of the redundant RNA occurred randomly. The only common feature among all revertants appeared to be the potential to form a hairpin with a short loop, suggesting that the single-stranded RNA negatively affected the viability of the phage. To test this hypothesis, the XbaI fragments of 34 nucleotides were introduced which could form either a long stem with a small loop or a short stem with a large loop of 26 nucleotides. The base-paired inserts were perfectly maintained for many generations, whereas the unpaired versions were quickly trimmed back to reduce the size of the loop. These data confirmed that the single-stranded RNA adversely affected phage fitness and was strongly selected against. It was highly remarkable that the repair of the RNA genomes appeared as the result of the random recombination. Of the plethora of recombinants, only those able to adopt a base-paired structure survived. The frequency with which the inserts were removed

seemed higher than measured by others for small inserts in a reading frame in the Qβ RNA. To account for this higher frequency, the authors proposed a model in which the single-stranded nature of the inserts induced random recombination at the site of the insertion (Olsthoorn and van Duin 1996a).

Klovins et al. (1997b) checked whether phage RNA structure was under selective pressure by host ribonucleases; in particular, by ribonuclease III. Thus, a fully double-stranded hairpin of 17 base pairs, a ribonuclease III target, was inserted into a noncoding region of the MS2 RNA genome. In a ribonuclease III⁻ host these phages survived, but in wild-type bacteria they did not. Here, the stem underwent Darwinian evolution to a structure that was no longer a substrate for the ribonuclease III. This was achieved in three different ways: (i) the perfect stem was maintained but shortened by removing all or most of the insert; (ii) the stem acquired suppressor mutations that replaced Watson-Crick base pairs by mismatches; (iii) the stem acquired small deletions or insertions that created bulges. These insertions consisted of short stretches of non-templated A or U residues. Their origin was ascribed to polyadenylation at the site of the ribonuclease III cut in the plus or minus strand either by E. coli polyA polymerase or by idling MS2 replicase (Klovins et al. 1997b; van Meerten et al. 1999).

Licis et al. (2000) forced the retroevolution of the phage MS2. To do so, the repressor complex I hairpin was partially randomized to monitor alternative solutions that would evolve on the passaging of mutant phages. The evolutionary reconstruction of the operator failed in the majority of mutants. Instead, a poor imitation developed containing only some of the recognition signals for the coat protein. The three mutants were of particular interest in that they contained the double nonsense codons in the lysis reading frame that run through the operator hairpin. The simultaneous reversion of the two stop codons into the sense codons had a very low probability of occurring. Therefore, the phage solved the problem by deleting the nonsense signals and, in fact, the complete operator, except for the initiation codon of the replicase gene. Several revertants were isolated with activities ranging from 1%–20% of the wild type. Surprisingly, the operator, long thought to be a critical regulator, now appeared to be a dispensable element. The results indicated how RNA viruses can be forced to step back to an attenuated form (Licis et al. 2000).

Furthermore, van Meerten et al. (2002) generated hybrids between different species or genera of the RNA phages by in vivo recombination. Thus, the MS2 cDNA located on a plasmid and lacking part of its 5′ untranslated leader was complemented with another plasmid carrying the 5′ half of the fr genome, a species of the group I, or of KU1, a species of the group II, with low sequence similarity. When the two plasmids were present in the same cell there was spontaneous production of hybrid phages. Interestingly, these hybrids did not arise by a double or single crossover that would replace the missing MS2 sequences with those of the fr or KU1. Rather, the hybrids arose by attaching the complete 5′ untranslated region of the fr or KU1 to the 5′ terminus of the defective phage MS2. Several elements of the 5′ untranslated region then occurred twice, one from KU1 (or fr) and the other from MS2. These redundant elements were in most cases deleted upon evolution of the hybrids. As a result, the 5′ untranslated region of KU1 (or fr) then replaced that of the MS2. However, when hybrids were competed against wild type, they were quickly outgrown, probably explaining their absence from natural isolates (van Meerten et al. 2002).

Much later, René C.L. Olsthoorn (2014) used the infectious MS2 cDNA system (Olsthoorn et al. 1994) to identify functionally important nucleotides in a self-cleaving ribozyme. To do this, the hammerhead ribozyme of satellite tobacco ringspot virus was inserted into the MS2 genome. The sequencing of the surviving phages revealed that the majority had acquired single base-substitutions that apparently inactivated the hammerhead ribozyme. The positions of these substitutions exactly matched that of the previously determined core residues of the ribozyme. Thus, the natural selection against a ribozyme in the MS2 genome functioned as a quick method to identify nucleotides required for self-cleavage.

Concerning further elucidation of the recombination by the Qβ-related parasitic sequences, Avota et al. (1998) found that the replicable RNA isolated from the natural 6S RNA found in Qβ-infected cells was a recombination product of Qβ RNA with cellular 23S rRNA and 10Sa RNA.

Finally, Chetverin et al. (1997) demonstrated the nonhomologous character of the RNA recombination that occurred between the 5′ and 3′ fragments of the replicable RNA in a cell-free system composed of the pure Qβ replicase and ribonucleoside triphosphates, providing direct evidence for the ability of RNAs to recombine without DNA intermediates and in the absence of host cell proteins. The recombination events were revealed by the molecular colony technique that allowed single RNA molecules to be cloned *in vitro*. The observed non-homologous recombination was entirely dependent on the 3′-hydroxyl group of the 5′ fragment and were due to a splicing-like reaction in which the RNA secondary structure guided the attack of this 3′-hydroxyl on phosphoester bonds within the 3′ fragment. Therefore, clear evidence for the transesterification mechanism guided by the RNA secondary structure was obtained. Moreover, Chetverina et al. (1999) discovered another mechanism of the RNA recombination, namely, the ability of RNAs to undergo spontaneous rearrangements of their sequences under physiological conditions without any protein. It was due to the sensitive molecular colony technique, which allowed the single RNA molecules to be detected. The rearrangements were Mg^{2+}-dependent, sequence-nonspecific, and occurred both *in trans* and *in cis* at a rate of 10^{-9} h^{-1} per site. These results suggested that the mechanism of the spontaneous RNA rearrangements differed from the transesterification reactions observed earlier in the presence of the Qβ replicase.

Later, Chetverin et al. (2005) explored the potential of the two RNA-directed RNA polymerases, namely, the Qβ replicase and the poliovirus 3Dpol protein, to promote the RNA recombination through a primer extension mechanism. The substrates of the recombination were fragments of complementary strands of a Qβ phage-derived RNA such

that if aligned at complementary 3′ termini and extended using one another as a template, they would produce replicable molecules detectable as RNA colonies grown in the Qβ replicase-containing agarose. The results showed that while 3Dpol efficiently extended the aligned fragments to produce the expected homologous recombinant sequences, only nonhomologous recombinants were generated by Qβ replicase at a much lower yield and through a mechanism not involving the extension of RNA primers. It was concluded that the mechanisms of RNA recombination by poliovirus and Qβ replicases were quite different. These data favored the RNA transesterification mechanism of the reaction catalyzed by a conformation acquired by the Qβ replicase during RNA synthesis and provided a likely explanation for the very low frequency of the homologous recombination in the phage Qβ (Chetverin et al. 2005).

The fine mechanism of this splicing-type RNA recombination performed by the Qβ replicase was presented in the numerous detailed reviews written later by Alexander B. Chetverin (1997, 1999, 2004).

As noted briefly in Chapters 2 and 6, by their studies on the RNA phage detection with the ecological aims, Vinjé et al. (2004) identified the new *Levivirus* group JS that included strains having more than 40% nucleotide sequence diversity with the known *Levivirus* groups I and II represented by the phages MS2 and GA, respectively. According to the authors' statement, further genomic sequence and serological data were needed to confirm that the group JS belongs to a novel subgroup or genotype, or whether these strains were a result of recombination or rearrangement events. After 8 years, Friedman et al. (2012) isolated two novel JS strains, DL52 and DL54. As in the Vinjé et al. (2004) study, the DL52 and DL54 strains were isolated from separate coastal waters. The complete genomic sequencing suggested a natural recombination event in the formation of a genogroup I subgroup JS-like levivirus, represented by the DL52 and DL54 strains. There was high nucleotide and amino acid identity in the three genes, the maturation, capsid, and lysis genes (≥95%) but a lack of similarity in the replicase gene (84–85%) when the JS strains were compared to the group I MS2-like strains. The four different recombination programs demonstrated one or two breakpoint regions in the replicase gene, signifying a recombination event. The recombination event occurred downstream of the replicase catalytic site, thereby maintaining viral integrity and replication function. However, after phylogenetic tree analysis, the authors concluded that the JS strains were not a unique *Levivirus* genogroup and proposed classification of the recombinant JS strains as the JS-like subgroup of the group I phages (Friedman et al. 2012). After 50 years of RNA phage studies, this was the first description of the recombinant viruses from natural isolates in the *Leviviridae* family.

In the above-characterized evolutionary study of Bansho et al. (2012), the parasitic RNA seemed to be derived from the sources in the cell-free system by the recombination events. The most possible source was the recombination within the β subunit-encoding RNA vector because the latter was originally constructed by inserting the β subunit gene of the Qβ replicase into the MDV-poly vector, a derivative of the classical MDV-1 RNA. Bansho et al. (2012) concluded that the non-homologous recombination of the β subunit-encoding RNA vector at appropriate sites was responsible for the production of the parasitic MDV-1 derivatives. The results of the mathematical modeling indicated that the size dependency of the appearance of the parasitic RNA could be well explained when the rate of appearance was set to 10^{-8}/hr per vector RNA molecule. This value was close to that reported previously: 10^{-7}/hr and 10^{-9}/hr (Chetverin et al. 1997; Chetverina et al. 1999).

Bansho et al. (2012) hypothesized that in the RNA World, such recombination might have produced parasitic replicators by deleting the replication ribozyme encoded on the genome but maintaining the recognition sequence for the ribozyme. Once this type of parasite appeared, it would have replicated much faster than the genome because of its small size and eventually inhibit replication of the genome. Therefore, the frequency of the appearance of such parasites must generally be low to achieve long-lasting self-replication of the genome. This restricted the possible conditions under which primitive replication may have emerged on the ancient Earth.

QUASISPECIES AND EVOLUTION

The elucidation of the RNA phage quasispecies traced its origin to the site-directed mutagenesis of the phage Qβ in the mid-1970s (Flavell et al. 1974, 1975), as described in the Mutants of the phages Qβ, SP, and FI section of Chapter 8. Thus, Esteban Domingo et al. (1976) described experiments in which the phage Qβ with an A → G mutation in the 40th nucleotide from the 3′ terminus, either pure or mixed with an equal number of plaque-forming units of the wild type phage Qβ, were propagated serially on *E. coli*. The infection was carried out at an overall multiplicity of 20 so as to favor intracellular competition of the phages. The lysate resulting from the first infectious cycle was used to initiate a second one under the same conditions as before. This process was continued for 10 or more cycles, and the ratio of the mutant to wild-type phage was determined by analyzing the ^{32}P-labeled viral RNA in regard to the mutation at the position 40. It was found that if the experiment was initiated with the cloned mutant phage, the 3%–8% of the phage was the wild type after three infectious cycles, and that the proportion of the wild type increased rapidly with the successive propagations. If the experiment was started with an equal mixture of the mutant and wild-type phage, the proportion of mutant dropped to about 20% after 2, and to less than 0.5% after 10 transfers. These results could be explained by assuming that the wild-type phages (i) arose at a significant rate by reversion and (ii) outgrew the mutant phage. To solve this discrepancy, a mathematical model was formulated that made it possible to describe the composition of the phage population, i.e., the proportion of the revertant and mutant phages, as a function of reversion and growth rate in the population (Batschelet et al. 1976). The resulting model curves were fitted to the

experimental data obtained with the Qβ phage mutant 40 and its wild-type counterpart. The growth rate of the mutant relative to wild type, under competitive conditions, was 0.25, and the reversion rate of the mutant was estimated to be about 10^{-4} per doubling.

Furthermore, Domingo et al. (1978) determined the nucleotide sequence heterogeneity of an RNA phage population. The nucleotide sequence of ^{32}P-RNA from Qβ phage clones was subjected to the ribonuclease T1 fingerprinting. About 15% of the clones derived from a multiply passaged Qβ population showed fingerprint patterns which deviated from that of the RNA from the total population. All deviations examined could be attributed to one and, less frequently, to two or more nucleotide transitions. It was estimated that each viable phage genome in a multiply passaged population differed in one to two positions from the average sequence of the parental population. The several deviant clones were tested by growth competition against a wild-type population; after 10–20 generations, the resulting phage showed the wild-type T1 fingerprint pattern. This study resulted in the important finding that a Qβ phage population resided in a dynamic equilibrium, with viable mutants arising at a high rate (Batschelet et al. 1976; Domingo et al. 1976), on the one hand, and being strongly selected against on the other (Domingo et al. 1978). It led to the revolutionary conclusion that the Qβ genome, just as other RNA phage genomes, cannot be described as a defined unique structure, but rather as a weighted average of a large number of different individual sequences. The historical Domingo et al. (1978) paper is considered now to mark the beginning of experimental studies on the viral *quasispecies*.

Later, the same fundamental conclusion was repeated by Esteban Domingo's team in their study on genetic heterogeneity generated upon passage of the foot-and-mouth disease virus in cell cultures (Sobrino et al. 1983).

The term *quasispecies*, defining the clouds of related elements that behave almost, or *quasi*, like a single type of molecule, or *species*, was used for the first time in the exhaustive review written by Domingo et al. (1985). The authors emphasized that cloned or non-cloned populations of most RNA viruses did not consist of a single genome species of defined sequence, but rather of heterogeneous mixtures of related genomes, or quasispecies. Due to very high mutation rates, the genomes of a quasispecies virus population shared a consensus sequence but differed from each other and from the consensus sequence by one, several, or many mutations. Needless to say, the quasispecies structure of the RNA virus populations should have many important theoretical and practical implications, because mutations at only one or a few sites might alter the phenotype of an RNA virus. It is noteworthy that the classical review of Esteban Domingo and John J. Holland (1988) was reprinted after 30 years by CRC Press (Domingo and Holland 2018).

John W. Drake (1993) was the first to estimate the rates of spontaneous mutation among RNA viruses. He found a clear central tendency for lytic RNA viruses, such as the phage Qβ and poliomyelitis, vesicular stomatitis, and influenza A viruses, to display rates of spontaneous mutation of approximately 1 per genome per replication. This rate was some 300-fold higher than previously reported for the DNA-based microbes. Therefore, the lytic RNA viruses mutated at a rate close to the maximum value compatible with the viability. The Qβ mutation rates were fixed later as 1.4×10^{-3} per base and 6.5 per genome (Drake et al. 1998; Gros 2008).

The effect of deleterious mutation-accumulation on the fitness of the phage MS2 was determined experimentally by de la Peña et al. (2000). Thus, the fitness declined by the accumulation of deleterious mutations, and the rate of fitness decline was estimated to be as high as 16% per bottleneck transfer.

A series of investigations were devoted to the adaptation issue. Bacher et al. (2003) studied the evolution of the Qβ phage in the presence of an amino acid analog 6-fluorotryptophan. The phage Qβ initially grew poorly, but after 25 serial passages the fitness of the phage on the analog was substantially increased. There was no loss of fitness when the evolved phage was passaged in the presence of tryptophan. Seven mutations were fixed throughout the phage in the two independent lines of descent, while none of the mutations changed a tryptophan residue. It was concluded that a relatively small number of mutations allowed an unnatural amino acid to be functionally incorporated into a highly interdependent set of proteins. These results seemed to support the ambitious "ambiguous intermediate" hypothesis for the emergence of divergent genetic codes in which the adoption of a new genetic code was preceded by the evolution of proteins that can simultaneously accommodate more than one amino acid at a given codon.

The limits of phage adaptation were measured when four pairs of phages were adapted for rapid growth under similar conditions to compare their evolved endpoint fitnesses (Bull et al. 2004). The first pair was MS2 and Qβ, while other three consisted of the pairwise related DNA phages. The fitness was measured as absolute growth rate per hour under the same conditions used for adaptation. The phages T7 and T3 achieved the highest fitnesses, able to increase by 13 billionfold and three-quarters billionfold per hour, respectively. In contrast, the RNA phages achieved low fitness maxima, with growth rates approximately 400-fold and 4000-fold per hour. It appeared that the highest fitness limits were not attributable to high mutation rates or small genome size, even though both traits were expected to enhance adaptation for fast growth. The authors suggested that the major differences in fitness limits stem from different "global" constraints, determined by the organization and composition of the phage genome affecting whether and how it overcame potentially rate-limiting host processes, such as transcription, translation, and replication.

The first explicit investigation into the effect of spatial heterogeneity on the stability of the host-phage coexistence in the absence of permanent spatial refuges was performed with the phage PP7 (Brockhurst et al. 2006a). The populations of PP7 and *P. aeruginosa* were propagated in homogeneous, i.e., by constant mixing, or shaken, and heterogeneous, i.e., unmixed, or static, microcosms. To prevent the formation of permanent spatial refuges that persisted for the whole experiment, an aliquot of each population was transferred to a fresh microcosm daily.

To test for coexistence, the population densities of the bacteria and phage were measured every 4 days. To ascertain the mode of coexistence, i.e., persistent coevolution or the maintenance of a sensitive subpopulation of bacteria, the changes in bacterial resistance and phage infectivity were measured through time. To determine the effect of the environmental regimes on parasite dispersal, the transmission rates were measured in both homogeneous and heterogeneous microcosms. Finally, the fitness of resistant bacteria was measured in both environments. The spatial heterogeneity promoted stable coexistence of host and parasite, while coexistence was significantly less stable in the homogeneous environment. It was concluded that the spatial heterogeneity would lead to more stable coexistence of host and parasite through providing ephemeral refuges by reducing dispersal (Brockhurst et al. 2006a).

Moreover, using the same phage-host pair, Brockhurst et al. (2006b) studied the impact of phages on interspecific competition in experimental populations of bacteria and found that the phage altered competitive interactions between bacterial species in a way that was consistent with the maintenance of coexistence. However, the stability of coexistence was likely to depend upon the nature of the constituent host-phage interactions and environmental conditions. Later, the library of the PP7-resistant *P. aeruginosa* variants was investigated for the effect of phage resistance on cytotoxicity of host populations toward cultured mammalian cells (Hosseinidoust et al. 2013b). Remarkably, the phage-resistant variants exhibited a greater ability to impede metabolic action of cultured human keratinocytes and have a greater tendency to cause membrane damage even though they could not invade the cells in large numbers. These results clearly indicated a significant change in the *in vitro* virulence of *P. aeruginosa* following phage predation and highlighted the need for caution in the selection and design of phages and phage cocktails for therapeutic use. All aspects of the understanding and exploitation of the *Pseudomonas* predators, including the phage PP7, were presented in the recent great review (De Smet et al. 2017).

The phages MS2 and R17, among 16 most characterized phages, were involved in the great study on the phage "life history," a tradeoff between survival and reproduction among phages (De Paepe and Taddei 2006). The main life history traits of phages were the multiplication rate in the host, the survivorship of virions in the external environment, and their mode of transmission. By comparing the life history traits of 16 phages in *E. coli*, the authors showed that their mortality rate was constant with time and positively correlated to their multiplication rate in the bacterial host. Even though these viruses did not age, this result was in line with the tradeoff between survival and reproduction previously observed in numerous aging organisms. Furthermore, a multiple regression showed that the combined effects of two physical parameters, namely, the capsid thickness and the density of the packaged genome, accounted for 82% of the variation in the mortality rate. The correlations between life history traits and physical characteristics of virions therefore provided a mechanistic explanation of this tradeoff. The fact that this tradeoff was present in this very simple biological situation suggested that it might be a fundamental property of evolving entities produced under constraints. Moreover, such a positive correlation between the mortality and multiplication revealed an underexplored tradeoff in host-parasite interactions (De Paepe and Taddei 2006).

Bollback and Huelsenbeck (2007) found that the clonal interference was alleviated by high mutation rates in large populations of the phage MS2. The phage MS2 was the most qualified subject for this investigation, since it demonstrated high mutation rate of 10^{-3} per nucleotide/replication (Drake 1993) and low recombination rate of 4.22×10^{-11} per nucleotide in multiply infected cells (Palasingam and Shaklee 1992). When a beneficial mutation was fixed in a population with very low recombination, the genetic background linked to that mutation was fixed. As a result, the beneficial mutations on different backgrounds experienced competition, or "clonal interference," that could cause asexual populations to evolve more slowly than their sexual counterparts. Factors such as a large population size (N) and high mutation rates (μ) increased the number of competing beneficial mutations, and hence were expected to increase the intensity of clonal interference. However, the existing theory suggested that with very large values of $N\mu$, the severity of clonal interference might instead decline. The reason was that with large $N\mu$, the genomes, including both beneficial mutations, were rapidly created by recurrent mutation, obviating the need for recombination. Bollback and Huelsenbeck (2007) found that in the non-recombining MS2 populations with very large $N\mu$, the recurrent mutation did appear to ameliorate this cost of asexuality. In other words, this study selected populations of MS2 for growth at elevated temperatures and the frequency of a number of beneficial mutations was determined throughout the time course of the experiment (every ~10 passages). These MS2 experimental data were used by the development of a new method for estimating effective population sizes, N_e, and selection coefficients, s (Bollback et al. 2008). In this study, the method was applied, after MS2, to estimate selection coefficients acting on the CCR5-Δ32 mutation on the basis of published samples of contemporary and ancient human DNA. Next, Bollback and Huelsenbeck (2009) turned to the parallel genetic evolution within and between the phage species of varying degrees of divergence when the three phage species, MS2, fr, and NL95, were adapted to a novel high-temperature environment. By adapting the three phage species to a novel environment, the authors found (i) a high rate of parallel genetic evolution at orthologous nucleotide and amino acid residues within species, (ii) parallel beneficial mutations did not occur in a common order in which they fix or appear in an evolving population, (iii) low rates of parallel evolution and convergent evolution between species, and (iv) the probability of parallel and convergent evolution between species was strongly effected by divergence. It is notable that the experimental MS2 adaptation data (Bollback and Huelsenbeck 2007) were used later in the population genetics by the elaboration of an alternative approximation to the mutant-frequency distribution that did not make any assumptions about the magnitude of selection or mutation and

was much more computationally efficient than the standard diffusion approximation (Lacerda and Seoighe 2014).

The quasispecies of the phage Qβ were studied by the quantification of the effect of an increase in the replication error rate when the mutagenic nucleoside analog 5-azacytidine was used (Cases-González et al. 2008; Lázaro 2008; Lázaro and Arribas 2010; Arribas et al. 2011). The evolution of the Qβ population was estimated both in liquid medium and in semisolid agar. The Qβ populations replicating in liquid medium in the presence of the 5-azacytidine were extinguished after a variable number of transfers. In contrast, 5-azacytidine had no negative effects in the development of clonal Qβ populations in semisolid agar. The 5-azacytidine increased the replicative ability of the individual components of some virus clones that had experienced a large number of plaque-to-plaque transfers. Moreover, the individual virus clones, when isolated from mutagenized populations that were doomed to extinction, could replicate under 5-azacytidine mutagenesis during a number of plaque-to-plaque transfers. The response to mutagenesis was interpreted in the light of features of plaque development versus infections by free-moving virus particles and the distance to a mutation-selection equilibrium. The results suggested that the clonal phage populations away from equilibrium derived replicative benefits from increased mutation rates. This was relevant to the application of lethal mutagenesis *in vivo*, in the case of viruses that encountered changing environments and were transmitted from cell to cell under conditions of limited diffusion that mimicked the events taking place during plaque development. Therefore, the results documented an important effect of population bottlenecks in the response of viruses to lethal mutagenesis (Cases-González et al. 2008).

Then, the possible mechanisms providing resistance to 5-azacytidine in the phage Qβ were evaluated (Cabanillas et al. 2013). To further study whether the interference among beneficial mutations operated during the Qβ evolution at increased error rate, the authors analyzed how polymorphic mutations were distributed in individual virus genomes isolated at different points of the evolutionary series. This approach has made it possible to identify several competing lines carrying different combinations of polymorphic mutations which differed in their fitness values and in their ability to fix when present in a simpler mutant spectrum. The presence of additional mutations accompanying the polymorphic ones, the high frequency of recurrent mutations, and the occurrence of epistatic interactions contributed to generate a highly complex dynamic (Cabanillas et al. 2013).

The mutational fitness of random mutations was measured in the phages Qβ, MS2, and SP, in parallel with some single-stranded DNA phages (Domingo-Calap et al. 2009). The phages were subjected to plaque-to-plaque transfers and chemical mutagenesis. The genome sequencing and growth assays indicated that the average fitness effect of the accumulated mutations was similar in the RNA and DNA phages. Furthermore, the site-directed mutagenesis was used to obtain 42 clones of the phage Qβ carrying random single-nucleotide substitutions, which were assayed for fitness. In Qβ, 29% of such mutations were lethal, whereas viable ones reduced fitness by 10% on average. It therefore seemed that high mutational sensitivity was a general property of viruses with small genomes. These data on the mutational fitness effects revealed by the site-directed mutagenesis studies on the RNA phages were reviewed and commented on exhaustively (Sanjuán 2010, 2012; Combe and Sanjuán 2014).

Furthermore, Domingo-Calap et al. (2010) examined the correlation between mutational robustness and thermostability in the experimental Qβ populations. Thus, the thermostable viruses evolved after only six serial passages in the presence of heat shocks, and genome sequencing suggested that thermostability can be conferred by several alternative mutations. To test whether thermostable viruses have increased mutational robustness, additional passages in the presence of nitrous acid were performed. Whereas in control lines this treatment produced the expected reduction in growth rate caused by the accumulation of deleterious mutations, the thermostable viruses showed no such reduction, indicating that they were more resistant to mutagenesis. This suggested that the selection for thermostability could lead to the emergence of mutational robustness driven by plastogenetic congruence (Domingo-Calap et al. 2010).

Domingo-Calap and Sanjuán (2011) examined the proposition that the RNA phages are the fastest evolving entities in nature. They have compared the rates of adaptation and molecular evolution of the phages Qβ, SP, and MS2 with those of the single-stranded DNA phages φX174, G4, and f1. It appeared that the RNA phages evolved faster than the DNA phages under strong selective pressure, and that their extremely high mutation rates appeared to be optimal for this kind of scenario. However, their performance became similar to that of the DNA phages over the longer term or when the population was moderately well-adapted. Remarkably, the roughly 100-fold difference between the mutation rates of the RNA and DNA phages yielded less than a fivefold difference in the adaptation and nucleotide substitution rates. The results were therefore consistent with the observation that, despite their lower mutation rates, the single-stranded DNA viruses sometimes matched the evolvability of the RNA viruses (Domingo-Calap and Sanjuán 2011). Furthermore, Cuevas et al. (2012) checked the fitness effects of synonymous mutations in the DNA and RNA viruses. Thus, 53 single random synonymous substitution mutants were constructed in the phages Qβ and φX174 by site-directed mutagenesis, and their fitness was assayed. The selection at synonymous sites was found stronger in the Qβ than in the φX174. This type of selection contributed approximately 18% of the overall mutational fitness effects in the Qβ, and the random synonymous substitutions had a 5% chance of being lethal to the phage, whereas in the φX174, these figures dropped to 1.4% and 0%, respectively. In contrast, the effects of nonsynonymous substitutions appeared to be similar in the Qβ and φX174 (Cuevas et al. 2012).

At that time, Bradwell et al. (2013) refined the mutation rate of the phage Qβ. The *amber* reversion-based Luria-Delbrück fluctuation tests combined with the mutant sequencing gave an estimate of 1.4×10^{-4} substitutions per nucleotide per round of copying, the highest mutation rate reported for any virus using this method. This estimate was confirmed using a direct plaque sequencing approach and after reanalysis of previously published estimates for the phage Qβ. The comparison with other riboviruses, namely, all RNA viruses except retroviruses, provided statistical support for a negative correlation between the mutation rates and genome size. It was suggested that the mutation rates of the RNA viruses might be optimized for maximal adaptability and that the value of this optimum might in turn depend inversely on the genome size (Bradwell et al. 2013).

In parallel, the adaptive response of the phage MS2 to a novel environment was studied by Betancourt (2009). Thus, the three MS2 lines were adapted to rapid growth and lysis at cold temperature for a minimum of 50 phage generations and subjected to whole-genome sequencing. The adaptive substitutions were identified, the changes in frequency of adaptive mutations through the course of the experiment were monitored, and the effect on phage growth rate of each substitution was measured. All three lines showed a substantial increase in fitness, namely, a two- to threefold increase in growth rate, due to a modest number of substitutions, three to four. The data showed some evidence that the substitutions occurring early in the experiment had larger beneficial effects than later ones, in accordance with the expected diminishing returned relationship between the fitness effects of a mutation and its order of substitution. The patterns of the molecular evolution suggested an abundant supply of the beneficial mutations in this system. Nevertheless, some beneficial mutations appeared to have been lost, possibly due to accumulation of the beneficial mutations on other genetic backgrounds, clonal interference, and negatively epistatic interactions with other beneficial mutations (Betancourt 2009).

Furthermore, Betancourt (2010) used the material generated in the previous study (Betancourt 2009) to explore the role played by sign epistasis in adaptively evolving populations of the phage MS2. It was shown that the two mutations that previously occurred in different genetic backgrounds were beneficial on either background. This result suggested that the sign epistasis, in which a mutation was beneficial on one background but deleterious on another, was not the cause of different evolutionary trajectories observed in the Betancourt (2009) experiment. However, they could be explained by either magnitude epistasis, in which mutations have stronger or weaker beneficial effects depending on the background, or by the simultaneous fixation of multiple beneficial mutations. Surprisingly, large populations of the previous experiment showed less parallel evolution than the small populations of this experiment, which lends support to the fixation of multiple beneficial mutations contributing to the patterns seen in both experiments (Betancourt 2010). The evolutionary experiments on the phage MS2 were reviewed, among other phage models, in a great report on the experimental evolution and loss-of-function mutations (Behe 2010).

Kashiwagi and Yomo (2011) examined the ongoing changes driven by host-parasite interactions in a coevolution model consisting of the phage Qβ and *E. coli*. In the arms race coevolution through 54 daily copropagations (equivalent to 163–165 replication generations of Qβ) of the parasite and its host, *E. coli* first evolved partial resistance to infection and later accelerated its specific growth rate, while the phage counter-adapted by improving release efficiency with a change in host specificity and a decrease in virulence. The whole-genome analysis of *E. coli* and Qβ revealed that, despite the large difference in mutability of their genomes, approximately one to three orders of magnitude difference, a host with larger genome size, 4.6 Mbp, and a lower spontaneous mutation rate, 5.4×10^{-10} per base pair per replication, and a parasite with a smaller genome size, 4217 bases, and a higher mutation rate, 1.5×10^{-3} to 1.5×10^{-5} per base per replication, were capable of changing their phenotypes to coexist in the arms race. The phage accumulated 7.5 mutations, mainly in the A2 gene, 3.4-fold faster than in the Qβ propagated alone. The *E. coli* showed fixation of the two mutations (in *traQ* and *csdA*) faster than in the sole *E. coli* experimental evolution (Kashiwagi and Yomo 2011). Then, Kashiwagi et al. (2014) turned to the thermal adaptation of the Qβ mutations that did not alter the amino acid sequence. By the thermal adaptation experiment, the culture temperature was increased from 37.2° to 41.2°C and finally to an inhibitory temperature of 43.6°C in a stepwise manner in three independent lines. In fact, the occurring mutations contributed to increases in fitness of the phage. The whole-genome analysis revealed 31 mutations, including 14 mutations that did not result in amino acid sequence alterations. Eight of the 31 mutations were observed in all three lines. The adaptation effect was achieved by synonymous mutations or mutation in the untranslated region. Moreover, these mutations provided a genetic background for subsequent mutations to alter the fitness contribution from deleterious to beneficial (Kashiwagi et al. 2014). Moreover, Kashiwagi et al. (2018) introduced each mutation observed in these three end-point populations into the cDNA of the Qβ genome, and prepared three different mutants. The quantitative analysis showed that they tended to increase their fitness by increasing the adsorption rate to their host, shortening their latent period (i.e., the duration between phage infection and progeny release), and increasing the burst size (i.e., the number of progeny phages per infected cell), but all three mutants decreased their thermal stability (Kashiwagi et al. 2018).

In order to study the three faces of riboviral spontaneous mutation, namely, spectrum, mode of genome replication, and mutation rate, García-Villada and Drake (2012) used the phage Qβ replication in *E. coli* as a model ribovirus system. The Qβ readthrough gene was used as a mutation reporter. To reduce uncertainties in mutation frequencies due to selection, the experimental Qβ populations were established after a single cycle of infection and selection against readthrough-negative mutants during phage growth was ameliorated by the plasmid-based readthrough complementation *in trans*. A total of 32 readthrough mutants were detected among 7517 Qβ isolates. The sequencing analysis of the 45 readthrough mutations revealed a spectrum dominated by 39 transitions, plus 4 transversions and 2 indels. The average mutation rate per base replication was $\approx 9.1 \times 10^{-6}$ for base substitutions and $\approx 2.3 \times 10^{-7}$ for indels. The estimated mutation rate per genome replication, μ_g, was ≈ 0.04 (or, per phage generation, ≈ 0.08). These results, combined with other information about riboviral mutagenesis, depicted a ribovirus mutation spectrum largely dominated by transitions, a predominantly linear mode of genome replication, and a mutation rate per genome replication on the order of 0.04 for phages and plant viruses but perhaps an order of magnitude higher for mammalian riboviruses (García-Villada and Drake 2012). Furthermore, García-Villada and Drake (2013) used the Qβ model to test for an anticipated tradeoff between fecundity and lifespan in the coliphages through a selection experiment. First, the three independent wild-type Qβ populations were adapted to display an increased multiplication rate. Next, their capability to survive outside the host cell was determined and compared with that of the ancestral wild-type population. Further, it was determined which Qβ life-history traits were involved in the evolution of the multiplication rate, together with the genetic changes associated with the observed phenotypic changes. The results revealed that the short-term evolution of increased fecundity in the Qβ was associated with the decreased viability of phage virions. This tradeoff apparently arose because the fecundity increased at the expense of reducing the amount of resources, mainly time, invested per produced virion. Thus, the results also indicated that the Qβ fecundity might be enhanced through increases in the rates of adsorption to the host and progeny production. Finally, the genomic sequencing of the evolved populations pinpointed sequences likely to be involved in the evolution of the Qβ fecundity (García-Villada and Drake 2013).

Arribas et al. (2014) have chosen the fluctuating temperatures as a factor of the selective pressure. The authors propagated a heterogeneous population of the phage Qβ at its optimal growth temperature, 37°C, and at two new temperatures, one above the optimal value (43°C) and another one below (30°C). To compare the virus populations in constant versus deterministically fluctuating environments, the evolution experiments were carried out in which the temperature was kept invariable throughout the process, together with other ones in which the temperature alternated periodically among the three values assayed, using two patterns of change. The analysis of the evolved populations made it possible to determine the evolutionary pathway followed by the virus under the different propagation conditions. The genomic changes to be responsible for the adaptation were followed. It was concluded that (i) under periodically changing temperature conditions, the Qβ evolution was driven by the most stringent selective pressure, (ii) there was a high degree of evolutionary convergence between replicated populations and also among populations evolved at different temperatures, (iii) there were mutations specific of a particular condition, and (iv) adaptation to high temperatures in populations differing in their preexistent genetic diversity took place through the selection of a common set of mutations (Arribas et al. 2014). Furthermore, Arribas et al. (2016) combined selective temperature pressure with the 5-azacytidine treatment that increased the error rate as described above. The main finding was that the propagation at 42°C of the previously mutagenized Qβ populations led to the fixation of a set of mutations that were not present in the absence of the mutagenic treatment, showing that the sustained replication at increased error rate below the extinction threshold could strongly influence the future Qβ evolution. By further virus fitness studies, Arribas et al. (2018) addressed the competition dynamics between the wild-type virus and the two Qβ mutants that were isolated upon evolution at increased temperatures (Arribas et al. 2014). Since these mutants, which possessed an amino acid change in the replicase protein, have never been detected during evolution of the virus at standard temperature, it was expected that they would be dominant under restrictive conditions at 43°C and selected against under permissive conditions at 37°C. However, it appeared that the outcome of the competitions under permissive conditions depended on the previous evolutionary history of both competitors. The authors implemented a theoretical population dynamics model that described the evolution of a heterogeneous population of individuals, each characterized by a fitness value, subjected to subsequent cycles of replication and mutation. The experimental results were explained in the framework of the theoretical model under the two non-excluding, likely complementary assumptions: (i) the relative advantage of both competitors changed as populations approach mutation-selection equilibrium, as a consequence of differences in their growth rates, and (ii) one of the competitors was more robust to mutations than the other. The main conclusion was that the nearness of an RNA virus population to the mutation-selection equilibrium was a key factor determining the fate of particular mutants arising during replication (Arribas et al. 2018). Lázaro et al. (2018) showed that Qβ can optimize its survivability when exposed to short-term high-temperature extracellular heat shocks, as well as its replicative ability at higher-than-optimal temperature. The mutations responsible for simultaneous adaptation were the same as those selected when adaptation to each condition proceeded separately, showing the absence of important tradeoffs between survival and reproduction in this virus (Lázaro et al. 2018; Somovilla et al. 2019).

Remarkably, the disinfection was regarded as a specific selection pressure on the RNA phage evolution (Rachmadi et al. 2018).

The 30th anniversary of the quasispecies of the classical Domingo et al. (1978) paper was celebrated in a special meeting on the past, present, and future of quasispecies (Domingo and Wain-Hobson 2009) and marked by the special review written by Esteban Domingo (2009). The 40 years since the publication of "Nucleotide sequence heterogeneity of an RNA phage population" in *Cell* (Domingo et al. 1978) was celebrated by an interview in *Nature* with Esteban Domingo (Pariente 2018).

A short but sound history of the viral quasispecies was published by Esteban Domingo and Celia Perales (2016). Furthermore, the novel reviews on the quasispecies and virus (Domingo and Perales 2018) and on the history of viral fitness (Domingo et al. 2019) were published recently.

Vadim I. Agol and Anatoly P. Gmyl (2018) published an impressive review on the relationships among mutational robustness, resilience, and evolvability of viral RNA genomes.

Recently, Hartman et al. (2019) performed an interesting study on the protein evolution. The authors generated and characterized a 6615-member library of all two-amino acid combinations in the highly mutable FG loop of the MS2 particle. In addition to characterizing the effect of all double mutants on assembly, thermostability, and acid stability, many instances of epistasis were observed, where combinations of mutations were either more deleterious or more beneficial than expected. These results were used to generate rules governing the effects of multiple mutations on the self-assembly of the virus-like particles (Hartman et al. 2019). This prospective approach was reviewed operatively in detail by Hartman and Tullman-Ercek (2019).

19 Expression

The secret of getting ahead is getting started.

Mark Twain

My work is a game, a very serious game.

M.C. Escher

ERA OF GENE ENGINEERING

The point when gene engineering will rule the RNA phage studies was prophesied in the 1970s by Charles Weissmann's team (Flavell et al. 1974). The authors acknowledged in the paper on site-directed mutagenesis that the "expression of modified DNA *in vivo* or *in vitro* might ultimately allow the synthesis of proteins with predetermined amino acid substitutions." At that time, the Qβ RNA was systematically prepared for the cloning and expression techniques. As described in the By-products and direct consequences of the MS2 sequencing section of Chapter 10, the Qβ RNA was extended 3′-terminally with a polyA segment and shown to retain full infectivity in spheroplasts (Gilvarg et al. 1974, 1975a). Then, the DNA copy of the full-length Qβ RNA was prepared with reverse transcriptase by means of the poly(dA)•poly(dT) tailing procedure and the first plasmids carrying Qβ genome were obtained (Mekler et al. 1977). Finally, Weissmann's team showed that the complete Qβ DNA copy inserted in either orientation in the low-copy pCR1 plasmid, which was popular at that time, gave rise to the phage Qβ formation in the transfected host cells (Taniguchi et al. 1978a,b). To do this, the Qβ RNA extracted from the plaque-purified phage was used as a template for the synthesis of Qβ minus strands by the Qβ replicase. Both Qβ RNA plus and minus strands were elongated 3′-terminally by polyA polymerase and both used as templates for reverse transcriptase, with oligo(dT) as primer. Then, the plus and minus strands were hybridized, and the incomplete duplex was repaired with DNA polymerase I. The pCR1 plasmid was cleaved with *Eco*RI, its ends were trimmed back with 5′-exonuclease, the double-stranded Qβ DNA and the pCR1 DNA were elongated with dAMP and dTMP residues, respectively, using terminal nucleotidyl transferase. The two components were hybridized and transfected into *E. coli* HB101 (Taniguchi et al. 1978b).

In parallel, Walter Fiers' team performed the cloning of the apparently full-length DNA copy, lacking only 5′-terminal 14 nucleotides, of the MS2 RNA in a plasmid (Devos et al. 1979b). The history and methodology of this cloning were described in detail in the By-products and direct consequences of the MS2 sequencing section of Chapter 10. In a few words, the complementary DNA copy of the *in vitro* polyadenylated MS2 RNA was synthesized with avian myeloblastosis virus RNA-dependent DNA polymerase. After the MS2 RNA template was removed from the complementary

DNA strand with T1 and pancreatic ribonuclease digestion, the complementary DNA became a good template for the synthesis of the double-stranded MS2 DNA with *E. coli* DNA polymerase I. The double-stranded MS2 DNA was inserted into the *Pst*I restriction site of the classical pBR322 plasmid by means of the poly(dA)•poly(dT) tailing procedure. An *E. coli* transformant carrying a plasmid with the nearly full-length MS2 DNA insertion was named pMS2-7 and used further as the traditional source of the MS2 genes for the cloning and expression experiments. Remarkably, the restriction sites within the pMS2-7 plasmid correlated well with the predicted ones from the primary MS2 RNA structure. As mentioned in Chapter 10, the pMS2-7 plasmid contained an extra DNA insertion that was identified as the translocatable element IS1 (Devos et al. 1979a), a constant and imminent threat by the cloning procedures in future.

Meanwhile, Fiers' team introduced the strong leftward promoter of the phage λ, namely, the promoter P$_L$, for the expression of the cloned genes (Remaut et al. 1981, 1983b). The P$_L$ promoter was closed at 28°C by a thermosensitive repressor encoded by the host chromosome. At 42°C, the repressor was unstable, and the promoter was turned on. Remaut et al. (1981) described for the first time the famous plasmid pPLc24. Under the P$_L$ promoter control, this plasmid carried the ribosome binding site and the first 98 amino acid residues of the replicase protein of the phage MS2. This fragment was taken from the above-mentioned pMS2-7 (Devos et al. 1979b). The plasmid pLc24 was used by the historical expression of the human fibroblast interferon gene as a fusion to the 98 N-terminal amino acid residues of the MS2 replicase (Derynck et al. 1981; Fiers et al. 1982). Later, this N-terminal replicase fragment, together with the efficient ribosome initiation site, will be used by numerous expressions that are listed below in the Replicase section of this chapter.

Fiers' team expressed for the first time the three major MS2 genes, namely, maturation, coat, and replicase, by the individual insertion into the thermoinducible expression plasmids under control of the phage λ P$_L$ promoter (Remaut et al. 1982). Figure 19.1 presents the scheme how the individual MS2 genes were inserted into the corresponding P$_L$ expression vectors.

The phage-coded proteins were synthesized at high efficiency, up to 35% of the *de novo* protein synthesis, as early as 30 min after induction. It is highly remarkable that the induced cultures specifically complemented superinfecting MS2 *amber* mutants. Thus, the regulatory mechanisms operating during the natural infection cycle of the phage MS2 were reproduced by the plasmid expression system (Remaut et al. 1982). The same P$_L$ vectors were used to investigate the intriguing mode of regulated synthesis of the fourth phage MS2 gene coding for the lysis protein that was reported just at

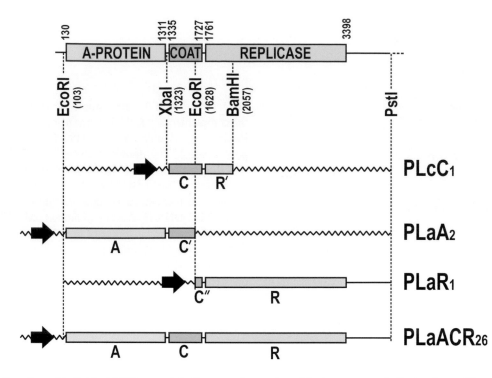

FIGURE 19.1 Insertion of individual MS2 genes into P_L expression vectors. The heavy arrow indicates the position of the P_L promoter. The boxed areas represent phage MS2 genes. The numbers refer to the positions on the linear cDNA map of cloned MS2 RNA (Devos et al. 1979b). A, A protein; C, coat protein; R, replicase; R', part of the replicase; C' and C'', parts of the coat protein. The wavy line indicates the acceptor plasmid. (Redrawn in color with permission of EMBO Press from Remaut E, Waele PD, Marmenout A, Stanssens P, Fiers W. *EMBO J.* 1982;1:205–209.)

that time (Kastelein et al. 1982). An analogous procedure with the cloning of the Qβ genes under the same P_L promoter was performed 1 year later by Karnik and Billeter (1983).

The experience with the expression of the individual MS2 genes made it possible to achieve the inducible high-level synthesis of the mature human fibroblast interferon in *E. coli* by Fiers' team (Remaut et al. 1983a). With this purpose in mind, a special vector was designed to allow easy coupling of a DNA coding region to the initiator AUG of the MS2 replicase gene cloned downstream of the P_L promoter. The activity of the promoter was regulated by temperature, while the translation of the inserted gene was controlled by the well-known ribosome binding site of the MS2 replicase gene. As a result, the induced cells accumulated the interferon up to 4% of the total cellular protein (Remaut et al. 1983a). Furthermore, the ribosome binding site of the MS2 replicase was used by the high-level expression of human interferon gamma (IFN-γ) in *E. coli* under the P_L promoter, although the *E. coli* tryptophan attenuator region functioned a bit better (Simons et al. 1984). Nevertheless, under both conditions, the IFN-γ levels of up to 25% of the total cellular protein were achieved. The highest levels were obtained when a terminator of transcription was cloned downstream from the IFN-γ sequence. It is remarkable that the simultaneous inducible expression of the MS2 lysis gene was tried in order to improve the solubility of IFN-γ that was almost entirely found in the initial pellet fraction and not in soluble extracts. Nevertheless, the presence of the lysis gene did not enhance the biological activity of the IFN-γ present in the supernatant fraction. The fine methodology of the

P_L-driven expressions using the ribosome binding site of the MS2 replicase was unveiled by Remaut et al. (1986). One of the expression plasmids constructed by Remaut et al. (1981) was used further by Jan van Duin's team by the expression of the rat interferon α1 gene, where the first four nucleotides of the MS2 coat were connected to the second nucleotide of the mature IFN-α_1 sequence (Spanjaard et al. 1989b).

Moreover, the rat interferon α_1 gene expression contributed to the generation of novel vectors based on the gene coupling and using the ribosome frameshift, as mentioned in the Ribosomal frameshifting and re-initiation sections of Chapter 16, Thus, the sequence 5'-AAG-GAG-GU-3', which was complementary to the 3' terminus of *E. coli* 16S rRNA, was inserted in a reading frame into the fused MS2 coat-IFN-α_1 gene just before the interferon part (Spanjaard and van Duin 1988). The translation over the inserted sequence yielded a 50% ribosomal frameshift if the reading phase was A-AGG-AGG-U and led to the synthesis of the desired rat IFN-α_1. Furthermore, the fine mechanism of the translational re-initiation was studied by this two-gene system, the upstream MS2 coat gene and the downstream rat IFN-α_1 gene, where expression of the downstream gene was dependent on the translation of an upstream reading frame (Spanjaard and van Duin 1989; Spanjaard et al. 1989a). As noted in Chapter 16, this dependence was almost absolute, since the upstream translation increased expression of the downstream gene by two to three orders of magnitude. Furthermore, the MS2 coat and replicase fusions were used by the first productive bacterial production of nonglycosylated human transferrin (de Smit et al. 1995).

The well-studied initiation signals of the RNA phages with their well-established Shine-Dalgarno sequences (for detail, see Chapter 16) were tested for their efficiency with heterologous genes. Thus, Looman et al. (1985) employed the ribosome binding sites of the MS2 and Qβ maturation proteins by the studies on the transcription and translation of the *lacZ* gene of *E. coli*. The highest β-galactosidase production was found in the clone with the Qβ A2 initiation site and the lowest amount in those containing the initiation of the MS2 maturation protein, although the latter possessed the longest Shine-Dalgarno sequence but contained GUG instead of AUG as the initiation codon. Changing the GUG into AUG resulted in a threefold increase in the expression level (Looman and van Knippenberg 1986). No strict correlation was found between the level of the *lac* mRNA and β-galactosidase production, and the translation, but not the transcription, appeared to be most efficient when the homologous *lacZ* initiation signal was present when the outcome was calculated per molecule of the *lac* mRNA (Looman et al. 1985). When more different ribosome binding sites were used for the expression of the model *lacZ* gene in parallel with the above-mentioned ones, namely, those of the Qβ replicase, MS2 replicase, Qβ coat, and MS2 coat, Looman et al. (1986) concluded that the secondary structure was the primary determinant of their efficiency in *E. coli*. The Shine-Dalgarno sequence and the length and sequence of the spacer region up to the initiation codon alone were not able to explain the observed translation efficiency. The deletion mutations created in the Qβ replicase ribosome binding site revealed a complex pattern of control of expression, probably involving the use of a "false" initiation site (Looman et al. 1986).

The large collection of the mutant ribosome binding sites from the MS2 coat protein gene, which was generated by van Duin's team (de Smit and van Duin 1990b, 1994a, 2003) and described in the Secondary and tertiary structure section of Chapter 16, was employed by the interesting investigation of the role of cryptic Rho-dependent transcription terminators in the *E. coli* transcriptional polarity (de Smit and van Duin 2008). The transcriptional polarity resulted in decreased synthesis of inefficiently translated mRNAs and therefore in decreased expression not only of downstream genes in the same operon (*intercistronic polarity*) but also of the gene in which termination occurred (*intracistronic polarity*). To study the relevance of the intracistronic polarity for the expression regulation *in vivo*, the polarity-prone *lacZ* reporter gene was fused to a range of the mutated ribosome binding sites, with a wide range of translational efficiencies, and repressed to different degrees by local RNA structure. The quantitative analysis of the protein and mRNA syntheses showed that the polarity occurred on functionally active mRNA molecules and affected expression of the gene carrying the terminator, thus enhancing the effect of translational repression (de Smit and van Duin 2008; de Smit et al. 2009).

GENERATION OF MUTANTS

The gene engineering methodology brought novel unique perspectives for the elucidation of the infection process and generation of recombinant phage mutants, since for the first time after the RNA phage discovery, their nucleotide sequences, as well as their proteins, could be easily manipulated and examined both *in vitro* and *in vivo*. Many manipulations that were difficult or absolutely impossible at the RNA level appeared simple and easy at the DNA level.

This very promising approach was started with the dazzling separation of the replicase gene and its RNA substrate on different plasmids (Mills 1988). It was the first successful replication of an active messenger RNA by the freely and independently expressed Qβ replicase, as narrated in the Mapping of Qβ replicase section of Chapter 18. Furthermore, the famous pQβml00 plasmid was generated, where full-length wild-type phage Qβ minus strand RNA was constitutively expressed *in vivo* from a modified tryptophan promoter (Mills et al. 1990). When these Qβ minus strands were replicated to generate Qβ plus RNA, an infectious cycle was initiated during which the infectious Qβ particles were produced. This system allowed an extensive mutational analysis of the Qβ replicase gene and engineering a series of the site-directed mutations and/or deletions into the replicase gene cDNA within the pQβml00 plasmid, when the active Qβ replicase was supplied *in trans* (Mills et al. 1989, 1990). These experiments have yielded unique information as to how these *cis* and *trans*-acting regions of the Qβ replicase gene were integrated and packaged into the structural organization of the phage RNA. At last, Donald R. Mills' team succeeded to dissect nearly all of the Qβ viral genome (Priano et al. 1995). Thus, the two new Qβ protein expression plasmids have been developed: one that generated Qβ readthrough protein and the other that generated Qβ maturation protein, both under the control of bacteria-based promoters. Moreover, Priano et al. (1995) have developed a second series of analogous expression plasmids that generated the maturation protein, readthrough protein, and replicase protein of the group IV phage SP. As a result, the authors achieved the complementation of Qβ genomes that were defective in the synthesis of functional proteins by supplying one or more of the Qβ or SP helper plasmids *in trans*. In this complementation system, the Qβ minus strand RNAs were constitutively transcribed from plasmid cDNA by *E. coli* RNA polymerase. The replication of these minus strands resulted in the synthesis of the Qβ plus RNA, thereby triggering the Qβ infectious cycle. The genetically engineered Qβ genome mutations that resulted in defective viral proteins were complemented *in trans* by the products of one or more Qβ helper plasmids that expressed either: (i) Qβ maturation protein, which complemented defects in the Qβ maturation gene; (ii) Qβ readthrough protein, which complemented defects in the readthrough gene; or (iii) Qβ replicase, which complemented defects in the replicase gene. Each plasmid component of this system contained a unique origin of replication and carried a different antibiotic gene, thereby enabling all combinations of these plasmids to coexist in the same host. Each of the corresponding SP helper proteins complemented respective defects within the Qβ genome with efficiencies similar to those observed for the Qβ helper proteins (Priano et al. 1995). Using this *in vivo* complementation system, a thorough mutational analysis of the Qβ readthrough gene was conducted (Arora et al. 1996). Thus, in the Qβ cDNA-containing pQβm100 plasmid, the six defined Qβ deletion cDNA genomes, each missing between 86 and 447 nucleotides from within the readthrough gene, were generated. These deletion plasmids were introduced into host cells that were constitutively supplied with the Qβ readthrough protein *in trans*. Under these conditions, all six deletion genomes spontaneously generated phage particles, each exhibiting a characteristic plaque phenotype and virus forming potential. The isolated readthrough-defective phage particles were subsequently used to infect host cells that carried helper readthrough protein. The passaged viruses yielded both larger plaques and higher titers compared with those of the parent phages. The sequence analysis revealed that the genomes of the passaged phages had deleted additional regions of the readthrough RNA sequence. These biologically significant results were interpreted in light of the possibilities that (i) the disruption of a well-defined structural domain in the Qβ RNA was selectively disadvantageous to phage infection, and that (ii) the evolved viral populations were selected by virtue of their ability to restore critical integrity of short- and/or long-range nucleotide interactions within this region of Qβ RNA. By the felicitous authors' expression, "fold as you please, but fold you must!!" (Arora et al. 1996). Moreover, the studies on the Qβ deletion variants located within the readthrough domain led to discovery that the co-translational regulation of the viral coat and replicase genes could be uncoupled in viral genomes carrying a deletion and then it could be restored by an additional deletion occurring in the naturally evolved pseudorevertant, as described in more detail in the Repressor complex II section of Chapter 16 (Jacobson et al. 1998).

In parallel with Mills' team, the plasmids containing the full-length cDNA copies of the Qβ RNA genome under flanking T7 promoters were constructed by Paul Kaesberg's team (Shaklee et al. 1988). The plus strand Qβ RNA, generated by the *in vitro* transcription of these plasmids with T7 RNA polymerase, was infectious to *E. coli* spheroplasts, while the minus strand RNA was infectious to *E. coli* carrying a plasmid that allowed expression of the Qβ replicase. Furthermore, Shaklee (1990) constructed plasmids which allowed preparation of the positive- and negative-strand MS2 RNA by *in vitro* transcription and which expressed the MS2 replicase. If the transcribed plus strand MS2 RNA was infectious to *E. coli* spheroplasts, the transcribed minus strand MS2 RNA was infectious to spheroplasts containing either MS2 replicase or the Qβ replicase. The results clearly showed that the Qβ replicase was quite effective in copying or replicating the MS2 minus strand RNA. Needless to say, these results were absolutely surprising, since they contradicted all existing traditions on the interrelation of the leviviruses and alloleviviruses starting from the first direct *in vitro* indication of Feix et al. (1968) that the Qβ replicase did not recognize the MS2 RNA minus strands. Furthermore, Shaklee (1991) found that the MS2 and Qβ minus strands, prepared by *in vitro* transcription, were each separated into two forms on non-denaturing

agarose gels. Only a single form of RNA was observed for the minus strands denatured with glyoxal prior to agarose analysis, demonstrating that the two forms were conformers in each case. The truncated and elongated versions of these RNAs, analyzed under non-denaturing conditions, demonstrated that a region of the RNA complementary to the maturation protein gene was at least partially responsible for the presence of these conformers in both MS2 and Qβ minus strands (Shaklee 1991). Moreover, using the gene engineering methodology, Palasingam and Shaklee (1992) disclosed the recombination as a mechanism responsible for the reversion of the Qβ mutants *in vivo*, as demonstrated in the Recombination section of Chapter 18.

The huge amount of work on the genetically engineered RNA phage mutants was performed by van Duin's team. The striking experiments of this team in this area were described in detail in the Recombination section of Chapter 18. The mutants concerning the translation process were described in the Secondary and Tertiary Structure section of Chapter 16. The recombinant phage display structures generated by van Duin's team are described below.

CHIMERIC PHAGES

After the numerous revolutionary topics on the gene-engineered reconstruction of the phage genome, van Duin's team succeeded by the generation of the peptide display on the live MS2 phage (van Meerten et al. 2001b). The live recombinant MS2 phages represented one of the first examples of the specific branch in the huge field of the virus-like particles (VLPs). These were divided into the two large branches: (i) noninfectious, or classical, VLPs and (ii) infectious VLPs, and those were divided further into the replication-competent and replication-noncompetent chimeric viruses (Pumpens and Grens 2002). The live recombinant MS2 phages generated by van Meerten et al. (2001b) belonged to the rare examples of the replication-competent infectious VLPs. Van Duin's group was therefore the first to try to accommodate and display the extra amino acids on the surface of the live MS2 particle, in spite of the strong restrictions at the RNA genome level. Clearly, if one could display peptides on the surface of live phages, this would potentially expand the technique with selection from libraries or with natural *in vivo* selection. To test the feasibility of this approach, van Meerten et al. (2001b) took a noninfectious VLP mutant generated earlier by Elmārs Grēns' team (Pushko et al. 1993), in which 5 amino acids, namely, ASISI, were inserted at position 1 of the coat protein of the phage fr, a close relative of MS2. This mutant was chosen because its capacity to form capsids, in this case so-called empty VLPs that did not contain genome inside, had remained intact (for more detail, see complete descriptions below in the sections devoted to the generation of the noninfectious RNA phage VLPs). In the MS2 crystal structure, this part of the coat protein formed a loop that extended from the outer surface of the icosahedral virion, while at the RNA level, the insert formed a large loop at the top of an existing hairpin. According to the van Meerten et al. (2001b) considerations, the *in vivo* approach faced several additional difficulties when compared with these *ex vivo* studies on the VLP formation. Most important, the peptide insertion must be compatible with the formation of an infectious phage. This required the correct incorporation of the maturation protein in the now-modified virion. Furthermore, the new coat protein must still be able to act as a translational repressor by binding the replicase operator. Another problem facing peptide display on live phages was that the extra RNA encoding the peptide could come under selection pressure for various reasons. If it could not adopt a proper secondary structure, it might fall prey to *E. coli* endonucleases, as was shown clearly at that time (Arora et al. 1996; Olsthoorn and van Duin 1996b; Klovins et al. 1997b; van Meerten et al. 1999). On the other hand, the 15 extra nucleotides encoding the 5 amino acids were likely to change the local RNA secondary structure at the coat start, and this might affect translational yield significantly (de Smit and van Duin 1990b). The degeneracy of the genetic code allowed the authors to make various mutants, all encoding the same ASISI insert in the coat protein. Several of these mutants were unstable and suffered deletions that reduced the insert to two or three amino acids, while others underwent adaptation by base substitutions. Some mutants were fully stable. Whether or not an insert was stable appeared to be determined by the choice of the nucleic acid sequence used to encode the extra peptide. This effect was not caused by differential translation, because the coat-protein synthesis was equal in wild-type and mutants. It was concluded that the stability of the insert depended on the structure of the large RNA hairpin loop, as demonstrated by the fact that a single substitution could convert an unstable loop into a stable one (van Meerten et al. 2001b).

Continuing the story of the evolutionary optimization of the RNA phages presented in Chapter 18, van Duin's team destroyed the most intriguing regulatory structure, namely, the target of the repressor complex I (Licis and van Duin 2006). Thus, the MS2 RNA was disabled by a deletion of 4 nucleotides in the 36-nucleotide-long intergenic region separating the coat and replicase genes. In fact, this piece of the genome was the focus of several RNA structures conferring high fitness: first, the operator hairpin, which, in the course of infection, bound the coat-protein dimer, thereby precluding further replicase synthesis and initiating encapsidation; second, a structure ensuring the polarity phenomenon and controlling replicase expression. Although the 4-nucleotide deletion did not directly affect the operator hairpin, it disrupted the polarity mechanism. The third effect was a frame shift in the overlapping lysis gene. Therefore, the mutation also disrupted the regulation of the replicase synthesis and provoked the uncontrolled replicase synthesis, which led to a reduced output of functional virus, depriving the phage of one of its high-fitness features. Licis and van Duin (2006) observed the large variety of ways in which the phage MS2 has repaired the damage inflicted by the deletion. In fact, four different solutions were obtained by sequencing 40 plaques. The three had cured the frame shift in the lysis gene by inserting one nucleotide in the loop of the operator hairpin causing its inactivation. Yet these

low-fitness revertants could further improve themselves when evolved. The inactivated operator was replaced by a substitute and thereafter these revertants found several ways to restore control over the replicase gene. To allow for the evolutionary enrichment of low-probability but high-fitness revertants, the lysate samples were passaged before plating. The revertants obtained in this way also restored the frame shift, but not at the expense of the operator. By taking larger and larger lysate samples for such bulk evolution, ever higher-fitness and lower-frequency revertants surfaced. Only one made it back to the wild type. As a rule, however, the revertants moved further and further away from the wild-type sequence because restorative mutations were, in the majority of cases, selected for their capacity to improve the phenotype by optimizing one of several potential alternative RNA foldings that emerged as a result of the initial deletion. This illustrated the role of the structural constraints which limited the path of subsequent restorative mutations. Therefore, the first step was always the restoration of the lysis reading frame, often through sacrificing the operator. Thereafter, the evolutionary efforts were directed to repair the operator hairpin. Finally, the control of the replicase gene expression was pursued. The authors concluded that it could be the general order in which these three critical features contributed to the phage fitness (Licis and van Duin 2006).

MATURATION GENE

After the first expression of the MS2 maturation gene in *E. coli* and demonstration of its ability to complement the corresponding mutant gene *in vivo* (Remaut et al. 1982), the interest was turned to the Qβ maturation, or A2, protein. When Karnik and Billeter (1983) cloned the full-length and partial Qβ cDNA sequences in the expression plasmid provided with the strong P_L promoter of Remaut et al. (1981), the A2 gene was found to cause cell lysis. The alterations of the A2 gene, leading to proteins either devoid of ~20% of the C-terminal region or of six internal amino acids, abolished the lysis function. The expression of other genes, in addition to the A2 gene, did not enhance the host lysis. Thus, the A2 protein appeared to trigger the cell lysis besides its maturation function. This was in flagrant contradiction with the situation in the group I phages, such as f2 and MS2, where a small lysis polypeptide was coded for by a region overlapping the end of the coat gene and the beginning of the replicase gene. This intriguing obstacle initiated the cloning interest to the Qβ maturation gene, as well as the further active elucidation of its role in the cell lysis, the full account of which is given in Chapter 11. In brief, Winter and Gold (1983) confirmed the cell lysis by the A2 protein after cloning of the A2 gene under control of the *lac* promoter/operator. Remarkably, the induction of the A2 clone was lethal to the host, and the lethality correlated with the A2 synthesis and was provoked by the cell lysis. As in the MS2 case, the plasmid-derived A2 protein specifically complemented the Qβ A2 *amber* mutants. The cell lysis activity was abolished in clones that synthesized truncated or internally deleted A2 polypeptides ranging from

10%–95% of the length of wild-type protein. Furthermore, the bacterial peptidoglycan was identified as a target of the A2 protein (Bernhardt et al. 2001, 2002) and the fine mechanism of its action was clarified (Reed 2012; Reed et al. 2012, 2013). Finally, thanks to the high-level production and thorough purification ensured by the gene engineering techniques, the structure of the Qβ complex with UDP-N-acetylglucosamine enolpyruvyltransferase (MurA) was resolved by single-particle cryoelectron microscopy at 6.1 Å resolution, where the outer surface of the β-region in A2 was identified as the MurA-binding interface (Cui et al. 2017).

The A2 gene of the group IV phage SP was expressed by Nishihara (2002). As in the case of the Qβ phage, the A2 gene product induced the cell lysis. The transmission electron micrographs of *E. coli* cells induced by the cloned SP A2 gene and by the lysis gene of the group II phage GA revealed various morphological aspects of the intermediates of lysing cells. The cells induced by the SP A2 gene became stretched and also tapered in shape; fragmentation of parts of the cells had also occurred. The cells induced by the GA lysis gene showed many ballooning structures on the cell surfaces, and others leaked material through the cell wall. However, some balloon-like structures also appeared on the surfaces of cells induced by the SP A2 gene, where material also appeared to be leaking through the cell wall in the photographs (Nishihara 2002).

Kazaks et al. (2011) overexpressed the A protein gene of the caulophage φCB5 in *E. coli*, but this protein did not cause cell lysis.

LYSIS GENE

In contrast to the three basic RNA phage genes, the clear evidence of the lysis gene function was obtained in full measure thanks to the gene engineering methodology. This follows from the exciting story of the lysis gene, which was outlined in Chapters 8 and 14. In a few words, the MS2 lysis gene was expressed, in its natural tandem with the coat protein, under the transcriptional P_L promoter control by Fiers' team (Kastelein et al. 1982). To do this, the P_L-driven expression plasmids depicted on Figure 19.1 were used (Remaut et al. 1981, 1982). The expression of the lysis gene under the P_L promoter was used in the further studies on the fine mechanisms of lysis gene functioning, which were performed now under the Jan van Duin's guidance (Kastelein et al. 1983b). Using the same P_L promoter-driven expression, Berkhout et al. (1985a) mapped the MS2 lysis protein and showed that the first 40 amino acids of the lysis protein were dispensable for its function. Then, Berkhout et al. (1987) studied how translation of the coat gene activated the start site of the lysis gene in the MS2 genome. The MS2 lysis gene was cloned under the SP6 promoter control when the RNA secondary structure around the lysis gene initiation region was examined (Schmidt et al. 1987).

The plasmids expressing the MS2 lysis gene under the P_L promoter were used by the elucidation of the cell lysis mechanism (Höltje et al. 1988; Walderich et al. 1988; Ursinus-Wössner and Höltje 1989; Markiewicz and Höltje 1992), as described in detail in the Lysis mechanism section of Chapter 14.

By analogy with the phage MS2, Adhin and van Duin (1989) cloned the coat-lysis gene tandem of the phage fr under the P_L promoter control and investigated the translational coupling between the coat and lysis genes. It was especially important in this case, since the UUG initiation codon of the fr lysis gene was nonfunctional in the *de novo* initiation but was activated by translational termination at the overlapping coat gene. The replacement of the UUG by either GUG or AUG resulted in the loss of genetic control of the lysis gene (Adhin and van Duin 1989). The P_L promoter-driven expression of the fr lysis gene was used by the generation of the scanning model for the translational re-initiation in *E. coli* (Adhin and van Duin 1990), which was described in the Natural expression of the lysis gene section of Chapter 14.

In the case of the group II phage JP34, the lysis gene was expressed within the region of the JP34 cDNA that contained the complete nucleotide sequences for the maturation, coat, and lysis genes under the P_L promoter control (Adhin et al. 1989). The thermoinduction of the P_L promoter in this plasmid led to the cell lysis. However, when the plasmid was deprived of the coat translation initiation region, it failed to lyse the host, indicating that the lysis expression depended on the coat gene translation, as in the group I phages. The P_L-driven expression was employed also to demonstrate the properties of the lysis gene of the group II phage KU1, which had an insertion of 18 nucleotides at the start codon, but nevertheless was coupled to the termination of the coat gene translation (Groeneveld et al. 1996).

The lysis gene of the group II phage GA, in tandem with the coat gene, was cloned and expressed under the P_L promoter control, first of all, in order to study the role of ribosome recycling factor RRF in the translational coupling (Inokuchi et al. 2000). Then, the P_L promoter-driven expression was used by the excellent electron microscopy investigation of the *E. coli* lysis process, which demonstrated clear differences when the GA lysis gene or the SP A2 gene (Nishihara 2002), or the MS2 lysis gene (Nishihara 2003) were expressed.

As presented in the Natural expression of the lysis gene section of Chapter 14, the lysis gene of the pseudomonaphage PP7 was described in detail after the sequencing of the PP7 genome, when it was cloned in expression plasmids (Olsthoorn et al. 1995a).

The function of the lysis gene of the acinetophage AP205, which was found at the unusual place at the 5′ end of the phage genome, was confirmed by the expression under the same P_L promoter control in the Remaut et al. (1981) plasmids (Klovins et al. 2002).

Then, the cloning and expression under the P_L promoter revealed the cryptic Rg-lysis gene, which was found near the start of the Qβ replicase gene in the +1 frame and caused cell lysis (Nishihara et al. 2004). To purify the Rg-lysis protein, the authors have constructed and used by the expression the hexahistidine-tagged version of the Rg-lysis gene. Furthermore, Nishihara et al. (2006) have expressed the lysis genes, in tandem with the coat genes, of the group I phages ZR and BO1 and of the group II phages SD, TH1, JP500, and TL2.

To confirm the lysis gene function of the pseudomonaphage PRR1, the putative gene was inserted into an expression vector, with the start codon changed to AUG, and the cell lysis was demonstrated upon promoter induction (Ruokoranta et al. 2006). In a similar manner, the lysis function was confirmed by the cloning and expression of the putative lysis genes in the caulophage φCB5 (Kazaks et al. 2011), the phages C-1 and HgaII (Kannoly et al. 2012), and the IncM pili-dependent phage M (Rumnieks and Tars 2012). Moreover, as described in the Natural expression of the lysis gene section of Chapter 14, both potential transmembrane helices of the φCB5 lysis protein were cloned and expressed separately (Kazaks et al. 2011).

Harkness and Lubitz (1987) constructed and expressed under the P_L promoter control the first chimeric lysis gene E-L encompassing 32 N-terminal amino acid sequence of the E protein, a lysis protein of the DNA phage φX174, and the 51 C-terminal amino acid sequence of the MS2 lysis protein. As stated in the Lysis mechanism section of Chapter 14, the chimeric gene product demonstrated very high lysis-inducing activity. Later, an entire family of the chimeric E-L genes was expressed under the P_L promoter (Witte et al. 1998). These expression systems involving the MS2 lysis gene were used further for the bacterial ghost technology (Szostak et al. 1990; Jechlinger et al. 2005; Mayrhofer et al. 2005).

The expressed versions of the phage MS2 and M lysis genes were used by the fine mapping and by the thorough elucidation of their lysogenic action mechanism in the capacity of a protein antibiotic (Chamakura et al. 2017a,b,c). More particularly, the hexahistidine-tagged variant of the phage M lysis gene was constructed (Chamakura et al. 2017b).

At the same time, the MS2 lysis gene was involved in the complicated genetic circuits aimed to increase the efficiency of the DNA transfer from E. coli to B. subtilis (Juhas et al. 2016; Juhas and Ajioka 2017). The details of the exquisite technique, such as bacterial ghosts, protein antibiotics, and DNA transfer were discussed in more detail in the Other Applications section of Chapter 14.

At last, as mentioned in the Other applications section of Chapter 14, the MS2 lysis gene was fused C-terminally with the GFP gene, in order to track protein localization within the cell and to provide a "handle" for future pull-down experiments (Rasefske and Piefer 2017).

REPLICASE GENE

Biebricher et al. (1986) were the first to prepare Qβ replicase by the standard purification method (Sumper and Luce 1975) but from uninfected E. coli containing a plasmid carrying the structural gene of the viral replicase, during their debates with Hill and Blumenthal (1983) about the template-free synthesis. Such replicase preparations were used further by the classical studies of the autocatalytic Qβ amplification (Biebricher 1987; Biebricher and Luce 1992, 1993, 1996; Rohde et al. 1995; Schober et al. 1995; Zamora et al. 1995; Preuß et al. 1997; Wettich and Biebricher 2001). For the full story, see the De novo synthesis? section of Chapter 18.

The detailed purification protocol of the Qβ enzyme with the plasmid-expressed β subunit was published by Moody et al. (1994). This protocol allowed the production of gram quantities of the highly purified Qβ replicase, without any troubles with the phage infection. The Qβ enzyme prepared according to Moody et al. (1994) was used by many functional Qβ replicase investigations (Brown and Gold 1995a,b, 1996; Chen et al. 1996; Yoshinari et al. 2000; Yoshinari and Dreher 2000; Tretheway et al. 2001), as well as by the Qβ amplification diagnostics (Stefano et al. 1997) and by the creation of diverse libraries of single-domain antibody fragments (Kopsidas et al. 2006, 2007), as discussed in the corresponding sections of Chapters 13 and 18.

To improve the purification efficiency, the His-tagging of the β subunit was applied for the high-quality purification of the greater quantities of the Qβ replicase enzyme (Nakaishi et al. 2000, 2002). The purified His-tagged enzyme assumed almost the same template specificity as the wild type purified by a conventional method when the MDV-poly(+) RNA or Qβ RNA were used as templates. An alternate system, where a wild-type E. coli strain was transfected with two plasmids, one encoding the β-subunit and another encoding EF-Tu with a His tag, was generated by Mathu et al. (2003). This system made it possible to study the role of the EF-Tu mutations within the Qβ enzyme.

Then, Wang (2004) coexpressed the β-subunit with the other three subunits in order to improve the molecular ratio of the subunits within the cell. Moreover, Kita et al. (2006) have fused genetically the β-subunit with EF-Tu and EF-Ts and the fused protein, a single-chain α-less Qβ replicase, was readily purified. Nevertheless, the two-plasmid systems encoding separately the β-subunit and the EF-Tu and EF-Ts were also used for the Qβ purification (Kita et al. 2006). Later, by using this two-plasmid approach, the E. coli elongation factor EF-Ts within the Qβ replicase molecule was replaced by its homolog from the thermophilic bacterium Thermus thermophilus (Vasiliev et al. 2010) and the role of the S1 within the enzyme was studied (Vasilyev et al. 2013). The Qβ replicase purification method of Kita et al. (2006) was used in other studies (Hosoda et al. 2007; Ichihashi et al. 2010), where Ichihashi et al. (2010) employed the two-plasmid approach with separated expression of the genes expressing (i) the TuTs fusion and (ii) the β replicase subunit or the cysteine mutants of the latter, as described in the some special features of the Qβ replicase enzyme section of Chapter 13. The same purification method was employed also by Tomita et al. (2015).

Concerning the crystal structures of the Qβ replicase complex, which were described in detail in the Fine structure of the Qβ replicase section of Chapter 13, Kidmose et al. (2010) operated with the enzyme purified from the E. coli strain carrying the single Ts-Tu-β expression vector, according to Kita et al. (2006). The latter was modified to insert a tobacco etch virus (TEV) protease cleavage site in the fusion protein EFTs–EFTu–TEV–βS–6xHis. The E. coli BL21 (DE3) strain was transformed with the resulting plasmid, grown in LB medium, and induced with L-arabinose. The fusion protein was purified by Ni²⁺ chelate chromatography and subsequently digested by

TEV protease. The cleaved fusion protein was further purified by hydrophobic interaction chromatography and gel filtration. The complex of the subunit β, or II, and EF-TuTs was resolved then at 2.5 Å resolution (Kidmose et al. 2010).

Takeshita and Tomita (2010) employed, by their crystal structure, the single-chain Qβ enzyme expression in the same *E. coli* BL21 (DE3) strain, using the same pBAD33-Ts-Tu-β-3 plasmid of Kita et al. (2006), but without the TEV protease cleavage site. The protein was also purified on the Ni²⁺ column with further additional chromatography steps, and the crystal structure of the core Qβ replicase was determined at 2.8 Å resolution (Takeshita and Tomita 2010). The same strategy of the Qβ enzyme purification was also used in the subsequent brilliant studies of this team (Takeshita and Tomita 2012; Takeshita et al. 2012). Takeshita et al. (2014) added the His-tagged expression variants of the protein S1 to the single-chained β subunit, EF-Tu, and EF-Ts complex. This study revealed that the two N-terminal OB-folds of S1 anchored S1 onto the β subunit, and the third OB-fold was mobile and protruded beyond the surface of the β subunit (Tomita 2014; Tomita and Takeshita 2014), as described in detail in the Fine structure of the Qβ replicase section of Chapter 13.

Gytz et al. (2015) used the same single-chain EFTs–EFTu–TEV–βS–6×His fusion protein, according to Kidmose et al. (2010), namely, with the TEV protease cleavage site.

In order to get the Qβ replicase enzyme with the definite exchanges within the critical sites (Inokuchi and Hirashima 1987; Inokuchi et al. 1994) or with C-terminal deletion (Inokuchi and Hirashima 1990; Inokuchi and Kajitani 1997), the expression plasmid was constructed, where the replicase subunit was under the *lac* promoter control. Then, the mutant plasmids were generated with the desired substitutions or deletions within the replicase subunit gene and tested for the complementation with the mutant phage Qβsus51, defective in the β subunit gene, or for the inhibitory effect on the replication of the wild-type phages, or for the *in vitro* activity. The results of this mapping were presented in the Some special features of the Qβ replicase enzyme section of Chapter 13.

Greater possibilities were opened by the combination within the bacterial cell of the plasmid expressing the replicase gene with an independent plasmid expressing the desired subject of the study. First, as mentioned in the first section of this chapter, this approach allowed separation of the replicase gene and its RNA substrate on different plasmids (Mills 1988). The *in trans* supply of the replicase subunit allowed the extensive mapping of the Qβ replicase gene (Mills et al. 1989, 1990) and the mutual Qβ and SP complementation (Priano et al. 1995). The mapping of the Qβ replicase performed by Mills' team (Mills 1988; Mills et al. 1989, 1990) was described in detail in the Mapping of Qβ replicase section of Chapter 18. The complementation experiments are described above in the Generation of mutants section of this chapter.

The Qβ replicase expression plasmid of Mills (1988) was used by the GENE-TRAK Systems company for the preparation of the pure enzyme and its use by the elaboration of the Qβ amplification diagnostics (Pritchard and Stefano 1990, 1991; Shah et al. 1994; An et al. 1995a,b; Chernesky and Mahony

1999). In the diagnostics context, it is noteworthy to mention that the purification of the Qβ replicase enzyme from *E. coli* cells carrying the appropriate expression plasmid was filed in 1989 and granted with an US patent in 1992 (DiFrancesco and Borcherts 1997). The references on the patent are present in the further papers of the GENE-TRAK company (Burg et al. 1996). The GENE-TRAK-produced Qβ replicase was used also by Fred Russell Kramer's team (Wu et al. 1992). The principles and application of the Qβ amplification-based diagnostics were described in detail in the Diagnostics section of Chapter 18.

In parallel, Shaklee et al. (1988) constructed full-length cDNA copies of the Qβ RNA with flanking T7 promoters. The Qβ replicase gene from the cloned DNA was subcloned in *E. coli* cells within the expression plasmid under the control of the temperature-inducible P_R promoter of the phage λ. The full-length, negative-strand Qβ transcripts were infectious when transfected into spheroplasts containing the induced *in trans* replicase gene. The same system was established with the cloned MS2 replicase (Shaklee 1990). The Qβ replicase-expressing vector of Shaklee et al. (1988) was employed later in further studies (Palasingam and Shaklee 1992; Ugarov et al. 2003; Chetverin et al. 2005; Ugarov and Chetverin 2008), as described in the corresponding sections of Chapter 18.

The gene engineering-based strategies resulted in the cell-free mRNA amplification system in the rabbit reticulocytes when the replicase gene was inserted into a small replicable RNA and mRNAs encoding the phage-coded replicase subunit, as well as EF-Tu and EF-Ts, were co-translated in the rabbit reticulocyte cell-free system (Fukano et al. 2002). This idea was described in the "EF Tu-Ts" section of Chapter 13 and in more detail in the "RNA World" section of Chapter 18. Furthermore, the simplified life model was constructed by Kita et al. (2008). As described in detail in the section "RNA World" in Chapter 18, the self-encoding system consisted of the liposome-encapsulated macromolecules, where genetic information was replicated by the self-encoded Qβ replicase. This unique evolutionary application of the Qβ replicase was developed further in detail (Ichihashi et al. 2008, 2013; Urabe et al. 2010; Bansho et al. 2012; Usui et al. 2013, 2014, 2015; Mizuuchi et al. 2014, 2016; Tomita et al. 2015; Ichihashi and Yomo 2016; Yoshiyama et al. 2016; Yumura et al. 2017).

Finally, Gunasekaran et al. (2013) elaborated a novel improved method for the facile production and rapid purification of the functional recombinant Qβ replicase heterotetramer. As followed from this study, the successful expression of the soluble Qβ enzyme depended on the EF-Ts and EF-Tu subunits being coexpressed prior to the expression of the β subunit. Efficient coexpression required two different inducible operons to coordinate the expression of the heterotrimer. The complete heterotetramer enzyme complex was achieved by production of the recombinant S1 protein in a separate host. This approach represented a facile way of producing and purifying large amounts of the soluble and active recombinant Qβ replicase tetramer without the necessity of a His-tag for purification.

REPLICASE FUSIONS

The above-mentioned plasmid pPLc24 (Remaut et al. 1981) carried the *EcoR*I-*Bam*HI fragment from the pMS2-7 plasmid harboring the quite complete MS2 genome (Devos et al. 1979b). The location of the two restriction sites within the cloned MS2 genome was exposed in Figure 19.1. The pPLc24 plasmid carried therefore, under the thermoinducible P_L promoter, the ribosome binding site and the first 98 amino acid residues of the replicase gene split at the *Bam*HI site. This plasmid served as one of the first and most important tools for the high-level expression of foreign, both eukaryotic and prokaryotic, genes in the 1980s–1990s. The triumphant procession of this approach was started with the historical expression of the human fibroblast interferon gene (Derynck et al. 1981; Fiers et al. 1982). Table 19.1 presents the rather full list of the fusioned genes. The real number of the N-terminal amino acid residues of the MS2 replicase varied in different fusions from 97–100, depending on the construction scheme used.

Not long after, Remaut et al. (1983a) constructed the plasmid vector that allowed easy insertion of any coding region downstream of the P_L promoter and the in-phase MS2 replicase initiator, without the 98 N-terminal MS2 replicase amino acid residues. It is also noteworthy that the pPLc24-derived plasmid was used by the historical expression of the hepatitis B virus core gene in *Bacillus subtilis* (Hardy et al. 1981). A set of MS2 replicase-based expression vectors was generated as derivatives of the vector pPLc24 by Strebel et al. (1986). Most of the listed MS2 replicase fusions were applied not only for purely scientific, but also for diagnostic purposes, especially by the obtaining of specific antibodies or detection of antibodies in clinical specimens. For example, the specific antibodies were generated against pre-S segment of the envelope protein of duck hepatitis B virus (DHBV), which made it possible to precipitate viral particles from serum and to characterize the two pre-S proteins of DHBV in sera and liver of the DHBV-infected ducks (Schlicht et al. 1987). Moreover, the MS2 replicase fusions with fragments of the S, C, and X genes of hepatitis B virus were employed for the search of the corresponding antibodies in the patient sera (Hess et al. 1988; Liang et al. 1988; Abraham et al. 1989; Vitvitski-Trépo et al. 1990). With the corresponding MS2 replicase fusions, a great amount of human sera was tested for antibodies against the HPV16 early proteins E4 and E7, in order to correlate them with virus replication or cervical cancer (Jochmus-Kudielka et al. 1989; Jochmus et al. 1992). The MS2 replicase fusions increased sensitivity and specificity of serological testing for tuberculosis (Amicosante et al. 1995).

The MS2 replicase fusions were used for the fine mapping of the immunological B-cell epitopes within the putative infection markers, such as, for example, the protein X of hepatitis B virus (Stemler et al. 1990) or the E7 protein of HPV18 (Selvey et al. 1990). Furthermore, antibodies to the HPV16 and HPV18 E7 proteins, which were tested with the corresponding MS2 replicase fusions, were used as a diagnostic marker for cervical cancer (Müller et al. 1992). The MS2 replicase fusions were used frequently for the generation of not only polyclonal antibodies against the desired fusioned parts, but also of the monoclonal antibodies, as in the case, for example, of the nucleocapsid protein of equine arteritis virus (Weiland et al. 2000).

Some fusion proteins explained good potencies as putative vaccine candidates. Thus, for example, immunization with the MS2 replicase fusion to the extremely conserved serine-rich and alanine-histidine rich proteins protected *Aotus* monkeys, or owl monkeys, from a severe experimental *P. falciparum* infection (Knapp et al. 1988, 1992; Enders et al. 1992a,b). Then, the MS2 replicase fusion with *K-fgf/hst* oncogene-encoded growth factor induced high titers of the specific antibodies and protected mice against tumor growth (Talarico et al. 1990).

Furthermore, the MS2 replicase fusion technique was employed for the bacterial production of short hormones encoded by synthetic genes when the hormones in question, such as aprotinin (Auerswald et al. 1988; Brinkmann and Tschesche 1990), stefin B (Thiele et al. 1988), and cystatin (Auerswald et al. 1989), were deliberated from the long fusion proteins by the cyanogen bromide cleavage and renaturation.

The application of the MS2 replicase fusion technique reached its absolute maximum in 1990 and 1991 with 20 and 19 papers per year, respectively. The activity and success of Hans Will's team must be specially emphasized. It is noteworthy that the last application of the MS2 replicase fusion was fixed in 2009 by the purification of herpes simplex virus 1 helicase-primase subcomplex (Schreiner et al. 2009).

The ribosome binding site and first codons of the MS2 replicase gene were used in the pioneering approach elaborated by David S. Peabody (1990) who constructed a genetic two-plasmid system that placed the synthesis of a hybrid replicase-β-galactosidase enzyme under translational control of coat protein. The MS2 replicase-β-galactosidase hybrid contained the first three codons of replicase, which were fused to the *lacZ* at codon 10. This approach was described in detail in the Repressor complex I: MS2 section of Chapter 16.

Pohlner et al. (1993) gradually reconstructed the traditional MS2 replicase-based plasmids for the expression of fused genes. The N-terminal fragment of the MS2 replicase was shortened from about 100 amino acid residues to 11 amino acids plus His_6 and 4 additional Gln residues in the novel pEV41 vector. This shortening did not affect the production and stability of the recombinant proteins but allowed convenient affinity purification by means of the His_6 peptide. The use of the novel plasmid was exemplified by the production of the variable heavy and light chain domains of a monoclonal antibody, where the fusion proteins were specifically processed with IgA protease. The chemical cross-linking of the processed VH and VL domains resulted in a recombinant antibody Fv fragment that bound specifically to its antigen. Furthermore, this vector was employed by the fusion of the short MS2 replicase fragment with the vacuolating cytotoxin encoded by the *vacA* gene, which demonstrated structural similarities with the IgA protease type of exported protein (Schmitt and Haas 1994), and the CagA protein of *Helicobacter pylori*, associated with severe gastritis,

TABLE 19.1

Genes Expressed in *E. coli* in the Form of Fusions to the 98 N-Terminal Amino Acid Residues of the MS2 Replicase

Subjects	Genes/Proteins	References
Human genes	Human fibroblast interferon β	Derynck et al. (1981)
	The parts of the human (U1) ribonucleoprotein specific 68-kDa autoantigen	Guldner et al. (1988, 1990); Netter et al. (1990)
	The protease nexin-II, a secreted form of the Alzheimer's amyloid precursor protein	Oltersdorf et al. (1989)
	The gene encoding a human nuclear antigen recognized by autoantibodies from patients with primary biliary cirrhosis	Szostecki et al. (1990)
	The human α-fetoprotein gene	Giuliani et al. (1989)
	The human hepatic interleukin-6 receptor gene	Schooltink et al. (1991); Pietzko et al. (1993)
	The human gene of casein kinase II subunit β	Lorenz et al. (1993)
	The interferon-γ-inducible protein p16 and the prevalence of anti-p16 in the systemic lupus erythematosus patients	Seelig et al. (1994)
	The human interleukin-11 gene	Miao et al. (1995)
	The nuclear dot protein 52	Sternsdorf et al. (1995, 1997)
	The nuclear dot-associated Sp100 protein	Guldner et al. (1999)
Animal genes	The HLA-DRw6 β chain and the murine Ia-associated invariant chain	Koch et al. (1987)
	The intracisternal A-particle retrotransposon *env* genes in rodents	Reuss (1992)
	The mouse simple repeated sequence	Di Carlo et al. (1994)
	The microphthalmia-associated basic helix-loop-helix leucine zipper transcription factor Mi	Planque et al. (1999)
	The tyrosine kinase receptors in mice	Menn et al. (2000)
Avian genes	The chicken *limb deformity* (*ld*) gene	Trumpp et al. (1992)
	The Pax-QNR genes from quail neuroretina	Carriere et al. (1993); Turque et al. (1994, 1996); Casado et al. (1996)
	The *en-2* homeodomain from chicken	Vincent et al. (1996)
Plant genes	The class I chitinase genes of rice	Pan et al. (1996)
Oncogenes	The viral *myc* gene	Bunte et al. (1984)
	The human cellular *myc* gene	Benter et al. (1985)
	The *erbB* gene of avian erythroblastosis virus	Heimann et al. (1985)
	The *ets* oncogene of avian retrovirus E26	Ghysdael et al. (1986)
	The *K-fgf/hst* oncogene encoding a fibroblast growth factor in mice	Talarico et al. (1990)
	The *c-fos*, *fra-1*, and *junD* genes	Pompéia et al. (1997)
	The MafA transcription factor	Benkhelifa et al. (2001)
	The Rev-erbα and Rev-erbβ orphan receptors	Chopin-Delannoy et al. (2003)
	The microphthalmia transcription factors in posterior uveal melanomas	Mouriaux et al. (2003)
	The human midkine gene fragment	Zhang et al. (2005)
Hormones and short synthetic genes	The bovine hypothalamic hormone oxytocin precursor	Nörenberg et al. (1988)
	The [Arg15,Glu52] aprotinin gene	Auerswald et al. (1988)
	The [Phe15,17, Glu52] aprotinin gene	Brinkmann and Tschesche (1990)
	The aprotinin homolog with new inhibitory specificity for cathepsin G	Brinkmann et al. (1991)
	The gene of stefin B, a human intracellular cysteine proteinase inhibitor	Thiele et al. (1988)

(Continued)

TABLE 19.1 (Continued)
Genes Expressed in *E. coli* in the Form of Fusions to the 98 N-Terminal Amino Acid Residues of the MS2 Replicase

Subjects	Genes/Proteins	References
Insect genes	The gene of chicken cystatin, a cysteine proteinase inhibitor	Auerswald et al. (1989)
	The rat trypstatin gene	Dolinar and Auerswald (1991)
	The nicotinic acetylcholine receptor genes of *Drosophila*	Schloss et al. (1991)
	The heterogeneous ribonucleoprotein particles of *D. melanogaster*	Hovemann et al. (2000)
Genes of blood flukes	The Sm31/Sm32 proteins of *Schistosoma mansoni*	Felleisen et al. (1988); Klinkert et al. (1988, 1989, 1991); El-Sayed et al. (1998)
	The cDNA library of *S. mansoni*	Moser et al. (1990, 1992); Richter and Harn (1993); Richter et al. (1993); Ghoneim and Klinkert (1995); Michel et al. (1995)
Genes of nematodes	An antigen of *Onchocerca volvulus*, a cause of river blindness, cross-reactive with a component of the retinal pigment epithelium	Braun et al. (1991)
	A gene of *O. volvulus* encoding an immunologically reactive polypeptide	Seeber et al. (1993)
Genes of amoeba	The tail fragment of *Dictyostelium discoideum* myosin	Wagle et al. (1988)
Unicellular eukaryotes	Merozoite stage-specific genes of *Plasmodium falciparum*	McGarvey et al. (1984)
	The histidine alanine rich protein, an analogue of the malarial histidine-rich protein, or HRPII, of *P. falciparum*	Knapp et al. (1988)
	The SERP (serine-stretch protein) and HRPII (histidine-alanine rich protein II) gene-based vaccines against *P. falciparum*	Enders et al. (1992a,b); Knapp et al. (1992)
	The heat shock protein of *Theileria parva*	Gerhards et al. (1994)
Yeast	A DNA binding protein from killer system of *Kluyveromyces lactis*	Tommasino et al. (1991)
	The enolase and heat shock protein 90 of *Candida albicans*	Peterson et al. (1996)
Bacteria	Genomic library of *Mycoplasma hyopneumoniae*	Klinkert et al. (1985)
	The P1 operon of *Mycoplasma pneumoniae*	Sperker et al. (1991)
	The *ompV* gene of *Vibrio cholerae*	Pohlner et al. (1986)
	Five subunits of pertussis toxin of *Bordetella pertussis*	Nicosia et al. (1987); Pizza et al. (1988); De Magistris et al. (1989)
	The filamentous hemagglutinin gene of *B. pertussis*	Guzmán et al. (1992)
	Mapping of the immunodominant epitopes of the filamentous hemagglutinin of *B. pertussis*	Piatti (1999)
	The RNA polymerase genes from *Methanobacterium thermoautotrophicum*	Schallenberg et al. (1988)
	The 34-Kda protein of *Mycobacterium tuberculosis*	Vismara et al. (1990, 1991); Amicosante et al. (1995)
	The toxin gene of *Bacillus thuringiensis*	Rhim et al. (1990)
	The *rolB* gene of *Agrobacterium rhizogenes*	Trovato et al. (1990)
	The *traI* gene encoding DNA helicase I of the *E. coli* F factor	Benz and Müller (1990)
	The fibronectin-binding protein gene of *Streptococcus pyogenes*	Talay et al. (1992); Valentin-Weigand et al. (1993)
	The *xylS* genes of *Pseudomonas putida*	Michán et al. (1992)
	The *xylR* gene of *P. putida*	Fernández et al. (1995)
	The flagellin gene *flaA* of *Helicobacter pylori*	Leying et al. (1992)
	A plasmid gene of *Chlamydia trachomatis* encoding 28 kDa antigen (pgp3)	Comanducci et al. (1993)
	The *comA* gene of *Neisseria gonorrhoeae*	Facius and Meyer (1993)
Bacteriophages	The *com* gene of phage Mu	Kahmann et al. (1985)
Foot and mouth disease virus	The VP1 gene of foot-and-mouth disease virus	Küpper et al. (1981)
	Non-structural genes of foot-and-mouth disease virus	Klump et al. (1984a,b)

(Continued)

TABLE 19.1 (Continued)

Genes Expressed in *E. coli* in the Form of Fusions to the 98 N-Terminal Amino Acid Residues of the MS2 Replicase

Subjects	Genes/Proteins	References
Hepadnaviruses	Segments of the cloned foot-and-mouth disease virus genome	Strebel et al. (1986)
	The L gene of foot-and-mouth disease virus	Strebel and Beck (1986)
	The pre-S segment of duck hepatitis B virus envelope gene	Schlicht et al. (1987); Lambert et al. (1990)
	The X gene of hepatitis B virus	E. Pfaff and H. Schaller, cited by Pugh et al. (1986)
	The parts of the S, C, and X genes of hepatitis B virus	Will et al. (1986); Hess et al. (1988); Liang et al. (1988); Abraham et al. (1989)
	Mapping of B cell epitopes of the hepatitis B virus X protein	Stemler et al. (1990); Kay et al. (1991); zu Putlitz et al. (1998)
	Detection of the X and/or anti-X in hepatitis B virus infection	Vitvitski-Trépo et al. (1990); Jung et al. (1991)
	The parts of the gene P of hepatitis B virus	Weimer et al. (1989)
	The preS1 fragment of hepatitis B virus	Nassal (1992)
	The S and preS2 regions of woodchuck hepatitis virus	Shamoon et al. (1991)
	The C genes of hepatitis B virus and ground squirrel hepatitis virus	Bichko et al. (1993)
	The preS fragment of hepatitis B virus	Chen Z et al. (1996)
Hepatitis C virus	The P450 2D6 gene encoding cytochrome P450 monooxygenase for the detection of anti-liver-kidney microsome antibodies by hepatitis C virus infection	Seelig et al. (1993)
	The core antigen of hepatitis C virus	Goeser et al. (1994)
	The NS3, NS4, and NS5 antigens of hepatitis C virus	Cao et al. (1996)
Hepatitis delta	The immunodominant region of hepatitis delta antigen	Saldanha et al. (1990)
	The mass spectrometry of the recombinant hepatitis delta antigen	Tecce et al. (1992)
Human papilloma viruses	The early genes of the human papilloma viruses HPV16 and HPV18 in cervical carcinoma cells	Bernard et al. (1987); Seedorf et al. (1987)
	The E6 and E6 genes of HPV18	Schneider-Gädicke et al. (1988)
	The E4 and E7 genes of HPV16	Jochmus-Kudielka et al. (1989); Krchnák et al. (1990); Suchánková et al. (1991); Jochmus et al. (1992); Müller et al. (1992); Nemecková et al. (2002)
	Mapping of B-epitopes of the E7 protein of HPV18	Selvey et al. (1990)
	Refolding and purification of the MS2 replicase-E7 fusion protein	Suttnar et al. (1994)
	The full-length E7 gene of HPV16	M. Pawlita, cited by Bodenbach et al. (1994)
Rotavirus	The VP3 and VP4 genes of rotavirus SA11	Arias et al. (1987); Lizano et al. (1991)
Caliciviruses	The genes of rabbit hemorrhagic disease virus	Meyers et al. (1991); Boniotti et al. (1994); Wirblich et al. (1994, 1995, 1996)
Coxsackievirus	The capsid proteins of coxsackievirus B3	Werner et al. (1988); Hofschneider et al. (1990)
Retroviruses	The core and envelope genes of bovine leukemia virus	Ulrich et al. (1990, 1991)
	The *vif* and *nef* genes of human immunodeficiency virus 1	Wieland et al. (1990, 1991)
	The *env* and *bel* genes of human foamy retrovirus	Mahnke et al. (1990); Keller et al. (1991); Löchelt et al. (1991)
Pestiviruses	The genomic regions p23 and p14 of classical swine fever virus	Stark et al. (1993)
Asfiviruses	The p54 gene of African swine fever virus	Alcaraz et al. (1995)
Arteriviruses	The nucleocapsid protein of equine arteritis virus	Weiland et al. (2000)
Herpesviruses	The amino acid residues 845–882 of the DNA helicase of herpes simplex virus 1	Schreiner et al. (2009)

Note: The data are presented in chronological order within the subject groups and subgroups.

duodenal ulceration, and gastric adenocarcinoma and lymphoma (Eck et al. 1997).

The independent out-of-phase initiation after translation of the 100-amino acid residues long N-terminal part of the MS2 replicase was used by the optimized production of the nonglycosylated human transferrin and its half-molecules in *E. coli* (de Smit et al. 1995).

COAT GENE

As described above and presented in Figure 19.1, the coat gene, namely, the MS2 coat, was expressed for the first time by Fiers' team (Remaut et al. 1982). The highly efficient coat gene expression occurred in *E. coli* under the P_L promoter control and led to the 20% of total *de novo* protein synthesis. Kastelein et al. (1983a) examined the translational efficiency of the MS2 coat gene *in vivo* with respect to neighboring sequences by gradual shortening of the upstream region. They showed that the removal of the 3′-terminal sequences had little influence, but the gradual removal of the 5′-terminal region profoundly reduced translation. Long before the ribosome binding site of the coat gene was reached, the yield of the coat protein had dropped by one order of magnitude. At the same time, the functional half-lives of the various messengers were found not to be significantly different. The secondary structure of the ribosome binding site in the native and shortened MS2 RNA was also identical. It seemed that the important determinants for ribosome recognition lay 5′ to the ribosome binding region of the MS2 coat gene (Kastelein et al. 1983a). Although the synthesized MS2 coat was tested immunologically by polyacrylamide gel electrophoresis, the question of its self-assembly remained untouched in this paper.

For the first time, Tatyana Kozlovska from Grēns' team demonstrated that the expression in *E. coli* of an RNA phage coat gene, in this case, of the phage fr, another representative of the group I, led to the highly-efficient formation of virus-like particles (VLPs) that were undistinguishable in electron microscopy from the original phage (Kozlovskaia et al. 1986). This historically important picture obtained by electron microscopist Velta Ose is presented in Figure 19.2.

The high-level synthesis of the fr coat protein occurred under the control of the strong promoter of the *E. coli* tryptophan operon P_{trp}. The fr VLPs did not differ from the native fr phage also by the double radial immunodiffusion test according to Ouchterlony (1965). At the same time, Kozlovska performed a huge work on the influence of the 5′-terminal shortening on the fr coat gene expression, in order to confirm or disprove the above-mentioned data of Kastelein (1983a), but with the phage fr, but not MS2, as a model. This large study was never published, but it seemed generally that the high-level synthesis of the coat protein was also possible for the maximally shortened upstream regions, still to the ribosome binding site.

Next, the coat gene of the phage JP34, an intermediate between the groups I and II, was sequenced and expressed by van Duin's team (Adhin et al. 1989). Again, the question of the coat state after expression, in the form of VLPs or not, was not touched on in this study.

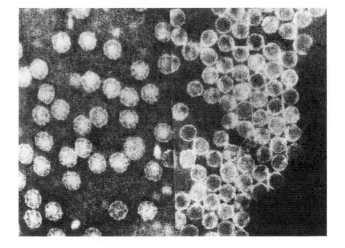

FIGURE 19.2 Electron microscopy of the phage fr particles (left) and of the purified phage fr VLPs (right), negatively stained with phosphotungstic acid. Multiplication 400,000×. (Reprinted with permission of Russian Academy of Sciences from Kozlovskaia TM, Pumpen PP, Dreilinia DE, Tsimanis AIu, Ose VP, Tsibinogin VV, Gren EJ. *Dokl Akad Nauk SSSR*. 1986;287:452–455.)

Then, Peabody (1990) constructed a genetic two-plasmid system that placed the synthesis of a hybrid replicase-β-galactosidase enzyme under the translational control of the MS2 coat, as described in the Repressor complex I: MS2 section of Chapter 16. In parallel, an analogous two-plasmid system was constructed for the GA coat protein (Peabody 1990) and a bit later for the Qβ coat protein (Lim et al. 1996). This pioneering system permitted the straightforward isolation of mutations that affected the repressor function of coat protein and functioned brilliantly *in vivo*, but the self-assembly problem of the coat products remained in the shadows.

Thanks again to the skillful mastery of Tatyana Kozlovska, Grēns' team expressed the coat gene of the phage Qβ, a representative of the group III, in *E. coli* (Kozlovska et al. 1993). In this case, the Qβ coat gene was amplified from native Qβ RNA using a reverse transcription-PCR technique. The amplified gene contained not only the sequence coding for the 133-amino acid residues of the essential Qβ coat, but also the sequence encoding the additional 196-amino acid residues of the 329-aa readthrough protein, or protein AI, and separated from the proper coat by an opal UGA stop codon, as described in Chapter 12. This obstacle allowed further progress in the elaboration of the original VLP display vectors, as described below. The cloning was achieved by using primers that ensured the natural environment for the Qβ coat gene, especially within the ribosome binding site, and supplying the gene with unique restriction sites at both ends. As a result, an amplified 1062-bp PCR fragment was positioned under the control of the strong *E. coli* P_{trp} promoter. The synthesis of the Qβ coat was confirmed first electrophoretically and immunologically. Moreover, the Ouchterlony double radial immunodiffusion test with antibodies against the phage Qβ and the electron microscopy evaluation performed by Velta Ose showed that when expressed, the Qβ coat gene was responsible for the high-level synthesis and correct self-assembly of

Qβ CP monomers into the VLPs indistinguishable morphologically and immunologically from the phage Qβ particles.

Figure 19.3 presents electron micrographs of the Qβ VLPs within the sliced *E. coli* cells, as well as the Qβ VLPs after their purification. The proper Qβ VLPs are compared in this historical illustration with the chimeric ones that will be described in Chapter 20.

Priano et al. (1995) expressed in *E. coli* the coat of the phage SP of the group IV, among other genes of this phage, in their substantial work on the mutual Qβ and SP complementation,

as outlined above in the Generation of mutants section of this chapter.

The expression of the pseudomonaphage PP7 coat gene in *E. coli* was achieved by David S. Peabody's team when they generated the genetic two-plasmid system, by analogy with that elaborated for the phages MS2, GA, and Qβ (Lim et al. 2001), as described in detail in the Repressor complex I: PP7 section of Chapter 16. Furthermore, Caldeira et al. (2007) tested the stability of the PP7 VLPs and regarded the latter as the putative subjects of growing nanobiotechnology. The

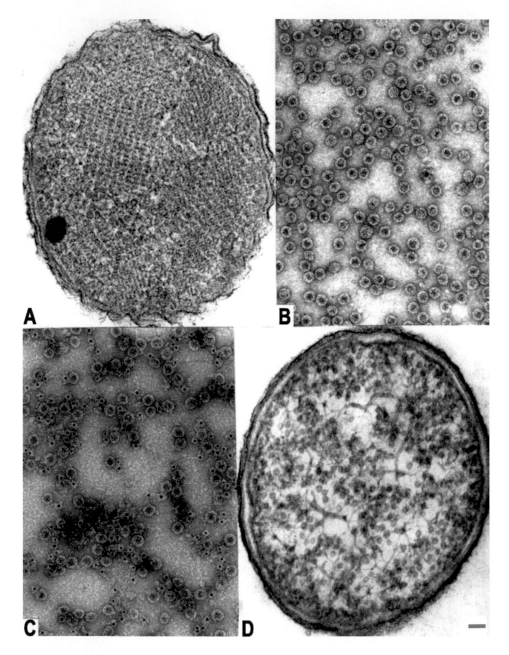

FIGURE 19.3 Electron micrographs of RNA phage coats as VLP carriers. Slices of *E. coli* cells filled with paracrystals of authentic Qβ particles (A) or with mosaic Qβ particles harboring 39 aa residues of the preS1 segment (D). Purified Qβ particles formed by authentic Qβ coat and A1 proteins (B), or by authentic Qβ coat as a helper and a readthrough A1 derivative carrying the same preS1 segment of 39 aa residues; preS1 epitopes exposed on the particles are labeled with monoclonal anti-preS1 antibody MA18/7 by immunogold technique (C). Scale bar, 50 nm. Micrographs are the generous gift of Velta Ose. (Reprinted with permission of Taylor & Francis Group from Pumpens P, Grens E. In: *Artificial DNA. Methods and Applications*. Khudyakov YE, Fields HA (Eds). Boca Raton: CRC Press LLC, 2002, pp. 249–327.)

authors specifically addressed the crosslinking of the PP7 VLPs by disulfide bonds between coat protein dimers at its fivefold and quasi-sixfold symmetry axes. This study determined, first of all, the effects of these disulfides on the thermal stability of the PP7 VLPs.

The further expressions in *E. coli* of the coat protein genes of the phage GA from the group II (Ni et al. 1996), of the pseudomonaphage PRR1 (Persson et al. 2008), and of the caulophage φCB5 (Plevka et al. 2009) were connected directly with determination of the three-dimensional structure, as will be presented in Chapter 21. The expression of the coat protein gene of the unique acinetophage AP205 in *E. coli* appeared as a result of the long-term and fruitful collaboration of Martin F. Bachmann's and Elmārs Grēns' teams (Spohn et al. 2010; Tissot et al. 2010) and also led to the high resolution structure of the AP205 VLPs (Shishovs et al. 2016).

In contrast to other RNA phage genes, the nanotechnological aims directed the highly efficient expression of the coat genes in yeast. Thus, in order to adapt the RNA phage VLPs to gene delivery, Legendre and Fastrez (2005) achieved for the first time the production of the MS2 VLPs in *Saccharomyces cerevisiae* and showed capability of the latter to package functional heterologous mRNAs. In parallel, the Qβ coat gene was expressed in both *S. cerevisiae* and *Pichia pastoris* by Andris Kazāks' group (Freivalds et al. 2006). In this study, the Qβ VLP production reached a high level of 3–4 mg/1 g of wet cells for *S. cerevisiae* and 4–6 mg for *P. pastoris*, which was about 15%–20% and 20%–30% of the *E. coli* expression level, respectively. The yeast-produced Qβ VLPs were easily purified by size-exclusion chromatography in both cases and contained some nucleic acid, shown by native agarose gel electrophoresis. The obtained particles were highly immunogenic in mice and the resulting sera recognized both *E. coli*- and yeast-derived Qβ VLPs equally well (Freivalds et al. 2006). Next, the phage GA VLPs were obtained in *S. cerevisiae* and *P. pastoris* (Freivalds et al. 2008). To optimize growth conditions, the three expression systems have been explored: GAL1 and GAL10 promoter-directed expression in *S. cerevisiae*, as well as AOX1 promoter-directed expression in *P. pastoris*. The formation of the GA VLPs was observed in all three cases. The GA VLPs were purified by a single size-exclusion chromatography step to 80%–90% of homogeneity. The final outcome of the purified VLPs varied between 0.6–2.0 mg per 1 g of cells for *S. cerevisiae*, while expression in *P. pastoris* resulted in the GA VLP yields of up to 3 mg from the same number of cells (Freivalds et al. 2008). At last, due to the strong efforts of Andris Kazāks and his group, the

S. cerevisiae- and *P. pastoris*-produced VLPs were obtained for the coliphage fr, as well as for the acinetophage AP205, pseudomonaphage PP7, and caulophage φCB5 (Freivalds et al. 2014). Surprisingly, all attempts to prepare the VLPs of the coliphage SP were unsuccessful in both *S. cerevisiae* and *P. pastoris* (Freivalds et al. 2014). In this excellent technological study, the VLP outcome varied from 0.2–8 mg from 1 g of wet cells. It was found that the φCB5 VLPs were easily dissociated into coat protein dimers when applied to strong anion exchangers. Upon salt removal and the addition of nucleic acid or its mimics and calcium ions, the dimers reassembled into VLPs with high efficiency.

Berkhout and Jeang (1990) expressed the MS2 coat gene in mammalian COS cells in their experiments by downmodulation of HIV-1 gene expression, as one of the first examples of the gene therapy approaches that will be presented in the Gene therapy section of Chapter 24. In parallel, Selby and Peterlin (1990) expressed the first chimeric MS2 coat derivative, namely, the coat fused to the HIV-1 Tat, in eukaryotic cells. This study paved the long way for the efficient tethering applications, as will be presented in the Tethering section of Chapter 24. In the same chapter, the two other historically important MS2 coat fusions will be presented: the MS2 fusion with the LexA protein that started the so-called three-hybrid approach and the MS2 fusion with the green fluorescence protein (GFP) that gave onset to the pioneering visualization, or imaging, studies in living cells. In parallel, the MS2 coat fusions with the target protein fragments were used in eukaryotic cells to identify specific functions of the latter, for example, the presence of nuclear export and localization signals (Hiriart et al. 2003), or mRNA stabilizing (Kong et al. 2003), or processing (Szymczyna et al. 2003) activity. By the tethering studies in eukaryotic cells (for detail, see Chapter 24), the MS2 coat was provided with the HA tag, an epitope 98–106 from human influenza hemagglutinin (Kim et al. 2005). Then, the SNAP domain of O-6-alkylguaninalkyltranferase was fused to the MS2 coat with an idea to link the MS2-SNAP fusion protein to DNA when the latter would be labeled with the SNAP substrate benzylguanine (Paul et al. 2013).

The MS2 and Qβ coats were expressed in plants, when they were used for the translational repression of transgene expression (Cerny et al. 2003). The MS2 coat fused to target proteins was overexpressed also in the baculovirus-infected Sf9 cells (Graveley et al. 1998; Sciabica and Hertel 2006).

The expression of the RNA phage coat genes has led to the emergence of a special branch of nanotechnology, which will be described in the next chapters of this book.

20 Chimeric VLPs

You see things and you say, "Why?" But I see things that never were; and I say, "Why not?"

George Bernard Shaw
Back to Methuselah, Part I, Act I

Everything should be made as simple as possible. But not simpler.

Albert Einstein

PROTEIN ENGINEERING AND VLP NANOTECHNOLOGY

The designed genetic reconstruction of proteins started in the early 1980s when Alan R. Fersht and his colleagues conceived the basic idea of the protein engineering, or mutational intervention, in the structure of proteins, which was based on spatial knowledge. Early protein engineering was oriented toward the creation of some kind of artificially improved proteins with desired functions (Winter et al. 1982; Wilkinson et al. 1983, 1984; Leatherbarrow and Fersht 1986). Since then, protein engineering as a specific branch of gene engineering has achieved much success, constructing new forms of enzymes and their inhibitors and elucidating the basic rules of protein folding and changing their specific activities (Fersht and Winter 1992).

Concurrent with the protein engineering of enzymes, i.e., monomeric or oligomeric proteins, a demand arose for the creation of artificial multimeric structures, which could be considered the primary tools for improving immunological activity of specific immunological epitopes from desired proteins. Success in the field was ensured by the highly impressive progress in the two major methods of attack were set in motion simultaneously in early 1980s. The first direction led to comprehension of the significance of the specific epitopes as antigenic regions, which were necessary and sufficient for (i) the induction of antigen-specific immune response and (ii) the recognition of antigens by monoclonal or polyclonal antibodies. The idea gained general acceptance, and many epitope sequences were mapped. The fine mapping of the immunological epitopes resulted, first, in the determination of a minimal epitope sequence, which could be defined as the shortest stretch within a protein molecule capable of inducing an immunological response and binding to the appropriate antibody. The next level of recognition within the epitope was defined as an antigenic determinant, namely, the exact side groups of amino acid residues involved in the epitope recognition by the paratope sequences of antibodies. The mapping techniques, based in general on the examination of libraries of short overlapping synthetic peptides or fusion proteins, revealed the existence of linear and conformational forms of the epitopes. The linear epitopes were defined as the

antigenic determinants that were localized in the short stretch of amino acid residues, usually from 4 to 10, and their recognition was possible within the polypeptides deprived of specific folding, for example, within the short synthetic peptides and SDS-denatured fusion proteins. By contrast, the conformational epitopes presupposed dispersion of the antigenic determinants in the more distant polypeptide stretches and were strongly dependent on the specific protein folding. In the 1980s, the peak of the mapping powers was reached thanks to gene engineering, which offered experimental resources to construct DNA copies of both linear and conformational epitopes, to define their length and composition at a single amino acid resolution, to combine them in a different order, and to introduce them into carrier genes, first of all, into the genes encoding the self-assembly competent monomers capable of assembling into virus-like particles (VLPs).

The increasing manipulations on the VLPs were reinforced by the simultaneous development of the second direction of the two major directions mentioned above. This direction was ensured by the fullest flower of structural knowledge about viral structural units by the use of high-resolution techniques. Thus, x-ray crystallography revealed high-resolution structures of such putative VLP carriers as the RNA phages (Valegård et al. 1990; Liljas et al. 1994; Golmohammadi et al. 1996; Tars et al. 1997, 2000a,b), mainly, due to the efforts of Lars Liljas' team in Uppsala and some contributions by Elmārs Grēns' team in Riga, as will be described in Chapter 21. Electron cryomicroscopy and x-ray crystallography also revealed in 1980–1990s the three-dimensional structures of some other prospective VLP carriers, such as rhinovirus (Rossmann et al. 1985), poliovirus (Hogle et al. 1985), tobacco mosaic virus (Namba and Stubbs 1986; Namba et al. 1989), bluetongue virus (Grimes et al. 1997), and hepatitis B cores (HBc) as one of the most popular, together with the RNA phages, VLP carriers (Crowther et al. 1994; Böttcher et al. 1997; Conway et al. 1997; Wynne et al. 1999).

Therefore, the monomeric constituents of capsids, envelopes, and rods of, in most cases, viral origin were regarded as fruitful carriers by VLP protein engineering and construction of the tailored VLPs. The first thorough review article on this novel field was published by Gren and Pumpen (1988). Those recombinant particles were called chimeric VLPs (Krüger et al. 1999) after *Chimaera*, a fire-breathing monster of Lycia in Asia Minor, composed of the parts of more than one animal, usually depicted as a lion with the head of a goat arising from its back, and a tail that might end with a snake's head.

In fact, the VLPs, as supramolecular structures built symmetrically from hundreds of subunits of one or more types, seemed to be the most efficient molecular carriers to the regular arrangement of epitope chains on the desired positions of the outer surface of the carrier. In contrast to the monomeric and oligomeric protein carriers, the VLP carriers were able to provide not only a high density of introduced foreign oligopeptides per particle, but also a distinctive three-dimensional conformation, which could be especially important for the presentation of the conformational epitopes. Therefore, the regular, repetitive pattern and correct conformation of inserted epitopes remained the factors that encouraged the persistent work on the functional activity of the tailored chimeric VLPs in inducing immunologically protective response.

The striking success in structural investigations of the putative carrier moieties and of the insertion-intended sequences, as well as the development of protein structure prediction techniques, completed VLP protein engineering as a specific branch of molecular biology and biotechnology until 2000. VLP protein engineering was defined at that time as a discipline working on knowledge-based approaches to the theoretical design and experimental construction of recombinant genomes and genes that might enable the efficient synthesis and correct self-assembly of the chimeric VLPs with programmed structural and functional properties (Pumpens and Grens 2002). More artistically, as stated above, the VLPs were defined as descendants of the monstrous *Chimaera* (Ulrich et al. 1996).

According to Pushko et al. (2013), the VLPs were defined as nanodimensional structures that (i) were built from one or several viral structural constituents in the form of recombinant proteins synthesized in efficient homologous or primarily heterologous expression systems (bacteria, yeast, or eukaryotic cell culture); (ii) were identical or closely related by their three-dimensional architecture and immunochemical characteristics to naturally occurring viral structures, and (iii) lacked genomes or infectivity. The term chimeric VLPs was accepted therefore for the structures in which the original structural proteins were covalently modified by the addition or substitution of foreign polypeptide stretches with desired functional properties, such as immunological epitopes, or cell-targeting, or encapsidation signals. The covalent integration into the chimeric VLPs could be achieved either by the expression of the VLP monomer genes containing the appropriate insertions encoding the desired protein stretches (Grens and Pumpen 1988) or by the chemical coupling of peptides, proteins, or other molecules to the VLPs (Bachmann and Dyer 2004).

Chimeric VLPs therefore targeted the three main functional applications: (i) presentation of foreign epitopes leading to novel immunological content and subsequently to the creation of novel vaccines; (ii) encapsidation of various therapeutic or diagnostic agents, such as nucleic acids as adjuvants for vaccines or gene therapy tools, proteins, or mRNAs for diagnostic or therapeutic purposes, or low molecular-mass drugs to be delivered to specific cells; and (iii) specific targeting of desired organs, tissues, or cells.

The early candidates for VLP protein engineering appeared in the mid-1980s and covered at once the three main structural forms of the putative VLP candidates. First, the filamentous phage f1 came from the *Inoviridae* family of rod-shaped

phages (Smith 1985). This revolutionary paper paved the way to the enormous field of the phage display methodology. It is our special pride that George Pearson Smith was rewarded with the Nobel Prize 2018 in Chemistry, together with Frances H. Arnold and Sir Gregory P. Winter, "for the phage display of peptides and antibodies." The rod-shaped viruses were also represented in the early works by tobacco mosaic virus as a representative of *Tobamovirus* (Haynes et al. 1986). Then, the outer envelope, or surface, or HBs, proteins of the enveloped hepatitis B virus from the *Hepadnaviridae* family were used to expose foreign sequences on the so-called 22-nm particles formed by the HBs (Valenzuela et al. 1985; Delpeyroux et al. 1986). Finally, the regular icosahedral particles were employed, namely, the RNA phage coats (Borisova et al. 1987; Gren et al. 1987; Kozlovskaia et al. 1988; Kozlovska et al. 1993), nucleocapsids, or cores, of hepatitis B virus (Borisova et al. 1987; Clarke et al. 1987; Newton et al. 1987), the VLPs encoded by yeast retrotransposon Ty1 (Adams et al. 1987), and nonenveloped capsids of picornaviruses (Burke et al. 1988). VLPs based on the hepatitis B virus products: surface and core antigens were reviewed in detail by the Grēns' team (Pumpens and Grens 1999, 2001, 2016; Pumpens et al. 2008). Remarkably, the enveloped VLPs from complex enveloped viruses, such as, for example, influenza viruses (Pushko et al. 2007, 2008) or the Ebola virus (Warfield et al. 2003) were only introduced after more than 20 years of intensive studies on the chimeric VLPs.

The first attempts were extremely fruitful not only from the viewpoint of the VLP candidate symmetry, but also in laying down the course of development of the two main directions of VLP engineering based on infectious and noninfectious structures. The noninfectious structures that did not contain any genomic material capable of productive infection or replication were represented by the RNA phage coats, hepatitis B virus envelopes and capsids, and Ty1 VLPs. They formed the basis of a long list of further replication-noncompetent models. As for the infectious structures, the chimeric derivatives of the filamentous DNA phage f1 and poliovirus initiated the line of the replication-competent viruses employed as the VLP models. The phage f1, a close neighbor of the phage f2 by the discovery time and place, initiated construction of the DNA phage-display vectors, not only classical, based on the single-stranded DNA viruses, but also novel, employing the double-stranded DNA-based entities. Furthermore, more poliovirus, rhinovirus, and plant (tobacco and cowpea mosaic) viruses were developed later as favorite replication-competent models. The latter could be divided again into the two main categories. First, there were replication-competent vectors, which carried full-length infectious viral genomes with hybrid genes, allowing the production of chimeric virus progeny. The second type represented the replication-noncompetent vectors, which were infectious but gave rise to chimeric progeny only in special replication conditions, e.g., in the presence of special helpers. Therefore, the genomes enveloped by such chimeric progeny were unable to give rise to chimeric particles. If full-length genomes and long genome fragments were usually obtained by PCR multiplication using specific synthetic oligonucleotides as primers, the individual

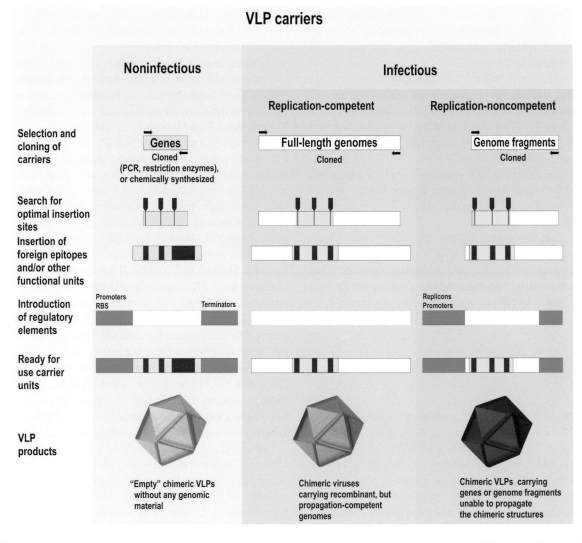

FIGURE 20.1 Major principles of VLP protein engineering, which comprise general constructing schemes of the appropriate carrier genes and genomes. VLP products are shown conditionally as icosahedrons, since the latter are prevalent in the viral world. Nevertheless, rods as well as isometric- and bacilliform-enveloped capsids are also possible VLP types. (Reprinted with permission of Taylor & Francis Group from Pumpens P, Grens E. In: *Artificial DNA. Methods and Applications.* Khudyakov YE, Fields HA (Eds). Boca Raton: CRC Press LLC, 2002, pp. 249–327.)

carrier genes could also be generated by total chemical synthesis. Thus, the first fully synthetic genes were constructed: the tobacco mosaic virus coat, constructed by Joel R. Haynes et al. (1986) and the hepatitis B virus core gene, generated by Michael Nassal (1988).

Figure 20.1 presents the general scheme taken from one of the first global reviews on VLP protein engineering (Pumpens and Grens 2002), demonstrating the general directions of the VLP technologies, one of which, namely, the noninfectious VLPs based on the icosahedral RNA phage coats, is the subject of the present chapter.

The DNA copies of the foreign oligopeptide sequences, which could be selected for the insertion into the VLP carrier genes, comprised further components of the protein engineering technology. Since these sequences might differ in length from a few to several hundred amino acid residues, the appropriate DNA copies were prepared by total chemical synthesis or by PCR multiplication with the use of specific synthetic primers.

These DNA copies were adjusted at that time to specific insertion sites on carrier genes, which must be selected and proven theoretically and experimentally for each VLP candidate.

The next of the most crucial steps for the successful realization of the chimeric VLP production consisted of the choice of the suitable expression system among prokaryotic bacterial cells and eukaryotic yeast, plant, insect, or mammalian cells. The choice of the expression system determined the nature of the regulatory elements, which needed to be added to the genomes or genes for the optimal expression. These regulatory elements (promoters, ribosome binding sites, transcription terminators, etc.) were prepared most often by chemical synthesis and comprised, after their introduction into the appropriate positions within the expression cartridge, the essential parts of the sophisticated vector arrangement that could be responsible for the desired high-level production of the complete VLPs. As presented in the preceding chapters, the RNA phages gave rise to many functional elements necessary for high-level production. The classical application of the VLP nanotechnology for the presentation of foreign immunological epitopes and designing of vaccines will be reviewed in detail in Chapter 22.

The latest advances in the VLP field showed, however, that the chimeric VLPs could be capable of presenting not only the immunological epitopes but also other functional protein motifs such as DNA or RNA binding and packaging sites, receptors and receptor binding sequences, immunoglobulins, elements recognizing low molecular mass substrates, etc. This inevitably broadened the VLP ideology from the conventional area of vaccines and diagnostic tools to gene therapy applications. This novel set of VLP applications will be explained in Chapter 23.

In this global context, Rother et al. (2016) presented a meaningful overview of the popular VLP models, or, in other words, natural protein nanocages, in comparison with potential artificial protein nanocages by their possible fruitful symbiosis for drug delivery applications, bionanotechnology, and materials science.

GENETIC FUSIONS OF COATS

fr

The first chimeric RNA phage coats were constructed for the phage fr by Grēns' team (Borisova et al. 1987; Gren et al. 1987; Gren and Pumpen 1988; Kozlovskaia et al. 1988; Pushko et al. 1993) and for the phage MS2 by Peter G. Stockley' team (Mastico et al. 1993). In both cases, the main idea consisted of the presentation of the foreign immunological epitopes on the VLP surface by genetic fusion at the specially chosen sites on the VLP carrier. To identify such coat regions that could tolerate the foreign insertions in the case of the fr VLPs, the synthetic oligonucleotide linkers encoding short amino acid sequences and containing convenient restriction sites were inserted into different regions of the fr coat gene (Borisova et al. 1987; Gren et al. 1987; Gren and Pumpen 1988; Kozlovskaia et al. 1988). Grēns' team performed this work in parallel and with the same oligonucleotides on two putative VLP carriers, namely, the fr coat and the HBc. Remarkably, this work was based on the computer predictions of the spatial structure of both VLP carriers and was conducted before the first crystal structure of an RNA phage, namely that of the homologous MS2 capsid, had been resolved (Valegård et al. 1986, 1990, 1991; Liljas and Valegård 1990; Golmohammadi et al. 1993) and the first electron cryomicroscopy and crystal structure of the HBc was achieved (Crowther et al. 1994; Wynne et al. 1999).

The first fr chimeras were constructed by Tatyana Kozlovska and Peter Pushko in Riga. The chimeric fr coats containing 2–12 amino acid residue long additions to their N- or C-termini or insertions at position 50 in the RNA binding region were capable of self-assembly, but this was not the case at positions 97–111 in the αA-helix (Kozlovskaia et al. 1988). The majority of the other fr coat mutants demonstrated reduced self-assembly capabilities and formed either coat dimers, as by the amino acid exchanges at residues 2, 10, 63, or 129, or both dimer and capsid structures, as at the residues 2 or 69 (Pushko et al. 1993).

As noted in the Chimeric phages section of Chapter 19, the five amino acid ASISI insertion from the Pushko et al. (1993) study was employed by Jan van Duin's team to accommodate and display the genetic fusion stretch on the surface of the live MS2 particles (van Meerten et al. 2001b).

The short epitope 31-DPAFRA-36 (DPAFR, or still DPAF), a hepatocyte-binding domain from the hepatitis B virus preS1 region, served as the first real immunological marker in the first VLP mapping studies, and further in the numerous elucidations of different VLP carriers, such as, for example, the hamster polyomavirus major capsid protein VP1 (Gedvilaite et al. 2000; Zvirbliene et al. 2006). This linear epitope was recognized strongly by the classical anti-preS1 monoclonal antibody MA18/7 generated by the famous Wolfram H. Gerlich's team (Heermann et al. 1984) and mapped to the four minimal DPAF amino acid residues by Grēns' team (Sominskaya et al. 1992b) and then still to the three letters D, P, and F by the famous Ken Murray's team (Germaschewski and Murray 1995). When the preS1 epitope DPAFR was inserted as the standard immunological marker at positions 2, 10, and 129 of the fr coat, it appeared in all cases on the particle surface (Borisova et al. 1987). In parallel, Grēns' team used a longer immunological marker, namely, the well-known and widely used V3 loop of the envelope gp120 protein of human immunodeficiency virus 1, in this case, of the MN subtype (Goudsmit et al. 1988). This model epitope was 39 or 40 amino acid residues long. Unfortunately, all attempts to tolerate this 40-amino acid-long V3 loop into the N-terminus, at positions 10, 12, and 15, or within the FG loop led to unassembled and mostly insoluble products (Kaspars Tārs, unpublished data).

The FG loop of the fr coat was also tested as a potential target for insertions/replacements, and was initially modified by a 4-aa-long deletion (Axblom et al. 1998). The deletion variant retained the ability to form capsids, although the VLP products displayed significantly reduced thermal stability. Furthermore, the three-dimensional structures of the mutant capsids revealed that the modified loops were disordered near the fivefold axis of symmetry and were too short to interact with each other (Axblom et al. 1998). Because of the high importance of the FG loop in capsid stability, further development of the chimeric fr VLPs based on the implementation of the FG loops was not pursued.

The fr coat vectors with the polyepitope linkers at position 2 (Borisova et al. 1987; Kozlovskaia et al. 1988; Pushko et al. 1993) were employed for the cloning of foreign sequences in all possible reading frames, with an aim of the fine mapping of linear immunological epitopes. This method did not expect VLP formation but was managed with immunoblotting of the polyacrylamide gel distributions of the corresponding fr coat fusion libraries. Thus, the minimal length of the preS1 epitope DPAFR (Sominskaya et al. 1992b), the preS2 epitope QDPR (Sominskaya et al. 1992a), and many other hepatitis B virus-derived epitopes (Sällberg et al. 1993; Meisel et al.

1994; Sobotta et al. 2000; Sominskaya et al. 2002) was established. Later, this approach was adapted to the immunological screening of the epitope libraries and described within the reliable protocol collection (Sominskaya and Tars 1998). It is noteworthy that the fr coat epitope selection approach was used by the establishing and mapping of the structural proteins of hamster polyoma virus (Siray et al. 1999, 2000). Moreover, the fr-based expression vectors were used for the efficient expression in *E. coli* of the hamster polyoma virus VP1 gene fragments (Voronkova et al. 2007). It is necessary to note here that the powerful hamster polyoma virus direction was developed due to the efforts of Rainer G. Ulrich's and Kęstutis Sasnauskas' teams in Berlin and Vilnius, respectively, with a definite participation of the young Riga researchers.

In parallel, the fr coat was used by Grēns' team as a leader sequence, by analogy with the above-described application of the replicase subunit, as presented in the Replicase fusions section of Chapter 19. Thus, the high level expression of α-human atrial natriuretic factor (ANP) in *E. coli* was achieved in the form of the fusion polypeptide with the fr coat protein that fulfilled in this case the function to protect the short peptide from proteolytic degradation in *E. coli* cells (Berzins et al. 1993). In this study, the synthetic DNA sequence encoding the 28 amino acid residues of α-human atrial natriuretic factor with the N-terminal linker tripeptide Ile-Asp-Lys was fusioned to the fr coat gene after the last C-terminal amino acid residue 130. This hybrid gene was placed under the control of the inducible *E. coli* tryptophan promoter and ribosome binding site of the fr coat gene. The expressed fusion protein did not self-assemble but accounted for at least 10% of the total cellular protein. By improving cultivation conditions, the yield of the fusion protein exceeded 45 mg per g of cell dry weight. The fusion protein from solubilized *E. coli* cells was purified to homogeneity by ion exchange chromatography on a set of different columns (Berzins et al. 1993). The fusion protein was used first to raise polyclonal antibodies and to develop of immunoassay of ANP on isolated rabbit aortic strips in order to examine hypotensive, diuretic, and natriuretic activity, as well as renal creatinine clearance, in an *in vivo* rat model (Baumanis et al. 1997; Mandrika et al. 1997). Surprisingly, the natural ANP and the recombinant fusion protein demonstrated identical activity in the functional tests. The physiological activity of the fusion protein has made it possible to predict its conformation and the method by which fusion protein interacted with the specific ANP receptors. Since the ANP did not contain any lysine, tryptophan, proline, histidine, threonine, glutamic acid, or valine residues, the authors planned to introduce the appropriate linking molecules and apply the selective cleavage methods to isolate ANP from the fusion protein (Baumanis et al. 1997).

In parallel, in an attempt to develop chimeric VLPs as an experimental vaccine against human papilloma virus (HPV)-induced tumors, the fr coat was tested as a putative VLP carrier of the relatively long stretch of the HPV16 E7 oncoprotein epitopes spanning amino acid residues 35–54 and 35–98 (Pumpens et al. 2002). These HPV E7 fragments were expressed in parallel on the three popular VLP carriers: fr coat and hepatitis B virus surface and core proteins. When inserted at position 2/3 of the fr coat, the E7 fragment 35–98 prevented, however, the VLP formation. Nevertheless, the purified fr-E7(35–98) induced the Th1 and Th2 subsets of T helper cells and elicited equally high antibody responses to both E7 and fr coat carrier (Pumpens et al. 2002). The fr coat also failed to form VLPs following the insertion of long HBV preS1 sequences, longer than the above-mentioned DPAFR epitope (Pumpens et al. unpublished).

Conversely, the fr coats showed unusually high capacities as VLP vectors by the addition of long segments of hamster polyomavirus VP1, a major capsid protein (Voronkova et al. 2002). Thus, the fr VLPs tolerated the N-terminal fusion of the immunodominant epitopes located at the C-terminal positions 333–384, 351–384, 351–374, and 364–384 of the hamster polyomavirus VP1. The efficient induction of the VP1-specific antibodies in rabbits and mice by immunization with chimeric virus-like particles harboring amino acids 333–384, 351–384, and 364–384 of the VP1 suggested the immunodominant nature of the C-terminal region of VP1 (Voronkova et al. 2002). When discovered, these findings markedly enhanced interest in using RNA phage coats as potent vaccine candidates.

Later, the purified fr VLPs carrying the hamster polyomavirus VP1 epitopes 333–384 and 351–384 were employed by the study on lymphoma outbreak in a GASH:Sal colony of Syrian golden hamsters *Mesocricetus auratus* (Muñoz et al. 2013). It is necessary to note here that the genetic audiogenic seizure hamster (GASH:Sal), inbred at the University of Salamanca, constituted a new strain which exhibited a genetic, spontaneous pathology with a predisposition to undergo convulsive crises of an audiogenic origin.

MS2

As mentioned above, the chimeric MS2 VLPs were first generated by Stockley's team. Thus, Mastico et al. (1993) employed for the first time the N-terminal β-hairpin exposed at the surface of the MS2 capsid, namely, at the site between amino acid residues Gly15 and Thr16, or between positions Gly14 and Thr15 if the N-terminal methionine is not considered, as preferable for toleration of foreign insertions. The insertion of DNA oligonucleotides at this site allowed the production of the chimeric MS2 VLPs carrying foreign peptide sequences expressed at the central part of the hairpin. The foreign sequences for the insertion were chosen because of their known antigenic properties and resulted in a number of different peptides up to 24 aa residues in length. The chimeric coat proteins self-assembled into largely RNA-free chimeric VLPs in *E. coli* and were easily disassembled and reassembled *in vitro*. The foreign epitopes exposed on these chimeric VLPs were found to be immunogenic in mice (Mastico et al. 1993). The first short outline of these data appeared in the review papers of Hill and Stockley (1996) and Lomonossoff and Johnson (1996). The technical details, such as the immunoaffinity chromatography, of the constructions in question were described in detail in the corresponding methodological paper (Stockley and Mastico 2000). The full list of these chimeric VLPs is given in Table 22.1 of Chapter 22, since they possessed all necessary properties of the putative vaccine candidates.

The putatively protective epitope T1, spanning 24 amino acid residues in length and derived from the immunodominant liver stage antigen-1 (LSA-1) antigen of the malaria parasite *Plasmodium falciparum* must be distinguished here. It was inserted at the tip of the N-terminal β-hairpin, between positions 15 and 16, of the MS2 coat (Heal et al. 2000). The chimeric VLPs carrying the LSA-1 epitope elicited both humoral and cellular immune responses in BALB/c mice with significant upregulation of interferon-γ, a finding which corroborated naturally acquired resistance to liver stage malaria (Heal et al. 2000). This study was regarded at that time as an important step to the development of the putative LSA-1 vaccine (Kurtis et al. 2001).

In parallel, David S. Peabody (1997b) has chosen so-called Flag as a standard immunological marker. The Flag octapeptide DYKDDDDK was proposed as a specific marker by Hopp et al. (1988). The Flag peptide was engineered onto the N-terminus of a variety of recombinant lymphokines for the purpose of aiding in their detection and purification from yeast supernatants or *E. coli* extracts. An antibody specific for the first four amino acid residues of this sequence was used as a detection reagent and for the affinity purification of products under mild conditions, by analogy with the above-mentioned monoclonal anti-preS1 antibody MA18/7. The authors thought that the proteins would remain unaffected by the Flag presence and would retain their biological activity, because of the small size of the peptide moiety and its hydrophilic nature. Moreover, it seemed possible to remove the Flag peptide by an enzymatic cleavage using enterokinase (Hopp et al. 1988).

Peabody et al. (1997b) studied the consequences of the Flag insertion on the translational repressor and capsid assembly functions of the chimeric MS2 coat. When the Flag epitope was added to the N-terminus in a way that is explained in Table 22.1, the chimeric coat folded properly into the form that recognized RNA, i.e., the dimer form, but was defective for capsid assembly. On the other hand, a chimeric coat which was expected to display the Flag insertion as a surface loop did not fold correctly and, as a consequence, was proteolytically degraded. The genetic fusion of the coat dimer resulted in a protein considerably more tolerant of these structural perturbations and mostly corrected the defects accompanying the Flag peptide insertion. Thus, the genetic fusion of the subunits reversed the defects of the AB loop insertions when either half of the fused dimer contained the wild-type sequence. However, the fusion of subunits was insufficient to compensate fully for the structure defects imposed by the presence of the Flag in both AB loops of the fused dimer. It did, nevertheless, prevent degradation of the protein and allowed its accumulation within inclusion bodies. The increased resistance of the single-chain coat protein to the urea denaturation indicated that the fused dimer was substantially more stable than the wild-type one. The covalent joining of subunits was

supposed to be a general strategy for the engineering of the MS2 chimeras with increased protein stability (Peabody et al. 1997b).

The histidine-tagged MS2 VLPs were constructed by the introduction of the His$_6$ linker between coat codons 15 and 16 (Cheng et al. 2006). The His$_6$ linker simplified purification of the VLP-covered RNAs, since this study was oriented mostly to generation of the so-called armored RNA tools that will be discussed in Chapter 23.

Qβ

The relatively low tolerance of the fr- and MS2-based VLP vectors to long foreign insertions provoked a definite interest in the generation of so-called mosaic particles. In contrast to the homogenously formed VLPs, the mosaic particles were intended to be composed of wild-type, or *helper*, and epitope-carrying, or *chimeric*, subunits. This was supposed to improve the VLP capacity for the foreign insertions and was used for the first time in the case of hepatitis B virus surface antigen as a carrier (Delpeyroux et al. 1988).

After successful cloning of the full-length A1 gene of the phage Qβ in *E. coli* (Kozlovska et al. 1993), Tatyana Kozlovska et al. (1996, 1997) recognized the 195 amino acid residues long readthrough extension of the phage Qβ coat as a promising site for foreign insertions by the generation of the mixed Qβ VLP particles.

The readthrough A1 extension was proposed to contain elements that could protrude as the spike-like structures on the Qβ VLP surface. This assumption was achieved by the sensitive homology programs elaborated at that time by Indulis Gusārs in Riga. He detected some unexpected similarity when the readthrough Qβ A1 extension was compared with the protruding preS part of the hepatitis B virus surface antigen (Kozlovska et al. 1996). Remarkably, the real three-dimensional structure of the Qβ A1 readthrough domain was revealed at high resolution by x-ray crystallography after 15 years in Riga by Jānis Rūmnieks and Kaspars Tārs (Rumnieks and Tars 2011). The fold was found unique among all proteins in the protein data bank. However, the real fold of the protruding hepatitis B virus preS1 region remains unresolved.

A next argument to the applicability of the readthrough extension was provided by the self-assembly capabilities of the capsids with mutually exchanged extensions of the Qβ and SP A1 proteins, which were demonstrated experimentally at that time (Priano et al. 1995).

Figure 20.2 presents the general outline of the planned approach (Kozlovska et al. 1996).

Thus, in order to realize the idea of the mosaic Qβ coat/A1 VLPs carrying foreign epitopes inserted into the readthrough extension part of A1, the two possible strategies were chosen. First, it was the expression of the original A1 gene in conditions of the enhanced UGA suppression. Second, it was the separated expression of the proper coat gene and of the full-length A1 gene after mutational changing of the UGA stop codon to UAA stop or GGA sense codons, respectively. By the second strategy, the expression

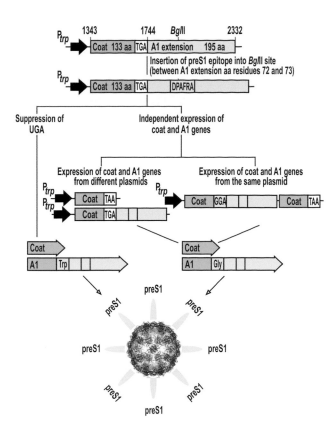

FIGURE 20.2 General scheme of expression in *E. coli* of the Qβ coat gene and its application for the exposure of foreign epitopes. The start and end points of the Qβ gene are indicated in accordance with the full-length Qβ genome sequence (Mekler 1981). The "immunomarker" sequence DPAFRA, which was introduced as an example of the epitope cloning, represents a fragment from HBV preS1 sequence recognized by monoclonal anti-preS1 antibody MA18/7 (Heermann et al. 1984). (Adapted from Kozlovska TM et al. *Intervirology.* 1996;39:9–15, with permission of S. Karger AG, Basel.)

of the coat and full-length A1 genes was achievable (i) from a single plasmid harboring the appropriate genes within a cassette under the control of a strong promoter, and (ii) from the two different plasmids conveying the two antibiotic resistances to the *E. coli* cells.

The above-presented preS1 epitope DPAFRA was employed as a standard foreign epitope, short enough and easy to test with well-characterized antibody MA18/7, as described above. The 39 or 40 amino acid residues long V3 loop of the envelope gp120 protein of human immunodeficiency virus 1 was used as a more complicated immunological marker that would represent epitopes longer than 4−6 amino acid residues and possessing some definite conformational features.

Concerning the direction based on the enhanced UGA suppression, Hofstetter et al. (1974) were the first to find out that the content of A1 within the Qβ phage particles could be elevated in conditions of the UGA suppression. Following this indication, a plasmid harboring cloned gene of *opal* suppressor tRNA which recognized the UGA codon as the Trp codon (Smiley and Minion 1993) was introduced into *E. coli* K802

cells bearing the appropriate Qβ expression plasmids. As a result, the synthesis of the A1, or A1 carrying the DPAFRA, or A1 carrying the V3 loop was enhanced to up to 50% of the Qβ coat synthesis (Kozlovska 1996, 1997).

Irrespective of the method by which the wild-type Qβ helper subunits were supplied, the chimeric Qβ subunits with additions at their C-termini, replacing the readthrough part, acquired the ability to be included into the chimeric particles in the cases when the chimeric subunits alone failed to form homogeneous particles from identical subunits (Kozlovska 1996, 1997).

Tatyana Kozlovska's group carefully mapped the potentially tolerant sites of the readthrough extension by the consistent insertion of the two above-mentioned model epitopes (Vasiljeva et al. 1998). In conditions of enhanced UGA suppression, the chimeric VLPs were always observed, although the proportion of the A1-extended to the short coats in mosaic particles dropped from 48% to 14%, with an increase of the length of the A1 extension. Nevertheless, the model preS1 epitope, DPAFR in this case, was found on the surface of the mosaic Qβ particles and ensured their specific antigenicity and immunogenicity in mice (Vasiljeva et al. 1998). The antibody response to the preS1 epitope was clearly higher for the self-assembled Qβ-preS1 VLPs than for a nonassembled Qβ-preS1 fusion variant (Fehr et al. 1998). When the Qβ coat was modified to carry long hepatitis B virus preS insertions, namely, full-length preS, preS1 alone, or preS2 alone, instead of the A1 extension, the mosaic Qβ particles were formed that possessed the surface-exposed preS chains, while regular VLPs did not form without the presence of the Qβ coat as a helper (Indulis Cielēns and Regīna Renhofa, unpublished data).

Further development of the idea of the mosaic Qβ coat particles was accomplished by the famous M.G. Finn's team, who performed at that time impressive studies on the functionalization of the Qβ VLPs, as described below in the Functionalization: Qβ section of the present chapter. Concerning specifically the genetic fusions, Udit et al. (2008) coexpressed two plasmids in *E. coli*, where the first, carbenicillin-resistant expression vector, carried the wild-type Qβ coat gene, but the other, spectinomycin-resistant vector, encoded the chimeric Qβ coat gene with the His-tag GSGSGHHHHHH coding sequence appended to the C-terminus of the Qβ coat gene. This resulted in the polyvalent display of approximately 37 His$_6$ tags per particle and allowed immobilization of the Qβ VLPs on metal-derivatized surfaces, without any observed change in the VLP stability compared to the wild-type VLPs.

By further engineering of metal- and metallocycle-binding sites on the Qβ VLP scaffolds, the three motifs, His$_6$, His$_6$-His$_6$, and Cys-His$_6$, were compared by incorporation into the capsid via a coexpression methodology at ratios of 1.1 : 1, 1.1 : 1, and 2.3 : 1 for wild-type to chimeric coat protein (Udit et al. 2010b). The size-exclusion chromatography yielded elution profiles identical to wild-type particles, while Ni-NTA affinity chromatography resulted in retention times that increased according to Qβ-His$_6$ < Qβ-Cys-His$_6$ < Qβ-His$_6$-His$_6$. In addition to interacting with metal-derivatized surfaces, the

Qβ-Cys-His$_6$ and Qβ-His$_6$-His$_6$ VLPs bound heme (Udit et al. 2010b).

Furthermore, instead of relying on the differential readthrough of a stop codon as in the natural virus, Brown et al. (2009) employed again the compatible plasmids coding for the wild-type Qβ coat and for the chimeric Qβ with the C-terminal extension of the 58 amino-acid Z domain derived from *S. aureus* protein A and fused over an octapeptide spacer. As known, the Z domain binds to th6e CH2-CH3 hinge region of a group of IgG subtypes and can be employed in numerous nanotechnological applications. The fusion plasmid yielded copious quantities of the Qβ-Z fusion protein, but no intact particles, consistent with the expectation that the extended subunit was incapable of assembling into a particle on its own. In contrast, when *E. coli* cells were transformed with both wild-type and fusion plasmids, the mosaic Qβ particles were isolated in high yields, approximately 50 mg per liter of culture, and approximately 20 Z domains were incorporated per particle. However, electron cryomicroscopic reconstruction of the chimeric Qβ-Z particles did not show protein density other than that of the standard Qβ capsid, presumably due to the sparse and presumably irregular positioning of the Z domains. Nevertheless, the hybrid VLPs were found to bind strongly to immobilized IgG in an ELISA assay, whereas native Qβ VLPs showed no interaction. Since the chimeric Qβ-Z VLPs did not disassemble under the conditions used in the ELISA, the result demonstrated that the fused domains were accessible on the exterior surface of the particle. It was found striking that the number of incorporated Qβ-Z fusion proteins dropped to approximately six per particle by changing growth media. Therefore, it was the first indication to the incorporation of the two different forms of a coat protein into VLPs in cells in response to the environmental factors. The authors suspected that the increase in osmotic pressure was at least partially responsible for this result and supposed variation in the VLP composition by manipulation of the conditions of cellular expression during practical generation of the functional nanoparticles (Brown et al. 2009).

Recently, Liao et al. (2017) constructed the mosaic Qβ VLPs by expressing two proteins, namely, the original Qβ coat and the Qβ-GFP fusion protein, where GFP was fused to the C-terminus of the Qβ coat, linked in tandem by an internal ribosome entry site sequence, in *E. coli*. The two proteins were expressed in roughly equal amounts, but the Qβ-GFP fusion protein was incorporated into the Qβ VLPs less well, at a ratio ~1:10 relative to the original Qβ protein. The particle size and geometric symmetry of the VLPs containing the Qβ-GFP fusion were similar to that of the Qβ-VLPs by electron microscopy examination (Liao et al. 2017).

Finally, after 20 years from the first experiments on the mosaic Qβ particles, Skamel et al. (2014) generated the Qβ display system via engineering of the A1 protein and suggested the phage Qβ as a favorable alternative to M13 for *in vitro* evolution of displayed peptides and proteins due to high mutagenesis rates in the Qβ RNA replication that could better simulate the affinity maturation processes of the immune response. As a result, Skamel et al. (2014) constructed the chimeric Qβ phages bearing G-H loop peptide of the VP1 protein of foot-and-mouth disease virus (FMDV). Thus, the DNA fragment encoding the G-H loop was fused to the Qβ A1 protein in a replication-competent hybrid phage that efficiently displayed the FMDV peptide. The surface-localized FMDV VP1 G-H loop cross-reacted with the anti-FMDV monoclonal antibody and was found to decorate the corners of the Qβ icosahedral shell by electron microscopy. The evolution of the Qβ-displayed peptides, starting from fully degenerate coding sequences corresponding to the immunodominant region of VP1, allowed rapid *in vitro* affinity maturation to the monoclonal antibody. The phage Qβ selected under evolutionary pressure revealed a non-canonical but essential epitope for the monoclonal antibody recognition. Finally, the selected chimeric Qβ phages induced polyclonal antibodies in guinea pigs with good affinity to both FMDV and hybrid Qβ-G-H loop, validating the requirement of the tandem pair epitope. Therefore, the A1 protein-based Qβ-display emerged as a novel framework for the rapid *in vitro* evolution by affinity maturation to the molecular targets (Skamel et al. 2014).

PP7

After successful application of the MS2 VLPs as a platform for the peptide display (Peabody et al. 2008), as described above in the Genetic fusions of coats: MS2 section, David S. Peabody and Bryce Chackerian turned to the VLPs generated by the pseudomonaphage PP7. By analogy with the MS2 coat, the AB loop of the PP7 coat and of the PP7 coat single-chain dimer, between positions 11 and 12, was employed first (Caldeira et al. 2010). In this first paper, the broadly cross-type neutralizing human papillomavirus L2 epitope of 15 amino acid residues served as the most important model. This and the two other models, namely, the Flag peptide and the V3 loop of HIV, are listed in Table 22.1 presenting the fusion-based VLP vaccines in Chapter 22. The single-chain dimer was also highly tolerant of the random 6-, 8-, and 10-amino acid insertions, as in the case of the MS2 VLPs.

As Caldeira et al. (2010) noted, the PP7 VLPs offered several potential advantages and improvements over the MS2 VLPs. First, the particles were more stable thermodynamically, because of the presence of the stabilizing inter-subunit disulfide bonds (Caldeira and Peabody 2007). Second, the PP7 VLPs were not cross-reactive immunologically with those of the MS2. This could be important in applications where serial administration of the VLPs would be necessary. Third, the authors anticipated that the correct folding and assembly of the PP7 VLPs might be more resistant to the destabilizing effects of peptide insertion, or that the PP7 VLPs might at least show tolerance of some peptides not tolerated in the MS2 VLPs.

As the first result, the single-chain dimer of the PP7 coat demonstrated broad tolerance to random and specific peptide insertions, displayed peptides on the VLP surface, ensured high immunogenicity of the inserted epitopes, and packaged the RNA that directed their synthesis. The two prospective vaccine candidates were generated. The first specific PP7 VLP vaccine successfully induced antibodies against a peptide derived from the V3 loop of HIV that was the target of neutralizing antibodies. The second vaccine induced antibodies against a broadly cross-neutralizing epitope from the minor capsid protein L2 of HPV16. It was highly important, since the neutralizing antibodies that targeted the minor capsid protein, L2, were broadly cross-neutralizing, suggesting that the neutralizing epitopes on L2 were conserved across HPV types, unlike the L1-specific neutralizing antibodies that were largely HPV type-specific. Caldeira et al. (2010) showed that the PP7 VLP vaccine displaying an L2 epitope induced antibody responses that protected mice from genital challenge with homologous (HPV16) and heterologous (HPV45) pseudovirus.

Moreover, Chackerian and Peabody's team advertised successful generation of a pan-HPV vaccine based on the PP7 VLPs displaying the broadly cross-neutralizing L2 epitope, in contrast to the classical HPV vaccines that were based on the VLPs of the major HPV capsid protein L1 (Tumban et al. 2011). As listed in detail in Table 22.1, the eight L2 epitopes from diverse HPV types were displayed successfully on the PP7 VLPs. These chimeric PP7-L2 VLPs, both individually and in combination, elicited high-titer anti-L2 IgG serum antibodies and protected the immunized mice from high dose infection with HPV pseudovirus. Thus, the mice immunized with 16L2 PP7 VLPs or 18L2 PP7 VLPs were nearly completely protected from both pseudovirus 16 and 18 challenge. The mice immunized with the mixture of the eight L2 VLPs were strongly protected from genital challenge with pseudovirus representing eight diverse HPV types and cutaneous challenge with HPV5 pseudovirus. It is noteworthy that Tumban et al. (2011) have removed RNA, for the first time in the history of the chimeric VLPs, from the inside of the L2-PP7 VLPs by alkaline treatment, in accordance with the procedure that was applied earlier on the non-chimeric MS2 VLPs (Hooker et al. 2004) and demonstrated its high efficiency later on the VLPs based on hepatitis B virus cores (Strods et al. 2015).

Then, Hunter et al. (2011) demonstrated successful induction of local mucosal and systemic antibody responses after direct immunization of the mouse genital tract by the L2-PP7 VLP vaccine in the aerosol form. It is noteworthy that the phage Qβ was used as a control in this study. The intravaginal aerosol administration of the VLPs induced high titer IgG and IgA antibodies in the female genital tract, as well as IgG in the sera, and protected mice from genital infection with a pseudovirus.

Tumban et al. (2013) performed a solid immunological study and showed that the L2-PP7 VLPs, although being monovalent vaccines, ensured long-lasting protective antibody responses in mice, persisting over 18 months after vaccination and protecting strongly against infection. Remarkably, this study unveiled the possible role of the encapsidated RNA as an endogenous adjuvant. Surprisingly, the exogenous and endogenous adjuvants, namely, LPS and encapsidated single-stranded RNA, had minor effects on antibody titers. The immunization with VLPs containing the encapsidated single-stranded RNA predominantly shifted the response to a Th1,

rather than a Th2-like pathway. Importantly, the immunization with the L2-VLPs without endogenous and exogenous adjuvants, but in the presence of alum hydroxide, elicited a robust antibody response (Tumban et al. 2013).

In parallel, Tumban et al. (2012) reported the generation of the L2 HPV vaccine on the MS2 VLP background, by analogy with the previous PP7 VLP experience, but with insertion of the L2 epitope at the N-terminus of the first MS2 coat copy within the single-chain dimer. The reason for this study was the fact that the authors were somewhat concerned about the context of the L2 peptide after its insertion into the AB loop of the PP7 coat. Therefore, they explored a different display site at the N-terminus of the coat protein in the alternate MS2 VLP display platform. The epitopes of different length, representing amino acids 20–29, 17–31, 14–40, and 14–65 of HPV16 L2, were inserted at the N-terminus of the MS2 single-chain dimer. Out of the four recombinant 16L2 MS2 coat proteins that were expressed, three of them, namely, L2(20–29), L2(17–31), and L2(14–40), assembled into VLPs. The coat protein displaying L2 amino acids 14–65 failed to form VLPs. Somewhat surprisingly, the immunization with MS2 VLPs displaying an HPV16 L2 peptide at the N-terminus of the coat protein elicited much broader cross-reactive antibodies compared to the same epitope displayed on the AB-loop of the PP7 VLPs. Moreover, the mice immunized with the MS2 VLPs displaying the HPV16 L2 peptide were strongly protected against vaginal and intradermal challenge with HPV pseudovirions representing diverse types. These data demonstrated that a single recombinant VLP displaying an L2 peptide could provide broad protection against HPV infection (Tumban et al. 2012).

The conclusive success of the PP7 and MS2-based L2 HPV vaccines was self-reviewed at that time by Tyler et al. (2014a).

Furthermore, Chackerian and Peabody's team targeted other potentially cross-reactive epitopes derived from the L2 protein (Tyler et al. 2014b). In this study, the authors used the two different VLP display methodologies, not only the PP7 VLP-based fusion technology but also Qβ phage-based chemical coupling technology, which will be explained in next sections of this chapter. As presented in Table 22.1, the L2 epitopes at amino acid positions 17–31, 35–50, 51–65, 65–79, and 65–85 of HPV16 were employed for the insertion into the AB loop of the PP7 coat as well as for the cross-linking to the phage Qβ surface. All variants invariably led to a high-titer antibody response against the peptide, but the *in vivo* protection observed upon vaginal challenge with HPV pseudovirus was quite varied. Thus, the vaccination with a VLP displaying HPV16 L2 epitope 65–85 induced strong homologous protection against the appropriate pseudovirus, but little to no cross-protection against heterologous HPV pseudovirus types. This obstacle was surmounted by immunizing with VLPs displaying an L2 peptide representing the amino acid 65–85 consensus sequence of high-risk HPV types. The sera from mice immunized with VLPs displaying the consensus sequence peptide were able to effectively neutralize heterologous high-risk HPV pseudovirus (Tyler et al. 2014b).

Then, Tyler et al. (2014c) developed a brilliant strategy to display the two different peptides of the L2 protein on a single, hybrid VLP in a multivalent, highly immunogenic fashion. The authors designed tricky plasmids that carried the two identical expression cassettes of the PP7 or MS2 coat, each containing a T7 promoter, an open reading frame encoding the single-chain dimer version of either PP7 or MS2 ready to display an epitope at the N-terminus or in the AB loop, and a transcriptional terminator. This enabled the production of hybrid VLPs that displayed the two different epitopes on the same particle in the same highly immunogenic display context. In general, the hybrid VLPs elicited high-titer antibody responses against both targets. The immunization with hybrid particles elicited antibodies that were able to neutralize heterologous HPV types at higher titers than those elicited by particles displaying one epitope alone, indicating that the hybrid VLP approach might be an effective technique to target epitopes that undergo antigenic variation (Tyler et al. 2014c).

Finally, the PP7 and MS2 VLPs displaying the L2 epitope 17–31, as presented in the vaccine Table 22.1, were subjected to thorough preclinical refinements (Tumban et al. 2015). As shown above, the immunization with the vaccine candidate VLPs carrying the L2 epitope of HPV16, or, in other words, 16L2 VLPs, elicited high titer and broadly cross-reactive and cross-neutralizing antibodies against diverse HPV types. The two refinements were introduced for the candidate vaccines, with an eye toward enhancing efficacy and clinical applicability in the developing world. First, the role of antigen dose and boosting on immunogenicity was addressed. The 16L2-MS2 VLPs at doses ranging from 2–25 μg with or without alum demonstrated high immunogenicity in mice at all doses. Alum appeared to have an adjuvant effect at the lowest dose. Although boosting enhanced antibody titers, even a single immunization could elicit strong and long-lasting antibody responses. Second, a method to enhance the vaccine stability was elaborated, generating dry powder vaccine formulations by a spray dry apparatus and protein stabilizers.

After the HPV story, Peabody's team turned to a putative contraceptive vaccine targeting human chorionic gonadotropin (hCG) (Caldeira et al. 2015). The peptides derived from several locations in the hCG sequence (see Table 22.1) were inserted into the AB-loop of the PP7 coat single-chain dimer gene. Some peptides were taken from the disordered C-terminal tail of the hormone, another came from an internal loop, and yet another was an epitope mimic produced by affinity-selection on an hCG-neutralizing antibody target. As a result, the immunization of mice with some VLPs yielded antisera that bound the hormone and neutralized its biological activity (Caldeira et al. 2015).

Recently, the PP7 coat single-chain dimer platform was used to express a peptide of prostatic acid phosphatase, namely PAP$_{114-128}$ (Sun and Sun 2016). As a result, the PAP$_{114-128}$ epitope was displayed on the surface of the chimeric PP7 VLPs and stimulated induction of high titer antibodies. Furthermore, the PP7 VLPs were provided with externally displayed cell-penetrating peptide and encapsidated GFP mRNA (Sun Y et al. 2016). Therefore, both display and packaging capabilities of the PP7 VLPs were involved to construct a targeted delivery vector for both peptides and mRNA. More detail of this construct is given in Table 23.1 of Chapter 23. Furthermore, Yi et al. (2018) exposed the most immunogenic peptide 91–110 of the 16 kD constituent of the *Mycobacterium tuberculosis* on the surface of the PP7 VLPs and evaluated their diagnostic value in tuberculosis.

AP205

Comparing to the above-described RNA phage coats, the AP205 coat demonstrated unique capabilities by the toleration of foreign insertions. An impressive collection of the chimeric AP205 VLPs was generated by the fruitful collaboration of Bachmann's and Grēns' teams (Tissot et al. 2010). The technical description of the gained chimeric AP205 VLPs is given in Table 22.1, since all structures were designed and tested as putative vaccine candidates. Generally, the six different peptides of (i) angiotensin II, (ii) outer membrane protein of *S. typhi*, (iii) CXCR4 receptor, (iv) HIV1 Nef, (v) gonadotropin-releasing hormone, and (vi) influenza A M2-protein were fused to either N- or C-terminus of the AP205 coat protein. The AP205-peptide fusions assembled well into VLPs, and the peptides displayed on the VLP surface were highly immunogenic in mice. Therefore, the AP205 VLPs appeared as a new and highly efficient vaccine system, suitable for complex and long epitopes of up to at least 55 amino acid residues in length. The AP205 VLPs conferred a high immunogenicity to displayed epitopes, as shown by inhibition of endogenous gonadotropin-releasing hormone and protective immunity against influenza infection in mice.

Due to the strong efforts of Regīna Renhofa's group, a set of the preS1-carrying AP205 VLPs was constructed, mainly, with the addressing/targeting/delivery, but not vaccination purposes (Kalniņš et al. 2013). For this reason, the data are included in Table 23.1 of Chapter 23. The whole preS1 region of hepatitis B virus, subtype *ayw*, corresponding to positions 2–108, as well as its fragments 9–108, 20–58, and 10–36, were added to the N- or C-terminus of the AP205 coat via flexible GTAGGGSG or SGTAGGGSGS linkers, respectively. The AP205 VLPs tolerated these extremely long insertions well in the case of the C-terminal fusions. Since the N-terminal fusions led to imperfect particles, unstable or tightly stuck to bacterial membranes, the mosaic technique was applied in this case. These mosaic preS1-carrying VLPs were regarded as a first priority of the study, since they mimicked precisely the natural exposure of the preS1 on the surface of hepatitis B virions or 22-nm HBsAg particles and could be therefore preferential for the correct recognition of hepatitis B virus receptors on the susceptible cells. A drawback of the mosaic approach was the relatively small amount of the recombinant preS1 fusions incorporated into the VLPs, when compared to the homogeneous chimeric VLPs (Kalniņš et al. 2013).

Then, Cielens et al. (2014) generated the mosaic AP205 VLPs carrying the virus-neutralizing domain III (DIII) of the West Nile virus (WNV) glycoprotein E of 111 amino acid residues. The DIII sequence was added to the C-terminus of the AP205 coat protein via *amber* or *opal* termination codons, and mosaic AP205-DIII VLPs were generated by cultivation in *amber*- or *opal*-suppressing *E. coli* strains. After extensive purification to 95% homogeneity, the mosaic AP205-DIII VLPs retained up to 11%–16% monomers carrying the DIII domains. Remarkably, the DIII domains on the VLP surface were fully accessible to anti-DIII antibodies. The immunization of BALB/c mice with the AP205-DIII VLPs resulted in the induction of specific anti-DIII antibodies, the level of which was comparable to that of the anti-AP205 antibodies generated against the VLP carrier. The AP205-DIII-induced anti-DIII response was represented by a significant fraction of IgG2 isotype antibodies, in contrast to parallel immunization with the DIII oligopeptide, which failed to induce IgG2 isotype antibodies. The formulation of the AP-205-DIII VLPs in alum adjuvant stimulated the level of the anti-DIII response but did not alter the fraction of the IgG2 isotype antibodies. These data made it possible to recognize the mosaic AP205-DIII VLPs as a promising prototype of a putative West Nile vaccine (Cielens et al. 2014).

The His-tagged AP205-His$_6$ VLPs were generated recently and compared with the analogous HBc-His$_6$ VLPs (Gillam et al 2018).

GA

The chimeric GA VLP derivatives were constructed by Renhofa's group with the addressing/targeting/delivery purposes. Thus, a 61-amino acid residues-long fragment of Z$_{HER2:342}$ affibody, potentially capable of recognizing HER2 receptor on cancer cells, was added to the C-terminus of the GA coat. The stromal cell-derived factor SDF1 of 41- or 19-amino acid residues, able to recognize CXCR4 receptor on leukocytes, was added at the N-terminus of the first or second single-chain dimer copy. These derivatives formed excellent particles, as discovered by electron microscopy (Arnis Strods and Regīna Renhofa, unpublished). These structures are included into Table 23.1.

Figure 20.3, prepared by Velta Ose, a brilliant expert of electron microscopy in Riga, illustrates the quality of the chimeric VLPs obtained by the gene fusion technology. This picture includes derivatives of the two representative models, namely, the group II phage GA and acinetophage AP205 and compares the RNA phage virions, recombinant VLPs, and chimeric VLPs carrying long foreign insertions. It is remarkable that in these high-quality electron micrographs the surfaces of the long insertions-carrying VLPs differ considerably from the surfaces of unmodified VLPs by the appearance of

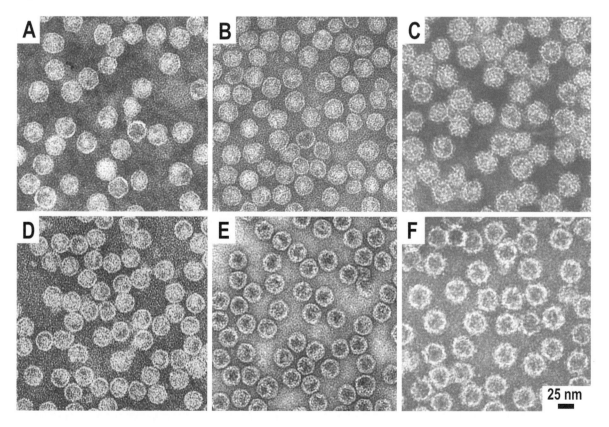

FIGURE 20.3 Electron micrographs of negatively stained AP205 virions (A), recombinant AP205 VLPs (B), chimeric AP205 VLPs carrying 151 aa residues of human interleukin-1β at the C-terminus (C), GA virions (D), recombinant GA VLPs (E), and chimeric GA VLPs carrying a 61-aa-long Z HER2: 342 affibody at the C-terminus (F). VLPs are purified from the appropriate gene-expressing *E. coli* cells. The electron microscopy was performed by Velta Ose. For electron microscopy, the grids with the adsorbed particles were stained with aqueous solutions of 1% uranyl acetate (pH 4.5) or 2% phosphotungstic acid (pH 7.0) and examined with JEM-100C or JEM-1230 electron microscopes (Jeol Ltd., Tokyo, Japan) at 100 kV. Well-ordered knobs are clearly visible on the surfaces of the chimeric VLPs. (Reprinted with permission of S. Karger AG, Basel from Pumpens P et al. *Intervirology*. 2016;59:74–110.)

distinct knobs that are presumably formed by the inserted sequences.

PEPTIDE DISPLAY

As described in the Chimeric phages section of Chapter 19, van Meerten et al. (2001b) found that the phage MS2 was able to accept foreign peptides and therefore suitable for peptide display. More specifically, the viable MS2 virions accommodated some pentapeptide sequences added to the N-terminus of its coat subunit. Following the same peptide display principle, David S. Peabody and Bryce Chackerian, together with their colleagues, proposed the immunogenic display of diverse peptides on the MS2 VLPs, but not on the viable phages (Peabody et al. 2008). This novel approach served not only for the tailored display of the specific peptides but was also adapted to the generation of random peptide libraries, where desired binding activities could be recovered by affinity selection. The first specific peptides were represented by the V3 loop of HIV gp120 and the ECL2 loop of the HIV coreceptor, CCR5, which were inserted into a surface loop of the MS2 coat protein, as described in the summarizing Table 22.1 of Chapter 22. Both insertions disrupted the VLP assembly, apparently by interfering with the correct protein folding, but these defects were suppressed efficiently by the genetical fusion of the two coat polypeptides into a single-chain dimer, as described above. The resulting VLPs displayed the V3 and ECL2 peptides on their surfaces, where the inserted epitopes showed the potent immunogenicity that was the hallmark of the VLP-displayed antigens. The experiments with random-sequence peptide libraries acknowledged the high tolerance of the MS2 single-chain dimer to the insertion of six, eight, and ten amino acid stretches. In fact, the MS2 VLPs supported the display of a wide diversity of peptides in a highly immunogenic format. Moreover, the MS2 VLPs encapsidated the mRNAs that directed their own synthesis, thus establishing the genotype/phenotype linkage necessary for the recovery of the affinity-selected sequences. Thus, the single-chain MS2 VLPs united in a single structural platform the selective power of the phage display with the high immunogenicity of the tailored VLPs (Peabody et al. 2008).

Furthermore, the specialized plasmid vectors were described that facilitated the construction of high-complexity random sequence peptide libraries on the MS2 VLPs and that allowed control of the stringency of affinity selection through the manipulation of display valency (Chackerian et al. 2011). The system was used to identify epitopes for several previously characterized monoclonal antibody targets and showed that the MS2 VLPs thus obtained elicited antibodies in mice whose activities mimicked those of the selected antibodies. The advantage of the MS2 VLP-based display system over the well-known filamentous phage display libraries consisted of the fact that the MS2-displayed peptides were much more immunogenic than the same epitopes displayed on the filamentous phages. The latter were usually not very immunogenic because the epitopes were presented on them at low densities that were not sufficient for efficacious B-cell activation. In contrast, the proposed MS2 VLP peptide display system combined the high immunogenicity of the MS2 VLPs with the powerful affinity selection capabilities of other phage display systems (Chackerian et al. 2011). The affinity selection in the MS2 VLP platform was performed successfully by the development of a mimotope vaccine targeting the *Staphylococcus aureus* quorum-sensing pathway (O'Rourke et al. 2014), described in detail in a special review article (O'Rourke et al. 2015).

The MS2 VLP peptide display was used for the presentation of the mimotope that mimicked a conserved epitope of the *Plasmodium falciparum* blood stage antigen AMA1 (Crossey et al. 2015b).

Then, the plasma-derived IgG from a pool of five patients with advanced ovarian cancer was subjected to iterative biopanning, using a library of the MS2 VLPs displaying diverse short random peptides (Frietze et al. 2016b). After two rounds of biopanning, the selectant population of the MS2 VLPs was analyzed by deep sequencing. One of the top 25 most abundant peptides identified, namely, DISGTNTSRA, had sequence similarity to the cancer antigen 125 (CA125/MUC16), a well-known ovarian cancer–associated antigen. The mice immunized with the chimeric MS2-DISGTNTSRA VLPs generated antibodies that cross-reacted with the purified soluble CA125 from ovarian cancer cells but not with the membrane-bound CA125, indicating that the DISGTNTSRA peptide was a CA125/MUC16 peptide mimic of the soluble CA125. Looking for anti-DISGTNTSRA, anti-CA125, and CA125 markers in preoperative ovarian cancer patient plasma, it was concluded that the deep sequence-coupled biopanning was applicable for the identification of the auto-antibody responses against tumor-associated antigens in the cancer patient plasma (Frietze et al. 2016b).

Furthermore, the pathogen-specific antigen fragment library displayed on the MS2 VLPs, in combination with deep sequence-coupled biopanning, was used for the detailed mapping of the linear epitopes targeted by antibody responses to secondary Dengue virus infection in humans (Frietze et al. 2017). Although there was considerable variation in the responses of individuals, several epitopes within the envelope glycoprotein and nonstructural protein 1 were commonly enriched. This study therefore established a novel approach for the characterization of the pathogen-specific antibody responses in human sera (Frietze et al. 2017).

Furthermore, the four different functional single-chain variable fragments (scFvs) were displayed successfully on the surface of the MS2 VLPs (Lino et al. 2017). Each scFv was validated both for its presence on the surface of the MS2 VLPs and for its ability to bind its cognate antigen. This prospective paper showed clearly how the scFv libraries could be displayed in a similar manner on the VLP surface and could then be biopanned in order to discover the novel scFv targets (Lino et al. 2017).

Detailed methodology of the affinity selection using the MS2 VLP display platform was published recently by Frietze et al (2020).

Recently, the predominant epitope 131–160 of VP1 of foot-and-mouth disease virus was displayed on the top of the MS2 coat (Wang G et al. 2018), as presented in Table 22.1. Despite the impressive length of the epitope, the chimeric protein self-assembled successfully into chimeric nanoparticles of 25–30 nm in diameter, when expressed in *E. coli*. The chimeric nanoparticles stimulated the antibody levels in mice, which were similar to that induced by the commercial synthetic peptide vaccine PepVac, and were significantly more immunogenic than the control peptide. Moreover, the results from the specific interferon-γ responses and lymphocyte proliferation tests indicated that the nanoparticle-immunized mice exhibited significantly enhanced cellular immune response. This made it possible to regard the MS2-derived nanoparticles as a potential alternative vaccine for the future FMDV control (Wang G et al. 2018).

CHEMICAL COUPLING

MARTIN F. BACHMANN'S TEAM

The chemical coupling of foreign oligopeptides to the surface of VLPs was developed as an alternative method to the genetic fusion of epitope-encoding sequences. For the first time, it was applied to Qβ VLPs by the famous highly enthusiastic and productive Martin F. Bachmann's team in Zürich (Jegerlehner et al. 2002a). They used an approach that was initially applied in their laboratory for another broadly used recombinant VLP model, namely, hepatitis B virus core antigen (Bachmann and Kopf 2002; Jegerlehner et al. 2002a,b, 2004; Jennings and Bachmann 2002; Ruedl et al. 2002, 2005; Storni et al. 2002, 2003; Storni and Bachmann 2003). For the first time with the Qβ VLPs, the model peptide D2 of 15 amino acid residues with N-terminal CGG linker (for the source and sequence of the D2 epitope see Table 22.2 in Chapter 22) containing therefore a free cysteine residue at the N-terminus was coupled to a naturally exposed lysine residue on the Qβ VLP surface using the hetero-bifunctional cross-linker maleimidobenzoic acid sulfosuccinimidyl ester (Jegerlehner et al. 2002a). The modified VLPs showed efficient induction of oligopeptide-specific antibodies in mice. Moreover, it was verified experimentally that the density of epitopes displayed on the VLPs was a key parameter for efficient antibody response (Jegerlehner et al. 2002a).

This chemical coupling approach initiated a novel era of therapeutic vaccination and led to the development of a panel of experimental therapeutic vaccines by Bachmann's team. Thus, Bachmann and Dyer (2004) announced the therapeutic vaccination for chronic diseases as a new field for the VLP technology and presented an impressive list of vaccine candidates to be generated. The therapeutic vaccination was proclaimed as a second vaccine revolution for the new epidemics of the twenty-first century (Dyer et al. 2006).

It is noteworthy, however, that the fruitful combination of the chemical coupling with the VLP approach was forecasted earlier by Bachmann and Jennings (2001). In turn, the idea of the therapeutic vaccines was based on the assumption that the VLP carriers could present surface-displayed self-antigens, overcome the natural tolerance of the immune system toward self-proteins, and induce high levels of specific autoantibodies (Bachmann et al. 1993). This classical paper was prepared by young Martin

F. Bachmann under the guidance of Rolf M. Zinkernagel, the 1996 Nobelist in physiology or medicine for the discovery of how the immune system recognizes virus-infected cells.

The novel direction of the therapeutic vaccination was imposed with the introduction of the Qβ VLPs as a major vaccine carrier by Bachmann's team and the ambitious and prospective Cytos Biotechnology Ltd. Starting from 2002, the Qβ VLPs were the focus of attention. In this connection, Figure 20.4 presents the general idea of the Qβ VLP-based chemical coupling approach to the generation of a long list of prospective vaccine candidates proposed by Bachmann at Cytos.

All further general reviews of Bachmann's team were targeted to the VLP application in vaccinology and to the role of the Qβ VLP model as the most successful methodologically and functionally by the chemical coupling approach (Jennings and Bachmann 2007, 2008, 2009; Spohn and Bachmann 2008; Bachmann and Jennings 2010, 2011; Rebeaud and Bachmann 2012b; Bachmann and Whitehead 2013; Zabel et al. 2013; Bachmann and Zabel 2015; Gomes et al. 2017a,b; Mohsen et al. 2017b; El-Turabi and Bachmann 2018). Recently, the chemical coupling on the Qβ VLPs was distinguished as still one of the central routes for the generation of novel vaccines (Engeroff and Bachmann 2019). The full list of the chemically coupled Qβ VLP-based vaccine candidates generated by Bachmann's team and Cytos is available in Table 22.2 of Chapter 22.

BRYCE CHACKERIAN AND DAVID S. PEABODY

At first, the phage Qβ, but not the Qβ VLPs, in parallel with human papillomavirus VLPs, was used to generate a vaccine candidate targeting the amyloid-beta (Aβ) peptide as a promising potential immunotherapy for the Alzheimer's disease patients (Chackerian et al. 2006). Thus, a twelve-amino acid peptide that contained the N-terminal nine amino acids of Aβ provided with additional C-terminal GGC tripeptide (for the sequence of the peptide see Table 22.2 of Chapter 22) was directly linked to the phage Qβ. As described in the previous section, the bifunctional SMPH cross-linker with amine- and sulfhydryl-reactive arms was used, where the amine-reactive arm of SMPH was linked again to surface-exposed lysines on Qβ, and then the Aβ-GGC peptide was linked to Qβ-SMPH by virtue of the exposed sulfhydryl residue on the C-terminal cysteine residue of the peptide. The immunization with the Aβ peptide conjugated to the phage Qβ or to the HPV VLPs elicited anti-Aβ antibody responses at low doses and without the use of adjuvants. Remarkably, both papillomavirus VLP- and Qβ phage-based Aβ vaccines induced weak or negligible T cell responses against Aβ, while T cell responses were largely directed against linked viral epitopes.

After classical work on the human papillomavirus VLPs and creation of the worldwide accepted HPV VLP vaccine, John T. Schiller's team, and personally Bryce Chackerian, exhibited a purely immunological interest in the phage Qβ as a promising display system (Chackerian et al. 2008). By using the phage Qβ that was supplied by David S. Peabody, in parallel with bovine and human papillomavirus VLPs, Chackerian et al. (2008) studied the ability to distinguish between self

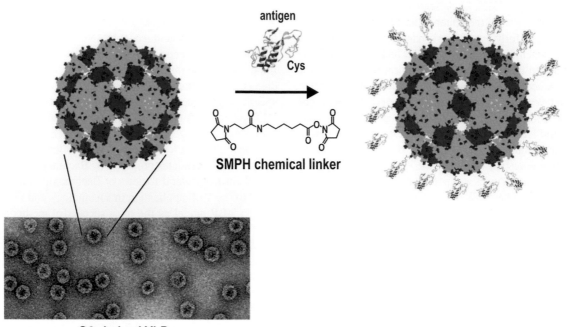

Qβ-derived VLPs

FIGURE 20.4 VLPs and the chemical coupling approach. The figure depicts a structural representation of a Qβ VLP obtained from the crystal structure of the phage Qβ. It also illustrates the modular approach to vaccine production whereby the protein antigen and VLP are separately produced and then covalently linked using a heterobifunctional chemical cross-linker such as succinimidyl-6-[(β-maleimidopropionamido) hexanoate], or SMPH. The resultant conjugate vaccine displays the target antigen in an ordered and highly repetitive fashion. Shown too is an electron-micrograph of 25–30 nm diameter icosahedral Qβ VLPs. (Bachmann MF, Jennings GT. Therapeutic vaccines for chronic diseases: Successes and technical challenges. *Philos Trans R Soc Lond B Biol Sci.* 2011;366:2815–2822. Reproduced by permission of the Royal Society of Chemistry.)

and foreign antigens in a B cell receptor transgenic mouse model that expressed a soluble form of hen egg lysozyme (HEL). The HEL-conjugated VLPs were generated by linking biotinylated HEL to biotinylated VLPs using streptavidin, whereas VLPs were biotinylated by incubation with commercial N-hydroxysuccinimidyl-long-chain-biotin. Immunization with multivalent VLP-arrayed HEL, but not a trivalent form of HEL, induced high-titer antibody responses against HEL in both soluble HEL transgenic mice and double transgenic mice that also expressed a monoclonal HEL-specific B cell receptor. In contrast to some previous data showing the T-independent induction of antibodies to foreign epitopes on VLPs, the ability of HEL-conjugated VLPs to induce anti-HEL antibodies in tolerant mice was dependent on the presence of CD4+ Th cells and could be enhanced by the presence of preexisting cognate T cells. In the *in vitro* studies, the VLP-conjugated HEL was more potent than trivalent HEL in upregulating surface activation markers on purified anergic B cells. Moreover, immunization with VLP-HEL reversed B cell anergy *in vivo* in an adoptive transfer model. Thus, it was shown convincingly that the antigen multivalency and T help cooperated to reverse B cell anergy, a major mechanism of B cell tolerance (Chackerian et al. 2008).

As mentioned above in the Genetic fusions of coats: PP7 section, the chemical coupling of the HPV16 L2 epitopes to the phage Qβ, but not to the Qβ VLPs, was used in parallel to the PP7 VLP-based fusion technology, where the same peptide

sequences were applied in both arms of the study (Tyler et al. 2014b). The technical data of the products are presented in Table 22.2 of Chapter 22. The data of both displaying capacities were generally in good concordance. It was concluded generally that displaying a consensus peptide on the phage Qβ can broaden the neutralizing specificity of a region of L2 that normally only elicited type-specific neutralizing antibodies (Tyler et al. 2014b).

FUNCTIONALIZATION

MS2

An alternate chemical coupling approach was developed on the phage MS2 by David S. Peabody (2003). Thus, the surface cysteines were introduced onto the MS2 coat. It is noteworthy that cysteines were among the most useful functional groups found in proteins, as they could bind a variety of metals and react with a large collection of organic reagents and were therefore obvious targets for the protein modification. The two cysteine residues that were present in the natural MS2 were internally located and therefore relatively unreactive. Therefore, the introduction of the cysteine residues onto the MS2 coat surface was the first step in the technology that could be regarded as a classical example of the rational functionalization of VLPs. After such rational design, the modified MS2 VLPs displayed a reactive thiol on the VLP surface as a result of the T15C substitution in the MS2 coat. It is necessary to note that the amino acid positions are given here as in the original paper, where the initial methionine was not taken into account. According to the genetic map of the MS2 RNA (see Chapter 7), the T15 and other positions mentioned below would appear as T16, and so on. In fact, the five different amino acids of MS2 coat protein were selected initially for the cysteine substitution, based on their accessibility on the surface of the viral capsid. The location of these residue candidates is presented in Figure 20.5.

Three of the five, namely, G13, G14, and T15, were located in the AB-loop, a short β-turn that connected the A and B β-strands of coat protein. The other two, D114 and G115, resided in a loop connecting the two coat protein α-helices (for more structural detail, see Chapter 21). Each of these five amino acids was converted to cysteine by site-directed mutagenesis, and the mutant genes were cloned and expressed. In the four of the five mutant cases, namely, G13C, G14C, D114C, and G115C, no VLPs were detected and the coat proteins were found predominantly in the insoluble fraction of cell lysates. An attempt to suppress the effects of these mutations on the MS2 coat folding/stability by incorporating them into the single-chain MS2 coat dimers was unsuccessful. In contrast to the destabilizing substitutions, the T15C mutant produced significant quantities of soluble coat protein that assembled into particles with the same electrophoretic mobility as the wild-type virus. The accessibility of the new cysteine was further illustrated by its active reaction with thiol-specific chemical reagents (Peabody 2003). This study therefore generated for the first time the RNA phage scaffold, to which a variety of substances could be chemically attached in definite geometric patterns. Moreover, the reaction with fluorescein-5-maleimide imparted green fluorescence to the mutant particle.

Cheng et al. (2005) performed similar site-directed mutagenesis at the codon 15, or 16 by the genetic map, of the MS2 coat gene. The thiolated MS2 Cys-VLPs were then chemically modified with fluorescein-5'-maleimide and the fluorescent nanoparticles were obtained.

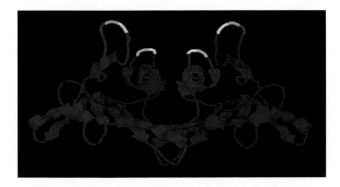

FIGURE 20.5 A view of the MS2 coat protein dimer with its two polypeptide chains shown as blue and red ribbons. The positions of amino acids altered in this study by site-directed mutagenesis are shown as yellow (glycine 13), green (glycine 14), magenta (threonine 15), cyan (glycine 113), and white (aspartic acid 114). (Reprinted from Peabody DS. *J Nanobiotechnology.* 2003;1:5.)

As mentioned briefly in the Dyed phage particles section of Chapter 6, the phage MS2, but not MS2 VLPs, was labeled at that time with (i) fluorescein-5-isothiocyanate (FITC) (ii) fluorescein, (iii) 5-(4,6-dichlorotriazinyl)aminofluorescein (5-DTAF), and (iv) rhodamine B (Gitis et al. 2002a). The FITC and DTAF were used for the conjugation of lysine residues. The rhodamine B and fluorescein labeling were performed by using 1-[3-(Dimethylamino)propyl]-3-ethylcarbodiimide hydrochloride (DEC) together with the dye. This procedure resulted in permanent attachment without covalent conjugation of the dye, probably due to DEC-assisted caging of the fluorescent labels in the hydrophobic environment of the dye. The fluorescent phage MS2 was introduced into ecological studies as a new tracer for the investigation of pathogen transport in porous media (Gitis et al. 2002b).

In parallel, Wu et al. (2002) set in motion another MS2 functionalization approach. They conjugated the MS2 coat to transferrin. To do this, the MS2 VLPs were reacted with a thiol-introducing reagent SATA, while transferrin was then reacted with a maleimide modification reagent SMCC, and the SATA-derivative of the coat protein was activated by hydroxylamine treatment. Finally, the two proteins were mixed and made it possible to form covalent conjugates at room temperature for 30 min. As a result, the cross-linked transferrin was found in the MS2 VLPs (Wu et al. 2002).

Furthermore, the MS2 VLPs were converted into an efficient nanoparticle magnetic resonance imaging (MRI) contrast agent (Anderson et al. 2006). Thus, more than 500 gadolinium chelate groups were conjugated onto a viral capsid. The MS2(Gd-DTPA-ITC)m contrast agent was synthesized through the conjugation of pre-metalated Gd-DTPA-ITC to lysine residues. Figure 20.6 presents the location of the potentially reactive lysine residues in the MS2 particle. This illustration is also helpful in the case of all other functionalization approaches that have used reactivity of the MS2 coat lysine residues, for example, that performed by Kang and Yoon (2004). The latter, as well as the work of Anderson et al. (2006) are described in more technical detail in Table 23.2 of Chapter 23, together with other RNA phage-based imaging applications.

The further strong progress in the functionalization field is connected with the long-term successful activities of Matthew B. Francis' team. They were the first that started to use the surface functionalization methodology not only to the outer VLP surface but also to the interiors of the MS2 VLPs by modification of tyrosine residues via a recently developed hetero-Diels-Alder bioconjugation reaction (Hooker et al. 2004). Using diazonium-coupling reactions, the tyrosine 85 was modified with high efficiency and selectivity. The virtually complete coupling was achieved in 15 min through exposure to five equivalents of a diazonium salt (Hooker et al. 2004). This pioneering approach was applied successfully for the dual-surface modification of the tobacco mosaic virus (Schlick et al. 2005).

A new protein modification reaction was based on a palladium-catalyzed allylic alkylation of tyrosine residues using π-allylpalladium complexes (Tilley and Francis

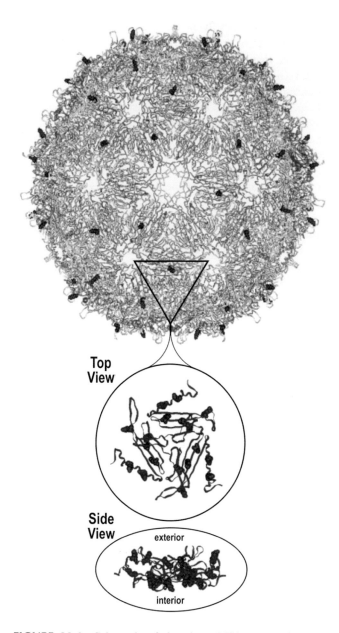

Top View

Side View

exterior

interior

FIGURE 20.6 Schematic of the phage MS2 nanoparticle. Top panel, the 27.4-nm-diameter capsid surface. The insets depict a trimer of coat proteins showing 18 lysine residues (6 per monomer). A subset of the 1080 lysine residues that are anticipated to be most reactive are highlighted in red. Structures are based on the 2MS2 pdb coordinates, with the full capsid structure rendered using VIPERdb (Shepherd et al. 2006) and iMol (http://www.pirx.com/iMol/). (Reprinted with permission from Anderson EA et al. Viral nanoparticles donning a paramagnetic coat: Conjugation of MRI contrast agents to the MS2 capsid. *Nano Lett*, 1160–1164. Copyright 2006, American Chemical Society.)

2006). This technique employed electrophilic π-allyl intermediates derived from allylic acetate and carbamate precursors and was used to modify the phage MS2 in aqueous solution at room temperature. To facilitate the detection of modified proteins, a fluorescent allyl acetate was synthesized and coupled to the phage MS2, while the tyrosine selectivity of the reaction was confirmed through trypsin digest analysis.

As described later in detail in a thorough review of Algar et al. (2011), Francis' team has specifically modified tyrosine residues native to the interior surfaces of the MS2 phage particles with nitrophenyl diazonium salts (Hooker et al. 2007, 2008; Kovacs et al. 2007; Datta et al. 2008). The nitrophenyl diazonium salts were linked to other functional groups that enabled the attachment of reporters such as organic dyes, MRI contrast agents, and radiolabels for positron emission tomography (PET), as listed in Table 23.2. The diazonium reaction also allowed selective bioconjugation to the interior of the MS2 particles, while allowing lysine residues native to the exterior surface to be modified in orthogonal reactions. The latter included functionalization of the MS2 particles with succinimidyl esters of PEG and targeting ligands (Kovacs et al. 2007). Although a nitrophenyl diazonium salt derivative of a reporter might be prepared directly (e.g., fluorescein), the greater versatility was achieved through the use of nitrophenyl diazonium salts that could introduce a bioorthogonal functional group at tyrosine residues.

For example, Francis' team labeled MS2 tyrosine residues with aldehyde-functionalized nitrophenyl diazonium salts for subsequent oxime ligations (Kovacs et al. 2007) and developed a four-step hetero-Diels-Alder reaction that allowed further chemical modification.

McFarland and Francis (2005) were the first to begin to explore the possibilities of "green chemistry," resulting in new tools for the modification of the phage MS2 particles. This resulted in the application of iridium-catalyzed transfer hydrogenation by the reductive alkylation of lysine residues of the phage MS2, which contained altogether six lysine residues, at room temperature and neutral pH.

The next functionalization approach was achieved by Francis' team by oxidative coupling to aniline-containing side chains (Carrico et al. 2008; Stephanopoulos et al. 2009; Tong et al. 2009). Thus, the one-to-one oxidative coupling of aniline with *N,N*-diethyl-*N'*-acylphenylene diamine using sodium periodate was achieved for the attachment of peptides to the surface of the MS2 VLPs (Carrico et al. 2008). The aniline functionality was installed into the MS2 capsid through the incorporation of an unnatural amino acid residue, namely, *para*-amino-L-phenylalanine (*p*AF) by *amber* suppression methods. To optimize the surface accessible *p*AF incorporation, the five stop codon variants were constructed: Q6TAG, D11TAG, T15TAG, D17TAG, and T19TAG, and the appropriate suppression technique was applied, where the observable MS2 coat protein expression was found only in the presence of *p*AF. As the MS2 mutant with *p*AF at position T19 produced the highest yield and was found to be the robust oxidative coupling scaffold, it was used for all subsequent experiments. Three peptides known to target specific tissues were attached by the oxidative coupling via *N,N*-diethyl-*N'*-acylphenylene diamine, which was placed at the N-terminus of each peptide during synthesis on the solid phase. The oxidative coupling to aniline was also used to decorate the exterior surface of the MS2 particles with zinc porphyrins for photocatalysis (Stephanopoulos et al. 2009) and a DNA aptamer for cellular delivery (Tong et al. 2009). Furthermore, a highly efficient protein bioconjugation method involving addition of anilines to *o*-aminophenols in the presence of sodium periodate was used (Behrens et al. 2011). The impressive progress in the dual-surface modification of MS2 capsids and generation of novel nanomaterials was summarized at that time by Witus and Francis (2011). The detailed protocol on how to conjugate antibodies to the MS2T19*p*AF VLPs was published recently by ElSohly et al. (2018). The practical results of the pioneering studies of Francis' team are presented in the summarizing Table 23.1 of Chapter 23, together with other achievements by the addressing/targeting/delivery applications of the RNA phage VLPs.

Furthermore, Francis' team mutagenically introduced cysteine at position 87 by the N87C exchange and targeted this interior residue with maleimide reagents to introduce drug molecules (Wu et al. 2009; Stephanopoulos et al. 2010b) or imaging agents (Meldrum et al. 2010) to the interior surface of the capsids with high efficiency approaching 100%. On the basis of this success, this site was also chosen for modification with Gd(III)-hydroxypyridonate complexes through the use of linking groups that possessed a minimal number of rotatable bonds (Garimella et al. 2011). Udit et al. (2010a) reviewed at that time the striking success of the MS2 VLP-based chemically tailored multivalent virus platforms that found their broad application from drug delivery to catalysis.

Seim et al. (2011) described an unprecedented oxidative coupling strategy that selectively modified tyrosine and tryptophan residues using cerium(IV) ammonium nitrate as a one-electron oxidant. This oxidative method was found to couple electron-rich aniline derivatives directly to tyrosine and tryptophan residues. The new tyrosine and tryptophan residues, namely, T15W, T19W, T15Y, and T19Y, were introduced to the external surface of the MS2 VLPs. Thus, this method was adapted for the chemoselective modification of the MS2 VLPs and was optimized to proceed with high yields at neutral pH and low substrate concentrations. The strategy has been used to modify both native and introduced residues on proteins with polyethylene glycol and small peptides. The reaction was also used in conjunction with cysteine alkylation to doubly modify the MS2 capsids with both targeting and imaging functionalities (Seim et al. 2011).

Then, the modification strategy using *o*-aminophenols or *o*-catechols that were oxidized to active coupling species *in situ* using potassium ferricyanide was tested, among other approaches, with the aniline side chains of the *p*AF residues (Obermeyer et al. 2014b) introduced into the MS2 coat, as described above.

The modification of aniline-containing MS2 VLPs was performed with ortho-azidophenols under photochemical conditions (El Muslemany et al. 2014). This technology was developed as a facile approach to the biomolecular photopatterning in order to yield useful platforms for drug screening, synthetic biology applications, diagnostics, and immobilization of live cells. With this aim, an easily accessible and operationally simple photoinitiated reaction was proposed, which involved the photolysis of 2-azidophenols to generate iminoquinone intermediates that coupled rapidly to aniline groups. As a specific application, the reaction was adapted for the photolithographic patterning of azidophenol DNA on aniline glass substrates (El Muslemany et al. 2014).

Meanwhile, ElSohly and Francis (2015) published a thorough account of the development of oxidative coupling strategies for site-selective protein modification, while ElSohly et al. (2015) presented how the synthetically modified MS2 VLPs could be used as versatile carriers by antibody-based cell targeting. Finally, ElSohly et al. (2017) identified *o*-methoxyphenols as molecules that undergo efficient oxidative couplings with anilines in the presence of periodate as oxidant. This approach was used to link epidermal growth factor to the MS2 VLPs bearing aniline moieties, therefore affording nanoscale delivery vectors that could target a variety of cancer cell types. The most recent account on the principles and methodology of the dual-surface modification of the MS2 VLPs for the nanomaterial fabrication and delivery applications was published by Aanei and Francis (2018).

In 2009, the phage MS2 was used for the first time as a biotemplate for semiconductor nanoparticle synthesis (Cohen et al. 2009). Thus, the synthesis of cadmium sulfide nanoparticles was demonstrated in aqueous solution with the phage MS2 as a result of the bionanofabrication approach that offered unique opportunities for nanomaterials synthesis, where choice of biotemplate could influence the size, shape, and function of the produced material. The phage MS2 was chosen as a potential template system for the bionanofabrication of nanoscale materials due to its capability to be (i) genetically modified to incorporate functional groups and sequences and (ii) devoid of the inner RNA by degradation or replacement with other oligonucleotide sequences. Remarkably, the cadmium sulfide nanoparticles possessed predicated fluorescent properties when formed in the presence of the phage MS2, and remained stable in aqueous solutions. Surprisingly, the MS2-templated CdS nanoparticles were stable at room temperature for several weeks.

A novel functionalization approach was achieved in the cell-free protein synthesis platform by the production of MS2 and Qβ VLPs with surface-exposed methionine analogs: azidohomoalanine and homopropargylglycine containing azide and alkyne side chains (Patel and Swartz 2011). Such VLPs could be used for one-step, direct conjugation schemes to display multiple ligands of interest. Using this technology, the proteins including an antibody fragment and granulocyte-macrophage colony-stimulating factor, as well as nucleic acids and poly(ethylene glycol) chains, were displayed on the VLP surface using Cu(I) catalyzed click chemistry (Patel and Swartz 2011). The practical outcome of this study is presented in Table 23.1 of Chapter 23.

In 2011, the phage MS2 was chosen to meet the growing requirements for the quality of bio-artificial tracers for the inline measurement of virus retention in membrane processes (for more detail, see Chapter 6). With this aim, the surface of the phage MS2 was modified by the grafting of enzymes, namely, horseradish peroxidase (HRP) (Soussan et al. 2011a,b). The HRP catalyzed a highly specific and rapid reaction, permitting its detection, and could be operated in drinking water without inhibition by the molecules present in solution. This tracer was thus built to enable direct detection of its induced enzymatic activity, notably by an amperometric method. The generation of the MS2 tracer was started with covalent binding of activated biotin to the lysine residues of the phage. This was not the case for bigger and less hydrophilic activated molecules, such as the activated HRP enzymes that were tested at the very beginning for the direct grafting method. For this reason, the neutravidin–HRP conjugates were chosen as enzymatic probes for the second labeling on biotin molecules, notably because of the highly specific, strong interaction between neutravidin and biotin. Moreover, a spacer present before the terminal reactive group of the chosen activated biotin allowed the steric hindrance between enzymatic probes and capsid proteins to be reduced (Soussan et al. 2011a). The novel MS2-based tracer was found as representative as possible of native pathogenic viruses (Soussan et al. 2011b).

Recently, Francis' team proposed ortho-methoxyphenols as convenient oxidative bioconjugation reagents and applied them to conjugate epidermal growth factor to the MS2 VLPs (ElSohly et al. 2017).

The conjugated MS2 VLPs were used to study parvovirus B19, which has an extraordinary narrow tissue tropism, showing only productive infection in erythroid precursor cells in the bone marrow. The VP1u, a unique region of viral protein 1 that mediated the B19 uptake into cells of the erythroid lineage, was chemically coupled to the MS2 capsid and tested for the internalization capacity on permissive cells (Leisi et al. 2016). The MS2-VP1u bioconjugate mimicked the specific internalization of the native B19 V into erythroid precursor cells, which further coincided with the restricted infection profile.

The MS2 functionalization was employed by the development of a nanoparticle system for blood–brain barrier (BBB) targeting and delivery (Curley and Cady 2018; Curley et al. 2018). To do this, the thiolated angiopep-2-FAM (AP2), a synthetic peptide that has been shown to have the greatest transcytosis efficiency across the BBB over any other targeting ligand, was conjugated to the MS2 capsid surface. The injection of the MS2-AP2 particles into the tail vein of rats was successful in allowing the particles to cross into the brain.

Recently, the MS2 VLPs were used as a mass spectrometry signal multiplier (Yuan et al. 2019). Thus, the dibenzylcyclooctynepoly(ethylene glycol)-folate (DBCO-PEG-FA), and DOTA-Eu complex tag-modified (FA-PEG)$_{69}$-MS2-(DOTA-Eu)$_{965}$ targeted the folate receptor on the human nasopharyngeal carcinoma cells and made it possible to quantify at least five cancer cells by ^{153}Eu-species unspecific isotope dilution inductively coupled plasma mass spectrometry.

Qβ

As just noted above in the Genetic fusion of coats: Qβ section, M.G. Finn's team struck the keynote of the Qβ functionalization. The practical outcome of these studies is included systematically in Tables 23.1 and 23.2 of Chapter 23.

At the beginning, following the above-mentioned idea of Anderson et al. (2006) who worked with the phage MS2, Prasuhn et al. (2007) turned to the generation of viral MRI contrast agents, but on the basis of the Qβ VLPs, in parallel with another classical viral model, namely cowpea mosaic virus (CPMV). Thus, the labeling of the Qβ VLPs with Gd^{3+} was reported in the two novel ways, by (i) conjugation of a monoalkylated 1,4,7,10-tetraazacyclododecanetetraacetic acid (DOTA) analog using the Cu^I-mediated azide–alkyne cycloaddition (CuAAC) reaction, and by a natural high affinity interaction of aqueous Gd^{3+} ions with the polynucleotide encapsulated by a virus particle. Remarkably, the lysine knockout mutant K16M of the Qβ VLPs was employed in this study. Such Qβ VLPs were derivatized at the available amine functional groups by N-hydroxysuccinimide ester to give the polyvalent azides and then the attachment of the Gd(DOTA) complex to the VLP azides was accomplished with the aid of the Cu-bathophenanthroline complex (Prasuhn et al. 2007).

The Qβ VLPs decorated with gadolinium complexes using the CuAAC reaction were used to determine the plasma clearance and tissue distribution of the particles in mice (Prasuhn et al. 2008). In this study, two point Qβ mutant forms were generated: K16M and T93M. The K16M chimera removed the most highly accessible lysine on the external surface, leaving two others, K2 and K12, plus the N-terminus, which was somewhat hindered, as active amine nucleophiles. The reactivity of lysines on the interior surface was strongly passivated by association with packaged polynucleotide. The K16M and T93M substitutions also allowed genetic incorporation of unnatural amino acids in place of methionine on the external and internal surfaces, respectively. Thus, the unnatural amino acid azidohomoalanine (AHA) was incorporated into the Qβ coat to provide structure T93M Qβ-AHA. The latter was then subjected to the CuAAC reaction. Figure 20.7 presents a cross-sectional view of the Qβ capsid showing the potentially reactive amino acid residues on the outer and inner surface of the particle.

The incorporation of the unnatural amino acid residue AHA, as well as of the homopropargyl glycine, into the Qβ VLPs was described for the first time by Strable (2008). Remarkably, this pioneering labeling with azide- or alkyne-containing unnatural amino acids by expression in a methionine auxotrophic strain of E. coli, which was followed by the Cu^I-catalyzed cycloaddition, was performed in parallel with both Qβ VLPs and hepatitis B virus core particles, i.e., in the two most popular VLP carriers that have been applied together in numerous classical VLP studies. Figure 20.8, from Strable et al. (2008), is summoned to compare three-dimensional structures of the two famous VLP vectors.

In contrast to the hepatitis B virus cores, the Qβ VLPs suffered no instability and were not decomposed by the formation of triazole linkages. Therefore, the marriage of these well-known techniques of the sense-codon reassignment and bioorthogonal chemical coupling provided the capability to construct polyvalent particles displaying a wide variety of functional groups with near-perfect control of spacing (Strable et al. 2008).

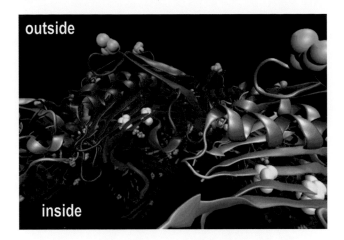

FIGURE 20.7 Cross-sectional view of the structure of the phage Qβ showing Lys16 on the outer surface in green and Thr93 on the inner surface in yellow. Capsid subunits are shown in ribbon representation in gray, blue, and red. Coordinates for the capsid structure were generated using the VIPER database (http://viperdb.scripps.edu/) (Reddy et al. 2001). (Reprinted with permission from Prasuhn DE Jr et al. Plasma clearance of bacteriophage Qβ particles as a function of surface charge. *J Am Chem Soc*, 1328–1334. Copyright 2008, American Chemical Society.)

Recently, Bhushan et al. (2018) proposed *S*-allylhomocysteine as an unnatural genetically encodable methionine analog for the Qβ VLPs. This analog is processed by translational cellular machinery and is also a privileged olefin cross-metathesis reaction tag in proteins. It was used for efficient Met-codon reassignment in a Met-auxotrophic strain of *E. coli*.

The next success of Finn's team was the display of polyvalent glycan ligands for the cell-surface receptors on the Qβ VLPs, again in parallel with the CPMV model (Kaltgrad et al. 2008). As the authors noted, the cell-surface receptors that recognize glycans as ligands mediated a vast range of events in cellular biology through low-affinity, multivalent binding interactions. The development of ligand-based probes of these receptors must therefore address both the challenges of glycan synthesis and the need for the multivalent platforms. Kaltgrad et al. (2008) have immobilized glycan reaction precursors on the Qβ VLPs and allowed a "building out" from such surface to make polyvalent and precisely placed oligosaccharides. In such a way, the exterior surface of the Qβ VLPs was labeled, taking advantage of the compatibility of the particles with glycosyltransferase enzymes and chemoenzymatic reaction conditions. Thus, the glycans arrayed on the exterior of the Qβ VLPs were used as substrates for glycosyltransferase reactions to build di- and trisaccharides from the VLP surface. The resulting particles exhibited tight and specific associations with cognate receptors on beads and cells, in one example defeating *in cis* cell-surface interactions in a manner characteristic of polyvalent binding. This methodology therefore provided a convenient and powerful way to prepare complex carbohydrate ligands for the clustered receptors.

Furthermore, the polyvalent sugar-coated Qβ VLPs were used by the elaboration of the back-scattering interferometry method for the quantitative determination of glycan-lectin interactions, one of the most important classes of interactions in biochemistry (Kussrow et al. 2009).

Astronomo et al. (2010) tried to breach the *glycan shield* defense of HIV, since it was known that the broadly neutralizing antibody 2G12 recognized a conserved cluster of high-mannose glycans on the surface envelope spike of HIV and could appear that the oligomannosides would serve as a vaccine target. In an attempt to recreate features of the glycan shield semisynthetically, the corresponding oligomannosides were coupled to the surface lysines on the Qβ VLPs and CPMV. In contrast to the CPMV, the Qβ VLP glycoconjugates presented oligomannose clusters that bound the antibody 2G12 with high affinity. However, the antibodies against these 2G12 epitopes were not detected in immunized rabbits. Rather, the alternative oligomannose epitopes on the conjugates were immunodominant and elicited high titers of anti-mannose antibodies that did not cross-react with the HIV envelope. Nevertheless, this pioneering study presented the highly important design considerations for the putative carbohydrate-based vaccine component against HIV and possibly other infectious agents.

Meanwhile, Hong et al. (2008) improved the CuAAC reaction by the use of an electrochemical potential to maintain catalysts in the active Cu^I oxidation state in the presence of air. This simple procedure efficiently achieved excellent yields of the CuAAC products without the use of potentially damaging chemical reducing agents. The electrochemically protected bioconjugations in air were performed with the Qβ VLPs that were derivatized with azide moieties at surface lysine residues, and complete derivatization of more than 600 reactive sites per particle was demonstrated (Hong et al. 2008). The optimized CuAAC reaction was reviewed under the headline "How to Click with Biomolecules" as the most widely recognized example of the *click chemistry* applications (Hong et al. 2009). At the same, Strable and Finn (2009) reviewed the principles of the chemical modification of viruses and virus-like particles. The principles and approach of the CuAAC reaction in connection with the Qβ VLPs, among other substrates, were reviewed further in detail (Lallana et al. 2011, 2012; Such et al. 2012; Levine et al. 2013; Kim H et al. 2015).

The Qβ VLPs were used to display the human iron-transfer protein transferrin, a high-affinity ligand for the receptors upregulated in a variety of cancers (Banerjee et al. 2010). The selective oxidation of the sialic acid residues on the glycan chains of transferrin was followed by the introduction of a terminal alkyne functionality through an oxime linkage. The attachment of the protein to the azide-functionalized Qβ VLPs in an orientation allowing access to the receptor binding site was accomplished by the CuAAC reaction. The transferrin conjugation to the Qβ VLPs allowed specific recognition by transferrin receptors and cellular internalization through clathrin-mediated endocytosis. By testing the Qβ VLPs bearing different numbers of transferrin molecules, it was demonstrated that the cellular uptake was proportional to the ligand density, but the internalization was inhibited by equivalent concentrations of free transferrin. These results paved the way for cell targeting with transferrin.

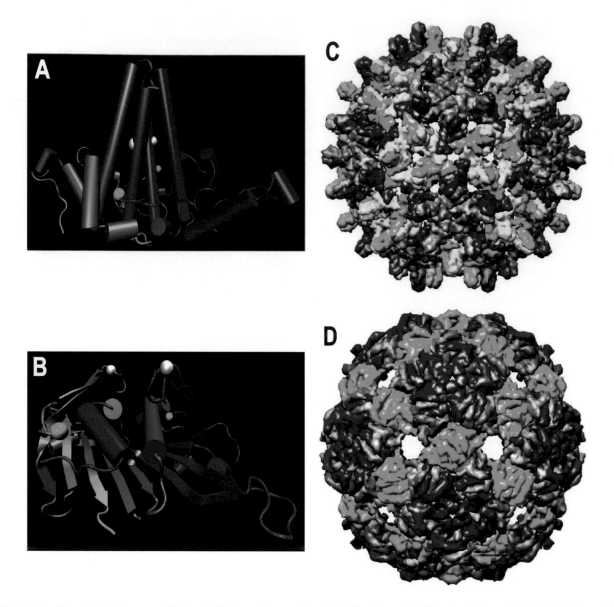

FIGURE 20.8   Hepatitis B virus core (HBc) and Qβ virus-like particles. (A) HBc dimer; (B) HBc virus-like particle; (C) Qβ dimer; (D) Qβ virus-like particle (Golmohammadi et al. 1996; Wynne et al. 1999; Shepherd et al. 2006). Representations (A) and (C) look obliquely down onto the outside capsid surface, showing the 4-helix bundle for HBV and intertwined R-helix and loop segments over adjacent β-sheets for Qβ. The spheres mark the locations of methionine residues: green = N-termini, white = Met66 in HBc and K16 M in Qβ, purple = Qβ T93M. (Reprinted with permission from Strable E et al. Unnatural amino acid incorporation into virus-like particles. *Bioconjug Chem*, 866–875. Copyright 2008, American Chemical Society.)

The CuAAC reaction was used by the covering of the Qβ VLP surface by polymer chains, to which multiple connections were made, thereby crosslinking protein cage subunits (Manzenrieder et al. 2011). Such polymers could extend *in vivo* circulation lifetime, diminishing nonspecific adsorption or passivating the immune response. The poly(2-oxazoline)s were chosen for the role of such polymers because of their advantageous properties of versatile controlled syntheses by means of living cationic polymerization, aqueous-phase solubility, and chemical stability that made them attractive for a variety of biomedical and materials applications (Manzenrieder et al. 2011). The polymerization of oligo(ethylene glycol)-methacrylate and

its azido-functionalized analog was performed directly from the outer surface of the Qβ VLPs by atom transfer radical polymerization (ATRP) (Pokorski et al. 2011a). The introduction of chemically reactive monomers during polymerization therefore provided a robust platform for post-synthetic modification via the CuAAC reaction. The ATRP methodology was developed further by Hovlid et al. (2014) and reviewed in an excellent minireview of Wallat et al. (2014), who specifically highlighted the concept of the *grafting-from* proteins as the new and efficient chemistry to polymerize directly from protein substrates in aqueous media, with the Qβ VLPs as the most successful story. The authors demonstrated efficient post-polymerization modification via the CuAAC reaction to introduce fluorophores, Gd-based contrast agents, and doxorubicin, a common anticancer drug. Then, Isarov et al. (2016) generated the *graft-to* protein/polymer conjugates using polynorbornene block copolymers, which were prepared via ring-opening metathesis polymerization (ROMP). The ROMP technique afforded low-dispersity polymers and allowed for the strict control over polymer molecular mass and architecture. Such polymers consisted of a large block of PEGylated monoester norbornene and were capped with a short block of norbornene dicarboxylic anhydride. This cap served as a reactive linker that facilitated attachment of the polymer to lysine residues under mildly alkaline conditions. Thus, the multivalent polynorbornene-modified Qβ VLPs were constructed. The conjugated nanoparticles showed no cytotoxicity to NIH 3T3 murine fibroblast cells. This study markedly expanded the toolbox for the protein bioconjugates (Isarov et al. 2016).

Lee PW et al. (2017a) described in detail the conformation of several biocompatible polymers conjugated to the Qβ VLPs. It is widely known that the covalent conjugation of water-soluble polymers to proteins was critical for evading immune surveillance in the field of biopharmaceuticals. Thus, the Qβ polymer conjugates were synthesized using the *grafting-to* approach with end-functionalized amino-reactive polymers. The Qβ-polymer conjugates were prepared with linear mPEG, poly(oligo(ethylene glycol) methyl ether acrylate) (POEGMEA), and anhydride-end-capped poly(norbornene-(oligo(ethylene glycol) ester)) (PNB), each with a target mass of 10 kDa, corresponding to degrees of polymerization of ~220, ~30, and ~18, respectively (Lee PW et al. 2017a). Crooke et al. (2018) described the assembly and characterization of a series of polymer conjugates of the Qβ VLPs, using (poly(oligo(ethylene glycol)methacrylate), poly(methacrylamido glucopyranose), and PEG, and investigated their ability to shield the protein from antibody recognition as a function of polymer loading density, chain length, architecture, and conjugation site.

It is highly important that Nicole F. Steinmetz' team recently raised a question of the bio-inspired shielding strategies for nanoparticle drug delivery applications (Gulati et al. 2018). Since nanoparticle surface coatings were often required to reduce immune clearance and thereby increase circulation times allowing the carriers to reach their target site, the origin of the coating remained highly important, whether of synthetic or biologic origin. To this end, polyethylene glycol has long been used, with several PEGylated products reaching clinical use. Unfortunately, the growing use of PEG in consumer products has led to an increasing prevalence of PEG-specific antibodies in the human population, which in turn has fueled the search for alternative coating strategies. The exhaustive review of Gulati et al. (2018) highlighted alternative bio-inspired nanoparticle shielding strategies that may be more beneficial moving forward than PEG and other synthetic polymer coatings.

Lee PW et al. (2017c) utilized melt-encapsulation to fabricate the Qβ VLP-laden polymeric materials that effectively delivered intact particles and generated carrier-specific antibodies *in vivo*. The dispersion of particles within the poly(lactic-*co*-glycolic acid) (PLGA) matrix was studied to develop processing windows for the scale-up and the creation of more complex materials. In fact, melt-processing was recommended as an emerging manufacturing method to create polymeric materials laden with proteins as an alternative to typical solvent-based production methods. The melt-processing was advantageous because it is continuous, solvent-free, and 100% of the therapeutic protein appeared encapsulated. The effect of mild and aggressive ball milling of the solid Qβ VLPs manufactured via hot melt extrusion (HME) in the PLGA matrix was studied by Lee PW et al. (2017b).

As an alternate approach to the click chemistry, Finn's team employed genetic engineering and chemical conjugation to construct the Qβ VLPs with cationic amino acid motifs on the surface (Udit et al. 2009). Thus, the application of the point mutation strategies made it possible to generate Qβ VLPs bearing the K16M, T18R, N10R, or D14R mutations. The mutants therefore provided a spectrum of particles differing in surface charge by as much as +540 units (K16M *versus* D14R). Whereas larger poly-Arg insertions, for example, C-terminal Arg$_8$, did not yield intact virions, it was possible to append chemically synthesized oligo-Arg peptides to the stable Qβ wild-type and K16M platforms. These particles were applied as inhibitors of the anticoagulant action of heparin, which was a common anticlotting agent subject to clinical overdose. Moreover, the engineered cationic Qβ VLPs retained their ability to inhibit heparin at high concentrations and showed no anticlotting activity of the kind that limited the utility of antiheparin polycationic agents that were traditionally in clinical use (Udit et al. 2009). Then, Gale et al. (2011) investigated the clinical heparin reversal function of one of these active Qβ VLPs, namely, the T18R mutant, in which the solvent-exposed residue Thr18 of the coat protein was mutated to Arg, thereby giving the particle substantially greater positive surface charge. The K16M mutant served as a negative control since its positive charge was diminished relative to the wild-type structure and it did not bind or reverse heparin *in vitro*. The plasma specimens were collected from patients who were treated with either therapeutic heparin or with high doses of heparin while undergoing cardiac catheterization procedures. The reversal of heparin anticoagulation in these patient plasma samples was investigated in comparison to protamine. As a result, the T18R particles showed significantly more consistent heparin reversal activity, with much greater potency in terms of particle concentration *versus* protamine concentration (Gale et al. 2011). This pioneering work was based generally on the three successful approaches to modify the Qβ VLPs by (i) chemically appending poly-Arg peptides, (ii) point mutations to Arg on the virus capsid, and (iii) incorporation of heparin-binding peptides displayed externally on the virus surface. Each approach generated particles with good heparin antagonist activity with none of the toxic side effects of protamine that remained nevertheless the only currently FDA-approved drug for clinical use as the heparin antagonist (Udit 2013). Finally, Cheong et al. (2017) presented further strong arguments in favor of the potential clinical use of the Qβ T18R VLPs as effective, nontoxic heparin antagonists.

To further develop the heparin antagonist properties of the chimeric Qβ VLPs, Choi JM et al. (2018) backed off from the Qβ point mutants and the CuAAC reaction and turned to the good old two-plasmid coexpression methodology that was first described by Tatyana Kozlovska's group (Vasiljeva et al. 1998) and further developed by Brown et al. (2009), as narrated above in the Genetic fusions of coats: Qβ section. The two-plasmid system was utilized to generate the Qβ VLPs that contained both the wild-type coat protein and a second coat protein with either a C- or N-terminal cationic peptide extension of 4–28 amino acids in length. The incorporation of the modified coat proteins varied from 8%–31%. The particles with the highest incorporation rate and best anti-heparin activity therefore displayed the C-terminal peptide ARK_2A_2KA, which corresponded to the Cardin-Weintraub consensus sequence for binding to glycosaminoglycans.

In parallel with heparin antagonists, Mead et al. (2014) generated the sulfated phage Qβ VLPs that elicited heparin-like anticoagulant activity. The sulfate groups were appended to the Qβ VLPs by the synthesis of single- and triple-sulfated ligands that also contained azide groups. Following conversion of the VLP surface lysine groups to alkynes, the sulfated ligands were attached to the VLP via the CuAAC reaction. The clotting activity analyzed by activated partial thrombin time assay showed that the sulfated particles were able to perturb coagulation, with the VLPs displaying the triple-sulfated ligand approximately as effective as heparin on a per-mole basis, and this activity could be partially reversed by protamine (Mead et al. 2014). The sulfated Qβ VLPs were examined thoroughly for the binding kinetics to cationic peptides (Groner et al. 2015).

The chemistry of the VLP functionalization, including both MS2 and Qβ VLPs, was reviewed by Schoonen and van Hest (2014). The methodology used to generate the hybrid Qβ particles was described in detail by Brown (2018).

A new direction in the Qβ VLP applications was opened by Cigler et al. (2010) by the generation of novel nanomaterials. In this study, the organic Qβ VLPs were combined with inorganic gold nanoparticles. The authors argued that both nanoparticles were radially symmetric building blocks of well-controlled size, although their properties were very different. The modification of these core particles with the same non-interacting DNA sequences allowed particle assembly to be induced by Watson-Crick hybridization using the appropriate linker sequences, creating a binary system. The distance of the DNA *sticky ends* from the center of the nanoparticle was relatively well defined, mainly as a consequence of the rigidity of the double helical DNA and the dense loading of DNA strands on the surface. The effective radii of the DNA-linked nanoparticles, a crucial parameter in determining the crystalline structure, included the DNA corona and could therefore be controlled by changing the DNA length. Therefore, a non-compact lattice was created by the DNA-programmed crystallization using the Qβ VLPs and gold nanoparticles, engineered to have similar effective radii (Cigler et al. 2010).

Pokorski et al. (2011b) ensured the cell targeting with the Qβ VLPs displaying epidermal growth factor (EGF). The latter was fused C-terminally to the Qβ coat. The co-assembly of the wild-type Qβ and EGF-modified subunits resulted in the structurally homogeneous nanoparticles displaying between 5–12 copies of EGF on their exterior surface. The particles were found to be amenable to bioconjugation by standard methods as well as the high-fidelity CuAAC reaction. Remarkably, such chemical derivatization did not impair the ability of the particles to specifically interact with the EGF receptor. In addition, the particle-displayed EGF remained biologically active when promoting autophosphorylation of the EGF receptor and apoptosis of A431 cells. Therefore, the mosaic Qβ-EGF VLPs were proposed as useful vehicles for the targeted delivery of imaging and/or therapeutic agents.

The colorful Qβ VLPs were prepared by encapsulation of multiple copies of fluorescent proteins (Rhee et al. 2011). The encapsidated proteins were nearly identical in photochemical properties to monomeric analogs, were more stable toward thermal degradation, and were protected from proteolytic cleavage. It is important that the residues on the outer capsid surface could be chemically derivatized by the CuAAC reaction without affecting the fluorescence properties of the packaged proteins. A high-affinity carbohydrate-based ligand of the CD22 receptor was thereby attached, and specific cell labeling by the particles was successfully detected by flow cytometry and confocal laser microscopy (Rhee et al. 2011). Moreover, the simultaneous modification of the Qβ VLPs with the metalloporphyrin derivative for photodynamic therapy and a glycan ligand for specific targeting of cells bearing the CD22 receptor was proposed for a nonsurgical method of cancer treatment based on photosensitizer molecules that produced toxic concentrations of singlet oxygen and other reactive oxygen species upon illumination (Rhee et al. 2012).

The CuAAC click chemistry reaction allowed conjugation of C_{60}, or *buckyball*, possessing high electron affinity and charge transport capabilities that have made derivatives of C_{60} and carbon nanotubes particularly attractive for the next-generation photovoltaic and electrical energy storage devices. First, the grafting to the Qβ VLPs made it possible to surmount the insolubility of C_{60} in water (Steinmetz et al. 2009). Second, it led to the serious investigation of photodynamic activity of the Qβ VLPs conjugated with C_{60} (Wen et al. 2012). Moreover, the cell uptake and cell killing using white light therapy and a prostate cancer cell line was demonstrated (Wen et al. 2012).

Furthermore, the CuAAC reaction also allowed conjugation of tumor-associated carbohydrate antigens (TACAs) to the Qβ VLPs (Yin et al. 2013). In fact, the TACAs were overexpressed on a variety of cancer cell surfaces and therefore presented tempting targets for the anticancer vaccine development. Since such carbohydrates were often poorly immunogenic, it was intriguing to display a very weak TACA, the monomeric Tn antigen, or GalNAc-α-O-Ser/Thr, on the Qβ VLPs. The local density of antigen rather than the total amount of antigen administered was found to be crucial for the induction of high Tn-specific IgG titers. The ability to display antigens in an organized and high-density manner was therefore a key advantage of the Qβ VLPs as vaccine carriers.

Glycan microarray analysis showed that the antibodies generated were highly selective toward Tn antigens. Furthermore, the Qβ-based VLPs elicited much higher levels of IgG antibodies than other types of VLPs, and the produced IgG antibodies reacted strongly with the native Tn antigens on human leukemia cells (Yin et al. 2013).

Finn's team took a next imperative step to the carbohydrate-based antitumor vaccine (Yin et al. 2016). They focused on the ganglioside GM2 TACA, which was overexpressed in a wide range of tumor cells. The GM2 was synthesized chemically and conjugated to the Qβ VLPs. Although the CuAAC reaction efficiently introduced 237 copies of GM2 per Qβ, this construct failed to induce significant amounts of anti-GM2 antibodies compared to the Qβ control. In contrast, the GM2 immobilized on the Qβ VLPs through a thiourea linker elicited high titers of IgG antibodies, which recognized GM2-positive tumor cells and effectively induced cell lysis through complement-mediated cytotoxicity. Thus, the Qβ VLPs remained a suitable platform to boost antibody responses toward GM2, a representative member of an important class of TACA: the ganglioside (Yin et al. 2016).

An excellent global review on the VLPs as definite platform technologies for modern vaccines, which included the newest data on numerous Qβ and HBc VLP-based vaccine candidates on the general background of other VLP carriers, was published at that time by Lee et al. (2016).

After putative vaccines using carbohydrate epitopes, the Qβ VLPs assisted in the elaboration of an excellent diagnostic tool of Chagas disease (Brito et al. 2016). Thus, the α-Gal antigen [Galα(1,3)Galβ(1,4)GlcNAcα] that was an immunodominant epitope displayed by infective trypomastigote forms of *Trypanosoma cruzi*, the causative agent of Chagas disease, was displayed on the Qβ VLPs by the CuAAC reaction. The Qβ VLPs displaying a high density of α-Gal were found to be a superior reagent for the ELISA-based serological diagnosis of Chagas disease and the assessment of treatment effectiveness. A panel of sera from patients chronically infected with *T. cruzi*, both untreated and benznidazole-treated, was compared with sera from patients with leishmaniasis and from healthy donors. The nanoparticle-α-Gal construct allowed for perfect discrimination between Chagas patients and the others, avoiding false negative and false positive results obtained with current state-of-the-art reagents (Brito et al. 2016). Next, such Qβ VLPs displaying approximately 540 α-Gal molecules were used to assess the protective effect of anti-α-Gal responses in falciparum malaria (Coelho et al 2019).

The CuAAC reaction was also used recently to construct a putative anticancer vaccine candidate (Yin Z et al. 2018). Thus, the mucin-1, one of the top ranked tumor associated antigens, was displayed on the Qβ VLPs. The mucin-1 peptides and glycopeptides were covalently conjugated to the Qβ VLPs. The immunization of mice with these constructs led to highly potent antibody responses with IgG titers over one million, which were among the highest anti-mucin-1 IgG titers reported to date. The high IgG antibody levels persisted for more than 6 months. The constructs also elicited mucin-1 specific cytotoxic T cells, which selectively killed mucin-1 positive tumor cells. The unique abilities of the Qβ-mucin-1 conjugates to powerfully induce both antibody and cytotoxic T cell immunity targeting tumor cells therefore bode well for the future translation of the constructs as anticancer vaccines (Yin Z et al. 2018).

The CuAAC reaction was used for the generation of a specific class of glycoproteins, namely, fluoroglycoproteins, as powerful substrates in NMR, MRI, and PET techniques (Boutureira et al. 2010). This approach exploited the incorporation of L-homopropargylglycine as non-natural amino acid and triazole formation chemistry. The 1,2,3-triazoles, formed from reaction of a fluoro-labeled glycosyl azide with an alkynic-amino acid partner, might be considered hydrolytically stable bioisosteres of the amide bond found in natural Asn-linked glycoproteins. Thus, the achieved introduction of the fluoro-sugars represented a variant named FGlyco-CCHC that expanded the previous CuAAC reaction. The FGlyco-CCHC approach was applied to the Qβ VLPs, among other models. The reaction of L-homopropargylglycine-modified Qβ VLPs with a fluoroglucosyl azide afforded F-glycoprotein in >95% of positions and a particle contained therefore 180 site-selectively positioned fluoroglycans (Boutureira et al. 2010). This pioneering work was reviewed and commented later by Boutureira and Bernardes (2015). The growing impact of the possible anti-glycan vaccines was accessed recently (Krumm and Doores 2018; Polonskaya et al. 2019).

A highly specific approach for the further development of VLP vectors was achieved by the asymmetrization of the Qβ VLPs after the introduction of a single copy of the maturation protein A2, which allowed the production of VLPs with a single unique modification (Smith et al. 2012). This approach was driven by the estimations that, by bioconjugation described above, the theoretical maximum of the Qβ VLPs having a single molecule attached could be 37%, according to the Poisson probability distribution. The remaining VLP population would contain zero (37%) or multiple (26%) molecules attached to each VLP and would not be easily separated from the VLPs with one molecule. The *E. coli*-based cell-free protein synthesis system was employed for the coexpression of the cytotoxic A2 protein and the coat of the phage Qβ to form a nearly monodispersed population of novel VLPs. The cell-free protein synthesis allowed for direct access and optimization of both protein synthesis and VLP self-assembly. The A2 was shown to be incorporated at high efficiency, approaching a theoretical maximum of one A2 per VLP. This work demonstrated for the first time the *de novo* production of a novel VLP, which contained a unique site that could have the potential for future nanometric engineering applications (Smith et al. 2012).

A novel pioneering approach to the bioconjugation, brilliant fluorescent functionalization across quaternary structure of the Qβ VLPs, was elaborated by Jeremiah J. Gassensmith's team (Chen Z et al. 2016a, 2017). Thus, the dibromomaleimide moiety was employed to break and re-bridge the exposed and structurally important disulfides in good yields. Not only was the stability of the quaternary structure retained after the reaction, the newly functionalized Qβ particles became brightly fluorescent and could be tracked *in vitro* using a commercially available filter set. Therefore, this highly efficient bioconjugation reaction not only introduced a new functional handle *between* the disulfides of VLPs, without compromising their thermal stability, but it was used to create a fluorescent probe (Chen Z et al. 2017). Figure 20.9 presents the location of cysteines along the pores, while Figure 20.10 list the dibromomaleimide compounds used for the bioconjugation.

At that time, the role of the chemical functionalization of viral architectures in order to create new technologies was reviewed by Chen Z et al. (2016b). Chen Z et al. (2018) further presented detailed methodology of how to prepare the fluorescent PEGylated Qβ VLPs by dibromomaleimide-disulfide chemistry and obtain a bright yellow fluorophore.

Moreover, Gassensmith's team elaborated a fundamentally novel composite nanomaterial (Benjamin et al. 2018). Thus, the surface of the Qβ VLPs was equipped with natural ligands for the synthesis of small gold nanoparticles. By exploiting disulfides in the protein secondary structure and the geometry formed from the capsid quaternary structure, regularly arrayed patterns of ~6 nm gold nanoparticles across the surface of the Qβ VLP were produced. It was further shown that the entrapped genetic material held upward of 500 molecules of the anticancer drug doxorubicin without leaking and without interfering with the synthesis of the gold nanoparticles. This direct nucleation of nanoparticles on the capsid allowed for exceptional conduction of photothermal energy upon

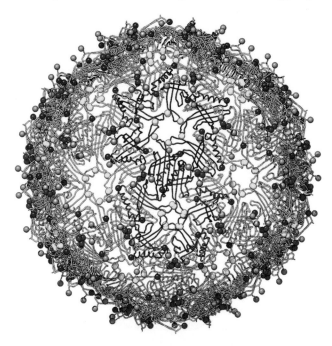

FIGURE 20.9 Qβ VLP is composed of 180 identical coat proteins, each with four solvent exposed primary amines, N-termini = orange, K2 = blue, K13 = red, K16 = green. These monomers are connected by disulfides (yellow) to form several interlinked subunits of either five (the pentameric subunit colored purple) or six (the hexameric subunit colored brown) coat proteins. (Chen Z et al.: Dual functionalized bacteriophage Qβ as a photocaged drug carrier. *Small*. 2016a. 12. 4563–4571. Copyright Wiley-VCH Verlag GmbH & Co. KGaA. Reproduced with permission.)

nanosecond laser irradiation. As a proof of principle, it was demonstrated that this energy was capable of rapidly releasing the drug from the capsid without heating the bulk solution, allowing for highly targeted cell killing *in vitro* (Benjamin et al. 2018).

Finally, Lee et al. (2018) regulated the uptake of the Qβ VLPs in macrophage and cancer cells via a pH switch. The cellular uptake by macrophages and cancer cells of the Qβ-derived nanoparticles was inhibited by conjugating negatively charged terminal hexanoic acid moieties onto the Qβ VLP surface. When hydrazone linkers were installed between the surface of the Qβ VLPs and the terminal hexanoic acid moieties, this resulted in a pH-responsive conjugate that, in acidic conditions, released the terminal hexanoic acid moiety and allowed for the uptake of the Qβ nanoparticle. The installation of the pH switch did not change the structure-function properties of the hexanoic acid moiety and the uptake of the Qβ VLP conjugates by macrophages (Lee et al. 2018).

The functionalization of the RNA phage capsids was reviewed recently in terms of a growing novel discipline that could be classified as *physical, chemical, and synthetic virology*, with a special aim to create reprogrammed virus particles as controllable nanodevices (Chen MY et al. 2019).

PLUG-AND-DISPLAY: AP205

In 2010, Bachmann's team involved the AP205 VLPs in the chemical coupling procedures. Thus, following the lysine-cysteine oligopeptide coupling methodology that was described above in the Chemical coupling: Martin F. Bachmann's team and Qβ section, an experimental WNV vaccine was constructed on the AP205 VLP platform (Spohn et al. 2010). Since its first appearance in the United States in 1999, WNV has spread in the Western hemisphere and continued to represent an important public health concern. In the absence of effective treatment, there was a strong medical need for the development of a safe and efficient vaccine. Live attenuated WNV vaccines have shown promise in preclinical and clinical studies but might carry inherent risks due to the possibility of reversion to more virulent forms. For this reason, the first conjugate vaccine candidate against WNV was generated. This vaccine consisted of the recombinantly expressed domain III of the WNV E glycoprotein chemically cross-linked to the AP205 VLPs. In contrast to the isolated DIII protein, which required three administrations to induce detectable antibody titers in mice, the high titers of the DIII-specific antibodies were induced after a single injection of the conjugate AP205 VLP vaccine. These antibodies were able to neutralize the virus *in vitro* and provided partial protection from a challenge with a lethal dose of WNV. The three injections of the vaccine induced high titers of virus-neutralizing antibodies, and completely protected mice from WNV infection (Spohn et al. 2010).

At the same time, the AP205 and Qβ VLP conjugates with the model D2 peptide from outer membrane protein *Salmonella typhi* were used by the solid immunological investigation performed by Bachmann's team on a process termed carrier induced epitopic suppression (CIES) (Jegerlehner et al. 2010).

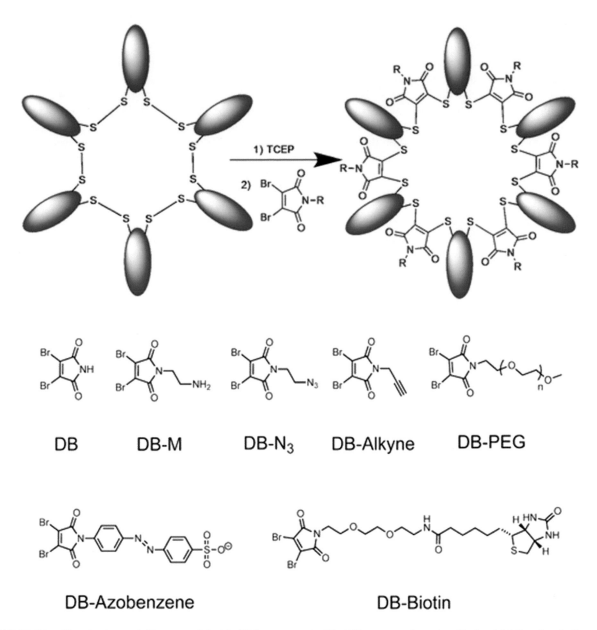

DB DB-M DB-N$_3$ DB-Alkyne DB-PEG

DB-Azobenzene DB-Biotin

FIGURE 20.10 Qβ reduction and dibromomaleimide (DB) compounds. The DB compound reacts with the thiol function in the reduced Qβ to form the corresponding Qβ-maleimide conjugates. (Reprinted with permission from Chen Z et al. Fluorescent functionalization across quaternary structure in a virus-like particle. *Bioconjug Chem*, 2277–2283. Copyright 2017, American Chemical Society.)

The preexisting immunity against the VLP carrier proteins has been reported to inhibit the immune response against antigens conjugated to the same carrier by the CIES process. Hence understanding the phenomenon of CIES was of major importance for the further development of conjugate vaccines. Thus, the impact of a preexisting VLP-specific immune response on the development of antibody responses against a conjugated model peptide was estimated after primary, secondary, and tertiary immunization. Although the VLP-specific immune responses led to reduced peptide-specific antibody titers, the CIES against peptide-VLP conjugates could be overcome by high coupling densities, repeated injections, and/or higher doses of conjugate vaccine. Then, the fluorescent Alexa 647-conjugated AP205 VLPs and Alexa 488-conjugated Qβ

VLPs were employed in an excellent immunological study showing that intranasal administration of VLPs resulted in the splenic B-cell responses with strong local germinal-center formation (Bessa et al. 2012). Surprisingly, the VLPs were not transported from the lung to the spleen in a free form but by B cells. The interaction between VLPs and B cells was initiated in the lung and occurred independently of complement receptor 2 and Fcγ receptors but was dependent upon B-cell receptors. Thus, the B cells passing through the lungs bound VLPs via their B-cell receptors and delivered them to local B cells within the splenic B-cell follicle. Therefore, this process was fundamentally different from the delivery of blood- or lymph-borne particulate antigens, which were transported into B cell follicles by binding to complement receptors on

B cells (Bessa et al. 2012). These and other chemical coupling investigations on the AP205 platform are included in the summarizing vaccine Table 22.2 of Chapter 22.

Recently, a novel superglue, or plug-and-display, system was established for the modular RNA phage VLP functionalization via further decoration of VLPs with peptides of interest (Brune et al. 2016; Thrane et al. 2016). The plug-and-display decoration of VLPs was based on the so-called bacterial superglue approach, namely on the ability of a peptide termed SpyTag and a protein termed SpyCatcher to form spontaneous covalent isopeptide bonds between lysine and aspartic acid residues under physiological conditions (Zakeri et al. 2012). The SpyTag and SpyCatcher were split units of the *Streptococcus pyogenes* fibronectin-binding protein FbaB and formed a highly stable amide bond by an irreversible reaction that occurred within minutes (Veggiani et al. 2014). The SpyTag or SpyCatcher sequences were genetically fused to the N- and/or C-terminus of the AP205 coat. After mixing the modified AP205 VLPs with the correspondingly linked peptides, the quantitative covalent coupling of the peptides to the VLPs was observed (Brune et al. 2016; Thrane et al. 2016). This approach resulted in the promising malaria vaccine candidate comprised of the full-length circumsporozoite protein (CSP) antigen that was presented on the AP205 VLPs using the SpyTag/SpyCatcher system (Janitzek et al. 2016). In this study, the full-length CSP was genetically fused at the C-terminus to SpyCatcher. The CSP-SpyCatcher antigen was then covalently attached, via the SpyTag/SpyCatcher interaction, to the AP205 VLPs, which displayed one SpyTag per VLP subunit. Moreover, the malaria transmission-blocking activity was improved by introduction of the *P. falciparum* protein 48/45 into the SpyTag/SpyCatcher-mediated display on the AP205 VLPs (Singh et al. 2017).

Furthermore, three platforms were compared directly for the displaying Pfs25 target (Leneghan et al. 2017). These platforms comprised the three important routes to antigen-scaffold linkage: the plug-and-display SpyTag/SpyCatcher, chemical crosslinking, and genetic fusion. Thus, the SpyCatcher-AP205 allowed covalent conjugation of antigen to the VLPs by the formation of a spontaneous isopeptide bond between SpyCatcher and its partner SpyTag, as described above. It was highly important that the fusion of SpyCatcher to the VLPs and SpyTag to the antigen allowed heterologous expression of antigen and VLP in the system most suited to each and avoided the typical downfalls of antigen-fusion VLPs such as poor expression levels, poor solubility, and misfolding. The Qβ VLPs were used for the chemical crosslinking. For the genetic fusion, the chosen multimerizing coiled-coil protein IMX313 was a multimerization domain based on a hybrid version of the chicken complement inhibitor C4b-binding protein. It was shown before that the IMX313 when fused to various antigens increased antibody titer and avidity, as well as improved cellular immunogenicity. It was intriguing that the chemically-conjugated Qβ VLPs elicited the highest quantity of antibodies, while the SpyCatcher-AP205-VLPs elicited the highest quality anti-Pfs25 antibodies for transmission blocking upon mosquito feeding (Leneghan et al. 2017). The full account of the numerous vaccine candidates generated by this novel technology on the AP205 VLPs is presented in Table 22.2.

Developing this fruitful novel technique, the dual plug-and-display system was elaborated (Brune et al. 2017). Thus, a dually addressable synthetic nanoparticle was generated by engineering the IMX313 and two orthogonally reactive split proteins. The SpyCatcher protein formed an isopeptide bond with the SpyTag peptide through spontaneous amidation, while the SnoopCatcher formed an isopeptide bond with the SnoopTag peptide through transamidation. The SpyCatcher-IMX-SnoopCatcher provided a modular platform, whereby SpyTag-antigen and SnoopTag-antigen could be multimerized on opposite faces of the particle simply upon mixing.

The *stick*, *click*, and *glue* routes were compared in the recent exciting review published by Karl D. Brune and Mark Howarth (2018). The authors concentrated on the potential vaccine candidates that used non-covalent assembly methods, or *stick*, unnatural amino acids for bioorthogonal chemistry, or *click*, and spontaneous isopeptide bond formation by SpyTag/SpyCatcher, or *glue*. The impressive applications of these methods were outlined in detail and critically considered, with particular insight on the novel Tag/Catcher plug-and-display decoration (Brune and Howarth 2018). The catching a SPY, or using the SpyCatcher-SpyTag approach, was reviewed recently by Hatlem et al. (2019).

STABILITY

The novel discipline was named virus engineering, by analogy to gene and protein engineering, and included, mainly, the functionalization and stabilization of VLPs for different nanotechnological applications. The MS2 and Qβ VLPs, together with cowpea chlorotic mottle virus (CCMV), were among the main subjects of the practical recommendations that were reviewed exhaustively by Mauricio G. Mateu (Mateu 2011, 2013, 2016). As he rightly noted (Mateu 2011), for many modern nanotechnological applications, improvement of the physical stability of the viruses, as well their VLPs, might be critical to adequately meet the demanding physicochemical conditions that they could encounter during production, storage, and medical or industrial use.

As it was known before, the natural RNA phage VLPs differed markedly by their stability. Thus, for example, the Qβ VLPs were more stable than the MS2 VLPs due to inter-subunit crosslinking by disulfide bonds (Ashcroft et al. 2005), although the phage Qβ was less stable than the phage MS2 in natural environment, as described in Chapter 6. The mutational mapping revealed residues that were responsible for the special inter- and intramolecular contacts and therefore for the greater thermal stability of the MS2 capsids (Stonehouse and Stockley 1993).

Efforts to improve the stability of the natural RNA phage VLPs were started with the development of a methodology that allowed the screening of bacteria for the synthesis of the mutant MS2 coats with altered assembly properties (Peabody and Al-Bitar 2001) and the selection of the MS2 coat D11N variant that formed virions more stable than the wild-type

MS2 coat (Lima et al. 2004). The introduction of interdisulfide bonds into the fivefold axis of symmetry to crosslink the MS2 VLPs improved their thermal stability to the level of that seen in the Qβ VLPs, which possessed natural intersubunit disulfide bonds (Ashcroft et al. 2005). In contrast, the crosslinking at the threefold axis of symmetry resulted in variant coats that were unable to self-assemble (Ashcroft et al. 2005). The development of an *E. coli*-based cell-free protein synthesis system opened a direct avenue for studying the role of disulfide bond formation in the stability of the mutant MS2 VLPs in comparison to the Qβ and HBc VLPs (Bundy and Swartz 2011). Through the construction of a set of the Qβ coat mutants, it was found that the disulfide linkages were the most important stabilizing elements in the VLPs and that interdimer interactions were less important than intradimer interactions for the Qβ VLP assembly (Fiedler et al. 2012).

The elucidation of the thermal stability of the foreign epitope-carrying VLPs, such as the chimeric MS2 VLPs formed by single-chain dimers, in comparison to the natural disulfide crosslinked PP7 VLPs (Caldeira and Peabody 2007, 2011) was of great importance for the further development of the stability principles.

The genetic fusions of two copies of the MS2 (Peabody et al. 2008), PP7 (Caldeira et al. 2010), and GA or Qβ (Induls Cielēns and Arnis Strods, unpublished) coats resulted in the self-assembly competent single-chain dimer that not only increased thermodynamic stability but also considerably improved tolerance to the foreign insertions within the AB-loop, as presented above. The resultant correctly assembled VLPs encapsidated mostly their "own" mRNAs encoded by the coat genes and expressing the coat proteins.

The stability of the RNA phage VLPs against the attack by antibodies was achieved by the PEGylation of the VLPs (Kovacs et al. 2007), as described above in the Functionalization: MS2 section. In fact, the ability of the polymer coating to block the access of polyclonal antibodies to the capsid surface was probed using a sandwich ELISA, which indicated a 90% reduction in binding.

By the above-described generation of the magnetic resonance contrast agents (Hooker et al. 2007), the MS2 capsids sequestering the gadolinium-chelates on the interior surface, when attached through tyrosine residues, not only provided higher relaxivities than their exterior functionalized counterparts, which relied on lysine modification, but also exhibited improved water solubility and capsid stability. Therefore, the attachment of a functional cargo to the interior surface was envisioned to minimize its influences on biodistribution, yielding significant advantages for the tissue targeting by additional groups attached to the capsid exterior (Hooker et al. 2007).

Bundy and Swartz (2011) demonstrated the ability to control the disulfide bond formation in VLPs by directly controlling the redox potential during or after production and assembly of VLPs in the open cell-free protein synthesis environment. Remarkably, the cell-free system produced VLPs at yields comparable to or greater than the traditional *in vivo* technologies. The optimal conditions for the disulfide bond formation were found to be VLP-dependent, and a cooperative effect in the formation of such bonds was observed. This study used the three most popular VLP platforms: the Qβ VLPs, the VLPs of an MS2 mutant which formed disulfide bonds (Ashcroft et al. 2005), and HBc VLPs. The Qβ VLPs formed disulfide bonds at high efficiency after exposure to a reductase-free extracellular aerobic environment. Similar to the MS2 and HBc VLPs, the endoplasmic reticulum redox conditions with a glutathione redox buffer were not sufficient to stimulate complete formation of the disulfide bonds in the Qβ VLPs. By incubating with the extremely oxidizing but bio-incompatible hydrogen peroxide, close to 100% of the Qβ VLP disulfide bonds were formed (Bundy and Swartz 2011).

When Finn's team engineered a set of mutations in the Qβ VLPs, it was generally concluded that the disulfide linkages were the most important stabilizing elements and that acidic conditions significantly enhanced the resistance of VLPs to thermal degradation (Fiedler et al. 2012). The interdimer interactions were found to be less important for the VLP assembly than the intradimer interactions. The special sections on the general factors influencing stability of the MS2 VLPs including both physicochemical and structural observations were included in the extensive review on the MS2-based delivery platform (Fu Yu and Li 2016).

Recently, Francis' team performed the first study on the MS2 VLPs, where the chemical modification conditions were used as a selection factor for the protein fitness and stability (Brauer et al. 2019; Maza et al. 2019).

PACKAGING

The efficient encapsidation of the desired materials, together with the precise assessment of the VLP innage, are playing the central role by the development of the two major biomedicinal goals of the VLP platforms consisting, in a few words, of the immunological display and targeted drug delivery. The materials that may need delivery in diagnostics, vaccines, and therapeutic modalities are very different and may include RNA and DNA fragments, proteins, and peptides, as well as very different low-molecular-mass molecules and inorganic nanomaterials.

Thomas Hohn and Barbara Hohn were probably the first to present systematically the role of nonspecific polyribonucleotides by the RNA phage particle assembly (Hohn and Hohn 1970), as described in the Reconstitution section of Chapter 9. Other studies at that time also denoted the possibility of unspecific RNA packaging. For example, the defective MS2 particles, which appeared after treatment with fluorouracil, packaged one to three molecules of 6S stable RNA and a similar amount of tRNA derived from *E. coli* together with a broken viral RNA molecule (Shiba and Saigo 1982).

David S. Peabody was the first to start the practical application of the translation complex I properties for putative RNA delivery. Thus, the ability of the MS2 translational operator to stimulate encapsidation of the desired RNAs *in vivo* demonstrated for the first time the real potential of the RNA phages as potential gene delivery vectors (Pickett and Peabody 1993).

In parallel, the Qβ coat residues responsible for RNA recognition were mapped (Lim et al. 1996; Spingola and Peabody 1997). This was very helpful for the knowledge-based development of the packaging and gene transfer technologies based on the Qβ phage model.

Stockley's team was the first to announce the cell-specific delivery by the MS2 VLPs encapsidated with the desired drug. They performed covalent coupling of deglycosylated ricin A chain to the MS2 translational operator and encapsidated the toxin into the MS2 VLPs (Wu et al. 1995). The resultant particles were directed to specific cells by receptor-mediated endocytosis of complexes formed with anti-MS2 coat antibodies or by further covalent modification of the capsids by addition of human transferrin molecules. It was the first example showing that the encapsidation and targeting could be an efficient way to deliver toxins and other molecules, such as antisense reagents, in a cell-specific fashion. The targeted delivery strategy was further developed by using soluble recombinant human CD4 molecules or anti-human immunodeficiency virus antibodies as ligands (Wu et al. 1996). The fruitful combination of the drug delivery options with the epitope presentation was described in detail by Brown et al. (2002).

It is noteworthy that an attempt was made to perform transduction of foreign RNA by noninfectious MS2 virions (Kniazhev et al. 2002). For such model extra genes, aminoglycoside-3′-phosphotransferase, or kanamycin resistance, gene and a sequence directed against hepatitis C virus were employed.

The packaging exercises in the immunological applications of the VLPs were stimulated strongly as a result of the epochal discovery that explained the role of the CpG motifs in DNA by the B-cell activation (Krieg et al. 1995). It was still hypothesized that the immunostimulatory CpG oligonucleotides might trigger changes that could make the cellular microenvironment unfavorable for HIV replication (Schlaepfer et al. 2004).

The famous CpG oligonucleotides were packaged successfully into the model VLPs by Bachmann's team, first, into the HBc VLPs (Storni et al. 2002, 2003; Schwarz et al. 2003; Storni and Bachmann 2003). The CpG packaging into the chimeric Qβ VLPs provided with the famous LCMV gp33 epitope followed (Bachmann et al. 2004b; Storni et al. 2004). In parallel, an extensive elucidation of the LCMV gp33 epitope occurred without any VLP carriers (Bachmann et al. 2004a). Storni et al. (2004) showed that the packaging of CpGs into the HBc or Qβ VLPs was not only simple but could reduce the two major problems of the CpG usage, namely, their unfavorable pharmacokinetics and systemic side effects, including splenomegaly. Thus, it was demonstrated that the CpGs packaged into VLPs were resistant to the DNase I digestion. In contrast to free CpGs, the packaged CpGs prevented splenomegaly in mice without affecting their immunostimulatory capacity. In fact, vaccination with the CpG-loaded VLPs was able to induce high frequencies of peptide-specific CD8⁺ T cells. It protected from infection with recombinant vaccinia viruses, and eradicated established solid fibrosarcoma tumors. It was concluded therefore for the first time that packaging

CpGs into the Qβ VLPs improved both their immunogenicity and pharmacodynamics (Storni et al. 2004). This study paved the long way for the development of the CpG-containing VLP vaccine candidates, especially in the allergy vaccines elaborated by Bachmann's team (Senti et al. 2009; Klimek et al. 2014), as will be presented in detail in Chapter 22.

Francis' team elaborated a simple and highly efficient alkali treatment method to remove the encapsidated RNA (Hooker et al. 2004) that was employed later by other VLP technologies, such as HBc VLPs, for the specific packaging (Strods et al. 2015).

Legendre and Fastrez (2005) produced the MS2 VLPs in *S. cerevisiae* with an aim to package functional heterologous mRNAs containing the MS2 packaging sequence and to adapt the MS2 VLPs therefore to the gene delivery. For instance, linking the MS2 translational operator to the human growth hormone mRNA enabled the packaging of this particular mRNA into the MS2 VLPs (Legendre and Fastrez 2005). A similar approach was employed by the expression of the phage GA VLPs and packaging of the functional heterologous mRNA *in vivo* during expression in *S. cerevisiae* by Renhofa's group (Rūmnieks et al. 2008; Strods et al. 2012; Ārgule 2013; Strods 2015; Ārgule et al. 2017).

By the above-mentioned prospective studies on the immunogenic display of foreign epitopes on the MS2 VLPs, Peabody and Chackerian's team found that the covalent single-chain MS2 coat dimers not only tolerated the six-, eight-, and ten-amino acid insertions, but also encapsidated the mRNAs that directed their synthesis, thus establishing the genotype/phenotype linkage necessary for the recovery of affinity-selected sequences (Peabody et al. 2008).

The embedding of other than RNA cargos into the RNA phage VLPs was stimulated by the rapid development of the click chemistry approaches, as described in detail in the preceding sections. Thus, fullerene buckyballs met the Qβ VLPs (Steinmetz et al. 2009). The porphyrins were attached to the MS2 VLPs (Stephanopoulos et al. 2009). After this covalent bioconjugation of porphyrins, Cohen et al. (2011, 2012) employed the purified capsids of the phage MS2 as a delivery vessel of porphyrins that were bound by "nucleotide-driven packaging." In other words, the MS2 capsids were capable of efficient packaging of cationic porphyrins by the RNA present within the MS2 capsid. The removal of the interior RNA significantly diminished the porphyrin interaction with the MS2 capsid. This interaction could likely occur through binding of the positively charged porphyrin with the negatively charged RNA phosphate backbone, as opposed to an intercalating base-stacking mechanism. The porphyrin, when packaged inside MS2 capsid, was capable of singlet oxygen photogeneration in solution when illuminated with light at 632 nm. These results demonstrated for the first time the generation of reactive oxygen species by self-packaged porphyrins contained within a nanoscale vessel (Cohen et al. 2012). Moreover, Cohen and Bergkvist (2013) were the first to perform targeted photodynamic therapy *in vitro* by the porphyrin-loaded MS2 capsids that were targeted to the cancer cells by a DNA aptamer exposed on the capsid surface.

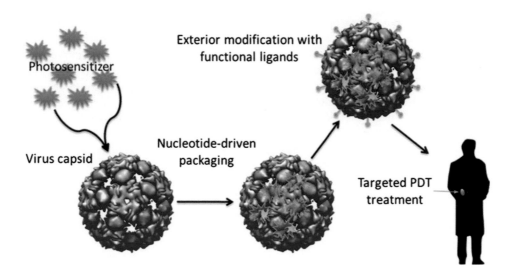

FIGURE 20.11 Schematic illustration of the capsid loading and targeting concept. Small functional molecules (in this case TMAP, or meso-tetra-(4-N,N,N,-trimethylanilinium)-porphine) infiltrate the capsid and bind to RNA/DNA. The exterior surface can subsequently be modified with specific cell recognition ligands enabling targeted treatment. (Reprinted from *J Photochem Photobiol B*, 121, Cohen BA, Bergkvist M, Targeted in vitro photodynamic therapy via aptamer-labeled, porphyrin-loaded virus capsids, 67–74, Copyright 2013, with permission from Elsevier.)

Figure 20.11 presents the concept of the capsid loading and targeting concept. More details are given in Table 23.1 of Chapter 23.

The gold-nanoshell encapsulated quantum dots were synthesized by attaching single quantum dots to hairpin RNA that directed the formation of the MS2 VLPs (Miao et al. 2010).

Finn's team was the first to achieve the *in vivo* packaging of proteins, which are listed in Table 23.1, into the Qβ VLPs (Fiedler et al. 2010). To facilitate this RNA-directed encapsidation, the two binding domains were introduced to the coat protein mRNA: an RNA aptamer developed by *in vitro* selection to bind an arginine-rich peptide HIV-1-derived peptide Rev and the sequence of the Qβ packaging hairpin. The cargo enzymes were N-terminally tagged with the Rev peptide and inserted into another compatible group plasmid. The transformation with both plasmids and expression in *E. coli* yielded VLPs encapsidating the Rev-tagged protein (Fiedler et al. 2010).

Another method of protein encapsidation was demonstrated by Francis' team with the MS2 capsids (Glasgow et al. 2012). Thus, the two improved methods for the encapsulation of heterologous molecules inside the MS2 capsids were elaborated. First, attaching DNA oligomers to a molecule of interest and incubating it with MS2 coat protein dimers yielded reassembled capsids that packaged the tagged molecules. The addition of a protein-stabilizing osmolyte, trimethylamine-N-oxide, significantly increased the yields of reassembly. Second, it was found that the expressed proteins with genetically-encoded negatively-charged peptide tags could also induce capsid reassembly, resulting in high yields of reassembled capsids containing the protein. This second method was used to encapsulate alkaline phosphatase tagged with a 16-amino acid peptide. The purified encapsulated enzyme was found to have the same K_m value and a slightly lower k_{cat} value than the free enzyme, indicating that this method of encapsidation had a minimal effect on enzyme kinetics (Glasgow et al. 2012).

Furthermore, Finn's team embedded a high affinity RNA aptamer against a heteroaryldihydropyrimidine structure, which represented a drug-like molecule with no cross-reactivity with mammalian or bacterial cells, as well as an aptamer against theophylline, in a longer RNA sequence that was encapsidated inside the Qβ VLP by the above-described expression technique (Lau et al. 2011). The system therefore comprised a general approach to the production and sequestration of functional RNA molecules, which was used further for the generation of armored RNAs, as it will be described in Chapter 23. The general use of the Qβ VLP-protected aptamers for the delivery of quantum dots, gold, silver, and silica nanoparticles was reviewed by Wu et al. (2011). A good general account on the specific aptamers was given at that time by Özalp and Schäfer (2012).

Then, Francis' team presented the controlled integration of gold nanoparticles and organic fluorophores into the MS2 VLPs (Capehart et al. 2013). They started with the idea that the placement of fluorophores in close proximity to metal nanoparticle surfaces would enhance photophysical properties of the dyes, potentially leading to improved quantum yields and decreased photobleaching. It was difficult in practice, however, to establish and maintain the nanoscale distances that were required to maximize these effects. The type of metal, size, and shape of the nanoparticle, the physical distance separating the metal nanoparticle from the organic dye, and the spectral properties of the fluorophore itself were all proposed to influence the quantum yield and lifetime. The MS2 VLPs were chosen as a system that made it possible to explore these effects while physically preventing the fluorophores from contacting the nanoparticle surfaces. Thus, the MS2 VLPs were used to house gold particles within its interior volume. The exterior surface of each capsid was then modified with Alexa Fluor 488-labeled DNA strands. By placing the dye at distances of 3, 12, and 24 bp from the surface of capsids containing 10 nm gold nanoparticles, fluorescence intensity enhancements of 2.2, 1.2, and 1.0, respectively, were observed. A corresponding decrease in fluorescence lifetime was observed for each distance. Because of its well-defined and modular nature, this architecture allowed the rapid exploration of the many variables involved in metal-controlled fluorescence (Capehart et al. 2013). This intense investigation was included as an example in the general review on the use of biomolecular scaffolds for assembling multistep light harvesting and energy transfer devices (Spillmann and Medintz 2015).

The MS2 VLPs were used to deliver the packaged siRNA that was targeted against the anti-apoptotic factor *BCL2* into HeLa cells (Galaway and Stockley 2013). The specific siRNA was delivered by the MS2 VLPs to hepatocellular carcinoma cells (Ashley et al. 2011a). Then, the MS2-mediated RNAi delivery was carried out *in vivo* when MS2 VLPs were packaged with a pre-miRNA, pre-miR-146a, and surface-decorated with HIV Tat peptides for cell penetration, while no cell-targeting ligands were used (Pan et al. 2012a,b). Upon intravenous delivery to mice, these VLPs displayed widespread biodistribution in the plasma, lung, spleen, and kidney, where high levels of mature miR-146a were detected and knockdown of known targets of miR-146a was observed. Furthermore, the MS2 VLPs were used to deliver mRNA vaccines, which successfully protected mice from prostate cancer (Li et al. 2014). The general potential of the targeted RNAi-based anticancer therapy including the MS2 VLP-based examples were reviewed at that time by Yan et al. (2014). Figure 20.12 presents an outline of the MS2 VLP construction for siRNA delivery.

Fang et al. (2017) demonstrated a method for production of a novel RNAi scaffold, packaged within the Qβ VLPs. The RNAi scaffold was a general utility chimera that contained a functional RNA duplex with paired silencing and carrier sequences stabilized by a miR-30 stem-loop. The Qβ RNA hairpin on the 5′ end conferred affinity for the Qβ coat protein. The silencing sequences could include mature miRNAs and siRNAs and could target essentially any desired mRNA. The VLP-RNAi assembled upon coexpression of the Qβ coat and the RNAi scaffold in *E. coli*. The annealing of the scaffold to form functional RNAs was intramolecular and was therefore robust and concentration independent. The dose- and time-dependent inhibition of GFP expression in human cells with VLP-RNAi was demonstrated in this study. Then, the 3′UTR of oncogenic Ras mRNA was targeted, and the Pan-Ras expression was suppressed, attenuating cell proliferation and promoting mortality of brain tumor cells (Fang et al. 2017). This approach is illustrated in Figure 20.13.

Furthermore, Fang PY et al. (2018) explored factors that could influence the stability and quantity of the target RNA that was packaged *in vivo* within the Qβ VLPs. Thus, it was shown clearly that the VLP packaging chemically protected RNA from small diffusible chemicals such as hydroxyl radicals and divalent cations that could readily penetrate the VLPs. It appeared that the packaging chemically stabilized RNA by mechanisms beyond direct exclusion of reactive species from proximity to the RNA. The extent of unmediated cleavage, in the absence of reactive species, was the same for RNA that was free or packaged within VLPs, and was very slow. Remarkably, the *in vivo* packaging of RNA within VLPs appeared to be more efficient for intrinsically compact RNAs, such as rRNA, and less efficient for unstructured, elongated RNAs, such as mRNA. The packaging efficiency was reduced by addition of the ribosome binding site to a target RNA. Remarkably, the Qβ hairpin was necessary but not sufficient for the efficient packaging (Fang PY et al. 2018).

The above-described SpyTag/SpyCatcher approach was used for the construction of a catalytic nanoreactor based on

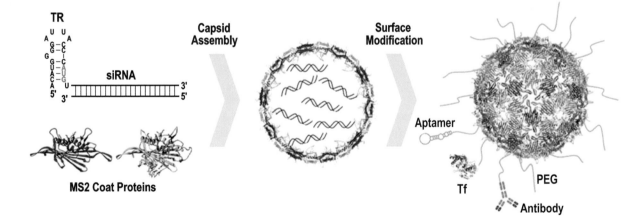

FIGURE 20.12 Construction of an MS2 VLP-based siRNA delivery system. MS2 coat proteins associate as non-covalent dimers (magenta, PDB code 1MSC). Capsid assembly from these dimers can be triggered at neutral pH by addition of the MS2 translational operator (TR), a 19-nt RNA stem-loop (Rolfsson et al 2008; Borodavka et al 2012). TR binding induces a conformation change in the coat protein dimer (green/blue PDB code 1ZDI), which can then act as an assembly initiation complex (Valegård et al. 1994b; Stockley et al. 2007). Any siRNA conjugated to TR will be packaged inside the assembled capsids (Galaway and Stockley 2013). The capsid surface can then be chemically modified to display ligands for immunomasking, endosomal escape, or tumor targeting. The examples presented include PEG, a nucleic acid aptamer, transferrin (Tf) and an antibody. These are not drawn to scale. (Reprinted with permission of Portland Press from Yan R et al. *Biochem J.* 2014;461:1–13.)

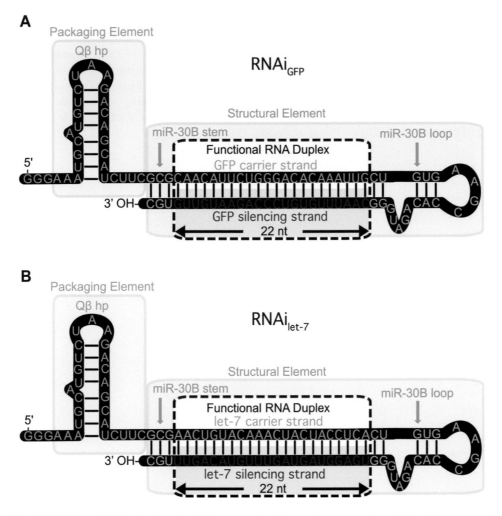

FIGURE 20.13 The Qβ RNAi scaffold is designed to fold into two hairpins and co-assemble with the Qβ coat *in vivo*. A single RNA molecule forms the Qβ hairpin (blue) linked to a miR-30B stem loop (green), which contains the functional RNA duplex (yellow/red). The functional RNA duplex can encode miRNA or siRNA. The scaffold can incorporate any 22 base pair sequence, depending on the desired target. (A) An RNAi scaffold that targets GFP mRNA. (B) An RNAi scaffold that targets the Pan-Ras mRNA 3′ UTR. (Reprinted from Fang PY et al. *Nucleic Acids Res.* 2017;45:3519–3527.)

the *in vivo* encapsulation of multiple enzymes into the MS2 VLPs (Giessen and Silver 2016a). Thus, a new functional nanocompartment in *E. coli* was generated by engineering the MS2 VLPs to encapsulate multiple cargo proteins. The sequestration of multiple proteins in the MS2 VLPs was achieved by the SpyTag/SpyCatcher protein fusions that covalently crosslinked with the interior surface of the capsid. Further, the functional two-enzyme indigo biosynthetic pathway was targeted to the engineered capsids, leading to a 60% increase in the indigo production *in vivo*. The enzyme-loaded particles were purified in their active form and showed enhanced long-term stability *in vitro*, about 95% activity after 7 days, compared with free enzymes that demonstrated only about 5% activity after 7 days. It was concluded that this engineered *in vivo* encapsulation system provided a simple and versatile way for generating highly stable multi-enzyme nanoreactors for the *in vivo* and *in vitro* applications (Giessen and Silver 2016a). Encapsulation as a prospective strategy for the design of biological compartmentalization was reviewed by Giessen and Silver (2016b).

In parallel, Finn's team packaged some technologically and medicinally important enzymes into the Qβ VLPs (Fiedler 2018). The same Rev tagging approach mentioned above and described earlier by Fiedler (2010) was applied to the four chosen enzymes that are listed in Table 23.1 of Chapter 23. The simultaneous expression in *E. coli* cells of the Qβ VLP protein and protein "cargo" tagged with the positively charged Rev peptide led to the spontaneous self-assembly of VLPs with multiple copies of the cargo inside. The captured enzymes were active while inside the nanoparticle shell, and were protected from environmental conditions that led to the free-enzyme destruction. The technology was extended to create, via self-assembly, the VLPs that simultaneously displayed protein ligands on the exterior and contained enzymes within. The displayed ligands are listed in Table 23.1. The inverse relationships were observed between the size of both the packaged and externally displayed protein or domains and nanoparticle yield (Fiedler et al. 2018).

Finally, Francis' team presented a solid methodological report on their success by the encapsulation of proteins with tags comprising anionic amino acids or DNA and gold nanoparticles with negative surface charges into the MS2 VLPs (Aanei et al. 2018a).

Meanwhile, due to strong efforts of the Andris Kazāks' group in yeast expression, Grēns' team presented a wide variety of novel VLP models, namely, derived from the phages fr, SP, AP205, PP7, and φCB5, with the special aim of nanomaterial packaging (Freivalds et al. 2014). It was stressed particularly that a variety of compounds, including RNA, DNA, and gold nanoparticles, could be packaged efficiently inside novel φCB5 VLPs. The ease with which the phage φCB5 coat protein dimers have been purified in high quantities and reassembled into VLPs made them especially attractive for the packaging of nanomaterials and the chemical coupling of peptides of interest on the surface (Freivalds et al. 2014).

21 3D Particles

"Sometimes," said Pooh, "the smallest things take up the most room in your heart."

Alan Alexander Milne

The soul never thinks without a picture.

Aristotle

VIRUSES AND/OR VLPs?

The current chapter is simply the logical continuation of Chapter 9. The long gap between these two chapters was necessary to present first two general subjects, namely, the expression of the RNA phage genes and the generation of chimeric VLPs; the knowledge of both is necessary to understand this chapter. Thus, this chapter appears after the genetic engineering of the RNA phages, which was explained in Chapter 19 and the general outline of the chimeric VLPs, which was presented in Chapter 20. As described in the Symmetry section of Chapter 9, the general idea of the spatial structure of the icosahedral RNA phages was clarified in the mid-1970s. The study of high-resolution structures began in the early 1990s with x-ray crystallography and is continued up to now. The first crystals were obtained for the phage preparations, while further investigations quite often preferred the VLPs that were obtained after expression of the RNA phage coat gene in *E. coli* or yeast. Moreover, since the active studies of the chimeric VLPs began in the mid-1990s, the determination of the three-dimensional (3D) structure was tightly connected with the urgent problem of how to display foreign peptides on the particle surface. To learn the global history of viral architecture, a review article written by the famous crystallographer Michael G. Rossmann (2013) is recommended.

Figure 21.1 demonstrates the x-ray crystallography structures of the resolved RNA phage representatives. The fine 3D structures are in fact very similar, despite the marked diversity in the primary structures of their coats, as described in Chapter 12.

The first resolved 3D structure of the RNA phages was for the phage MS2. At that time, this structure showed no similarity to any other known viruses or proteins of any origin. The MS2 virion structure was first determined at a resolution of 3.3 Å (Valegård et al. 1986, 1990, 1991; Liljas and Valegård 1990) and then refined to the 2.8 Å resolution (Golmohammadi et al. 1993). Next, the crystal structure of the MS2 VLPs with amino acid exchanges in the FG loop was resolved (Stonehouse et al. 1996a,b; Tars 1999). It is noteworthy from a historical point of view that the first phage MS2 crystals and the first preliminary x-ray examination data were obtained in Walter Fiers' lab still in 1979 (Min Jou et al. 1979). Walter Fiers and Bror Strandberg were great enthusiasts and initiators of the spatial MS2 structure, while the

brilliant professional realization of the idea belonged to Karin Valegård and Lars Liljas and later to Kaspars Tārs.

According to the 3D structure of MS2, the 180 coat subunits were arranged in dimers as initial building blocks and formed a lattice with a T = 3 triangulation number. The coat subunit consisted of a five-stranded β-sheet facing the inside of the particle and a hairpin and two α-helices on the outside. This particular hairpin was used as a favorite target for the genetic insertions and/or chemical coupling of the desired foreign sequences, as explained in detail in Chapter 20.

Later, the structure of a recombinant capsid, i.e., VLPs, of the RNA phage fr was determined by x-ray crystallography at a resolution of 3.5 Å and was shown to be identical to the protein shell of the native phage fr (Bundule et al. 1993; Liljas et al. 1994). Again, this study was performed by Lars Liljas' team in Uppsala, but with the active participation of Elmārs Grēns' team in Riga, where Maija Bundule appeared as the first enthusiast of the fine structure determination.

As noted in the Genetic Fusions of Coats: fr section of Chapter 20, the FG loop of the fr coat was tried as a potential target for the insertions/replacements and was initially modified by a 4-amino acid-long deletion. The crystal structure of the FG deletion mutant, which retained, however, the ability to self-assemble, revealed that the modified loops were disordered near the fivefold axis of symmetry and were too flexible and too short to interact with each other (Axblom et al. 1998). This seemed incompatible with a potential role of the FG loop in the regulation of the capsid size.

The structures of both virions and VLPs of the phage Qβ were resolved at a resolution of 3.5 Å (Valegård et al. 1994; Golmohammadi et al. 1996). These structures differed from the previously determined RNA phage MS2 and fr structures by the presence of stabilizing disulfide bonds on each side of the flexible FG loop, which covalently linked the coat dimers. The nanotechnological preferences of this linkage were explained in the Functionalization: Qβ section of Chapter 20 and outlined in Figure 20.9. A profound comparison with the structure of the related phage MS2 showed that, although the fold of the Qβ coat was very similar, the details of the protein–protein interactions were completely different (Golmohammadi et al. 1996).

The first MS2, Qβ, and fr structures were followed by the determination of the structure of the group II phage GA, which showed some structural differences compared to the MS2 and fr virions or VLPs, especially in the N- and C-terminal regions (Tars et al. 1997; Tars 1999). This fine GA structure initiated the powerful carrier of Kaspars Tārs, a versatile and successful crystallographer, presently working in Riga after 11 years in Uppsala.

The knowledge of the 3D structure attracted interest in the fine mechanisms of the RNA phage morphogenesis and

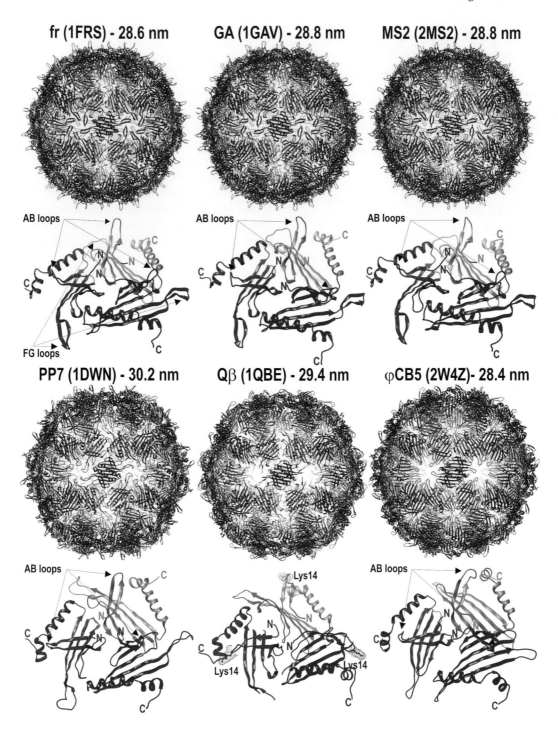

FIGURE 21.1 Crystal structures of the RNA phages that have been used as VLP carriers. The structures are presented in alphabetical order with their protein data bank IDs shown in parentheses and the outer diameters indicated for each species. The AB loops are exposed on the full-capsid surfaces. Also shown are the corresponding trimeric asymmetric units with the indicated N- and C-termini. The outer surface is oriented toward the reader. The AB loops are indicated by arrows. The Qβ AB loops are distinguished by Lys14 residues, which are indicated by the shaded areas. The coat chains A, B, and C are indicated in red, green, and blue, respectively. The structural data are compiled from the VIPERdb (http://viperdb.scripps.edu) database (Carrillo-Tripp et al. 2009) and are visualized using Chimera software (Pettersen et al. 2004). (Reprinted with permission of S. Karger AG, Basel from Pumpens P et al. *Intervirology*. 2016;59:74–110.)

virus assembly. Highly qualified reviews on the phage design-ing subject were published at that time (Stockley et al. 1994; Liljas 1999; Stockley and Stonehouse 1999; Rowlands and Stockley 2001).

The crystal structure of the pseudomonaphage PP7 was resolved at a resolution of 3.7 Å (Tars 1999; Tars et al. 2000a,b).

As in the case of the MS2 and Qβ coats, the RNA recognition site on the PP7 coat was determined at that time (Lim and Peabody 2002). It is remarkable that a detailed biophysical study of the PP7 virions, including the effects of pH and salt concentration on the charge transition, from the net-positive to the net-negative, was undertaken recently (Nap et al. 2014).

The crystal structure of another pseudomonaphage, PRR1, was resolved at a resolution of 3.5 Å and exhibited a binding site for a calcium ion close to the quasi-threefold axis (Persson et al. 2008).

The crystal structure of a very distant RNA phage, the caulophage φCB5, was resolved at 3.6 Å, while the structure of the φCB5 VLPs was resolved at 2.9 Å (Plevka et al. 2009a). The structures appeared to be nearly identical, with some differences in the average density of RNA. Unlike other phages, the φCB5 capsids were significantly stabilized by calcium ions, similarly to some plant viruses (Hull 1977). The disassembly of these capsids occurred when the calcium ions were chelated with EDTA and/or there was a reduction in the surrounding salt concentration. Another unique feature of the φCB5 capsids was the involvement of RNA bases in the stabilization of its interdimer contacts (Plevka et al. 2009a).

After 10 years of trials by Kaspars Tārs' group on the acinetophage AP205, the crystal structure of the AP205 coat dimer was solved at a resolution of 1.7 Å and then fitted into the 6.6-Å resolution electron cryomicroscopy map (Shishovs et al. 2016). Figure 21.2 presents this very interesting and highly important structure.

In fact, the structure of the AP205 coat dimer could be regarded as a circular permutant relative to the structures of

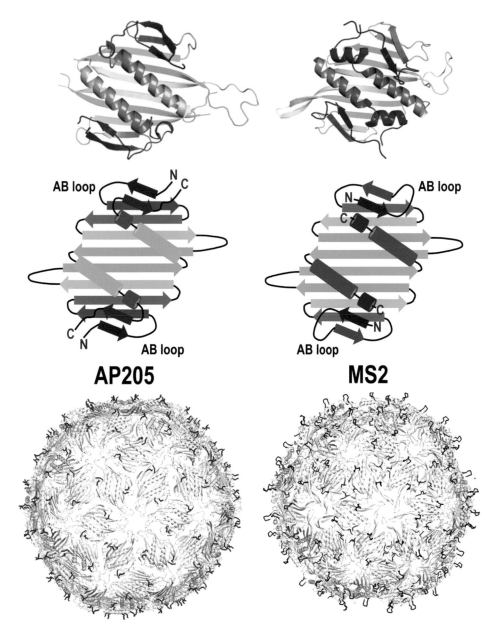

FIGURE 21.2 Differences in the coat structures between the AP205 and MS2 phages. The overall folds of the coat dimers (top) are similar for both phages. The coat monomers are shown in rainbow coloring, from the blue N-terminus to the red C-terminus. From the more schematic picture of secondary structure elements (middle), it is obvious that the AP205 coat has N- and C-termini that are located in roughly the same place as the AB loops from MS2. As a result, the N- and C-termini in the AP205 coat are well exposed on the capsid surface (blue and red, respectively), similarly to the AB loops (black) in the MS2 coat. (Reprinted with permission of S. Karger AG, Basel from Pumpens P et al. *Intervirology.* 2016;59:74–110.)

the phage MS2 and other family members. This feature was made possible by the fact that the N-terminus of one monomer in the dimer is in close proximity to the C-terminus of the other monomer.

Compared to the MS2 and other phages with known structures, the AP205 coat was missing one β-strand in its N-terminus, but it had an extra β-strand in its C-terminus. However, when considered from a 3D perspective, the position of the β-strand was essentially the same in both cases. As a consequence, the AP205 structure demonstrated the N- and C-termini in the same locations as those occupied by the surface-exposed AB loops in other phage capsids. It is notable that the N- and C-termini in other phages were not well exposed on the surface and were clustered around the quasi-threefold axes. This explained the previous observations narrated in the Genetic fusions of coats: AP205 section of Chapter 20 that the AP205 coat, in contrast to other phages, tolerated long additions at its N- and C-termini without compromising the capsid assembly (Tissot et al. 2010).

In addition to the structures of the self-assembled RNA phage capsids, the crystal structures of unassembled mutant coat dimers of MS2 (Ni et al. 1995a,b) and GA (Ni et al. 1996) were also resolved. These structures showed only minor differences in comparison to their self-assembled counterparts.

As noted in the previous chapters, the GA and MS2 coat proteins differed in sequence at 49 of 129 amino acid residues. The sequence differences that contributed to distinct immunological and physical properties of both proteins were found at the surface of intact phages in the AB and FG loops. There were six differences in potential RNA contact residues within the RNA-binding site located in an antiparallel β-sheet across the dimer interface. The three differences involved residues in the center of this concave site: Lys/Arg 83, Ser/Asn 87, and Asp/Glu 89. The residue 87 was shown by molecular genetics to define the RNA-binding specificity by the GA or MS2 coat proteins (Lim et al. 1994). This sequence difference reflected recognition of the nucleotide at position -5 in the unpaired loop of the translational operators bound by these coat proteins. In the GA structure, the nucleotide at this position was a purine, whereas it was a pyrimidine in the MS2 structure (Ni et al. 1996). The structure of this GA RNA hairpin was determined using heteronuclear NMR spectroscopy to an overall precision of 2.0 Å (Smith and Nikonowicz 1998).

As discussed in Chapter 20, the MS2 coat sustained a genetic fusion, resulting in a duplicated coat that folded normally and functioned as a translational repressor due to its physical proximity to the N- and C-termini of the CP (Peabody and Lim 1996).

Furthermore, the crystal structure of an icosahedral MS2 capsid that was assembled from covalently joined dimers, or so-called single-chain dimers, was resolved at 4.7 Å (Plevka et al. 2009b). The structure resembled the wild-type MS2 virion except for the intersubunit linker regions, but a fraction of the capsids was unstable in phosphate buffer because of the assembly defects (Plevka et al. 2009b). Moreover, the organization of the MS2 single-chain dimers into crystals may have resulted in an arrangement of subunits that corresponded to

the T = 3 octahedral particles (Plevka et al. 2008). In this case, the arrangement of dimers was somewhat similar to that in the normal T = 3 icosahedral particles, except that the four FG loops interacted near the fourfold axis of symmetry on an octahedron rather than five FG loops interacting near the fivefold axis of symmetry on an icosahedron. However, when the MS2 coat dimers were not crystallized in the F cubic crystal form, they were assembled into the T = 3 icosahedral capsids that were indistinguishable from the wild-type particles (Plevka et al. 2008).

In the pure structural connection, it is noteworthy that the interface of the SARS-CoV N protein dimer appeared as a four-stranded β-sheet, superposed by two long α-helices, i.e., this topology closely resembled that of the RNA phage capsids (Chang et al. 2005). The sequence alignment and secondary structure prediction suggested that other coronavirus N proteins also adopted a similar dimerization mechanism.

Recently, the phage MS2 structure was resolved by a novel approach in frame of the global single particle imaging (SPI) initiative launched in December 2014 at the Linac Coherent Light Source, SLAC National Accelerator Laboratory, USA (Sun et al. 2018). The all-atom molecular dynamics computer model was built for the phage MS2 particle without its genome, using high-resolution electron cryomicroscopy measurements for initial conformation (Farafonov and Nerukh 2019).

RNA APTAMERS

As explained in detail in the numerous repressor complex I sections of Chapter 16, x-ray crystallography provoked the real breakthrough in understanding the protein-RNA interactions that were occurring during the RNA phage translational repression and genome encapsidation. This breakthrough was particularly apparent after the first crystal structure of a complex of recombinant MS2 capsids with the 19-nucleotide RNA operator was resolved at 2.7 Å (Valegård et al. 1994b, 1997; Stockley et al. 1995). These studies were the result of the tight and fruitful cooperation of both the prospective and successful Lars Liljas' and Peter G. Stockley's teams. The residues responsible for the protein-RNA interactions were localized by analysis of the amino acid exchanges at positions 45 and 59 (Lago et al. 1998; van den Worm et al. 1998; Peabody and Chakerian 1999). Moreover, numerous other mutations responsible for altering the specificity (Lim et al. 1994) and efficiency (Lim and Peabody 1994) of the translational operator complexes were identified. The crystal structures of the MS2 VLPs complexed with the RNA aptamers, which differed by their secondary structure from wild-type RNA (Convery et al. 1998; Rowsell et al. 1998; Grahn et al. 1999) or involved the presence of 2'-deoxy-2-aminopurine at the critical -10 position (Horn et al. 2004) were resolved.

The structure of the coat protein complexed with the operator RNA fragments were also solved for the phages PP7 (Chao et al. 2008), PRR1 (Persson 2010; Persson et al. 2013), and Qβ (Rumnieks and Tars 2014). The coat dimers formed by the RNA phages studied to date all recognized the stem-loop sequence around the replicase start codon. As explained in detail in Chapters 16 and 17, this coat-RNA interaction served as a mechanism for the translational repression and self-genome recognition during virion assembly. Although the overall binding mode of the stem-loop to the coat was similar in all the studied cases, the details were surprisingly different among different viruses. A number of nucleotides formed sequence-specific and sequence-unspecific interactions with coats. As demonstrated in Chapter 16, the most important nucleotides for the coat recognition were two adenosines, one located in the loop region and another in a stem bulge. In the MS2, PP7, and Qβ operators, these adenosines formed quite different interactions with the coat dimer, as presented in Figure 16.11 of Chapter 16. It should be noted here that all attempts to identify analogous coat-RNA interactions in the acinetophage AP205 and caulophage φCB5 failed until now (Kaspars Tārs, unpublished observations), suggesting that mechanisms of genome recognition and translational repression might differ significantly among the distant *Leviviridae* family members.

It is remarkable that an unexpected 3D homology was found to the PP7 coat fold. This was N-terminus of the coat morphogenetic protein SpoVID of *Bacillus subtilis*, which was necessary for spore encasement in a tough protein shell known as the coat, which consists of at least 70 different proteins (Wang KH et al. 2009). It was suggested that similar mechanisms may be operating during the assembly of spore and of the phage external protective structures.

The electron cryomicroscopy studies have shown that, in addition to the operator, many other RNA sequences in the MS2 genome were able to bind to the coat dimer (Koning et al. 2003).

Such studies have allowed the 3D visualization of the icosahedrally averaged genomic MS2 RNA at a resolution of 9 Å (Toropova et al. 2008). Recently, direct evidence for the packaging signal-mediated assembly of the MS2 phage was presented, when based on crosslinking studies of peptides and oligonucleotides at the interfaces between the capsid proteins and the genomic RNA of this phage (Rolfsson et al. 2016; Stockley et al. 2016). Remarkably, the same coat-RNA and maturation protein-RNA interfaces were identified in every viral particle.

Surprisingly, the icosahedral Qβ VLPs were converted into rods after modification of the FG loop structure (Cielens et al. 2000). Since the Qβ VLPs were stabilized by disulfide bonds of cysteine residues 74 and 80 within the FG loop, the stability of capsids was mutationally reduced by converting the amino acid stretch 76-ANGSCD-81 within the FG loop into the 76-VGGVEL-81 sequence. It led to production in *E. coli* cells of aberrant rod-like Qβ VLPs, along with normal icosahedral capsids. The length of the rod-like particles exceeded 4–30 times the diameter of icosahedral Qβ VLPs (Cielens et al. 2000).

As mentioned in the Reconstitution section of Chapter 9, the appearance of alternate VLP forms of RNA phages was further confirmed by the presence of the rod-like structures in the case of the co-assembly of the phage fr and GA coats (Rūmnieks 2006; Rumnieks et al. 2009). Figure 21.3 demonstrates electron micrographs of the rod-like mixed particles, while Figure 21.4 shows how such particles may appear.

Figure 21.5 compares the inter-dimer interactions around the threefold axis in the wild-type fr and GA capsids and the modeled mixed particles.

Using a chromatographic selection for the VLP assembly, Asensio et al. (2016) further established that a single amino acid mutation S37P led to the smaller MS2 VLPs. This mutation mediated a uniform switch in particle geometry from the T = 3 to T = 1 icosahedral symmetry. The resulting smaller particle retained the ability to be disassembled and reassembled *in vitro* and to be chemically modified to load cargo into its interior cavity. Remarkably, the pair of 27 and 17 nm MS2 particles could allow direct examination of the effect of size on function in established applications of VLPs, including drug delivery and imaging (Asensio et al. 2016).

Recently, de Martín Garrido et al. (2019) have assembled MS2 VLPs with non-genomic RNA containing the capsid incorporation sequence and revealed by electron cryomicroscopy not only traditional T = 3, but also T = 4 and mixed capsids between these two triangulation numbers to 4 Å and 6 Å, respectively.

PACKING OF GENOMIC RNA

The electron cryomicroscopy approach was used, first of all, to evaluate the structure of the genomic RNA within the phage particles. As noted above, the first electron cryomicroscopy visualization was achieved for the phage MS2 by Koning et al. (2003). It showed that the RNA density was present inside the capsids, suggesting many coat protein-binding sites in addition to the hairpin on which the assembly and packaging process was initiated. Next, the electron cryomicroscopy structures of the phages Qβ, PP7, and AP205 were solved (van den Worm et al. 2006). These reconstructions demonstrated the expected icosahedral capsid enclosing the genomic RNA. The general RNA organization inside the capsids indeed resembled that found in MS2, giving an important clue to the existence of the secondary coat binding sites (Koning et al. 2003). The details of the RNA densities inside the capsids of PP7 and AP205, however, differed significantly from the MS2 and Qβ patterns, which were nearly indistinguishable. Moreover, the characteristic combination of triangles and pentagons was not observed in other virus structures determined so far, where the overall RNA pattern was formed by either pentagons building a dodecahedron, such as in pariacoto virus or nodavirus, or triangles building an icosahedron, as in satellite tobacco necrosis virus or satellite tobacco mosaic virus. Thus, the combined structure in the *Leviviridae* seemed to be unique (van den Worm 2006). Figure 21.6 presents the exteriors and interiors of the virions by the electron cryomicroscopy reconstructions.

It is noteworthy that the cryo-electron micrographs of the phage MS2 contributed markedly at that time to the elaboration of the TYSON, a suitable program for automatic particle selection from electron micrographs (Plaisier et al. 2004).

Stockley's team concentrated on the role of the 19 nucleotide RNA stem-loop by the direction of the pathway toward a T = 3 capsid (Stockley et al. 2005). Using mass spectrometry, Stockley et al. (2007) detected both assembly intermediates

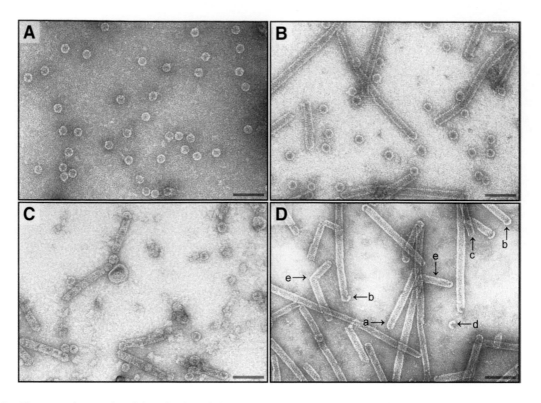

FIGURE 21.3 Electron micrographs of the mixed particles. *In vitro* reconstructed particles from mixtures of (A) fr and MS2 and (B) fr and GA coat proteins. (C) Lysozyme lysate of *E. coli* cells expressing both fr and GA coat proteins. (D) Complete and aberrant rod-like particles found in reassembly reactions of fr and GA CPs: (a) a rod-like particle with a completely capped end; (b) "oversized" caps and caps attached to only one side of the tube; (c) a rod-like particle with an open end; (d) separated cap; (e) bent and L-shaped tubules. Bar, 100 nm. (Reprinted from *Virology*, 391, Rumnieks J et al. Assembly of mixed rod-like and spherical particles from group I and II RNA bacteriophage coat proteins, 187–194, Copyright 2009, with permission from Elsevier.)

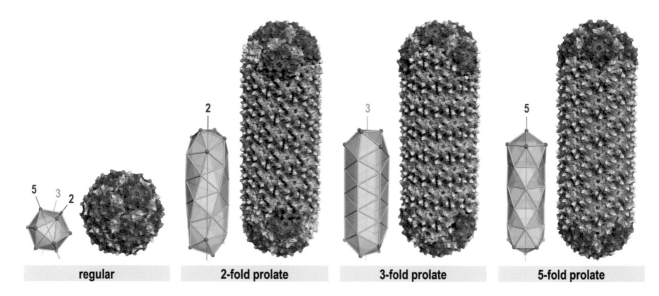

regular   2-fold prolate   3-fold prolate   5-fold prolate

FIGURE 21.4 Three-dimensional models of the RNA phage rod-like particles. Prolate icosahedrons can be constructed by splitting a regular icosahedron into two halves containing six vertices each and connecting those with a hexagonal tubular lattice. Three possibilities exist corresponding to prolates along the twofold, threefold and fivefold symmetry axes. In absence of experimental high-resolution structures of the rod-like particles, their three-dimensional models can be constructed in accordance with the geometric principles of prolate icosahedrons. 2-fold, 3-fold, and 5-fold prolate structures were built using coordinates of the MS2 coat protein and are presented here as their solvent-accessible surface models. A regular icosahedral RNA phage particle is shown for comparison. A smaller polyhedral representation adjacent to each molecular model is provided to visualize the underlying geometry of the particles, with the fivefold (5), threefold (3), and twofold (2) symmetry axes indicated. In the models, the coat dimers involved in pentameric interactions are shaded brown and the corresponding vertices in the polyhedral models are marked as brown spheres. (Generated by Jānis Rūmnieks. Courtesy of Jānis Rūmnieks.)

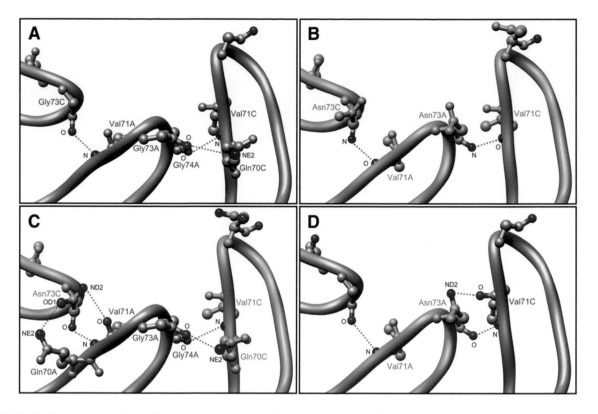

FIGURE 21.5 Comparison of inter-dimer interactions around the threefold axis in wild-type fr and GA capsids and the modeled mixed particles. Wild-type fr (A) and GA (B) capsid models are compared to those built from fr AB + GA CC dimers (C) and GA AB + fr CC dimers (D). For a better fit, the models have interchanged OD1 and ND2 atoms in Asn73 compared to the published GA structure. For clarity, the FG loops are shown looking from inside the capsid; residues not involved in contacts are hidden. The interacting atoms are indicated. The numbering of residues is after fr, the letter following the residue name designates the subunit. In all images, fr coat protein molecules are shown in blue and GA in green. (Reprinted from *Virology*, 391, Rumnieks J, Ose V, Tars K, Dislers A, Strods A, Cielens I, Renhofa R, Assembly of mixed rod-like and spherical particles from group I and II RNA bacteriophage coat proteins, 187–194, Copyright 2009, with permission from Elsevier.)

and the final product, the T = 3 viral capsid, during reassembly of the phage MS2. The assembly was only efficient when both types of quasi-equivalent coat protein dimers seen in the final capsid were present in solution. The NMR studies confirmed that the interconversion of these conformers was allosterically regulated by sequence-specific binding of the short RNA stem-loop. The isotope pulse-chase experiments demonstrated that all intermediates observed were competent for further coat protein addition, i.e., they were all on the pathway to capsid formation, and that the unit of capsid growth was a coat protein dimer. The major intermediate species were dominated by stoichiometries derived from formation of the particle threefold axis, implying that there was a defined pathway toward the T = 3 shell. These results provided the first experimental evidence for a detailed mechanistic explanation of the regulation of quasi-equivalent capsid assembly. Therefore, a direct role for the encapsidated RNA in the *in vivo* assembly was suggested (Stockley et al. 2007). The use of mass spectrometry to probe virus structures was substantiated thoroughly at that time by Morton et al. (2008). This review also provides the readers with the full list of intact viruses and assembly intermediates studied by mass spectrometry.

Furthermore, using a combination of biochemistry, mass spectrometry, NMR spectroscopy, and electron cryomicroscopy,

the control of quasi-equivalent conformer switching in the MS2 coat by a sequence-specific protein interaction with the short RNA stem-loop was confirmed (Rolfsson et al. 2008; Toropova et al. 2008). In fact, the binding of the RNA component of this complex resulted in the conversion of the coat dimer from a symmetrical conformation to an asymmetric one. Only when both symmetrical and asymmetrical dimers were present in solution, the T = 3 phage capsid assembly was efficient. This implied that the conformers characterized by NMR corresponded to the two distinct quasi-equivalent conformers seen in the 3D structure of the virion. Moreover, a single-particle electron cryomicroscopy reconstruction of the wild-type phage to ~9 Å resolution has revealed the icosahedrally ordered density encompassing up to 90% of the single-stranded RNA genome. The RNA was seen with a novel arrangement of two concentric shells, with connections between them along the fivefold symmetry axes. The RNA in the outer shell interacted with each of the 90 CP dimers in the T = 3 capsid and, although the density was icosahedrally averaged, there appeared to be a different average contact at the different quasi-equivalent protein dimers: precisely the result that would be expected if the protein conformer switching was RNA-mediated throughout the assembly pathway (Rolfsson et al. 2008; Toropova et al. 2008).

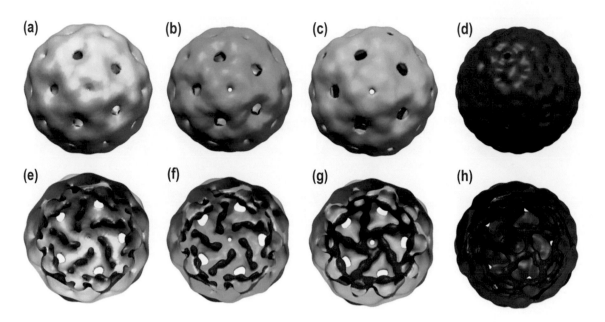

FIGURE 21.6 The three-dimensional structures of the phages MS2, Qβ, PP7, and AP205 obtained by electron cryomicroscopy and icosahedral object reconstruction. (a)−(d) Exterior of the virions and (e)−(h) interior of a hemisphere of the electron cryomicroscopy reconstructions of the phages (a) MS2, (b) Qβ, (c) PP7, and (d) AP205 viewed along the fivefold axis (red, RNA). Clearly visible are the holes at the threefold axes in the capsids of all four phages. Perturbations on the inside of the capsids, representing RNA, are located at all coat dimer interfaces for MS2 and Qβ. In AP205, the RNA density forms a continuous network, covering all coat dimer interfaces. The RNA density in PP7 is of a form intermediate between the coliphages and AP205. (Reprinted from *J Mol Biol*, 363, van den Worm SHE, Koning RI, Warmenhoven HJ, Koerten HK, van Duin J, Cryo electron microscopy reconstructions of the *Leviviridae* unveil the densest icosahedral RNA packing possible, 858–865, Copyright 2006, with permission from Elsevier.)

Figure 21.7 presents the gained organization of the genomic RNA in the MS2 virion by using electron cryomicroscopy, single-particle image processing, and three-dimensional reconstruction with icosahedral averaging.

Further studies developed this approach and showed that the RNA sequences which were not the short RNA stem-loop triggered the conformational change in the MS2 coat dimer required to initiate efficient assembly and led to the conclusion that sequences adjacent to the stem-loop in the genome were playing roles in directing the assembly pathway toward the T = 3 shell (Basnak et al. 2010). It therefore appeared clearly that the genomic RNA contributed actively to its own encapsidation.

The impact of viral RNA on assembly pathway selection was specified by the kinetic studies of the assembly intermediate formation and by the generation of a kinetic model of the assembly reaction (Morton et al. 2010b). Using a combination of mass spectrometry and kinetic modeling, this study

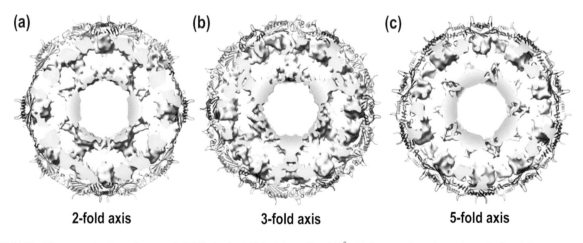

FIGURE 21.7 The organization of genomic RNA in the MS2 virion. The 40 Å thick central sections through the virion are shown, perpendicular to the (a) twofold, (b) threefold, and (c) fivefold icosahedral symmetry axes. The crystallographic coordinates for the MS2 protein shell are shown in each view with regions of the electron cryomicroscopy map at lower radii shown as radially colored density (from pale blue ($r \sim 108$ Å) to pale red ($r \sim 42$ Å)). The cut surface of the electron microscopy density is shown in the same color scheme. (Reprinted from *J Mol Biol*, 375, Toropova K, Basnak G, Twarock R, Stockley PG, Ranson NA, The three-dimensional structure of genomic RNA in bacteriophage MS2: Implications for assembly, 824–836, Copyright 2008, with permission from Elsevier.)

made it possible to determine the dominant assembly pathways explicitly, as well as to estimate the effect of the RNA on the free energy of association between the assembling protein subunits. Generally, the results revealed only a small number of dominant assembly pathways, which varied depending on the relative ratios of RNA and protein.

Using hydrogen/deuterium exchange-mass spectrometry (HDX-MS), Morton et al. (2010a) located regions of the MS2 coat dimer that exhibited conformational/dynamical changes, and hence changes in their HDX kinetics, upon binding to a genomic RNA stem-loop known to trigger the assembly initiation. These results demonstrated the potential utility of the HDX-MS to probe conformational and dynamical changes within the non-covalently bound protein-RNA complexes (Morton et al. 2010a).

Furthermore, Rolfsson et al. (2010) assayed the ability of the MS2 coat to package large, defined fragments of its genomic RNA. Thus, the efficiency of packaging into a T = 3 capsid *in vitro* was inversely proportional to the RNA length, implying that there was a free-energy barrier to be overcome during assembly. All the RNAs examined had greater solution persistence lengths than the internal diameter of the capsid into which they became packaged, suggesting that the protein-mediated RNA compaction must occur during assembly. The electron cryomicroscopy structures of the capsids assembled in these experiments with the sub-genomic RNAs showed a layer of RNA density beneath the coat protein shell but lacked density for the inner RNA shell seen in the wild-type virion. The inner layer was restored when full-length virion RNA was used in the assembly reaction, implying that it became ordered only when the capsid was filled, presumably because of the effects of steric and/or electrostatic repulsions. The data were consistent with mutual chaperoning of both RNA and coat protein conformations, partially explaining the ability of such viruses to assemble so rapidly and accurately (Rolfsson et al. 2010).

Since the 19-nucleotide RNA stem-loop acted as an allosteric effector and its binding to the symmetric coat protein dimers resulted in conformational changes, principally at the FG loop, yielding an asymmetric structure, it was intriguing to check unrelated stem-loops for this activity (Dykeman and Twarock 2010; Dykeman et al. 2010). It appeared that the assembly was triggered by a number of the stem-loops unrelated to the original stem-loop in sequence and detailed secondary structure, suggesting that there was little sequence specificity to the allosteric effect. While the stem-loop binding site on the coat protein dimer was distal to the FG-loops, the mechanism of this switching effect needed to be investigated. Thus, the vibrational modes of both stem-loop-bound and RNA-free coat protein dimers were studied using an all-atom normal-mode analysis. The results suggested that asymmetric contacts between the A-duplex RNA phosphodiester backbone and the EF-loop in one coat protein subunit resulted in the FG-loop of that subunit becoming more dynamic, while the equivalent loop on the other monomer decreased its mobility. The increased dynamic behavior occurred in the FG-loop of the subunit required to undergo the largest conformational change when adopting the quasi-equivalent B conformation.

The free energy barrier on the pathway to form this new structure would consequently be reduced compared to the unbound subunit. In fact, the allosteric effect was independent of the base sequence of the bound stem-loop (Dykeman and Twarock 2010; Dykeman et al. 2010).

In order to quantify the impact of RNA on the association rates at the dimer interfaces, the diffusional association of the MS2 dimers was measured in the presence and absence of the RNA stem-loop, using Brownian dynamics simulation (ElSawy et al. 2010). Moreover, an occupancy landscape was constructed and used for investigating the preferred configurations of the dimer-dimer diffusional encounter complexes. Briefly, the stem-loop binding resulted in the self-association of AB dimers being inhibited, while association of AB with CC dimers was greatly enhanced. This provided an explanation for experimental results in which an alternating assembly pattern of AB and CC dimer addition to the growing assembly intermediate has been observed to be the dominant mode of assembly. The presence of the RNA hence dramatically decreased the number of dominant assembly pathways and thereby reduced the complexity of the self-assembly process of these viruses (ElSawy et al. 2010). This conclusion was especially important in light of the knowledge that the coat self-assembly may appear without any RNA participation.

It is noteworthy that at that time the ion mobility spectrometry-mass spectrometry was combined with tandem mass spectrometry to characterize the MS2 virion intermediates in terms of mass, shape, and stability in a single experiment (Knapman et al. 2010).

ElSawy et al. (2011) concentrated on more general questions of the origin of order in the genome organization of single-stranded RNA viruses. Starting from the electron cryomicroscopy and x-ray crystallography images of the phage MS2, the relative effect of different energetic contributions, as well as the role of confinement, on the genome packaging was modeled via a series of biomolecular simulations in which different energy terms were systematically switched off. It appeared that the bimodal radial density profile of the packaged genome was a consequence of the RNA self-repulsion in confinement, suggesting that it should be similar for all single-stranded RNA viruses with a comparable ratio of capsid size/genome length. In contrast, the detailed structure of the outer shell of the RNA density depended crucially on the steric contributions from the capsid inner surface topography, implying that the various different polyhedral RNA cages observed in experiment were largely due to the differences in the inner surface topography of the capsid (ElSawy et al. 2011).

Bleckley and Schroeder (2012) turned to the structure of the RNA stem-loop. Since the RNA hairpin motif appeared as energetically unfavorable and the free energy predictions were biased against this motif, the authors used computer programs called Crumple, Sliding Windows, and Assembly to predict the viral RNA secondary structures when the traditional assumptions of the RNA structure prediction by free energy minimization might not apply. These methods allowed incorporation of global features of the RNA fold and motifs that were difficult to include directly in minimum free energy predictions. For example, for the MS2 RNA, the experimental data from SELEX experiments, crystallography, and theoretical calculations of the path for the series of hairpins were incorporated into the RNA structure prediction, and thus the influence of the free energy considerations was modulated. This approach thoroughly explored conformational space and generated an ensemble of secondary structures (Bleckley and Schroeder 2012).

Further progress in the RNA packing evaluation came after experimental corroboration of the nonuniform packing of the RNA genome due to the presence of a single copy of the maturation protein.

MATURATION PROTEIN

As described in Chapters 9 and 11, the RNA phage particles contain a single copy of the maturation protein, also known as the A protein, or as the A2 protein in the group III and IV phages, which binds to genomic RNA and is absolutely necessary for the infectivity of the RNA phages by their attachment to the bacterial pili. It was logical to assume that the maturation protein must be exposed at both the inner and outer surfaces of the coat shell. Because the capsids of the RNA phages contained holes at the fivefold and threefold axes of symmetry that were large enough for the diffusion of RNA fragments, it was long believed that the maturation protein was bound in proximity to these axes. However, it was recently shown by electron cryomicroscopy that the A protein actually replaces a single-coat dimer (Dent et al. 2013; Koning et al. 2016). It appears therefore that the RNA phage virion actually contains 178 coat monomers, but not 180, as was asserted in textbooks for 50 years.

It is remarkable that the definite changes in the particle structure after embodiment of the maturation protein were noticed 10 years before this discovery. To arrange it, the small-angle neutron scattering (SANS) approach was introduced into the phage MS2 studies in order to extend the structural examination of the physical MS2 characteristics in solution (Kuzmanovic et al. 2003). Specifically, the contrast variation technique was employed to determine the molecular mass of the individual components of the MS2 virion, namely, of the protein shell and of the genomic RNA, and the spatial relationship of the genomic RNA to its protein shell. Remarkably, one of the consequences of this work was the evaluation of a novel particle-counting instrument, the Integrated Virus Detection System (IVDS), which was mentioned in Chapters 5 and 9. The IVDS, in combination with SANS, demonstrated the definite potential to provide rapid quantitative physical characterization of unidentified viruses (Kuzmanovic et al. 2003). Moreover, the comparative investigation of the MS2 VLPs lacking the maturation protein and of the MS2 virions containing a copy of the maturation protein by the SANS method revealed the apparent differences, particularly the presence of "thin," or preinfection, and "thick," or postinfection, capsids (Kuzmanovic et al. 2006b). These differences were not seen by crystallography. The authors speculated that the role of the maturation protein during virus assembly might involve

the accumulation of tension that could be later used to eject the genomic RNA together with the maturation protein into a host cell. At the same time, the small-angle x-ray scattering (SAXS) method was introduced for the determination of the overall particle size and for the quantification of RNA within the MS2-like virions in solution (Kuzmanovic et al. 2006a). Further application of the small-angle scattering techniques, like SAXS and SANS, was tested with the phage MS2 as a biological size standard (Kuzmanovic et al. 2008).

In 2011, Stockley's team visualized the MS2 RNA genome poised for release from its receptor complex by electron cryomicroscopy (Toropova et al. 2011). The corresponding reconstruction of the phage MS2 bound to the bacterial F pilus was achieved at $\sim$20 Å resolution. The site of interaction was identified between virus and pilus as occurring at a viral fivefold vertex, thus localizing at least the surface-accessible portion of the single copy of the pilin-binding maturation protein to this position. As a result, the structure of the virus-receptor complex was calculated with fivefold rotational symmetry averaging, revealing for the first time in any virus the nonuniform distribution of the packaged RNA genome. Thus, the RNA density was unevenly distributed within the capsid, with distinct occupancies at different distances from the pilus. Strikingly, at the vertex that contacted the pilus, a rod of density that might include contributions from both genome and maturation protein sat above a channel that went through the capsid to the outside. This density was reminiscent of the DNA density observed in the exit channel of the double-stranded DNA phages, suggesting that the RNA-maturation protein complex was poised to leave the capsid as the first step of the infection process. The data on the analogous Qβ complex were most consistent with the same location for the Qβ maturation protein as that seen for MS2, albeit with perhaps a more flexible interaction with the pilus (Toropova et al. 2011).

After the electron cryomicroscopy structure of the phage MS2 was determined with only fivefold symmetry averaging, revealing the asymmetric distribution of its encapsidated genome, Dykeman et al. (2011) showed by mathematical modeling that this RNA distribution was consistent with an assembly mechanism that followed two simple rules derived from the experiment: (i) the binding of the MS2 maturation protein to the RNA constrains its conformation into a loop, and (ii) the capsid must be built in an energetically favorable way.

Finally, Dent et al. (2013) demonstrated the asymmetric structure of the phage MS2, attached to its receptor, the F pilus, by electron cryotomography. The subtomographic averaging of such complexes resulted in a structure containing clear density for the packaged genome, implying that the conformation of the genome was the same in each virus particle. The data also suggested for the first time that the single-copy viral maturation protein broke the symmetry of the capsid, occupying a position that would be filled by a coat protein dimer in an icosahedral shell. This capsomere could thus fulfill its known biological roles in receptor and genome binding and suggested an exit route for the genome during infection (Dent et al. 2013). Figure 21.8 demonstrates how the maturation protein replaces the coat dimer in the MS2 virion.

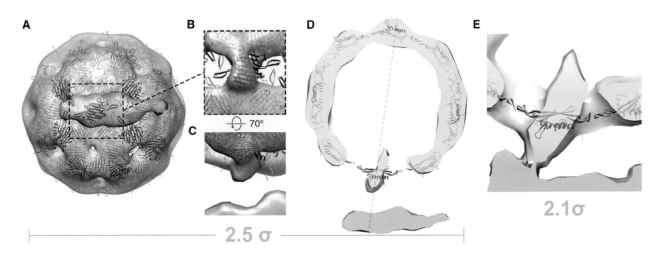

FIGURE 21.8 The maturation protein replaces a coat protein dimer. (A) A view through the pilus on to the capsid beneath at 2.5 σ. (B) A close-up from an identical viewpoint, but with the pilus density clipped away, showing the point of interaction. This position is occupied by a C/C-type dimer on a twofold axis in the icosahedrally averaged structure. (C) An oblique view of the same point from the side, showing that density at this point is different from all other 29 equivalent positions. (D) A 25-Å-thick section through the complex showing the coordinates for a C/C dimer and their poor fit to density that was ascribed to maturation protein. The disruption of the capsid structure around this point is indicated by the large pores at either side. (E) A close-up view of the interaction between virus and pilus. The section is thicker (35 Å) and at a lower contour level (2.1 σ) than in (D) to show connected density between pilus and virus. (Reprinted with permission of Cell Press from Dent KC et al. The asymmetric structure of an icosahedral virus bound to its receptor suggests a mechanism for genome release. *Structure*. 2013;21:1225–1234.)

Figure 21.9 is invented to show high potential of the electron cryotomography approach. It demonstrates fantastic quality of the image of the MS2-decorated F pili (Dent et al. 2013). Summarizing, this study made it possible for the first time to overcome the traditional oversimplified presentation of the RNA phage as a perfect icosahedral assembly.

In parallel, Roman I. Koning's team demonstrated the potential capabilities on how to increase the resolution in electron cryotomography up to 2–5 nm (Diebolder et al. 2012). It would make it possible to link the electron cryotomography to the high-resolution techniques and contribute to the molecular mapping of whole cells.

Since the resolution of the asymmetric tomographic reconstruction remained relatively low, about 39 Å only, and the molecular details were still unclear, the asymmetric MS2 genome organization was revealed further via a new analysis method, graph-theoretical analysis of tomographic data (Geraets et al. 2015). This approach opened unprecedented opportunities to analyze viral genomes, revealing conserved structural features and mechanisms that could be targeted in antiviral drug design (Geraets et al. 2015).

Koning's team improved the asymmetric structure of the phage MS2 to the resolution of 8.7 Å by single-particle electron cryomicroscopy (Koning et al. 2016). Figure 21.10 demonstrates this highly impressive reconstruction.

The map clearly outlined the maturation protein, which replaced one coat dimer. Moreover, it showed an ordered genome that was shaped as a branched network of connected

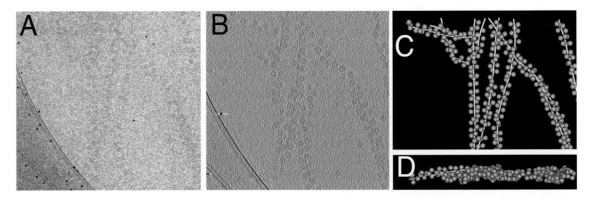

FIGURE 21.9 The electron tomography of MS2-decorated F pili. (A) The untilted (0°) image from a tilt series of an F pilus heavily decorated with MS2. To prevent radiation damage, very low doses are used (∼1e⁻/Å 2 per tilt); thus, contrast is very poor. The dark black dots are 10 nm gold particles used for alignment. (B) A section through the tomographic reconstruction of the sample in (A). (C and D) Three-dimensional models of the pili (tubes) and MS2 particles (spheres) in (A) and (B). (Reprinted with permission of Cell Press from Dent KC et al. *Structure*. 2013;21:1225–1234.)

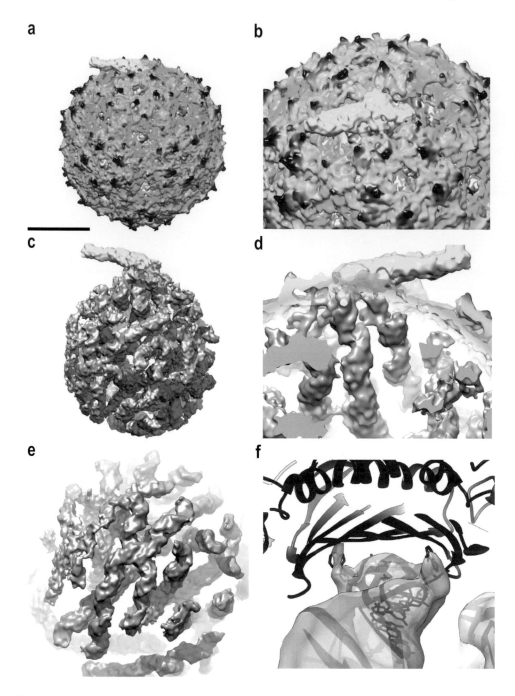

FIGURE 21.10 The asymmetric reconstruction of bacteriophage MS2. Asymmetric structure of bacteriophage MS2 (green–blue radially colored) shows the maturation protein (yellow) (a), which replaces one coat dimer (b). Inside the protein capsid a structured genome (gray) is present (c) that is connected to the maturation protein (d). The reconstruction shows the double-stranded helices in the stem loop structures (e). At some positions individual NAs connecting to the capsid are resolved, as shown by fitting of the x-ray structure of the 19-nucleotide translational operator TR (magenta) bound to the capsid (blue) (pdb: 1ZDH) in the electron microscopy density (gray) (f). Scale bar is 100 Å. (Reprinted from Koning RI et al. *Nat Commun.* 2016;7:12524.)

RNA stem-loops, the majority of which interacted with the inside of the capsid. Remarkably, the RNA-coat interactions were primarily located on one side of the capsid, which might have consequences for the genome packaging and virion assembly.

Furthermore, the MS2 structure was revealed by electron cryomicroscopy and asymmetric reconstruction at 3.6 Å resolution (Dai et al. 2017). Due to high resolution, approximately 80% of the backbone of the viral genome was traced,

the atomic models for 16 RNA stem-loops were built, and three conserved motifs of the RNA-coat protein interactions were identified among 15 of these stem-loops with diverse sequences. The stem-loop at the 3′ end of the genome interacted extensively with the maturation protein, which, with just a six-helix bundle and a six-stranded β-sheet, formed a genome-delivery apparatus and joined 89 coat protein dimers to form a capsid. This atomic description of the genome-capsid interactions in a spherical single-stranded RNA virus

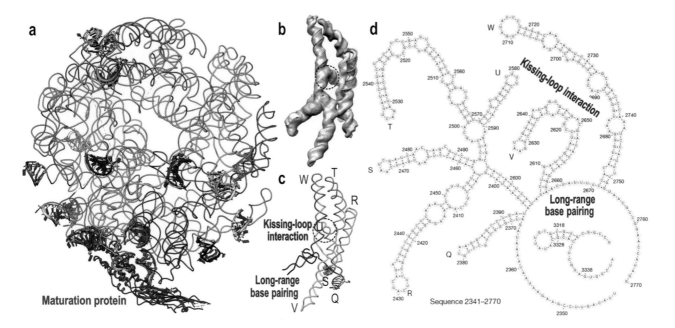

FIGURE 21.11 Modeling the single-stranded RNA genome. (a) Backbone structure of the genome (wire) and non-uniform distribution of the high-resolution stem-loops (ribbons). Backbone is rainbow-colored (blue to red) from 5′ to 3′. (b–d) Examples of tracing RNA backbone. Part of the genome density (gray in b) is segmented out and superimposed with its backbone model (rainbow-colored wire, blue to red from 5′ to 3′; b, c). For each of the two high-resolution stem-loops (ribbons in c) contained in this segment, a degenerate sequence was derived on the basis of the resolved bases and used to search against the genome to identify sequence candidates. Each of these short sequence candidates was expanded in both directions to include about 500 bases for secondary structure prediction. The predicted secondary structure was then correlated with the backbone obtained in (b) and only one of these sequence candidates yielded the correct sequence registration of individual stem-loops (indicated by letters Q–W in c, d). The backbone model reveals kissing-loop and long-range base-pairing interactions as indicated. (Reprinted by permission from Springer Nature. *Nature.* Dai X et al. Copyright 2017.)

provided the first detailed insight into the genome delivery via the host sex pilus and mechanisms underlying the RNA-capsid co-assembly (Dai et al. 2017). Figure 21.11 presents the successful modeling of the MS2 genome in question.

The outstanding success of the MS2 structure, among other novel insights into the molecular architecture of viruses, bacteria, and parasites of microorganisms that have been gained by the 3D electron microscopy, was reviewed by Cyrklaff et al. (2017). Almeida et al. (2018) included the resolution of the MS2 asymmetric structure as one of the important steps in their brief history of phage imaging.

Finally, Gorzelnik et al. (2016) resolved the electron cryomicroscopy structures of the phage Qβ with and without symmetry applied. The icosahedral structure, at 3.7 Å resolution, resolved loops not previously seen in the published x-ray structure, whereas the asymmetric structure, at 7 Å resolution, revealed maturation, or A2, protein and the genomic RNA. The A2 protein contained a bundle of α-helices and replaced one dimer of coat proteins at a twofold axis. The helix bundle bound genomic RNA, causing denser packing of RNA in its proximity, which asymmetrically expanded the surrounding coat protein shell to potentially facilitate RNA release during infection. A fixed pattern of the genomic RNA organization was observed among all viral particles, with the major and minor grooves of RNA helices clearly visible. A single layer of RNA directly contacted every copy of the coat protein, with one-third of the interactions occurring at operator-like RNA

hairpins. These RNA-coat interactions stabilized the tertiary structure of the genomic RNA within the virion, which could further provide a roadmap for the capsid assembly (Gorzelnik et al. 2016). The astonishing importance of breaking the symmetry of a viral capsid was commented on by Morais (2016).

Furthermore, Cui et al. (2017) used electron cryomicroscopy to reveal structures of Qβ virions, Qβ VLPs, and the Qβ-MurA complex at 4.7, 3.3, and 6.1 Å resolutions, respectively. The outer surface of the β-region in the A2 protein was identified as the MurA-binding interface, as described in the Lysis section of Chapter 11. Figure 11.5 illustrates these notable findings.

A detailed protocol for the evaluation of phages, with the phage MS2 as one of the experimental models, by electron cryomicroscopy was published recently (Cuervo and Carrascosa 2018). The amazing capabilities of the modern atomic electron cryomicroscopy by the resolution of virus structure were reviewed recently (Jiang and Tang 2017; Kaelber et al. 2017; Stass et al. 2018; de Ruiter et al. 2019). With the phage MS2 as a model, Zhang J et al. (2019) elaborated PIXER, an automated particle-selection method for the electron cryomicroscopy. The PIXER is based on segmentation using a deep neural network.

Finally, electron cryomicroscopy made it possible to reveal the fine structure of the phage MS2 with the *E. coli* F pilus, showing a network of hydrophobic and electrostatic interactions at the interface of the F pilus and maturation protein

(Meng et al. 2019). This study demonstrated clearly that the binding of the F pilus induced slight orientational variations of the maturation protein relative to the rest of the phage capsid, priming the maturation protein-connected genomic RNA for its release from the virions. The authors found that the exposed tip of the attached maturation protein pointed opposite to the direction of the pilus retraction, which might facilitate the translocation of the genomic RNA from the capsid into the host cytosol (Meng et al. 2019). A brief overview of the methods used to investigate non-symmetric capsid features, including the phage MS2, was published recently by Conley and Bhella (2019).

SELF-ASSEMBLY

From the very beginning, the phage MS2 became a traditional prototype by exploring the paths of viral assembly. In the 2010s, the great calculation and experimental capabilities gave a powerful incentive to the assembly studies. Thus, the highly prospective Adam Zlotnick's team have chosen a hypothetical icosahedral 30-mer as a model (Moisant et al. 2010). In the context of this capsid, each dimer-like subunit was tetravalent and had twofold symmetry. The subunits were analogous to those found in many viruses, including MS2, hepatitis B virus core, cowpea chlorotic mottle virus, brome mosaic virus. The authors examined the complete set of intermediates available for the assembly of a hypothetical VLP and the connectivity between these intermediates in a graph-theory-inspired study. Using a buildup procedure, assuming ideal geometry, the complete set of 2,423,313 species for formation of an icosahedron from 30 dimeric subunits was enumerated. The stability of each n-subunit intermediate was defined by the number of contacts between subunits, but the probability of forming an intermediate was based on the number of paths to it from its predecessors. When defining population subsets predicted to have the greatest impact on assembly, both stability- and probability-based criteria selected a small group of compact and degenerate species; ergo, only a few hundred intermediates made a measurable contribution to assembly (Moisant et al. 2010). The comprehensive reviews of Zlotnick's team always demonstrated remarkable interest to the RNA phage assembly (Zlotnick 2005; Katen and Zlotnick 2009; Zlotnick and Mukhopadhyay 2011; Zlotnick 2015; Kondylis et al. 2018; Wang JC et al. 2018).

In parallel, the modern experimental approaches were introduced actively by the assembly studies in RNA phages. Thus, the *in situ* grazing-incidence small-angle x-ray scattering (GISAXS) was used to follow the self-assembly process of the MS2 and Qβ VLPs in real time (Ashley et al. 2011b). In contrast to electron microscopy or optical methods, the GISAXS approach enabled the *in situ* characterization of nanostructure during dynamic assembly processes in real time under ambient environments and over large areas. In the GISAXS, an x-ray beam was incident upon a sample at an angle greater than the critical angle of the film but less than that of the substrate, thus maximizing the scattering volume inside the film. Coupled with the high photon flux obtained at a synchrotron source, it enabled the investigation of fast (on the time scale of seconds) self-assembly phenomena of films as thin as one monolayer. By this approach, the rapid assembly of the icosahedral MS2 and Qβ VLPs into highly ordered (domain size >600 nm), oriented 2D superlattices was demonstrated directly onto a solid substrate using convective coating. From water, the GISAXS data were consistent with a transport-limited assembly process where convective flow directed assembly of the VLPs into a lattice oriented with respect to the water drying line. The addition of a nonvolatile solvent, e.g., glycerol, modified this assembly pathway, resulting in non-oriented superlattices with improved long-range order. The modification of electrostatic conditions, such as solution ionic strength, substrate charge, also altered assembly behavior (Ashley et al. 2011b).

Stockley's team employed single-molecule fluorescence correlation spectroscopy (smFCS) to selectively monitor coat protein or viral RNA components in the *in vitro* reassembly reactions (Borodavka et al. 2012). Such assays were carried out at low concentrations, ≤1 μM, allowing observation of mechanistic features that were not dominated by high coat concentrations. The smFCS assays were applied in parallel to the two viral model systems, namely, the phage MS2 and satellite tobacco necrosis virus, which represented a large number of plant RNA viruses. The differences in the virus architecture made it possible to identify the conserved and distinct mechanistic processes. The SmFCS technique revealed that there was no simple correlation between RNA length and hydrodynamic radius. The viral RNAs were larger in the absence of their coats than the capsids into which they must eventually fit. Remarkably, instead of a steady condensation of the RNA, which would be expected by a charge neutralization mechanism, the addition of coats to their cognate RNAs resulted in a rapid collapse (<1 min) in the solution conformation. The collapse depended on the protein–protein interactions and did not occur on nonviral RNA controls or with the noncognate viral RNA, showing that it depended on the specific RNA-coat interactions mediated by the sequence and structure of each genome. The collapsed state was smaller than the capsid and appeared to consist of complexes with substoichiometric amounts of coat proteins with respect to capsids, but with roughly the correct shell curvature. The full complement of coats was recruited in a second slower stage of assembly. The nonviral RNAs supported assembly inefficiently and with much lower fidelity. Therefore, the RNA packaging occurred as a two-stage process (Borodavka et al. 2012). The two-stage mechanism of viral RNA compaction revealed by the SmFCS approach was presented in further detail by Borodavka et al. (2013).

In a comprehensive review at this stage of the investigations, Stockley et al. (2013a) opposed the traditional view on the coats as the dominant components by the self-assembly. In fact, this view was reinforced because in many systems the coats formed VLPs *in vitro* in the absence of RNA, or in the presence of noncognate RNA, polyanions, or even nanoparticles. The studies of Stockley's team led to the conclusion that the genomic RNAs could in fact play many active and cooperative roles during virion assembly. These facilitated rapid and faithful assembly *in vivo*, conferring selective advantages that would not arise if the RNA was merely a passive scaffold. Such roles were irrelevant or undetectable in many *in vitro* studies but have major impacts on viral evolution (Stockley et al. 2013a).

It is noteworthy that the hydrodynamic radii of a wide range of biologically relevant long RNA molecules including MS2 RNA were determined at that time at low nanomolar concentration using fluorescence correlation spectroscopy (Borodavka et al. 2016).

In parallel, Dykeman et al. (2013a) constructed a generalized framework for modeling assembly that incorporated the regulatory functions provided by cognate protein-nucleic acid interactions between capsid proteins and segments of the genomic RNA, called packaging signals, into the model. The yield strongly depended on the distribution and nature of the packaging signals, highlighting the importance of the crucial roles of the RNA in this process. Moreover, Dykeman et al. (2013b) combined the conserved assembly mechanism and geometry of the packaged genome with the multiple RNA packaging signals. The authors determined the existence of such signal within viral genomes, which facilitated assembly by binding coat proteins in such a way that they promoted the protein–protein contacts needed to build the capsid. The binding energy from these interactions enabled the confinement or compaction of the genomic RNAs. The location of the packaging signals was determined using Hamiltonian paths, a concept from graph theory, in combination with bioinformatics and structural studies. Concerning their sequence, the packaging signals had a common secondary structure motif but distinct consensus sequences and positions within the studied MS2 and satellite tobacco necrosis virus genomes. Despite these differences, the distributions of the packaging signals in both viruses implied defined conformations for the packaged RNA genomes in contact with the protein shell in the capsid, consistent with a recent asymmetric structure determination of the MS2 virion (Dykeman et al. 2013b). The features of the MS2 assembly were compared further with the sequence-specific RNA-coat interactions in the assembly of satellite tobacco necrosis virus, when different RNA aptamers were employed (Ford et al. 2013). In a thorough review, Stockley et al. (2013b) opposed the previous theory that the assembly process could be explained entirely by electrostatics. The authors presented an alternative theory which recognized the important cooperative roles played by the RNA-coat protein interactions at the packaging signal sites. The hypothesis was formulated in a way that multiple copies of packaging signals, repeated according to capsid symmetry, aided formation of the required capsid protein conformers at defined positions, resulting in significantly enhanced assembly efficiency. These insights paved the way to a new antiviral therapy, reducing capsid assembly efficiency by targeting of the vital roles of the packaging signals, and opened up new avenues for the efficient construction of protein nanocontainers in bionanotechnology (Stockley et al. 2013b). The direction to a unique antiviral strategy was strengthened by a successful attempt

to solve a Levinthal's paradox (Dykeman et al. 2014). The authors formulated the viral equivalent of the well-known Levinthal's paradox as the selection of the assembly intermediates and pathways best suited to enhance viral load and the navigation among the vast numbers of possible pathways and potential kinetic traps in a limited amount of time. For the phage MS2, there were on the order of 10^{15} different possible intermediate species. If each of these were explored, the capsid assembly would be excessively slow, or even worse, potentially result in kinetic traps where capsid proteins formed stable off-pathway intermediates. This contrasted with what was observed in nature: the assembly of capsids was highly efficient and robust to kinetic trapping. Using an *in silico* assembly model, Dykeman et al. (2014) showed that a vital step in resolving the viral assembly paradox lay in the timing of viral coat production, the protein "ramp," i.e., the accumulation of coat that would naturally occur in the *in vivo* process. The protein ramp, or in other words the increasing coat concentration that occurred in infected cells, played unexpected and vital roles in avoiding potential kinetic assembly traps, significantly reducing the number of assembly pathways and assembly initiation sites and resulting in enhanced assembly efficiency and genome packaging specificity (Dykeman et al. 2014). Because capsid assembly is a vital determinant of the overall fitness of a virus in the infection process, it was concluded that these insights have important consequences for the understanding of how selection impacts on the evolution of viral quasispecies. Moreover, these results suggested novel strategies for optimizing the production of protein nanocontainers for drug delivery and of VLPs for vaccination. The *in silico* modeling showed that drugs targeting the specific RNA-capsid protein contacts could delay assembly, reduce viral load, and lead to an increase of misencapsidation of cellular RNAs, hence opening up unique avenues for antiviral therapy (Dykeman et al. 2014).

At the same time, Michael F. Hagan and colleagues performed an impressive line of atomic simulations on the phage MS2 (Perkett et al. 2012, 2016; Hagan 2014; Perlmutter and Hagan 2015). When Harvey et al. (2013) viewed the icosahedral RNA virus as a grotto, the phage MS2 seemed to be organizing its genome into a stalagmite, but not stalactite, structure inside the mature virus. Kelly et al. (2016) developed a Flory mean-field theory to evaluate the sequence dependence of viral RNA encapsidation and the genomic Qβ RNA was used as a model. This approach extended the current RNA folding algorithms to include interactions between different sections of the secondary structure. The theory was applied to the sequence-selective RNA encapsidation.

To the general search for the prospective universal building block of self-assembly in coats, Polles et al. (2013) elaborated a general computational scheme for identifying the stable domains of a given viral capsid. The method was based on elastic network models and quasi-rigid domain decomposition and was first applied to a heterogeneous set of well-characterized viruses, such as MS2, cowpea chlorotic mottle virus, satellite tobacco necrosis virus, and satellite tobacco mosaic virus, for which the known mechanical or assembly domains were correctly identified. The validated method was next applied to other viral particles such as L-A, Pariacoto, and polyoma viruses, whose fundamental functional domains were still unknown or debated, and for which verifiable predictions were formulated. As a result of this study, the numerical code implementing the domain decomposition strategy was made freely available (Polles et al. 2013).

In order to directly confirm the presence of the multiple packaging signals within the MS2 genome, the contact points between the proteins and RNA components of intact virions were determined in a great study of Rolfsson et al. (2016). Using crosslinking coupled to matrix-assisted laser desorption/ionization mass spectrometry and CLIP-Seq sequencing, the peptide and oligonucleotide sequences were determined at the interfaces between the capsid proteins and the genomic RNA of the phage MS2. The results suggested that the same coat protein-RNA and maturation protein-RNA interfaces were used in every viral particle. The portions of the viral RNA in contact with coat subunits span the genome, consistent with a large number of discrete and similar contacts within each particle. Moreover, the previous predictions of the packaging signal sites in contact with the phage coat were remarkably accurate, unambiguously confirming the packaging signal-mediated assembly of this virion. The chemical RNA footprinting was used to compare the secondary structures of protein-free genomic fragments and the RNA in the virion. Some packaging signals were partially present in protein-free RNA, but others would need to refold from their dominant solution conformations to form the contacts identified in the virion. Figure 21.12 presents distribution of the RNA sequences in the MS2 virion that contact the capsid proteins.

The data gained also made it possible to assign the functions of the single-copy maturation protein to separate domains, the N-terminal domain seemingly encompassing the RNA-binding function, while the pilus-binding function lies within the C-terminal portion. The sequence comparisons with other phage maturation proteins suggested that the C-terminal portion was a conserved arrangement. Figure 21.13 shows the maturation sequences homologies in connection with the role in the RNA contacts.

In addition, the RNA structure probing was used to monitor the presence of the packaging signal sites in the coat-free genome transcripts or the virion, revealing that the packaging signals fell into at least two classes, those that were present at least some of the time in the protein-free RNA and those that must refold as assembly proceeded. An ordered series of induced-fit interactions, similar to those occurring during ribosome assembly, likely accounted for the MS2 virion formation and presumably for assembly of other single-stranded RNA viruses using the packaging signal-mediated mechanism (Rolfsson et al. 2016).

The full assembly instruction manual for the capsid, as it was encoded by the MS2 RNA and deciphered by Stockley's team, was reviewed in detail and compared with the packaging signal-mediated assembly of the single-stranded RNA plant virus, namely, the satellite tobacco necrosis virus (Stockley et al. 2016). Figure 21.14 presents an asymmetric tomographic MS2 structure as a result of the packaging signal-mediated assembly, while Figure 21.15 shows the full cycle of the MS2 assembly. The latter figure is taken from ElSawy (2017), who turned to the thermodynamics and calculated the impact of the MS2 RNA on the association free energies of coat protein assembly.

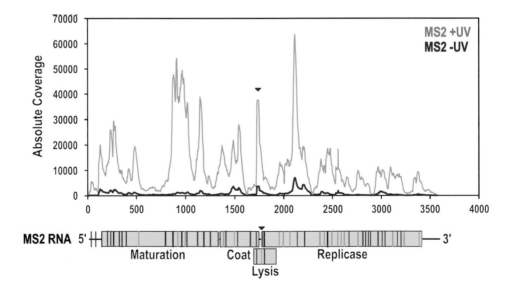

FIGURE 21.12 RNA sequences in the MS2 virion that contact the capsid proteins. Abundances of cDNA fragments that co-purified with the MS2 coat protein. The sequences for the cDNAs identified by Illumina DNA sequencing were aligned with the MS2 genome, and histograms were produced denoting the frequency of particular sequences within the datasets for irradiated (green) and control (blue) samples. The peak that corresponds to the MS2 operator hairpin is identified by an inverted blue triangle. A schematic of the MS2 genome is shown below the graph to allow identification the approximate locations of the most significant peaks. Lines in blue highlight those sites matching previous packaging signal predictions by Dykeman et al (2013b); those in orange, similarly for Bleckley and Schroeder (2012); and green, for matches to both predictions. Black lines indicate the 10 peaks not predicted previously. (Reprinted from Rolfsson Ó et al. *J Mol Biol.* 2016;428:431–448.)

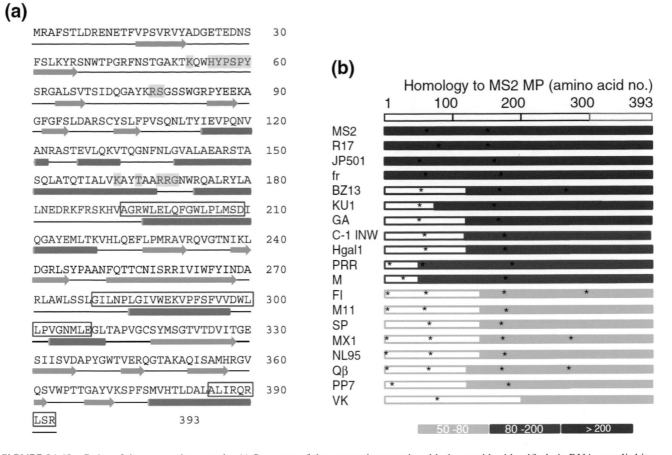

FIGURE 21.13 Roles of the maturation protein. (a) Sequence of the maturation protein with the peptides identified via RNA crosslinking and peptide (RCAP) mapping assay highlighted in red. The secondary structure elements of the maturation protein are predicted by a Jpred3 algorithm. The predicted RNA-binding residues (RNABindRPlus) are highlighted in orange. Peptides that are highly conserved across both *Levivirus* and *Allolevivirus* phage maturation proteins are boxed (motif 1, amino acids 193–210, 51% average identity over 18 amino acids; motif 2, amino acids 279–308, 56% average identity over 30 amino acids; motif 3, amino acids 385–393, average 52% identity over 9 amino acids). These regions may therefore be part of the pilin-binding site. (b) Sequence homology of the *Leviviridae* maturation protein sequences. Filled bars represent regions homologous to the MS2 maturation protein and open boxes represent non-homologous regions. These bars cover the full length of each protein, without indicating the short gaps required to accommodate the alignment. Most proteins in this comparison are ~400 amino acids long. Colors represent the homology score using the BLOSUM62 matrix. Potential RNA-binding sites predicted by RNABindRPlus are represented by asterisks. (Reprinted from Rolfsson Ó et al. *J Mol Biol*. 2016;428:431–448.)

The triumphant march of the packaging signal-mediated self-assembly, the classics of which were ensured by the phage MS2, was continued further by the corresponding structures of the above-mentioned satellite tobacco necrosis virus (Patel et al. 2017) and human parechovirus, a representative of the *Picornaviridae* family (Shakeel et al. 2017). The idea of the general RNA-mediated instruction manual and the packaging signals as conservative parts of genome were interpreted in light of virus evolution and further development of antiviral therapy (Bingham et al. 2017). Thus, a quasispecies-based model of a viral infection was introduced. This model incorporated structural and mechanistic knowledge of the packaging signal function in assembly to construct a phenotype-fitness map, capturing the impact of this RNA code on assembly yield and efficiency. The details of viral replication and assembly inside an infected host cell were coupled with a population model of a viral infection, allowing the occurrence of therapy resistance to be assessed in response to drugs

inhibiting packaging signal recognition. More specifically, the stochastic simulations of viral quasispecies evolution in chronic HCV infection under drug action and/or immune clearance revealed that the drugs targeting all RNA signals in the assembly code collectively had a high barrier to drug resistance, even though each packaging signal in isolation had a lower barrier than conventional drugs. This MS2-initiated study suggested that the drugs targeting the RNA signals in the assembly code could be promising routes for exploitation in the antiviral drug design (Bingham et al. 2017).

The large-scale *ab initio* computation centered on critical aspects of the consensus MS2 protein-RNA interactions recognition motif led to the localization of the exact binding sites with the strongest hydrogen bonding (Poudel et al. 2017; Poudel and Ching 2018). This interesting study was commented on in detail in the Repressor complex I: MS2 section of Chapter 16.

Eric Charles Dykeman (2017) generated the stochastic assembly model that incorporated explicit nucleotide

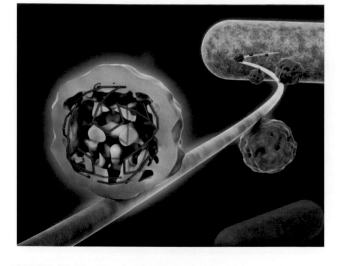

FIGURE 21.14 The implications of packaging signal-mediated assembly for the earliest stages of infection. Shown is the asymmetric tomographic structure of MS2, at low resolution, bound to its initial cellular receptor, the *E. coli* F pilus. This structure shows extensive density for the viral RNA, confirming the idea that packaging signal-mediated assembly leads to almost identical RNA conformations within each viral particle. The contact to the pilus is mediated by the maturation protein, MP, which also binds to specific sites close to either end of the viral genome. Only the MP-RNA complex enters the bacterial cell, leaving the coat shell largely intact. MP replaces a C/C capsomere in an otherwise icosahedral protein shell. (Reprinted from Stockley PG et al. *Bacteriophage.* 2016;6:e1157666.)

sequence information as well as simple aspects of the RNA folding that would be occurring during the RNA/capsid co-assembly process. Applying this paradigm to a dodecahedral viral capsid, a computer-derived nucleotide sequence was evolved *de novo* that was optimal for packaging the RNA into capsids, while also containing capacity for the coding of a viral protein (Dykeman 2017).

Recently, Reidun Twarock et al. (2018b) highlighted the role of the Hamiltonian path analysis of viral genomes. In fact, the Hamiltonian path analysis is a mathematical abstraction of virus assembly pathways, simultaneously encoding the order in which capsomers are recruited to the growing capsid shell along different assembly pathways. It captured geometric constraints on the packaging signal positions in the linear genomic sequence that arose from the relative positions of the RNA-coat-binding sites in the inner capsid surface (Twarock et al. 2018b). The modeling for the RNA virus assembly was reviewed by Twarock et al. (2018a) and by Twarock and Stockley (2019).

Chechetkin and Lobzin (2019) examined the genome packaging and large-scale segmentation in viral genomic sequences of the phage MS2 in parallel with other icosahedral viruses carrying not only single-stranded RNA, but also single- and double-stranded DNA genomes. Thus, combining discrete direct and double Fourier transforms, the significant quasi-regular segmentation of genomic sequences related to the virion assembly and the genome packaging within icosahedral capsid. The calculations corresponded well to the available electron cryomicroscopy data on the capsid

structures and genome packaging in these viruses. The complex story of the packaging of genomic RNA in the positive-sense single-stranded RNA viruses was reviewed recently by Comas-Garcia (2019).

Concerning the charge of the RNA phage particles, Lošdorfer Božič and Podgornik (2017) studied the pH dependence of charge multipole moments in the MS2 particles together with three other model proteins. The detailed representation of the charges on the proteins was in this way replaced by the magnitudes and orientations of the multipole moments of varying order. Focusing on the three lowest-order multipoles, namely, the total charge, dipole, and quadrupole moment, it appeared that the value of pH influenced not only their magnitudes, but more notably also the spatial orientation of their principal axes. Moreover, Lošdorfer Božič and Podgornik (2018) examined the pH-dependent interplay of charge on both N-terminal tails and whole coats of the icosahedral, positive-sense single-stranded RNA viruses, including RNA phages. They concluded that, in contrast to the charge on the coats, the net positive charge on the N-tails persisted even to very basic pH values. Moreover, Lošdorfer Božič et al. (2018) turned to the robustness of the sequence-structure interplay involved in determining the size of RNA folds by mutating the wild-type genomes of the phage MS2 and brome mosaic virus using three different mutation schemes. The authors mutated either continuous stretches of nucleotides, spaced at regular intervals along the genome, or stretches of nucleotides sharing a similar distance relationship from the central branching hub of the wild-type fold. The dispersed mutations picked uniformly along the genome were also considered to provide a comparison with existing studies. As a result, mutating local stretches of the genomes was less disruptive to their compactness compared to inducing randomly distributed mutations. Therefore, a mechanism for the conservation of compactness was encoded on a global scale of the genomes (Lošdorfer Božič et al. 2018). Recently, Wang B et al. (2019) proposed the first multiscale model to simulate the assembling process of coat proteins in the phage MS2.

LATTICES

Aloysio Janner has explored the use of lattices to explain the features of the three-dimensional geometry and genome organization in viruses, including the phage MS2 as a pet subject (Janner 2006, 2008, 2010, 2011a,b, 2013, 2016). Reidun Twarock and Peter G. Stockley and their colleagues introduced a method that focused on the symmetry group of the underlying lattice and worked with quasi-lattices, i.e., structures with long-range order lacking periodicity (Keef et al. 2013). They considered Janner's lattices as approximations of quasi-lattices with icosahedral symmetry and described the structure of the phage MS2 as one of the classical models. This new concept in virus biology provided for the first time predictive information on the structural constraints on coat protein and genome topography and revealed a previously unrecognized structural interdependence of the shapes and sizes of different viral components (Keef et al. 2013). Moreover, this

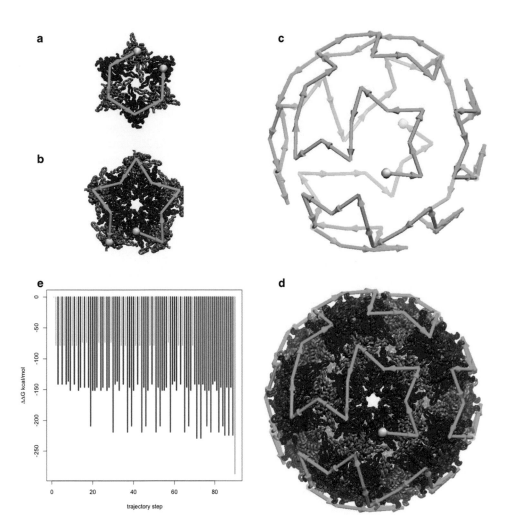

FIGURE 21.15 Implications for MS2 virus assembly. (a, b) The three- and fivefold symmetric intermediates that are detected experimentally during MS2 virus capsid reassembly. (c) The computed full-capsid dimer-dimer assembly trajectory. (d) The dimer-dimer assembly trajectory superposed on the MS2 capsid. (e) The corresponding stepwise intermediate stabilization ($\Delta\Delta G$). In (a–d), the assembly trajectory is shown in green (a white sphere marks the trajectory start and a yellow sphere marks the trajectory end) and the capsid proteins A, B, and C are shown in blue, red, and gray. In (e), trajectory steps that involve formation of one, two, three, and four interfaces are shown in gray, blue, red, and green, respectively. (Reprinted by permission from Springer Nature from *J Mol Model.*, ElSawy KM. Copyright 2017.)

study opened up the possibility of distinguishing the structures of different viruses with the same T-number, suggesting a refined viral structure classification scheme. The relations between viruses and geometry, with a sight on the antiviral therapy, were reviewed briefly by Reidun Twarock (2016).

In parallel, Matthew B. Francis' team used DNA origami as a scaffold and performed the one-dimensional arrangement of virus capsids with nanoscale precision (Stephanopoulos et al. 2010a). To do this, the interior surface of the MS2 VLPs was modified with fluorescent dyes as a model cargo, but the unnatural amino acid on the external surface (for detail, see the Functionalization: MS2 section of Chapter 20) was then coupled to DNA strands that were complementary to those extending from origami tiles, while the two different geometries of DNA tiles, rectangular and triangular, were used. The capsids associated with tiles of both geometries with virtually 100% efficiency under mild annealing conditions, and the location of capsid immobilization on the tile could be controlled by the position of the probe strands. The rectangular

tiles and capsids could then be arranged into one-dimensional arrays by adding DNA strands linking the corners of the tiles. The resulting structures consisted of multiple capsids with even spacing of approximately 100 nm. This hierarchical self-assembly made it possible to position the MS2 particles with unprecedented control and allowed the future construction of integrated multicomponent systems from biological scaffolds using the power of rationally engineered DNA nanostructures (Stephanopoulos et al. 2010a). Furthermore, Wang et al. (2014) presented hierarchical assembly of plasmonic nanostructures using MS2 scaffolds on DNA origami templates. These specific nanostructures were used as programmable scaffolds that provided molecular-level control over the distribution of fluorophores and nanometer-scale control over their distance from a gold nanoparticle antenna. While previous research using DNA origami to assemble plasmonic nanostructures focused on determining the distance-dependent response of single fluorophores, here the authors addressed the challenge of constructing hybrid nanostructures that presented an organized ensemble of fluorophores and then investigated the plasmonic response. By combining finite-difference time-domain numerical simulations with atomic force microscopy and correlated scanning confocal fluorescence microscopy, it was found that the use of the scaffold kept the majority of the fluorophores out of the quenching zone, leading to increased fluorescence intensity and mild levels of enhancement. Therefore, these bioinspired plasmonic nanostructures provided a flexible design for manipulating photonic excitation and photoemission (Wang et al. 2014). The full and lavishly illustrated path of polyhedra building with many references on the phage MS2 was presented by Kaplan et al. (2014).

Recently, Shih et al. (2015) proposed a novel mathematical approach to the Platonic solid structure of MS2 particles. Since the overall shape of MS2 is spherical, it has long been known that it is difficult to see the three-dimensional figure in electron microscopy two-dimensional images. In this connection, the authors introduced a mathematical method to predict solid 3D figures before performing 3D reconstruction. As a result, the MS2 particle was confirmed to be icosahedron, but not dodecahedron or a pentakis dodecahedron.

A computational framework, *de novo* RNP modeling in real-space through assembly of fragments together with electron density in rosetta (DRRAFTER), was elaborated which made it possible to build a full-atom model of 1508 resolved nucleotides of the packaged MS2 genome (Kappel et al. 2018).

AP205

The first 3D structure of the most enigmatic RNA phage AP205 was achieved by electron cryomicroscopy at relatively low resolution estimated to be between 17 Å and 24 Å (van den Worm et al. 2006). This structure is presented in Figure 21.6. Figure 21.2 presents x-ray crystal structure that was resolved to the AP205 coat dimer at a resolution of 1.7 Å and then fitted into a 6.6-Å resolution map obtained by electron cryomicroscopy (Shishovs et al. 2016).

It is noteworthy, however, that the phage AP205, due to the initiative of Kaspars Tārs' group, served as a model by the pioneering 3D studies using solid-state nuclear magnetic resonance under ultra-fast magic-angle spinning (MAS) (Barbet-Massin et al. 2013, 2014; Andreas et al. 2015). As an appropriate result, the phage AP205 contributed to the first examples of protein structure determination by the MAS NMR (Andreas et al. 2016; Stanek et al. 2016; Cala-De Paepe et al. 2017). Finally, dynamic nuclear polarization (DNP) made it possible to overcome the sensitivity limitation of the MAS NMR experiments (Jaudzems et al. 2018). Thus, the high-quality DNP-enhanced NMR spectra were obtained for the phage AP205 by combining high magnetic field (800 MHz) and fast MAS (40 kHz). These conditions yielded enhanced resolution and long coherence lifetimes, allowing the acquisition of resolved 2D correlation spectra and of previously

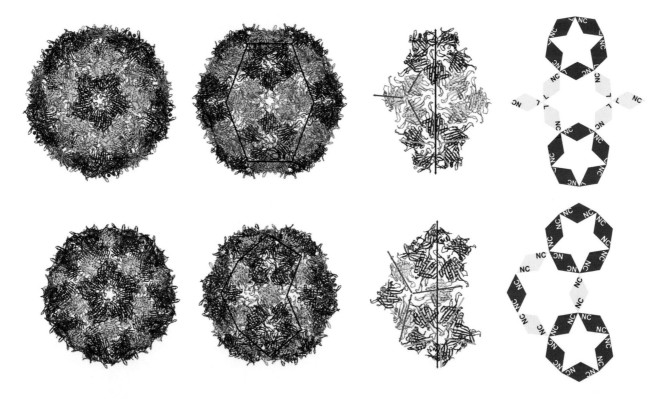

FIGURE 21.16 The electron cryomicroscopy structure of the PP7-PP7 virus-like particle [top, T = 4 (h = 2; k = 0) structure, PDB ID 6N4 V] compared to the previously published structure of the PP7 particle [bottom, T = 3 (h = 1; k = 1) subunits colored the same way, PDB ID 1DWN]. (left) View down the fivefold symmetry axes; (middle) view down the twofold symmetry axis; (right) a cartoon representation showing the organization of the coat-protein dimers in the T = 4 structure. The linker loop is labeled "L," and the termini are labeled "NC." (Reprinted with permission from Zhao L, Kopylov M, Potter CS, Carragher B, Finn MG. Engineering the PP7 virus capsid as a peptide display platform. *ACS Nano*;13, 4443–4454. Copyright 2019 American Chemical Society.)

unfeasible scalar-based experiments. This enabled the assignment of aromatic resonances of the AP205 coat protein and its packaged RNA, as well as the detection of long-range contacts (Jaudzems et al. 2018).

The pioneering role of the phage AP205 model by the development of the MAS NMR methodology of proteins was acknowledged in several exhaustive reviews (Quinn et al. 2015, 2018; Quinn and Polenova 2017; Linser 2017).

The AP205 capsid was selected as the first protein target in the Critical Assessment of Structure Prediction (CASP12) program, which was intended to present modeling efforts for targets with no obvious templates of high sequence/structure similarity in the Protein Data Bank (PDB) (Abriata et al. 2018; Kryshtafovych et al. 2018; Mishra et al. 2018).

CHIMERIC VLPs

The first 3D images for the chimeric VLPs were obtained by M.G. Finn's team (Zhao et al. 2019). They generated numerous chimeric derivatives of the PP7 VLPs as one of the most promising peptide display platforms. Generally, the PP7 VLPs were shown to tolerate the display of sequences from 1 kDa, a cell penetrating peptide, to 14 kDa, the Fc-binding double Z-domain, on its exterior surface as C-terminal

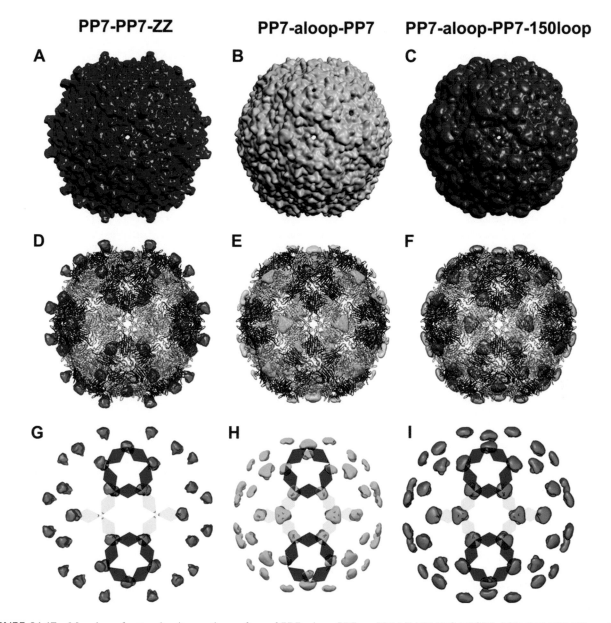

FIGURE 21.17 Mapping of extra density on the surface of PP7-a-loop-PP7, or PP7-LEAEMDGAKGRL-PP7, PP7-PP7-ZZ, and PP7-a-loop-PP7-150-loop, or PP7- LEAEMDGAKGRL-PP7-NDTGHETDEN, particles. Overall surface density maps: (A) PP7-PP7-ZZ, (B) PP7-a-loop-PP7, and (C) PP7-a-loop-PP7-150-loop. Extra densities highlighted: (D) PP7-PP7-ZZ, (E) PP7-a-loop-PP7, and (F) PP7-a-loop-PP7-150-loop, comparing to PP7-PP7. (G) PP7-PP7-ZZ, (H) PP7-a-loop-PP7, and (I) PP7-a-loop-PP7-150-loop, showing the same loop insertions and C-terminal extensions as in (D−F). Subunit organization in (G−I) is indicated as in Figure 21.16. (Reprinted with permission from Zhao L, Kopylov M, Potter CS, Carragher B, Finn MG. Engineering the PP7 virus capsid as a peptide display platform. *ACS Nano*;13, 4443–4454. Copyright 2019 American Chemical Society.)

genetic fusions to the coat protein. In addition, a dimeric construct allowed the presentation of exogenous loops between capsid monomers and the simultaneous presentation of two different peptides at different positions on the icosahedral structure. The PP7 particle was thereby significantly more tolerant of these types of polypeptide additions than Qβ and MS2, which remained in common use. Moreover, for the first time, the appearance of a T = 4 structure was shown for the *Leviviridae*-derived capsids, adopted unexpectedly by particles self-assembled from coat dimers. (Zhao et al. 2019).

First, to explore the parameters of exterior surface display, the authors have chosen the Z-domain, a 58-amino acid analog of the IgG-binding domain B in Staphylococcal Protein A, or a dimeric version ZZ joined by a four-amino acid linker. As described in Chapter 20, Finn's team previously fused these polypeptides to the Qβ coat as C-terminal extensions and were able to prepare hybrid particles bearing 24 Z-domains (Brown et al. 2009) or more than 30 ZZ-domains (Fiedler et al. 2018) per particle when expressed with the wild-type capsid protein. The self-assembling properties of the PP7 capsid were tested by appending Fc-binding domains in three different ways: ZZ-domains (14.2 kDa) added to the N-terminus (designated ZZ-PP7), single Z-domains at both the N- and C-termini of the PP7 CP (Z-PP7-Z), and ZZ-domains fused to the C-terminus (PP7-ZZ). When coexpressed with the wild-type coat, all three constructs were found to successfully self-assemble into icosahedral hybrid particles bearing significant numbers of polypeptide extensions, isolated in good yields.

Moreover, with an eight-amino acid linker to connect the ZZ-domain extension to the coat, it was possible to express and isolate intact ZZ-PP7 and PP7-ZZ particles, each with 180 copies of the ZZ-domain and each exhibiting similar hydrodynamic radii as their hybrid versions of ~22 nm. The difference between PP7 and Qβ in this regard was striking: the authors found no extended Qβ capsid protein that would be able to assemble into a discrete particle.

To decrease the number of the extension incorporated onto the particle, the single-chain PP7 dimer was applied, in which the N-terminus of one coat was linked to the C-terminus of another, by a four-amino acid sequence AYGG instead of the dipeptide YG employed by Caldeira and Peabody (2007). The ZZ-domain was fused onto the C-terminus of the dimer with an eight-amino acid linker GGPSESGA to generate a protein that assembled in high yield into particles designated PP7-PP7-ZZ. All of the Z-domain-bearing particles described above were found to bind to an antibody Fc domain.

Several other functional peptides were then added to the C-termini of the PP7 coat and its dimeric analog, in each case after a common eight-amino acid spacer sequence GGASESGA. The authors were able to successfully produce intact PP7 constructs from both proteins alone (in other words, without the truncated capsid protein) bearing the following peptide extensions: a transferrin recognition peptide HAIYPRH, an EGFR recognition peptide NPVVGYIGE RPQYRDL, and standard ovalbumin SIINFEKL and ISQ AVHAAHAEINEAGR epitopes. In addition, two other functionally interesting peptides—a repeat of the NANP motif from *Plasmodium falciparum* circumsporozoite protein NANPNVDPNANPNANPNANPNANP and a *Trypanosoma* trans-sialidase epitope ATIENRDVM— were successfully displayed on the single-chain PP7 dimer particle.

For the dual display, the four Zika virus epitopes from the loops of the antigenic domain III of the envelope protein were selected. These epitopes are listed in Table 22.1 of Chapter 22. To simultaneously present two epitopes, the NDTGHETDEN sequence was added to the C-termini of the PP7- LEAEMDGAKGRL-PP7 and PP7- VEFKDAH AKRQTVVV-PP7 capsid proteins.

The thermal stability studies showed that the PP7 platform, regardless of how it was manipulated by dimerization, extension, or loop insertion, reliably maintained stability to ~80°C if it formed particles at all. Appending ZZ domains to either terminus of the PP7 coat monomer, or to the N-terminus of the PP7-PP7 coat dimer, gave equivalently stable particles.

When the PP7-PP7, PP7-PP7-ZZ, and PP7- LEAEMDGA KGRL-PP7 particles were resolved by electron cryomicroscopy at ~3 Å resolution for each, the surprising finding arose that all of the dimer-based structures, including the simple PP7-PP7 dimer, formed T = 4 capsids. This stood in contrast to the T = 3 structure reported for the monomeric PP7 particle and confirmed for the ZZ-PP7 structure. Figure 21.16 presents these breakthrough unexpected data.

Zhao et al. (2019) visualized excellently the extra peptides on the PP7-PP7-ZZ and PP7- LEAEMDGAKGRL-PP7 particles using low-pass filtered maps displayed at low threshold. Figure 21.17 presents the first 3D visualization of the chimeric particles.

The PP7-PP7-ZZ particle, composed of 120 proteins, showed 60 doublet protrusions, while the extra densities of the PP7- LEAEMDGAKGRL-PP7 particle had 20 trileaves that were formed by the clustering of three loops at each of 20 threefold axis sites and 60 monoleaves that were formed by single loops. These great images therefore provided the first insight as to how the covalent dimeric coats are organized and the extra peptides are displayed. Examination of the dual-epitope presenting PP7 particles showed the two types of display to be compatible with each other.

Finally, Zhao et al. (2019) performed the simultaneous construction of a bifunctional particle with both targeting and catalytic function. To do this, they expressed the thermostable cytosine deaminase tagged with the Rev peptide at the N-terminus with each of the PP7-PP7-OVA1 and PP7-PP7-OVA2 VLP constructs. The resulting particles contained an average of 15−25 cytosine deaminase enzymes per particle.

22 VLP Vaccines

Necessity is the mother of invention.

Plato
The Republic

There is nothing impossible to him who will try.

Alexander The Great

GENERAL

The vaccine candidates represent the most advanced field of the RNA phage VLP applications due to the excellent and well-established scaffold properties and structural tolerance to decoration by foreign immunogenic sequences. Such decoration can be performed both genetically and chemically, and the VLP scaffold may provide foreign epitopes with a strong T cell response. Moreover, the RNA phage VLPs can serve as nanocontainers that can encapsulate specific adjuvants, such as the immunostimulatory CpG oligodeoxynucleotides, known as specific TLR9 ligands (Temizoz et al. 2016). The RNA phage VLPs can also be packaged with single-stranded or double-stranded RNA fragments, known as TLR7 and TLR3 ligands, respectively (Brencicova and Diebold 2013). Moreover, the recombinant RNA phage VLPs contain encapsulated bacterial RNA, which may act as an adjuvant.

The RNA phage VLPs are one of the most popular carriers among other VLP vectors. By number of constructed vaccine candidates, the RNA phage VLPs can be compared only with hepatitis B core (HBc) VLPs.

The numerous reviews on the global VLP application by the generation of vaccine candidates and their mode of action present the RNA phage VLPs among other popular VLP candidates (Pumpens and Grens 2002; Bachmann and Dyer 2004; Garcea and Gissmann 2004; Dyer et al. 2006; Reichel et al. 2006; Chackerian 2007; Jennings and Bachmann 2007, 2008; Ramqvist et al. 2007; Peek et al. 2008; Bachmann and Jennings 2010, 2011; Federico 2010, 2012; Roldão et al. 2010, 2011; Buonaguro et al. 2011; Plummer and Manchester 2011; Kushnir et al. 2012; Bárcena and Blanco 2013; Pushko et al. 2013; Ungaro et al. 2013; Zeltins 2013, 2018; Al-Barwani et al. 2014; Lua et al. 2014; Tan and Jiang 2014; van Kan-Davelaar et al. 2014; Zhao et al. 2014; Bachmann and Zabel 2015; Diederich et al. 2015; López-Sagaseta et al. 2015; Lundstrom 2015; Naskalska and Pyrć 2015; Yan et al. 2015; Frietze et al. 2016a, 2018; Karch and Burkhard 2016; Lee KL et al. 2016; Shirbaghaee and Bolhassani 2016; Charlton Hume and Lua 2017; Cimica and Galarza 2017; Jeong and Seong 2017; Kalkhoran 2017; Kelly et al. 2017; Moradpour and Sardari 2017; Negahdaripour et al. 2017; Sunderland et al. 2017; Tagliamonte et al. 2017; Bao et al. 2018; Donaldson et al. 2018; Fogarty and Swartz 2018; Goes and Fuhrmann 2018; He et al. 2018; Bao et al. 2019; Charlton Hume et al.

2019; El-Sayed and Kamel 2018; Kelly et al. 2019; Lei et al. 2019; Syomin and Ilyin 2019; Tsoras and Champion 2019; Wallis et al. 2019; Xu Z and Kulp 2019). Some articles were focused either on the RNA phage VLPs in total (Chackerian and Peabody 2015; Pumpens et al. 2016) or on the use of a specific RNA phage species as vaccines, such as the MS2 VLPs (Fu Yu and Li 2016). The remarkable advances of the VLP vectors, including the Qβ and PP7 VLPs, by mucosal immunization were reviewed by Vacher (2013).

VLP technology gave birth to two popular human vaccines to date. First, the prophylactic hepatitis B vaccine of the 22-nm particles of hepatitis B virus surface (HBs) antigen produced in yeast and applied since 1986 in human healthcare. Second, the cervical cancer vaccine that is composed from the human papillomavirus VLPs produced in yeast or baculovirus expression systems, which went on the market in 2006 and 2007, respectively. After these two global vaccines, the VLP-based hepatitis E vaccine was approved in 2011. It is necessary to call special attention to the fact that these vaccines are based on the recombinant but not chimeric VLPs. The same is true for the animal vaccines, where non-chimeric circovirus and parvovirus VLPs were accepted as vaccines against infections in pigs and dogs. Other animal vaccines were generated against calicivirus (RHDV), papillomavirus (BPV and CRPV), reovirus (BTV), and birnavirus (AHSV) infections by using the appropriate non-chimeric VLPs. The introduction of the chimeric VLPs as vaccines is forthcoming.

BASIC IMMUNOLOGY

In parallel with the applied goals by the generation of vaccine candidates, the RNA phage VLPs, mostly Qβ VLPs, contributed greatly to the basic immunological knowledge. These systematic investigations were performed by Martin F. Bachmann's team that grew out of the Nobelist 1996 Rolf M. Zinkernagel's school. Bachmann's team started in the early 2000s with the HBc VLPs but soon switched to the Qβ VLPs, and later broadened their interest also to the AP205 VLPs. The pioneering role of this team was presented in the Chemical coupling: Martin F. Bachmann's team and Qβ section of Chapter 20.

First, Bachmann's team performed deep immunological studies with the Qβ VLPs decorated by the attractive model epitopes: p33, which was derived from the lymphocytic choriomeningitis virus (LCMV) glycoprotein (Bachmann et al. 2004b, 2005a; Storni et al. 2004; Schwarz et al. 2005; Agnellini et al. 2008; Bessa et al. 2008; Keller et al. 2010a,b), OVA peptide from ovalbumin (Jegerlehner et al. 2007), and D2 epitope derived from *Salmonella* (Jegerlehner et al. 2010). These three models have played a central role in the elucidation of the fine immunological mechanisms that governed responses to the chimeric VLPs. Second, Bachmann's team was the first to display a small antigen, namely, nicotine, on the Qβ VLPs (Maurer et al. 2005; Beerli et al. 2008; Cornuz et al. 2008; Lang R et al. 2009). This was a novel approach that enabled the generation of strong immunological responses against nonpeptide antigens and paved the way for the development of experimental vaccines against nicotine addiction. Moreover, it led to the determination, in collaboration with Kaspars Tārs' group, of the 3D structure for the human antibody-nicotine complex and oriented future studies to increasing the antibody affinity to nicotine (Tars et al. 2012). The structure of the Fab fragment of the human anti-nicotine antibody that was induced by the Qβ VLPs decorated with nicotine molecules was presented at 1.85 Å resolution in complex with nicotine and, in addition, in complex with the nicotine-linker derivative nicotine-11-yl-methyl-(4-ethyl-amino-4-oxo)-butanoate, which was used for the conjugation of nicotine to the Qβ VLPs, at 2.1 Å resolution (Tars et al. 2012). Later, M.G. Finn's team displayed another small antigen, a carbohydrate moiety, on the Qβ VLPs, and tested them as a prospective cancer vaccine (Yin et al. 2013, 2016).

Third, Bachmann's team succeeded by the packaging of the CpG sequences into the Qβ VLPs and elaborated the allergy vaccine candidate CYT003-QbG10, which did not carry any attached epitopes but contained the encapsulated CpG sequence, so-called QbG10 (Senti et al. 2009; Klimek et al. 2011, 2013b; Beeh et al. 2013; Casale et al. 2015).

In 2002, Bachmann's team published their first self-review, which presented the two VLP models, namely, HBc and Qβ, as molecular assembly systems that rendered antigens of choice highly repetitive and were able to induce efficient antibody responses in the absence of adjuvants and provide protection from viral infection and allergic reactions (Lechner et al. 2002).

Moreover, the Qβ VLPs as a standard antigen contributed markedly to the solution of basic immunological problems. Thus, the *in vivo* response of marginal zone and follicular B cells to the Qβ VLPs was compared (Gatto et al. 2004). The Qβ VLPs were T cell-independent and induced IgM responses in the absence of T help because of multimeric interactions with cognate B cell receptors that induced a strong activation signal in B cells. It appeared that the marginal zone B cells responded with slightly faster kinetics, but numerically, follicular B cells dominated the response, while both marginal zone and follicular B cells underwent isotype switching, with marginal zone B cells again exhibiting faster kinetics (Gatto et al. 2004; Gatto and Bachmann 2005). The role of the CD21-CD35 complement receptors in the generation of the B cell memory was elucidated by the immunization of the appropriate deficient mice with the Qβ VLPs (Gatto et al. 2005). Using the Qβ VLPs, it was shown that the early B cell proliferation and development of B cell memory in mice were highly antigen-dependent, whereas persisting antigen was not essential for the maintenance of B cell and antibody memory in the late phase of the response (Gatto et al. 2007b). Furthermore, the heterogeneous antibody repertoire of the marginal zone B cells was evaluated after immunization of

mice with the two different VLPs, namely, Qβ and AP205 (Gatto et al. 2007a). In this study, the direct comparison of the heavy chain variable region sequences from murine marginal zone and follicular B cells indicated that the response of the marginal zone B cells to the VLPs was clonotypically heterogeneous and suggested that the marginal B cell compartment was capable of generating variable and diverse antibody responses.

The Qβ VLPs decorated with the p33 epitope and packaged with the CpG adjuvant were used for evaluation of the recall proliferation potential of the memory CD8+ T cells by antiviral protection (Bachmann et al. 2005b). The p33-Qβ VLPs later played an important role by explaining VSIG4, a B7 family-related protein, as a negative regulator of the T cell activation (Vogt et al. 2006). Furthermore, Bachmann et al. (2006a) compared memory CD8+ T cell development after infection with live LCMV or after vaccination in mice with the p33-Qβ VLPs, where prolonged but balanced antigenic exposure was identified for up to 9 days due to viral replication or repetitive vaccination as a pivotal factor driving the generation of long-lived CD8+ T cell memory. The p33-decorated and CpG-packaged VLPs were employed by the estimation of the differential role of IL-2R signaling for the CD8+ T cell responses in acute and chronic viral infections (Bachmann et al. 2007). In an extensive review, Hinton et al. (2008) explained how viruses and VLPs triggered Toll-like receptors (TLRs), which, in addition to increasing overall antibody levels, drove the switch to the IgG2a isotype that was more efficient in viral and bacterial clearance and activated complement, which in turn lowered the threshold of the B cell receptor activation. This conclusion was very helpful in the vaccine design, demonstrating that the safe recombinant vaccines could still remain as effective as a virus in inducing B cell responses.

The Qβ VLPs labeled with Alexa-488 were employed for the trafficking of nanoparticles *in vivo*, in parallel with 20-, 500- and 1000-nm polystyrene fluorescent nanoparticles, when injected into footpads of mice (Manolova et al. 2008). It appeared that nanoparticles trafficked to the draining lymph nodes in a size-dependent manner. Whereas large particles of 500–2000 nm were mostly associated with dendritic cells from the injection site, small nanoparticles of 20–200 nm in diameter, as well as the Qβ VLPs of 30 nm, were also found in lymph node—resident dendritic cells and macrophages, suggesting free drainage of these particles to the lymph nodes. These data provided clear evidence that the particle size determined the mechanism of trafficking, and that only small nanoparticles could specifically target the lymph node—resident cells (Manolova et al. 2008).

The Qβ and AP205 VLPs carrying single-stranded RNA of bacterial origin were used to assess the role of TLR7 signaling in driving IgA responses, both at the systemic level and in the respiratory tract mucosa of mice (Bessa et al. 2009). Thus, depending on the site of the antigen sampling, the TLR7 signaling was required either to lung dendritic cells and alveolar macrophages or to B cells, and resulted in T cell-dependent *versus* T cell-independent IgA responses, respectively (Bessa et al. 2009; Bessa and Bachmann 2010). Furthermore, the Qβ VLPs loaded with bacterial RNA or devoid of this RNA, therefore allowing investigation of B cell responses against the same antigen in the presence or absence of TLR7/8 ligands, were used to demonstrate that IL-21 acted directly on B cells, cooperated with TLR signaling, and was crucially required for the VLP-specific IgG responses (Bessa et al. 2010). The presence of the TLR7/8 ligands largely overcame this dependency, demonstrating that B cells integrated innate and Th cell-derived signals for optimal IgG responses. Moreover, the key roles of IL-2 in inhibiting IL-21 production by CD4+ T cells and of IL-21 in negatively regulating marginal zone B cell survival and antibody production were highlighted by using the Qβ VLPs loaded with the bacterial RNA (Tortola et al. 2010).

By exhaustive investigation on the mechanisms by which Th cells promote CD8+ T cell responses, the Qβ VLPs were used to display interleukin-15 by chemical coupling and induce the corresponding antibodies in mice (Wiesel et al. 2010). The Qβ VLPs displaying the cytokines IL-1α and IL-1β, which are listed in Table 22.2, were used to neutralize the endogenous cytokines in mice by the evaluation of how oxidative stress and metabolic danger signals converge and mutually perpetuate the chronic vascular inflammation that drives atherosclerosis (Freigang et al. 2011).

The immunization of mice with the Qβ VLPs that were packaged with CpG oligodeoxynucleotides led to the qualified conclusion that the CpG adjuvant functioned as the TLR9 agonist and strongly promoted the T cell-dependent antibody response (Hou et al. 2011). Remarkably, the immune-stimulating effect of the CpG molecules incorporated inside the Qβ VLPs depended on the adaptor MyD88 signaling, because the IgG response to the Qβ VLPs was almost completely ablated in Myd88-/- mice (Hou et al. 2011). Later, the Qβ VLPs carrying TLR ligands were employed by the extensive study of the fine MyD88 signaling mechanisms (Tian et al. 2018). It was demonstrated that the B cell MyD88 signaling was important for both T cell-independent and T cell-dependent phases of the Qβ VLP-induced antibody response. The role of the B cell MyD88 in the germinal center response was related mainly to its effect in promoting early germinal center B cell precursors. Finally, the B cell-intrinsic MyD88 signaling was also required for the Qβ VLP-induced T-bet expression, explaining the biased Ig isotype switch in mice deleted of the B cell MyD88 (Tian et al. 2018).

Link et al. (2012) performed a highly important study that demonstrated directly the major immunological difference of the Qβ VLPs and their dimeric subunits. Thus, the same antigenic determinants either as VLPs or dimers were compared. It was demonstrated that the VLPs were transported efficiently to murine splenic follicular dendritic cells in vivo in the absence of prior immunity. The natural IgM antibodies and complement were required and sufficient to mediate capture and transport of the VLPs by noncognate B cells. In contrast, the Qβ dimer was only deposited on follicular dendritic cells in the presence of specifically induced IgM or IgG antibodies. Unexpectedly, the IgG antibodies had the opposite effect on viral particles and inhibited follicular dendritic cells deposition. For the first time, these findings clearly identified the size and repetitive structure of the Qβ VLPs as critical factors for the efficient antigen presentation to B cells and highlighted important differences between viral particles and their subunits (Link et al. 2012).

Using the Qβ VLPs, Zabel et al. (2014) demonstrated rapid differentiation of the memory B cells to powerful secondary plasma cells, whereas the secondary pool of the memory B cells was to a large extent derived from naive B cells, allowing plasticity of the memory B cell repertoire upon multiple antigenic exposures. To do this, a system based on the Qβ VLPs was established that allowed tracking of the VLP-specific B cells by flow cytometry as well as histology. Using allotype markers, it was possible to adoptively transfer the memory B cells into a naive mouse and track responses of naive and memory B cells in the same mouse under physiological conditions (Zabel et al. 2014). Furthermore, using the Qβ VLPs, Zabel et al. (2017) demonstrated that the VLP-specific memory B cell responses exhibited a hierarchical dependence on Th cells. To acknowledge this, the memory B cells were generated in a primary host and adoptively transferred into different secondary hosts globally deficient in Th cells (MHC II-deficient) or specifically lacking CD40L or IL-21 receptor (IL-21R), as a model for the follicular Th cell deficiency. It was observed that proliferation of VLP-specific memory B cells showed a strong Th cell dependence, and furthermore, required CD40L and IL-21R signaling. In contrast, the differentiation of the Qβ-specific memory B cells into plasma cells and fully mature secondary plasma cells was only strongly reduced in the global absence of Th cells, whereas CD40L and IL-21R were dispensable to a large degree (Zabel et al. 2017).

Concerning the role of the packaged E. coli RNA, the Qβ VLPs were used recently to show that the RNA and the TLR7-signaling in B cells synergize for the regulation of the secondary plasma cell response (Krueger et al. 2019). The absence of the RNA or the TLR7-signaling resulted in complete failure to generate the memory B cells competent of forming secondary plasma cells. Moreover, the stimulation of the memory B cells generated in the presence of the RNA also failed to result in secondary plasma cell induction in the absence of TLR7-signaling during recall. Therefore, it was clearly demonstrated that the generation of secondary plasma cells is regulated by the RNA and TLR7-signaling at multiple levels (Krueger et al. 2019).

The Qβ VLPs carrying endogenous bacterial RNA were used by the extensive studies on the indispensable role of the thioredoxin-1 (Trx1) and the glutathione (GSH)/glutaredoxin-1 (Grx1) systems for the development and functionality of the marginal zone B cells in mice (Muri et al. 2019). It appeared that the redox capacity driving proliferation was more robust and flexible in B cells than in T cells, which might have profound implications for the therapy of B and T cell neoplasms (Muri et al. 2019).

Using the CpG-loaded Qβ VLPs and E7 protein of HPV, Bachmann's team brought out clearly that the physical association of the VLP carrier with the vaccine protein was more critical for B than T cell responses, in the case when the vaccine protein possessed great size comparable with that of the VLP carrier (Gomes et al. 2017a). Thus, the HPV E7 protein spontaneously formed oligomers with molecular mass ranging from 158 kDa to 10 MDa at an average size of 50 nm. When the E7 oligomers were either chemically linked or simply mixed with the CpG-loaded Qβ VLPs, the E7-specific IgG responses were strongly enhanced if the antigen was linked to the VLPs. In contrast, both CD4+ and CD8+ T cell responses as well as T cell-mediated protection against tumor growth were comparable for linked and mixed antigen formulations. The mixing free E7$_{49-57}$ peptide with CpG-loaded Qβ VLPs, however, failed to induce strong T cell responses, suggesting that the adjusted particle size may be sufficient to co-deliver antigen and adjuvants to the same dendritic cells for optimal T cell induction, eliminating the requirement for the linkage of the two entities. Therefore, the B cell but not T cell responses required antigen linkage to the carrier and adjuvant for optimal vaccination outcome by the use of both particulate carriers and antigens of similar size. In this case, the necessity of conjugating the two entities together could be avoided. This would be especially important for the development of patient-specific vaccines, where VLPs could remain an attractive platform for personalized vaccines considering the convenience of production and low cost (Gomes et al. 2017a).

Moreover, using the Qβ VLPs, which displayed the model LCMV p33 peptide, Mohsen et al. (2017a) demonstrated that the CpGs functioned efficiently as adjuvants when they were packaged in separate VLPs and mixed with the VLPs displaying the CTL epitopes prior to administration *in vivo*. This novel method generated results comparable to the standard method where the CpG adjuvants and CTL epitopes were linked to the same VLP. Thus, packaging adjuvants into the carrier VLPs eliminated the need for physical linkage to the antigenic VLPs (Mohsen et al. 2017a).

The Qβ VLPs decorated with foreign epitopes were used to generate monoclonal antibodies targeted to the epitope in question, as for example the Qβ VLPs carrying the major cat allergen Fel d 1 (Zha et al. 2018). These antibodies were used to further elucidate basic immunological problems, such as capability of low-affinity antibodies to inhibit mast cell activation but not to neutralize the allergen. This indicated that the allergen-specific immunotherapy generated a larger protective umbrella of inhibitory IgG antibodies than previously appreciated (Zha et al. 2018).

The basic problems of the interaction of the VLPs with the innate immune system were analyzed recently in an extensive review from Bachmann's team (Mohsen et al. 2018). Further important immunological details by the immunization with the Qβ VLPs were revealed in the recent papers of Bachmann's team (Gomes et al. 2019a; Krüger et al. 2019).

In the context of the CpG efficiency, it is noteworthy that the Qβ VLPs encapsulated with CpG demonstrated remarkable therapeutic potential against peritoneal carcinomatosis in a murine model (Miller AM et al. 2019). The robust anti-Qβ immune response likely contributed to the enhanced survival

and decreased disease progression in these mice. The authors of this recent investigation concluded that the promising preclinical results suggested that the CpG-packed Qβ VLPs may have potential as an immunotherapy for the treatment of patients with peritoneal carcinomatosis and are worthy of further evaluation (Miller AM et al. 2019).

Due to David Klatzmann's vigorous activity, the AP205 and Qβ vaccine vectors were included in the impressive list of platforms by the elaboration of a methodology that used transcriptomic data in dendritic cells to predict the adaptive immune responses induced by large sets of vaccine vectors of different classes, ranging from infectious particles to VLPs and DNA (Dérian et al. 2016; Tsitoura et al. 2019). Thus, the 41 vector, classified in 13 categories of vaccines and all expressing the same standard LCMV p33 epitope, were evaluated and compared for their ability to induce an adaptive T cell immune response after vaccination. The model was based on the analysis of transcriptomic data, obtained 6 hours after vaccination, that could predict the antigen-specific immune responses induced at the peak of the response, 5–10 days later. This model, developed in mice, successfully predicted vaccine-induced responses from literature-mined human datasets. Remarkably, the RNA phage-derived VLPs induced the maximal CD8+ T cell expansion (Dérian et al. 2016).

Liao et al. (2017) developed methods to label and enrich the Qβ-VLP-specific B cells and followed these cells in immunized mice for 1 year. They found that, as expected for a T cell independent response, Qβ+ B cells expanded much more quickly after immunization than had been previously observed for the B cells specific for T cell-dependent antigens. The initiation of germinal center B cell differentiation was T cell-independent upon the Qβ VLP immunization. More interestingly, the Qβ-VLP-induced T cell-independent response generated memory B cells, some of which were class-switched and somatically hypermutated, indicating the efficient induction of activation-induced cytidine deaminase in the absence of T cells. Then, the T cell dependent response induced the long-lived Qβ-specific memory B cells, which were mainly IgG+. Although IgM+ memory cells were generated during the early immune response to the Qβ VLPs, they disappeared within a few months. This was quite different from previous reports for other T cell-dependent antigens, in which IgM+ and IgG+ memory cells were generated in similar numbers, and IgM+ memory cells were stably maintained for a very long time (Liao et al. 2017).

Hong et al. (2018) acknowledged that the antigen-specific B cells were the dominant antigen-presenting cells that initiated naive CD4+ T cell activation by immunization of mice with the Qβ VLPs. The B cells were sufficient to induce T follicular helper cell development in the absence of dendritic cells. The Qβ-specific B cells promoted CD4+ T cell proliferation and differentiation via cognate interactions and through the TLR signaling-mediated cytokine production (Hong et al. 2018).

Raso et al. (2018) used the Qβ VLPs for the identification of the regulation mechanism of the germinal center B cell TLR signaling, mediated by α_v integrins and noncanonical autophagy. Furthermore, the Qβ VLPs were employed in a substantive study that utilized autoimmune models to assess the impact of TYK2, a JAK family member, variants on T cell subsets and cytokine signaling and on normal and autoimmune responses *in vivo* (Gorman et al. 2019).

GENETIC FUSIONS

Table 22.1 presents a detailed list of the vaccine candidates that have been constructed from the RNA phage VLPs using genetic fusion methodology. First, the success attended the experimental human papilloma virus (HPV) vaccine elaborated by Bryce Chackerian and David S. Peabody's team and based on the PP7 single-chain-dimer VLPs (Caldeira et al. 2010; Hunter et al. 2011; Tumban et al. 2011, 2013; Tyler et al. 2012, 2014b,c) developed to preclinical studies (Tumban et al. 2015). A similar HPV vaccine candidate was then constructed by the same team on the MS2 single-chain-dimer VLPs and tested in preclinical studies (Tumban et al. 2012, 2015; Saboo et al. 2016; Peabody J et al. 2017). Both PP7- and MS2 VLP-based vaccines were immunogenic, but the MS2-L2 VLPs induced a broader HPV-neutralizing antibody response. This was likely because of the structural context of L2 display on the VLPs, since L2 was displayed on the AB-loop of the PP7 coat but at the N-terminus of the MS2 coat (Caldeira et al. 2010). These studies were extensively reviewed (Tyler et al. 2014a). A broader review on HPV vaccine candidates, including the RNA phage VLP-based vaccines, was published later (Jiang RT et al. 2016). Recently, a novel MS2 VLP-based vaccine was tested and demonstrated in mice the protective power against cervicovaginal infection with HPV pseudoviruses 16, 18, 31, 33, 45, and 58 at levels similar to mice immunized with the standard Gardasil-9 vaccine (Zhai et al. 2017). The vaccine candidates were represented by the MS2 VLPs displaying the tandem HPV31/16L2 peptide 17–31 or by a mixture of VLPs displaying either the tandem peptide or the consensus peptides 69–86 from HPV L2. Moreover, the MS2-L2 VLPs were active by oral immunization and demonstrated remarkable storage potential (Zhai et al. 2019). The high potential of the RNA phage VLPs by the generation of a pan-HPV vaccine was recognized in a review dealing with the role of VLPs by the prevention of HPV-associated malignancies (Wang JW et al. 2013). The remarkable role of the MS2 and PP7 VLPs in the development of the L2-based human HPV vaccines was reviewed by Schellenbacher et al. (2017).

A malaria vaccine based on the MS2 VLPs has also been reported as very promising (Ord et al. 2014; Crossey et al. 2015b).

A huge work was performed recently with the predicted Zika virus epitopes (Basu et al. 2018). Because of its broad character and complexity, this study is not referenced in Tables 22.1 and 22.2. Thus, the identified potential B cell epitopes on the Zika virus envelope protein, namely, aa 241–259 from the domain II and aa 294–315, 317–327, 346–361, 377–388, and 421–437 from the domain III were displayed on the three VLP platforms: MS2, PP7, and Qβ, and their immunogenicity in mice was assessed. When Zika virus epitopes could not be successfully displayed by the genetic insertion on the MS2 or PP7 VLPs due to the failure of the recombinant coat proteins to assemble into VLPs, they were displayed on the Qβ VLPs by chemical conjugation. In fact, mice immunized with a mixture of VLPs displaying Zika virus epitopes elicited anti-ZIKV antibodies. However, immunized mice were not protected against a high challenge dose of Zika virus, but sera—albeit at low titers—from immunized mice neutralized *in vitro* a low dose of Zika virus. Taken together, these results showed that these epitopes were B cell epitopes and they were immunogenic when displayed on the Qβ VLP platform. The results also showed that immunization with the VLPs displaying a single B cell epitope minimally reduced Zika virus infection, whereas immunization with a mixture of VLPs displaying a combination of the B cell epitopes neutralized Zika virus infection (Basu et al. 2018).

The MS2 (O'Rourke et al. 2014) and PP7 (Daly et al. 2017) VLPs were used by Chackerian and Peabody's team to generate promising vaccine candidates against methicillin-resistant *Staphylococcus aureus* (MRSA), a cause of the growing incidence of skin and soft tissue infections (SSTI). Compared to antibiotic-susceptible strains, the MRSA SSTI treatment failure required added interventions, with associated increases in human suffering and medical costs. Therefore, it was of high importance that the PP7 VLPs carrying a short amino acid sequence, as described in Table 22.1, were able to demonstrate efficiency in a murine SSTI challenge model with a highly virulent MRSA isolate (Daly et al. 2017).

A new concept, the "epitope-RNA VLP vaccine," was introduced by Dong et al. (2016), who generated a foot-and-mouth disease vaccine that combined exposure of an epitope on the MS2 VLP surface with the packaged antisense RNA. This complex was expected to have the virtues of both the VLP and RNAi vaccines and was obtained by co-transformation of the two plasmids into bacteria. Thus, the antisense RNA against the *3D* genes of foot-and-mouth disease virus (FMDV) was packaged into the MS2 VLPs carrying the VP1 epitope 141–160 presented on the VLP surface. As indicated in Table 22.1, the vaccine demonstrated definite potency against FMDV infection in mice and guinea pigs (Dong et al. 2016).

Bolli et al. (2018) demonstrated inhibition of the progression of metastatic breast cancer *in vivo* after immunization of mice with the MS2 VLPs carrying an epitope from the cystine-glutamate antiporter protein xCT that regulates cystine intake, conversion to cysteine, and subsequent glutathione synthesis, protecting cells against oxidative and chemical insults. The MS2 VLPs carrying the xCT epitope elicited a strong antibody response against xCT including high levels of IgG2a antibody. The IgG antibodies isolated from the MS2 VLP-treated mice bound to tumorspheres, inhibited xCT function as assessed by reactive oxygen species generation, and decreased breast cancer stem cell growth and self-renewal (Bolli et al. 2018).

It is noteworthy that the MS2 VLPs carrying matrix protein 2 ectodomain (M2e) of influenza A virus were used as a positive control of the M2e display by the elaboration of a novel category of vaccines ensuring the simultaneous surface display and cargo loading of encapsulin nanocompartments (Lagoutte et al. 2018). Some considerations on the prospective

TABLE 22.1

Vaccines Constructed on the RNA Phage VLPs and Viable Virions by Using Genetic Fusion Methodology

Vaccine Target	Source of Epitope	Epitope length, aa	Position of Insertion or Addition	Comments	References
AP205					
Acquired immunodeficiency: human immunodeficiency virus (HIV)	Co-receptor CCR5, ECL2 loops: mini-loop CRSQKEGLHYTC and full-length loop CRSQK…QTLKC	12 / 33	C-terminus	VLPs carrying mini-loop are formed with GTAGGGSG, but not with GSG linker. VLPs carrying full-length loop are formed with both linkers.	Indulis Cielēns and Regīna Renhofa, unpublished
	Co-receptor CXCR4, mini-loop: CNVSEADDRYIC	12	C-terminus	VLPs carrying mini-loop are formed with GTAGGGSG, but not with GSG linker.	
	Co-receptor CXCR4, aa 1–39	39	N-terminus	The epitope is definitely displayed on the VLP surface.	Tissot et al. (2010)
	Nef protein, aa 66–100 and 132–151	55	C-terminus	The Nef T cell epitopes are displayed on the VLP surface.	
Autoimmune arthritis	Interleukin-1β, human	17 kDa	C-terminus	Chimeric VLPs display knobs on the surface (see Figure 20.3).	Juris Jansons, Irina Sominskaya, and Regīna Renhofa, unpublished
	Interleukin-1α, murine Interleukin-1β, murine and human	157 / 152 / 154		Mosaic particles are formed by (i) suppression of C-terminal amber or opal codons or (ii) coexpression of the chimeric gene with the helper coat gene. The GSG or GSGG linkers were used.	Indulis Cielēns and Regīna Renhofa, unpublished
Cancer, prostate	Gonadotropin releasing hormone (GnRH), aa 1–10	10	N-terminus C-terminus	The VLPs carrying C-terminal fusion induce strong antibody response that inhibits GnRH function *in vivo*.	Tissot et al. (2010)
Chikungunya virus (ChikV)	Glycoprotein E1, the virus-neutralising domain III (DIII); Glycoprotein E2, full-length, Domain A, Domain B, Domains A+B	84 / 361 / 131 / 60 / 298	C-terminus	Mosaic particles formed by suppression of the C-terminal amber codon. The GSG linker was used in all cases.	Indulis Cielēns and Regīna Renhofa, unpublished
Hepatitis B virus (HBV)	preS1, aa 21–47	27	N-terminus	The SGTAGGGSGS linker was more preferable for the outcome than the SGG linker.	Tissot et al. (2010)
	preS1, aa 21–47, 20–58, 20–119	27 / 39 / 100	C-terminus	The GTAGGGSG linker was used.	
Hepatitis C virus (HCV)	E2 protein, genotype 1α, HVR sequence, aa 384–411; a "consensus" HVR sequence	28 / 31	N-terminus	The SGTAGGGSGS linker was used in both cases.	
Hypertension	Angiotensin II, aa 1–8	8	N-terminus C-terminus	Both variants display angiotensin II epitope on the VLP surface.	Tissot et al. (2010)
Influenza virus	M2e protein, N-terminal ectodomain, consensus sequence, aa 2–24	23	N-terminus	Strong M2e-specific antibody response was achieved upon immunization in mice: protection of 100% mice from a lethal influenza infection.	Tissot et al. (2010); Schmitz et al. (2012)

(Continued)

TABLE 22.1 (Continued)
Vaccines Constructed on the RNA Phage VLPs and Viable Virions by Using Genetic Fusion Methodology

Vaccine Target	Source of Epitope	Epitope length, aa	Position of Insertion or Addition	Comments	References
Lymphocytic choriomeningitis virus (LCMV)	Glycoprotein, peptide p33: KAVYNFATM	9	C-terminus	Chimeric VLPs were formed also when different linkers and another epitope variant KAVYNFATMA have been used.	Indulis Cielēns and Regīna Renhofa, unpublished
Obesity	Ghrelin, aa 24–31: GSSFLSPE	8		The SGTAGGGSGS linker was more preferable for the outcome than the SGG linker.	Tissot et al. (2010)
	Gastric inhibitory peptide (GIP), aa 1–15 of mature GIP (42 aa): YAEGTFISDYSIAMD	15			
Salmonella typhi	Outer membrane protein, D2 peptide, aa 266–280	15	N-terminus C-terminus	The epitope is displayed on the VLP surface.	Cielēns et al. (2014)
West Nile virus (WNV)	Glycoprotein E, the virus-neutralizing domain III (DIII), aa 296–406	111	C-terminus	Mosaic particles. Immunization of mice resulted in the induction of IgG2 isotype anti-DIII. antibodies.	
fr					
Hamster polyomavirus (HaPV)	VP1, aa 364–384, 351–374, 351–384, 333–384	21 24 34 52	N-terminus	Induction of anti-VP1 antibodies in rabbits and mice.	Voronkova et al. (2002)
GA					
Human immunodeficiency virus (HIV)	Co-receptor CCR5, ECL1 loop YAAAQWDFGNTMCQ	14	FG loop	A shorter insertion without Cys residue GYAAAQWDFGNTG did not result in the self-assembled VLPs.	Arnis Strods and Regīna Renhofa, unpublished
	Co-receptor CCR5, ECL2a loop QKEGLHYTG	9	AB loop, aa 14/15		Indulis Cielēns and Regīna Renhofa, unpublished
	Co-receptor CCR5, N-terminus, aa 1–27 aa 1–31	27 31	N-terminus of the second coat copy	Mosaic particles with the GA coat as a helper were formed in both cases.	
	Co-receptor CXCR4, aa 1–40	40	N-terminus	Four variants with different linkers and surrounding sequences formed VLPs.	
Influenza virus	M2e protein, N-terminal ectodomain consensus sequence, aa 2–24 with C-terminally added G residue	24	N-terminus C-terminus	The GSGS (GSRS) and GSG linkers were used for N-terminal and C-terminal insertions, respectively.	
	M2e protein, N-terminal ectodomain consensus sequence, aa 2–24 with C-terminally added G	24	N-terminus of the first coat copy of the coat single-chain dimer	The GSG linker was used.	
Salmonella typhi	Outer membrane protein, D2 peptide, aa 266–280	15	N-terminus		

(Continued)

TABLE 22.1 (Continued)
Vaccines Constructed on the RNA Phage VLPs and Viable Virions by Using Genetic Fusion Methodology

Vaccine Target	Source of Epitope	Epitope length, aa	Position of Insertion or Addition	Comments	References
		MS2			
Cancer, breast	xCT protein, ECD6 peptide LYSDPFST derived from the 6th loop	8	AB loop of the coat single-chain dimer "15/14"	Dosing of BALB/c mice with the VLPs generated high titer antibodies that bound to xCT expressing breast cancer stem cells and these antibodies inhibited their function *in vitro*. The mice immunized with the VLPs had significantly less pulmonary metastases compared to controls in syngeneic models of breast cancer.	Bolli et al. (2018)
Cancer, ovarian	A peptide mimicking the cancer-associated antigen 125 (CA125/MUC16)	10	AB loop of the coat single-chain dimer, aa "15/14"	The MS2-DISGTNTSRA VLPs induced murine antibodies that cross-reacted with CA125 from ovarian cancer cells. Preoperative ovarian cancer patient plasma was assessed for anti-DISGTNTSRA.	Frietze et al. (2016b)
Flag peptide	DYKDDDDK	11	N-terminus	The defect caused by the Flag insertion was tolerated by the genetic fusion of the two subunits into the coat dimer, independently of the position of the chimeric half. When both AB loops of the dimer carried the Flag epitope, the self-assembly was not tolerated.	Peabody (1997b)
		9	AB loop, between D12 and N13		
Foot-and-mouth disease virus (FMDV)	VP1, aa 141–160 serotype O/OZK	20	AB loop	The effective immune response in mice and protection of guinea pigs and swine against FMDV were achieved.	Dong et al. (2015)
				The antisense RNA against the 3D genes of FMDV was packaged into VLP with the 141–160 epitope presented on the surface. The potency studies indicated that the vaccine protected 40% of suckling mice and 85% (17/20) of guinea pigs from FMDV.	Dong et al. (2016)
	VP1, aa 131–160, serotype O	30		The chimeric nanoparticles of 25–30 nm in diameter were significantly more immunogenic than the control epitope tandem peptide.	Wang Guoqiang et al. (2018)
Human immunodeficiency virus (HIV)	gp120 protein, V3 loop	10	AB loop of the coat single-chain dimer, aa "15/14"	The V3 insertion disrupted self-assembly but was tolerated by the coat fusion into the single-chain dimer. High immunogenicity in mice. The ability to pack their own mRNA was demonstrated.	Mastico et al. (1993); Stockley and Mastico (2000); Peabody et al. (2008)
	Co-receptor CCR5, ECL2 loop	10		The ECL2 insertion disrupted self-assembly but was tolerated by the coat fusion into the single-chain dimer. High immunogenicity in mice. Ability to pack their own mRNA.	
Human papillomavirus (HPV)	L1 protein L2 protein	20 20	AB loop, aa "15/14"		Mastico et al. (1993); Stockley and Mastico (2000)
	L2 protein of HPV16, HPV18, HPV31, aa 17–31	15	N-terminus and AB loop of the coat single-chain dimer	Two different epitopes were displayed on the same particle. The strong protection of mice from genital infection with HPV pseudoviruses representing 11 diverse HPV types was demonstrated. The spray-dried vaccine was stable after storage at room temperature for 34 months. The longevity of protection reached 2 years after immunization.	Tumban et al. (2012, 2015); Tyler et al. (2014c); Saboo et al. (2016); Peabody J et al. (2017)

(Continued)

TABLE 22.1 (Continued)
Vaccines Constructed on the RNA Phage VLPs and Viable Virions by Using Genetic Fusion Methodology

Vaccine Target	Source of Epitope	Epitope length, aa	Position of Insertion or Addition	Comments	References
	Tandem HPV31/16 L2 peptide, aa 17–31, consensus peptides from L2, aa 69–86 or 108–122, multivalent epitope consensus L2(108–122)/ HPV31L2(20–31)/HPV16L2(17–31)	27 18 15 42	N-terminus and AB loop of the coat single-chain dimer	Mice immunized with the MS2-L2 VLPs displaying the tandem peptide or immunized with a mixture of VLPs (displaying the tandem peptide and consensus peptide 69–86) elicited high titer antibodies against individual L2 epitopes. The vaccinated mice were protected from cervicovaginal infection with HPV pseudoviruses 16, 18, 31, 33, 45, and 58 at levels similar to mice immunized with Gardasil-9. The VLPs also have the potential to protect, orally, against the same oncogenic HPVs. Mixed MS2-L2 VLPs were thermostable at room temperature for up to 60 days after spray-freeze drying and they were protective against oral HPV infection.	Zhai et al. (2017, 2019)
Hypercholesterolemia	Proprotein convertase subtilisin/kexin type 9 (PCSK9), aa 153–163, 188–200, 208–222, 368–381	11 13 17 14	N-terminus of the coat single-chain dimer	There were not as dramatic reductions in total cholesterol in mice immunized with recombinant MS2-PCSK9 VLPs, in comparison to chimeric Qβ VLPs (see Table 22.2).	Crossey et al. (2015a)
Influenza virus	Hemagglutinin, epitope YPYDVPDYA	9	AB loop, aa "15/14"		Mastico et al. (1993); Stockley and Mastico (2000)
	M2e protein, conserved epitope EVETPIRNE	9	AB loop of the coat single-chain dimer, aa "15/14"	The scalable purification protocol for the potential veterinary vaccine application was elaborated. This construct was used further as a reference by the elaboration of the simultaneous surface display and cargo loading of the encapsulin nanocompartments, a novel approach to the rational vaccine design.	Lagoutte et al. (2016, 2018)
Malaria	*Plasmodium falciparum*, liver stage antigen-1 (LSA-1), T1 epitope	24	AB loop, aa "15/14"	The LSA-1-carrying VLPs stimulated a type 1-polarized response, with significant upregulation of interferon-γ, a finding which corroborates naturally acquired resistance to liver stage malaria	Mastico et al. (1993); Heal et al. (2000); Stockley and Mastico (2000)
	Plasmodium falciparum, RH5 protein, a peptide SAIKKPVT mimicking a linear epitope	8	AB loop of the coat single- dimer, aa "15/14"	The chimeric VLPs elicited antibodies that inhibit parasite invasion and could form the basis of an effective vaccine against malaria.	Ord et al. (2014)
	Plasmodium falciparum, blood stage antigen AMA1 (apical membrane antigen-1)	10		The mimotope identified by VLP-peptide display induced murine antibodies that cross-reacted with AMA1.	Crossey et al. (2015b)
Staphylococcus aureus	*agr*IV quorum-sensing operon, mimotope peptides that immunologically mimic the autoinducing peptide 4 (AIP4)	6 7		The first report of an efficacious active vaccine targeting the secreted autoinducing peptides of the *S. aureus agr* quorum-sensing system.	O'Rourke et al. (2014)

(Continued)

TABLE 22.1 (Continued)
Vaccines Constructed on the RNA Phage VLPs and Viable Virions by Using Genetic Fusion Methodology

Vaccine Target	Source of Epitope	Epitope length, aa	Position of Insertion or Addition	Comments	References
			PP7		
Flag peptide	DYKDDDDK	8	AB loop of the coat and of the coat single-chain dimer, aa 11/12	The VLPs were highly immunogenic in mice and packaged their own mRNA.	Caldeira et al. (2010)
Human immunodeficiency virus (HIV)	gp120 protein, V3 loop, a peptide IQRGPGRAPV	10		The VLPs were highly immunogenic in mice and packaged their own mRNA.	Caldeira et al. (2010)
Human papillomavirus (HPV)	L2 protein of HPV16, aa 17–31: QLYKTCKQAGTCPPD	15		The VLPs were highly immunogenic in mice and packaged their own mRNA. The intravaginal immunization protected mice from genital infection with HPV16 pseudovirions.	Caldeira et al. (2010); Hunter et al. (2011)
	L2 protein of HPV1, 5, 6, 11, 16, 18, 45, 58, aa 17–31	15	AB loop of the coat single-chain dimer, aa 11/12	Mice immunized with the mixture of eight L2 VLPs were strongly protected from genital challenge with pseudovirions representing 8 diverse HPV types. The anti-L2 antibodies persisted over 18 months and vaccinated mice retained protection over a year after immunization. For the first time in the generation of chimeric VLPs, RNA was removed from the VLP inside by alkaline treatment, as it has been performed before on non-chimeric MS2 VLPs (Hooker et al. 2004).	Tumban et al. (2011, 2013)
	L2 protein of HPV16, aa 17–31, 35–50, 51–65, 65–79, 65–85, Consensus 65–85	21 (for 65–85 variant)		Insertion of HPV16 L2 aa 35–50 and aa 51–65 was compatible with VLP assembly, but insertion of aa 65–79 was not. The VLPs displaying the 65–85 consensus peptide of high-risk HPV types induced murine sera that neutralized heterologous high-risk HPV pseudovirions.	Tyler et al. (2014b)
	L2 protein of HPV1, HPV16, HPV18, aa 17–31	15	N-terminus and AB loop of the coat single-chain dimer	Two different epitopes were displayed on the same particle. The strong protection of mice from genital infection with HPV pseudoviruses representing 11 diverse HPV types was detected.	Tyler et al. (2014c)
Influenza	Hemagglutinin, long alpha helix, or LAH, domain, strain H1N1/09, with a GSG linker	58	N-terminus	Although formation of the correct PP7-LAH VLPs failed, expression of this chimeric construct resulted in formation of soluble protein aggregates that were capable to induce strong seroconversion to both group 1 and group 2 hemagglutinin antigens in mice.	Lu et al. (2018)
Pregnancy	Human chorionic gonadotropin (hCG), aa 39–56, 45–55. 66–81, 69–80, 111–120, 116–125, 121–130, 131–140, 136–145	8–16	AB loop of the coat single-chain dimer, aa 11/12	Preincubating the hormone with antiserum elicited by VLPs displaying peptides 121–130 and 136–145 failed to inhibit hCG activity, while sera generated by immunization with VLPs displaying peptides 116–125, 126–135, and 131–140 dramatically inhibited uterine weight gain.	Caldeira et al. (2015)

(Continued)

TABLE 22.1 (Continued)

Vaccines Constructed on the RNA Phage VLPs and Viable Virions by Using Genetic Fusion Methodology

Vaccine Target	Source of Epitope	Epitope length, aa	Position of Insertion or Addition	Comments	References
Prostate cancer	Prostatic acid phosphatase, 114–128	15	AB loop of the coat single-chain dimer, aa 11/12	The epitope induced high levels of antibodies.	Sun and Sun (2016)
Staphylococcus aureus	accessory gene regulator operon *agr*, AIP1S peptide, YSTSDFIM, where original C in the AIP1 peptide was changed to S (red)	8	AB loop of the coat single-chain dimer, aa 11/12	The PP7-AIP1S elicited antibodies which recognized the original AIP1 *in vitro* and provided protection in a murine model of *S. aureus* dermonecrosis upon challenge with a highly virulent MRSA *agr* type I isolate.	Daly et al. (2017)
Zika virus	Envelope protein, Domain III LEAEMDGAKGRL VEFKDAHAKRQTVVV LVDRGWGNGCGLFGKG C-terminally: NDTGHETDEN	12 15 16 10	AB loop of the coat single-chain dimer, aa 11/12, and C-terminus	The chimeric particles demonstrated excellent quality and were resolved by electron cryomicroscopy. The immunological investigation was not performed.	Zhao et al. (2019)
			Qβ		
Foot-and-mouth disease virus (FMDV)	VP1 protein, G-H loop peptide	14	C-terminally added to the shortened A1 protein gene within the viable Qβ genome	A replication-competent hybrid phage that efficiently displayed the FMDV peptide was achieved. The surface-localized FMDV VP1 G-H loop cross-reacted with the anti-FMDV monoclonal antibody SD6 and was found by electron microscopy to decorate the corners of the Qβ icosahedral shell. The hybrid phages induced polyclonal antibodies in guinea pigs with good affinity to both FMDV and hybrid Qβ-G-H loop.	Skamel et al. (2014)
Hepatitis B virus (HBV)	preS1 epitope: 31-DPAFR-35 31-DPAFRA-36	5 6	A1 protein, C-terminal extension, aa 72/73, after aa 3, 6, 13, 19; as well as instead of the C-terminal extension	Mosaic particles were formed by A1-derived chimeras and Qβ coat helper via (i) suppression of leaky UGA stop codon of the coat gene and (ii) simultaneous expression of coat helper and A1-derived genes obtained after the changing of coat-terminating UGA to strong UAA stop codon or sense GGA codon, respectively. The proportion of A1-extended to short coat in mosaic particles varied from 48% to 14% after increase of the length of A1 extension. The preS1 epitope ensured specific immunogenicity in mice.	Kozlovska et al. (1996, 1997); Vasiljeva et al. (1998)
	Full-length preS1 and preS2; preS1, aa 20–47, 20–58, or 31–58	28–163	A1 protein, C-terminal extension, aa 18 or after leaky termination codon instead of the C-terminal extension	Mosaic particles were formed by (i) suppression of terminal *amber* or *opal* codons or (ii) coexpression of the chimeric gene with the coat helper gene. In the second case, it was necessary to place the chimeric gene first within the plasmid.	Indulis Cielēns and Regīna Renhofa, unpublished

(Continued)

TABLE 22.1 (Continued)

Vaccines Constructed on the RNA Phage VLPs and Viable Virions by Using Genetic Fusion Methodology

Vaccine Target	Source of Epitope	Epitope length, aa	Position of Insertion or Addition	Comments	References
Human immunodeficiency virus (HIV)	gp120 protein, V3 loop, aa 299–337	39	A1 protein, C-terminal extension, aa 72/73; after aa 19; instead of the C-terminal extension	Mosaic particles were formed.	Kozlovska et al. (1996, 1997)
	gp120 protein, epitopes b122a: N(410–423)-NG-(435–449)-NG-(291–341)-GSAGSAGSA-(365–392)C; OD$_{EC}$, and b122a1-b		C-terminal	The epitopes were individually cloned with an 8-amino acid hydrophilic spacer separating the two domains. Sera raised against these particles in 6-month long rabbit immunization studies could neutralize Tier1 viruses across different subtypes.	Purwar et al. (2018)
Paroxysmal nocturnal hemoglobinuria	C5 protein, epitope ASYKPSKEESTSGS linked with the PADRE peptide AKFVAAWTLKAAA via GSG linker and N-terminal sequence AYGG	34		Mosaic particles were formed, in which an average of 50 copies of the extension were incorporated per capsid. The C5 epitope was fused with the tolerance breaking PADRE peptide. Immunizing mice with the mosaic VLPs elicited strong humoral responses against recombinant mouse C5, reduced hemolytic activity, and protected the mice from complement-mediated intravascular hemolysis in a model of paroxysmal nocturnal hemoglobinuria.	Zhang L et al. (2017)

Source: The original table from Pumpens P et al. *Intervirology.* 2016;59:74–110, with permission of S. Karger AG, Basel, was extended with recent data.

Note: The data are given in alphabetical order of the (i) phage model and (ii) vaccine target for the same phage model. When the precise number of the amino acid residues within the epitope was difficult to assess, the size of the epitope was expressed in kDa.

direction of the encapsulin nanocompartments will be presented in Chapter 25.

It was attempted to use the PP7 VLPs for the generation of chimeras that would reproduce on their surface the natural trimer of the long alpha helical (LAH) domain of the influenza virus hemagglutinin (Lu et al. 2018). Unfortunately, the native self-assembly did not occur, but the aggregates demonstrated good immunological capabilities. It is remarkable that Andris Kazāks and Kaspars Tārs' team succeeded by the self-assembly of analogous chimeras based on the hepatitis B core platform (Kazaks et al. 2017). The attempt to present the trimeric hemagglutinin on the Qβ VLPs was reported at that time by Lauster et al. (2018).

Furthermore, mosaic Qβ particles were generated as a putative vaccine antagonizing the pathological effects of IL-13 in severe human diseases (Bai et al. 2018). Thus, the human IL-13 peptide was fused to the C-terminus of the Qβ coat, which was then coexpressed with the native Qβ coat and formed the mosaic Qβ-IL-13 particles.

Finally, concerning the selection of optimal VLP carriers for generation of genetically fused vaccine candidates, the AP205 VLPs have demonstrated a high capacity and tolerance to foreign insertions (Tissot et al. 2010). Moreover, the AP205 VLPs appeared as good candidates for construction of the mosaic VLPs (Cielens et al. 2014).

The exceptional capabilities of the PP7 platform, in comparison with the traditional RNA phage platforms, were announced by Zhao L and Finn (2017). Finally, Zhao et al. (2019) demonstrated unique properties of the chimeric PP7 VLPs, as described in the Chimeric VLPs section of Chapter 21.

CHEMICAL COUPLING

Table 22.2 presents a detailed list of the vaccine candidates generated by the chemical coupling approach. As noted in the Chemical coupling: Martin F. Bachmann's team section of Chapter 20, Bachmann's team moved as pioneers to the forefront of this approach, mainly, in the field of therapeutic vaccination against non-infectious diseases.

The principal validity of the chemical coupling approach was approved by a long line of experimental therapeutic vaccines. The impressive list of the latter was first proposed by Bachmann and Dyer (2004), but their generation was followed further by the numerous reviews of the team, as were referenced in the Chemical coupling: Martin F. Bachmann's team section of Chapter 20. The general idea of the therapeutic vaccination with self-antigens was presented in the same section of Chapter 20.

Mainly, this approach was planned as a method to replace the host-specific monoclonal antibodies in the treatment of acute and chronic diseases, starting with noninfectious diseases (Bachmann and Dyer 2004). The proposed transition from passive administration of monoclonal antibodies to active vaccination against self-antigens was a logical step in the drug development, focusing on affordable medicines and broader patient acceptance and regulatory compliance. The

induction of autoantibodies might be beneficial under certain physiological conditions in order to remove unwanted excess of a particular self-antigen, such as, for example, angiotensin in the case of hypertension.

In order to be maximally informative, Table 22.2 includes the RNA phage VLPs carrying the model epitopes p33, D2, and OVA, although these chimeric VLPs were not designed as actual vaccine candidates. The well-known Qβ VLP-based allergy vaccine CYT003-QbG10 elaborated by Bachmann's team is also included in Table 22.2, since this potentially successful vaccine was originally initiated by the epitope-coupling methodology.

There are excellent reviews not only on the novel vaccine candidates constructed on the RNA phage VLPs in comparison with other VLP platforms, which are listed above in the General section, but also special reviews on the VLP-based vaccines against allergies (Moingeon 2012, 2015; Klimek and Pfaar 2013; Klimek et al. 2013b, 2014, 2018; van Hage and Pauli 2014; Aryan and Rezaei 2015; Kündig et al. 2015; Sandrini et al. 2015; Creticos 2016; Bachmann and Kündig 2017; Ricketti et al. 2017; Anzaghe et al. 2018; Givens et al. 2018; Satitsuksanoa et al. 2018; Scheiblhofer et al. 2018; Zahirović and Lunder 2018), Alzheimer's disease (Chackerian 2010; Fettelschoss et al. 2014; Liu E and Ryan 2016; Sterner et al. 2016; Gallardo and Holtzman 2017; Marciani 2017; Mo et al. 2017; Bachmann et al. 2019; Kaminaka and Nozaki 2019), atherosclerosis (Govea-Alonso et al. 2017; Amirfakhryan 2019; Kobiyama et al. 2019), cancer (Grippin et al. 2017; Ong et al. 2017; Qiu et al. 2017; Wei et al. 2018; Mohsen et al. 2019b), hypercholesterolemia (Chackerian and Remaley 2016), hypertension (Ready 2005; Bachmann et al. 2006b; Gradman and Pinto 2008; Miller et al. 2008; Campbell 2009, 2012; Phisitkul 2009; Maurer and Bachmann 2010; Bairwa et al. 2014; Nakagami et al. 2014; Tamargo and Tamargo 2017; Nakagami and Morishita 2018; Azegami and Itoh 2019), influenza (Sączyńska 2014; Pushko and Tumpey 2015; Kolpe et al. 2016; Quan et al. 2016; Wong and Ross 2016), malaria (Reyes-Sandoval and Bachmann 2013; Wu Yimin et al. 2015; Noe et al. 2016; Draper et al. 2018; Huang et al. 2018), nicotine addiction (Cerny 2005; Heading 2007; Maurer and Bachmann 2006, 2007; Brown 2008; Didilescu 2009; Crain and Bhat 2010; Hartmann-Boyce et al. 2012; Kitchens and Foster 2012; Raupach et al. 2012; Fahim et al. 2013; Lieber and Millum 2013; Wang and Zhu 2013; Pentel and LeSage 2014; Collins and Janda 2016; Ilyinskii and Johnston 2016; Kalnik 2016; Kosten et al. 2017; Zhao Z et al. 2017), and obesity (Monteiro 2014; Na et al. 2014). The potential Qβ VLP-based vaccines for metabolic diseases were reviewed by Morais et al. (2014). Rynda-Apple et al. (2014) concentrated on the capability of the VLPs to induce protective immune responses in the lung. The Qβ-IgE VLP-driven therapeutic induction of anti-IgE antibodies to trigger the human immune system was reviewed by Licari et al. (2017). Recently, the role of the Qβ VLPs, among other protein-based nanoparticles, in the development of cancer vaccines was reviewed by Neek et al. (2019).

TABLE 22.2

Vaccines Constructed on the RNA Phage VLPs by Chemical Coupling or Plug-and-Display Methodology

Vaccine Target	Source of Epitope	Epitope Length, aa	Coupling Site	Comments	References
			AP205		
Asthma/allergy	IL-5, murine	33 kDa	SpyTag or SpyCatcher: genetically fused to the N-terminus and/or C-terminus	The VLP display led to the efficient breaking of self-tolerance.	Thrane et al. (2016)
Cancer	Human telomerase reverse transcriptase, the mutant Telo epitope biotin-GAHIVMVDAYKPTREARPALLTSRLRFIPK	30	SpyCatcher genetically fused to the N-terminus	SpyCatcher is a genetically-encoded protein designed to spontaneously form a covalent bond to its peptide-partner SpyTag carrying desired epitope.	Brune et al. (2016)
	Human epidermal growth factor receptor (EGFR) from glioblastoma, fusion junction epitope LEEKKGNYVVTDHGAHIVMVDAYKPTK–biotin	27			
	Murine proteins involved in cancer (CTLA-4, PD-L1, Survivin and HER2)	kDa: 15, 27, 30, 83	SpyTag or SpyCatcher: genetically fused to the N-terminus and/or C-terminus	The VLP display led to the efficient breaking of self-tolerance.	Thrane et al. (2016)
	Full HER2 extracellular domain including subdomains I–IV, aa 23–652, genetically fused with SpyCatcher, aa 23–139, at the N-terminus, via GGS linker, and His_6 at the C-terminus; the SpyTag peptide sequence AHIVMVDAYKPTK was separated from the coat by a flexible linker GSGTAGGGSGS at the N-terminus and GTASGGSGGSG at the C-terminus	656	SpyTag genetically fused to both the N- and C-termini	The stable, non-aggregated VLPs were each coated with an average of 183 SpyCatcher-HER2 antigens. The therapeutically potent anti-HER2 autoantibody response was induced. Vaccination reduced spontaneous development of mammary carcinomas by 50%–100% in human HER2 transgenic mice and inhibited the growth of HER2-positive tumors implanted in wild-type mice.	Palladini et al. (2018)
Cardiovascular disease	PCSK9, murine	84 kDa			Thrane et al. (2016)
Human immunodeficiency virus (HIV)	The RC1 epitope that facilitates the recognition of the V3-glycan patch on the envelope protein of HIV-1		C-terminal SpyTag sequence (13 residues) was added to the RC1-4fill epitope. The SpyCatcher-AP205 was prepared by Brune et al. (2016) and purified by Tissot et al (2010)	Immunization of mice, rabbits and rhesus macaques with the AP205-RC1 VLPs elicited serological responses that targeted the V3-glycan patch and expanded clones of B cells that carried the anti-V3-glycan antibodies, which resembled precursors of human broadly neutralizing antibodies.	Escolano et al. (2019)

(Continued)

TABLE 22.2 (Continued)
Vaccines Constructed on the RNA Phage VLPs by Chemical Coupling or Plug-and-Display Methodology

Vaccine Target	Source of Epitope	Epitope Length, aa	Coupling Site	Comments	References
	Six peptides covering the alpha-helical regions of gp41: 3–13, 3–17, 3–20, 3–24, P1, P8, provided with a C-terminal Cys residue	12 16 19 23 40 66	Lysine residues	The gp41 peptides were coupled to AP205 VLPs through a bifunctional cross-linker SMPH. The chimeric VLPs elicited high titers of gp41-specific antibodies. 1, 2, or 3 peptide copies were coupled to each AP205 subunit.	Pastori et al. (2012)
Hypertension	ATR-AP205–001 vaccine: ATR-001 peptide AFHYESQ corresponding to an epitope of the ECL2 of human AT1R, a G protein coupling receptor, with N-terminally added cysteine	8		The vaccine is constructed by analogy to the earlier ATRQβ-001 vaccine (see below). In mice, the vaccine induced potent humoral immunity through collaboration of B cells, follicular dendritic cells and follicular helper T cells, providing an effective and safe intervention for hypertension.	Hu et al. (2017)
Lymphocytic choriomeningitis virus (LCMV)	Glycoprotein, the p33 peptide KAVYNFATM	9		Chimeric VLPs were packaged by CpG and induced effective CTL response in mice.	Schwarz et al. (2005)
Malaria	*Plasmodium falciparum*, membrane protein 1 (PfEMP1) containing the complex lysine and cysteine-rich inter-domain region (CIDR). *Plasmodium falciparum*, Pfs25 protein	27 kDa	SpyCatcher genetically fused to the N-terminus	A pair of SpyTag and SpyCatcher as a functional unit were used (see above). Injecting SpyCatcher-VLPs decorated with a malarial antigen efficiently induced antibody responses after only a single immunization.	Brune et al. (2016)
	Plasmodium falciparum, CSP, CIDR, VAR2CSA, and Pfs25 proteins	kDa: 53, 32, 118, 40	SpyTag or SpyCatcher: to the N-terminus and/ or C-terminus	The Pfs25 and VAR2CSA vaccines showed efficacy that is comparable with the efficacy of the best existing analogs. An average of 112 CSP molecules were displayed per VLP. Vaccination of mice with the CSP Spy-VLP vaccine resulted in significantly increased antibody titers over a course of 7 months.	Thrane et al. (2016); Janitzek et al. (2016);
	Plasmodium falciparum, Pfs48/45 protein, region 6C. The SpyTag sequence AHIVMVDAYKPTK genetically fused along with flexible linkers GSGTAGGGSGS, N-terminus and GGSGly, C-terminus. The DNA fragment encoding amino acids 24–139 of the SpyCatcher domain sequence	~130 kDa, ~27 kDa	SpyCatcher-R0.6C and SpyCatcher-6C to the both Spy-tagged N- and C-terminus	Two different recombinant proteins (SpyCatcher-R0.6C and SpyCatcher-6C), comprising the Pfs48/45–6C region, were covalently attached to the surface of Spy-tagged *Acinetobacter* phage AP205 VLPs. Resulting Pfs48/45-VLP complexes appeared as non-aggregated particles of 30 nm, each displaying an average of 216 (R0.6C) or 291 (6C) copies of the antigens.	Singh et al. (2017)
	Plasmodium falciparum, Pfs48/45 and Pfs230 fusion proteins		SpyTag was fused to the N-terminus of AP205	Multicomponent hybrid protein containing both Pf s48/45 and Pfs230 holds the potential to lower the required threshold of functional antibodies and to reduce the risk of escape mutations.	Singh et al. (2019)
	Plasmodium falciparum, Pfs25	40 kDa	SpyTag-Pfs25 to the SpyCatcher-VLPs	The SpyCatcher-AP205-VLPs elicited the highest quality anti-Pfs25 antibodies for transmission blocking upon mosquito feeding.	Leneghan et al. (2017)

(Continued)

TABLE 22.2 (Continued)
Vaccines Constructed on the RNA Phage VLPs by Chemical Coupling or Plug-and-Display Methodology

Vaccine Target	Source of Epitope	Epitope Length, aa	Coupling Site	Comments	References
Salmonella typhi	Outer membrane protein, D2 peptide, aa 266–280: TSNGSNPSTSYGFAN with N-terminal CGG linker	15	Lysine residues	The vaccine preparations of 13, 56, 94, 142, and 293 peptides per VLP were used. A phenomenon of carrier-induced epitopic suppression (CIES) could be overcome by high coupling densities, repeated injections and/or higher doses of conjugate vaccine.	Jegerlehner et al. (2010)
Tuberculosis	*Mycobacterium tuberculosis*, Ag85A protein	48 kDa	SpyTag or SpyCatcher: genetically fused to the N-terminus and/or C-terminus		Thrane et al. (2016)
West Nile virus (WNV)	Glycoprotein E, the virus-neutralizing domain III (DIII), aa 582–696 of the WNV polyprotein precursor	115	Lysine residues	Domain III was engineered to comprise a His$_6$ tag and a Cys-containing linker at its C-terminus. The antibodies induced in mice were able to neutralize virus *in vitro* and provided partial protection from a challenge with a lethal dose of WNV.	Spohn et al. (2010)
			Qβ		
Acquired immunodeficiency: feline immunodeficiency virus (FIV)	Transmembrane (TM) glycoprotein, a peptide containing tryptophan-rich motif, aa 767–786: LQKWEDWVGWIGNIPQYLKG	20	Lysine residues	Immunized cats developed antibodies that reacted with the epitope but failed to recognize whole FIV. The coupling efficiency was never higher than 20%.	Freer et al. (2004)
Acquired immunodeficiency: human immunodeficiency virus (HIV)	Co-receptor CCR5, N-terminal ECL domain, circularized	20		Immunized mice and rabbits generated antibodies that recognized native CCR5 and inhibited entry of pseudotype viruses bearing envelope glycoproteins from diverse primary strains *in vitro*.	Huber et al. (2008); Sommerfelt (2009)
Acquired immunodeficiency: human immunodeficiency virus (HIV)/simian immunodeficiency virus (SIV)	Co-receptor CCR5, macaque, EC1, N-terminal aa MDYQVSSPTYDIDYYTSEPC; ECL2, a cyclic peptide, aa 168–177 DRSQREGLHYTG linked through an DG dipeptide spacer	20 / 10		Immunization of mice and rats induced anti-CCR5 antibodies that recognized native CCR5 and inhibited SIV infection *in vitro*. Equal amounts of both constructs were mixed to formulate the vaccine that was protective in macaques. The coupling efficiency reached 90 EC1 or 270 ECL2 peptides per particle.	Hunter et al. (2009, 2010); van Rompay et al. (2014)
Allergy	Cysteine protease, the major fecal allergen of the house dust mite *D. pteronyssinus*, Der p 1 peptide, aa 117–133: CGIYPPNANKIREALAQTHSA	21		The vaccine was administered without adjuvants and found safe and immunogenic in humans after evaluation of different doses and routes of immunization.	Lechner et al. (2002); Kündig et al. (2006)

(Continued)

TABLE 22.2 (Continued)

Vaccines Constructed on the RNA Phage VLPs by Chemical Coupling or Plug-and-Display Methodology

Vaccine Target	Source of Epitope	Epitope Length, aa	Coupling Site	Comments	References
Allergy, cat	Fel d1 protein, major cat allergen, a covalent dimer of chain 2 and chain 1 of Fel d1 spaced by a 15 aa-linker (GGGGS)x3 and added to the coding sequence for LEHHHHHHGGC at the C- terminus	23 kDa	Lysine residues	A single vaccination by Qβ-Fel d1 was sufficient to induce protection against type I allergic reactions in mice. Moreover, Qβ-Fel d1 did not induce degranulation of basophils derived from human volunteers with cat allergies. The coupling density was 40%, or 70 covalent Fel d1 dimers per VLP. The allergen displayed on VLPs failed to cause mast cell activation.	Schmitz et al. (2009); Uermösi et al. (2010); Engeroff et al. (2018); Zha et al. (2018)
Allergy, rhinoconjunctivitis, rhinitis	A-type CpG (QbG10) was encapsulated as a TLR9 ligand. Mixed with house dust mite (HDM) extract	-	No coupling	A phase I study. All patients achieved practically complete alleviation of allergy symptoms after 10 weeks of immunotherapy.	Senti et al. (2009)
	CYT003-QbG10: CpG QbG10 was encapsulated	-		A phase IIb study. Treatment with high-dose CYT003-QbG10 improved disease symptoms. The QbG10 content was approximately 20% of the total mass, corresponding to ~60 molecules per VLP.	Klimek et al. (2011, 2013a)
Allergy, asthma	CYT003-QbG10	-		A successful phase IIb study	Beeh et al. (2013); Casale et al. (2015)
Allergy, asthma and rhinitis	Two IgE peptides, different loops of the C3 domain: ADSNPRGVSAYLSRPSPGGC and YQCRVTHPHLPRALMRS	20 16	Lysine residues	The vaccine induced high titers of anti-human IgE antibodies by preclinical studies in mice and non-human primates.	Champion et al. (2014); Akache et al. (2016); Weeratna et al. (2016)
Allergy to red meat	α-1,3-galactosyl transferase, Galα3LN epitope. For conjugation, α-Gal trisaccharide and glucose were converted to their respective alkyne derivatives. Each alkyne was attached by a two-step procedure in which the protein nanoparticle was first acylated with an azide-terminated N- hydroxysuccinimide ester and then addressed by copper-catalysed azide-alkyne cycloaddition	Carbohydrates are conjugated		Search for the presence of α-Gal-containing epitopes in the saliva of *Amblyomma sculptum*. The bites from the *A. sculptum* tick may be associated with the allergic reactions to red meat in Brazil.	Araujo et al. (2016)
Alzheimer's disease	Aβ peptide (1–9)-GGC: DAEFRHDSGGGC	12		The phage Qβ, but not Qβ VLPs was used in this case. This Qβ phage-Aβ complex elicited anti-Aβ antibody responses at low doses and without the use of adjuvants.	Chackerian et al. (2006)
	Aβ peptide: N-terminal Aβ(1–9) or C-terminal Aβ (28–40)	12 16		Both of these immunogens produced significant antibody titers without use of additional adjuvants and reduced Aβ levels when tissues were examined 8 months after the first inoculation.	Li QY et al. (2010)

(Continued)

TABLE 22.2 (Continued)
Vaccines Constructed on the RNA Phage VLPs by Chemical Coupling or Plug-and-Display Methodology

Vaccine Target	Source of Epitope	Epitope Length, aa	Coupling Site	Comments	References
	CAD106 vaccine: Aβ peptide 1–6 DAEFRH plus a GGC spacer	9	Lysine residues	CAD106 avoided activation of Aβ-specific T cells and was efficacious in reducing the amyloid accumulation in transgenic mice without evidence of unwanted side effects and is currently tested in patients by phase II study. Each VLP contained ~350–550 Aβ peptides. The first clinical trial demonstrated a favorable safety profile and acceptable antibody response in patients with Alzheimer's disease.	Wiessner et al. (2011); Winblad et al. (2012); Farlow et al. (2015); Vandenberghe et al. (2017)
Atherosclerosis	Interleukin-1α, full length, containing aa linker at C-terminus	17 kDa		Immunization of mice reduced both the inflammatory reaction in the plaque as well as plaque progression.	Tissot et al. (2013)
Autoimmune arthritis	Interleukin-1α, murine, aa 117–270 of IL-1α precursor Interleukin-1β, murine, aa 119–269 of IL-1β precursor, both provided with aa linkers at C-termini	17 kDa		Immunization of mice elicited a rapid and long-lasting auto antibody response. In the collagen-induced arthritis model, both vaccines strongly protected mice from inflammation and degradation of bone and cartilage. In the T and B cell-independent collagen Ab transfer model, immunization with the IL-1β vaccine strongly protected from arthritis, whereas immunization with the IL-1α vaccine had no effect. The coupling efficiency was about 20% in the case of IL-1α and 28% in the case of IL-1β, or 36 molecules of IL-1α or 50 molecules of IL-1β per VLP.	Spohn et al. (2008); Guler et al. (2011)
Autoimmune arthritis, encephalomyelitis, and myocarditis	Interleukin 17, murine, aa 26–158, with a C-terminal linker GGGGC	32 kDa		Immunization induced high levels of autoantibodies in mice and was effective in ameliorating disease symptoms in animal models of autoimmunity. The coupling efficiency was about 10%, or 18 IL-17 homodimers per VLP.	Röhn et al. (2006); Sonderegger et al. (2006); Dallenbach et al. (2015)
Cancer, induction of antitumor antibodies	Tumor associated carbohydrate antigens (TACAs): monomeric Tn antigen (GalNAc-α-O-Ser/Thr) that is overexpressed on the surface of a variety of cancer cells including breast, colon, and prostate cancer and is involved in aggressive growth and lymphatic metastasis of cancers	Carbohydrates are conjugated by CuAAC	Lysine residues and N-terminus	An average of 340 copies of Tn was attached per Qβ capsid through a triazole linker (termed internal triazole), as well as 200 copies of triazole without Tn (external triazole). The antibodies generated in mice were highly selective toward Tn antigens and reacted strongly with the native Tn antigens on human leukemia cells.	Yin Z et al. (2013, 2014, 2015)
	Tumor associated carbohydrate antigens (TACAs): ganglioside GM2, a tetrasaccharide, overexpressed in a wide range of tumor cells			GM2 immobilized on VLPs through a thiourea linker elicited high titers of IgG antibodies that recognized GM2-positive tumor cells and effectively induced cell lysis through complement-mediated cytotoxicity.	Yin et al. (2016)

(Continued)

TABLE 22.2 (Continued)

Vaccines Constructed on the RNA Phage VLPs by Chemical Coupling or Plug-and-Display Methodology

Vaccine Target	Source of Epitope	Epitope Length, aa	Coupling Site	Comments	References
Cancer, induction of antitumor antibodies	Tumor associated mucin-1 peptides and glycopeptides. The (glyco) peptides containing 20–22 aa residues as the backbone covering one full length of the tandem repeat region with the sequence PDTRPAPGSTAPPAHGVTSA	20–22	Lysine residues and N-terminus	Mucin-1 displayed on VLPs by the CuAAC reaction induced highly potent antibody and cytotoxic T cell responses. From 30 to 140 copies of mucin-1 per VLP were exposed.	Yin Z et al. (2018)
	Mucin-1 peptide SAPDT*RPAP, where * denotes glycosylation	9		The critical protective epitope of MUC1 was identified to be conjugated to the Qβ VLPs. The antibodies elicited exhibited high tumor binding and killing activities, as well as excellent recognition of human breast cancer over normal breast tissues.	Wu Xuanjun et al. (2018)
Chronic inflammatory disorders: rheumatoid arthritis, psoriasis, Crohn's disease	TNF-α, murine, aa 80–235 of the transmembrane form, aa 4–23 peptide: CGGSSQNSSDKPVAHVVANHQVE	156 20		Immunization by Qβ-TNFα(4–23) showed a potential to become an effective and safe therapy against inflammatory disorders. The coupling efficiency was about 60 TNFα 1–156 or ~340 TNFα 4–23 molecules per VLP.	Spohn et al. (2007b)
Chronic inflammatory illnesses: type 2 diabetes mellitus	Interleukin-1β, murine, rhesus monkey and human, two muteins: mIL-1β (D143 K), mIL-1β (D143 K) with strongly reduced inflammatory activity.	153		As a potential therapy for the prevention and long-term treatment of type 2 diabetes, the vaccine showed good tolerability in mice and non-human primates and induced long-lasting, but reversible, IL-1β-neutralizing antibody titers, improved glucose tolerance, and enhanced insulin secretion in a mouse model of diet-induced diabetes. The preclinical study was followed by a phase I/II clinical trial in patients with type 2 diabetes mellitus using the human version of the vaccine hIL1βQβ and the neutralizing IL-1β-specific antibody response was registered. The epitope density was about 0.5 IL-1β molecules per Qβ monomer corresponding to a total of 90 IL-1β molecules per Qβ VLP.	Spohn et al. (2014); Cavelti-Weder et al. (2016)
	Interleukin-1β, human, a set of site-specific mutations to mimic modifications occurring during inflammatory responses and possibly leading to alterations or masking of certain cytokine sites			Although antibody responses generated to one variant were potently inhibiting IL-1β, antibody responses induced by the other variant even potentiated the *in vivo* effects of IL-1β; the latter led to enhanced morbidity in two different IL-1β-mediated mouse models, including a model of inflammatory bowel disease and an inflammatory arthritis model.	Spohn et al. (2017)
Clostridium difficile	Receptor-binding domain of the toxin B			The antibodies neutralized the activity of two distantly related toxins B with similar efficacy, suggesting that a polyclonal antibody response induced by active vaccination gives protection against different *C. difficile* strains.	Rebeaud and Bachmann (2012a)

(Continued)

TABLE 22.2 (Continued)

Vaccines Constructed on the RNA Phage VLPs by Chemical Coupling or Plug-and-Display Methodology

Vaccine Target	Source of Epitope	Epitope Length, aa	Coupling Site	Comments	References
Eosinophilia	Recombinant interleukin-5; recombinant eotaxin, both murine	17 and 8 kDa	Lysine residues and N-terminus	Both vaccines overcame self-tolerance and induced high antibody titers against the corresponding self-molecules in mice. Immunization with either of the two vaccines reduced eosinophilic inflammation of the lung in an ovalbumin-based mouse model of allergic airway inflammation. The coupling efficiency was about 47% or 15% Qβ monomers crosslinked to rIL-5 or r-eotaxin, respectively, or about 80–90 rIL-5 and 25–30 r-eotaxin molecules per VLP.	Zou et al. (2010)
Hen egg lysozyme (HEL), as a model for overcoming self-tolerance	Full-length	129		HEL-conjugated Qβ phage was generated by linking biotinylated HEL to biotinylated VLPs using streptavidin. The Qβ-HEL complex elicited high-titer IgG responses against HEL in tolerant HEL transgenic mice in the absence of exogenous adjuvants. Therefore, the presentation of HEL onto Qβ phage allowed overcoming the effects of anergy in the HEL mouse model.	Chackerian et al. (2008)
Human immunodeficiency virus (HIV)	gp41 protein, cluster I, a conserved immunodominant loop connecting the heptad repeat 1 and heptad repeat 2 of the HIV-1 envelope glycoprotein	20		The antibodies to the cluster 1 region of gp41 were elicited and possessed properties similar to the known monoclonal antibodies.	Sharma et al. (2012)
	gp120 protein, several conserved epitopes of broadly neutralizing antibodies			The epitopes were built in through surface engineering. Over 87% of seropositive participants showed specific antibody responses to the epitopes displayed on the Qβ VLPs.	Nchinda et al. (2017)
	gp120 protein, variants of the b122a epitope—see Table 22.1 for the structure			The conjugation was performed by the CuAAC method. Sera raised against these particles in six-month long rabbit immunization studies could neutralize Tier1 viruses across different subtypes.	Purwar et al. (2018)
Human papilloma virus (HPV)	L2 protein, N-terminal HPV16 peptides: aa 34–52, 49–71, 65–85, 108–120, consensus 65–85 peptide	19 23 21 13		The phage Qβ, but not VLPs, was used. Although all VLP-displayed peptides elicited high-titer anti-peptide antibody responses, only a 65–85 peptide-induced response strongly neutralized HPV16 pseudovirion infection. To overcome HPV type specificity, a consensus 65–85 peptide was introduced and broad neutralizing activity against all of the HPV types tested was achieved.	Tyler et al. (2014b)
	E7 protein; E7 peptide 49–57: RAHYNIVTFGGC	17 kDa 12		The E7 oligomers were either chemically linked or simply mixed with CpG-loaded VLPs. The E7-specific IgG responses were strongly enhanced if the antigen was linked to the VLPs. In contrast, both CD4+ and CD8+ T cell responses as well as T cell-mediated protection against tumor growth were comparable for linked and mixed antigen formulations.	Gomes et al. (2017a)

(Continued)

TABLE 22.2 (Continued)

Vaccines Constructed on the RNA Phage VLPs by Chemical Coupling or Plug-and-Display Methodology

Vaccine Target	Source of Epitope	Epitope Length, aa	Coupling Site	Comments	References
Hypercholesterolemia	Proprotein convertase subtilisin/kexin type 9 (PCSK9), human, aa 68–76, 153–163, 207–223	9 11 17	Lysine residues and N-terminus	Vaccination of mice and macaques led to significant reductions in total cholesterol, free cholesterol, phospholipids, and triglycerides (see Table 22.1).	Crossey et al. (2015a)
	Proprotein convertase subtilisin/kexin type 9 (PCSK9), human, aa 150–157, 161–170, 236–243, 273–281, and 577–585	8 10 8 9 9		Mice vaccinated with VLP-PCSK9 peptide vaccines, especially PCSK9Qβ-003 displaying the 150–157 epitope, developed high titer IgG antibodies against PCSK9. The PCSK9Qβ-003 vaccine decreased plasma total cholesterol in mice and up-regulated LDLR expression in liver.	Pan et al. (2017)
Hypertension	CYT006-AngQb vaccine: Angiotensin II-derived peptide CGGDRVYIHPF where CGG is a linker	11		Successful preclinical trials in mice and rats and phase I and IIa clinical trials were performed.	Ambühl et al. (2007); Tissot et al. (2008)
Hypertension, diabetic nephropathy, atherosclerosis	ATRQβ-001 vaccine: ATR-001 peptide AFHYESQ corresponding to an epitope of the ECL2 of human AT1R, a G protein coupling receptor, with N-terminally added cysteine	8		A successful preclinical trial: the ATRQβ-001 vaccine decreased the blood pressure of Ang II-induced hypertensive mice and spontaneously hypertensive rats. The vaccine provided a promising method to treat diabetic nephropathy in rats and atherosclerosis in mice.	Chen X et al. (2013); Ding D et al. (2016); Zhou Y et al. (2016); Pan et al. (2019)
Inflammatory hyperalgesia: potential long-term therapy for chronic pain	Nerve growth factor (NGF), murine, aa 19–241 of pro-NGFβ and an additional 9 aa extension at the C terminus comprising a His₆ tag and GGC sequence	223		Vaccination with NGFQβ substantially reduced hyperalgesia in collagen-induced arthritis or postinjection of zymosan A, two models of inflammatory pain in mice. The coupling efficiency was about 60 NGF molecules per VLP.	Röhn et al. (2011)
Influenza virus	M2 protein, N-terminal extracellular domain	23		Intranasal immunization of mice with Qβ VLP-M2 resulted in strong M2-specific antibody responses as well as anti-viral protection.	Bessa et al. (2008)
	M2e peptide SLLTEVETPIRNEWGC-azide coupled through the copper-free click chemistry crosslinker DBCO	16		VLPs packing prokaryotic RNA must be preferred whenever a response dominated by IgG2 is desired, while eukaryotic RNA should be employed in order to induce a response dominated by IgG1.	Gomes AC et al. (2019b)
	Hemagglutinin, globular head domain (gH1), A/California/07/2009 (H1N1) strain, aa 49–325 and C-terminal extension GGGCG	281		Preclinical studies of a set of gH1 variants from mouse-adapted virus and 2009 H1N1 virus were performed. During phase I trial, non-adjuvated gH1-Qβ showed similar antibody mediated immunogenicity and a comparable safety profile in healthy humans to commercially available vaccines.	Skibinski et al. (2013, 2018); Jegerlehner et al. (2013); Low et al. (2014)
	Two hemagglutinin stalk peptides of subtypes H1 and H3 overlapping at the C terminus	57 72		Immunization of mice with chimeric particles induced robust cross-reactive antibody response to hemagglutinin by both peptides 1 and 2 but was unable to induce higher protection against influenza virus infection than a peptide vaccine.	Kiršteina (2017)

(Continued)

TABLE 22.2 (Continued)

Vaccines Constructed on the RNA Phage VLPs by Chemical Coupling or Plug-and-Display Methodology

Vaccine Target	Source of Epitope	Epitope Length, aa	Coupling Site	Comments	References
Leishmaniasis	α-Gal trisaccharide epitope, *Leishmania infantum*, *Leishmania amazonensis*		Lysine residues and N-terminus	The α-Gal-Qβ VLPs were tested against *Leishmania* infection in a C57BL/6 α-galactosyltransferase knockout mouse model, which mimics human hosts in producing high titers of anti-α-Gal antibodies. Vaccination protected the knockout mice against *Leishmania* challenge.	Moura et al. (2017)
Lyme disease	modified outer surface protein CspZ of *Borrelia burgdorferi*	~25 kDa		The Qβ VLPs carrying modified CspZ with eliminated binding of FH, or complement regulator factor H, demonstrated greater bactericidal antibody titers in mice and cleared spirochete infection.	Marcinkiewicz et al. (2018)
Lymphocytic choriomeningitis virus (LCMV)	Glycoprotein, peptides: p33, KAVYNFATM; p13, GLNGPDIYKGVYQFKSVEFD; p33-gp61, CKSLKAVYNFATMGLNGPDI YKGVYQFKSVEF with a GGC linker added to the C-terminus	9 20 32		The LCMV peptides played crucial role in the studies of the CTL induction by the CpG packaging. The average presence of about two p33 peptides per subunit was found, indicating that about 360 peptides were displayed per particle.	Bachmann et al. (2004b, 2005a); Storni et al. (2004); Schwarz et al. (2005); Bessa et al. (2008); Agnellini et al. (2008); Keller et al. (2010a,b); Mohsen et al. (2017a)
Malaria	*Plasmodium falciparum*, circumsporozoite protein (CSP), an almost full-length CSP consisting of 19 NANP and 3 NVDP repeats and the majority of the N- and C-terminal regions (residues 26_{Tyr}–127_{Asp} linked to 207_{Pro}–383_{Ser})	45 kDa		By immunization of mice, the Qβ-CSP induced higher anti-NANP repeat titers, higher levels of cytophilic IgG2b/c antibodies and a trend towards higher protection against transgenic parasite challenge as compared to soluble CSP formulated in the same adjuvant. The coupling efficiency was about 12%–25%, or average of ~30 CSP molecules per VLP. The vaccine was efficient also in the rhesus monkey model.	Khan et al. (2015); Phares et al. (2017)
	Plasmodium falciparum, Pfs25	40 kDa		The highest quantity of antibodies was elicited when compared to antigen displayed by AP205 VLPs or fused to IMX313.	Leneghan et al. (2017)
	Plasmodium vivax, cell-traversal protein for ookinetes and sporozoites (*P. vivax* CelTOS)	21 kDa		Four different vaccine platforms were used. No immunization regimens provided significant protection against a novel chimeric rodent *P. berghei* parasite *Pb-PvCelTOS*.	Alves et al. (2017)
Melanoma	Mel-QbG10 vaccine: Melan-A/Mart-1 A27L variant peptide CGHGHSYTTAEELAGIGILTV Packaging with QbG10 CpG: GGGGGGG GGGGACGATCGTCGGGGGGGGGG	20		Promising results were obtained by phase I and IIa clinical trials on stage II-IV melanoma patients. The coupling density reached 160 peptide copies per VLP.	Speiser et al. (2010); Braun et al. (2012); Goldinger et al. (2012)

(Continued)

TABLE 22.2 (Continued)
Vaccines Constructed on the RNA Phage VLPs by Chemical Coupling or Plug-and-Display Methodology

Vaccine Target	Source of Epitope	Epitope Length, aa	Coupling Site	Comments	References
	The vaccines were produced with germline epitopes, germline-multi target vaccine (GL-MTV) or mutated epitopes (Mutated-MTV) or a combination of both (Mix-MTV) in mice transplanted with aggressive B16F10 melanoma tumors	8 9 10	Lysine residues and N-terminus	Personalized multi-target vaccine was based on identified germline and predicted mutated peptides coupled to the TLR 9-ligand loaded Qβ VLPs by the non-toxic bio-orthogonal Cu-free click chemistry. The generated vaccine candidates induced effective CD8+ T cell responses.	Mohsen et al. (2019c)
Nicotine addiction	CYT002-Nic-Qb vaccine, or NIC002 (formerly known as Nicotine-Qβ or Nic-Qβ); nicotine was covalently coupled to Qβ VLPs via a succinimate linker.	-		Preclinical studies in mice and successful phase I and II clinical trials were performed. Stable vaccine formulations enabling storage were developed. An experimental comparison to a NIC-CRM vaccine by Pfizer was achieved. The coupling density was about 3.25 nicotine molecules per coat monomer, or 585 nicotine molecules per VLP.	Maurer et al. (2005); Beerli et al. (2008); Cornuz et al. (2008); Lang R et al. (2009); McCluskie et al. (2015)
Obesity	Qβ-GIP vaccine: gastric inhibitory peptide (GIP, also known as glucose-dependent insulinotropic polypeptide), aa 1–15 YAEGTFISDYSIAMD of mature GIP (42 aa) with C-terminally added linker GC.	15		Immunization of mice with Qβ-GIP vaccine induced high titers of specific antibodies and efficiently reduced body weight gain in animals fed a high fat diet. Moreover, increased weight loss was observed in obese mice vaccinated with VLP-GIP. The coupling density was about 1.6–2.2 GIP molecules per coat monomer.	Fulurija et al. (2008)
Osteoporosis	Qβ-TRANCE/RANKL vaccine: TNF-related activation-induced cytokine (TRANCE), also known as receptor activator of NF-kappaB ligand (RANKL), aa 158–316 (extracellular domain) of the mature form of murine TRANCE/RANKL with a Cys-linker and His$_6$ tag	159		Immunization of mice with Qβ-TRANCE/RANKL VLPs overcame the natural tolerance of the immune system toward self- proteins and produced high levels of specific antibodies without the addition of any adjuvant. Serum antibodies of immunized mice neutralized TRANCE/RANKL activity in vitro and were highly active in preventing bone loss in a mouse model of osteoporosis. The coupling density was about 14%, indicating that about one of seven coat monomers was covalently attached to a peptide. As TRANCE/RANKL forms a trimer, ~25 C-TRANCE trimers were displayed per VLP.	Spohn et al. (2005); Spohn and Bachmann (2007a)
OVA peptide	Ovalbumin, CSSAESLKISQAVHAAHAEINEAGR	25		TLR9 signaling in B cells determines class switch recombination to IgG2a.	Jegerlehner et al. (2007)
Parkinson's disease	Alpha-synuclein, peptides: CGGKNEEGAPQ (PD1), MDVFMKGLGGC (PD2), CGGEGYQDYEPEA (PD3)	11 11 13		The antibodies induced by vaccination in mice recognized Lewy bodies and toxic oligomeric a-syn species, while recognition of monomeric a-syn was essentially absent. The antibodies were unable however to treat symptoms of Parkinson's disease in the mouse model.	Doucet et al. (2017)

(Continued)

TABLE 22.2 (Continued)

Vaccines Constructed on the RNA Phage VLPs by Chemical Coupling or Plug-and-Display Methodology

Vaccine Target	Source of Epitope	Epitope Length, aa	Coupling Site	Comments	References
Salmonella typhi	Outer membrane protein, D2 peptide, aa 266–280 TSNGSNPSTSYGFAN with N-terminal CGG linker	18	Lysine residues and N-terminus	The vaccine preparations of 13, 56, 94, 142, and 293 peptides per VLP were used. It was shown that the phenomenon of carrier induced epitopic suppression (CIES) could be overcome by high coupling densities, repeated injections and/or higher doses of conjugate vaccine (see above for AP205).	Jegerlehner et al. (2010)
Salmonella enteritidis	Synthetic tetrasaccharide antigen			High levels of specific and long-lasting anti-glycan IgG antibodies were induced by the conjugate, which completely protected mice from lethal bacterial challenge.	Huo et al. (2019)
Streptococcus pneumoniae	Synthetic glycans, TS3, a linear repeat of a disaccharide Glcβ1–3GlcAβ1–4Glcβ1–3GlcAβ1, TS14, Galβ1-4Glcβ1-6[Galβ1-4]GlcNAcβ1, a branched tetrasaccharide			The CuAAC reaction was used with loading of ~20–200 tetrasaccharides per VLP for both TS3 and TS14. Protection was due to the induction of very-high-affinity antiglycan antibodies in a completely CD4 T cell–dependent manner.	Polonskaya et al. (2017)
Tauopathy	pT181, tau peptide [175]TPPAPKpTPPSSGEGGC[190], phosphorylated at Thr181 and modified with two glycine and one cysteine spacer sequence for conjugation	13		Vaccination of mice with pT181-Qβ reduced both soluble and insoluble species of hyperphosphorylated pTau in the hippocampus and cortex, avoided a Th1-mediated pro-inflammatory cell response and rescued cognitive dysfunction in a mouse model of frontotemporal dementia. The study provided a premise for the development of VLP-based immunotherapy to target pTau and potentially prevent Alzheimer's diseases and related tauopathies.	Maphis et al. (2019)

Source: The original table from Pumpens P et al. *Intervirology.* 2016;59:74–110, with permission of S. Karger AG, Basel was extended with recent data.

Note: The order of data presentation is the same as in Table 22.1. When the precise number of the amino acid residues within the epitope was difficult to assess, the size of the epitope was expressed in kDa.

The numerous anticytokine vaccine candidates including a full list of the Qβ VLP-based vaccines have been reviewed (Delavallée et al. 2008; Link and Bachmann 2010; Uyttenhove and Van Snick 2012). More generally, the vaccines against chronic non-communicable diseases were analyzed by Röhn and Bachmann (2010). The Qβ VLP-based vaccine against the signaling molecule IL-17 was described as a safe and effective tool in the common skin disease psoriasis (Foerster and Bachmann 2015). It was specifically emphasized that the vaccination against IL-17 could be capable of replacing the costly manufacture of antibodies currently in clinical use, with huge implications for treatment availability and health economics. The prospects of an active vaccine against asthma, targeting IL-5, were outlined by Bachmann et al. (2018).

Bachmann's team elaborated the CAD106 vaccine against Alzheimer's disease (Wiessner et al. 2011). The CAD106 first-in-human study demonstrated a favorable safety profile and promising antibody response (Winblad et al. 2012). Further in-human investigations supported the CAD106 favorable tolerability profile after repeated CAD106 injections (Farlow et al. 2015; Vandenberghe et al. 2017). The CAD106 vaccine, as well as the Qβ VLP nicotine vaccine, were reviewed exhaustively in a special paper devoted to the vaccines in psychiatry (Kuppili et al. 2018).

The existing noninfectious disease vaccine candidates are thoroughly examined in a recent review by El-Turabi and Bachmann (2018). The major problem of immunogenicity and immunodominance in antibody responses against the epitopes displayed on the VLPs was analyzed recently by Vogel and Bachmann (2019).

Concerning especially the infectious disease vaccines, the influenza gH1-Qβ vaccine, which is presented in Table 22.2, induced influenza-specific CD4$^+$ and CD8$^+$ T cell responses and a number of influenza-specific cytokines including antiviral IFN-γ, in both non-adjuvanted and adjuvanted formulations (Skibinski et al. 2018). Moreover, the transcriptome analysis revealed gene signatures that correlated with the vaccine response. The authors believed that further studies, including a planned human challenge study with the gH1-Qβ VLP vaccine, will reveal any protective capability against influenza infection, as well as the role of T cell responses in protection, and allow validation of the identified gene signatures (Skibinski et al. 2018).

An interesting effort was attempted to display a complex conformational HIV1 gp120 epitope, namely, the epitope b122a, recognized by the CD4 binding site-directed broadly neutralizing antibody b12, on the Qβ VLPs (Purwar et al. 2018). The artificially designed b122a polypeptide was generated earlier as a recombinant protein carrying a substantive part of the b12 epitope (Bhattacharyya et al. 2013). Then, this recombinant protein, in parallel with some other epitopes, was displayed on the Qβ VLPs by both fusion and coupling technologies, as presented in Tables 22.1 and 22.2.

A prospective approach for developing vaccines against bacterial infectious agents was demonstrated recently by the putative Lyme disease vaccine, where outer surface protein CspZ of *Borrelia burgdorferi* was modified by eliminating ability to bind the complement regulator factor H and then coupled to the Qβ VLPs (Marcinkiewicz et al. 2018). This is the first example of the promising strategy by conjugating a bacterial antigen to a VLP and eliminating binding to the target ligand.

The Qβ VLPs were used to display high-mannose glycans that were identified on the HIV-1 envelope glycoprotein, gp120, as a target for neutralizing antibody 2G12, the first HIV-1 antiglycan neutralizing antibody described (Doores et al. 2013). This study contributed markedly to the putative HIV carbohydrate vaccine design strategies. The Qβ VLP-based glycovaccines were reviewed by Restuccia et al. (2016). Furthermore, a third-generation anti-glycan vaccine was generated against two *Streptococcus pneumoniae* pathogenic serotypes, where long-term protective immunity in mice was elicited with exquisite specificity (Polonskaya et al. 2017). Furthermore, the α-Gal trisaccharide epitope was identified on the surface of the protozoan parasites *Leishmania infantum* and *L. amazonensis*, the etiological agents of visceral and cutaneous leishmaniasis, respectively, and displayed on the Qβ VLPs (Moura et al. 2017). The putative vaccine demonstrated high efficiency in a C57BL/6 α-galactosyltransferase knock-out mouse model and protected animals against *Leishmania* challenge, eliminating the infection and proliferation of parasites in the liver and spleen as probed by qPCR. The α-Gal epitope might therefore be considered as a vaccine candidate to block human cutaneous and visceral leishmaniasis (Moura et al. 2017). A detailed protocol for the conjugation of a prototypical tumor-associated carbohydrate antigen (TACA), the Tn antigen, to the Qβ VLPs was published by Sungsuwan et al. (2017).

The plug-and-display methodology, principles of which were described in the Chemical coupling and plug-and-display: AP205 section of Chapter 20, led to the generation of an impressive number of vaccine candidates, which are listed in Table 22.2. The first place in this list belongs, without a doubt, to the putative malaria vaccines (Brune et al. 2016; Janitzek et al. 2016; Thrane et al. 2016; Leneghan et al. 2017; Singh et al. 2017). The potential importance of the AP205 VLPs displaying VAR2CSA epitopes was reviewed by Pehrson et al. (2017). It is noteworthy that the VAR2CSA-based vaccines were targeted against placental malaria appearing during pregnancy, since the *var2csa* gene is expressed on the surface of parasite isolates from placental tissue and by parasites selected to bind to chondroitin sulfate A (CSA).

In parallel, high-density display of human HER2 on the surface of the AP205 VLPs allowed overcoming B cell tolerance and induced high-titer therapeutically potent anti-HER2 IgG antibodies (Palladini et al. 2018). Moreover, prophylactic vaccination with the AP205-HER2 VLPs prevented (i) tumor growth in wild-type FVB mice grafted with mammary carcinoma cells expressing human HER2 and (ii) spontaneous development of human HER2-positive mammary carcinomas in tolerant transgenic mice. Therefore, the AP205-HER2 VLPs demonstrated definite potential to become a tool for treatment and prevention of HER2-positive cancers, while the AP205 platform appeared as a good choice for development of vaccines against other noncommunicable diseases (Palladini et al. 2018).

The AP205 platform contributed by the generation of novel protein cages to be employed by the plug-and-display SpyTag/SpyCatcher methodology (Bruun et al. 2018). Thus, the i301 nanocage, which mimicked the structure of VLPs, was based on the 2-keto-3-deoxy-phosphogluconate aldolase from the Entner–Doudoroff pathway of the hyperthermophilic bacterium *Thermotoga maritima*. The i301 had five mutations that altered the interface between the wild-type protein trimer, promoting assembly into a higher-order dodecahedral 60-mer. The novel modular antigen was designed by fusing SpyCatcher to the N-terminus of the protein and compared with the AP205 VLPs for the ruggedness and immunogenicity of the scaffold, when different transmission-blocking and blood-stage malaria antigens were displayed (Bruun et al. 2018).

Finally, the AP205 platform was employed by the construction of the first combinatorial vaccine (Janitzek et al. 2019). Since cervical cancer and placental malaria are major public health concerns, for example, in Africa, and the target population for vaccination against both diseases, adolescent girls, would be overlapping, the authors decided to combine both vaccines by displaying the appropriate antigens on the AP205 VLPs by the plug-and-display technique. Therefore, the proof-of-concept for a combinatorial vaccine was demonstrated by simultaneous display of the two clinically relevant antigens, namely, the human papillomavirus HPV RG1 epitope and the placental malaria VAR2CSA antigen. The three distinct combinatorial VLPs were produced displaying one, two, or five concatenated RG1 epitopes without obstructing the VLP capacity to form. The co-display of VAR2CSA was achieved through a split-protein Tag/Catcher interaction without hampering the vaccine stability. Vaccination with the combinatorial vaccines was able to reduce HPV infection *in vivo* and induced anti-VAR2CSA IgG antibodies, which inhibited binding between native VAR2CSA expressed on infected red blood cells and chondroitin sulfate A in an *in vitro* binding inhibition assay (Janitzek et al. 2019). This is the first successful attempt to use the AP205 plug-and-display system to make a combinatorial vaccine capable of eliciting antibodies with dual specificity.

CHIMERIC PHAGES

As presented in the Genetic fusions of coats: Qβ section of Chapter 20, the live Qβ phage was employed as a display system via engineering of the A1 protein (Skamel et al. 2014). This unique system employed the high Qβ capability to mutate and adopt to the sequence changes. Thus, as

described in Chapter 20, the chimeric Qβ phages bearing G-H loop peptide of the VP1 protein of FMDV were constructed. Developing this approach further, the membrane proximal external region (MPER) of HIV-1 envelope glycoprotein-41 (gp41) was selected for the presentation on the Qβ phage (Waffo et al. 2017). Thus, a fragment representing the 50-amino acids consensus region within the HIV-1 gp41 MPER was fused in frame with the A1 protein of the phage Qβ. The three variant MPER expression cassettes were obtained with the MPER cDNA in frame with the A1 gene, including pQβMPER, pQβMPERHis with an additional C-terminal hexa-histidine tag and QβMPERN with a C-terminal not1 site. The expression cassettes were used for the production of QβMPER recombinant phages after transformation of *E. coli* HB101 strain. The engineered Qβ phages displayed 12 molecules of MPER per phage particle on the particle surface. The antigenicity of the chimeric phages was assessed with plasma from long-standing antiretroviral naive HIV-1 infected persons, while immunogenicity studies were done in mice. The fusion of MPER and Qβ genes was confirmed by RT-PCR followed by gel electrophoresis and sequencing. The novel recombinant QβMPER phages were proposed, first, to monitor MPER-specific immune responses in HIV-1 exposed or infected people. Second, it was concluded that the recombinant QβMPER phages could be used as immunogens for the induction of the MPER-specific immunity against HIV-1, as a component of the possible HIV-1 vaccines (Waffo et al. 2017).

The detailed methodology of the Qβ phage engineering by the generation of peptide libraries for the FMDV G-H loop and the HIV-1 MPER was published by Singleton et al. (2018).

Finally, Waffo et al. (2018) generated chimeric phages QβMSP3, QβUB05, and a combined variant of the two QβUB05-MSP3, that displayed asexual blood stage antigens UB05 and merozoite surface protein 3 (MSP3) of *Plasmodium falciparum*. Since naturally acquired immune responses to the MSP3 and UB05 were implicated in semi-immunity in populations living in malaria-endemic areas, it would be reasonable to use them by designing malaria vaccine candidates displaying the epitopes upon the surface of the phage in their native form. In fact, the chimeric phages differentially detected blood stage antigen targeting antibodies in children living in a high malaria transmission region of Cameroon. As a result, the chimeric phage QβUB05-MSP3 was validated as an appropriate antigen for tracking immunity to malaria. Furthermore, the comparative analysis of IgG responses to the QβUB05-MSP3 vaccine was performed in dual HIV-malaria infected adults living in areas differing in malaria transmission intensities (Lissom et al. 2018; Nchinda et al. 2018).

23 Non-Vaccine VLPs

At its heart, engineering is about using science to find creative, practical solutions. It is a noble profession.

Queen Elizabeth II

Ever tried. Ever failed. No matter. Try again. Fail again. Fail better.

Samuel Beckett

ARMORED RNA

The *armored* nucleic acids were implemented as useful noninfectious, easily available reagents for quality control in the routine diagnosis of pathogenic viruses. The classical armored RNA technology is based on the MS2 VLPs, where *in vivo* encapsidation of a desired RNA is performed by including the MS2 operator sequence into the RNA molecule in question to enable its packaging. The armored RNA is therefore protected from ribonuclease digestion and can be used as stable, well-characterized controls and standards in routine clinical runs.

For the first time, the versatile armored RNA technology was elaborated by Brittan L. Pasloske and colleagues (1998) from Ambion, Inc, Austin, Texas, and applied to the human immunodeficiency virus type 1 (HIV-1) assay. Thus, the DNAs encoding the (i) MS2 coat protein, (ii) target RNA sequence represented by a 172-base consensus sequence from a portion of the HIV-1 *gag* gene, and (iii) MS2 operator sequence were cloned downstream of an inducible *lac* promoter. As the MS2 coat gene was translated, it was bound to the operator sequence at the 3′ end of the recombinant RNA, initiating the encapsidation of the recombinant RNA to produce pseudoviral particles. The capacity of the armored RNA to accept the desired foreign RNA was measured by a lambda fragment of different length, and estimated as approximately 500 nucleotides. Unlike the phage MS2, which was released into the spent medium by lysing *E. coli*, the armored RNA particles were localized in the cytoplasmic fraction of the *E. coli* cells. After production and purification, the resulting HIV-1 armored RNA particles were shown to be totally resistant to degradation in human plasma and produced reproducible results in the Amplicor HIV-1 Monitor assay for 180 days when stored at −20°C, or for 60 days at 4°C. It was important that the armored RNA preparations were homogeneous and noninfectious. Thus, the straightforward manufacturing process and reliable performance made the armored RNA technology ideal to produce the RNA controls and standards for clinical diagnostics (Pasloske et al. 1998).

Next, the same team generated the armored RNA control for the hepatitis C virus (HCV) genotyping assay (WalkerPeach et al. 1999). To do this, a consensus 412-base sequence from the 5′-noncoding region/core of HCV subtype 2b was synthesized *de novo*, cloned, and expressed by the armored RNA approach. The resulting HCV-2b armored RNA contained the complete HCV-2b consensus RNA sequence encapsidated within the protective MS2 VLPs. The armored HCV RNA was fully recoverable from human plasma incubated at 4°C for >300 days. When added into seronegative, nonviremic plasma, the armored HCV RNA showed reproducible signals and linear dilutions in the three different clinical assay formats (WalkerPeach et al. 1999). The practical methodology of the armored RNAs was presented at that time by DuBois et al. (1999).

Furthermore, Drosten et al. (2001) employed the armored RNA prepared by Pasloske et al. (1998) method in their technically reliable HIV-1 assay for high-throughput blood donor screening. The MS2 VLP-based armored HCV control was then used by Forman and Valsamakis (2004).

Beld et al. (2004) elaborated a highly sensitive assay for the detection of enterovirus in clinical specimens by RT-PCR with the armored RNA internal control, where the armored enterovirus RNA was represented by a part of the 5′-noncoding region, namely, nucleotides 428–691, packaged into the MS2 VLPs. Eisler et al. (2004) used the MS2 VLP-based armored RNA to detect West Nile virus RNA in mosquito pools. The commercially available armored RNA from Ambion was employed by the elaboration of the real-time RT-PCR assays for detection of severe acute respiratory syndrome coronavirus (SARS-CoV) (Bressler and Nolte 2004) and hepatitis C virus (HCV) (Konnick et al. 2005). Donia et al. (2005) used the Ambion Diagnostic kit with armored enterovirus RNA sequence, which included 263 nucleotides of highly conserved 5′-UTR region, to construct a calibration curve for the real-time RT-PCR. Moreover, at that time, the MS2 VLP-based armored RNAs were commercially available from Ambion to be used as reference material in assays for HIV, hepatitis A, C, and G, Dengue, and Norwalk virus, among others (Schaldach et al. 2006).

Hietala and Crossley (2006) adopted the MS2 VLP-based armored RNA approach to the detection of the four high-consequence animal pathogens: classical swine fever virus, foot-and-mouth disease virus, vesicular stomatitis virus of New Jersey serogroup, and vesicular stomatitis virus of Indiana serogroup. The armored RNA spiked into oral swab fluid specimens mimicked virus-positive clinical material through all stages of the RT-PCR testing process, including the RNA recovery by the four different commercial extraction procedures, reverse transcription, PCR amplification, and real-time detection at target concentrations consistent with the dynamic ranges of the existing real-time PCR assays. The armored RNA concentrations spiked into the oral swab fluid specimens were stable under storage conditions selected to approximate the extremes of time and temperature expected for shipping and handling of proficiency panel samples, including 24 h at 37°C and 2 weeks at temperatures ranging from ambient room temperature to −70°C (Hietala and Crossley 2006).

Cheng et al. (2006) prepared the first His-tagged MS2 armored RNA-carrying VLPs for RT-PCR detection of SARS-CoV. In order to facilitate armored RNA purification, a His_6 tag was introduced into the loop region of the MS2 coat protein, which allowed the exposure of multiple His tags on the surface during armored RNA assembly. The His-tagged armored RNA particles were purified to homogeneity and verified to be free of DNA contamination in a single run of affinity chromatography. The armored RNA control was applied for the quantification of viral loads in clinical samples (Cheng et al. 2006).

In parallel, the triumphant march of the methodology was continued by the preparation of the MS2 VLP-based armored RNAs containing the M gene of influenza H3N2 (Yu et al. 2007), the integrase region of the HIV-1 polymerase gene for diverse subtypes in the Abbott RealTime™ HIV-1 assay (Tang et al. 2007) and other HIV-1 sequences (Huang et al. 2008), the hepatitis E virus ORF3 (Zhao et al. 2007), the rabies virus sequences (Wang et al. 2008), the SARS-CoV sequences (Stevenson et al. 2008), the influenza A, influenza B, and SARS-CoV sequences altogether (Yu et al. 2008), and the rotavirus NSP3 gene (Chang et al. 2012). The Abbott RealTime HIV-1 assay, with the same armored RNA as a standard, was adopted later to the detection of HIV-1 in dried blood spots and plasma (Huang et al. 2011).

Summarizing up to that time, the armored RNA contained approximately 1.7 kb of MS2 RNA encoding the maturation and coat proteins, and the operator, or pac, site. Taking into account the MS2 genome length of 3569 nucleotides at most, 1.9 kb of the non-phage RNA sequence might be encapsulated by this method. Practically, the packaging of 500 bases of RNA has been demonstrated to be very efficient; however, packaging of 1- and 1.5-kb amounts of RNA was inefficient (Pasloske et al. 1998). Huang et al. (2006) used armored RNA technology to package a 1200-nucleotide foreign RNA sequence in their RT-PCR assay for HCV, HIV-1, SARS-CoV1, and SARS-CoV2 by deleting some disposable sequences between the multiple cloning site and the transcription terminator.

To enhance the capacity capabilities of the approach, Wei Y et al. (2008) generated a method for packaging long, more than 2000 nucleotides, RNA sequences, which was referred to as armored L-RNA technology. To do this, they applied the two-plasmid coexpression system, in which the maturation and coat proteins were expressed from one plasmid and the target RNA sequence with modified MS2 stem-loop, or pac site, was transcribed from another plasmid. It made it possible to pack the 3V armored L-RNA of 2248 bases containing six gene fragments: hepatitis C virus, severe acute respiratory syndrome coronavirus SARS-CoV1, SARS-CoV2, and SARS-CoV3, avian influenza virus matrix gene, and H5N1 avian influenza virus. The 3V armored L-RNA functioned successfully as a calibrator for multiple virus assays (Wei Y et al. 2008). In parallel, Wei B et al. (2008) improved the MS2 VLP-based armored RNA by increasing the number and

affinity of the pac site in exogenous RNA and of the sequence encoding coat protein. In this obstacle, the one-plasmid expression system allowed the extension of the length of the armored RNA to 1891 bases, in the case when two C-variant pac sites were used. The C-variant pac site differed from that of the wild type by the substitution of U^{-5} with C, which yielded a significantly higher affinity for coat protein than wild-type pac site in vitro, as described in Chapter 16. Therefore, the armored RNA with the 1891 bases target RNA was expressed successfully by the one-plasmid expression system with two C-variant pac sites, while for one pac site, no matter whether the affinity was changed or not, only the 1200 bases target RNA was packaged. The armored RNA contained SARS-CoV1, SARS-CoV2, SARS-CoV3, HCV, and influenza H5N1 fragments (Wei B et al. 2008). The armored L-RNA technology with the improved pac site was used by the elaboration of the branched DNA assay by the HIV-1 detection (Zhan et al. 2009). This laboratory also presented protocols for the MS2 VLP-based armored RNA detection for enterovirus 71 and coxsackievirus A16 (Song et al. 2011) and avian influenza H7N9 virus (Sun et al. 2013).

In parallel, the traditional MS2 VLP-based armored RNA was constructed in full accordance with the original protocol (Pasloske et al. 1998) and applied in an HCV diagnostic using the novel target in the HCV genome, namely, 3′-X-tail, which was highly conserved across genotypes (Drexler et al. 2009).

Then, the MS2 VLP-based armored RNA standard was applied to measure microcystin synthetase E gene (mcyE) expression in toxic Microcystis sp. (Rueckert and Cary 2009). It was highly important, since microcystin was a secondary metabolic peptide toxin known to cause hepatotoxicosis and carcinogenicity in vertebrates. It was produced by various bloom-forming cyanobacterial species constituting a serious threat to the quality of freshwater reservoirs worldwide. This important study indicated a growth phase−dependent expression of the mcyE gene, with the maximal level of expression observed during mid-log growth. With progressing age of the cultures, the mcyE was gradually down-regulated, with expression levels having declined by more than three orders of magnitude during the stationary growth phase (Rueckert and Cary 2009; Wood et al. 2011).

Okello et al. (2010) employed the MS2 VLP-based armored HIV-1 RNA by quantitative assessment of the sensitivity and reproducibility of 11 commercially available reverse transcriptases.

Monjure et al. (2014) were the first to apply the armored RNA technique to optimize PCR for quantification of simian immunodeficiency virus (SIV) genomic RNA in the plasma of rhesus macaques. Remarkably, this quantification procedure was optimized by inclusion of the MS2 VLP-based HCV RNA because HCV did not infect non-human primates to confound data interpretation. Then, the two novel control systems for RT-qPCR were established by the MS2 VLP-based armored RNAs of hantavirus and Crimean-Congo hemorrhagic fever virus, where both target-specific gene cassettes were combined with two internal C-variant pac-sites and a multiple cloning site (Felder and Wölfel 2014).

The armored RNA technology contributed markedly to a set of original methods directed to the efficient *in vivo* expression of large amounts of homogeneous RNA (Ponchon et al. 2013). The use of specific RNA scaffolds was extended by the addition of the MS2 operator to the original chimeric tRNA-mRNA sequence, in order to be encapsulated into the MS2 VLPs. The genes encoding the desired RNA and the MS2 coat were positioned on the two different plasmids. After cell lysis and VLP purification by size exclusion chromatography, the desired RNA was recovered by heat treatment, without the need for phenol-extraction steps, and the strategy was amenable to simple 96-well format parallelization and could be conceived by novel combinatorial approaches that might focus on RNA, for example, genomic library expression, or *in vivo* mutagenesis and evolution of RNA structures, followed by screening. It is remarkable that the addition of His_6 tag into the AB loop region of the MS2 coat resulted in the formation of the MS2 dimers instead of the VLPs. Nevertheless, these dimers retained ability to bind the desired RNA and allowed easy purification of the latter in the form acceptable for the crystallographic studies (Ponchon et al. 2013). Later, Zhang Jin et al. (2015) got ribonuclease-resistant His_6-tagged MS2 VLPs carrying armored influenza RNA.

The armored double-stranded DNAs were encapsulated for the first time into the MS2 VLPs by Zhang Lei et al. (2015). To do this, the MS2 VLPs were dissociated into coat dimers first, and then the latter were reassembled into the VLPs in the presence of *pac* site in the RNA or DNA form. The armored DNAs were obtained for 1.3-, 3-, 3.5-, and 6.5-kb HBV or HPV DNA sequences, when they were coupled to the RNA or DNA *pac* site. It appeared therefore that the *pac* DNA also triggered the spontaneous reassembly of the MS2 dimers. Thus, this was the first successful encapsulation of the double-stranded DNA by the RNA phage VLPs, which remained indistinguishable from native MS2 capsids in size and morphology (Zhang L et al. 2015).

The 2013–2015 Ebola virus disease humanitarian crisis provoked the development of the laboratory-free, point-of-care nucleic acid testing solutions, realized by the EbolaCheck, an international consortium to deliver the clinical molecular diagnostic standard-of-care testing suitable for the West African milieu within 12 months (Moschos 2015). One of the solutions was represented by the internal assay standards developed as the MS2 VLP-based armored RNA. Next, the MS2 VLP-based armored RNA was used successfully by the elaboration of the promising one-step RT FRET-PCR assay for differential detection of five Ebola virus species (Lu et al. 2015), as well as for the evaluation of seven commercial Ebola virus RT-PCR detection kits (Wang G et al. 2015). The MS2 VLP-based armored RNA was used also in other Ebola virus kits (Dedkov et al. 2016). Furthermore, an extensive evaluation of the MS2 VLP-based armored RNAs in the context of the Ebola virus diagnostics was performed by Shah (2017). It is remarkable in this connection that the phage MS2 itself was sometimes used as a positive control for the sample preparation by the Ebola virus detection directly from blood (Benzine

et al. 2016). A detailed protocol of Ebola virus detection with the armored RNA, as well as with the phage MS2 as controls, was published by Grolla (2017).

In parallel, the MS2 VLP-based armored RNA was employed by the external quality assessment program of the RT-PCR detection of measles virus (Zhang D et al. 2015) and by the rapid detection of HCV genotypes 1a, 1b, 2a, 3a, 3b, and 6a, where the apropriate armored RNAs were produced for each of six HCV genotypes (Athar et al. 2015).

At that time, Mikel et al. (2015) published an extensive review on the MS2 VLP-based armored RNA with a special aim to teach how to prepare and use the armored tools by the RT-PCR and qRT-PCR detection of RNA viruses not only in clinical specimens, but also in food matrices. To monitor the sample analysis, Mikel et al. (2016) generated two sorts of the MS2 VLPs with armored RNAs derived from mitochondrial DNA sequences of two extinct species, namely, thylacine *Thylacinus cynocephalus* and the moa bird *Dinornis struthoides*. Due to the fact that the sequences were derived from mitochondrial DNA of extinct species, its natural occurrence in the analyzed samples was highly unlikely. Next, Mikel et al. (2017, 2019) adopted His-tag system for rapid production and easy purification of MS2 VLPs. This was the first case when the single-chain version of the MS2 coat dimer containing the His-tag and capable to form intact His-tagged MS2 VLPs was supposed for the armored RNA technology.

In 2015, the appearance of Middle East respiratory syndrome coronavirus (MERS-CoV) initiated rapid construction of the external quality assessment systems, which were based again on the MS2 VLPs encapsulating specific RNA sequences of MERS-CoV as positive specimens (Zhang L et al. 2016).

Recently, the MS2 VLP-based armored RNA technology was applied to the detection of hepatitis E virus (Wang Shen et al. 2016), a set of mosquito-borne viruses and parasites (Zhang Y et al. 2017), Dengue virus serotypes (Fu et al. 2017), Zika virus (Lin G et al. 2017), HCV (Zambenedetti et al. 2017), a broad set of influenza viruses (Zhang D et al. 2018), rabies virus (Dedkov et al. 2018a), filoviruses (Dedkov et al. 2018b), HIV-1 (Gholami et al. 2018a,b), and Lassa virus (Dedkov et al. 2019). Petrov et al. (2018) adapted the armored MS2 VLPs for the detection of Ebola, Marburg, Lassa, Machupo, Venezuelan encephalitis equine (VEE), Rift Valley fever, and rabies viruses. The MS2 VLP-based armored RNA technology was also employed by the p210 *BCR-ABL1* testing, as a proof of common cytogenetic abnormalities in leukemia (Fu et al. 2019). Recently, the armored MS2 technique was applied to generate standard RNA for measuring hepatitis B virus (Gao M et al. 2019).

After the most popular MS2 VLPs, the Qβ VLPs were introduced for the first time in the role of the armored RNA tools by the HCV detection (Villanova et al. 2007). Then, the Qβ VLP-based armored RNA technology was applied to the norovirus detection (Zhang Q et al. 2018). At last, the armored Qβ VLPs were found more stable than the armored MS2 VLPs for packaging human norovirus RNA, by incubation and storage at different temperatures (Yao L et al. 2019).

NANOCONTAINERS: DRUG DELIVERY

Table 23.1 summarizes the RNA phage VLP-based experimental approaches that could be regarded as the VLP packaging and targeting methodologies. Historically, the idea of encapsulation/targeting by the RNA phage VLPs appeared in the early 1990s, only a few years after the vaccine/epitope display approaches described above. The first attempt of the RNA phage nanocontainer-targeted drug delivery demonstrated encapsulation of the deglycosylated ricin A chain coupled to the RNA operator stem-loop and decoration of the MS2 VLPs by transferrin (Wu et al. 1995, 1996, 1997). Such structures, called *synthetic virions* by the authors, demonstrated high toxicity to leukemia cells carrying transferrin receptor. A detailed review of these and similar pioneering experiments was published at that time (Brown et al. 2002).

The next step in development of the VLP packaging/targeting technology included the chemical coupling of putative cargo and targeting molecules to the outer and inner surfaces of VLPs, respectively (Hooker et al. 2004; Kovacs et al. 2007; Carrico et al. 2008; Stephanopoulos et al. 2009, 2010b; Tong et al. 2009; Wu et al. 2009). It was demonstrated that the MS2 VLPs could be conjugated to peptides recognizing human hepatocellular carcinoma cells and could be loaded with vastly different types of cargo, including low molecular mass chemotherapeutic drugs, siRNA cocktails, protein toxins, and nanoparticles, resulting in the selective killing of target cells (Ashley et al. 2011a). The first review declaring the MS2 VLPs as an element of synthetic biology for nanotechnology and presenting the pioneering data of Matthew B. Francis' team was published by Philip Ball (2005). From the early reviews, the MS2 and Qβ VLPs occupied the stable place, as potentially important nanocages, among a long set of other nanoscale viruses and non-viral protein cage candidates involved in adaptations for medical applications, mainly for drug delivery (Lee and Wang 2006; MaHam et al. 2009; Seow and Wood 2009; Grasso and Santi 2010; Koudelka and Manchester 2010; Stephanopoulos and Francis 2011; Yildiz et al. 2011; Yoo et al. 2011).

The packaging of the Qβ VLPs with immunostimulatory CpG sequences led not only to the development of potential allergy vaccines, as described in the Chemical coupling section of Chapter 22, but also strongly contributed to the general understanding of the mechanics of oligodeoxynucleotide-induced stimulation (Bachmann et al. 2004b, 2005a; Storni et al. 2004; Schwarz et al. 2005; Agnellini et al. 2008; Bessa et al. 2008; Senti et al. 2009; Keller et al. 2010a, 2010b; Hou et al. 2011; Klimek et al. 2011, 2013a; Link et al. 2012; Beeh et al. 2013; Casale et al. 2015), as narrated in the Basic immunology section of Chapter 22. The Qβ VLP vaccines carrying packaged CpG were reviewed in the global context of the CpG oligodeoxynucleotide nanomedicines for the prophylaxis or treatment of cancers, infectious diseases, and allergies (Hanagata 2017). Furthermore, the Qβ VLP vaccines were ranked on a par with the total *E. coli*–derived VLPs in vaccine development (Huang et al. 2017).

Numerous reviews concentrated at that time on the general problems of the VLP cages as the most attractive nanocarrier platforms for drug packaging and targeted delivery, always with special emphasis on the MS2 and Qβ VLPs, mentioning, less often, also the PP7, AP205, and φCB5 VLPs (Ma Y et al. 2012; Doll et al. 2013; Steinmetz 2013; Wu Y et al. 2013; Deshayes and Gref 2014; Galdiero et al. 2014; Hong and Nam 2014; Jin et al. 2014; Molino and Wang 2014; Schoonen and van Hest 2014; Shi et al. 2014; Vannucci et al. 2014; Bajaj and Banerjee 2015; Ferrer-Miralles et al. 2015; Putri et al. 2015; Schwarz and Douglas 2015; Varshosaz and Farzan 2015; Yan et al. 2015; Aljabali 2016, 2018; Luo et al. 2016; Zdanowicz and Chroboczek 2016; Bhaskar and Lim 2017; Czapar and Steinmetz 2017; Gardiner Heddle et al. 2017; Karimi et al. 2017; Meghani et al. 2017; Parodi et al. 2017; Rohovie et al. 2017; Roudi et al. 2017; Sunderland et al. 2017; Tashima 2017; Aljabali et al. 2018; Alshaer et al. 2018; Aumiller et al. 2018; Choi B et al. 2018; Diaz et al. 2018; Hill et al. 2018; Kawano et al. 2018; Kelemen et al. 2018; Lam and Steinmetz 2018; Lemire et al. 2018; Ngandeu Neubi et al. 2018; Casanova et al. 2019; Jeevanandam et al. 2019; Li B et al. 2019; Pottash et al. 2019; Sokullu et al. 2019).

The special importance of the cell-type-specific aptamer-functionalized VLPs, including those of MS2, for targeted disease therapy was highlighted in the review published by Zhou J and Rossi (2014). In this context, they revived Paul Ehrlich's 110-years-old concept of the *magic bullet*. This concept presumed that a therapeutic agent would only kill the specific cells it targeted. The RNA phage VLPs provided with the specific aptamers were considered therefore as promising tools of such magic bullet, or, in other words, *targeted therapy*. The convincing examples of the aptamer-targeted therapy are presented in fact in Table 23.1. Another detailed review devoted to anticancer perspectives of nanoparticles filled with siRNA was published at that time (Yan et al. 2014). Later, the aptamers in connection with the MS2 VLPs were reviewed by Vorobyeva et al. (2016).

The remarkable potential of the MS2 VLPs by the treatment of systemic lupus erythematosus was reviewed by Guo et al. (2014). The logical connection of the specific delivery with the targeted immunomodulation using antigen-conjugated nanoparticles was highlighted by McCarthy et al. (2014).

Concerning the current success in the cargo delivery, an extensive investigation on the microRNA delivery was published recently (Wang G et al. 2016). This study was based on the MS2 VLPs displaying the cell-penetrating peptide TAT and aimed to treat hepatocellular carcinoma. Previously, this highly prominent team developed the MS2 VLP-based microRNA-146a delivery system, when the TAT peptide was conjugated chemically to the VLPs (Wei et al. 2009). These MS2-miR-146a VLPs effectively eliminated autoantibodies, thus ameliorating systemic lupus erythematosus progression in lupus-prone mice (Pan et al. 2012b) and suppressed osteoclast differentiation (Pan et al. 2012a). The MS2 VLPs were used to encapsulate the long noncoding MEG3 lncRNA and were crosslinked with the GE11 polypeptide, proving beneficial to cancer therapy (Chang et al. 2016). Nevertheless, the technical problems of crosslinking provoked the team to switch to the phage display technique (Wang G et al. 2016).

TABLE 23.1

RNA Phage VLPs as Models for Nanocontainer Packaging and Decoration with Addressing/Targeting/Delivery Purposes

Interior: Packaged by	Exterior: Decorated by	Addressed to	Supposed to Treat	Comments	References
			AP205		
Random *E. coli* RNA, without any specific packaging	HBV, full length preS1, aa 13–119	Eukaryotic cell lines HepG2, Hek293, Jurkat, Namalwa, and BHK-21	HBV infection	The uptake took place, but seemed as the cell-line unspecific and not highly efficient.	Kalniņš et al. (2013)
			fr		
No specific packaging	IgG-binding Z domain at the C-terminus of the coat coexpressed with native coat as a helper	Rather unspecific targeting	Potentially broad applications in diagnostics	This is a first example of the generation of mosaic VLPs carrying the IgG-binding Z domain that could be targeted to antibodies displayed on the cell surface and used in diagnostics.	Juris Jansons and Irina Sominskaya, unpublished
			GA		
GA operator-specified mRNA *in vivo*	No decoration	No cell targeting	Broad spectrum of potential targets	Packaging of different mRNAs encoding GA coat protein, ENA-78, and GFP *in vivo* in yeast *Saccharomyces cerevisiae.*	Rūmnieks et al. (2008)
	Cell-penetrating peptide TAT, HIV-1, 48–60, or WNV E protein DIII domain	Human peripheral blood mononuclear cells (PBMCs)	PBMC-derived failure	Production of mosaic VLPs that are packaged by mRNAs encoding IL2 or GFP was achieved *in vivo* in yeast *Saccharomyces cerevisiae.*	Strods et al. (2012)
MS2 operator-specified mRNA *in vivo*	No decoration	No cell targeting	Broad spectrum of potential targets	Introduction of single (S87N, K55N, R43K) and double (S87N + K55N and S87N + R43K) aa exchanges into the GA coat allowed self-assembly and packaging of MS2 operator-carrying mRNAs *in vivo* in yeast *Saccharomyces cerevisiae.*	Ārgule et al. (2017)
No specific packaging	Z$_{HER2:342}$ affibody, 61 aa in length	HER2 receptor on cancer cells	Diagnostics/therapy	Ability of GA-$_{ZHER2:342}$ –VLPs to recognize selectively and enter the HER2 receptor-bearing cells was demonstrated. Electron microscopy of this construction is presented in Figure 20.3.	Arnis Strods and Regīna Renhofa, unpublished
	Stromal cell-derived factor (SDF1), aa 1–41 and 1–19	CXCR4 receptor on leukocytes	Diagnostics/therapy	Insertions at the N-terminus of the first or second single-chain dimer copy were performed. Some members of the family recognized specific cells.	
			MS2		
Ricin A coupled to RNA operator; 5-fluorouridine coupled to RNA operator	Transferrin, anti-MS2 antibodies, anti-DF3 antibodies	Cells of the immune system; breast carcinoma and leukemia cells	Broad spectrum of potential targets	Classical attempt to use RNA operator as a carrier of the desired material.	Wu et al. (1995, 1997)
Antisense ODNs targeted to human nucleolar protein p120 mRNA	Covalent decoration with transferrin on the VLP surface	Promyelocytic leukemia cell line	Acute myelogenous leukemia	The ODNs were synthesized as covalent extensions to the 19-nt-long operator sequence.	Wu M et al. (2005)

(Continued)

TABLE 23.1 (Continued)

RNA Phage VLPs as Models for Nanocontainer Packaging and Decoration with Addressing/Targeting/Delivery Purposes

Interior: Packaged by	Exterior: Decorated by	Addressed to	Supposed to Treat	Comments	References
50–70 Fluorescein molecules conjugated to the interior of VLPs	180 PEG-2000 or PEG-5000 chains on the exterior of VLPs	Tumor cells	Solid tumors	An early attempt to construct an addressed nanocontainer for the potential delivery of therapeutic cargo.	Kovacs et al. (2007)
Fluorescent dye conjugated to the VLP interior	A 41-nt operator-containing sequence covalently bound to the VLP surface	Tyrosine kinase receptor on the Jurkat T cells	Leukemia	Colocalization experiments using confocal microscopy indicated that the operator-labeled capsids were endocytosed and trafficked to lysosomes for degradation that could allow the targeted drug delivery of acid-labile prodrug.	Tong et al. (2009)
Antisense RNA against the 5′-UTR and IRES of HCV	Cell-penetrating peptide TAT, HIV-1, 47-YGRKKRRQRRR-57	Huh-7 cells containing an HCV reporter system	Chronic hepatitis C	The TAT peptide was conjugated chemically to VLPs. The packaged antisense RNA showed an inhibitory effect on the translation of HCV genome.	Wei et al. (2009)
Fluorescent dyes as donor chromophores	Zinc porphyrins capable of electron transfer	No cell targeting	No definite target disease specified	Specific positioning allowed energy transfer and sensitization of the porphyrin at previously unusable wavelengths, as demonstrated by the system's ability to effect a photocatalytic reduction reaction at multiple excitation wavelengths.	Stephanopoulos et al. (2009)
Coupling of 180 porphyrins capable of generating cytotoxic singlet oxygen upon illumination	~20 copies of a Jurkat-specific aptamer	Jurkat leukemia T cells	Leukemia	The doubly modified VLPs were able to target and kill Jurkat cells selectively even when mixed with erythrocytes.	Stephanopoulos et al. (2010b)
Porphyrin: meso-tetra-(4-N,N,N,-trimethylanilinium)-porphine-tetrachloride, or TMAP. ~ 250 TMAP were loaded by nucleotide-driven packaging	Cancer cells-targeting DNA aptamers via chemical conjugation	MCF-7 human breast cancer cells	Breast cancer upon photoactivation	TMAP interacted with the RNA present within the MS2 capsid. Removal of the interior RNA led to significantly lower porphyrin interaction with the MS2 capsid. The MCF-7 cells incubated with targeted, porphyrin-loaded virus capsids exhibited cell death. The strategy offered an approach for efficient targeted delivery of photoactive compounds for site-specific photodynamic cancer therapy.	Cohen et al. (2011, 2012); Cohen and Bergkvist (2013)
Porphyrin	Angiopep-2	Midbrain	Intractable brain conditions, such as tinnitus	Angiopep-2 facilitates the transport of the MS2 VLPs across the blood-brain-barrier.	Apawu et al. (2018); Cacace et al. (2018)
Nanoparticles, chemotherapeutic drugs, siRNA cocktails, and protein toxins	SP94 peptide that binds human hepatocellular carcinoma (HCC) cells	Hep3B cells	HCC	The targeted VLPs loaded with doxorubicin, cisplatin, and 5-fluorouracil selectively killed the HCC cell line, Hep3B. Encapsidation of a siRNA cocktail induced growth arrest and apoptosis of Hep3B cells. Loading of VLPs with ricin toxin A-chain killed the entire population of Hep3B cells.	Ashley et al. (2011a)
MicroRNA: pre-miR 146a	Cell-penetrating peptide TAT, HIV-1, 47-YGRKKRRQRRR-57	Human peripheral blood mononuclear cells (PBMCs)	Systemic lupus erythematosus, osteoclastogenesis	The TAT peptide was conjugated chemically to VLPs. Restoring the loss of miR-146a was effective in eliminating the production of autoantibodies.	Pan et al. (2012a,b); Yao et al. (2015)

(Continued)

TABLE 23.1 (Continued)

RNA Phage VLPs as Models for Nanocontainer Packaging and Decoration with Addressing/Targeting/Delivery Purposes

Interior: Packaged by	Exterior: Decorated by	Addressed to	Supposed to Treat	Comments	References
RNA conjugate encompassing a siRNA and the operator sequence	Covalent attachment of human transferrin	HeLa cells	Potentially broad applications	The VLPs entered cells via receptor-mediated endocytosis and produced siRNA effects better than by traditional lipid transfection route.	Galaway and Stockley (2013)
Gold nanoparticles	Alexa Fluor 488 (AF 488) labeled DNA strands	No cell targeting	No definite target disease specified	The VLP architecture by placing the dye at distances of 3, 12, and 24 bp from the surface of VLPs bearing 10-mm gold nanoparticles allowed the rapid exploration of many variables involved in metal-controlled fluorescence.	Capehart et al. (2013)
Various reporters to be used by fluorescence-based flow cytometry, confocal microscopy, and mass cytometry	Antibodies using a rapid oxidative coupling strategy	Receptors on human breast cancer cell lines	Breast cancer	The broad set of conjugates with various reporters on the interior of VLPs may lead to many clinically relevant applications, including drug delivery and *in vivo* diagnostics.	ElSohly et al. (2015)
^{64}Cu labeling	Anti-EGFR antibodies	Receptors on tumor xenografts in mice	Breast cancer	The targeting antibodies did not lead to increased uptake *in vivo* despite *in vitro* enhancements.	Aanei et al. (2016); Aanei and Francis (2018)
Proteins with tags comprising anionic amino acids or DNA and gold nanoparticles with negative surface charges	No decoration	No cell targeting	No definite target disease specified	The possible application in biocatalysis, protein stabilization, vaccine development, and drug delivery.	Aanei et al. (2018a)
AlexaFluor680	Antibodies specific to VCAM	VCAM1, vascular cell adhesion molecule	Atherosclerosis	The VCAM-targeted MS2 VLPs were used for the detection of plaques in the early stages of atherosclerosis development.	Aanei et al. (2018b)
Doxorubicin	No decoration	Convection-enhanced delivery	Glioblastoma	The MS2 VLPs conjugated to doxorubicin, as well as tobacco mosaic virus disks, showed the best response in mice, in comparison with phage filamentous rods.	Finbloom et al. (2018)
Long noncoding RNA (lncRNA): MEG3 RNA	GE11, a dodecapeptide YHWYGYTPQNVI, ligand of epidermal growth factor receptor (EGRF), chemically coupled	EGFR receptors	HCC cell line	The targeted delivery was dependent on clathrin-mediated endocytosis and MEG3 RNA suppressed tumor growth mainly via increasing the expression of p53 and its downstream gene GDF15, but decreasing the expression of MDM2.	Chang et al. (2016)
MicroRNA: miR-122	Cell-penetrating peptide TAT, HIV-1, 47-YGRKKRRQRRR-57	Hep3B, HepG2, and Huh7 cells and Hep3B related animal models	Hepatocellular carcinoma	The TAT peptide was displayed on the MS2 VLPs by genetic fusion, instead of being chemically crosslinked. The MS2 VLPs displaying TAT penetrated effectively the cytomembrane and delivered miR-122. The inhibitory effect was shown on the hepatocellular carcinoma model cells.	Wang G et al. (2016)
Thallium (I) ions from TlNO3	iRGD peptide	Endotheliocytes of the tumor tissue neovasculature and certain tumor cells	Breast cancer	iRGD peptide-conjugated MS2 VLPs filled with Tl+ caused cell death in two types of cultivated human breast cancer cells and effected necrosis of these tumor xenografts in mice.	Kolesanova et al. (2019)

(Continued)

TABLE 23.1 (Continued)

RNA Phage VLPs as Models for Nanocontainer Packaging and Decoration with Addressing/Targeting/Delivery Purposes

Interior: Packaged by	Exterior: Decorated by	Addressed to	Supposed to Treat	Comments	References
LacZ RNA fused to RNA operator	No decoration	No cell targeting	Potentially broad applications	This is a classical attempt to ensure operator-specific packaging *in vivo*.	Pickett and Peabody (1993)
MS2 operator sequence linked to the human growth hormone (hGH) mRNA for *in vivo* packaging in *S. cerevisiae*			No definite target disease specified	This is a sort of application belonging to the "armored RNA" technology. Functionality of packaged mRNA was confirmed by translation of mRNAs purified from VLPs.	Legendre and Fastrez (2005)
Taxol			Breast, lung, and ovarian cancers	Aa exchange N87C on the interior of VLP allowed use of sulfhydryl groups for the attachment of taxol, a potent chemotherapeutic, as a cargo.	Wu et al. (2009)
HIV-1 gag mRNAs (1544 bases) produced in *S. cerevisiae*			HIV/AIDS: as a vaccine	The HIV antigen-specific antibody responses were elicited by immunization of Balb/C mice.	Sun et al. (2011)
Alkaline phosphatase tagged with a 16 aa peptide			No definite target disease specified	This is a first attempt to encapsulate enzymes where the encapsulated enzyme had the same K(m) value and a slightly lower k(cat) value than the free enzyme.	Glasgow et al. (2012)
mRNA joined to 19-nt operator/aptamer		Macrophages	Prostate cancer	The packaged mRNA ensured strong humoral and cellular immune responses and protected mice as a therapeutic vaccine against prostate cancer.	Li J et al. (2014)
No specific packaging	Tumor cell-specific peptides	Tumor cells	Solid tumors	Three peptides known to target specific tissues: (i) neuroblastoma and breast cancer cell lines, (ii) matrix metalloproteinases, (iii) kidney were chosen as attachment models.	Carrico et al. (2008)
	Conjugation of azide- and alkyne-containing proteins (an antibody fragment and the granulocyte-macrophage colony stimulating factor), nucleic acids, and PEG	Broad spectrum of potential targeting	Tumors and other possible targets	This is a universal approach based on the inclusion of surface exposed methionine analogues (azidohomoalanine and homopropargylglycine) containing azide and alkyne side chains by cell-free synthesis technology.	Patel and Swartz (2011)
	PEG, small peptides	Broad spectrum of potential targeting	Tumors and other possible targets	The oxidative coupling to tyrosine and tryptophan residues using cerium(IV) ammonium nitrate as an oxidant. Six MS2 viral capsids were generated with tryptophan and tyrosine residues on the exterior surface of each monomer. Two of these capsids were also expressed with a cysteine on the interior surface.	Seim et al. (2011)
	PEG, small peptides	Broad spectrum of potential targeting	Tumors and other possible targets	A substantially more efficient version of addition of anilines to o-aminophenols.	Behrens et al. (2011)
	Epidermal growth factor	Variety of cancer cell types	Tumors and other possible targets	The o-methoxyphenols were identified as air-stable, commercially available derivatives that undergo efficient oxidative couplings with anilines in the presence of periodate as oxidant.	ElSohly et al. (2017)

(Continued)

TABLE 23.1 (*Continued*)
RNA Phage VLPs as Models for Nanocontainer Packaging and Decoration with Addressing/Targeting/Delivery Purposes

Interior: Packaged by	Exterior: Decorated by	Addressed to	Supposed to Treat	Comments	References
			PP7		
mRNA encoding GFP	Low molecular weight protamine VSRRRRRRGGRRRR	Cell-penetrating ability to mouse prostate cancer cells	Demonstration of principle	The chimeric PP7 VLPs carrying GFP mRNA penetrated the mouse prostate cancer cells RM-1 after 24 h incubation due to the displayed cell-penetrating peptide.	Sun Y et al. (2016)
MicroRNA: pre-miR-23b	Cell-penetrating peptide TAT, HIV-1, 47-YGRKKRRQRRR-57	Hepatoma SK-HEP-1 cells	Demonstration of principle	The TAT peptide was displayed on the PP7 VLPs by the insertion into the AB loop of the single-chain PP7 coat dimer.	Sun Y et al. (2017)
Cytosine deaminase with Rev peptide	ZZ domain OVA1 OVA2		Potential applications in medicine and chemical manufacturing	All of the Z-domain-bearing particles recognized antibody Fc domain. For details see the Chimeric particles section of Chapter 21. An average of 15—25 cytosine deaminase molecules were packaged.	Zhao et al. (2019)
			Qβ		
CpG oligodeoxynucleotides (ODNs)	With and without specific addressing	Cells of immune system	Potentially broad applications	This is a classical example of the ODN encapsulation for the broad clinical applications as prophylactic and/or therapeutic vaccines.	Bachmann et al. (2004d, 2005a); Storni et al. (2004d); Schwarz et al. (2005); Agnellini et al. (2008); Bessa et al. (2008); Senti et al. (2009); Keller et al. (2010a, 2010b); Klimek et al. (2011, 2013b); Beeh et al. (2013); Casale et al (2015)
~60 Alexa Fluor 568 fluorophores per VLP	Fullerene C$_{60}$ and PEG	HeLa cell line	Cancer	This approach overcame the insolubility of C$_{60}$ in water and opened the door for the applications in photoactivated tumor therapy.	Steinmetz et al. (2009)
Metalloporphyrin derivative for photodynamic therapy	Glycan ligand for specific targeting of cells bearing the CD22 receptor	CD22 receptor		This approach benefited from the presence of the strong targeting function and the delivery of a high local concentration of singlet oxygen-generating payload.	Rhee et al. (2011, 2012)
Aptamers embedded in a longer RNA sequence with the Qβ coat operator	No decoration	No cell targeting	Potentially broad applications in therapy	The VLPs ensured the delivery of the encapsulated aptamers that were protected from degradation and retained ability to bind their small-molecule ligands.	Lau et al. (2011); Wu et al. (2011)
Positively charged synthetic polymer by atom transfer radical polymerization (ATRP) methodology			Potentially broad applications in therapy and imaging diagnostics	This is a robust method for removing encapsidated RNA from VLPs and the use of the empty interior space for site-specific, "graft-from" ATRP reactions.	Hovlid et al. (2014)

(*Continued*)

TABLE 23.1 (Continued)

RNA Phage VLPs as Models for Nanocontainer Packaging and Decoration with Addressing/Targeting/Delivery Purposes

Interior: Packaged by	Exterior: Decorated by	Addressed to	Supposed to Treat	Comments	References
The Rev tagged proteins: 25-kDa N-terminal aspartate dipeptidase, peptidase E, 62-kDa firefly luciferase (Luc), a thermostable mutant of Luc			No definite target disease specified	The encapsulated enzymes were stabilized against thermal degradation, protease attack, and hydrophobic adsorption.	Fiedler et al. (2010)
The Rev tagged proteins: 2-deoxyribose-5-phosphate aldolase, superoxide dismutase, cytosine deaminase, purine-nucleoside phosphorylase	ZZ domain; epidermal growth factor EGF; peptide GE7 which binds the EGF receptor; peptide F56 which binds the subtype 1 of the vascular EGF receptor		Potential applications in medicine and chemical manufacturing	The captured enzymes were active while inside the nanoparticle shell and were protected from environmental conditions that led to free-enzyme destruction.	Fiedler et al. (2018)
Universal RNAi scaffold that could target any desired mRNA	No decoration	Specific messengers	Potentially broad applications in therapy	The dose- and time-dependent inhibition of GFP expression in human cells and of Pan-Ras expression in brain tumor cells.	Fang et al. (2017)
Doxorubicin	Gold nanoparticles	No cell targeting	Cancer	The highly selective drug release and cell killing of macrophage and cancer cells *in vitro* exclusively within the laser path while cells outside the path, even though they were in the same culture, showed no drug release or death.	Benjamin et al. (2018)
EGFP, enhanced green fluorescent protein, or CD, cytosine deaminase	IgG-binding ZZ domains	Undifferentiated cells	Cancer	After labeling with antibodies against the hPSC-specific surface glycan SSEA-5, the EGFP-containing particles were shown to specifically bind undifferentiated cells in culture, and the CD-containing particles were able to eliminate undifferentiated hPSCs with virtually no cytotoxicity to differentiated cells upon treatment with the prodrug 5-fluorocytosine.	Rampoldi et al. (2018)
RNAi; luciferase mRNA	Cell-penetrating peptide and apolipoprotein E peptide	Cells of murine glioblastoma models	Brain tumors: glioblastoma	The modified Qβ VLPs crossed blood-brain-barrier and acted synergistically with temozolomide for promoting clinical chemotherapy.	Pang et al. (2019)
No specific packaging	Copperᴵ-catalyzed azide-alkyne cycloaddition reaction	Variety of cells	Potentially broad applications	Bioconjugations of Qβ VLPs derivatized with azide moieties at surface lysine residues was performed. Complete derivatization of more than 600 reactive sites per particle was achieved.	Hong et al. (2008)
	Glycans used as substrates for glycosyltransferase reactions to build di- and trisaccharides	Cognate receptors on the appropriate beads and cells	Potentially broad applications	The elaborated methodology provided a convenient and powerful way to prepare complex carbohydrate ligands for clustered receptors.	Kaltgrad et al. (2008); Kussrow et al. (2009)
	Cationic aa motifs by genetic engineering or chemical conjugation	Inhibiting the anticoagulant action of heparin	Clinical overdose of heparin	The polycationic motifs displayed on the mutated Qβ VLPs acted as heparin antagonists.	Udit et al. (2009); Udit (2013); Gale et al. (2011); Cheong et al. (2017)

(Continued)

TABLE 23.1 (Continued)

RNA Phage VLPs as Models for Nanocontainer Packaging and Decoration with Addressing/Targeting/Delivery Purposes

Interior: Packaged by	Exterior: Decorated by	Addressed to	Supposed to Treat	Comments	References
	Cationic aa motifs by genetic engineering	Inhibiting the anticoagulant action of heparin	Clinical overdose of heparin	The two-plasmid system was employed. The highest incorporation rate and best anti-heparin activity was achieved by the mosaic particle that displayed the C-terminal peptide ARK$_2$A$_3$KA.	Choi JM et al. (2018)
	IgG-binding Z domain at the C-terminus of the coat coexpressed with native coat as a helper	Rather unspecific targeting	Potentially broad applications in diagnostics	This presents generation of mosaic VLPs carrying the IgG-binding Z domain that could be targeted to antibodies displayed on the cell surface or used in diagnostics.	Brown SD et al. (2009)
	Oligomannosides that are modeling the "glycan shield" of HIV envelope	No cell targeting	HIV/AIDS as a model	The oligomannose clusters were recognized by monoclonal anti-HIV antibody, but did not induce antibodies against the HIV epitopes by immunization of rabbits.	Astronomo et al. (2010)
	Transferrin	Transferrin receptors	Potentially broad applications in diagnostics and therapy	This approach allowed cellular internalization of chimeric VLPs through clathrin-mediated endocytosis.	Banerjee et al. (2010)
	DNA	No cell targeting	Potentially broad applications in nanotechnology	A non-compact lattice was created by DNA-programmed crystallization using surface-modified Qβ VLPs and gold nanoparticles, engineered to have similar effective radii.	Cigler et al. (2010)
	Poly(2-oxazoline)	No cell targeting	Potentially broad applications in diagnostics and therapy	This showed that the size and content of VLP–polymer constructs could be controlled by changing polymer chain length and attachment density. The system is universal because of the convenient click chemistry applications.	Manzenrieder et al. (2011)
	Conjugation of azide- and alkyne-containing proteins	Broad spectrum of potential targeting	Tumors and other possible targets	See above; the same for MS2.	Patel and Swartz (2011)
	Modification via the copper-catalyzed azide-alkyne cycloaddition reaction	No definite cell targeting specified	Potentially broad applications in therapy and imaging diagnostics	This is a methodology to use VLPs as multivalent macroinitiators for atom transfer radical polymerization (ATRP)	Pokorski et al. (2011a)
	Human epidermal growth factor (EGF) as a C-terminal fusion to the Qβ coat	EGF receptor	Therapy based on interactions with cells over-expressing their cognate receptor	Mosaic particles with an approximately 5–12 EGF molecules on the VLP surface were obtained. The particles were found to be amenable to bioconjugation by standard methods as well as the high-fidelity copper-catalyzed azide-alkyne cycloaddition reaction (CuAAC).	Pokorski et al. (2011b)

(Continued)

TABLE 23.1 (Continued)
RNA Phage VLPs as Models for Nanocontainer Packaging and Decoration with Addressing/Targeting/Delivery Purposes

Interior: Packaged by	Exterior: Decorated by	Addressed to	Supposed to Treat	Comments	References
	Sulfate groups that elicit heparin-like anticoagulant activity	No cell targeting	Blood coagulation - as heparin-like drugs	Following conversion of VLP surface lysine groups to alkynes, the sulfated ligands were attached to the VLP via copper-catalyzed azide-alkyne cycloaddition (CuAAC). 3–6 attachment points per coat monomer were modified via CuAAC. The sulfated VLPs were able to perturb coagulation.	Mead et al. (2014); Groner et al. (2015)
	Polynorbornene block copolymers	No cell targeting	Potentially broad applications in therapy and imaging diagnostics	Poly(norbornene-PEG)-b-poly(norbornene anhydride) of three molecular mass: 5, 10, and 15 kDa were added to lysine residues.	Isarov et al. (2016)
	Poly(ethylene glycol)	No cell targeting	Potentially broad applications in therapy and imaging diagnostics	The polymer conformation of poly-(ethylene glycol) was compared with those of water-soluble polyacrylate and polynorbornene when attached to the Qβ VLPs.	Lee PW et al. (2017a)
	Poly(lactic-co-glycolic acid), or PLGA	No cell targeting	Potentially broad applications in therapy and imaging diagnostics	The melt-encapsulation was found to be an effective method to produce composite materials that can deliver viral nanoparticles over an extended period and elicit an immune response comparable to typical administration schedules.	Lee PW et al. (2017b,c)
	Low molecular mass inhibitors of the GCPII	Glutamate carboxypeptidase II, or GCPII	Recognition, imaging, and delivery of treatments to prostate cancer cells	The GCPII is a membrane protease overexpressed by prostate cancer cells and detected in the neovasculature of most solid tumors.	Neburkova et al. (2018)
φCB5					
Gold nanoparticles, tRNA, diphteria toxin, mRNA, CpG	No decoration	No cell targeting	Potentially broad applications in therapy	The ease with which the φCB5 coat dimers can be purified and re-assembled into VLPs makes them attractive for the internal packaging of nanomaterials and the chemical coupling of peptides.	Freivalds et al. (2014)

Source: The original table from Pumpens P et al. *Intervirology.* 2016;59:74–110, with permission of S. Karger AG, Basel, was extended with recent data.
Note: The data are presented in alphabetical order of the phage model. For each model, the studies are ordered chronologically, but presenting first the studies describing both packaging and decoration, then data with packaging only, and then data with decoration only.

This was the first study in which the cell-penetrating peptides were displayed on the MS2 VLP surface by genetic fusion. Therefore, the MS2 VLP-based microRNA delivery system was combined for the first time with the phage surface display platform. It is necessary to mention that the HIV TAT peptide was the first reported cell penetrating peptide, since the TAT region 47–57 functioned as the clear protein transduction domain and possessed efficient transport capacity. Therefore, the TAT 47–57 peptide was displayed on the MS2 VLPs by genetic fusion instead of chemical crosslinking. As a result, the TAT peptide was connected to the N-terminus of the single-chain dimer of the MS2 coat, whereas the pre-miR122 was designed with a *pac* site to facilitate packaging. The study acknowledged not only the cell-penetrating capability of the MS2 VLPs displaying TAT peptide, but also the inhibitory effect of the encapsulated miR-122 on the hepatocellular carcinoma cells (Wang G et al. 2016). It seems worth mentioning that a solid article on the cell-specific delivery of mRNA and microRNA by MS2 VLPs carrying cell-penetrating peptide (Sun and Yin 2015) was retracted at that time.

Furthermore, Sun Y et al. (2017) inserted the TAT peptide into the AB loop of the PP7 coat single-chain dimer, while the pre-miR-23b microRNA was packaged inside the PP7 VLPs. The PP7 VLPs carrying the cell-penetrating TAT peptide and packaged microRNA were efficiently expressed in *E. coli* using the one-plasmid double expression system, penetrated hepatoma SK-HEP-1 cells, and delivered the pre-microRNA-23b, which was processed into a mature product within 24 hours and inhibited the migration of hepatoma cells by downregulating liver-intestinal cadherin (Sun Y et al. 2017).

Recently, the MS2 VLPs were used to identify a novel lncRNA, lncRNA-m433s1, as an intergenic lncRNA located in the cytoplasm (Han DX et al. 2019). The authors employed the MS2-RIP, or RNA immunoprecipitation, assays in order to demonstrate functional interactions of the discovered lncRNA-m433s1.

Concerning the employment of the Qβ VLPs, Neburkova et al. (2018) involved them in a large comparative study of different VLPs, namely, polymer-coated nanodiamonds, mouse polyomavirus VLPs, and polymeric poly(HPMA) nanoparticles, for their ability to display low-molecular-mass inhibitors of glutamate carboxypeptidase II (GCPII), a membrane protease that is overexpressed by prostate cancer cells. Regardless of the diversity of the investigated nanosystems, they all strongly interacted with the GCPII and effectively targeted the GCPII-expressing cells. Recently, the Qβ VLPs were checked for lung tissue delivery mediated by macrolide antibiotics (Crooke et al. 2019). It was established that azithromycin was able to direct the Qβ VLPs to the lungs in mice, with significant accumulation within 2 h of systemic injection. These results suggested that this new class of bioconjugate could serve as an effective platform for intracellular drug delivery in the context of pulmonary infections (Crooke et al. 2019).

As presented in detail in Chapter 20, the orthogonal chemistries were elaborated, which allowed modification of the exterior and interior surfaces of the RNA phage VLPs. The capsid exterior was endowed with cell-specific targeting capabilities, particularly via the appendage of receptor-specific antibodies on the external surface. Given the potential of the MS2 VLPs as a drug delivery and imaging system and its success in *in vitro* experiments (Ashley et al. 2011a; ElSohly et al. 2015), the *in vivo* behavior of constructs based on the MS2 scaffold warranted investigation. Thus, Aanei et al. (2016) conjugated the anti-EGFR antibodies to the MS2 VLPs, in order to target them toward receptors overexpressed on breast cancer cells. First, the specific binding of the MS2 VLPs to the targeted receptors was achieved *in vitro*. Then, the MS2 VLPs radiolabeled with ^{64}Cu isotopes were injected into mice possessing tumor xenografts, and both positron emission tomography-computed tomography (PET/CT) and scintillation counting of the organs *ex vivo* were used to determine the localization of the agents. The capsids exhibited surprisingly long circulation times, 10%–15% ID/g in blood at 24 h, and moderate tumor uptake, 2%–5% ID/g. However, the targeting antibodies did not lead to the increased uptake *in vivo* despite *in vitro* enhancements, suggesting that extravasation was a limiting factor for delivery to tumors by these particles (Aanei et al. 2016).

In parallel, the epidermal growth factor itself was conjugated to the MS2 VLPs in order to target them directly to the epidermal growth factor receptor on the cancer cells (ElSohly et al. 2017).

It is worth mentioning also some patents that claimed at that time the MS2 VLPs as tools to "enclose and subsequently isolate and purify target cargo molecules of interest including nucleic acids such as siRNAs and shRNAs, miRNAs, messenger RNAs, small peptides and bioactive molecules" (Oates et al. 2016; Arhancet et al. 2018).

Contributing structurally to the nanotechnological packaging and delivery ideology, Hartman et al. (2018) elaborated an impressive strategy, termed systematic mutation and assembled particle selection (SyMAPS), which was used to a library generation and single-step selection, in order to study the VLP self-assembly. This selection did not rely on infectivity, clinical abundance, or serum stability, and therefore enabled experimental characterization of the assembly competency of all single amino acid variants of the MS2 coat protein and challenged some conventional assumptions of protein engineering. The resulting high-resolution fitness landscape (AFL) presented a fundamental roadmap to altering the MS2 coat to achieve tunable chemical and physical properties. After recapitulating the results of many previous investigations in a single experiment (Peabody 2003; Carrico et al. 2008; Wu et al. 2009), Hartman et al. (2018) calculated the effect of ten physical properties on the AFL and evaluated the validity of several common protein engineering assumptions. An additional round of selection identified a previously unknown variant, the MS2 coat T71H variant, that exhibited acid-sensitive properties that were promising for the engineering controlled endosomal release of cargo by targeted drug delivery. The library of the MS2 variants can be subjected to future selections to address any number of additional engineering goals. Moreover, Hartman et al. (2018) recommended the SyMAPS as a straightforward approach that could be applied more broadly to assess the fitness landscapes of the coat proteins of clinically relevant pathogens, including hepatitis B and human papillomavirus virions.

Figure 23.1 presents the quantitative AFL, which contains an apparent fitness score (AFS), for every variant at every residue across the backbone of the self-assembling MS2 coat protein.

Contributing to the theoretical background of bionanoparticle delivery in cancer treatment, Shah et al. (2018) employed the MS2 VLPs in the Brownian dynamics simulations and showed how to interpret these results for real applications.

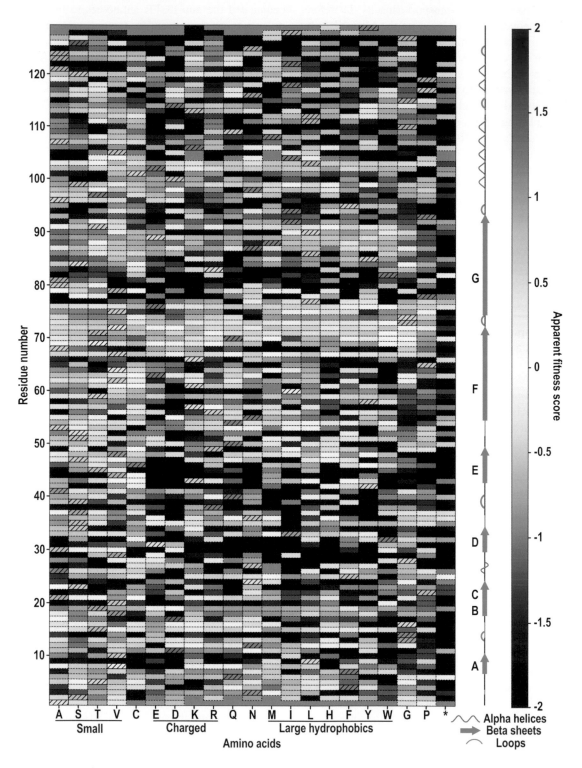

FIGURE 23.1 Apparent fitness scores (AFS, $n=3$) for all single amino acid variants of the MS2 coat protein. Wild-type residues are indicated with hatches, and missing values are green. Dark red variants were sequenced before selection but absent following selection. (Reprinted from Hartman EC et al. *Nat Commun*. 2018;9:1385.)

Recently, the MS2 VLPs were used to elaborate a general cargo-compatible approach to encapsulate guest materials based on the apparent critical assembly concentration (CAC_{app}) of the viral nanoparticles (Li L et al. 2019). The new method drove the reassembly of the latter to encapsulate cargoes by simply concentrating an adequately diluted mixture of the VLP building blocks and cargoes to a concentration above the CAC_{app}. Moreover, the MS2 VLPs devoid of genomic RNA were used as a model in theoretical calculations for their pH-dependent electrostatic interactions with metal nanoparticles (Phan and Hoang 2019). The authors found that the outer surface charge was fully responsible for the variations in the electrostatic interaction of the MS2 VLPs with the metal nanoparticle. The charge of the outer surface changed sign monotonically from positive to negative values at $pH = 4$, which led to the repulsive-attractive transition seen at this pH (Phan and Hoang 2019).

The MS2 VLPs, among other cargo models including popular hepatitis B virus core VLPs, were tested recently for the nuclear import kinetics (Paci and Lemke 2019). This study revealed the key molecular determinants on large cargo import kinetics into the cells.

NANOMATERIALS AND IMAGING

The RNA phage VLP-based applications that have been developed as novel nanomaterials and/or imaging tools are compiled in Table 23.2.

Mainly, the use of imaging agents in combination with RNA phage VLPs has contributed to the high-resolution and noninvasive visualization of these particles, as well as for the potential treatment of diseases. Thus, the first studies on the generation of nanoparticles for magnetic resonance imaging applications and the first comparisons of interior *versus* exterior cargo strategies appeared in the mid-2000s (Anderson et al. 2006; Hooker et al. 2007). Following these reports, the MS2 VLPs were loaded with positron emission tomography radiolabels (Hooker et al. 2008). Currently, major attention is focused on the use of the RNA phage-based nanoparticles as potential scaffolds for novel biomaterials and as subjects for nanoscale engineering applications involving exposure to various chemical compounds. As can be seen from Table 23.2, the MS2 phage and VLPs are playing to date a leading role in the novel material and bioimaging studies.

The specific application of the RNA phages and their VLPs in the generation of novel nanomaterials and imaging tools remains a favorite subject for numerous review articles (Manchester and Singh 2006; Yin 2007; Lee LA et al. 2009, 2011; Cormode et al. 2010; Hooker 2010; Li K et al. 2010; Rodríguez-Carmona and Villaverde 2010; Steinmetz 2010; Zourob and Ripp 2010; Chung et al. 2011; Dedeo et al. 2011; Jutz and Böker 2011; Pokorski and Steinmetz 2011; Veerapandian and Yun 2011; Ng et al. 2012; Bittner et al. 2013; Bruckman et al. 2013; Smith et al. 2013; Wen et al. 2013, 2015; Wojta-Stremayr and Pickl 2013; Farr et al. 2014; Glasgow and Tullman-Ercek 2014; Li F and Wang 2014; Nie et al. 2014; Tawil et al. 2014; Capek 2015; Koudelka et al. 2015; Krejcova et al. 2015; Shukla and Steinmetz 2015; Tsvetkova and Dragnea 2015; Chakraborty et al. 2016; Karimi et al. 2016; Lee EJ et al. 2016; Maassen et al. 2016; Peltomaa et al. 2016; Raeeszadeh-Sarmazdeh et al. 2016; Samanta and Medintz 2016; Wen and Steinmetz 2016; Zhang Y et al. 2016; Glidden et al. 2017; Nagamune 2017; Sun H et al. 2017; Zhang W et al. 2017; Almeida-Marrero et al. 2018).

In the context of the RNA phage VLPs as templates for nanomaterials and as carriers for vaccines, bioimaging labels or drug delivery tools, Machida and Imataka (2015) published a comprehensive summary of the most applicable production methods for viral particles. Yeast was then evaluated as a prospective expression system for the production of VLPs including MS2 and Qβ VLPs (Kim HJ and Kim HJ 2017). It is noteworthy that the stability of the phage MS2 was enhanced by the formulation with poly-γ-glutamic acid (PGLA), a biodegradable polymer which demonstrated remarkable potential protection of beneficial viruses (Khalil et al. 2016).

NANOREACTORS

The idea that the MS2 VLPs could be used as a nanoreactor was published first by de la Escosura et al. (2009). The authors relied on the two key studies performed by Francis' team. First, on the two orthogonal modification strategies to decorate the exterior surface of MS2 capsids with PEG chains, while installing 50–70 copies of a fluorescent dye inside as a drug cargo mimic (Kovacs et al. 2007). Second, on the solubilization of the phage MS2 in organic solvents, subsequent removal of water, and resolubilization of MS2 in a solvent of choice (Johnson et al. 2007). The resolubilized MS2 was then derivatized with stearic acid in chloroform, illustrating that the bioconjugation reactions could be performed with reagents that were completely insoluble in water. Therefore, the extended range of potential chemical modifications and the enhanced thermal stability of these MS2 cages in organic solvents opened the way for their use as organic nanoreactors (de la Escosura et al. 2009). In their review on smart nanocontainers and nanoreactors, Kim KT et al. (2010) adduced the same experimental proofs for the use of MS2 as a nanoreactor, with an additional argument that the MS2 VLPs were modified on the exterior with aptamers to direct uptake of the capsid drug carrier by specific cells, according to Tong et al. (2009). The MS2 cages were also listed among the possible nanoreactor candidates by Bode et al. (2011). Then, Cardinale et al. (2012) included both MS2 and Qβ in the list of virus scaffolds as potential enzyme nano-carriers, using the papers of Ashley et al. (2011a) and Strable et al. (2008), respectively, as the experimental arguments. Cardinale et al. (2015) extended the arguments by references on the experimental studies of Wu et al. (1995), in the case of MS2, and of Fiedler et al. (2010) and Rhee et al. (2011, 2012), in the case of Qβ.

Thus, Finn's team reported packaging of the 25-kDa N-terminal aspartate dipeptidase peptidase E (PepE), 62-kDa firefly luciferase (Luc), and a thermostable mutant of luciferase (tsLuc), inside Qβ VLPs (Fiedler et al. 2010). The average number of the encapsidated cargo proteins was controlled by

TABLE 23.2

Bioimaging Agents on the Basis of RNA Phage VLPs and Virions

Contrast Agent	Comments	References
	MS2	
MS2 phage was labeled with the succinimidyl ester of [Ru(2,2'-bipyridine)2(4,4'-dicarboxy-2,2'-bipyridine)]2+ (RuBDc) which is a very photostable probe that possesses favorable photophysical properties including long lifetime, high quantum yield, large Stokes' shift, and highly polarized emission. The RuBDc luminophore attacks lysine residues.	The intensity and anisotropy decays of RuBDc when conjugated to RNA phage MS2 were examined using frequency domain fluorometry with a high-intensity, blue light-emitting diode (LED) as the modulated light source. The results showed that RuBDc can be useful for studying rotational diffusion of biological macromolecules.	Kang and Yoon (2004)
The MS2(Gd-DTPA-ITC)m contrast agent was synthesized through the conjugation of pre-metalated Gd-DTPA-ITC (2-(4-isothiocyanatobenzyl)-diethylenetriaminepentaacetic acid) to the lysine residues of MS2 VLPs (not phage) on the VLP exterior.	A magnetic resonance imaging (MRI) contrast agent was developed by conjugation of more than 500 Gd chelate groups onto a VLP. The high density of paramagnetic centers and slow tumbling rate of modified VLPs provided enhanced T1 relaxivities up to 7200 mM-1s-1 per particle. A bimodal imaging agent was generated by sequential conjugation of fluorescein and Gd3+ chelate.	Anderson et al. (2006)
In order to maximize the relaxivity of each Gd3+ complex attached to the phage scaffold, hydroxypyridonate (HOPO)-based contrast agents were used. The interior (Tyr85) and exterior (Lys106, Lys113, and the N-terminus) surfaces of "empty" phage were targeted independently through the appropriate choice of reagents.	The phage capsids sequestering the Gd-chelates on the interior surface (attached through tyrosine residues) not only provided higher relaxivities than their exterior functionalized counterparts (which relied on lysine modification) but also exhibited improved water solubility and capsid stability. There are strong advantages of using the internal surface for contrast agent attachment, leaving the exterior surface available for the installation of tissue targeting groups.	Hooker et al. (2007); Datta et al. (2008)
The "empty" shell of phage was labeled on its inside surface with 18F-fluorobenzaldehyde through a multistep bioconjugation strategy. An aldehyde functional group was first attached to interior tyrosine residues through a diazonium coupling reaction. The aldehyde was further elaborated to an alkoxyamine functional group, which was then condensed with 18F-fluorobenzaldehyde.	This is a first example of the positron emission tomography (PET) radiolabels that have been developed on RNA phages. Relative to fluorobenzaldehyde, the fluorine-18-labeled MS2 exhibited prolonged blood circulation time and a significantly altered excretion profile.	Hooker et al. (2008)
Approximately 125 xenon MRI sensor molecules were incorporated in the interior of an MS2 VLPs, conferring multivalency and other properties of the VLP to the sensor molecule.	The resulting signal amplification facilitated the detection of sensor at 0.7 pM, the lowest to that date for any molecular imaging agent used in magnetic resonance. This amplification promised the detection of chemical targets at much lower concentrations than would be possible without the VLP scaffold.	Meldrum et al. (2010)
Multivalent, high-relaxivity MRI contrast agents using rigid cysteine-reactive gadolinium complexes.	Greater contrast enhancements were seen for MRI agents that were attached via rigid linkers, validating the design concept and outlining a path for future improvements of nanoscale MRI contrast agents.	Garimella et al. (2011)
The PET imaging characteristics were improved by the usage of PEG chains added to MS2 VLPs. The MS2- and MS2-PEG VLPs possessing interior DOTA chelators and labelled with 64Cu were compared by injecting intravenously into mice possessing tumor xenografts.	The biodistribution and circulation properties of the VLP-based PET imaging agents were investigated carefully, in order to realize the potential of such agents for the future use in in vivo applications.	Farkas et al. (2013)

(Continued)

TABLE 23.2 (Continued)
Bioimaging Agents on the Basis of RNA Phage VLPs and Virions

Contrast Agent	Comments	References
The VLPs were modified using an oxidative coupling reaction, conjugating ~90 copies of a fibrin targeting peptide to the exterior of each protein shell. The installation of near-infrared fluorophores on the interior surface of the capsids enabled optical detection of binding to fibrin clots. The targeted capsids bound to fibrin, exhibiting higher signal-to-background than control, non-targeted VLP-based nanoagents.	The chemically functionalized and specifically targeted VLPs were used for fibrin imaging. The modified capsids outperformed the free peptides and were shown to inhibit clot formation at effective concentrations over tenfold lower than the monomeric peptide alone. The *in vitro* assessment of the capsids suggests that fibrin-targeted VLPs could be used as delivery agents to thrombi for diagnostic or therapeutic applications.	Obermeyer et al. (2014a)
The targeted, selective, and highly sensitive ^{129}Xe NMR nanoscale biosensors were synthesized using MS2 VLPs, Cryptophane A, or CryA, molecules, and DNA aptamers. The MS2 VLPs were modified with CryA-maleimide at ~110 N87C positions on the interior surface of the capsid. This was followed by the attachment of aptamers on the exterior surface.	The biosensor showed strong binding specificity toward targeted lymphoma cell line. This work provided a strong basis for the continued exploration of targeted cancer cell imaging agents in NMR and MRI.	Jeong et al. (2016)
PP7		
Technetium-99m (^{99m}Tc) labeling of PP7 phage was achieved by a HYNIC bifunctional agent. Radiochemical purity higher than 90% was obtained.	The labeled PP7 phage was used as a specific tracer of *Pseudomonas aeruginosa* infection in animals.	Cardoso et al. (2016); Elena et al. (2016)
Qβ		
The VLPs were decorated with gadolinium complexes using the Cu^I-mediated azide-alkyne cycloaddition reaction. The interior surface labeling was engineered by the introduction of an azide-containing unnatural amino acid into the coat.	The circulation life-time, plasma clearance, and distribution in major organs were studied in mice for contrast agents based on original VLPs and VLPs with mutated surface aa residues.	Prasuhn et al. (2007, 2008)
The FGlyco-CCHC technique, as a subset of the CuAAC reaction, was used for the generation of fluoroglycoproteins on the basis of the Qβ VLPs as potential substrates in NMR, MRI, and PET techniques.	The reaction of L-homopropargylglycine-modified Qβ VLPs with a fluoroglucosyl azide led to occupation of more than 95% of positions with F-glycoprotein. The Qβ VLP displayed finally 180 site-selectively positioned fluoroglycans.	Boutureira et al. (2010); Boutureira and Bernardes (2015)
R17		
The metal-ligand complex, [Ru(2,2′-bipyridine)$_2$(4,4′-dicarboxy-2,2′-bipyridine)]$^{2+}$ (RuBDc) was used as a spectroscopic probe for studying hydrodynamics of biological macromolecules.	The combination of the use of long-lifetime Ru(II) metal-ligand complexes with blue LED excitation made it possible to perform time-resolved intensity and anisotropy decay measurements with a simpler and lower-cost instrument.	Kim MS et al. (2010)

Source: The original table from Pumpens P et al. *Intervirology.* 2016;59:74–110, with permission of S. Karger AG, Basel, was extended with recent data.

changing expression conditions or by removing interaction elements from the plasmids. In this way, the PepE incorporation was varied reproducibly between 2 and 18 per particle. Fewer copies of Luc proteins were packaged: 4–8 copies per particle were found for most conditions, whereas the number of the packaged tsLuc molecules varied between 2 and 11 per VLP. The encapsulated enzymes demonstrated normal functional capabilities. All production and assembly steps occurred *in vivo* within the bacterial cell, with indirect control of amount of packaged cargo possible by simply changing the expression media or the nature of the components of the packaging system. The VLPs were produced in high yields and were purified by a convenient standard procedure, independent of the protein packaged inside (Fiedler et al. 2010).

Furthermore, Francis' team encapsulated enzymes inside the MS2 capsids, using a new osmolyte-based method (Glasgow et al. 2012). Attaching DNA oligomers to a molecule of interest and incubating it with the MS2 coat protein dimers *in vitro* yielded reassembled capsids that packaged the tagged molecules. The addition of a protein-stabilizing osmolyte, trimethylamine-N-oxide, significantly increased the yields of

reassembly. Second, the expressed proteins with genetically encoded negatively-charged peptide tags could also induce capsid reassembly, resulting in high yields of reassembled capsids containing the protein. This second method was used to encapsulate alkaline phosphatase tagged with a 16-amino acid peptide. The purified encapsulated enzyme was found to have the same K_m value and a slightly lower k_{cat} value than the free enzyme, indicating that this method of encapsulation had a minimal effect on the enzyme kinetics. This method therefore provided a practical and potentially scalable way of studying the complex effects of encapsulating enzymes in protein-based compartments (Glasgow et al. 2012).

The functioning of this enzymatic nanoreactor was studied further to explore pore-structure effects on substrate and product flux during the catalyzed reaction (Glasgow et al. 2015). The capsids mutated in the FG loop were able to encapsulate alkaline phosphatase enzymes bearing a localization tag. Figure 23.2 presents the idea of the experiment.

When these enzymes were inside capsids with pores having no charge or a charge opposite that of the substrate, the apparent K_M ($K_{M,app}$) was similar to that of the free enzyme;

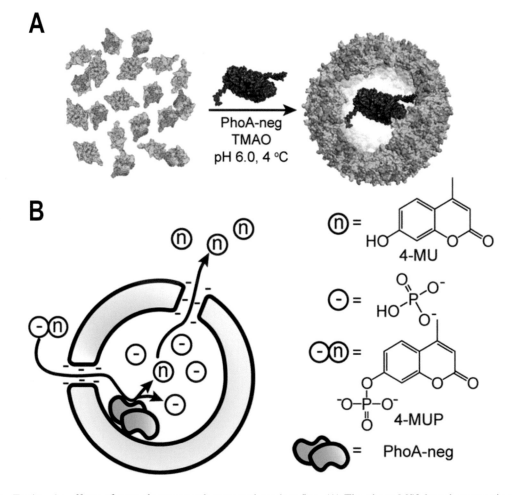

FIGURE 23.2 Testing the effects of pore charge on substrate and product flux. (A) The phage MS2-based enzymatic nanoreactors are created through self-assembly of the capsid around negatively charged alkaline phosphatase (PhoA-neg) variants. (B) Capsid pore charge is varied by mutagenesis, and reactions of negatively charged phosphatase substrates are monitored. (Reprinted with permission from Glasgow JE, Asensio MA, Jakobson CM, Francis MB, Tullman-Ercek D. Influence of electrostatics on small molecule flux through a protein nanoreactor. *ACS Synth Biol.* 2015;4, 1011–1019. Copyright 2015, American Chemical Society.)

however, when pore and substrate charge were the same, the $K_{M,app}$ of the enzyme increased significantly. The kinetic modeling suggested this could be caused primarily by reduced product efflux, leading to inhibition of the enzyme. These experiments represented the first step in creating selective nanoreactors with the potential either to protect cargo by inhibiting entrance of interfering molecules or to enhance multistep pathways by concentrating internal intermediates. These findings also lend support to the hypothesis that protein compartments can modulate the transport of small molecules and thus influence metabolic reactions and catalysis *in vitro* (Glasgow et al. 2015).

The construction of a catalytic nanoreactor based on *in vivo* encapsulation of multiple enzymes into the MS2 VLPs using the SpyTag/SpyCatcher approach (Giessen and Silver 2016a) was described in the Packaging section of Chapter 20. Briefly, the sequestration of two enzymes from the indigo biosynthetic pathway in the MS2 VLPs was achieved by the SpyTag/SpyCatcher protein fusions that covalently crosslinked with the interior surface of the capsid. The resultant nanoparticles were catalytically active, and the enzyme retained higher activity than free enzyme 1 week later (Giessen and Silver 2016a).

The success of the RNA phage-derived nanoreactors has been reviewed extensively (Giessen and Silver 2016b; Koyani et al. 2017; Quin et al. 2017; Schwarz B et al. 2017; Banerjee and Howarth 2018; Timmermans and van Hest 2018; Wilkerson et al. 2018; Ren et al. 2019).

24 Non-VLPs

Be braver—you can't cross a chasm in two small jumps.

David Lloyd George

Any sufficiently advanced technology is indistinguishable from magic.

Arthur C. Clarke
Profiles of the Future: An Inquiry
into the Limits of the Possible

GENE THERAPY

Sanford and Johnston (1985) were the first to discuss the concept of parasite-derived resistance by supplying resistance genes from the phage's own genome. The authors used the phage Qβ to illustrate specifically how parasite-derived resistance might be engineered. The given theoretical examples included the expression of the coat and replicase genes encoding potential repressors as well as the expression of ribosome binding and antisense sequences. Remarkably, the maturation gene was also included in this list, since it would induce premature lysis of a host cell upon initial infection by Qβ, constituting on the population level a form of hypersensitivity.

In addition, the RNA phages were behind the first experimental example of gene therapy applications, namely, of the antisense-based gene therapy that was termed by the authors the "mRNA-interfering complementary RNA (micRNA) immune system" and was originally designed to prevent proliferation of the group IV phage SP (Coleman et al. 1985; Hirashima et al. 1986, 1989). The first report announced that the induction of the micRNAs directed against the coat protein and/or the replicase genes of the phage SP prevented its replication (Coleman et al. 1985). To show this, the two DNA fragments of the cloned phage SP genome were used to construct the micRNA immune system: fragment A consisted of the 247-base pairs encompassing the Shine-Dalgarno sequence and the initiation codon of the coat gene, while fragment B consisted of the 159-base pairs encompassing the Shine-Dalgarno sequence and the initiation codon of the replicase gene. The control fragment C consisted of ~690 bp: 411 bp from the C-terminal coding region of the replicase gene, 107 bp from the 3′-noncoding region, ~60 bp from the poly(A) tail, and 111 bp from the pBR322 vector. The transcription of the fragments was controlled by the strong *lpp* promoter and the *lac* promoter-operator, so that the production of a large amount of RNA transcript could be induced in the presence of *lac* inducer, such as isopropyl-β-D-thiogalactoside (IPTG). Those plasmids in which the fragments A, B, and C appeared in the antisense transcription direction with respect to the *lpp* and *lac* promoters were used to test the micRNA immune system. To clear gene dosage effects, the plasmids carrying two or three micRNA genes were constructed. The induction of the micRNAs in the plasmid carrying the A, B, and C fragments inhibited phage proliferation. The partial inhibition was observed even without the addition of IPTG, a result of incomplete repression of the micRNA gene transcription in the cells. The plaque formation was also inhibited when the cells carrying the plasmids with the A, B, and C fragments were used as indicator bacteria in the presence of IPTG.

In the case of single micRNA genes, the *micB* and *micA* appeared to be more inhibitory to the phage SP in the presence of IPTG than the *micC*. Although the latter did not encompass the gene initiation regions, it was thought to prevent replicase from transcribing the positive RNA strand to produce the negative RNA strand, by forming a hybrid at the 3′ end of the phage RNA.

The authors of this revolutionary paper speculated about the possible application of the micRNA immune system to both plant and animal cells, establishing, for example, transgenic animals carrying a micRNA immune system(s) directed against a specific virus, such as poliovirus or foot-and-mouth disease virus, by using micRNA genes that would be stably integrated into the desired host genome. Similarly, the micRNA immune systems directed against plant viruses could be introduced into the genomes of plant protoplasts which could be subsequently regenerated into mature fertile plants (Coleman et al. 1985). This study provoked at that time numerous papers on the function of complementary RNAs in bacterial, insect, and animal cells.

In the next paper (Hirashima et al. 1986), the construct that more effectively conferred immunity to the phage SP was designed, and the specificity of the micRNA immune system was evaluated by involvement of other related RNA phages. Thus, the gene for the maturation protein was found to be the best target for this type of immune system; the mRNA-interfering complementary RNAs specific to the genes for coat protein and replicase were less effective in preventing infection. The greatest inhibitory effect was observed with a 240-base sequence encompassing the 24-base noncoding region of the maturation gene plus the 216-base coding sequence. Significantly, even a 19-base sequence covering only the Shine-Dalgarno sequence without the coding region exerted a strong inhibitory effect on phage proliferation. In contrast to the highly specific action against the phage SP exhibited by the longer mRNA-interfering complementary RNA, the specificity with the shorter mRNA-interfering complementary RNA was broadened to the phages Qβ and GA as well as SP, all of which were classified in the different groups of RNA coliphages. It was thereby shown that this type of antiviral reagent could have a particular breadth of specificity, thus increasing its value in various research and possibly clinical applications (Hirashima et al. 1986).

Hirashima et al. (1989) attempted to identify the most effective micRNA sequences for antiviral activity within the 5′-terminal noncoding region of 54 nucleotides. It was found that a 30-nucleotide micRNA against the sequence from base 32 to 61 of the SP genome exhibited nearly complete inhibition of phage production. Upon further dissection of this sequence, it was concluded that the most effective micRNA against the phage SP production should contain the sequences complementary to the Shine-Dalgarno sequence of the maturation gene and its 13-nt upstream sequence. The addition of downstream sequences had little effect, while the addition of further upstream sequences had a significant negative effect.

Then, a ribozyme designed to cleave the maturation gene of the phage SP, when transcribed from a plasmid in *E. coli*, caused failure of the proliferation of progeny phage, while inactive ribozymes with altered catalytic sequences did not affect phage growth (Inokuchi et al. 1994b). It was indicated that this was mainly the catalytic activity of the ribozyme and not its function as an antisense molecule that could be responsible for suppressing the proliferation of the RNA phage. Moreover, an analysis based on numbers of plaque-forming units and the function of the maturation protein indicated that the antisense RNA may successfully compete with ribosomes in targeting mRNA, while ribozymes did not compete with ribosomes in the naturally occurring bacterial transcription/translation-coupled systems (Inokuchi et al. 1994b).

Later, the optimism about the possible antiviral application of the micRNA technique was lowered by the mutational considerations. Thus, Bull et al. (1998) propagated the phage SP for several generations on a host expressing a 240-base antisense RNA complementary to the region 31–270 of the SP genome, an efficient micRNA suggested earlier by Hirashima et al. (1986). As a result, phages evolved that escaped inhibition. Typically, these escape mutants contained three or four base substitutions, but different sequences were observed among different isolates. The mutations were located within three different types of structural features within the predicted secondary structure of SP genomic RNA: (i) hairpin loops, (ii) hairpin stems, and (iii) the 5′ region of the phage genome complementary to the antisense molecule. Computer modeling of the mutant genomic RNAs showed that all of the substitutions within hairpin stems improved the Watson-Crick pairing of the stem. No major structural rearrangements were predicted for any of the mutant genomes, and most substitutions in coding regions did not alter the amino acid sequence. Although the evolved phage populations were polymorphic for substitutions, many substitutions appeared independently in two selected lines. The creation of a new, perfect, antisense RNA against an escape mutant resulted in the inhibition of that mutant but not of other escape mutants nor of the ancestral, unevolved phage. Thus, at least in this system, a population of viruses that evolved to escape from a single antisense RNA would require a cocktail of several antisense RNAs for inhibition (Bull et al. 1998).

Another variant of the antiviral gene therapy predicted by Sanford and Johnston (1985), the expression of the Qβ repressor coat gene before phage infection, was evaluated by Lindemann et al. (2002). The effect of this type of antiviral gene therapy was studied over the course of more than 100 generations. In 13 experiments carried out in parallel, 12 phage populations became resistant and 1 became extinct. The sequence analysis revealed that only 2 distinct phage mutants emerged in the 12 surviving phage populations. For both escape mutants, sequence variations located in the repressor binding site of the viral genomic RNA, which decrease affinity for the repressor protein, conferred resistance to translational repression (Lindemann et al. 2002).

A bit earlier, the MS2 coat was used as a translational repressor by down-modulation of the HIV-1 gene (Berkhout and Jeang 1990). When the coat protein binding site was attached to the HIV-1 5′ leader RNA, the HIV-1 LTR-directed expression was inhibited *in trans* by the coat protein. It was shown therefore for the first time that the prokaryotic RNA-binding proteins were useful as genetic modulators in eukaryotic cells.

In parallel, Selby and Peterlin (1990) employed the MS2 coat in their studies of the HIV-1 trans-activator Tat and its target TAR, an RNA stem-loop located at the 5′ end of all HIV-1 transcripts. The authors fused Tat to the MS2 coat and replaced TAR in the HIV-1 LTR by the MS2 operator stem-loop. The hybrid Tat-coat protein transactivated HIV-1 LTRs containing either TAR or operator sequences. The mutations in the operator that weakened binding of the MS2 coat *in vitro* led to decreased levels of transactivation *in vivo* (Selby and Peterlin 1990).

The non-encapsidation activities of positive-stranded RNA viruses, including the RNA phages, were reviewed thoroughly by Ni and Kao (2013). Masayori Inouye (2016) later narrated the exciting history how the RNA interference to inhibit mRNA function—the first gene therapy attempt—was demonstrated for the first time on the RNA phages.

TETHERING

Chapter 16 presented in detail the mechanism by which the RNA phage coats recognize and bind the natural and modified short RNA operator hairpins and therefore perform their repressor function. The existence of the repressor complex I was demonstrated experimentally for all RNA phage groups with the exception of the acinetophage AP205. In the latter case, such data are still missing.

This highly specific ability of the RNA phage coats to recognize the corresponding operator stem-loop led to the development of a set of efficient methodologies based on the *tethering* of the coat to the coat-operator-tagged RNAs. Using these techniques, mRNAs that were tagged with the operator sequence, were highly specifically recognized by the phage coats, which were fused to fluorescent or other functional probes.

The tethering methodology mainly exploited the coat-operator composition from the phage MS2. The closely related composition of the phage R17 was also employed, for example, in the original paper on the tethering approach (Bardwell and Wickens 1990). The appropriate PP7-based methodology

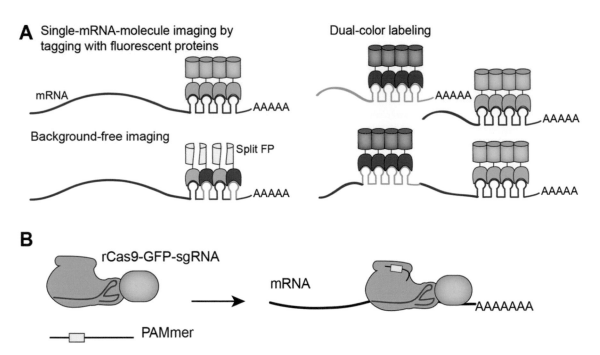

FIGURE 24.1 RNA binding proteins fused to fluorescent reporters. (A) MS2 and related hairpin binding proteins fused to fluorescent proteins or split fluorescent proteins for the visualization of a specific RNA. (B) rCas9-GFP targeting of natural RNA without the need to encode recognition elements. (Reprinted with permission from Alexander SC, Devaraj NK. Developing a fluorescent toolbox to shed light on the mysteries of RNA. *Biochemistry.* 2017;56, 5185–5193. Copyright 2017 American Chemical Society.)

was developed later and led to the generation of the simultaneous MS2 and PP7 two-color labeling of live cells. These studies will be described in detail in the New challenges: PP7 section of the present chapter. Figure 24.1 presents a general view of the current approach in the tethering technology.

The high potential of the RNA phages as a protein-RNA tethering system was ensured by the properties of the complex. Thus, the MS2 coat dimer bound to the MS2 RNA stem-loop with a K_d of $(1–3) \times 10^{-9}$ M, but most current applications employ a high-affinity mutant of the MS2 RNA stem-loop that binds the coat protein with a K_d of $(2–6) \times 10^{-10}$ M (Johansson et al. 1998; Helgstrand et al. 2002; Horn et al. 2004) more strongly than the native operator, as described in detail in Chapter 16. Most current applications are employing the single-chain dimer of the MS2 coat, which is more stable than the monomer, as explained in Chapters 20 and 22. The tethering methodology also employs the MS2 coat mutants designed by Peabody's team and characterized by reduced ability for self-assembly. To improve binding, the number of the inserted MS2 stem-loops grew steadily from 2 to 4, 6, 8, 24 tandem copies, and sometimes reached more than 100 repeats; 128, for example (Antolović et al. 2019).

Historically, the tethering technique was first applied as a chromatography method for the affinity purification of the RNAs bearing a short, 21-nucleotides hairpin binding motif of the phage R17 (Bardwell and Wickens 1990). This method was useful to isolate specific RNAs and RNA-protein complexes formed *in vivo* or *in vitro*. The RNAs containing that hairpin bound to the R17 coat protein that has been covalently attached to a solid support. The bound RNA-protein

complexes were eluted with excess of the R17 stem-loops. It was demonstrated that binding to the column-immobilized coat protein was highly specific and enabled one to separate an RNA of interest from a large excess of other RNAs in a single step. However, binding of an RNA containing non-R17 sequences to the support required two recognition sites in tandem; a single site was insufficient. The potential utility of this novel approach in purifying RNA-protein complexes was illustrated by the demonstration that a U1 snRNP formed *in vivo* on an RNA containing tandem hairpins was selectively retained by the coat protein support (Bardwell and Wickens 1990).

In parallel, Selby and Peterlin (1990) fused the HIV-1 trans-activator Tat protein to the MS2 coat and replaced the TAR sequence in the HIV-1 LTR by the MS2 operator hairpin. The hybrid Tat-coat protein transactivated HIV-l LTRs containing either TAR or MS2 operator sequences. The mutations in the operator that weakened binding of the coat protein *in vitro* led to decreased levels of transactivation *in vivo*. The authors explained in this pioneering paper that the coat proteins of MS2 and R17 were identical, recognized the same operator sequence, and had the same functions (Selby and Peterlin 1990). However, when the next paper of this team on the lentivirus subject appeared, the phage R17 name was used (Modesti et al. 1991). The same was true when the R17 coat fusions with the Tat proteins of HIV-1 and equine infectious anemia virus, another lentivirus, were generated (Derse et al. 1991; Carroll et al. 1992). The contribution of the R17-based tethering made it possible to clear up the optimal interactions between Tat and TAR of lentiviruses (Luo and Peterlin 1993). It is remarkable that only one copy of the hairpin ensured the tethering mechanism in these studies. However, in a subsequent study, Luo et al. (1993, 1994) not only turned to the MS2 name but also used four copies of the MS2 hairpin in their constructs. This way, they followed to the recommendations that the target RNA would contain two tandem copies of the MS2 hairpin to efficiently bind the MS2 coat that was fused to the C-terminus of the HIV-1 Rev or human T-cell leukemia virus type I Rex proteins (McDonald et al. 1992; Venkatesan et al. 1992).

Although the MS2 name was still dominant in the 1990s, some studies retained the traditional early R17 name in the tethering studies at that time. Thus, Han et al. (1992) and Newstein et al. (1993) reported that they constructed chimeric HIV-1 TAR sequences that contained the R17 RNA sequence and a fusion of the Tat protein with the MS2 coat. The R17 platform was used when the gel retardation assays were performed to compare the R17 coat-RNA and the HIV Tat-TAR interactions (Pearson et al. 1994). The R17 hairpin was linked to the ribozyme helix 4, which lies at the periphery of the catalytic domain (Sargueil et al. 1995). In the absence of the MS2 coat, the incorporation of the R17 hairpin increased the catalytic efficiency of the hairpin ribozyme by 2-fold for the cleavage reaction and 16-fold for the ligation reaction. The kinetics of cleavage and ligation reactions were not altered by the presence of saturating concentrations of coat protein, while the coat protein remained bound to the ribozyme throughout the catalytic cycle (Sargueil et al. 1995).

The R17 name was employed by the generation of the one-hybrid system for detecting RNA-protein interactions (Wilhelm and Vale 1996). The specific R17/MS2 designation was then used for the same platform by the development of the improved method for selecting RNA-binding activities *in vivo* (Fouts and Celander 1996). Here, the RNA challenge phages were modified versions of the DNA phage P22 that encoded the operator sequence and underwent lysogenic development in bacterial cells that expressed the R17/MS2 coat (Fouts and Celander 1996, 1998). The R17 coat fusion with glutathione S-transferase (GST) moiety, a popular tool that enabled inducible, high-level protein expression and purification from bacterial cell lysates, was used in the investigation of mammalian pre-mRNAs (Chen and Wilusz 1998). For the last time, the R17 name appeared in connection with the tethering approach in the study on pharmacological switches that could control gene expression at the translational level (Boutonnet et al. 2004).

At the turn of the century, the MS2 tethering platform grew steadily by the number of different applications, first of all, for the fine mechanisms of translational regulation of mRNAs in eukaryotic cells (Stripecke et al. 1994; Coller et al. 1998; Paraskeva et al. 1998; Coburn et al. 2001; Kruse et al. 2002; Sànchez and Marzluff 2002; Xu et al. 2002; Cunningham and Collins 2005; Brandt et al. 2007), as well as in bacterial cells (Sawata et al. 2004; Vieu and Rahmouni 2004; McGarry et al. 2005; Ali et al. 2006).

Among the numerous MS2-based tethering studies, the contribution of the latter to the unveiling of the mRNA polyadenylation and deadenylation mechanisms must be emphasized (Lykke-Andersen et al. 2000; Ruiz-Echevarría and Peltz 2000; Minshall et al. 2001, 2010; Ko and Gunderson 2002; Melo et al. 2003; Kwak et al. 2004; Minshall and Standart 2004; Hall-Pogar et al. 2007; Maciolek et al. 2008; Hockert et al. 2010; Clement et al. 2011; Hosoda et al. 2011; Sandler et al. 2011; Darmon and Lutz 2012; Ezzeddine et al. 2012; Stubbs et al. 2012; Shankarling and MacDonald 2013; Zekri et al. 2013; Weidmann et al. 2014; Bresson et al. 2015; Alhusaini and Coller 2016; Rambout et al. 2016; Yamagishi et al. 2016; Smith et al. 2017; Grozdanov et al. 2018).

The MS2-based tethering platform contributed greatly to the elucidation of the splicing machinery (Graveley and Maniatis 1998; Graveley et al. 1998, 2001; Del Gatto-Konczak et al. 1999, 2000; Hertel and Maniatis 1999; Dauksaite and Akusjärvi 2002; Jurica et al. 2002; Lam et al. 2003; McCracken et al. 2003; Philipps et al. 2003; Villemaire et al. 2003; Abruzzi et al. 2004; Expert-Bezançon et al. 2004; Spingola et al. 2004; Fushimi et al. 2005; Spellman et al. 2005; Tange et al. 2005; Deckert et al. 2006; Wang et al. 2006; Gesnel et al. 2007; Tintaru et al. 2007; Singh et al. 2010; Coltri et al. 2011; Ray et al. 2011; Hoskins and Moore 2012; Huang SC et al. 2012; Samatov et al. 2012; Shen and Mattox 2012; Sun et al. 2012; Voelker et al. 2012; Edge et al. 2013; Mo et al. 2013; Rahman et al. 2013; Shankarling and Lynch 2013; Ilagan and Jurica 2014; Kumar et al. 2014; Becerra et al. 2015; Naftelberg et al. 2015; Rahman et al. 2015; Boesler et al. 2016; Wongpalee et al. 2016; Wu X et al. 2017; Ahsan et al. 2017; Ying et al. 2017; Erkelenz et al. 2018; Wan and Larson 2018).

It is notable in this context that the MS2-based tethering approach was used for the identification of the splicing repressor domain in polypyrimidine tract-binding protein, a splicing regulator, and for the further elucidation of its role in the splicing process (Wagner and Garcia-Blanco 2002; Gromak et al. 2003; Robinson and Smith 2006; Spellman and Smith 2006; Sharma et al. 2008; Kafasla et al. 2012; Martin et al. 2013).

The MS2-based tethering approach also contributed markedly to the deciphering of the nonsense-mediated mRNA decay as a surveillance mechanism to accelerate the degradation of mRNAs containing premature translation termination codons in yeast (Kervestin et al. 2012) and mammals (Cho et al. 2013b; Fatscher et al. 2014).

The MS2-based tethering approach provided suitable information on the virus replication mechanisms for retroviruses: HIV-1 (Jeong et al. 2000; Reddy et al. 2000; Nam et al. 2001; Yi et al. 2002a,b; Molle et al. 2007; Blissenbach et al. 2010; Kellner et al. 2011; D'Orso et al. 2012; Kenyon et al. 2015; Smyth et al. 2015; Kennedy et al. 2016; Knoener et al. 2017), bovine immunodeficiency virus (Wakiyama et al. 2012), Moloney murine leukemia virus (Green et al. 2012), and foamy virus (Moschall et al. 2017), as well as for poliovirus (Spear et al. 2008; Ogram et al. 2010; Flather et al. 2016). The MS2-based tethering was subsequently applied in tombusviruses: tomato bushy stunt virus (Pathak et al. 2012) and red clover necrotic mosaic virus (Iwakawa et al. 2012). Shi et al. (2005) showed that the MS2-based tethering approach functioned correctly in *Trypanosoma brucei*.

Using MS2-based tethering, the so-called three-hybrid system was elaborated (SenGupta et al. 1996, 1999; Harrington et al. 1997). The three-hybrid system was based on the well-known yeast two-hybrid assay that relied on the ability of a protein such as the bacterial repressor LexA to tether another protein to DNA and on a protein–protein interaction to bring a transcriptional activation domain hybrid into close proximity with the DNA-binding domain hybrid. This three-hybrid assay used the MS2 coat to tether an RNA containing the MS2 stem-loop, while the MS2 coat was fused to LexA. Therefore, this hybrid protein should bring the RNA to which it was bound to a reporter gene regulated by the LexA binding sites. In addition, the RNA was bifunctional in that it also contained a binding site for a second RNA-binding protein. This second RNA-binding protein was present as a fusion to a transcriptional activation domain. Thus, the interaction of the RNA with both RNA-binding domains should result in the activation domain being present at the promoter of the LexA-regulated gene, and so should lead to transcriptional activation of the reporter. The three-hybrid system was used successfully in a number of studies (Martin et al. 1997; Bacharach and Goff 1998; Bieniasz et al. 1998; Vidal and Legrain 1999; Kickhoefer et al. 1999; Sonoda and Wharton 1999; Fagegaltier et al. 2000; Le et al. 2000; Matter et al. 2000; Zhang et al. 2000; Dickson et al. 2001; Fraldi et al. 2001; Al-Maghrebi et al. 2002; Ostrowski et al. 2002; Saha et al. 2003; Sokolowski et al. 2003; Evans et al. 2004; Lee and Linial 2004; Li B et al. 2004; Hook et al. 2005; Lucke et al. 2005; Poderycki et al. 2005; Riley et al. 2006; König et al. 2007; Vollmeister et al. 2009; Wurster and Maher 2010; Liu Y et al. 2015; Wallis et al. 2018; Stefanovic and Stefanovic 2019).

It is noteworthy that the three-hybrid system revealed meanwhile the interaction of the hepatitis C virus 3′-terminal nucleotides with human ribosomal proteins (Wood et al. 2001). Furthermore, Piganeau and Schroeder (2006), using the MS2-based tethering principle, elaborated a variant of the yeast RNA hybrid system for the identification and detection of RNA-RNA interactions. The methodological aspects by the analysis of RNA-protein interactions using the yeast three-hybrid system were described in detail by Stumpf et al. (2008).

The application of the three-hybrid system was reviewed first by Frederickson (1998) and subsequently illustrated (Serebriiskii et al. 2001; Bernstein et al. 2002). Franck Martin (2012) reviewed the three-hybrid system at the 15-year anniversary of its application. Koh and Wickens (2014a,b,c) later published extensive reviews on the three-hybrid approach. The use of the three-hybrid system in plants was recently described in detail by Cho SK and Hannapel (2018).

The MS2-based tethering allowed functional identification of the following important protein players: Y14, a component of the exon junction complex, which underwent post-translational modifications (Hsu et al. 2005, Chuang et al. 2013); the *Xenopus laevis* and human DAZ-associated protein 1 (DAZAP1), an RNA-binding protein required for normal growth, development, and fertility (Smith et al.

2011); Ki-1/57 protein from Hodgkin lymphoma (de Almeida Gonçalves et al. 2011); a quadruple BEN domain-containing protein (BEND3) that was highly conserved among vertebrates (Sathyan et al. 2011); Staufen 1 (Kim et al. 2005; Cho et al. 2013a) and Staufen2 (Miki et al. 2011), double-stranded RNA-binding proteins and mammalian orthologs of the Staufen protein in *Drosophila melanogaster,* the DEAD-Box Protein Dhh1 in yeast (Sweat et al. 2012); Ewing sarcoma protein (Huang L et al. 2012); AUF1, an AU-rich element binding factor heterogeneous nuclear ribonucleoprotein D (1/hnRNP D), as an interacting protein of Epstein-Barr virus—induced noncoding RNA EBER1 (Lee N et al. 2012); and PARP12, a member of a large family of ADP-ribosyl transferases, encoded by an interferon-induced gene (Welsby et al. 2014).

Using the MS2-based tethering approach, a novel long noncoding RNA, treRNA, was identified as an element playing critical role in various processes ranging from normal development to human diseases such as cancer progression (Gumireddy et al. 2013). The authors also identified a novel ribonucleoprotein complex consisting of a set of RNA-binding proteins that are required for treRNA functions. The isolation of protein complexes associated with long noncoding RNAs, which employed the MS2 coat fusion with maltose-binding protein, was described in detail (Gumireddy et al. 2016; Liu L et al. 2018).

The MS2-based tethering contributed strongly to the evaluation of the mechanism that ensured adaptation to iron deficiency and iron supplementation in yeast (Martínez-Pastor et al. 2013).

The MS2-based tethering approach enabled elaboration of efficient methods for the identification, isolation, and purification of the desired RNA-protein complexes of various origins (Bachler et al. 1999; Zhou and Reed 2003; Ji et al. 2004; Youngman et al. 2004; Willingham et al. 2005; Youngman and Green 2005; Hogg and Collins 2007; Batey and Kieft 2007; Beach and Keene 2008; Ortega et al. 2008; Baillat and Shiekhattar 2009; Dienstbier et al. 2009; Fayet-Lebaron et al. 2009; Grozdanov et al. 2009; Juillard et al. 2009; Keryer-Bibens et al. 2009; Said et al. 2009; Wang and Cambi 2009; Yoshimoto et al. 2009; Zychlinski et al. 2009; Barrett and Chin 2010; Sahin et al. 2010; Slobodin and Gerst 2010, 2011; Desai et al. 2011; Iioka et al. 2011; Tsai et al. 2011; Desnoyers and Massé 2012; Gong et al. 2012; Gummadi et al. 2012; Hoareau-Aveilla et al. 2012; Yoon et al. 2012; Tacheny et al. 2013; Braun et al. 2014; Choe et al. 2014a,b; Collier et al. 2014; Gupta N and Culver 2014; Han C et al. 2014; Joncourt et al. 2014; Kennedy et al. 2014; McHugh et al. 2014; Sesto et al. 2014; Valdés et al. 2014; Chen et al. 2015; Coelho et al. 2015, 2016; Cook et al. 2015; El Khouri et al. 2015; Fimiani et al. 2015, 2016; Gong and Maquat 2015; Lalaouna and Massé 2015; Lalaouna et al. 2015a,b, 2017, 2018a,b, 2019; Lebo et al. 2015; Liu S et al. 2015; Liu XY et al. 2015; Merchante et al. 2015; Peng et al. 2015; Reineke et al. 2015; Rigo et al. 2015; Samra et al. 2015; Scarola et al. 2015; Wang X et al. 2015; Carrier et al. 2016; Chai et al. 2016; Chen CK et al. 2016; Chen L et al. 2016; Guo et al. 2016; Hung and Leonard 2016;

Jang and Paek 2016; Karagiannis et al. 2016; Kim et al. 2016; Liu LX et al. 2016; Luan et al. 2016, 2018; Matsuguchi and Blackburn 2016; Park OH et al. 2016; Radhakrishnan et al. 2016; Samra and Arava 2016; Schmidt et al. 2016; Yoon and Gorospe 2016; Cook and Charlesworth 2017; Kassem et al. 2017; Lee HC et al. 2017; Li G et al. 2017; Pendleton et al. 2017; Portoso et al. 2017; Schmidt et al. 2017; Tomasini et al. 2017; Vazquez-Anderson et al. 2017; Wang X and Shi 2017; Abshire et al. 2018; Cheetham and Brand 2018; Garre et al. 2018; Liu Y et al. 2018; Ma P et al. 2018; MacRae et al. 2018, 2019; Nicholson et al. 2018; Ryan et al. 2018; Smith et al. 2018; Wang J et al. 2018; Zeidan et al. 2018; Zhang L et al. 2018; Zhitnyuk et al. 2018; Aughey et al. 2019; Chen Xing et al. 2019; Dahiya and Atreya 2019; Fukushima et al. 2019; Gatti da Silva et al. 2019; Huang Lin et al. 2019; Liu L et al. 2019; Mukherjee et al. 2019; Park et al. 2019; Silva IJ et al. 2019; Tan et al. 2019; Woo et al. 2019; Xiong et al. 2019; Zhang S et al. 2019). Flores et al. (2014) presented a detailed account of the contribution of the MS2 tethering to the preparation and purification of the RNA and RNA-protein complexes for crystallographic studies.

Using MS2-based tethering, Goodson et al. (2012) identified small RNAs that activated transcription of a target gene in *E. coli* to varying degrees. MS2 tethering was used to identify RNA-protein interactions in *Helicobacter pylori* (Rieder et al. 2012) and *Sinorhizobium meliloti*, a genetically tractable nitrogen-fixing plant symbiotic bacterium (Robledo et al. 2018).

Corcoran et al. (2012) presented a valuable step-by-step protocol for the use of aptamer tagging to identify *in vivo* protein binding partners of small regulatory RNAs in bacteria, starting from the Hfq-associated small RNA (Said et al. 2009).

The MS2 and Qβ coats and operators were used to develop the tethering approach in plants, when the phage-derived system functioned by the translational repression of reporter genes, thereby broadening approaches for transgene regulation in plants (Cerny et al. 2003).

The MS2 tethering approach contributed strongly to the development of a novel packaging and delivery vaults to deliver mRNA for translation *in vivo* (Pupols et al. 2010). These vaults were ubiquitous intracellular components found in most eukaryotes, including humans, as large, 13 MDa ribonucleoprotein particles (72.5×41 nm) that were found at 10^4–10^5 particles per cell. Their interior cavity (5×10^7 Å^3) seemed large enough to encapsulate hundreds of proteins (Pupols et al. 2010).

The MS2-based tethering approach also ensured involvement of near-infrared dyes for the highly sensitive localization of specific RNAs (Köhn et al. 2010).

Using MS2-based tethering, Kashida et al. (2012) demonstrated a three-dimensional structure-based molecular design principle for constructing short hairpin RNA-mediated genetic information converters. These converters responded to specific proteins and triggered the desired gene expression by modulating the function of the RNA-processing enzyme Dicer.

The original MS2 coat, without any fusions, was used together with the MS2 operators that were inserted in the 5′-untranslated region of an mRNA, by the elaboration of a method for quantitatively tuning the expression of multiple transgenes in mammalian cells (Endo et al. 2013).

The MS2 tethering function was used by the generation of a novel retrovirus-based RNA delivery instrument in cell or gene therapies (Prel et al. 2015). To improve retroviral transfer, a dimerization-independent MS2-driven RNA packaging system was designed that used MS2 coat-HIV-1 chimeras. The engineered chimeric HIV-1 particles promoted effective packaging of several types of RNAs provided with the MS2 stem-loops. Systemic injection of high-titer particles led to gene expression in mouse liver, transferring therapeutic mRNAs, and editing of the desired locus. It is noteworthy in this context that the first experiments in which HIV-1 nucleocapsid was replaced by the MS2 coat were performed by Zhang et al. (1998).

Recently, a first example was presented of engineering the MS2 tethering system to design an RNA editing enzyme complex capable of targeting specific point mutations and restoring the genetic code (Azad et al. 2017, 2019; Bhakta et al. 2018). The authors engineered the deaminase domain of adenosine deaminase acting on RNA (ADAR1) and the MS2 system to target specific adenosines, with the goal of correcting G-to-A mutations at the RNA level. The ADAR1 deaminase domain was fused downstream of the MS2 coat, and the guide RNAs complementary to target RNAs were designed. The guide RNAs directed the ADAR1 deaminase to the desired editing site, where it converted adenosine to inosine. This approach could facilitate the rational use of the ADAR variants for genetic restoration and treatment of genetic diseases. The ADAR recruitment by the MS2 tagging system was reviewed recently by Chen G et al. (2019).

The exhaustive methodological surveys how to perform the MS2 tethering were presented (Coller and Wickens 2007; Clement and Lykke-Andersen 2008; Gaur 2008; Gredell et al. 2012; Carrier et al. 2018a,b).

It is noteworthy that the MS2 tethering approach was used by the generation of the synthetic mRNA-delivered circuits with RNA-binding proteins for the logic computation in mammalian cells (Matsuura et al. 2018).

Generally, the MS2-based tethering approach and methodology was described sufficiently in numerous reviews (Graveley et al. 1999; Patel 1999; Fashena et al. 2000; Graveley 2004; Gehring et al. 2008; Keryer-Bibens et al. 2008; Stynen et al. 2012; Chen Y and Varani 2013; Andries et al. 2015; Ferrè et al. 2016; Marchese et al. 2016; Barra and Leucci 2017; Köster et al. 2017; Lybecker and Samuels 2017; Arora and Kaul 2018; d'Aquino et al. 2018; Wilkinson et al. 2018; Wong J et al. 2018; Kanwal and Lu 2019; Ramanathan et al. 2019; Tretbar et al. 2019; van den Ameele et al. 2019).

IMAGING

The MS2 tethering approach was pioneered as a revolutionary imaging tool by the Robert H. Singer team (Bertrand et al.

1998; Chartrand et al. 1999; Long et al. 2000). This team succeeded for the first time to image and track RNAs in live yeast by fusing the MS2 coat to green fluorescence protein (GFP). Figure 24.2 presents the first scheme for the visualization of reporter mRNAs in live cells. As it follows from the illustration, the MS2 dlFG coat variant was used in the constructs. It should be noted here that the MS2 dlFG variant was constructed by Peabody and Ely (1992) by deletion of 13 amino acid residues within the FG loop, in order to prevent contacts between dimers with further self-assembly.

Thus, the system was comprised of two components, the reporter mRNA and the MS2-GFP fusion protein. The

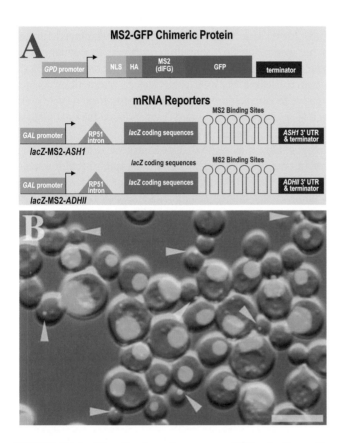

FIGURE 24.2 Visualization of reporter mRNAs in live cells. (A) Schematic describing the constructs used in this approach. The system is comprised of two components, a reporter mRNA and a GFP-MS2 fusion protein. The GFP-MS2 was expressed under the control of the constitutive *GPD* promoter, while the reporter mRNA was under the control of the *GAL* promoter. The reporter mRNA contains six binding sites for the coat protein of the bacterial phage MS2. To avoid possible interference with translation and the function of the 3′UTR, the MS2-binding sites were introduced immediately after a translation termination codon. The 3′UTRs were either from the *ASH1* gene, to induce mRNA localization at the bud tip, or from the *ADHII* gene, as control. In addition, a nuclear localization signal (NLS) followed by an HA tag was introduced at the N terminus of the fusion protein, so that that only the GFP protein that is bound to its target mRNA would be present in the cytoplasm. (B) Live cells expressing the GFP-MS2 fusion protein and the *lacZ-MS2-ASH1* reporter mRNA. Arrows indicate some of the particles, usually in the bud. Bar, 5 μm. (Reprinted with permission of Cell Press from Bertrand E et al. *Mol Cell*. 1998;2:437–445.)

MS2-GFP gene was expressed under the control of the constitutive *GPD* promoter, while the reporter mRNA was under the control of the *GAL* promoter. The reporter *ASH1* mRNA contained a tandem repeat of the six 19-nucleotide recognition stem-loops within the 3′UTR. To avoid possible interference with translation and the function of the 3′UTR, the MS2-binding sites were introduced immediately after the translation termination codon. The 3′UTRs were either from the *ASH1* gene, to induce mRNA localization at the bud tip, or from the *ADHII* gene, as a control. In addition, a nuclear localization signal (NLS) followed by an HA tag was introduced at the N terminus of the fusion protein, so that that only the GFP protein that was bound to its target mRNA would be present in the cytoplasm (Bertrand et al. 1998). Therefore, Singer and coworkers were able to track the movement of RNA within live yeast. The first review commenting on this success was published by Arn and Macdonald (1998). A bit later, a similar system of the localization and anchoring of mRNA in budding yeast was published by Beach et al. (1999). The next paper of this team contained nice video material of the visualization process (Beach and Bloom 2001). The same MS2-GFP approach was then used in other studies on the direct visualization of the RNA localization in living yeast cells (Corral-Debrinski et al. 2000).

The MS2-GFP tagging system was also applied for the visualization of mRNA translocation in living neurons, where eight tandem copies of the MS2 stem-loop were inserted into the mRNA in question (Rook et al. 2000). A methodological paper that addressed the identification of proteins affecting mRNA localization in living cells was published by Brodsky and Silver (2002), while the classical "tethering function" of the MS2 coat was analyzed methodologically by Coller and Wickens (2002). At the same time, MS2-GFP imaging contributed markedly to studies on the mechanisms of the RNA movements and targeting within the cell nucleus (Verheggen et al. 2001; Boulon et al. 2002).

The MS2-based imaging led to a reasonable inducible system to visualize gene expression in real-time at the levels of DNA, RNA, and protein in single living cells (Janicki et al. 2004). The system was composed of a 200-copy transgene array integrated into a euchromatic region of chromosome 1 in human U2OS cells. This system made it possible to correlate changes in chromatin structure with the progression of transcriptional activation, presenting a real-time integrative view of gene expression. This straightforward success was commented operatively by Henikoff (2004).

Moreover, the Singer team visualized for the first time the discrete "pulses" of gene activity that turned on and off at irregular intervals (Chubb et al. 2006). Surprisingly, the length and height of these pulses were consistent throughout development. The cells were more likely to re-express than to initiate new expression, indicating the existence of a transcriptional memory. The principles of the eukaryotic transcription were reviewed at that time by Golding and Cox (2006).

The first live visualization of endogenous RNA in *Drosophila* oocytes and embryos with the MS2-GFP approach was achieved by Forrest and Gavis (2003). This success earned a special review by Singer (2003) on the real-time visualization of RNA localization by the MS2-based imaging approach. Later, the impressive MS2-based imaging, as well as tethering studies in *D. melanogaster*, were performed (Ashraf et al. 2006; Brechbiel and Gavis 2008; Estes et al. 2008; Jaramillo et al. 2008; van Gemert et al. 2009; Belaya and St Johnston 2011; Lim et al. 2011).

The number of MS2-based visualization studies grew rapidly. Sylvestre et al. (2003) compared the mRNA transport machinery in yeast and human cells. The technique was used to localize mRNAs intracellularly in whole plants (Hamada et al. 2003; Zhang and Simon 2003). The transport of mRNAs was evaluated in neurons (Bi et al. 2003, 2006, 2007; Dynes and Steward 2007; Andreassi et al. 2018; Bauer KE et al. 2019).

The Singer team further improved the MS2-based imaging approach and reported a method to visualize single mRNA molecules in living mammalian cells, while, regardless of any specific cytoplasmic distribution, individual mRNA molecules exhibited rapid and directional movements on microtubules (Fusco et al. 2003). Furthermore, this team monitored mRNA expression to characterize the movement in real time of single mRNA-protein complexes in the nucleus of living mammalian cells (Shav-Tal et al. 2004).

Basyuk et al. (2005) employed at that time as many as 24 MS2 stem-loop copies by the RNA tagging in their study on the visualization of transport steps by the replication of Moloney murine leukemia virus.

Finally, Golding and Cox (2004) elaborated the MS2-based imaging method for tracking RNA molecules in *E. coli*, which was sensitive to single copies of mRNA and followed individual RNA molecules for many hours in living cells. The observed distinct characteristic dynamics of RNA molecules were fully consistent with the known life history of RNA in prokaryotes; namely, (i) localized motion consistent with the Brownian motion of an RNA polymer tethered to its template DNA, (ii) free diffusion, and (iii) a few examples of polymer chain dynamics that appeared to be a combination of chain fluctuation and chain elongation attributable to RNA transcription (Golding and Cox 2004). Le et al. (2005) further demonstrated that the MS2-based imaging in combination with fluorescence correlation spectroscopy allowed direct real-time measuring of the activity of any specific promoter in prokaryotes. It appeared that induced RNA levels within a single *E. coli* bacterium exhibited a pulsating profile in response to a steady input of inducer. The imaging technique therefore made possible the identification of relationships between genotypes and transcriptional dynamics that would be accessible only at the level of the single cell (Le et al. 2005). Furthermore, Golding et al. (2005) measured mRNA partitioning at cell division and correlated mRNA and protein levels in single *E. coli* cells. Next, Le et al. (2006) measured in real time within a single bacterium the transcription activity at the RNA level of the *acrAB-TolC* multidrug efflux pump system. Using fluorescence correlation spectroscopy, Guet et al. (2008) determined in a single living bacterial cell both the absolute mRNA concentration and the associated protein expressed from an inducible plasmid. This method allowed the real-time monitoring to approximately 40 nM mRNA, i.e., roughly two transcripts per volume of detection (Guet et al. 2008). Smolander et al. (2011) measured the transcriptional activity of a tetracycline-inducible promoter controlling the transcription of an RNA with 96 binding sites for MS2-GFP. So et al. (2011) quantified copy-number statistics of mRNA from 20 *E. coli* promoters.

Furthermore, the *in vivo* kinetics of transcription initiation in *E. coli* were measured (Kandhavelu et al. 2011, 2012a,b; Mäkelä et al. 2011; Chowdhury et al. 2012; Lloyd-Price et al. 2012; Muthukrishnan et al. 2012; Gupta A et al. 2014; Häkkinen et al. 2014; Viswanathan et al. 2014; Häkkinen and Ribeiro 2015, 2016; Sala et al. 2015; Anufrieva et al. 2016; Lloyd-Price et al. 2016; Lampo et al. 2017; Oliveira et al. 2018, 2019). The MS2-GFP imaging was employed further by the evaluation how temperature affects the dynamics of the repressilator in *E. coli*, where repressilator was a synthetic circuit consisting of three genes whose interactions form a negative feedback loop, i.e., the three genes form a cycle, with each gene expressing a protein that repressed the next gene in the cycle (Chandraseelan et al. 2013).

Stylianidou et al. (2014) mapped the generic motion of large protein complexes in the bacterial cytoplasm through quantitative analysis of thousands of complete cell-cycle trajectories of fluorescently tagged ectopic MS2-mRNA complexes. The motion of these complexes in the cytoplasm was strongly dependent on their spatial position along the long axis of the cell, and that their dynamics were consistent with a quantitative model that required only nucleoid exclusion and membrane confinement. This analysis also revealed that the nucleoid increased the mobility of MS2-mRNA complexes, resulting in a fourfold increase in diffusion coefficients between regions of the lowest and highest nucleoid density (Stylianidou et al. 2014).

The MS2-GFP platform contributed markedly to the single-cell evaluation of the phage λ, which resulted in the complete narrative for the λ life cycle (Golding 2016). Using the MS2 imaging technique, Sterk et al. (2018) revisited the MS2 coat protein system and assessed the translation efficiency from its sequestered ribosome-binding site by introducing standby mutations, in accordance with the standby model of de Smit and van Duin (2003a,b), which was described in the Secondary and tertiary structure section of Chapter 16. It is highly remarkable that the use of novel technologies by *in vivo* and *in vitro* examination confirmed the pioneering de Smit and van Duin's conclusions concerning the complexity of the natural standby site.

Gelderman et al. (2015) used the MS2 GFP-based imaging to elucidate the role of small RNAs (sRNAs). The authors used *in vivo* the trimolecular fluorescence complementation (TriFC) assay in order to detect and visualize the central regulatory sRNA-protein interaction of the carbon storage regulatory system in *E. coli*. The carbon storage regulator (Csr) consists primarily of an RNA-binding protein, CsrA, that alters the activity of mRNA targets, and of an sRNA, CsrB, that modulates the activity of CsrA. As a result, the fluorescence complementation system detected the interactions between CsrB and CsrA. This was the first attempt to use the

MS2-based imaging technique to study the sRNA-protein interaction directly in bacteria (Gelderman et al. 2015).

The RNA and protein localization and metabolism in single live bacterial cells were reviewed substantially (Broude 2011; Nevo-Dinur et al. 2012; Buskila et al. 2014; Kannaiah and Amster-Choder 2014; Kapanidis et al. 2018). Ribeiro (2016) reviewed recent quantitative assessments of delays in transcription events that most affect the dynamics of RNA production in *E. coli*. It was concluded that time delays in transcription could be powerful, evolvable regulators of the kinetics of gene expression in prokaryotes. Engl (2019) described specifically the noise in bacterial gene expression as a genome-wide phenomenon that arises from the stochastic nature of the biochemical reactions that take place during gene expression and the relatively low abundance of the molecules involved.

In parallel, the MS2-GFP-based imaging was used for the specific mRNA localization in *B. subtilis* (Tiago dos Santos et al. 2012). Later, MS2-GFP-based imaging provided a time-resolved single-cell view on coupled defects in transcription, translation, and growth during expression of heterologous membrane proteins in *Lactococcus lactis* (van Gijtenbeek et al. 2016). Moreover, van Gijtenbeek and Kok (2017) published an extensive overview of the RNA visualization in bacteria.

To make the MS2-based mRNA localization more universal, Haim et al. (2007) developed a simple PCR-based genomic-tagging strategy, termed m-TAG, that used homologous recombination to insert hairpins for the MS2 coat between the coding region and 3′ untranslated region of any yeast gene. By coexpression with the MS2-GFP fusion, localization of a set of endogenous mRNAs was followed in living yeast. The m-TAG approach was elaborated upon in detail by further development (Haim-Vilmovsky and Gerst 2009, 2011). Gadir et al. (2011) used the m-TAG approach for the sustained visualization of endogenous mRNAs in live yeast to localize mitochondrial proteins. Haim-Vilmovsky et al. (2011) described the advanced mp-TAG method that allowed the simultaneous visualization of both endogenously expressed mRNAs and their translation products in living yeast. For this reason, homologous recombination was used to insert the mCherry gene and MS2 coat binding sites downstream from any open reading frame, in order to localize protein and mRNA, respectively. The mTAG methodology was used further in the advanced studies on the fate of mRNAs encoding secreted/membrane proteins in yeast (Kraut-Cohen et al. 2013).

The MS2-based visualization in eukaryotic cells was used successfully by many nicely illustrated investigations in yeast (Entelis et al. 2006; Lange et al. 2008; Zipor et al. 2009, 2011; Heuck et al. 2010; Youk et al. 2010; Barberis et al. 2011; Gallardo and Chartrand 2011; Gallardo et al. 2011; Kilchert and Spang 2011; Casolari et al. 2012; Fundakowski et al. 2012; Gelin-Licht et al. 2012; Cusanelli et al. 2013; Genz et al. 2013; Hermesh et al. 2014; Leung et al. 2014; Lui et al. 2014; Shichino et al. 2014; Singer-Krüger and Jansen 2014; Garcia and Parker 2015, 2016; Jin et al. 2015, 2017; Haimovich et al.

2016; Mehta et al. 2018), bacteriovores (Chubb et al. 2013; Wang W et al. 2018; Antolović et al. 2019), nematodes (Yan et al. 2009; Scharf et al. 2011; Lee C et al. 2019), *Drosophila* (Catrina et al. 2012; Weil et al. 2012; Bothma and Levine 2013; Garcia et al. 2013; Lucas et al. 2013; Apte et al. 2014; Bothma et al. 2014; Gregor et al. 2014; Arib et al. 2015; Bor et al. 2015; Kim G et al. 2015; Desponds et al. 2016; Esposito et al. 2016; Ferraro et al. 2016a; Lo et al. 2016; Misra et al. 2016; Lim et al. 2017; Berrocal et al. 2018; Blazquez et al. 2018; Kordyukova et al. 2018; Lucas et al. 2018; Mir et al. 2018; Falo-Sanjuan et al. 2019; Scholes et al. 2019; Yamada et al. 2019), plants (Campalans et al. 2004; Sambade et al. 2006; Christensen et al. 2010; Komarova et al. 2010; Peña and Heinlein 2012; Schönberger et al. 2012; Cheng SF et al. 2013; Göhring et al. 2014; Michaud et al. 2014; Rosa et al. 2014; Tian and Okita 2014; Yagi et al. 2014; Peña et al. 2015; Tilsner 2015; Peña and Heinlein 2016; Scarpin et al. 2017; Seo and Chua 2017, 2019; Boavida 2018; Kinoshita et al. 2018; Luo et al. 2018; Mazzoni-Putman and Stepanova 2018), *X. laevis* oocytes (Gagnon et al. 2013; Powrie et al. 2016; Ciocanel et al. 2017), zebrafish (Campbell et al. 2015; Kaufman and Marlow 2016), birds (Tyagi and Alsmadi 2004; Yamagishi et al. 2009), and mammals (Bannai et al. 2004; Rackham and Brown 2004; Darzacq et al. 2007; Jonkers et al. 2008; Mili et al. 2008; Ruault et al. 2008; Trinkle-Mulcahy and Lamond 2008; Hautbergue et al. 2009; Grünwald and Singer 2010; Naarmann et al. 2010; Oh et al. 2010; Barakat et al. 2011; Basu et al. 2011; de Turris et al. 2011; Hodge et al. 2011; Jo and Kim 2011; Karimian Pour and Adeli 2011; Lee HY et al. 2011; Ng et al. 2011; Querido et al. 2011; Schmidt et al. 2011; Shevtsov and Dundr 2011; Wei and Loh 2011; Dynes et al. 2012; Han et al. 2012; Newhart et al. 2012; Tripathi et al. 2012; Kochanek and Wells 2013; Aizer et al. 2014; Gao et al. 2014; Kim et al. 2014; Negishi et al. 2014; Newhart and Janicki 2014; Rino et al. 2014; Battich et al. 2015; Berkovits and Mayr 2015; Lai et al. 2015; Paek et al. 2015; Rothé et al. 2015; Steward et al. 2015; Tsuboi et al. 2015; Zhang P et al. 2015; Corrigan et al. 2016; Kajita et al. 2016; Katz et al. 2016; Pankert et al. 2017; Avogaro et al. 2018, 2019; Ha et al. 2018; Tang et al. 2018). Kim SH et al. (2019a) elucidated the degradation pattern of the MS2-GFP-labeled mRNA in mammalian cells and tissues and found that accumulation of the MS2 binding site RNA decay fragments did not always occur, depending on the mRNA species and the model organisms used.

MS2-based imaging also contributed to deciphering the role of Huntington disease protein (Ma B et al. 2010, 2011; Savas et al. 2010; Culver et al. 2016).

MS2-based single cell imaging was used to visualize transcriptional dynamics of transgenes at high expressing retroviral integration sites in murine embryonic stem cells (Lo et al. 2012). MS2-based tethering and imaging contributed to the functional elucidation of the retroviruses and retrotransposons, namely, retrotransposon Ty3 (Beliakova-Bethell et al. 2006; Larsen et al. 2008; Clemens et al. 2013), retrotransposon LINE-1 (Doucet et al. 2010, 2016; Goodier et al. 2015), short interspersed degenerate retroposons (SIDERs), in *Leishmania* (Azizi et al. 2017a,b), human T-cell leukemia virus (Younis et al. 2006), Rous sarcoma virus (McNally et al. 2006), and bovine foamy virus (Wang W et al. 2010). As for other viruses, MS2-GFP-based imaging was employed by the localization of the replicative complex elements of the tombusviruses: cucumber necrosis virus (Panavas et al. 2005) and *Cymbidium* ringspot virus (Rubino et al. 2007) in yeast cells, as well as the cucumber mosaic virus (Choi et al. 2012, Chaturvedi et al. 2014).

Boireau et al. (2007) visualized the transcriptional cycle of HIV-1 in real-time and live cells. MS2-based imaging also contributed to the elucidation of replication process of retroviruses: HIV-1 (Chen et al. 2009; Jouvenet et al. 2009, 2011a,b; Molle et al. 2009; Baumgärtel et al. 2011; Kula et al. 2011; Maiuri et al. 2011a,b; Ni et al. 2011; Ghoujal et al. 2012; Kula and Marcello 2012; Nakamura et al. 2012; Phalora et al. 2012; Roesch et al. 2012; Nikolaitchik et al. 2013; Xu et al. 2013; Yin J et al. 2013; Woods et al. 2014; Boons et al. 2015; Dirk et al. 2016; Ferrer et al. 2016a,b; Itano et al. 2016; Pocock et al. 2016; Zhang XE et al. 2016; Becker et al. 2017; Behrens et al. 2017; Akiyama et al. 2018; Barajas et al. 2018; Itano et al. 2018), HIV-2 (Dilley et al. 2011), feline immunodeficiency virus (Kemler et al. 2010; Jouvenet et al. 2011a; Elinav et al. 2012), and Mason-Pfizer monkey virus (Pocock et al. 2016). MS2-based imaging was then applied to the trafficking of hepadnaviruses: hepatitis B virus and woodchuck hepatitis virus (Pocock et al. 2017) and flaviviruses: tickborne encephalitis virus (Hoenninger et al. 2008; Miorin et al. 2008, 2012, 2013, 2016), Semliki forest virus (Kallio et al. 2015), and hepatitis C virus (Fiches et al. 2016). The methodology of viral RNA imaging was described by Alonas et al. (2016).

MS2-based imaging contributed to the evaluation of the role of microRNAs (miRNAs) in plant organogenesis and hormone action (Fujioka et al. 2007) and in mammalian cells (Portal 2011; Srikantan et al. 2011; Tominaga et al. 2011; Bloom et al. 2014; Eades et al. 2015; Tuccoli et al. 2016; Steinkraus et al. 2016; Rehfeld et al. 2018).

MS2-based imaging paved the way for a novel approach to study and visualize the association of an RNA-binding protein with its native RNA target *in situ* by fluorescence resonance energy transfer (FRET) (Lorenz 2009). The MS2 coat was tagged with a yellow variant of GFP, and the RNA stained with SytoxOrange. The RNA binding resulted in a decrease of the fluorescence lifetime of YFP due to FRET, which could be measured by fluorescence lifetime imaging microscopy (FLIM). The binding produced an RNA-specific FRET signal (Lorenz 2009). In parallel, the MS2-based FRET approach was proposed by Huranová et al. (2009) and reviewed extensively by Šimková and Staněk (2012).

To overcome the restriction of MS2-based imaging to artificial cell lines and reporter genes, Singer's team developed an approach that enabled live-cell imaging of single endogenous labeled mRNA molecules transcribed in primary mammalian cells and tissue (Lionnet et al. 2011). To do this, the authors generated a knock-in mouse line with an MS2 binding site (MBS), cassette targeted to the 3′ untranslated region of the essential β-actin gene. As β-actin-MBS was ubiquitously expressed, the regulation of endogenous mRNA could be uniquely addressed in any tissue or cell type. Moreover, Brody and Shav-Tal (2011) measured the synthesis rate of the newly made mRNAs, using the MS2-YFP platform.

Using the MS2-based imaging platform, Maiuri et al. (2011b) estimated transcription rate of RNA polymerase II in live mammalian cells at above 50 kb/min. All available data on the estimated transcription elongation rates of RNA polymerase II were collected and analyzed (Cannon and Chubb 2011; Marcello 2012).

Park HY et al. (2014) visualized dynamics of single endogenous mRNA in live transgenic mouse with the MS2-GFP-based technique, where all β-actin mRNA was fluorescently labeled and provided insight into its dynamic regulation within the context of the cellular and tissue microenvironment. Later, an extended MS2-XFP-based method for visualizing mRNAs in living mouse was presented by Nwokafor et al. (2019).

Using the MS2-based imaging and ribozyme switching techniques, Bloom et al. (2015) presented a platform that combined a ligand-responsive ribozyme switch and synthetic miRNA regulators to create an OFF genetic control device based on RNA interference (RNAi). This novel genetic device increased the level of silencing from a miRNA in the presence of a ligand of interest, effectively creating an RNAi-based OFF control. The OFF switch platform had the flexibility to be used to respond to both small molecule and protein ligands.

To further improve the quantitative single-molecule assays by studies of when and where molecules were interacting inside living cells and where enzymes were active, English and Singer (2015) employed a three-camera imaging microscope for high-speed single-molecule tracking and super-resolution imaging in living cells. Singer's team also applied Bayesian model selection to hidden Markov modeling to infer transient transport states from trajectories of mRNA-protein complexes in live mouse hippocampal neurons and metaphase kinetochores in dividing human cells and provided the appropriate software to all persons concerned (Monnier et al. 2015).

Singer's team originally published detailed methodological protocols for the MS2-based imaging in yeast and mammalian cells (Wells et al. 2007a,b; Zenklusen et al. 2007). Numerous other detailed papers on the highly-sensitive imaging methodology followed (Dahm et al. 2008; Long and Urbinati 2008; Querido and Chartrand 2008; Santangelo et al. 2012; Kalo et al. 2013; Yunger et al. 2013; Corrigan and Chubb 2015; Rino et al. 2015, 2016, 2017; Bahar Halpern and Itzkovitz 2016; Cho et al. 2016; Chubb 2016; Martin RM et al. 2016; Moon and Park 2016; Park HY and Song 2016; Tantale et al. 2016; Tsanov et al. 2016; Chen Mingming et al. 2017; Dacheux et al. 2017; Germier et al. 2017; Haimovich et al. 2017; Ramat et al. 2017; Tamburino et al. 2017; Perez-Romero et al. 2018; Tutucci et al. 2018b). Shanbhag and Greenberg (2013) presented a detailed protocol on how to create multiple nuclease-induced DNA double-strand breaks at a known distance from an inducible and visualizable transcriptional reporter and methods to study both the DNA damage repair and its effects on local transcription. A segmentation method termed maximum likelihood estimation (MAMLE) was developed for MS2 visualization-based detection of cells within dense clusters (Chowdhury et al. 2013). Dundr (2013) described how to use MS2-based imaging to probe the contribution of specific RNAs in the formation of nuclear bodies.

Singer's team also developed a light-activation technology to turn on a single steroid-responsive gene and follow dynamic synthesis of RNA from the activated locus (Larson et al. 2013).

The MS2-GFP imaging approach was reviewed extensively, alone and in comparison with other imaging technologies (Rafalska-Metcalf and Janicki 2007; Rodriguez et al. 2007; Tyagi 2007, 2009; Fusco et al. 2008; Darzacq et al. 2009; Larson et al. 2009; Carmo-Fonseca 2010; Lionnet and Singer 2010; Park et al. 2010; Weil et al. 2010; Carmo-Fonseca and Rino 2011; Dange et al. 2011; Doerr 2011; Filipovska and Rackham 2011; Hocine and Singer 2011; Itzkovitz and van Oudenaarden 2011; Kalisky et al. 2011; Li GW and Xie 2011; Urbinati and Long 2011; Wang F and Greene 2011; Bakstad et al. 2012; Bann and Parent 2012; Chao et al. 2012; Chumakov et al. 2012; Dictenberg 2012; Lease et al. 2012; Lionnet and Singer 2012; Londoño-Vallejo and Wellinger 2012; Oeffinger and Zenklusen 2012; Palangat and Larson 2012; Selimkhanov et al. 2012; Ausländer and Fussenegger 2013; Cody et al. 2013; Hayashi and Okamoto 2013; Sheinberger and Shav-Tal 2013; Wang Y et al. 2013; Carmo-Fonseca and Kirchhausen 2014; Paszek 2014; Widom et al. 2014; Cui and Irudayaraj 2015; Liao et al. 2015; Shi et al. 2015; You and Jaffrey 2015; Dolgosheina and Unrau 2016; Ehses et al. 2016; Hövelmann and Seitz 2016; Mannack et al. 2016; McFadden and Hargrove 2016; Schieweck et al. 2016; Wang W and Chen 2016; Bigley et al. 2017; Donlin-Asp et al. 2017; Hochberg et al. 2017; Liu Xiaohui et al. 2017; Ma Z et al. 2017; Nicolas et al. 2017; Pauff et al. 2017; Salehi et al. 2017; Sheinberger and Shav-Tal 2017; Specht et al. 2017; George et al. 2018; Kato M and McKnight 2018; Li et al. 2018; Lim 2018; Urban and Johnston 2018; Van Driesche and Martin 2018; Schulz and Harrison 2019).

After MS2-GFP imaging, fluorescent oligodeoxynucleotide probes based on MS2 and cyclin D1, a known breast cancer mRNA marker, were used to detect the cyclin D1 mRNA in breast cancer (Segal et al. 2013). Such ds-nucleoside with intrinsic fluorescence (ds-NIF) probes were prepared using 5-((4-methoxy-phenyl)-trans-vinyl)-2′-deoxy-uridine as a fluorescent uridine analog exhibiting a 3000-fold higher quantum yield and maximum emission at 478 nm, which is 170 nm red-shifted as compared to uridine. In parallel, a strategy was proposed for how to image metabolites using genetically encoded fluorescent sensors composed of RNA (Song et al. 2013). This method involved fusing RNA aptamers to Spinach, a 98-nucleotide RNA that switches on the fluorescence of 3,5-difluoro-4-hydroxybenzylidene imidazolinone (DFHBI), an otherwise nonfluorescent small molecule. The metabolite-binding aptamer is fused therefore via a stem required for Spinach-DFHBI fluorescence. The stem does not form a stable duplex at the imaging temperature. Only after metabolite binding does the aptamer fold, bringing the strands of the stem in proximity, which results in a Spinach structure that can bind DFHBI. Song et al. (2013) generated a sensor for the MS2 coat protein, where the MS2 stem-loop was fused to

Spinach with various transducer stems and generated a sensor that exhibited a 41.7-fold increase in fluorescence upon addition of the MS2 coat. The optimal transducer module for the MS2 coat sensor was composed of a truncated stem found in the MS2 hairpin. The Spinach-based MS2 sensor was used in this study to monitor infection kinetics of *E. coli* by MS2. Remarkably, this investigation revealed cell-to-cell variability in infection kinetics that would be masked in population measurements of the MS2 coat synthesis after infection, highlighting the strength of these molecules as sensors *in vivo*. The MS2-Spinach approach was reviewed operatively by the authors (Strack and Jaffrey 2013).

Cao et al. (2013) generated a genetically encoded optogenetic system that activated mRNA translation in mammalian cells in response to light. Blue light induced the reconstitution of an RNA binding domain and a translation initiation domain, thereby activating target mRNA translation downstream of the binding sites. To do this, the MS2 coat was fused to the CIBN domain, an NLS-deficient truncated version of the CIB1 protein. The latter was the basic-helix–loop–helix 1 polypeptide interacting with the *Arabidopsis thaliana* Cryptochrome 2 (CRY2), protein upon light illumination.

Carrocci and Hoskins (2014) developed novel small molecule-based RNA tags by fusion of the MS2 coat to either the *E. coli* dihydrofolate reductase (MS2-eDHFR) or the so-called SNAP tag (MS2-SNAP), where the SNAP is a domain of human O^6-alkylguanine-DNA alkyltransferase that can covalently couple to benzylguanine and its derivatives. These two fusions were used to image RNAs in live *S. cerevisiae* cells with a diverse range of fluorophores. The use of MS2-eDHFR or MS2-SNAP fusions significantly expands the repertoire of fluorophores available for imaging RNAs in live cells. The eDHFR tag tightly bound fluorescent analogs of trimethoprim, while the SNAP tag reacted with fluorescent benzyl guanine or benzylchloropyrimidine derivatives to form a covalent adduct to the protein. The reporter mRNA in this case encoded six copies of the MS2 stem-loop (Carrocci and Hoskins 2014).

The MS2-based platform contributed to the development of novel imaging tools, such as far-red mNeptune-based bimolecular fluorescence complementation (BiFC), and TriFC systems with excitation and emission above 600 nm in the "tissue optical window" for imaging of protein–protein and RNA-protein interactions in live cells and mice (Han Y et al. 2014).

Martins et al. (2018) engineered a software, namely, SCIP, to analyze multimodal, multiprocess, time-lapse microscopy morphological and functional images. The SCIP software was capable of automatic and manually corrected segmentation of cells and lineages, automatic alignment of different microscopy channels, and was able to detect, count, and characterize fluorescent spots, such as RNA tagged by the MS2-GFP imaging. The RNA2DMut web tool was generated for the design and analysis of RNA structure mutations in order to find the optimal MS2 aptamer sequences (Moss 2018). The MS2 coat-RNA operator complex measurements were used by the development of a novel test for the RNA-protein binding affinity prediction (Kappel et al. 2019). In addition, the fully automated RiboLogic method was used to design riboswitches that could modulate their affinity to the MS2 coat protein upon binding of flavin mononucleotide, tryptophan, theophylline, and microRNA miR-208a (Wu MJ et al. 2019).

In a recent review, the RNA sequence, structure, and function, including those of the RNA phage aptamers, were linked to the revolutionizing capacities of high-throughput sequencing methods in order to catalog the diversity of RNAs and RNA-protein interactions that exist in cells (Denny and Greenleaf 2019).

NEW CHALLENGE: PP7

Novel opportunities were opened by the involvement of the novel actor, the PP7 platform, into the tethering and imaging approaches. As stated *Nature* (Kaganman 2008), Jeffrey Chao, a postdoc in Singer's lab, said "...the drawback of this technique [*the MS2 coat protein fused to fluorescent proteins to label RNA for imaging in living cells* – PP] is that you can only look at one RNA at a time." Singer's team therefore needed a second system with a high affinity orthogonal interaction similar to that of MS2 but with a different specificity, and they found this in PP7 (Chao et al. 2008).

The PP7 coat protein–RNA interaction was first adopted by Kathleen Collins' team, designing an aptamer for affinity purification of ribonucleoproteins (Hogg and Collins 2007; Hogg and Goff 2010). The first PP7 tethering study was performed by Yamasaki et al. (2007) who used the system of the expressed PP7 coat and six tandem PP7 operator copies inserted into the target RNA. The MS2 hairpins within the latter were not recognized by the PP7 coat. The combined use of MS2 and PP7 coat fusions by the tethering approach was achieved by Gesnel et al. (2009).

Alcid and Jurica (2008) constructed a special labeling protein, termed Beta-PP7, by fusion of the β-subunit of *E. coli* DNA polymerase III to the PP7 coat, which recognized the corresponding 24-nucleotide RNA hairpin target. This fusion was used to label spliceosomes assembled on a pre-mRNA that contained the target sequence in the exons. The label was clearly visible in electron microscopy images of the spliceosome, and subsequent image processing with averaging showed that the exons sat close to each other in the complex (Alcid and Jurica 2008).

The involvement of the PP7 platform, in parallel with the MS2, into the imaging strategy, was reviewed at that time by Schifferer and Griesbeck (2009).

The PP7 tethering was then applied in a study on the identification of a protein tightly associated with the deadenylation complex in human cells (Ozgur et al. 2010). As a result, the human protein Pat1b was characterized as a central component of the RNA decay machinery by physically connecting deadenylation with decapping. The FLAG-tagged PP7 coat was a subject to immunoprecipitation with anti-FLAG antibody in a study on ASBEL, an antisense transcript of the *ANA/BTG3* gene that was required for tumorigenicity of ovarian carcinoma (Yanagida et al. 2013). The MS2 and PP7 tethering was compared directly by the isolation of the desired ribonucleoprotein complexes (Leppek and Stoecklin 2014).

The MS2 and PP7 aptamer domains were involved in construction of the rationally designed RNA assemblies to be used as functional architecture *in vivo* (Delebecque et al. 2011). Thus, the synthetic RNA modules carrying the MS2 and PP7 aptamer domains were assembled into functional discrete 1D and 2D scaffolds *in vivo* and used to control the spatial organization of bacterial metabolism, namely, a hydrogen-producing pathway.

By analogy with MS2 tethering, PP7-based tethering was employed by unveiling of the splicing mechanisms (Kotzer-Nevo et al. 2014; Nasrin et al. 2014; Shefer et al. 2014; Ge et al. 2016; Nazim et al. 2017; Sperling R 2017; Sperling J and Sperling R 2017).

The PP7 platform was also used for the affinity purification of the desired complexes, as in the case of the nonsense-mediated mRNA decay (Baker and Hogg 2017).

The imaging potential of both MS2 and PP7 platforms was compared directly by Singer's team (Wu et al. 2012). Using fluorescence fluctuation spectroscopy to quantitatively characterize the imaging of single mRNAs directly in live cells, it first appeared that the single-chain tandem MS2 dimer increased significantly the uniformity and sensitivity of mRNA labeling. Second, the PP7 system performed better for RNA labeling. It is noteworthy that 24×MS2, 24×PP7, and 6×PP7 binding sites inserted into the 3′ untranslated region of mRNA were used in this study.

Singer's team was the first to include the MS2- and PP7-based dual-color imaging in live yeast (Hocine et al. 2013). Figure 24.3 presents an original illustration of dual-color imaging, when two-color yeast strains were generated by mating *MDN1-24PP7* and *MDN1-24MS2* haploids and expressed PP7 coat-2yEGFP and MS2 coat-mCherry fusions such that transcription of each individually tagged *MDN1* allele resulted in differentially labeled populations of mRNAs.

The *MDN1* is the longest yeast gene of 14.7 kb and encodes an essential ATPase involved in ribosome assembly. The *MDN1* gene is constitutively expressed at low steady-state numbers and remains transcriptionally active in nearly all cells. The dual-color methodology was used, among other goals, to measure intrinsic noise in mRNA levels and RNA polymerase II kinetics at a single gene (Hocine et al. 2013). The MS2- and PP7-based dual-color technique was employed further by the observation of the Nanog expression variability in mouse embryonic stem cells (Ochiai et al. 2014).

However, high background due to unbound MS2-GFP and the low abundance of RNA targets relative to protein probes severely limited the widespread use of this method, especially in deep tissue imaging applications. This limitation was addressed by harnessing fluorescent complementation using alternating MS2 and PP7 binding domains on the target RNA. Because YFP was split in two halves across the MS2 and PP7 coats, the fluorescence was reconstituted, without background, only when the two protein fusions were brought into proximity at the alternating multi-hairpin binding site within the target RNA and restored the split fluorescence protein together to image single reporter mRNAs (Wu B et al. 2014).

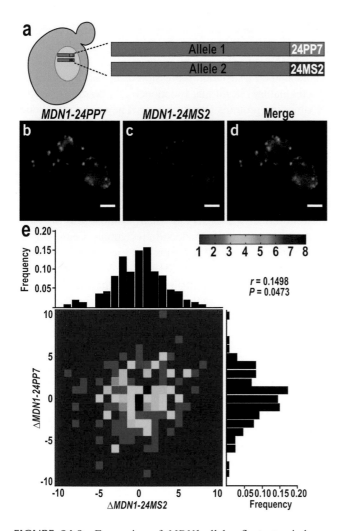

FIGURE 24.3 Expression of *MDN1* alleles fluctuates independently over time. (a) A diploid strain carries one PP7-tagged *MDN1* allele and one MS2-tagged *MDN1* allele. Coexpression of PCP-2yEGFP and MCP-mCherry results in allele-specific labeling of mRNA. (b,c) Transcription of the indicated allele produces green or magenta fluorescent spots. (d) Overlay of mRNA fluorescence for both alleles. (e) *MDN1* mRNAs are counted in single cells every 20 min. Fluctuations were determined as the change in steady-state levels for a given allele between time points, and a heat map was generated to represent these fluctuations. Peripheral histograms represent the distribution of fluctuations for each allele. Pearson's correlation coefficient (*r*) is generated from the data set ($P = 0.0473$, $n = 210$ time points in 33 cells). Scale bars, 3 μm. (Reprinted by permission from Springer Nature, *Nat Methods*, Hocine S et al., Copyright 2013.)

Meanwhile, Coulon et al. (2014) constructed the dual-color imaging platform, where 24 copies of the MS2 and PP7 stem-loops were used, but the PP7 coat was tagged with mCherry and MS coat was tagged with GFP. The platform was used to investigate kinetic competition during the transcription cycle in living human cells by the observation of a complete kinetic profile of transcription and splicing of the β-globin gene. Remarkably, the splicing of the terminal intron occurred stochastically both before and after transcript release, indicating there was not a strict quality control checkpoint (Coulon et al. 2014).

To further improve the visualization of translation from single mRNAs, Singer's team developed an RNA biosensor, termed translating RNA imaging by coat protein knockoff (TRICK), that could distinguish untranslated mRNAs from those that have undergone at least one round of translation (Halstead et al. 2015, 2016; Voigt et al. 2018). The MS2 and PP7 were used to label mRNA with two different fluorescent proteins. Simultaneous expression of the PP7 coat fused with GFP, and the MS2 coat fused with red fluorescent protein (RFP), resulted in yellow fluorescent signals from untranslated mRNAs. During the first round of translation, the PP7-GFP was removed from the mRNA as a ribosome traversed the coding region that contained the PP7 stem-loops. Thus, translated mRNAs were labeled only with MS2-RFP bound to the MS2 stem-loops in the 3′UTR. The pioneering TRICK approach was announced briefly by Popp and Maquat (2015). The TRICK technique was then illustrated plainly in a review by Lee BH et al. (2016).

As Lee BH et al. (2016) stated in this excellent review, while the TRICK technique provided information on the first round of mRNA translation, four independent groups (Morisaki et al. 2016; Wang C et al. 2016; Wu B et al. 2016; Yan et al. 2016) have demonstrated two-color imaging of mRNA and its nascent polypeptides. These four studies used the MS2 or PP7 systems to label mRNA and the SunTag (Tanenbaum et al. 2014; Pichon et al. 2016) or FLAG tag (Viswanathan et al. 2015) systems to label nascent proteins. The fluorescence-tagged MS2 or PP7 capsid proteins were expressed in cells, providing a strong fluorescence signal from each mRNA. The SunTag and FLAG tag systems used fluorescently labeled antibody fragments that bound to multiple copies of a short epitope on each polypeptide. An mRNA undergoing active translation was identified by the colocalization of two fluorescent signals from the mRNA and nascent proteins.

At the same time, both PP7 and MS2 platforms were employed by the generation of a technique based on two-photon fluctuation analysis that enabled the measurement of the protein-RNA interaction in live cells at the single-molecule level (Wu B et al. 2015a,b).

In addition, by colocalization of two-color labels, Singer's team developed a robust super registration methodology that corrected the chromatic aberration across the entire image field to within 10 nm, which was capable of determining whether two molecules were physically interacting or simply in proximity by random chance (Eliscovich et al. 2017).

In parallel, the PP7- and MS2-tagged RNAs were employed by the multicolor fluorescence *in situ* hybridization (FISH) experiments in yeast, where 10 and 24 copies of the MS2 and PP7 stem-loops, respectively, were used (Bajon et al. 2015). Moreover, in order to immunopurify the desired ribonucleoproteins, Bajon et al. (2015) involved Protein A-MS2 fusion protein that could bind with high affinity and high specificity to the MS2 RNA tag sequence and was expressed from an inducible promoter in the same yeast cells. Thus, immunoprecipitation of the MS2-tagged RNA was performed by immunoglobulin G-Sepharose beads.

Using both PP7 and MS2 platforms, Shcherbakova et al. (2016) proposed novel prospective tags, namely, three bright and spectrally distinct monomeric near-infrared fluorescent proteins (miRFPs) engineered from bacterial phytochrome, which were used as easily as GFP-like fluorescent proteins. The miRFPs were in fact two- to fivefold brighter in mammalian cells than other monomeric near-infrared fluorescent proteins and performed well in protein fusions, allowing multicolor structured illumination microscopy.

Using the PP7- and MS2-based approach, Horvathova et al. (2017) developed a fluorescent biosensor based on dual-color, single-molecule RNA imaging that allowed intact transcripts to be distinguished from stabilized degradation intermediates. Using this method, the mRNA decay was measured in single cells and it appeared that individual degradation events occurred independently within the cytosol and were not enriched within processing bodies. This method, termed (three)′-RNA end accumulation during turnover (TREAT 3), allowed the entire life of mRNAs from birth to death to be visualized (Horvathova et al. 2017).

Using the PP7 platform, Roszyk et al. (2017) were the first to use the Spinach, a genetically encodable RNA aptamer that starts to fluoresce upon binding of an organic molecule, with the FRET assembly. The authors described how Spinach was quenched when close to acceptors. The RNA-DNA hybridization was used to bring quenchers or red organic dyes in close proximity to the Spinach. The RNA-protein interactions were examined quantitatively on the PP7 coat and its interacting *pp7*-RNA. Therefore, this work represented a direct method to analyze RNA-protein interactions by quenching the Spinach aptamer (Roszyk et al. 2017).

Using the PP7-based approach, Hoek et al. (2019) developed an assay to visualize nonsense-mediated decay of individual mRNA molecules in real time. This made it possible to uncover real-time dynamics of NMD and revealed key mechanisms that influence the efficiency of the nonsense-mediated decay.

In eukaryotic cells, the PP7-based tethering and imaging, alone or together with the MS2 one, was performed successfully by investigations in yeast (Larson et al. 2011; Guet et al. 2015; Lenstra et al. 2015; Smith et al. 2015; Bensidoun et al. 2016; Lenstra and Larson 2016; Harlen et al. 2017; Heinrich et al. 2017b; Laprade et al. 2017; Rullan et al. 2017, 2018; Tutucci et al. 2018c; Donovan et al. 2019), *Ustilago maydis* fungus (Zander et al. 2018), bacteriovores (Chubb et al. 2013), *Drosophila* (Horan et al. 2015; Abbaszadeh and Gavis 2016; Ferraro et al. 2016b; Fukaya et al. 2016, 2017; Lefebvre and Lécuyer 2018; Lim et al. 2018a,b; Yan and Großhans 2018), and mammals (Day et al. 2016; Meredith et al. 2016; Vaňková Hausnerová and Lanctôt 2017; Balas et al. 2018; Braselmann et al. 2018; Das et al. 2018; Suzuki et al. 2018; Flores et al. 2019; Vítor et al. 2019).

As in the case of the MS2 platform, the PP7-based tethering and imaging was applied also for advanced *E. coli* studies (Sachdeva et al. 2014; Zhou Y et al. 2014; Zhang Jichuan et al. 2015; Kannaiah and Amster-Choder 2016; Berry and Hochschild 2018; Silva JPN et al. 2019).

Finally, the interesting *in vivo* binding assays based on repression of a reporter gene in bacteria were performed with a set of RNA phage coats by Katz et al. (2018, 2019). The authors studied translation repression in *E. coli* by engineering a regulatory circuit that functioned as a binding assay for RNA-binding proteins (RBP) *in vivo*. They did so by inducing expression of a fluorescent protein-RBP chimera, together with encoding its binding site at various positions within the ribosomal initiation region (+11–13 nt from the AUG) of a reporter module. When bound by their cognate RBPs, the PP7 and Qβ coats demonstrated strong repression for all hairpin positions within the initiation region. However, a sharp transition to no-effect was observed when positioned in the elongation region at a single-nucleotide resolution. Employing *in vivo* selective 2′-hydroxyl acylation analyzed by primer extension followed by sequencing (SHAPE-seq) for a representative construct, it was established that in the translationally active state the mRNA molecule was nonstructured, while in the repressed state a structured signature was detected. This regulatory phenomenon was utilized to quantify the binding affinity of the MS2, PP7, GA, and Qβ coats to 14 cognate and noncognate binding sites *in vivo*. Using the generated circuit, the qualitative differences appeared between *in vitro* to *in vivo* binding characteristics for various variants when comparing to past studies. By introducing a simple mutation to the loop region for the Qβ-wild type site, the binding of MS2 coat was abolished, creating the first high-affinity Qβ coat site that was completely orthogonal to the MS2 coat (Katz et al. 2018, 2019). This study therefore paved the way for the introduction of novel RNA phage-based platforms into the tethering and imaging methodology.

The PP7 platform, alone or together with the MS2 ones, contributed to the deciphering of the fine virus replication mechanisms, as in the case of HIV-1 (Chen J et al. 2016) and hepatitis C virus (Chopra et al. 2019).

The PP7 tethering and imaging investigations were referenced in numerous specialized and general reviews, almost in parallel with the MS2 tethering and imaging approach (Oeffinger 2012; Ferguson and Larson 2013; Ishikawa et al. 2013; Michaelis and Treutlein 2013; Dean and Palmer 2014; Faoro and Ataide 2014; Pitchiaya et al. 2014; Urbanek et al. 2014; Weng and Xiao 2014; Buxbaum et al. 2015a,b; Dacheux et al. 2015; Gaspar and Ephrussi 2015; Munsky and Neuert 2015; Rath and Rentmeister 2015; Spille and Kubitscheck 2015; ZJU_China Team and Chen 2015; Bos et al. 2016; Chekulaeva and Landthaler 2016; Chen H and Larson 2016; DeHaven et al. 2016; Lampasona and Czaplinski 2016; Lenstra et al. 2016; Li C et al. 2016; Moon et al. 2016; Schoonen and van Hest 2016; Bauer et al. 2017; Dermit et al. 2017; Eliscovich and Singer 2017; Glock et al. 2017; Heinrich et al. 2017a; Lyon and Stasevich 2017; Xia et al. 2017; Chao and Lionnet 2018; Fei and Sharma 2018; Morisaki and Stasevich 2018; Ohno et al. 2018; Pichon et al. 2018; Ruijtenberg et al. 2018; Tutucci et al. 2018a; Biswas et al. 2019; Elf and Barkefors 2019; Kim SH et al. 2019b; Latallo et al. 2019; Neubacher and Hennig 2019; Taliaferro 2019; Volkov and Johansson 2019; Wang Yaolai et al. 2019; Wilbertz et al. 2019). Recently, the MS2- and

PP7-based imaging techniques were described exhaustively in the second edition of the book *Imaging Gene Expression: Methods and Protocols* (Shav-Tal 2019). The first edition of this notable book appeared in 2013.

CRISPR-Cas9

The MS2-based tethering was applied by the generation of the CRISPR-Cas system-inspired RNA-guided transcriptional activation system (Mali et al. 2013). To generate aptamer-modified sgRNA, i.e., single-guide RNA, tethers capable of recruiting activation domains, the two copies of the MS2 stem-loop were appended to the 3' end of the sgRNA and these chimeric sgRNAs were expressed together with the Cas9N_, a nuclease-deactivated Cas9 (dCas9), and the MS2-VP64 fusion protein, where VP64 was a synthetic transcription activation domain. Figure 24.4a presents this original construction. This pioneering study was reviewed in the context of the revolutionary CRISPR-Cas methodology by Sander and Joung (2014).

To further improve the potency of the Cas9-mediated gene activation, Konermann et al. (2015) considered how transcriptional activation could be achieved in natural contexts, where endogenous transcription factors generally act in synergy with cofactors. Thus, the authors hypothesized that combining VP64 with additional, distinct activation domains could improve activation efficiency. They have chosen the NF-κB transactivating subunit p65 that, while sharing some common cofactors with VP64, recruited a distinct subset of transcription factors and chromatin remodeling complexes. Figure 24.4b presents the original structure of the MS2-p65-HSF1 fusion, where HSF1 is the activation domain from the human heat-shock transcription factor 1. After variation of the effector domain fused to dCas9 or MS2, the combination of dCas9-VP64 and MS2-p65-HSF1, termed the synergistic activation mediator (SAM), exhibited particularly robust transcriptional activation. The authors used these engineered Cas9 activation complexes to investigate sgRNA targeting rules for effective transcriptional activation, to demonstrate multiplexed activation of ten genes simultaneously, and to upregulate long intergenic noncoding RNA transcripts (Konermann et al. 2015).

Zalatan et al. (2015) were the first to turn the sgRNA into a scaffold to recruit different types of effectors including the MS2 and PP7 platforms. The original structures are presented in Figure 24.5.

In this way, the authors tried to overcome the disadvantage of the direct fusion of effectors to dCas9 that only one type of perturbation could occur within a given cell: dCas9 either activated or repressed genes but did not do both. Using RNA-binding proteins from different phages which recognized their specific RNA hairpin structures, the authors succeeded in different combinations of effectors with the RNA-binding proteins. This strategy was used in both yeast and mammalian cells to regulate genes in orthogonal directions simultaneously (Zalatan et al. 2015). Further, both MS2 and PP7 fusions were used by Shechner et al. (2015) and contributed to the elaboration of the novel CRISPRainbow system for labeling DNA in living cells (Ma H et al. 2016), and enabled

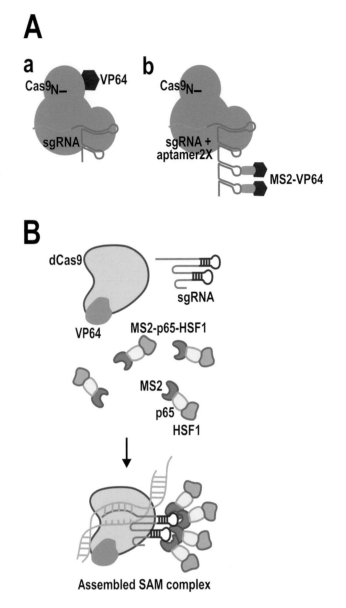

FIGURE 24.4 The MS2-based tethering and CRISPR-Cas system. (A) RNA-guided transcriptional activation: (a) to generate a Cas9N-fusion protein capable of transcriptional activation, the VP64 activation domain was directly tethered to the C terminus of Cas9_N_; (b) to generate sgRNA tethers capable of recruiting activation domains, two copies of the MS2 phage coat protein-binding RNA stem-loop were appended to the 3' end of the sgRNA and these chimeric sgRNAs were expressed together with Cas9_N_ and the MS2-VP64 fusion protein. (B) Structure-guided design and optimization of an RNA-guided transcription activation complex. Schematic of the three-component synergistic activation mediator system (SAM) that is the combination of sgRNA2.0, NLS-dCas9-VP64 and MS2-p65-HSF1. ([a] Reprinted by permission from Springer Nature, *Nat Biotechnol.*, Mali et al., Copyright 2013; [b] Reprinted by permission from Springer Nature, *Nature*, Konermann S et al., Copyright 2015.)

low-background visualization of target loci (Fu Yi et al. 2016; Shao et al. 2016; Wang Siyuan et al. 2016; Qin et al. 2017).

The MS2-based tethering contributed to the resolution of the crystal structure of the *S. aureus* Cas9, in complex with a sgRNA and its double-stranded DNA targets (Nishimasu

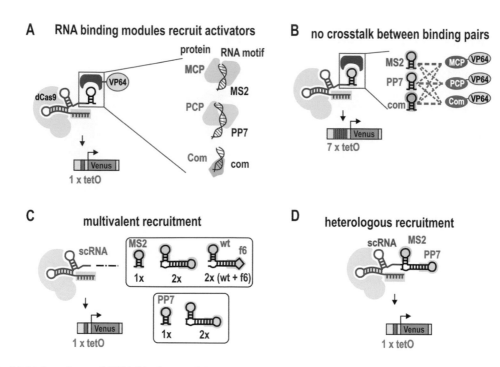

FIGURE 24.5 Multiple orthogonal RNA-binding modules can be used to construct CRISPR scaffolding RNAs. (A) scRNA constructs with MS2, PP7, or com RNA hairpins recruit their cognate RNA-binding proteins fused to VP64 to activate reporter gene expression in yeast. The MS2 and PP7 RNA hairpins bind at a dimer interface on their corresponding MCP- and PCP-binding partner proteins (Chao et al 2008), potentially recruiting two VP64 effectors to each RNA hairpin. (B) There is no significant crosstalk between mismatched pairs of scRNA sequences and non-cognate binding proteins. (C) Multivalent recruitment with two RNA hairpins connected by a double-stranded linker produces stronger reporter gene activation compared to single RNA hairpin recruitment domains. (D) A mixed MS2-PP7 scRNA constructed using the 23 double-stranded linker architecture recruits both MCP and PCP. (Adapted with permission of Cell Press from Zalatan JG et al. *Cell*. 2015;160:339–350.)

et al. 2015). Furthermore, Dahlman et al. (2015) developed a CRISPR-based method that used catalytically active Cas9 and distinct sgRNA constructs to knock out and activate different genes in the same cell. These sgRNAs, with 14- to 15-bp target sequences and MS2 binding loops, activated gene expression using an active *Streptococcus pyogenes* Cas9 nuclease, without inducing double-stranded breaks. Therefore, the MS2 stem-loops were involved as parts of the "dead RNAs" to perform orthogonal gene knockout and transcriptional activation in human cells.

The CRISPR/guide RNA technique, which involved screening of 16 MS2-mediated single-guide RNAs, led to a highly effective and target-specific Cas9 system that could serve as a novel HIV-latency-reversing therapeutic tool for the permanent elimination of HIV-1 latent reservoirs (Zhang Y et al. 2015). Furthermore, the MS2 coat fusions together with the MS2 stem-loops were applied by specific induction of endogenous viral restriction factors using CRISPR/Cas-derived transcriptional activators (Bogerd et al. 2015). The RNA-guided dCas9-VP64 reactivated HIV-1 in all latency models tested and did not lead to nonspecific global gene expression or adverse cellular toxicities, therefore providing an exciting new avenue for targeted reactivation of latent HIV (Bialek et al. 2016; Limsirichai et al. 2016; Saayman et al. 2016; Zhang Y et al. 2018). The successful application of the CRISPR-Cas9 approach against HIV-1 was reviewed in detail (Darcis et al. 2018; Wang Gang et al. 2018). Furthermore, the

CRISPR-mediated activation of endogenous BST-2/tetherin expression was shown to inhibit the wild-type HIV-1 production (Zhang Y et al. 2019).

Nelles et al. (2016) employed a catalytically inactive form of Cas9 to target endogenous RNA and, by way of fusion to GFP, provided a means of visualizing specific RNAs without the need to encode a recognition element.

The MS2-based tethering was employed by the generation of the CRISPR-Cas9-based "signal conductors" that regulated transcription of endogenous genes in response to external or internal signals of interest (Liu Y et al. 2016a,b). Furthermore, the CRISPR-Cas9 system was used to manipulate mammalian prion protein (Kaczmarczyk et al. 2016).

Concerning the immune activity of the CRISPR system against RNA phages, a breakthrough investigation was performed by Virginijus Šikšnys' excellent team (Tamulaitis et al. 2014). By their investigation of nucleic acid specificity and mechanism of CRISPR interference for the *Streptococcus thermophilus* complex, the authors found that the type III-A StCsm complex targeted RNA and not DNA. When expressed in *E. coli*, the StCsm complex restricted the phage MS2 in a Csm3 nuclease-dependent manner. It was stressed by the authors that the phage MS2 is a preferable model to investigate RNA targeting by the CRISPR-Cas system *in vivo* because no DNA intermediate is formed during the life cycle of this phage. The pCRISPR_MS2 plasmid carried the synthetic CRISPR array of five repeats interspaced by four

36-nucleotide spacers targeting correspondingly the maturation, lysis, coat, and replicase sequences of MS2 RNA. The data demonstrated clearly that the StCsm complex conveyed *in vivo* resistance to the RNA phage MS2 in the heterologous *E. coli* host (Tamulaitis et al. 2014).

Abudayyeh et al. (2016) further showed that the novel CRISPR-Cas effector C2c2, now classified as Cas13a (Shmakov et al. 2017), which was developed from the type-VI CRISPR/Cas adaptive immune system of the bacterium *Leptotrichia shahii*, demonstrated RNA-guided ribonuclease function. The C2c2 could therefore provide interference against RNA phage. The *in vitro* biochemical analysis showed that the C2c2 was guided by a single crRNA, a CRISPR RNA, and could be programmed to cleave single-stranded RNA targets carrying complementary protospacers. Moreover, coexpression of the C2c2 and a crRNA containing a 28-nucleotide spacer targeting the phage MS2 RNA conferred viral resistance to *E. coli*. To validate the interference activity of the enriched spacers, the four top-enriched spacers were cloned individually into the pLshC2c2 CRISPR arrays and observed a 3- to 4-$\log_{10}$ reduction in plaque formation. The 16 guides targeting distinct regions of the MS2 maturation gene were cloned. All 16 crRNAs mediated MS2 interference, indicating that the C2c2 can be effectively retargeted in a crRNA-dependent fashion to sites within the MS2 genome (Abudayyeh et al. 2016). This remarkable finding was reviewed operatively by Zhang Dandan et al. (2016).

It is noteworthy that the fine molecular architecture of the Cas13a, former C2c2, was deciphered recently (Liu L et al. 2017a,b).

The Cas13b, another type VI-B CRISPR-associated RNA-guided ribonuclease, was shown to achieve the MS2 resistance when heterologously expressed in *E. coli* (Smargon et al. 2017). The protection against MS2 infection was achieved further by the expression in *E. coli* of the Cas9 genes from both subtypes II-A and II-C, while guides targeted not only single-stranded, accessible regions but also those that formed apparently stable secondary structures in MS2 RNA (Strutt et al. 2018).

The MS2 and PP7 platforms participated in a set of further experimental CRISPR-Cas9 applications (Chavez et al. 2016; Hess et al. 2016; Hosogane et al. 2016; Kushwaha et al. 2016; Lebar and Jerala 2016; Xu et al. 2016; Braun et al. 2017; Ferry et al. 2017; Hass and Zappulla 2017; Koirala et al. 2017; Lundh et al. 2017; Park JJ et al. 2017; Powell et al. 2017; Schwarz KA et al. 2017; Sekiba et al. 2017; Xiong et al. 2017; Zhang T et al. 2017; Ade et al. 2018; Kunii et al. 2018; Lin D et al. 2018; Pyzocha and Chen 2018; Shrimp et al. 2018; Strother et al. 2018; Cai et al. 2019; Chen Xutao et al. 2019; Cunningham-Bryant et al. 2019; Devilder et al. 2019; Franklin et al. 2019; Han SW et al. 2019; Huang H et al. 2019; Malzahn et al. 2019; O'Geen et al. 2019; Román et al. 2019; Spille et al. 2019; Tran et al. 2019; Yu et al. 2019).

The participation of the MS2- and PP7-based tethering by the repurposing CRISPR-Cas9 for precision genome regulation and interrogation has been analyzed in detail in numerous reviews (Fujita and Fujii 2015; Heckl and Charpentier 2015; La Russa and Qi 2015; Moore JD 2015; Nelles et al. 2015; Saayman et al. 2015; Cano-Rodriguez and Rots 2016; Didovyk et al. 2016; Ding Y et al. 2016; Dominguez et al. 2016; Du and Qi 2016; Enríquez 2016; Genga et al. 2016; Jazurek et al. 2016; Lee HB et al. 2016; Nowak et al. 2016; Puchta 2016; Vera et al. 2016; Vora et al. 2016; Wang Haifeng et al. 2016; Alexander and Devaraj 2017; Brown A et al. 2017; Canver et al. 2017; Chira et al. 2017; Czapiński et al. 2017; Czaplinski 2017; Chen Meng and Qi 2017; Doetschman and Georgieva 2017; Joung et al. 2017; Komor et al. 2017; Liu Xuejun et al. 2017; Lo and Qi 2017; Marchisio and Huang 2017; Mitsunobu et al. 2017; Pineda et al. 2017; Puschnik et al. 2017; Richter et al. 2017; Shang et al. 2017; Adli 2018; Chen S et al. 2018; Plummer et al. 2018; Waryah et al. 2018; Zhou L et al. 2018; Zhu et al. 2018; O'Connell 2019; Pandey et al. 2019; Tadić et al. 2019; Wang F et al. 2019; Wu X et al. 2019; Xu X et al. 2019).

25 Epilogue

The greatest reward for doing is the opportunity to do more.

Jonas Salk

Everything is possible. The impossible just takes longer.

Dan Brown
Digital Fortress

STATISTICS

Figure 25.1 presents a rough estimate of the publication activity regarding the single-stranded RNA phages over 59 years, from their discovery in 1961 up to June 2019. These are not solely Medline-referenced papers, but printed products that have been collected during lifelong follow-up, as well as papers that have been found via Scholar or other internet resources.

The collected papers are devoted directly to the single-stranded RNA phage matter or are reviewing and referencing the original papers in question. The list of papers behind Figure 25.1 is far from complete. In addition, not all of the papers in this list have been referenced in the present book. Nevertheless, the list seems to properly reflect the global pattern of the *Leviviridae* phage expansion in scientific publications. It is striking that the total number of papers has grown markedly in the years from the early 2000s, after a strongly pronounced peak in the late 1960s–early 1970s and a following depression in the 1980–2000 period. Figure 25.1 shows that the early spurt in publishing activity in the 1960s was first connected with the rapid development of the classical *molecular biology* field. Then, the diagram specifies a special discipline called *gene engineering* at the beginning of the gene engineering era. This block of papers is constituted mostly from the applications of the mysterious Qβ replicase enzyme in the diagnostic tools (see the Diagnostics section in Chapter 18) and of the N-terminal part of the MS2 replicase gene in the generation of fused genes for the efficient expression in *E. coli*, with further purification and characterization of a large number of different proteins, both homologous and heterologous by their origin, as described in the Replicase fusions section of Chapter 19.

The permanent increase of the number of papers in the *ecology* and *nanotechnology* categories is in good agreement with the general observation concerning the total number of papers bearing the term "phage" (Henry and Debarbieux 2012). This estimation led to the conclusion that, in the past decades, two main fields have revitalized the interest in the phages-based approaches. First, the advent of nanotechnology that showed many promises including the rapid development of the virus-like particle (VLP) applications, as described in Chapters 20, 22, and 23. Second, the interest in the structural analysis of environmental samples, which is provoked not

only by the impressive reduction in the cost of DNA sequencing, but also by the broad involvement of the RNA phages in the environmental studies in their traditional role of favorite markers, models, and surrogates of biocontrol, in full accordance with the huge amount of data collected in Chapter 6. However, the RNA phages now offer a third active component by the growth of the total number of phage-derived papers; namely, the *imaging* field. As follows from Chapter 24, the MS2- and PP7-based tethering and imaging platforms contributed markedly to modern knowledge not only of the private cell life, but also of the mRNA metabolism in live biological subjects. As follows from Figure 25.1, the three identified directions of *ecology*, *nanotechnology*, and *imaging* are ensuring the total growth of the number of papers on the RNA phages in the last several years.

GOAL ACHIEVEMENTS

In 1965, on the fifth anniversary of the great discovery of the RNA phages, Norton D. Zinder, the "father of RNA phages," acknowledged that "it is certain that the whole story of these phages will be encompassed and, with this, there will then be one independent bit of genetic material about which we can say—we know and we understand" (Zinder 1965). Moreover, he hypothesized that "if one had to speculate upon which biological entity might be synthesized *de novo*, it would be a prime candidate." Despite their small size, the RNA phages offered a broad assortment of novel knowledge, first, on the astonishing combination of genetic functions with template functions and on the regulation of the phage-drawn events in space, time, and efficiency. It became true. although difficult to consider, that all regulative network is encoded by the same tiny genome, the level of knowledge about which seemed rather satisfying for the scientific community of those days in the rapidly growing molecular biology. Therefore, the RNA phage history appears as real history of molecular biology. In respect to how phages shaped modern biology, a review of Eric C. Keen (2015) is recommended. Now, when the finest details of the three-dimensional structures are resolved not only for the numerous RNA phages themselves, but also for the bacterial components, such as F pili, involved directly into the phage-host interplay, the metagenomic era is affording rather unexpected challenges from the RNA phage world.

For precisely this reason, the personal experiences of the real RNA phage fate-makers are of special interest and importance. Robert L. Sinsheimer, one of the great founders of the RNA phage field, narrated his story on the very beginning of this field (Sinsheimer 2015). Harvey F. Lodish (2013) presented his impressions from his groundbreaking time in the RNA phage studies. The great contribution of Norton N. Zinder

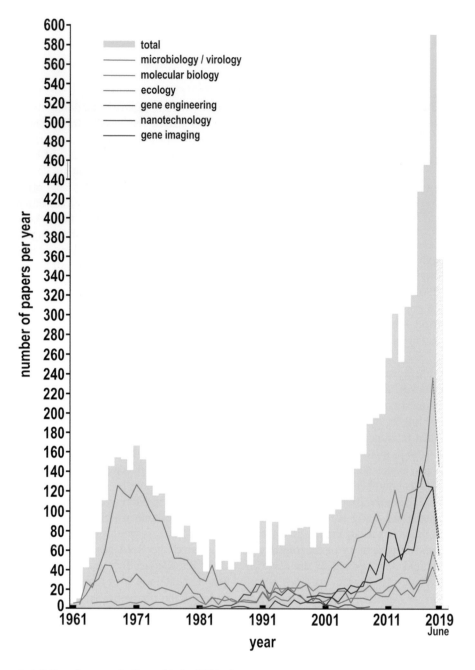

FIGURE 25.1   General statistics of papers dealing with the RNA phages.

was evaluated at that time by Kresge et al. (2011) and Harvey Lodish and Nina Fedoroff (2012). Concerning the historical role of the RNA phages in the global history of RNA viruses, a review of Daniel Kolakofsky (2015), one of the pioneers of the RNA phage studies, would be highly recommended. Witold Filipowicz (2017) published his reflections on the RNA phage studies in 1960s. And finally, Elmārs Grēns (2015), the "father of molecular biology" in Latvia, published his version of the early years in the RNA phage history.

In this context, Frederick M. Ausubel (2018) recently recalled Sol Spiegelman's seminar on phage Qβ replication *in vitro* as a major influence in his decision to apply to graduate programs in molecular biology rather than chemistry. A paper devoted to a "giant in medicine," Lucy Shapiro, one of

the authors of the first RNA-dependent RNA polymerase, or RNA replicase appeared recently (Neill 2019). In addition, Laura L. Mays Hoopes (2019) published a book about two great women of the RNA world, a dual biography of Joan Steitz and Jennifer Doudna.

VIROME

David E. Bradley (1965b) was one of the first or, possibly the first, to claim the idea that the RNA phages are the only phages with the genome and morphology close to that of non-bacterial viruses, a *rara avis* in the bacterial world. Richard M. Franklin's team were the next to write that the phage R17 "is the first bacteriophage to be assigned to this class [*of morphology arranged in icosahedral symmetry—PP*] which already includes one RNA animal virus, poliomyelitis virus; one RNA plant virus, TYMV, and one DNA animal virus, K rat virus" (Vasquez et al. 1966). This surprising evolutionary moment retains its intriguing character up to now. The capacities of the metagenomic era seem especially promising in the sense of the eternal question of why the single-stranded RNAs are much more popular in the nonbacterial than in the bacterial world.

The actual place of the RNA viruses, including the RNA phages, in the virosphere was defined systematically (Koonin et al. 2015; Krishnamurthy et al. 2016; Shi et al. 2016; Novik et al. 2018). Currently, the *Leviviridae* and *Cystoviridae* families remain the only two recognized families of prokaryotic RNA viruses. Siddharth R. Krishnamurthy and David Wang (2018) hypothesized recently that picobirnaviruses, which are bisegmented double-stranded RNA viruses commonly found in animal stool samples, are in fact prokaryotic RNA viruses, since they have never been propagated in cell culture or in an animal model. The authors identified and analyzed the genomes of 38 novel picobirnaviruses and determined that a classical bacterial sequence motif, the ribosomal binding site, is present in the 5′ untranslated regions of all of the novel as well as all previously published picobirnavirus sequences. Among all viruses, enrichment of the ribosome binding site motif was only observed in viral families that infect prokaryotes and not in eukaryotic infecting viral families (Krishnamurthy and Wang 2018). The hypothesis that picobirnaviruses are phages that belong to a novel RNA phage family with a high level of genomic diversity was supported experimentally by Adrianssens et al. (2018).

For the first time, the *Leviviridae* phages appeared as an intrinsic component of the viral metagenomic studies in the breakthrough paper of Edwards and Rohwer (2005). Later, it was generally acknowledged that the present metagenomic surveys of natural populations in different environments underestimated the RNA phage population (Rosario et al. 2009; Kristensen et al. 2010; Hyman and Abedon 2012; Abeles and Pride 2014; Goldhill and Turner 2014; Salmond and Fineran 2015; Temmam et al. 2015; Rastrojo and Alcamí 2017; Duerkop 2018; Kropinski 2018; Zhang YZ et al. 2018; Starr et al. 2019).

As explained in detail in the Metaviromics section of Chapter 2, Krishnamurthy et al. (2016) were the first to show that the *Leviviridae*-like RNA phages can be highly divergent from each other and from the previously described RNA phage prototypes. At the same time, Mang Shi et al. (2016) redefined at the invertebrate RNA virosphere and added 67 additional partial genomes to the collection of the levi-like RNA phages.

Nevertheless, it is highly important that, in parallel with the growing metagenomic survey, a search for the novel infectious RNA phages still continued. Thus, the novel *Leviviridae*-like phages with untypical appearance, namely, with 41–43 nm in size, were also found by the traditional phage identification methods, in piggery effluents tested on isolates of *Enterococcus faecalis*, *E. faecium*, and *E. gallinarum* for lytic phages (Mazaheri Nezhad Fard et al. 2010).

Therefore, the list of the RNA phages, as presented in Table 1.1 of Chapter 1, appears in the novel light as greatly limited, and the brilliant metagenomic methodology provokes fresh interest in the advanced search of the RNA phage representatives in ocean, soil, and animal ecosystems. Krishnamurthy and Wang (2017) presented thereupon a systematic roadmap for the investigation and accurate classification of viral dark matter, namely, the metagenomic sequences that originate from viruses but do not align to any reference virus sequences. This paper paved the first steps in the direction to the comprehensive defining of the virome with the single-stranded RNA phages as an essential part of it. The authors estimated the viral dark matter between 40% and 90% of the metagenomic sequences and identified three factors that contribute to the existence of viral dark matter: the divergence and length of virus sequences, the limitations of alignment-based classification, and limited representation of viruses in reference sequence databases. Thus, reducing the extent of viral dark matter, the knowledge of the single-stranded RNA phages will progress inevitably. Moreover, high capabilities of the novel sequencing approaches, such as single-cell RNA sequencing (Saliba et al. 2017) and nanopore sequencing (Viehweger et al. 2018; Harel et al. 2019; Wongsurawat et al. 2019) are inspiring scientists to perform total refreshing and resequencing of older data. In this connection, the general history of DNA sequencing, or sequence of sequencers, would be recommended (Heather and Chain 2016). The growing interest resulted in a special recent review on the RNA phages in the global metagenomic era (Callanan et al. 2018). Moreover, the modern metagenomics protocols are adjusted especially to the accurate isolation of the putative RNA phage genomes (Grasis 2018). Nevertheless, it is necessary to take into account real difficulties with the RNA phage detection by metagenomic analysis, which could appear with the RNA virus-search targeted protocols. Thus, Shkoporov et al. (2018) acknowledged that the attempt to artificially spike fecal samples with the phage Qβ did not result in recovery of any reads aligning to its genome. This suggested that although some larger rod-shaped plant RNA viruses of the *Virgaviridae* family were detected using the protocol, the latter failed to quantitatively recover the smaller RNA phage Qβ. Further, Shkoporov and Hill (2019) acknowledged that they were unable to detect the RNA phages in gut phage communities, likely because of low viral loads and unsuitable nucleic acid extraction procedures.

Concerning the basic virome problems, it is important to take into account that the advent of new biology produces tons of sequences in the terabyte range every day from sequencing centers (Bursteinas et al. 2016; Biji CL and Achuthsankar 2017). As the sequencing cost is reducing, the majority of life science laboratories can generate genomic data in terabyte scale and access a huge volume of data from public repositories, such as the European Bioinformatics Institute (EBI) or National Center for Biotechnology Information (NCBI). Therefore, the problems of data storage, compression, evaluation, and analysis are of great importance nowadays and provide additional insights to the biologist.

In this context, it is remarkable that the phage identification tools are elaborated, which uses supervised learning to classify metagenomic contigs as phage or non-phage sequences (Roux et al. 2015; Deaton et al. 2017; Kleiner et al. 2017; Meier-Kolthoff and Göker 2017; Ren J et al. 2017; Thannesberger et al. 2017; Zhao G et al. 2017; Amgarten et al. 2018; Zheng et al. 2019).

It should be noted that the RNA phages, among other viruses, remain an intrinsic subject of the endless debate as to whether or not "viruses are alive" (Koonin and Starokadomskyy 2016).

NOVEL VLP MODELS

As noted in the Metaviromics section of Chapter 2, Kaspars Tārs' team expressed a long list of the coat genes in *E. coli* of very different levi-like RNA genomes (Liekniņa et al. 2019). The authors performed the advanced BLAST analysis that revealed approximately 14 distinct levi-like RNA phage coat types, which was a noticeable increase from the three coat types known before. The authors selected 110 coat sequences from the metagenomic data to cover all coat groups and expressed in *E. coli* using a T7 promoter-driven system. The vast majority of the coats were produced in the expected high levels and only in very few cases was no expression detected. However, only about 60% of the coats were at least partially soluble, while the rest were found in inclusion bodies. In an effort to mitigate the issue, the insoluble proteins were propagated at 15°C that indeed rendered 85% of the previously insoluble coats at least partially soluble, and only six remained in inclusion bodies also at the lower temperature. In total, 80, or approximately 72%, of the soluble coats assembled into VLPs as confirmed by electron microscopy. In the majority of cases, the VLP morphology resembled that of the previously characterized RNA phage VLPs with an apparent spherical shape 28–30 nm in diameter that corresponded to a $T = 3$ icosahedral particle. However, notable deviations from the standard particle size and shape were not uncommon.

Some VLPs were noticeably bigger, reaching 35–40 nm in diameter, which could correspond to a $T = 4$ icosahedral particle. The two coats assembled into small particles approximately 18 nm in diameter with a presumed $T = 1$ icosahedral symmetry. A sizeable proportion of one of the coats appeared to have an elongated shape, while two coats demonstrated a mixture of $T = 3$ and $T = 1$ particles. The authors addressed the possible disulfide bonds, thermal stability, and potential RNA-protein interactions as the parameters crucial for the further nanotechnological application (Liekniņa et al. 2019).

This highly remarkable study demonstrated conclusively for the first time that the environmental viral sequences uncovered in metagenomic studies can be successfully employed to reconstruct virus-like particles in a laboratory setting with further nanotechnological goals. The 80 novel VLP platforms present a great contribution to the growing RNA phage applications.

On the other hand, Liu X et al. (2019) performed a powerful attempt to combine the RNA phage VLP approach with exosomes by the generation of novel cell-targeting and drug delivery devices. As known, the exosomes denote a naturally secreted nanoparticle family that carry RNA and protein cargos. Liu X et al. (2019) repurposed the exosomes for targeting peptide screening. To do this, the signal peptide region of Lamp2b, a membrane protein on the exosomes, was fused in the N-terminus with 10 aa-long random peptides, while the C-terminus of Lamp2b was fused with the MS2 coat protein. Then, the whole Lamp2b-MS2 coat open reading frame was further engineered to harbor a 3'UTR sequence of MS2 RNA. The resultant exosomes from engineered Lamp2b-MS2-coat expressing cells displayed the 10 aa peptides on the outside while containing the genetic information inside. By proof-of-principle experiments, the authors showed that the exosomes with different peptides could be preferentially distributed to different tissues. Furthermore, the target sequences for different tissues were enriched by some rounds of selection. This approach was termed *exosome display* and recommended for displaying and screening targeted peptides for the cells outside the capillary with condense barriers, like neurons in the brain (Liu X et al. 2019).

THREE-DIMENSIONAL STRUCTURE

The global impact of the structural elucidation of viruses is traditionally high (Reddy and Sansom 2016). Concerning the RNA phages, the three-dimensional interplay of all phage-derived components was excellently presented in a recent review by Rūmnieks and Tārs (2018). In addition to putative novel phages, the existing classical representatives have revealed unexpected capabilities of their capsids, such as their ability to transform into $T = 1$ (Freivalds et al. 2014), $T = 4$ (de Martín Garrido et al. 2019), or rod-like (Cielens et al. 2000) structures. Such studies have contributed worthwhile information toward understanding protein folding and virus assembly (Dykeman et al. 2014). Additional important directions for future studies involved the elucidation of the VLP surface properties in the context of the presence of internal RNA, as reported previously (Dika et al. 2014), and/or the external display of foreign sequences of different origins, such as oligonucleotides, peptides, sugars, and metal ions. Such studies would be helpful for the prediction of VLP characteristics in the context of aqueous and nonaqueous media, as well as under different biological conditions, which should facilitate the use of these particles as vaccines and/or gene therapy tools. The growing impact of the NMR in the high-resolution studies, in addition to x-ray crystallography and electron cryomicroscopy, must be also emphasized, where RNA phages always played a pioneering role (Cala-De Paepe et al. 2017; Gupta and Polenova 2018; Jaudzems et al. 2018, 2019; Kryshtafovych et al. 2018; Martial et al. 2018; Goldbourt 2019; Gupta et al. 2019; Loquet et al. 2019). This is also the time when novel ideas and approaches are appearing in the field of the macromolecule self-assembly (Cingil et al. 2017).

The icosahedral objects from the *Leviviridae* family, together with the representatives of the *Picobirnaviridae*, *Flaviviridae*, and *Picornaviridae* families, generated particular interest among theoreticians in anomalous small viral shells and simplest polyhedra with icosahedral symmetry (Pimonov et al.

2019). The role of the structural studies by the classification of virosphere, which is somehow neglected now, was accented by Nasir and Caetano-Anollés (2017) and highlighted recently by Jens H. Kuhn et al. (2019). The above-presented study of Tārs' team (Liekniņa et al. 2019), which resulted in an impressive set of novel RNA phage VLP carriers, paves the way for further unexpected discoveries in the three-dimensional field.

IMMEDIATE FUTURE

The ongoing tasks of the RNA phage VLP-based activities in the field of bionanotechnology are presented in Figure 25.2. The closest milestones provide the combination of both (i) decoration of the VLP scaffolds with molecules of interest and (ii) packaging of foreign material into VLP nanocontainers. It is noteworthy that Thomas and Barbara Hohn prophesied 50 years ago, in distant 1970: "Phage assembly might open an interesting possibility for biotechnology. It is possible to pack heterologous RNA into the phage capsid. Once the techniques for producing infective particles are better controlled this might become an interesting tool for providing a cell with mRNA at the experimentators' will" (Hohn and Hohn 1970).

Without doubt, vaccines must remain the most productive area for the application of the RNA phage VLPs. The packaging and delivery methodologies will progress in parallel with the development of other promising VLP carriers, such as hepatitis B core antigen and plant virus VLPs.

It is also noteworthy in this context that the RNA phages offered nowadays are an unexpected but highly favorable alternative to filamentous phages for the peptide display.

It would be intriguing to cheer the RNA phage VLPs as a tool for the future generation of nanomachines, understanding the nanomachines as self-maintained nanodevices ranging in size from 1 nm to 1 μm. The real perspectives of putative nanoscale engineering on the RNA phage VLPs as natural prefabricated scaffolds to contain molecules in precisely defined arrays were discussed in the previous nanotechnological chapters. The future also looks pretty good for the RNA-phage VLP-derived materials, including chemical strategies and bioconjugation technologies, which are applicable to change both the interior and exterior surfaces of capsids for the unlimited incorporation of organic and inorganic compounds.

Further development of VLPs with defined antigenic and immunogenic properties, as well as VLPs with improved packaging and targeting capabilities, will doubtless create novel viral nanotechnology applications and lead to construction of nanomachines with rationally designed characteristics. As described in the Lattices section of Chapter 21, DNA may play a remarkable role of a specific scaffold to arrange the RNA phage VLPs into one-dimensional arrays with precise nanoscale positioning. Another sort of nanomachine could exploit functional RNAs that explicitly take advantage of cotranscriptionally generated transient or metastable structures (Kobori et al. 2019).

The RNA phage VLPs paved the way for other viral models to gain the powerful role in successful nanotechnological

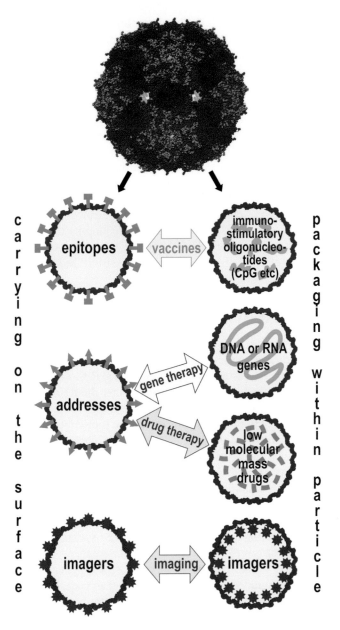

FIGURE 25.2 General milestones of the development of RNA phage VLPs as nanotools. The three-dimensional structure of the MS2 phage is presented at the top. (Reprinted with permission of S. Karger AG, Basel from Pumpens P et al. *Intervirology.* 2016;59:74–110.)

applications, as, for example, for the VLPs of cucumber mosaic virus (Zeltins et al. 2017; Cabral-Miranda et al. 2018; Mohsen et al. 2019a,b; Thoms et al. 2019). The RNA phages provoked much more complicated phages, such as the phage T4, to try the role of the vaccine delivery nanovehicle (Tao et al. 2019).

Numerous modern conjugation methods have been initiated by the RNA phage pioneering impulse (Zhang Lin et al. 2015; Sonawane and Nimse 2016; Sciore and Marsh 2017; Steele et al. 2017; Donaldson et al. 2018; Andersson et al. 2019).

The excellent book of Nicole F. Steinmetz and Marianne Manchester (2011) introduced readers to the role of VLPs in future nanotechnologies. This book was published by CRC

Press/ Taylor & Francis Group as an eBook in 2019. Recently, Steinmetz (2019) presented novel arguments for the future development of the VLP-based nanoengineering.

The persistent role of the RNA phages in the ecological field of science and practice should be highlighted. The systematic modeling of the behavior of different viruses, including RNA phages, in media filtration and flocculation is highly encouraging (Baltus et al. 2017). The marked contribution to ecology is confirmed by the generation of nanostructured electrochemical biosensors capable of the label-free detection of water- and foodborne pathogens including the RNA phages (Ertürk and Lood 2018; Reta et al. 2018). It is noteworthy that the phage MS2 remains as a permanent member of laboratory tests, modeling and statistical evaluation by the search of extraterrestrial life. Thus, the phage MS2 was used in the program elaborated for the transfer of unsterilized material from Mars to Phobos (Patel et al 2019).

BACTERIAL IMMUNITY

As described in the CRISPR-Cas9 section of Chapter 24, the MS2 tethering system contributed markedly to the experimental approaches that involved the CRISPR-Cas methodology. The logical question arose of whether the RNA phages themselves would be the subject of the CRISPR-Cas immunity action. After identification of reverse transcriptase activity in prokaryotes, Kojima and Kanehisa (2008) surmised the possible capability of the CRISPR-Cas system to provide immunity against RNA phages via cDNA synthesis. To investigate this possibility, the authors searched spacers derived from the RNA phages but without any access, and arrived at the conclusion that there were little sequence data on the phages infecting bacteria with CRISPR-Cas system in whose genomes the reverse transcriptase genes were found.

In their exhaustive survey of the CRISPR system as the small RNA-guided defense in bacteria and archaea, Karginov and Hannon (2010) established the fact that the corresponding spacers were detected only from phages with DNA genomes, but again emphasized that any conclusions based upon this observation must be tempered by the relative scarcity of RNA phage sequences.

The principal ability of the CRISPR-Cas immunity to restrict the phage MS2 was demonstrated experimentally (Tamulaitis et al. 2014; Abudayyeh et al. 2016; Kazlauskiene et al. 2016). However, although the *Streptococcus thermophilus* type III-A CRISPR-Cas system restricted the phage MS2 *in vivo* and cut RNA *in vitro,* the original CRISPR array spacers matched DNA and not RNA phages. It was concluded that the CRISPR-Cas is an efficient fail-safe mechanism for degradation of phage mRNA and transcriptionally coupled DNA while maintaining host genome integrity (Kazlauskiene et al. 2016).

The examination of bacterial genomes has revealed a class of the CRISPR-associated coding regions with Cas1 fused to a putative reverse transcriptase and raised the possibility of a concerted mechanism of spacer acquisition involving reverse transcription of RNA to DNA: a potentially host-beneficial

mechanism for RNA-to-DNA information flow and incorporation of new spacers directly from RNA (Silas et al. 2016, 2017). This finding made it possible to speculate that the CRISPR system goes retro (Marraffini and Sontheimer 2010; Sontheimer and Marraffini 2016). Again, due to limited knowledge on the abundance and distribution of RNA phages and other RNA parasites, with the vast majority restricted to the *Escherichia* and *Pseudomonas* genera, the solution of the role of spacers in the reverse transcriptase-associated CRISPR loci among natural populations of bacteria and their environment was left for further massive search for RNA phages (Silas et al. 2016). Up to now, the limited number of known RNA phages appear rather as key actors waiting for their appearance on the stage by the evaluation of the bacterial immune system.

Using the phage MS2 infection, the CRISPR spacer acquisition was demonstrated as a specific ability of *E. coli* cells to capture and convert intracellular RNAs into DNA, enabling DNA-based storage of transcriptional information (Schmidt et al. 2018). Thus, upon MS2 infection of the FsRT-Cas1-Cas2-expressing *E. coli* cells, the novel MS2-aligning spacers sampled from throughout the MS2 genome were observed readily, whereas the MS2-aligning spacers shared no sequence similarity with the plasmid or host genome, confirming their specificity. Therefore, the FsRT-Cas1-Cas2 enabled spacer acquisition directly from a foreign RNA, thereby providing a molecular memory of an invading phage.

The ability of the CRISPR-Cas system to protect bacteria against the RNA phages inspired a generation of plants resistant to RNA viruses: tomato yellow leaf curl virus (Tashkandi et al. 2018) and potato virus Y (Zhan et al. 2019).

Recently, the phage MS2 coat-operator complex was used for the generation of non-integrating murine leukemia virus (MLV)-based CRISPR-Cas9 all-in-one particles for targeted gene knockout (Knopp et al. 2018). The authors achieved efficient codelivery of nonviral *SpCas9* mRNA and sgRNA transcripts into target cells by replacement of retroviral components with analogous parts of the MS2 phage packaging machinery. The proof of concept was obtained by efficient knockout of a surrogate CRISPR-Cas9 reporter gene in human and murine cell lines, and other successful applications followed. The same MS2 coat-operator interaction was employed by the construction of a system able to package up to 100 copies of *Staphylococcus aureus Cas9* (*SaCas9*) mRNA in each lentivirus-like bionanoparticle (Lu et al. 2019). The *SaCas9* lentivirus VLPs mediated transient *SaCas9* expression and achieved highly efficient genome editing in the presence of guide RNA.

The diversity, modularity, and efficacy of the CRISPR-Cas systems were recommended recently as global driving forces of the current biotechnological revolution (Knott and Doudna 2018; Lau 2018; Lau and Suh 2018; Liu H et al. 2018; Lone et al. 2018; Zezulin and Musunuru 2018; Koonin 2019; Liu B et al. 2019).

It is noteworthy that the phage Qβ infection was used by the primary characterization of the *E. coli* BREX system, a poorly characterized prokaryote defense system, after well-known restriction-modification and CRISPR-Cas immunity systems (Gordeeva et al. 2019).

The dark-matter space of the CRISPR spacers that have been found for a large majority of reverse transcriptase-Cas1-derived spacers suggested the diversity of the RNA phages and RNA mobile genetic elements that remain to be discovered (McGinn and Marraffini 2019).

PHAGE THERAPY

Attempts to use lytic phages as a treatment to suppress bacterial infections have long history. However, this approach was rapidly overlooked by the emergence of antibiotics. In the twenty-first century, when antibiotic resistance in pathogenic bacteria has become a growing problem in medicine, interest in the putative phage therapy applications was renewed. Although the RNA phage lysins were proclaimed in due time as novel phage antibiotics, as described in the Protein antibiotics section of Chapter 14, direct use of the RNA phages as phage therapy tools might raise many serious questions because of their narrow host list, limited lytic capacities, and rapid emergence of resistant bacteria.

Nevertheless, Hosseinidoust et al. (2011) tried the phage MS2, among five different phage candidates also including the phages PRD1, P22, PR772, and T4, by coating of antimicrobial model surfaces and further characterization of the gained results with the most current x-ray photoelectron spectroscopy and atomic force microscopy methods. Remarkably, symmetric phages were found to be a better choice for antibacterial surfaces compared to more asymmetric tailed phages. The immobilized phages were found to disrupt the membranes of attached bacteria and were thus proposed as possible candidates for the coating of antimicrobial surfaces. Thus, the coated MS2 was more effective than T4 and PR772 among the tested *E. coli* phages. However, when free phages were tested, both tailed phages were more effective in lysis of bacteria. It was concluded that the phage MS2 exhibits superior capture efficiencies for their host bacteria, while coated on the surface.

In his review on phage therapy in modern medicine, John Evans (2012) emphasized the importance of studies on the ability of phages to respond to bacterial mutations that could give them favorable resistance for the growing phage therapy field, and cited a remarkable previous study (Kashiwagi and Yomo 2011). According to the latter, both the bacteria and phage went through multiple mutations throughout the observation. The *E. coli* cells mutated first to become partially resistant to Qβ. However, the phage Qβ was able then quickly to mutate advantageously, allowing it to better act against *E. coli*. This arms race, or *Red Queen hypothesis*, HAS caused these two organisms to continue to adapt to each other in order to improve their situation (Evans 2012).

The extensive review on the growing use of phage therapy in aquaculture (Oliveira et al. 2012) stated that the male-specific RNA phages have already been associated with shellfish, but their potential as therapeutic agents to control bacteria that caused infections in bivalves or that contaminated this food product has not yet been explored.

Theoretically, the phage therapy approach could be used not only to prevent bacterial infection, but also in virus wars, i.e., using one virus to block the spread of another. Paff et al. (2016) modeled this running start using the phage Qβ as a lytic virus and the phage f1 as a therapeutic virus.

In the most recent review on phage therapy, Chen et al. (2019) concluded reasonably that phage therapy, together with phage nanoparticle vaccination, could be the two promising strategies to address the emergence of multi-antibiotic-resistant bacterial pathogens and their continuing spread in the population. There are, however, limitations to the naturally occurring phages but, fortunately, recent progress in phage genome engineering promises to overcome these limits, such as expanding phage host range to facilitate phage therapy, and disrupting the immunodominant epitope of phage capsids to eliminate immune response against phages (Chen et al. 2019).

Some specific advantages of the RNA phages were recalled in phage therapy reviews. When phage therapy for respiratory infections was addressed (Chang et al. 2018), it was noted that the phage MS2 was the most robust during aerosolization. These data were acquired earlier by Turgeon et al. (2014, 2016), as mentioned in the Desiccation, humidity, and aerosols section of Chapter 6. Although not directly pointing to the RNA phages, the recent general phage therapy reviews did not exclude the involvement of leviviruses in the growing theoretical and possibly pharmacological studies in this direction (Forde and Hill 2018; Krut and Bekeredjian-Ding 2018; Huh et al. 2019). Moreover, in their extensive review on phage-derived antibacterials, Kim BO et al. (2019) directly named the phages MS2 and PP7 as putative targets of investigation in the phage therapy field because of the very high level of knowledge about them.

The phage PP7 was employed by the elaboration of a straightforward protocol to assess the antibacterial efficacy of the phages toward the opportunistic human pathogen *Pseudomonas aeruginosa*, which was based on the systemic infection model using the fruit fly, *Drosophila melanogaster* (Jang et al. 2019).

On the other hand, the connection of human diseases with phages is attracting more attention. The suggestion that bacterial viruses have different ways to directly and indirectly interact with eukaryotic cells and proteins, leading to human diseases, were thoroughly reviewed (Tetz and Tetz 2018). In this connection, Lawrence et al. (2019) urged to look for more appropriate techniques by the search of RNA viruses, including RNA phages. In addition, Vandamme and Mortelmans (2019) published an exciting story about the century of phage research and applications, including phage therapy.

PHAGE: THERAPY, APPLICATIONS, AND RESEARCH

The humble optimism about the forthcoming flourishing of the phage field was strengthened suddenly by the announcement of the novel specialized peer-reviewed journal *PHAGE: Therapy, Applications, and Research* on March 14, 2019. Applications in medicine, agriculture, aquaculture, veterinary applications, animal production, food safety, and food production are welcome. Per the publisher, Mary Ann Liebert, Inc., *PHAGE* "will be published quarterly online and in print and will serve as the cornerstone publication of phage research with the goal of solidifying, growing, and unifying its emerging community of innovative researchers and clinicians. A preview issue of the journal was planned for summer 2019." Among other goals, the journal planned to publish investigations on the life cycle of phages, capsid structure and morphology, genome organization and content, phage transcriptional takeover and regulation, roles of phages in natural ecosystems, impacts of phages on bacterial/eukaryotic physiology, immune cell modulation by phages, genetic engineering of phages, impact and exploitation of CRISPR-Cas systems, phages as vectors for gene delivery, vaccine development, phage-based diagnostics, phage proteins and enzymes, phage lysins, phages in food production and safety, nanotechnology and materials science, drug, gene, and antibody delivery, and clinical and animal trials. These are the topics that are also the subjects of the current book.

At last, by saying *Goodbye*! it remains to recommend to our indulgent reader to think for a short time like a phage, in accordance with the advice of the brilliant Merry Youle (2017) for *Thinking Like a Phage*. If you succeed, try to take part in the *secret social lives of viruses*, following the recommendation of Elie Dolgin (2019).

Author Index

Scientific research is one of the most exciting and rewarding of occupations.

Frederick Sanger

A

Aanei IL, 452, 466, 527
Aanes KJ, 116
Aas PA, 234
Abbaszadegan M, 133, 138, 148, 153, 548, 162
Abd-Elmaksoud S, 127
Abdelmohsen K, 311, 545
Abdelzaher AM, 107
Abe H, 130
Abebe LS, 139
Abedon ST, 29, 555
Abel K, 544
Abin M, 144
Aboubakr HA, 162
Aboul-ela F, 313
Abraham AK, 343, 427
Abrahams JP, 471
Abrams WR, 367
Abriata LA, 487
Abruzzi KC, 538
Abu-Bakar A, 149
Abudayyeh OO, 549, 550, 551, 558
Acca FE, 304
Accornero N, 542, 545
Achame EM, 543
Achtman M, 51
Achtman M, 51, 52, 61, 74, 98
Achuthsankar SN, 555
Acosta E, 418
Adair BD, 64
Adam V, 529
Adams CJ, 354, 359, 470, 518, 537
Adams JM, 1, 17, 19, 21, 81, 175, 186, 192, 211,
    214, 218, 220, 221, 237, 238, 249,
    319, 323
Adams MJ, 29, 30, 31
Adams SE, 436
Adams TD, 235
Adamson A, 545
Ade CM, 551
Adegoke AA, 132
Adelberg EA, 1, 53
Adeleke R, 106
Adeli K, 543
Ades EW, 165
Adesina OA, 234
Adeyemo A, 235
Adham SS, 120, 156
Adhikari A, 144, 163
Adhikari M, 311, 312
Adhin MR, 20, 40, 225, 227, 228, 294, 295, 299,
    310, 339, 424, 431
Adin A, 145, 449
Adler K, 280
Adli M, 551
Adolphsen J, 156
Aeppli M, 148
Afanasieva E, 543
Afonina E, 325

Afrooz ARMN, 153
Agabian N, 64
Agabian-Keshishian N, 55, 64, 67
Agafonov AP, 517
Agafonov DE, 538
Agans KN, 517
Agarwal A, 202
Agboluaje M, 96
Agnellini P, 489, 501, 518
Aguilar M, 117
Aguirre J, 418
Agulló-Barceló M, 142
Ahmad F, 113, 558
Ahmad V, 517
Ahmadian R, 149
Ahmed W, 164
Ahn C, 310
Ahn SH, 539
Ahn SJ, 188
Aho S, 160
Ahola T, 544
Ahsan KB, 538
Aigner S, 548
Aime S, 450
Air GM, 291
Aizer A, 543
Ajabali AAA, 197
Ajamian L, 544
Ajioka JW, 299, 425
Akaike Y, 543
Akbari M, 234
Åkerman L, 235
Akers TG, 143
Akhavan O, 131
Akhter M, 130
Akiba M, 104, 117
Akiyama H, 544
Akopjana I, 42, 73, 192, 243, 250, 476, 477, 486,
    556, 557
Akusjärvi G, 538
Akutagawa M, 129
Al-Barwani F, 489
Al-Bitar L, 460
Al-Maghrebi M, 539
Albernaz FP, 197
Albers M, 316
Albert T, 141
Alberti A, 203
Albihn A, 137
Alblas J, 227, 228
Albrecht J, 314
Albrecht R, 487, 556
Albuquerque SC, 197
Alcazar-Roman AR, 543
Alcid EA, 546
Alcock R, 163
Aldaye FA, 546
Alderson R, 306
Aleshkina LA, 217
Alexander JM, 545

Alexander L, 235
Alexander SC, 551
Algranati ID, 343
Alhmidi H, 164, 165
Alhusaini N, 538
Ali IK, 538
Ali M, 164
Ali Azam T, 280
Aligeti M, 544
Alimbarova LM, 378
Alimov AP, 339, 400
Aljabali AAA, 518
Alkawareek MY, 132
Allard AK, 157
Allen C, 539
Allende A, 121
Allenspach EJ, 493
Allison DP, 54
Allman AM, 356
Allmann E, 117
Allwood PB, 120, 159, 161
Almeida A, 130
Almeida DO, 166
Almeida FC, 197
Almeida-Marrero V, 529
Almo SC, 360, 361, 470, 546
Almond JW, 436
Aloia AL, 544
Alonso JM, 529
Alonso MC, 163
Alper T, 132
Alshaer W, 518
Alshraiedeh NH, 132
Altamirano J, 202
Althauser M, 15
Althof N, 162
Altintas Z, 108, 109, 312
Altman GG, 160
Alum A, 148, 153
Alvarez ME, 117
Alvarez P, 130
Alvarez PJ, 127, 130, 131
Álvarez-Benedicto E, 193, 207
Álvarez de Eulate E, 133
Alvarez-Lorenzo C, 131
Alvord WG, 543
Amarasiri M, 113
Amati P, 130
Ambroise J, 108
Ambu S, 117
Ameling S, 235
Ames BN, 176
Amgarten D, 556
Amicosante M, 427
Amir N, 139
Amirfakhryan H, 501
Amit R, 234
Amitai S, 85, 183, 253
Ammann J, 23, 367
Ammer G, 542

Ammer S, 55
Amorós I, 166
Amos LA, 17, 192
Amour C, 203
Amoussouvi A, 543
Ampenberger F, 491
Ampuero M, 203
Amster-Choder O, 543, 548
An Q, 397, 398
Anaya-Plaza E, 529
Anderegg JW, 189, 190
Anderer FA, 306
Anders R, 118, 151
Andersen ES, 548
Andersen GR, 268, 426
Andersen JA, 284
Anderson C, 15
Anderson CW, 19, 20, 176, 183, 184, 291, 294,
        327, 338
Anderson DG, 456
Anderson DJ, 164
Anderson EA, 529
Anderson EL, 234
Anderson ES, 50, 81
Anderson GP, 304
Anderson JS, 323
Anderson MA, 150, 153, 292, 463
Anderson N, 106
Anderson NG, 42
Anderson P, 313
Anderson TF, 47, 54, 70
Anderson WB, 153
Andersson AC, 514, 557
Ando A, 13, 23, 37, 39, 40, 99, 101, 102, 103,
        104, 112, 123, 145, 163, 187, 302
Andrade RC, 116
Andrade RV, 116
Andradottir HO, 152
Andreas LB, 486, 556
Andreasen N, 513
Andreassi C, 542
Andreola F, 234
Andrews CJ, 356
Andrews JO, 544
Andries O, 540
Androshchuk IA, 379
Androsov VV, 67
Androulaki B, 134
Androvic P, 235
Andryshak D, 159
Ang CYL, 114
Angelescu DG, 199
Angelin A, 157
Angenent LT, 40, 114, 144
Angénieux C, 234
Angus JC, 191
Ankoudinova I, 234
Anobom CD, 197
Anokhina MM, 538
Anthony KG, 50
Antoine MD, 198
Antolović V, 537, 543
Antonenko SV, 378
Antony A, 155
Anufrieva O, 542
Anwar D, 144
Anzaghe M, 501
Aoi T, 13, 19, 35, 99, 101, 102, 167, 184, 186, 187,
        253, 302
Aoki H, 517
Apgar J, 211

Apirion D, 85, 210, 313
Appert J, 144
Appleton H, 159
Appleyard G, 436
Apte MS, 543
Arai H, 404
Aranha-Creado H, 165
Araud E, 202
Araujo R, 115, 116
Araujo RM, 113
Araujo RN, 113
Aravantinou AF, 149, 151
Araya CL, 356
Arber W, 66
Arenas-Huertero C, 543
Arendt H, 513
Arentzen R, 324
Arfin SM, 317
Argetsinger JE, 1, 17, 19, 71, 175, 178, 185, 186,
        192, 193, 237
Argetsinger-Steitz JE, 18, 191, 195, 221, 238
Argos P, 285
Ãrgule D, 462
Argyrakis MP, 82
Argyrakis-Vomvoyannis MP, 82
Arhancet JP, 169, 527
Arhancet JPH, 169, 527
Arib G, 543
Aridi S, 138
Arkatkar T, 493
Arkhangelsky E, 153, 154
Arluison V, 280, 281, 545
Armanious A, 148
Armbruster MA, 226
Armentrout SA, 318
Armon R, 112, 113, 115, 120, 136, 148
Armstrong AM, 133
Armstrong GD, 14, 59
Armstrong-Major J, 331
Arnal C, 145, 154
Arndt H, 140
Arneodo AJ, 544
Arnold AJ, 132
Arnold S, 109, 200
Arnold SLM, 203
Arnon R, 309
Aronino R, 153
Arora M, 540
Arora R, 224, 227, 422, 423
Arredondo-Hernandez LJ, 116
Arriaga EA, 200
Arribas M, 396, 415, 417, 418
Arroyo-Solera I, 18
Arruda I, 153
Artman M, 372
Artus-Revel CG, 548
Arvai AS, 64
Aryan Z, 501
Asahina A, 124
Asami M, 104, 117
Asami T, 121
Asano Y, 538
Asensio MA, 463, 471, 532
Asghari A, 125
Ashbolt NJ, 12, 115
Ashcroft AE, 460, 461, 474
Ashe MP, 543
Ashley CE, 122, 464, 480, 518, 527
Ashraf SI, 542
Ashton T, 198
Ashwood ER, 515

Assavasilavasukul P, 539
Assfalg-Machleidt I, 427
Assier E, 513
Assis ASF, 166
Asteriadis GT, 226
Astier-Manifacier S, 286
Ataide SF, 548
Atala A, 324
Atala ZP, 324
Atar O, 234
Athanikar JN, 543
Athar MA, 517
Atherton JG, 148
Atkin SL, 235
Atkins J, 252, 294, 296
Atkins JF, 19, 20, 176, 181, 183, 184, 222, 291,
        294, 327, 338
Atmar RL, 539
Attinti R, 148
Auchincloss A, 555
Audoly G, 233
Auerswald EA, 427
Aufreiter E, 50, 98
August JT, 1, 13, 23, 29, 178, 212, 257, 258, 260,
        262, 263, 272, 273, 274, 275, 276,
        286, 315, 365, 366, 383, 386
Ault J, 232, 207
Aumiller WM, 518
Ausländer S, 545
Averell PM, 455
Avisar D, 129
Aviv H, 318
Aviv O, 136, 137, 139
Avogaro L, 543
Avota E, 394
Avots A, 225
Avots A, 225, 294
Avots AI, 225
Aw TG, 18, 19, 189, 205
Awasthi LP, 89
Axblom C, 438, 467
Aydt CM, 433, 540
Ayginin AA, 517
Ayliffe GA, 163
Ayres PA, 112
Azad MTA, 540
Azad TA, 540
Azegami T, 501
Azevedo MA, 457
Azizi H, 543
Aznar R, 162
Azzalin CM, 548

B

Ba T, 127
Baan RA, 314
Babb JR, 163
Babic A, 61
Babichenko S, 109
Babitzke P, 352
Babkina GT, 334
Bacharach E, 539
Bacher JM, 169
Bächi T, 458
Bachmann M, 501, 513
Bachmann MF, 235, 435, 436, 447, 459, 489,
        460, 462, 489, 490, 491, 492, 501,
        512, 513, 518
Bachrach U, 138
Backendorf C, 328

Backlund MP, 542
Bader DLV, 461
Badgett MR, 414, 536
Badgley BD, 113
Badiaga S, 203
Badireddy AR, 130, 131, 558
Badran AH, 551
Badr CE, 543
Bae EH, 234
Bae HW, 188, 234, 559
Bae J, 120
Bae KS, 125
Bae SW, 223
Baesi K, 517
Baggi F, 120
Baggs E, 538
Bahar Halpern K, 544
Bahlman JA, 153
Bahnfleth W, 127
Bahramian MB, 328
Bahrami S, 518, 529
Bai HM, 501
Bai Y, 539, 543
Baidya N, 354, 355
Bailey D, 203
Bailey ES, 106
Bailey KL, 543
Bailey M, 165
Bailey RD, 234
Bairwa M, 501
Bajaj S, 518
Bajew S, 543
Bajon E, 547
Bąk M, 42
Bakanga F, 154
Baker CS, 352
Baker M, 202
Baker SL, 546
Bakhshayeshi M, 166
Bakshi A, 538
Bakstad D, 545
Bal A, 42
Balas MM, 548
Bałazy A, 163
Balbinder E, 53
Bales RC, 120
Baliban SM, 131, 140
Balkema-Buschmann A, 202
Balkin H, 137
Balklava Z, 356, 411, 423
Ball CS, 201
Ball DA, 548
Ball LA, 22, 25, 179, 218, 237, 249, 319, 331, 339
Ball LM, 130
Ballantyne S, 538
Ballesté E, 116
Balluff C, 164
Balluz SA, 119
Balmori E, 234
Baltimore D, 23, 29
Baltus RE, 558
Balyozova D, 136
Balzer D, 54
Banas S, 113, 146, 159
Bancaud A, 544
Bancroft JB, 195, 197
Bandín I, 131
Bandle E, 261
Bandle EF, 187, 413, 419
Banerjee A, 533

Banerjee AK, 23, 258, 260, 262, 261, 267, 272, 274, 286, 365, 383, 386
Banerjee D, 453
Banerjee M, 201, 518
Banerji S, 136
Bangala Y, 490
Bange A, 310
Bann DV, 545
Bannai H, 543
Bannert H, 518
Bansho Y, 426
Bansode RR, 161
Bao L, 517
Bao Q, 489
Barajas BC, 544
Barakat TS, 543
Barbeau B, 136, 202
Barberis M, 543
Barbet-Massin E, 486, 556
Barbieri L, 314
Barbot C, 145
Barbre C, 200
Barceló Culleres D, 121
Barcena C, 549
Bárcena J, 489
Barclay L, 202
Bardwell VJ, 536, 537
Barinskii IF, 377, 378
Barkallah I, 132
Barkefors I, 548
Barker AM, 73, 243, 476, 477
Barmada SJ, 548
Barnett B, 310
Barnini S, 427
Baron AJ, 354, 359
Baron CS, 53
Barouch D, 42, 105, 163, 171, 407, 555
Barr JN, 73, 243, 476, 477
Barra J, 540
Barrell BG, 22, 176, 211, 212, 218, 220, 291, 315, 326
Barrera I, 279
Barrett M, 121
Barrios ME, 113, 166
Barron AR, 128
Barron LD, 113, 193
Barrs VR, 42
Barry KH, 543
Barta A, 543
Bartel H, 123
Barthel A, 130, 218
Bartocha W, 123
Bartsch C, 160
Bartz JA, 154
Barutello G, 493
Basak AK, 435
Baslé A, 487, 556
Basnak G, 474
Bass MB, 153
Bassel BA, 343, 344
Bassel BA Jr, 210, 212
Bassell GJ, 545
Bassik MC, 551
Bastien M, 203
Bastos RK, 116
Basu R, 493
Basu SK, 543
Basus VJ, 193
Basyuk E, 155, 541, 543
Batan D, 548
Bateman BW, 112, 113

Bathe M, 548
Batschelet E, 413
Batten K, 234
Battich N, 543
Battigelli DA, 114
Batzer G, 137
Bauer C, 235
Bauer DE, 551
Bauer GJ, 400
Bauer K, 325
Bauer KE, 542, 548
Bauer M, 459, 460, 489, 490
Baulcombe D, 285
Bauman V, 25, 76, 77, 80, 85, 91, 92, 93, 371, 372, 383
Bauman VR, 76
Baumanis V, 76, 91, 439, 441
Baumann S, 539
Baumann T, 153
Baumann U, 342
Baumanns S, 116
Baumgärtel V, 544
Baumgarten H, 438
Bausum HT, 143
Bavor HJ, 147
Bayer M, 54
Bazos I, 139
Beach DL, 541
Beall SG, 203
Bean CL, 137
Beard JP, 85, 295
Beard MR, 544
Beaudouin J, 538
Becerra S, 538
Bechanko R, 138
Beck C, 543
Beck CM, 64
Beck DL, 234
Beck E, 490, 518
Beck ET, 202
Beck NK, 107
Beck SE, 124, 129
Becker A, 283
Becker JT, 544
Becker S, 203
Beckett D, 115, 348
Beckham C, 543
Beeh KM, 490, 518
Beekwilder J, 40, 114, 169, 228, 335
Beekwilder MJ, 40, 113, 114, 169, 228, 335
Beemon KL, 546
Beer H, 1, 47, 74, 75
Beerendonk EF, 125
Beerli RR, 490, 501
Beeson J, 184
Beg AA, 548
Behe MJ, 416
Behl R, 202
Behlen LS, 354, 355
Behnke J, 143
Behrens CR, 451, 527
Behrens RT, 544
Behrsing O, 438
Behzadnia N, 538
Beich-Frandsen M, 281
Beigelman L, 470
Beisel CL, 97
Bekele A, 144
Bekele AZ, 131
Bekeredjian-Ding I, 559
Belanche-Muñoz L, 113, 116

Belancio VP, 551
Beland FA, 234
Belaya K, 542
BelBruno JJ, 109
Beld M, 210
Belfort M, 19, 85, 183, 192, 194, 253
Belgrader P, 202
Beliakova-Bethell N, 543
Belkina AC, 544
Bell K, 138
Bell SS, 148
Bellamy K, 163
Bellett-Travers MD, 105
Bellier B, 492
Belliot G, 160
Bellou MI, 151
Belousov YS, 234
Belov K, 42
Beltrán-López J, 501
Bely B, 555
Ben Ammar A, 113
Ben Said M, 127
Ben-Shaul A, 232, 481
Ben-Yishay R, 543
Bencsics CE, 548
Bendall SC, 452, 527
Bendis I, 16, 76
Benett W, 202
Bengoa-Vergniory N, 543
Benjamin CE, 131, 458
Benne R, 324
Bennett A, 164
Bennett GN, 279
Bennett HB, 107
Benoit F, 146
Bensadoun P, 202
Bensidoun P, 548
Bensimon M, 134
Benson D, 76
Bentahir M, 108, 109
Benurwar M, 235
Benyahya M, 118
Benz EJ Jr, 538
Benzer S, 175
Benzine JW, 517
Benzinger R, 67
Beraldi R, 234
Beranek A, 54
Berardi A, 518
Beremand MN, 19, 20, 176, 184, 291, 294
Berenger BM, 202
Berestowskaya NH, 268
Berg G, 105
Bergamini N, 89
Bergen K, 234
Berger EM, 340
Berger S, 120
Bergkvist M, 451, 452, 462
Bergmann JE, 318
Bergquist PL, 316, 341
Bergstrom DE, 127
Berindan-Neagoe I, 551
Berissi H, 23, 263, 324
Berkholz R, 166
Berkhout B, 20, 219, 252, 293, 294, 295, 296,
    420, 424, 431, 536, 550
Berkovits BD, 543
Berman D, 105, 120
Bernacki J, 84
Bernardes GJ, 457
Bernardi A, 221, 320

Bernat S, 538
Bernat X, 544
Bernhardt L, 135
Bernhardt TG, 167, 243, 298, 424, 425
Bernstein D, 539
Bernstein DS, 539
Berriman JA, 435, 438
Berrocal A, 543
Berry KE, 450, 461, 518, 529, 548
Bertarello A, 486, 556
Bertrand E, 155, 540, 541, 542, 545
Bertrand I, 120, 141, 145, 201, 540, 541
Bertz PD, 114
Berzin V, 83, 225, 294, 322, 333, 349, 357, 394
Berzin VM, 83, 219, 225, 283, 333, 349, 358
Berzins V, 394, 411, 439
Bespalova IA, 98, 185
Bessa J, 447, 459, 460, 489, 490, 491, 518
Bessis N, 513
Bestervelt LL, 138
Betancourt AJ, 416
Betancourt WQ, 202
Bettin N, 543
Betz-Stablein B, 544
Beulke I, 1, 19
Bevilacqua PD, 116
Bezuidenhout CC, 103
Bhaduri P, 113
Bhakta S, 540
Bhardwaj J, 109, 405, 543
Bharucha T, 202
Bhaskar S, 518
Bhat A, 501
Bhatnagar R, 233
Bhatta D, 310
Bhattacharyya S, 201, 513
Bhullar M, 126
Bhullar MS, 126, 161
Bi J, 542
Bialek JK, 550
Bibby K, 203
Bibi E, 543
Bibis SS, 283
Bichai F, 136
Biebricher C, 161
Biebricher CK, 61, 77, 285, 387, 388, 389, 394,
    396, 405, 410, 425
Bieniasz PD, 539, 544
Bigley RB, 545
Biji CL, 555
Bijkerk P, 132
Bilanchone V, 543
Bilek G, 200
Bilimoria SL, 377
Bilinska K, 551
Billard L, 42
Billeter M, 167, 243
Billeter MA, 19, 176, 210, 222, 223, 224, 248,
    238, 315, 327, 367
Bingham RJ, 483
Bingle WH, 63
Biondaro G, 556
Bird RE, 234
Birkbeck TH, 306
Birnbaum J, 377
Bischel HN, 164
Bischof K, 54
Bisha B, 108
Bishop DHL, 15, 249, 258, 261, 301, 367, 381
Bishop GA, 462
Bishop J, 316

Bishop JM, 262, 368, 371
Bispo FC, 166
Bisrat Y, 427
Biswas J, 548
Biswas P, 143, 144, 199
Biswas S, 460, 513
Bitsura JA, 120
Bittner AM, 529
Bitton G, 114, 117, 125
Bixby RL, 147
Bjørås M, 234
Blaas D, 200
Black EP, 158
Blackall PJ, 155
Blackbeard J, 155
Blaha B, 20, 40, 226, 295, 344, 406, 425
Blaisdell BE, 226
Blaise-Boisseau S, 160
Blanc R, 118
Blanch AR, 113, 116, 151
Blanch EW, 113, 193
Blanchard JM, 541, 542, 545
Blanco E, 489
Blanco Fernández MD, 113, 166
Blasdel BG, 414
Bläsi U, 281
Blatchley ER 3rd, 127
Blatter N, 234
Blazquez L, 543
Bleck M, 544
Bleckley S, 476
Blessing J, 19, 71, 178, 193, 237
Blinov VM, 406
Blinova EA, 517
Blissenbach M, 538
Block JC, 145
Block R, 314
Blom T, 294, 295
Blomquist CL, 398, 426
Blondel B, 436, 440
Bloom K, 541
Bloom RJ, 544
Bloomfield RA, 65
Blugeon C, 541
Blum-Emerique L, 94
Blumberg BM, 326
Blumenthal T, 19, 20, 23, 176, 184, 213, 262,
    264, 265, 266, 278, 279, 284, 285,
    291, 294, 363, 389, 550
Boavida LC, 543
Bobst AM, 193, 205
Bobst EV, 193
Boccard F, 285
Böck A, 314
Bock CT, 234
Bock WJ, 109
Boczek LA, 129
Bode W, 427
Bodlaender J, 421
Bodner K, 540
Bodnev SA, 517
Bodzek M, 157
Boedtker H, 18, 19, 189, 190, 193, 205, 208, 210,
    211, 314, 320
Boehm AB, 113, 131, 141, 153
Boehm V, 538
Boehme RE, 234
Boehringer D, 538
Boers R, 543

Boesler C, 538
Bofill-Mas S, 126
Bogans J, 42, 250, 445, 501, 556, 557
Bogatikov GV, 378
Bogdanov A, 325
Bogdanova S, 325
Bogerd HP, 538, 539, 550
Bohatier J, 118
Bohl F, 538
Böhm T, 162
Bohra L, 128
Boijoux O, 538
Boime I, 318
Bois JS, 543
Boissier MC, 513
Böker A, 529
Bolhassani A, 489
Bolinger H, 202
Bollback JP, 40, 169, 226, 405, 415
Bolle A, 175
Bolli E, 493
Bollum FJ, 223, 224, 419
Bolton PH, 193
Bonadonna L, 106
Bondeson K, 234
Bonhoeffer F, 261
Boni IV, 123, 267, 329, 364
Bonjoch X, 116
Bonné P, 148
Boom R, 210
Boon T, 314
Boons E, 544
Booth DS, 538
Boque M, 116
Bor B, 543
Borchardt MA, 114, 121, 162
Borcherts MK, 426
Borecký L, 377
Borek B, 551
Borelli IA, 190
Borén T, 341
Borer PN, 348, 349
Borges K, 113
Borgia PT, 529
Borisevich SV, 379
Borisova G, 436, 438
Borisova GP, 349, 435, 438
Borisova OF, 217, 438
Borjihan Q, 137
Bornhop DJ, 453
Borodavka A, 232, 207, 480, 481
Borrego JJ, 112, 163
Borst P, 23, 288, 367
Bosanac L, 545
Bosch A, 544
Bosch L, 85, 132, 208, 266, 314
Bose ME, 202
Bösl M, 148
Bos TJ, 548
Bosselaar A, 314
Bothma J, 543
Bothma JP, 543
Bothma L, 103
Bothma T, 16, 58, 59, 60, 122
Botstein D, 187
Botta M, 443, 449, 450, 462, 518
Böttcher B, 435, 438
Bottomley MJ, 489
Botzenhart K, 149
Bouabe H, 280
Bouadloun F, 317

Boudaud N, 141, 146, 154, 159, 201
Boujnan M, 202
Boulon S, 541
Bourgeois CF, 538
Boutonnet C, 538
Boutureira O, 457
Bouwes Bavinck JN, 427
Bouwknegt M, 108
Bowen MD, 202, 543
Bowman CM, 314
Bowman JC, 464
Bowman R, 138
Boyd CA, 65
Boyd SD, 458
Boyle DS, 203
Boz EB, 556
Bozkurt H, 162
Bozkurt H, 162
Bracker CE, 195, 197
Bradley A, 216, 221
Bradley CR, 163
Bradley D, 234
Bradley DE, 1, 13, 15, 16, 17, 37, 53, 55, 56, 57,
        58, 59, 60, 62, 63, 76, 79, 80, 97, 122,
        192, 249, 301, 302
Bradley I, 14, 151
Bradley ML, 154
Bradwell K, 396, 416
Brady-Estévez AS, 131
Braga ÉM, 457
Braga LPP, 556
Brakier-Gingras L, 313
Branch A, 115, 138, 155, 156
Brandão ML, 166
Brandina I, 543
Brandsma SR, 161
Brandt S, 538
Brandwein H, 165
Braselmann E, 548
Brassard J, 41, 161
Bratu DP, 543
Bräuchle C, 543
Brauer DD, 461
Braun G, 429
Braun J, 539
Braun M, 512
Braun PG, 141
Braun R, 427
Braun SMG, 551
Braunstein JL, 124
Braustein HE, 311
Braustein IE, 311
Braverman B, 325
Brawerman G, 324
Bray SJ, 543
Braymen C, 161
Brazos B, 136
Breakefield XO, 543
Breathnach R, 538, 551
Brechbiel JL, 542
Brencicova E, 489
Brenner A, 153, 154
Brenner R, 117, 148
Brenner S, 27, 84
Brenowitz J, 322
Bressler AM, 515
Bresson SM, 538
Bretscher MS, 332
Brewer DN, 543
Brewster L, 95
Brezesinski T, 130
Brian IJ, 460

Bricio SM, 166
Brié A, 141, 201
Briggs JA, 197
Brigham MD, 549, 551
Bright KR, 108, 155, 162
Brigneti G, 285
Brillen AL, 538
Brinkman NE, 156
Brinkmann T, 427
Brinton CC Jr, 1, 45, 46, 47, 49, 51, 55, 56, 61,
        66, 74, 75, 98
Brion G, 154
Brion GM, 113, 118, 136, 145
Brito CR, 113
Brito CRN, 457
Britto R, 555
Brochot E, 202
Brock TD, 87
Brockhurst MA, 66, 414
Brod F, 513
Brody Y, 543, 545
Broeze RJ, 319
Brombacher F, 491, 518
Bromberg DJ, 131
Bromberg L, 131
Bromenshenk JJ, 149
Bromley PA, 224
Brooks CL 3rd, 190
Brooks JP, 143
Brooks KM, 202
Brooks RR, 284
Brorson K, 165, 166, 234
Brorson KA, 166, 199
Broude NE, 334, 543
Brouqui P, 203
Brouwer-Hanzens AJ, 124, 153
Brown A, 551
Brown CM, 543
Brown D, 285, 425
Brown F, 1, 31, 436
Brown GA, 105
Brown J, 442
Brown JG, 343
Brown KM, 517
Brown L, 280
Brown MJ, 501
Brown MR, 166
Brown PO, 543
Brown S, 265, 266
Brown SD, 455, 461, 463, 488, 529, 532
Brown WL, 518
Brownlee GG, 212
Brownstein BL, 87, 149, 313
Bruckman MA, 529
Bruder K, 42
Brudo I, 90, 91
Bruenn JA, 406
Bruinsma R, 383
Brulé H, 539
Brune KD, 460, 513
Bruner R, 209
Bruno G, 235
Bruno JG, 356
Bruun TUJ, 514
Brūvere R, 378
Bryant EL, 128
Bryant S, 152
Bucardo F, 234
Buchanan JH, 342
Buchanan N, 131, 150
Buchanan TM, 63

Buckingham R, 184, 538
Buck KW, 267
Buck M, 455
Buckling A, 66
Budarz JF, 131
Budowsky EI, 98, 123, 135, 138, 185, 267, 329, 334, 364
Buenrostro JD, 356
Buhr DL, 304
Bui QT, 201
Buiķis A, 378
Bukata LA, 377
Bukhari Z, 124
Buldun CM, 513, 557
Bule P, 487, 556
Buley MD, 464, 518, 527
Bull JJ, 169, 414, 536
Bullock M, 235
Bulmer M, 218, 341
Bundule M, 431, 435, 436, 438, 440, 467
Bundy BC, 97, 339, 461
Buonaguro FM, 489
Buonaguro L, 489
Burbery L, 149
Burch EJ, 233
Burdick RC, 548
Burdon RH, 367
Burg JL, 398, 426
Burge WD, 112, 137, 140
Burger C, 107
Burhenne J, 538
Burkacky O, 543
Burke KL, 436
Burkhard P, 489, 518
Burkhardt J, 156
Burkhardt S, 162, 202
Burkhardt W 3rd, 113, 157
Bürki K, 447
Burlandy FM, 203
Burlingame AL, 538
Burman LG, 50
Burnham CA, 202
Burnham CD, 165
Burnham MS, 166
Burns AT, 529
Burns DJ, 316, 341
Burns S, 113
Burns SM, 161
Burrell CJ, 12
Burrows LL, 64, 66
Bursteinas B, 555
Burton DR, 513
Burton RE, 202
Burutarán L, 166
Busch B, 539
Buser RB, 490
Buskila AA, 543
Bustos J, 444
Buter N, 539
Butkus MA, 125
Butler AL, 235
Butler M, 119, 120, 135
Butler SS, 458
Butner J, 117
Butot S, 161
Buxbaum AR, 548
Buxton D, 398
Buyanova NI, 54
Buzard RL, 125
Bychkova EN, 377
Bystricky K, 544

C

Cabaj A, 132
Cabanillas L, 396, 415, 417, 418
Cabelli VJ, 120, 136, 161
Cabral-Miranda G, 447, 492, 557
Cabrera JV, 283
Cadena-Nava RD, 533
Cadnum JL, 164, 165
Cady NC, 452
Cahill J, 299
Cahill K, 501
Cahill PB, 398, 426
Cai JC, 493
Cai M, 130
Cai X, 136
Cai Y, 551
Caillou MS, 203
Cairns J, 261
Cala-De Paepe D, 486, 556
Calarco JP, 551
Calci K, 159
Calci KR, 39, 113, 150, 157, 331
Caldeira J, 444
Caldeira JC, 442, 446, 461
Caldeira Jdo C, 442, 446, 461, 493
Calderón E, 106, 117
Callahan K, 107
Callahan KM, 117
Callahan R III, 325
Callanan J, 555
Calogero RA, 329
Calvo JS, 458
Calvo M, 166
Camara J, 517
Cambillau C, 233
Camerini-Otero RD, 189, 190, 191
Cameron JM, 234
Cameron V, 347, 386
Cammarata RV, 113, 166
Campalans A, 543
Campbell A, 175
Campbell AM, 226
Campbell C, 317
Campbell DJ, 501
Campbell MG, 465, 488
Campbell PD, 543
Campbell S, 235
Campbell SG, 543
Campos C, 106, 113, 117, 136
Canary JW, 529
Canh VD, 42
Cannon D, 543, 544, 548
Cannon JL, 202
Cano-Rodriguez D, 551
Canonaco MA, 329
Cantera J, 203
Cantergiani F, 161
Cantor CR, 267
Cantwell RE, 125
Canver MC, 551
Cao H, 151
Cao Z, 139
Capecchi MR, 1, 17, 19, 21, 25, 175, 179, 186, 192, 237, 249, 317, 318, 319, 323, 330, 331, 332, 342, 344
Capehart SL, 463, 464, 466, 532, 533
Capek I, 529
Caporini M, 556
Caprais MP, 113, 116
Caputo A, 513

Caputo AT, 487, 556
Caraglia M, 518
Carattoli A, 58
Carazo JM, 73, 192, 243, 476, 477
Carberry TP, 453
Carbon P, 539
Cárdenas M, 106
Cárdenas-Youngs Y, 166
Carding SR, 42
Cardoso-Oliveira GP, 457
Care A, 518
Carey J, 347, 386
Cargile BJ, 250
Carl A, 162
Carlsen PHR, 514
Carman RK, 399, 425
Carmichael GG, 266, 278
Carmo-Fonseca M, 545
Carnahan J, 1, 46, 47, 66
Carnes EC, 464, 480, 518, 527
Caro LG, 16, 46, 54
Carpenter M, 63
Carrascosa JL, 479
Carratalà A, 134, 142
Carret G, 95
Carrico ZM, 450, 518, 527
Carrillo-Tripp M, 190
Carrocci TJ, 545
Carroll AR, 436
Carroll R, 537
Carrondo MJT, 489
Carsana A, 367
Carteret C, 161
Cartwright GA, 310
Carubelli R, 133
Carvajal G, 138, 156
Carvalho AL, 487, 556
Carvalho M, 537
Carvalho-Costa FA, 166
Casale TB, 490, 518
Casalegno JS, 42
Casanova I, 518
Casanova L, 41, 163, 169
Casanova LM, 106, 133, 164
Casas R, 235
Casas S, 544
Cascarino J, 158
Caserta S, 235
Cases-González C, 396, 415
Cashdollar JL, 129
Caskey CT, 332
Casolari JM, 543
Caspar DLD, 192
Cass C, 165
Casson LW, 203
Casteel MJ, 114, 130, 160
Castel A, 314
Castelain S, 202
Castelli LM, 543
Castello AA, 203
Castiglioni P, 314
Castillejos A, 465, 488
Castillo G, 153
Castillo-Rojas G, 116
Castracane J, 452
Catão RM, 116
Cates EL, 128
Cath TY, 156
Cathcart AL, 538
Catrina IE, 543
Cauchie HM, 149

Caudrelier Y, 160
Causse SZ, 545
Cavaleiro JA, 130
Cavallo F, 493
Caves LS, 475
Cazeaux C, 160
Ceballos BS, 116
Cech I, 117
Cech TR, 400
Cedergren RJ, 167, 218, 340
Celander DW, 400, 538
Celebanska A, 109
Celen P, 219
Celma ML, 313
Ceppellini M, 189
Cerdá-Olmedo E, 50, 85
Cernohorský I, 96
Cerny RE, 433, 540
Cerny T, 501
Cerretti DP, 440
Cerutti PA, 123
Céspedes MV, 18
Cetó X, 133
Chackerian B, 442, 444, 446, 447, 448, 461, 489, 493, 501, 514
Chae SR, 131
Chae YB, 324
Chain B, 555
Chakerian A, 351, 470
Chakrabarti S, 85
Chakraborti S, 518
Chakraborty K, 539
Chakraborty PR, 324
Chalkley RJ, 538
Chamakura K, 299
Chamakura KR, 298, 299, 425
Chamberlain AH, 136
Chambliss GH, 318
Chamontin C, 155, 156
Champe SP, 175
Champion LM, 543
Champney WS, 139, 331
Chan HK, 559
Chan HT, 202
Chan TS, 21, 331
Chan YS, 102
Chander Y, 144, 517
Chandler JC, 108, 144
Chandler-Bostock R, 139
Chandraseelan JG, 542
Chaney WG, 331
Chang A, 234
Chang AT, 359
Chang CF, 470
Chang CK, 470
Chang CT, 543
Chang H, 489
Chang HY, 356
Chang JS, 132
Chang JY, 167, 245, 424
Chang L, 516, 518
Chang LT, 114
Chang RYK, 559
Chao JA, 360, 361, 470, 543, 544, 545, 546, 547, 548
Chao KL, 487, 556
Chao SW, 202
Chapman S, 285
Chappell JD, 202
Chari R, 551
Charles KJ, 124

Charlesworth B, 414
Charlesworth D, 414
Charlton Hume HK, 489
Charni-Ben-Tabassi N, 145
Charnock C, 165
Charpentier E, 551
Chartrand P, 540, 541, 543, 544, 548
Chassaing M, 120
Chatterjee A, 518
Chattopadhyay S, 120
Chaturvedi P, 108, 312
Chatzinotas A, 152
Chau HC, 105, 111
Chaubey B, 267
Chaudhari SR, 556
Chaudhry RM, 156
Chaudhuri M, 147
Chau WS, 102
Chauvin C, 191
Chávez A, 139
Chavez A, 551
Chedid L, 309
Chekulaeva M, 548
Chelchessa B, 272
Chellam S, 130, 131, 133, 558
Chelliah R, 543
Chen C, 545
Chen DK, 117
Chen DR, 199
Chen F, 280, 538
Chen G, 136, 464
Chen H, 109, 136, 159, 162, 235, 425, 548
Chen J, 160, 221, 280, 330, 442, 544, 548
Chen KCS, 63
Chen L, 442, 457
Chen LH, 543
Chen LL, 551
Chen MY, 458
Chen NT, 144
Chen Q, 136, 453
Chen R, 492
Chen S, 551
Chen W, 453, 458
Chen WH, 311
Chen WM, 235
Chen X, 124, 131, 139
Chen Y, 202, 540
Chen Z, 131, 153, 457, 458, 539
Chen ZB, 132, 136
Chenciner N, 436
Cheng C, 134
Cheng D, 160
Cheng N, 435
Cheng R, 128, 132, 134
Cheng SF, 543
Cheng T, 551
Cheng X, 143
Cheng Y, 304, 440, 516
Chernesky MA, 397
Cherny D, 538
Cheroutre H, 20, 167, 181, 183, 216
Chetverin AB, 268, 339, 390, 391, 395, 400, 405, 412, 413, 426
Chetverina EV, 390, 391
Chetverina HV, 390, 405, 391, 412, 413, 426
Cheung LE, 543
Chevalley R, 175
Chi JY, 343
Chiang YC, 470
Chibani CM, 42
Chichkova N, 325

Chidlow G, 202
Chilukoti RK, 235
Ching WY, 356, 483
Chiniforooshan Y, 109
Chinivasagam HN, 155
Chira S, 551
Chircus LM, 356
Chiu W, 435, 479
Cho H, 298, 425, 538, 539
Cho HH, 156
Cho J, 267, 425
Cho M, 125, 127, 128, 130, 131, 133
Cho WK, 544
Cho YH, 111, 112, 188, 234, 559
Choe J, 538
Choe JK, 136
Choi B, 518
Choi DS, 529
Choi H, 518
Choi HJ, 150
Choi SB, 542
Choi SH, 117, 544
Choi SS, 156, 538
Choi W, 160, 233
Choi WS, 141
Chon H, 217
Chong R, 42
Choo QL, 234
Choobtashani M, 131
Chopra A, 544, 548
Choquet CG, 319
Chory EJ, 551
Chory J, 322
Chou YJ, 117
Chow CC, 547, 548
Chow M, 435
Chowdhury K, 203
Chowdhury S, 542, 545
Christen JM, 493
Christensen E, 42, 134
Christensen JR, 54
Christensen NM, 543
Christie PJ, 64, 120
Christoffels E, 116
Chroboczek J, 70, 91, 346
Chromiński M, 548
Chrysikopoulos CV, 118, 128, 143, 149, 151
Chu J, 135
Chu W, 114
Chu XN, 150
Chu Y, 120, 153, 543
Chu YG, 267
Chua NH, 543
Chuan YP, 489
Chuang DY, 279
Chuang TW, 161
Chubb JR, 537, 541, 543, 544, 548
Chui L, 202
Chumakov KM, 406
Chumakov PM, 545
Chumakov SP, 545
Chung DG, 333
Chung H, 106, 120, 157
Chung S, 234
Chung WJ, 529
Churchman LS, 548
Chutiraka K, 544
Cielens I, 333, 349, 357, 431, 433, 436, 440, 441, 444, 445, 462, 471, 501, 556
Ciesiolka A, 551
Cigler P, 456

Cimica V, 489
Cimprich KA, 551
Cingil HE, 556
Ciocanel MV, 543
Cioroch M, 543
Cirino PC, 540
Cirkel G, 136
Citovsky V, 543
Clancy J, 124
Clancy JL, 120
Clark AJ, 51, 74, 98
Clark BFC, 21, 323
Clark DS, 105
Clark G, 234
Clark KJ, 551
Clark RM, 145, 154
Clark VL, 325
Clarke BE, 436
Clasen S, 542
Clasen T, 155
Clauss M, 127
Clawson G, 193
Claybrook JR, 258, 261, 381
Cléard F, 543
Clemens K, 543
Clément B, 202
Clement SL, 538, 540
Clements MO, 234
Clémot M, 548
Clerté C, 155
Cleveland PH, 196
Clevenger T, 136
Clifford E, 121
Clifford K, 114
Cloninger MJ, 453
Close ED, 210
Clough C, 493
Coburn GA, 538
Cocito C, 313
Cock I, 140
Cock IE, 140
Cody JDM, 241, 293
Cody NA, 545
Coelho M, 538
Coelho ZBA, 457
Coetzee JN, 15, 16, 37, 58, 59, 60, 122, 302
Coetzee WF, 14, 58
Coffey LL, 201
Coffin JM, 548
Coffman RL, 235
Cohen BA, 451, 462
Cohen DR, 57
Cohen R, 234
Cohen SS, 84
Coia G, 399, 425
Colacicco D, 112, 140
Colantoni A, 540
Cole CN, 543
Cole D, 113, 114, 118
Cole J, 490, 518
Cole KD, 166, 199
Cole PE, 266, 267, 384, 385, 392
Cole RD, 88, 196, 197, 250
Coleman J, 252, 294, 296, 535
Colford Jr JM, 113
Colina R, 166, 203
Coller J, 538, 540
Coller JM, 538
Collier JH, 489
Collins K, 538, 546
Collins KC, 501

Collins KE, 151
Coltri P, 538
Colvin VL, 128
Combe M, 396, 416
Combs J, 165
Comeau AM, 203
Communal PY, 113
Comninellis C, 134
Concheiro A, 131
Conden-Hansson AC, 157
Confaloni A, 235
Conlon P, 543
Conlon PJ, 440
Conner TW, 433, 540
Connor V, 429
Connors NK, 489
Conrad JC, 312
Conrad NK, 538
Content J, 219, 419, 427
Conti F, 218
Conti L, 493
Conti SF, 15
Contreras A, 493
Contreras L, 543
Contreras R, 20, 167, 181, 183, 211, 212, 213, 214,
    216, 219, 238, 323, 339, 365, 419, 427
Contreras RD, 164
Contreras-Coll N, 116
Contzen M, 162
Converse B, 517
Convery MA, 352, 354, 470, 537
Conway JF, 435
Conway TW, 241, 293, 324, 325
Conway W, 544
Cook K, 210
Cook N, 142
Cooney JJ, 111
Cooper A, 42, 513
Cooper EM, 41, 169, 331
Cooper S, 11, 23, 82, 86, 87, 97, 178, 257, 258,
    286, 366
Coote BG, 115
Coppey M, 543, 548
Corapcioglu MY, 120, 150
Corash L, 123
Corchero JL, 518
Corcuera E, 137
Cordelier P, 551
Core SB, 446
Corman A, 95
Cormican M, 121
Cormier J, 81, 107
Cormode DP, 529
Cornax R, 112, 163
Cornetta K, 234
Cornish TJ, 198
Cornuet P, 286
Cornuz J, 490, 501
Corral-Debrinski M, 541
Corrigan AM, 543, 544, 548
Corsini V, 427
Corti E, 313
Cory S, 214, 220, 221, 238
Coskun A, 544
Cosson B, 540
Costa L, 130
Costán-Longares A, 113
Cotter RI, 193
Coubrough P, 112, 113
Coudron L, 109
Couix C, 95

Couladis M, 139
Coulanges P, 102
Coulliette AD, 18, 19, 136, 164, 189, 205
Coulon A, 547, 548
Coulson AR, 176, 220, 224, 286
Coulter CG, 96, 138
Cox CS, 89
Cox D, 306
Cox DB, 551, 558
Cox EC, 542
Cox G, 234
Coyle MP, 463, 464
Crabtree GR, 551
Craggs RJ, 141
Craig L, 64
Crain D, 501
Crainic R, 436, 440
Cramer JH, 87, 367, 368, 372
Cramer WN, 112, 135, 137, 140
Crandall PG, 162
Crawford EM, 1, 16, 46
Crépin M, 325
Crerar H, 542
Crescente V, 20, 40, 226, 295, 344, 406, 425
Crespi M, 543
Crestini A, 235
Creticos PS, 501
Cricca M, 107
Crick FHC, 16
Crissy K, 234
Crittenden J, 138
Cromar N, 131, 150
Cronin AA, 151
Crooke SN, 527
Crosbie ND, 202
Crossey E, 446, 493
Crossland L, 433, 540
Crossley BM, 516
Crothers DM, 221
Croucher D, 155
Crow JF, 414
Crowgey EL, 226
Crowther J, 152
Crowther RA, 17, 192, 435, 438
Cruz C, 538
Cruz MC, 166
Csík M, 53
Csuk R, 130
Cuervo A, 479
Cuevas JM, 169, 415, 416
Cui S, 551
Cui X, 235
Cui Y, 131, 545
Cui Z, 167, 245, 424, 479
Cuillel M, 234
Cullen BR, 518, 538, 539, 550
Culley A, 105
Culley AI, 42, 151
Culver BP, 543
Cunha A, 130
Cunha MÂ, 113
Cunningham C, 33
Cunningham DD, 538
Cunningham J, 436
Cunningham-Bryant D, 551
Curley SM, 452
Curran MD, 202, 326
Curry ME, 343
Curtiss LK, 241, 307
Curtiss R III, 45, 47, 54, 74
Cusanelli E, 543

Cuzin F, 50
Cvitkovitch DG, 64, 66
Czapar AE, 518
Czapiński J, 551
Czaplinski K, 544, 548, 551
Czernilofsky AP, 325
Czub S, 431
Czworkowski J, 329

D

D'Aquino AE, 540
D'Ávila DA, 457
D'Hooge F, 457
D'Orso I, 538
D'Souza DH, 1, 11, 12, 15, 139, 140, 143, 160, 161, 162
Da Poian AT, 143
da Rosa E Silva ML, 166
da Silva AM, 556
da Silva AV, 202
da Silva EE, 203
da Silva MF, 166
da Silva TP, 166
da Silva Felício MT, 121
da Silva Monteiro DC, 202
Dabbs ER, 331
Dacheux E, 544, 548
Daehnel K, 61
Dahlberg AE, 328
Dahlberg JE, 211, 212, 222, 223, 238, 327, 328
Dahling DR, 105
Dahlman JE, 550
Dahlson N, 234
Dahm R, 544
Dai RH, 136
Dai X, 192, 243, 478, 479
Dal Peraro M, 487
Dalachi M, 543
Dale T, 355
Dalgarno L, 22, 121
Dalrymple B, 64
Daly SM, 493
Damen M, 210
Damit B, 163
Dandekar S, 202
Dange T, 545
Danis-Wlodarczyk K, 414
Dannull J, 235
Danquah MK, 518
Dantham VR, 200
Danyluck GM, 234
Danziger RE, 69, 70
Dao TL, 203
Darby JL, 124
Darcis G, 550
Darga P, 202
Darimani HS, 141
Darmon SK, 538
Darzacq X, 543, 544, 545
Das AT, 550
Das R, 486, 545
Das S, 548
Das T, 267
Dasilva R, 543
Date T, 61, 70
Datta A, 443, 449, 450, 451, 461, 462, 518, 529
Datta N, 48, 49, 50, 56, 57, 58, 98
Dauksaite V, 538
Dave PK, 113
Davern CI, 12, 15, 19, 87, 93, 175, 301

David F, 154
David Z, 126
Davidi D, 78, 81
Davidson M, 144
Davidson PM, 1, 12, 15, 139, 140, 162
Davies CM, 147
Davies DR, 352
Davies J, 313
Davies JG, 163
Davies JW, 318
Davies-Colley RJ, 141
Davis BD, 87, 313, 324, 492
Davis BG, 457
Davis BK, 384, 400
Davis J, 67
Davis JE, 18, 23, 67
Davis N, 42
Davis R, 1, 12, 15, 139
Davis RM, 298, 425
Davong V, 202
Davydova AA, 377, 378
Dawson DJ, 159
Dawson KM, 436
Dawson SR, 113, 163
Day CR, 548
Day LA, 189
De S, 162, 311
de Abreu Corrêa A, 166
de Aceituno AF, 153
de Assis MR, 166
de Ávila AI, 418
De Boever JG, 19, 185, 273
De Clercq E, 219, 378, 419, 427
de Cortalezzi MM, 166
De Fernandez MT, 262
de Giorgi C, 223
De Graaf FK, 314
de Gruyter M, 421
De Guzman RN, 352
De Haseth PL, 278, 346, 347, 386
de Jong JC, 143
de la Cruz F, 58
de la Escosura A, 529
de la Salle H, 234
De La Tour EB, 175
De Las Heras Chanes J, 548
De Lay N, 283
de Lima Alves F, 546
De Luca G, 107, 136, 156
de Moura MO, 156
de Oliveira AC, 143, 461
De Paepe M, 414, 415
De Petris S, 78
de Roda Husman AM, 108, 132
De Smet J, 414
de Smit MH, 294, 295, 334, 335, 337, 421, 423
de Souza VC, 202
de Turris V, 543, 544, 545, 547
de Vicente A, 112
de Vries R, 556
De Wachter R, 40, 167, 211, 212, 213, 216, 217, 223, 241, 339, 419, 427
Dea M, 154
Dean KM, 548
Deaton J, 556
Debarbieux L, 553
Deboosère N, 107, 160
Debord JD, 427
Debreceni N, 33
Debusschere BJ, 200
Decarolis JF, 156

Dechant PP, 356
Decker R, 234
Deckert J, 538
DeClercq J, 543
Decrey L, 137, 164
Dedeo MT, 529
Dedkov VG, 517
DeFranco AL, 235
Degrave W, 20, 167, 181, 183, 216
DeGruson E, 136
Deharo D, 551
DeHaven AC, 548
Deinhardt F, 113
Dekel L, 19, 85, 86, 167, 183, 192, 194, 232, 253
Del Gatto-Konczak F, 538
Delabre K, 145
Delaney K, 169, 527
Delaney SF, 64
Delattre JM, 114
Delavallée L, 513
Delavari A, 558
Delbos F, 551
Delebecque CJ, 546
Delekar SD, 131
Delitheos A, 139
Delius H, 23, 367
Della-Negra S, 427
Dellagi K, 203
Dellunde J, 118
Delobel A, 160
DeLoyde JL, 153
Delpeyroux F, 436, 440
Demain AL, 377
Demarigny Y, 139
Demarta A, 120
Demidenko AA, 390, 412, 413, 426
Demirev PA, 198
Denaro M, 313
Deng J, 545
Deng L, 140
Deng W, 516
Deng X, 517
Deng Y, 124, 131, 549
Denhardt GH, 175
Denisova LI, 225
Denk H, 234
Denk J, 235
Dennehy JJ, 166
Dennis WH, 135
Dennison S, 66
Denny SK, 546
Dent KC, 73, 243, 476, 477
Denti MA, 235
Denys A, 513
Derbes RS, 551
Dergel I, 137
Dérian N, 492
DeRisi JL, 15, 42, 105, 170, 226
Dermit M, 548
Derrer CP, 548
Derynck R, 219, 419, 427
Dersch P, 280
Derse D, 537
Dertinger D, 355
Deshayes S, 518
Deshmukh SP, 131
Desjardins R, 136, 202
Desnoues N, 279
Desponds J, 543, 548
DeStefano J, 513
Dettori R, 189

Deutscher MP, 86, 475
Devane ML, 113, 115
Devaraj NK, 551
Devare SG, 234
Deviatkin AA, 517
Devilder MC, 551
Devos R, 20, 167, 181, 183, 216, 219, 224, 274, 284, 419, 427
Dey R, 518
Dhaese P, 249
Dhamane S, 311, 312
Dhara D, 131, 140
Dharmarwardana M, 131
Dharmasena M, 202
Dhillon EK, 102, 105, 111
Dhillon TS, 102, 105, 111
Di Natale P, 342
Dias E, 113, 118
Diaz D, 518
Diaz M, 199
Diaz MH, 203
Diaz-Avalos C, 116
Dibarrart F, 153
DiCaprio E, 202
DiCapua E, 234
Dichtelmüller H, 123
Dickman MJ, 210, 543
Dickson KS, 539
Dickson SE, 132
Dickson-Anderson SE, 133
Dictenberg J, 545
Didilescu C, 501
DiDonato GT, 115
Didovyk A, 551
Diebold SS, 489
Diebolder CA, 477
Diederich S, 489
Diederichs K, 234
Diekmann S, 388
Diener C, 543
Diener TO, 394
Dietmeier K, 489
Dietrich S, 42
DiFrancesco RA, 426
DiGiano FA, 153
Dijkstra A, 54
Dika C, 200, 201
Dikeakos JD, 544
Dikler S, 427
Dill V, 234
Dilley KA, 544
Dimitrov G, 235
Diner EJ, 64
Ding D, 280
Ding Y, 551
Dinh-Thanh M, 160
Dionisio Calado A, 142
Dipple A, 135
Dirk BS, 544
Dishler A, 25, 76, 77, 80, 85, 91, 92, 93, 371, 372, 383
Dishler AV, 77, 98, 175, 185, 371, 374, 375
Dishlers A, 501
Dislers A, 431, 433, 436, 440, 441
Ditta IB, 127
Dittmer DP, 130
Dizer H, 147
Djikeng A, 42
Djinović-Carugo K, 281
Djordjevic M, 356
Dlugy C, 145, 153, 449

Do WG, 117
do Nascimento VA, 202
Doane FW, 106
Dobkin C, 385
Dobric N, 399, 425
Dodel M, 548
Doerfler W, 316
Doerr A, 545
Doetschman T, 551
Dohme F, 325
Doi RH, 22, 23, 97, 210, 366
Dolgalev I, 543
Dolgosheina EV, 545
Dolgova AS, 517
Dolja VV, 406, 409, 555
Doll TAPF, 518
Dolman F, 272
Dolník V, 377
Domb AJ, 136, 137, 139
Domingo E, 19, 187, 188, 396, 413, 418, 415
Domingo-Calap P, 169, 396, 415, 416
Dominguez AA, 551
Donahue JP, 319
Donaldson B, 489, 557
Donath A, 152
Donchenko AP, 406
Dong A, 137
Dong C, 234
Dong Q, 235
Dong Y, 518
Dong YM, 493
Donia D, 493
Donis RO, 14
Donlin-Asp PG, 545
Donnat S, 493
Donnison AM, 115, 141
Donofrio RS, 138
Donovan BT, 548
Donskey CJ, 164, 165
Doolittle MM, 108
Doores KJ, 457, 513
Dore W, 121, 158
Doré WJ, 157
Dorigo B, 490
Dornbusch AJ, 65
Dornhelm P, 314
Dorokhov YL, 543
Dorsett B, 135
dos Passos RA, 202
Dos Santos IAL, 166
Doskočil I, 205, 377
Doskočil J, 88, 194, 377
Dostatni N, 543, 548
Dotan A, 161
Dotto GP, 86
Doucet AJ, 543
Doucet M, 543
Doudna JA, 558
Douglas S, 235
Douglas T, 518
Douglass J, 265
Doumatey AP, 235
Dovom MR, 162
Dowd SE, 120, 150
Doyle-Cooper C, 513
Dragnea B, 529
Drake JW, 176, 405, 414, 415, 417
Drake KL, 131
Draper DE, 267, 364
Draper LA, 555
Draper SJ, 460, 501, 514, 557

Drees KP, 133
Dreier J, 202
Dreilina D, 436, 440, 441
Dreilinia DE, 431, 436, 438
Dreimane A, 436, 438
Dreizin E, 144
Drewes JE, 138
Drexler JF, 516
Driessen AJ, 294, 295, 296
Drlica K, 86
Drulak M, 135
Drumond BP, 166
Dryden SK, 40, 114
Du D, 551
Du Y, 192, 243, 280, 398, 426, 478, 479
Du Toit L, 15, 16, 37, 59, 60, 122
Dubberke ER, 165
Dube SK, 94, 325
Dubnoff JS, 324
Dubochet J, 61
Dubois E, 107
Dubovi EJ, 143
DuBow MS, 278
Duchaine C, 144, 155, 203
Ducoste JJ, 129
Dudekula DB, 162
Dudley MA, 303, 305
Duek A, 154
Duerinck F, 24, 212, 288, 319, 323, 365
Duerkop BA, 555
Duffy PE, 440
Duffy S, 29
Duggan J, 203
Duguid JP, 46
Duijfjes JJ, 314
Duim H, 404
Duits AJ, 202
Duizer E, 132
Düker EM, 61
Dūks A, 378
Dular M, 139
Dumas C, 543
Dumas K, 234
Dunagin MC, 543
Dunay GA, 550
Dunbar S, 202
Duncan JN, 551
Duncanson R, 120, 136
Dundr M, 543
Dungeni M, 115
Dunker AK, 17, 18, 192, 241
Dunkin N, 96, 138
Dunlap BE, 322
Dunn G, 436
Dunphy DR, 480
DuPont HL, 120, 155
Duran AE, 117, 122, 136
Durbin GW, 114
Durchschlag H, 210
Durfee MR, 447, 448
Durfee P, 493
Durfee PN, 464, 518, 527
Durosay P, 538
Durso FT, 164
Dürwald H, 1, 15, 19, 97, 372
Dusing S, 143
Dussoix D, 66
Dutilh BE, 29, 30, 31
Dutka BJ, 117
Duval JF, 81, 153, 154, 200, 201
Duverlie G, 202

Dvornyk AS, 98
Dyer MR, 436, 447, 489, 501
Dykeman EC, 475, 477, 481, 483, 556
Dymshits GM, 203
Dynes JL, 542
Dzhagatspanian NG, 378
Dziewatkoski M, 310

E

Eades G, 544
Easingwood R, 200
Ebdon J, 113, 116, 118
Eboigbodin KE, 202
Ecelberger SA, 198
Eck M, 431
Ecker CM, 543
Eckerson HW, 54
Eckhardt H, 328
Eckly A, 234
Economou AE, 323
Economy J, 134
Edberg S, 114
Edelbluth C, 52
Edelbrock JF, 529
Edelman GM, 302, 303, 304
Eden JS, 202
Ederhof T, 401
Edgar RS, 175
Edge C, 538
Edmonds-Wilson SL, 164
Edmunds WJ, 202
Edouard S, 203, 233
Edwards DD, 117
Edwards GB, 298, 425
Edwards JR, 164
Edwards SF, 163
Efanova TN, 379
Eferl R, 54, 234
Effenberger K, 538
Eggan E, 543
Eggen K, 22, 25, 180, 237, 258, 317, 318
Eglinger J, 548
Ehara Y, 279
Ehlers MM, 113
Ehrenberg L, 138
Ehrenberg M, 313
Ehses J, 545
Eichler D, 211
Eiden M, 202
Eigen M, 77, 285, 387, 388, 389, 401
Eiji S, 266, 426
Eikhom TS, 258, 262, 278
Eilat D, 304, 342
Eisen MB, 543
Eisenstadt J, 249
Eisenstadt JM, 324
Eisenstark A, 125
Ekinci FY, 538
El Karoui M, 543
El-Khoury SS, 113
El Muslemany KM, 451
El-Sayed A, 489
El-Turabi A, 447, 501, 543
Elashvili I, 243, 476
Elf J, 548
Elf S, 202
ElHadidy AM, 156
Elias DJ, 455
Elias P, 341
Elimelech M, 131

Elinav H, 544
Eliscovich C, 547, 548
Elizaquível P, 162
Elkins BV, 117, 148
Ellenberg J, 538, 544
Ellington AD, 169, 352, 470
Elliott FB, 19, 186
Elliott JF, 317
Elliott M, 153
Elliott T, 280
Ellis DB, 23, 257, 367
Ellis J, 543
Ellis JS, 202, 326
Elmerich C, 279
Elmore BO, 493
Elsässer D, 108
Elsäßer D, 109
ElSawy KM, 475
ElSohly AM, 451, 452, 527
Elton RA, 218
Elving J, 137
Ely B, 64, 267
Ely KR, 350
Ely RW, 310
Emanuelli B, 551
Embrey SS, 114
Emechebe U, 161
Emelko MB, 121, 132
Emmerson GD, 310
Emmett W, 543
Emmoth E, 137
Enders B, 427
Endo H, 85, 183, 186, 265
Endo K, 540
Enea V, 86
Engel A, 61
Engelberg H, 79, 90, 91, 372
Engelberg-Kulka H, 19, 51, 85, 86, 167, 183, 192,
    194, 210, 232, 253, 343
Engelbrecht F, 429
Engelbrecht RS, 147
Engelhardt DL, 1, 19, 21, 25, 67, 176, 179, 178,
    183, 184, 193, 237, 317, 318, 319, 323,
    331, 367, 368
Enger KS, 18, 19, 189, 205
Enger MD, 15, 137, 247, 307
Engeroff P, 447
Englande AJ, 410
English BP, 544, 548
Englund JA, 202
Eninger RM, 144, 163
Enosi Tuipulotu D, 202
Enríquez P, 551
Entelis N, 543
Eoyang L, 1, 23, 29, 258, 260, 262, 263, 272,
    275, 286, 365
Ephrussi A, 547, 548
Epstein RH, 175
Erdmann I, 458
Erdos GA, 148
Eregno FE, 118
Erickson JW, 160, 367
Erickson SB, 518
Erikson E, 23, 367
Erikson RL, 23, 122, 287, 367, 368
Erkelenz S, 538
Ernberg I, 84
Ershov FI, 377, 378
Ertürk G, 558
Erukunuakpor K, 164
Escarmis C, 223

Escarnis C, 223, 224, 248
Eschweiler H, 132
Eskridge RW, 82
Esona MD, 202
Espinoza F, 234
Esposito E, 543
Esteban R, 356
Estes PS, 542
Estienney M, 160
Ettayebi M, 118
Ettinger CR, 544
Evans A, 162
Evans DJ, 518
Evans H, 135
Evans L, 203
Evans MJ, 539
Evans-Strickfaden TT, 165
Evdokimov YM, 205, 377
Everett GA, 211
Evison LM, 117
Ewald JC, 234
Ewen-Campen B, 551
Exner M, 116
Expert-Bezançon A, 538
Eyre NS, 544
Ezekiel DH, 89

F

Fabbri BJ, 433, 540
Faddy HM, 202
Fadrus P, 235
Fagegaltier D, 539
Fagnant CS, 156
Fahim RE, 501
Fahnestock S, 331
Fairbairn N, 119
Fairley CK, 113
Fajri JA, 117
Falahati H, 543
Falanga A, 518
Falconer RA, 149
Faldina VN, 379
Falgarone G, 513
Fallavolita D, 33
Fallgren PH, 136
Fallon KS, 120, 124
Fallowfield H, 131
Fallowfield HJ, 142, 149, 150
Falman JC, 156
Falnes PO, 234
Falo-Sanjuan J, 543
Famurewa O, 234
Fan L, 544
Fan M, 356
Fan S, 304
Fan X, 109
Fan YC, 235
Fang H, 113, 114
Fang JL, 234
Fang PY, 464
Fang Q, 136, 464
Faoro C, 548
Farabaugh PJ, 341
Farafonov VS, 470
Farah S, 137, 139
Faraway R, 543
Farinelle S, 20, 40, 140, 226, 295, 344, 406, 425
Farjot G, 433
Farkas K, 200

Farkas ME, 452, 527
Farlow MR, 513
Farner Budarz J, 131
Farr A, 42
Farr R, 529
Farrah S, 144
Farrah SR, 96, 114, 117, 125, 147, 148, 154, 155, 159
Farsund O, 108
Farzan M, 518
Fashena SJ, 540
Fassbender K, 235
Fastrez J, 462
Fatemizadeh SS, 162
Fatscher T, 538
Fattal B, 161
Fattal E, 518
Fauquet CM, 31
Faust TB, 538
Fausti ME, 233
Faustino MA, 130
Fauvel B, 149
Favelukes G, 316
Favor AH, 193, 207, 418
Favre G, 538
Fay JP, 159
Fayad ZA, 529
Faye JC, 538
Feary TW, 13, 15, 16, 37, 66, 71, 76, 79, 97, 122, 192, 302
Federico M, 489
Fedorcsák I, 138
Fedoroff NV, 23, 213, 286, 287, 373
Fedoseeva LA, 203
Fehlhaber K, 141
Fehr T, 441
Fei J, 548
Feichtmayer J, 140
Feig M, 352
Feinberg J, 538
Feix G, 210, 258, 283, 284, 288, 365, 368
Feldbrügge M, 539
Felder E, 516
Feldmane G, 378
Feliciano JR, 280
Fenaux H, 120
Feng C, 517
Feng YY, 125, 140, 145, 150
Feng Z, 125
Fenner F, 31
Fenton A, 414
Fenwick ML, 23, 122, 367
Ferdats A, 378
Ferguson AN, 164
Ferguson M, 436
Ferguson ML, 115, 547, 548
Fernández A, 389
Fernandez B, 551
Fernandes da Costa C, 202
Fernández-González M, 457
Fernández-Ibáñez P, 142
Fernandez-Lima FA, 427
Fernandez-Lopez R, 58
Fernandez-Megia E, 453
Fernández-Moya SM, 545
Fernández-Recio J, 361
Fernandez-Trillo F, 453
Ferraro T, 543, 548
Ferrè F, 540
Ferré V, 107
Ferrer M, 155, 156

Ferrer O, 544
Ferrer-Miralles N, 518
Ferreyra-Reyes L, 202
Ferring D, 538
Ferry QR, 551
Fersht A, 435
Fersht AR, 435
Fetka I, 543
Fettelschoss A, 501
Fettelschoss-Gabriel A, 501
Fiches GN, 544
Ficner R, 538
Fidy J, 132
Fiedler JD, 461, 463, 465, 488, 529, 532
Fiedler W, 296, 424
Field AK, 375, 376
Fiering S, 489
Fiers W, 1, 19, 20, 29, 40, 167, 181, 183, 185, 210, 211, 212, 213, 214, 216, 217, 219, 223, 224, 241, 273, 284, 288, 293, 319, 323, 337, 339, 365, 368, 373, 419, 424, 427
Figueiredo HT, 234
Figueroa R, 218
Fiksdal L, 154
Filipovska A, 545
Filipowicz W, 320
Fillatre A, 202
Filman DJ, 435
Filter M, 160
Finbloom JA, 538
Finch GR, 119, 138
Finch JT, 17, 192
Fineran PC, 555
Fink G, 187
Finlay BB, 52, 61, 64
Finley DT, 529
Finn MG, 453, 455, 456, 457, 461, 463, 488, 492, 527, 529, 532
Finnegan D, 51
Fioretti JM, 166
Firczuk H, 548
Fischbach FA, 189, 190
Fischer W, 342
Fischer WA 2nd, 164
Fisher E Jr, 13, 15, 16, 37, 66, 71, 76, 79, 97, 122, 192, 302
Fisher EM, 164
Fisher MB, 141, 163
Fisher RA, 163, 384
Fisher TN, 13, 15, 16, 37, 66, 71, 76, 79, 97, 122, 192, 302
Fischetti VA, 243
Fishman MR, 461
Fister S, 139
Fitch WM, 218, 341
Fitzgibbon JE, 136
Fitzhenry K, 121
Fives-Taylor P, 49, 91
Flace A, 492, 447
Flamm C, 231
Flanders JR, 233
Flandrois JP, 95
Flannery J, 142, 158
Flather D, 538
Flavell RA, 19, 187, 188, 413, 419
Fleming J, 16, 37, 59, 122, 302
Flores BN, 548
Floridi A, 216
Floyd RA, 130, 133
Flügel RM, 518

Fock C, 234
Fodor K, 317
Foerster J, 513
Foft JW, 539
Fogarty JA, 489
Folkhard W, 61, 63, 66
Folkmann AW, 543
Fomichev IuK, 66
Fomina AN, 377
Fong TT, 113
Fonseca SA, 352, 470, 518
Fontes CJF, 457
Forchhammer J, 317
Forde A, 559
Forest KT, 64
Formiga-Cruz M, 157
Forni M, 493
Forster RL, 234
Fortas E, 281
Foss TR, 202
Foster HA, 127
Foster SL, 501
Fouace J, 1, 12, 67
Fougereau M, 303
Fountain A, 117
Fout GS, 121
Fouts DE, 538
Foux M, 130
Fowler MJ, 324
Fowst G, 89
Fox R, 156
Fradin C, 543
Fraenkel-Conrat H, 29, 216, 221, 247, 249
Fraldi A, 539
Franceschi VR, 542
Francis MB, 105, 193, 207, 418, 443, 449, 450, 451, 452, 461, 462, 463, 464, 466, 518, 527, 529, 532, 533, 538
Francis MJ, 436
Franciszkowicz MJ, 97, 339
Francius G, 200
Francke B, 19, 20, 176, 260, 267, 367
Francki RIB, 31
Franco E, 106
Franco PF, 113
François C, 202
Franek F, 235
Frank H, 26, 196, 306
Franke C, 116
Frankel AD, 538
Franklin RM, 17, 23, 76, 78, 82, 83, 122, 189, 191, 287, 316, 344, 365, 367, 368, 369, 551
Franklin S, 161
Franks WT, 486
Franze de Fernandez MT, 23, 262, 272, 273, 274, 275, 276, 286, 365
Franzreb M, 148
Fraser AR, 334
Fraser VJ, 165
Frauchiger D, 155
Frazier CS, 150
Frei CS, 540
Freigang S, 491
Freivalds J, 433, 466, 556
French MJ, 191
French R, 14
French RH, 529
Frésard L, 551
Fréval-Le Bourdonnec A, 154
Frey DD, 166

Fridborg K, 435, 438, 467, 470
Friedman H, 15, 156
Friedman RM, 367
Friedman SD, 41, 42, 113, 169, 331, 412
Friedmann A, 232
Friedrich R, 210
Friesen JD, 84
Frietze K, 446, 489, 493
Frietze KM, 446, 493
Frist RH, 220
Fritzsch C, 545
Frohnert A, 137
Frolova LY, 219
Frolova OY, 543
Fromageot HPM, 90, 176, 315
Frömmel C, 438
Frost LS, 14, 49, 50, 52, 59, 61, 62, 63, 64, 98
Froussard P, 234
Fruetel JA, 200
Frydman A, 50
Fu X, 543, 548
Fu Y, 517
Fu Yi, 461, 489, 549
Fuchsberger N, 377
Fuerst JA, 55
Fuhrman JA, 118
Fuhrmann G, 489
Fujii H, 551
Fujiki K, 14, 226
Fujimura RK, 88
Fujimura T, 356
Fujino S, 116
Fujioka R, 114, 124
Fujioka Y, 544
Fujishima A, 127
Fujita T, 551
Fujita Y, 285
Fukami H, 320, 389
Fukano H, 266, 426
Fukatsu K, 543
Fukaya T, 548
Fukazawa T, 551
Fuke M, 211, 216, 223
Fukuda A, 14, 55, 226
Fukuma I, 84
Fukuma Y, 138
Fulga TA, 551
Fulmer A, 125
Fumasoli A, 164
Fumian TM, 166, 203
Funakoshi Y, 538
Funamizu N, 137, 141
Funayama R, 551
Fundakowski J, 543
Funderburg SW, 112
Furia A, 367
Furst KE, 137
Furumai H, 42, 104
Furuse K, 13, 19, 23, 35, 37, 39, 40, 99, 101, 102,
        103, 104, 105, 106, 111, 112, 123, 145,
        163, 186, 187, 302, 310, 535
Furuta H, 548
Fusco D, 542, 545
Fushimi K, 538
Fussenegger M, 545
Füzik T, 197

G

Gaboriaud F, 81, 153, 154
Gachet C, 234

Gadir N, 543
Gaggero A, 203
Gagné MJ, 41
Gagnon JA, 543
Gaj T, 550
Gala JL, 108, 109
Gałan W, 42
Galarza JM, 489
Galaway FA, 464
Galdiero M, 518
Galdiero S, 518
Gale AJ, 455
Gallandat K, 165
Gallarda J, 203
Gallardo F, 543, 544
Gallardo G, 501
Gallego I, 418
Gallerani R, 23, 263, 264, 265, 287, 363
Galli A, 544
Gally JA, 302, 303
Galofré B, 142, 544
Galvañ C, 544
Gamage S, 128, 138
Gandhi V, 125
Ganesan U, 233
Ganesh A, 113, 115
Ganime AC, 166
Ganoza MC, 313, 333, 334, 517, 538
Gantzer C, 41, 81, 106, 113, 116, 117, 120, 134,
        141, 145, 146, 149, 151, 153, 154, 159,
        161, 200, 201
Gantzer CJ, 151
Gao J, 161
Gao M, 517
Gao T, 137
Gao X, 543
Garamella J, 97, 339
Garcea RL, 489
Garcia A, 338
Garcia HG, 543
Garcia JF, 543
García M, 166, 203
García-Aljaro C, 113, 142
Garcia-Blanco MA, 538
García-Fernández A, 58
García-Fruitós E, 518
García-García L, 202
Garcia-Valles M, 126
García-Villada L, 417
Garcillan-Barcia MP, 58
Gardarsson SM, 152
Gardiner WC, 77
Gardiner WC Jr, 285, 388
Gardiner Heddle J, 518
Gardner EA, 155
Gardner T, 164
Garen A, 175
Garenne D, 97
Garges S, 268, 281
Garimella PD, 451
Garland J, 156
Garnier L, 202
Garrido A, 139, 141
Gartner TK, 184
Garvey G, 311
Garvin RT, 436
Garwes D, 20, 22, 167, 180, 252
Gary GW, 120
Gasc AM, 184
Gaspar I, 542, 548
Gaspar LP, 143

Gasparian MG, 378
Gasparini M, 235
Gassensmith JJ, 457, 458
Gast M, 109
Gatto D, 489, 490
Gaudin JC, 202
Gaur RK, 356, 540
Gauss-Müller V, 113
Gautam A, 235
Gautam R, 202, 543
Gautier C, 339
Gautreau-Rolland L, 551
Gautret P, 203
Gavis ER, 542, 548
Gawler A, 116
Gazit E, 78, 81
Ge L, 543
Ge S, 144
Ge WP, 543
Ge Z, 546
Geary SM, 501
Gebinoga M, 77
Gedvilaite A, 438, 489
Geeti A, 410
Gehr R, 137
Gehring NH, 538, 540
Gehringer P, 132
Gehrke CW, 341
Geier M, 67
Geis M, 231
Geisbert TW, 517
Gelbart WM, 232, 481
Gelderman G, 543
Gelin-Licht R, 543
Gelinas AD, 352
Gell C, 354
Gemski P Jr, 1, 46, 47, 53, 66
Gendron L, 144, 155
Gendronneau M, 116
Genenger G, 427
Genga RM, 551
Gengenbach TR, 133
Genthner FJ, 41, 42, 113, 169, 331, 412
Gentile DM, 138
Gentile GJ, 166
Gentilomi GA, 107
Gentry J, 42, 113
Gentry-Shields J, 120, 162
Genz C, 543
George L, 545
Georgiades K, 233
Georgiadis MM, 538
Georgieva T, 551
Geraets JA, 477
Gerba C, 155
Gerba CP, 57, 96, 113, 117, 120, 127,
        143, 150, 152, 155, 164,
        202, 391
Gerhardts A, 164
Gerlich WH, 438, 492
Germaschewski V, 438
Germer JJ, 234
Germier T, 544
Gerrity D, 128, 138
Gersberg RM, 117, 148
Gershon PD, 538
Gerst JE, 543, 544
Gervais P, 160
Gervelmeyer A, 121
Gesnel MC, 538, 551
Gessert SF, 202

Gesteland RF, 1, 16, 18, 19, 20, 46, 176, 181, 187, 189, 190, 193, 205, 208, 210, 211, 212, 220, 222, 227, 291, 315, 320, 327, 333, 338, 370
Getzoff ED, 64
Gevorkian RA, 378
Ghaderi E, 131
Ghadirzad S, 162
Ghanem N, 152
Ghasemi A, 518
Ghazvini M, 543
Ghernaout B, 149
Ghernaout D, 149
Gheysen D, 20, 167, 181, 183, 216
Ghezzi P, 235
Gholami M, 517
Ghosh M, 435
Ghosh RP, 551
Ghoshal D, 538
Ghoujal B, 544
Giammar DE, 40, 114
Giannakis S, 134, 142
Gibbons GH, 235
Gibson GR, 105
Gibson GW, 137
Gibson KE, 152, 162
Giddings JC, 96
Giddings TH Jr, 543
Gielen J, 249
Gierer A, 261
Giessen TW, 465, 533
Gilbert CE, 429
Gilbert N, 543
Gilchrist KH, 311
Gilham PT, 216, 221, 223, 226, 458
Gillam F, 445
Gillerman C, 151
Gillerman L, 151
Gilling DH, 162
Gillis E, 19, 185, 219, 224, 273, 274, 284, 384
Gillis EM, 219
Gillman L, 203
Gilmore BF, 132
Gilpin BJ, 113, 115
Gilvarg C, 223, 224, 419
Gimenez G, 233
Gin KY, 113, 114
Gingery M, 280
Gino E, 113, 116
Ginocchio CC, 203
Ginoza W, 46, 73, 132
Gintnere L, 225
Ginuino A, 166
Giordano A, 539
Giorgetti L, 548
Gipson CL, 164
Girard M, 148
Girón-Callejas A, 234
Girones R, 126, 142, 157
Gironés Llop R, 121
Gissmann L, 427, 489
Gitis V, 125, 145, 152, 153, 154, 449
Gittens M, 108, 109, 312
Givens BE, 501
Glaser SJ, 161, 396
Glasgow J, 529
Glasgow JE, 463, 464, 466, 471, 532, 533
Glasier LM, 63
Gleba YY, 543
Glenn EP, 150
Glick BR, 538

Glickman BW, 85, 132
Glidden MD, 529
Glitz DG, 211, 216, 221
Globa LI, 147
Glock C, 548
Gmurek M, 492
Göçerler H, 235
Godiska R, 517
Godson GN, 316, 365, 368, 369
Goeller LJ, 199
Goes A, 489
Goessens WH, 294, 295, 296
Goff SP, 333, 539, 546
Goh MY, 117
Goh SG, 113, 114
Gohr A, 543
Göhring J, 543
Göker M, 556
Gokulan K, 131
Gold A, 130
Gold L, 253, 285, 425
Goldberg E, 74
Goldberg IH, 313
Goldberg ML, 322
Goldberg S, 234
Goldbourt A, 556
Goldfarb P, 223
Goldhill DH, 555
Golding I, 542
Goldman E, 83, 94, 317, 325, 330, 341
Goldman ER, 304
Goldmann U, 438
Goldstein BP, 313
Goldstein J, 332
Golemis EA, 540
Goller KV, 234
Golmohammadi R, 435, 438, 467
Golubkov VI, 67
Gomes AC, 492, 447
Gomes J, 492
Gomez-Blanco J, 73, 192, 243, 476, 477
Gómez Ramos LM, 464
Gomila M, 126
Goncalves N, 542
Gong L, 442
Gonggrijp R, 377
Gonsalves S, 202
Gonzalez N, 117
Gonzales-Gustavson E, 156, 166
Gonzalez-Ibeas D, 103
Gonzalez-Jaen F, 163
Gonzalez-Orta M, 165
González-Rivera C, 64
Goodhouse J, 542
Goodier JL, 543
Gooding C, 538
Goodman H, 223
Goodman HM, 222, 223, 327, 238
Goodman LB, 203
Goodridge L, 107
Goodridge LD, 108, 113, 144
Goodwin K, 141
Goonetilleke A, 164
Gootenberg JS, 550, 551, 558
Gopalakrishna S, 356
Gophna U, 555
Gopisetty VV, 161
Gopmandal PP, 201
Gorbalenia AE, 406
Gorbalenya AE, 29, 30, 31
Gordeeva J, 559

Gordon C, 122
Gordon JA, 23, 367
Gordon MP, 29, 385
Gordon NR, 318
Göringer HU, 539
Gorini L, 85
Gorini P, 314
Gorman JA, 493
Gorman SP, 132
Gorospe M, 311, 545
Gorzelnik KV, 167, 245, 424, 479
Górzny ML, 529
Gosert R, 202
Göthe G, 130
Gott JM, 348, 349, 355
Gottesman S, 280, 283
Götze B, 544
Gou L, 109
Gough NM, 234
Gould JL, 489
Goumballa N, 203
Gourmelon M, 113, 116
Gouy M, 339, 341
Govea-Alonso DO, 501
Govindan K, 157
Gowda K, 517
Goyal SM, 96, 120, 144, 155, 159, 161, 162
Grabau E, 138
Grabovska V, 378
Grabovskaia KB, 67
Grabow WO, 113
Grabow WOK, 112, 113
Gradman AH, 501
Graef G, 203
Graf A, 513
Graf PCF, 202
Graham AF, 12, 69, 70, 82, 87
Graham ME, 75, 79, 235
Graham RC, 150
Graham WG, 132
Grahn E, 354, 359, 470, 537
Gralla J, 221
Grammatikakis I, 311
Gramstad R, 139
Granboulan N, 17, 76, 78, 82, 226, 368
Granovskii NN, 378
Grantham R, 339
Grasis JA, 555
Grasso M, 235
Grasso S, 518
Grathwohl G, 156
Gratz J, 203
Gratzer WB, 193
Graudiņa Ž, 378
Graule T, 200
Graveley BR, 433, 538, 540
Gray NK, 538, 539
Gray NM, 234
Grayson NE, 481, 556
Grdina TA, 539
Grechko VV, 217, 438
Gredell JA, 540
Greef CH, 355
Green L, 333
Green PJ, 535
Green RH, 538
Green RM, 106, 333
Greenaway PJ, 12
Greenbaum NL, 359
Greene EC, 545
Greenleaf WJ, 356, 486, 546

Gref R, 518
Grego S, 311
Gregor J, 149
Gregor T, 543
Greider CW, 542
Greiner A, 431
Gren E, 77, 83, 85, 91, 93, 288, 371, 422, 436, 437, 489
Gren E, 91
Gren EI, 76
Gren EJ, 25, 76, 77, 80, 83, 85, 91, 92, 93, 98, 175, 185, 194, 219, 283, 315, 322, 333, 349, 357, 371, 372, 374, 375, 383, 431, 436, 438
Greninger AL, 15, 42, 105, 170, 226
Grēns E, 1, 76, 316, 394, 436
Grewe B, 538
Gribanov V, 322
Gribanov VA, 349
Gribencha SV, 377, 378
Gribnau J, 543
Griebler C, 140
Grieco P, 518
Griffin DW, 40, 113, 114, 118
Griffin MT, 198
Griffith JF, 113
Griffiths WD, 164
Grigorieff N, 538
Grilo AM, 280
Grimes JM, 435
Grimm JB, 544
Grinshpun SA, 144, 163
Grinsted J, 56
Grinstein E, 436
Grinstein EE, 436, 438
Grippin AJ, 501
Grisafi P, 187
Grishina EV, 54
Groeneveld H, 40, 20, 225, 228, 294, 295, 335, 424
Grohmann A, 123
Grollman AP, 314
Gromak N, 538
Groman NB, 88, 93, 94, 95, 297
Grøndahl-Rosado RC, 116
Grondin GH, 120
Groner M, 264
Groner Y, 23, 263, 324, 456
Grootegoed JA, 543
Gros C, 414
Gros F, 313, 324, 325
Grosberg AY, 383
Groschup MH, 202
Grosjean H, 167, 218, 340
Grosse F, 161
Grossi G, 548
Grosveld F, 543
Grotto RM, 516
Grozdanov PN, 538
Gruau G, 113
Gruber CJ, 54, 79
Grubman MJ, 323
Grueber CE, 42
Gruffat H, 433
Grunberg-Manago M, 324
Grunert A, 137
Grünwald D, 543, 545, 548
Grunwald T, 538, 539
Grywna K, 516
Gschwender HH, 96
Gu J, 235

Gu W, 540
Gu X, 113, 114
Gualerzi C, 334
Gualerzi CO, 329
Gualtieri A, 234
Guan D, 39, 150, 158
Guan H, 145
Guan Y, 156, 517
Guarino H, 144
Gubbens J, 324
Guérit D, 548
Guerra CE, 392, 396
Guerra J, 544
Guerreiro A, 108, 109, 312
Guerreiro SI, 280
Guerrero A, 106
Guerrero-Latorre L, 126, 156
Guerrier-Takada C, 266, 267
Guessous G, 543
Guet CC, 54
Guet D, 548
Guha S, 199
Guillier L, 160
Guillou A, 543
Guindy YS, 341
Guisez Y, 337
Gulati NM, 455
Gulei D, 551
Gull K, 436
Gumireddy K, 539
Gummuluru S, 544
Gun J, 145, 154, 449
Gunderson SI, 538
Gunhanlar N, 543
Gunji Y, 279
Gunnarsdottir MJ, 152
Gunter-Ward DM, 126
Guo C, 304
Guo D, 551
Guo K, 133
Guo L, 109
Guo X, 153
Gupta A, 542
Gupta AK, 267
Gupta N, 19
Gupta P, 214
Gupta R, 556
Gupta SL, 221, 330
Gupta V, 501
Gurgo C, 313
Gusars I, 440
Gussin G, 19, 24, 177, 178, 186, 237, 344
Gussin GN, 1, 17, 19, 21, 175, 178, 186, 192, 237, 319, 331, 344
Gusti V, 356
Guthrie GD, 67
Gutierrez L, 131, 134
Gutierrez M, 107
Gutiérrez-Aguirre I, 139
Guttler T, 538
Guzmán C, 151
Guzmán MR, 280
Gwerder M, 490
Gygi SR, 538
Gytz H, 272

H

Ha N, 543
Haas B, 234, 544
Haas CN, 143, 203

Haas R, 427, 431
Habenstein B, 556
Haberman N, 543
Habibi Najafi MB, 162
Habteselassie M, 199
Hackett J Jr, 234
Haddad F, 109
Hadley D, 202
Hadravová R, 197
Haegeman G, 24, 167, 211, 212, 214, 216, 288, 319, 323, 339, 365, 419, 427
Hafner EW, 85
Hagan MF, 481
Hagen B, 234
Hagervall TG, 342
Häggström J, 136
Haghiri B, 202
Hagström AE, 311, 312
Hague LK, 538
Hahn H, 63
Hahn T, 149
Haider T, 132
Haig DA, 132
Haim-Vilmovsky L, 543
Haimovich G, 543, 544, 548
Haisma HJ, 558
Hajenian H, 120, 135
Hajitou A, 551
Hajnsdorf E, 283
Hake H, 284
Hakim H, 129
Häkkinen A, 542
Håkonsen T, 134
Halasa NB, 202
Hale EM, 306
Hall BD, 82
Hall KB, 169, 527
Hall PR, 493
Hall-Pogar T, 538
Haller C, 299, 425
Haller W, 96
Halsall HB, 310
Halstead JM, 547
Hamada S, 542
Hamaguchi K, 156
Hamasaki H, 19, 82, 186
Hamblin MR, 518, 529
Hambsch B, 148
Hamdi M, 108, 113, 132
Hamil B, 138
Hamilton KA, 164
Hamilton MT, 164
Hammamieh R, 235
Hammann-Haenni A, 490, 512, 518
Hammer TR, 164
Hämmerle H, 281
Hammond RW, 394
Hamon Y, 67
Han C, 150
Han CH, 405, 543
Han DX, 527
Han G, 489
Han J, 201
Han JC, 136
Han K, 551
Han S, 453, 538
Han SW, 551
Han Y, 517, 518, 545
Hanagata N, 518
Hanamoto S, 104
Handa H, 518

Handzel TR, 106
Hänninen ML, 154
Hansen CJ, 202
Hansen IA, 108, 312
Hansen JJ, 137
Hansen KC, 548
Hansmann B, 166
Happe M, 258
Happel R, 400
Haque M, 166
Haque SJ, 551
Hara Y, 551
Harada K, 88
Haramoto E, 104, 105, 113, 116, 117, 147
Hardesty B, 329, 337
Hardie JD, 197
Hardiman T, 234
Harding AS, 141
Hardy K, 427
Harel N, 555
Hargreaves M, 144
Hargrove AE, 545
Hargy TM, 120, 124
Harigai H, 35, 39, 99, 101, 102, 181, 187, 302,
        310, 535
Harik O, 136
Harkness RE, 299
Harlen KM, 548
Harley R, 202
Harm W, 123
Harms H, 152
Harmsen DJ, 125
Harnett G, 202
Harrach B, 29, 30, 31
Harrington GW, 153, 292, 463, 539
Harrington L, 153, 539
Harris D, 267
Harris JM, 82, 86
Harris KS, 235
Harris R, 61
Harris SA, 354
Harrison CJ, 202
Harrison MM, 545
Harrison PM, 189, 190
Hart GJ, 234
Hart RD, 235
Hartard C, 113, 146, 159
Harte F, 143, 160
Harte FM, 140
Hartikainen J, 235
Hartman EC, 193, 207, 418, 461, 471
Hartman KA, 207
Hartman PE, 176
Hartman PS, 125
Hartmann G, 396
Hartmann GR, 161, 396
Hartmann-Boyce J, 501
Hartmuth K, 538
Hartung J, 127
Haruki M, 217
Haruna I, 23, 38, 178, 257, 258, 261, 262, 288,
        315, 389
Harwood VJ, 113
Hasegawa M, 218
Hasegawa S, 87, 262
Hasegawa T, 129, 387
Haselkorn R, 315
Haseltine WA, 265, 314
Hashem E, 310
Hashimoto A, 129
Hashimoto K, 127

Hashsham SA, 113
Hasman H, 58
Hass EP, 551
Hassan F, 202
Hassanizadeh SM, 120
Hassen A, 127
Hasty J, 551
Hata A, 104, 106, 108, 127
Hatfield GW, 341, 543
Hatsukami D, 501
Hattier L, 410
Hattman S, 94
Hatton TA, 131
Hattori N, 551
Hauber J, 550
Haubova S, 197
Hauchman FS, 136
Haught RC, 145, 154
Hausen P, 29, 210
Hautbergue GM, 543
Havelaar A, 116
Havelaar AH, 40, 103, 105, 106, 111, 112, 114,
        119, 120, 123, 163, 169
Hawkins MA, 124
Hawley AL, 142
Hayasaka H, 285
Hayashi G, 545
Hayashi K, 540
Hayder Z, 517
Hayes CS, 64
Haynes JR, 436
Hayward AC, 55
Hayward WS, 262, 273, 274, 276
Haywood AM, 23, 67, 80, 82, 83, 86, 237
Hazan R, 85, 183, 253
He A, 539
He F, 235
He J, 543
He M, 250
He Q, 489
He WJ, 235
He X, 150
He Y, 543
He YH, 356
Heading CE, 501
Heal KG, 440, 518
Heaney C, 120
Heath MD, 557
Heather JM, 555
Hébrant M, 145
Hecht L, 113, 193
Hecht R, 400
Heck AJ, 199
Heckl D, 551
Hecquet D, 202
Hedberg CW, 120, 159, 161
Hedges RW, 15, 16, 37, 56, 57, 58, 59, 60, 66,
        122, 302
Heekin RD, 501
Heer S, 491
Heermann KH, 438
Heesemann J, 280
Heffron J, 134
Heger Z, 529
Heguy A, 543
Hehlmann R, 367
Heimbuch BK, 144
Heineman WR, 310
Heininger U, 202
Heinlein M, 543
Heinonen-Tanski H, 106, 124, 139, 144, 163

Heinrich S, 548
Heisele O, 378
Heisenberg M, 19, 71, 178, 185, 186, 193, 237
Heistad A, 118, 134
Hekmatmanesh A, 518
Helgstrand C, 354, 359, 537
Hellard ME, 113
Heller KJ, 105
Heller RC, 234
Helling A, 166
Hellingwerf KJ, 65
Hellström K, 544
Helm M, 538
Helmer RD, 138
Helmer-Citterich M, 540
Helmi K, 145
Helmuth R, 51, 61
Hendler RW, 315
Hendricks A, 398
Hendrix H, 414
Hengartner H, 447
Hengge-Aronis R, 280
Henkens RW, 193, 303
Hennechart C, 107
Hennechart-Collette C, 160
Hennig S, 548
Henriet S, 156
Henry M, 553
Henshilwood K, 157
Hentze MW, 538, 540
Herath G, 120, 121, 154
Herbold-Paschke K, 149
Herbst RH, 543
Hermann T, 352
Hermesh O, 543
Hermodson MA, 63
Hermoso JM, 263, 325
Hernandez M, 127
Hernández-Delgado EA, 112
Hernández-Munain C, 538
Hernroth BE, 157
Herold WH, 130
Heroven AK, 280
Herrlich P, 52
Herrmann A, 543
Herrmann R, 26, 195, 196
Hersberger M, 491
Herschlag D, 486
Hertel KJ, 433, 538, 540
Herzberg M, 501
Herzog AB, 113
Hess GT, 551
Heuck A, 543
Heumüller M, 548
Hewat E, 234
Hewitt J, 155, 202
Hicks DT, 39, 150, 158
Hidaka T, 42, 105
Hiebert E, 195, 197
Hierowski M, 96
Hietala SK, 516
Higgins CA, 544
Higgins MK, 501
Higginson C, 465, 488
Highfield PE, 436
Highsmith AK, 165
Higuchi T, 143
Hijikata N, 137
Hijnen WA, 124, 125, 148, 153
Hill BD, 352
Hill C, 555, 559

Hill D, 279, 389
Hill HR, 440, 518
Hillaireau H, 518
Hillebrand F, 538
Hilleman MR, 375, 376
Hillman BI, 394
Hillyard DR, 515
Hilner C, 113
Himes RH, 318
Himsworth D, 202
Hincapie R, 457
Hindennach I, 258, 316
Hindiyeh MY, 203
Hindley J, 19, 176, 210, 222, 223, 238, 315, 327
Hindocha DM, 210
Hinds L, 130
Hink T, 165
Hinton HJ, 447, 459, 460, 489, 490, 491, 518
Hippeläinen M, 235
Hiraga S, 82
Hirakata A, 8
Hiraku Y, 491, 518
Hirani ZM, 156
Hirano RA, 154
Hirao I, 352, 470
Hirasawa M, 538
Hirashima A, 13, 23, 35, 39, 40, 99, 101, 102, 103, 104, 112, 169, 181, 187, 224, 225, 228, 294, 302, 310, 331, 424, 426, 431, 535, 536
Hirata AA, 306
Hirata T, 129
Hirayama J, 130
Hiriart E, 433
Hirokawa G, 333
Hiron TK, 551
Hirooka K, 153
Hirose M, 143
Hirose S, 19, 82, 186
Hirose T, 40, 169, 225, 228
Hirota K, 8
Hirota Y, 46, 54, 235
Hirsch HH, 202
Hirsh D, 253
Hiscox JA, 73, 243, 476, 477
Hmaied F, 108, 113, 132
Ho DN, 538
Ho J, 109
Hoad VC, 202
Hoang VT, 203
Hobbs M, 64
Hochberg H, 545
Hochschild A, 548
Hocine S, 546
Hockert JA, 538
Hodge CA, 543
Hodgson CJ, 150
Hodgson KR, 159
Hoefer D, 164
Hoenninger VM, 544
Hofacker IL, 231
Hoff JC, 120
Hoffler E, 283
Hoffman R, 234
Hoffmann B, 538
Hoffmann V, 144
Hoffmann-Berling H, 1, 12, 15, 16, 18, 19, 23, 75, 79, 97, 189, 205, 367
Hofmann A, 299, 425

Hofmann R, 125, 140
Hofschneider PH, 1, 12, 15, 16, 19, 20, 23, 29, 67, 72, 75, 79, 85, 86, 91, 94, 96, 176, 190, 241, 260, 301, 367, 368
Hofstetter H, 19, 167, 192, 253
Hofstra KR, 234
Hogan CJ Jr, 143, 144, 163, 199
Hogeboom WM, 103, 112, 106, 112, 163
Högenauer G, 54, 314
Hogg C, 42
Hogg JR, 546
Hogle JM, 435
Hohn B, 1, 17, 25, 26, 45, 47, 49, 56, 74, 79, 178, 192, 195, 196, 237, 461, 557
Hohn T, 1, 17, 25, 26, 79, 178, 192, 195, 196, 237, 461, 557
Hokajärvi AM, 118
Hoke A, 235
Holady JC, 128
Holder PG, 450, 461, 518, 529
Holguin SY, 464
Holland AF, 115
Holland EG, 304
Holland J, 138, 410
Holland JJ, 413
Holler S, 109, 200
Holley RW, 211
Hollingdale MR, 440
Holloway BW, 55, 56, 98
Holloway RW, 156
Holmes EC, 42
Holmes WM, 341, 539
Holmes ZE, 548
Holmquist R, 406
Holt DM, 136
Höltje JV, 296, 297, 424
Holton J, 143
Holtz G, 342
Holtz J, 233
Holtzman DM, 501
Holub JM, 453
Homann M, 539
Honda T, 144, 163
Honeyman G, 153
Hong CA, 518
Hong S, 405, 453
Hong SS, 518
Hong SW, 133
Hong V, 492
Hong X, 131
Honikel KO, 396
Hoogendam BW, 314
Hook B, 539
Hook EC, 210
Hooker JM, 105, 443, 449, 450, 451, 461, 462, 518, 529
Hoover DG, 39, 150, 158
Höper D, 160
Hopp TP, 440
Horan L, 548
Hori K, 23, 85, 88, 258, 260, 261, 262, 265, 272, 286, 365, 368, 548
Hori M, 313
Horimoto H, 138
Horiuchi K, 1, 15, 19, 53, 175, 176, 177, 178, 180, 181, 183, 184, 187, 237, 252, 315, 331, 370, 410
Hörlein D, 427
Horm KM, 140, 161
Hörman A, 154
Hormann MP, 111, 112

Horn WT, 354, 359, 460, 461, 470, 537
Hornstra LM, 136, 152
Horodyski F, 138
Horovitz I, 129
Horvathova I, 548
Hoser M, 202
Hoshino S, 538
Hoskins AA, 538, 545, 548
Hosoda K, 285, 401, 402, 425, 426
Hosoda N, 538
Hosogane M, 551
Hosseini SR, 162
Hosseinidoust Z, 166, 414
Hotham-Iglewski B, 83, 344, 365, 369
Hottin J, 155
Hotz G, 122, 132
Hotze EM, 130
Hou B, 235, 491, 492, 518
Hou C, 543
Hou G, 556
Hou JJ, 539
Hou Y, 551
Houba-Herin N, 40
Houck-Loomis B, 333
Houde A, 115
Houghton M, 234
House WL, 134
Hövelmann F, 545
Hovlid ML, 465, 488
Howard G Jr, 165
Howarth M, 460, 513, 514, 533, 557
Howe TG, 85, 295
Hoyle NP, 543
Hoyles L, 42, 105
Hryc CF, 479
Hryckiewicz K, 544
Hsiao C, 464
Hsiao HH, 470
Hsieh CH, 235
Hsieh CM, 470
Hsieh K, 234
Hsieh YS, 235
Hsu FC, 40, 82, 107, 113, 114
Hsu HY, 162
Hsu IW, 161
Hsu M, 161
Hsu NY, 144
Hsu PD, 549
Hsu WT, 88, 135, 160, 325, 539
Hsu WY, 154
Hsu YC, 135, 325
Hsu YH, 543
Hu B, 120
Hu CY, 410
Hu F, 517
Hu H, 120, 161
Hu HY, 165
Hu JY, 125, 140, 145, 150
Hu L, 134
Hu TL, 61
Hu WS, 544, 548
Hu X, 124, 131, 542
Hu Y, 131, 544
Hua Z, 492
Huang C, 202
Huang GN, 545
Huang H, 155, 156, 551
Huang HY, 132
Huang J, 433, 440, 516, 540
Huang L, 161, 501, 518, 539
Huang NC, 518

Huang Q, 199, 539
Huang S, 433, 540
Huang SC, 538
Huang SH, 202
Huang T, 550
Huang TH, 470
Huang W, 543
Huang WW, 501
Huang X, 109, 131, 140, 155
Huang YJ, 527
Huang YP, 543
Huang Z, 551
Huber M, 513
Hubert N, 539
Huck PM, 153
Hudson JB, 82
Hudson R, 163
Huelsenbeck JP, 40, 169, 226, 405, 415
Huffman D, 136, 202
Huggins CJ, 543
Hughes JB, 131
Hughes SH, 342
Hughes VM, 15, 16, 37, 57, 59, 60, 122
Hügler M, 148
Huguet L, 161
Huh H, 559
Hukari K, 200
Hull NM, 127, 130
Hull R, 1, 469
Hulme AE, 543
Hüls D, 543
Humble PJ, 151
Humphrey TJ, 157
Humphries B, 149
Hundesa A, 126, 142, 156, 166
Hundhausen C, 493
Hundt E, 427
Hung ML, 543
Hung PP, 26, 39, 83, 94, 195, 306
Hung ST, 548
Hunn BHM, 543
Hunt D, 83, 84
Hunt SC, 235
Hunter AC, 538
Hunter OV, 538
Hunter T, 368
Hunter Z, 447, 493
Hunziker L, 462
Huo CX, 131, 140
Huppert J, 1, 12, 67, 94
Hur HG, 111, 112, 117
Huret C, 492
Hurst CJ, 117, 120
Hurst NJ, 203
Hurwitz J, 219
Husimi Y, 95, 404
Husmann D, 551
Hussain I, 306
Hussain SS, 517
Hutchins JE, 89
Hutchison III CA, 291
Hüttelmaier S, 538, 539
Hutten S, 542
Huynh A, 548
Huynh T, 466
Hwang CC, 202
Hwang HJ, 162
Hwang J, 133, 144
Hwang JY, 163
Hwang TM, 125
Hyman P, 29, 555

Hynek D, 529
Hyun J, 133

I

Iampietro C, 545
Iansone IV, 225
Iba H, 55
Ibáñez PF, 134
Ibarluzea JM, 116
Iberer R, 54
Ichida K, 105
Ichihashi N, 285, 401, 402, 404, 425, 426
Ichikawa T, 267, 425
Iemura S, 543
Igarashi K, 19, 82, 186, 333
Igarashi SJ, 19, 91, 186, 212, 289, 317
Iglewski WJ, 72, 73, 82, 83, 94, 186, 367
Ihara M, 154
Iida H, 55
Iijima T, 54
IJpelaar GF, 125
Ikawa Y, 548
Ikebuchi K, 130
Ikeda H, 130
Ikeda K, 88
Ikeda Y, 86, 122, 301, 367, 365
Ikegami T, 543
Ikehata K, 137
Ikemura T, 227
Ikner LA, 108, 155
Ilagan JO, 538
Illig C, 542
Ilyas A, 152
Ilyin YV, 320, 489
Ilyinskii PO, 501
Im D, 138
Imahori K, 320
Imai T, 143
Inal JM, 298
Inayama S, 40, 169, 225, 228
Indah S, 153
Indig FE, 545
Indik S, 234
Indikova I, 234
Ingram LC, 56
Inoko H, 39, 99, 102, 105, 163, 102
Inokuchi H, 333
Inokuchi Y, 35, 39, 40, 99, 101, 102, 105, 163,
            169, 187, 224, 225, 228, 279, 302, 331,
            424, 426, 535, 536
Inomata T, 285
Inoue K, 42
Inoue M, 8
Inoue T, 540, 543, 544
Inouye H, 23, 263, 325
Inouye M, 40, 232, 233, 252, 294, 296, 331, 535
Inuzuka M, 54, 61
Inuzuka N, 54
Ippen KA, 46, 47, 49, 53, 69, 70, 94
Ippen-Ihler K, 49, 51, 61, 98
Iranpour R, 154
Irenge L, 108
Iriarte M, 163
Irorere VU, 529
Irudayaraj J, 545
Irving WL, 234
Isaacman S, 529
Isaev A, 559
Isaeva DM, 329, 364
Isaksson LA, 317

Isarov SA, 455
Isenberg H, 193
Iserentant D, 181, 217, 288
Ishak M, 163
Ishibashi M, 46, 48
Ishida C, 138
Ishida H, 8
Ishida Y, 232
Ishigami M, 387
Ishiguro H, 127
Ishihama A, 279, 280
Ishikawa J, 548
Ishikura H, 387
Ishinaga H, 235, 491, 518
Ishiyama K, 542
Ishizuka AS, 460
Isidor MS, 551
Islam MS, 160
Isono K, 322
Isono S, 322
Israeli-Reches M, 19, 85, 86, 90, 91, 167, 183,
            192, 194, 232, 253
Itano MS, 544
Ito M, 538
Ito R, 137, 141
Ito Y, 262
Ito YH, 23, 178, 257, 288, 315
Itoh H, 501
Iturriza-Gomara M, 234
Itzkovitz S, 544, 545
Iushmanov SV, 406
Ivan C, 150
Ivanchenko S, 544
Ivics Z, 551
Iwakawa HO, 538
Iwasaki K, 20, 518
Iwata A, 280

J

Jacak J, 543
Jacak R, 148
Jacangelo JG, 96, 120, 138, 155, 156
Jackevica L, 445, 501
Jackson NB, 166
Jackson TC, 283
Jackson WS, 550
Jacob A, 543
Jacob F, 50
Jacob P, 145
Jacobson A, 49, 74, 227, 536, 538
Jacobson AB, 40, 169, 224, 225, 227, 228, 364,
            422, 423
Jacobson LA, 372
Jacobson LS, 135
Jacobson SC, 480
Jacquemyn M, 544
Jacrot B, 191
Jadas-Hécart A, 113
Jaedicke A, 538
Jaeger A, 203
Jafari S, 165
Jaffrey SR, 539, 545
Jaffrezic MP, 154
Jafry HR, 128
Jager A, 452, 527
Jäger P, 489, 518
Jain R, 84
Jakana J, 167, 245, 424, 435, 479
Jakes K, 22
Jakobson CM, 193, 207, 463, 471, 532

Jaktar K, 155
Jakubowska M, 42
Jakubowski H, 83
Jalinska A, 76
James TL, 193
Jan LY, 543
Jan YN, 543
Janaki L, 203
Janda KD, 501
Jandrig B, 438
Jandus C, 512
Janeczko R, 202
Janenisch R, 67
Janes M, 81, 107
Jang GM, 538
Jang HJ, 559
Jang J, 109, 144, 405, 543
Jang KL, 117
Jang KM, 233
Janicki SM, 541, 545
Janik M, 109
Janitzek CM, 460, 513, 514
Janjic N, 352
Janner A, 484
Janosi L, 224, 536
Janowski AB, 42, 105, 163, 171, 407, 555
Jansen RP, 538, 543
Jansone I, 225, 322, 333, 349, 357, 439
Jansone IV, 219, 349
Jansons J, 42, 250, 556, 557
Janssen GMC, 267
Jantsch MF, 543
Janzen EM, 117
Japhet MO, 234
Jaramillo A, 551
Jaramillo AM, 542
Jardé E, 113
Jarke C, 141
Jarmoskaite I, 486
Jarrold MF, 199
Jarvis CI, 202
Jarzyna PA, 529
Jasiecki J, 86
Jaspars EM, 197
Jasper MN, 129
Jaudzems K, 486, 487, 556
Javidi-Parsijani P, 324
Jay G, 263, 325, 367
Jayant L, 337
Jayanth N, 544
Jayaraj K, 130
Jayaraman K, 329
Jaykus LA, 157, 162
Jazurek M, 551
Jean J, 161
Jean MN, 233
Jeang KT, 536
Jeanis KM, 129
Jeanneau L, 113
Jeantet C, 313, 325
Jebri S, 108, 113, 132
Jechlinger W, 299, 425
Jeevanandam J, 518
Jegerlehner A, 447, 458, 459, 460, 489, 490, 491, 518
Jencson A, 164
Jencson AL, 165
Jenkins ST, 85, 295
Jennings GT, 447, 489, 490, 501
Jennings PA, 55
Jenny RM, 129

Jensen GJ, 64
Jensen SL, 120
Jeong EY, 117
Jeong H, 489
Jeong K, 538
Jeppesen PGN, 19, 176, 211, 214, 216, 218, 220, 315, 320, 370
Jerala R, 551
Jett M, 235
Jetté LP, 163
Jewett MC, 540
Jeyakumar V, 235
Ji D, 235
Ji M, 235
Ji X, 433
Jia J, 235
Jia T, 518
Jiang M, 391
Jiang RT, 493
Jiang SC, 114, 155
Jiang W, 479
Jiang Z, 480
Jiménez B, 139
Jin HE, 518
Jin J, 460, 513
Jin L, 120, 150
Jin M, 108, 153
Jin S, 136
Jin SE, 518
Jin Y, 120, 148, 153, 201, 543
Jo A, 543
Jochmus I, 427
Jochmus-Kudielka I, 427
Jockusch H, 22, 223, 224, 237, 258
Jockusch S, 130
Joerger RD, 150, 588
Joe YH, 144
Jofre J, 40, 106, 113, 115, 116, 117, 118, 122, 132, 136, 142, 151, 157, 544
Johannes GJ, 539
Johansen CA, 234
Johansson E, 144
Johansson HE, 352, 355, 537
Johansson M, 548
John A, 165
Johne R, 160
Johns MW, 41, 107, 115, 161
Johnson AE, 266
Johnson AM, 548
Johnson B, 324
Johnson CM, 14
Johnson DS, 544
Johnson GK, 233
Johnson HR, 105
Johnson JE, 190
Johnson K, 304
Johnson LA, 161
Johnson N, 489
Johnson PC, 120
Johnson PF, 543
Johnson RC, 64, 267
Johnson SA, 166
Johnston ID, 109
Johnston LP, 501
Johnston MI, 342
Johnston RJ Jr, 545
Johnston TC, 529
Jokitalo E, 544
Jolis D, 124, 154
Jones D, 138
Jones DL, 142

Jones HH, 119
Jones MV, 163
Jones R, 543
Jones TH, 41, 107, 115, 161
Jonkers I, 543
Jordan FL, 155
Joseph A, 545
Joseph O, 126
Joseph S, 538
Joshi A, 538
Jossent J, 145, 154
Joung J, 549, 550, 551, 558
Jouvenet N, 544
Joyce E, 151
Joyner JA, 493
Jubinville E, 203
Juffras AM, 398, 426
Juhas M, 299, 425
Julius C, 539
Jung BK, 551
Jung G, 235
Jung JH, 199
Jung SH, 162
Jung YT, 234
Jungi WF, 490
Jurač K, 96
Juretschko S, 202
Jurgens LA, 202
Jurica MS, 538, 546
Jutz G, 529

K

Kaas L, 155
Kabir MS, 234
Kabirov ShK, 377
Kacian DL, 383, 384, 386
Kacprowski T, 235
Kaczmarczyk L, 550
Kadoya T, 169, 417
Kadykov VA, 205, 377
Kadyrova AA, 377
Kaelber JT, 479
Kaempfer R, 263, 324, 325, 367
Kaerner HC, 12, 23, 25, 73, 75, 195, 241, 367
Kaesberg P, 1, 19, 22, 25, 35, 91, 137, 167, 179, 187, 194, 195, 196, 220, 226, 237, 249, 253, 302, 318, 319, 331
Kafasla P, 538
Kafri P, 543, 544
Kaganman I, 546
Kahrs C, 166
Kaido M, 538
Kailasa SK, 538
Kaisers W, 538
Kaji A, 224, 313, 316, 332, 333, 536
Kaji H, 333
Kajioka J, 127
Kajita K, 543
Kajitani M, 40, 279, 285, 426
Kakegawa T, 19, 82, 186
Kakoschke SC, 280
Kakoschke TK, 280
Kałafut J, 551
Kalashnikova JI, 217
Kalb R, 139
Kaledin LA, 131, 149, 155
Kaledin TG, 149, 155
Kaleni P, 163
Kalis J, 436
Kalisky T, 545

Kalkhoran BF, 489
Kallies R, 152
Kallio K, 544
Kalmakoff J, 377
Kalmykova A, 543
Kalnik MW, 501
Kalniņš G, 42, 250, 444, 556, 557
Kalnins P, 439
Kalnoky M, 203
Kalo A, 543, 544
Kaloyeros AE, 451
Kalt FR, 140
Kaltgrad E, 453
Kamalian LA, 378
Kamarasu P, 162
Kamel M, 489
Kamen R, 23, 212, 252, 258, 263, 264, 265, 278,
    287, 363, 386, 456
Kamenski P, 543
Kamer G, 285
Kamiko N, 121
Kaminski PA, 279
Kamiya T, 85, 265
Kamiyama M, 87
Kanaya S, 217
Kandavalli V, 542
Kandhavelu M, 542, 545
Kandolf R, 367
Kang CW, 127
Kang D, 162, 538
Kang DH, 129, 160
Kang JH, 117
Kang M, 128, 132, 134
Kang P, 458
Kang S, 131, 518
Kang ST, 235
Kang Y, 538
Kannaiah S, 543, 548
Kanniess F, 490, 518
Kannoly S, 40, 58, 59, 60, 71, 295, 425
Kano S, 543
Kantzas A, 152
Kanwal F, 540
Kanwar N, 202
Kapanidis AN, 543
Kaplan S, 322
Kappel K, 486, 545
Kapuscinski RB, 112, 141
Kar A, 311
Karathanasis SK, 139
Karch CP, 489
Karch F, 543
Karczag A, 132
Karim MR, 121, 150
Karimi M, 518, 529
Karimian Pour N, 543
Karna SKL, 558
Karnik S, 167, 243
Karpov OV, 378
Karpova GG, 334
Karpova T, 139
Karunanithy G, 197
Kashiwagi A, 52, 169, 285, 417, 559
Kashiwagi K, 19, 82, 186
Kastelein RA, 20, 219, 252, 293, 294, 296, 420,
    424, 431
Kastelic KA, 358
Kastner B, 538
Kasuga I, 42, 104
Katanaev VL, 97, 339
Katayama H, 42, 104, 106, 108, 117, 121, 156

Katayama Y, 543
Katen S, 480
Kato A, 279
Kato M, 545
Kato R, 121
Kato T, 156
Kato Y, 138
Kator HI, 106
Katsuki M, 23, 39, 40, 99, 102, 103, 104, 112
Katz BD, 137
Katz EM, 451, 543
Katz N, 234
Katz ZB, 548
Katzameyer MJ, 161
Katze J, 186, 249, 250, 307
Kaudewitz F, 19, 50, 134, 186
Kaufman OH, 543
Kaufmann B, 234
Kaul D, 540
Kauling J, 125
Kauppinen A, 116, 118
Kaushik AM, 234
Kaushik N, 203
Kavishe R, 460, 513
Kawai N, 542
Kawamura A, 129
Kawano M, 518
Kawashiro J, 35, 302
Kawata K, 135, 137
Kay AC, 324
Kay BK, 304
Kay D, 152
Kaye AM, 257, 315
Kaye G, 162
Kazaks A, 20, 40, 73, 140, 192, 226, 243, 295,
    344, 406, 425, 431, 433, 436, 440,
    445, 466, 476, 477, 489, 501, 556
Kazama S, 137
Kazazi D, 492
Kazazian HH Jr, 543
Kaziro Y, 87
Kazlauskiene M, 558
Kazuta Y, 404, 426
Kearns NA, 551
Keaveney S, 121, 158
Kedziora A, 131
Keef T, 484
Keegan A, 150, 156
Kehl SC, 202
Keifer DZ, 199
Keil TU, 85, 368
Keim EK, 156
Keino H, 8
Kekez MM, 138
Kelemen RE, 518
Kellenberger E, 175
Keller A, 151, 518
Keller E, 435, 491
Keller SA, 489
Kelley JJ, 220
Kelley R, 543
Kellner S, 538
Kellner T, 202
Kelly HG, 367
Kelly J, 383
Kelly RB, 23, 257, 367, 489
Kelly SH, 489
Kemler I, 544
Kempf C, 452
Keng D, 200
Kenndler E, 200

Kennedy EM, 538, 550
Kenner L, 234
Kent SJ, 367
Kenter A, 543
Kenyon JC, 538
Kenyon KF, 143
Keppie N, 203
Kerekes L, 53
Kern F, 235
Kervestin S, 538
Kerwick MI, 136
Keryer-Bibens C, 540
Kesawat MS, 548
Kessler PD, 501
Keswick BH, 120, 155
Ketratanakul A, 147
Kettleson EM, 144, 199
Keuckelaere A, 548
Keweloh HC, 95
Kézdy FJ, 326
Khalil IR, 529
Khalil SM, 141
Khan SJ, 138, 156
Khan SR, 84
Khara P, 120
Khare S, 131
Kharrat A, 538
Khatri M, 118
Khazaie K, 342
Khechara MP, 529
Khera E, 352
Khorana HG, 211
Ki SJ, 117
Kibiki G, 203
Kickhoefer VA, 539
Kidder L, 203
Kidmose RT, 268, 426
Kiebler MA, 544, 545, 548
Kieke BA, 121
Kiel JL, 356
Kiel MC, 313
Kiełbus M, 551
Kiesel B, 152
Kikuchi Y, 224
Kikumoto T, 279
Kilanzo-Nthenge A, 126
Kilchert C, 543
Kilonzo-Nthenge A, 126, 161
Kilwinski J, 162
Kim CK, 138
Kim DK, 129
Kim DS, 540
Kim ES, 188, 234
Kim EY, 543
Kim G, 543
Kim H, 78, 95, 107, 453, 518
Kim HE, 132, 133
Kim HJ, 529, 548
Kim HK, 543
Kim IS, 529
Kim J, 132, 144, 311, 312
Kim JH, 117, 125, 127, 128, 130, 131, 142, 148
Kim JJ, 64
Kim JY, 134
Kim KH, 538
Kim KM, 539
Kim KS, 117
Kim KW, 156
Kim MS, 133
Kim S, 281
Kim SB, 148

Kim SG, 233
Kim SH, 548
Kim SJ, 129, 141
Kim SS, 160, 199
Kim SY, 108, 117
Kim T, 133
Kim YK, 538, 539
Kimberley M, 149
Kimble J, 538, 543
Kimitsuna W, 266, 426
Kimmig R, 427
Kimmitt PT, 164
Kimmitt RT, 234
Kimura H, 285
Kindler P, 85, 368
Kines RC, 442, 461, 493
King AMQ, 29, 30, 31
King CR, 501
King DW, 303
King W, 398
Kingsley DH, 136, 159
Kingsman AJ, 436
Kingsman SM, 436
Kinjo M, 387
Kinor N, 543
Kinoshita N, 543
Kinouchi Y, 129
Kinsman M, 493
Kinzirskii AS, 378
Kinzler MG, 489
Kirchberger PC, 79
Kirchhausen T, 545
Kirk C, 87
Kirkland JG, 551
Kirs M, 41
Kirsebom LA, 317
Kirshenbaum K, 449, 453, 529
Kirsteina A, 140
Kirtikar DM, 316
Kiselev NA, 435, 438
Kish AZ, 76
Kishida N, 104, 117
Kisielow J, 491
Kislukhin AA, 465, 488
Kiss MM, 304
Kisselev L, 330
Kisselev LL, 219
Kistemann T, 116
Kister L, 538
Kita H, 267, 285, 401, 425
Kitada T, 540
Kitajima M, 104, 105, 106, 108, 113, 117, 147, 164
Kitamura H, 52
Kitano T, 1, 53
Kitchell BB, 197
Kitchens CM, 501
Kitis M, 154
Kitt D, 153
Kizek R, 529
Kjeldgaard NO, 84
Klabunde JS, 128
Klabunde KJ, 128, 136
Klämbt D, 233
Klatzmann D, 492
Klebanova LM, 98, 135, 185
Kleesiek K, 202
Klein A, 314
Klein EJ, 202
Klein HA, 332
Kleiner M, 556

Klem EB, 325
Klepacki D, 313
Klestil T, 235
Kleva D, 544
Kley N, 202
Klimek L, 462, 490, 501, 518
Klimke WA, 50
Klinck R, 313
Klingel K, 367
Klingler K, 490
Klinman DM, 462
Klinman NR, 303
Klipp E, 543
Klita S, 313
Kloiber K, 281
Klovins J, 16, 20, 40, 170, 225, 228, 231, 292, 294, 295, 356, 364, 405, 406, 411, 423, 424
Klug A, 17, 192
Klug C, 536
Kluge S, 89
Knapman TW, 475
Knapp B, 427
Knauber DC, 33
Knezevich A, 544
Kniazeva VF, 377
Kniazhev VA, 462
Kniel K, 39, 148
Kniel KE, 150, 158
Knight AC, 234
Knippers R, 12, 75
Knoener RA, 538
Knol AH, 125
Knolle P, 1, 11, 12, 19, 25, 46, 47, 53, 67, 69, 72, 81, 82, 84, 86, 97, 134, 186, 195, 367
Knopp Y, 558
Knoroz MIu, 378
Knorr AL, 84
Knott GJ, 558
Knudsen CR, 268, 426
Knudson DL, 31
Knüsel F, 396
Ko B, 538
Ko G, 108, 111, 112, 117, 127, 141, 144
Koay ES, 234
Kobayashi H, 343
Kobiyama K, 501
Kobori S, 557
Kocanova S, 544
Koch G, 371
Koch TH, 348, 355
Kock MM, 113
Kochanek DM, 54
Koedrith P, 162
Koenig JA, 195, 250
Koeppe JR, 457
Koerten H, 471
Kofler RM, 544
Koganti S, 164, 165
Koh CY, 201
Köhler E, 223
Kohlstaedt LA, 548
Köhn M, 141, 540
Kohn T, 125, 134, 137, 142, 148, 164
Koirala P, 551
Koizumi Y, 127
Kojanian N, 202
Kojima K, 104
Kokki H, 235
Kokkinos PA, 151
Kolakofsky D, 223, 258, 456

Kolb A, 325
Kolchenko V, 200
Koldovsky U, 427
Kolieva MK, 378
Koller T, 363, 364
Kolosova NG, 203
Kolpe A, 501
Komarov PA, 543
Komarova TV, 543
Komlenic R, 155
Komor AC, 551
Komor AT, 137
Konarev PV, 281
Kondo M, 23, 138, 261, 263, 264, 265, 287, 363, 386
Kondorosi A, 138, 543
Kondylis P, 480
Konermann S, 549, 550, 551, 558
Kong J, 433
Kong MG, 109
Konieczny A, 314
Konieczny K, 157
Konietzny R, 538
Konig A, 116
König H, 166
König J, 539
Konigsberg W, 186, 247, 249, 250, 307, 544
Koning RI, 73, 192, 243, 471, 476, 477
Konings RN, 19, 20, 176, 267
Konnick EQ, 515
Kontsek P, 377
Koo A, 154
Koonin EV, 32, 406, 409, 555, 556, 558
Koontz SW, 83
Kooti W, 123
Kopein DS, 412, 426
Koper OB, 128
Kopera HC, 543
Kopf M, 447, 462, 490, 491
Kopkova A, 235
Koppel DE, 191
Koppelman MHGM, 202
Kopsidas G, 399, 425
Koraimann G, 54
Korajkic A, 113
Kordium VA, 67, 186
Kordyukova M, 543
Koretzky GA, 462
Kornepati AV, 538, 550
Kornhuber J, 235
Korshin GV, 138
Kos J, 427
Kos M, 551
Kosel J, 139
Kosiakova NP, 378
Kosiol P, 166
Kossik AL, 156
Kost AA, 123, 138
Kostal V, 200
Kosten TR, 501
Koster AJ, 73, 192, 243, 476, 477
Koster J, 119, 123
Köster T, 540
Kostiuk G, 558
Kostiuk GV, 123
Kostyuk GV, 123, 138
Kotelovica S, 433, 486
Kothapalli R, 234
Kotrys AV, 548
Kott Y, 112, 113, 115, 148
Kotzer-Nevo H, 546

Koudelka J, 205, 377
Koudelka KJ, 518, 529
Kourentzi K, 311, 312
Kourouma F, 517
Kouznetsov MY, 151
Kovàcs AL, 218
Kovacs EW, 450, 461, 518, 529
Koval'chuk AV, 379
Kovanen S, 118
Kowalczuk M, 529
Kowalski W, 127
Koyani R, 533
Kozak M, 73, 241, 313, 327
Kozlovska T, 471, 556
Kozlovska TM, 431, 436, 440, 441
Kozlovskaia TM, 431, 436, 438
Kozlovskaya T, 436, 438
Kraal B, 314
Kraase M, 234
Kraft CS, 164
Kraft DJ, 556
Krahn PA, 69
Krahn PM, 69, 71, 73, 74, 241, 317
Kramer FR, 224, 384, 385, 386, 392, 393, 396, 397
Kramer G, 337
Krasheninnikov IA, 543
Kraus BL, 318
Krauss S, 140
Kraut-Cohen J, 543
Kravchenko AV, 391
Kravchenko IuE, 545
Kreiling JA, 543
Kreißel K, 148
Krejcova L, 529
Kremser L, 200
Krieg AM, 462
Krijnen S, 125
Krishnamurthy SR, 31, 42, 105, 163, 171, 407, 555
Krishnan Y, 529
Kristensen DM, 555
Krivisky A, 98, 134, 185
Krivisky AS, 98, 135, 185
Krivokhatskaya LD, 378
Krizkova S, 529
Křížová I, 197
Krohn K, 539
Krokan HE, 234
Krol A, 539
Kroll S, 156
Kropinski AM, 555
Krueger CC, 491
Krueger RG, 26, 35, 194, 196, 241, 307
Krueger S, 243, 476
Krug M, 346, 347
Krüger CC, 435
Krüger DH, 492
Krumm SA, 457
Krunic N, 202
Kruse C, 538
Kruse CW, 135
Krush R, 154
Krut O, 559
Kryshtafovych A, 487, 556
Krzych U, 440
Krzyzosiak WJ, 551
Kubista M, 235
Kubitscheck U, 548
Kubota K, 417, 418
Kubota Y, 127

Kudla G, 543
Kudlicki W, 337
Kuechler E, 325, 330
Kuehn TH, 144
Kuersten S, 550
Kuhlenschmidt MS, 134
Kuhlenschmidt TB, 134
Kukkonen L, 140
Kula A, 544
Kulozik AE, 538, 540
Kulpa DA, 543
Kumagai T, 169, 417
Kumagai Y, 99
Kumar A, 233
Kumar H, 364
Kumar PP, 161
Kumar R, 235
Kumar S, 53
Kumar Rai P, 538
Kumasaka N, 169, 417
Kumpin'sh (Kumpiņš) VK, 333, 358
Kunakh VA, 98
Kündig T, 458
Kündig TM, 447, 459, 460, 501
Kunduru KR, 136
Kunes S, 542
Kunii A, 551
Kunikane S, 104, 117
Kunin (Koonin) EV, 406
Kunze A, 108
Kuo CH, 122, 261, 262, 275, 365
Kuo D, 156
Kuo G, 234
Kuo J, 122
Kuo JF, 276
Kuo KC, 341
Kuo WN, 233
Kuosmanen SM, 235
Kupfer B, 516
Küpper H, 427
Küppers B, 283, 284
Kuppili PP, 513
Kuramitz H, 310
Kurgat EK, 164
Kuriyel R, 166
Kurland CG, 84, 325
Kurnasov OV, 97
Kurn N, 55
Kurokawa K, 543
Kurosawa K, 35, 99, 101, 102, 302
Kurtis JD, 440
Kurucz N, 234
Kus JV, 64, 66
Kushmaro A, 113
Kushner DJ, 319
Kushnir N, 489
Kushwaha M, 551
Kusova KS, 334
Kussova KS, 334
Kussrow A, 453
Kutlubaeva Z, 272
Kutter M, 398, 426
Kuwahara I, 501, 518, 539
Kuwano M, 85, 88, 183, 186, 210, 265
Kuwano Y, 543
Kuzmanovic DA, 243, 476
Kuzmich L, 235
Kuzmickas R, 42
Kuznetsov M, 151
Kuznetsov YG, 197
Kuznetsova NV, 217, 438

Kvitsand HM, 152
Kwak JE, 538
Kwok LW, 202
Kwon EE, 538
Kwon JH, 165
Kwon SJ, 544
Kyeremeh AG, 279

L

La Russa MF, 551
Labadz JC, 150
Labare MP, 125
Laborda S, 139
Labzo SS, 377
Lacadie S, 538
Lacerda M, 415
Lackovic V, 377
Laduron F, 108
Lagenaur C, 64
Lagerkvist U, 341
Lagha M, 543
Lago BD, 377
Lago H, 352, 354, 359, 460, 461, 470
Lagoutte P, 493
LaGue E, 202
Lahti K, 154
Lai CP, 543
Lai LT, 543
Lai M, 192, 243, 478, 479
Lai SK, 543
Lai TP, 234
Laîné JM, 120
Lainé S, 155, 156, 544
Lallana E, 453
Lalli D, 486
Lalonde M, 548
Lalucat J, 126
Lam BJ, 538
Lam P, 518
Lamb DC, 543, 544
Lammers NC, 543
Lamond AI, 543
Lampasona AA, 548
Lampo TJ, 542
Lampson GP, 375, 376
Lamsa A, 543
Lanar DE, 440
Lancaster JH, 13
Lancaster L, 538
Lanctôt C, 548
Landau JV, 319
Lander G, 190
Landers TA, 23, 213, 262, 264, 265, 278, 363
Landini P, 313
Landthaler M, 548
Lane TW, 200
Lane WS, 235
Lane-Bell PM, 63
Lang AS, 42, 151
Lang KM, 65
Lang NL, 105
Lang R, 490
Lang S, 54, 74, 79
Langbeheim H, 309
Lange S, 543
Lange UC, 550
Langlais C, 167, 245, 424
Langlais CL, 243
Langlet J, 81, 153, 154, 155
Langley M, 95

Langridge R, 367
Lania L, 539
Lanka E, 54
Lannan JE, 184
Lantagne D, 165
Lanza F, 234
Lanzardo S, 493
Laout N, 136, 137, 139
Lapen DR, 161
Laprade H, 548
Larason TC, 124
Larburu K, 116
Lari A, 548
Larsen A, 550
Larsen KP, 545
Larsen LS, 543
Larson DR, 538, 545, 547, 548
Lasobras J, 113, 116, 118
Lassen C, 490, 518
László VG, 53
Latallo MJ, 548
Lateef Z, 489, 557
Laterreur N, 543, 547
Latham MP, 538
Lau BL, 153, 292, 463
Lau CH, 558
Lau J, 461
Lau JL, 463, 529, 532
Laval A, 121
Lavelle C, 545
Laveran H, 118
Lavigne R, 414, 555
Lavin K, 47, 48, 98
Lavoie B, 542, 545
Lavrukhina LA, 378
Lawal OR, 129
Lawley PD, 134
Lawn AM, 16, 48, 49, 50, 55
Lawrence DS, 545
Lawton WD, 53
Layton CJ, 356
Lazar P, 281
Lazarenko AA, 378
Lázaro E, 396, 415, 417, 418
Le KM, 513
Le S, 542
Le Blanc Smith PM, 163
Le Calvé M, 107
Le-Clech P, 115, 155
Le Corre M, 203
Le Guiner C, 538
Le Guyader FS, 539
Le Marchand T, 486
Le Mennec C, 116
Leahy D, 234
Leak ES, 18, 19, 189, 205
Leary S, 202
Lease RA, 545
Leatherbarrow RJ, 435
Lebar T, 551
Lebarbenchon C, 203
Lebleu B, 323
Lebtag I, 135
Lecaer JP, 538
Lecatsas G, 15, 16, 37, 59, 60, 122, 302
LeChevallier M, 124
Lechner F, 458, 490
Leclerc D, 313
Leclerc H, 114
Leclerc M, 159
Lécuyer E, 545, 548

LeCuyer KA, 348, 354, 355
Leder P, 20, 38, 318, 319, 323, 325, 345
Lee A, 538
Lee AB, 304
Lee AHF, 202
Lee BH, 223
Lee C, 132, 133, 134, 163, 543
Lee CH, 551
Lee CK, 234
Lee CN, 144
Lee D, 130, 156
Lee EG, 539
Lee EJ, 529
Lee GU, 250
Lee H, 20, 111, 112, 117, 131, 132, 133
Lee HB, 551
Lee HC, 162
Lee HJ, 133
Lee HK, 234
Lee HS, 126
Lee HY, 543
Lee I, 130
Lee J, 127, 131, 457, 538
Lee JC, 216, 221, 458
Lee JE, 111, 112, 117, 127, 141, 199
Lee JH, 144, 446
Lee JS, 52, 61
Lee JV, 113, 163
Lee JW, 233
Lee JY, 188
Lee KL, 489
Lee KW, 281
Lee LA, 518, 529
Lee LH, 132
Lee LM, 235
Lee LY, 125
Lee MH, 143, 144
Lee N, 539
Lee NK, 529
Lee PW, 455
Lee S, 117, 125, 127, 147, 154, 156, 543
Lee SG, 133
Lee SH, 156
Lee SJ, 470
Lee SP, 117
Lee SW, 529
Lee Y, 138, 156, 281
Lee YD, 163
Lee YG, 117
Lee YJ, 234
Lee YS, 117
Lee-Huang S, 20, 94, 324
Leemhuis T, 234
Lees DN, 157
Lefebvre FA, 548
Leffler S, 96, 343
Lefkowitz EJ, 29, 30, 31
Legagneux V, 540
Legault-Demare L, 313, 318
Legeay O, 107
Legendre D, 462
Legrain P, 539
Lehman C, 161
Lehman G, 156
Lei L, 235
Lei Y, 489
Leibovici J, 138
Leibowitz D, 234
Leiknes T, 154
Leipold B, 72, 241
Leis JP, 219

Leisi R, 452
Leitão JH, 280
Leite JP, 166, 203
Leite JPG, 166, 203
Lelong JC, 325
Lemire S, 518
Lemke EA, 529
Lemuth K, 234
Lenaerts A, 249
Lenaerts F, 249
Leneghan DB, 460, 513
Lénès D, 145
Lengyel P, 221, 249, 330
Lenn T, 537, 543
Lennick M, 436
Lenstra TL, 548
Lentzen G, 313
Leonard EF, 130
Leonard KR, 61
Leone CM, 202
Leong CG, 65
Leoni E, 136, 156
Leoratti FMS, 491
Leow A, 156
Lepage C, 135
Lepore A, 543
Lepoutre L, 40, 181, 211, 216, 241
Leppek K, 546
Lerner TJ, 52, 79
Lesage A, 486, 487
LeSage MG, 501
Leschine SB, 372
Leslie G, 155
Leslie RA, 164
Lessl M, 54
Leth H, 132
Leucci E, 540
Leung E, 543
Leung RLC, 197
Leung SSY, 559
Lev O, 145, 154, 449
Levanova A, 234
Levashev VS, 67
Levchenko A, 543
Lever AM, 538
Levesque RC, 203
Levican A, 203
Levican J, 203
Levin JZ, 544
Levine M, 543, 548
Levine PM, 453
Levintow L, 262, 368
Levisohn R, 261, 382
Levitz R, 86
Levonen AL, 235
Levy A, 202
Levy R, 166
Lewin R, 391
Lewis GD, 115, 149
Lewis JB, 327
Li A, 539, 551
Li B, 518, 539
Li C, 52, 108, 142, 143, 144, 161, 538, 548
Li CS, 138, 144
Li D, 117, 130, 548
Li F, 52, 117, 153, 517, 529
Li G, 134
Li GW, 330, 545
Li H, 551
Li HY, 127, 543
Li J, 162, 166, 517, 539

Li JM, 516
Li K, 529
Li L, 52, 117, 551
Li M, 120, 165, 235
Li N, 457, 458
Li Q, 128, 137, 544
Li R, 136
Li S, 120, 489
Li W, 545
Li WK, 105, 111
Li X, 130, 131, 134, 326, 489, 539, 548, 551
Li Y, 131, 137, 460, 489, 513, 518, 543
Li Z, 192, 235, 243, 478, 479, 539
Lian Y, 131
Liang CK, 427
Liang L, 113
Liang S, 199, 538
Liang SY, 165
Liao CT, 235
Liao G, 545
Liao L, 453
Liao ML, 202
Liao N, 160, 442
Liao W, 492
Liao Y, 517
Libby RT, 440
Liberti R, 106
Libonati M, 216, 222, 367, 368
Licciardo P, 539
Licis N, 356, 411, 423
Licklider LJ, 538
Liebana E, 121
Lieber L, 65
Lieber SR, 501
Liebhaber SA, 433
Liebscher V, 235
Liekniņa I, 42, 250, 556, 557
Lielausis A, 175
Liepa S, 439
Liesegang H, 42
Liga MV, 128
Light YK, 201
Lighthart B, 149
Lihavainen E, 542
Lijek R, 446
Liljas L, 190, 197, 350, 352, 354, 435, 438, 467, 468, 470, 537
Lillis L, 203
Lim A, 436
Lim B, 543, 545, 548
Lim F, 470, 360, 468
Lim MY, 117
Lim S, 518
Lim WA, 551
Lim YA, 117
Lima A, 161
Lima JBP, 202
Lima SM, 143, 461
Limsawat S, 40
Limsirichai P, 550
Lin A, 543
Lin CF, 61
Lin CH, 227
Lin D, 551
Lin G, 517
Lin G, 517
Lin H, 199
Lin J, 113, 115
Lin JY, 247
Lin K, 144
Lin L, 71, 123, 492

Lin MY, 144
Lin RI, 161
Lin S, 113, 131
Lin XT, 144
Lin Y, 348, 349, 559
Lin YF, 202
Lin YP, 542
Lindahl L, 317
Lindemann BF, 394, 536
Linden KG, 124, 127, 129, 130
Linden Y, 156
Lindner AB, 61, 546
Lindner AJ, 161, 396
Ling CM, 39, 94, 195
Ling V, 214, 221
Lingappa JR, 544
Linial ML, 539
Link A, 459, 460, 491, 513, 518
Linker SB, 551
Lino CA, 442, 446, 461, 464, 493, 518, 527
Linse P, 199
Linser R, 487
Linville A, 446
Lionnet T, 544, 545, 547, 548
Liphardt JT, 551
Lipmann FR, 23, 264
Lipowsky G, 489
Lipp P, 148
Lipp S, 543
Liquier J, 280
Lisle JT, 117
Lissom A, 514
Listello JJ, 433, 540
Litsis (Līcis) NG, 333, 358
Little SC, 543
Liu AP, 453
Liu B, 558
Liu C, 131, 492
Liu CB, 501
Liu D, 109
Liu DR, 551
Liu E, 501
Liu F, 65
Liu G, 545
Liu H, 453, 558
Liu J, 203, 397, 543
Liu JL, 304
Liu L, 128, 539, 551
Liu M, 156
Liu N, 131
Liu P, 134
Liu S, 134
Liu S, 545
Liu W, 108, 137
Liu WT, 125
Liu X, 136, 556
Liu Xiaohui, 545
Liu Xuejun, 551
Liu Y, 136, 150, 539, 548, 550
Liu YP, 128
Liu Z, 109, 200, 310
Livingston DM, 323
Livingston NM, 548
Lizardi PM, 392, 393, 396, 397
Lizasoain A, 166, 203
Lizzio E, 234
Llano-Sotelo B, 313
Llewelyn MJ, 235
Llorian M, 538
Lloveres CP, 50, 85
Lloyd SJ, 64

Lloyd-Price J, 542
Lo A, 551
Lo MY, 543
Lobba MJ, 193, 207, 418
Löchelt M, 518
Lockett S, 548
Lockwood AH, 324
Lodder W, 132
Lodder WJ, 108
Lodish HF, 1, 11, 19, 20, 22, 23, 24, 25, 38, 94, 147, 175, 176, 177, 178, 180, 183, 184, 237, 317, 318, 319, 320, 322, 323, 323, 325, 327, 344, 345, 367, 368, 370
Loeb S, 142
Loeb T, 11, 16, 75, 79, 81, 95, 122, 190, 248, 301
Loge FJ, 124
Loginova NS, 378
Loginova SI, 378, 379
Logue CH, 203
Loh HH, 542
Loh TP, 234
Lomax TD, 149
Lombard B, 107
Lomeli H, 392, 393, 396, 397
Lomonossoff GP, 193
Londoño-Vallejo JA, 545
Lone BA, 558
Long CA, 460, 501, 513
Long G, 202
Long RM, 540, 541, 544, 545
Long SC, 113, 114, 118
Longstaff M, 285
Lood R, 558
Looman AC, 421
Lopez GU, 164
López MI, 134
Lopez-Jones M, 544
Lopez-Pila JM, 123
López-Sagaseta J, 489
Lopez-Vidal Y, 116
Loquet A, 556
Lord JM, 518
Lorenz M, 544
Lorenzoni K, 54
Lorimer E, 540
Lory S, 64
Lošdorfer Božič A, 484
Lostroh CP, 65
Lou F, 162
Lou X, 442
Lough TJ, 234
Loughran TP Jr, 234
Loutreul J, 146, 159, 160
Love DC, 41, 134, 141, 158, 159, 160
Lovelace G, 113, 157
Lovelace GL, 141, 158
Lovmar L, 234
Lovmar M, 313
Low DA, 64
Lowary PT, 347, 386
Lowrie RJ, 341
Lowry CV, 323
Loyd B, 318
Loža V, 378
Lozzi L, 129
Lu B, 324
Lu C, 540
Lu CH, 61
Lu G, 326
Lu HY, 542
Lu IN, 20, 40, 140, 226, 295, 344, 406, 425

Lu J, 517
Lu M, 556
Lu P, 15, 156
Lu R, 125, 501
Lu TK, 518
Lu W, 155
Lu X, 150, 529
Lu Y, 117, 558
Lua LH, 489
Lua LHL, 489
Lubitz W, 299, 425
Lucas-Lenard JE, 23, 264
Lucas T, 543, 548
Lucas WJ, 234
Lucena F, 106, 108, 113, 115, 116, 117, 118, 122,
    136, 142, 151, 544
Luce R, 284, 387, 388, 394, 405, 410, 425
Luciani F, 544
Lucier KJ, 133
Lucke S, 539
Ludányi M, 53
Ludlow AT, 234
Ludvigsson J, 235
Lugari A, 493
Lührmann R, 328, 538
Lui J, 543
Luisi BF, 283
Lukasik J, 154, 159
Lukowiak A, 131
Lun JH, 166, 202
Lund O, 58
Lundgren DA, 144
Lundh M, 551
Lundstrom K, 489
Lunt MR, 87, 383
Luo KR, 518
Luo L, 551
Luo Q, 543
Luo VM, 549
Luo X, 161
Luo Y, 537, 558
Lupker JH, 85, 132
Luria SE, 66, 175
Luster E, 129
Lustig F, 341
Lute S, 165, 166, 199
Luther K, 114
Luty AJ, 440
Lutz CS, 538
Luxbacher T, 129
Luzi S, 142
Luzzatto L, 313
Ly TDA, 203
Ly-Chatain HM, 201
Ly-Chatain MH, 139
Lybecker MC, 540
Lycheva IA, 378
Lykke-Andersen J, 139, 538, 540
Lynch KW, 538
Lyon K, 548
Lyon SR, 117, 148
Lyons B, 197
Lytton-Jean AK, 456
Lyu P, 324
Lyutova R, 551

M

Ma B, 138, 543
Ma C, 492
Ma H, 155, 157, 549

Ma HF, 489
Ma J, 539
Ma L, 108
Ma M, 539
Ma X, 545
Ma Y, 202, 518, 529, 551
Ma Z, 134, 545
Maas W, 556
Maaløe O, 84
Maassen SJ, 529
Macbeth MR, 314
Macchi P, 544
MacColl R, 196
MacDonald CC, 538
Macdonald J, 109
Macdonald PM, 543
Macedo JM, 460, 461
Machida CA, 539
Machinal C, 154
Maciejak A, 235, 538
Maciorowski KG, 111
Mack AK, 233
Mackay P, 12
Mackey ED, 120
Mackeyev Y, 127, 130, 131
Mackie C, 306
Mackie M, 157
MacMorris M, 550
Macrina FL, 53
Maddera L, 61
Mäde D, 160, 202
Madec A, 155
Madi M, 234
Madison JT, 211
Madonna AJ, 109, 198
Maeda RK, 543
Maehr R, 551
Magassouba NF, 517
Maggi N, 89
Magri ME, 137
Maguire KM, 120
Maguire-Boyle SJ, 128, 156
Magzamen SL, 144
Mahajan N, 226
Mahan DE, 398, 426
Maher E, 134
Maherchandani S, 159, 161
Mahmoudabadi G, 78
Mahoney W, 234
Mahony J, 202
Mahony JB, 397
Mai L, 131
Maier RM, 133
Maiga AH, 141
Maillard JY, 137
Maita T, 186, 247, 249, 250, 307, 544
Maitra U, 324
Maître B, 234
Maiztegi MJ, 116
Majello B, 539
Majev SP, 217
Majiya H, 130, 139, 243, 312
Makani V, 324
Mäkelä J, 542
Makki FM, 235
Maksymiuk GM, 304
Malabirade A, 281, 283
Maldonado Y, 202
Maleev VV, 517
Malekzad H, 518
Malham SK, 142

Mali P, 549
Malik AM, 548
Malik R, 543
Malik YS, 120, 159, 161
Malito E, 489
Malki K, 42
Malpiece Y, 436
Malsch C, 235
Malsey S, 61
Maltais F, 203
Malys N, 544
Malyuta SS, 98
Mamais D, 154
Mamane H, 125, 129, 551
Mamedova SA, 378
Mana TC, 165
Mana TSC, 165
Manak M, 143
Manariotis ID, 143
Manchak J, 50, 61
Manche L, 378
Manchester M, 489, 518, 529
Mancini L, 543
Mandal NC, 52, 120, 85
Mandl CW, 231, 544
Mandrekar JN, 234
Mandrika I, 439
Mandrika IK, 439
Manet E, 433
Mangiarotti G, 324
Mangues R, 18
Mangus LM, 234
Manhart M, 54
Maniatis T, 433, 538, 540
Manifacier SA, 286
Mankin AS, 313
Mannack LV, 545
Mannerström H, 542
Manning PA, 52
Manohar H, 513
Manolova V, 447, 489, 490, 492, 518, 462
Manor Y, 203
Manrique P, 163
Manzenrieder F, 454, 465, 488
Mao C, 489
Mao H, 442
Mao L, 232
Mapp L, 114
Mar V, 153
Maraccini P, 14, 151
Maraccini PA, 131
Marcello A, 544
March CJ, 440
Marchalonis JJ, 304
Marchin G, 136
Marchin GL, 128, 136, 145
Marchisio MA, 551
Marciani DJ, 501
Marcker KA, 21
Marco S, 280, 281
Mardakheh FK, 548
Marets N, 156
Margeat E, 155
Margolin AB, 137
Marie J, 538
Marie V, 115
Mariella R Jr, 202
Marin VA, 166
Mariñas BJ, 134
Marino P, 90, 91
Marjanovic M, 142

Marjoshi D, 200
Markazi S, 14, 151
Markiewicz Z, 297, 424
Markuševiča V, 76
Markushevich V, 91
Markushevich VN, 76
Marlow FL, 543
Maro A, 203
Marondedze C, 540
Marques AF, 457
Marques IJ, 542
Marques MV, 55
Marquisee M, 211
Marr LC, 144
Marraffini LA, 558, 559
Marras SA, 543
Marshall JB, 538, 550
Marshall R, 97, 339
Marti Villalba M, 310
Martin F, 539, 538
Martin J, 166, 332
Martin K, 157
Martin KC, 545
Martin MJ, 555
Martin P, 234
Martin RM, 544
Martin RP, 543
Martin S, 489, 518, 462
Martin SW, 490
Martin TD, 139
Martinez J, 548
Martinez S, 126
Martínez-Hernández R, 235
Martínez-Manzanares E, 112
Martínez-Pastor M, 539
Martins R, 492
Marvin DA, 15, 16, 18, 19, 45, 47, 49, 56, 74, 61,
      63, 66, 97, 189, 205
Marx A, 234
Marziali F, 234
Mas Marques A, 234
Masachessi G, 203
Masago Y, 104
Masahiro O, 127
Maschler R, 325
Mason JL, 529
Masoumzadeh E, 538
Mast J, 493
Mastico RA, 438, 439, 440, 518
Masuda A, 538
Masuda K, 543
Matassova N, 313
Mateu MG, 460
Mathies RA, 451
Mathieu L, 145
Mathu SG, 266
Mathur M, 267
Matikka V, 118
Matlashov M, 559
Matondo S, 460, 513
Matos A, 492
Matson S, 462
Matsuhashi S, 19, 175, 176, 177, 178, 180, 181,
      187, 237, 252, 315, 331, 370
Matsui M, 518
Matsui Y, 121, 148
Matsumoto K, 501, 518, 539
Matsunaga T, 304
Matsuo I, 130
Matsushita T, 107, 121, 148
Matsuura S, 540

Matsuura T, 267, 285, 401, 402, 404, 425, 426
Matsuzaki S, 19, 82, 186
Mattei E, 234
Matter N, 539
Matthaei JH, 20, 21
Matthews HR, 26, 31, 196, 226
Matthews KR, 161
Matthews KS, 88, 196, 197, 250
Matthews RE, 96, 197
Mattick JS, 64
Mattle MJ, 125, 142, 202
Mattox W, 538
Matulic-Adamic J, 470
Maudru T, 234
Maul A, 159
Maunula L, 154
Maurer P, 447, 490, 501, 512
Maurice CF, 163
Mavromara P, 492
Mawatari K, 129
Mawhinney DB, 128
Maxwell SL, 164
May T, 145
Mayer BK, 134, 148, 377
Mayer J, 234
Mayer V, 129
Mayotte JM, 122, 149
Mayr C, 543
Mayrhofer P, 299, 425
Maza JC, 461
Mazaheri Nezhad Fard R, 555
Mazari-Hiriart M, 116
Mazé R, 79
Mazumder R, 324
Mazzoni EO, 549
Mazzoni-Putman SM, 543
Mbayed VA, 113, 166
Mbonimpa EG, 127
McAlister M, 165
McArdell CS, 164
McBride MT, 202
McBurnett SR, 356
McCaffrey RL, 202
McCalla JI, 212
McCarthy JE, 544, 548
McCartney AL, 105
McCaskill JS, 389, 400
McClellen RE, 83
McClenahan SD, 235
McCloskey JA, 342
McClure CP, 234
McCluskey DK, 109
McColl IH, 113, 193
McConnell MM, 15, 16, 60, 122
McCracken S, 538
McDermid B, 134
McDonel JL, 319
McDonnell MB, 109, 310
McFadden EJ, 545
McFeters GA, 117
McGahey C, 154
McGarry KG, 538
McGill E, 200
McGinn J, 559
McGregor WC, 351
McIntire FC, 306
McKay CS, 113, 457
McKay LD, 150
McKay R, 144, 163
McKechnie NM, 429
McKinlay R, 235

McKnight SL, 545
McLaughlin MR, 118
McLellan NL, 108
McLoon AL, 542
McLuckey SA, 198, 250
McMenemy P, 159
McMinn BR, 108, 113, 121
McMurry LM, 343
McNally LM, 544
McNamara PJ, 134
McNaughton P, 96
McNeill K, 141
McNicholas PM, 313
McParland K, 329
McPhail T, 153
McPherson A, 197
McQuarrie J, 138
Mduma E, 203
Mead DA, 517
Mead G, 219, 456
Meade GK, 136
Meagher RJ, 201
Means GE, 250
Medema GJ, 124, 125, 133, 136, 152, 153, 163
Meder F, 156
Medford A, 442, 461, 493
Mediannikov O, 233
Medintz IL, 464
Medvedeva NI, 334
Meehan A, 544
Mège JL, 233
Meghani NM, 518
Megnekou R, 514
Mehdizadeh F, 518
Mehl RA, 450, 518, 527
Mehraein-Ghomi F, 324
Mehta A, 166
Mehta GD, 543
Meier D, 91
Meier JL, 548
Meier-Kolthoff JP, 556
Meierhofer R, 142
Meijerink E, 489, 518, 462
Meinecke F, 42
Meir M, 555
Meisenheimer KM, 356
Meissner SM, 117
Mejia GL, 458
Mekler P, 35, 40, 224, 405, 419
Melançon P, 313
Meldrum T, 451
Melgaço FG, 166
Meller VH, 543
Mellits KH, 139
Mello-Grand M, 235
Meltzer RH, 202
Mempel TR, 543
Mena MP, 153
Menard-Szczebara F, 145
Mende Y, 550
Mendelson E, 203
Mendes P, 544
Méndez J, 113
Mendez J, 132
Méndez X, 106, 122
Meng J, 234
Meng QS, 120
Meng R, 480
Meng S, 516
Meng X, 544, 545
Meng XH, 540

Mengozzi M, 235
Menon DU, 543
Menon V, 513
Menzel G, 81, 88, 89
Meo M, 201
Merante F, 202
Meriläinen P, 118
Merle G, 107, 160
Merlin C, 201
Merregaert J, 211, 212, 213, 214, 216, 288, 319, 323, 339, 365
Merrick WC, 14
Merril CR, 67
Merrill SH, 211
Merryman AE, 131
Merz ZN, 461
Merzlyak A, 529
Mesa J, 544
Meschke JS, 107, 113, 118, 120, 150, 160
Mesquita MMF, 121
Messens E, 167, 419, 427
Mestdagh P, 235
Mester P, 139
Metafora S, 221, 314
Metelev VG, 221, 216
Metelyev VG, 219
Metlitskaya AZ, 98, 135, 185
Metz DH, 125
Meulemans CC, 119, 123
Meunier SM, 126
Meyer F, 25, 264, 362
Meyer K, 540
Meyer PL, 383, 384
Meyn T, 148
Meynell E, 48, 49, 50, 58, 60
Meynell GG, 48, 50, 98
Meyvisch C, 76
Miagostovich MP, 166, 203
Miao J, 463
Michaelis J, 548
Michalek P, 529
Michaud M, 543
Micheel B, 438
Michel AA, 310
Michel P, 138
Micheletti C, 484
Michen B, 148, 200
Michot L, 161
Mickleburgh I, 538
Middelberg APJ, 489
Middendorf M, 231
Middlebrook JL, 193
Mido T, 109
Midorikawa K, 491, 518
Mielecki D, 86
Mienie C, 103
Miermont A, 537, 543
Mieszkin S, 116, 158
Miettinen IT, 116, 118
Mignon C, 493
Miguel MG, 140
Mihelcic JR, 142
Mijatovic-Rustempasic S, 202
Miki T, 51, 539
Mikoshiba K, 543
Mikulic P, 109
Milanovich F, 202
Miles SL, 155
Milev MP, 544
Mili S, 543
Militello V, 281

Miller AM, 492
Miller CM, 544
Miller DL, 266
Miller M, 163
Miller MJ, 20, 111, 263, 324, 328, 501
Miller N, 123
Miller P, 329
Miller S, 203
Milligan JF, 348
Mills DR, 224, 227, 258, 285, 337, 343, 344, 365, 381, 383, 384, 385, 386, 392, 393, 396, 405, 421, 422, 423, 426
Millum J, 501
Milman G, 332
Milot J, 203
Milton D, 210
Min JH, 155
Min L, 298, 425
Min Jou W, 24, 40, 167, 210, 211, 212, 213, 214, 216, 218, 220, 238, 241, 247, 286, 288, 319, 320, 323, 339, 340, 365, 419, 427, 467
Mindich L, 29
Miner TA, 341
Mingle L, 545
Minion FC, 441
Minnaar R, 210
Minney-Smith CA, 234
Minor PD, 436
Minoshima M, 127, 133
Minshall N, 538
Miorin L, 544
Miossec L, 157
Mir M, 543
Miranda G, 268, 279
Miranda JA, 234
Mire CE, 517
Mironova A, 543
Mirshekari H, 529
Mirskaia EE, 378
Miryuta NY, 98
Mirzaei MK, 163
Mise K, 538
Misiak D, 539
Misra AK, 131, 140
Misra M, 487, 543
Mitch WA, 136, 137
Mitchell DA, 501
Mitchell R, 112, 141
Mitra RD, 19, 208, 341
Mitra S, 137
Mitsui K, 61
Mitsui Y, 317
Mitsunari Y, 262
Mitsunobu H, 551
Miura A, 19, 82, 186
Miura K, 460, 513
Miyagawa R, 538
Miyakawa K, 14, 16, 37, 55, 67, 81, 302
Miyake T, 13, 23, 35, 37, 38, 39, 72, 76, 98, 99, 101, 102, 103, 178, 187, 195, 257, 288, 302, 308, 315
Miyakoshi M, 280
Miyata T, 218
Miyazaki K, 138
Mizaikoff B, 109
Mizuno A, 89, 132, 313
Mizuno T, 40
Mizutani A, 543
Mizutani T, 210
Mizuuchi R, 404, 426

Mo JJ, 538
Mo S, 501
Mocé-Llivina L, 106, 136, 140, 154
Modak MJ, 12, 203
Model P, 19, 20, 175, 176, 177, 178, 183, 184, 291, 331, 338
Modesti N, 537
Modolell J, 313, 492
Moelling K, 235
Moen RC, 234
Moerner WE, 543
Moghoofei M, 518, 529
Mohi-El-Din H, 543
Mohr D, 272
Mohraz M, 517
Mohsen M, 447
Mohsen MO, 447, 492, 501, 557
Moilanen K, 202
Moineau S, 144, 155
Moingeon P, 501
Moisant P, 480
Moldovan JB, 543
Molineux IJ, 414, 536
Molino NM, 518
Moll DM, 147
Moll I, 280
Molle D, 538, 544
Møller Aarestrup F, 58
Møller T, 205, 280, 281, 316
Molnar DM, 53
Molugu SK, 455
Momba MN, 115, 163
Momma T, 280
Monastyrskyy B, 487
Mondal PK, 141, 152
Monis P, 150, 156
Monkhorst K, 543
Monnier N, 544, 548
Monroe SS, 410
Monstein HJ, 19, 167, 192, 253, 263
Monteiro MP, 501
Monteiro P, 148, 15
Montemayor M, 113, 126, 151
Montes M, 538
Montgomery SB, 551
Montpetit B, 548
Moody MD, 390
Moohr JW, 88, 160
Mooijman K, 116
Mooijman KA, 106
Moon AM, 161
Moon HC, 544, 548
Moon K, 125
Moor KJ, 131, 162
Moore A, 234
Moore B, 161
Moore CH, 147, 253, 315, 333
Moore G, 164
Moore JD, 551
Moore MD, 162
Moore MJ, 538
Moosavi Basri SM, 529
Mor SK, 144
Moradi Moghadam S, 162
Moradpour S, 489
Moraes MR, 116
Morales I, 152
Morales S, 559
Morales-Morales HA, 165
Moran VA, 543
Moran-Gilad J, 203
Morel Y, 202

Moreland V, 124
Morella NM, 471
Morelli G, 518
Moreno B, 116, 446
Moreno Y, 166
Morgan M, 166
Morgunova V, 543
Mori H, 265
Moriarty EM, 113, 115
Morikawa M, 217
Morin T, 138, 160, 540, 541
Morinigo MA, 112
Moriñigo MA, 163
Morisaki T, 547, 548
Morishita R, 501
Morita H, 105, 147
Moroe I, 166
Morón A, 106, 117
Morozov IY, 399
Morozov SI, 258
Morozova N, 559
Morrell L, 202
Morris AJ, 331
Morris JM, 136
Morrison TG, 318
Morse DE, 184
Morse JW, 86, 475
Morton VL, 474
Mosberg JA, 136
Moschall R, 538
Moschos SA, 517
Moser M, 161
Moser MJ, 234
Mosig H, 161
Moss T, 354, 537
Moss WN, 545
Mostafavi ST, 153
Motorin Y, 538
Mougel M, 155, 156, 544
Mouland AJ, 544
Mourão LC, 457
Moussaoui S, 139
Mowry KL, 543
Moyon M, 551
Mozejko JH, 216
Mozhylevs'ka LP, 98
Mu L, 234
Mucha H, 164
Muehlhauser V, 161
Mueller JP, 114
Mujeriego R, 113
Mujtic M, 89
Mukherjee J, 233
Mukhopadhyay S, 480
Mulder WJ, 529
Mullani SB, 131
Müller A, 154
Müller B, 544
Muller CP, 140
Müller L, 427, 538
Muller S, 513
Müller-Hermelink HK, 431
Müller-McNicoll M, 543
Mullers WJ, 377
Mullon CJ, 96
Mumma JM, 164
Munakata N, 122
Munch HK, 538
Münch M, 148
Muniain-Mujika I, 157
Muniesa M, 113, 122, 136, 146

Munishkin AV, 391, 405, 410
Muñoz LJ, 439
Munoz R, 116
Munro I, 109
Munsky B, 548
Munson M, 543
Muntwiler S, 489, 490
Mura C, 280, 281
Muralikrishna P, 329
Muramoto T, 543, 548
Murata A, 138, 145
Murata M, 235, 491, 518
Murchie AI, 313
Murén E, 317
Muri J, 491
Murphy B, 33
Murphy E, 86
Murphy J, 141
Murphy ST, 527
Murphy TM, 29
Murray JB, 352, 470, 518
Murray K, 12, 438
Murray KW, 458
Murray R, 156
Muscillo M, 540, 541
Mushegian AR, 29, 30, 31, 555
Musychenko ML, 329, 364
Muthukrishnan AB, 542
Mutyam SK, 281
Mwakalinga SB, 460, 513
Myers CA, 202
Myers K, 120
Myers MN, 96
Myers MP, 544
Mylon SE, 199
Myrmel M, 42, 116, 118, 134, 158
Mysore C, 136, 202

N

Na HN, 326, 501
Naaktgeboren N, 324
Naarmann IS, 543
Nabergoj D, 96
Naceur MW, 149
Naftelberg S, 538
Nagai K, 352
Nagai S, 95
Nagamune T, 529
Nagano K, 387
Nagao T, 35, 302
Nagaoka K, 38
Nagashima K, 543
Nagel JH, 218
Nagels JA, 141
Nagels JW, 141
Naimski P, 91
Nair S, 356
Najafi-Shoushtari SH, 235
Najm HN, 200
Nakada D, 25, 180, 323, 369
Nakada N, 138
Nakagami H, 501
Nakagawa N, 137
Nakai Y, 501, 518, 539
Nakaishi T, 267, 285, 425
Nakamoto T, 88, 25, 186, 323, 326, 333
Nakamura A, 543
Nakamura M, 88, 154, 544
Nakamura S, 54, 234, 491, 518
Nakano R, 127

Nakao H, 279
Nakayama K, 551
Nam YS, 518, 538
Namanda-Vanderbeken A, 540
Namba K, 435
Nanassy OZ, 202
Nangmenyi G, 134
Nap RJ, 468
Nappier SP, 107, 113, 140, 145, 158
Naraginti S, 131
Naranjo J, 155
Naranjo JE, 127
Nariya H, 232
Narum DL, 446, 493
Nasarabadi S, 202
Nascimento S, 202
Nasrin F, 546
Nasser A, 118, 145, 449
Nasser AM, 148, 153
Nastri HG, 343
Natarajan P, 190
Nathans D, 19, 20, 21, 22, 73, 87, 176, 180, 185, 237, 241, 247, 258, 313, 315, 316, 317, 318, 344
Natsume T, 543
Naudts I, 490, 518
Nava G, 78
Naveca FG, 202
Navet B, 551
Navidad J, 202
Naydenova EV, 517
Nazim M, 538, 546
Nchinda G, 514
Neal KR, 113, 163
Nederlof MM, 153
Neeli-Venkata R, 542
Neeman I, 148
Neetoo H, 162
Negahdaripour M, 489
Neggers JE, 544
Negishi M, 543
Nehrer S, 235
Neill US, 554
Neilson T, 329, 334
Neiman AM, 120, 150
Neiman LA, 198
Nelles DA, 551
Nelson KL, 134, 141, 156, 159, 163
Nelson RW, 198
Nelson W, 202
Nemazee D, 513
Nemes MM, 375, 376
Németh Z, 155
Nemoto N, 404
Nepomnyaschaya NM, 139
Nerukh D, 470
Nerva L, 410
Netirojjanakul C, 452, 527, 538
Neto MF, 203
Netongo PM, 514
Netzler NE, 202
Neubacher S, 548
Neubauer Z, 93, 122
Neuert G, 548
Neve H, 105
Neves MG, 130
Nevo-Dinur K, 543
Newbury SF, 235
Newby MI, 359
Newhart A, 543
Newton A, 55, 71

Newton SE, 436
Ng HY, 121
Ng K, 529, 543
Ng S, 543
Ng T, 264
Ng WJ, 125, 140, 145
Ngandeu Neubi GM, 518
Ngo T, 538
Ngo TD, 201
Ngoh AA, 514
Ngu LL, 514
Nguyen HG, 529
Nguyen HT, 201
Nguyen K, 543
Nguyen MT, 133, 142
Nguyen MV, 433
Nguyen TH, 14, 113, 125, 131, 134, 151, 501
Nguyen TT, 132, 201
Ni CZ, 470
Ni N, 433, 470, 544, 548
Nichol S, 138
Nichols JL, 211, 218, 220, 222, 315, 345, 545
Nicorescu I, 160
Niculescu-Morzsa E, 235
Nie L, 529
Nie Y, 517
Nielsen CM, 501
Nielsen K, 113, 193
Nielsen MA, 460, 513, 514
Nielsen R, 415
Nielsen SO, 458
Niemi RM, 151
Niemira BA, 159
Niendorf S, 234
Nierhaus KH, 325
Niessing D, 543
Niessner R, 108, 109
Nieto-Juarez JI, 134
Nieuwenhuizen R, 40, 114, 169, 228, 335
Nieuwkoop AJ, 486
Nieuwstad TJ, 119, 120, 123
Nikitina TT, 333, 349, 358
Nikolaitchik O, 548
Nikolaitchik OA, 544, 548
Nikonowicz EP, 359, 470
Nikovskaia GN, 122
Nilsen V, 134, 156
Nimse SB, 557
Ninham B, 141
Ninham BW, 139, 141
Ninove L, 203
Nirenberg M, 38
Nirenberg MW, 20, 21
Nishida K, 113, 147, 543
Nishigaki K, 404
Nishihara T, 20, 38, 39, 40, 189, 243, 247, 249,
            278, 291, 297, 384, 385, 386, 392,
            406, 424
Nishikawa S, 426
Nishimura A, 280
Nishimura K, 129
Nishimura Y, 46, 49, 235
Niu F, 130
Niu J, 440, 516
Niveleau A, 278
Nkwe KI, 115
Noble KN, 543
Noble RT, 113, 118, 141
Noble-Wang J, 164
Nogueira TC, 544
Noh JH, 311

Noiges R, 54
Noireaux V, 97, 339
Nojima T, 339
Nolan GP, 452, 527
Nolf F, 242
Noll M, 324
Noller HF, 23, 538
Nolte FS, 515
Nolte O, 234
Nomellini JF, 63
Nomura M, 93, 314, 323
Nomura S, 135
Nonoyama M, 13, 86, 122, 185, 301, 367, 365,
            367
Norden IS, 548
Nordin A, 141, 142
Norton G, 107
Nosach LN, 379
Nosik NN, 377, 378
Nosova LY, 54
Noss CI, 136
Notani GW, 12, 19, 21, 24, 176, 178, 183, 331
Nout MJ, 161
Novak U, 219
Novak G, 555
Novik G, 555
Novokhatskii AS, 377
Novotny CP, 47, 48, 49, 53, 54, 91, 98
Nowak CM, 131
Nowak L, 233, 551
Nozu K, 261
Nshama R, 203
Nüesch J, 396
Nuk M, 54
Nunes DF, 457
Nupen EM, 112
Nusbaum C, 544
Nuss AM, 280
Nussbacher JK, 548
Nussberger J, 501
Nussinov R, 218
Nuttall SD, 399, 425
Nwachuku N, 18
Nwokeoji AO, 210
Nyanachendram D, 117
Nykovskaya GN, 147

O

O'Banion NB, 136, 145
O'Brien DJ, 147
O'Callaghan CA, 551
O'Callaghan RJ, 49, 69, 73, 74, 192,
            196, 241
O'Connell C, 243, 476
O'Connell KP, 107
O'Connell MR, 551
O'Dell HD, 107
O'Donnell CP, 130
O'Donovan C, 555
O'Flaherty V, 121, 158
O'Geen H, 551
O'Hara PJ, 385
O'Neil JP, 305, 306, 450, 527, 529
O'Reilly MK, 453
O'Rourke JP, 446, 493
O'Shea M, 542
O'Sullivan L, 162
O'Toole JS, 165
Oates E, 169, 527
Obel J, 303, 304
Oberhauser F, 235

Oberholzer T, 400
Obermeyer AC, 451
Oberti S, 154
Obinata M, 284
Obregon KA, 281, 283
Obregon-Perko V, 203
Ochiai H, 138
Ochiai T, 546
Ochoa S, 1, 20, 22, 94, 167, 180, 252, 258, 324,
            367, 368
Ockerman B, 156
Odegaard C, 127
Odermatt B, 490
Odom OW, 328, 329, 337
Oeffinger M, 545, 548
Oehlenschläger F, 401
Oeschger MP, 22, 87, 176, 237, 258, 317, 318
Ogami K, 538
Ogata N, 136
Ogawa K, 332
Ogle JM, 313
Ogonah O, 20, 40, 226, 295, 344, 406, 425
Ogorzaly L, 41, 116, 122, 149
Ogram SA, 538
Oguma K, 117, 126, 129, 130
Oh HM, 117
Oh JY, 543
Oh S, 155
Ohba Y, 544
Ohgaki S, 40, 117, 121, 147, 154
Ohkawa J, 426
Ohkawara B, 538
Ohki K, 261
Ohnishi K, 317
Ohno H, 548
Ohno K, 538
Ohsawa H, 334
Ohshima H, 201
Ohsuka S, 163
Ohtaka Y, 19, 247, 261, 344
Ohtsubo E, 51, 98
Ohyama K, 85
Oikawa S, 491, 518
Oikonomou CM, 64
Oishi W, 137
Öjstedt U, 139
Ok YS, 538
Okabe S, 113, 120, 137
Okada Y, 55, 14, 226, 261
Okamoto A, 545
Okeke MI, 514
Okita TW, 542, 543
Okuno T, 538
Okuyama A, 313
Olalemi A, 159
Olenkina OM, 543
Oliinyk A, 118
Olins DE, 302, 303
Olive DM, 398
Olive M, 538
Oliveira AC, 143, 197
Oliveira J, 542, 559
Oliveira SM, 542
Olivieri VP, 135, 136, 154
Olovnikov I, 543
Olsen RH, 13, 14, 16, 37, 38, 56, 58, 59, 66, 71,
            76, 98, 122, 302
Olsthoorn RCL, 29, 40, 55, 170, 225, 228, 294,
            335, 344, 356, 406, 411, 412, 423, 424
Olszewski J, 165
Ong HK, 501

Ong SL, 125, 140, 145
Ono M, 85, 265
Ooi SS, 117
Opaleye OO, 234
Oparka KJ, 543
Opgenorth A, 52, 61
Oppenheimer JA, 124
Opsomer C, 384
Orange N, 160
Ord RL, 493
Orias E, 184
Oriel PJ, 73, 140, 185, 186, 250
Orkin SH, 551
Orlinger KK, 544
Orna L, 438, 467
Orom UA, 539
Oron G, 151, 153
Orr DC, 234
Ortega J, 200
Ortiz PJ, 257, 315
Osaman TAM, 267
Osawa H, 42
Osawa S, 13, 19, 35, 39, 99, 101, 102, 103, 104,
        105, 111, 163, 186, 238, 302
Osborne HB, 540
Oschkinat H, 486, 487
Ose V, 378, 433, 435, 438, 440, 441, 445, 466,
        467, 471, 501, 556
Ose VP, 431
Oshima K, 165
Oshima KH, 165
Osman H, 127
Oss Pegorar C, 543
Østerhus SW, 152
Osterman-Golkar S, 138
Osterwalder J, 490
Ostrove JM, 234
Ostrowski J, 539
Osumi N, 538
Ota N, 129
Otagiri M, 105, 116, 147
Otaka T, 313
Otaki M, 121, 137, 147, 154
Otenio MH, 166
Ott G, 235
Otten H, 400
Otterlei M, 234
Otto S, 404
Ottoson J, 106, 116, 137, 540, 541
Ottoson JR, 137
Ou AC, 538
Ou JT, 47, 54, 70
Ouambo HF, 514
Oudejans SJG, 113
Oudot F, 335
Ouenzar F, 543
Oulton R, 153
Ouwens RN, 153
Ouyang H, 545
Ouyang T, 545
Overbeek GP, 20, 225, 252, 293, 294, 296, 328,
        420, 424
Overby L, 234
Overby LR, 16, 18, 19, 26, 35, 39, 81, 83, 94, 189,
        195, 306, 307
Ovryn B, 548
Owen A, 65
Owens RA, 394
Oxenius A, 462, 489, 490, 501, 518
Oyelere AK, 527
Ozaki M, 84, 85

Ozawa A, 35, 302
Ozel M, 438
Özel Duygan BD, 164
Ozgur S, 546

P

Paar J 3rd, 107, 108
Pace NR, 67, 258, 261, 262, 365, 381
Pachepsky YA, 151
Paci G, 529
Pacitti C, 218
Padilla DP, 464, 518, 527
Padkina M, 539
Paek KY, 543
Page MA, 134
Pai CI, 543
Paigen K, 82
Paish A, 159
Pal K, 518
Palangat M, 545, 547
Palasingam K, 405, 411, 415, 426
Paleček E, 377
Paliwal S, 543
Pallin R, 106
Palm P, 316
Palmateer GA, 115, 117
Palmen R, 65
Palmer AE, 548
Palmer CJ, 107
Palmieri M, 222
Pan C, 126, 161
Pan H, 143
Pan L, 117
Pan M, 144
Pan T, 200, 348
Pan X, 132
Pan Y, 518
Pan YC, 205
Pan YL, 464
Pana' A, 493
Panach L, 235
Panavas T, 544
Pandey PK, 551
Pandey VN, 203, 267
Pang HH, 115, 152
Pang L, 200
Pankert T, 543
Pannier C, 202
Panning M, 516
Pantophlet R, 513
Panyanouvong P, 202
Paone G, 427
Papadopoulou B, 543
Paparrodopoulos SC, 151
Paranchych W, 1, 12, 14, 17, 18, 23, 51, 52, 59,
        61, 62, 63, 64, 66, 67, 69, 70, 71, 72,
        73, 74, 75, 82, 192, 241, 257, 317, 367
Parant M, 309
Pardo CG, 543
Parent LJ, 545
Pariente N, 418
Park C, 188
Park DH, 144
Park E, 131, 163, 545
Park GH, 117
Park GW, 126, 128, 165
Park HY, 223, 544, 548
Park J, 141, 538
Park JA, 134, 148, 149, 152, 156
Park JH, 163, 232, 233

Park JJ, 551
Park KT, 108
Park S, 131
Park SG, 538
Park SH, 234, 551
Park SY, 223
Parker J, 342, 529
Parker R, 543
Parks S, 164
Parpari L, 161
Parrott AM, 354
Parshina OV, 378
Parsons B, 202
Partin KM, 518
Pascale JM, 446
Pascalis H, 203
Pascente C, 311
Pasceri P, 543
Pashley R, 141
Pashley RM, 139, 141
Pashneva NN, 135, 185
Pasloske BL, 64, 515, 516
Passent J, 91
Pastuszak AW, 538
Paszek P, 545
Pászti J, 53
Patel DJ, 352, 483, 540, 558
Patel HP, 548
Patel KG, 451
Pathak KB, 538
Pathak VK, 548
Patil SM, 131
Patitucci T, 202
Patkar A, 283
Patras A, 126, 160, 161
Patras AS, 126
Patskovsky Y, 360, 361, 470, 546
Pattinson DJ, 557
Pauff S, 545
Paul AV, 235
Paul JH, 40, 113, 114, 118
Paul S, 433
Paulson JC, 453
Pavé A, 95
Pavlovic M, 203
Pawliszyn J, 200, 310
Payán A, 113
Payment P, 106, 137
Payumo AY, 545
Peabody D, 352
Peabody D, 489
Peabody DS, 143, 197, 349, 350, 351, 352, 354,
        359, 360, 431, 440, 442, 444, 446,
        447, 449, 460, 461, 462, 468, 470,
        489, 493, 527, 529
Peabody J, 444, 446, 493, 514
Pearson L, 538
Pease LF 3rd, 166, 199
Pecson BM, 107, 137, 141
Peden K, 234
Pedersen LB, 234
Pedley S, 151
Peduzzi R, 120
Peek LJ, 489
Pei L, 107, 108
Peillon N, 440
Pekárek J, 377
Pekhov AP, 54
Pekonen P, 139
Pelczer I, 348, 349
Peldszus S, 156

Pelham SJ, 489
Pelkmans L, 543
Pell AJ, 486, 556
Pellegrini O, 280
Pelleïeux S, 145
Pellicioli E, 490
Peltomaa R, 529
Peltz SW, 538
Peña EJ, 543
Peña M, 156, 543
Pendyala B, 161
Peng WP, 199
Peng Y, 281
Penn CR, 234
Penrod SL, 120, 189
Penswick JR, 211
Pentel PR, 501
Pepper IL, 143
Perales C, 418
Pereira GC, 543
Pereira-Gómez M, 416
Perelle S, 160
Pererva TP, 98
Perez DR, 14
Pérez JA, 134
Pérez JS, 166
Pérez-Cano L, 361
Pérez-Méndez A, 108, 116, 144
Perez-Pinera P, 551
Pérez-Robles J, 533
Perez Romero C, 543
Perez Romero CA, 543, 544
Pericle F, 493
Perkett M, 481
Perkins EA, 310
Perkins J, 150
Perlmutter JD, 481
Permogorov VI, 377
Peron Y, 67
Perreault J, 109
Perrin A, 200
Perrott P, 144, 202
Perry KA, 164
Person MD, 543
Persson M, 344, 433, 469, 470
Pesaro F, 122
Pessel-Vivares L, 544
Petereit A, 141
Peterlin BM, 536, 537
Petersen N, 516
Petersen PSS, 551
Peterson BM, 141
Peterson LA, 136
Petre J, 23, 263, 325
Petri B, 127
Petrovska R, 378
Petrovskis I, 471, 556
Petsev DN, 480
Pettitt BM, 352
Pezeshki P, 162
Pfeifer D, 67
Pfister T, 490
Pfleger S, 132
Phair RD, 543
Phalora PK, 544
Pham HP, 492
Pham M, 148, 201
Phan AD, 201, 529
Phillips DJ, 193, 205
Phillips GK, 464, 518, 527
Phillips KM, 202

Phillips LA, 186, 344, 369
Phillips SE, 352, 354, 470, 537
Phillips SL, 332, 333, 538
Phisitkul S, 501
Phuong DJ, 544
Piatkovsky M, 501
Pica A, 556
Picard C, 42
Piccirilli F, 281
Píchalová R, 197
Pichon M, 42
Pichon X, 547, 548
Pickering LK, 155
Pieczenik G, 218
Piefer AJ, 299, 425
Piepenbrink KH, 65
Pierre G, 155
Pierrel J, 219, 222
Pierz V, 116
Pierzo V, 114
Pietrzak M, 346
Piette AS, 109
Pietzner M, 235
Piffaretti JC, 248, 308
Pilania M, 501
Piletsky S, 108, 109, 312
Pillai SD, 120, 145, 150
Pilmane M, 378
Pimienta G, 539
Pimonov VV, 556
Pineda JMB, 543
Pineda M, 551
Pines A, 538
Pinheiro MD, 140
Pinkett MO, 149, 313
Pinon A, 160
Pintacuda G, 486, 487, 556
Pinto D, 235
Pinto F, 202
Pinto R, 501
Piorkowski G, 202
Pique ME, 64
Pisanic N, 120
Pisano MB, 203
Pisarenko A, 138
Piscopo P, 235
Pistsov MN, 379
Pitchiaya S, 548
Pitek AS, 529
Pitkänen T, 116, 118
Pitol AK, 153, 163
Pitt P, 154
Pitt PA, 154
Pitt TL, 56
Pittoggi C, 234
Pitton JS, 50, 81, 248, 308
Pivarnik LF, 39, 150, 158
Piyatigorskaya TL, 205, 377
Pizzinga M, 543
Plaisier JR, 471
Planta RJ, 314
Platt RJ, 551
Platten III WE, 156
Platts-Mills JA, 203
Plaza-Rodriguez C, 160
Pleij CW, 218, 338
Pleiss MG, 123
Plevka P, 433, 469, 470
Plinston CA, 341
Plisov S, 548
Pluciñska K, 551

Plummer EM, 489
Plummer JD, 113
Plummer RJ, 551
Pocock GM, 544
Pocock J, 108, 109, 312
Poderycki MJ, 539
Podgornik A, 96
Podgornik R, 484
Poeschla EM, 544
Poindexter JS, 96
Poirier MG, 548
Poitras E, 115
Pokharel B, 126, 160, 161
Pokharel YR, 558
Pokhrel S, 493
Pokorski JH, 529
Pokorski JK, 455, 529
Polakovic M, 166
Polashock JJ, 394
Polaski JT, 548
Polenova T, 486, 487, 556
Poleshchuk EM, 517
Poling BC, 538
Poliquin PG, 203
Politi AZ, 544
Polivtsev OF, 205, 377
Pollack Y, 23, 94, 263, 324, 325
Polles G, 482
Pollet R, 258, 288
Polo-López MI, 134, 142
Polonskaya Z, 455, 457, 513
Polozov AI, 378
Poluektova L, 378
Poma HR, 166, 203
Pomerantz SC, 342
Pommepuy M, 158
Pommier de Santi V, 203
Pon CL, 329
Poncet D, 540
Ponchon L, 517
Ponta H, 123
Pontes-Braz L, 399, 425
Ponthoreau C, 116
Ponting CP, 543
Pooi CK, 121
Poole F, 189
Poon J, 125
Poot R, 169, 228, 335
Popa LM, 208
Pope DH, 319
Poppell CF, 112
Poranen MM, 234
Porman AM, 548
Portal MM, 544
Porter AG, 223, 327
Poryvkina L, 109
Posevaia TA, 378
Pot R, 103, 111, 123, 163
Pot-Hogeboom WM, 111, 112, 119, 123
Potapov I, 542
Potekaev NS, 378
Potgieter N, 136
Pothier P, 160
Potschka M, 96
Pottage T, 165
Potter K, 83
Poudel L, 356, 483
Pouillot R, 159
Pourcher AM, 113
Powell D, 164
Powell J, 156

Powell SK, 148
Powell T, 551
Powelson DK, 152
Powrie EA, 543
Prado T, 166, 203
Praharaj I, 202
Prakash S, 551
Prangishvili D, 555
Prasad BV, 435
Prasanth KV, 541
Prasuhn DE Jr, 452
Prat GS, 136
Pratt CW, 267, 364
Precup J, 342
Presolski SI, 492
Preston DR, 147, 148, 155
Preston WD, 203
Prestwood LJ, 538
Preuss A, 79
Preuß R, 399, 425
Prévost M, 136, 202
Priano C, 224, 227, 337, 421, 422, 423, 426, 440
Pribil W, 132
Price M, 106
Price VL, 440
Prickett KS, 440
Prier KR, 149
Prieto-Simón B, 133, 311
Prikhod'ko AV, 378
Primakoff P, 184
Primrose SB, 105
Pritchard CG, 393, 396, 397
Prives C, 383, 386
Prokofyev MA, 219
Prokulevich VA, 66
Propst-Ricciuti B, 80
Propst Ricciuti C, 80
Prouty WF, 88
Prozesky OW, 113
Pruitt BW, 551
Prusa J, 234
Ptok J, 538
Puchta H, 551
Puglisi EV, 545
Puglisi JD, 545
Puig A, 113
Pulgarin C, 134, 142
Pulikowska J, 267
Puls RW, 120
Pumpen P, 25, 76, 77, 78, 80, 81, 83, 85, 91, 92, 93, 179, 184, 185, 187, 288, 371, 372, 373, 383, 422, 436, 437, 438, 439, 489, 441
Pumpen PP, 77, 98, 175, 185, 371, 374, 375, 431
Pumpens P, 433, 435, 438, 440, 441, 445, 467, 471, 489, 490, 491, 501, 556
Pupols M, 540
Purushothaman P, 544
Purwar M, 513
Pusey PN, 190, 191
Pushko P, 422, 436, 438, 489, 501
Pushko PM, 436, 438
Pustoshilova NM, 225
Putallaz T, 161
Putonti C, 42
Putri RM, 518
Pyankov OV, 517
Pye QN, 133
Pyne MT, 203
Pyra H, 234
Pyzocha NK, 551

Q

Qi L, 551
Qi LS, 551
Qi Y, 433, 540
Qiao M, 235
Qiao Z, 123
Qinjian Zhao Q, 109
Qin P, 549
Qin Z, 458
Qiu H, 134
Qiu X, 501
Qiu Z, 153, 280
Qiu ZG, 108
Qu H, 137
Quadt R, 267
Quake S, 556
Quake SR, 545
Quan FS, 501
Quaye O, 202
Quek BL, 546
Quek PH, 150
Querido E, 543, 544
Quezada E, 538
Quin MB, 533
Quinlan ME, 543
Quinn CM, 487
Quinn LY, 487
Quiñones O, 128
Quiñones OA, 543
Quinta-Ferreira R, 492
Qulsum U, 540

R

Raab C, 88
Rabinowitz JC, 318
Rackham O, 543, 545
Rački N, 139
Radecka I, 529
Radloff RJ, 187, 194
Radman M, 61
Raeeszadeh-Sarmazdeh M, 453, 529
Raeymakers A, 24, 212, 319, 323, 365
Rafalska-Metcalf IU, 545
Raffl S, 54, 79
Ragheb K, 127
Raghunathan G, 234
Rahaman MS, 155
Rahi H, 326
Rahman A, 304
Rahman MA, 115, 538
Rahman R, 538
Rahman SA, 548
Rahmouni AR, 538
Rai M, 132
Raicevic A, 399, 425
Rainey PB, 66, 414
Raj A, 543
Raj R, 197
Raja B, 311
Rajal VB, 166
Rajala-Mustonen RL, 106, 124
Rajala RL, 153
Rajamohan F, 234
RajBhandary UL, 342
Rajca M, 157
Rajendra VK, 54
Raji I, 527
Rajko-Nenow P, 142, 158
Rajkovic A, 137

Rajkowitsch L, 281
Rajpurohit R, 233
Rakariyatham K, 126
Ram D, 203
Ramachandran V, 235, 544
Raman S, 518
Ramanathan M, 540
Ramasundaram S, 129
Ramaswami B, 40, 114, 144, 199
Rambout X, 538
Rames E, 109
Ramirez A, 20, 40, 226, 295, 344, 406, 425
Ramm M, 543
Ramon C, 126
Ramón-Núñez LA, 235
Ramos Bordajandi L, 121
Ramqvist T, 489
Ramsay WJ, 123
Rangarajan M, 157
Rangel M, 447
Ranque S, 203
Ranson NA, 73, 243, 476, 477, 481, 484, 556
Rao A, 202
Rao AL, 544
Rao G, 310
Rao RS, 128
Raoufmoghaddam S, 197
Raoult D, 203, 233
Rapala J, 154
Rapaport D, 543
Rappaport I, 70, 75, 122, 307
Rappsilber J, 546
Rappuoli R, 489
Raschke C, 130
Rascon O, 117
Rasefske KA, 299, 425
Rashid A, 354
Rasolofonirina R, 102
Rassi E, 54
Rathore U, 513
Ratmanova KI, 219
Ratner S, 136, 137, 139
Ratnesar-Shumate S, 138
Rattanakul S, 126, 130
Rattanavong S, 202
Rattner BP, 543
Raupach T, 501
Ravanshad M, 517
Ravantti JJ, 84
Ravensbergen CJ, 328
Ravi R, 161
Raviram R, 549
Ravnikar M, 139
Ravva SV, 115, 118
Rawa ARA, 162
Rawlinson WD, 202
Ray P, 538
Raymond KN, 443, 449, 450, 451, 462, 518
Raymond P, 546, 548
Rayner JC, 501
Raynor PC, 144
Razzell WE, 211
Read CA Jr, 427
Ready T, 501
Rebeaud F, 447
Rebillat I, 202
Rechenburg A, 116
Redaschi N, 555
Redder P, 325
Reddy SM, 136
Reddy T, 556

Reddy TR, 538
Reddy VS, 190, 453
Redman JA, 120, 121, 152
Redondo S, 58
Redway KF, 164
Redzej A, 79
Reed CA, 167, 244, 245, 424, 479
Reed JC, 544
Rees JC, 198
Reeve P, 156
Reeve PJ, 149
Refardt D, 148
Regan M, 527
Regel R, 156
Régnier P, 280
Rehfeld F, 544
Rehnstam-Holm AS, 157
Reid BR, 19, 20, 291, 338
Reid GE, 250
Reid PJ, 184
Reimers RS, 410
Rein A, 235
Reinhardt G, 427
Reissmann S, 130
Rekosh DM, 314
Remaley A, 501
Remaut E, 20, 167, 181, 183, 185, 212, 214, 216,
219, 293, 315, 344, 337, 365, 373, 419,
423, 424, 427, 431
Remsen JF, 123
Ren H, 144, 533
Ren J, 556
Ren L, 545
Renhof R, 194, 210, 333, 349, 358
Renhof RF, 349
Renhofa R, 433, 444, 445, 462, 466, 471, 501,
556
Renkhof (Renhof) RF, 333, 358
Renner WA, 447, 489, 490, 501
Renshaw RW, 202
Rensing U, 23, 214, 220, 260, 262, 272, 274, 286,
365, 383, 386
Rensing UFE, 212, 220, 365
Rentmeester E, 543
Renzi RF, 200
Reponen L, 144
Reponen T, 144, 163
Resch S, 299, 425
Reske KA, 165
Reta N, 108, 558
Retel JS, 486
Reuss M, 234, 339
Reusser F, 313
Revel M, 23, 263, 324, 325, 456
Reynard AM, 54
Reynolds KA, 164
Reynolds S, 69, 72, 241
Reynolds SJ, 144
Rezaei N, 501
Rezaeinejad S, 120
Rezwan K, 156
Rhee JK, 456
Rhodes MW, 106
Ribeiro AS, 542, 545
Ribitsch G, 210
Ricca DM, 111
Ricci L, 235
Riccio A, 542
Rice RH, 250
Rich A, 15, 156, 325, 330, 331
Richards DH, 136

Richards EG, 226
Richards H, 57
Richards J, 202
Richelson E, 73, 241
Richmond MH, 56
Richter F, 551
Ricke SC, 162
Ricketti PA, 501
Rideau A, 538
Ried T, 541
Rieder R, 540
Rigby MH, 235
Rigo N, 538
Rigobello V, 139
Riguera R, 453
Rikta SY, 132
Riley JJ, 155
Riley KJ, 539
Riley MR, 199
Rimhanen-Finne R, 154
Rincé A, 146
Ringuette L, 163
Rino J, 543, 544, 545
Rio DC, 548
Riordan W, 166, 200
Ripp S, 529
Rippey SR, 113, 157
Rise ML, 151
Rissmann M, 202
Riva S, 91
Rival-Gervier S, 543
Rivero-Müller A, 551
Rivet R, 113, 159
Riviere ME, 513
Rivoire C, 555
Roach A, 143, 160
Robakis NK, 334
Robbens J, 337
Roberts AS, 399, 425
Roberts BN, 15
Roberts JW, 18, 19, 178, 191, 195, 238, 319
Roberts RC, 118
Robertson HD, 19, 21, 23, 25, 176, 180, 183, 214,
216, 217, 221, 222, 287, 322, 323, 331,
344, 345, 365, 367, 368, 369, 370, 371
Robertson L, 121
Robertson LJ, 116, 134
Robin JD, 234
Robinson A, 233
Robinson BA, 544
Robinson BJ, 115
Robinson F, 538
Robinson JP, 127
Robinson JW, 185, 186, 216, 221, 316
Robinson K, 138
Robinson L, 398
Robinson MDM, 197
Robinson MO, 153, 539
Robinson SA, 418
Robledo M, 540
Robson B, 113
Rocha MS, 166
Rocha PP, 549
Roden RB, 446, 493
Rodrigo MM, 529
Rodrigues BCM, 457
Rodrigues C, 113
Rodrigues H, 113
Rodrigues JS, 166
Rodrigues SM, 516
Rodriguez J, 548

Rodriguez RA, 124, 141, 545
Rodriguez SD, 108, 312
Rodríguez-Carmona E, 518, 529
Rodriguez-Manzano J, 142
Roesch F, 544
Roesti ES, 557
Rogers H, 544
Rogerson DL Jr, 95
Roggenbuck MW, 348, 349
Rogovskaya SI, 379
Rohde N, 394, 425
Rohloff H, 223
Rohlova E, 235
Rohovie MJ, 518
Rohrer UH, 447
Rohrmann GF, 26, 194, 196, 307
Rojiani M, 83
Rokjer D, 127
Rokutan K, 543
Roldão A, 489
Rolfe KJ, 202
Rolfsson Ó, 471, 473, 474, 475, 482
Rolih V, 493
Román I, 418
Romanini DW, 450, 451, 461, 518, 527, 529
Romaniuk PJ, 347, 351
Romanyuk N, 235
Rome LH, 539
Romeo P, 235
Romeo T, 352
Römer W, 264, 363, 386
Romero CA, 543
Romero LC, 166
Romero OC, 141
Romero P, 112, 512
Romond C, 135
Rontó G, 132
Rook MS, 541
Roos RP, 234
Rörsch A, 85, 132
Ros C, 452
Rosa S, 543
Rosado-Lausell SL, 142
Rosales-Mendoza S, 501
Rosario K, 555
Rosay M, 556
Rosbash M, 538
Röschenthaler R, 88
Rose JB, 18, 19, 40, 113, 114, 118, 136, 164, 189,
205
Rose RJ, 543
Rosemeyer V, 234
Rosenberg H, 221, 223, 226
Rosenberg M, 221, 223, 226
Rosenberger RF, 342
Rosenblum JS, 129
Rosenthal G, 83, 349
Rosenthal GF, 83, 95, 349
Roser DJ, 138, 156
Rosiles-González G, 116
Rosina G, 156
Rosman M, 332
Ross CM, 115, 141
Ross RP, 555
Rossier O, 280
Rossmanith P, 139
Rossmann MG, 435
Rossoll W, 545
Rostain W, 551
Roszyk L, 361, 548
Rotering H, 296, 424

Roth JR, 187
Rothenbacher FP, 232
Rothman-Denes LB, 315
Rothwell JD, 91
Rotimi CN, 235
Rots MG, 551
Rottman F, 322
Roudi NE, 518
Roufa DJ, 319, 330, 345
Rouha H, 544
Roulston B, 414
Roush DJ, 166
Roussel C, 202
Roussis V, 139
Roux S, 556
Rouzbahani NH, 517
Rovenský J, 377
Rovira J, 137
Rowan N, 121
Rowlands DJ, 29, 303, 305, 436, 468
Rowlands DT Jr, 303
Rowsell S, 352, 356, 470
Rowson-Baldwin A, 543
Roy P, 435
Roy-Engel AM, 551
Rozenboom W, 314
Rozentāls G, 316
Rtimi S, 134, 142
Ruault M, 543
Rubino FA, 298, 544
Ruda VM, 545
Rudland PS, 94, 324, 325
Rudolph U, 26, 196
Ruedl C, 447, 490
Rueedi J, 151
Ruenphet S, 129
Ruigrok RW, 234
Ruijtenberg S, 548
Ruiz N, 298, 425
Ruiz-Echevarría MJ, 538
Ruiz-Ruiz F, 311
Rullan M, 548
Rumlová M, 197
Ruml T, 197
Rūmnieks J, 40, 42, 94, 169, 170, 195, 243, 250,
        255, 295, 406, 425, 433, 440, 462,
        466, 470, 471, 501, 556, 557
Ruokoranta TM, 40, 58, 226, 295, 425
Rupasov VV, 258
Rusch KA, 151
Rushizky GW, 95, 216
Rusiñol M, 126, 142
Ruskin RH, 117
Russell E, 316
Russell GJ, 218
Russell RJ, 313
Russell SC, 199
Rustad G, 108
Rustad M, 97, 339
Rutala WA, 163, 164
Rutjes SA, 108
Ruža L, 76
Ryabova LA, 334, 399
Ryan FJ, 385
Ryan JM, 501
Ryan JN, 120
Ryan T, 278
Rybalko SL, 378
Rychlik W, 318
Rymar' SE, 67, 186
Ryoji M, 186, 332

Ryu H, 116, 129, 138, 148, 153
Ryu KY, 551
Ryu TH, 163

S

Saari B, 266
Saavedra ME, 108
Saayman SM, 550, 551
Saber A, 558
Sabir T, 354
Sabo D, 413, 418
Sabo DL, 187, 188, 413, 419
Sabol S, 20, 324, 493
Saboo S, 548
Sacchetti R, 107, 136, 156
Sachidanandam R, 541
Sachsenröder J, 203
Sack BK, 501
Sączyńska V, 501
Sade D, 78, 81
Saeidi N, 108, 113
Saenz D, 117
Safonova MV, 517
Sagredo S, 548
Sagripanti JL, 136
Saha R, 138
Saha S, 539
Sahar S, 356
Sahinovic E, 543
Said N, 540
Saigo K, 461
Sainsbury F, 489
Saint-Antoine P, 163
Saito H, 540
Saito Y, 83
Sakaguchi K, 224
Sakaki Y, 82
Sakamoto G, 124
Sakamoto T, 156
Sakchaisri K, 543
Sakellaris M, 117
Sakharova NK, 217, 438
Sakuma T, 551
Sakurai T, 13, 23, 35, 37, 39, 40, 98, 99, 101, 102,
        103, 104, 105, 112, 163, 302
Sala A, 542
Sala L, 142
Salanti A, 460, 513, 514
Salazar-González JA, 501
Saleh NB, 131
Salehi S, 545
Salem AK, 501
Salesse R, 538
Salghetti SE, 541
Salman MA, 557
Salmond GP, 555
Salter RS, 114
Salveson A, 138
Salzman J, 543
Saman E, 219, 419, 427
Samatov TR, 390, 538
Samiee SM, 517
Samoto M, 138
Samsonoff WA, 15
Samuels DS, 540
Samuelson G, 196
Samuelsson T, 341
San Millan MJ, 314
Sana J, 235
Sanchez C, 106

Sánchez G, 162, 540, 541
Sánchez-Gonzalez L, 156
Sánchez-Navarro M, 457
Sander AF, 460, 513, 514
Sander M, 148
Sanders NN, 540
Sanderson JD, 105
Sanderson KE, 52, 54, 98
Sandhu SK, 113
Sandler H, 538
Sandmeyer S, 543
Sands J, 155
Sandstede B, 543
Sanger DM, 115
Sanger F, 19, 22, 176, 211, 212, 218, 220, 224,
        286, 315
Sangsanont J, 117
Sanhueza CA, 113
Sanhueza-Chavez C, 457
Sanjana NE, 551
Sanjuán R, 169, 396, 415, 416
Sankarakumar N, 165
Sankaran B, 471
Sankaran S, 202
Sankoff D, 167, 218, 340
Sano D, 104, 113, 137
Sano H, 283, 284
Sano S, 104
Sano Tsushima F, 52, 417
Sansom MS, 556
Santa Marina L, 116
Santangelo PJ, 544
Santarpia JL, 138
Santi L, 518
Santiago-Frangos A, 283
Santiago-Rodriguez TM, 42
Santillan CA, 132
Santo Domingo J, 116
Santos LC, 113
Santos LCB, 457
Sarapuu T, 334
Sardari S, 489
Sardo L, 548
Sargeant K, 95
Sarhan WM, 324
Sarkar D, 132
Sarkar NK, 372
Sarnquist C, 202
Sasges M, 126, 160, 161
Sasidharan S, 149
Sasnauskas K, 433, 438, 489
Sassi HP, 141, 164, 165
Sassoubre LM, 113, 131
Sastry P, 224, 248
Sastry PA, 14, 59, 64, 223
Satake Y, 543
Sathyan KM, 539
Satitsuksanoa P, 501
Sato K, 543
Sato S, 234
Sato T, 127, 166
Sato-Asano K, 211
Satoh K, 129
Sattar SA, 138
Satterwhite TK, 120
Saucedo NM, 109
Saudan P, 235, 447, 459, 460, 489, 490, 491, 492,
        518
Saulquin X, 551
Saunders GC, 303
Saunders JR, 56

Sautrey G, 200
Sauvageau D, 96
Savas JN, 543
Savichtcheva O, 120
Savin FA, 98, 123, 138, 185
Savoye P, 124
Sawaki S, 40, 331
Sawamura S, 35, 302
Sawata SY, 538
Sawchyn I, 267
Sayour EJ, 501
Scadden AD, 538
Scarcelli JJ, 543
Scarpin MR, 543
Schaal H, 538
Schaar H, 138
Schabikowski M, 132
Schaefer DW, 191
Schaefer K, 138
Schaefer L, 221, 330
Schaefer M, 540, 541
Schaeffer JW, 144
Schäfer C, 550
Schaffer DV, 550
Schaffer S, 145
Schaffner W, 284, 386, 387, 393, 405
Schandel KA, 52
Schaper M, 40, 113, 115, 137, 140, 145
Schaub SA, 143
Schauer A, 54
Schechter AN, 304
Scheckel C, 538
Schedl PD, 325
Scheible OK, 127
Scheiblhofer S, 501
Schellenbacher C, 493
Schendzielorz P, 501
Scheps R, 456
Scherneck S, 438
Scherrer K, 226
Schertenleib A, 164
Scheulen M, 331
Scheurer S, 501
Scheuss V, 542
Schieweck R, 545
Schifano JM, 232
Schiff KC, 113
Schijven JF, 108, 120, 123, 136, 148, 150, 151,
    153, 540, 541
Schildbach JF, 54
Schilder C, 490, 518
Schiller JT, 442, 447, 448, 461, 493
Schimer J, 527
Schimmele B, 490
Schindler J, 1, 87, 88
Schink T, 134
Schlesinger S, 410
Schlessinger D, 85, 210, 313
Schlicht HJ, 427
Schlick TL, 449
Schlicksup CJ, 480
Schlissel G, 543
Schlumberger HD, 306
Schmausser B, 431
Schmid M, 489, 543
Schmid SL, 453
Schmidt CE, 160
Schmidt F, 294
Schmidt J, 71
Schmidt JM, 11, 15, 16, 37, 55, 66, 67, 71, 76,
    81, 302

Schmidt K, 424, 543
Schmidt S, 125
Schmidt U, 558
Schmitt M, 328
Schmitt W, 427
Schmitz N, 447, 459, 460, 489, 518
Schnabel E, 427
Schnarr M, 234
Schnegg B, 50
Schneider A, 427
Schneider C, 543
Schneider D, 130, 233
Schneider JE Jr, 130, 133
Schneider MC, 258, 368
Schneider P, 490, 491
Schnetzler Y, 491, 518
Schneweis KE, 427
Schnoor MH, 131
Schnös M, 16, 46
Schober A, 190, 394, 425, 543
Schoenfeld TW, 234
Schoenitz M, 144
Schoenmakers JGG, 220
Schönberger J, 543
Schoonen L, 518, 548
Schoulaker R, 51
Schoulaker-Schwarz R, 51
Schrader C, 160
Schreiber G, 543
Schreier MH, 318
Schreil W, 54
Schreiner U, 427
Schriewer A, 141
Schröder W, 427
Schroeder SJ, 476
Schroeter ML, 235
Schu DJ, 283
Schubert D, 26, 196, 250
Schuchalter S, 78, 81
Schuech R, 141
Schuldiner M, 543
Schülke S, 501
Schulz KN, 545
Schulze-Makuch D, 145, 149, 151
Schumacher MA, 280
Schumacher R, 143
Schuman EM, 548
Schupbach T, 542
Schuppli D, 279, 364, 386
Schur FK, 197
Schuster D, 234
Schuster P, 232, 400
Schuster-Wallace CJ, 133
Schvoerer E, 120
Schwab KJ, 96, 120, 138, 141, 156
Schwartz AM, 543
Schwartz FM, 17, 78
Schwartz JH, 19, 75, 141, 167, 313, 314, 316
Schwartz W, 438
Schwarz B, 518, 533
Schwarz K, 162, 462, 489, 490, 501, 512, 518
Schwarz KA, 551
Schweet R, 316
Schweikert EA, 427
Schwertz H, 234
Schwienhorst A, 394, 536
Sciabica KS, 433
Sciamanna I, 234
Sclarsic SM, 542
Scolnick E, 332
Scott DW, 15, 35, 301

Scott JW, 83, 94
Scott TM, 107, 153, 154, 155, 159
Seaman R, 155
Sebastian T, 543
Sebbel P, 458, 490
Secor SL, 120
Secundo F, 52
Sedat JW, 210
Seder RA, 501
Sedillo JL, 165
Sedláček J, 218
Sedlak DL, 134
Seeberg E, 234
Seed CR, 202
Seedorf K, 427
Seephonelee M, 202
Segal M, 545
Segard C, 202
Segre AL, 218
Segura I, 542, 548
Seidel K, 123
Seidel M, 108, 109
Seidu-Larry S, 538
Seifer M, 349
Seiler F, 166
Seiler S, 165
Seim KL, 451
Seitz O, 545
Sekiba K, 551
Sekine M, 143
Sekine Y, 224, 536
Sela M, 309
Selby MJ, 536, 537
Selimkhanov J, 545
Selinka HC, 137
Seliskar CJ, 310
Sellou H, 544
Selvaraju SB, 203
Selvey LA, 427
Semenova TB, 378
Semler BL, 538
Sen S, 233
Sena M, 529
Senaud J, 118
Senear AW, 278
Senecal J, 137
Sengvilaipaseuth O, 202
Sensi P, 89
Senti G, 462, 490, 501, 518
Seo JK, 544
Seo JS, 543
Seo K, 141
Seo Y, 466, 527
Seoighe C, 415
Seong BL, 342, 489
Seow Y, 518
Sepúlveda A, 218
Sepúlveda-Robles O, 14
Serafino A, 234
Serafino A, 234
Serebriiskii I, 540
Serebriiskii IG, 539
Sergeant A, 433
Sergeev AA, 517
Sergeev NV, 202
Serrano E, 116
Seth P, 202
Setiyawan AS, 117
Seto E, 115, 303
Setoyama T, 539
Setubal JC, 556
Seurinck-Opsomer C, 219, 224, 274, 284

Seweryn P, 272
Sexton DJ, 164
Sexton JD, 164, 165
Seyffarth T, 438
Seyfried PL, 106
Seymour IJ, 159
Seyrig G, 113
Shabalina NV, 378
Shaban HA, 544
Shabarova ZA, 219
Shade L, 160
Shaffer R, 163
Shaffer RE, 164
Shafranski P, 346
Shah J, 398
Shah JS, 397, 398
Shah S, 235
Shah SA, 134, 491, 518
Shahi N, 558
Shahid M, 141
Shaikhet L, 375
Shakarisaz D, 311
Shakeel S, 483
Shaklee PN, 223, 405, 411, 415, 422, 426
Shaku M, 233
Shallcross J, 203
Sham LT, 298, 425
Shamshiev AT, 491
Shan J, 153
Shandalov S, 153
Shang C, 136, 464
Shang H, 250
Shang W, 126, 551
Shankarling G, 538
Shao J, 489
Shao S, 549
Shao Y, 40, 58, 59, 60, 71, 295, 425
Shapiro JW, 42
Shapiro K, 23
Shapiro L, 16, 23, 55, 64, 67, 76, 166, 178, 257,
          262, 272, 273, 258, 286, 365, 366
Shapiro R, 325
Sharipova MR, 73
Sharma A, 109, 133, 543
Sharma CM, 548
Sharma M, 117, 143, 538
Sharp JD, 233
Sharp PM, 219
Shash K, 234
Shatalov M, 129
Shav-Tal Y, 541, 542, 543, 544, 545, 549
Shaw EJ, 56
Shchelkanov MY, 517
Shchipkov VP, 50, 54
Shefer K, 546
Sheinberger J, 545
Sheldon J, 418
Shelton MB, 165
Shen C, 127
Shen H, 165, 166
Shen M, 538
Shen SH, 436
Shen T, 391
Shen Y, 210
Shen Z, 153
Shen ZP, 128
Shenoy SM, 540, 541, 542, 543, 545, 547
Shepherd CM, 190, 218
Sherban TP, 98, 122, 134, 135, 175, 185
Sherchan SP, 126
Sherer NM, 544

Shevelev OB, 203
Shevtsov SP, 543
Shi C, 151
Shi D, 326
Shi H, 42, 518, 545
Shi L, 128, 132, 134
Shi M, 42, 555
Shi Q, 551
Shi W, 304
Shi X, 160, 279, 543
Shi Y, 446, 551
Shiba T, 23, 37, 39, 72, 73, 74, 99, 101, 102,
          103, 178, 187, 195, 257, 288, 308,
          315, 461
Shibahara S, 387
Shichino Y, 543
Shieh YC, 157
Shieh YS, 40, 82, 113, 114
Shields PA, 96
Shigemura H, 156
Shih NC, 314
Shim J, 128, 160
Shim JH, 117
Shim JJ, 223
Shima Y, 267, 425
Shimura Y, 87, 185, 194, 333
Shin DS, 64
Shin G, 107
Shin GA, 120, 124, 125, 136
Shin H, 543
Shin IS, 199
Shine J, 22, 121
Shiozaki A, 285
Shipkov VP, 54
Shipley P, 13, 14, 37, 56, 66, 302
Shipulin GA, 517
Shirasaka Y, 104
Shirasaki N, 121, 134, 148, 149
Shirbaghaee Z, 489
Shirota M, 551
Shishovs M, 433, 469, 486
Shkoporov A, 555
Shmakov S, 551, 558
Shokair I, 200
Sholtes KA, 129
Shooter KV, 134, 135
Shores LS, 489
Shorter D, 501
Shouse PJ, 150
Shraga A, 543
Shrimp JH, 551
Shtatland T, 356
Shu MD, 139
Shu S, 192, 243, 478, 479
Shuai D, 131
Shubladze AK, 377
Shukla D, 529
Shukla S, 455, 529
Shulman LM, 107, 203
Shupe A, 165
Shuval HI, 161
Shwartzbrood L, 106
Si H, 161, 551
Si J, 141
Sible E, 42
Sicard AM, 184
Sickbert-Bennett EE, 163, 164
Siddiqi O, 175
Sidhu JP, 122, 141
Sidikaro J, 314
Sidler CL, 548

Sidorov GN, 517
Siegel J, 84
Siemann M, 339
Siemann-Herzberg M, 234
Sierakowska H, 233
Sierka RA, 127
Sierra ML, 112
Sierro N, 559
Sifuentes LY, 164
Sigafoos AN, 551
Sigman DS, 221
Sigstam T, 134, 136, 146, 147, 148
Siksnys V, 558
Silas S, 558
Sillanpää M, 126
Sillero A, 20, 22, 167, 180, 252
Silva JL, 143, 197, 461
Silva JPN, 548
Silva-Beltrán NP, 140
Silvennoinen O, 543
Silver PA, 465, 533, 546
Silver R, 234
Silverman A, 159
Silverman AI, 50, 52, 69, 141
Silverman P, 50, 51, 52, 61, 142
Silverman PM, 52, 69, 72, 73, 85, 120, 262, 276,
          277, 383, 386
Silverstein J, 136
Silvertand LH, 200
Siminoff P, 89
Simmel FC, 548
Simmons OD, 129
Simon R, 131, 140
Simon SM, 544
Simonet J, 145
Simonova EG, 517
Simons G, 420
Simper SC, 235
Simpson CE, 543
Simpson JC, 518
Sims JT, 148
Simukova NA, 123
Šimůnek J, 151
Sinclair MI, 113
Sinclair R, 120, 121
Sinclair TR, 156
Singaram SW, 232, 481
Singer MF, 20
Singer RE, 324, 325
Singer RH, 360, 361, 470, 538, 540, 541, 542,
          543, 544, 545, 546, 547, 548
Singer-Krüger B, 543
Singh J, 548
Singh KK, 538
Singh M, 64, 115, 427
Singh P, 513, 529
Singh R, 235, 311
Singh S, 542
Singh SK, 460, 513
Singh SN, 57, 155, 391
Singh T, 166, 199
Sinha NK, 19, 210, 220, 226, 339
Sinibaldi-Vallebona P, 234
Sinkunas T, 559
Sinsheimer R, 67
Sinsheimer RL, 15, 16, 18, 19, 23, 67, 82, 86, 87,
          189, 205, 237, 257, 365, 367, 368, 372,
          383, 489, 553
Sinton LW, 106, 113, 141, 151
Sinzel M, 543
Siomos MA, 399, 425

Siray H, 439
Sirgel FA, 14, 16, 58, 59
Sirivithayapakorn S, 151
Šišovs M, 42, 250, 556, 557
Sisson SA, 138, 156
Sitabkhan A, 202
Siu KH, 453
Sivakumar A, 543
Sivasubramani SK, 163
Sjöblom B, 281
Sjogren JC, 127
Skali-Lami S, 145
Skaltsa H, 89, 139
Skangals A, 394, 439
Skerker JM, 64
Skibinski DAG, 156, 513
Skiniotis G, 545
Skinner MA, 557
Skoblov YM, 123
Skogan G, 108
Skogerson L, 223, 345
Skogerson LS, 325
Skoglund C, 235
Skok JA, 549
Sköld O, 84
Skorpen F, 234
Skraber S, 106
Skraber S, 116, 117, 145
Skrastina D, 433, 441
Skripkin EA, 224, 227
Skurray RA, 49, 61, 98
Slaby O, 235
Sladek M, 538
Slavcev R, 559
Slay E, 65
Slaymaker IM, 551, 558
Sleep BE, 152
Slegers H, 217
Slobin LI, 324
Slobodin B, 543, 544
Slor H, 258, 365
Slupphaug G, 234
Small MJ, 143
Smeets PW, 152
Smiley BK, 441
Smirnov SN, 108, 312
Smirnov VD, 219
Smit J, 63, 64
Smith CW, 538
Smith D, 265
Smith DA, 354
Smith DC, 41
Smith DH, 87
Smith DW, 202
Smith GP, 436
Smith JH, 234, 539
Smith JJ, 117
Smith JS, 359, 470
Smith MT, 457, 529
Smith RW, 538, 548
Smith SJ, 543
Smith SO, 276
Smith SR, 105
Smith WP, 319
Smits FJC, 136
Smolke CD, 544
Smyth H, 493
Smyth RP, 538
Snead MC, 112
Snellgrove WC, 42, 169, 412
Snijders AP, 543

Sniker DI, 436, 438
Snow SD, 131
Snyder MP, 356
Snyder S, 138
Snyder SA, 128, 138
Snyder-Cappione JE, 544
Soares GC, 143
Soares NE, 116
Sobek H, 109
Sober HA, 216
Sobrero P, 283
Sobrino F, 413
Sobsey M, 113
Sobsey MD, 40, 41, 42, 82, 106, 107, 113, 114,
        117, 118, 120, 126, 130, 133, 136, 141,
        150, 157, 158, 160, 163, 164, 165,
        169, 442
Sobura JE, 325
Socolovschi C, 233
Sodoyer R, 493
Sofer D, 203
Sogo JM, 279
Sokolova TM, 378
Sokolowski M, 539
Sol C, 210
Solecki O, 113
Solis JJ, 126
Söll D, 184, 249
Solo-Gabriele HM, 107
Solomon CJ, 319
Solomon O, 234
Solymosy F, 138
Sominskaya I, 438, 439
Sommer J, 139
Sommer JM, 55, 71
Sommer M, 202
Sommer R, 124, 132, 138, 156, 162
Somovilla P, 418
Sonawane MD, 557
Song B, 516
Song K, 545
Song LF, 125
Song M, 544
Song S, 551
Song W, 129
Song Y, 551, 556
Sonnleitner E, 279, 280
Sonoda J, 539
Sontheimer EJ, 558
Sood AK, 150
Sorber CA, 112
Sorial G, 156
Šorm F, 88, 194
Soropogui B, 517
Soroushian F, 136
Soto MA, 218
Soto-Beltran M, 108
Soudry E, 79, 90, 91
Sousa SA, 280, 452
Sousa-Herves A, 453
Soussan L, 451, 452
Southby J, 538
Souza TL, 143
Sovago J, 513
Spadafora C, 234
Spahr PF, 19, 25, 176, 210, 211, 212, 214, 220,
        221, 227, 238, 250, 315, 320, 333, 345,
        370, 422
Spakowitz AJ, 542
Spang A, 543
Spangler VL, 82, 186

Spanjaard RA, 294, 295, 338, 339, 420
Sparrow IJ, 310
Specht EA, 130
Specht KG, 545
Spector DL, 541
Speed DJ, 141
Speers D, 202
Speight S, 164
Speiser DE, 512
Spellman R, 538
Spencer SK, 114
Sperling J, 546
Sperling R, 546
Spiegelman S, 1, 15, 19, 22, 23, 38, 67, 82, 97,
        210, 212, 219, 247, 257, 258, 261, 262,
        278, 344, 365, 366, 381, 382, 383,
        384, 385, 386, 392
Spille JH, 548, 551
Spiller DG, 545
Spillmann CM, 464
Spillmann SK, 140
Spindler K, 138
Spingola M, 352, 354, 360, 462, 538
Spirin AS, 96, 97, 339
Spiro DJ, 42
Spivak NY, 378
Spohn G, 447, 458, 491
Spohn M, 550
Sprakel J, 556
Spremulli LL, 318
Springman R, 414
Springorum AC, 127
Springthorpe SV, 120
Sprinzl M, 342
Sproul OJ, 138
Spuhler D, 134
Sreenivas V, 202
Srikantan S, 544
Srivastava R, 84
Srivastava S, 53
Srour EF, 234
St Jean J, 559
St Johnston D, 542
Stabentheiner E, 54
Stachler E, 203
Staczek J, 78
Stadden E, 310
Stadler MB, 548
Stadler PF, 231, 400
Stadthagen G, 493
Staffell LM, 159
Stahl S, 427
Stahl SJ, 435
Staiger D, 540
Staley C, 113
Stallcup MR, 318
Stallions DR, 54
Standart N, 538
Standridge JH, 539
Standring DN, 349
Stanek J, 486, 556
Stanier RY, 11, 15, 16, 37, 66, 71, 76, 81, 302
Stanisich VA, 15, 57
Staniulis J, 438
Stankevich EI, 333, 358, 436, 438
Stanley WM Jr, 20, 324
Stano M, 35
Stanssens P, 219, 419, 427
Staples DH, 19, 176, 210, 222, 223, 238, 315, 327
Stapleton CM, 152
Stapleton JA, 113, 540

Starega-Roslan J, 551
Starikova I, 235
Stark H, 538
Starke JA, 125
Starokadomskyy P, 556
Startceva S, 542
Stasevich TJ, 548
Stass R, 479
Stauber CE, 153
Stavis RL, 1, 262
Stedtfeld RD, 113
Steele A, 127, 198, 557
Stefano JE, 397, 398, 425, 426
Stefanovic B, 539
Stefanovic L, 539
Stein H, 210
Stein KE, 234
Steinberg AD, 304
Steinberg CM, 175
Steinbergs J, 304
Steiner T, 113
Steinmetz N, 529
Steinmetz NF, 455, 456, 462, 489, 518, 529
Steitz JA, 18, 19, 20, 22, 24, 71, 139, 176, 183,
        184, 191, 194, 211, 214, 218, 220, 221,
        238, 278, 286, 288, 291, 294, 315,
        319, 320, 322, 325, 326, 327, 338, 345,
        370, 539
Stemler M, 427
Stenström TA, 132
Stent GS, 84
Stenz E, 88, 89
Stepanova AN, 543
Stepanova OB, 216
Stephan W, 123
Stephanopoulos N, 450, 462, 485, 451, 518
Stephen AG, 539
Stephenson JL Jr, 250
Stepulak A, 551
Stern A, 555
Sterner RM, 501
Sternglanz R, 120, 150, 542
Stetler RE, 106
Steven AC, 435
Stévenin J, 538
Stevense M, 543, 548
Stevenson IM, 115
Stevenson J, 516
Steward GF, 151, 234
Steward O, 542, 543
Stewart EJ, 61
Stewart II WE, 378
Stewart J, 120
Stewart JR, 40, 113, 115
Stewart ML, 313, 314
Stewart PL, 455
Stewart TA, 148
Stewart TS, 314, 319
Stewart-Pullaro J, 113
Sthiannopkao S, 117
Stiegler P, 391
Stirpe F, 314
Stockdale SR, 555
Stocker BA, 52, 98
Stockley DJ, 262, 278
Stockley PG, 29, 63, 73, 207, 232, 243, 351, 352,
        354, 440, 460, 461, 464, 468, 470,
        471, 473, 474, 475, 476, 477, 480, 481,
        482, 484, 518, 537, 556
Stoeckel DM, 113
Stoecklin G, 538, 546

Stoeger T, 543
Stöffler G, 325
Stoimenov P, 128
Stokes R, 40, 113, 114, 118
Stollar BD, 342
Stoltz KP, 234
Stone HO Jr, 82, 185, 315
Stonehouse NJ, 351, 352, 354, 460, 461, 467,
        468, 470, 474, 518, 537
Stoner BR, 311
Störmer M, 202
Storni FL, 557
Storni T, 447, 462, 489, 518
Storz G, 283
Stott R, 141
Stotter C, 235
Stouthamer AH, 314
Strack RL, 545
Strand M, 46, 49, 69, 70, 71, 183, 237
Strande L, 164
Strathmann TJ, 134
Stratton P, 202
Straub A, 14, 151
Straub TM, 117, 133
Straub U, 149
Strauss J, 67
Strauss JH Jr, 16, 18, 19, 23, 67, 187, 189, 205
Streatfield SJ, 489
Streeck RE, 436, 440
Streeton CL, 113
Strek W, 131
Strelnikova A, 431, 436, 440, 471, 556
Streltsov VA, 399, 425
Stringfellow L, 266
Stripecke R, 538
Strods A, 443, 445, 462, 501
Strohalmová K, 197
Stroheker T, 161
Strom MS, 64
Struck DK, 167, 243, 298, 424
Strunk G, 401
Strutt SC, 551
Strych U, 311, 312
Stuart DI, 435
Stubbs EA, 137
Stubbs G, 435
Stubbs SH, 538
Stucka R, 550
Stuckey D, 156
Stulberg MP, 334
Stumpf C, 539
Stumpf F, 202
Stumpp S, 314
Stumptner C, 234
Stutt EL, 399, 425
Stuyfzand PJ, 136
Stylianidou S, 542
Su C, 543
Su HJ, 144
Su Q, 139, 140, 143
Su X, 139, 140, 143, 160, 161
Subak-Sharpe JH, 218
Subramanian AR, 268, 324, 329, 331
Such GK, 453
Sue SC, 470
Suetina IA, 378
Sugawara R, 417
Sugino Y, 82
Sugiyama T, 19, 25, 82, 178, 180, 185, 195, 237,
        250, 315, 344, 345, 369
Suh J, 458

Suh Y, 558
Suhre K, 235
Sukhodolets MV, 268, 281, 283
Sukias JPS, 141
Sukumar M, 165
Sullivan R, 120
Summers MF, 352
Summers N, 169, 527
Sumper M, 283, 284, 387, 425
Sun D, 324
Sun H, 529
Sun J, 538, 551
Sun JE, 455
Sun L, 160, 442, 545
Sun P, 125
Sun Q, 453
Sun R, 192, 243, 478, 479
Sun S, 516, 538
Sun SM, 102
Sun TP, 53
Sun W, 136, 556
Sun WJ, 501
Sun XL, 527
Sun Y, 139, 444, 527
Sun Z, 470
Sunada K, 127
Sunami T, 285, 401, 404, 425
Sundarrajan S, 203
Sundberg BN, 551
Sunderland KS, 489, 518
Sundheim O, 234
Sundler F, 378
Sundram A, 113, 115
Suñé C, 538
Sunna A, 518
Sunun L, 398
Sureau A, 538
Surman SB, 113, 163
Surovoy A, 235
Suryanarayana T, 329
Susman M, 175
Suttle CA, 42
Suwa M, 156
Suylen GM, 153
Suzukake K, 313
Suzuki G, 88, 93, 94, 95, 297
Suzuki H, 235, 426
Suzuki M, 404
Suzuki S, 108
Suzuki Y, 72, 73, 74, 548
Sverdlov ED, 98, 135, 185
Svobodová J, 377
Swan D, 323
Swann PG, 234
Swanson P, 234
Swartz JR, 97, 339, 451, 461, 489
Swatkoski S, 198
Sy TL, 203
Sykes RB, 56
Sylla B, 517
Sylvain S, 155
Syme CD, 193
Syngouna VI, 128, 143, 151
Syomin BV, 320, 489
Syred AD, 436
Syvänen AC, 234
Szabo K, 160
Szafrański P, 313
Szekely M, 324
Szekeres GP, 156
Szer W, 96, 233, 263, 313, 322, 325, 343

Szewzyk R, 137
Szostak MP, 299, 425

T

Taboga OA, 166
Táborský I, 377
Taddei F, 414, 415
Tadić V, 551
Taghbalout A, 281
Tagliamonte M, 489
Tahmasebi S, 315
Tahmasebi Nick S, 131, 140
Tai PC, 317
Tainer JA, 64
Taira K, 426
Tajima Y, 538
Takahara Y, 279
Takahashi A, 129
Takahashi R, 40, 169, 228
Takamura Y, 235
Takeda JI, 538
Takeda M, 323
Takehara K, 129
Takenaga M, 551
Takeshita D, 269, 270, 426
Taketo A, 67, 86
Takeuchi K, 235, 491, 518
Takizawa S, 42
Talarico G, 235
Talbot M, 125
Talbot SJ, 351
Talebi Amiri M, 134
Taliaferro JM, 548
Tam KI, 202
Tamargo J, 501
Tamargo M, 501
Tameike N, 137
Tamimi AH, 164
Tamò GE, 487
Tamotsu Z, 266, 426
Tamulaitis G, 550, 551, 558
Tamura M, 129
Tan CKL, 109
Tan L, 489
Tan TM, 117
Tan TW, 125
Tan XL, 140, 145, 150
Tan Z, 131, 140
Tanaka H, 104, 127, 138, 153, 154
Tanaka M, 544
Tanaka N, 313, 330
Tanaka R, 35, 302
Tanenbaum ME, 547
Tanese N, 543
Tang C, 202
Tang JW, 234
Tang L, 479
Tang N, 516
Tang WZ, 126
Tang X, 231, 517, 543
Tange TØ, 538
Tanida S, 88
Taniguchi H, 538
Taniguchi I, 267, 425
Taniguchi T, 98, 188, 224, 328, 413, 418, 419, 538
Tanner BD, 143, 165
Tanneru CT, 133
Tannous BA, 543
Tansey WP, 541

Tantale K, 544
Tanzer A, 231
Tao P, 557
Tapia L, 337
Tarassov I, 543
Tarján I, 132
Tarlov MJ, 199
Tarn WY, 161
Tarr PI, 202
Tärs K, 40, 42, 73, 140, 169, 170, 192, 243, 250, 255, 295, 354, 359, 406, 425, 433, 435, 438, 439, 440, 467, 468, 470, 476, 477, 486, 490, 501, 537, 556, 557
Tartoni PL, 427
Tashima T, 518
Tashkandi M, 558
Tauchert MJ, 538
Taurins V, 91
Tavernier J, 419, 427
Tavi P, 235
Tavira B, 235
Tawil N, 529
Taya M, 127
Taylor DJ, 117
Taylor DW, 429
Taylor H, 113, 118
Taylor IJ, 460, 513
Taylor MJ, 66, 120, 150
Taylor RK, 64
Taylor SE, 450, 529
Taylor SM, 235
Taylor Murphy S, 527
Taylor-Robinson AW, 440, 518
Tchobanoglous G, 124
Tchorsh Y, 161
Tchou J, 539
Teal LJ, 164
Teasdale R, 462
Teeka J, 143
Teferle K, 54
Teirstein PS, 455
Teitelbaum D, 309
Tejedor I, 153, 292, 463
Tellam JH, 151
Temin HM, 23
Temmam S, 555, 160
Temmerman J, 323
Templeton M, 113, 125, 126, 127, 153
Tener GM, 211
Tenson T, 313
Tepper F, 131, 149, 155, 200
Ter-Ovanesyan D, 551
Terrazas WCM, 202
Teruyuki N, 266, 426
Tetz G, 559
Tetz V, 559
Teuchy H, 76
Teutsch G, 149
Tezuka R, 137
Thach RE, 192, 324, 330, 368
Thach SS, 210
Thammakarn C, 129
Thannesberger J, 556
Thapar R, 352
Theander TG, 460, 513, 514
Theiss S, 538
Theoleyre S, 538
Therezien M, 131
Thiele U, 427
Thimon K, 40, 20, 225, 228, 294, 295, 335, 424
Thirion JP, 220, 226

Thom V, 166
Thomas CM, 58, 193, 198, 250, 329
Thomas DD, 13, 14, 16, 37, 38, 56, 58, 59, 71, 76, 98, 122, 302
Thomas GJ Jr, 207
Thomas JH, 202, 310
Thomas JJ, 198
Thompson BC, 115
Thompson HA, 319
Thompson KA, 108, 109, 312
Thompson MA, 543
Thompson R, 73, 243, 476, 477
Thompson SR, 539
Thompson SS, 120
Thoms F, 435, 491, 557
Thorsen M, 235
Thrane S, 460, 513, 514
Thurner C, 231
Thurston-Enriquez JA, 120
Tiago dos Santos V, 543
Tian L, 543
Tian M, 453, 491
Tian W, 551
Tian Y, 548
Tiehm A, 109
Tikchonenko TI, 1, 16
Tikhonenko AS, 98, 185
Tikhonov M, 543
Tiligada E, 139
Tilsner J, 543
Timchak E, 125
Timmermans SB, 533
Timmins PA, 234
Timokhina GI, 217, 438
Tinoco I Jr, 221
Tischendorf GW, 325
Tissot A, 458
Tissot AC, 433, 444, 470, 490, 501
Tissot-Dupont H, 203
Titcombe Lee M, 144
Tito MA, 250
Titz A, 491, 518
Tiwari BK, 130
Tocchini-Valentini GP, 90
Toh RJ, 133
Tohá J, 218
Toivola M, 163
Tolchard J, 556
Toles M, 156, 164
Tomari Y, 538
Tomas ME, 164
Tomé AC, 130
Tomé JP, 130
Tominaga K, 544
Tomita K, 269, 270, 272, 426
Tomlinson RV, 211
Tomoeda M, 51, 52, 54, 61, 98
Tompkins R, 332
Tong GJ, 148, 450, 518, 527
Tong YW, 165
Tooze J, 1, 17, 19, 24, 175, 176, 177, 186, 192, 237, 319, 370
Topp E, 161
Toranzos GA, 112, 148
Torii S, 156
Tornesello ML, 489
Torok VA, 159
Toropova K, 137, 471, 473, 476, 481, 556
Torres C, 113, 166
Torres T, 529
Torrey JR, 104

Tort LF, 166
Tort LFL, 203
Tortola L, 491
Tosto G, 235
Tóth I, 235
Toth K, 122
Tothill IE, 108, 109, 312
Touré IM, 88
Toze S, 122
Tracey MC, 109
Tran EJ, 543
Tran H, 542, 543
Tran JS, 298, 299, 425
Tran M, 359
Tran NH, 113
Tran NT, 551
Tran TV, 202
Traoré O, 107
Traub F, 140, 233, 263, 323
Trcek T, 541
Tree JA, 112, 120, 132
Trenholm RA, 128, 138
Trépout S, 281
Tretbar US, 540
Treutlein B, 548
Trevino AE, 549
Tricot G, 234
Trinkle-Mulcahy L, 543
Tripathi V, 543
Triplett KD, 493
Trites JR, 235
Trojnar E, 160, 162
Trokan L, 490, 518
Trouwborst T, 139, 143
Trown PW, 383, 384
Truden JL, 316, 369
True HL, 538
Trunov M, 144
Truyen U, 141
Tryland I, 116, 118
Tsai AY, 88, 160
Tsai CH, 543
Tsai CK, 470
Tsai DH, 166, 199
Tsai FT, 151
Tsai MH, 202
Tsai NP, 542
Tsang AH, 105, 111
Tsang YF, 538
Tsanov N, 544
Tsao NH, 156
Tschesche H, 427
Tselepi MA, 151
Tsen KT, 133
Tsen SW, 133
Tseng CC, 138, 144
Tseng YH, 199
Tsibinogin V, 436, 438
Tsibinogin VV, 431
Tsielens IE, 334
Tsimanis A, 225
Tsimanis AI, 225, 349, 431
Tsimring L, 551
Tsitoura E, 492
Tsuchida N, 13, 19, 94, 185
Tsui HC, 279
Tsujimoto M, 538
Tsujimura M, 129
Tsukada K, 78, 285, 401, 425
Tsukahara T, 235, 538, 540
Tsukamoto T, 541

Tsunaka Y, 217
Tsushima FS, 169, 417
Tsvetkova EA, 139
Tsvetkova I, 529
Tsvetkova K, 559
Tu J, 310
Tubiana L, 484
Tuck A, 543
Tufenkji N, 166, 414
Tuite MF, 319
Tullis E, 64, 66
Tullman-Ercek D, 193, 207, 418, 461, 463, 471,
    529, 532, 533
Tuma R, 232, 207, 480, 481
Tumban E, 443, 444, 493
Tumpey TM, 501
Tung-Thompson G, 164
Tunnacliffe E, 543
Tuong L, 131
Turabi AE, 492, 447
Turchinsky MF, 98, 123, 135, 185, 193
Turgeon N, 144
Turnage NL, 152
Turnbull AR, 159
Turner A, 156
Turner PE, 29, 555
Turner RB, 352
Turner SJ, 115
Turnowsky F, 314
Turri M, 122, 134
Turro NJ, 130
Tussie-Luna I, 392, 396
Tuttle M, 551
Tutucci E, 192, 544, 548
Twa D, 95
Twarock R, 475, 477, 481, 483, 484, 556
Tweehuysen M, 16, 57, 60, 98, 122
Twilt JC, 314
Twite AA, 451
Twyman RM, 489
Tyagi P, 113
Tyagi S, 393, 396, 397, 399, 543, 545
Tyler M, 443, 444, 448, 493
Tymokhina GI, 217
Tyrrell SA, 120
Tytell AA, 375, 376
Tzareva NV, 329, 364

U

Uberla K, 538, 539
Uchida M, 518
Udert KM, 137, 164
Udit AK, 264, 441, 442, 455
Ueda K, 404
Ueda S, 280
Uehara Y, 313
Ueta M, 331
Uetrecht C, 199
Ugarov VI, 390, 395, 399, 412, 413, 426
Uhlenbeck OC, 115, 278, 346, 347, 348, 349,
    351, 352, 354, 355, 386, 537
Uike M, 138
Ulbrich P, 197
Ulbricht M, 166, 501
Ulrich R, 202, 436, 438, 492
Ulrich RG, 489
Umbraško J, 378
Umesha KR, 158
Ungaro F, 489
Unnithan VV, 107, 150

Unrau PJ, 545
Unzueta U, 18
Urabe H, 402, 426
Urabe I, 267, 401, 425
Urase T, 120, 121, 154
Urban EA, 545
Urbanek MO, 548
Urbinati CR, 544, 545
Urdal DL, 440
Urdzikova-Machova L, 235
Urlaub H, 538
Usachev EV, 311
Ushijima K, 137
Ushijima T, 551
Usui K, 426, 402, 404
Utagawa E, 121
Utsumi M, 544
Uyttendaele M, 548
Uyttenhove C, 319, 513
Uzawa T, 319

V

Vacher G, 343
Vacher J, 184
Vaculovicova M, 529
Vågbø CB, 234
Vagias C, 139
Vagner S, 538
Vaidyanathan PP, 486
Vaill A, 304
Vajpayee M, 202
Valbonesi P, 314
Valegård K, 190, 197, 350, 351, 352, 354, 355,
    356, 435, 438, 467, 470
Valente AP, 197
Valentin-Hansen P, 280
Valentine RC, 1, 19, 46, 47, 49, 53, 69, 70, 71,
    72, 73, 84, 94, 175, 177, 183, 184, 237,
    301, 316
Valenzuela P, 436
Valero F, 122
Valette M, 42
Valihrach L, 235
Valverde C, 283
Van Assche W, 19, 185
Van Boom JH, 328
van Boven CP, 377
Van Breda A, 210
Van Cuyk S, 153
van de Ven TG, 166, 414
Van De Water L, 545
van de Winckel E, 529
van den Akker B, 156
Van Den Berg H, 132
van den Berg HH, 108
van den Born E, 234
van den Oord J, 544
van den Worm S, 352, 470, 471, 486
van der Avoort H, 210
Van Der Hoek K, 544
Van Der Marel G, 328
van Der Meide PH, 266
van Der Merwe RR, 115
Van der Plas J, 328
van der Schoot P, 556
Van der Stede Y, 234
Van der Werf S, 436
Van Dieijen G, 267, 325
Van Dolah RF, 115
Van Doren JM, 159

Van Driesche SJ, 545
Van Duin J, 1, 20, 29, 40, 113, 114, 123, 169, 219, 225, 227, 228, 252, 263, 293, 294, 295, 296, 299, 310, 325, 328, 334, 335, 337, 338, 339, 356, 411, 420, 421, 423, 424, 431, 471
Van Dyke MI, 153, 156
van Emmelo J, 167, 219, 224, 273, 274, 284, 419, 427
van Gijtenbeek LA, 543
van Hest JC, 518, 533, 548
van Hoorebeke C, 202
van Knippenberg PH, 314, 421
van Leerdam E, 314
van Leerdam RC, 125
van Melle G, 490
Van Montagu M, 19, 167, 181, 185, 217, 242, 249, 419, 427
van Nguyen S, 163
Van Nynatten LR, 544
Van Olphen M, 119
van Oudenaarden A, 545
van Ravenswaay Claasen JC, 314
Van Regenmortel MHV, 30
van Reis R, 166
van Sinderen D, 105
Van Snick J, 319, 513
Van Styvendaele B, 40, 167, 212, 216, 288, 419, 427
Van Voorthuizen EM, 120
Vandamme E, 219
Vandamme EJ, 19, 185, 258
Vandekerckhove J, 19, 185, 238, 241, 242
Vandekerckhove JS, 249
Vandenberge V, 234
Vandenberghe A, 40, 167, 181, 211, 212, 213, 214, 216, 339, 365, 368, 419, 427, 513
Vandenbussche P, 211
Vandendriessche L, 40, 181, 216, 241
VandePol S, 138
Vandermeer J, 33
VanderNoot VA, 200
Vaňková Hausnerová V, 548
Vannucci L, 518
Vanstreels E, 544
Vantarakis A, 151
Vantarakis AC, 116, 158
Varani G, 540
Vargas J, 73, 192, 243, 476, 477
Vargas-Morales J, 501
Varghese S, 127
Varsani A, 200
Varshavsky YM, 205, 377
Varshosaz J, 518
Vasiliev NN, 266, 268, 269, 425, 426
Vasiliev VD, 268
Vasiljeva I, 431, 436, 440, 441, 456
Vasilyev NN, 425
Vassilenko SK, 217
Vasuhdivan V, 148
Vaz AC, 143
Vazquez D, 313
Vázquez E, 518
Vazquez-Duhalt R, 533
Večerek B, 281
Vecitis CD, 131
Veeneman G, 328
Veerapandian M, 529
Veerasubramanian P, 137
Veetil AT, 539
Vega E, 159, 160

Veggiani G, 460
Veillette M, 144, 155, 203
Veinalde R, 378
Ven'iaminova AG, 225
Venclovas Č, 558
Venezia R, 162
Venkataraman S, 31, 190
Venkatesan S, 538
Venkstern T, 330
Venuto AP, 457
Vepsäläinen A, 118
Vera A, 139
Vera M, 551
Vera-Garcia M, 398
Verbraeken E, 194, 241
Verbyla ME, 142
Vercruysse T, 544
Vergara GG, 118, 120
Vergara GGR, 113, 114
Vergne MJ, 160, 161, 162
Verhassel JP, 212, 223
Verheggen C, 541
Verhoef NJ, 314
Verkhoturov SV, 427
Verma H, 144
Verma S, 233
Vermeer C, 314, 324
Vermeire J, 234
Vermeire J, 234
Verplancke H, 288
Verreault D, 144, 155
Verschoor GH, 85
Versoza M, 144
Verweij JJ, 203
Verykokidou E, 139
Vestal JR, 147
Vetrak J, 377
Vetrák J, 377
Vicente MS, 166
Victoria M, 166, 203
Vidal M, 539
Vidigal J, 489
Vieira M, 548
Viennet E, 202
Vieritz A, 164
Vieu E, 538
Vijayavel K, 113
Villa L, 58
Villanova GV, 517
Villaverde A, 18, 518, 529
Villemaire J, 538
Villems R, 334
Vincentelli R, 233
Vinetskii IuP, 132
Vingerhoed JP, 217
Vinjé J, 15, 40, 42, 113, 113, 141, 202
Vinnerås B, 137
Vinograd J, 209
Vinograd NA, 378
Viñuela E, 20, 22, 176, 180, 185, 237, 238, 258, 315, 317, 323, 367, 368
Vishnevsky J, 394
Vishnevsky Y, 394
Visser A, 148
Viswanathan S, 542, 547
Vitaliti A, 513
Vithanage G, 115
Vitiello M, 518
Vitvitski-Trépo L, 427
Vladimirov SN, 334
Vladimirova IN, 334
Vlassiouk I, 108, 312

Vo HT, 143, 163
Voelcker NH, 133
Voelker RB, 538
Vogel J, 283
Vogel M, 435, 501
Vogel MA, 491
Vogelaar A, 421
Vögeli G, 223
Vogelmeier CF, 490, 518
Voges M, 550
Vogl B, 88
Vogt L, 490
Voigt F, 547, 548
Volckaert G, 212, 214, 216, 217, 238, 365, 419, 427
Voldby Larsen M, 58
Volk AE, 235
Volkov AA, 268
Volkov IL, 548
Volkova (Kozlovska) TM, 349
Vollenweider HJ, 25, 362, 363, 364
Volpetti V, 193
Volráte Ā, 378
Volterra L, 106
von Allmen CE, 489
von Bonsdorff CH, 154
von Gunten U, 138
von Hippel PH, 267, 364
Von Nordheim M, 452
Von Seefried A, 436
Vondrejs V, 96
Vongsouvath M, 202
Voorhees KJ, 198
Voorma HO, 314, 318, 324
Vora S, 551
Voráčková I, 197
Voronkova T, 433, 439, 501
Voss NR, 453
Vostiar I, 513
Votaw NL, 489
Vought K, 161
Voumard M, 142
Vournakis JN, 284
Vulic M, 61

W

Waade A, 136
Wada A, 279
Wade TJ, 113
Waffo AB, 514
Wagner D, 513
Wagner EGH, 163, 283
Wagner EJ, 538
Wagner F, 490, 518
Wagner M, 137, 139
Wagner SJ, 130, 163
Wagstaff BJ, 551
Wah MJ, 117
Wahba AJ, 20, 23, 263, 278, 324, 325
Wahl GM, 342
Wain-Hobson S, 418
Waite TD, 134
Wakiyama M, 538
Walczak AM, 543, 548
Walderich B, 296, 297, 424
Waldschmidt TJ, 462
Walker CM, 144
Walker DC, 141
Walker EM, 165
Walker GF, 489, 557

Walker J, 108, 109, 312
Wall K, 151
Wallace MA, 165
Wallace RL, 127, 139, 313
Wallat JD, 455
Wallbank AM, 135
Waller JP, 249
Wallin M, 559
Wallis J, 539, 489
Walls HJ, 144
Walongo T, 203
Walotka L, 538
Walser R, 122
Walsh EC, 343
Walsh MJ, 543
Walshe GE, 152
Walther N, 544
Walton AG, 191
Walton S, 490
Wan G, 150
Wan T, 551
Wan Y, 538
Wan Z, 200
Wander J, 144
Wang B, 202
Wang C, 326, 542, 547
Wang CJ, 527
Wang D, 31, 42, 105, 131, 136, 163, 171, 407, 555
Wang EY, 529
Wang F, 64, 545, 551
Wang G, 447, 517, 518, 527
Wang Gang, 550
Wang GB, 501
Wang GQ, 136
Wang GX, 543
Wang H, 559
Wang Haifeng, 551
Wang Hanting, 153
Wang IN, 40, 58, 59, 60, 71, 167, 243, 295, 298,
        424, 425
Wang J, 132, 134, 539
Wang JC, 480
Wang JF, 108
Wang JL, 128
Wang JR, 144
Wang JW, 493
Wang K, 356
Wang KH, 471
Wang KS, 234
Wang L, 130, 166, 199, 538, 551, 558
Wang LN, 516
Wang M, 154, 280, 539
Wang Q, 518, 529
Wang R, 543
Wang S, 117, 120, 132, 150, 348, 349, 545
Wang Shen, 517
Wang Siyuan, 549
Wang SK, 513
Wang SW, 518
Wang TH, 234
Wang W, 264, 453, 543, 544, 545
Wang WH, 527
Wang WX, 235
Wang X, 109
Wang Y, 139, 539, 545
Wang Yaolai, 548
Wang YJ, 137
Wang Yulin, 127
Wang Z, 126
Wanjugi P, 115
Ward P, 115

Ward R, 19, 20, 176, 267
Ward RL, 25, 122, 132
Ward S, 113, 163
Ward VK, 489, 557
Warden PS, 137
Wardman JP, 484
Warfield KL, 436
Warner RC, 367
Warriner K, 162
Warriner SL, 470
Wassarman KM, 280
Watanabe I, 1, 13, 19, 23, 35, 37, 38, 39, 40, 76,
        80, 82, 83, 88, 98, 99, 101, 102, 103,
        104, 105, 112, 123, 145, 163, 178, 181,
        184, 186, 187, 189, 224, 257, 288, 301,
        302, 310, 315, 424, 535
Watanabe K, 35, 99, 101, 102, 302
Watanabe M, 13, 83, 84, 212, 223
Watanabe Y, 544
Waterson J, 313
Watkins SC, 42
Watkins WD, 113, 157
Watson JD, 1, 16, 17, 19, 84, 175, 186, 192, 237,
        330
Watts TH, 63, 66
Weaver L, 149
Webb AR, 143
Webb J, 538
Webb SJ, 143
Webber BD, 137
Weber D, 164, 186, 211, 213, 214, 222, 238, 250,
        252, 253, 264, 307, 330, 346, 362, 365
Weber DJ, 163, 164
Weber G, 21, 24, 143
Weber H, 223, 264, 279, 330, 331, 362
Weber K, 1, 17, 19, 23, 24, 25, 167, 175, 176, 177,
        186, 192, 213, 223, 237, 238, 249,
        250, 253, 258, 264, 265, 278, 286,
        307, 315, 319, 333, 362, 363, 370
Weber P, 162
Webster RE, 19, 20, 21, 53, 176, 178, 179, 180,
        181, 184, 221, 249, 252, 315, 317, 318,
        319, 323, 331, 332, 367, 368
Webster-Brown J, 113, 115
Wechsler ME, 490, 518
Wedel H, 47, 69, 72
Weel J, 210
Weerathunge P, 109
Wege C, 529
Wegelin M, 141
Wegmann M, 134
Wegrzyn G, 86
Wehrli W, 89
Wei B, 516, 518
Wei J, 148
Wei LN, 542, 543
Wei M, 543, 551
Wei MM, 501
Wei Y, 516
Weiche B, 538
Weidhaas J, 115, 116
Weigert MG, 176
Weil TT, 542, 543, 545
Weinbauer MG, 42
Weiner A, 234
Weiner AM, 19, 167, 192, 223, 238, 253, 258,
        286, 315, 333
Weiner MP, 304
Weinfeld H, 82
Weis K, 548
Weisberg SB, 113

Weisberger AS, 318
Weisbrod N, 152
Weiss R, 540
Weiss SB, 88, 160, 325, 539
Weiss VU, 198
Weissleder R, 543
Weissman SM, 221, 330
Weissmann C, 1, 19, 20, 21, 23, 29, 82, 97, 167,
        176, 178, 187, 188, 192, 210, 222, 223,
        224, 238, 248, 253, 257, 258, 261, 262,
        263, 264, 265, 287, 288, 315, 327, 328,
        363, 365, 367, 368, 386, 413, 418, 419
Weith HL, 221, 223, 458
Welch RP, 131, 458
Weller SA, 203
Wellinger RJ, 543, 545, 547
Wells AL, 544
Wells DG, 54
Welsby I, 539
Welte W, 234
Wemmer DE, 538
Wen AM, 456, 529
Wen F, 352
Wen J, 131
Wendt L, 46, 49, 69
Weng S, 96, 138
Weng X, 548
Wenger SL, 160, 163
Wenk J, 142
Wennberg A, 118
Wente SR, 543
Weppelman RM, 56, 67, 71, 76
Werbin H, 122
Werle B, 493
Werner J, 156
Werner M, 386
Wert EC, 138
Wertheim-van Dillen P, 210
Wéry N, 113
West JA, 200
West SE, 493
Westbye MP, 234
Westwood J, 354
Weyrauch K, 54
Wharton RP, 539
Wheatley AK, 367
Wheeler ML, 235
Wheeler SS, 201
Whelan J, 53
Whitaker N, 64
Whitchurch CB, 64
White EJ, 202
White H, 203
White HB III, 221
White MR, 545
White PA, 166, 202
White TB, 551
Whitehead P, 447
Whitman RL, 164
Whitworth JA, 429
Wibowo N, 489
Wick C, 106, 190, 198, 243, 476
Wick LY, 152
Wickens M, 536, 537, 538, 539, 540
Wickens MP, 538
Widén F, 137
Widmer G, 137
Widom JR, 545
Wiegand HL, 538
Wieland M, 234

Wierzba AJ, 548
Wiesehahn G, 123
Wiesel M, 489, 491, 518
Wiesner MR, 130, 131
Wiggins PA, 542
Wigginton KR, 123, 128, 129
Wilbertz JH, 547, 548
Wilches Pérez D, 137
Wilderer C, 164
Wildy P, 31
Wilharm G, 65
Wilhelm JE, 538
Wilhelm LJ, 348, 355
Wilhelm RC, 19, 179, 184, 317, 318, 319
Wilk HE, 233
Wilkerson JW, 533
Wilkes G, 161
Wilkinson AJ, 435
Wilkinson JF, 46
Wilkinson ME, 540
Will CL, 280, 538
Will H, 428, 430
Will WR, 280
Willers J, 490, 518
Willetts N, 51, 52, 74, 98
Williams FP, 106
Williams FP Jr, 106
Williams LD, 464
Williams P, 198
Williams S, 202
Williams SM, 515
Willieme S, 140
Willis MC, 348, 355
Willkomm M, 116
Willson CS, 201
Willson RC, 311, 312
Wilschut J, 294, 295, 296
Wilson B, 165
Wilson BA, 458
Wilson BM, 165
Wilson CJ, 136
Wilson LJ, 127, 130, 131
Wilson SA, 543
Wiltfang J, 235
Wilusz J, 538
Wimmer E, 235
Winblad B, 513
Winchell JM, 203
Winey M, 543
Wingfield PT, 435
Winkler SM, 544
Winkler U, 122
Winterbourn JB, 142
Winter G, 435, 490
Winter RB, 167, 243
Wippich F, 547
Witherell GW, 348
Witte AK, 139
Witte ON, 296, 425
Wittmann HG, 247
Wittmann-Liebold B, 247
Witz J, 191
Wlotzka W, 543
Wodnar A, 320
Wodnar-Filipowicz A, 320
Woessner WW, 150
Wohl DA, 164
Wohlhieter JA, 53
Wojta-Stremayr D, 529
Wolf S, 41, 42, 158
Wolf YI, 410

Wolfe RL, 138
Wölfel R, 516
Wolfenden ML, 453
Wolfinger MT, 218, 231
Wolfner M, 187
Wolfson B, 544
Wolin SL, 544
Wolint P, 489, 490, 501, 518, 462
Wong C, 299, 425
Wong CH, 513
Wong DK, 310
Wong J, 210, 540
Wong K, 69, 73, 152, 241
Wong MS, 234
Wong RS, 202
Wong S, 559
Wongpalee SP, 538
Wongsurawat T, 555
Woo HH, 144, 155
Woo SY, 543
Wood D, 113, 115
Wood H, 210
Wood J, 539, 516
Wood MJ, 518
Wood NB, 267
Wood TR, 543
Woods J, 159
Woods MW, 544
Woods WS, 551
Woodward WE, 155
Woody MA, 76, 77, 112
Wool IG, 314
Worley-Morse T, 113
Wray GW, 88
Wright A, 548
Wright CC, 113
Wright HB, 120, 124, 125
Wu B, 156, 358, 399, 426, 462, 518, 527, 545, 546, 547, 548
Wu C, 558
Wu CY, 144, 163
Wu H, 124, 131, 539
Wu HN, 115, 144
Wu J, 118
Wu LT, 202
Wu M, 518
Wu MJ, 546
Wu Q, 125, 517, 556
Wu QQ, 489
Wu WJ, 470
Wu X, 538, 551
Wu Xueyin, 142
Wu XZ, 310
Wu Y, 136, 160, 398, 426, 518, 544
Wu Yan, 132
Wu Yimin, 501
Wurster SE, 539
Wyer MD, 152
Wyn-Jones P, 540, 541
Wynne SA, 435, 438
Wynyard S, 235
Wysocki KM, 144

X

Xagoraraki I, 539
Xia J, 235
Xia N, 109
Xia T, 548
Xia Y, 144
Xiang F, 107, 250

Xiang W, 235
Xiao H, 126, 160, 161
Xiao Y, 131
Xiao ZH, 108
Xie C, 551
Xie F, 356
Xie J, 108, 202, 234, 517
Xie JH, 516
Xie K, 551
Xie MH, 202
Xie X, 131, 517
Xie XS, 545
Xing X, 551
Xiong K, 551
Xiong X, 131
Xoconostle-Cázares B, 234
Xu C, 52
Xu H, 538
Xu J, 160
Xu L, 544
Xu R, 109, 131
Xu X, 136, 162, 551
Xu Y, 150, 517
Xu Z, 489, 517
Xue B, 153
Xue J, 122
Xue X, 134
Xue XY, 128
Xue Y, 304

Y

Yada Y, 138
Yadav K, 501
Yadav SK, 314
Yager TD, 202
Yaghoubian S, 202
Yagi Y, 543
Yahya M, 108, 113, 115, 133
Yahya MT, 120
Yakhini Z, 234
Yamabe S, 89
Yamada K, 113
Yamada S, 313, 325, 543
Yamada T, 117
Yamada Y, 38
Yamagishi M, 543
Yamagishi R, 538
Yamaguchi Y, 127, 142, 232, 233
Yamahara KM, 141
Yamaji Y, 267
Yamamoto K, 120, 121, 154
Yamamoto KR, 96
Yamamoto M, 85, 265, 491
Yamamoto N, 133, 134
Yamamoto T, 551
Yamane K, 23, 178, 257, 261, 288, 315
Yamashita N, 104, 127, 154
Yamazaki H, 83, 84, 89, 91
Yan D, 518, 543
Yan F, 543
Yan GJH, 202
Yan J, 539
Yan R, 489
Yan S, 280, 548
Yan W, 59
Yan X, 304
Yanagawa Y, 35, 302
Yanagida S, 546
Yanazaki Y, 265
Yandell M, 161

Yang CM, 516
Yang D, 108, 130
Yang FJ, 96
Yang G, 304, 538, 556
Yang HL, 314
Yang HW, 464
Yang J, 137, 543, 551
Yang M, 489
Yang PE, 235
Yang S, 124, 131, 202, 280
Yang X, 311, 356, 501, 517, 543, 551, 556
Yang Y, 107, 118
Yang Z, 326
Yanko WA, 150
Yannam S, 161
Yannam SK, 126
Yannitsaros A, 139
Yao J, 545
Yao JD, 234
Yao L, 107, 517
Yao Y, 127, 235, 375, 544, 551
Yao Z, 543
Yarbrough ML, 164
Yarosh E, 94
Yarzabal A, 116
Yasuda S, 235
Yasuhara JC, 548
Yasui N, 154
Yasunaga T, 218
Yasuo K, 86
Yates MV, 85, 113, 117, 119, 120, 150, 153
Yavarmanesh M, 160, 162
Yazdi M, 160
Ye Y, 107, 118, 129
Yeager M, 64
Yee SYF, 117
Yeh HJ, 538
Yehl KM, 518
Yen S, 126
Yeo GW, 548
Yeo RG, 95
Yermakov M, 144
Yi AK, 462
Yi Vanessa SP, 543
Yildiz I, 518
Yin G, 527
Yin J, 78, 95, 456, 457, 490, 529, 544
Yin S, 304
Yin Z, 457
Ying R, 517
Ying Y, 538
Yli-Harja O, 542, 545
Yofe I, 543
Yoklic MR, 155
Yomo T, 169, 417
Yomo T, 267, 285, 401, 402, 404, 417, 425, 426, 559
Yonesaki T, 39, 288, 289
Yong SF, 117
Yoo J, 202, 518
Yoo KW, 324
Yoon J, 132, 134
Yoon MH, 233
Yoon SH, 156
Yoon YJ, 545, 548
York TL, 415
Yosef I, 543
Yoshida M, 324, 325
Yoshimura Y, 272
Yoshinaga K, 96
Yoshinari S, 284, 425

Yoshio M, 156
Yoshiyama T, 426
You J, 129, 128, 134, 153, 545
You L, 539
Youk H, 543
Young DV, 85, 265, 295, 297, 324
Young P, 150
Young R, 167, 243, 245, 298, 299, 424, 425, 479
Young RA, 266, 278
Young SL, 489, 557
Young TA, 155, 156
Younis I, 127, 544
Yousefi Z, 106, 147
Yousri NA, 235
Ysebaert M, 210, 211, 212, 213, 214, 216, 238, 365
Yu B, 538
Yu C, 155
Yu F, 538, 556
Yu H, 516
Yu MS, 543
Yu TH, 470
Yu TS, 518
Yu X, 529, 551
Yu XF, 144
Yu Z, 551
Yu ZW, 527
Yuan B, 125, 202
Yuan BL, 489
Yuan L, 234, 556
Yuan R, 452
Yuan Z, 40, 114, 480
Yueh A, 333
Yumura M, 404, 426
Yun HC, 117
Yun K, 529
Yun SG, 130, 202
Yunger S, 544
Yusibov V, 489
Yuyama N, 426

Z

Zabel F, 447, 459, 460, 489, 491, 492, 501, 518, 543
Zabetakis D, 304
Zabezhinsky D, 544
Zachariah MR, 199
Zacharias M, 218, 352
Zagórska L, 320, 342, 346
Zagórski W, 318, 331, 346
Zagrebel'nyy SN, 225
Zagury JF, 513
Zak A, 352
Zalatan JG, 549, 551
Zalesskaya MA, 138
Zambenedetti MR, 517
Zambriski JA, 203
Zamir A, 211
Zamora H, 394, 425
Zander S, 548
Zanetti F, 107, 136, 137, 155, 156
Zangabad PS, 518
Zangger K, 79
Zankari E, 58
Zappulla DC, 551
Zare H, 518
Zariņa J, 378
Zarnescu DC, 542
Zatloukal K, 234
Závada V, 122
Zavada V, 93, 122

Závada V, 96
Zawadzki P, 202
Zdanowicz M, 518
Zechner EL, 54, 74, 79
Zeevi M, 86, 183, 253
Zehner ZE, 539
Zeighami R, 202
Zeitelhofer M, 544
Zeka F, 235
Zekri L, 538
Zelazo PO, 26, 196
Zellnig G, 54, 79
Zeltins A, 489, 557
Zeng F, 304
Zeng X, 131
Zenklusen D, 545, 546, 548
Zerda KS, 96, 147, 155
Zettl M, 54
Zeuzem C, 280
Zevnik B, 550
Zha L, 492
Zhai L, 493
Zhan S, 131, 516
Zhan Y, 539, 548
Zhang B, 539
Zhang C, 131, 326, 445, 501
Zhang CW, 132, 136
Zhang D, 517
Zhang F, 550, 551
Zhang G, 130
Zhang H, 124, 131, 304, 556
Zhang J, 109, 167, 245, 326, 424, 455, 479
Zhang Jichuan, 548
Zhang K, 120, 150, 516, 517, 518
Zhang L, 109, 161, 517, 518
Zhang Lin, 557
Zhang M, 543
Zhang N, 140
Zhang P, 543
Zhang Q, 136, 517
Zhang R, 160, 442, 516, 517, 551
Zhang S, 144
Zhang T, 42, 105, 128, 551
Zhang W, 52, 126, 131, 529
Zhang X, 148, 150, 165, 518, 539, 545
Zhang XE, 52, 544
Zhang XF, 131
Zhang Y, 107, 136, 160, 440, 442, 489, 516, 517, 529, 543, 544, 550
Zhang YZ, 555
Zhang Z, 267, 453, 545, 551
Zhang ZP, 52
Zhao C, 202, 487, 488, 489, 516
Zhao F, 489
Zhao G, 42, 105, 163, 171, 407, 555, 556
Zhao H, 551
Zhao L, 150
Zhao Q, 109
Zhao R, 545
Zhao W, 131
Zhao Y, 109
Zhao Z, 501
Zhelyabovskaya OB, 299
Zhen Y, 543
Zheng S, 356
Zheng T, 128
Zheng X, 128, 132, 134, 556
Zheng Y, 527
Zholobak NM, 378
Zhong L, 551
Zhong Q, 136

Zhou A, 538
Zhou J, 235, 551
Zhou K, 152
Zhou L, 551
Zhou NA, 156
Zhou Q, 544
Zhou W, 153
Zhou Y, 124, 131, 548
Zhou Z, 125, 160, 161
Zhou ZH, 192, 243, 478, 479, 545
Zhou ZM, 489
Zhu CJ, 501
Zhu M, 136
Zhu X, 310, 453
Zhu Y, 489, 517
Zhuang J, 120, 134, 151
Zhuang S, 132
Zielina M, 153
Ziemba C, 140
Zillig W, 316

Zimmerman B, 156
Zimmerman RK, 202
Zimta AA, 551
Zinder ND, 1, 11, 15, 16, 17, 18, 19, 20, 21, 23, 24, 25, 38, 45, 52, 75, 73, 78, 79, 81, 82, 86, 87, 90, 93, 95, 97, 122, 147, 167, 175, 176, 177, 178, 180, 183, 184, 189, 190, 191, 205, 221, 237, 248, 257, 258, 286, 287, 291, 296, 301, 315, 316, 319, 323, 331, 339, 344, 365, 366, 367, 368, 370, 405, 553
Zinkernagel RM, 441, 447, 462
Zipor G, 543
Zipori P, 324
Zipper P, 26, 190, 196, 210
Ziprin H, 135
Zivanovic S, 1, 12, 15, 139
Zlatkin IV, 267
Zlotnick A, 435, 480
Zodrow K, 156

Zolotareva GN, 378
Zorzini V, 232
Zou S, 160
Zou X, 551
Zourob M, 529
Zuaretz-Peled S, 161
Zubairov MM, 139
Zubay G, 1, 84
Zuber S, 161
Zuidema D, 197
Zuker M, 224, 227, 364, 391
Zunder RM, 543
Zuo Z, 144
Zurcher A, 490
Zvereva AS, 543
Zvirbliene A, 438, 489
Zwahlen S, 490
Zwart PH, 471
Zyara AM, 126, 129
Zydney AL, 166

Subject Index

It is never too late to be wise.

Daniel Defoe, *Robinson Crusoe*

A

A1 protein, 19
A2 protein, 237; *see also* Maturation protein
AAA, *see* Amino acid analysis
Aβ peptide, *see* Amyloid-beta peptide
Acinetophage AP205, 295
Adsorption, 69–71; *see also* RNA phage physiology and growth
Advanced oxidation processes (AOPs), 142
Aerosol OT (AOT), 199
AFS, *see* Apparent fitness score
Água-mel, 140
AHA, *see* Azidohomoalanine
Allolevivirus genome, 167
AMA1, *see* Apical membrane antigen 1
Ambiguous intermediate hypothesis, 414; *see also* Evolution
Amino acid analysis (AAA), 166
AMV, *see* Avian myeloblastosis virus
Amyloid-beta peptide (Aβ peptide), 447
Anaerobic membrane bioreactors (AnMBRs), 156
Angiopep-2-FAM (AP2), 452
AnMBRs, *see* Anaerobic membrane bioreactors
ANP, *see* Atrial natriuretic factor
Antibiotic (-) rugulosin, 88
Antibodies, 15; *see also* Immunology
phylogeny of, 304–305
target of, 306–309
Antigenic determinant, 435
AOPs, *see* Advanced oxidation processes
AOT, *see* Aerosol OT
AP2, *see* Angiopep-2-FAM
AP205 phages, 486–487; *see also* Chimeric VLPs; 3D particles
AP205 VLPs, 458–460
and MS2 phage coat structures, 469
3D structures of, 474
Apical membrane antigen 1 (AMA1), 399
Apparent fitness score (AFS), 528
A protein, 73, 237, 238; *see also* Maturation protein
Aptamer selection express (ASExp), 356
Armored RNA, 515–517; *see also* Non-vaccine VLPs
ASExp, *see* Aptamer selection express
ASV, *see* Avian sarcoma virus
ATA, *see* Aurintricarboxylic acid
Atom transfer radical polymerization (ATRP), 454
Atrial natriuretic factor (ANP), 439
ATRP, *see* Atom transfer radical polymerization
Attachment, 69–71; *see also* RNA phage physiology and growth
Aurintricarboxylic acid (ATA), 262, 314
Avian myeloblastosis virus (AMV), 219
Avian sarcoma virus (ASV), 219
Azidohomoalanine (AHA), 452
Azure mutants, 181, 183; *see also* Mutations

B

BAC, *see* Bacterial artificial chromosome
Bacterial artificial chromosome (BAC), 299
Bacterial immunity, 558–559
Bacterial lysis by RNA phage, 77
BBB, *see* Blood–brain barrier
Bdellovibrio bacteriovirus, 35
Beet western yellows virus (BWYV), 338
Beta-PP7, 546
BiFC, *see* Bimolecular fluorescence complementation
Bimolecular fluorescence complementation (BiFC), 545
Biodosimetry, 127
Biofilms, 144–145; *see also* Phage ecology
Bioimaging agents, 530–531; *see also* Non-vaccine VLPs
Biological agents, 310
Biological weighting functions (BWFs), 142
Blood–brain barrier (BBB), 452
Butterfly model of bacteriophage RNA synthesis, 370
BWFs, *see* Biological weighting functions
BWYV, *see* Beet western yellows virus

C

CAC$_{app}$, *see* Critical assembly concentration
Canavanine, 232
CANC, *see* Capsid and nucleocapsid domains of Gag
Capillary isoelectric focusing with liquid-core waveguide laser-induced fluorescence whole-column imaging detection (CIEF-LCW-LIF-WCID), 200, 310
Capsid and nucleocapsid domains of Gag (CANC), 197
Carbon nanotubes (CNTs), 311
Carbon storage regulator (Csr), 542
Carrier induced epitopic suppression (CIES), 458
CASP12, *see* Critical Assessment of Structure Prediction
Cationic conjugated polyelectrolytes (CPE), 139
Cationic organic molecules, 202
Caulobacteraceae, 66–67
pili of, 64
CCBA, *see* Coordinate and chemical bonding and adsorption
CCL-3, *see* Contaminant Candidate List
CCMV, *see* Cowpea chlorotic mottle virus
CDR, *see* Complementarity-determining region
Cell-free synthesis, 96–97; *see also* RNA phage physiology and growth
Cellstat, 95
CFD, *see* Computational fluid dynamics
CGE, *see* Chip gel electrophoresis
Chemical coupling, 447, 513–514; *see also* Chimeric VLPs

of foreign oligopeptides to surface of VLPs, 447
HEL-conjugated VLPs, 448
papillomavirus VLP- and Qβ phage-based Aβ vaccines, 447
vaccines constructed by, 501–512
VLPs and, 448
Chimeric phages, 422–423, 514
Chimeric VLPs, 435; *see also* Phage MS2; Phage Qβ functionalization; Virus-like particles
antigenic determinant, 435
AP205 VLPs, 458–460
chemical coupling, 447–448
functionalization, 436, 448–458
genetic fusions of coats, 438–446
nucleotide-driven packaging, 461–466
peptide display, 446–447
protein engineering and VLP nanotechnology, 435–438
stability, 460–461
3D particles, 487–488
Chip gel electrophoresis (CGE), 200
Chitosan, 139
Chloramphenicol, 87
Chondroitin sulfate A (CSA), 513
Chronic obstructive pulmonary disease (COPD), 203
CIEF-LCW-LIF-WCID, *see* Capillary isoelectric focusing with liquid-core waveguide laser-induced fluorescence whole-column imaging detection
CIES, *see* Carrier induced epitopic suppression
Circumsporozoite protein (CSP), 460
CLAT method, 41
CNTs, *see* Carbon nanotubes
Coat gene, 431–433; *see also* Expression
Coat protein, 247
aggregation, 249–250
homology, 250–251
isolation, 249
lysis, 251–252
molecular mass of viral proteins, 248, 253
N-terminal modification, 249
physical properties, 250
readthrough protein, 252–255
regions of readthrough domains, 255
sequences comparison, 251
sequencing, 247–249
single-stranded RNA phage coat similarity groups, 252
spatial structure, 251
structure of readthrough domain, 254
Cold Spring Harbor Symposium, 220
Colicine E2, 88
Coliphages, 12–13
Complementarity-determining region (CDR), 304
Complementation, 177; *see also* Mutations
Computational fluid dynamics (CFD), 127, 129

Conditionally lethal mutations, 175; *see also* Mutations
Conduction theory, 45
Contact-dependent stimulation of T4SS, 75
Contaminant Candidate List (CCL-3), 161
Coordinate and chemical bonding and adsorption (CCBA), 148
COPD, *see* Chronic obstructive pulmonary disease
Counterion release, 201
Coupled replication-translation, 399–400; *see also* Evolution
Cowpea chlorotic mottle virus (CCMV), 195, 197, 460
Cowpea mosaic virus (CPMV), 452
CPE, *see* Cationic conjugated polyelectrolytes
CPMV, *see* Cowpea mosaic virus
CRISPR-Cas9, 549–551; *see also* Non-VLPs
    C2c2, 551
    Cas13b, 551
    CRISPR/guide RNA technique, 550
    MS2-based tethering, 549
    multiple orthogonal RNA-binding modules, 550
CRISPR-Cas system, 558
Critical assembly concentration (CAC$_{app}$), 529
Critical Assessment of Structure Prediction (CASP12), 487
CRY2, *see* Cryptochrome 2
Cryptochrome 2 (CRY2), 545
Crystalline inclusions, 78–79; *see also* RNA phage physiology and growth
CSA, *see* Chondroitin sulfate A
CSP, *see* Circumsporozoite protein
Csr, *see* Carbon storage regulator
CuAAC reaction, *see* CuI-mediated azide–alkyne cycloaddition reaction
CuI-mediated azide–alkyne cycloaddition reaction (CuAAC reaction), 452, 457
Cyclin D1, 545

D

Darwinian selection experiments, 381–383; *see also* Evolution
DB, *see* Dibromomaleimide
DEC, *see* 1-[3-(Dimethylamino) propyl]-3-ethylcarbodiimide hydrochloride
De novo RNP modeling, 486; *see also* 3D particles
De novo synthesis, 387–390; *see also* Evolution
Density functional theory (DFT), 356
DESI-MS, *see* Desorption electrospray ionization-mass spectrometry
Desorption electrospray ionization-mass spectrometry (DESI-MS), 199
DFT, *see* Density functional theory
DHBV, *see* Duck hepatitis B virus
Dibromomaleimide (DB), 459
5-(4,6-dichlorotriazinyl) aminofluorescein (5-DTAF), 449
Digital PCR (dPCR), 109
DIII, *see* Domain III
1-[3-(Dimethylamino) propyl]-3-ethylcarbodiimide hydrochloride (DEC), 449
7,12-dimethylbenz[a]anthracene (7,12-DMBA), 88
Dinitrophenol (DNP), 241
Dissolved organic matter (DOM), 142
Distribution, 99

enumeration, 105–109
geography, 99–102
human beings, 105
RNA coliphages in animal sources, 103
of RNA phages, 100, 101, 102
source, 102–105
7,12-DMBA, *see* 7,12-dimethylbenz[a] anthracene
DNA viruses, 203
DNP, *see* Dinitrophenol
DOM, *see* Dissolved organic matter
Domain III (DIII), 444
DOTA, *see* 1,4,7,10-tetraazacyclododecanetet-raacetic acid
Double agar layer method, 81
Double-stranded DNA (dsDNA), 30
dPCR, *see* Digital PCR
Drug delivery, 518, 527–529; *see also* Non-vaccine VLPs
dsDNA, *see* Double-stranded DNA
ds-NIF, *see* ds-nucleoside with intrinsic fluorescence
ds-nucleoside with intrinsic fluorescence (ds-NIF), 545
5-DTAF, *see* 5-(4,6-dichlorotriazinyl) aminofluorescein
Dual-color imaging, 546, 547
Duck hepatitis B virus (DHBV), 427
Dyed phage particles, 145; *see also* Phage ecology

E

EBI, *see* European Bioinformatics Institute
Eclipse, 71–73; *see also* RNA phage physiology and growth
EC-MWNT, *see* Electrochemical multiwalled carbon nanotube
E. coli dihydrofolate reductase (eDHFR), 545
E. coli MazE-MazF system, 232
eDHFR, *see* E. coli dihydrofolate reductase
EF Tu-Ts, *see* Elongation factor Tu-Ts complex
EGF, *see* Epidermal growth factor
Ejection, 71–73; *see also* RNA phage physiology and growth
Electrochemical biosensors, 311
Electrochemical multiwalled carbon nanotube (EC-MWNT), 155
Electrocoagulation, 149
Electronegative membrane-vortex (EMV), 104
Electrophiles, 138
Electrospray differential mobility analysis (ES-DMA), 166, 199
Electrospray ionization–ion trap mass spectrometry (ESI-MS), 107, 198
Electrospray technique, 199
ELISAs, 310
Elongation factor Tu-Ts complex (EF Tu-Ts), 265–267; *see also* Replicase
    structural differences, 271
E-L protein, 296
eLS (Encyclopedia of Life Sciences), 30
EMV, *see* Electronegative membrane-vortex
Endogenous RNA, 541–542
Energy dot plot, 224
Enolpyruvyl (EP), 244
Enterobacteriaceae, 66
Enzymatic treatment (ET), 107
EP, *see* Enolpyruvyl
Epidermal growth factor (EGF), 456
Epitope-RNA VLP vaccine, 493

Escherichia coli, 78
    phage MS2 RNA synthesis and growth in rifamycin-treated, 93
    R17-infected *E. coli* cells, 79
    superinfection of, 94
ES-DMA, *see* Electrospray differential mobility analysis
ESI-MS, *see* Electrospray ionization–ion trap mass spectrometry
ET, *see* Enzymatic treatment
European Bioinformatics Institute (EBI), 555
Evolution, 381
    ambiguous intermediate hypothesis, 414
    artificial cell, 402
    autocatalytic synthesis of heterologous RNAs, 391–393
    coupled replication-translation, 399–400
    Darwinian experiments, 381–383
    de novo synthesis, 387–390
    diagnostics, 396–399
    gene contraction and expansion hypothesis, 405
    gene hypothesis, 405
    hybridized binary probes, 398
    mapping of Qβ replicase, 395–396
    MDV-1, 384–386
    micro-and nanovariants, 386–387
    molecular colonies, 390–391
    molecular parasites, 390
    mutagenesis system, 399
    of narnavirus-like RNA viruses, 408
    nucleotide sequence of mutant MDV-1 RNA viruses, 391
    nucleotide sequences of recombinant transcripts, 392
    nutritional mutants, 382
    phylogeny, 404–406
    quasispecies and, 413–418
    recombination, 410–413
    relationship between viruses and hosts, 409
    replicating complex structure, 395
    replicative complexes, 381
    reversible target capture, 397
    RNA-dependent polymerases, 406–410
    RNA self-replication system, 403
    RNA world, 400–404
    RQ variants, 393–395
    self-encoding system, 401
    6S RNA, 383–384
    V-1 variant, 381–382
    viral RNA dependent polymerase phylogenetic tree, 407
Exclusion phenomenon, 94
Expression, 419
    chimeric phages, 422–423
    coat gene, 431–433
    gene engineering, 419–421
    of genes in *E. coli*, 428–430
    heterotetramer enzyme complex, 426
    lysis gene, 424–425
    maturation gene, 423–424
    MS2 gene insertion in PL vectors, 420
    MS2 replicase fusion technique, 427
    mutant generations, 421–422
    phage-coded proteins, 419
    phage fr particles and of purified phage fr VLPs, 431
    replicase fusions, 427–431
    replicase gene, 425–426
    RNA phage coats as VLP carriers, 432

F

f2 coat-RNA complexes, 346; *see also* Repressor complex I
f2 replicase, 286–287; *see also* Replicase
f2 sequencing, 221–222; *see also* RNA
FCS, *see* Fluorescence correlation spectroscopy
Fecal pollution indicators, 113
F-E method, *see* Filtration-elution method
Fertility factor (FP), 55
FESEM, *see* Field emission scanning electron microscopy
FESEM of donor strain DT1944, 60
FET, *see* Field-effect transistor
FFE, *see* Forced-flow electrophoresis
FFF, *see* Flow field-flow fractionation
Field-effect transistor (FET), 108
Field emission scanning electron microscopy (FESEM), 60
FILTER, *see* Filtration and Irrigated cropping for Land Treatment and Effluent Reuse
Filtration and Irrigated cropping for Land Treatment and Effluent Reuse (FILTER), 155
Filtration-elution method (F-E method), 38
Fine mapping
    of F pili, 61
    of polar pili, 61–64
FISH, *see* Fluorescence *in situ* hybridization
FITC, *see* Fluorescein-5-isothiocyanate
FLIM, *see* Fluorescence lifetime imaging microscopy
Flower model, 214, 215
Flow field-flow fractionation (FFF), 96
Fluorescein-5-isothiocyanate (FITC), 166, 449
Fluorescence correlation spectroscopy (FCS), 354
Fluorescence *in situ* hybridization (FISH), 547
Fluorescence lifetime imaging microscopy (FLIM), 544
Fluorescence resonance energy transfer (FRET), 544
Fluorophenylalanine, 88
5-fluorouracil (5-FU), 87
FMDV, *see* Foot-and-mouth disease virus
fMet-tRNA$_f$, *see* N-formyl-methionine tRNA
Foot-and-mouth disease virus (FMDV), 442, 493
Forced-flow electrophoresis (FFE), 96
Förster resonance energy transfer (FRET), 361
FP, *see* Fertility factor
F pili, 45–47; *see also* Pili
    in *Enterobacteriaceae* other than *E. coli*, 53
    fine mapping of, 61
    F pili–RNA phage complexes, 48
    genetics of, 50–52
    non-F-encoded pili, 55–57
    packing geometry of, 61
    production, 47
    R factors, 49–50
    and RNA phages, 47
    role of, 47
Franklin's structure (FS), 261, 367
fr coat-RNA complex I, 356–358; *see also* Repressor complex I
FRET, *see* Fluorescence resonance energy transfer; Förster resonance energy transfer
FRNA phages, *see* F-specific RNA phages
FS, *see* Franklin's structure
F-specific RNA phages (FRNA phages), 55, 105, 111–112; *see also* Phage ecology

distribution of subgroups of, 146
impact of, 163
survival of prototype strains of, 119
5-FU, *see* 5-fluorouracil
Full width at half maximum (FWHM), 198
FWHM, *see* Full width at half maximum

G

GA phage coat protein, 358–359; *see also* Repressor complex I
GA phage replicase, 288–289; *see also* Replicase
GaRV-MS2, *see* *Gremmeniella abietina* RNA virus MS2
Gas-phase electrophoretic mobility molecular analyzer (GEMMA), 198
GCPII, *see* Glutamate carboxypeptidase II
Gel electrophoresis, 208
GEMMA, *see* Gas-phase electrophoretic mobility molecular analyzer
Gene; *see also* Evolution; Mutations
    contraction hypothesis, 405
    engineering, 553
    expansion hypothesis, 405
    order, 176
Generalized likelihood uncertainty estimation (GLUE), 122
Gene therapy, non-VLPs, 535–536
Genetic fusions, 493, 501
    vaccines constructed using, 494–500
Genetic fusions of coats, 438; *see also* Chimeric VLPs
    AP205, 444–445
    fr, 438–439
    GA, 445–446
    MS2, 439–440
    PP7, 442–444
    Qβ, 440–442
Genogrouping, 40–42; *see also* Grouping
Genome, 167
    full-length RNA phage, 167, 168–169
    full-length sequences, 167–171
    genetic makeup, 167
    IncM plasmid-dependent RNA phage M, 169
    peculiarities of, 171–173
    of RNA phages, 171–172
    sequence comparison of RNA phages, 173
    similarities and dissimilarities, 173–174
    small RNA phage, 167
Genomic RNA packing, 471–476; *see also* 3D particles
Genotyping of RNA phages, 15
GFP, *see* Green fluorescence protein
GISAXS, *see* Grazing-incidence small-angle x-ray scattering
GLUE, *see* Generalized likelihood uncertainty estimation
Glutamate carboxypeptidase II (GCPII), 527
Glutathione S-transferase (GST), 538
GMP-PCP, *see* Guanylyl-5′-methylene-diphosphonate
Grazing-incidence small-angle x-ray scattering (GISAXS), 480
Green fluorescence protein (GFP), 433, 541
Gremmeniella abietina RNA virus MS2 (GaRV-MS2), 1
Grouping, 29; *see also* Taxonomy
    affinities to millipore filters, 37–38
    of *E. coli* RNA phages, 37
    genogrouping, 40–42

hybrid particles, 39
metaviromics, 42–44
phylogenetic analyses of RNA bacteriophages, 43
physicochemical parameters, 39–40
serogrouping, 35–37
template specificity of replicases, 38–39
unrooted phylogenetic analysis of full-length nucleotide sequences, 41
GST, *see* Glutathione S-transferase
Guanylyl-5′-methylene-diphosphonate (GMP-PCP), 330

H

haRPA, *see* Heterogeneous asymmetric recombinase polymerase amplification
HBc, *see* Hepatitis B virus core
HBCE, *see* Hot bubble column evaporator
HBs, *see* Hepatitis B virus surface
hCG, *see* Human chorionic gonadotropin
HCV, *see* Hepatitis C virus
HDTMA-Br, *see* Hexadecyltrimethylammonium bromide
HDX-MS, *see* Hydrogen/deuterium exchange-mass spectrometry
Healthy Workplace Project (HWP), 164
HEL, *see* Hen egg lysozyme
Hen egg lysozyme (HEL), 448
HEPA, *see* High-efficiency particulate air
Hepatitis B virus core (HBc), 299, 435, 489
Hepatitis B virus surface (HBs), 489
Hepatitis C virus (HCV), 515
Heterogeneous asymmetric recombinase polymerase amplification (haRPA), 109
Heterogeneous nuclear ribonucleoprotein particles (hrRNPs), 233
Hexadecyltrimethylammonium bromide (HDTMA-Br), 149
HFI, *see* Host factor I
HFII, *see* Host factor II
hfq gene, 279–280; *see also* Replicase
Hfq protein, 280–281, 283; *see also* Replicase
    from *E. coli*, 281
    structural model of fibrillar, 282
Hibernation promoting factor (HPF), 331
Hidden Markov models (HMMs), 44
High-efficiency particulate air (HEPA), 144
HiPOX reactor, 138
HIV-1, *see* Human immunodeficiency virus type 1
HME, *see* Hot melt extrusion
HMMs, *see* Hidden Markov models
Hofschneider's structures (HS), 260, 367
Horseradish peroxidase (HRP), 451
Host factor I (HFI), 273, 278–279; *see also* Replicase
Host factor II (HFII), 273
Host factors, 272; *see also* Replicase
Host mutants, 84–86; *see also* RNA phage physiology and growth
Hot bubble column evaporator (HBCE), 141
Hot melt extrusion (HME), 455
HPF, *see* Hibernation promoting factor
HPV, *see* Human papilloma virus
HRP, *see* Horseradish peroxidase
hrRNPs, *see* Heterogeneous nuclear ribonucleoprotein particles
HS, *see* Hofschneider's structures

Human chorionic gonadotropin (hCG), 444
Human immunodeficiency virus type 1 (HIV-1), 515
Human papilloma virus (HPV), 439, 493
Human virus surrogates, 118–122; *see also* Phage ecology
HWP, *see* Healthy Workplace Project
Hybridization, 210; *see also* RNA
Hybridized binary probes, 398; *see also* Evolution
Hydrogen/deuterium exchange-mass spectrometry (HDX-MS), 475

I

Icosahedra, 16–17
ICTV, *see* International Committee on Taxonomy of Viruses
ICTV taxonomy, 29–31; *see also* Taxonomy
IFN-γ, *see* Interferon gamma
IIa protein, 237; *see also* Maturation protein
IL-21R, *see* IL-21 receptor
IL-21 receptor (IL-21R), 491
Imaging, non-VLPs, 540–546
    evaluation of role of microRNAs, 544
    OFF genetic control device, 544
    live visualization of endogenous RNA, 541–542
    MS2-based visualization studies, 542, 543
    MS2-GFP imaging, 541, 542, 543
    m-TAG, 543
    reporter mRNAs in live cells, 541
    SNAP tag, 545
    Spinach-based MS2 sensor, 545
Immunology, 301
    antiserum for phage selection, 301
    immunodetection, 310–312
    immunoglobulins, 302–304
    mapping, 309–310
    phylogeny of antibodies, 304–305
    relationship between bacteriophages R17 and M12, 301
    serogrouping, 301–302
    serological criteria, 301
    target of antibodies, 306–309
IMS, *see* Integrated membrane system
Inc groups, *see* Incompatibility groups
IncM plasmid-dependent RNA phage M, 169
Incompatibility groups (Inc groups), 57; *see also* Pili
    and RNA phages, 58–61
Infection processes, 69; *see also* RNA phage physiology and growth
Integrated membrane system (IMS), 153
Integrated virus detection system (IVDS), 106, 198, 476
Interferon gamma (IFN-γ), 420
International Committee on Taxonomy of Viruses (ICTV), 29
International Organization for Standardization (ISO), 106
IPTG, *see* Isopropyl-β-D-thiogalactoside
ISO, *see* International Organization for Standardization
Isopropyl-β-D-thiogalactoside (IPTG), 535
IVDS, *see* Integrated virus detection system

K

Kolakofsky-Weissmann model, 362

L

LAH, *see* Long alpha helical
LAPS, *see* Light-addressable potentiometric sensor
Larifan, 378–379; *see also* Replication
Lateral flow immunoassay (LFA), 311
Lattices, 484–486; *see also* 3D particles
Layered double hydroxide (LDH), 148, 153
LCMV, *see* Lymphocytic choriomeningitis virus
LDH, *see* Layered double hydroxide
LEDs, *see* Light-emitting diodes
Lethal mutations, conditionally, 175; *see also* Mutations
Levallorphan, 88
Levi-like viruses, 44
Levinthal's paradox, 481
Leviviridae phages, 169
Levorphanol, 88
LFA, *see* Lateral flow immunoassay
Light-addressable potentiometric sensor (LAPS), 310
Light-emitting diodes (LEDs), 124
Liver stage antigen-1 (LSA-1), 440
Long alpha helical (LAH), 501
LSA-1, *see* Liver stage antigen-1
Lyme disease vaccine, 513; *see also* VLP vaccines
Lymphocytic choriomeningitis virus (LCMV), 489
Lysis gene, 167, 424–425; *see also* Expression
Lysis protein, 291
    applications, 299
    cloned MS2 lysis protein, 296
    comparison of lysine sequences, 293
    discovery, 291
    E-L protein, 296
    gene regulation, 294
    homology, 291–292
    lysis mechanism, 295–298
    molecular mass of lysins, 292
    mutational analysis of lysis gene, 298
    natural expression of lysis gene, 293–295
    open reading frames, 291
    of phage M, 296
    protein antibiotics, 298–299
    scanning model, 295
Lysogeny, 97–98; *see also* RNA phage physiology and growth

M

Magic-angle spinning (MAS), 486
Magnetic hybrid colloids (MHC), 131
Magnetic resonance imaging (MRI), 449
MALDI, *see* Matrix-assisted laser desorption/ ionization
MALDI-MS, *see* Matrix-assisted laser desorption ionization mass spectrometry
MALDI-TOF mass spectrometry (MALDI-TOF-MS), 123
MALDI-TOF-MS, *see* MALDI-TOF mass spectrometry
Male specificity, 1
MAMLE, *see* Maximum likelihood estimation
Mapping, 309–310; *see also* Immunology
MAS, *see* Magic-angle spinning
Matrix-assisted laser desorption/ionization (MALDI), 198

Matrix-assisted laser desorption ionization mass spectrometry (MALDI-MS), 146
Maturase, 237; *see also* Maturation protein
Maturation gene, 423–424; *see also* Expression
Maturation protein, 237, 476–480; *see also* 3D particles
    chemical studies, 238
    conserved regions of RNA phage, 244
    fate of, 238, 241
    gel electrophoresis, 237–238
    genomic RNA binding of Qβ A2 protein, 244
    homology, 242
    isolation, 238
    lysis, 243–245
    molecular mass of, 239–240
    mutant allusions, 237
    Qβ A2 protein, 243
    replacing coat protein dimer, 477
    roles of, 483
    sequence comparison, 242
    sequencing, 241–242
    spatial structure, 242–243
    structural mechanism of A2-mediated host lysis, 245
    structure of Qβ bound with MurA, 245
Maximum likelihood estimation (MAMLE), 544
MazF toxin, 232
MBR, *see* Membrane bioreactor
MBS, *see* MS2 binding site
MDP, *see* N-acetylmuramyl-L-alanyl-D-isoglutamine
MDV-1, 384–386; *see also* Evolution
Measure of pollution, 111–112
Medical Research Council (MRC), 26
Medical Subject Headings (MeSH), 45
Membrane bioreactor (MBR), 156
Membrane filtration, 154–157; *see also* Phage ecology
Membrane proximal external region (MPER), 514
Merozoite surface protein 3 (MSP3), 514
MERS-CoV, *see* Middle East respiratory syndrome coronavirus
MeSH, *see* Medical Subject Headings
Metagenomic sequencing, 226; *see also* RNA
Metaupon, 89
Metaviromics, 42–44; *see also* Grouping
Methicillin-resistant *Staphylococcus aureus* (MRSA), 493
Methylglyoxal bis(guanylhydrazone) (MGBG), 317
MGBG, *see* Methylglyoxal bis(guanylhydrazone)
MGEs, *see* Mobile genetic elements
MHC, *see* Magnetic hybrid colloids
micRNA, *see* mRNA-interfering complementary RNA
Micro- and nanovariants, 386–387; *see also* Evolution
Microbial source tracking (MST), 113, 158
MicroRNAs (miRNAs), 544
Middle East respiratory syndrome coronavirus (MERS-CoV), 517
Minivariants (MNV), 388; *see also* Evolution
MIPs, *see* Molecularly imprinted polymers
Miracil D, 87
miRFPs, *see* Monomeric near-infrared fluorescent proteins
miRNAs, *see* MicroRNAs
Mixed infections, 93–95; *see also* RNA phage physiology and growth
MLV, *see* Murine leukemia virus
MMDB, *see* Molecular Modeling Database

MNV, *see* Minivariants
Mobile genetic elements (MGEs), 58
MOI, *see* Multiplicity of infection
Molecular colonies, 390–391; *see also* Evolution
Molecularly imprinted polymers (MIPs), 108, 312
Molecular Modeling Database (MMDB), 33
Molecular parasites, 390; *see also* Evolution
Monomeric near-infrared fluorescent proteins
    (miRFPs), 548
Moraxellaceae, 67
MPER, *see* Membrane proximal external region
MRC, *see* Medical Research Council
MRI, *see* Magnetic resonance imaging
mRNA-interfering complementary RNA
    (micRNA), 535
MRSA, *see* Methicillin-resistant *Staphylococcus
    aureus*
MS2 binding site (MBS), 544; *see also* Phage MS2
MS2 phage, *see* Phage MS2
MS/MS, *see* Tandem MS
MSP3, *see* Merozoite surface protein 3
MST, *see* Microbial source tracking
m-TAG, 543
Multiplicity of infection (MOI), 54
Multiwalled carbon nanotube (MWNT), 155
Murine leukemia virus (MLV), 558
Mutant, 351; *see also* Expression; Mutations
    allusions, 237
    azure, 181, 183
    gene function, 177–181
    generation of, 421–422
    genes, 177–181
    op-3 mutant, 183
    of phage f2, 183–184
    of phage fr, 186
    of phage GA, 186–187
    of phage MS2, 184–186
    of phage PP7, 188
    of phage R17, 186
    of phages Qβ, SP, and FI, 187
    replicase, 178
    synthesis rate of RNA phage coat amber, 179
Mutations, 175
    complementation, 177
    conditionally lethal, 175
    frequencies of, 176–177
    frequency of nonsense, 177
    gene order, 176
    MS2 RNA 3′-terminal segment model, 182
    polar, 176
    polarity effect, 179
    stimulation of *in vitro* synthesis of IIb, 180
    streptomycin suppression, 183
    super-repressor, 350
MWNT, *see* Multiwalled carbon nanotube

N

N-acetylmuramyl-L-alanyl-D-isoglutamine
    (MDP), 309
NALDI, *see* Nano-assisted laser desorption/
    ionization
Nano-assisted laser desorption/ionization
    (NALDI), 199
Nanocontainers, 518, 527–529; *see also*
    Non-vaccine VLPs
Nanofiltration method, 108
Nanomaterials and imaging, 529, 532–533;
    see also Non-vaccine VLPs
Nanoreactors, 529; *see also* Non-vaccine VLPs
NanoSight (NS), 81

National Center for Biotechnology Information
    (NCBI), 32, 555; *see also* Taxonomy
    classification, 32–35
    Leviviridae in, 34
    RNA phage entries at, 36
NCBI, *see* National Center for Biotechnology
    Information
N-formyl-methionine tRNA (fMet-tRNA$_f$), 323
NLM, *see* U.S. National Library of Medicine
NLS, *see* Nuclear localization signal
Non-*Allolevivirus* replicases, 286; *see also*
    Replicase
Non-coli phages, 13–14
Non-vaccine VLPs, 515
    AFS for single amino acid variants of MS2
       coat protein, 528
    armored RNA, 515–517
    bioimaging agents, 530–531
    drug delivery, 518, 527–529
    effects of pore charge on substrate and
       product flux, 532
    MS2 VLP-based armored RNA
       approach, 515
    nanomaterials and imaging, 529, 532–533
    nanoreactors, 529
    RNA phage VLPs, 519–526
    VLP packaging technology, 518
Non-VLPs, 535
    CRISPR-Cas9, 549–551
    gene therapy, 535–536
    imaging, 540–546
    PP7, 546–549
    tethering, 536–540
Novobiocin, 87
NS, *see* NanoSight
Nuclear localization signal (NLS), 541
Nucleic acid pump, 47, 50
Nucleotide-driven packaging, 461; *see also*
    Chimeric VLPs
    antigenic determinant, 435
    capsid loading and targeting concept, 463
    MS2 VLP-based siRNA delivery system, 464
    novel VLP models, 466
    Qβ RNAi scaffold, 465
    RNA-directed encapsidation, 463
    siRNA delivery, 464
    SpyTag/SpyCatcher approach, 464
Nutritional mutants, 382; *see also* Evolution

O

OFF genetic control device, 544
Office wellness intervention (OWI), 164
Oligomeric conjugated polyelectrolytes
    (OPEs), 139
Oligo-phenylene ethynylenes (OPE), 139
op-3 mutant, 183; *see also* Mutations
OPE, *see* Oligo-phenylene ethynylenes
Open reading frames (ORFs), 170, 225
OPEs, *see* Oligomeric conjugated
    polyelectrolytes
OPRA, *see* Optimal protein-RNA area
Optimal protein-RNA area (OPRA), 361
ORFs, *see* Open reading frames
OWI, *see* Office wellness intervention

P

PAA, *see* Peracetic acid
Packaging signals, 481
*p*AF, *see* *para*-amino-L-phenylalanine

p-aminophenol (PAP), 310
p-aminophenyl galactopyranoside (PAPG), 310
PAP, *see* p-aminophenol; Pokeweed antiviral
    protein
PAPG, *see* p-aminophenyl galactopyranoside
PAP I, *see* Poly(A) polymerase I
Papillomavirus VLP- and Qβ phage-based Aβ
    vaccines, 447
para-amino-L-phenylalanine (*p*AF), 450
Parenteral Drug Association (PDA), 166
Particles, 189; *see also* 3D particles
    aggregation, 201–202
    coat-RNA contact, 193
    composition, 191–192
    defective particles, 193–195
    diameters of RNA phages, 191
    molecular mass, 189–190
    monitoring, 198–201
    morphology, 192
    reconstitution, 195–197
    size, 190–191
    standards and internal controls, 202–203
    symmetry, 192
PDA, *see* Parenteral Drug Association
PDB, *see* Protein Data Bank
PEG, *see* Polyethylene glycol
PEI, *see* Polyethyleneimine
Penetration, 73–75; *see also* RNA phage
    physiology and growth
PEP, *see* Phosphoenolpyruvate
PepE, *see* Peptidase E
Pepper mild mottle virus (PMMoV), 116
Peptidase E (PepE), 529
PepVac, 447
Peracetic acid (PAA), 137
PES, *see* Polyether sulfone
PET, *see* Polyethylene terephthalate; Positron
    emission tomography
PET/CT, *see* Positron emission tomography-
    computed tomography
Phage; *see also* AP205 phages; Coat protein;
    Expression; Repressor complex I;
    RNA phage physiology and growth;
    3D particles
    bacterial lysis by RNA, 77
    chimeric, 422–423, 514
    -coded proteins, 419
    effect on host, 81–84
    f1 mutants, 187
    f2 mutants, 183–184
    FI, 436
    fr mutants, 186
    GA mutants, 186–187
    growth inhibition, 86–89
    of *Microviridae* family, 15
    Phage 7s, 302
    -pili attachment, 70
    PP7 mutants, 188
    Qβ, coat of group III, 359–360
    Qβ mutants, 187
    R17 mutants, 186
    -specific replicase subunits, 259
    SP mutants, 187
    therapy, 559
Phage ecology, 105, 111
    adsorption and coagulation, 147–150
    aerosols, 143–144
    alkaline stabilization, 137
    alkylating agents, 134–135
    ammonia, 137
    biofilms, 144–145

Phage ecology (*Continued*)
  bisulfite, 135
  chlorine, iodine, and bromine, 135–137
  CO₂, 143
  desiccation, 143–144
  different resistance to inactivation, 145–147
  different survival in water, 117–118
  distribution of FRNA phages, 146
  dyed phage particles, 145
  fecal pollution indicators, 113
  filtration by soil and sand columns,
      152–153
  food total, 161–162
  fracture flow and transport, 150–152
  FRNA coliphages from feces, 113
  FRNA phages, 111–112
  FRNA prototype strain survival in surface
      waters, 119
  fruits and vegetables, 159–161
  graphene and carbon nanomaterials, 131
  heat, 140–141
  high pressure, 143
  humidity, 143–144
  hydroxylamine, 135
  inactivating agents, 138–140
  inactivation, 122
  ionizing radiation, 132
  magnetic hybrid colloids, 131
  measure of pollution, 111–112
  membrane filtration, 154–157
  metal compounds, 133–134
  nanoalumina fibers, 131
  nitrous acid, 134
  oysters and mussels, 157–159
  ozone, 138
  patient care, 163–165
  peracetic acid, 137–138
  photocatalytic disinfection, 127–129
  photosensitization of fullerol nanoparticles,
      130–131
  photosensitizing agents, 130
  plasma, 132–133
  polycationic superparamagnetic core-shell
      nanoparticles, 131
  pseudomonaphage PP7, 165
  selective photonic disinfection, 133
  silver nanowire-carbon fiber cloth
      nanocomposites, 131
  sunlight, 141–143
  surrogates of human viruses, 118–122
  ultrasound, 143
  ultraviolet irradiation, 122–127
  UV-LEDs, 129–130
  water quality, 112
  water surveillance, 113, 114–117
Phage genomic RNAs; *see also* RNA
  molecular mass of, 206–207
  nucleotides ratio in, 209
Phage MS2, 12, 109, 114, 189, 349–356,
      446–447; *see also* Chimeric VLPs;
      Non-vaccine VLPs; Particles; Phage
      ecology Repressor complex I;
      Tethering; 3D particles
  aggregates, 202
  -based tethering, 538–540
  -based visualization studies, 542, 543
  as biologic model for assessing virus
      removal, 154
  as biotemplate, 451–452
  coat protein dimer, 449
  -decorated F pili, 477

  differences in coat structures between
      AP205 and, 469
  dyed phage MS2, 145
  functionalization, 448
  genome, 167
  -GFP imaging, 541, 542, 543
  immunological characteristics of, 301
  implications for MS2 virus assembly, 485
  LP-based armored RNA approach, 515
  lysis protein, 296
  modification of aniline-containing MS2
      VLPs, 451
  mutants, 184–186
  nanoparticle, 450
  organization of genomic RNA in MS2
      virion, 474
  Ouchterlony immunodiffusion assay, 308
  photoinactivation of, 131
  solar photocatalytic inactivation, 128
  3D structures of, 474
  UV inactivation of, 125
  -VP1u bioconjugate, 452
Phage MS2 replicase, 288; *see also* Expression;
      Replicase
  fusion technique, 427
Phage MS2 RNA; *see also* RNA
  cleavage assay, 232
  as control, 234–235
  and programmed cell death, 232–233
  as substrate and interaction probe, 233–234
Phage MS2 sequencing, 211–217; *see also* RNA
  by-products and direct consequences of,
      217–219
  preparations for, 210–211
Phage Qβ
  protein gene sequence, 222
  sequencing, 222–224
  3D structures of, 474
  VLPs, 458, 490
Phage Qβ A2 protein, 243; *see also* Maturation
      protein
  gene expression in *E. coli* of, 441
  genomic RNA binding of, 244
Phage Qβ functionalization, 452; *see also*
      Chimeric VLPs
  approaches to modify Qβ VLPs, 455
  ATRP methodology, 454
  cellular uptake of Qβ, 458
  CuAAC reaction, 457
  hepatitis B virus core and Qβ virus-like
      particles, 454
  Qβ reduction and DB compounds, 459
  Qβ showing Lys16, 453
  Qβ VLPs, 453, 456, 458
  ROMP technique, 455
Phage Qβ replicase, 258–262, 283–286; *see also*
      Replicase
  core complex, 277
  intermolecular contacts within, 270
  mapping, 395–396
  properties of subunits of, 263
  RNA elongation model, 274
  structure, 268–272
  subunits of, 262–265
Phage Qβ RNA; *see also* RNA
  first 860 nucleotides of, 230
  proposed secondary structure for 3′
      replicase, 229
  structure proposal, 230
Phage R17; *see also* Replicase; Repressor
      complex I; RNA; Tethering

  -based tethering, 537
  complex I, 346–349
  replicase, 289
  sequencing, 219–221
Phage RNA
  secondary structure of, 336
  sequencing, 224–226
Phleomycin, 88
Phosphoenolpyruvate (PEP), 244
Pili, 14, 45
  Caulobacter and *Acinetobacter* pili, 64–66
  chromosomal mutations, 52–53
  conduction theory, 45
  conjugation and RNA phages, 53–54
  electron cryomicroscopy reconstruction and
      filament model, 65
  elimination of sex factors, 54–55
  FESEM of donor strain DT1944, 60
  fine mapping of polar pili, 61–64
  global sequencing stage of pili studies, 61
  hypothetical phylogenetic tree of ssRNA
      phages, 60
  incompatibility groups, 57–58
  incompatibility groups and RNA phages,
      58–61
  non-F-encoded pili, 55–57
  physical parameters of, 63
  Pseudomonas aeruginosa strain 1 with
      PP7 phage virions adsorbed to polar
      pili, 56
  retraction hypothesis, 49
  retraction theory, 45
  RNA phage hosts, 66–67
  and RNA phages, 62
  spheroplasts, 67
  theories regarding operation mode of, 45
  working mechanism, 47–49
Pili of *Caulobacter* and *Acinetobacter*, 64–66
PI protein, *see* Polyamine-induced protein
³²P-labeled phage assay, 47
Plaque assay, 81; *see also* RNA phage physiology
      and growth
PLGA, *see* Poly(lactic-*co*-glycolic acid)
Plug-and-display, 460, 513
PMMoV, *see* Pepper mild mottle virus
PNB, *see* Poly(norbornene-(oligo(ethylene
      glycol) ester))
POEGMEA, *see* Poly(oligo(ethylene glycol)
      methyl ether acrylate)
Pokeweed antiviral protein (PAP), 234
Polarity effect, 179; *see also* Mutations
Polar mutations, 176; *see also* Mutations
Polyamine-induced protein (PI protein), 317
Poly(A) polymerase I (PAP I), 86
Polyether sulfone (PES), 154
Polyethylene glycol (PEG), 96
Polyethyleneimine (PEI), 131, 156
Polyethylene terephthalate (PET), 141
Poly(lactic-*co*-glycolic acid) (PLGA), 455
Poly(norbornene-(oligo(ethylene glycol) ester))
      (PNB), 455
Poly(oligo(ethylene glycol) methyl ether acrylate)
      (POEGMEA), 455
Poly(phenylene ethynylene) (PPE), 139
Polyvinylidene fluoride (PVDF), 154
Positron emission tomography (PET), 450
Positron emission tomography-computed
      tomography (PET/CT), 527
Potato spindle tuber viroid (PSTV), 284
PP7 phage, 360–361, 546; *see also* Non-VLPs;
      Repressor complex I

dual-color imaging, 546, 547
PP7-based tethering, 546
structure of PP7-PP7 virus-like particle, 486
tethering and imaging investigations, 548–549
3D structures of, 474
TRICK technique, 547
PPE, *see* Poly(phenylene ethynylene)
PQI, *see* p-quinone imine
p-quinone imine (PQI), 310
Preparative growth and purification, 95–96;
  see also RNA phage physiology and
  growth
Protein 5, 291
antibiotics, 298–299
ramp, 481
S1, 263
Protein Data Bank (PDB), 487
Protein engineering, 435
candidates for VLP, 436
principles of VLP, 437
PRR1 complex I, 361–362; *see also* Repressor
  complex I
Pseudolysogenization, 98, 187
Pseudomonadaceae, 66
Pseudomonaphage, 97–98; *see also* Phage
  ecology
PP7, 40, 165
Pseudomonas aeruginosa
filament model of, 65
with PP7 phage virions adsorbed to polar
  pili, 56
after RNA phage PP7 infection, 80
pSIP, *see* Self-immobilizing plasmid DNA
PSTV, *see* Potato spindle tuber viroid
Pulse-chase experiments, 261
PVDF, *see* Polyvinylidene fluoride
Py-GC-IMS, *see* Pyrolysis-gas chromatography-
  ion mobility spectrometry
Pyrolysis-gas chromatography-ion mobility
  spectrometry (Py-GC-IMS), 198

Q

QMRA, *see* Quantitative microbial risk
  assessment
QPCR, *see* Quantitative real-time polymerase
  chain reaction
Quantitative microbial risk assessment
  (QMRA), 115
Quantitative real-time polymerase chain reaction
  (QPCR), 81
Quasispecies, 393, 413–418; *see also* Evolution
QUASR, *see* Quenching of Unincorporated
  Amplification Signal Reporters
Quenching of Unincorporated Amplification
  Signal Reporters (QUASR), 201
Qβ replicase, 258–262, 263, 271
core complex, 269
structure, 268–272

R

Rabbit's ears, 76
Rara avis, 554
RBP, *see* RNA-binding proteins
RdRps, *see* RNA-dependent RNA polymerases
RE, *see* Replicative ensemble
Readthrough protein, 252–255
Reaferon, 378; *see also* Replication
Recognition mutants, 350
Recombinant norovirus VLPs (rNV-VLPs), 121

Recombinant Norwalk virus (rNV), 120
RED, *see* Reduction equivalent dose
Red fluorescent protein (RFP), 547
Reduced graphene oxide (RGO), 311
Reduction equivalent dose (RED), 125
Reduction equivalent fluence (REF), 124
REF, *see* Reduction equivalent fluence
Regulatory elements, 437
Related to Qβ sequences (RQ sequences), 393;
  see also Evolution
Release, 79–81; *see also* RNA phage physiology
  and growth
Replicase, 257
activity, 257–258
EF Tu-Ts, 265–267
f2 replicase, 286–287
fusions, 427–431
GA replicase, 288–289
gene, 425–426
Hfq, 272–278, 278–279
hfq gene, 279–280
Hfq structure, 280–281, 283
homology, 258
initiation of negative strand Qβ RNA
  synthesis, 276
intermolecular contacts in Qβ, 270
molecular mass of phage-specific replicase
  subunits, 259
MS2 replicase, 288
mutants, 178
non-*Allolevivirus* replicases, 286
R17 replicase, 289
replicase subunit, 258
ribosomal protein S1, 267–268
RNA binding and surface properties of
  β-subunit, 271
RNA elongation model by Qβ replicase, 274
SDS-polyacrylamide gel pattern of Oβ RNA
  replicase, 262
sequences comparison, 260
special features of Qβ replicase enzyme,
  283–286
structural differences between EF-Tu and –
  Ts, 271
structural model of fibrillar Hfq from EM
  and image analysis, 282
structure of Hfq protein from *E. coli*, 281
structure of initiation stage, 275
structure of monomeric form of Qβ replicase
  core complex, 277
structure of β-subunit, 273
subunits of Qβ replicase, 262–265
subunit sequences, 260
Replicating complex structure, 395
Replication, 365
bacteriophage RNA synthesis model, 370
early studies, 366–367
essentials, 365–366
induction of interferon, 375–378
Larifan, 378–379
nascent strains, 365
normalized cycles of MS2 and Qβ RNA
  synthesis, 372
protein inhibition effect on rate of phage
  RNA synthesis, 376
regulation of replication cycle, 371–375
replicative intermediate, 367–371
RNA phage replication, 366, 377
Replication cycle, 75–78; *see also* RNA phage
  physiology and growth
Replicative complexes, 381; *see also* Evolution

Replicative ensemble (RE), 287
Replicative form (RF), 366
Replicative intermediate (RI), 323, 367–371
Reporter mRNAs, 541
Repressor complex I, 344; *see also* Translation
binding of operator stem-loop sequences to
  coat dimers, 361
classes of mutants, 351
f2 coat-RNA complexes, 346
fr coat-RNA complex I, 356–358
GA coat protein, 358–359
group III phage Qβ coat, 359–360
MS2, 349–356
MS2 coat-operator interactions, 356
MS2 CP-RNA complex, 357
MS2 genome containing translational
  repression site, 345
PP7 ΔFG, 360
PRR1 complex I, 361–362
Qβ coat, 350
R17 complex I, 346–349
recognition mutants, 350
replicase-operator hairpins in RNA
  phages, 358
secondary structure of Qβ RNA, 346
soaking technique, 351–352
super-repressor mutations, 350
3D structures of, 353
two-plasmid system, 349
Repressor complex II, 362–364; *see also*
  Translation
Qβ RNA from polysome to replicating
  complex, 362
Qβ RNA with Qβ replicase, 363
Requinomycin, 88
Resistance factors (R factors), 49
Respiratory virus panel (RVP), 202
Retraction
hypothesis, 49
theory, 45
Reverse line blot hybridization (RLB), 15
Reverse quantitative PCR (RT-qPCR), 40
Reverse-transcription loop-mediated isothermal
  amplification (RT-LAMP), 201
Reversible target capture, 397; *see also*
  Evolution
RF, *see* Replicative form
R factors, *see* Resistance factors
RFP, *see* Red fluorescent protein
Rg-lysis gene, 291
RGO, *see* Reduced graphene oxide
RI, *see* Replicative intermediate
Ribonuclease A, 211
Ribonuclease H, 208; *see also* RNA
Ribosomal
delivery mechanism, 299
hibernation, 331
protein S1, 267–268
Ribosome-inactivating proteins (RIPs), 314
Ribosome modulation factor (RMF), 331
Ribosome releasing or recycling factor (RRF),
  332–333
Rice yellow mottle virus (RYMV), 198
Rifamycin, 87, 89; *see also* RNA phage
  physiology and growth
naturally occurring, 90
phage MS2 RNA synthesis rate and single-
  step growth, 93
on phage replication, 91
RNA synthesis rate and single-step growth
  curves, 92

Ring-opening metathesis polymerization (ROMP), 455
RIPs, *see* Ribosome-inactivating proteins
RLB, *see* Reverse line blot hybridization
RMF, *see* Ribosome modulation factor
RNA, 205, 208–210; *see also* Evolution; 3D particles
 aptamers, 470–471
 biophysics, 205–208
 by-products and consequences of MS2 sequencing, 217–219
 comparison of RNA secondary structures, 231
 -dependent polymerases, 406–410
 electron microscopy, 226–227
 energy dot plot, 224
 f2 sequencing, 221–222
 first 860 nucleotides of Qβ RNA, 230
 flower model, 214, 215
 gel electrophoresis, 208
 as genome, 1, 11
 genomic phage RNA mass, 206–207
 hybridization, 210
 metagenomic sequencing, 226
 MS2 RNA and programmed cell death, 232–233
 MS2 RNA as control, 234–235
 MS2 RNA as substrate, 233–234
 MS2 RNA cleavage assay, 232
 MS2 sequencing, 211–217
 plasmids, 394
 -protecting protein, 237
 Qβ coat protein gene sequence, 222
 Qβ sequencing, 222–224
 R17 sequencing, 219–221
 ratio of nucleotides in phage genomic RNAs, 209
 relation of size to sedimentation constant, 208
 ribonuclease H, 208
 RNA phage parasitism, 218
 secondary structure, 227–228, 231–232
 sequencing of phage RNAs, 224–226
 sequencing, 210–211
 6S RNA, 383–384
 small RNAs, 542
 ssRNA phage phylogenetic tree, 60
 synthetase research team, 21
 3′ replicase domain of Qβ RNA, 229
 2DMut web tool, 545
 viruses, 203
 world, 400–404
RNA-binding proteins (RBP), 548
RNA-dependent RNA polymerases (RdRps), 267
RNAi, *see* RNA interference
RNA interference (RNAi), 544
RNA phage, 1; *see also* 3D particles
 antibodies, 15
 applications and research, 559–560
 attachment and penetration, 69
 bacterial immunity, 558–559
 classical reviews, 1
 coats as VLP carriers, 432
 coliphages, 12–13
 composition, 18–19
 CRISPR-Cas system, 558
 crystal structures of, 468
 discovery of, 1, 553
 diversity of pili, 14
 F pili–RNA phage complexes, 48
 genome, 19–20
 growth cycle, 11–12

 immediate future, 557–558
 in Germany, 12
 incompatibility groups and, 58–61
 male specificity, 1
 modern genotyping, 15
 MS2 virion sequences, 482
 non-coli phages, 13–14
 novel VLP models, 556
 parasitism, 218
 phage of *Microviridae* family, 15
 phage tales, 26–27
 phage therapy, 559
 pili and, 62
 portraits of icosahedra, 16–17
 reconstitution, 25–26
 regulation, 24–25
 replication scheme, 366
 replication, 22–24
 RNA as genome, 1, 11
 RNA synthetase research team, 21
 rod-like particles, 472
 scientific mode of life, 2–10
 statistics, 553, 554
 three-dimensional structure, 556–557
 translation, 20–22
 virome, 554–556
 worldwide, 13
RNA phage hosts, 66; *see also* Pili
 Caulobacteraceae, 66–67
 Enterobacteriaceae, 66
 Moraxellaceae, 67
 Pseudomonadaceae, 66
RNA phage physiology and growth, 69
 A protein, 73
 attachment, 69–71
 attachment of phage to *E. coli*, 72
 cell-free synthesis, 96–97
 contact-dependent stimulation of T4SS model, 75
 crystalline inclusions, 78–79
 ejection, 71–73
 exclusion phenomenon among RNA phages, 94
 growth inhibition, 86–89
 host mutants, 84–86
 infection processes, 69
 lysis of bacteria, 77
 lysogeny, 97–98
 mixed infections and superinfections, 93–95
 penetration, 73–75
 phage effect on host, 81–84
 phage-pili attachment, 70
 plaque assay, 81
 preparative growth and purification, 95–96
 Pseudomonas aeruginosa after RNA phage infection, 80
 release, 79–81
 replication cycle, 75–78
 rifamycin, 89–93
RNA phage VLPs, 519–526; *see also* Non-vaccine VLPs
 as nanotools, 557
rNV, *see* Recombinant Norwalk virus
rNV-VLPs, *see* Recombinant norovirus VLPs
ROMP, *see* Ring-opening metathesis polymerization
RQ sequences, *see* Related to Qβ sequences
RQ variants, 393–395; *see also* Evolution
RRF, *see* Ribosome releasing or recycling factor
RT-LAMP, *see* Reverse-transcription loop-mediated isothermal amplification

RT-qPCR, *see* Reverse quantitative PCR
RVP, *see* Respiratory virus panel
RYMV, *see* Rice yellow mottle virus

S

SAM, *see* Synergistic activation mediator
SaMazF toxin, 232
SANS, *see* Small-angle neutron scattering
SARS-CoV, *see* Severe acute respiratory syndrome-associated coronavirus
SAXS, *see* Small-angle x-ray scattering
Scanning model, 299
scFvs, *see* Single-chain variable fragments
sdAbs, *see* Single-domain anti-MS2 antibodies
SDS, *see* Sodium dodecyl sulfate
SD sequences, *see* Shine-Dalgarno sequences
Secondary ion mass spectrometry (SIMS), 199
Selective photonic disinfection (SEPHODIS), 133
SELEX, *see* Systematic evolution of ligands by exponential
Self-assembly, 3D particles, 480–484; *see also* 3D particles
Self-encoding system, 401; *see also* Evolution
Self-immobilizing plasmid DNA (pSIP), 299
Self-induced switch, 218
Self-sustained sequence replication system (3SR system), 401
Semipreparative MS2 propagation method, 95
Sensu lato species, 35
SEPHODIS, *see* Selective photonic disinfection
Serogrouping, 35–37, 301–302; *see also* Grouping; Immunology
 inactivation of RNA phages by various antisera, 37
 of RNA phages, 38
SERS, *see* Surface-enhanced Raman spectroscopy
Serum-blocking assay, 48
Severe acute respiratory syndrome-associated coronavirus (SARS-CoV), 136, 515
sgRNA, *see* Single-guide RNA
SHAPE-seq, 548
Shine-Dalgarno sequences (SD sequences), 313
Short interspersed degenerate retroposons (SIDERs), 543
SIDERs, *see* Short interspersed degenerate retroposons
Signal-mediated assembly, packaging, 484; *see also* 3D particles
Simian immunodeficiency virus (SIV), 516
SIMS, *see* Secondary ion mass spectrometry
Single-chain variable fragments (scFvs), 446
Single-domain anti-MS2 antibodies (sdAbs), 304
Single-guide RNA (sgRNA), 549
Single jellyroll capsid proteins (SJR-CPs), 410
Single-molecule fluorescence correlation spectroscopy (smFCS), 480
Single-molecule fluorescence resonance energy transfer (SM-FRET), 354
Single-particle fluorescence spectrometer (SPFS), 200
Single particle imaging (SPI), 470
Single-stranded DNA (ssDNA), 30
Single-stranded RNA genome modeling, 479; *see also* 3D particles
Single-walled CNTs (SWCNTs), 311
SIV, *see* Simian immunodeficiency virus
SJR-CPs, *see* Single jellyroll capsid proteins
Skin and soft tissue infections (SSTI), 493
Small-angle neutron scattering (SANS), 476

Small angle spectroscopy, *see* Small-angle x-ray scattering
Small-angle x-ray scattering (SAXS), 200, 476
Small nuclear ribonucleoproteins (snRNPs), 280
Small RNAs (sRNAs), 542
smFCS, *see* Single-molecule fluorescence correlation spectroscopy
SM-FRET, *see* Single-molecule fluorescence resonance energy transfer
SMPs, *see* Soluble microbial products
SNAP tag, 545
snRNPs, *see* Small nuclear ribonucleoproteins
Soaking technique, 351–352
SODIS, *see* Solar disinfection system of water
Sodium dodecyl sulfate (SDS), 54, 200
Soft agar overlay method, 81
Solar disinfection system of water (SODIS), 141
Soluble microbial products (SMPs), 125
Spermine, 87–88
SPFS, *see* Single-particle fluorescence spectrometer
Spheroplasts, 67; *see also* Pili
SPI, *see* Single particle imaging
Spiegelman, Sol, 401
Spinach-based MS2 sensor, 545
SpyCatcher protein, 460
SpyTag/SpyCatcher, 460
sRNAs, *see* Small RNAs
ssDNA, *see* Single-stranded DNA
ssRNA phage phylogenetic tree, 60
SSTI, *see* Skin and soft tissue infections
Streptomycin suppression, 183; *see also* Mutations
Superinfections, 93–95; *see also* RNA phage physiology and growth
Super-repressor mutations, 350
Surface-enhanced Raman spectroscopy (SERS), 199
Surface plasmon resonance–based immunosensor, 311
SWCNTs, *see* Single-walled CNTs
SyMAPS, *see* Systematic mutation and assembled particle selection
Synergistic activation mediator (SAM), 549
Systematic evolution of ligands by exponential (SELEX), 355
Systematic mutation and assembled particle selection (SyMAPS), 527

T

T4CPs, *see* Type IV coupling proteins
T4SSs, *see* Type IV secretion systems
TACAs, *see* Tumor-associated carbohydrate antigens
TACB, *see* Threo amino chlorobutyric acid
Tailing phenomenon, 125
Tandem MS (MS/MS), 107
Taxonomy, 29; *see also* Grouping
  entries in books of reference and encyclopedias, 29
  NCBI classification, 32–35
  necessity for improvements in, 44
  official ICTV taxonomy, 29–31
  phylogenetic tree, 32
  phylogeny of viral RNA dependent replicases, 34
  relatedness to other taxa, 31–32
  virus taxa infecting bacteria and archaea, 33
Tethering, 536
  MS2-based, 538–540
  R17-based, 537

RNA binding proteins fused to fluorescent reporters, 537
1,4,7,10-tetraazacyclododecanetetraacetic acid (DOTA), 452
15,20-tetrakis (1-methyl-4-pyridinio) porphyrintetra p-toluenesulfonate (TMPyP), 130
TEV, *see* Tobacco etch virus
3D particles, 467; *see also* Particles
  AP205 phage, 486–487
  chimeric VLPs, 487–488
  comparison of inter-dimer interactions, 473
  crystal structures of RNA phages, 468
  de novo RNP modeling, 486
  genomic RNA in MS2 virion, 474
  implications for MS2 virus assembly, 485
  implications of packaging signal-mediated assembly, 484
  lattices, 484–486
  mapping of extra density on surface of PP7-a-loop-PP7, 487
  maturation protein, 476–480, 483
  maturation protein replacing coat protein dimer, 477
  mixed particles, 472
  MS2-decorated F pili, 477
  MS2 phage reconstruction, 478
  packing of genomic RNA, 471–476
  phage coat structures, 469
  RNA aptamers, 470–471
  RNA sequences in MS2 virion, 482
  self-assembly, 480–484
  single-stranded RNA genome modeling, 479
  structure of PP7-PP7 virus-like particle, 486
  3D models of RNA phage, 472
  3D structures of phages, 474
  viruses and VLPs, 467–470
(three)′-RNA end accumulation during turnover (TREAT 3), 548
3SR system, *see* Self-sustained sequence replication system
Threo amino chlorobutyric acid (TACB), 88
Thymectomized, irradiated, and bone marrow–reconstituted (TXBM), 309
Time-of-flight (TOF), 198
TLRs, *see* Toll-like receptors
TMPyP, *see* 15,20-tetrakis (1-methyl-4-pyridinio) porphyrintetra p-toluenesulfonate
TMV, *see* Tobacco mosaic virus
Tobacco etch virus (TEV), 425
Tobacco mosaic virus (TMV), 197, 211, 261
TOF, *see* Time-of-flight
Toll-like receptors (TLRs), 490
Transcription-translation system (TXTL system), 97
Translating RNA imaging by coat protein knockoff (TRICK), 547
Translation, 313; *see also* Repressor complex I
  analogous phenomenon, 338
  codons in MS2 coat, 340
  elongation, 330–331
  essentials, 313–315
  genetic code, 339–343
  heterologous translation *in vitro*, 318–319
  initiation, 323–327
  intergenic sequences, 333
  in vitro, 316–318, 334
  in vivo, 315–316
  large-scale translation, 339
  MS2 RNA-programmed *in vitro* synthesis, 313

mutations introduced in coat initiator hairpin, 334
nascent chains, 322–323
non-coliphages, 343–344
polarity, 319
Qβ RNA translation, 318, 337
reinitiation, 338–339
repressor complex II, 362–364
ribosomal frameshifting, 338
RNA fragments as messengers, 333
secondary and tertiary structure, 334–337
secondary structure of phage RNAs, 336
secondary structures for RNA phage initiator regions, 321
Shine-Dalgarno, 327–330
termination, 331–333
Translation-coupled RNA self-replication system, 403; *see also* Evolution
TREAT 3, *see* (three)′-RNA end accumulation during turnover
TRICK, *see* Translating RNA imaging by coat protein knockoff
TriFC, *see* Trimolecular fluorescence complementation
Trimolecular fluorescence complementation (TriFC), 542
Tubercidin triphosphate (TuTP), 382
Tumor-associated carbohydrate antigens (TACAs), 456, 513
Tunable Wavelength Interrogated Sensor Technology (TWIST), 311
Turnip yellow mosaic virus (TYMV), 197
TuTP, *see* Tubercidin triphosphate
TWIST, *see* Tunable Wavelength Interrogated Sensor Technology
Two-plasmid system, 349
TXBM, *see* Thymectomized, irradiated, and bone marrow–reconstituted
TXTL system, *see* Transcription-translation system
TYMV, *see* Turnip yellow mosaic virus
Type IV coupling proteins (T4CPs), 74
Type IV secretion systems (T4SSs), 64

U

UDP-NAG, *see* Uridine 5′-diphosphate-N-acetylglucosamine
Ultraviolet (UV), 122
Uridine 5′-diphosphate-N-acetylglucosamine (UDP-NAG), 244
Uridine photohydrates, 122
U.S. Environmental Protection Agency (USEPA), 106, 114
USEPA, *see* U.S. Environmental Protection Agency
U.S. National Library of Medicine (NLM), 32
UV, *see* Ultraviolet
UV-LEDs, *see* UV light-emitting diodes
UV light-emitting diodes (UV-LEDs), 129–130

V

V-1 variant, 381–382; *see also* Evolution
VapBC, *see* Virulence-associate proteins
Variable lymphocyte receptors (VLRs), 304
VEE, *see* Venezuelan encephalitis equine
Venezuelan encephalitis equine (VEE), 517
VIRADEN procedure, 106
Virologists, 29
Virome, 554–556

Virulence-associate proteins (VapBC), 233
Virus; *see also* Particles
    aggregation, 201–202
    species, 30
    taxa infecting bacteria and archaea, 33
Virus-like particles (VLPs), 25, 97, 195,
        219, 309, 422, 431, 553; *see also*
        Chimeric VLPs
    carriers, 438
    and chemical coupling approach, 448
    HEL-conjugated VLPs, 448
    models, 556
    Qβ VLP, 458
    RNA phage coats as VLP carriers, 432, 436
    as supramolecular structures, 435
Virus Sensitivity Index (VSI), 126
Virustat, 95
VLP packaging/targeting technology, 518;
        see also Non-vaccine VLPs

VLPs, *see* Virus-like particles
VLP technology, 489
VLP vaccines, 489
    basic immunology, 489–493
    chemical coupling, 501–514
    by chemical coupling methodology, 501–512
    chimeric phages, 514
    epitope-RNA VLP vaccine, 493
    by genetic fusion methodology, 494–500
    genetic fusions, 493–501
    Lyme disease vaccine, 513
    Qβ VLPs, 490
VLRs, *see* Variable lymphocyte receptors
VSI, *see* Virus Sensitivity Index

W

Water quality, 112
Water surveillance, 114; *see also* Phage ecology

    in Africa, 115
    in Antarctica, 117
    in Canada, 115
    in Central and South America, 116
    in China, 117
    in Europe, 116
    in India, 117
    in Israel, 115–116
    in Japan, 116–117
    in New Zealand and Australia, 115
    in Southeast Asia Countries, 117
    in the United States, 114–115
    of ground and surface waters, 113
West Nile virus (WNV), 444
WGM-h, *see* Whispering gallery
        mode−nanoshell hybrid
Whispering gallery mode−nanoshell hybrid
        (WGM-h), 200
WNV, *see* West Nile virus